Modification and Editing of RNA

Modification and Editing of RNA

EDITED BY

Henri Grosjean
Laboratoire d'Enzymologie et Biochimie Structurales
Centre National de la Recherche Scientifique
Gif-sur-Yvette, France

Rob Benne
Department of Biochemistry
Academic Medical Center
University of Amsterdam
Amsterdam, The Netherlands

ASM PRESS   WASHINGTON, D.C.

Copyright © 1998 American Society for Microbiology
1325 Massachusetts Avenue NW
Washington, DC 20005-4171

Library of Congress Cataloging-in-Publication Data

Modification and editing of RNA / edited by Henri Grosjean and Rob Benne.
   p.   cm.
Includes bibliographical references and index.
ISBN 1-55581-133-7
   1. RNA editing. I. Grosjean, Henri. II. Benne, Rob, 1945–   .
QH450.25.M63 1998
572.8'845—dc21   97-47125
                                                                                                                                                                      CIP

All Rights Reserved
Printed in the United States of America

Front cover figure: *Escherichia coli* cytidine deaminase (ECCDA) provides a model for both APOBEC and its minimal RNA substrate. (Left) The ECCDA dimer, showing the locations of two peptides (GAPs) that are deleted in APOBEC sequences. Monomers in the dimer are purple and red; deleted peptides in the two monomers are green and gold, respectively. (Right) Remodeling of ECCDA to give a model for APOBEC-1, the editing subunit of the apolipoprotein B editosome. Color coding for the dimeric enzyme is the same as in the ECCDA on the left; zinc atoms are teal, and zinc ligands are yellow. The GAP peptides from both monomers (green and gold) together with the transition state analog inhibitors form a peptide mimic of the RNA substrate, which exposes the targeted C and a downstream U with approximate twofold symmetry. This mimic represents single-stranded 5' and 3' regions and a central region with an unspecified tertiary structure. Vertical arrows show the fit to the two active sites. See Chapter 20. (Courtesy of Charles W. Carter, Jr. [University of North Carolina].)

Back cover figure: Model of the interaction between prokaryotic tRNA-guanine transglycosylase (TGT) and a cognate tRNA substrate. The electrostatic surface potential of TGT shows a bipolar distribution with positively charged regions (blue), near and at the zinc-binding subdomain, which interact with the anticodon stem-loop phosphate backbone (phosphates are green). The negatively charged pockets (red) at the center of the C-terminal face of the $(\beta/\alpha)_8$ barrel specifically recognize the $U_{33}G_{34}U_{35}$ sequence of the anticodon loop. The wobble guanine is indicated as G_{34}. The preQ$_1$-binding pocket is found underneath the surface indicated as the preQ$_1$ site. See Chapter 9. (Courtesy of Christophe Romier and Dietrich Suck [European Molecular Biology Laboratory] and Ralf Ficner [Philipps-Universität Marburg].)

CONTENTS

Contributors • vii
Preface • xi

1. Historical Perspectives on RNA Nucleoside Modifications • 1
Byron G. Lane

2. RNA-Modifying and RNA-Editing Enzymes: Methods for Their Identification • 21
Henri Grosjean, Yuri Motorin, and Annie Morin

3. Detection and Structure Analysis of Modified Nucleosides in RNA by Mass Spectrometry • 47
Pamela F. Crain

4. Incorporation of Modified Nucleotides into RNA for Studies on RNA Structure, Function, and Intermolecular Interactions • 59
Robert A. Zimmermann, Michael J. Gait, and Melissa J. Moore

5. Biophysical and Conformational Properties of Modified Nucleosides in RNA (Nuclear Magnetic Resonance Studies) • 85
Darrell R. Davis

6. Effects of Pseudouridylation on tRNA Hydration and Dynamics: a Theoretical Approach • 103
Pascal Auffinger and Eric Westhof

7. Modulation Role of Modified Nucleotides in RNA Loop–Loop Interaction • 113
Henri Grosjean, Claude Houssier, Pascale Romby, and Roland Marquet

8. Mechanisms of RNA-Modifying and -Editing Enzymes • 135
George A. Garcia and DeeAnne M. Goodenough-Lashua

9. Structural Basis of Base Exchange by tRNA-Guanine Transglycosylases • 169
Christophe Romier, Ralf Ficner, and Dietrich Suck

10. Biosynthesis and Functions of Modified Nucleosides in Eukaryotic mRNA • 183
Joseph A. Bokar and Fritz M. Rottman

11. Posttranscriptional Modifications in the U Small Nuclear RNAs • 201
Séverine Massenet, Annie Mougin, and Christiane Branlant

12. The Pseudouridine Residues of rRNA: Number, Location, Biosynthesis, and Function • 229
James Ofengand and Maurille J. Fournier

13. Small Nucleolar RNAs Guide the Ribose Methylations of Eukaryotic rRNAs • 255
Jean-Pierre Bachellerie and Jérôme Cavaillé

14. Functional Aspects of the Three Modified Nucleotides in Yeast Mitochondrial Large-Subunit rRNA • 273
Thomas L. Mason

15. Regulatory Aspects of rRNA Modifications and Pre-rRNA Processing • 281
Denis L. J. Lafontaine and David Tollervey

16. Editing of tRNA • 289
David H. Price and Michael W. Gray

17. RNA Editing by Base Conversion in Plant Organellar RNAs • 307
Anita Marchfelder, Stefan Binder, Axel Brennicke, and Volker Knoop

18. Apolipoprotein B mRNA Editing • 325
Benny Hung-Junn Chang, Paul P. Lau, and Lawrence Chan

19. Adenosine-to-Inosine Conversion in mRNA • 343
Susan M. Rueter and Ronald B. Emeson

20. Nucleoside Deaminases for Cytidine and Adenosine: Comparison with Deaminases Acting on RNA • 363
Charles W. Carter, Jr.

21. Mitochondrial mRNA Editing in Kinetoplastid Protozoa • 377
Stephen L. Hajduk and Robert S. Sabatini

22. RNA Editing in *Physarum* Mitochondria • 395
Jonatha M. Gott and Linda M. Visomirski-Robic

23. Contranscriptional Paramyxovirus mRNA Editing: a Contradiction in Terms? • 413
Daniel Kolakofsky and Stéphane Hausmann

24. Intracellular Locations of RNA-Modifying Enzymes • 421
B. Edward H. Maden

25. Genetics and Regulation of Base Modification in the tRNA and rRNA of Prokaryotes and Eukaryotes • 441
Malcolm E. Winkler

26. Links between tRNA Modification and Metabolism and Modified Nucleosides as Tumor Markers • 471
Glenn R. Björk and Torgny Rasmuson

27. Modified Nucleosides in Translation • 493
James F. Curran

28. Importance of Modified Nucleotides in Replication of Retroviruses, Plant Pararetroviruses, and Retrotransposons • 517
Roland Marquet

29. Modified Nucleotides Always Were: an Evolutionary Model • 535
Nicolas Cermakian and Robert Cedergren

Appendix 1. Chemical Structures and Classification of Postranscriptionally Modified Nucleosides in RNA • 543
Yuri Motorin and Henri Grosjean

Appendix 2. RNA Editing Types and Characteristics • 551
Rob Benne and Dave Speijer

Appendix 3. General Properties of RNA-Modifying and -Editing Enzymes • 555
George A. Garcia and DeeAnne M. Goodenough-Lashua

Appendix 4. Genetic Locations and Database Accession Numbers of RNA-Modifying and -Editing Enzymes • 561
Hon-Chiu Eastwood Leung, Tord G. Hagervall, Glenn R. Björk, and Malcolm E. Winkler

Appendix 5. Location and Distribution of Modified Nucleotides in tRNA • 569
Pascal Auffinger and Eric Westhof

Appendix 6. Modified Nucleosides at Positions 34 and 37 of tRNAs and Their Predicted Coding Capacities • 577
Glenn R. Björk

Index • 583

CONTRIBUTORS

Pascal Auffinger
Institut de Biologie Moléculaire et Cellulaire, CNRS, 15, rue René Descartes, F-67084 Strasbourg Cedex, France

Jean-Pierre Bachellerie
Laboratoire de Biologie Moléculaire du CNRS, Université Paul-Sabatier, 118, route de Narbonne, F-31062 Toulouse Cedex, France

Rob Benne
Department of Biochemistry, Academic Medical Center, University of Amsterdam, Meibergdreef 15, 1105 AZ Amsterdam, The Netherlands

Stefan Binder
Allgemeine Botanik, Universität Ulm, Albert-Einstein-Allee, D-89069 Ulm, Germany

Glenn R. Björk
Department of Microbiology, Umeå University, S-90187 Umeå, Sweden

Joseph A. Bokar
Department of Medicine, Case Western Reserve University, Cleveland, OH 44106

Christiane Branlant
Laboratoire d'Enzymologie et de Génie Génétique, URA-CNRS 457, Faculté des Sciences, Université de Nancy I, F-54506 Vandoeuvre-lès-Nancy Cedex, France

Axel Brennicke
Allgemeine Botanik, Universität Ulm, Albert-Einstein-Allee, D-89069 Ulm, Germany

Charles W. Carter, Jr.
Department of Biochemistry and Biophysics, CB 7260, University of North Carolina, Chapel Hill, NC 27599-7260

Jérôme Cavaillé
Laboratoire de Biologie Moléculaire du CNRS, Université Paul-Sabatier, 118, route de Narbonne, F-31062 Toulouse Cedex, France

Robert Cedergren
Département de Biochimie, Université de Montréal, Montréal, Québec H3C 3J7, Canada

Nicolas Cermakian
Département de Biochimie, Université de Montréal, Montréal, Québec H3C 3J7, Canada

Lawrence Chan
Departments of Cell Biology and Medicine, Baylor College of Medicine, One Baylor Plaza, Houston, TX 77030

Benny Hung-Junn Chang
Departments of Cell Biology and Medicine, Baylor College of Medicine, Houston, TX 77030

Pamela F. Crain
Department of Medicinal Chemistry, University of Utah, Salt Lake City, UT 84112

James F. Curran
Department of Biology, Wake Forest University, P.O. Box 7325, Winston-Salem, NC 27109

Darrell R. Davis
Department of Medicinal Chemistry, University of Utah, Salt Lake City, UT 84112

Ronald B. Emeson
Department of Pharmacology, Vanderbilt University, 460 Medical Research Bldg. 2, Nashville, TN 37232-6600

Ralf Ficner
Institut für Molekularbiologie und Tumorforschung, Philipps-Universität Marburg, Emil-Mannkopff-Strasse 2, D-35037 Marburg, Germany

Maurille J. Fournier
Department of Biochemistry and Molecular Biology, University of Massachusetts, Amherst, MA 01003

Michael J. Gait
MRC Laboratory of Molecular Biology, Hills Rd., Cambridge CB2 2QH, United Kingdom

George A. Garcia
Interdepartmental Program in Medicinal Chemistry, College of Pharmacy, University of Michigan, Ann Arbor, MI 48109-1065

DeeAnne M. Goodenough-Lashua
Interdepartmental Program in Medicinal Chemistry, College of Pharmacy, University of Michigan, Ann Arbor, MI 48109-1065

Jonatha M. Gott
Center for RNA Molecular Biology, Department of Molecular Biology and Microbiology, Case Western Reserve University, Cleveland, OH 44106

Michael W. Gray
Department of Biochemistry, Dalhousie University, Halifax, Nova Scotia B3H 4H7, Canada

Henri Grosjean
Laboratoire d'Enzymologie et Biochimie Structurales, Centre National de la Recherche Scientifique, F-91198 Gif-sur-Yvette, France

Tord G. Hagervall
Department of Microbiology, University of Umeå, S-90187 Umeå, Sweden

Stephen L. Hajduk
Department of Biochemistry and Molecular Genetics, University of Alabama at Birmingham School of Medicine, Birmingham, AL 35294

Stéphane Hausmann
Department of Genetics and Microbiology, University of Geneva School of Medicine, CMU, 9, ave. de Champel, CH-1211 Geneva, Switzerland

Claude Houssier
Laboratorire de Chimie Macromoléculaire et Chimie Physique, Université de Liège, Sart-Tilman (B6), B-4000 Liège, Belgium

Volker Knoop
Allgemeine Botanik, Universität Ulm, Albert-Einstein-Allee, D-89069 Ulm, Germany

Daniel Kolakofsky
Department of Genetics and Microbiology, University of Geneva School of Medicine, CMU, 9, ave. de Champel, CH-1211 Geneva, Switzerland

Denis L. J. Lafontaine
Institute of Cell and Molecular Biology, University of Edinburgh, Mayfield Rd., Edinburgh EH9 3JR, United Kingdom

Byron Lane
Biochemistry Department, University of Toronto, Toronto, Ontario M5S 1A8, Canada

Paul P. Lau
Departments of Cell Biology and Medicine, Baylor College of Medicine, Houston, TX 77030

Hon-Chiu Eastwood Leung
Department of Microbiology and Molecular Genetics, University of Texas Houston Medical School, Houston, TX 77030-1501

B. Edward H. Maden
School of Biological Sciences, University of Liverpool, Crown St., Liverpool L69 7ZB, United Kingdom

Anita Marchfelder
Allgemeine Botanik, Universität Ulm, Albert-Einstein-Allee, D-89069 Ulm, Germany

Roland Marquet
Institut de Biologie Moléculaire et Cellulaire, CNRS, 15, rue René Descartes, F-67084 Strasbourg, France

Thomas L. Mason
Department of Biochemistry and Molecular Biology, University of Massachusetts, Amherst, MA 01003

Séverine Massenet
Laboratoire d'Enzymologie et de Génie Génétique, URA-CNRS 457, Faculté des Sciences, Université de Nancy I, F-54506 Vandoeuvre-lès-Nancy Cedex, France

Melissa J. Moore
Howard Hughes Medical Institute, Department of Biochemistry, Brandeis University, Waltham, MA 02254-9110

PREFACE

RNA MODIFICATION AND EDITING: TWO SIDES OF THE SAME COIN

The intricate series of events which produces functional, fully mature RNA from primary transcripts of genes is called RNA maturation. The simplicity of this term largely misrepresents the complexity of the underlying processes, which involve numerous steps and the participation of a large number of distinct catalytic macromolecules (enzymes and ribozymes) and of cofactors. Two basically distinct modes of RNA maturation exist: one in which phosphodiester bonds are broken and/or formed (e.g., splicing, polyadenylation and polyuridylation, 5' and 3' trimming, capping, addition or deletion of internal nucleotides, etc.) and one in which bases and/or ribose moieties are chemically modified while the ribose-phosphate backbone is left intact. From the impressive amount of energy (and genes) that cells invest in these processes, it must be concluded that they are important for cellular metabolism and function. The challenge is to understand how and why these posttranscriptional RNA alteration processes have evolved: do they represent relics of the hypothetical prebiotic RNA world or acquisitions of vital new functions (or both)?

The first evidence for the presence of modified nucleosides in RNA was obtained in 1951 (pseudouridine, the fifth nucleoside), while the first RNA modification enzyme was identified in 1962 (*Escherichia coli* tRNA-uracil methylase). To date, 95 chemically modified nucleosides have been identified. This is probably not an upper limit, and sequencing of more RNAs (not only tRNAs but also rRNAs, mRNAs, and snRNAs) from various organisms may still raise this number, especially now that tools for easier identification exist. Some are simple modifications, such as base or ribose methylation and base isomerization, reduction, thiolation, or deamination. Others are more complex hypermodifications whose biosynthesis involves several enzymes. Among the different types of RNA, tRNA is generally the most heavily modified: in some tRNAs from higher eukaryotes, up to 25% of the nucleotides can be modified in one way or the other. However, despite an impressive amount of research, the effect of most of the modifications on RNA structure and function is still far from clear.

In 1986 the concept of RNA editing was introduced, initially to describe the process of posttranscriptional insertion of nongenomically encoded uridylate residues within the coding region of mitochondrial mRNAs of kinetoplastid protozoa. At that point in time, RNA editing was another addition to the list of processes that involve phosphodiester bond cleavage or ligation. Things became more complicated when in the following years the term "RNA editing" was used to describe any process other than splicing and capping that changes the sequence of an RNA transcript (mRNA, tRNA, or rRNA) from that encoded by the corresponding gene. Although the precise mechanism is not yet known in many cases, some of the RNA-editing processes again appeared to involve cleavage or ligation reactions. However, others were highly reminiscent of well-established RNA modification reactions, which not only create nucleosides that are chemically distinct from the four canonical nucleosides but also convert one canonical nucleoside into another one (e.g., C's to U's, U's to C's, etc.). However, the effect of base conversion editing appeared more obvious than that of many nucleoside modifications, since editing directly changes base pairing potential and thus (for example) the genetic content of mRNAs or the secondary and/or tertiary structure of tRNAs and rRNAs. Therefore, it seemed logical also to use the term "RNA editing" to describe the conversion by deamination of adenosines to inosines in certain mammalian mRNAs, given that the I's are read as C's by the translational machinery. As a result, the distinction between RNA base conversion editing and RNA nucleoside modification has become fuzzy, because A-to-I deaminations in tRNAs and rRNAs have, since their discovery more than 25 years ago, always been known as classical examples of RNA modification (but not RNA editing). In addition, readers of this book will realize that there are other examples where a clear distinction between RNA modification and RNA editing is not obvious. For example, the formation of lysidine (a C-modified nucleoside that decodes A instead of G) in the wobble position of a tRNA^{Ile} of several organisms clearly affects base pairing potential during translation. As for anticipating further developments, it does not seem too far-fetched to assume that other RNA modifications will qualify as RNA editing and that, on the other hand, the mechanism of some of the currently poorly characterized RNA-editing reactions will turn out to be some form of base modification. Evidently, RNA nucleoside modification and RNA base conversion editing are two sides of the same coin. This is strikingly illustrated by the recent finding that the deaminase domain of the yeast tRNA:A$_{37}$ deaminase HRA400 (recently renamed YGL243w) has a clear evolutionary relationship to those of the mammalian mRNA A deaminases ADAR1 and ADAR2.

SCOPE AND PURPOSE OF THIS BOOK

The purpose of this book is to cover all aspects of nucleoside modification and editing in different classes of RNA in different organisms. This includes the description of the different types of modification and editing; the identification of the corresponding enzymes; and discussion of *cis*- and *trans*-acting sequences, machineries, mechanisms, cellular locations, biological roles of individual modified and/or edited bases, involvement in diseases like cancer and AIDS, and possible evolutionary connections between different processes. Apart from covering base conversion editing, we decided to include cleavage-and-ligation RNA editing, not only because certain systems such as *Physarum polycephalum* mitochondria possess both forms of editing, but also because clear mechanistic parallels may exist between them. For example, editing-site selection is mediated by small guide RNAs in both U insertion-deletion editing in kinetoplastid mitochondria and 2'-O-ribose methylation and pseudouridylation in rRNA of eukaryotic cells. Last but not least, in many cases insufficient data are available to classify a particular editing event: certain apparent RNA base conversions may actually be carried out by cleavage-ligation reactions and vice versa.

The only previous thorough coverage of nucleoside modifications in RNA was published 27 years ago (Ross H. Hall, *The Modified Nucleosides in Nucleic Acids*, Columbia University Press, 1971). Since then, no book has given enough credit to the modified nucleosides in RNA. Even recent authoritative multiauthor volumes, such as *The Ribosome: Structure, Function, and Evolution* (ASM Press, 1990), *The RNA World* (Cold Spring Harbor Laboratory Press, 1993), and *RNA Structure and Function* (Cold Spring Harbor Laboratory Press, 1998), exhibit little or no recognition of the possibility that modified nucleosides have an important—even essential—role in RNA and/or protein biosynthesis. However, the book *tRNA: Structure, Biosynthesis and Function* (ASM Press, 1995; edited by D. Söll and U. L. RajBhandary) contains a few excellent chapters on tRNA modification. As far as RNA editing is concerned, the only book was put together 5 years ago by one of us (*RNA Editing: the Alteration of Protein Sequences of RNA* [Ellis Horwood Limited, 1993]).

In the last few years, new biochemical and physicochemical techniques have been developed, such as chemical and enzymatic mapping of RNA sequence and conformation, mass spectrometry, nuclear magnetic resonance, and fast kinetic relaxation techniques. Synthetic substrates can now easily be produced enzymatically or chemically in vitro. Several modifying and editing enzymes and their corresponding genes have been identified and characterized. Several bacterial and eukaryotic mutants defective in RNA modification are now available. These advances have generated a wealth of new data, further underlining the importance of modified and edited nucleosides for the structure and function of RNA. Given the rapid progress of the research in the different areas, it seems timely to have all of the information together in an unifying general framework.

The different chapters of the book have been written by leading researchers (geneticists, biochemists, physical chemists, organic chemists, and medical doctors) who are currently working on various aspects of nucleoside modification and editing in tRNA, mRNA, rRNA, and/or snRNA. The book also provides a historical perspective and contains descriptions of the techniques used for the detection and identification of modified nucleosides in RNA, for the measurement of their physicochemical properties, and for site-directed incorporation. Last but not least, it includes six appendixes which give complete lists of the currently known nucleoside modifications and their location and distribution in tRNAs, the different types of editing, and the properties of modifying and editing enzymes and their genomic locations. The book is well referenced up to the end of 1997 and contains numerous citations of research that will be published in 1998. We hope that this book will serve as an invaluable reference tool for researchers, teachers, and students alike and that it will be used for the

Annie Morin
Laboratoire d'Enzymologie et Biochimie Structurales, Centre National de la Recherche Scientifique, F-91198 Gif-sur-Yvette, France

Yuri Motorin
Laboratoire d'Enzymologie et Biochimie Structurales, Centre National de la Recherche Scientifique, F-91198 Gif-sur-Yvette, France

Annie Mougin
Laboratoire d'Enzymologie et de Génie Génétique, URA-CNRS 457, Faculté des Sciences, Université de Nancy I, F-54506 Vandoeuvre-lès-Nancy Cedex, France

James Ofengand
Department of Biochemistry and Molecular Biology, University of Miami School of Medicine, Miami, FL 33101-6129

David H. Price
Department of Biochemistry, Dalhousie University, Halifax, Nova Scotia B3H 4H7, Canada

Torgny Rasmuson
Department of Oncology, Umeå University, S-90187 Umeå, Sweden

Pascale Romby
Institut de Biologie Moléculaire et Cellulaire, CNRS, 15 rue René Descartes, F-67084 Strasbourg, France

Christophe Romier
Structural Biology Programme, European Molecular Biology Laboratory, Meyerhofstrasse 1, D-69117 Heidelberg, Germany

Fritz M. Rottman
Department of Molecular Biology and Microbiology, Case Western Reserve University, Cleveland, OH 44106-4960

Susan M. Rueter
Department of Pharmacology, 460 Medical Research Building 2, Vanderbilt University, Nashville, TN 37332-6600

Robert S. Sabatini
Department of Biochemistry and Molecular Genetics, University of Alabama at Birmingham School of Medicine, Birmingham, AL 35294

Dave Speijer
Department of Biochemistry, Academic Medical Center, University of Amsterdam, Meibergdreef 15, 1105 AZ Amsterdam, The Netherlands

Dietrich Suck
Structural Biology Programme, European Molecular Biology Laboratory, Meyerhofstrasse 1, D-69117 Heidelberg, Germany

David Tollervey
Institute of Cell and Molecular Biology, University of Edinburgh, Mayfield Rd., Edinburgh EH9 3JR, United Kingdom

Linda M. Visomirski-Robic
Department of Biological Sciences, Carnegie Mellon University, 4400 Fifth Ave., Pittsburgh, PA 15213

Eric Westhof
Institut de Biologie Moléculaire et Cellulaire, CNRS, 15, rue René Descartes, F-67084 Strasbourg Cedex, France

Malcolm E. Winkler
Department of Microbiology and Molecular Genetics, University of Texas Houston Medical School, Houston, TX 77030-1501

Robert A. Zimmermann
Department of Biochemistry and Molecular Biology, University of Massachusetts, Amherst, MA 01003-4505

Chapter 1

Historical Perspectives on RNA Nucleoside Modifications

BYRON G. LANE

A retrogressive consequence of the swift progress in RNA research over the past 40 years has been the reduction of its history to episodic accounts of "exciting" developments. The history of the continuous development of our knowledge of RNA biochemistry might yet become as difficult to uncover as the biochemical evolution of RNA itself. The principal intention of this article is to sketch the historical roots of current avenues of research on RNA nucleoside modifications by describing some functional sites, structural environments, and biosynthetic strategies with which the modifications are associated. Consonant with my working experience, I shall describe selected examples of nucleoside modifications in ribosomal, transfer, messenger, and small nuclear RNAs, and in keeping with my career-long interests, I shall give special emphasis to a conspicuous association of the most abundant nucleoside modifications in eukaryotic rRNA with the peptidyl transfer site in the ribosome. Beforehand, I shall briefly recount the historical and chronological development of general knowledge about RNA structure and function.

AN ABBREVIATED HISTORY OF RNA RESEARCH

Before the advent of molecular biology, there was a distinguished epoch when research on RNA and DNA was a domain of specialists in the organic chemistry of natural products. Typical of the best of his generation, one such investigator was G. C. Butler, my Ph.D. supervisor. Having obtained his doctorate in the 1930s in the laboratory of Guy Marrian (who first assigned the correct tetracyclic structure to estrogenic hormones), Gordon Butler sought new adventure during the late 1940s in the still-uncharted field of chromatin structure. The many germinal contributions of his laboratory, such as the pioneering use of sodium dodecyl sulfate (SDS) to prepare cellular DNA (Marko and Butler, 1951), have been recounted elsewhere (Lane, 1979a). First isolated as parts of nucleoprotein complexes approximately 130 years ago (see Allen, 1962), the nucleic acids were only the second class of phosphorus-containing organic substances found in living cells. By the early part of the present century (see Levene and Bass, 1931), studies of the organic constituents of RNA and DNA had extended the natural occurrence of some substances and uncovered others that were not previously known even as products of laboratory synthesis: two sugars, the O-heterocycles, D-ribofuranose and 2-deoxy-D-ribofuranose, and five bases, the N-heterocycles, adenine, guanine, cytosine, uracil, and thymine (see Chargaff and Davidson, 1955). For historical reasons, RNA and DNA are called nucleic acids, and uracil and thymine are called bases; however, under physiological conditions, RNA and DNA are nucleate salts, and uracil and thymine are closely related to barbituric acids. These chemical distinctions were critical for arriving at the polymeric structures of DNA and RNA.

Before 1956: Covalent Bonds in the Classical Mononucleotides

By 1956, investigations of the preceding 25 years had shown how the bases and sugars in the classical nucleosides of RNA and deoxynucleosides of DNA are united through β N-glycosyl bonds. Investigations of this period also showed that most if not all nucleosides in RNA and DNA are linked through $3'$-$5'$ phosphodiester bridges. An era of research on the natural polynucleotides ended when the methods of classical organic chemistry adduced definitive covalent structures for the four classical nucleotide constituents in each of RNA and DNA. This achievement

Byron G. Lane • Biochemistry Department, University of Toronto, Toronto, Ontario, Canada M5S 1A8.

was marked by the appearance of the first of a three-volume series edited by Chargaff and Davidson (1955). *The Nucleic Acids* surveyed research from the latter part of a classical era and the early part of a modern era that began about 50 years ago with the application of powerful new chromatographic techniques for separating hydrolysis products derived from RNA and DNA. These new separation methods soon showed that cellular RNA, which was abbreviated as PNA (pentose nucleic acid) in *The Nucleic Acids*, contains a "modified" sugar, 2'-O-methylribose, that can replace 0.1–2% of the ribose in each classical nucleoside, adenosine, guanosine, cytidine, and uridine (A, G, C, U), to give corresponding 2'-O-methylnucleosides (Am, Gm, Cm, Um) (Smith and Dunn, 1959a).

After 1956: Hydrogen Bonds between Polynucleotides

Resistance to the idea that DNA (not protein) is the genetic substance had gradually given way to Avery, MacLeod, and McCarty's report in 1944 that a bacterial transforming factor was DNA (see Dubos, 1976). With increasing acceptance of the notion that DNA structure was not "too simple" to encode the very complex array of amino acid sequences that Sanger and his colleagues were contemporaneously finding in insulin (Sanger and Tuppy, 1951), attention focused on more ambitious template (Dounce, 1956) and catalytic (Binkley, 1951) roles for the "other" nucleate—RNA. Cytological studies in the 1930s and 1940s strongly implicated RNA in the biosynthesis of proteins (see Brachet, 1955; Chantrenne, 1961) but the biochemistry of RNA was a "black box" until the 1950s. As recently recollected (Santer and Dahlberg, 1996), the period between 1953 and 1960 gave a distinctive new meaning to "molecular biology." In earlier times, progress was propelled by the determination of covalent-bond structures in the nucleic acids, but in this new era biological function would be deduced from putative hydrogen-bond relations, first between DNA and later between RNA molecules.

Analyses of DNA hydrolysates (see Chargaff, 1955) revealed the A=T and G=C equivalences which, when used to evaluate the relative merits of competing X-ray diffraction models of DNA structure, had dictated, in 1953, that the Watson-Crick "double helix" was an irresistibly attractive choice (Watson and Crick, 1953). By extending this molecular biological approach, in which a vision of the storage and replication properties of the genetic material had been implicit in the structure of the DNA double helix, Crick suggested how RNA might mediate protein biosynthesis, and in so doing bridged a conceptual gulf between nucleic acid templates and protein biogenesis. In 1955, when the singularly pervasive engagements of RNA with genome (transcription) and ribosome (translation) were only dimly perceived, Crick advanced an RNA "adaptor hypothesis" to explain how RNA structure might be "translated" into protein structure (see Hoagland, 1960).

Although "natural" adaptors proved to be much larger than those originally envisioned by Crick (Judson, 1979), the fundamental correctness of the idea was soon realized when Zamecnik's laboratory isolated "soluble" or "transfer" RNA (tRNA) (Hoagland et al., 1958). After a tRNA, carrying a specific activated amino acid (Holley, 1957), is transferred from the soluble cytoplasm to an RNA-rich particle aptly named the ribosome (Dintzis, 1958), the anticodon-containing tRNA was assumed to adapt to (H-bond with) a complementary (triplet) codon in a protein-encoding RNA template (mRNA), the activated amino acid becoming bonded to another amino acid at a peptidyl transfer center in the ribosome. Working in the Lipmann laboratory in 1961, Francois Chapeville and his coworkers proved the "adaptor hypothesis" (Chapeville et al., 1962). Because nucleoside modifications in tRNA were expertly reviewed very recently (Bjork, 1995; Yokoyama and Nishimura, 1995), and research of the past 3 years has revolutionized our understanding of nucleoside modifications in rRNA, the emphasis in this account is on rRNA modification, which was last comprehensively reviewed 8 years ago (Maden, 1990).

The Dawn of RNA Biochemistry: Phenolic Extraction and Salt Precipitation of RNA

Before 1956, the status of cellular RNA as a biopolymer was in serious doubt. There was still persistent and lingering belief in a time-honored notion that the RNA of living cells might consist of aggregates of small oligonucleotides (Levene and Bass, 1931; Magasanik, 1955). In the mid-1960s, mean degrees of polymerization were <20 for commercial specimens of bulk (yeast) RNA (Singh and Lane, 1964a). Shortly after publication of the first volume of the three-volume series by Chargaff and Davidson (1955), Colter and Brown (1956) disrupted Ehrlich ascites cells in aqueous phenol and precipitated their RNA by a new method that Allen's laboratory had just introduced to precipitate "hot SDS" RNA from yeast: aqueous 1 M NaCl at 0°C (Crestfield et al., 1955).

By preliminary partitioning (not inactivation) of the bulk of the RNases (which enter the phenol phase) from RNA (which enters the aqueous phase),

planning of future research to answer the many remaining questions. It is meant not only for our colleagues who are already working in the field but indeed for all those wanting to know (more) about RNA modification and editing. We hope that it will act as a catalytic factor in convincing researchers to join the "RNA modification-editing club."

A book of this size is the result of the efforts of many different people. First of all, we would like to give the credit where it belongs: to our colleagues who contributed the chapters and found the time in their busy academic and scientific lives to produce such high-quality products in only a few months and to patiently respond to our queries and our frequent reminders of deadlines. Certainly, the use of electronic mail has allowed rapid exchange of information and the successive versions of each manuscript between authors and editors. The production of this book would have been impossible without the help and advice of several persons at ASM Press in Washington: Gregory Payne and Susan Birch at the early stages of the preparations for the book and Ken April during the final production stage. We greatly appreciate their advice and tireless efforts. We also acknowledge Ed Atkeson for the cover design. Finally, we apologize to our colleagues in Gif-sur-Yvette and Amsterdam for any inconvenience that may have arisen from our spending much more time in preparing this book than initially anticipated.

Henri Grosjean
Rob Benne

and secondary partitioning of residual RNase activity, proteins, DNA, and polysaccharides (which remain soluble) from the bulk of the cellular RNA (which is precipitated from aqueous 1 M NaCl solution at 0°C), Colter and Brown (1956) were the first to see the discrete "18S" (small-subunit [SSU]) and "26S" (large-subunit [LSU]) RNAs. This set new benchmarks for degrees of nucleotide polymerization (>1,000) in cellular RNA, and studies of such phenol-extracted or SDS-stabilized (Kurland, 1960) RNA specimens led Doty's laboratory to propose their fundamental loop/hairpin model of RNA structure (Fresco et al., 1960).

In the years that immediately followed introduction of the "phenol method," it was widely reported that the integrity of the global superstructure in isolated eukaryotic rRNA specimens could be easily disrupted, possibly owing to the presence of numerous "hidden breaks" in bihelical regions (Hall and Doty, 1959; Brown et al., 1960; Otaka et al., 1961). An "interrupted strand" model, envisioning staggered discontinuities in the two strands of the double helix, had once been proposed to explain the cell-free properties of DNA specimens (Dekker and Schachman, 1954). Similar notions about eukaryotic rRNA were eliminated when studies of chain termini in phenol-extracted, NaCl-precipitated wheat rRNA (Lane, 1965; Diemer et al., 1966) definitively disqualified the "subunit hypotheses." The same studies had nonetheless detected a small number of supernumerary termini in SSU and/or LSU high-molecular-weight rRNAs. The cryptic termini corresponded to about one break per mole of LSU RNA in bulk wheat rRNA (Lane, 1965), and they were localized to the LSU RNA in L-cell rRNA (Lane and Tamaoki, 1967). These findings presaged discovery, soon afterward, of the single hidden break in LSU RNA that gives rise to 5.8S rRNA in eukaryotes (Pene et al., 1968; see Nazar, 1984).

After a primary structure and a possible "cloverleaf" secondary structure for yeast tRNA^{Ala} were reported (Holley et al., 1965), almost 10 years passed before independent X-ray crystallographic studies in two laboratories established an L-shaped tertiary structure for tRNA^{Phe} (Suddath et al., 1974; Robertus et al., 1974). Most tRNAs of known primary structure conform to a full or truncated form of cloverleaf secondary structure (see Appendix 1 by Motorin and Grosjean). Only recently, X-ray crystallography has provided a partial tertiary structure for *E. coli* 5S rRNA (Correll et al., 1997). Primary and secondary structural paradigms for high-molecular-weight rRNAs, with some tertiary interactions that stem largely from evolutionary comparisons (Woese and Pace, 1993), are regularly updated (Gutell et al., 1993; Gutell, 1994).

THE FIRST AND MOST ABUNDANT MODIFIED NUCLEOSIDE DETECTED IN RNA IS AN ISOMER OF URIDINE, PSEUDOURIDINE (ψ)

By mid-century, it had been definitively shown that a substance thought to be 5-methylcytosine (Hotchkiss, 1948) is a constituent of DNA (Wyatt, 1950). During the first half of the century there had been sporadic reports of the presence of nonclassical bases in RNA, but such claims had been relinquished, contradicted, or dismissed by 1957 (see Chargaff and Davidson, 1955), when Allen's laboratory demonstrated (Davis and Allen, 1957) that a salt-soluble RNA fraction (mainly tRNA) contained much more of a "fifth nucleotide" than did salt-insoluble RNA (mainly ribosomal RNA). The "fifth nucleoside" accounted for ~4 mol% of the constituent nucleosides in yeast tRNA and was identified as 5-ribosyluracil, a C-glycosyl isomer of the classical N-glycosyl nucleoside, uridine (1-ribosyluracil) (Yu and Allen, 1959; Scannell et al., 1959; Cohn, 1959, 1960). The novel nucleoside, assigned the name pseudouridine and abbreviation ψ by Michelson (see Cohn, 1959), contains the only C-C "base-sugar bond" in RNA (or DNA) (see Chambers, 1966).

The early identifications of modified nucleosides relied heavily on application of the same "new" approaches (anion-exchange and filter-paper chromatography and UV spectrophotometry) that were described in considerable detail in the first volume of *The Nucleic Acids*. They had just been used with definitive effect to refine the structures that had been deduced for the classical nucleotides by classical methods of organic chemistry. The first book solely devoted to the modified nucleosides in nucleic acids appeared about a decade after the third and final volume of the series edited by Chargaff and Davidson (1960): *The Modified Nucleosides In Nucleic Acids*, written by Ross Hall (1971), an important early contributor to this field of research. It dealt heavily with the idiosyncratic approaches then used to isolate modified nucleosides, and it provided a compendium of UV spectral data, as did a section by David Dunn and Ross Hall in a CRC Handbook edited by Fasman, which also included a wealth of anecdotal information about the discovery and characterization of individual modified nucleosides (Dunn and Hall, 1975). This comprehensive collection of UV spectral data followed the pristine form used by David Shugar and Jack Fox (Shugar and Fox, 1952) in their pioneering systematic studies of pH-dependent changes in the UV spectra of classical bases and nucleosides. Such data were invaluable in establishing the identities of the most abundant modified nucleosides. For example, the UV maximum of ψ (Davis and Allen, 1957), like that for uracil, but unlike that for uridine, is strongly shifted to higher wavelengths above pH 10, and this suggested that, like the base (uracil), but unlike the nucleoside (uridine), with its N-1-ribosyl substituent, ψ has a dissociable proton at N1.

The Influence of Frank Allen's Laboratory

The way Frank Allen's laboratory in Berkeley, and later at the San Francisco Medical Center, contributed to the sudden transition from the classical to the modern era of RNA biochemistry, and to the almost simultaneous inauguration of the field of RNA nucleoside modifications, circa 1956, is not likely familiar to most and deserves some explication. Historical influence, not chronological precedence, has guided my account in important ways. Having been a pioneer in the purification (Crestfield et al., 1955; Davis and Allen, 1957; Allen, 1962) and structural analysis (Crestfield and Allen, 1956, 1958) of RNA, Allen felt his laboratory had just made significant strides toward overcoming what had long been the most formidable limitation to progress in RNA biochemistry: nuclease-induced degradation during RNA isolation (Bacher and Allen, 1950; Allen, 1954). Frank Allen had no doubts about the "credentials" of any unusual nucleotide in his NaCl-insoluble RNA or salmine-precipitated NaCl-soluble RNA (rich in a "fifth nucleotide"), both being virtually devoid of adventitious contaminants. Cohn's laboratory had detected an unknown compound in RNA hydrolysates (designated "?") (Cohn and Volkin, 1951), but did not call it a "fifth nucleotide" and did not publish about it for 8 years thereafter (Cohn, 1959). Only after Davis and Allen (1957) reported intriguing UV spectral characteristics for their "fifth nucleotide" was there galvanization of the search for and discovery of novel RNA nucleotides in other laboratories.

Ambiguity, then posed by the fact that the so-called (4S) "subunits" of rRNA (Otaka et al., 1961) were not larger than 4S tRNA, could be circumvented by determining the amount of ψ per mole of 3′-hydroxyl terminal nucleoside in "biosynthetic units." This led to a definitive demonstration that ψ is a bona fide constituent of eukaryotic (wheat) rRNA and left no doubt about a role for ψ in the biochemistry of rRNA (Lane and Allen, 1961). As an Allen alumnus, Tom Yu (Yu and Allen, 1959) was welcome in Zamecnik's laboratory, where he initiated one of the earliest studies of a possible role for ψ in the secondary structure of tRNA (see Zamecnik, 1962). Likewise, as an Allen alumnus, I was welcome in the laboratory of Nobel laureate Fritz Lipmann, where I intended to initiate exploration of a possible role for rRNA-ψ in peptide-bond synthesis. This intention did not materialize for 32 years.

"Pressing" RNA into Familiar Protein Functions: a Motive Force in RNA Research

The notion that RNA catalysis had its roots, in the mid-1960s, in fertile evolutionary reflections (see Orgel and Crick, 1993) about what Gilbert called the "RNA World" (Gilbert, 1986), and/or in watershed discoveries of RNA catalysis (see Cech, 1990; Altman et al., 1995), is understandable. However, to those "ancients" among us, the notion of RNA as an enzyme had common currency at least a decade before these justly celebrated landmarks. The earliest literature accounts of a possible role for "RNA enzymes" in protein synthesis antedate the paramount position of ribosomes in studies of cell-free protein biosynthesis. "RNA enzymes" had currency at a time when many distinguished workers (e.g., Hanes, 1953) thought peptide bonds in proteins might be formed through the action of peptidases. Conjectural and experimental support for the view that peptidases might be "RNA enzymes" (Binkley, 1951) proved wanting (Semenza, 1957; Matheson and Hanes, 1959), but the ideas were espoused with the same sense of excitement (see Binkley 1951) that was more recently invested in the possibility that protein-free, rRNA-catalyzed, cell-free synthesis of peptide bonds might be at hand (see Waldrop, 1992; Noller et al., 1992; Joyce and Orgel, 1993; Noller, 1993).

The notion that peptidases might be "RNA enzymes" was justly influential because it responded to an apprehended need to address a vital, if too-simplistically framed, issue: if enzymes are proteins and proteins catalyze cellular reactions, what catalyzes protein synthesis? This was yet another (albeit less-sophisticated) version of "the paradox of the chicken and the egg" (Joyce and Orgel, 1993). I soon had a personal encounter with this recurrent tendency to "press" familiar protein function on RNA when I joined Fritz Lipmann's laboratory to study a possible biochemical role for ψ in protein biosynthesis (see Lane et al., 1995). Unexpectedly, but not unpleasantly, when I arrived in the Lipmann laboratory I was asked to work on what was then thought to be the RNA whose full sequence would most likely be the first one to be established.

Believed to be an intermediate in glutathione synthesis, tRNA$^{Cys-\gamma Glu}$ was said to be easily purified in a large quantity. This was 5 years before Bob Holley's laboratory reported a structure for tRNA^{Ala}, when the specimens I was then shown were still blackened by the charcoal being used in their purification. By contrast, large white flakes of pure tRNA$^{Cys-\gamma Glu}$ fluttered in suspension during what was said to be a copious formation of oxidation-induced disulfide dimer, the result of linking terminally attached Cys residues in two molecules of tRNA$^{Cys-\gamma Glu}$. Alas, for reasons described by others (Bennett, 1988; Edsall, 1995), tRNA$^{Cys-\gamma Glu}$ was not to be, and the biogenesis of glutathione reverted to its status as an RNA-independent reaction that is still catalyzed by two protein enzymes (Lane and Lipmann, 1961).

During the long hiatus that had to be devoted to negotiating a decent burial for tRNA$^{Cys-\gamma Glu}$, I "moonlighted" the preparation of massive amounts of *Escherichia coli* rRNA, which were then being discarded during Geoffrey Zubay's preparation of *E. coli* tRNA^{Leu} in the Lipmann laboratory. To establish or dismiss a general role for ψ in the biochemistry of rRNA, gram quantities of rRNA would be needed to detect any trace amounts of ψ in the bacterial rRNA, something that soon became less likely after David Dunn, an impeccable analyst, reported that he could not find ψ in *E. coli* rRNA (Dunn et al., 1963). Using the gram quantities of rRNA then needed for controlled experimentation, which were likely unavailable to David Dunn and most others outside the Lipmann laboratory, we detected ψ in *E. coli* rRNA (Nichols and Lane, 1967) soon after my return to Canada, in 1961, to assume my first academic position in John Colter's department in Edmonton.

As suspected, the amount of ψ was very much (~10-fold) smaller in bacterial rRNA than in wheat rRNA. However, until the powerful cloning technologies developed in the 1970s (e.g., Messing, 1988) were used to obtain unmodified transcripts of *E. coli* high-molecular-weight rRNA genes (Kryzosiak et al., 1987; Weitzmann et al., 1990), there was little scope for studying a relation between rRNA-ψ and peptidyl transfer. Realization of this pursuit became possible when another Lipmann alumnus, Jim Ofengand, agreed to pool resources for such an initiative in 1992.

ψ Is Localized to the Peptidyl Transfer Center (PTC) in LSU rRNA

Selective clustering of modified nucleosides (e.g., UmGmψ) (Lane, 1965) suggested that rRNAs, like tRNAs and proteins, might contain "executive sites" (e.g., acceptor termini in tRNA) that comprise only a small part of the rRNA molecules. For instance, the idiosyncratic N1 position in ψ, as in uracil (Spector and Keller, 1958), might serve as a site of peptidyl transfer, and/or the tandem of peptide bonds in ψ (see Lane et al., 1995), if present in the PTC of the ribosome, might engage in β sheet type intermolecular interactions with the peptide backbone during polypeptide formation, and thereby help direct exit of the nascent polypeptide (or mRNA) from the ribosome. With such limitless possibilities in mind, and not immune to the attractions of "pressing" familiar protein functions on RNA, we (Lane et al., 1992) decided to determine if—as seemed possible based on data then available for *E. coli* (Branlant et al., 1981), mitochondrial (Klootwijk et al., 1975; Dubin and Taylor, 1978), and eukaryotic (Veldman et al., 1981; Maden, 1988) LSU RNAs—most or all ψ residues in LSU RNA are concentrated in the PTC of the ribosome.

The speculation proved to be productive. There are 9 ψs in *E. coli* LSU RNA, all in or near the PTC (Bakin and Ofengand, 1993), and surprisingly, all 30 ψs in yeast LSU RNA (Bakin et al., 1994) and all 55 ψs in human LSU RNA are also located in or near the PTC (Ofengand et al., 1995; Ofengand and Bakin, 1997). The single ψ in each of yeast (Klootwijk et al., 1975) and mammalian (Dubin and Taylor, 1978) mitochondrial LSU RNA is at a position homologous with one of the 9 ψs in the PTC of *E. coli* LSU RNA (ψ2580, *E. coli* numbering) (Bakin et al., 1994). In the long term, Jim Ofengand's laboratory will study the effects of disrupting, individually and in combination, genes for enzymes (e.g., Wrzesinski et al., 1995) that can generate all 9 of the ψ residues in the PTC of *E. coli* LSU RNA (see Chapter 12 by Ofengand and Fournier).

Coclustering of Different Types of Modification in the PTC of All Ribosomes

More than 30 years ago, it was shown that the levels of ψ and Nm in eubacterial rRNAs (Nichols and Lane, 1966a, 1966b) are vastly lower than in eukaryotic rRNAs (Singh and Lane, 1964b). From this, it was plain that ribosome function, in vivo, is not dependent on the relatively high levels of rRNA nucleoside modifications that are found in eukaryotes. Even so, it is now known that ψ and Nm often cocluster in the functional centers of all types of ribosome. In *E. coli* SSU RNA, the single ψ (Nichols and Lane, 1967) at position 516 (Ofengand et al., 1995) and the single Nm (m^{4}Cm) (Nichols and Lane, 1966a) at position 1402 (Fellner, 1969; Brosius et al., 1978) are in the decoding site of the ribosome (O'Connor et al., 1997). In *E. coli* LSU RNA, all 9 ψ residues (Bakin and Ofengand, 1993) and all 3 Nm residues in the GmG, CmC, and UmG sequences (Nichols and Lane, 1966a) at positions 2251, 2498, and 2552 (Branlant et al., 1981), respectively, are in the PTC, as are most of the 30–50 ψ (Bakin et al., 1994) and 40–60 Nm (Veldman et al., 1981; Maden, 1988) residues in eukaryotic LSU RNA (Ofengand et al., 1995). Firm conclusions are not possible, but available evidence indicates there has not, as in high-molecular-weight rRNA, been a large increase in levels of modification in ribosomal proteins of eukaryotes, relative to prokaryotes (Alix, 1988; Vladimirov et al., 1996).

The realization that ψ, mN, and Nm are selectively located in the PTC of prokaryotic and eukaryotic ribosomes constitutes significant progress (Maden, 1990; Lane et al., 1992; Smith et al., 1992; Brimacombe et al., 1993; Bakin et al., 1994), but biochemical roles for these rRNA nucleoside modifications, known as "mods" in the jargon, remain entirely speculative (see Lane et al., 1995).

Biochemical roles are still unknown for the single ψ in the 1,559-residue LSU RNA of the human mitochondrial ribosome and for 55 ψ residues in the 5,025-residue LSU RNA of the human cytoplasmic ribosome. Contributions of ψ to the assembly, transport, catalysis, fidelity, stability, and durability of the ribosome (see Lane et al., 1995) will likely emerge in current gene-defect studies with prokaryotes (see above) and eukaryotes (see below). Reasons were once given (Lane, 1988) for thinking that modified nucleosides generally exert modulator and not effector influences on the activity of RNA molecules—they modulate (attenuate/amplify) pre-existing structure/function but are not sine qua non elements of structure/function (Kwong and Lane, 1975). In keeping with this, only a minority of tRNA transcripts (see Bjork, 1995), when they are totally devoid of modifications, are found to have less than the full aminoacyl acceptor activity of their naturally modified counterparts (see Samuelsson et al., 1988; Perret et al., 1990).

BASE-METHYLATED (mN) AND SUGAR-METHYLATED (Nm) NUCLEOSIDES

Shortly after the reported existence of a "fifth nucleoside" (ψ) in RNA, Gutman's laboratory extended its own and earlier (see Kruger and Salomon, 1898) studies of methylated bases in urine to commercially accessible RNA and found several methylated bases (Adler et al., 1958). In the same year, David Dunn, John Smith, and their coworkers submitted the first in a highly influential series of papers on the isolation of a host of base-methylated (mN) and sugar-methylated (Nm) nucleosides that, cumulatively, became the second most abundant class of modified nucleosides (Littlefield and Dunn, 1958; Smith and Dunn, 1959a, 1959b). Dunn (1959) found that, like ψ, mN is present in greater quantity in eukaryotic (liver) tRNA than rRNA, although Nm is present in greater quantity in eukaryotic (wheat) rRNA (Lane, 1965) than tRNA (Hudson et al., 1965). Methyl groups are bonded to endocyclic carbon (e.g., 2-methyladenosine), endocyclic nitrogen (e.g., 1-methyladenosine), and exocyclic nitrogen (e.g., N^6-methyladenosine) in the four classical bases, but only to the exocyclic $O^{2'}$-oxygen (e.g., 2'-O-methyladenosine) in the case of ribose. Methyl substitution usually imparts increased hydrophobicity, but may impart increased polarity, as in positively charged 1-methyladenosine (at physiological pH) and 7-methylguanosine (at all pH values). For this discussion, emphasis is given to three of the base- and sugar-methylated nucleosides with which I have had some working experience: N^2-dimethylguanosine (m^2_2G), a base-methylated nucleoside found mainly in eukaryotic tRNA; N^6-dimethyladenosine (m^6_2A), a base-methylated nucleoside found mainly in prokaryotic and eukaryotic SSU RNA; and $N^4,2'$-O-dimethylcytidine (m^4Cm), a base- and sugar-methylated nucleoside found only in bacterial SSU RNA.

Nucleoside Modifications (mN, Nm, and ψ) Arise Posttranscriptionally

In Ross Hall's monograph (1971), progress could already be reported about the biogenesis of base- and sugar-methylated nucleosides. Using RC^{Rel} mutants of *E. coli*, which accumulate "undermodified" RNA during amino acid starvation, Borek's laboratory showed that AdoMet-dependent enzymes can mediate methylation, in vitro, of specific bases (Mandel and Borek, 1963; Fleissner and Borek, 1963) in macromolecular precursors of RNA, i.e., posttranscriptionally after polynucleotide synthesis. Building on this, our studies of RNA from RC^{Rel} mutants showed there is much rarer, but equally specific, AdoMet-dependent enzymatic methylation, in vitro, of specific sugars in polynucleotide precursors of tRNA and rRNA (Nichols and Lane, 1968a, 1968b, 1969). It was soon shown that ψ also arises by posttranscriptional RNA modification (Johnson and Soll, 1970). Just as modifications of the 20 canonical amino acids in proteins occur posttranslationally, modifications of the 4 canonical nucleosides in RNA occur posttranscriptionally.

RNA Architecture as a Determinant of tRNA Modifications

When it was found that bulk (Lane and Butler, 1959a), ribosomal (Lane and Allen, 1961; Lane, 1965), and transfer (Lane and Allen, 1961; Hudson et al., 1965) RNAs have 5'-monophosphate termini, not 5'-triphosphate termini, it was clear that most "mature" cellular RNAs arise by processing, perhaps endonucleolysis of polynucleotide precursors (see Lane, 1965). Tertiary structure is now known to be needed (see Lee et al., 1997) for the formation of 5'-monophosphate termini by RNase P-catalyzed endonucleolysis of precursor tRNAs (Robertson et al., 1972). The importance of tertiary structure for precise tRNA modifications (aminoacylations) was cogently articulated at some length by Loftfield as early as 1972, and we pursued the same notion in the context of tRNA methylations (Streeter and Lane, 1970; Kwong and Lane, 1975; Kennedy et al., 1977). Based on a study of the methylation, by wheat AdoMet-tRNA methyltransferases, of some selected *E. coli* tRNAs and yeast tRNA^{Asp}, all "naturally" devoid of m$^2_2$G, it was concluded that secondary and tertiary structures play central roles in generating m$^2_2$G at the only site (position 26) where m$^2_2$G occurs naturally in eukaryotic tRNA—between the stems of the "dihydrouridine" and "anticodon" loops.

Nishikura and de Robertis (1981) found that some RNA nucleoside modifications (e.g., m^{5}U) take place in the "large" tRNA precursors (>100 nucleotides) whereas others (e.g., m$^2_2$G) occur only after tRNA has been processed to its "mature" size (~80 nucleotides). These pioneering observations have recently assumed a new dimension (see Grosjean et al., 1996). For the recent studies, yeast [^{32}P]tRNA^{Asp} transcripts (devoid of the modifications which, typical of tRNAs, are present in the central "half" of naturally modified tRNA^{Asp}) were injected, with or without site-specific mutagenesis, into the cytosol of *Xenopus* oocytes. Without mutagenesis, modifications are introduced (0.2–1 per mole of tRNA) at all 8 sites that are naturally modified in yeast tRNA^{Asp}, at 2 additional sites that are modified in *Xenopus* tRNA^{Asp} (74% homologous with yeast tRNA^{Asp}), and at 3 more sites that are not modified in either yeast or *Xenopus* tRNA^{Asp}. Because fragmented tRNA can be modified at sites that are not naturally modified (Kuchino et al., 1971), it was important that 60–80% of the yeast [^{32}P]tRNA^{Asp} was stable for as long as 63 h postinjection and that only the intact [^{32}P]tRNA^{Asp} molecules were analyzed.

Consonant with the unique importance of the U8-A14 base pair in tRNA, which is predicted to be the solitary tertiary interaction essential for sustaining the (L form) tertiary structure of tRNA^{Phe}, the pleiotropic effects of mutagenically disrupting the U8-A14 base pair distinguish between two discrete groups of nucleoside modifications. The formation of one group, including m^{5}C49, m^{5}U54, ψ55 and m^{1}A58 in the amino acid acceptor domain of the L form, is independent of the disruption of U8-A14, whereas formation of the second group, mostly on the "inner" side of the anticodon branch of the L form, including ψ13, m^{2}G26, manQ34, m^{1}G37 and ψ40, depends on the maintenance of U8-A14 base pairing. It was therefore proposed that group 1 modifying enzymes "chaperone" (promote and/or stabilize) the folding of tRNA precursors to give a tertiary structure that is only then recognized by group 2 modifying enzymes (Grosjean et al., 1996). This formulation of how an idiosyncratically modified three-dimensional structure for tRNA^{Asp} is generated in vivo is fundamental to the broader issue of how a tRNA adaptor is uniquely recognized by its cognate aminoacyl-tRNA synthase (Hagervall et al., 1990; Perret et al., 1990). The multistep maturation of pre-tRNA to mature tRNA, with particular emphasis on the intron-dependent formation of some modified nucleosides, has been recently reviewed (Grosjean et al., 1997).

Modifications Sometimes Occur at or Proximal to 3' Termini in tRNA and rRNA

When 2'- and 3'-O-aminoacyladenosines were prepared from the aminoacylated termini of tRNAs, they were among the first modified nucleosides to be identified as RNA constituents (Zachau et al., 1958). Aminoacyladenosines, corresponding to the 20 (unmodified) amino acids found in proteins, have never previously been classified as modified nucleosides. As recently recounted (Wood et al., 1995), my laboratory once treated 3' termini in rRNA as potential group-transfer sites in the ribosome, a possibility that gradually faded as aminoacyladenosines were not detected in rRNA. However, excepting a rare ψ that is antepenultimate to the 3' terminus in *Crithidia fasciculata* SSU RNA (Schnare and Gray, 1981), a "quasi-universal" sequence (m$^6_2$Am$^6_2$A), first found in *E. coli* rRNA (Nichols and Lane, 1966b), is closer to the 3' terminus than any other nucleoside modification.

Alkali stability is used to quasi-identify sugar-methylation sites (as gaps) in sequencing ladders, but it is less well known that base methylation can strongly affect the stability of internucleoside 3'-5'-phosphodiester bridges in RNA. The strongest stabilization is observed in tetramethylated ApA (Lane and Butler, 1959b); small amounts of m$^6_2$Am$^6_2$A were recovered even after *E. coli* rRNA was incubated in 1 M NaOH for 100h (Nichols and Lane, 1966b). The extreme hydrophobicity responsible for such remarkable stability has been the subject of comprehensive study (Tazawa et al., 1980). The $m^6_2Am^6_2A$ sequence surmounts a "hairpin" and occurs in *E. coli* SSU RNA at positions 1518 and 1519, 23 residues removed from the 3' terminus (A1542) (Ehresmann et al., 1971; Brosius et al., 1978), and in human SSU RNA at positions 1850 and 1851, 18 nucleotides from the 3' terminus (A1869) (see Bachellerie et al., 1995b). Interestingly, a form of SSU RNA that is known to be peculiar to malaria pathogenesis has striking deletions in the decoding region and does not contain $m^6_2Am^6_2A$ at the customary "hairpin" site (Li et al., 1997).

A very rare nucleoside, the first one found to have methyl substituents in the base and sugar of the same nucleoside (m^4Cm), was shown to be the sole site of sugar methylation in *E. coli* SSU RNA (Nichols and Lane, 1966a), and was later shown to be proximal to the 3' terminus in *E. coli* SSU RNA, at position 1402 (Fellner, 1969; Brosius et al., 1978). When Zimmermann's laboratory (Prince et al., 1982) showed that the anticodon in P site-bound tRNA can be cross-linked to C1400 in *E. coli* SSU RNA, it was an early demonstration of the proximity of a modified nucleoside to an executive site in SSU RNA. The $m^4Cm1402$ residue in *E. coli* SSU RNA is also immediately adjacent to G1401, which forms a functionally important tertiary base pair (to C1501) between two highly conserved, single-strand regions in SSU RNA (Cunningham et al., 1992, 1993).

Modifications Sometimes Occur at 5' Termini in tRNA, snRNA, and mRNA

Shortly after pGp was shown to be the dominant nucleoside diphosphate in hydrolysates of bulk cellular RNA (Lane and Butler, 1959a), the dominance was found to be because of a preponderant presence of guanosine at the 5' termini in the most copious class of low-molecular-weight cellular RNAs—tRNA (Singer and Cantoni, 1960; Lane and Allen, 1961). More comprehensive study soon showed that much smaller amounts of pAp, pCp, and pUp also derive from 5' termini of tRNAs (Hudson et al., 1965; Gray and Lane, 1967). Because the chain termini in tRNA comprise a relatively "high" proportion of the total nucleosides (~1.3%), it is not surprising that the first 5'-terminal modified nucleoside found in an RNA molecule was found in the course of studies of bulk yeast tRNA. RajBhandary et al. (1964) reported that a modified nucleoside could be detected in a 5'-terminal position in yeast tRNA, but its identity could not be established. Some years later, we showed (Gray and Lane, 1967) that it was ψ, which is released as pψp when bulk yeast tRNA is hydrolyzed in alkali; subsequently, ψ was found at the 5' terminus of yeast tRNA^{Lys} (Madison and Boguslawski, 1976). Similarly comprehensive study failed to detect modified nucleosides at 5' or 3' termini in ribosomal RNAs, in which the chain termini account for <0.1% of the total constituent nucleosides (Lane, 1965; Oakden and Lane, 1976).

Again, because their chain termini comprise a relatively "high" proportion (~0.5%) of their total nucleosides, the so-called small nuclear RNA (snRNA) molecules were the second class of cellular RNA molecules found to have a 5'-terminal modified nucleoside, in this case $N^2,N^2,7$-trimethylguanosine ($m^{2,2,7}G$), which was previously shown to be a constituent of snRNA (Saponara and Enger, 1969). Remarkably, in the only exception to the standard 3'-5'-phosphodiester bridge between adjacent nucleosides in cellular RNA molecules, this modified-nucleoside terminus was shown to be bonded to the penultimate nucleoside via the 5'-5'-triphosphate bridge $m^{2,2,7}GpppN$ (Shibata et al., 1975). This same seminal study also showed that some types of mammalian snRNA (e.g., U2 snRNA) are among the most densely modified RNA molecules in nature—19 of 48 residues in the 5'-terminal region of liver U2 snRNA are modified.

After a method was introduced for separating poly(A)-containing mRNA from bulk RNA, Perry's laboratory detected methylation in L-cell mRNA (Perry and Kelley, 1974). Part of the methylation in eukaryotic (not prokaryotic) mRNA was at internal sites (m^6A), and another part was in terminal "cap" structures of the kind previously found in snRNA (Perry et al., 1975) (see Chapter 10 by Bokar and Rottman). Linkage of the terminal modified nucleoside to the penultimate nucleoside is through the 5'-5'-triphosphate bridge m^7GpppN, N often being a 2'-O-methylnucleoside in snRNA and mRNA of animal (Shibata et al., 1975; Perry et al., 1975) but not plant (Haffner et al., 1978; Kennedy and Lane, 1979) cells. In mRNA, the "cap" plays a part in initiating protein synthesis by mediating mRNA binding to the ribosome (Rhoads, 1991). In mRNA and snRNA, it prevents 5'-exonucleolysis in the cell nucleus (Nevins, 1983; Cavaille and Bachellerie, 1996). Striking turnover of RNA in the nucleus of mammalian cells was discovered in the 1950s when Harris' laboratory found that most "rapidly labeled" (newly synthesized) RNA in cultured animal cells, cryptically, was degraded and never exited the nucleus (see Harris, 1974).

TARGETING SITES OF SCISSION AND MODIFICATION BY COMPLEXING BETWEEN "ADAPTOR" RNAs AND PRECURSOR RNAs IN EUKARYOTES: snRNA AND SMALL NUCLEOLAR RNA (snoRNA)

The precept that the "biosynthetic units" in NaCl-insoluble RNA, even after appreciable cell-free degradation (Lane, 1962; Lane et al., 1963; Diemer et al., 1966) or biodegradation (Lane, 1965), are measureable as nonphosphorylated, 3′-hydroxyl terminal nucleosides, was again serviceable for analyzing "rapidly labeled" (newly synthesized) polynucleotides. The mean chain length of "rapidly labeled" NaCl-insoluble RNA from L cells was found to be very much shorter (~10-fold) (Tamaoki and Lane, 1967) than that of its principal ("45S") mass component, as measured by sedimentation equilibrium (McConkey and Hopkins, 1969). An unexpectedly low degree of (mean) polymerization and an inexplicably rapid turnover of a large part of the "rapidly labeled" RNA in the nucleus of mammalian cells in culture (see Harris, 1974) were harbingers of unimagined degrees of "small" RNA-mediated, nonconservative processing and modification of "large" mRNA and rRNA precursors in eukaryotes. Not only does a large part (~40%) of the "rapidly labeled" 45S RNA (Scherrer and Darnell, 1962) in the cell nucleus (Tamaoki and Mueller, 1962), which is composed of rRNA precursors, lose ~50% of its total sequence during conversion to mature cytoplasmic rRNA (Weinberg et al., 1967; Vaughan et al., 1967), but a larger part (~60%) (Brandhorst and McConkey, 1974) of the rapidly labeled 45S RNA, composed of the heterogeneous nuclear RNA (hnRNA), often loses >75% of its total sequence, in the form of introns, as it is spliced to yield cytoplasmic mRNA (Breathnach et al., 1977; Dugaiczyk et al. 1978; Roop et al., 1978; Tilghman et al., 1978).

"Rapidly labeled" NaCl-insoluble RNA from exponentially growing yeast, like that from L cells, was also found to contain many cryptic supernumerary chain termini (Oakden and Lane, 1973). The same study showed that without aqueous denaturation (Oakden et al., 1972) of yeast "rapidly labeled" RNA, a mass of cryptic RNAs cosedimented with yeast "35S" pre-rRNA/pre-mRNA in sucrose-density gradients; however, after aqueous denaturation, the cryptic RNAs sedimented behind 26S rRNA and well ahead of pre-tRNAs (Rosbash and Penman, 1972; Kennedy and Lane, 1984). Like the cryptic RNAs in yeast pre-rRNA/pre-mRNA, snRNAs (and mRNAs) are NaCl-insoluble (Haffner et al., 1978), are rapidly labeled in exponentially growing cells (see Busch et al., 1982), and without denaturation cosediment with pre-rRNAs in sucrose-density gradients (Tollervey, 1987).

Complexing of snRNAs with Pre-mRNAs and Complexing of snoRNAs with Pre-rRNAs

The nature of snRNA involvement in scission/splicing of eukaryotic pre-mRNA was a subject of productive conjecture (Lerner et al., 1980; Rogers and Wall, 1980). Several snRNP particles, each composed of proteins and one or two snRNAs (U1, U2, U4/6, U5), form a spliceosome that cuts and splices pre-mRNA (Moore et al., 1993). Sites that are operative in targeting and modulating the splicing activity of snRNAs are present in 5′ regions that are often heavily modified (mostly ψ and Nm as in eukaryotic rRNA) (see Chapter 11 by Massenet et al.).

More recently, complexing between pre-rRNA and snoRNAs (reviewed by Maxwell and Fournier, 1995) has been implicated in the processing and modification of eukaryotic pre-rRNA (see Tollervey, 1987, and Tollervey et al., 1993). Bachellerie, Fournier, and their coworkers suggested (Bachellerie et al., 1995a) that members of a large subfamily of 50–100 fibrillarin-associated snoRNAs, each containing a specific tract of 12–21 nucleotides that is complementary to a region in rRNA, act as "guide sequences" in directing most of the sugar methylations in eukaryotic rRNA (see Chapter 13 by Bachellerie and Cavaillé). This was later proved for vertebrate and yeast cells. Even more recently, a separate class of snoRNAs, associated with the Gar1 protein rather than fibrillarin, has been found to direct, by site-specific annealing to short stretches of flanking nucleotides on the 3′ and 5′ sides of solitary uridines in pre-rRNA, most of the 50–100 pseudouridylations in eukaryotic rRNAs (see Chapter 12 by Ofengand and Fournier). Many snoRNAs are "capped" (see Chapter 11 by Massenet et al.), but snoRNAs involved with site-directed 2′-O-methylations have 5′-phosphate (and 3′-hydroxyl) termini (Cavaille and Bachellerie, 1996), and thus, like "capped" snRNAs and mRNAs, may have contributed to the 3′-hydroxyl termini; however, unlike "capped" RNAs, they could not have contributed to the slow emergence of steady-state levels of 5′-phosphate termini in "rapidly labeled" RNA (see Tamaoki and Lane, 1967).

HYPERMODIFIED NUCLEOSIDES

Between the mid-1950s and mid-1960s, about 25 different modified nucleosides were identified as constituents of bulk tRNA and/or rRNA specimens.

These so-called "simple" modified nucleosides (ψ, mN, and Nm) were identified by using idiosyncratic mobilities in maximum-resolution anion-exchange (Cohn and Volkin, 1953) and filter-paper (Lane, 1963) chromatographic separations as a supplement to UV spectra for bases, and specific color reactions for sugars, while relying on CD spectra of the parent nucleosides to establish anomeric configurations (e.g., Nichols and Lane, 1968c). Analyses of these simple modifications were especially useful for "fingerprinting" bulk tRNA (Gray and Lane, 1967) and bulk rRNA (Singh and Lane, 1964b; Lane, 1965; Lane and Tamaoki, 1969; Lau et al., 1974; Gray, 1974) from different organisms.

Many of these "simple" modifications had been localized to specific sites when the inaugural structure of a natural RNA molecule was reported—tRNA^{Ala} (Holley et al., 1965). Holley's laboratory soon found that sites of modification in tRNA^{Phe} from a multicellular organism (wheat) (Dudock et al., 1969) were widely conserved relative to those previously reported for the tRNA^{Phe} from a unicellular organism (yeast) (RajBhandary et al., 1967). One modified nucleoside (Y) was found (in variant forms) next to the 3' ends of anticodons in both tRNA^{Phe} molecules, but it was unidentified. Developments between 1965 and 1968 laid the groundwork for identifying Y and a panoply of other "hypermodified" nucleosides so that, by 1994, 93 modified nucleosides had been identified as constituents of natural RNAs (see Appendix 1 by Motorin and Grosjean).

Most Hypermodified Nucleosides Are Found in tRNA

In a global survey (Limbach et al., 1994), 79 of a total of 93 modified nucleosides in cellular RNA were found in tRNA (cf. 28 in rRNA, 12 in mRNA and 11 in snRNA). Remarkably, 59 of them are peculiar to tRNA, whereas only 7 are peculiar to rRNA, 5 to mRNA, and 2 to snRNA. About 80% of the 59 modified nucleosides that are peculiar to tRNA (Limbach et al. 1994) are what Hall called "hypermodified" nucleosides, and most of these are located either in the "wobble" position of the anticodon or in the position immediately adjacent to the 3' end of the anticodon. By hypermodification, Ross Hall did not refer to the enhanced modification of simple modified nucleosides, as when m^{6}A is further modified to m$^6_2$A or Cm is further modified to m^{4}Cm. Rather, he noted (Hall, 1971) that "most of the modified nucleosides detected in nucleic acids are relatively simple modifications of the various major nucleosides" whereas "some of the modified nucleosides recently detected in tRNA possess a more elaborate structure resulting from the attachment of a complex side chain. Not only does the presence of the larger side chain create bulk but also such side chains contain functional groups (organic chemical definition)." Hall (1971) concluded that "six such nucleosides have now been detected in tRNA": isopentenylated and carbothreonylated derivatives of adenosine, and carboxymethylated derivatives of uridine.

Carboxylate Salts, Esters, and Amides Are Commonly Found in and near Anticodons

The first hypermodified nucleoside to be reported and later correctly identified was described in a report from my laboratory in 1965. Our research had been geared to detecting, in gram quantities of undegraded RNA, modified sequences that occur in only unimolecular quantity in wheat LSU RNA (e.g., UmGmψ, now known to be the most densely modified sequence in the PTC of LSU RNA). Because tRNA is ~50-fold smaller than LSU RNA, the same procedures could uncover a unimolecular amount of a modified nucleoside in gram quantities of bulk tRNA, even if the "mod" were present in only one or a few (e.g., 1 in 50) of the isoacceptor species in bulk tRNA.

Found as an unidentified nucleotide (~1.5 mg) in the dinucleotide fraction of alkali hydrolysates of gram quantities of wheat tRNA (Hudson et al., 1965), and called T? because of its UV spectral resemblance to thymidine, the corresponding nucleoside was later shown to be an anionic salt at neutral pH—uridine 5-acetic acid (cm^{5}U) (Gray and Lane, 1967). We suggested that a neutral counterpart of cm^{5}U in enzymatic hydrolysates of tRNA was an ester (Gray and Lane, 1968). The neutral methylester (mcm^{5}U) was found in enzymatic hydrolysates of yeast but not wheat tRNA (Tumaitis and Lane, 1970). David Dunn's laboratory later showed that the neutral nucleoside in enzymic hydrolysates of wheat tRNA is an amide (ncm^{5}U) (Dunn and Trigg, 1975). It was shown that mcm^{5}U occupies the "wobble" position in yeast tRNA$^{Arg}_3$ (Kuntzel et al., 1975), and as with bulk yeast tRNA (Bronskill et al., 1972), selective ^{14}C-methyl esterification of mcm^{5}U can be achieved by incubating saponified tRNA$^{Arg}_3$ in Ado[^{14}C-methyl]Met-supplemented yeast extract.

High-resolution mass spectrometry was used to identify "natural" i^{6}A (Biemann et al., 1966) in the course of determining the structure of tRNA^{Ser} in Zachau's laboratory, and was again used by Biemann, in collaboration with Baczynskyj and Ross Hall, to characterize s^{2}mcm^{5}U (Baczynskyj et al., 1968). Combined with liquid chromatography, mass spectrometry initiated a new era of increased sensitivity in nucleoside analysis, and these methods are now standard for characterizing nanogram amounts of modified nucleosides in microgram amounts of RNA (see Chapter 3 by Crain).

The 5-substituted, hypermodified uridines, beginning with cm⁵U and s²mcm⁵U, comprise one of the largest families of modified nucleosides, accounting for 17 of the 79 modified nucleosides in tRNA. All occupy the "wobble" position. Including the 12 carboxyl derivatives among these 5-substituted uridines, an even larger family of 26 hypermodified nucleosides, accounting for nearly one-third of the modified nucleosides in tRNA, contains carboxylate, carboxyl ester, or carboxyl amide groups. This family constitutes ~50% of those modified nucleosides peculiar to tRNA, and each member causes a carboxylate salt, ester, or amide group to be near or part of an anticodon.

A productive interface between RNA nucleoside hypermodification and the health sciences is discussed in depth in Chapter 28 by Marquet. In this instance, host-cell tRNA$^{Lys}_3$ is known to be needed to prime reverse transcription of human immunodeficiency virus (HIV) RNA and, surprisingly, the hypermodified "wobble" nucleoside (s²mcm⁵U) in tRNA$^{Lys}_3$, which is not part of the primer sequence, is responsible for a "spectacular" enhancement of transcription initiation.

GENERAL PERSPECTIVES AND PROSPECTS

All known cells contain tens to hundreds of genes that encode RNA-modifying enzymes, many of which have a requirement for S-adenosylmethionine (AdoMet), the synthesis of one mole of which requires about 12 moles of ATP (see Bakin et al., 1994). For some, such impressive energy investments have left no doubt about the biochemical importance of RNA nucleoside modifications (e.g., Borek, 1980). However, as a plant biologist (Lane, 1988, 1991, 1994), I have to be more-than-usually impressed by the energy that biological systems sometimes expend in the manufacture of what, in an allusion to the wonderfully complex world of plant alkaloids, was once called "flotsam thrown up on a [metabolic] beach" (Tschirch, 1923). In a further reference to alkaloid-bearing plants, which grow equally well if depleted of their alkaloids, it was observed that putative protective strategies are often illusory and it was concluded that, in the highly anabolic synthesis of plant alkaloids, "we are witnesses to the wealth of possibility in a half-way stage of metabolic evolution" (James, 1950).

Certainly, in the case of what is by far the most abundant RNA nucleoside modification, thermodynamic arguments about demands on cellular energy lose some force. Biosynthesis of ψ has no direct energy requirement: U-to-ψ conversion in RNA is spontaneous (Nurse et al., 1995). When the precepts of Gibbs' chemical thermodynamics are applied to biochemical systems (Lane, 1979b), an overriding concern must surely be to assess factor-dependent change in the potential of a system (e.g., a cell with and without ψ). Like thermodynamicists, biochemists and geneticists have preferred, when possible, to formulate their convictions on the basis of biochemical differences (e.g., Rose's biochemical definition of the dietary amino acids needed for normal growth of young animals) and genetic differences (e.g., the growth of bacterial auxotrophs).

Accordingly, I shall briefly sketch contexts in which "differential strategies" have long been and continue to be active and attractive (see Chapter 25 by Winkler for in-depth coverage). It merits emphasis that many of the gene-defect experiments described below do not address the "long-term" implications (relative to RNA half-life) that some mutations may have for viability and stability of function in nondividing cells (e.g., quiescent embryos in higher plants), or for the superiority of wild types in mixed-population (competition) experiments (Björk and Neidhardt, 1975).

Gene Mutations That Affect tRNA Modifications

N^2-Dimethylguanosine (m²₂G) is absent from *E. coli* tRNAs (Nichols and Lane, 1968b) and its customary presence in about 50% of eukaryotic (yeast and wheat) tRNAs (Gray and Lane, 1967) is not needed for normal growth. A mutant yeast can grow normally even though its tRNA is totally devoid of m²₂G (Phillips and Kjellin-Straby, 1967). As assumed in studies of factors that affect the cell-free synthesis of m²₂G (Kwong and Lane, 1975; Edqvist et al., 1995), a single enzyme is likely responsible for mono- and dimethylation at G26 since the transfection of *E. coli* with a single wild-type yeast gene that encodes the tRNA(m²₂G)methyltransferase leads to the appearance of m²₂G in the tRNAs of the transformed *E. coli* cells (Ellis et al., 1986).

Just as the absence of m²₂G from the tRNA of a mutant yeast scarcely affects the growth rate of a eukaryote that normally contains m²₂G in about one-half of its tRNAs (Phillips and Kjellin-Straby, 1967), complete absence of m⁵U from the tRNA of an *E. coli* mutant (Björk and Isaksson, 1970) scarcely affects the growth rate (Björk and Neidhardt, 1975) of a prokaryote that ordinarily contains m⁵U in all of its tRNAs. Björk's laboratory has since shown, paradoxically, that total absence of m^5U54 from *E. coli* tRNA scarcely affects growth rate, but disruption of the gene for tRNA(m^5U54)methyltransferase is lethal (Persson et al., 1992). This stresses, in an unexpected way, why it can be wrong to conclude that because a cell grows at a near-normal rate in the absence of a specific nucleoside modification in its tRNA, disrupting the gene encoding an enzyme responsible for the modification is harmless.

More generally, it would also be wrong to dismiss as unimportant the gene for a nucleoside-modifying enzyme, the disruption of which does not lead to diminished cell growth. For example, neoplasia in higher organisms is believed to be the consequence of a small number of events, each of which is individually tolerable, but which in concert can lead to malignancy. Recently, a model for multifactorial lethality (in yeast) has fused the issue of increased levels of RNA modification in eukaryotes with nuclear-pore proteins (Simos et al., 1996). Mutations in three genes, one encoding a nuclear-pore protein (a nucleoporin) (Nsp1p), another encoding a nucleoporin that affects pre-tRNA splicing (Los1p), and a third that encodes a pseudouridine ($\psi 27,28,34,35,36$) tRNA synthase (Grosjean et al., 1997) (Pus1p), combine to produce a lethal phenotype, i.e., the combination of these three proteins is required to sustain the biogenesis of functional tRNA.

Information about organisms which contain mutations that affect tRNA modifications has been catalogued (Björk, 1995). In general, such mutations are not lethal individually, although the slowed-growth effect that results from mutation of the *Salmonella typhimurium hisT* gene (which encodes a $\psi 38,39,40$ synthase) has been described as "catastrophic" (Tsui et al., 1991). Briefs for the importance of tRNA modifications have sometimes become "overheated" (see Borek, 1980) but the evidence in favor of the evolutionary impact of nucleoside modifications is difficult to gainsay. As Björk (1995) has noted, the approximate proportion of the *E. coli* genome directed to the manufacture of 79 tRNAs (~0.25%) is only about 25% as great as that devoted to manufacture of (at least) 45 tRNA-modifying enzymes. Even in the smallest organisms (intracellular parasites), the mycoplasmas, the genomes of which are ~25% as large as the *E. coli* genome, there are 13 modified nucleosides in tRNA, including hypermodified t^6A, and most were likely present in the tRNA of the progenote of prokaryotes and eukaryotes (Björk, 1995).

Biochemical and Genetic Changes That Affect rRNA Modification

Darnell's laboratory (Vaughan et al., 1967) showed that methionine deprivation blocks the growth of cultured HeLa cells. Such cells do not synthesize complete ribosomes and they process pre-rRNA atypically. There is ~20% of normal sugar methylation in 45S pre-rRNA, which gives rise to SSU, LSU, and 5.8S rRNAs, and ~65% of normal sugar methylation in 32S pre-rRNA, which yields LSU and 5.8S RNAs. The study also showed that during methionine deprivation, methylation of 45S pre-rRNA does not have to be completed before its cleavage to 32S pre-rRNA; the precursor to SSU RNA, made during cleavage of 45S pre-rRNA to 32S pre-rRNA, is lost; and processing of 32S pre-rRNA to LSU RNA of the large ribosomal subunit is blocked.

Our related investigation showed that if L cells are cultured in a puromycin-containing medium, a striking concomitant of the resulting decrease in sugar methylation is a pronounced increase in the proportion of pre-rRNA in "rapidly labeled" (newly synthesized) RNA. Retarded maturation of pre-rRNA and reduced synthesis of mature rRNA in puromycin-treated L cells are not allied with conspicuous changes in the relative amounts of different sugar-methylated sequences (Nm-N), but are allied with selective reduction in the relative amount of one base-methylated nucleoside, m^6_2A, part of the 3'-proximal $m^6_2Am^6_2A$ sequence in SSU RNA. It was therefore noted that m^6_2A may arise at a "late" stage in the processing of pre-rRNA in animal cells (Tamaoki and Lane, 1968).

This notion of "late" methylations in eukaryotic pre-rRNAs was expanded by Brand et al. (1977), who showed that $m^6_2Am^6_2A$ is formed during final maturation of the SSU RNA precursor in the yeast cytoplasm. The yeast SSU RNA(m^6_2A)methyltransferase gene is essential (Lafontaine et al., 1994), but notably, the corresponding m^6_2A modification in SSU RNA is not essential for the normal growth of yeast (Lafontaine et al., 1997) or *E. coli* (van Knippenberg, 1986), just as the *E. coli* tRNA(m^5U)methyltransferase gene is essential, but m^5U in tRNA is not needed for normal *E. coli* growth (Persson et al., 1992). Synthesis of (ribosomal A-site) UmGmψ also occurs "late" during maturation of the yeast LSU RNA precursor (Brand et al., 1977; Eladari et al., 1977). While $\psi 2919$ is ordinarily absent in *E. coli* LSU RNA (with its "undermodified" UmGU homolog) (Bakin and Ofengand, 1993), its absence (in UmGmψ) in yeast LSU RNA leads to a cold-sensitive, slow-growth phenotype in which the rate of ribosome assembly is likely reduced (Ni et al., 1997).

The two sugar-methylation sites in yeast mitochondrial LSU RNA (Dubin and Taylor, 1978; Sirum-Connolly et al., 1995) are homologous with two of the three sugar-methylation sites in *E. coli* LSU RNA (Nichols and Lane, 1966a, 1966b). An enzyme (Pet56p) encoded by a nuclear LSU RNA(Gm2270)methyltransferase gene (*PET56*) catalyzes synthesis, in yeast mitochondrial LSU RNA, of the *E. coli* Gm2251G homolog, and although it is ordinarily needed to form functional ribosomes (Sirum-Connolly and Mason, 1993), a weak suppressor of *pet56* loss-of-function mutations exists (Mason et al., 1996) (see Chapter 14 by Mason). In a related context, mutation of G2251 in *E. coli* LSU RNA leads to severe impairment of translation, in vivo (Green et al., 1997a), but sugar methylation of G2251 is not needed to assemble LSUs active in the "fragment reaction" (Maden et al., 1968) in vitro (Green and Noller, 1996).

What History Can Tell Us about Where We May Be Going

The availability of unmodified tRNA transcripts made it possible to prove a biphasic, incremental pathway (Nishikura and de Robertis, 1981) of pre-tRNA modification (see Grosjean et al., 1996, 1997) and to demonstrate that posttranscriptional modifications can be indispensable for ensuring the fidelity of tRNA aminoacylations in small RNP particles (tRNA/aminoacyl-tRNA synthase) in vitro and for stabilizing mature tRNAs in vivo (Hagervall et al., 1990; Perret et al., 1990). Likewise, the availability of unmodified transcripts of *E. coli* SSU and LSU RNAs (Kryzosiak et al., 1987; Weitzmann et al., 1990) has made it possible to study localization and biogenesis of ψ in eubacterial rRNAs (Ofengand et al., 1995; Ofengand and Bakin, 1997), and should soon make it possible to determine if, as in "small" RNP particles, posttranscriptional modifications in rRNA (decoding site in SSU RNA and PTC in LSU RNA) ensure fidelity of peptide α-aminoacylation and stability of the resident rRNA in "large" RNP particles (ribosomes) in vivo.

When extended to eukaryotic cells (Bakin et al., 1994; Lane et al., 1995; Ofengand et al., 1995), the studies of ψ in eubacterial rRNA blended in a timely way with advances in research on eukaryotic pre-rRNA modification. The recent surge in research on the 100–200 snoRNAs involved with nucleolar sugar methylation and pseudouridylation in eukaryotic pre-rRNA has been aptly dubbed a "sno storm" (Smith and Steitz, 1997). Record "sno storms" in 1996 and 1997 have had much in common with the discovery of splicing in eukaryotic pre-mRNA more than 20 years ago. Both phenomena are peculiar to eukaryotes, both originated in areas of research that lay dormant for decades despite early observations pregnant with cryptic promise, and both are mediated by "small" nuclear RNAs—nucleoplasmic snRNAs and nucleolar snoRNAs.

The hypothesis that ψ in the PTC of LSU RNA may direct peptidyl transfer between P and A site-bound tRNAs during peptide-bond synthesis in the ribosome (Lane et al., 1992) is now amenable to testing, in vivo, by depleting individual snoRNA-directed ψ residues in the eukaryotic PTC. It has already been shown that, individually, each of 5 ψ residues in domain V (Ni et al, 1997) and, collectively, 3 ψ residues in domain II and 1 ψ in domain IV (Ganot et al., 1997) in or near the yeast PTC (Bakin et al., 1994), are not essential for yeast growth or, therefore, protein biosynthesis. Further research of this kind will yield a wealth of insight into the likely role of rRNA modifications in sustaining normal eukaryotic cell growth. Extended study of even the nonessential modifications may follow a nonessential plant-alkaloid paradigm by spawning felicitous serendipities for the health sciences; that is, the potential for site- and tissue-specific, snoRNA-directed mRNA modification (see Cavaille et al., 1996) could have untold and far-reaching impact in antioncogenic strategies.

In *E. coli* ribosomes, 2 of the 3 sugar methylations in LSU RNA (Nichols and Lane, 1966a, 1966b) are in the immediate vicinity of peptidyl transfer: Gm2251 is near the 3' end of P site-bound tRNA (Samaha et al., 1995), and Um2552 (part of UmGmψ in nucleocytoplasmic LSU RNA) (Lane, 1965) is near the 3' end of A site-bound tRNA (Green et al., 1997b). Their proximity to the site of peptidyl transfer suggests Gm2251 and Um2552 may stabilize key internucleoside phosphodiester bridges against OH^- ions liberated during a metal ion-catalyzed aminoacyl transfer (Yarus, 1993) and/or prevent errant H bonding (e.g., with ψ) during stereostructural modeling of the PTC. Found in all of the (limited number of) LSU RNAs analyzed (Schnare and Gray, unpublished data), Gm2251 and Um2552, as well as the variably methylated 5.8S-rRNA Um14 implicated in cell differentiation (Munholland and Nazar, 1987), may be signally important sugar-methylation sites.

No ψ site in LSU RNA is "universally" conserved, but it is noteworthy that ψ1915 and ψ1917 (*E. coli* numbering) have been found in all except yeast and mammalian mitochondrial LSU RNAs (Ofengand and Bakin, 1997). Regardless of its site, ψ may have a peculiar "slippage potential," akin to topoisomerase effects on DNA, that allows (perhaps in concert with [ribosomal] proteins) a form of stereostructural modeling that is needed to adjust and spontaneously (Nurse et al., 1995) stabilize executive sites—the decoding site (SSU RNA) and PTC (LSU RNA). This may be why ψ could not be fully eliminated in the important reductionist experiments of Green and Noller (1996). Alternatively, owing to the special properties of the N1 hydrogen in ψ (Spector and Keller, 1958; Lane et al., 1992; Wood et al., 1995), an appositely placed ψ near the aminoacyl acceptor site, such as ψ2580 in the PTC of *E. coli* ribosomes (Porse and Garrett, 1995), might neutralize OH⁻ ions and/or form dative bonds with metal ions (see Yarus, 1993) in the PTC of the ribosome.

Some of the intriguing questions for ongoing studies of RNA nucleoside modifications follow.

If methylations, by promoting efficient transport of pre-rRNA to the cytoplasm, are essential for normal growth and protein synthesis in eukaryotes (Caboche and Bachellerie, 1977), does highly selective localization of methylations/pseudouridylations in or near the PTC indicate that they also play a significant part in modeling/stabilizing a functional, fully active PTC in LSU RNA in vivo?

Will selected or global deletions of snoRNA-directed formation of ψ and Nm residues in the PTC of LSU RNA uncover structural or functional synergisms that underlie a long-recognized (analytical) linkage between ψ and Nm in rRNAs (Nichols and Lane, 1966a, 1967)?

Do 50-100 snoRNA "adaptors" direct one "master" pseudouridylating (or sugar-methylating) enzyme (see Eichler, 1994)? Do snoRNAs direct all ψ and Nm modifications in eukaryotic rRNA, and if so, do "guide sequences" exist for eubacterial rRNA (ψ and Nm) modifications?

Are the lethal consequences of disrupting genes that encode synthases for nonessential rRNA (m^6_2A) (Lafontaine et al., 1997) and tRNA (m^5U) (Persson et al., 1992) modifications related, in common, to scattered observations (Alberty et al., 1978; Persson et al. 1992) and speculations (Lane, 1985) that can be interpreted to suggest there may be a central role for the 3′-terminal region of SSU RNA in the coordinated control of ribosome function and turnover (see also van Knippenberg, 1986)?

In conclusion, it seems fitting to remark that the direct functional consequence (frameshifting) of the first reported RNA editing event (Benne et al., 1986) was instantly transparent, whereas the functional consequence of the first reported, and by far the most abundant, RNA modification (ψ), has for more than 40 years remained all but opaque. Happily, it now seems certain that by using "differential strategies," the dividends of persistent research over these past forty years will soon define important roles for ψ and other modifications in catalyzing/chaperoning/stabilizing the assembly and transport of functional RNP particles (e.g., ribosomes, spliceosomes), and in catalyzing/mediating/modulating biochemical processes (e.g., protein synthesis, and splicing), in vivo.

Acknowledgment. It is a profound pleasure to thank the Medical Research Council of Canada, a solitary source of uninterrupted research funding (grant MRC-MT-1226) over a period of 37 years, 1962-1999.

REFERENCES

Adler, M., B. Weissmann, and A. B. Gutman. 1958. Occurrence of methylated purine bases in yeast ribonucleic acid. *J. Biol. Chem.* 230:717-723.

Alberty, H., M. Raba, and H. J. Gross. 1978. Isolation from rat liver and sequence of a RNA fragment containing 32 nucleotides from position 5 to 36 from the 3′ end of ribosomal 18S RNA. *Nucleic Acids Res.* 5:425-434.

Allen, F. W. 1954. Nucleic acids. *Annu. Rev. Biochem.* 23:99-124.

Allen, F. W. 1962. *Ribonucleoproteins and Ribonucleic Acids.* Elsevier Publishing Co., Amsterdam, The Netherlands.

Alix, J.-H. 1988. Post-translational methylations of ribosomal proteins. *Adv. Exp. Med. Biol.* 231:371-385.

Altman, S., L. Kirsebom, and S. Talbot. 1995. Recent studies of RNase P, p. 67-78. *In* D. Söll and U. L. RajBhandary (ed.), *tRNA: Structure, Biosynthesis, and Function.* ASM Press, Washington, D.C.

Bachellerie, J.-P., B. Michot, M. Nicoloso, A. Balakin, J. Ni, and M. J. Fournier. 1995a. Antisense snoRNAs: a family of nucleolar RNAs with long complementarities to rRNA. *Trends Biochem. Sci.* 20:261-264.

Bachellerie, J.-P., M. Nicoloso, L.-H. Qu, B. Michot, M. Caizergues-Ferrer, J. Cavaille, and M.-H. Renalier. 1995b. Novel intron-encoded small nucleolar RNAs with long sequence complementarities to mature rRNAs involved in ribosome biogenesis. *Biochem. Cell Biol.* 73:835-843.

Bacher, J. E., and F. W. Allen. 1950. A comparison of pentose nucleic acid from pancreas with ribonucleic acid from yeast. *J. Biol. Chem.* 183:641-645.

Baczynskyj, L., K. Biemann, and R. H. Hall. 1968. Sulfur-containing nucleoside from yeast transfer ribonucleic acid: 2-thio-5(or 6)-uridine acetic acid methyl ester. *Science* 159:1481-1483.

Bakin, A., B. G. Lane, and J. Ofengand. 1994. Clustering of pseudouridine residues around the peptidyltransferase center of yeast cytoplasmic and mitochondrial ribosomes. *Biochemistry* 33:13475-13483.

Bakin, A., and J. Ofengand. 1993. Four newly located pseudouridylate residues in *Escherichia coli* 23S ribosomal RNA are all at the peptidyltransferase center: analysis by the application of a new sequencing technique. *Biochemistry* 32:9754-9762.

Benne, R., J. Van Den Burg, J. P. J. Brakenhoff, P. Sloof, J. H. Van Boom, and M. C. Tromp. 1986. Major transcript of the frameshifted *coxII* gene from trypanosome mitochondria contains four nucleotides that are not encoded in the DNA. *Cell* 46:819-826.

Bennett, T. P. 1988. Lipmann and "Not Strictly Biochemistry," p. 85-93. *In* H. Kleinkauf, H. von Dohren, and L. Jaenicke (ed.), *The Roots of Modern Biochemistry.* Walter de Gruyter, Berlin, Germany.

Biemann, K., S. Tsunakawa, J. Sonnenbichler, H. Feldmann, D. Dutting, and H. G. Zachau. 1966. Structure of an odd nucleoside from serine-specific transfer ribonucleic acid. *Angew. Chem. Intl. Ed.* 5:590-591.

Binkley, F. 1951. Metabolism of glutathione. *Nature* (London) 167:888-889.

Björk, G. R. 1995. Biosynthesis and function of modified nucleosides, p. 165-205. *In* D. Söll and U. L. RajBhandary (ed.), *tRNA: Structure, Biosynthesis, and Function.* ASM Press, Washington, D.C.

Björk, G. R., and L. A. Isaksson. 1970. Isolation of mutants of *Escherichia coli* lacking 5-methyluracil in transfer ribonucleic acid or 1-methylguanine in ribosomal RNA. *J. Mol. Biol.* 51: 83–100.

Björk, G. R., and F. C. Neidhardt. 1975. Physiological and biochemical studies on the function of 5-methyluridine in the transfer ribonucleic acid of *Escherichia coli*. *J. Bacteriol.* 124:99–111.

Borck, E. 1980. Transfer RNA and its by-products as tumor markers, p. 445–462. *In* S. Sell (ed.), *Cancer Markers: Diagnostic and Developmental Significance*. Humana Press, Clifton, N.J.

Brachet, J. 1955. The biological role of pentose nucleic acids, p. 475–519. *In* E. Chargaff and J. N. Davidson (ed.), *The Nucleic Acids: Chemistry and Biology*, vol. 2. Academic Press Inc., New York, N.Y.

Brand, R. C., J. Klootwijk, T. J. M. Van Steenbergen, A. J. De Kok, and R. J. Planta. 1977. Secondary methylation of yeast ribosomal precursor RNA. *Eur. J. Biochem.* 75:311–318.

Brandhorst, B. P., and E. H. McConkey. 1974. Stability of nuclear RNA in mammalian cells. *J. Mol. Biol.* 85:451–463.

Branlant, C., A. Krol, M. A. Machatt, J. Pouyet, and J.-P. Ebel. 1981. Primary and secondary structures of *Escherichia coli* MRE 600 23S ribosomal RNA. Comparison with models of secondary structure for maize chloroplast 23S rRNA and for large portions of mouse and human 16S mitochondrial rRNAs. *Nucleic Acids Res.* 9:4303–4324.

Breathnach, R., J. L. Mandel, and P. Chambon. 1977. Ovalbumin gene is split in chicken DNA. *Nature* (London) 270:314–319.

Brimacombe, R., P. Mitchell, M. Osswald, K. Stade, and D. Bochkariov. 1993. Clustering of modified nucleotides at the functional center of bacterial ribosomal RNA. *FASEB J.* 7:161–167.

Bronskill, P., T. D. Kennedy, and B. G. Lane. 1972. Cell-free enzymic esterification of 5-carboxymethyluridine residues in bulk yeast transfer RNA. *Biochim. Biophys. Acta* 262:275–282.

Brosius, J., M. L. Palmer, P. J. Kennedy, and H. F. Noller. 1978. Complete nucleotide sequence of a 16S ribosomal RNA gene from *Escherichia coli*. *Proc. Natl. Acad. Sci. USA* 75:4801–4805.

Brown, R. A., K. A. O. Ellem, and J. S. Colter. 1960. Reversible dissociation of ribonucleic acid. *Nature* (London) 187:509–511.

Busch, H., R. Reddy, L. Rothblum, and Y. C. Choi. 1982. SnRNAs, snRNPs, and RNA processing. *Annu. Rev. Biochem.* 51: 617–654.

Caboche, M., and J.-P. Bachellerie. 1977. RNA methylation and control of eukaryotic RNA biosynthesis. Effects of cycloleucine, a specific inhibitor of methylation, on ribosomal RNA maturation. *Eur. J Biochem.* 74:19–29.

Cavaille, J., and J.-P. Bachellerie. 1996. Processing of fibrillarin-associated snoRNAs from pre-mRNA introns: an exonucleolytic process exclusively directed by the common stem-box terminal structure. *Biochimie* 78:443–456.

Cavaille, J., M. Nicoloso, and J.-P. Bachellerie. 1996. Targeted ribose methylation of RNA *in vivo* directed by tailored antisense RNA guides. *Nature* (London) 383:732–735.

Cech, T. R. 1990. Self-splicing of Group I introns. *Annu. Rev. Biochem.* 59:543–568.

Chambers, R. W. 1966. The chemistry of pseudouridine. *Prog. Nucleic Acid Res. Mol. Biol.* 5:349–397.

Chantrenne, H. 1961. *International Series of Monographs on Pure and Applied Biology. Modern Trends in Physiological Sciences. The Biosynthesis of Proteins.* Pergamon Press, New York, N.Y.

Chapeville, F., F. Lipmann, G. Von Ehrenstein, B. Weisblum, W. J. Ray, Jr., and S. Benzer. 1962. On the role of soluble ribonucleic acid in coding for amino acids. *Proc. Natl. Acad. Sci. USA* 48:1086–1092.

Chargaff, E. 1955. Isolation and composition of the deoxypentose nucleic acids and of the corresponding nucleoproteins, p. 307–371. *In* E. Chargaff and J. N. Davidson (ed.), *The Nucleic Acids: Chemistry and Biology*, vol. 1. Academic Press Inc., New York, N.Y.

Chargaff, E., and J. N. Davidson (ed.). 1955. *The Nucleic Acids: Chemistry and Biology*, vol. 1. Academic Press Inc., New York, N.Y.

Chargaff, E., and J. N. Davidson (ed.). 1960. *The Nucleic Acids*, vol. 3. Academic Press Inc., New York, N.Y.

Cohn, W. E. 1959. 5-Ribosyl uracil, a carbon-carbon ribofuranosyl nucleoside in ribonucleic acids. *Biochim. Biophys. Acta* 32: 569–571.

Cohn, W. E. 1960. Pseudouridine, a carbon-carbon linked ribonucleoside in ribonucleic acids: isolation, structure, and chemical characteristics. *J. Biol. Chem.* 235:1488–1498.

Cohn, W. E., and E. Volkin. 1951. Nucleoside-5'-phosphates from ribonucleic acid. *Nature* (London) 167:483–484.

Cohn, W. E., and E. Volkin. 1953. On the structure of ribonucleic acids. I. Degradation with snake venom diesterase and the isolation of pyrimidine diphosphates. *J. Biol. Chem.* 203: 319–332.

Colter, J. S., and R. A. Brown. 1956. Preparation of nucleic acids from Ehrlich ascites tumor cells. *Science* 124:1077–1078.

Correll, C. C., B. Freeborn, P. B. Moore, and T. A. Steitz. 1997. Metals, motifs and recognition: 5S rRNA domain crystal structure. *The Second Annual Meeting of the RNA Society Abstracts*, p. 46.

Crestfield, A. M., and F. W. Allen. 1956. Studies on the enzymatic liberation of diphosphonucleosides from the ribonucleic acids of yeast. *J. Biol. Chem.* 219:103–110.

Crestfield, A. M., and F. W. Allen. 1958. Terminal components of ribonucleic acids. *Arch. Biochem. Biophys.* 78:334–337.

Crestfield, A. M., K. C. Smith, and F. W. Allen. 1955. The preparation and characterization of ribonucleic acids from yeast. *J. Biol. Chem.* 216:185–193.

Cunningham, P. R., K. Nurse, A. Bakin, C. J. Weitzmann, M. Pflumm, and J. Ofengand. 1992. Interaction between the two conserved single-stranded regions at the decoding site of small subunit ribosomal RNA is essential for ribosome function. *Biochemistry* 31:12012–12022.

Cunningham, P. R., K. Nurse, C. J. Weitzmann, and J. Ofengand. 1993. Functional effects of base changes which further define the decoding center of *Escherichia coli* 16S ribosomal RNA: mutation of C1404, G1405, C1496, G1497, and U1498. *Biochemistry* 32:7172–7180.

Davis, F. F., and F. W. Allen. 1957. Ribonucleic acids from yeast which contain a fifth nucleotide. *J. Biol. Chem.* 227:907–915.

Dekker, C. A., and H. K. Schachman. 1954. On the macromolecular structure of deoxyribonucleic acid: an interrupted two-strand model. *Proc. Natl. Acad. Sci. USA* 40:894–909.

Diemer, J., B. McLennan, and B. G. Lane. 1966. Studies of the primary structure of 18-S + 28-S ribonucleates. *Biochim. Biophys. Acta* 114:191–194.

Dintzis, H. 1958. Cited by R. B. Roberts. 1967. Ribosome synthesis, p. 441. *In* H. J. Vogel, J. O. Lampen, and V. Bryson (ed.), *Organizational Biosynthesis*. Academic Press, New York, N.Y.

Dounce, A. 1956. Nucleoproteins. Round-table discussion. *J. Cell. Comp. Physiol.* 47(Suppl. 1):103–112.

Dubin, D. T., and R. H. Taylor. 1978. Modification of mitochondrial ribosomal RNA from hamster cells: the presence of GmG and late-methylated UmGmU in the large subunit (17S) RNA. *J. Mol. Biol.* 121:523–540.

Dubos, R. J. 1976. *The Professor, the Institute, and DNA. Oswald T. Avery, His Life and Scientific Achievements.* The Rockefeller University Press, New York, N.Y.

Dudock, B. S., G. Katz, E. K. Taylor, and R. W. Holley. 1969. Primary structure of wheat germ phenylalanine transfer RNA. *Proc. Natl. Acad. Sci. USA* 62:941–945.

Dugaiczyk, A., S. L. C. Woo, E. C. Lai, M. L. Mace, Jr., L. McReynolds, and B. W. O'Malley. 1978. The natural ovalbumin gene contains seven intervening sequences. *Nature* (London) **274:**328–333.

Dunn, D. B. 1959. Additional components in ribonucleic acid of rat-liver fractions. *Biochim. Biophys. Acta* **34:**286–288.

Dunn, D. B., and R. H. Hall. 1975. Purines, pyrimidines, nucleosides, and nucleotides: physical constants and spectral properties, and natural occurrence of the modified nucleosides, p. 65–250. *In* G. D. Fasman (ed.), *CRC Handbook of Biochemistry and Molecular Biology, Nucleic Acids*, 3rd ed., vol. 1. CRC Press, Cleveland, Ohio.

Dunn, D. B., J. H. Hitchborn, and A. R. Trim. 1963. Studies on the ribonucleic acid from soluble and particulate fractions of plant leaves. *Biochem. J.* **88:**34P.

Dunn, D. B. and M. D. M. Trigg. 1975. 5-Carbamoylmethyluridine: a new minor nucleoside of transfer ribonucleic acid. *Biochem. Soc. Trans.* **3:**656–659.

Edqvist, J., K. B. Straby, and H. Grosjean. 1995. Enzymatic formation of N^2,N^2-dimethylguanosine in eukaryotic tRNA: importance of the tRNA architecture. *Biochimie* **77:**54–61.

Edsall, J. T. 1995. On the hazards of whistleblowers and on some problems of young biomedical scientists in our time. *Sci. Eng. Ethics* **1:**329–340.

Eichler, D. C. 1994. Characterization of a nucleolar 2'-O-methyltransferase and its involvement in the methylation of mouse precursor ribosomal RNA. *Biochimie* **76:**1115–1122.

Ehresmann, C., P. Fellner, and J.-P. Ebel. 1971. The 3'-terminal nucleotide sequence of the 16S ribosomal RNA from *Escherichia coli*. *FEBS Lett.* **13:**325–328.

Eladari, M.-E., A. Hampe, and F. Galibert. 1977. Nucleotide sequence neighbouring a late modified guanylic residue within the 28S ribosomal RNA of several eukaryotic cells. *Nucleic Acids Res.* **4:**1759–1767.

Ellis, S. R., M. J. Morales, J.-M. Li, A. K. Hopper, and N. C. Martin. 1986. Isolation and characterization of the *TRM*1 locus, a gene essential for the N^2,N^2-dimethylguanosine modification of both mitochondrial and cytoplasmic tRNA in *Saccharomyces cerevisiae*. *J. Biol. Chem.* **261:**9703–9709.

Fellner, P. 1969. Nucleotide sequences from specific areas of the 16S and 23S ribosomal RNAs of *E. coli*. *Eur. J. Biochem.* **11:**12–27.

Fleissner, E., and E. Borek. 1963. Studies on the enzymatic methylation of soluble RNA. I. Methylation of the s-RNA polymer. *Biochemistry* **2:**1093–1100.

Fresco, J. R., B. M. Alberts, and P. Doty. 1960. Some molecular details of the secondary structure of ribonucleic acid. *Nature* (London) **188:**98–101.

Ganot, P., M.-L. Bortolin, and T. Kiss. 1997. Site-specific pseudouridine formation in preribosomal RNA is guided by small nucleolar RNAs. *Cell* **89:**799–809.

Gilbert, W. 1986. The RNA world. *Nature* (London) **319:**618.

Gray, M. W. 1974. The presence of $O^{2'}$-methylpseudouridine in the 18S + 26S ribosomal ribonucleates of wheat embryo. *Biochemistry* **13:**5453–5463.

Gray M. W., and B. G. Lane. 1967. Studies of the sequence distribution of 2'-O-methylribose in yeast soluble ribonucleates. *Biochim. Biophys. Acta* **134:**243–257.

Gray, M. W., and B. G. Lane. 1968. 5-Carboxymethyluridine, a novel nucleoside derived from yeast and wheat embryo transfer ribonucleates. *Biochemistry* **7:**3441–3453.

Green, R., and H. F. Noller. 1996. In vitro complementation analysis localizes 23S rRNA posttranscriptional modifications that are required for *Escherichia coli* 50S ribosomal subunit assembly and function. *RNA* **2:**1011–1021.

Green, R., R. R. Samaha, and H. F. Noller. 1997a. Mutations at nucleotides G2251 and U2585 of 23S rRNA perturb the peptidyl transferase center of the ribosome. *J. Mol. Biol.* **266:**40–50.

Green, R., C. Switzer, and H. F. Noller. 1997b. Localization of the A site on the 50S subunit of the ribosome, p. 87. *In The Second Annual Meeting of the RNA Society Abstracts*.

Grosjean, H., J. Edqvist, K. B. Straby, and R. Giege. 1996. Enzymatic formation of modified nucleosides in tRNA: dependence on tRNA architecture. *J. Mol. Biol.* **255:**67–85.

Grosjean, H., Z. Szweykowska-Kulinska, Y. Motorin, F. Fasiolo, and G. Simos. 1997. Intron-dependent enzymatic formation of modified nucleosides in eukaryotic tRNAs: a review. *Biochimie* **79:**293–302.

Gutell, R. R. 1994. Collection of small subunit (16S- and 16S-like) ribosomal RNA structures: 1994. *Nucleic Acids Res.* **22:**3502–3507.

Gutell, R. R., M. W. Gray, and M. N. Schnare. 1993. A compilation of large subunit (23S and 23S-like) ribosomal RNA structures: 1993. *Nucleic Acids Res.* **21:**3055–3074.

Haffner, M. H., M. B. Chin, and B. G. Lane. 1978. Wheat embryo ribonucleates. XII. Formal characterization of terminal and penultimate nucleoside residues at the 5'-ends of 'capped' RNA from imbibing wheat embryos. *Can. J. Biochem.* **56:**729–733.

Hagervall, T. G., J. U. Ericson, K. B. Esberg, L. Ji-nong, and G. R. Bjork. 1990. Role of tRNA modification in translational fidelity. *Biochim. Biophys. Acta* **1050:**263–266.

Hall, B. D., and P. Doty. 1959. The preparation and physical chemical properties of ribonucleic acid from microsomal particles. *J. Mol. Biol.* **1:**111–126.

Hall, R. H. 1971. *The Modified Nucleosides in Nucleic Acids*. Columbia University Press, New York, N.Y.

Hanes, C. S. 1953. The formation of peptides in enzymatic reactions. *Br. Med. Bull.* **9:**131–134.

Harris, H. 1974. *Nucleus and Cytoplasm*. Clarendon Press, Oxford, United Kingdom.

Hoagland, M. B. 1960. The relationship of nucleic acid and protein synthesis as revealed by studies in cell-free systems, p. 349–408. *In* E. Chargaff and J. N. Davidson (ed.), *The Nucleic Acids*, vol. 3. Academic Press Inc., New York, N.Y.

Hoagland, M. B., M. L. Stephenson, J. F. Scott, L. I. Hecht, and P. C. Zamecnik. 1958. A soluble ribonucleic acid intermediate in protein synthesis. *J. Biol. Chem.* **231:**241–257.

Holley, R. W. 1957. An alanine-dependent, ribonuclease-inhibited conversion of AMP to ATP, and its possible relationship to protein synthesis. *J. Am. Chem. Soc.* **79:**658–662.

Holley, R. W., J. Apgar, G. A. Everett, J. T. Madison, M. Marquisee, S. H. Merrill, J. R. Penswick, and A. Zamir. 1965. Structure of a ribonucleic acid. *Science* **147:**1462–1465.

Hotchkiss, R. D. 1948. The quantitative separation of purines, pyrimidines, and nucleosides by paper chromatography. *J. Biol. Chem.* **175:**315–332.

Hudson, L., M. Gray, and B. G. Lane. 1965. The alkali-stable dinucleotide sequences and the chain termini in soluble ribonucleates from wheat germ. *Biochemistry* **4:**2009–2016.

James, W. O. 1950. Alkaloids in the plant, p. 15–90. *In* R. H. F. Manske and H. L. Holmes (ed.), *The Alkaloids. Chemistry and Physiology*, vol. 1. Academic Press Inc., New York, N.Y.

Johnson L., and D. Soll. 1970. In vitro biosynthesis of pseudouridine at the polynucleotide level by an enzyme extract from *Escherichia coli*. *Proc. Natl. Acad. Sci. USA* **67:**943–950.

Joyce, G. F., and L. E. Orgel. 1993. Prospects for understanding the origin of the RNA world, p. 1–25. *In* R. F. Gesteland and J. F. Atkins (ed.), *The RNA World*. Cold Spring Harbor Laboratory Press, Cold Spring Harbor, N.Y.

Judson, H. F. 1979. *The Eighth Day of Creation. Makers of the Revolution in Biology*, p. 327. Simon and Schuster, New York, N.Y.

Kennedy, T. D., T. C. Kwong, and B. G. Lane. 1977. Wheat embryo ribonucleates. IX. Generation of N^2-dimethylguanylate when bulk wheat embryo tRNA is used as substrate for wheat embryo S-adenosylmethionine-tRNA methyltransferases, in vitro. *Can. J. Biochem.* **55**:1039–1048.

Kennedy, T. D., and B. G. Lane. 1979. Wheat embryo ribonucleates. XIII. Methyl-substituted nucleoside constituents and 5′-terminal dinucleotide sequences in bulk poly(A)-rich RNA from imbibing wheat embryos. *Can. J. Biochem.* **57**:927–931.

Kennedy, T. D., and B. G. Lane. 1984. Wheat-embryo ribonucleates. XV. Characterization of a fraction of RNA subject to conspicuous terminal labeling during early germination of wheat embryos. *Can. J. Biochem. Cell Biol.* **62**:321–328.

Klootwijk, J., I. Klein, and L. A. Grivell. 1975. Minimal posttranscriptional modification of yeast mitochondrial ribosomal RNA. *J. Mol. Biol.* **97**:337–350.

Kruger, M., and G. Salomon. 1898. Epiguanin. *Hoppe-Seyler's Z. Physiol. Chem.* **26**:389–394.

Krzyzosiak, W., R. Denman, K., Nurse, W. Hellmann, M. Boublik, C. W. Gehrke, P. F. Agris, and J. Ofengand. 1987. In vitro synthesis of 16S ribosomal RNA containing single base changes and assembly into a functional 30S ribosome. *Biochemistry* **26**:2353–2364.

Kuchino, Y., T. Seno, and S. Nishimura. 1971. Fragmented *E. coli* methionine tRNA$_f$ as methyl acceptor for rat liver tRNA methylase: alteration of the site of methylation by the conformational change of tRNA structure resulting from fragmentation. *Biochem. Biophys. Res. Commun.* **43**:476–483.

Kuntzel, B., J. Weissenbach, R. E. Wolff, T. D. Tumaitis-Kennedy, B. G. Lane, and G. Dirheimer. 1975. Presence of the methylester of 5-carboxymethyl uridine in the wobble position of the anticodon of tRNA$^{Arg}_{III}$ from brewer's yeast. *Biochimie* **57**:61–70.

Kurland, C. G. 1960. Molecular characterization of ribonucleic acid from *Escherichia coli* ribosomes. Isolation and molecular weights. *J. Mol. Biol.* **2**:83–91.

Kwong, T. C., and B. G. Lane. 1975. Wheat embryo ribonucleates. V. Generation of N^2-dimethylguanylate when 'fully sequenced' homogeneous species of transfer RNA are used as substrates for wheat embryo methyltransferases. *Can. J. Biochem.* **53**:690–697.

Lafontaine, D., J. Delcour, A-L. Glasser, J. Desgres, and J. Vandenhaute. 1994. The *DIM1* gene responsible for the conserved $m^6_2Am^6_2A$ dimethylation in the 3′-terminal loop of 18S rRNA is essential in yeast. *J. Mol. Biol.* **241**:492–497.

Lafontaine, D. L. J., T. Preiss, and D. Tollervey. 1997. The yeast 18S rRNA dimethylase Dim1p; quality control in ribosome synthesis, p. 327. *In The Second Annual Meeting of the RNA Society Abstracts*.

Lane, B. G. 1962. Studies of the terminal residues of high molecular weight ribonucleates. *Can. J. Biochem. Physiol.* **40**:1071–1078.

Lane, B. G. 1963. The separation of adenosine, guanosine, cytidine and uridine by one-dimensional filter-paper chromatography. *Biochim. Biophys. Acta* **72**:110–112.

Lane, B. G. 1965. The alkali-stable trinucleotide sequences and the chain termini in 18S + 28S ribonucleates from wheat germ. *Biochemistry* **4**:212–219.

Lane, B. G. 1979a. Nucleate research in Toronto. *Trends Biochem. Sci.* **4**:N124–N125.

Lane, B. G. 1979b. The language of biochemistry. *Can. Res.* **12(4)**:31–34.

Lane, B. G. 1985. Sequences in the 3′-terminal coding regions of 5S and 18S rRNA genes may contribute to co-ordinated expression of 5S rRNA and pre(18S/5.8S/26S) rRNA genes. *FEBS Lett.* **186**:11–12.

Lane, B. G. 1988. The wheat embryo, then and now, p. 457–476. *In* H. Kleinkauf, H. von Dohren, and L. Jaenicke (ed.), *The Roots of Modern Biochemistry*. Walter de Gruyter, Berlin, Germany.

Lane, B. G. 1991. Cellular desiccation and hydration: developmentally regulated proteins, and the maturation and germination of seed embryos. *FASEB J.* **5**:2893–2901.

Lane, B. G. 1994. Oxalate, germin, and the extracellular matrix of higher plants. *FASEB J.* **8**:294–301.

Lane, B. G., and F. W. Allen. 1961. The terminal residues of wheat germ ribonucleates. *Biochim. Biophys. Acta* **47**:36–46.

Lane, B. G., and G. C. Butler. 1959a. The isolation, identification, and properties of dinucleotides from alkali hydrolyzates of ribonucleic acids. *Can. J. Biochem. Physiol.* **37**:1329–1350.

Lane, B. G., and G. C. Butler. 1959b. The exceptional resistance of certain oligoribonucleotides to alkaline degradation. *Biochim. Biophys. Acta* **33**:281–283.

Lane, B. G., J. Diemer, and C. A. Blashko. 1963. End group and sedimentation data on fragmented high molecular weight ribonucleates. *Can. J. Biochem. Physiol.* **41**:1927–1941.

Lane, B. G., and F. Lipmann. 1961. Nonparticipation of ribonucleic acid in glutathione and ophthalmic acid synthesis. *J. Biol. Chem.* **236**:PC80–PC81.

Lane B. G., J. Ofengand, and M. W. Gray. 1992. Pseudouridine in the large-subunit (23S-like) ribosomal RNA. The site of peptidyl transfer in the ribosome? *FEBS Lett.* **302**:1–4.

Lane, B. G., J. Ofengand, and M. W. Gray. 1995. Pseudouridine and $O^{2'}$-methylated nucleosides. Significance of their selective occurrence in rRNA domains that function in ribosome-catalyzed synthesis of the peptide bonds in proteins. *Biochimie* **77**:7–15.

Lane, B. G., and T. Tamaoki. 1967. Studies of the chain termini and alkali-stable dinucleotide sequences in 16S and 28S ribosomal RNA from L cells. *J. Mol. Biol.* **27**:335–348.

Lane, B. G., and T. Tamaoki. 1969. Methylated bases and sugars in 16-S and 28-S RNA from L cells. *Biochim. Biophys. Acta* **179**:332–340.

Lau, R. Y., T. D. Kennedy, and B. G. Lane. 1974. Wheat embryo ribonucleates. III. Modified nucleotide constituents in each of the 5.8S, 18S and 26S ribonucleates. *Can. J. Biochem.* **52**:1110–1123.

Lee, Y., D. W. Kindelberger, J.-Y. Lee, S. McClennen, J. Chamberlain, and D. R. Engelke. 1997. Nuclear pre-tRNA terminal structure and RNase P recognition. *RNA* **3**:175–185.

Lerner, M. R., J. A. Boyle, S. M. Mount, S. L. Wolin, and J. A. Steitz. 1980. Are snRNPs involved in splicing? *Nature* (London) **283**:220–224.

Levene, P. A., and L. W. Bass. 1931. *Nucleic acids*. The Chemical Catalogue Co., Inc., New York, N.Y.

Li, J., R. R. Gutell, S. H. Damberger, R. A. Wirtz, J. C. Kissinger, M. J. Rogers, J. Sattabongkot, and T. F. McCutchan. 1997. Regulation and trafficking of three distinct 18S ribosomal RNAs during development of the malaria parasite. *J. Mol. Biol.* **269**:203–213.

Limbach, P. A., P. F. Crain, and J. A. McCloskey. 1994. Summary: the modified nucleosides of RNA. *Nucleic Acids Res.* **22**:2183–2196.

Littlefield, J. W., and D. B. Dunn. 1958. The occurrence and distribution of thymine and three methylated-adenine bases in ribonucleic acids from several sources. *Biochem. J.* **70**:642–651.

Loftfield, R. B. 1972. The mechanism of aminoacylation of transfer RNA. *Prog. Nucleic Acid Res. Mol. Biol.* **12**:87–128.

Maden, B. E. H. 1988. Locations of methyl groups in 28S rRNA of *Xenopus laevis* and man. Clustering in the conserved core of molecule. *J. Mol. Biol.* **201**:289–314.

Maden, B. E. H. 1990. The numerous modified nucleotides in eukaryotic ribosomal RNA. *Prog. Nucleic Acid Res. Mol. Biol.* **39**:241–303.

Maden, B. E. H., R. R. Traut, and R. E. Monro. 1968. Ribosome-catalyzed peptidyl transfer: the polyphenylalanine system. *J. Mol. Biol.* 35:333–345.

Madison, J. T., and S. J. Boguslawski. 1976. Partial digestion of a yeast lysine transfer ribonucleic acid and reconstitution of the nucleotide sequence. *Biochemistry* 13:524–527.

Magasanik, B. 1955. Isolation and composition of the pentose nucleic acids and of the corresponding nucleoproteins, p. 373–407, 474–475. *In* E. Chargaff and J. N. Davidson (ed.), *The Nucleic Acids: Chemistry and Biology*, vol. 1. Academic Press Inc., New York, N.Y.

Mandel, L. R., and E. Borek. 1963. The nature of the RNA synthesized during conditions of unbalanced growth in *E coli* $K_{12}W$-6. *Biochemistry* 2:560–566.

Marko, A. M., and G. C. Butler. 1951. The isolation of sodium deoxyribonucleate with sodium dodecyl sulphate. *J. Biol. Chem.* 190:165–176.

Mason, T. L., C. Pan, M. E. Sanchirico, and K. Sirum-Connolly. 1996. Molecular genetics of the peptidyl transferase center and the unusual Var1 protein in yeast mitochondrial ribosomes. *Experientia* 52:1148–1157.

Matheson, A. T., and C. S. Hanes. 1959. The chemical nature of intracellular peptidases. *Biochim. Biophys. Acta* 33:292–294.

Maxwell, E. S., and M. J. Fournier. 1995. The small nucleolar RNAs. *Annu. Rev. Biochem.* 64:897–934.

McConkey, E. H., and J. W. Hopkins. 1969. Molecular weights of some HeLa ribosomal RNAs. *J. Mol. Biol.* 39:545–550.

Messing, J. 1988. M13, the universal primer and polylinker. *Focus* 10:21–26.

Moore, M. J., C. C. Query, and P. A. Sharp. 1993. Splicing of precursors to mRNA by the spliceosome, p. 303–357. *In* R. F. Gesteland and J. F. Atkins (ed.), *The RNA World*. Cold Spring Harbor Laboratory Press, Cold Spring Harbor, N.Y.

Munholland, J. M., and R. N. Nazar. 1987. Methylation of ribosomal RNA as a possible factor in cell differentiation. *Cancer Res.* 47:169–172.

Nazar, R. N. 1984. The ribosomal 5.8S RNA: eukaryotic adaptation or processing variant? *Can. J. Biochem. Cell Biol.* 62:311–320.

Nevins, J. R. 1983. The pathway of eukaryotic mRNA formation. *Annu. Rev. Biochem.* 52:441–466.

Ni, J., A. L. Tien, and M. J. Fournier. 1997. Small nucleolar RNAs direct site-specific synthesis of pseudouridine in ribosomal RNA. *Cell* 89:565–573.

Nichols, J. L., and B. G. Lane. 1966a. The characteristic alkali-stable dinucleotide sequences in each of the 16S and 23S components of ribosomal ribonucleates from *Escherichia coli*. *Can. J. Biochem.* 44:1633–1645.

Nichols, J. L., and B. G. Lane. 1966b. N^4-methyl-2'-O-methyl cytidine and other methyl-substituted nucleoside constituents of *Escherichia coli* ribosomal and soluble RNA. *Biochim. Biophys. Acta* 119:649–651.

Nichols, J. L., and B. G. Lane. 1967. *In vivo* incorporation of methyl groups into the ribose of *Escherichia coli* ribosomal RNA. *J. Mol. Biol.* 30:477–489.

Nichols, J. L., and B. G. Lane. 1968a. The in vitro $O^{2'}$-methylation of RNA in the ribonucleoprotein precursor-particles from a *Relaxed* mutant of *Escherichia coli*. *Can. J. Biochem.* 46:109–115.

Nichols, J. L., and B. G. Lane. 1968b. In vitro $O^{2'}$-methylation of sugars in *E. coli* RNA. II. Methylation of ribosomal and transfer RNA by homologous methylases in crude cell-free extracts and particulate suspensions from a *relaxed* mutant of *E. coli*. *Can. J. Biochem.* 46:1487–1495.

Nichols, J. L., and B. G. Lane. 1968c. Characterization of $N^4,O^{2'}$-dimethylcytidine, a rare nucleoside constituent of *Escherichia coli* 16-S RNA. *Biochim. Biophys. Acta* 166:605–615.

Nichols J. L., and B. G. Lane. 1969. In vitro $O^{2'}$-methylation of sugars in *E. coli* RNA. III. Methylation of *E. coli* transfer RNA by heterologous methylases in particulate-free extracts of wheat embryo. *Can. J. Biochem.* 47:863–869.

Nishikura, K., and E. M. de Robertis. 1981. RNA processing in microinjected *Xenopus* oocytes. Sequential addition of base modifications in a spliced transfer RNA. *J. Mol. Biol.* 145:405–420.

Noller, H. F. 1993. tRNA-rRNA interactions and peptidyl transferase. *FASEB J.* 7:87–89.

Noller, H. F., V. Hoffarth, and L. Zimniak. 1992. Unusual resistance of peptidyl transferase to protein extraction procedures. *Science* 256:1416–1419.

Nurse, K., J. Wrzesinski, A. Bakin, B. G. Lane, and J. Ofengand. 1995. Purification, cloning, and properties of the tRNA ψ55 synthase from *Escherichia coli*. *RNA* 1:102–112.

Oakden, K. M., A. A. Azad, R. Y. Lau, and B. G. Lane. 1972. Aqueous denaturation of wheat embryo ribonucleates. *Biochim. Biophys. Acta* 272:252–261.

Oakden, K. M., and B. G. Lane. 1973. Chain termini of the satellite RNA from yeast ribosomes. *Can. J. Biochem.* 51:520–528.

Oakden, K. M., and B. G. Lane. 1976. Wheat embryo ribonucleates. VI. Comparison of the 3'-hydroxyl termini in 'rapidly labeled' RNA from metabolizing wheat embryos with the corresponding termini in ribosomal RNA from differentiating embryos of wheat, barley, corn and pea. *Can. J. Biochem.* 54:261–271.

O'Connor, M., C. L. Thomas, R. A. Zimmermann, and A. E. Dahlberg. 1997. Decoding fidelity at the ribosomal A and P sites: influence of mutations in three different regions of the decoding domain in 16S rRNA. *Nucleic Acids Res.* 25:1185–1193.

Ofengand, J. and A. Bakin. 1997. Mapping to nucleotide resolution of pseudouridine residues in large subunit ribosomal RNAs from representative eukaryotes, prokaryotes, archaebacteria, mitochondria and chloroplasts. *J. Mol. Biol.* 266:246–268.

Ofengand, J., A. Bakin, J. Wrzesinski, K. Nurse, and B. G. Lane. 1995. The pseudouridine residues of ribosomal RNA. *Biochem. Cell Biol.* 73:915–924.

Orgel, L. E., and F. H. C. Crick. 1993. Anticipating an RNA world. Some past speculations on the origin of life: where are they to-day? *FASEB J.* 7:238–239.

Otaka, E., Y. Oota, and S. Osawa. 1961. Sub-unit of ribosomal ribonucleic acid from yeast. *Nature* (London) 191:598–599.

Pene, J. J., E. Knight, Jr., and J. E. Darnell, Jr. 1968. Characterization of a new low molecular weight RNA in HeLa cell ribosomes. *J. Mol. Biol.* 33:609–623.

Perret, V., A. Garcia, H. Grosjean, J.-P. Ebel, C. Florentz, and R. Giege. 1990. Relaxation of a transfer RNA specificity by removal of modified nucleotides. *Nature* (London) 344:787–789.

Perry, R. P., and D. E. Kelley. 1974. Existence of methylated messenger RNA in mouse L cells. *Cell* 1:37–42.

Perry, R. P., D. E. Kelley, K. Friderici, and F. Rottman. 1975. The methylated constituents of L cell messenger RNA: evidence for an unusual cluster at the 5' terminus. *Cell* 4:387–394.

Persson, B. C., C. Gustafsson, D. E. Berg, and G. R. Bjork. 1992. The gene for a tRNA modifying enzyme, m^5U54-methyltransferase, is essential for viability in *E. coli*. *Proc. Natl. Acad. Sci. USA* 89:3995–3998.

Phillips, J. H., and K. Kjellin-Straby. 1967. Studies on microbial ribonucleic acid. IV. Two mutants of *Saccharomyces cerevisiae* lacking N^2-dimethylguanosine in soluble ribonucleic acid. *J. Mol. Biol.* 26:509–518.

Porse, B. T., and Garrett, R. A. 1995. Mapping important nucleotides in the peptidyl transferase centre of 23S rRNA using a random mutagenesis approach. *J. Mol. Biol.* 249:1–10.

Prince, J. B., B. H. Taylor, D. L. Thurlow, J. Ofengand, and R. A. Zimmermann. 1982. Covalent crosslinking of $tRNA^{Val}_1$ to

16S RNA at the ribosomal P site: identification of cross-linked residues. *Proc. Natl. Acad. Sci. USA* **79:**5450–5454.

RajBhandary, U. L., S. H. Chang, A. Stuart, R. D. Faulkner, R. M. Hoskinson, and H. G. Khorana. 1967. Studies on polynucleotides. LXVIII. The primary structure of yeast phenylalanine transfer RNA. *Proc. Natl. Acad. Sci. USA* **57:**751–758.

RajBhandary, U. L., R. J. Young, and H. G. Khorana. 1964. Studies on polynucleotides. XXXII. The labeling of end groups in polynucleotide chains: the selective phosphorylation of phosphomonoester groups in amino acid acceptor ribonucleic acids. *J. Biol. Chem.* **239:**3875–3884.

Rhoads, R. E. 1991. Initiation: mRNA and 60S subunit binding, p. 109–148. *In* H. Trachsel (ed.), *Translation in Eukaryotes.* CRC Press, Boca Raton, Fla.

Robertson, H. D., S. Altman, and J. D. Smith. 1972. Purification and properties of a specific *Escherichia coli* ribonuclease which cleaves a tyrosine transfer ribonucleic acid precursor. *J. Biol. Chem.* **247:**5243–5251.

Robertus, J. D., J. E. Ladner, J. T. Finch, D. Rhodes, R. S. Brown, B. F. C. Clark, and A. Klug. 1974. Structure of yeast phenylalanine tRNA at 3 A resolution. *Nature* (London) **250:**546–551.

Rogers, J., and R. Wall. 1980. A mechanism for RNA splicing. *Proc. Natl. Acad. Sci. USA* **77:**1877–1879.

Roop, D. R., J. L. Nordstrom, S. Y. Tsai, M.-J. Tsai, and B. W. O'Malley. 1978. Transcription of structural and intervening sequences in the ovalbumin gene and identification of potential ovalbumin mRNA precursors. *Cell* **15:**671–685.

Rosbash, M., and S. Penman. 1972. The precipitation of precursor tRNA in high salt. *Biochem. Biophys. Res. Commun.* **46:**1469–1475.

Samaha, R. R., R. Green, and H. F. Noller. 1995. A base pair between tRNA and 23S rRNA in the peptidyl transferase centre of the ribosome. *Nature* (London) **377:**309–314.

Samuelsson, T., T. Boren, T.-I. Johansen, and F. Lustig. 1988. Properties of a transfer RNA lacking modified nucleosides. *J. Biol. Chem.* **263:**13692–13699.

Sanger, F., and H. Tuppy. 1951. The amino-acid sequence in the phenylalanyl chain of insulin. 2. The investigation of peptides from enzymic hydrolysates. *Biochem. J.* **49:**481–490.

Santer, M., and A. E. Dahlberg. 1996. Ribosomal RNA: an historical perspective, p. 3–20. *In* R. A. Zimmermann and A. E. Dahlberg (ed.), *Ribosomal RNA. Structure, Evolution, Processing, and Function in Protein Biosynthesis.* CRC Press, Boca Raton, Fla.

Saponara, A. G., and M. D. Enger. 1969. Occurrence of $N^2,N^2,7$-trimethylguanosine in minor RNA species of a mammalian cell line. *Nature* (London) **223:**1365–1366.

Scannell, J. P., A. M. Crestfield, and F. W. Allen. 1959. Methylation studies on various uracil derivatives and on an isomer of uridine isolated from ribonucleic acids. *Biochim. Biophys. Acta* **32:**406–412.

Scherrer, K., and J. E. Darnell. 1962. Sedimentation characteristics of rapidly labelled RNA from HeLa cells. *Biochem. Biophys. Res. Commun.* **7:**486–490.

Schnare, M. N., and M. W. Gray. 1981. 3′-Terminal nucleotide sequence of *Crithidia fasciculata* small ribosomal subunit RNA. *FEBS Lett.* **128:**298–304.

Schnare, M. N., and M. W. Gray. Unpublished data.

Semenza, G. 1957. Chromatographic purification of cysteinylglycinase. *Biochim. Biophys. Acta* **24:**401–413.

Shibata, H., T. S. Ro-Choi, R. Reddy, Y. C. Choi, D. Henning, and H. Busch. 1975. The primary nucleotide sequence of nuclear U-2 ribonucleic acid. *J. Biol. Chem.* **250:**3909–3920.

Shugar, D., and J. J. Fox. 1952. Spectrophotometric studies of nucleic acid derivatives and related compounds as a function of pH. I. Pyrimidines. *Biochim. Biophys. Acta* **9:**199–218.

Simos, G., H. Tekotte, H. Grosjean, A. Segref, K. Sharma, D. Tollervey, and E. C. Hurt. 1996. Nuclear pore proteins are involved in the biogenesis of functional tRNA. *EMBO J.* **15:**2270–2284.

Singer, M. F., and G. L. Cantoni. 1960. Studies on soluble ribonucleic acid of rabbit liver. Terminal groups and nucleotide composition. *Biochim. Biophys. Acta* **39:**182–183.

Singh, H., and B. G. Lane. 1964a. The separation, estimation, and characterization of alkali-stable oligonucleotides derived from commercial ribonucleate preparations. *Can. J. Biochem.* **42:**87–93.

Singh, H., and B. G. Lane. 1964b. The alkali-stable dinucleotide sequences in 18S + 28S ribonucleates from wheat germ. *Can. J. Biochem.* **42:**1011–1021.

Sirum-Connolly K., and T. L. Mason. 1993. Functional requirement of a site-specific ribose methylation in ribosomal RNA. *Science* **262:**1886–1889.

Sirum-Connolly, K., J. M. Peltier, P. F. Crain, J. A. McCloskey, and T. L. Mason. 1995. Implications of a functional large ribosomal RNA with only three modified nucleosides. *Biochimie* **77:**30–39.

Smith, C. M., and J. A. Steitz. 1997. Sno storm in the nucleolus: new roles for myriad small RNPs. *Cell* **89:**669–672.

Smith, J. E., B. S. Cooperman, and P. Mitchell. 1992. Methylation sites in *Escherichia coli* ribosomal RNA: localization and identification of 4 new sites of methylation in 23S rRNA. *Biochemistry* **31:**10825–10834.

Smith, J. D., and D. B. Dunn. 1959a. An additional sugar component of ribonucleic acids. *Biochim. Biophys. Acta* **31:**573–575.

Smith, J. D., and D. B. Dunn. 1959b. The occurrence of methylated guanines in ribonucleic acids from several sources. *Biochem. J.* **72:**294–301.

Spector, L. B., and E. B. Keller. 1958. Labile acetylated uracil derivatives. *J. Biol. Chem.* **232:**185–192.

Streeter, D. G., and B. G. Lane. 1970. Studies of the biogenesis of N^2-dimethylguanylate. I. Generation of N^2-dimethylguanylate when bulk *Escherichia coli* transfer RNA is used as a substrate for wheat embryo methyltransferases. *Biochim. Biophys. Acta* **199:**394–404.

Suddath, F. L., G. J. Quigley, A. McPherson, D. Sneden, J. J. Kim, S. H. Kim, and A. Rich. 1974. Three-dimensional structure of yeast phenylalanine tRNA at 3 A resolution. *Nature* (London) **248:**20–24.

Tamaoki, T., and B. G. Lane. 1967. The chain termini and alkali-stable dinucleotide sequences in rapidly labeled ribonucleates from L cells. *Biochemistry* **6:**3583–3591.

Tamaoki, T., and B. G. Lane. 1968. Methylation of sugars and bases in ribosomal and rapidly labeled ribonucleates from normal and puromycin-treated L cells. *Biochemistry* **7:**3431–3440.

Tamaoki, T., and G. C. Mueller. 1962. Synthesis of nuclear and cytoplasmic RNA of HeLa cells and the effect of actinomycin D. *Biochem. Biophys. Res. Commun.* **9:**451–454.

Tazawa, I., T. Koike, and Y. Inoue. 1980. Stacking properties of a highly hydrophobic dinucleotide sequence, N^6,N^6-dimethyladenylyl(3′-5′)N^6,N^6-dimethyladenosine, occurring in 16-18-S ribosomal RNA. *Eur. J. Biochem.* **109:**33–38.

Tilghman, S. M., P. J. Curtis, D. C. Tiemeier, P. Leder, and C. Weissmann. 1978. The intervening sequence of a mouse β-globin gene is transcribed within the 15S β-globin mRNA precursor. *Proc. Natl. Acad. Sci USA* **75:**1309–1313.

Tollervey, D. 1987. A yeast small nuclear RNA is required for normal processing of pre-ribosomal RNA. *EMBO J.* **6:**4169–4175.

Tollervey, D., H. Lehtonen, R. Jansen, H. Kern, and E. C. Hurt. 1993. Temperature-sensitive mutations demonstrate roles for yeast fibrillarin in pre-rRNA processing, pre-rRNA methylation, and ribosome assembly. *Cell* **72**:443–457.

Tschirch, A. 1923. Cited by W. O. James. 1950. Alkaloids in the plant, p. 82. *In* R. H. F. Manske and H. L. Holmes (ed.), *The Alkaloids. Chemistry and Physiology*, vol. 1. Academic Press Inc., New York.

Tsui, H.-C. T., P. J. Arps, D. M. Connolly, and M. E. Winkler. 1991. Absence of *hisT*-mediated tRNA pseudouridylation results in a uracil requirement that interferes with *Escherichia coli* K-12 cell division. *J. Bacteriol.* **173**:7395–7400.

Tumaitis, T. D., and B. G. Lane. 1970. Differential labelling of the carboxymethyl and methyl substituents of 5-carboxymethyluridine methyl ester, a trace nucleoside constituent of yeast transfer RNA. *Biochim. Biophys. Acta* **224**:391–403.

van Knippenberg, P. H. 1986. Structural and functional aspects of N^6,N^6-dimethyladenosines in 16S ribosomal RNA, p. 412–424. *In* B. Hardesty and G. Kramer (ed.), *Structure, Function, and Genetics of Ribosomes*. Springer-Verlag, Berlin, Germany.

Vaughan, M. H., Jr., R. Soeiro, J. R. Warner, and J. E. Darnell, Jr. 1967. The effects of methionine deprivation on ribosome synthesis in HeLa cells. *Proc. Natl. Acad. Sci. USA* **58**:1527–1534.

Veldman, G. M., J. Klootwijk, V. C. H. F. de Regt, R. J. Planta, C. Branlant, A. Krol, and J.-P. Ebel. 1981. The primary and secondary structure of yeast 26S rRNA. *Nucleic Acids Res.* **9**:6935–6952.

Vladimirov, S. N., A. V. Ivanov, G. G. Karpova, A. K. Musolyamov, T. A. Egorov, B. Thiede, B. Wittmann-Liebold, and A. Otto. 1996. Characterization of the human small-ribosomal-subunit proteins by N-terminal and internal sequencing, and mass spectrometry. *Eur. J. Biochem.* **239**:144–149.

Waldrop, M. M. 1992. Finding RNA makes proteins gives 'RNA World' a big boost. *Science* **256**:1396–1397.

Watson, J. D., and F. H. C. Crick. 1953. Molecular structure of nucleic acids. A structure for deoxyribose nucleic acid. *Nature* (London) **171**:737–738.

Weinberg, R. A., U. Loening, M. Willems, and S. Penman. 1967. Acrylamide gel electrophoresis of HeLa cell nucleolar RNA. *Proc. Natl. Acad. Sci. USA* **58**:1088–1095.

Weitzmann, C. J., P. R. Cunningham, and J. Ofengand. 1990. Cloning, *in vitro* transcription, and biological activity of *Escherichia coli* 23S ribosomal RNA. *Nucleic Acids Res.* **18**:3515–3520.

Woese, C. R., and N. R. Pace. 1993. Probing RNA structure, function, and history by comparative analysis, p. 91–117. *In* R. F. Gesteland and J. F. Atkins (ed.), *The RNA World*. Cold Spring Harbor Laboratory Press, Cold Spring Harbor, N.Y.

Wood, D. D., H. Pang, A. Hempel, N. Camerman, B. G. Lane, and M. A. Moscarello. 1995. Participation of acetylpseudouridine in the synthesis of a peptide bond *in vitro*. *J. Biol. Chem.* **270**:21040–21044.

Wrzesinski, J., K. Nurse, A. Bakin, B. G. Lane, and J. Ofengand. 1995. A dual-specificity pseudouridine synthase: an *Escherichia coli* synthase purified and cloned on the basis of its specificity for $\psi 746$ in 23S RNA is also specific for $\psi 32$ in tRNA^{Phe}. *RNA* **1**:437–448.

Wyatt, G. R. 1950. Occurrence of 5-methyl-cytosine in nucleic acids. *Nature* (London) **166**:237–238.

Yarus, M. 1993. How many catalytic RNAs? Ions and the Cheshire cat conjecture. *FASEB J.* **7**:31–39.

Yokoyama, S., and S. Nishimura. 1995. Modified nucleosides and codon recognition, p. 207–223. *In* D. Söll and U. L. RajBhandary (ed.), *tRNA: Structure, Biosynthesis, and Function*. ASM Press, Washington, D.C.

Yu, C.-T., and F. W. Allen. 1959. Studies on an isomer of uridine isolated from ribonucleic acids. *Biochim. Biophys. Acta* **32**:393–406.

Zachau, H. G., G. Acs, and F. Lipmann. 1958. Isolation of adenosine amino acid esters from a ribonuclease digest of soluble, liver ribonucleic acid. *Proc. Natl. Acad. Sci. USA* **44**:885–889.

Zamecnik, P. C. 1962. The First Jubilee Lecture: unsettled questions in the field of protein synthesis. *Biochem. J.* **85**:257–264.

Chapter 2

RNA-Modifying and RNA-Editing Enzymes: Methods for Their Identification

HENRI GROSJEAN, YURI MOTORIN, AND ANNIE MORIN

One of the most distinctive structural features of cellular RNAs is the presence of a significant proportion of modified nucleosides derived from adenosine, guanosine, cytidine, and uridine (see Appendix 1). The formation of each of them is catalyzed by specific enzymes that act concomitantly to the other RNA processing events, such as 5' and 3' trimming, intron splicing, and nucleotide addition at the 5' or 3' ends. Thus, RNA modification is an integral part of a complex, sequential, multienzymatic process of RNA maturation. In eukaryotic cells, the majority of these posttranscriptional events occur in the nucleus, but a few additional steps of RNA maturation take place in the cytoplasm or simultaneously with the transport of RNA through the nuclear pores. A distinct RNA maturation apparatus is also present in mitochondria and chloroplasts; however, all of the corresponding enzymes in these organelles are probably the products of nuclear genes. The recent discovery of RNA editing phenomena (by base conversion, or by addition or deletion of nucleotides in certain mRNAs or tRNAs [see Appendix 2]) has added a new dimension to the RNA maturation processes and to cellular gene expression in general.

HOW MANY MODIFICATION AND EDITING ENZYMES ARE THERE?

Only a very limited number of RNA modification and RNA-editing enzymes have been identified and characterized to date (see Appendices 2, 3, and 4). Most of the studied modification enzymes are from *Escherichia coli* and were shown to be specific for the target nucleotide within a given class of RNA (tRNA, rRNA, snRNA, or mRNA). Therefore, besides the expected variety of enzymes catalyzing different types of reactions (see Chapter 8 by Garcia and Goodenough-Lashua), a large number of distinct enzymes are present in the cell, catalyzing the same reaction but at different locations in an RNA molecule or in different classes of RNA.

One modification enzyme (*E. coli* RNA:pseudouridine synthase) was shown to act on different types of RNAs (at position 32 in tRNA and at position 746 in 23S rRNA ["dual-specificity"]) (Wrzesinski et al., 1995; also see Chapter 12 by Ofengand and Fournier). Also, a multisite specific yeast tRNA:pseudouridine synthase, catalyzing the formation of pseudouridines at at least 5 (possibly 8) distinct positions in different tRNA isoacceptors, was described (Simos et al., 1996; Motorin et al., unpublished data). Moreover, the formation of many pseudouridines present in eukaryotic rRNAs is probably catalyzed by only one enzyme, the specificity of which is guided by a large number of *trans*-acting factors (including small nucleolar RNAs [snoRNAs] [see Chapter 12 by Ofengand and Fournier and Chapter 15 by Lafontaine and Tollervey]). Likewise, a single rRNA:methyltransferase probably catalyzes 2'O-methylation of the majority of sites (if not all) in eukaryotic rRNAs (see Chapter 13 by Bachellerie and Cavaillé). More examples of that sort of dual or multisite specificity will probably emerge in the near future, thus reducing considerably the possible number of modification enzymes to be discovered.

On the other hand, multiple distinct enzymes are needed to catalyze the attachment of the various chemical groups to different atoms of the same nucleoside. Also, complex substituents like methylaminomethyl or *cis*-hydroxyisopentenyl (see Appendix 1) found respectively on carbon 5 of uridine at position

---

Henri Grosjean, Yuri Motorin, and Annie Morin • Laboratoire d'Enzymologie et Biochimie Structurales, Centre National de la Recherche Scientifique, 91198 Gif-sur-Yvette, France.

34 (mnm^5U_{34}) or on nitrogen 6 of adenosine at position 37 (io^6A_{37}) of several prokaryotic tRNAs are synthesized by the sequential action of several distinct enzymes (reviewed in Björk, 1995a, 1995b). In contrast, the formation of the complex hypermodified queuosine at position 34 in eukaryotic tRNAs requires only one enzyme, the tRNA:guanine transglycosylase, which exchanges the guanosine-34 with the free precursor queuine (a deazaguanine derivative). However in prokaryotes, at least three enzymes are required to accomplish the stepwise formation of the same queuosine-containing tRNAs (reviewed in Slany and Kersten, 1994) (also see Chapter 8 by Garcia and Goodenough-Lashua and Chapter 9 by Romier et al.).

As far as the editing enzymes are concerned, less information is available (see Appendix 2). Only one core enzyme may edit most, if not all, of the sites by base addition or deletion in mitochondrial RNAs (see Chapter 21 by Hajduk and Sabatini; also see Chapter 22 by Gott and Visomirski-Robic). Also, only one enzyme (APOBEC) has been found for the C to U conversion in eukaryotic mRNA (see Chapter 18 by Chang et al.). In contrast, for the adenosine to inosine conversion in both mRNA (editing) and tRNA (modification), several distinct RNA:adenosine deaminases are now being discovered (see Chapter 19 by Rueter and Emeson). Also, the different types of base conversions in mitochondrial and chloroplast RNAs may require distinct enzymes (see Chapter 16 by Price and Gray and Chapter 17 by Marchfelder et al.).

As a result, the total number of distinct enzymes accounting for all the posttranscriptionally modified or edited nucleotides in all of the cellular RNAs is almost impossible to estimate at present. This is especially the case for the higher eukaryotic cells, in which cytoplasmic RNAs are much more extensively modified than in prokaryotic organisms. Nevertheless, for a simple organism like *E. coli*, it has been estimated that at least 45 different enzymes are required to account for the 29 chemically different modified nucleotides present in the complete set of *E. coli* tRNAs (Björk, 1995a). A total of 79 tRNAs were identified in the *E. coli* genome by Komine et al. (1990), but not all may be expressed. Assuming an average molecular mass of about 30 kDa for each putative tRNA modification enzyme (which is in fact a lower limit estimation [see Appendix 3]), requiring a coding sequence of about 1 kb of DNA (not including the flanking sequences of each structural gene), the total genetic information required to account for all of these genes should correspond to no less than 1% (about 45 kb) of the total bacterial genome (about 4.6 Mb). This amount of genetic information is already four times higher than the part of the *E. coli* genome coding for a full set of tRNAs [79 tRNA genes of about 150 nucleotides each totaling about 12 kb, according to Komine et al. (1990)]. This evaluation does not even include the other numerous enzymes involved in nucleotide modifications in rRNA, snRNA and mRNA.

This multiplicity of these enzymes, although they are possibly expressed at low levels (reviewed in Chapter 25 by Winkler), evidently points out that the synthesis of the so-called "minor," "rare," or "unusual" nucleotides in RNA is far from being a minor cellular metabolic affair.

The basic problems impeding detection, identification, and characterization of the RNA modification and editing enzymes are manifold. Some of these are related to the intrinsic properties of the enzymatic machinery (instability, tendency to associate with other proteins, enzymes or subcellular structures, dependence on still unknown *trans*-acting factors acting on the enzyme or on the substrate), while the others are purely technical (lack of appropriate substrates, difficulties in identifying the products of enzymatic reactions).

In this chapter we review the methods that have been successfully used to date for the identification of RNA modification and editing enzymes. Some earlier techniques that were used in the 1960s and 1970s are discussed in Chapter 1 by Lane. The selected references cited here serve only to illustrate the techniques described; it is not our intention to provide extensive coverage of all of the literature.

HOW TO OBTAIN AN RNA SUBSTRATE

Naturally occurring, fully modified RNAs isolated from the normal growing cell are the end products of the whole maturation process and thus cannot serve as substrates, at least in homologous assay systems in which both the enzyme and the RNA substrate originate from the same cell. Also, except in certain cases (see, for example, Hopper et al., 1978; Harada et al., 1984), partially processed RNA intermediates, which in principle could be used as substrates for selected RNA modification enzymes, do not accumulate in a normally growing cell. Most probably, under the physiological conditions of a living cell, the product of a given maturation enzyme is immediately channeled and efficiently used (without lag time) as a substrate for the subsequent step. Alternatively, some modification or editing reactions may occur simultaneously at different sites, especially when a large RNA molecule is considered.

Fortunately, the development of recombinant DNA techniques now offers numerous alternatives to obtain synthetic or semisynthetic, unmodified or partially modified pre-RNA substrates suitable for the analysis of the enzymatic posttranscriptional RNA modifications both in vitro and in vivo.

Runoff Transcripts of Synthetic or Natural RNA Genes

In the pioneering work of Zeevi and Daniel (1976), E. coli RNA polymerase was successfully used to transcribe several E. coli tRNA genes in vitro in reaction mixtures containing [α-^{32}P]CTP, [α-^{32}P]UTP or [α-^{32}P]ATP. The synthesized radiolabeled RNA transcripts were then further processed by incubation with an E. coli S100 extract supplemented with several cofactors (ATP, S-adenosyl-L-methionine, isopentenyl-pyrophosphate, folinic acid, L-cysteine, L-threonine, pyridoxal phosphate). At the end of the incubation period, the resulting matured tRNAs were analyzed for the presence of modified nucleotides. This simple system allowed to detect the formation of several modified nucleotides (Ψ, m^{5}U, D, s^{4}U, i^{6}A, t^{6}A, Gm and m^{7}G) in tRNA transcripts. Similarly, in vitro transcription of [5-^{3}H]uridine-labelled tRNA^{Tyr} precursor from a bacteriophage DNA carrying the E. coli tRNA^{Tyr} gene was successfully used to study the biosynthesis of pseudouridines at both positions 39 and 55 in tRNAs (Schaefer et al., 1973; Ciampi et al., 1977).

Presently, techniques for gene cloning and automated DNA synthesis allow us to obtain natural genes or synthetic DNA oligomers at will. Also, E. coli RNA polymerase is now replaced by the more convenient SP6 or T7 bacteriophage RNA polymerases (Dreher et al., 1984; Melton et al., 1984). Thus, natural or synthetic DNAs are now routinely used as templates to produce in vitro a variety of so-called "runoff transcripts," lacking all of the modified nucleotides that are normally present in fully matured RNAs.

Figure 1 summarizes various pathways that are currently used to produce DNA templates carrying a T7 or SP6 bacteriophage promoter. When an RNA transcript of less than 35 nucleotides (nt) is required, simple antisense single-stranded DNA (ssDNA) downstream of the bacteriophage double-stranded DNA (dsDNA) promoter can be used (Milligan and Uhlenbeck, 1989). However, dsDNA should always be preferred because serious technical problems may arise during transcription of ssDNA templates (see Cazenave and Uhlenbeck, 1994; Zhou et al., 1995). Long RNA transcripts are usually produced from DNA templates assembled into a gene from synthetic DNA oligomers, or are obtained by cloning a natural RNA gene. The "cassette gene construction" easily allows a portion of the sequence to vary by replacing only 2 or 3 synthetic DNA oligomers, and is therefore extensively used to produce a set of RNA mutants. A particularly useful technique is based on polymerase chain reaction (PCR) amplification, which can pick up any gene of interest from a given genome. If appropriate primers are used for amplification, the resulting amplified product can be directly linked to the desired bacteriophage promoter and restriction sites for linearization (Mullis and Faloona, 1987). Coupled with other recombinant DNA techniques that allow the introduction of specific point mutations (for examples see Kunkel et al., 1987; Horton and Pease, 1991; Sayers et al., 1992; Zhao et al., 1993) or the amplification of a library of randomized sequences (the SELEX technique [see Szostak and Ellington, 1993; Joyce, 1994; Wright and Joyce, 1997]), it is now possible to produce a large variety of runoff transcripts in vitro.

The in vitro transcription of natural or synthetic genes is certainly the method of choice for producing RNA molecules of known nucleotide sequence. It allows the synthesis of transcripts radiolabeled with ^{32}P or ^{3}H, which greatly facilitates the subsequent detection and identification of even a tiny amount of modified or edited nucleotides. It also allows the production of large quantities of unlabeled RNAs (on the order of milligrams) for structural studies (Lowary et al., 1986; Hall et al., 1989; Doudna et al., 1993). However, for the large-scale production of RNA transcripts, optimization of experimental conditions is often required (Yin and Carter, 1996). Cotranscriptional incorporation of various chemically altered nucleotides (such as thio, bromo, or fluoro derivatives) in RNA transcripts by bacteriophage RNA polymerases has also been described (see Chapter 4 by Zimmermann et al.). Worthwhile to mention is the existence of T7 RNA polymerase mutants, which are able to use both ribo- and deoxyribonucleotide triphosphates, thus allowing synthesis of deoxyribose-containing RNAs (Kostyuk et al., 1995; Aphasizhev et al., 1997).

The methodologies previously described are now routinely used in many laboratories to produce RNA substrates for a large number of modification-processing enzymes acting on different types of RNAs (for examples see Krzyzosiak et al., 1987; Reyes and Abelson, 1987; Sampson et al., 1987; Narayan and Rottman, 1988; Samuelsson et al., 1988; Yisraeli and Melton, 1989; Giegé et al., 1990a, 1990b; Patton, 1991; Mueller and Slany, 1995). Radiolabeled T7 runoff transcripts for detecting the activity of several tRNA modification enzymes and for studying various

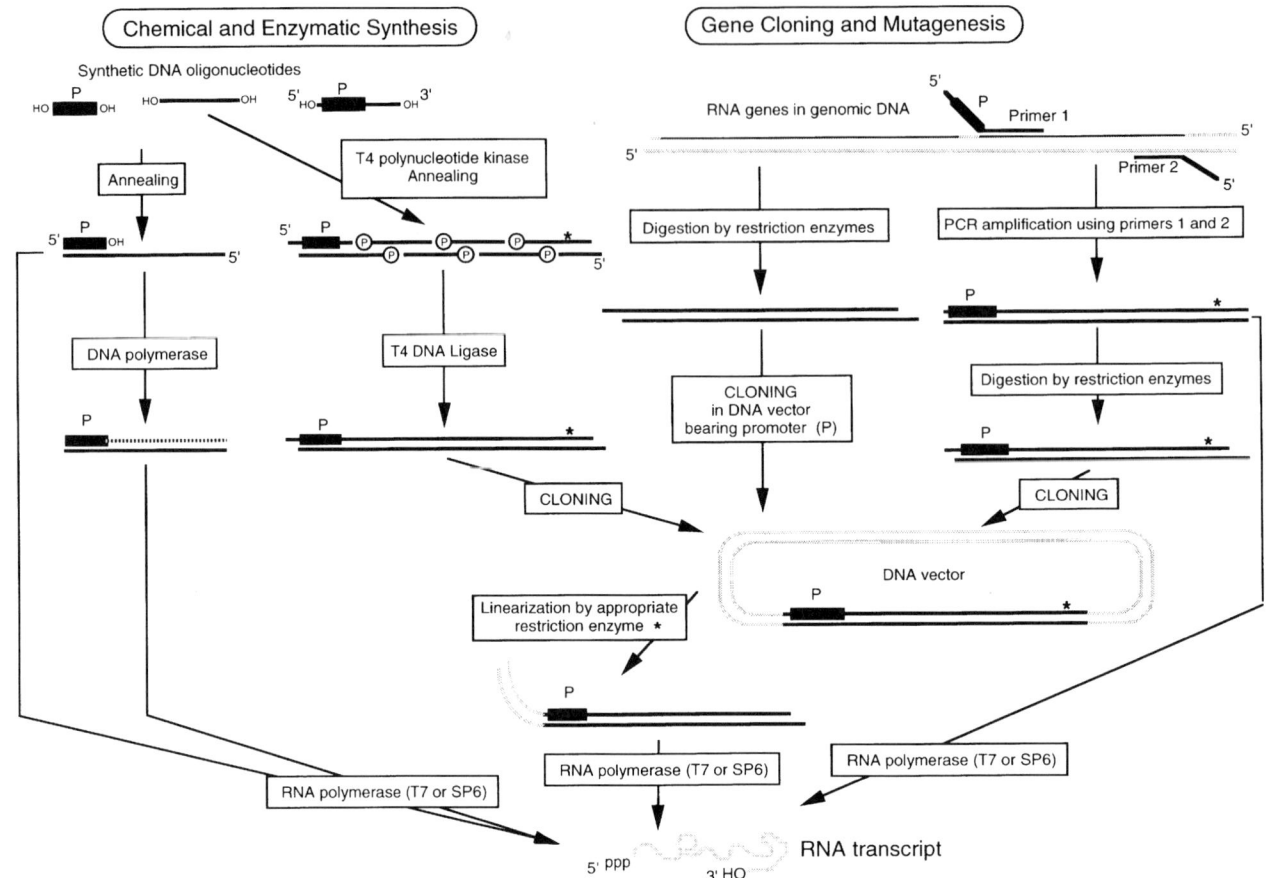

Figure 1. Schematic procedures for construction of DNA templates that can be used for in vitro transcription into RNA using specific DNA-dependant RNA polymerases. On the left side, the starting material corresponds to pure chemically synthesized oligodeoxynucleotides, while on the right, the starting material corresponds to natural RNA genes present in isolated genomic DNA. The different steps of the procedure are indicated in boxes. DNA is represented by bars. A thick bar stands for the DNA portion corresponding to a promoter (P) of bacteriophage T7 or SP6 RNA polymerase. The asterisk stands for the restriction site needed for the linearization of the plasmid, before the transcription. The resulting "runoff" RNA transcript is indicated by a curved gray line. It can be radiolabeled internally on one of the four nucleotides with either ^{32}P, ^{3}H, or any other isotope by adding the corresponding radiolabeled nucleotide triphosphate into the transcription mixture. Hydroxyl and phosphate terminal groups are indicated only when these are essential. Intermediate purification steps by electrophoresis on agarose or polyacrylamide gels are not indicated.

aspects of their mechanism and specificity have been extensively used in our laboratory (Grosjean et al., 1990, 1995a, 1996a, 1996b; Edqvist et al., 1992; Szweykowska-Kulinska et al., 1994; Auxilien et al., 1996; Becker et al., 1997b; Jiang et al., 1997; Motorin et al., 1997; Lecointe et al., 1998; Morin et al., 1998).

In Vitro Recombinant RNA Molecules

In the synthetic (or semisynthetic) approach described above, RNA substrates lack modified nucleotides and therefore may not be suitable for studying the latest events in a complex sequential RNA maturation process. Indeed, a few "early" modification events may facilitate the attainment of a proper pre-RNA conformation by preventing misfolding, by stabilizing a particular domain, or by allowing correct splicing events and other subsequent maturation steps (for examples, see Steinberg and Cedergren, 1995; Fossé et al., 1997; Helm et al., 1997; Lafontaine et al., 1997). The absence of these "early" modified and edited nucleotides may therefore compromise the action of some "late" modification enzymes that depend on such conformational parameters. In this case, the recombinant RNA methodology, as pioneered by Kaufmann, Ohtsuka, and Uhlenbeck and their coworkers (Kaufmann and Littauer, 1975; Ohtsuka et al., 1976; Uhlenbeck and Cameron, 1977; Meyhack et al., 1978), can be an alternative way to overcome at least part of these problems.

In this method, a naturally occurring RNA bearing all of its modified nucleotides is first selectively cleaved into pieces. However, in contrast to recombinant DNA technology, where a large set of restriction enzymes can be used, only few base-specific RNases or RNases with rather broad specificities are available. Thus, it is not an easy task to obtain sufficient amounts of selected, well-defined large RNA fragments from naturally occurring RNA. Fortunately, many "tricks" have been developed to overcome this difficulty; Fig. 2 and 3 illustrate some of them. For example, to select RNA fragments from large molecules, such as rRNA or mRNA, one can synthesize a portion of cDNA complementary to the region of interest or use a DNA fragment resulting from a restriction enzyme digestion, followed by annealing the complementary DNA to the RNA (Fig. 2). If stringent enough conditions are used for the RNA/DNA hybridization and if appropriate fragment-purification procedures are used, the target RNA does not have to be pure. RNA that is unprotected in the hybrid can then be degraded with nuclease S1, or preferably with RNase T_1 (see, for example, Miele et al., 1983; Draper et al., 1988). In the latter case, the protected portion of the RNA fragments will have a guanosine 3'-phosphate.

An alternative variant of this methodology allows site-specific cleavage of large RNA molecules into two pieces by taking advantage of RNase H and of chimeric 2'-O-methyloligodeoxyribonucleotides (Fig. 3) (see Inoue et al., 1987; Nakamura et al., 1991; Hayase et al., 1992; Lapham and Crothers, 1996; Lapham et al., 1997). However, when RNA has only to be reduced in size, the chimeric 2'-O-methyloligodeoxyribonucleotides can be replaced advantageously by less costly and more easily obtained

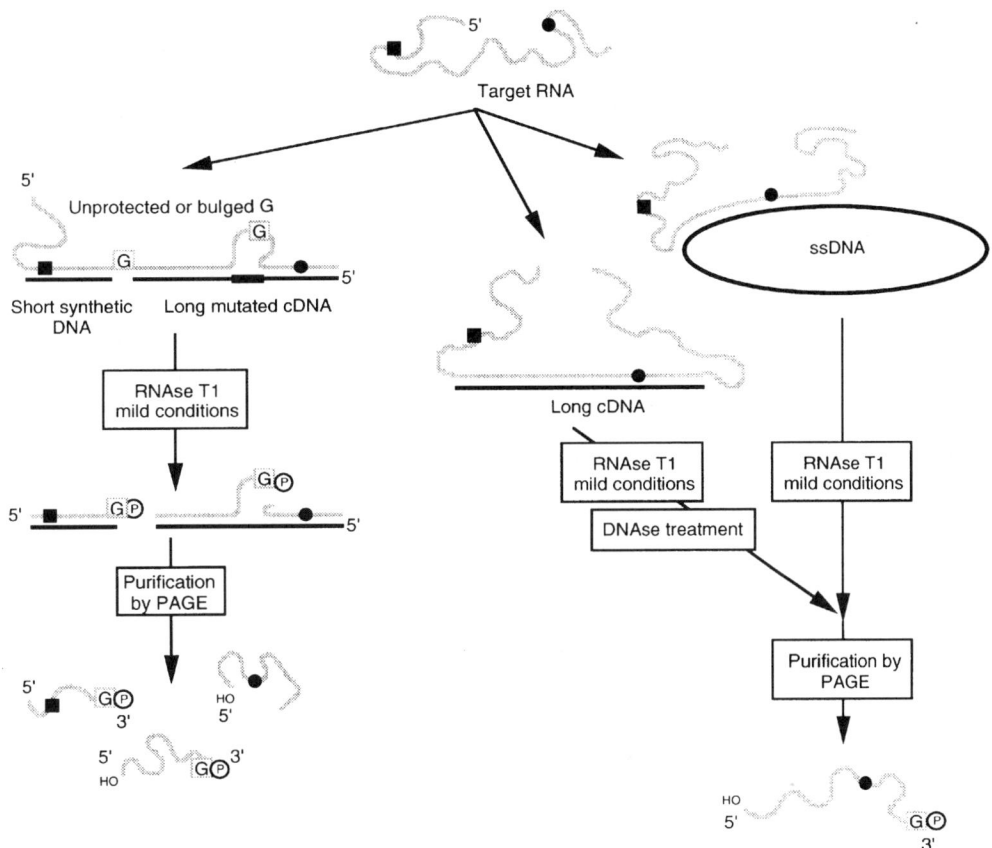

Figure 2. Site-specific cleavage of RNA. RNA fragments are prepared from long natural RNA by hybridization selection and addressed cleavages by RNase T_1 (G specific). Short synthetic DNA, long complementary DNA (cDNA) obtained by reverse transcription of a natural RNA, or single-stranded phage DNA (ssDNA) may be used to protect the desired region of the targeted RNA from nuclease digestion. Modified nucleotides present in the natural RNA are illustrated by a black dot and a black square. If present in the RNA portion to be protected, they should allow formation of a perfect Watson–Crick base pair. "Mild conditions" means that a low ratio of RNase T_1 to the amount of substrate to be digested is used; also, the digestion time is carefully controlled. After removal of the RNase T_1 by phenolization, the resulting RNA fragments are purified by gel electrophoresis or any other purification procedure. The purified fragments have a 5'-hydroxyl group and a 3'-phosphate, which are not suitable for subsequent RNA ligation (see Fig. 5 and 6).

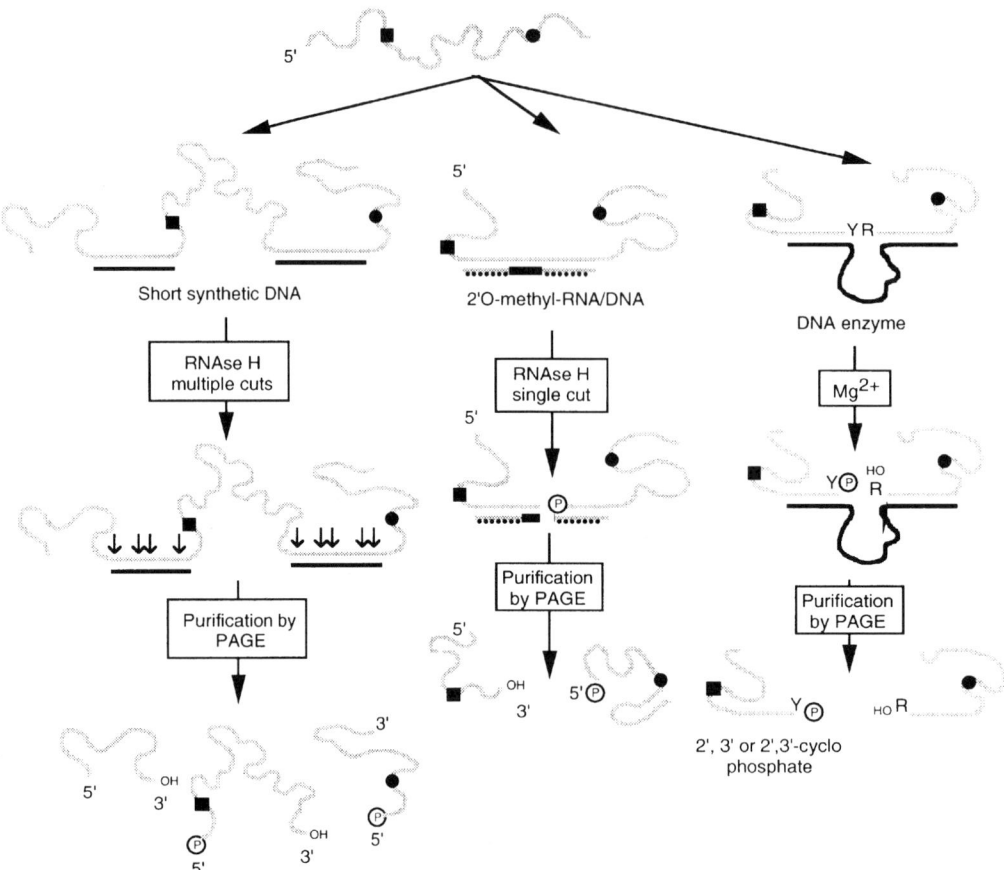

Figure 3. Isolation of large RNA fragments. RNA is cleaved within an RNA-DNA hybrid by RNase H or catalytic DNA enzyme. With complementary DNA 6 to 20 nucleotides long, multiple cuts (shown by arrows) arise in the hybrid region (left part of the figure). A site-specific single cut in RNA can be produced only when a short cDNA (4-mers) is flanked by two complementary 2′-O-methylated RNAs of 6 to 10 nucleotides on each side (central part of the figure). The DNA enzyme (DNAzyme) shown in the right part of the figure corresponds to a catalytic domain of 15 synthetic deoxynucleotides, flanked by two RNA-complementary domains of 7 to 8 deoxynucleotides on each side. The DNAzyme binds to the target RNA through Watson–Crick base pairs. The substrate is cleaved at a phosphodiester bond located between an unpaired purine and a paired pyrimidine residue. If modified nucleotides (illustrated by a black dot and a black square) are present in the target RNA, they should allow formation of a perfect Watson–Crick base pair in the RNA/DNA hybrid.

synthetic DNA. It might well be that in the near future, site-specific cleavage of long RNA molecules will be achieved with synthetic "deoxy-ribozymes" (artificial RNA restriction enzymes) (Santoro and Joyce, 1997). However, these RNase H-induced cleavage methods require the formation of a perfect Watson–Crick type of double helix and therefore cannot be applied when certain modified nucleotides (such as m^1G, m^1A, t^6A, or any bulky modified nucleotides) are present around the cleavage site. Also, because undesired cuts may arise during the cleavage procedure, the identity of the nucleotides at both the 5′ and the 3′ ends of the resulting fragment must always be verified. Depending on the method of cleavage, the resulting RNA fragment bears a phosphate at the 3′ or 5′ end that may not be suitable for the subsequent ligation of fragments. Therefore, additional enzymatic or chemical treatment of RNA fragments with alkaline phosphatase, T4 kinase, or periodate/aniline may be needed (Cedergren and Grosjean, 1987) (also see Chapter 4 by Zimmermann et al.). Purification of the fragments is usually performed by denaturing gel electrophoresis, or by anion exchange or adsorption chromatography. However, alternative methods, such as affinity/hybridization chromatography, can be used (Fig. 4) (Grabowski and Sharp, 1986; Mörl et al., 1994; Tsurui et al., 1994).

The next step consists of the ligation (recombination) of RNA fragments to the final chimeric product. This essential step is performed in vitro by T4 DNA ligase (which efficiently ligates RNA/DNA duplexes) using a DNA splint (Fig. 5A) (Moore and Sharp, 1992; reviewed in Moore and Query, 1997; see also Chapter 4 by Zimmermann et al.). Again, when a stable RNA/DNA hybrid cannot be formed

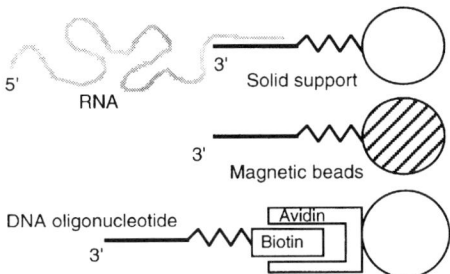

Figure 4. Simple one-step procedure for isolating of RNA or RNA fragments from bulk RNA by the affinity purification technique. A DNA (synthetic or natural, represented by a thick bar) bearing a tail complementary to the target RNA is covalently linked via an appropriate spacer (represented by a zigzag line) to a solid support (usually agarose beads, paramagnetic beads or diol-silica beads). Alternatively, synthetic DNA biotinylated at its 5' end is immobilized on avidin-coated agarose beads (which could also be paramagnetic). Bulk cellular RNA or fragmented RNA bearing the complementary region of the DNA tail is hybridized to the bound DNA. The beads are then collected by centrifugation or with a magnet and washed several times. The trapped RNA is recovered by using appropriate experimental conditions (use of low salt concentration, high temperature, proteinase K). After concentration, the isolated RNA is further purified by gel electrophoresis or by any other technique such as anion exchange or adsorption chromatography.

(for example, because of the presence of a bulky modified nucleotide nearby the cleavage site), then T4 RNA ligase has to be used. In the latter case, a reasonable yield can be obtained only if the two fragments form a stable stem-loop structure, allowing the two extremities to come close to one to another (Fig. 5B).

By using a combination of specific cleavage and recombination, chimeric RNA molecules can be assembled, having one selected modified nucleotide replaced by the originally encoded unmodified residue (Fig. 6). During the RNA remodeling, a ^{32}P label adjacent to the replaced nucleotide or a new base labeled with either ^{3}H, ^{14}C, ^{13}C, ^{35}S, or ^{15}N can be introduced. Elegant adaptations of these methodologies have been proposed by Wower et al. (1994) and Yu and Steitz (1997) for the introduction of various photoactivable nucleotides into specific positions in an RNA molecule (also see Chapter 4 by Zimmermann et al.). This strategy has also been applied for replacing a portion of a sequence with a different RNA sequence (Bruce and Uhlenbeck, 1982; Miele et al., 1983; Ohtsuka et al., 1983; Wang, 1984; Zagorska et al., 1984; Sampson et al., 1987; Kretz et al., 1990; Iwase et al., 1992; Sylvers et al., 1993).

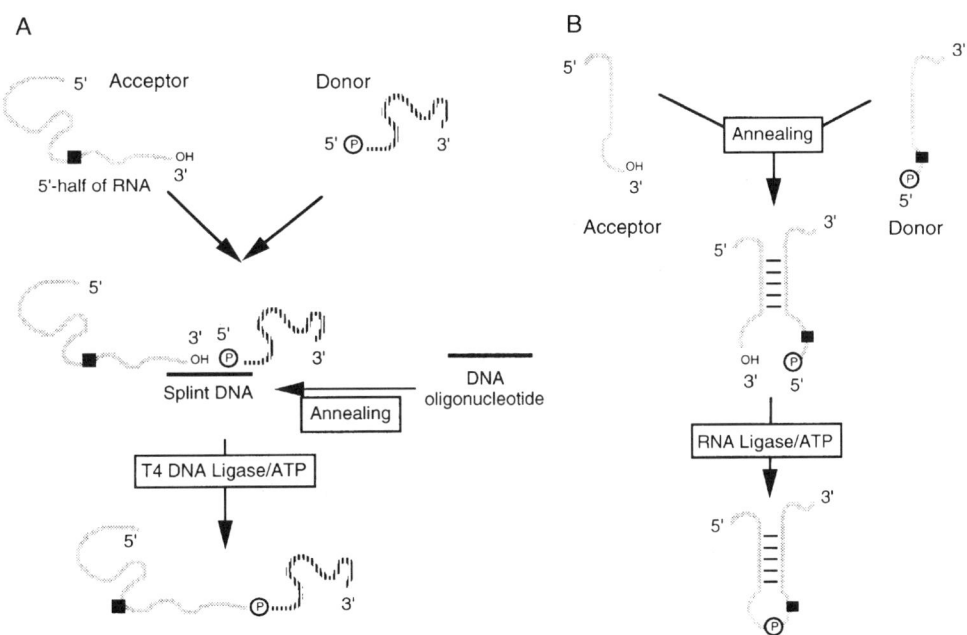

Figure 5. In vitro recombination of RNA fragments (natural or synthetic). The joining reactions can be catalyzed either by T4 DNA ligase or by T4 RNA ligase. With DNA ligase (A), the RNA acceptor and the RNA donor have to be bridged with a DNA splint (black thick line) complementary to the two RNA extremities, thus forming a sort of nicked duplex. With RNA ligase (B), the RNA donor and the RNA acceptor have to be self-complementary in the vicinity of the ligation sites, thus forming a stem-loop structure. RNA recombination can also occur with RNA partners that cannot form such stem-loop structures, but the yield of the ligation reaction is usually dramatically low. For both ligation reactions, the 3' end of the acceptor RNA should have free 2'- and 3'-hydroxyl groups, while the 5' end of the donor RNA molecule has to be phosphorylated. If this 5'-phosphate is radiolabeled with ^{32}P, the resulting recombinant RNA will contain a site-specific radiolabeled phosphate precisely at the ligation site. The black dot stands for naturally occurring modified or edited nucleotides.

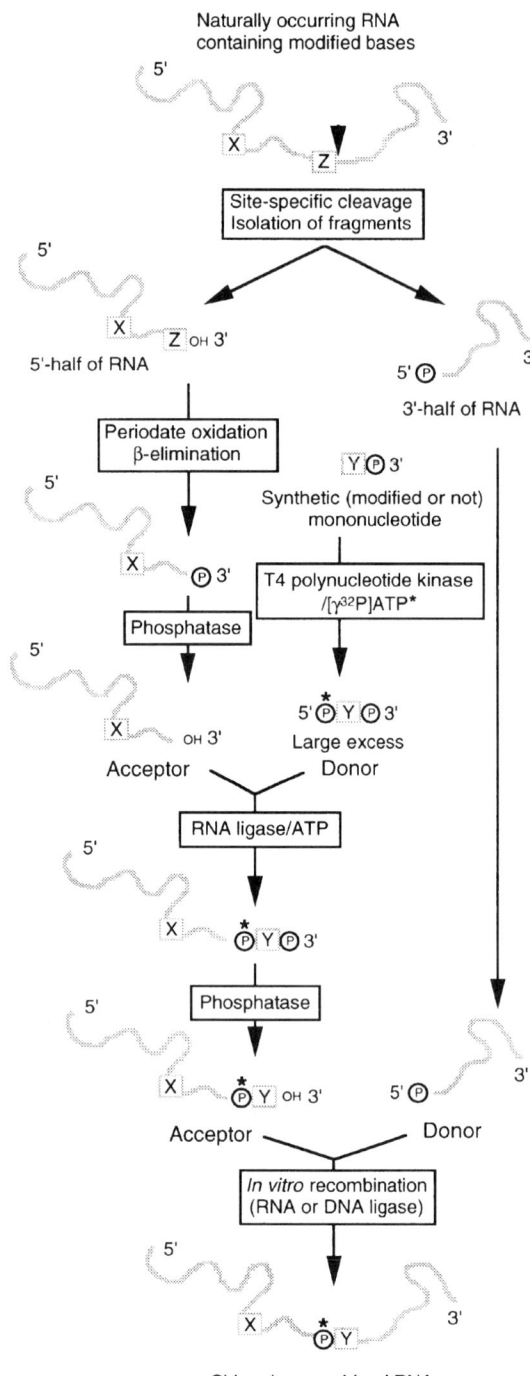

Figure 6. Strategies for site-specific replacement of a nucleotide, Z (modified or not), by another nucleotide, Y (modified or not), in a naturally occurring RNA. Site-specific cleavage (initial step) as well as in vitro recombination of fragments (last step) is performed by one of the methods outlined above (Fig. 2, 3, and 5). Terminal nucleotide Z of the 5'-half RNA fragment can be quantitatively removed after periodate oxidation of the 2'-3'-glycol end followed by a β-elimination of the nucleoside induced by either aniline or lysine. The resulting 3'-phosphate of the RNA fragment, now lacking nucleotide ZMP, is removed by phosphatase. The subsequent ligation with a large excess of nucleotide Y (in the form of 3'-5'-diphosphate) is catalyzed by T4 RNA ligase. The 5'-P of the newly incorporated donor (Y) dinucleotide phosphate can be

The recombinant RNA technology has now been used in many laboratories for site-specific replacement of nucleotides in various kinds of RNA, mainly in tRNA but also in rRNA, phage RNA, and mRNA (reviewed in Cedergren and Grosjean, 1987; Moore and Query, 1997; also see Chapter 4 by Zimmermann et al.). While the experimental details of each reconstruction may vary, the overall principle remains the same. Moreover, the association of recombination technology and in vitro T7 or SP6 transcription now allows construction of a large variety of long RNA molecules (even containing selectively modified nucleotides) that can be used as potential substrates for different RNA modification and editing enzymes (Sampson et al., 1987; Sylvers et al., 1993).

tRNA molecules obtained by recombination technologies were successfully used to detect selected nucleotide modifications such as inosine, Gm, manQ, mcm^{5}U, m^{1}G, m^{1}I, t^{6}A, i^{6}A and Wye at position 34 or 37 in the anticodon loop in various tRNAs (Carbon et al., 1983; Fournier et al., 1983; Haumont et al., 1984, 1987; Droogmans et al., 1986; Droogmans and Grosjean, 1987, 1991; Grosjean et al., 1987; Kretz et al., 1990).

Chemical Synthesis of RNA Substrates

Stepwise automated chemical synthesis of DNA oligonucleotides is widely used for producing starting material for gene construction, primers for PCR, DNA splints, or DNA probes for hybridization and deoxyribozymes. Recently, the synthesis of protected phosphoramidite derivatives has also enabled the development of automated chemical synthesis of RNA (Usman and Cedergren, 1992; Agris et al., 1995). However, for long RNA molecules (more than 100 nt), the yield is dramatically low. This is mainly because of the presence of a protecting group on the 2'-OH of ribonucleotides which considerably slows down the rate and yield of each coupling reaction. Other technical problems are related to incomplete deprotection of the synthesized RNA, and to partial degradation of the full-length product during the procedures. Nevertheless, oligoribonucleotides of about 40–50 nt in length, which is the size of some biologically active RNAs or RNA domains, can now

---

radiolabeled with ^{32}P (symbolized by an asterisk). Phosphatase treatment of this newly synthesized RNA molecule allows a new cycle of ligation reaction with an appropriate donor RNA molecule. The global yield of this stepwise synthesis of chimeric RNA is rather low (at best a few percent of the starting material). Alternative methods for site-specific introduction of modified nucleotides in RNA are detailed in Chapter 4 by Zimmermann et al. (see also Moore and Query, 1997).

be obtained with a reasonable yield but still at a rather high cost. It is noteworthy that despite the above-mentioned difficulties, small amounts of unmodified or partially modified RNAs of the length of a tRNA (77 nt) have been produced by chemical synthesis (Ohtsuka et al., 1981; Wang, 1984; Ogilvie et al., 1988; Bratty et al., 1990; Gasparutto et al., 1992). Recently, the production of milligram amounts of chemically synthesized RNA (tRNA, ribozymes) has been reported (Doudna et al., 1993; Goodwin et al., 1994; Pley et al., 1994; Scott et al., 1995).

The main advantage of chemical synthesis is that it allows a free choice in the design of polynucleotides. Provided that the starting synthons are available and stable enough during the different chemical steps of the automated synthesis, certain natural or synthetic modified nucleotides (containing base, ribose, or phosphate analogs) can be introduced at specific sites of the synthetic RNA molecule; also, mixed deoxy- and ribopolynucleotides can be obtained (Usman et al., 1992; Eckstein and Heidenreich, 1994) (for details see Chapter 4 by Zimmermann et al.).

The drawback is that the introduction of a radiolabeled nucleotide into the molecule is difficult because of the cost and nonavailability of the corresponding radioactively labeled synthons. However, the combination of the automated RNA synthesis and the enzymatic RNA recombination technology should in most cases solve this problem. Thus, chemical synthesis of RNA remains a useful technique for preparing RNA fragments that have to be incorporated into longer RNA molecules by recombination techniques. However, great care has to be taken to ensure that all the blocking groups have been removed and that no 2′-3′ phosphate isomerization has occurred during the final alkaline deprotection of the synthetic RNA.

RNA Substrates from Mutant Strains Defective for a Given Nucleotide Modification

Isolation of natural RNAs that specifically lack one type of modified nucleotide from mutant cells defective in the corresponding enzymatic activity is another way to obtain RNA substrates (see Appendix 4) (reviewed in Hopper, 1990; Björk, 1995b). However, the search for such mutant strains has always been hampered by the general lack of a defined phenotype or simple selectable markers. Also, because the functions of most of the nucleotide modifications are unknown, no logical screening strategy can be set up. Nevertheless, a few useful mutants have been isolated for *E. coli*, *Salmonella typhimurium*, and some fungi, either fortuitously or by exploiting different phenotypes, such as the deregulation of the *trp* or *his* operon [*miaA*/*hisT*], resistance to a particular antibiotic, inability to suppress a nonsense suppressor or induction of a translational frameshift phenotype. Most of these naturally isolated mutant strains are deficient in nucleotide modification in the anticodon loop of tRNA. Laborious methods of nonselective screening, such as a search for the alteration of the chromatographic profile for one or several tRNAs or differential migration of tRNA on polyacrylamide gels, have also been used (reviewed in Björk, 1995b) (also see Chapter 25 by Winkler and Appendix 4).

On the other hand, with the accumulation of sequence information, it now becomes possible to identify putative genes for RNA modification enzymes by homology search (for examples see Gustafsson et al., 1996; Koonin, 1996). With the proviso that the disrupted mutant strain is viable or that no side effect on other steps of the complex stepwise RNA maturation process arises, it should soon be possible to obtain a large collection of additional RNAs lacking just one or few types of modified nucleotides.

Successful disruption of genes coding for RNA modification enzymes has been performed with *E. coli* and yeast (see Chapter 12 by Ofengand and Fournier and Chapter 25 by Winkler) (for examples see Aström and Byström, 1994; Lafontaine et al., 1994; Simos et al., 1996; Becker et al., 1997a; Lecointe et al., 1997).

Use of Natural Heterologous RNAs

Modification patterns of RNAs vary greatly between organisms. As a rule, the higher the organism stands in the phylogenetic classification, the more modified its RNAs are. Variations involve both the type of modification and their location in the molecule. For example, m^2_2G is found at position 26 of most eukaryotic tRNAs, while the same G_{26} is not methylated in any prokaryotic tRNA sequenced so far. Assuming that there is no species specificity between an enzyme from one organism and the RNA substrate from another, these naturally "hypomodified" RNAs can be used to probe selected enzymatic activities present in extracts of other organisms.

This kind of natural substrate has been widely used in the past to detect various methyltransferases catalyzing the formation of methylated bases (m^2G_{26}, $m^2_2G_{26}$, $m^5C_{48/49}$, m^5U_{54} and m^1A_{58}) and two tRNA:Ψ synthases (specific for synthesis of Ψ_{39} and Ψ_{55}) in different heterologous enzyme/tRNA systems (for examples see Johnson et al., 1970; Streeter and Lane, 1970; Pegg, 1972; Wildenauer et al., 1974; Mullenbach et al., 1976; reviewed in Nau, 1976) (also see Chapter 1 by Lane).

Depending on how easily a natural RNA can be obtained in a pure form and in sufficient amounts, this kind of heterologous system may still be valuable. However, because the identity elements for RNA recognition by a given enzyme from different organisms are often not exactly the same, such heterologous RNAs may not necessarily work properly. This situation is reminiscent of that observed for the aminoacylation of the tRNAs by heterologous aminoacyl-tRNA synthetases (see Giegé et al., 1993).

HOW TO LOCALIZE THE MODIFIED OR EDITED NUCLEOTIDES IN RNA

Two distinct approaches are generally used to gain the information on the precise location of modification or editing sites in RNA molecules. The first is based on direct sequencing of pure RNA and is applied mostly for the analysis of rather short RNA molecules, while the other is based on the use of reverse transcription with unfractionated RNA mixtures and specific oligonucleotide primers. Structural characterization of modified nucleotides is usually done by comparison to reference compounds of known chemical structure, as determined in most cases by various mass spectrometry based methods (Pomerantz and McCloskey, 1990; McCloskey, 1991) (see Chapter 3 by Crain).

Direct Sequencing Approach

Direct sequencing of the RNA molecules provides the exact information about the presence (and in most cases about the type) of a modified nucleotide and its precise location in a given RNA molecule. The method proved to be extremely fruitful for the analysis of short RNA molecules (less than 150 nt in length; mostly tRNA, some snRNAs, and fragmented rRNA or mRNA molecules). However, the RNA to be sequenced has to be pure, which often poses technical problems for RNAs present in the cell at very low levels (such as cytoplasmic mRNAs or RNAs from organelles).

The very first methods of sequencing used a set of specific RNases and column and thin-layer chromatographies, and they required gram amounts of pure RNA ("cold" sequencing; Holley et al., 1965). Only the use of in vivo ^{32}P-labeling methods reduced the amount of the required starting material to the reasonable level ("hot" sequencing; reviewed in Branch et al., 1989). Currently, such uniformly labeled RNAs are not used for sequencing, although the ^{32}P/^{14}C RNA fingerprinting technique developed at that time is still a valuable method to analyze the RNase T$_1$ digestion products of large RNA molecules (see Chapter 22 by Gott and Visomirski-Robic) (Smith et al., 1992).

A major breakthrough in RNA sequence analysis came with the discovery of T4 polynucleotide kinase and its use for ^{32}P postlabeling of RNA fragments. Several postlabeling procedures for direct sequence analysis of RNA fragments resulting from enzymatic digestion or chemical degradation of as little as 0.5 μg of pure RNA were then developed (Donnis-Keller et al., 1977; Stanley and Vassilenko, 1978; Peattie and Gilbert, 1980; reviewed in Silberklang et al., 1979; Randerath and Randerath, 1983). Among these procedures, the ^{32}P-thin-layer readout technique proposed by Gupta and Randerath (1979) and Diamond and Dudock (1983) is certainly still a method of choice because it provides detailed information concerning the identity of each modified (or edited) nucleotide.

This procedure (Fig. 7) begins with random single-hit chemical cleavage of the RNA molecule. The resulting fragments, differing by one nucleotide in length, are then radiolabeled at their 5'-OH with ^{32}P by using T4 polynucleotide kinase, followed by size separation of the radiolabeled fragments on a denaturing polyacrylamide gel. The fragments present in each radioactive band of the "ladder" are digested in situ with RNase T$_2$ and then contact-transferred to a polyethyleneimine (PEI)-cellulose, anion-exchange, thin-layer sheet ("print" step). The resulting 3',5'-[^{32}P]diphosphates are then resolved by one-dimensional ascending chromatography on PEI-cellulose. After autoradiography of the chromatogram, the nucleotide sequence can be read directly from the spot pattern. In the alternative procedure, the fragments from the gel are eluted and digested by nuclease P1, and the resulting 5'-^{32}P mononucleotides are analyzed by two-dimensional (2D) thin-layer chromatography. In both approaches, the identification of modified nucleotides is usually done by comparison with reference chromatography maps (Rogg et al., 1976; Gupta and Randerath, 1979; Keith, 1995). 2'-O-methylated nucleotides present in RNA result in characteristic gaps in the "ladder" during separation by polyacrylamide gel electrophoresis (PAGE) and in the appearance of a dinucleotide in a T$_2$ hydrolysate when analyzed by thin-layer chromatography. However, when P1 hydrolysates (in the alternative procedure) are analyzed, 2'-O-methyl derivatives appear as 5'-monophosphate derivatives. One drawback of this method is that some modified nucleotides, such as m^{1}A, m^{7}G, m^{3}A, and S^{4}U, are unstable under the alkaline conditions used to generate the single-hit cleavages of the entire RNA. Also, not all of the modified nucleotides identified thus far in

RNA have a known migration pattern on cellulose thin-layer plates or PEI plates. The direct sequencing technique is also an excellent tool for the identification of the type and location of the edited bases in purified RNA.

The other modern ways for direct RNA sequencing (including the identification of modified or edited sites) are based on a combination of high-performance liquid chromatography and mass spectrometric analysis (Pomerantz and McCloskey, 1990) (see Chapter 3 by Crain).

The precise mapping of the modified nucleotides in long RNA molecules, such as mRNA or pre-rRNA, has always been a daunting task. Nevertheless, despite technical difficulties, the majority of modified nucleotides in rRNA have been localized by direct sequencing of purified rRNA fragments using in vivo labeling and postlabeling methods, followed by enzymatic or chemical degradation and resolution of the digestion products on a sequencing gel (reviewed in Maden et al., 1995).

Base Composition Analysis

When only analysis of the base composition of a pure RNA or RNA fragment is required, a simplified postlabeling procedure can be applied, as illustrated in Fig. 8 (Silberklang et al., 1979). The RNA (0.1 to 0.5 μg) is first completely hydrolyzed into 3'P-NMP by RNase T_2, followed by postlabeling of the free 5'-hydroxyl with [γ-^{32}P]ATP and T4 polynucleotide kinase. All 3'-phosphates are then removed by using the 3'-phosphatase activity of nuclease P1, and the resulting [5'-^{32}P]NMP is analyzed by 2D thin-layer chromatography. The method identifies the presence of a given modified nucleotide in RNA; however, due to differential reactivity of modified nucleotides toward enzymatic 5'-^{32}P labeling, only qualitative information can be obtained (for examples, see Xue et al., 1993; Paul and Bass, 1997).

Quantitative measurement of the modified nucleotide content can also be performed by reversed phase high-pressure liquid chromatography after

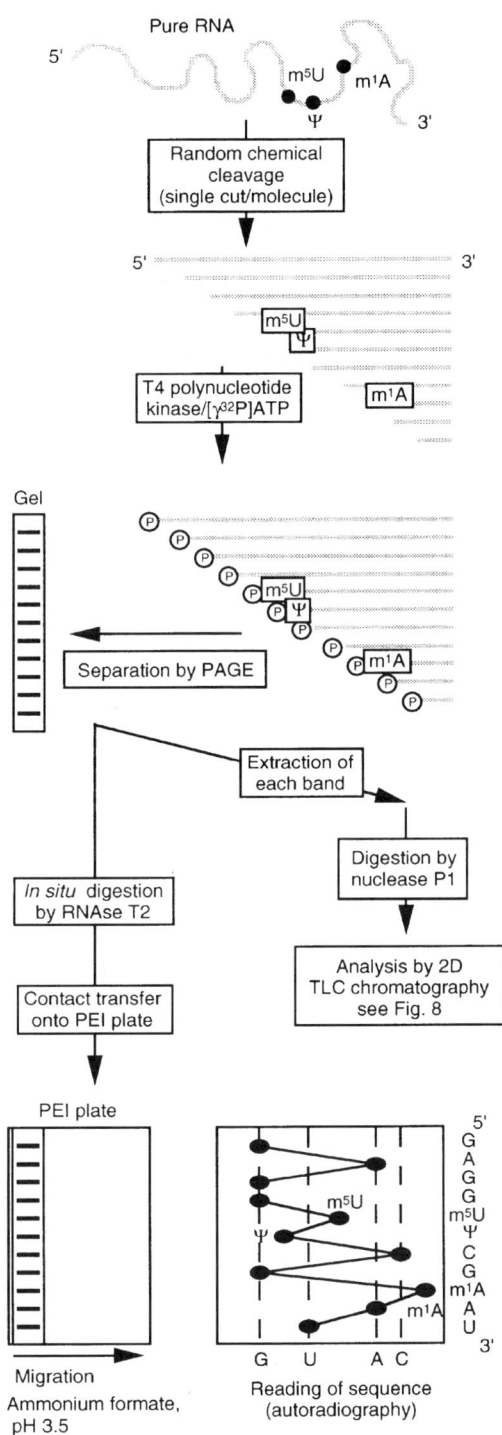

Figure 7. Schematic of a procedure for direct RNA sequencing of an RNA fragment containing modified nucleotides. The RNA is first cleaved randomly (single cut per molecule) under mild conditions. Each fragment is then radiolabeled with ^{32}P at the 5' end. The resulting 5'-radiolabeled fragments are separated by electrophoresis on a polyacrylamide gel. Hydrolysis of each fragment into mononucleotides is performed in situ with RNase T_2. Contact transfer of the resulting nucleotides is performed on a PEI-cellulose plate. After the chromatography, the plate is autoradiographed. Identification of each radiolabeled terminal 3',5'-dinucleotide phosphate on the plate is obtained from its position as compared with those of published reference maps (method of Gupta and Randerath, 1979). The presence of 2'-O-methylated nucleosides in RNA is signaled by "gaps" in the ladder on the gel. Alternatively, the RNA fragment from each band of the gel can be eluted and completely digested by nuclease P1, and the resulting 5'-nucleotide monophosphate can be analyzed by 2D thin-layer chromatography. Identification of each radiolabeled spot on the plate is performed by comparison with those of published reference maps (Keith, 1995). The example illustrates the sequencing of the tRNA region corresponding to the TΨ-loop. This method could also be used to detect base-conversion-type RNA editing processes.

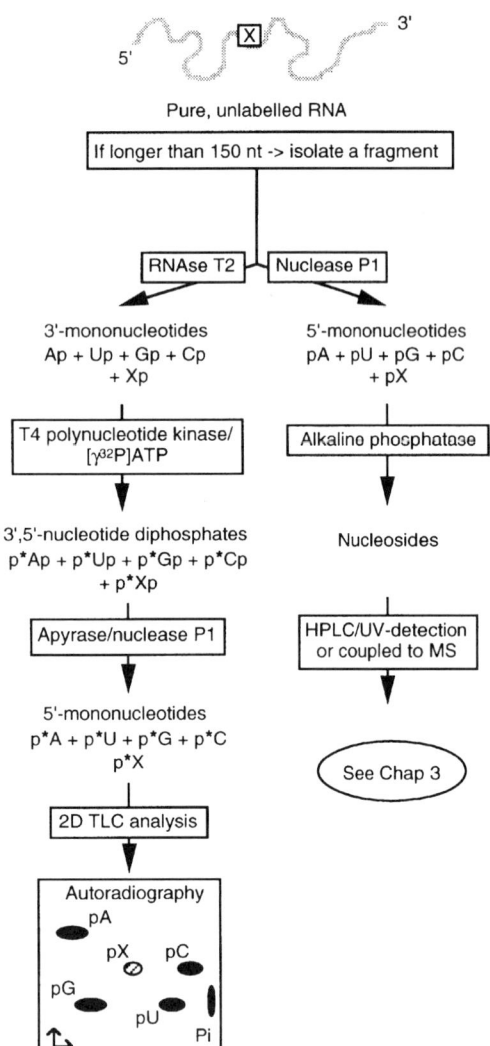

Figure 8. Experimental protocol for the determination of the base composition of an RNA fragment, including modified nucleotides (designated X in the example). The series of steps ends up in 2D thin-layer chromatographic analysis of the 5'-^{32}P-nucleotide monophosphate (qualitative test). Identification of each radiolabeled spot on the plate is performed by comparison with those in published reference maps (Keith, 1995). An alternative approach is based on quantitative analysis of the nucleoside composition by high-pressure liquid chromatography (HPLC) coupled to a diode array spectrophotometer (spectral analysis) and/or mass spectrometry (for details see Chapter 3 by Crain).

complete hydrolysis of the purified RNA (at least 20 μg) into 5'-NMP by nuclease P1 followed by dephosphorylation into nucleosides by alkaline phosphatase (Fig. 8). Each nucleoside component detected in the eluate can be rigorously identified by comparison of its UV spectrum and the retention time with ribonucleoside reference standards. Quantification can be performed using the internal standards and gives results expressed in moles of residue per mole of RNA analyzed (Buck et al., 1983; Gehrke and Kuo, 1990; Xue et al., 1993).

Analysis of the nucleosides in RNA can also be performed by the combined liquid chromatography–mass spectrometry technique. The main advantage of mass spectrometry is that it enables the precise structural characterization and identification of almost all of the modified nucleotides known so far in RNA (see Chapter 3 by Crain). It is also worth mentioning that formation of [^{18}O]inosine in dsRNA catalyzed by dsRAD and of [^{18}O]inosine-34 in tRNA catalyzed by tRNA:A$_{34}$ deaminase has been unambiguously demonstrated by high-pressure liquid chromatography and electrospray mass spectrometry (Polson et al., 1991; Auxilien et al., 1996).

Nearest-Neighbor Type of Analysis

In many cases, the sequence of the RNA substrate is already known and the task of the analysis is to detect the presence of a new modified/edited nucleotide within a defined region. In the case of tRNAs (74–92 nt in length), more than 500 different sequences have been reported (Sprinzl et al., 1998). One can often guess where a given modified nucleotide may appear in a given tRNA (Grosjean et al., 1995b) (see also Appendixes 5 and 6). Therefore, the identification by 2D thin-layer chromatography of a modified nucleotide (usually obtained from the analysis of 5'-^{32}P-nucleotides of a total tRNA nuclease P1 hydrolysate) and that of its 3'-adjacent nucleotide (as determined by the analysis of 3'-^{32}P-nucleotides after an RNase T$_2$ digestion), allows in most cases the unambiguous localization of the position of the modification within a known tRNA cloverleaf framework (Fig. 9). However, to identify all possible types of modified nucleotides and their corresponding nearest-neighbor nucleotides, four independent samples of the same RNA substrate, each radiolabeled with one of the four α-^{32}P-nucleoside triphosphates (NTPs), have to be analyzed. If doubt still persists concerning the precise location of one or a few identified modified nucleotides, partial or complete cleavage of the RNA by RNase T$_1$ or partial fragmentation of the RNA using the hybrid protection technique and RNase T$_1$ or RNase H (Fig. 2 and 3), followed by the analysis of the modified nucleotide content of the isolated fragments by one of the previously mentioned techniques, may solve the problem.

The base conversion type of RNA editing (such as C to U and A to I) can also be detected by this methodology. However, because 3'-mononucleotide mixtures resulting from RNase T$_2$ analysis always contain all four ^{32}P-nucleoside monophosphates (NMPs), only nuclease P1 analysis can provide some

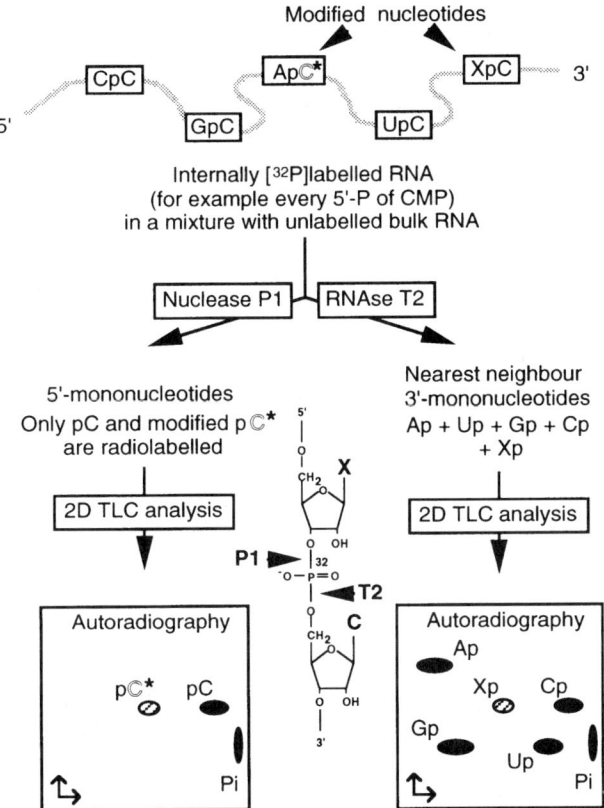

Figure 9. Nearest-neighbor analysis of modified nucleotides in RNA. The RNA fragment has to be internally labeled with ^{32}P at one of the four nucleotides (in this example, 5'-P of CMP). The 2D thin-layer analysis of the nuclease P1 hydrolysate (that retains the 5'-phosphate) allows the identification of the modified nucleotides corresponding to the parent radiolabeled nucleotide (in the example of the figure, a 5'-CMP derivative posttranscriptionally modified on the base or sugar represented by C*). The chromatographic analysis of the nucleotides obtained after complete hydrolysis of the same radiolabeled RNA but with RNase T_2 (leaving the 3'-phosphate) allows all of the modified nucleotides that are 3'-adjacent to a CMP in the original RNA chain (nearest neighbor) to be revealed.

useful information. Doubly labeled substrates, with ^{32}P on phosphate and ^{3}H on the base, were successfully used to demonstrate that the inosine formation in tRNA (Auxilien et al., 1996; reviewed in Grosjean et al., 1996a) and the "C to U" conversion in the mitochondrial system of mRNA editing (Araya et al., 1995; Yu and Schuster, 1995) both proceed by deamination.

The main drawbacks of the nearest-neighbor approach are the limited size of the RNA substrate that can be analyzed (not more than 150 nt) and the necessity to have prior indication of the location of the modification or editing site in the RNA molecule. The molar yield of modified [^{32}P]NMP can easily be calculated by measuring the radioactivity found in the corresponding spot of the thin-layer plate and normalizing this number to the theoretically calculated number of moles of the parent nucleotide, determined from the known primary sequence of the RNA or the RNA fragment.

2D thin-layer chromatographic analysis of complete nucleotide hydrolysates has also been applied to recombinant RNA molecules. With this type of substrate, the [^{32}P] label is generally present at only one position in the RNA substrate (Fig. 6). Therefore, very clear and unambiguous data are obtained. In this case, the accuracy of the method is optimal and the technique can be used irrespective of the size of the RNA substrate. Because only one nucleotide in the molecule bears a radioactive label, even a rather low level of conversion provides a reliably detectable signal for quantification.

For examples of the application of nearest-neighbor analysis to study several nucleotide modifications in tRNA transcripts or recombinant tRNAs, see Nishikura and De Robertis, 1981; Grosjean et al., 1996b; and Jiang et al., 1997.

Primer Extension Sequencing Analysis of RNA

For longer RNA molecules (mRNA or rRNA), the various techniques based on the reverse transcription reaction are certainly the most appropriate. They consist of sequencing of RNA molecules by using specific synthetic DNA primers, viral reverse transcriptase as the chain elongation enzyme, deoxynucleoside triphosphate (dNTP)-dideoxynucleoside triphosphate (ddNTP) mixtures for sequencing, or dNTPs for primer extension. The main advantage of the method is the use of unfractionated RNA as a starting material and the rapidity and selectivity of the sequencing, ensured by the use of a unique sequence-specific DNA probe as the transcription primer. Also, by using appropriate "walking" primers, a complete scan of a long RNA can be achieved (Boorstein and Craig, 1989; Hahn et al., 1989).

One of the most straightforward applications of reverse transcriptase-based RNA sequencing is the search for putative editing sites by comparing the RNA sequence obtained by reverse transcription with the corresponding genomic DNA sequence. This method allowed the identification of many editing sites in RNA altered by base insertion or base conversion editing (reviewed in Benne, 1996) (see Chapters 16–19, 21, and 22 and Appendix 2).

An interesting variation of the reverse transcriptase sequencing approach has been used to detect the base conversion editing (C → U) in mRNAs from organelles (Fig. 10). Here, a ^{32}P-radiolabeled oligonucleotide primer is hybridized to the edited and unedited RNA (both present in the bulk RNA), a few

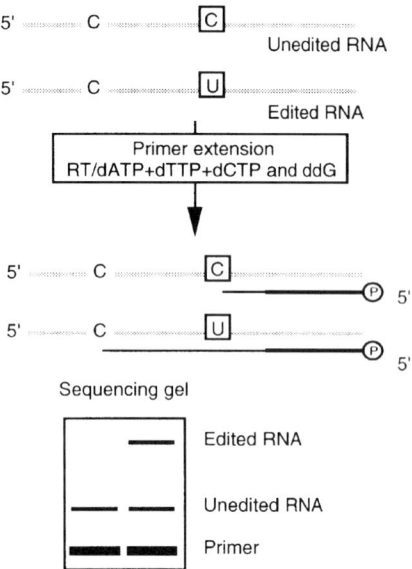

Figure 10. Identification of a C-to-U editing site by the primer extension assay. Here a 5'-^{32}P-labeled DNA primer (thick black line) is prepared so that it hybridizes close to the editing site, leaving no C residue between the 3' end of the primer and the first putative edited C. Reverse transcription is performed in the presence of dATP, dTTP, dCTP, and ddGTP. A "strong stop" will appear on the sequencing gel at unedited C, except if an edited U is present (Driscoll et al., 1989). The same analysis can be performed for testing A-to-I editing or modification sites in RNA. Indeed, inosine will be read as G during transcription. From such analysis, the ratio of edited to nonedited molecules at a given position in the RNA population can be evaluated.

nucleotides downstream of the known conversion site (C in this case). The primer is then extended by reverse transcriptase in the presence of dATP, dCTP, dTTP, and ddGTP (instead of dGTP), which generates a strong "stop" only at the unedited C site, while the polymerase reads through the edited U site until it reaches the next C (Driscoll and Casanova, 1990; Seiwert and Stuart, 1994).

A two-step procedure was developed for indirect sequencing of RNAs present in very small amounts in the cell. The cDNA synthesized by reverse transcription is further amplified by PCR and cloned in the appropriate vector, and individual clones are sequenced by the standard methods. This two-step approach was extensively used to study the editing phenomena in mitochondrial tRNAs from marsupials (Yokobori and Paabo, 1995) (also see Chapter 16 by Price and Gray and Chapter 22 by Gott and Visomirski-Robic). However, in contrast to the direct RNA sequencing by reverse transcription, which gives an image of the average sequence of an RNA population, in the PCR-based indirect approach each individual clone represents the image of a single RNA molecule. In addition, the possibility of mistranscription by the thermostable DNA polymerases during the PCR amplification should always be taken into account (Lundberg et al., 1991; Barnes, 1992).

Figure 11 illustrates some applications of primer extension analysis for RNA containing modified nucleotides. In all cases, the DNA primers are 5'-end labeled with [^{32}P]phosphate and the length of the radiolabeled transcription products is determined by electrophoresis of the reaction mixture on a denaturing sequencing gel followed by autoradiography.

Bulky modified nucleotides like wybutosine (yW) or threonylcarbamoyl adenosine (t^{6}A) in tRNA or N^6,N^6-dimethyladenosine (m$^6_2$A) or 1-methyl-3-(3-amino-3-carboxyprolyl) pseudouridine (m^{1}acp$^3\Psi$) in rRNA (see Appendix 1) can easily be detected by reverse transcription. Indeed, upon primer extension with reverse transcriptase, these bulky modified bases block the elongation of the growing DNA chain and consequently give rise to strong stops in the sequencing gel (Hagenbüchle et al., 1978; Wittig and Wittig, 1978; Youvan and Hearst, 1981; Ehresmann et al., 1987; Lafontaine et al., 1995; Ofengand and Bakin, 1997). The presence of N^2-methylguanine during reverse transcription of RNA induces only a pause detectable on gels (Youvan and Hearst, 1979). On the other hand, some other modified nucleotides, such as 1-methyladenosine (m^{1}A) or dihydrouridine (D), that are unable to form a canonical Watson–Crick base pair, not only slow down the rate of chain elongation, but in addition induce "mistranscription" (mutagenic insertion of noncomplementary deoxynucleotides) by the reverse transcriptase (Wittig and Wittig, 1978; Liu et al., 1995). The presence of such mutations or an ambiguous sequence of the cDNA may indicate the presence of modified nucleotides. However, the exact nature and chemical structure have to be determined by alternative methods.

Nonbulky modified nucleotides in RNA, like those from Watson–Crick base pairs (such as m^{5}C or s^{4}U), as well as those modified nucleotides (such as Ψ) which do not impair the incorporation of nucleotides into the growing cDNA chain, cannot be localized within the template RNA by simple primer extension. In such cases a prior chemical treatment changing the structure of the base has to be used. For mapping pseudouridine in RNA, the method takes the advantage of the enhanced alkaline stability of the bulky carbodiimide derivative of pseudouridine as compared to the same derivative formed by uridine (Ho and Gilham, 1971). Therefore, under suitable conditions, Ψ residues in RNA chain can be randomly chemically altered, giving rise to corresponding "strong stops" in the sequencing gel at each Ψ position (Fig. 11) (also see Chapter 12 by Ofengand and Fournier and Chapter 11 by Massenet et al.).

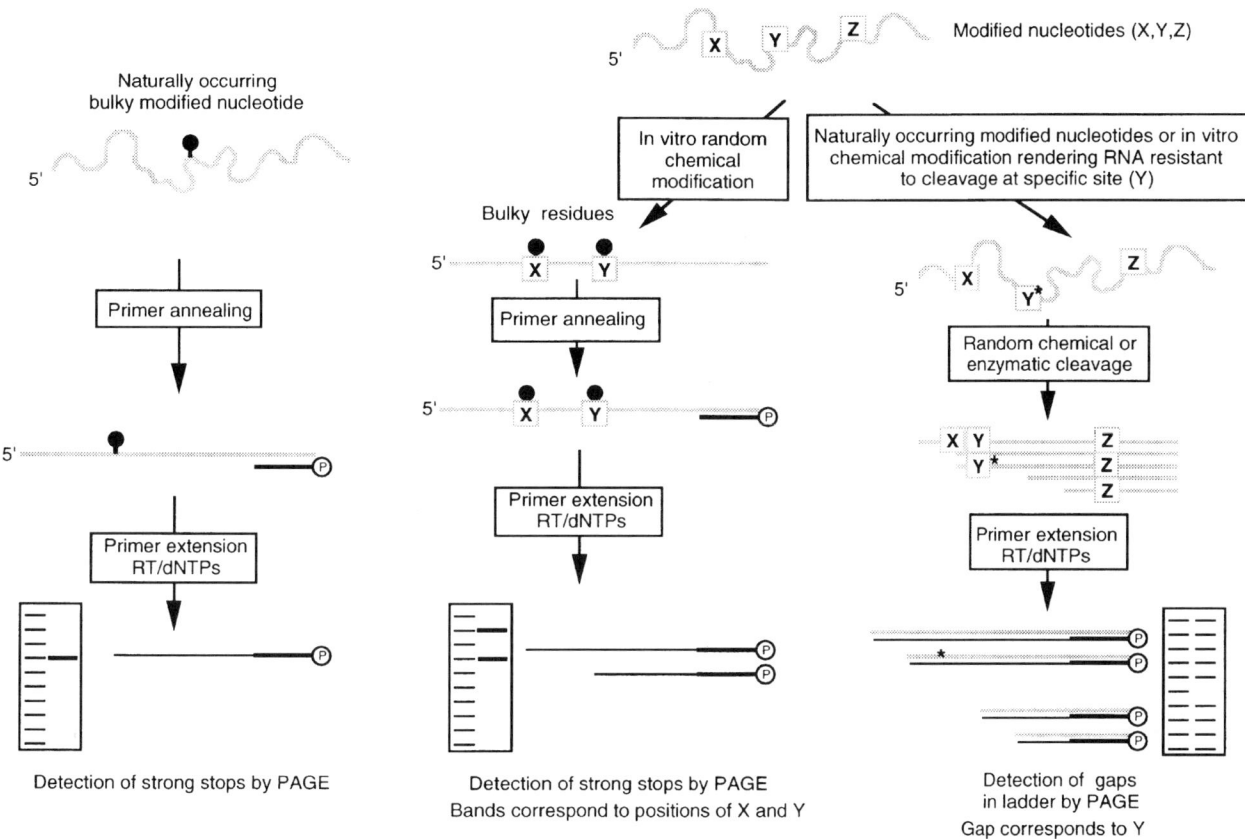

Figure 11. Detection of modified nucleotides in RNA by combined chemical and reverse transcription techniques. The cartoon illustrates different potential situations. In one case (left part) RNA contains a naturally occurring bulky residues. It will cause a strong stop upon reverse transcription that is easily detected on the sequencing gel after primer extension. In vitro chemical alteration of RNA [for example, with 1-cyclohexyl-3-(2-morpholinoethyl)-carbodiimide metho-p-toluenesulfonate (CMCT)] can produce bulky residues (middle part of the figure) that will also generate strong stops on the sequencing gel. However, if the in vitro chemical conditions are optimized for producing random (partial) chemical "hypermodification," several positions of the same RNA molecule will give the strong stop phenomenon. The right part of the figure illustrates the case where naturally occurring nucleotide modifications (like 2'-O-methylation of the ribose moiety) or an in vitro random chemical alteration (as with glyoxal on G) can render RNA resistant to enzymatic or hydrolytic cleavage at specific sites. These sites will be revealed on the sequencing gel by "gaps" in the radiolabeled RNA "ladder" after primer extension. This type of analysis gives only qualitative information, and cannot be used to quantify the relative amount of a modified nucleotide at a given position of an RNA population. The main advantages are that the RNA does not have to be pure, and that a long RNA can be explored with different primers complementary to different regions of the RNA molecule.

The development of such efficient mapping methods, based on prior alteration of nucleotides in RNA by chemical procedures, is now being explored in several laboratories (Stern et al., 1988; Rhee et al., 1995). One breakthrough in this direction is the recent work of Morse and Bass (1997) describing the approach based on the differential chemical reactivities of guanosine and inosine. The method uses the fact that glyoxal reacts specifically with guanosine, but not with inosine, despite their structural similarity. Moreover the glyoxalated guanosine is not recognized by RNase T_1, while this enzyme can cut, albeit inefficiently, after inosine. Primer extension with reverse transcriptase is therefore used to locate the remaining RNase T_1 cleaving sites that correspond to inosines in RNA. Using this method, Morse and Bass (1997) have successfully confirmed the presence of inosine at two known editing sites in the messenger RNA encoding a subunit of the glutamate receptor in rat (see Chapter 19 by Rueter and Emeson). While this method may be of general use to detect the presence of inosine in RNA, an alternative approach may be based on the recently discovered inosine-specific RNase (Scadden and Smith, 1997).

A difficult task is to locate the sites of 2'-O-methylation in RNA. One approach is based on the natural resistance of the 2'-O-ribose diphosphoester bond to alkaline hydrolysis or enzymatic cleavage (Fig. 11). This method was routinely used for mapping of 2'-O-ribose methylated sites in rRNA (see

Chapter 13 by Bachellerie and Cavaillé). Here, primer extension performed on partially digested RNAs results in the appearance of "gaps" in the reverse transcription profile. Similarly, the resistance to phosphodiester cleavage following hydrazinolysis was used to map the pseudouridine sites in rRNA and in snRNA (Fig. 11) (see Chapter 11 by Massenet et al.).

The other method is based on the observation that reverse transcriptase slows down at the sites of 2′-O-methylation, especially when a low dNTP concentration is used. This reveals "pauses" as bands with abnormal intensities on the autoradiogram of the sequencing gel, thus revealing positions of putative 2′-O-methylation sites. However, as previously pointed out, other modified nucleotides often give the same pausing phenomena and results have to be interpreted with extreme caution (discussed in Maden et al., 1995) (also see Chapter 13 by Bachellerie and Cavaillé). To circumvent the difficulties of unambiguous localization of 2′-O-methylated nucleotides in rRNA, a totally different new method has been recently presented. It is based on the use of a DNA splint and RNase H, which cannot cut the DNA/RNA hybrid at the methylation sites (Yu et al., 1997).

In a few cases, the edited nucleotides create (or destroy) the recognition sequence for a given restriction enzyme in double-stranded DNA synthesized from an RNA template. This principle has been exploited to develop a restriction enzyme assay for the identification and localization of edited nucleotides in mRNA after their transcription into dsDNA (Casey et al., 1992; Zheng et al., 1992) (for more details and other methods, see Chapter 22 by Gott and Visomirski-Robic).

All of the previously described methods based on primer reverse transcription provide extremely useful information on the location of modification or editing sites in long RNA molecules, but are not well adapted for quantification of modification or editing yields because the band intensity varies dramatically as a function of the nucleotide context or the efficiency of the chemical or enzymatic procedures used (Szkukalek et al., 1995) (also see Chapter 11 by Massenet et al.).

HOW TO DETECT ENZYME ACTIVITY

Many laboratories have now succeeded in testing a large variety of RNA modification and editing enzymes under in vitro conditions (see Chapter 8 and Appendix 3 by Garcia and Goodenough-Lashua). However, not all RNA modification and editing enzymes may be easily detectable in such experimental conditions, especially when the enzymes require unknown cofactors, such as for the Wye base in position 37 of eukaryotic tRNA^{Phe} (Droogmans and Grosjean, 1987), or belong to a more complex modification/editing machinery, such as the rRNA 2′-O-methylation and pseudouridylation machinery in the nucleolus of eukaryotic cells (see Chapter 12 by Ofengand and Fournier and Chapter 13 by Bachellerie and Cavaillé). The observation that the activity of RNA modification enzymes is often lost upon chromatographic fractionation might reflect the dependence on such cofactors (proteins, RNA, and other modification and editing enzymes). A good example is the methylation of mRNA, which requires several components that can be separated by chromatography (Bokar et al., 1997) (see Chapter 10 by Bokar and Rottman). Also, certain RNA modification and editing enzymes may require an even more complex "functional microenvironment," such as association with specific subcellular structures (membrane, nuclear pores, endoplasmic reticulum, or elements of the cytoskeleton), the exact optimal composition of which could be difficult to reproduce in vitro.

Moreover, the majority of the methods available for detecting a modified or edited nucleotide are cumbersome and time-consuming, thereby complicating the task of purification when several chromatographic steps are involved. In addition, some assay techniques provide only qualitative or semiquantitative information, making it difficult to evaluate properly the enzyme activity present in a test tube.

Figure 12 illustrates several strategies that can be used to detect the activities of enzymes catalyzing the formation of a modified or edited nucleotide, either in vitro or in vivo. The next section briefly reviews only those studies in which a foreign RNA substrate or its gene has been introduced into a living cell.

The Oocyte Microinjection Technique

The oocytes of the South African clawed toad, *Xenopus laevis*, are giant cells that are arrested at the second meiotic metaphase, waiting until fertilization occurs. At stages IV–VI the synthesis of DNA, tRNA, and rRNA has ceased, but the enzymatic machinery remains active, ready to work on exogenous microinjected DNA or RNA. The cells are preferred at stage VI for microinjection because of their larger size (about 0.8 to 1.2 mm in diameter). In addition to size, the hardiness of oocytes makes them ideally suited for microinjection of relatively large volumes of liquid (up to 20 nl into the nucleus and 50 nl into the cytoplasm; that is, about 2 to 5% of the total oocyte volume). Oocytes may even be injected twice;

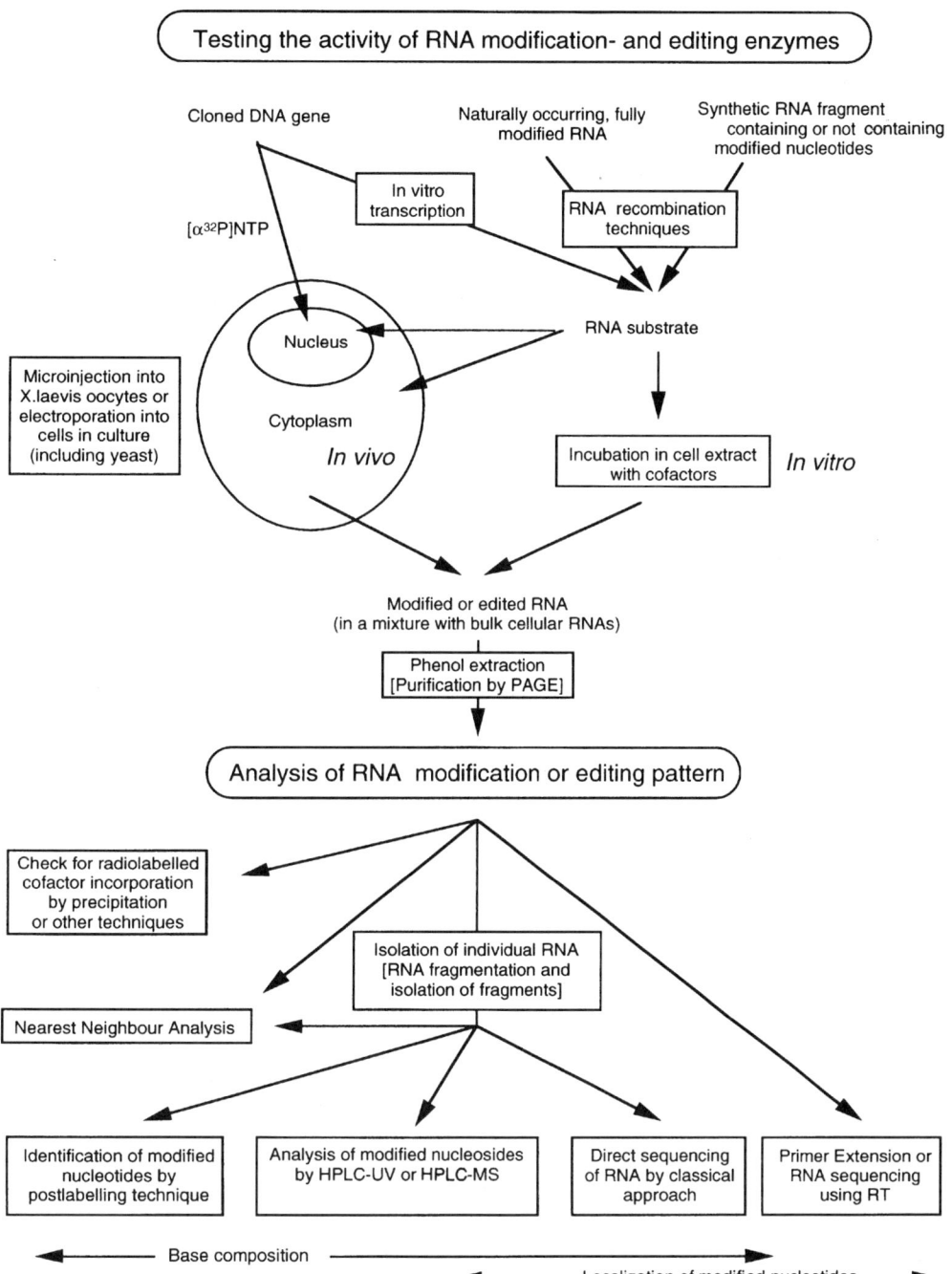

Figure 12. Alternative protocols for the identification of enzyme activities corresponding to selected RNA modification or editing phenomena. For certain types of analysis, the RNA used as substrates has to be radiolabeled with ^{32}P. This can be done either during in vitro or in vivo transcription of synthetic or natural genes, or during the recombination of RNA fragments. Details of the preparation of different substrates and RNA modification/editing analysis are described in the previous figures. HPLC, high-performance liquid chromatography; MS, mass spectrometry.

however, a few do not withstand the second injection. The microinjected oocytes may be kept alive outside the toad's body for days (usually up to 3 or 4 days) and procedures for maintaining them as functional cells are well settled. Advantageously, the supply of *Xenopus* oocytes is not restricted to a season as is the case for most other amphibians. This test system has been widely used to study various aspects of the cellular metabolism such as gene transcription, processing and maturation of various gene products (RNA and proteins), mRNA translation, stability and subcellular compartmentalization of injected macromolecules, and transport phenomena through nuclear pores (for more details consult Celis et al., 1986; Kay and Peng, 1991; Colman, 1984; Grosjean and Kubli, 1986; and Izaurralde et al., 1995).

For studying the RNA modification or editing machinery of *Xenopus laevis* oocytes, a plasmid containing a gene coding for a given RNA under the control of an appropriate promoter, together with one of the four α-^{32}P-NTPs at high specific radioactivity (3,000 Ci/mmol), can be microinjected into the nucleus. Alternatively, a ^{32}P-radiolabeled RNA precursor (T7 transcript or recombinant RNA molecule) can also be used and microinjected into either the nucleus or the cytoplasm. After a given period of incubation at 20°C (from 3 h up to 60 h), the bulk RNA from the oocytes is phenol-chloroform extracted and fractionated by electrophoresis on a denaturing polyacrylamide gel. Radioactive bands corresponding to the newly synthesized [^{32}P]RNA in the case of a microinjected gene, or the remaining radiolabeled pre-RNA transcript in the case of microinjected [^{32}P]pre-RNA, are then analyzed for their content of modified nucleotides. With this simple method, the enzymes catalyzing the formation of many of the modified nucleotides usually present in eukaryotic tRNAs have been detected, including several of the hypermodified nucleotides present in the anticodon loop (for examples see Nishikura and De Robertis, 1981; Bossi and Ciampi, 1983; Grosjean et al., 1987). Also, the existence of intermediate products in the stepwise formation of hypermodified nucleotides, such as m$^2_2$G, manQ, and Wye base, have been revealed (Droogmans and Grosjean, 1987; Haumont et al., 1987; Grosjean et al., 1990; Edqvist et al., 1992).

Microinjection of radiolabeled snRNA into the nucleus or the cytoplasm of *X. laevis* oocytes also revealed interesting problems related to the cellular trafficking of this class of RNA between the nucleus and the cytoplasm, a process that in the case of snRNA is mediated, among different proteins, by the cap-binding protein (reviewed in Görlich and Mattaj, 1996) (also see Chapter 11 by Massenet et al. and Chapter 24 by Maden).

Likewise, microinjection of antisense RNA, with the aim of blocking translation of selected mRNAs in the *Xenopus laevis* oocyte or fertilized frog eggs, revealed the existence of what was first identified as an unwinding RNA protein (Bass and Weintraub, 1987; Rebagliati and Melton, 1987). It turned out later that this protein is in fact a dsRNA deaminase, now called ADAR1 (see Chapter 19 by Rueter and Emeson).

Recently, Steitz and collaborators have used the oocyte microinjection technique to detect the involvement of specific snoRNA guides in 2'-O-methylation at a given site in *Xenopus* rRNA. To this end, the oocytes were microinjected with a short antisense DNA oligonucleotide designed to deplete the corresponding snoRNA guide in vivo. Several hours after injection, the newly synthesized rRNA was extracted from isolated oocyte nuclei and analyzed for the presence of a 2'-O-methyl group at the selected site. Detection was performed by random cleavage of rRNA under alkaline conditions followed by primer extension with reverse transcriptase (Fig. 11). The absence of 2'-O-methylation was reflected by the appearance of a stop at the unmethylated position (Tycowski et al., 1996).

To study the RNA modification and editing enzymes within eukaryotic cells smaller than the *X. laevis* oocytes, one can also use the microinjection technique (see Celis et al., 1986; also see Graessmann and Graessmann, 1983). In this case, only a tiny amount of liquid can be introduced into the cells (about 0.01 pl per cell on average; that is, about one million-fold less than into a *Xenopus* oocyte). In addition, this method requires much more laborious and time-consuming micromanipulations which seriously limit its use to study RNA modification and editing in vivo.

Cell Electroporation Technique

A better way to introduce ^{32}P-radiolabeled pre-RNA substrates into eukaryotic cells is electroporation. This technique is widely used to introduce DNA into mammalian cells in culture (Potter, 1988; Foster and Neumann, 1989; Bates, 1995; Morin et al., 1996a). Recently, it has also been used to introduce RNA (tRNA or plant viral RNA) into mammalian cells (Negrutskii and Deutscher, 1991; Tarassov and Entelis, 1992; Boyer et al., 1993).

This technique allows transient opening of small pores in the outer membrane of the cell by a controlled electric pulse of the exponential decay type. As a result, a few percent of the macromolecules present in the extracellular medium can be introduced into the cells. The cell density (about 10^7 cells per ml in case of DNA and 10^8 cells per ml in case of RNA), the pulse parameters and the time of cell incubation after the electric pulse (between 2 and 20 h) have to be adapted to obtain enough nucleic acid incorporated into the cell. Only 60% of the electroporated cells usually survive this treatment.

The main advantage of this versatile method is its applicability to a wide range of cell types. By using this technique, the enzymatic activities for a large variety of posttranscriptional modifications, including the more complex ones, were successfully detected in several pre-tRNAs (T7 transcripts) of various origins (human, *Bombyx mori*, yeast, *E. coli*) after their electroporation into human HeLa cells or *Xenopus* embryonic cells (XTC-2 and XB693T strains) (Morin et al., 1996b; Morin et al., unpublished data).

Biolistic Technique

Recent studies also report the use of biolistic methods for transfection of eukaryotic cells, mainly of insect or plant origin because of the difficulty in transfecting these cells by other methods (Mialhe and Miller, 1994). Biolistic techniques use DNA-coated particles, which are delivered into the cells at high velocity. Depending on the instrument used, three variants of the technique were described. They are referred to as "dry," "high-pressure aqueous," and "low-pressure aqueous" biolistic techniques, depending on how the DNA-coated particles are used at the time of the acceleration. Both aqueous methods offer the advantage of reproducibly delivering 100-fold larger amounts of DNA to the cells than the dry method. An important drawback of the biolistic technique, and of the electroporation technique previously described, is the lack of control of the amount and localization of the RNA substrate incorporated into the cell.

Transformed Cells

An ideal assay system for studying the activity of a given RNA modification or editing enzyme would be to analyze the formation of modified nucleotides or edited RNAs within a living cell with "normal" amounts of both the substrate and the enzyme of interest and at the right location in a living cell. Indeed, the various elements of the cellular machinery that normally processes a given RNA transcript may need to act in a temporal order. Also, interplay among different metabolisms (processing versus degradative enzymes, for example) may regulate the level of the various intermediate RNA molecules and ultimately the production of functional fully mature (normally modified or edited) RNA within the various compartments of an eukaryotic cell.

Such an ideal situation might exist in certain transformed cells in which a gene for a given RNA modification or editing enzyme, or one of its substrates, has been deleted, disrupted, or mutated. Depending on how essential the deleted enzyme or mutated substrate was, the transformed cells may be nonviable, slow growing, or almost perfectly normal (at least in appearance). Transformation and complementation of such mutant strains with plasmids carrying a gene for a new RNA modification or editing enzyme or a new RNA substrate, under the control of an appropriate promoter, should allow the testing of the activity of a foreign enzyme or an altered RNA molecule under physiological conditions that should be very close to those in the normal wild strain. Depending on how different the foreign enzyme or mutated RNA substrate is from the homologous one, nearly "normal" or artificial experimental situations may be created. The main problem with such types of experiments is the difficulty of detecting, among the bulk natural cellular RNAs, the presence or absence of a given modified or edited nucleotide(s). Such an in vivo system has been recently described for testing the structural requirements for the formation of 1-methylguanosine in tRNA^{Pro} of *Salmonella typhimurium* (Qian and Björk, 1997).

CONCLUSIONS AND PERSPECTIVES

The Substrate Problem

The lack of adequate substrates for testing the various modification enzymes in vitro was one of the major problems during the 30-year period from 1956 to 1986. Not only did it hinder the progress of the research on RNA maturation, but it also certainly discouraged many scientists working on the purification and characterization of the RNA modification enzymes. Fortunately, the development of recombinant DNA and RNA techniques and techniques for chemical synthesis of DNA and RNA now offers adequate alternatives for obtaining synthetic or semisynthetic unmodified or partially modified RNA substrates suitable for analyzing enzymatic posttranscriptional RNA modifications both in vivo and in vitro.

An unsolved problem concerns the adequacy of RNA substrates for testing RNA modification and editing enzymes. Indeed, natural or synthetic RNAs are usually isolated and purified under rather drastic denaturing conditions, such as extraction by phenol and chloroform, precipitation by ethanol, dehydration under vacuum, sometimes purification by electrophoresis in the presence of 8 M urea or formamide, and dialysis against buffers containing chelators to remove potential stabilizing metal ions. To what extent such denatured RNA molecules mimic the desired unique and biologically relevant conformation, even after renaturation procedures, is totally unknown. Also, certain RNAs may need a specific binding protein to generate or to maintain their active conformation. This remark is particularly important for the long RNA polymers, such as rRNA, mRNA, and a few snRNAs for which we know almost nothing about their "biologically relevant" conformation. The situation is more favorable for most tRNA molecules, for which these denaturation processes seem not to limit the ability to fold back to their 2D cloverleaf or their 3D L-shaped structure; however, even in this case, few examples of denatured tRNAs have been encountered (discussed in Uhlenbeck, 1995).

Identification of New RNA Modification and RNA Editing Enzymes

With the recent access to complete genome sequences, it has become possible to search for homologous genes in the different genomes, to clone them into suitable vectors after amplification by PCR, and to express them in appropriate host cells. In this way, it should soon be possible to identify, study and compare a large number of new RNA modification and editing enzymes. However, those gene products that require other cofactors or are active only under special conditions, such as associated with other proteins or certain cellular substructures, still remain difficult to study by this approach.

In Vitro Versus In Vivo Assay Systems

The in vitro approach to studying the activity and properties of the RNA modification and editing enzymes is still the most straightforward. When successfully reconstituted from isolated cellular components, such artificial systems readily allow the study of enzyme kinetics and specificity. The concentrations of enzyme and substrate are easily controlled and the experimental setup may be simple. Numerous examples of the in vitro approach are documented in the modern literature. However, such artificial in vitro systems have several drawbacks.

First, the exact nature of the indispensable components of the system is often not known, thus making the reconstitution difficult from a technical point of view. Also, substrates (like S-adenosylmethionine or isopentenyl-pyrophosphate) of the RNA modification enzymes may not be stable or may not be available commercially. As far as the macromolecular RNA substrate is concerned (especially large RNAs, such as mRNA or rRNA), their conformation in vitro is not well defined, or even shown to be different from that prevailing in vivo, where it is stabilized by interactions with other cellular proteins or other nucleic acids. Successful reconstitution of such complex ribonucleoprotein particles has been only rarely described.

Second, the enzyme (very often of recombinant origin) used in vitro is separated from its normal "physiological" cellular environment. This fact may not be of a great importance for most cytoplasmic "soluble" proteins, but may considerably influence the properties of the enzyme normally bound to subcellular structures (such as membranes or ribosomes) or incorporated into high-molecular-mass complexes with other proteins. Thus, the behavior of such an isolated protein in a test tube may not necessarily reflect its properties in the cell.

The third major complication in reconstituting an in vitro system concerns the choice of the "natural" substrate to be tested. Indeed, in addition to the existence of most of the cellular RNA in the form of ribonucleoprotein particles, one should take into account the particular spatial compartmentalization of the enzyme and its substrate, especially in the case of eukaryotic cells. In fact, most potential modification or editing sites present in cellular RNA molecules may not be accessible to the enzyme because of protection by tightly bound cellular components or because of the location of the enzyme and its potential target in the different cellular compartments.

In contrast, the in vivo approach to study the modification and editing of enzymes allows the detection of the enzymatic reaction under the "physiologically" relevant conditions. All components are present in their normal environment and in the presence of all of the necessary cofactors. However, depending on the method used to introduce the substrate or the enzyme into the cell (electroporation, microinjection, or expression from a plasmid), the fate of such artificially introduced RNA may be different from the real in vivo situation. Very little is yet known about how a microinjected or electroporated substrate diffuses within the cell and about its eventual entrapment by cellular substructures. Does it diffuse fast enough within the different compartments compared to the duration of the experiment? Does it have free access to the pool of RNA modification and editing enzymes, as do the endogenous RNA transcripts during the normal process of RNA maturation? At least in the case of aminoacyl-tRNA synthetases and other components of the protein synthesizing machinery, it was demonstrated that the intracellular pool of cytoplasmic tRNAs is sequestered and does not easily exchange with the exogenous tRNA introduced by electroporation (Negrutskii and Deutscher, 1992). These remarks also apply for the oocyte microinjection technique.

Moreover, the methods used for introducing the substrate into the living cells are not exempt from artifacts related to the cellular stress or possible cellular damage (Knight and Scrutton, 1986; van den Hoff et al., 1990; van den Hoff et al., 1992).

In actual fact the in vitro and in vivo approaches are complementary. In vitro tests are useful for detailed studies of enzyme mechanism and specificity, while in vivo approaches (including deletion mutants) give a better picture of enzyme behavior inside the cell and can provide information about its real function within the network of the complex cellular metabolism.

Despite these difficulties, and considering the exponentially increasing numbers of recent papers devoted to the problem of RNA modification or RNA editing, one can be optimistic and believe that most problems will be solved soon. At least we now have the tools to do so.

Acknowledgments. H.G. and A.M. are recipients of research grants from the Centre National de la Recherche Scientifique, the Actions de la Recherche sur le Cancer (1992–1997) and the Agence Nationale de Recherche sur le Sida (1996–1997). The salary of Y.M. was provided by an operating grant from the Medical Research Council of Canada (grant MRC-MT-1226) awarded to Dr. B. Lane, University of Toronto, Toronto, Canada. We acknowledge critical reading of the manuscript and useful comments by Dr. J.-P. Waller (CNRS, Gif-sur-Yvette, France).

REFERENCES

Agris, P. F., A. Malkiewicz, A. Kraszewski, K. Everett, B. Nawrot, E. Sochacka, J. Jankowska, and R. Guenther. 1995. Site-selected introduction of modified purine and pyrimidine ribonucleosides into RNA by automated phosphoramidite chemistry. *Biochimie* **77:**125–134.

Aphasizhev, R., B. Senger, and F. Fasiolo. 1997. Importance of structural features for tRNA^{Met} identity. *RNA* **3:**489–497.

Araya, A., V. Blanc, D. Begu, F. Crabier, A. Mouras, and S. Litvak. 1995. RNA editing in wheat mitochondria. *Biochimie* **77:**87–91.

Aström, S. U., and A. S. Byström. 1994. Rit1, a tRNA backbone-modifying enzyme that mediates initiator and elongator tRNA discrimination. *Cell* **79:**536–546.

Auxilien, S., P. F. Crain, R. W. Trewyn, and H. Grosjean. 1996. Mechanism, specificity and general properties of the yeast enzyme catalysing the formation of inosine 34 in the anticodon of transfer RNA. *J. Mol. Biol.* **262:**437–458.

Barnes, W. M. 1992. The fidelity of Taq polymerase catalyzing PCR is improved by an N-terminal deletion. *Gene* **112:**29–35.

Bass, B. L., and H. Weintraub. 1987. A developmentally regulated activity that unwinds RNA duplexes. *Cell* **48:**607–613.

Bates, G. W. 1995. Electroporation of plant protoplasts and tissues. *Methods Cell Biol.* **50:**363–373.

Becker, H. F., Y. Motorin, R. J. Planta, and H. Grosjean. 1997a. The yeast gene YNL292w encodes a pseudouridine synthase (Pus4) catalyzing the formation of Ψ55 in both mitochondrial and cytoplasmic tRNAs. *Nucleic Acids Res.* **25:**4493–4499.

Becker, H. F., Y. Motorin, M. Sissler, C. Florentz, and H. Grosjean. 1997b. Major identity determinants for enzymatic formation of ribothymidine and pseudouridine in the TΨ-loop of yeast tRNAs. *J. Mol. Biol.* **274:**505–518.

Benne, R. 1996. RNA editing: how a message is changed. *Curr. Opin. Genet. Dev.* **6:**221–231.

Björk, G. R. 1995a. Biosynthesis and function of modified nucleosides. p. 165–206. *In* D. Söll, and U. RajBhandary, (ed.), *tRNA: Structure, Biosynthesis and Function*, ASM Press, Washington.

Björk, G. R. 1995b. Genetic dissection of synthesis and function of modified nucleosides in bacterial transfer RNA. *Progr. Nucl. Acid Res. Mol. Biol.* **50:**263–338.

Bokar, J. A., M. E. Shambaugh, D. Polayes, A. G. Matera, and F. M. Rottman. 1995. Purification and cDNA cloning of the AdoMet-binding subunit of the human mRNA (N^6-adenosine)-methyltransferase. *RNA* **3:**1233–1247.

Boorstein, W. R., and E. A. Craig. 1989. Primer extension analysis of RNA. *Methods Enzymol.* **180:**347–369.

Bossi, L., and M. S. Ciampi. 1983. The expression of prokaryotic tRNA genes in frog oocytes. *Nucleic Acids Res.* **11:**3207–3226.

Boyer, J.-C., B. Zaccomer, and A.-L. Haenni. 1993. Electrotransfection of turnip yellow mosaic virus RNA into *Brassica* leaf protoplasts and detection of viral RNA products with a nonradioactive probe. *J. Gen. Virol.* **74:**1911–1917.

Branch, A. D., B. J. Bonenfeld, and H. D. Robertson. 1989. RNA fingerprinting. *Methods Enzymol.* **180:**130–154.

Bratty, J., T. F. Wu, K. Nicoghosian, K. K. Ogilvie, J. P. Perreault, G. Keith, and R. Cedergren. 1990. Characterization of a chemically synthesized RNA having the sequence of the yeast initiator tRNA^{Met}. *FEBS Lett.* **269:**60–64.

Bruce, A. G., and O. C. Uhlenbeck. 1982. Enzymatic replacement of the anticodon of yeast phenylalanine transfer ribonucleic acid. *Biochemistry* **21:**855–861.

Buck, M., M. Connick, and B. N. Ames. 1983. Complete analysis of tRNA-modified nucleosides by high-performance liquid chromatography: the 29 modified nucleosides of *Salmonella typhimurium* and *Escherichia coli* tRNA. *Anal. Biochem.* **129:**1–13.

Carbon, P., E. Haumont, M. Fournier, S. de Henau, and H. Grosjean. 1983. Site-directed *in vitro* replacement of nucleosides in the anticodon loop of tRNA: application to the study of structural requirements for queuine insertase activity. *EMBO J.* **2:**1093–1097.

Casey, J. L., K. F. Bergmann, T. L. Brown, and J. L. Gerin. 1992. Structural requirements for RNA editing in hepatitis delta virus: evidence for a uridine-to-cytidine editing mechanism. *Proc. Natl. Acad. Sci. USA* **89:**7149–7153.

Cazenave, C., and O. C. Uhlenbeck. 1994. RNA template-directed RNA synthesis by T7 RNA polymerase. *Proc. Natl. Acad. Sci. USA* **91:**6972–6976.

Cedergren, R., and H. Grosjean. 1987. RNA design by *in vitro* RNA recombination and synthesis. *Biochem. Cell. Biol.* **65:**677–692.

Celis, J. E., A. Graessmann, and A. Loyter (ed.). 1986. *Microinjection and Organelle Transplantation Techniques.* Academic Press, London, United Kingdom.

Ciampi, M. S., F. Arena, and R. Cortese. 1977. Biosynthesis of pseudouridine in the *in vitro* transcribed tRNA^{Tyr} precursor. *FEBS Lett.* **77:**75–82.

Colman, A. 1984. Translation of eukaryotic messenger RNA in *Xenopus* oocytes. *In* D. Rickwood and B. D. Hames (ed.), *Transcription and Translation: a Practical Approach*. IRL Press, Oxford, United Kingdom.

Diamond, A., and B. Dudock. 1983. Methods of RNA sequence analysis. *Methods Enzymol.* **100:**431–453.

Donnis-Keller, H., A. M. Maxam, and W. Gilbert. 1977. Mapping adenines, guanines, and pyrimidines in RNA. *Nucleic Acids Res.* **4:**2527–2538.

Doudna, J. A., C. Grosshans, A. Gooding, and C. E. Kundrot. 1993. Crystallization of ribozymes and small RNA motifs by a sparse matrix approach. *Proc. Natl. Acad. Sci. USA* **90:**7829–7833.

Draper, D. E., S. A. White, and J. M. Kean. 1988. Preparation of specific ribosomal RNA fragments. *Methods Enzymol.* **164:**221–237.

Dreher, T. W., J. J. Bujarski, and T. C. Hall. 1984. Mutant viral RNAs synthesized *in vitro* show altered aminoacylation and replicase template activities. *Nature* **311:**171–175.

Driscoll, D. M., and E. Casanova. 1990. Characterization of the apolipoprotein B mRNA editing activity in enterocyte extracts. *J. Biol. Chem.* **265:**21401–21403.

Driscoll, D. M., J. K. Wynne, S. C. Wallis, and J. Scott. 1989. An in vitro system for editing of apolipoprotein B mRNA. *Cell* **58:**519–525.

Droogmans, L., and H. Grosjean. 1987. Enzymatic conversion of guanosine 3' adjacent to the anticodon of yeast tRNA^{Phe} to N^1-methylguanosine and the Wye nucleoside: dependence on the anticodon sequence. *EMBO J.* 6:477–483.

Droogmans, L., and H. Grosjean. 1991. 2'-O-Methylation and inosine formation in the wobble position of anticodon-substituted transfer RNA^{Phe} in a homologous yeast *in vitro* system. *Biochimie* 73:1021–1025.

Droogmans, L., E. Haumont, S. de Henau, and H. Grosjean. 1986. Enzymatic 2'-O-methylation of the wobble nucleoside of eukaryotic tRNA^{Phe}: specificity depends on structural elements outside the anticodon loop. *EMBO J.* 5:1105–1109.

Eckstein, F., and O. Heidenreich. 1994. Synthesis of modified RNA: approaches and applications. *FASEB J.* 7:90–96.

Edqvist, J., H. Grosjean, and K. B. Straby. 1992. Identity elements for N^2-dimethylation of guanosine-26 in yeast tRNAs. *Nucleic Acids Res.* 20:6575–6581.

Ehresmann, C., F. Baudin, M. Mougel, P. Romby, J.-P. Ebel, and B. Ehresmann. 1987. Probing the structure of RNAs in solution. *Nucleic Acids Res.* 15:9109–9128.

Fossé, P., M. Mougel, G. Keith, E. Westhof, B. Ehresmann, and C. Ehresmann. 1998. Modified nucleotides of tRNA^{Pro} restrict interactions in the binary primer/template complex of M-MuLV. *J. Mol. Biol.* 275:731–746.

Foster, W., and E. Neumann. 1989. Gene transfer by electroporation, p. 299–318. *In* E. Neumann, A. E. Sovers, and C. A. Jordan (ed.), *Electroporation and Electrofusion in Cell Biology*. Plenum Press, New York, N.Y.

Fournier, M., E. Haumont, S. de Henau, J. Gangloff, and H. Grosjean. 1983. Posttranscriptional modification of the wobble nucleotide in anticodon-substituted yeast tRNA^{ArgII} after microinjection into *Xenopus laevis* oocytes. *Nucleic Acids Res.* 11:707–718.

Gasparutto, D., T. Livache, H. Bazin, A. M. Duplaa, A. Guy, A. Khorlin, D. Molko, A. Roget, and R. Teoule. 1992. Chemical synthesis of a biologically active natural tRNA with its minor bases. *Nucleic Acids Res.* 20:5159–5166.

Gehrke, C. W., and K. C. Kuo. 1990. Ribonucleoside analysis by reversed-phase high performance liquid chromatography, p. A3–A71. *In* C. W. Gehrke and K. C. Kuo (ed.), *Chromatography and Modifications of Nucleosides*, vol. 45A. Elsevier Press, Amsterdam, The Netherlands.

Giegé, R., C. Florentz, A. Garcia, H. Grosjean, V. Perret, J. Puglisi, A. Theobald-Dietrich, and J.-P. Ebel. 1990a. Exploring the aminoacylation function of transfer RNA by macromolecular engineering approaches. Involvement of conformational features in the charging process of yeast tRNA^{Asp}. *Biochimie* 72:453–461.

Giegé, R., J. Rudinger, T. Dreher, V. Perret, E. Westhof, C. Florentz, and J.-P. Ebel. 1990b. Search of essential parameters for the aminoacylation of viral tRNA-like molecules. Comparison with canonical transfer RNAs. *Biochim. Biophys. Acta* 1050:179–85.

Giegé, R., J. D. Puglisi, and C. Florentz. 1993. tRNA structure and aminoacylation efficiency. *Prog. Nucleic Acid Res. Mol. Biol.* 45:129–206.

Goodwin, J. T., W. A. Stanick, and G. D. Glick. 1994. Improved solid-phase synthesis of long oligoribonucleotides: application to tRNA^{Phe} and tRNA^{Gly}. *J. Org. Chem.* 59:7941–7943.

Görlich, D., and I. W. Mattaj. 1996. Nucleoplasmic transport. *Science* 271:1513–1518.

Grabowski, P. J., and P. H. Sharp. 1986. Affinity chromatography of splicing complex U2, U5 and U4+U6 small nuclear ribonucleoprotein particles in the spliceosome. *Science* 233:1294–1298.

Graessmann, M., and A. Graessmann. 1983. Microinjection of tissue culture cells. *Methods Enzymol.* 101:482–492.

Grosjean, H., and E. Kubli. 1986. Functional aspects of tRNA microinjected into *Xenopus laevis* oocytes: results and perspectives, p. 304–326. *In* J. E. Celis, A. Graessmann, and A. Loyter (ed.), *Microinjection and Organelle Transplantation Techniques*. Academic Press, London, United Kingdom.

Grosjean, H., E. Haumont, L. Droogmans, P. Carbon, M. Fournier, S. de Henau, T. Doi, G. Keith, J. Gangloff, K. Kretz, and R. Trewyn. 1987. A novel approach to the biosynthesis of modified nucleosides in the anticodon loops of eukaryotic tRNAs, p. 355–378. *In* K. Bruzik and W. Stec (ed.), *Biophosphates and Their Analogues: Synthesis, Structure, Metabolism and Activity*. Elsevier Science Publishers, Amsterdam, The Netherlands.

Grosjean, H., L. Droogmans, R. Giegé, and O. C. Uhlenbeck. 1990. Guanosine modifications in runoff transcripts of synthetic transfer RNA^{Phe} genes microinjected into *Xenopus* oocytes. *Biochim. Biophys. Acta* 1050:267–273.

Grosjean, H., F. Constantinesco, D. Foirct, and N. Benachenhou. 1995a. A novel enzymatic pathway leading to 1-methylinosine modification in *Haloferax volcanii* tRNA. *Nucleic Acids Res.* 23:4312–4319.

Grosjean, H., M. Sprinzl, and S. Steinberg. 1995b. Posttranscriptionally modified nucleosides in transfer RNA: their locations and frequencies. *Biochimie* 77:139–141.

Grosjean, H., S. Auxilien, F. Constantinesco, C. Simon, Y. Corda, H. F. Becker, D. Foiret, A. Morin, Y. X. Jin, M. Fournier, and J. L. Fourrey. 1996a. Enzymatic conversion of adenosine to inosine and to N-1-methylinosine in transfer RNAs: a review. *Biochimie* 78:488–501.

Grosjean, H., J. Edqvist, K. B. Straby, and R. Giegé. 1996b. Enzymatic formation of modified nucleosides in tRNA: dependence on tRNA architecture. *J. Mol. Biol.* 255:67–85.

Gupta, R. C., and K. Randerath. 1979. Rapid print-readout technique for sequencing of RNAs containing modified nucleotides. *Nucleic Acids Res.* 6:3443–3458.

Gustafsson, C., R. Reid, P. J. Greene, and D. V. Santi. 1996. Identification of new RNA modifying enzymes by iterative genome search using known modifying enzymes as probes. *Nucleic Acids Res.* 24:3756–3762.

Hagenbüchle, O., M. Santer, and J. A. Steitz. 1978. Conservation of the primary structure at the 3' end of 18S rRNA from eucaryotic cells. *Cell* 13:551–563.

Hahn, C. S., E. G. Strauss, and J. H. Strauss. 1989. Dideoxy sequencing of RNA using reverse transcriptase. *Methods Enzymol.* 180:121–130.

Hall, K. B., J. R. Sampson, O. C. Uhlenbeck, and A. G. Redfield. 1989. Structure of an unmodified tRNA molecule. *Biochemistry* 28:5794–5801.

Harada, F., M. Matsubara, and N. Kato. 1984. Stable tRNA precursors in HeLa cells. *Nucleic Acids Res.* 12:9263–9269.

Haumont, E., M. Fournier, S. de Henau, and H. Grosjean. 1984. Enzymatic conversion of adenosine to inosine in the wobble position of yeast tRNA^{Asp}: the dependence on the anticodon sequence. *Nucleic Acids Res.* 12:2705–2715.

Haumont, E., L. Droogmans, and H. Grosjean. 1987. Enzymatic formation of queuosine and of glycosyl queuosine in yeast tRNAs microinjected into *Xenopus laevis* oocytes. The effect of the anticodon loop sequence. *Eur. J. Biochem.* 168:219–225.

Hayase, Y., M. Jahn, M. J. Rogers, L. A. Sylvers, M. Koizumi, H. Inoue, E. Ohtsuka, and D. Söll. 1992. Recognition of bases in *E. coli* tRNA^{Gln} by glutaminyl-tRNA synthetase: a complete identity set. *EMBO J.* 11:4159–4165.

Helm, M., H. Brulé, C. Degoul, C. Cepanec, J.-P. Leroux, R. Giegé, and C. Florentz. 1997. The presence of a modified nucleoside is required for the cloverleaf folding of a human mitochondrial tRNA. *17th International tRNA Workshop, Kazusa Akademia Center, Japan*, p. 3–13.

Ho, N., and P. Gilham. 1971. Reaction of pseudouridine and inosine with N-cyclohexyl-N'-beta-(4-methylmorpholinium) ethylcarbodiimide. *Biochemistry* **10**:3651–3657.

Hollcy, R. W., J. Apgar, G. A. Everett, J. T. Madison, M. Marquise, S. H. Merril, J. R. Penswick, and R. Zamir. 1965. Structure of a ribonucleic acid. *Science* **147**:1462–1465.

Hopper, A. K. 1990. Genetic methods for study of trans-acting genes in processing of precursors to yeast cytoplasmic tRNAs. *Methods Enzymol.* **181**:400–421.

Hopper, A. K., F. Banks, and V. Evangelidis. 1978. A yeast mutant which accumulates precursor tRNAs. *Cell* **14**:211–219.

Horton, R. M., and L. R. Pease. 1991. Recombination and mutagenesis of DNA sequences using PCR, p. 217–247. In M. J. McPherson (ed.), *Directed Mutagenesis: a Practical Approach*. IRL Press, Oxford, United Kingdom.

Inoue, H., I. Hayase, A. Imura, K. Iwai, and E. Ohtsuka. 1987. Sequence-dependent hydrolysis or RNA using modified oligonucleotide splints and RNase H. *FEBS Lett.* **215**:327–330.

Iwase, R., M. Maeda, T. Fujiwara, M. Sekine, T. Hata, and K. I. Miura. 1992. Molecular design of a eukaryotic messenger RNA and its chemical synthesis. *Nucleic Acids Res.* **20**:1643–1648.

Izaurralde, E., J. Lewis, C. Gamberi, A. Jarmolowski, C. McGuigan, and I. Mattaj. 1995. A cap-binding protein complex mediating U snRNA export. *Nature* **376**:709–712.

Jiang, H.-Q., Y. A. Motorin, Y.-X. Jin, and H. Grosjean. 1997. Pleiotropic effects of intron removal on base modifications pattern of yeast tRNA^{Phe}: an *in vitro* study. *Nucleic Acids Res.* **25**:2694–2701.

Johnson, L., H. Hayashi, and D. Söll. 1970. Isolation and properties of a transfer ribonucleic acid deficient in ribothymidine. *Biochemistry* **9**:2823–2831.

Joyce, G. F. 1994. *In vitro* evolution of nucleic acids. *Curr. Opin. Struct. Biol.* **4**:331–336.

Kaufmann, G., and U. Z. Littauer. 1975. Covalent joining of phenylalanine transfer ribonucleic acid half-molecules by T4 RNA ligase. *Proc. Natl. Acad. Sci. USA* **71**:3741–3745.

Kay, B. K., and H. B. Peng (ed.). 1991. *Methods in Cell Biology*, vol. 36. *Xenopus laevis: Practical Uses in Cell and Molecular Biology*. Academic Press, New York, N.Y.

Keith, G. 1995. Mobilities of modified ribonucleotides on two-dimensional cellulose thin-layer chromatography. *Biochimie* **77**:142–144.

Knight, D. E., and M. C. Scrutton. 1986. Gaining access to the cytosol: the technique and some applications of electropermeabilization. *Biochem. J.* **234**:497–506.

Komine, Y., T. Adachi, H. Inokuchi, and H. Ozeki. 1990. Genomic organization and physical mapping of the transfer RNA genes in *Escherichia coli* K12. *J. Mol. Biol.* **212**:579–598.

Koonin, E. V. 1996. Pseudouridine synthases: four families of enzymes containing a putative uridine-binding motif also conserved in dUTPases and dCTP deaminases. *Nucleic Acids Res.* **24**:2411–2415.

Kostyuk, D. A., S. M. Dragan, D. L. Lyakhov, V. O. Rechinsky, V. L. Tunitskaya, B. K. Chernov, and S. N. Kochetkov. 1995. Mutants of T7 RNA polymerase that are able to synthesize both RNA and DNA. *FEBS Lett.* **369**:165–168.

Kretz, K. A., R. W. Trewyn, G. Keith, and H. Grosjean. 1990. Site directed replacement of nucleotides in the anticodon loop of tRNA: application to the study of inosine biosynthesis in yeast tRNA^{Ala}, p. B144–B171. In C. W. Gehrke and K. C. T. Kuo (ed.), *Chromatography and Modification of Nucleosides Part B: Biological Roles and Function of Modification. J. Chrom. Library*, vol. 45B. Elsevier Science Publishing, Amsterdam, The Netherlands.

Krzyzosiak, W., R. Denman, K. Nurse, W. Hellmann, M. Boublik, C. Gehrke, P. Agris, and J. Ofengand. 1987. *In vitro* synthesis of 16S ribosomal RNA containing single base changes and assembly into a functional 30S ribosome. *Biochemistry* **26**:2353–2364.

Kunkel, T. A., J. D. Robert, and R. A. Zakour. 1987. Rapid and efficient site-specific mutagenesis without phenotypic selection. *Methods Enzymol.* **154**:367–382.

Lafontaine, D., J. Delcour, A. L. Glasser, J. Desgres, and J. Vandenhaute. 1994. The DIM1 gene responsible for the conserved m$^6_2$Am$^6_2$A dimethylation in the 3'-terminal loop of 18 S rRNA is essential in yeast. *J. Mol. Biol.* **241**:492–497.

Lafontaine, D., J. Vandenhaute, and D. Tollervey. 1995. The 18S rRNA dimethylase Dim1p is required for pre-ribosomal RNA processing in yeast. *Genes and Devel.* **9**:2470–2481.

Lafontaine, D. L. J., C. Bousquet-Antonelli, Y. Henry, M. Caizergues-Ferrer, and D. Tollervey. 1998. The box H+ACA snoRNAs carry Cbf5p, the putative rRNA pseudouridine synthase. *Genes Dev.* **12**:527–537.

Lapham, J., and D. M. Crothers. 1996. RNase H cleavage for processing of *in vitro* transcribed RNA for NMR studies and RNA ligation. *RNA* **2**:289–296.

Lapham, J., Y.-T. Yu, M.-D. Shu, J. A. Steitz, and D. M. Crothers. 1997. The position of site-directed cleavage of RNA using RNAse H and 2'-O-methyl oligonucleotides is dependent from the enzyme source. *RNA* **3**:950–951.

Lecointe, F., G. Simos, A. Sauer, E. C. Hurt, Y. Motorin, and H. Grosjean. 1998. Identification of yeast protein Deg 1 as Pseudouridine synthase (Pus 3) catalysing the formation of Ψ_{38} and Ψ_{39} in tRNA anticodon loop. *J. Biol. Chem.* **273**:1316–1323.

Liu, J., W. Zhou, and P. W. Doetsch. 1995. RNA polymerase bypass at sites of dihydrouracil: implications for transcriptional mutagenesis. *Mol. Cell. Biol.* **15**:6729–6735.

Lowary, P., J. Sampson, J. Milligan, D. Groebe, and O. C. Uhlenbeck. 1986. A better way to make RNA for physical studies, p. 69–76. In P. H. van Knippenberg and C. W. Hilbers (ed.), *Structure and Dynamics of RNA*, vol. 110. Plenum Press, New York, N.Y.

Lundberg, K. S., D. D. Shoemaker, M. W. Adams, J. M. Short, J. A. Sorge, and E. J. Mathur. 1991. High-fidelity amplification using a thermostable DNA polymerase isolated from *Pyrococcus furiosus*. *Gene* **108**:1–6.

Maden, B. E. H., M. E. Corbett, P. A. Heeney, K. Pugh, and P. M. Ajuh. 1995. Classical and novel approaches to the detection and localization of the numerous modified nucleotides in eukaryotic ribosomal RNA. *Biochimie* **77**:22–29.

McCloskey, J. A. 1991. Structural characterization of natural nucleosides by mass spectrometry. *Acc. Chem. Res.* **24**:81–88.

Melton, D. A., P. A. Krieg, M. R. Rebagliati, T. Maniatis, K. Zinn, and M. R. Green. 1984. Efficient *in vitro* synthesis of biologically active RNA and RNA hybridization probes from plasmids containing a bacteriophage SP6 promoter. *Nucleic Acids Res.* **12**:7035–7056.

Meyhack, B., B. Pace, O. C. Uhlenbeck, and N. Pace. 1978. Use of T4-RNA ligase to construct model substrate for a ribosomal RNA maturation endonuclease. *Proc. Natl. Acad. Sci. USA* **75**:3045–3049.

Mialhe, E., and L. H. Miller. 1994. Biolistic techniques for transfection of mosquito embryos (*Anopheles gambiae*). *BioTechniques* **16**:924–931.

Miele, E. A., D. R. Mills, and F. R. Kramer. 1983. Autocatalytic replication of a recombinant RNA. *J. Mol. Biol.* **171**:281–295.

Milligan, J. F., and O. C. Uhlenbeck. 1989. Synthesis of small RNAs using T7 polymerase. *Methods Enzymol.* **180**:51–62.

Moore, M., and P. Sharp. 1992. Site-specific modification of pre-mRNA: the 2'OH groups at the splice sites. *Science* **256**:992–997.

Moore, M. J., and C. C. Query. 1998. Uses of specifically modified RNAs constructed by RNA ligation, p. 75–108. *In* C. Smith (ed.), *RNA-Protein Interactions: a Practical Approach*. IRL Press, Oxford, United Kingdom.

Morin, A., A. Belayew, J. A. Martial, C. Tougard, and A. Tixier-Vidal. 1996a. Expression and secretion of rat prolactin in transfected pituitary cells in culture. *Mol. Cell Endocrinol.* 117:59–73.

Morin, A., D. Foiret, and H. Grosjean. 1996b. Variation of tRNAs modifying enzymes activities during differentiation of normal cells and in tumoral cells. *6th International Congress on Cell Biology, December 7–11, San Francisco, California*, 493a.

Morin, A., S. Auxilien, B. Senger, R. Tewari, and H. Grosjean. 1998. Structural requirements for enzymatic formation of threonylcarbamoyladenosine (t^6A) in tRNA: an *in vivo* study with *Xenopus laevis* oocytes. *RNA* 4:24–37.

Morin, A., C. Simon, D. Foiret, and H. Grosjean. Unpublished data.

Mörl, M., M. Dörner, and S. Pääbo. 1994. Direct purification of tRNAs using oligonucleotides coupled to magnetic beads, p. 107–111. *In* M. Uhlén, E. Hornes, and O. Olsvik (ed.), *Advances in Biomagnetic Separation*. Eaton Publishers, Natick, Mass.

Morse, D. P., and B. L. Bass. 1997. Detection of inosine in messenger RNA by inosine-specific cleavage. *Biochemistry* 36:8429–8434.

Motorin, Y., G. Bec, R. Tewari, and H. Grosjean. 1997. Transfer RNA recognition by the *Escherichia coli* Δ^2-isopentenylpyrophosphate:tRNAΔ^2-isopentenyl transferase: dependence on the anticodon arm structure. *RNA* 3:721–733.

Motorin, Y., C. Simon, D. Foiret, G. Keith, G. Simos, E. Hurt, and H. Grosjean. Unpublished data.

Mueller, S. O., and R. K. Slany. 1995. Structural analysis of the interaction of the tRNA modifying enzymes Tgt and QueA with a substrate tRNA. *FEBS Lett.* 361:259–264.

Mullenbach, G. T., H. O. Kammen, and E. E. Penhoet. 1976. A heterologous system for detecting eukaryotic enzymes which synthesize pseudouridine in transfer ribonucleic acids. *J. Biol. Chem.* 251:4570–4578.

Mullis, K. B., and F. A. Faloona. 1987. Specific synthesis of DNA *in vitro* via a polymerase-catalyzed chain reaction. *Methods Enzymol.* 155:335–350.

Nakamura, H., Y. Oda, S. Iwai, H. Inoue, E. Ohtsuka, S. Kanaya, S. Kimura, C. Katsuda, K. Katayanagi, K. Morikawa, H. Miyashiro, and M. Ikehara. 1991. How does RNase H recognize a DNA*RNA hybrid? *Proc. Natl. Acad. Sci. USA* 88:11535–11539.

Narayan, P., and F. Rottman. 1988. An *in vitro* system for accurate methylation of internal adenosine residues in messenger RNA. *Science* 242:1159–1162.

Nau, F. 1976. The methylation of tRNA. *Biochimie* 58:629–645.

Negrutskii, B. S., and M. P. Deutscher. 1991. Channeling of aminoacyl-tRNA for protein synthesis *in vivo*. *Biochemistry* 30:4991–4995.

Negrutskii, B. S., and M. P. Deutscher. 1992. A sequestered pool of aminoacyl-transfer RNA in mammalian cells. *Proc. Natl. Acad. Sci. USA* 89:3601–3604.

Nishikura, L., and E. M. De Robertis. 1981. RNA processing in microinjected *Xenopus* oocytes. Sequential addition of base modification in a spliced transfer RNA. *J. Mol. Biol.* 154:405–420.

Ofengand, J., and A. Bakin. 1997. Mapping to nucleotide resolution of pseudouridine residues in large subunit ribosomal RNAs from representative eukaryotes, prokaryotes, archaebacteria, mitochondria and chloroplasts. *J. Mol. Biol.* 266:246–268.

Ogilvie, K. K., N. Usman, K. Nicoghosian, and R. J. Cedergren. 1988. Total chemical synthesis of a 77-nucleotide-long RNA sequence having methionine-acceptance activity. *Proc. Natl. Acad. Sci. USA* 85:5764–5768.

Ohtsuka, E., S. Nishikawa, M. Ikehara, and S. Takemura. 1976. Reconstitution of chemically synthesized ribooligonucleotides with naturally occurring tRNA fragments. *Eur. J. Biochem.* 66:251–255.

Ohtsuka, E., S. Tanaka, T. Tanaka, T. Miyake, A. F. Markham, E. Nakagawa, T. Wakabayashi, Y. Taniyama, S. Nishikawa, R. Fukumoto, H. Uemura, T. Doi, T. Tokunaga, and M. Ikehara. 1981. Total synthesis of a RNA molecule with sequence identical to that of *Escherichia coli* formylmethionine tRNA. *Proc. Natl. Acad. Sci. USA* 78:5493–5497.

Ohtsuka, E., T. Doi, R. Futumoto, H. Matsugi, and M. Ikehara. 1983. Modification of the anticodon triplet of *E. coli* tRNA^{fMet} by replacement with trimers complementary to non-sense codons UAG and UAA. *Nucleic Acids Res.* 11:3863–3872.

Patton, J. R. 1991. Pseudouridine modification of U5 RNA in ribonucleoprotein particles assembled *in vitro*. *Mol. Cell. Biol.* 11:5998–6006.

Paul, M. S., and B. L. Bass. 1997. Sensitive methods for the detection of inosine in messenger RNA. *Second RNA Meeting, 21–26 May, Banff, Canada*, p. 408.

Peattie, D. A., and W. Gilbert. 1980. Chemical probes for higher-order structure in RNA. *Proc. Natl. Acad. Sci. USA* 77:4679–4682.

Pegg, A. E. 1972. Methylation of yeast aspartic acid transfer RNA by rat liver extracts. *FEBS Lett.* 22:339–342.

Pley, H. W., K. M. Flaherty, and D. B. McKay. 1994. Three-dimensional structure of a hammerhead ribozyme. *Nature* 372:68–74.

Polson, A. G., P. F. Crain, S. C. Pomerantz, J. A. McCloskey, and B. L. Bass. 1991. The mechanism of adenosine to inosine conversion by the double-stranded RNA unwinding/modifying activity—a high-performance liquid chromatography-mass spectrometry analysis. *Biochemistry* 30:11507–11514.

Pomerantz, S. C., and J. A. McCloskey. 1990. Analysis of RNA hydrolyzates by LC/MS. *Methods Enzymol.* 193:796–824.

Potter, H. 1988. Electroporation in biology: methods applications and instrumentation. *Anal. Biochem.* 174:361–373.

Qian, Q., and G. R. Björk. 1997. Structural requirements for the formation of 1-methylguanosine *in vivo* in tRNA^{Pro}(GGG) of *Salmonella typhimurium*. *J. Mol. Biol.* 266:283–297.

Randerath, E., and K. Randerath. 1983. Selected postlabeling procedures for base composition and sequence analysis of nucleic acids, p. 169–233. *In* S. M. Weissman (ed.), *Methods of DNA and RNA sequencing*. Praeger Publishers, New York, N.Y.

Rebagliati, M. R., and D. A. Melton. 1987. Antisense RNA injections in fertilized frog eggs reveal an RNA duplex unwinding activity. *Cell* 48:599–605.

Reyes, V. M., and J. Abelson. 1987. A synthetic substrate for tRNA splicing. *Anal. Biochem.* 166:90–106.

Rhee, Y., M. R. Valentine, and J. Termini. 1995. Oxidative base damage in RNA detected by reverse transcriptase. *Nucleic Acids Res.* 23:3275–3282.

Rogg, H., R. Brambilla, G. Keith, and M. Staehelin. 1976. An improved method for the separation and quantitation of the modified nucleosides of transfer RNA. *Nucleic Acids Res.* 3:285–295.

Sampson, J., F. Sullivan, L. Behlen, A. DiRenzo, and O. C. Uhlenbeck. 1987. Characterization of two RNA-catalyzed RNA cleavage reactions. *Cold Spring Harbor Symp. Quant. Biol.* 52:267–275.

Samuelsson, T., T. Boren, T. I. Johansen, and F. Lustig. 1988. Properties of a transfer RNA lacking modified nucleosides. *J. Biol. Chem.* 263:13692–13699.

Santoro, S. W., and G. F. Joyce. 1997. A general purpose RNA-cleaving DNA enzyme. *Proc. Natl. Acad. Sci. USA* 94:4262–4266.

Sayers, J., C. Krekel, and F. Eckstein. 1992. Rapid high-efficiency site-directed mutagenesis by the phosphorothioate approach. *BioTechniques* 13:592–596.

Scadden, A. D. J., and C. W. J. Smith. 1997. A ribonuclease specific for inosine containing RNA: a potential role in antiviral defence? *EMBO J.* 16:2140–2149.

Schaefer, K. P., S. Altman, and D. Söll. 1973. Nucleotide modification *in vitro* of the precursor of transfer RNA of *Escherichia coli*. *Proc. Natl. Acad. Sci. USA* 70:3626–3630.

Scott, W. G., J. T. Finch, R. Grenfell, J. F. T. Smith, M. J. Gait, and A. Klug. 1995. Rapid crystallization of chemically synthesized hammerhead RNAs using a double screening procedure. *J. Mol. Biol.* 250:327–332.

Seiwert, S. D., and K. Stuart. 1994. RNA editing: transfer of genetic information from gRNA to precursor mRNA *in vitro*. *Science* 266:114–117.

Silberklang, M., A. M. Gillum, and U. L. RajBhandary. 1979. Use of *in vitro* ^{32}P labelling in the sequence analysis of nonradioactive tRNAs. *Methods Enzymol.* 59:58–109.

Simos, G., H. Tekotte, H. Grosjean, A. Segref, K. Sharma, D. Tollervey, and E. C. Hurt. 1996. Nuclear pore proteins are involved in the biogenesis of functional tRNA. *EMBO J.* 15:2270–2284.

Slany, R. K., and H. Kersten. 1994. Genes, enzymes and coenzymes of queuosine biosynthesis in procaryotes. *Biochimie* 76:1178–1182.

Smith, J. E., B. S. Cooperman, and P. Mitchell. 1992. Methylation sites in *Escherichia coli* ribosomal RNA: Localization and identification of four new sites of methylation in 23S rRNA. *Biochemistry* 31:10825–10834.

Sprinzl, M., C. Horn, M. Brown, A. Ioudovitch, and S. Steinberg. 1998. Compilation of tRNA sequences and sequences of tRNA genes. *Nucleic Acids Res.* 26:148–153.

Stanley, J., and S. Vassilenko. 1978. A different approach to RNA sequencing. *Nature* 274:87–89.

Steinberg, S., and R. Cedergren. 1995. A correlation between N^2-dimethylguanosine presence and alternate tRNA conformers. *RNA* 1:886–891.

Stern, S., D. Moazed, and H. Noller. 1988. Structural analysis of RNA using chemical and enzymatic probing monitored by primer extension. *Methods Enzymol.* 164:481–489.

Streeter, D. G., and B. G. Lane. 1970. Studies of the biogenesis of N^2-dimethylguanylate. I. Generation of N^2-dimethylguanylate when bulk *Escherichia coli* transfer RNA is used as a substrate for wheat embryo methyltransferases. *Biochim. Biophys. Acta* 199:394–404.

Sylvers, L. A., K. C. Rogers, M. Shimizu, E. Ohtsuka, and D. Soll. 1993. A 2-thiouridine derivative in tRNA^{Glu} is a positive determinant for aminoacylation by *Escherichia coli* glutamyl-tRNA synthetase. *Biochemistry* 32:3836–3841.

Szkukalek, A., E. Myslinski, A. Mougin, R. Luhrmann, and C. Branlant. 1995. Phylogenetic conservation of modified nucleotides in the terminal loop 1 of the spliceosomal U5 snRNA. *Biochimie* 77:16–21.

Szostak, J. W., and A. D. Ellington. 1993. In vitro selection of functional RNA sequences, p. 511–533. *In* R. F. Gesteland and J. F. Atkins (ed.), *The RNA World*. Cold Spring Harbor Laboratory Press, Cold Spring Harbor, N.Y.

Szweykowska-Kulinska, Z., B. Senger, G. Keith, F. Fasiolo, and H. Grosjean. 1994. Intron-dependent formation of pseudouridines in the anticodon of *Saccharomyces cerevisiae* minor tRNA^{Ile}. *EMBO J.* 13:4636–4644.

Tarassov, I. A., and N. S. Entelis. 1992. Mitochondrially imported cytoplasmic transfer RNA^{Lys}(CUU) of Saccharomyces cerevisiae: *in vivo* and *in vitro* targetting systems. *Nucleic Acids Res.* 20:1277–1281.

Tsurui, H., Y. Kumazawa, R. Sanokawa, Y. Watanabe, T. Kuroda, A. Wada, K. Watanabe, and T. Shirai. 1994. Batchwise purification of specific tRNAs by a solid-phase DNA probe. *Anal. Biochem.* 221:166–172.

Tycowski, K. T., C. M. Smith, M. D. Shu, and J. A. Steitz. 1996. A small nucleolar RNA requirement for site-specific ribose methylation of rRNA in *Xenopus*. *Proc. Natl. Acad. Sci. USA* 93:14480–14485.

Uhlenbeck, O. C., and V. Cameron. 1977. Equimolar addition of oligoribonucleotides with T4-RNA ligase. *Nucleic Acids Res.* 4:85–98.

Uhlenbeck, O. C. 1995. Keeping RNA happy. *RNA* 1:4–6.

Usman, N., and R. Cedergren. 1992. Exploiting the chemical synthesis of RNA. *Trends Biochem. Sci.* 17:334–339.

Usman, N., M. Egli, and A. Rich. 1992. Large scale chemical synthesis, purification and crystallization of RNA-DNA chimeras. *Nucleic Acids Res.* 20:6695–6699.

van den Hoff, M. J. B., W. T. Labruyère, A. F. M. Moorman, and W. H. Lamers. 1990. The osmolarity of the electroporation medium affects the transient expression of genes. *Nucleic Acids Res.* 18:6464.

van den Hoff, M. J. B., A. F. M. Moorman, and W. H. Lamers. 1992. Electroporation in 'intracellular' buffer increases cell survival. *Nucleic Acids Res.* 20:2902.

Wang, Y. 1984. A total synthesis of yeast alanine transfer RNA. *Acc. Chem. Res.* 17:393–397.

Wildenauer, D., H. J. Gross, and D. Riesner. 1974. Enzymatic methylations: III. Cadaverine-induced conformational changes of E. coli tRNA^{fMet} as evidenced by the availability of a specific adenosine and specific cytidine residue for methylation. *Nucleic Acids Res.* 1:1165–1182.

Wittig, B., and S. Wittig. 1978. Reverse transcription of tRNA. *Nucleic Acids Res.* 5:1165–1178.

Wower, J., K. V. Rosen, S. S. Hixson, and R. A. Zimmermann. 1994. Recombinant photoreactive tRNA molecules as probes for cross-linking studies. *Biochimie* 76:1235–1246.

Wright, M. C., and G. F. Joyce. 1997. Continuous *in vitro* evolution of catalytic function. *Science* 276:614–617.

Wrzesinski, J., K. Nurse, A. Bakin, B. G. Lane, and J. Ofengand. 1995. A dual-specificity pseudouridine synthase: an *Escherichia coli* synthase purified and cloned on the basis of its specificity for Ψ746 in 23S RNA is also specific for Ψ32 in tRNA^{Phe}. *RNA* 1:437–448.

Xue, H., A.-L. Glasser, J. Desgres, and H. Grosjean. 1993. Modified nucleotides in *Bacillus subtilis* tRNA^{Trp} hyperexpressed in *Escherichia coli*. *Nucleic Acids Res.* 21:2479–2486.

Yin, Y. H., and C. W. Carter. 1996. Incomplete factorial and response surface methods in experimental design: yield optimization of tRNA^{Trp} from *in vitro* T7 RNA polymerase transcription. *Nucleic Acids Res.* 24:1279–1286.

Yisraeli, J. K., and D. A. Melton. 1989. Synthesis of long, capped transcripts *in vitro* by SP6 and T7 RNA polymerases. *Methods Enzymol.* 180:42–50.

Yokobori, S., and S. Paabo. 1995. tRNA editing in metazoans. *Nature* 377:490.

Youvan, D. C., and J. E. Hearst. 1979. Reverse transcriptase pauses at N^2-methylguanine during in vitro transcription of *Escherichia coli* 16S ribosomal RNA. *Proc. Natl. Acad. Sci. USA* 76:3751–3754.

Youvan, D. C., and J. E. Hearst. 1981. A sequence from *Drosophila melanogaster* 18S rRNA bearing the conserved hypermodified nucleoside amΨ: analysis by reverse transcription and high-performance liquid chromatography. *Nucleic Acids Res.* 9:1723–1741.

Yu, W., and W. Schuster. 1995. Evidence for a site-specific cytidine deamination reaction involved in C to U RNA editing of plant mitochondria. *J. Biol. Chem.* 270:18227–18233.

Yu, Y. T., and J. A. Steitz. 1997. A new strategy for introducing photoactivatable 4-thiouridine (s⁴U) into specific positions in a long RNA molecule. *RNA* **3:**807–810.

Yu, Y.-T., M.-D. Shu, and J. A. Steitz. 1997. A new method for detecting sites of 2′-O-methylation in RNA molecules. *RNA* **3:**324–331.

Zagorska, L., J. Van Duin, H. Noller, B. Pace, K. Johnson, and N. Pace. 1984. The conserved 5s rRNA complement to tRNA is not required for translation of natural mRNA. *J. Biol. Chem.* **259:**2798–2802.

Zeevi, M., and V. Daniel. 1976. Aminoacylation and nucleoside modification of *in vitro* synthesised transfer RNA. *Nature* **260:**72–74.

Zhao, L.-J., Q. X. Zhang, and R. Padmanabhan. 1993. Polymerase chain reaction-based point mutagenesis protocol. *Methods Enzymol.* **217:**218–227.

Zheng, H., T. B. Fu, D. Lazinski, and J. Taylor. 1992. Editing on the genomic RNA of human hepatitis delta virus. *J. Virol.* **66:**4693–4697.

Zhou, W., D. Reines, and P. W. Doetsch. 1995. T7 polymerase bypass of large gaps on the template strand reveals a critical role of the non-template strand in elongation. *Cell* **82:**577–585.

Modification and Editing of RNA
Edited by Henri Grosjean and Rob Benne
© 1998 ASM Press, Washington, D.C.

Chapter 3

Detection and Structure Analysis of Modified Nucleosides in RNA by Mass Spectrometry

PAMELA F. CRAIN

There are presently 95 known modified nucleosides, and every type of cellular RNA is modified to different extents (McCloskey and Crain, 1998). tRNA is the most highly modified species; it presently contains 81 known modified nucleosides. The occurrence, biosynthesis, and function of nucleoside modification and editing in the various RNAs are the focus of this book. All of these aspects of nucleoside modification and editing require methods for their identification and placement within the RNA.

Conventional methods for analyzing modified nucleosides in RNA rely solely on chromatographic mobility. The location of a modified site within an RNA sequence is usually carried out by ^{32}P-postlabeling methods (Kuchino et al., 1987; Keith, 1990) (also see Chapter 2 by Grosjean et al.) In these protocols, the target RNA is subjected to limited digestion, and subsequently labeled on one end. The resulting labeled oligonucleotides are separated electrophoretically, and following the excision of the bands from the gel, the identity of the terminal labeled nucleotide is determined from the thin-layer chromatography (TLC) of a total digest of the band. This approach is relatively easy to implement and very sensitive, but requires a high level of skill to interpret the chromatograms and offers lower chromatographic resolution than high-performance liquid chromatography (HPLC).

Reversed-phase, high-performance liquid chromatography of total digests of RNA can likewise be used to reveal the modified nucleoside content. Two different gradient elution systems have been described. One of these systems utilizes phosphate buffers, which are UV transparent and permit the acquisition of full UV spectra concomitant with the analysis (Gehrke and Kuo, 1990). A second system utilizes an ammonium acetate-based volatile buffer system (Buck et al., 1983), and permits the recovery of isolated nucleosides by solvent evaporation. Both of these systems provide adequate resolution of most of the known modified nucleosides; however, given the increasing number of modified nucleosides being discovered in RNA, recognition requires that they be totally separated from closely eluting nucleosides. In addition, great care must be taken with the preparation of solvents and the implementation of the chromatographic fractionation to ensure reproducibility of elution times, the major and often sole parameter for nucleoside identity.

The elution system of Buck et al. (1983) is suitable for directly combined liquid chromatography/mass spectrometry (LC/MS) of RNA digests, and a method for analysis was devised in combination with thermospray ionization mass spectrometry (Pomerantz and McCloskey, 1990). Mass is a fundamental property of matter, and its measurement is not subject to systemic variability as is the determination of a chromatographic retention parameter. The combined measurement, therefore, offers a method that is more useful than either method used alone.

MASS SPECTROMETRY OF NUCLEIC ACID CONSTITUENTS

Mass spectrometry is a time-honored method for structure determination of modified nucleosides from nucleic acids, including a breadth of applications to naturally and xenobiotically modified species (see McCloskey, 1990, for references). Indeed, of the 95 known modified nucleosides in RNA (McCloskey and Crain, 1998), the structures of more than half have been characterized by some form of mass spectrometry. In most cases, the nucleoside was first iso-

---

Pamela F. Crain • Department of Medicinal Chemistry, University of Utah, Salt Lake City, Utah 84112.

lated, then subjected to mass spectrometry; however, LC/MS techniques are now assuming a greater role.

In addition to its use as a tool for structure studies, the measurement of mass allows the analysis of reaction products for which no radioactive label exists (such as oxygen and nitrogen). For example, $H_2^{18}O$ was used as a substrate to explore the details of enzymatic reactions for the A → I reactions associated with RNA editing (Polson et al., 1991) and modification of A-34 in tRNA (Auxilien et al., 1996). Mass spectrometry was used in both studies to determine the ^{18}O content of inosine following total digestion of the respective RNAs to nucleosides.

Finally, MS-based methods offer a means for the high-sensitivity quantification of modified nucleic acid constituents by using stable isotope dilution (Millard, 1979). The general approach consists of adding to the nucleic acid or to a digest a known amount of a stable-isotope-labeled (such as ^2H, ^{13}C, ^{15}N, or ^{18}O) analog of the nucleoside or base of interest. The ratio of ion signals for the (unlabeled) nucleic acid constituent and the labeled analog (which has a different mass) can be referred back to the amount of analog added to give the amount of natural nucleoside or base present. A recent example of this approach describes the quantification of dihydrouridine levels in tRNA from psychrophilic microorganisms, in which the amounts were found to be elevated in comparison with mesophiles (Dalluge et al., 1996, 1997).

NEW IONIZATION METHODS FOR MASS SPECTROMETRY OF NUCLEIC ACIDS AND THEIR CONSTITUENTS

The analysis of biopolymers by mass spectrometry has been revolutionized by the development of new methods for generating ions from proteins and nucleic acids. These two methods are electrospray ionization (ESI) (Fenn et al., 1989) and matrix-assisted laser desorption ionization (MALDI) (Karas and Hillenkamp, 1988). Both of these techniques have been applied to the analysis of nucleic acids, as recently reviewed (Fitzgerald and Smith, 1995; Nordhoff et al., 1996; Crain, 1997; Crain and McCloskey, in press). However, methods based on ESI currently hold the advantage, primarily because of greater accuracy of mass determination. This feature is essential, because the difference in mass between C and U is 1 Da, and useful measurements will require sufficient accuracy to distinguish compositions differing by C versus U, especially where the potential for RNA editing exists. In directly comparable analyses, for example, intact tRNAs were mass measured with errors less than 0.004% (within 1 Da at an M_r of 25,000) by ESI (Limbach et al., 1995; Mangroo et al., 1995), while the errors observed upon measurement using MALDI were on the order of 0.1% (23 Da at an M_r of 25,000) (Gruić-Sovulj et al., 1997)). Moreover, mass measurement of oligonucleotides from RNase digests can be carried out with about 0.01% accuracy for lengths up to at least the 19-mer level, while the utility of MALDI for analytically useful measurements is currently restricted to "quite small oligoribonucleotides" (Hahner et al., 1997).

It should be noted, however, that MALDI-based methods are presently receiving much attention for the mass measurement of small PCR fragments in genotyping and diagnostics applications, where high sample throughput and ease of automated sample handling are advantageous (Köster et al., 1996; Little et al., 1997), and lesser mass accuracy can be tolerated because of the need to measure mass only to within 9 Da (the smallest mass difference, for A versus T).

The determination of protein–nucleic acid crosslinks is emerging as an important area of study in which both MALDI and ESI are assuming prominence. The application of both ionization techniques to a model system provides valuable insights into the issues involved with each method (Jensen et al., 1996). Recent applications include the characterization of cross-linked products generated from E. coli 30S rRNA (Urlaub et al., 1997a; Urlaub et al., 1997b), and from Escherichia coli tRNA$_i^{Met}$ cross-linked to Met-tRNA formyltransferase (Gite and RajBhandary, 1997).

Another major advantage of nucleic acid analysis using electrospray ionization-based methods is the ease with which they can be implemented with chromatographic (Voyksner, 1997) or electrophoretic (Severs and Smith, 1997) sample introduction, thus providing retention time or mobility as an additional parameter for identification or structure characterization. In the case of RNA and its constituents, there are also a greater number of published applications utilizing ESI, permitting a more reliable evaluation of its utility. Consequently, this chapter will review applications of ESI to the analysis of RNA and its constituents.

ELECTROSPRAY IONIZATION MASS SPECTROMETRY OF NUCLEIC ACIDS AND THEIR CONSTITUENTS

Figure 1 shows a schematic representation of an electrospray ionization interface coupled to a triple quadrupole mass analyzer, which is the instrument

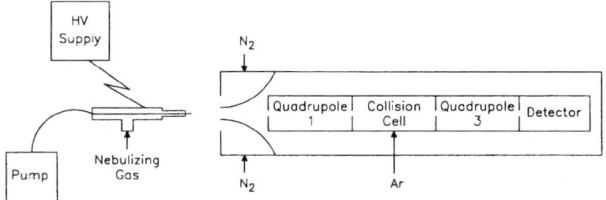

Figure 1. Schematic diagram of the sample inlet and ESI for a triple quadrupole mass spectrometer.

configuration currently used in our laboratory. Other mass analyzers can also be coupled with the ESI interface (Cole, 1997). The sample to be analyzed is pumped (by syringe pump or liquid chromatograph) through a fused silica or stainless steel capillary. Ionization of the analyte is accomplished by the application of a high voltage (several kilovolts) to the tip of the probe capillary through which the sample flows. A spray consisting of analyte ions and solvent is produced, and as the solvent is selectively removed by the application of heat and a drying gas (typically N_2), the resulting gas phase ions can then be mass analyzed. Very large ions can be produced by ESI (Cheng et al., 1996), and the molecular mass of intact 5S rRNA has been measured to within a 2-Da accuracy (Limbach et al., 1995). ESI is also useful for mass spectrometry of nucleosides derived from DNA and RNA (reviewed by Crain, 1997).

DETECTION OF MODIFIED NUCLEOSIDES IN RNA HYDROLYSATES BY LC/ESI-MS

Directly combined LC/MS with thermospray ionization (Edmonds et al., 1985) has been extensively used for surveying the modified nucleosides in tRNA (see Pomerantz and McCloskey, 1990; Edmonds et al., 1991; Takeda et al., 1991; Kowalak et al., 1994; Dallluge et al., 1997) and rRNA (see Kowalak et al., 1993; Bruenger et al., 1993; Kowalak et al., 1995). Although the use of thermospray ionization has been superseded by electrospray ionization, the strategies for data acquisition and interpretation (Pomerantz and McCloskey, 1990) remain unchanged. However the chromatographic conditions must be altered (Auxilien et al., 1996) for compatibility with ESI, which is conducted in low ionic strength buffers (for example, 5 mM ammonium acetate, pH 5.3).

In a typical analysis, the RNA of interest is digested completely to nucleosides using nuclease P_1, phosphodiesterase I, and alkaline phosphatase (Crain, 1990b). Dephosphorylated alkali-stable dinucleotides from RNase T_2 digestion are also amenable to LC/MS analysis (Takeda et al., 1991). The digest can be injected directly into the liquid chromatograph without the removal of salts or enzymes. For mixed tRNA, an A_{260} of 0.5 to 1 is a typical amount for screening, while isoaccepting tRNAs can generally be analyzed at an A_{260} of 0.05 to 0.1. A somewhat greater amount of material is required for screening rRNA, which is typically 10- to 20-fold less highly modified than tRNA.

Mass spectra are usually acquired over a mass range of 100–500, once per second, throughout the chromatographic fractionation. Electrospray mass spectra of nucleosides consist of two ion types. The first type represents the protonated intact molecule (MH^+), while the second type represents the protonated base moiety (BH_2^+), released by cleavage of the glycosidic bond concomitant with the transfer of a hydrogen from the sugar to the base. The sugar identity can be deduced from the difference in mass between the two ions: a difference of 132 indicates that the sugar is a normal ribose, while a 146 difference corresponds to the presence of 2′-O-methylribose.

Figure 2 shows the chromatogram from an LC/ESI-MS analysis of a nucleoside digest of *E. coli* MRE600 mixed tRNA, showing some of the most abundant modified nucleosides identified. The insert shows the summed mass spectrum from the broad peak eluting at 16.8 min; the solvent background has been subtracted. Three prominent ions are evident in this spectrum. In the initial assignment, the prominent m/z 298 ion is taken to be an MH^+ ion, from which either 132 or 146 is subtracted in the search for the accompanying BH_2^+ ion. In this example, ions resulting from both losses (m/z 166 and m/z 152, respectively) are present. Because only two ions are expected from most modified nucleosides, two possibilities exist: either a sugar-methylated nucleoside and a base-methylated nucleoside are coeluting, or only one nucleoside is present and the third ion results from a background contaminant that cannot be completely subtracted.

To distinguish between these possibilities, reconstructed ion chromatograms can be displayed, representing time versus abundance for the ions of interest. Figure 3 shows chromatograms for the three ions in question, along with the 254-nm UV detection trace. The m/z 298 channel exhibits a leading "shoulder," while traces for the candidate BH_2^+ ions (m/z 166 and m/z 152) form sharp peaks that are slightly displaced from each other. The spectrum can now be interpreted as representing a mixture of two nucleosides. If a contaminant were present it would not track with the MH^+ ion. The two nucleosides are assigned as 1-methylguanosine and 2′-O-methylguanosine, based on their relative retention times

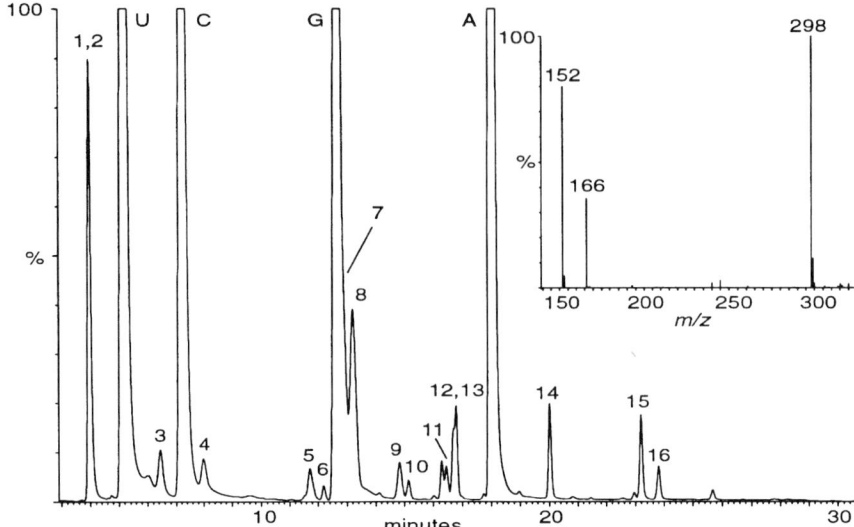

Figure 2. Chromatogram (detection at 254 nm) from a directly combined high-performance LC/ESI-MS analysis of nucleosides from a total digest of mixed tRNA from *E. coli* MRE600. Major modified nucleosides are as follows: 1, Ψ; 2, D; 3, acp^{3}U; 4, s^{2}C; 5, Cm; 6, I; 7, m^{7}G; 8, m^{5}U; 9, s^{4}U; 10, Um; 11, oQ; 12, m^{1}G; 13, Gm; 14, t^{6}A; 15, m^{2}A; 16, m^{6}A. (Inset) ESI mass spectrum of the peak at centered at 16.8 min. Symbols are defined in Appendix 1.

(Pomerantz and McCloskey, 1990). Figure 3 also illustrates the power of detection by mass. Note that there is a third component possessing coincident m/z 298 and m/z 166 ions, but it does not exhibit a distinct peak in the 254-nm trace. This component is, in fact, 7-methylguanosine, which elutes under the abundant guanosine peak and would not be detected by UV absorbance alone.

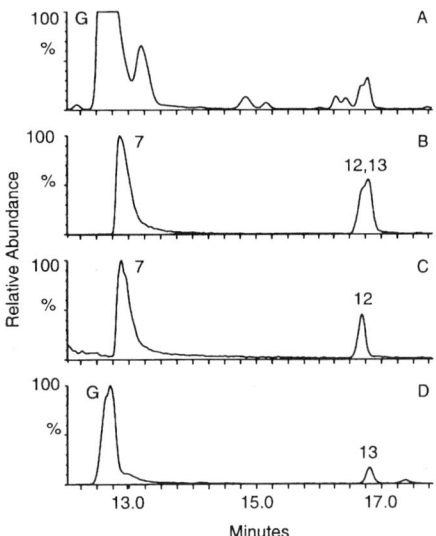

Figure 3. Expanded UV chromatogram (A) and reconstructed ion chromatograms for the MH$^+$ ion of methylguanosine (B) (m/z 298) and the BH$_2^+$ ions for methylguanine (C) (m/z 166) and guanine (D) (m/z 152), extracted from the analysis shown in Fig. 2. For clarity, ion abundances have been normalized to the most abundant peak in each panel.

STRUCTURE CHARACTERIZATION OF MODIFIED NUCLEOSIDES DIRECTLY FROM MIXTURES

By the approaches previously summarized, the presence of any of the 95 known modified RNA nucleosides (McCloskey and Crain, 1998) can be evaluated by examining the mass spectra of components giving peaks in the UV trace and, if necessary, generating ion traces for the calculated MH$^+$ and BH$_2^+$ ions for each. If a new nucleoside is present, it can usually be recognized because it will generate a set of ions that track together and differ from each other by 132 or 146 m/z units; other ions may also be present if the nucleoside is hypermodified (Pomerantz and McCloskey, 1990). Retention times should likewise be taken into account as a parameter for identification. If a diode array UV detector is used in series with the mass spectrometer, UV spectra of individual peaks can also be used, although the sensitivity is lower than that available from ion (mass) detection.

In favorable cases, new nucleosides identified in this way may be structurally characterized even when present in amounts too low to permit their isolation for characterization by conventional methods (McCloskey, 1990), either because they are a trace constituent (for example, present in a single low abundance isoacceptor in mixed tRNA) or because the RNA may be difficult to obtain in sufficient quantity. An ancillary method that provides structurally useful information is LC/MS using deuterated HPLC solvents (Edmonds et al., 1988), which establishes the number of replaceable hydrogens in the nucleoside. Consideration of mass spectral data (also discussed in the following section) may permit a likely structure candidate to be deduced, which is then synthesized by an unambiguous route, rigorously characterized, and used for chromatographic and mass spectral comparison with the natural nucleoside in consecutive analyses, including coinjection, if appropriate. Some examples of modified nucleosides discovered by LC/MS and characterized in this fashion include four sugar-methylated nucleosides in tRNA from hyperthermophilic archaea (Edmonds et al., 1987), 5-methylcarboxymethyl-2′-O-methyluridine in rat liver tRNA$^{Ser[Sec]*}$ (Diamond et al., 1993), and two sugar-methylated nucleosides in the cap4 subunit of trypanosomal mRNA (Bangs et al., 1992).

COLLISION-INDUCED DISSOCIATION OF MODIFIED NUCLEOSIDE IONS

"Soft" ionization methods such as electrospray (Crain, 1997) and fast atom bombardment (FAB) (Crow et al., 1984) have proven extremely useful for elucidation of nucleoside structure (Crain, 1990a) because of their ability to generate molecular ion species from these relatively polar compounds, as compared with electron ionization, which is one of the first ionization methods for nucleosides. The mass spectra, however, generally contain few informative fragment ions from which the structure of an unknown nucleoside may be deduced. Tandem mass spectrometry, also called mass spectrometry/mass spectrometry (MS/MS), can overcome this limitation by producing subsequent fragmentation of one of the two major nucleoside ions (MH$^+$ or BH$_2^+$).

There are many ways to implement the MS/MS experiment (Busch et al., 1988); Fig. 1 illustrates a triple quadrupole mass spectrometer, which is used in our laboratory. Briefly, an ion (precursor ion) resulting from the initial ionization event is selected in the first mass analyzer (quadrupole 1) and conducted into a quadrupole gas cell, where it collides with gas atoms (such as argon), causing it to dissociate, hence the term collision-induced dissociation (CID). (The gas cell is counted as quadrupole 2, although no mass analysis occurs.) Fragment ions produced from the precursor ion following collision with Ar are then directed from the collision cell into the third quadrupole, where they are mass analyzed to produce a conventional mass spectrum of the product ions.

Any ion produced in the initial ionization event can be selected as the precursor ion for CID; however, the BH$_2^+$ ion provides more structurally informative fragment ions than the MH$^+$ ion (Crain, 1990a). In the case of the base-related ion, the heterocycle ring itself fragments upon CID, while if the MH$^+$ ion is dissociated it produces primarily the BH$_2^+$ ion. The latter ion typically is already produced spontaneously from MH$^+$, and therefore little additional information generally results.

An example of isomer differentiation by MS/MS is shown in Fig. 4. Product ion spectra from CID of m/z 166 (the BH$_2^+$ ion for any base-methylated guanosine) were acquired directly from an LC/MS analysis of nucleosides from a total digest of mixed wheat germ tRNA (Micromass Quattro II instrument at 23 eV collision energy; chromatographic conditions as described in Auxilien et al., 1996). A comparison of the product ion spectra in Fig. 4 with the primary mass spectrum (see Fig. 2 inset) reveals the additional structural information produced from CID. The 1-, N^2-, and 7-methylguanosines give distinctive spectra, reflecting different sites of methyl substitution. (See Gregson and McCloskey, 1997, for a detailed interpretation of product ion spectra of methylguanosines).

It is evident that of MS/MS techniques can provide a powerful tool for modified nucleoside structure studies. CID was the key element in the structure

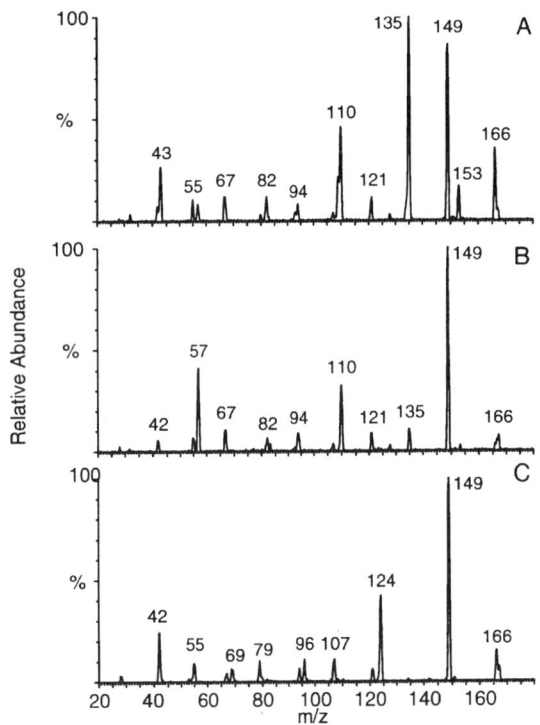

Figure 4. Product ion spectra from collision-induced dissociation of the BH$_2^+$ ions (m/z 166) characteristic for base-methylated guanosines, acquired from an LC/MS/MS analysis of a digest of wheat germ tRNA. (A) 1-Methylguanosine; (B) N^2-methylguanosine; (C) 7-methylguanosine.

elucidation of archaeosine, the modified G at position 15 of archaeal tRNAs (Gregson et al., 1993). This nucleoside was found to be not a guanosine derivative, but rather a 7-deazapurine nucleoside. Structure determination of two unusual nucleosides found in thermophilic bacteria and archaea, hn^6A and ms^2hn^6A, both of which are t^6A analogs in which the amino acid substituent is hydroxynorvaline instead of threonine, was also aided by MS/MS (Reddy et al., 1992). In the examples shown in Fig. 4 the nucleosides were analyzed in their free form; however, in the cases of the latter two hypermodified nucleosides, the most structurally informative spectra were produced by CID of derivatives prepared from isolated nucleosides. Archaeosine was analyzed as a permethylated derivative, while hn^6A was determined as a trimethylsilyl derivative; FAB ionization was used in both studies.

METHODS FOR MAPPING MODIFIED NUCLEOSIDES IN OLIGONUCLEOTIDES DERIVED FROM RNA

The capability of ESI/MS to produce and accurately mass measure ions from nucleic acids has provided a powerful tool for locating and characterizing modified nucleosides at the oligonucleotide level. Based on extensive calculations of molecular mass versus nucleoside composition, it was discovered that base composition of an oligonucleotide can be specified up to the 14-mer length, provided that the number of occurrences of any single nucleoside is known, and that mass can be measured to within ±0.01% (Pomerantz et al., 1993). Both objectives can readily be met in the case of oligonucleotides derived from RNase T_1, for which the number of guanosines is defined as one. A method has been implemented for mapping modified nucleosides in RNA, based on electrospray ionization mass spectrometry of RNase T_1-derived oligonucleotides (Kowalak et al., 1993). The basic approach consists essentially of measuring molecular weights of these oligonucleotides and comparing the values with those calculated from either a primary sequence of the RNA or its gene. Oligonucleotides that contain a modified nucleoside that produces an increase in mass will be recognized by the failure of their molecular weights to correspond to any of the calculated (unmodified) values. The U → Ψ modification reaction does not result in an increase in mass, and therefore it cannot be located to a T_1 fragment by this approach. A useful method for determining the location of Ψ residues in RNA is described in Chapter 12 by Ofengand and Fournier.

An outline of elements of the mapping strategy is shown in Fig. 5. The initial step is the analysis of a total nucleoside digest by LC/MS to determine the identities of the modified nucleosides in the RNA. These nucleosides must ultimately be accounted for upon completion of the map. The next step is the generation of a complete RNase T_1 digest of the RNA. In the initial account of the method, 16S rRNA from *E. coli* was the target molecule, from which RNase T_1 digestion is predicted to generate 131 different sequence isomers. A mixture of this complexity required a preliminary fractionation using anion exchange HPLC, and 42 pools were collected. A second LC/MS analysis of a total nucleoside digest of each pool was conducted to permit targeting those that contain modified nucleosides for subsequent analysis. Oligonucleotide pools selected for molecular weight determination were further fractionated by reversed-phase HPLC, and then manually mass measured by infusion of the collected peaks into the mass spectrometer.

Direct LC/MS analysis of oligonucleotide mixtures is difficult to implement because the volatile

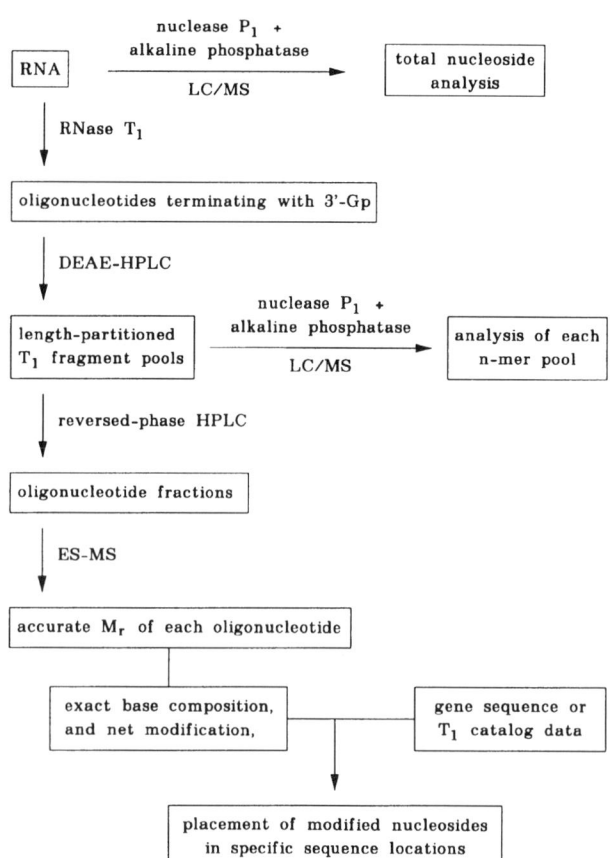

Figure 5. Scheme for mapping modified nucleosides in RNA by mass spectrometric analysis of RNase T_1-generated oligonucleotides.

buffers necessary for compatibility with ESI give poor separations (Crain, 1997). Very recently, however, a new buffer system based on 1,1,1,3,3,3-hexafluoro-2-propanol (HFIP) has been described for reversed-phase LC/MS of DNA oligonucleotides (Apffel et al., 1997). This buffer system is superior to those previously described, and we have found it also appropriate for LC/MS of RNA oligonucleotides. Figure 6 shows the chromatogram (UV detection) from LC/MS analysis of an RNase T$_1$ digest of 20 pmol (about 0.5 µg) of tRNA$_{QUA}^{Tyr}$ from *E. coli*. Table 1 shows the calculated and measured molecular masses for the oligonucleotides predicted from the sequence of this tRNA (entry RY1660 in the tRNA sequence database) (Sprinzl et al., 1998). The measured molecular masses were generally determined to within several tenths of a dalton, thus providing sufficient accuracy for the assignment of composition. Also evident in the chromatogram are several small peaks for which mass values could be obtained and likely compositions could be assigned; they undoubtedly arise from small amounts of a contaminating RNA. The ability to mass measure these small peaks suggests that oligonucleotides can be analyzed at levels lower than the 20 pmol of tRNA utilized in the present example.

The electrospray mass spectrum of the RNase T$_1$-derived 12-mer containing the anticodon sequence from *E. coli* tRNA$_{QUA}^{Tyr}$ is shown in Fig. 7. Unlike nucleosides, for which singly charged ions are produced by electrospray ionization, oligonucleotides are multiply charged. Each peak represents a population of ions that differ according to whether the individual phosphodiester groups are ionized or neutral, and each likewise serves as an independent measure of molecular mass. The molecular mass derived for this oligonucleotide from the mass spectrum is inset in Fig. 7. Note that the abscissa is mass, and not mass/charge, so this representation of the data is not formally a mass spectrum. The M_r 4,100.5 species represents the neutral molecule, while the other components present represent the monosodium (M_r, 4,122.5) and monopotassium (M_r, 4,138.4) salts. Chromatographic sample introduction permits analysis without removal of enzymes or buffer salts because the sample is effectively desalted on line, in contrast to MALDI-based techniques, for which sample cleanup is required before data acquisition (Nordhoff et al., 1996; Hahner et al., 1997).

The basic protocol outlined in Fig. 5 (Kowalak et al., 1993) and discussed in this section was modified for application to 23S rRNA from *E. coli* (2,904 nt), from which 210 sequence isomers are generated from 991 RNase T$_1$ cleavage sites. Because many modified nucleosides are clustered in the region of the peptidyl transferase center (domain V), this region was targeted for selective analysis (Kowalak et al., 1995). Two cDNAs, a 39-mer and a 51-mer, were used for hybridization to the regions of interest, and the remainder of the RNA was digested away with mung bean nuclease. The RNA 39-mer and 51-mer were then digested with RNase T$_1$ and analyzed as previously described.

MAPPING MODIFIED NUCLEOSIDES IN rRNAs

It was known from earlier studies that one 3-methyluridine was present in *E. coli* 16S rRNA, and that there was an unspecified modified uridine at position 1498. By the basic method previously described (Kowalak et al., 1993), 3-methyluridine was assigned as the nucleoside at position 1498. Dihydrouridine was assigned as the uncharacterized modified uridine at position 2449 of *E. coli* 23S rRNA, and certain errors and inconsistencies in earlier work were reconciled (Kowalak et al., 1995). These methods also allowed the characterization of a new nucleoside not previously known in RNA,

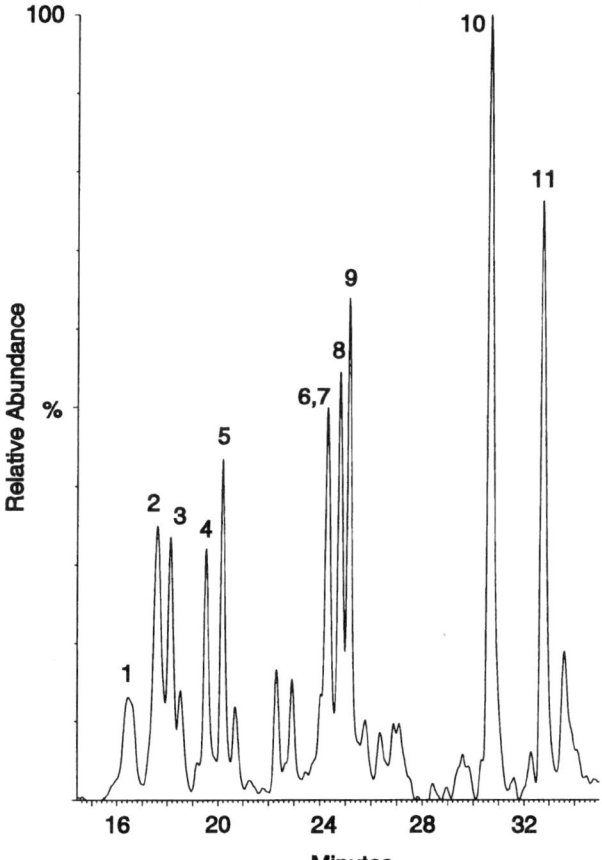

Figure 6. Chromatogram from directly combined high-performance LC/ESI-MS of an RNase T$_1$ digest of 20 pmol of tRNA$_{QUA}^{Tyr}$.

Table 1. Assignments of oligonucleotides from RNase T_1 digestion of *E. coli* tRNA$_{QUA}^{Tyr}$

Oligonucleotide (peak number)[a]	Sequence location[b]	Mass Calculated[c]	Measured
CCGp (1)	43–45	973.1	973.3
CAGp (2)	28–30	997.2	997.3
CGmGp (3)	16–19	1,027.2	1,027.2
AAGp (4)	50–52	1,021.2	1.021.3
TΨCGp (5)	54–57	1,294.2	1.294.5
s^{4}UUCCCGp (6)[d]	8–13	1,907.2	1,906.8
UCAUCGp (7)	45e12–45e23	1,915.1	1,914.9
ACUUCGp (8)	45a22–49	1,915.1	1,914.9
CCAAAGp (9)	18–23	1,961.2	1,961.0
ACUQUAms^{2}i^{6}AAΨCUGp (11)	31–42	4,100.7	4,100.5
AAUCCUUCCCCCACCACCA-OH (10)	58–76	5,859.6	5,859.6

[a] Peak numbers are those used in the chromatogram shown as Fig. 6.
[b] Sequence locations utilize the standard tRNA cloverleaf numbering system (Sprinzl et al., 1998).
[c] Calculated mass values are monoisotopic for trimers and tetramers and isotope-averaged for larger oligonucleotides.
[d] Oligonucleotides with mass values consistent with the presence of unmodified as well as dithiolated species are also present.

3-methylpseudouridine, in domain IV of *E. coli* 23S rRNA (Kowalak et al., 1996).

Nucleoside modification in 5S rRNA is rare (McCloskey and Crain, 1998); nonetheless, two modified nucleosides were identified in 5S rRNA from the extremely thermophilic archaeon *Pyrodictium occultum* (Bruenger et al., 1993). One of them, N^4-acetyl-2'-O-methylcytidine, was placed at position 35, within a 9-mer. N^4-acetylcytidine was placed within a (C,U)Gp sequence, but its exact location could not be assigned because of multiple occurrences of this trimer. An alkali-stable modified cytidine described in earlier studies of 5S rRNA from *Sulfolobus solfataricus*, and present in the site analogous to the location of ac^{4}C in *P. occultum*, was assigned as Cm in the same study (Bruenger et al., 1993). N^4-Acetyl-2'-O-methylcytidine exhibits extreme conformational rigidity and is predicted to aid thermal stabilization of RNA in hyperthermophiles (Kawai et al., 1992).

MAPPING MODIFIED NUCLEOSIDES IN tRNA

Although originally conceived as a means for mapping of RNAs for which a sequence is available, the mapping technique has been applied successfully to locate a modified nucleoside in tRNA (Kowalak et al., 1994). To examine levels of modified nucleosides in tRNA as a function of growth temperature in the thermophilic archaeon *Pyrococcus furiosus*, it was grown at 70, 85, and 100°C. Modified nucleosides were screened by LC/MS, and amounts of several modified nucleosides, including 5-methyl-2-thiouridine, were observed to increase with increasing culture temperature. Because this nucleoside had never been previously reported in archaeal tRNA, it was of interest to determine whether or not it was located in the T loop. Mixed tRNA from *P. furiosus* was accordingly digested with RNase T_1, and the tetramer pool was mass measured to determine whether the nucleoside would be found in the highly conserved loop sequence Gm^{5}UΨCG. Indeed, an oligonucleotide corresponding in mass to the sequence m^{5}UΨCG + 16 Da was found, and when the composition of the fragment was determined by LC/MS of a total digest, m^{5}s^{2}U and Ψ were both observed to be present. In this application, although no tRNA sequences were available, conservation of sequence (in the T loop) could be used as a guide for modified nucleoside placement.

This approach is potentially applicable to the sequencing of tRNA from hyperthermophilic micro-

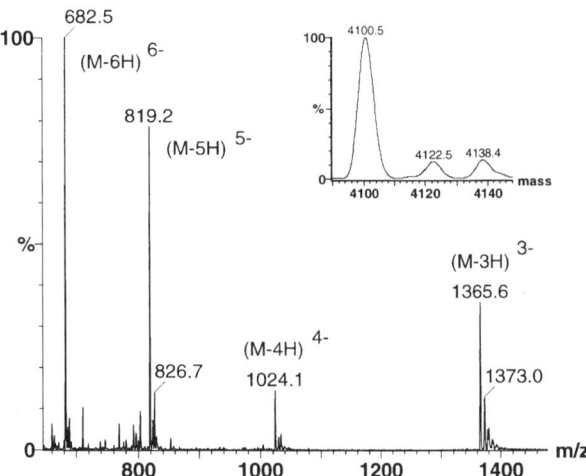

Figure 7. ESI mass spectrum of ACUoQUAms^{2}i^{6}AAΨCUGp (peak 10 in Fig. 6). (Inset) Oligonucleotide masses derived from the mass spectrum.

organisms, for which conventional ^{32}P-based postlabeling methods will most likely not succeed. Implementation of these methods (Kuchino et al., 1987; Keith, 1990) requires that the initial treatment to cleave the RNA before postlabeling be limited (less than one break per molecule) and completely random by site. An attempt to sequence tRNA$_i^{Met}$ from *Pyrodictium occultum* (optimal growth temperature 105°C) failed as a consequence of the extreme thermal stabilization effected by the nucleoside modifications themselves, which limited cleavages to only a few favored sites (Ushida et al., 1996).

OLIGONUCLEOTIDE SEQUENCING BY MASS SPECTROMETRY

In the foregoing applications, the placement of the modified nucleoside in question was achieved either because an unspecified modified nucleoside was known from earlier studies to be located at the specific position, or because the modification was confined to the only occurrence of the unmodified residue in the oligonucleotide. For example, 3-methyluridine could be placed at position 1498 of *E. coli* 16S rRNA because the T$_1$ fragment that was shifted in mass by +14 Da and in which m^{3}U was found to occur, $_{1498}$UAACAAG$_{1504}$, contained only one U. In another fortuitous occurrence, ac^{4}Cm was assigned to a 9-mer in the 5S rRNA of *P. occultum* because the nuclease-resistant dinucleotide ac^{4}CmpG was found in a total digest of the RNA (Bruenger et al., 1993). Since only one G is expected in a T$_1$ oligonucleotide in which there is no Gm, ac^{4}Cm could be placed adjacent to the 3'-terminal Gp. In these examples, there was no need to actually determine the sequence location because it could be predicted with certainty. In situations where the location of a modified nucleoside may not be so readily ascertained, it will be necessary to determine the sequence. Mass spectrometry may also be applied to this task.

There are several possible approaches that can be applied, as reviewed recently by Limbach (1996). Based on the work of McLuckey et al. (1992), who developed the framework for interpretation of the product ion spectrum from collision-induced dissociation of multiply charged DNA ions produced by electrospray ionization, a procedure for interpreting de novo the product ion spectrum of an oligonucleotide of unknown sequence has been developed (Ni et al., 1996). Mass ladders are constructed by assigning the ion types that indicate sequence in the 3'→5' and 5'→3' directions. The resulting product ion spectrum is complex in appearance, but sufficiently well understood to permit the placement of a modified residue by a shift in mass of any of the several product ion types produced upon CID of the oligonucleotide. One particular ion type, the so-called "a$_n$–B$_n$" series (McLuckey et al., 1992) consists of the loss of the base at the *n* position, followed by chain cleavage, and may be used to determine in the case of a methylated oligonucleotide whether the methyl group is in the sugar or the base. Combinatorial mixtures of oligonucleotides that contain certain conserved sequence features can be sequenced (Pomerantz et al., 1997), and the strategies presented can potentially be applied to sequencing of uncomplicated mixtures of T$_1$ oligonucleotides.

In favorable cases, sequences can be determined for oligonucleotide lengths up to about a 20-mer, provided that as a minimum requirement mass ladders built up from ions belonging to cleavage reactions from both ends extend more than halfway into the interior of the chain. This requirement may not always be met, and gaps will occur in the mass ladder. Another approach is use of ESI/MS to monitor the products of sequential nucleotide removal by exonuclease digestion of the oligonucleotide (Limbach et al., 1994). The oligonucleotide is treated with either phosphodiesterase I (for sequencing in the 3'→5' direction) or phosphodiesterase II (5'→3' direction), and the digest is infused into the mass spectrometer with a syringe pump. The identities of both the released terminal mononucleotide and the residual oligonucleotide can be determined in real time as the digestion proceeds, permitting the mass ladder to be constructed.

SUMMARY AND FUTURE PROSPECTS FOR NUCLEIC ACID CHARACTERIZATION BY MASS SPECTROMETRY

The strategies and examples of nucleic acid analysis using mass spectrometry presented in this chapter utilize ESI because of its successful implementation for the analysis of nucleic acids and their constituents from the monomer level in digests all the way up to intact tRNA and 5S rRNA. Although it is more complicated to implement than MALDI-based methods, it is generally more adaptable and has the benefit of generally yielding more accurate measurements. The merits of a new solvent system for LC/MS of oligonucleotides (Apffel et al., 1997) cannot be overstated. Although not discussed in detail, electrospray probes that utilize submicroliter flow rates (Crain, 1997) provide substantially reduced sample consumption and should facilitate the implementation of direct sequencing of small amounts of oligonucleotides. Improvements in instrumentation continue to evolve, and it can safely be said that to date the limit has not been defined.

Acknowledgment. Previously unpublished results and preparation of the manuscript were supported by NIH grant GM29812, which is gratefully acknowledged.

REFERENCES

Apffel, A., J. A. Chakel, S. Fischer, K. Lichtenwalter, and W. S. Hancock. 1997. Analysis of oligonucleotides by HPLC-electrospray ionization mass spectrometry. *Anal. Chem.* **69:** 1320–1325.

Auxilien, S., P. F. Crain, R. W. Trewyn, and H. Grosjean. 1996. Mechanism, specificity and general properties of the yeast enzyme catalyzing the formation of inosine-34 in the anticodon of transfer RNA. *J. Mol. Biol.* **262:**351–379.

Bangs, J. D., P. F. Crain, T. Hashizume, J. A. McCloskey, and J. C. Boothroyd. 1992. Mass spectrometry of mRNA cap 4 from trypanosomatids reveals two novel nucleosides. *J. Biol. Chem.* **267:**9805–9815.

Bruenger, E., J. A. Kowalak, Y. Kuchino, J. A. McCloskey, H. Mizushima, K. O. Stetter, and P. F. Crain. 1993. 5S rRNA modification in the hyperthermophilic archaea *Sulfolobus solfataricus* and *Pyrodictium occultum*. *FASEB J.* **7:**196–200.

Buck, M., M. Connick, and B. N. Ames. 1983. Complete analysis of tRNA-modified nucleosides by high-performance liquid chromatography: the 29 modified nucleosides of *Salmonella typhimurium* and *Escherichia coli* tRNAs. *Anal. Biochem.* **129:**1–13.

Busch, K. L., G. L. Glish, and S. A. McLuckey. 1988. *Mass Spectrometry/Mass Spectrometry. Techniques and Applications of Tandem Mass Spectrometry.* VCH Publishers, Inc., New York, N.Y.

Cheng, X., I. Camp, D. G., Q. Wu, R. Bakhtiar, D. L. Springer, B. J. Morris, J. E. Bruce, G. A. Anderson, C. G. Edmonds, and R. D. Smith. 1996. Molecular weight determination of plasmid DNA using electrospray ionization mass spectrometry. *Nucleic Acids Res.* **24:**2183–2189.

Cole, R. B. (ed.). 1997. *Electrospray Ionization Mass Spectrometry. Fundamentals, Instrumentation, and Applications.* Wiley-Interscience, New York, N.Y.

Crain, P. F. 1990a. Mass spectrometric techniques in nucleic acid research. *Mass Spectrom. Rev.* **9:**504–554.

Crain, P. F. 1990b. Preparation and enzymatic digestion of RNA and DNA for mass spectrometry. *Methods Enzymol.* **193:** 782–790.

Crain, P. F. 1997. Nucleic acids and their constituents, p. 421–427. *In* R. B. Cole (ed.), *Electrospray Mass Spectrometry. Fundamentals, Instrumentation, and Applications.* Wiley-Interscience, New York, N.Y.

Crain, P. F., and J. A. McCloskey. Applications of mass spectrometry to the characterization of oligonucleotides and nucleic acids. *Curr. Opin. Biotechnol.*, in press.

Crow, F. W., K. B. Tomer, M. L. Gross, J. A. McCloskey, and D. E. Bergstrom. 1984. Fast atom bombardment combined with tandem mass spectrometry for the determination of nucleosides. *Anal. Biochem.* **139:**243–262.

Dalluge, J. J., T. Hamamoto, K. Horikoshi, R. Y. Morita, K. O. Stetter, and J. A. McCloskey. 1997. Posttranscriptional modification of transfer RNA in psychrophilic bacteria. *J. Bacteriol.* **179:**1918–1923.

Dalluge, J. J., T. Hashizume, and J. A. McCloskey. 1996. Quantitative measurement of dihyrouridine in RNA using isotope dilution chromatography-mass spectrometry (LC/MS). *Nucleic Acids Res.* **24:**3242–3245.

Diamond, A., I. S. Choi, P. F. Crain, T. Hashizume, S. C. Pomerantz, R. Cruz, C. Steer, K. E. Hill, R. F. Burk, J. A. McCloskey, and D. Hatfield. 1993. Dietary selenium affects methylation of the wobble nucleoside in the anticodon of selenocysteine tRNA [Ser]Sec. *J. Biol. Chem.* **268:**14215–14223.

Edmonds, C. G., P. F. Crain, R. Gupta, T. Hashizume, C. H. Hocart, J. A. Kowalak, S. C. Pomerantz, K. O. Stetter, and J. A. McCloskey. 1991. Posttranscriptional modification of tRNA in thermophilic archaea (archaebacteria). *J. Bacteriol.* **173:** 3138–3148.

Edmonds, C. G., P. F. Crain, T. Hashizume, R. Gupta, K. O. Stetter, and J. A. McCloskey. 1987. Structural characterization of four ribose-methylated nucleosides from transfer RNA of extremely thermophilic archaebacteria. *J. Chem. Soc. Chem. Commun.* **173:**909–910.

Edmonds, C. G., S. C. Pomerantz, F. F. Hsu, and J. A. McCloskey. 1988. Thermospray liquid chromatography/mass spectrometry in deuterium oxide. *Anal. Chem.* **60:**2314–2317.

Edmonds, C. G., M. L. Vestal, and J. A. McCloskey. 1985. Thermospray liquid chromatography-mass spectrometry of nucleosides and enzymatic hydrolysates of nucleic acids. *Nucleic Acids Res.* **13:**8197–8206.

Fenn, J. B., M. Mann, C. K. Meng, S. F. Wong, and C. M. Whitehouse. 1989. Electrospray ionization for mass spectrometry of large biomolecules. *Science* **246:**64–71.

Fitzgerald, M. C., and L. M. Smith. 1995. Mass spectrometry of nucleic acids: the promise of matrix-assisted laser desorption-ionization (MALDI) mass spectrometry. *Annu. Rev. Biophys. Biomol. Struct.* **24:**117–140.

Gehrke, C. W., and K. C. Kuo. 1990. Ribonucleoside analysis by reversed-phase high performance liquid chromatography, p. A3–A64. *In* C. W. Gehrke and K. C. Kuo (ed.), *Chromatography and Identification of Nucleosides, Part A. Journal of Chromatography Library*, vol. 45A. Elsevier, New York, N.Y.

Gite, S., and U. L. RajBhandary. 1997. Lysine 207 as the site of cross-linking between the 3′-end of *Escherichia coli* initiator tRNA and methionyl-tRNA formyltransferase. *J. Biol. Chem.* **272:**5305–5312.

Gregson, J. M., P. F. Crain, C. G. Edmonds, R. Gupta, T. Hashizume, D. W. Phillipson, and J. A. McCloskey. 1993. Structure of the archaeal transfer RNA nucleoside G*-15: 2-amino-4,7-dihydro-4-oxo-7-β-D-ribofuranosyl-1H-pyrrolo[2,3-d]pyrimidine-5-carboximidamide (archaeosine). *J. Biol. Chem.* **268:** 10076–10086.

Gregson, J. M., and J. A. McCloskey. 1997. Collision-induced dissociation of protonated guanine. *Int. J. Mass Spectrom. Ion Processes* **165/166:**475–485.

Gruić-Sovulj, I., H.-C. Ludemann, F. Hillenkamp, I. Weygand-Durasević, Z. Kućan, and J. Peter-Katalinić. 1997. Matrix-assisted laser desorption/ionisation mass spectrometry of transfer ribonucleic acids isolated from yeast. *Nucleic Acids Res.* **25:** 1859–1861.

Hahner, S., H.-C. Lüdemann, F. Kirpekar, E. Nordhoff, P. Roepstorff, H.-J. Galla, and F. Hillenkamp. 1997. Matrix-assisted laser desorption/ionization mass spectrometry (MALDI) of endonuclease digests of RNA. *Nucleic Acids Res.* **25:**1957–1964.

Jensen, O. N., S. Kulkarni, J. V. Aldrich, and D. F. Barofsky. 1996. Characterization of peptide oligonucleotide heteroconjugates by mass spectrometry. *Nucleic Acids Res.* **24:**3866–3872.

Karas, M., and F. Hillenkamp. 1988. Laser desorption ionization of proteins with molecular masses exceeding 10,000 daltons. *Anal. Chem.* **60:**2299–2301.

Kawai, G., T. Hashizume, M. Yasuda, T. Miyazawa, J. A. McCloskey, and S. Yokoyama. 1992. Conformational rigidity of N^4-acetyl-2′-O-methylcytidine found in tRNA of extremely thermophilic archaebacteria (Archaea). *Nucleosides Nucleotides* **11**: 759–771.

Keith, G. 1990. Nucleic acid chromatographic isolation and sequence methods, p. A103–A140. *In* C. W. Gehrke and K. C. Kuo (ed.), *Chromatography and Identification of Nucleosides, Part A. Journal of Chromatography Library*, vol. 45A. Elsevier, New York, N.Y.

Köster, H., K. Tang, D.-J. Fu, A. Braun, D. van den Boom, C. L. Smith, R. J. Cotter, and C. R. Cantor. 1996. A strategy for rapid and efficient DNA sequencing by mass spectrometry. *Nat. Biotechnol.* **14**:1123–1128.

Kowalak, J. A., E. Bruenger, T. Hashizume, J. M. Peltier, J. Ofengand, and J. A. McCloskey. 1996. Structural characterization of U*-1915 in domain IV from *E. coli* 23S ribosomal RNA as 3-methylpseudouridine. *Nucleic Acids Res.* **24**:688–693.

Kowalak, J. A., E. Bruenger, and J. A. McCloskey. 1995. Posttranscriptional modification of the central loop of domain V in *E. coli* 23S ribosomal RNA. *J. Biol. Chem.* **270**:17758–17764.

Kowalak, J. A., J. J. Dalluge, J. A. McCloskey, and K. O. Stetter. 1994. The role of posttranscriptional modification in stabilization of transfer RNA from hyperthermophiles. *Biochemistry* **33**: 7869–7876.

Kowalak, J. A., S. C. Pomerantz, P. F. Crain, and J. A. McCloskey. 1993. A novel method for the determination of posttranscriptional modification in RNA by mass spectrometry. *Nucleic Acids Res.* **21**:4577–4585.

Kuchino, Y., N. Hanyu, and S. Nishimura. 1987. Analysis of modified nucleosides and nucleotide sequences of tRNA. *Methods Enzymol.* **155**:379–396.

Limbach, P. A. 1996. Indirect mass spectrometric methods for characterizing and sequencing oligonucleotides. *Mass Spectrom. Rev.* **15**:297–336.

Limbach, P. A., P. F. Crain, and J. A. McCloskey. 1994. Enzymatic sequencing of oligonucleotides with electrospray mass spectrometry. *Nucleic Acids Res. Symp. Ser.* **31**:127–128.

Limbach, P. A., P. F. Crain, and J. A. McCloskey. 1995. Molecular mass measurement of intact ribonucleic acids using electrospray ionization quadrupole mass spectrometry. *J. Am. Soc. Mass Spectrom.* **6**:27–39.

Little, D. P., A. Braun, M. J. O'Donnell, and H. Köster. 1998. Mass spectrometry from miniaturized arrays for full comparative DNA analysis. *Nat. Med.* **3**:1415–1416.

Mangroo, D., P. A. Limbach, J. A. McCloskey, and U. L. RajBhandary. 1995. An anticodon mutant of *Escherichia coli* initiator tRNA: importance of a newly acquired base modification next to the anticodon on its activity in initiation. *J. Bacteriol.* **177**:2858–2862.

McCloskey, J. A. 1990. Constituents of nucleic acids: overview and strategy. *Methods Enzymol.* **193**:771–781.

McCloskey, J. A., and P. F. Crain. 1998. The RNA modification database—1998. *Nucleic Acids Res.* **26**:198–200.

McLuckey, S. A., G. J. Van Berkel, and G. L. Glish. 1992. Tandem mass spectrometry of small, multiply charged oligonucleotides. *J. Am. Soc. Mass Spectrom.* **3**:60–70.

Millard, B. J. 1979. *Quantitative Mass Spectrometry*. Heyden & Son, Ltd., London, United Kingdom.

Ni, J., S. C. Pomerantz, J. Rozenski, Y. Zhang, and J. A. McCloskey. 1996. Interpretation of oligonucleotide mass spectra for determination of sequence using electrospray ionization and tandem mass spectrometry. *Anal. Chem.* **68**:1989–1999.

Nordhoff, E., F. Kirpekar, and P. Roepstorff. 1996. Mass spectrometry of nucleic acids. *Mass Spectrom. Rev.* **15**:67–138.

Polson, A. G., P. F. Crain, S. C. Pomerantz, J. A. McCloskey, and B. Bass. 1991. The mechanism of adenosine to inosine conversion by the double-stranded RNA unwinding/modifying activity: a high performance liquid chromatography-mass spectrometry analysis. *Biochemistry* **30**:11507–11514.

Pomerantz, S. C., J. A. Kowalak, and J. A. McCloskey. 1993. Determination of oligonucleotide composition from mass spectrometrically measured molecular weight. *J. Am. Soc. Mass Spectrom.* **4**:204–209.

Pomerantz, S. C., and J. A. McCloskey. 1990. Analysis of RNA hydrolyzates by LC/MS. *Methods Enzymol.* **193**:796–824.

Pomerantz, S. C., J. A. McCloskey, T. Tarasow, and B. E. Eaton. 1997. Deconvolution of combinatorial oligonucleotide libraries by electrospray ionization tandem mass spectrometry. *J. Am. Chem. Soc.* **119**:3861–3867.

Reddy, D. M., P. F. Crain, C. G. Edmonds, R. Gupta, T. Hashizume, K. O. Stetter, F. Widdel, and J. A. McCloskey. 1992. Structure determination of two new amino acid-containing derivatives of adenosine from tRNA of thermophilic bacteria and archaea. *Nucleic Acids Res.* **20**:5607–5615.

Severs, J. C., and R. D. Smith. 1997. Capillary electrophoresis-electrospray ionization mass spectrometry, p. 343–382. *In* R. B. Cole (ed.), *Electrospray Ionization Mass Spectrometry. Fundamentals, Instrumentation, and Applications.* Wiley-Interscience, New York, N.Y.

Sprinzl, M., C. Horn, M. Brown, A. Ioudovitch, and S. Steinberg. 1998. Compilation of tRNA sequences and sequences of tRNA genes. *Nucleic Acids Res.* **26**:148–153.

Takeda, N., S. C. Pomerantz, and J. A. McCloskey. 1991. Detection of ribose-methylated nucleotides in enzymatic hydrolysates of RNA by thermospray liquid chromatography-mass spectrometry. *J. Chromatogr.* **562**:225–235.

Urlaub, H., B. Thiede, E.-C. Müller, R. Brimacombe, and B. Wittmann-Liebold. 1997a. Identification and sequence analysis of contact sites between ribosomal proteins and rRNA in *Escherichia coli* 30 S subunits by a new approach using matrix-assisted laser desorption/ionization-mass spectrometry combined with N-terminal microsequencing. *J. Biol. Chem.* **272**: 14547–14555.

Urlaub, H., B. Thiede, E.-C. Müller, and B. Wittmann-Liebold. 1997b. Contact sites of peptide-oligoribonucleotide cross-links identified by a combination of peptide and nucleotide sequencing with MALDI MS. *J. Protein Chem.* **16**:375–383.

Ushida, C., T. Muramatsu, H. Mizushima, T. Ueda, K. Watanabe, K. O. Stetter, P. F. Crain, and J. A. McCloskey. 1996. Structural feature of the initiator tRNA gene from *Pyrodictium occultum* and the thermal stability of its gene product, tRNA^{Met}. *Biochemie* **768**:847-855.

Voyksner, R. D. 1997. Combining liquid chromatography with electrospray mass spectrometry, p. 323–341. *In* R. B. Cole (ed.), *Electrospray Ionization Mass Spectrometry. Fundamentals, Instrumentation, and Applications.* Wiley-Interscience, New York, N.Y.

Chapter 4

Incorporation of Modified Nucleotides into RNA for Studies on RNA Structure, Function and Intermolecular Interactions

ROBERT A. ZIMMERMANN, MICHAEL J. GAIT, AND MELISSA J. MOORE

This chapter focuses on techniques for intentionally incorporating both natural and unnatural modified nucleotides into RNA molecules. The number of uses to which modified RNAs have been put is almost as great as the number that have been prepared. The ability to vary individual functional groups of nucleotides by synthetic means extends the concept of mutagenesis to a new level. In contrast to the rather limited set of base changes allowed by conventional mutagenesis, the contributions of individual functional groups at particular positions can be probed singly or in any desired combination by using chemically modified nucleotide bases. This concept is sometimes referred to as "atomic mutagenesis." Moreover, synthetic approaches permit the incorporation into RNA of modified sugars and phosphates, something that is not possible with standard mutagenesis techniques. Modified ribose and phosphate moieties are particularly useful as mechanistic probes for RNA processing enzymes that catalyze phosphotransfer reactions. The incorporation of modified nucleotides can also facilitate the structural analysis of RNAs and RNA-containing complexes. RNAs labeled with heavy atom derivatives such as Br or I at single positions have been invaluable for solving X-ray crystal structures. Photocross-linking agents introduced at specific sites have been used to probe both protein-RNA and RNA-RNA interactions in a wide variety of systems, ranging from relatively simple complexes containing just one polypeptide and one RNA to large multicomponent particles like the ribosome and spliceosome. In addition, RNAs labeled with fluorescent groups can be used to investigate RNA structure and the kinetics of RNA folding with the aid of fluorescence energy transfer techniques. It is even possible to introduce disulfide cross-links to restrict an RNA structurally in order to assess the activity of a particular conformer.

Modified nucleotides can be introduced into RNA molecules either at specific positions within the polynucleotide chain, or randomly throughout the molecule. The former approach is most appropriate when information is sought about the role of a particular nucleotide or nucleotide sequence in an interaction, such as that of the tRNA anticodon with the 30S ribosomal subunit, of the 5' splice site of a eukaryotic pre-mRNA intron with components of the spliceosome, or of an RNA-RNA interaction in a ribozyme. If the experiment involves intermolecular cross-linking, it is also much simpler to characterize and interpret a cross-link if its position in one of the molecules is predetermined. Random incorporation is better suited for situations where little is known about the interacting surfaces. For instance, a quick way to identify positions in an RNA that are important for its structure or function is through modification interference studies where modifications are introduced randomly into the host molecule. After an effector nucleotide is identified by a random screen, the position through which the effect is transmitted can then itself be investigated by modification techniques.

When one considers how to use a synthetic approach in the study of an RNA or RNA complex, the most important decision is how to tailor the available technology to the specific needs of the system. This chapter is intended to give the reader a broad overview of the types of experiments that can be performed with chemically modified RNAs. We first describe the properties of particular modified nucleotides, then discuss techniques for their incorporation into polynucleotide chains. The last section pre-

---

Robert A. Zimmermann • Department of Biochemistry & Molecular Biology, University of Massachusetts, Amherst, Massachusetts 01003.   **Michael J. Gait** • MRC Laboratory of Molecular Biology, Hills Road, Cambridge CB2 2QH, United Kingdom.   **Melissa J. Moore** • Howard Hughes Medical Institute, Department of Biochemistry, Brandeis University, Waltham, Massachusetts 02254-9110.

sents a number of examples of the ways in which synthetic approaches have been used to investigate a variety of problems. It should be noted, however, that a comprehensive listing of all of the uses of artificially modified RNAs is beyond the scope of this chapter. Rather, we have chosen to focus on illustrations of the methods that pertain to our individual areas of expertise.

MODIFIED NUCLEOTIDES

Reagents for Affinity Labeling

The establishment of covalent cross-links between two interacting molecules provides an important avenue for relating specific structural features to molecular function. Short-range cross-links, on the order of 2 to 4 Å, are well suited to defining the immediate neighborhood of the interacting partners. For RNA, this can be achieved by incorporating reactive—especially photoreactive—nucleotides into the polynucleotide chain. Nucleotides that have proven particularly useful for this purpose fall into three categories, thionucleotides, halonucleotides, and azidonucleotides (Fig. 1), all of which can be photoactivated by irradiation with near UV light. Many of these modified nucleotides retain the structural properties of their unmodified counterparts and can be randomly incorporated into polynucleotide chains by enzymatic synthesis. Others, in which the modifications interfere with proper base pairing (Sylvers and Wower, 1993), must be introduced by chemical means. All of these reagents can be converted to reactive species by light with a wavelength of 300 to 350 nm. As this is well above the region of the spectrum that is absorbed by nucleotide bases and amino acids, irradiation inflicts little or no damage upon unmodified residues within RNA and protein.

Thionucleotides

Thionucleotides are among the most effective and widely used photoaffinity reagents for RNA because replacement of carbonyl oxygens within the base with sulfur atoms leads to minimal perturbation of the structure of the polynucleotide chains into which they are incorporated. 4-ThioU, the most frequently utilized member of this class of reagents, has been introduced into RNA molecules both randomly, by transcription with T7 RNA polymerase (Tanner et al. 1988), and in a site-specific fashion, by recombinant RNA methods (Wyatt et al., 1992; Sontheimer and Steitz, 1993). Another thionucleotide, 6-thioG, has only recently been put to use as a photoaffinity reagent for RNA (Sergiev et al., 1997). The 5′-triphosphates of 4-thioU and 6-thioG are incorporated by T7 RNA polymerase at roughly 25% and 10%, respectively, of the rate of their parent molecules (Favre and Fourrey, 1995; Sergiev et al., 1997). Irradiation of these photoprobes with light of 330 to 350 nm converts them to highly reactive species that can form stable, covalent bonds with molecules within a few angstroms of the modified residue. Model studies suggest that the mechanism of 4-thio-U cross-linking entails both cycloaddition and the participation of radical intermediates (Favre and Fourrey, 1995). Randomly substituted RNAs containing 4-thioU have been used for the study of systems as diverse as tRNA processing, tRNA- and

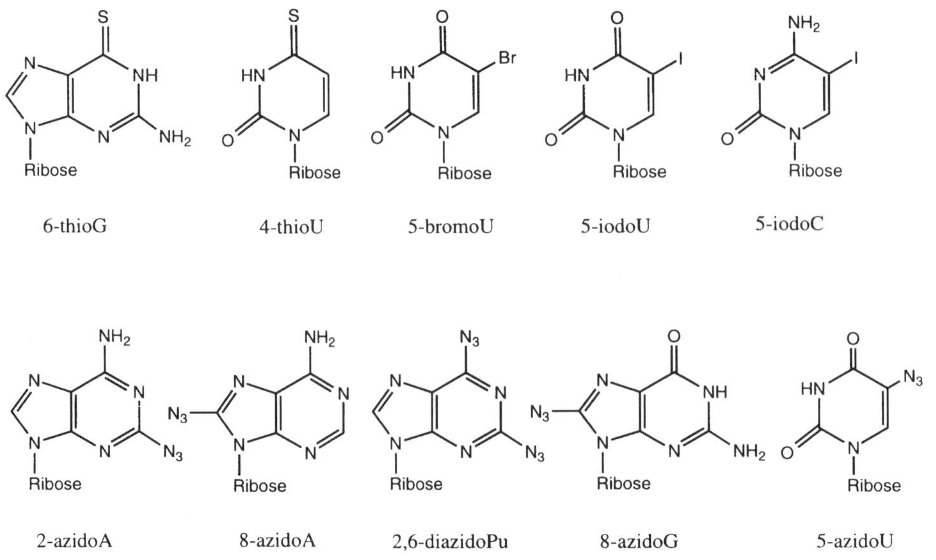

Figure 1. Modified nucleotides useful in RNA cross-linking experiments.

mRNA-ribosome interactions, and the folding of ribozymes (Tanner et al., 1988; Stade et al., 1988; Dontsova et al., 1992a; Rinke-Appel et al., 1993; Rosen et al., 1993; Rosen and Zimmermann, 1997; Favre and Fourrey, 1995). Analogs of mRNA substituted with 6-thioG have been used in delineating the pathway of mRNA through the ribosome (Sergiev et al., 1997). In addition to random incorporation, single 4-thioU residues can be introduced into RNA molecules at preselected sites. In one approach, the dinucleotide 4-thioUpG was used to initiate transcription by T7 RNA polymerase in vitro, yielding a transcript with 4-thioU at its 5' terminus. This molecule was then joined to the 3' end of an unmodified RNA molecule by T4 DNA ligase in the presence of a complementary oligodeoxyribonucleotide that bridges the two RNA fragments (Sontheimer and Steitz, 1993; Sontheimer, 1994).

Halonucleotides

The photoreactivity of halogenated pyrimidine derivatives has long been recognized and 5-bromodeoxyuridine was incorporated into DNA for use in protein-DNA cross-linking experiments more than two decades ago (Weintraub, 1973; Lin and Riggs, 1974). The usefulness of halopyrimidines for protein-RNA cross-linking has been demonstrated more recently, particularly as the 5'-triphosphates of 5-bromoU, 5-iodoU, and 5-iodoC, have the same hydrogen-bonding properties as their unmodified counterparts and can be efficiently incorporated into RNA molecules by in vitro transcription with T7 RNA polymerase (Tanner et al., 1988; Gott et al., 1991; Willis et al., 1993; Meisenheimer et al., 1996). During the past few years, 5-bromoU, 5-iodoU and 5-iodoC have all been introduced at the same position within the loop of a small RNA hairpin from either bacteriophage R17 (Gott et al., 1991; Willis et al., 1993) or the closely related bacteriophage MS2 (Meisenheimer et al., 1996), which comprises the binding site for the phage coat protein. The 5-bromoU-containing RNA became cross-linked to the coat protein at a 30 to 50% yield when complexes between them were irradiated with near UV light by using either a broad-spectrum transilluminator with a peak wavelength of 312 nm or a monochromatic xenon chloride laser at 308 nm (Gott et al., 1991). Similar complexes containing 5-iodoU- and 5-iodoC-substituted RNAs became cross-linked at even higher frequencies (75–95%) when irradiated with a helium-cadmium laser at 325 nm (Willis et al., 1993; Meisenheimer et al., 1996). At these wavelengths, photochemical damage to RNA and protein is minimal. In all three cases, cross-linking is believed to occur via a similar triplet-state mechanism (Willis et al., 1994) and photocoupling is apparently selective for electron-rich aromatic amino acids such as tyrosine, tryptophan, and histidine (Dietz and Koch, 1987; Willis et al., 1994; Meisenhemer et al., 1996). It has been suggested that cross-linking may entail stacking of the critical pyrimidine residue on the target amino acid (Meisenheimer et al., 1996). Given the preference of the halopyrimidines for specific amino acid side chains, they are not suitable for establishing RNA-RNA cross-links.

Azidonucleotides

Azidoadenosine triphosphates have been used extensively for labeling nucleotide binding sites in proteins (for a review, see Haley, 1983) but have been introduced into RNA molecules as photoaffinity probes only in the last decade. Although they are not substrates for DNA-dependent RNA polymerases (Sylvers and Wower, 1993), 8-azidoA, 2-azidoA and 2,6-diazidopurine riboside can be incorporated into RNA chains as 3',5'-bisphosphates by using T4 RNA ligase (Wower et al., 1988, 1994b; Sylvers et al., 1989). All three azidonucleosides have been placed at specific positions within tRNA molecules through the use of RNA reconstruction techniques and used to investigate the topography of tRNA binding sites in the the *Escherichia coli* ribosome (Wower et al., 1988, 1989, 1990, 1993b, 1994a; Sylvers et al., 1992). The preparation of $p8N_3Gp$ has also been described (Owens et al., 1987), but there are no reports to date of its incorporation into RNA molecules. The azidopurines yield highly reactive nitrenes when exposed to light in the vicinity of 300 nm and can insert into many different kinds of chemical bonds with little specificity and frequently with high yields (Sylvers and Wower, 1993). In contrast to the azidopurine nucleotides, 5-azido-UTP can serve as a substrate for DNA-dependent transcription by *E. coli* and T7 RNA polymerases (Evans and Haley, 1987; Woody et al., 1988; Dontsova et al., 1992b). This reagent, which has an absorption maximum at 288 nm, has been incorporated into mRNA analogs that were cross-linked to proteins of the 30S ribosomal subunit in and around the site of translation initiation (Dontsova et al., 1992b). 5-azidoU has not been widely adopted for cross-linking studies, perhaps because of its instability and problematic synthesis.

Convertible nucleotides

An alternative approach, which can be used to access intermolecular contacts at a greater distance from the modified nucleotide, entails the attachment of photoreactive moieties to modified bases randomly incorporated into the RNA chain by T7 RNA polymerase. These reagents have been dubbed "convertible" nucleotides because they are incorporated in one form and then derivatized with an additional functional group (MacMillan and Verdine, 1990; Moore and Query, 1998). Two strategies involving U analogs modified at the 5 position, 5-mercaptouridine and 5-aminomethyluridine, have been successfully deployed in recent years. Both of the analogs can be incorporated with roughly the same efficiency as unmodified U and allow total replacement of U with the modified U derivatives. In one case, 5-mercaptouridine was reacted posttranscriptionally with *p*-azidophenacyl bromide (Hixson and Hixson, 1975) to form the nitrene-generating residue 5-[4-(azidophenacyl)thio]-U (He et al., 1995; Hanna, 1996). In a second case, U was replaced by 5-aminomethyluridine and then converted to a photolabile, carbene-generating diazirine by reaction with the N-hydroxysuccinimide ester of 4-(trifluoromethyl diazirino)-benzoic acid (Bochkariov and Kogon, 1992; Sergiev et al., 1997). In both cases, the photoreactive groups are 10–11 Å from the base to which they are attached and are photoactivated by light of 300 and 350 nm, respectively, well away from the absorption maxima of protein and RNA. Other examples of "convertible" nucleotides include phosphorothioates, which have been derivatized with *p*-azidophenacyl bromide (Burgin and Pace, 1990; Musier-Forsyth and Schimmel, 1994), and adenosine containing an N^6-thioalkyl moiety that has been replaced with benzophenone and other photolabile reagents (MacMillan et al., 1994).

Ribose modifications

A further opportunity for the introduction of affinity reagents into RNA arises from the fact that the 2′-hydroxyl groups of nucleotides in duplex regions point away from the helix and are not often involved in hydrogen-bonding interactions. Thus, the 2′ positions of ribose sugars are suitable sites for the incorporation of chemical crosslinks between RNA domains or between RNA and protein. For example, alternative models of the hammerhead ribozyme were distinguished by the use of an aryl disulfide cross-link (Fig. 2a) spanning two different sets of residues in arms I and II of the hammerhead, only one set of which retained catalytic cleavage ability upon cross-linking (Sigurdsson et al., 1995). Similarly, the tolerance of aryl disulfide cross-linking and shorter alkyl disulfide cross-linking (Fig. 2b) has been used to aid molecular modeling of the two domains of the hairpin ribozyme (Earnshaw et al., 1997). The rates

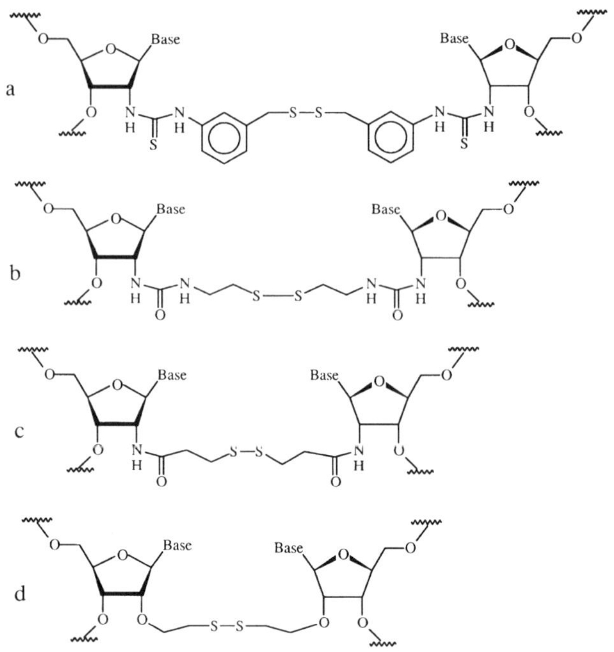

Figure 2. Structures of disulfide cross-links used in RNA structural studies.

of formation of a third type of disulfide cross-link (Fig. 2c) between distant regions of a large catalytic ribozyme derived from a group I intron were used to show unexpectedly large interdomain thermal motions (Cohen and Cech, 1997). In each of these three examples, the disulfide linkage was introduced by chemical synthesis of RNAs carrying 2′-amino groups at the desired locations followed by appropriate postsynthetic reactions. 2′-amino groups are also convenient for attaching site-specific labels (Sigurdsson and Eckstein, 1996). An alternative chemical cross-linking route is illustrated by the induction of 2′-O-ethyl disulfide cross-links (Fig. 2d) between residues located in different regions of yeast $tRNA^{Phe}$ (Goodwin et al., 1996).

Reagents for Functional Group Analysis

A number of base analogs are useful in RNA structure-function analysis (Fig. 3). For example, an exocyclic amino group is absolutely required for the base pairing of A and required in all but one type of base pair for G. Exocyclic amino group removal is effected by the replacement of guanosine by inosine (hypoxanthine riboside) or adenosine by nebularine (purine riboside), modifications that result in each case in minimal disruption to base stacking. Such substitutions are helpful in the prediction of base pairs within RNA internal loops and bulges. Inosine is also a good probe for structure-specific minor groove interactions between G amino groups in RNA duplex

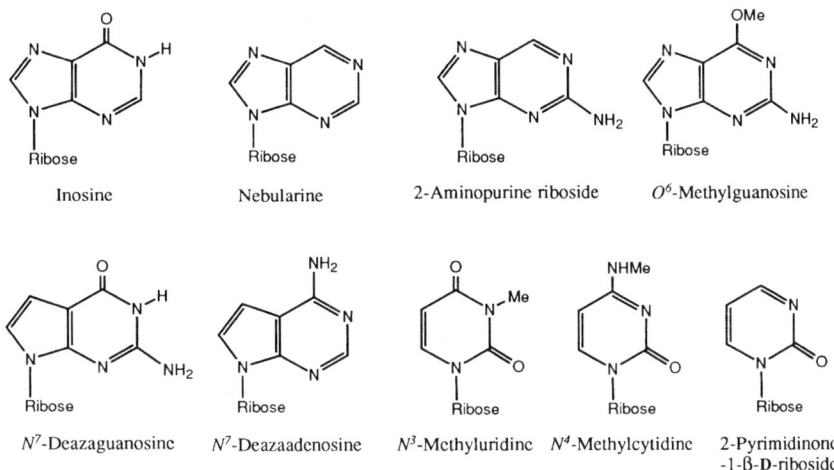

Figure 3. Base analogs useful in functional group analysis of RNA structures.

regions and proteins (Musier-Forsyth et al., 1991; Iwai et al., 1992; Hamy et al., 1993). Apart from nebularine (Slim and Gait, 1992; Fu et al., 1993), other useful purine nucleoside analogs are 2-aminoadenosine (Doudna et al., 1990; SantaLucia et al., 1991; Tuschl et al., 1993), O^6-methylguanosine (Grasby et al., 1993), N^7-deazaguanosine (Seela and Mersmann, 1992, 1993; Fu et al., 1993), and N^7-deazaadenosine (Fu and McLaughlin, 1992; Seela et al., 1993). The N^7-deazanucleosides are useful for probing the presence of Hoogsteen interactions, chelation sites for metal ions, and contacts to proteins.

Only a few pyrimidine ribonucleoside base analogs have been used for RNA structure-function analysis (Fig. 3). N^3-Methyluridine (Sumner-Smith et al., 1991; Iwai et al., 1992), N^4-methylcytidine (Grasby et al., 1995), and the fluorescent analog 2-pyrimidinone-1-β-D-riboside (Adams et al., 1994) are perhaps the most valuable. A more drastic modification of RNA is the complete removal of a base by use of an abasic analog (Fig. 4), for example in studies of the hammerhead and hairpin ribozymes (Beigelman et al., 1994; Schmidt et al., 1996). To assess the need for both sugar and base together at a particular position in an RNA, a propyl linker maintains the number and type of atoms between neighboring phosphate residues. Tolerance of propyl linker substitution shows that the residue plays no significant structural role (for example, in the hairpin ribozyme; Schmidt et al., 1996) or interactive role (for example, in protein-RNA interactions; Sumner-Smith et al., 1991; Iwai et al., 1992), and merely acts as a structural spacer element.

A single 2'-deoxyribonucleoside substitution in an RNA often has a minimal effect on overall RNA conformation. Accordingly, 2'-deoxynucleosides have been used as probes for the requirement of particular hydroxyl groups for ribozyme activity (Perreault et al., 1990, 1991; Chowrira et al., 1993). In the few cases where an RNA-like 3'-endo (N) configuration is essential, 2'-deoxy-2'-fluoro- (Pieken et al., 1991) or 2'-O-methyl nucleoside substitution allows the maintenance of an RNA-like conformation. In contrast to a 2'-deoxynucleoside, fluorosubstitution maintains the proton-accepting properties but not the proton-donating properties of a hydroxyl group. A 2'-deoxy-2'-mercapto group is an even closer hydroxyl mimic (Hamm and Piccirilli, 1997).

Two types of phosphate analog can be incorporated into RNA: a phosphorothioate and a 2'-deoxy-3'-methylphosphonate. Both types of analog are useful as probes of RNA contacts with proteins (Milligan and Uhlenbeck, 1989; Pritchard et al., 1994). An advantage of incorporating these compounds by chemical synthesis is that the R_p and S_p isomers can usually be separated by high-performance liquid chromatography (HPLC) (Slim and Gait, 1991; Pritchard et al., 1994). Thus, the effect of substitution of each oxygen atom can be assessed separately.

INCORPORATION OF MODIFIED NUCLEOTIDES INTO RNA BY SYNTHETIC METHODS

Chemical Synthesis

Recent improvements in methods of chemical synthesis now allow oligoribonucleotides up to about 50 residues in length to be made routinely by automated synthesis (Scaringe et al., 1990; Damha and Ogilvie, 1993; Davis, 1995; Sproat et al., 1995; Tsou et al., 1995b; Wincott et al., 1995). In most cases, the same techniques can be used to incorporate a

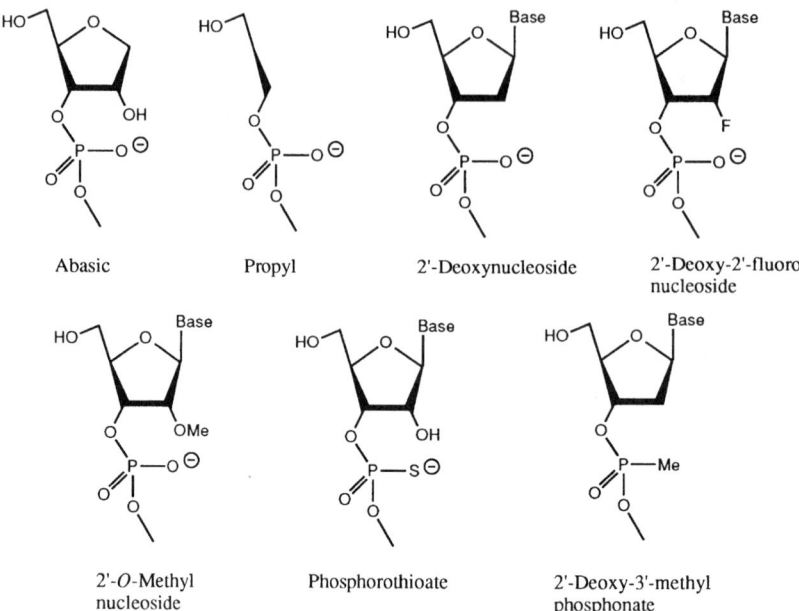

Figure 4. Sugar and phosphate analogs useful in analysis of RNA structures.

wide range of nucleotide analogs into RNA (Usman and Cedergren, 1992; Grasby and Gait, 1994; Eaton and Pieken, 1995; Gait et al., 1995). Practical methods of solid-phase oligoribonucleotide synthesis have been described previously (Gait et al., 1991; Damha and Ogilvie, 1993).

Assembly of oligoribonucleotides is carried out by sequential addition of protected ribonucleoside phosphoramidite monomers to a solid support (usually controlled pore glass or polystyrene) that carries a single nucleoside attached through its 2' or 3' position via an alkali-labile succinate linkage (Fig. 5). Almost all commercially supplied ribophosphoramidites contain the fluoride-labile t-butyldimethylsilyl (TBDMS) group (Ogilvie et al., 1974) for protection of the 2'-hydroxyl function. The main variable is the choice of heterocyclic base protecting groups. For many applications, especially those involving modified residues, it is important to select A, C, and G phosphoramidites having a compatible set of base protecting groups that can be removed under very mild ammoniacal conditions. Examples include the protection of A and G by phenoxyacetyl (Pac) groups (Chaix et al., 1989; Wu et al., 1989) or their more lipophilic p-alkyl derivatives (Sinha et al., 1993). For cytidine, standard N^4-protection is achieved by benzoyl groups or by the more labile acetyl group. Uridine requires no base protection.

A 1-μmol scale synthesis provides suffcent oligoribonucleotide (about 1 mg) for most biochemical

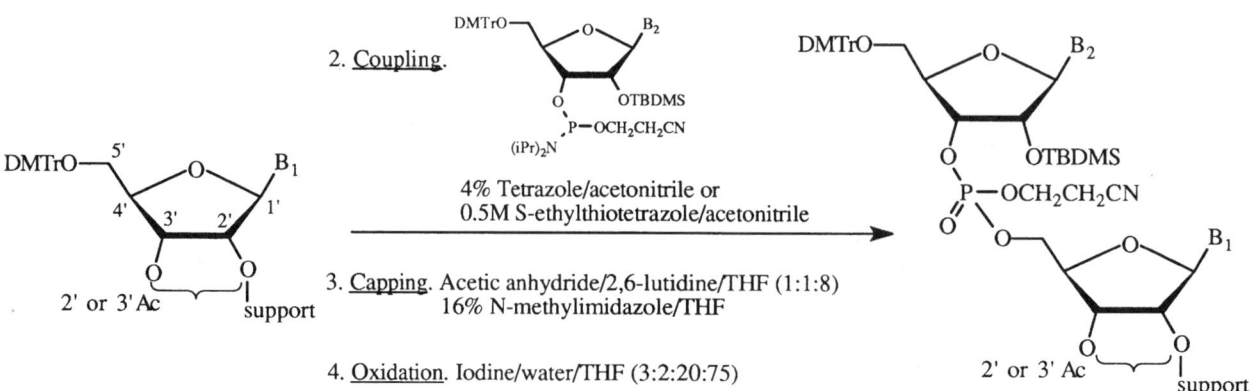

Figure 5. Steps involved in one cycle of solid-phase synthesis of oligoribonucleotides by the silyl-phosphoramidite method.

purposes, and three or four repeated syntheses are usually sufficient for crystallization trials. Each cycle of nucleotide addition requires four steps: (i) acidic deprotection of 5'-terminal dimethoxytrityl (DMTr) groups, (ii) coupling of a protected nucleoside derivative as its 3'-phosphoramidite, (iii) capping (acetylation) of unreacted 5'-hydroxyl groups, and (iv) oxidation of phosphite linkages to the corresponding phosphotriester. Each step is programmed on a DNA/RNA synthesizer with respect to delivery and wait time (time for reaction), and these steps are interspersed with a number of solvent washes. The precise details of these oligonucleotide assembly steps are dependent on the type of synthesizer.

Some alterations to the cycles are required for certain oligoribonucleotide modifications. For example, the insertion of a phosphorothioate is effected by replacing the oxidation step by a sulfurization step, for example, using 3H-1,2-benzodithiol-3-one 1,1-dioxide (the Beaucage reagent). Sulfurization generally results in 98–99% phosphorothioate incorporation, the remainder being phosphodiester. Another reagent reported to lead to even higher yields is 3-ethoxy-1,2,4-dithiazoline-5-one (EDITH) (Xu et al., 1996).

After the appropriate number of cycles of oligoribonucleotide synthesis, protecting groups are removed and the oligonucleotide is released into solution. First, the support is treated with concentrated aqueous ammonia-ethanol (3:1), anhydrous methanolic ammonia, or methylamine-concentrated ammonium hydroxide (1:1) (AMA) (Strobel et al., 1994; Reddy et al., 1995), depending on the type of base protection. This treatment removes base and phosphate protecting groups and cleaves the succinate linkage to the support. Second, 2'-O-TBDMS groups are removed by treatment with triethylamine trihydrofluoride (TEA.3HF) (Gasparutto et al., 1992) either as a 3:1 mixture with N,N-dimethylformamide (DMF) (Tsou et al., 1995a) or with N-methylpyrrolidinone as a cosolvent (Wincott et al., 1995). After deprotection, the oligoribonucleotide is desalted before purification, either by precipitation with butan-1-ol (Gait et al., 1991) or n-propanol (Sproat et al., 1995) or by gel filtration. Some slight alterations to these procedures are required when certain modified residues are incorporated.

It is possible to retain the terminal DMTr groups (TRITYL ON) during the procedures described above and carry out purification by the use of reversed-phase HPLC, but the resolution of contaminants of different lengths is often inadequate. A better procedure is to remove the DMTr groups on the synthesizer (TRITYL OFF) before commencing the remaining deprotection. The completely deprotected oligoribonucleotide is then resolved by polyacrylamide gel electrophoresis under denaturing conditions (Gait et al., 1991), followed by electroelution and dialysis against 1 M salt and then water (Scott et al., 1995). For optimal resolution of impurities, the use of anion-exchange HPLC, such as NucleoPac PA-100 (Dionex), is preferable. Both electrophoresis and anion-exchange HPLC exploit the unit charge difference on oligoribonucleotides, but separations on the NucleoPac column have a very slight hydrophobic element, so that impurities containing base modifications are also frequently resolved. After HPLC, salts are best removed by dialysis against water, and the product is isolated by lyophilization. If very high purity is required, or if there is a need to separate phosphorothioate or methylphosphonate diastereoisomers, it is usual to carry out an additional reversed-phase HPLC step (Slim and Gait, 1991; Pritchard et al., 1994).

It is essential to check the purity of an oligoribonucleotide and to determine that a desired modification is still present. Purity can be assessed by the use of capilliary electrophoresis. The separations are carried out in a thin capillary tube which requires only a very small amount of sample and results in a resolution that is far superior to that of a slab gel. However, this method gives no information as to the presence of a modified nucleotide. To obtain the molecular mass of the oligonucleotide, the method of choice is matrix-assisted laser desorption ionization–time of flight (MALDI-TOF) mass spectrometry. This procedure is extremely fast and requires only about 1 pmol of sample. The accuracy is usually sufficient to distinguish masses differing by only 3 or 4 Da. Nevertheless, mass spectrometry is not a quantitative technique and it cannot be used as a criterion of purity.

In some cases, mass spectrometry is not useful for the determination of product integrity. For example, the substitution of a 2'-amino group for a 2'-hydroxyl group in an oligoribonucleotide leads to a difference of only a few daltons in mass. Also, oligonucleotides of more than about 30 residues are less well resolved by MALDI-TOF mass spectrometry. In these cases, one can enzymatically digest the RNA to its constituent nucleosides and quantitate the base composition by separating the nucleosides on a reversed-phase HPLC column (Eadie et al., 1987).

Transcription with DNA-Dependent RNA Polymerases

The availability of highly purified DNA-dependent RNA polymerases from bacteriophages SP6, T7, and T3 has greatly contributed to the use of runoff transcription in vitro for the synthesis of RNAs tailored to serve a wide variety of purposes (Chamberlin and Ryan, 1982; Milligan and Uhlenbeck, 1988; Gurevich, 1996). These enzymes are well suited not only to the production of milligram quantities of unmodified RNAs for structural investigations, but also to the synthesis of small amounts of RNAs randomly substituted with modified nucleotides for analytical and functional studies. The most extensively used of these enzymes is T7 RNA polymerase, which can be obtained from numerous commercial sources. In addition, the recent construction of plasmids encoding T7, T3, and SP6 RNA polymerase variants with N-terminal His tags has made the isolation of these enzymes a convenient and simple task within the reach of any laboratory (He et al., 1997).

Short RNAs, on the order of 10–50 residues, can be readily transcribed from synthetic DNA templates in which only the 17-nucleotide T7 promoter need be double stranded (Milligan and Uhlenbeck, 1989). At the same time, RNAs of hundreds—or even thousands—of residues can be transcribed from double-stranded plasmid DNA templates into which the sequence of interest has been cloned downstream from a T7 promoter. Templates corresponding to defined sequences can also be prepared by PCR amplification of cloned genes or segments thereof. These approaches are schematized in Fig. 6. For efficient transcription, the initial templated nucleotide must be a G, although the yield of RNA also depends on the subsequent pentanucleotide sequence and other structural features of the template that are not well understood. As a rule, double-stranded templates are almost invariably better substrates for T7 RNA polymerase than partially duplex synthetic templates, perhaps because secondary structure can form in the single-stranded portions of the latter. Under optimal conditions, the best templates can produce up to several thousand copies of the RNA sequences they encode. The most important variables governing transcription yield are the concentrations of polymerase, template, nucleoside triphosphates, and divalent cations. Various additives have been reported to enhance the amount of RNA produced, including a number of different amines (Frugier et al., 1994) and pyrophosphatase, which removes inorganic pyrophosphate, a byproduct of the synthetic reaction that can inhibit transcription when present at high concentrations (Cunningham and Ofengand, 1990; Gurevich, 1996). In vitro transcription can lead to RNAs that are heterogeneous at both their 5' and 3' ends (Schenborn and Mierendorf, 1985; Milligan and Uhlenbeck, 1988; Moroney and Piccirilli, 1991), the most common manifestation of which is the addition

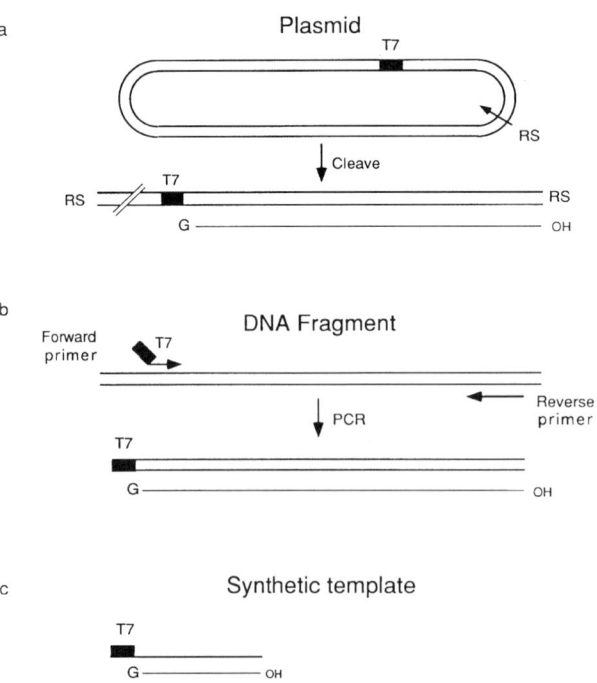

Figure 6. Transcription by T7 RNA polymerase of plasmid DNA (a), amplified DNA (b), and synthetic DNA (c) templates. Black rectangles denote the 17-bp T7 RNA polymerase promoter; RS signifies a restriction site.

of one or more untemplated nucleotides at the 3′ end. To overcome this problem, ribozymes can be incorporated into the structure of the transcripts, or used in *trans*, to cleave specific bonds at the 5′ and 3′ termini and thereby ensure the homogeneity of the products (for examples, see Price et al. 1995; Ferré-D'Amaré and Doudna, 1996).

In addition to unlabeled ATP, GTP, CTP, and UTP and their radioactively labeled counterparts, T7 RNA polymerase can utilize a number of modified nucleotides as long as they are able to form standard Watson–Crick base pairs with the template. For example, phosphorothioates, the halogenated pyrimidine nucleotides 5-bromoU, 5-iodoU and 5-iodoC, the thionucleotides 4-thioU and 6-thioG, the azidonucleotide 5-azidoU, and a number of "convertible" nucleotides can all be randomly incorporated into RNA chains by this enzyme (Schatz et al., 1991; Gott et al., 1991; Willis et al., 1993; Meisenheimer et al., 1996; Tanner et al., 1988; Sergiev et al., 1997; Dontsova et al., 1992b; He et al., 1995; Moore and Query, 1997), as can deoxyribonucleotides and 2′-O-methyl nucleotides if low concentrations of Mn^{2+} ions are present in the reaction mixture (Conrad et al., 1995). There are exceptions, however, such as 2- and 8-azidoA, which are not substrates for T7 polymerase owing to steric or conformational peculiarities (Sylvers and Wower, 1993). It is not possible, of course, to introduce modified nucleotides at specific sites in the RNA chain by in vitro transcription alone, and questions have arisen as to whether the absence of specific, naturally occurring modifications, such as those that typically occur in tRNA molecules, compromises the function of the resulting RNA molecules. In fact, tRNA transcripts lacking modified residues have proven to be good substrates for aminoacyl-tRNA synthetases and the partial reactions of protein synthesis (for examples, see Sampson and Uhlenbeck, 1988; Rosen et al., 1993). It is worthy of note that modified nucleotides can be introduced specifically at the 5′ end of in vitro transcripts because T7 RNA polymerase is relatively undiscriminating with regard to chemical groups adjoining the 5′ position of the initial guanosine residue. It is therefore possible to initiate transcription with GMP if it is desired to have a 5′-monophosphate for ligation or another purpose. T7 RNA polymerase will also initiate transcription with di- or trinucleotides as long as the 3′ residue is a G. In this way, it has been possible to incorporate deoxyribonucleotides, 2′-O-methyl nucleotides, phosphorothioates, 4-thioU and the pre-mRNA "cap" structure, G(5′)ppp(5′)G, into the 5′ end of T7 polymerase transcripts (see, Moore and Sharp, 1992, 1993; Sontheimer and Steitz, 1993; Wang et al., 1993; Sontheimer, 1994).

INCORPORATION OF MODIFIED NUCLEOTIDES INTO RNA BY RECOMBINANT RNA TECHNIQUES

In addition to the synthetic methods discussed previously, modified nucleotides can also be introduced into RNA chains in a site-specific fashion by recombinant RNA technology, a group of methods that encompasses the preparation of discrete RNA fragments and their reconstruction to form larger, biologically relevant RNA molecules (Cedergren and Grosjean, 1987).

Labeling the 3′ End

Perhaps the simplest method for introducing a modified nucleotide into an RNA molecule is its addition to the 3′ end of a polynucleotide chain as a 3′,5′-bisphosphate by using T4 RNA ligase. Because of the broad tolerance of RNA ligase for modified donor substrates, a wide variety of nucleotides with substituted base and sugar moieties can be employed for this purpose (Barrio et al., 1978). For example, this approach has been used to incorporate photoreactive azidoadenosines into the 3′ end of tRNA molecules or fragments derived therefrom (Wower et al., 1988, 1989, 1990; Sylvers et al. 1992).

Introduction of Modified Nucleotides at Internal Positions of RNA Chains

Modified nucleotides can be introduced at preselected internal positions of RNA molecules, given the availability of efficient techniques to specifically cleave and religate the RNA. These recombinant RNA methods encompass site-directed chemical and enzymatic cleavages, specialized synthetic techniques, and ligation with either RNA or DNA ligase (Fig. 7). Whereas some of them take advantage of structural peculiarities of molecules such as tRNA, others are of more general applicability.

Chemical cleavage methods

Reliable and efficient methods for the sequential truncation of RNA molecules from one or the other of their termini are rare. One such method, the Whitfeld degradation, permits the stepwise—and essentially quantitative—removal of residues from the 3′ end by a cyclic procedure that entails periodate oxidation of the 3′-terminal ribose, cleavage of the oxidized nucleoside with aniline, and removal of the 3′-terminal phosphate with alkaline phosphatase (Whitfeld and Markham, 1953; Paulsen and Wintermeyer, 1984). Modified nucleotides can then be li-

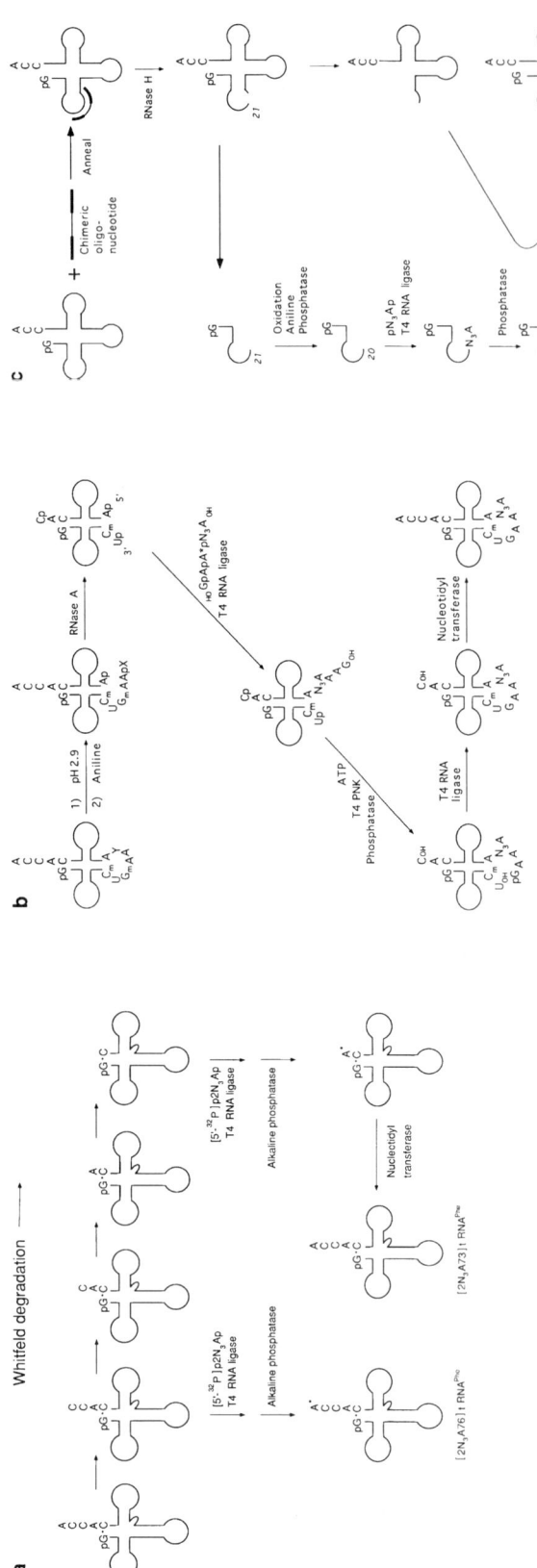
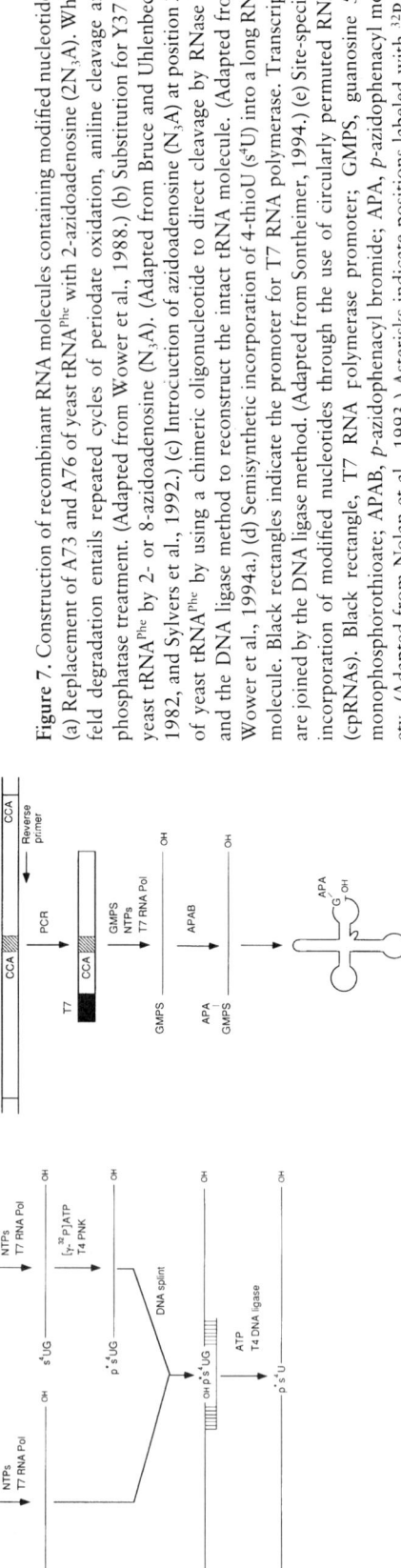

Figure 7. Construction of recombinant RNA molecules containing modified nucleotides. (a) Replacement of A73 and A76 of yeast tRNA^{Phe} with 2-azidoadenosine (2N$_3$A). Whitfeld degradation entails repeated cycles of periodate oxidation, aniline cleavage and phosphatase treatment. (Adapted from Wower et al., 1988.) (b) Substitution for Y37 of yeast tRNA^{Phe} by 2- or 8-azidoadenosine (N$_3$A). (Adapted from Bruce and Uhlenbeck, 1982, and Sylvers et al., 1992.) (c) Introduction of azidoadenosine (N$_3$A) at position 21 of yeast tRNA^{Phe} by using a chimeric oligonucleotide to direct cleavage by RNase H and the DNA ligase method to reconstruct the intact tRNA molecule. (Adapted from Wower et al., 1994a.) (d) Semisynthetic incorporation of 4-thioU (s^{4}U) into a long RNA molecule. Black rectangles indicate the promoter for T7 RNA polymerase. Transcripts are joined by the DNA ligase method. (Adapted from Sontheimer, 1994.) (e) Site-specific incorporation of modified nucleotides through the use of circularly permuted RNAs (cpRNAs). Black rectangle, T7 RNA polymerase promoter; GMPS, guanosine 5'-monophosphorothioate; APAB, p-azidophenacyl bromide; APA, p-azidophenacyl moiety. (Adapted from Nolan et al., 1993.) Asterisks indicate positions labeled with ^{32}P.

gated to the 3' end enzymatically (Fig. 7a). This approach has been used to remove either the 3'-terminal A or ACCA sequence from yeast tRNA^{Phe}, allowing the introduction of azidoadenosines at positions 76 and 73, respectively (Wower et al., 1988, 1989). In the latter case, the 3'-terminal CCA sequence was restored by the action of nucleotidyl transferase, a method of 3' end reconstruction specific to tRNA molecules.

The presence of modified residues in tRNAs sometimes provides unique opportunities for specific cleavage of the RNA chain (Fig. 7b). A classic example is the acid lability of the wyosine residue at position 37 of yeast tRNA^{Phe} (Philippsen et al., 1968), which permits the deletion or substitution of bases in the anticodon loop (Kaufmann and Littauer, 1974). The ability to cleave, reconstruct, and religate the anticodon loop was later exploited to evaluate the role of the anticodon in the interaction of yeast tRNA^{Phe} with Phe-tRNA synthetase (Bruce and Uhlenbeck, 1982) and to prepare derivatives of the same tRNA containing photolabile azidoadenosines adjacent to the anticodon loop (Sylvers et al., 1992). Although quite effective, this procedure again depends on the structural peculiarities of individual tRNAs.

Several attempts have been made to create artificial RNases targeted to defined RNA sequences by covalently attaching nucleases or nucleolytic chemical groups such as EDTA-Fe(II) or 1,10-phenanthroline-Cu(II) to complementary oligodeoxyribonucleotides (for examples, see Dreyer and Dervan, 1985; Chen and Sigman, 1986). Although both approaches proved feasible, they have not been used extensively for RNA reconstruction.

Enzymatic cleavage methods

During the past 15 years there have been numerous efforts to cleave RNA in a site-specific manner with the aid of nucleases. The most successful of these relied on the unusual and virtually unique susceptibility of the anticodon loop of tRNA to single-strand-specific nucleases such as RNases A, T$_1$, U$_2$, and S1. This approach proved quite valuable for replacing the anticodon of numerous tRNAs to be used in the analysis of features recognized by aminoacyl-tRNA synthetases (for examples, see Schulman and Pelka, 1983; Ohtsuka et al., 1983; Bare and Uhlenbeck, 1986; Beauchemin et al., 1986) and for generating substrates for modification enzymes (for examples, see Fournier et al., 1983; Haumont et al., 1984; also see Chapter 2 by Grosjean et al.).

More general strategies for site-specific RNA cleavage have emerged in recent years; in principle, they permit the introduction of modified nucleotides at any position within an intact RNA chain. One approach is based on the use of "addressed" cleavage with RNase H, an enzyme that cleaves the RNA strand within RNA/DNA heteroduplexes (Donis-Keller, 1979; Stepanova et al., 1979). In most cases, however, it was found that multiple scissions occurred within the heteroduplex region, producing fragments with heterogeneous ends. An ingenious method for overcoming this problem was developed by Ohtsuka and collaborators, who found that RNase H cleavage could be directed to a specific phosphodiester bond by annealing the RNA to a chimeric oligonucleotide consisting of a block of four deoxyribonucleotides flanked on both sides by short 2'-O-methyloligoribonucleotides (Inoue et al., 1987; Shibahara et al., 1987; Hayase et al., 1990). Chimeric oligonucleotides of this type target RNase H cleavage to the phosphodiester bond opposite the 5' end of the tetradeoxyribonucleotide block. At this point, the 3' nucleotide of the 5' fragment can be removed by one cycle of the Whitfeld degradation and replaced by a modified nucleotide by using the 3',5'-bisphosphate and RNA ligase if so desired. The 5' and 3' fragments are then resealed with RNA or DNA ligase as appropriate (Fig. 7c). This method has been employed to construct tRNA variants for use in the study of tRNA identity (Hayase et al., 1992; Sylvers et al., 1993) and in photochemical cross-linking experiments (Wower et al., 1994a). A similar method has recently been used to introduce 4-thioU at a specific position within a eukaryotic pre-mRNA (Yu and Steitz, 1997).

A new approach to site-directed RNA cleavage has recently been described and offers considerable promise with respect to generality, flexibility and convenience. Using in vitro selection techniques, Santoro and Joyce (1997) have isolated a sequence-specific endoribonuclease composed exclusively of DNA. The catalytic core of the DNA enzyme consists of 15 nucleotides and is flanked by two sequences of 7 to 8 bases that can pair with the substrate RNA on either side of the target nucleotide. After heteroduplex formation, cleavage occurs on the 3' side of the unpaired nucleotide. Although the specificity of the enzyme can be altered by changing the flanking sequences, it requires that the scissile bond lie between an unpaired purine and a paired pyrimidine. Nonetheless, the reaction is genuinely enzymatic in that it is characterized by multiple turnovers, and it may be possible to isolate a completely unselective variant of the endoribonuclease in the future.

Semisynthetic methods

In the "cut-and-paste" approaches described above, intact RNA molecules are cleaved chemically or enzymatically into 5' and 3' fragments, a modified nucleotide is introduced at the 3' end of the 5' fragment, and the two sequences are ligated either by T4 RNA ligase or by T4 DNA ligase with the aid of a "bridging" oligodeoxyribonucleotide complementary to the splice junction (see Moore and Sharp, 1992; Moore and Query, 1998). RNA molecules with modified nucleotides at predetermined sites can also be constructed by semisynthetic methods in which the modified residue is introduced at the 5' end of the 3' RNA fragment by transcription and then ligated to the 5' RNA fragment (Fig. 7d). This approach has been used extensively in the study of eukaryotic pre-mRNA splicing (Moore and Sharp, 1992, 1993; Wyatt et al., 1992; Sontheimer and Steitz, 1993; Sontheimer, 1994; Gaur et al., 1995; Newman et al., 1995; Reyes et al., 1996). The RNA sequence 5' to the site of insertion can be prepared by any of the chemical or enzymatic cleavage methods described above or by in vitro transcription. The 3' fragment is transcribed from an appropriate DNA template with a dinucleotide of the form XpG, where X is the modified nucleoside and G is the first templated nucleotide, to initiate the RNA chain. X may be a nucleotide analog, such as a 2'-deoxy or 2'-O-methyl nucleotide, for functional group analysis (Moore and Sharp, 1992) or contain a photolabile moiety, such as 4-thioU, for cross-linking experiments (Wyatt et al., 1992). Alternatively, the X-G bond may consist of a phosphorothioate which can be used for mechanistic studies (Moore and Sharp, 1993). The 5' end of the 3' fragment is then phosphorylated with T4 polynucleotide kinase and ligated to the 5' fragment with T4 DNA ligase by using a complementary DNA splint (Moore and Sharp, 1992). For affinity labeling studies, it is convenient to introduce ^{32}P at the phosphorylation step so as to have a radioactive label adjacent to the cross-linking site. The main limitation of this method is that there must be a G one residue 3' to the junction between the two fragments, a consequence of the requirement that G be the first templated nucleotide incorporated into the 3' fragment by T7 RNA polymerase.

A novel strategy for the site-specific incorporation of modified nucleotides makes use of circularly permuted RNAs (cpRNAs). Although originally developed for cross-linking tRNA to the RNA component of RNase P (Nolan et al., 1993), it is potentially applicable to a much wider range of RNA molecules. The method entails the synthesis of tRNAs in which the normally occurring 5' and 3' ends are connected by a short linker, and artificial termini are introduced within internal portions of the RNA sequence (Fig. 7e). The new 5' and 3' termini can then be used to attach affinity labels or other modified nucleotides.

DNA templates encoding full-length RNAs with displaced 5' and 3' termini are amplified by PCR from tandemly duplicated tRNA genes with primers that define the new ends. Incorporation of a promoter for T7 RNA polymerase into the forward primer permits the variant RNAs to be prepared by in vitro transcription. In the experiments of Nolan et al. (1993), transcription was initiated with guanosine 5'-monophosphorothioate, providing a unique 5' thiol to which the photochemical cross-linking reagent p-azidophenacyl bromide was attached. The tRNA analogs constructed in this way were found to be substrates for RNase P RNA and were subsequently cross-linked to the ribozyme to identify sequences within the tRNA binding site. This technique can be generalized to any RNA whose native 5' and 3' termini lie close to one another, such as at the end of a duplex stem. Moreover, modified bases can be incorporated at the 3' end of the cpRNAs as well as at the 5' end with nucleoside bisphosphates and T4 RNA ligase and, if necessary, the nonnative termini can be ligated to one another.

Covalent Joining of RNA Molecules by Enzymatic and Chemical Ligation

There are presently three enzymatic joining methods available for ligating RNA molecules; they use (i) T4 RNA ligase, (ii) T4 DNA ligase, and (iii) chemical ligation. Both T4 RNA ligase and T4 DNA ligase use ATP to drive the ligation of a downstream donor RNA having a 5'-terminal monophosphate to an upstream acceptor RNA terminating with a 3'-hydroxyl group (Fig. 8). The main difference between the two enzymes is that T4 RNA ligase is specific for single-stranded polynucleotides, whereas T4 DNA ligase joins nicks in double-stranded regions of nucleic acid molecules. These reactions generally are straightforward to perform by incubating the substrates together with the enzyme in the appropriate buffer containing ATP. However, each enzyme has both advantages and disadvantages for particular applications.

T4 RNA ligase

T4 RNA ligase has been successfully used to prepare both modified oligonucleotides (England and Uhlenbeck, 1978; Krug et al., 1982; Romaniuk and Uhlenbeck, 1983) and tRNAs with site-specific modifications in the anticodon loop (Bare et al., 1983; Wittenberg and Uhlenbeck, 1985) (see also Chapter 2 by Grosjean et al.). In fact, it appears that anticodon loops are the natural substrates for this enzyme (Uhlenbeck and Gumport, 1982). T4 RNA ligase is

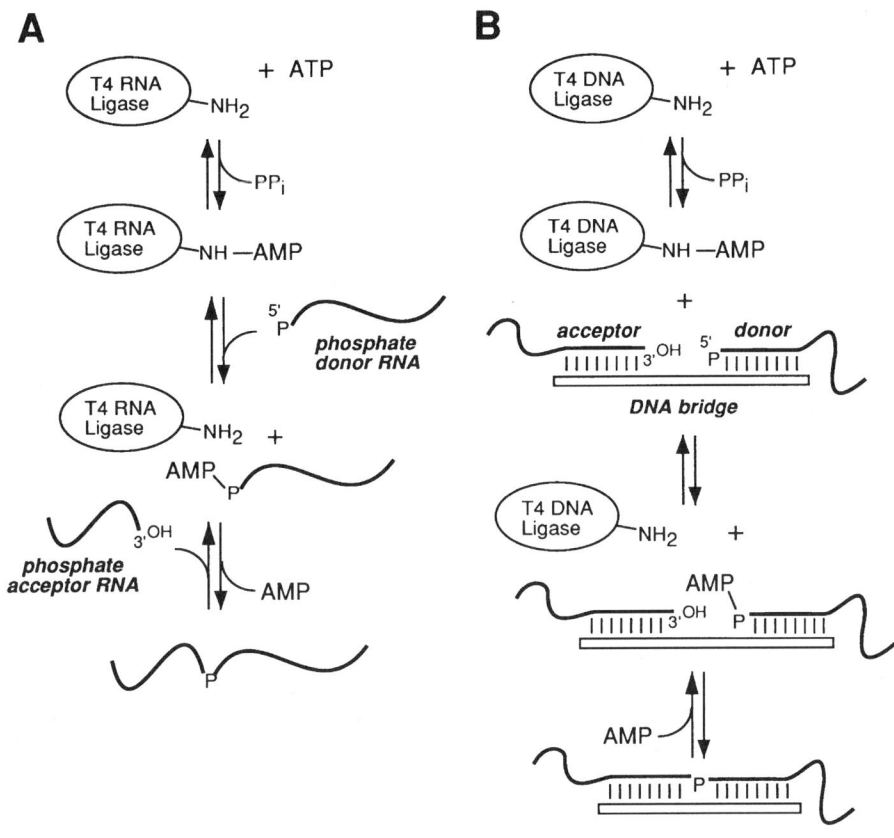

Figure 8. Mechanisms of RNA joining reactions catalyzed by T4 RNA ligase (A) and T4 DNA ligase (B).

also used widely for labeling the 3' ends of RNAs with [5'-^{32}P]pCp (Barrio et al., 1978; Romaniuk and Uhlenbeck, 1983; Enright and Sollner-Webb, 1994; Lamond and Sproat, 1994; Wahle and Keller, 1994). There are, however, a number of features of the enzyme that limit its usefulness as a general agent for joining long RNAs. First, its high K_m for polynucleotides (>1 mM; Uhlenbeck and Gumport, 1982) means that ligation reactions can proceed quite slowly at low RNA concentrations. Second, ligation efficiency is highly dependent on the sequences to either side of the ligation junction (England and Uhlenbeck, 1978; Romaniuk et al., 1982). Third, since the enzyme is specific for single-stranded regions, ligation reactions cannot be readily templated. Thus, significant amounts of side products, such as circular and oligomeric RNAs, can accumulate unless the 3'-OH of the phosphate donor RNA is protected or destroyed (Uhlenbeck and Gumport, 1982; Romaniuk and Uhlenbeck, 1983). For the same reason, T4 RNA ligase is unable to discriminate between "N" and "N+1" transcripts that are often present in RNA synthesized in vitro. This can result in the incorporation of a high percentage of extra, undesired nucleotides in the ligated product (Moore and Sharp, 1992).

T4 DNA ligase

T4 DNA ligase, most commonly used to join fragments of DNA, can also be used to ligate RNA (Higgins and Cozzarelli, 1979; Engler and Richardson, 1982; Moore and Sharp, 1992; Moore and Query, 1998). Because this enzyme has a strict requirement for nicks in double-stranded regions, T4 DNA ligase reactions must be templated; that is, the RNAs to be joined must first be annealed to a complementary DNA splint to form a nicked (RNA/RNA)·DNA duplex (Fig. 8). Templating eliminates many of the problems inherent in T4 RNA ligase reactions. In particular, T4 DNA ligase produces far fewer side products and it can select against "N+1" RNAs (Moore and Sharp, 1992). Additional advantages include the low K_m of the enzyme for double-stranded nucleic acid (in the submicromolar range; Engler and Richardson, 1982) and the fact that there is little or no dependence of the reaction on the sequence around the nick. For these reasons, T4 DNA ligase has become the enzyme of choice for ligating long RNAs of varying sequence (Moore and Sharp, 1992; Moore and Query, 1998). On the other hand, T4 DNA ligase does not work particularly well with very short substrates (<5 nucleotides; Suntharalingam et al., 1997), and highly structured substrates like tRNAs and large ribozymes can be extraordinarily difficult to anneal to the DNA template (see Strobel and Cech, 1993). In addition, because T4 DNA ligase does not turn over effectively on mixed RNA·DNA duplexes, stoichiometric quantities of the enzyme are required (Suntharalingam et al., 1997). For very large scale ligations this can result in a prohibitive cost, although a $(His)_6$-tagged version of the enzyme is now available (Strobel and Cech, 1995).

Template-dependent chemical ligation: the incorporation of trisubstituted pyrophophosphates

Template-dependent chemical ligation is a useful technique for the incorporation of pyrophosphate analogs in place of phosphodiester linkages (Kuznetsova et al., 1990). A trisubstituted pyrophosphate (tsp) derivative at a particular position in duplex DNA or RNA is particularly valuable in protein-nucleic acid recognition studies because it is relatively stable to hydrolysis at pH 5.5–7.5, but rapidly reacts with primary amines to form a phosphoamide derivative (Naryshkin et al., 1996). Reaction occurs at the most substituted phosphate of the tsp linkage. A tsp linkage can be incorporated into an oligoribonucleotide containing a single 2′-deoxyribonucleoside on the 5′ side of the tsp linkage by means of template-dependent chemical ligation using the water soluble carbodiimide EDC (Fig. 9). The ligated oligonucleotide tsp analog may be isolated by gel electrophoresis. A tsp linkage was incorporated into certain positions of an RNA stem-loop representing the TAR recognition sequence of the HIV-1 Tat protein and found to bind a Tat peptide indistinguishably from unmodified TAR RNA. Longer incubation resulted in the formation of a covalent phosphoamide derivative by reaction of the tsp linkage with a specific lysine residue of Tat in the bound state (Naryshkin et al., 1997).

THE USE OF NUCLEOTIDE MODIFICATION TECHNIQUES IN RNA RESEARCH

Functional Group Analysis

The manner in which functional groups conjoined with the base, sugar, and phosphate moieties of individual nucleotides contribute to the structural and functional properties of RNA molecules has provided insights into many different phenomena (Grasby and Gait, 1994). These include RNA catalysis, as exemplified by the hammerhead (McKay, 1996), hairpin (Schmidt et al., 1996) and RNase P (Loria and Pan, 1997) ribozymes, the splicing of group I introns (Strobel and Cech, 1995) and eukaryotic pre-mRNAs (Query et al., 1996), and a variety of protein-RNA interactions (Musier-Forsyth et al., 1991; Hayase et al., 1992; Iwai et al., 1992; Hamy et al., 1993; Baidya and Uhlenbeck, 1995; Bevilacqua and Cech, 1996). The use of functional group anal-

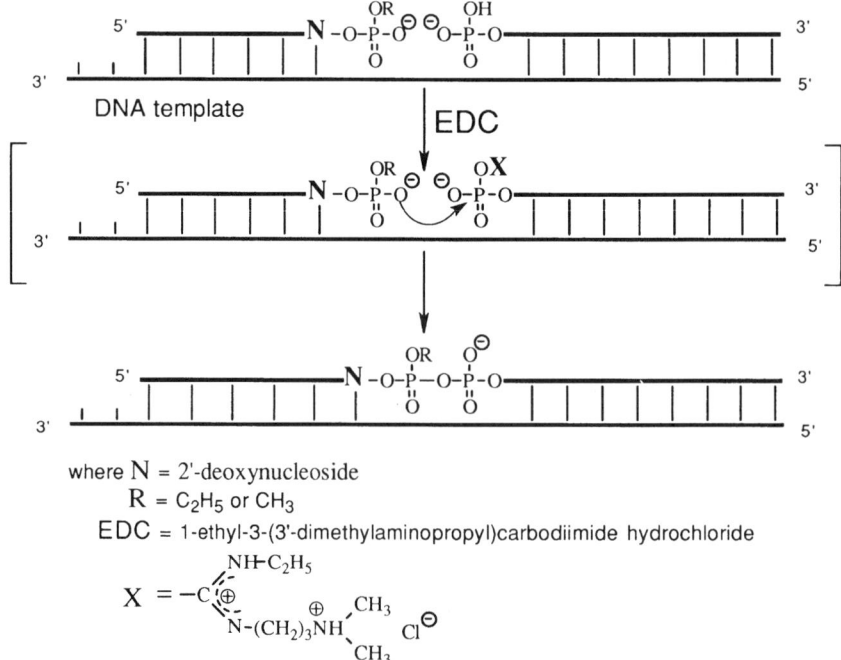

Figure 9. Template-dependent chemical ligation to form an oligoribonucleotide containing a trisubstituted pyrophosphate linkage.

ysis to investigate specific associations between tRNAs and their cognate aminoacyl-tRNA synthetases and between the human immunodeficiency virus type 1 (HIV-1) Tat protein and TAR RNA is described below.

Interaction of tRNAs with aminoacyl-tRNA synthetases

The tRNAs provide excellent models for the study of protein-RNA interaction owing to their small size and well-defined three-dimensional structure. The addition, deletion or alteration of functional groups can therefore be used to answer specific questions about the role of individual nucleotides in tRNA recognition and, in particular, to confirm the existence of molecular contacts suggested by crystallographic analysis of tRNA–aminoacyl-tRNA synthetase complexes. The latter is illustrated by the work of Hayase et al. (1992), who constructed a series of full-length $E.$ $coli$ tRNA^{Gln} derivatives by enzymatically ligating chemically synthesized $5'$-$\frac{1}{4}$ molecules, in which individual G residues had been replaced by I, to a common $3'$-$\frac{3}{4}$ molecule obtained from in vitro T7 transcription. With the aid of these recombinant tRNAs, they showed that the 2-amino groups of G2, G3, and G10 are necessary for optimal recognition by glutaminyl-tRNA synthetase, corroborating the inference from X-ray crystallographic analysis that these exocyclic amines interact with specific amino acid side chains in the enzyme.

In the case of $E.$ $coli$ tRNA^{Ala}, the principal feature recognized by alanine-tRNA synthetase is a wobble base pair between G3 and U70 within the acceptor arm (Hou and Schimmel, 1988; McClain and Foss, 1988). The discovery that short duplex RNAs mimicking the anticodon arm can serve as substrates for aminoacylation made it possible to determine the effects of nucleotide modifications on synthetase interaction and catalysis (Musier-Forsyth et al., 1991; Musier-Forsyth and Schimmel, 1992). Through a variety of substitutions at position 3, it was found that the unpaired, exocyclic 2-amino group of G3, which extends into the minor groove of the RNA helix, is required for aminoacylation and probably contacts the enzyme (Musier-Forsyth et al., 1991). Subsequently, the systematic replacement of individual ribonucleotides with their deoxyribonucleotide or $2'$-O-methyl counterparts established that the $2'$-OH groups of G4, U70, and C71 also contribute to interaction with the synthetase (Musier-Forsyth and Schimmel, 1992). These hydroxyls also lie in the minor groove of the helical stem and are situated within 5 Å of the 2-amino group of G3. Interestingly, association of the catalytic core of the Tetrahymena group I intron with a conserved G·U base pair at the $5'$ splice site is mediated by a similar cluster of minor groove functional groups (Strobel and Cech, 1995). This structural motif, consisting of the exocyclic amine of a G·U wobble pair and a subset of nearby $2'$-OH groups, can therefore be recognized specifically by both protein and RNA. In another example, recombinant RNA techniques were used to assemble a series of full-length human tRNA^{Ser} analogs from a transcript corresponding to the $5'$-$\frac{3}{4}$ fragment and chemically synthesized $3'$-$\frac{1}{4}$ fragment in which the "discriminator" base, G73, was replaced by either 2-aminopurine or inosine (Breitschopf and Gross, 1996). These studies established that both the 2-amino group and the 6-carbonyl oxygen of G73 are important for synthetase interaction and that the 2-amino group is required for aminoacylation.

Interaction of the HIV-1 Tat protein with TAR RNA

The interaction of Tat, the essential HIV-1 gene regulatory protein, with its RNA recognition sequence, the $trans$-activation response region (TAR), has been the center of much study because of the potential for the development of anti-HIV drugs (Gait and Karn, 1993; Karn et al., 1996). Preliminary nuclear magnetic resonance (NMR) analysis of Tat protein alone has provided some information on its possible conformation (Bayer et al., 1995), and NMR has been used to propose structural models of TAR RNA both free (Aboul-ela et al., 1996) and in the presence of a peptide ligand (Aboul-ela et al., 1995). Nonetheless, a full structural determination of the Tat-TAR complex has remained elusive. The complex has also defied all attempts at X-ray crystallographic analysis to date.

HIV-1 Tat is an 86-amino acid protein which can be subdivided into six regions. Of these, a central section rich in arginine and lysine residues is thought to to be particularly important for TAR recognition. Flanking hydrophobic core and glutamine-rich regions are proposed to help form a more pronounced structural domain for interaction of the basic region. TAR is a 59-residue RNA stem-loop that is found at the $5'$ end of all HIV transcripts. Essential residues of TAR required for RNA recognition were shown to be U23, one of the residues in a three-base bulge, and the two base pairs immediately above the bulge (Delling et al., 1992; Churcher et al., 1993) (Fig. 10). Functional group mapping of TAR proved very valuable for showing that in the complex the basic region of Tat lies in the major groove of the TAR RNA duplex, rather than in the minor groove. Thus, whereas replacement of G26 by inosine in a

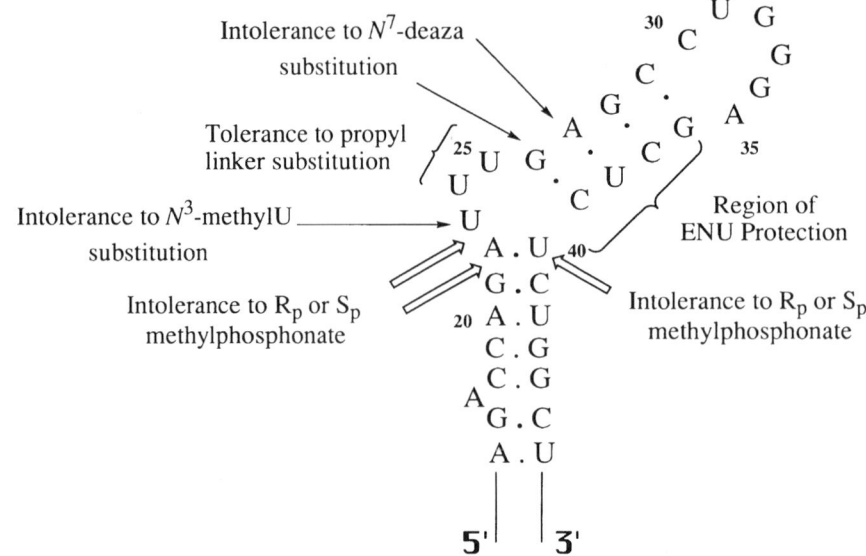

Figure 10. The apex region of the TAR RNA stem-loop, showing sites where chemical substitution or interference has been used to determine functionalities important in recognition by HIV-1 Tat.

model TAR RNA duplex had no effect on Tat binding, replacement by N^7-deazaG was highly deleterious (Hamy et al., 1993). Similarly, N^7-deazaA substitution at A27 reduced Tat binding. A further likely Tat-TAR contact was demonstrated at U23 because of its intolerance for substitution by N^3-methylU (Delling et al., 1992). By contrast, U24 and U25 could be replaced by propyl linkers without substantial loss of Tat binding (Sumner-Smith et al., 1991), suggesting that these residues merely act as flexible spacers, a finding later confirmed by NMR analysis (Aboul-ela et al., 1995).

In addition to bases, phosphate groups play an important role in Tat binding. Ethylation interference analysis showed two regions of TAR protection, two phosphates on one strand (P21 and P22) and five phosphates on the other strand (P36 to P40) (Fig. 10) (Calnan et al., 1991; Churcher et al., 1993), suggesting that Tat closely approaches these regions. Replacement of individual phosphates by R_p or S_p methylphosphonates at P21, P22, or P40 was in each case very detrimental to Tat binding (Pritchard et al., 1994), making these phosphates likely candidates for direct interaction with basic residues in Tat. Phosphate-arginine and phosphate-lysine contacts may make essential contributions to Tat-TAR recognition via "indirect readout" in that phosphates in the vicinity of the TAR bulge take up positions that are displaced from those they would occupy in a regular A-form duplex, allowing them to make structure-specific, and hence sequence-specific, contacts with Tat (Aboul-ela et al., 1996). Cross-linking studies using tsps in place of specific phosphate linkages indicate that Lys51 in the basic region of Tat reacts specifically with TAR replaced by tsp at P38 (Naryshkin et al., 1997). These and other cross-linking studies in progress should help considerably to distinguish models obtained from NMR studies of the Tat-TAR complex.

Biochemical Mechanisms

Site-specifically modified RNAs have been especially useful as mechanistic and structural probes of RNA splicing by both self-splicing introns and the spliceosome. In these cases, the RNAs involved usually are quite long, necessitating preparation of the substrates by the T4 DNA ligase approach. The best-studied group I and group II self-splicing introns encompass 413 and 887 nucleotides, respectively, whereas the spliceosome contains five snRNAs in the range of 85 to >1,000 nucleotides, a pre-mRNA of variable size, and myriad proteins. Because these RNAs are so large, detailed information about their three-dimensional structures at the subnanometer level is extremely limited. And, although significant progress is being made toward determining crystallographic structures of RNAs as large as group I introns (Doudna and Cate, 1997), the sheer size of the 60S spliceosome may prove to be an insurmountable challenge to atomic-level, three-dimensional structural analysis of such complexes in the foreseeable future. Thus, the ability to implement and analyze the effects of individual functional group alterations may be an even more valuable tool in large RNA-containing systems, such as the spliceosome, than it is for smaller RNAs and RNA-protein complexes for which structural information is more readily attained by other means.

The introduction of site-specific modifications into specific ribose and phosphate groups by the T4 DNA ligase approach has illuminated a number of key mechanistic and structural features both of pre-mRNA splicing by the spliceosome and of self-splicing introns. Placement of 2′-deoxy and arabinose (the C2′ epimer of ribose) sugars at the pre-mRNA branch site provided good evidence that the 2′-OH nucleophile for the first step of splicing must be in the bulged position of a short RNA duplex between the branch site and U2 snRNA in order to be used for lariat formation (Query et al., 1994). Incorporation of individual 2′-deoxy and 2′-O-methyl groups at the 5′ and 3′ splice sites showed distinct differences in the need for an adjacent 2′-OH group during the first and second chemical steps of splicing (Moore and Sharp, 1992). Further differences in the stereochemical details of the active sites were revealed by incorporation of chiral R_p and S_p phosphorothioates at the splice sites. Like most enzymes that catalyze phosphotransfer reactions, the spliceosome displays a distinct preference for one phosphorothioate diastereomer over the other. The pattern of stereospecificity in this case led to the hypothesis that the first and second steps of splicing are catalyzed by distinguishable active sites (Moore and Sharp, 1993). Stereochemical analysis of spliced products formed with the allowed phosphorothioate diastereomer at each site also permitted unambiguous determination of the chemical mechanisms for both splicing reactions (Maschhoff and Padgett, 1993; Moore and Sharp, 1993). Similar experiments subsequently carried out with the ai5g group II intron (Padgett et al., 1994; Podar et al., 1995) gave results that paralleled those obtained with the spliceosome. The long list of mechanistic similarities between group II introns and the spliceosome is taken by many to indicate a common ancestry for the two systems (Sharp, 1985; Cech, 1986). More recently, the incorporation of a 3′-sulfur atom at either splice site in a spliceosomal pre-mRNA revealed the interaction of a metal ion with the phosphodiester linkage at the 5′ splice site, but interestingly, no evidence was found for a similarly situated metal ion at the 3′ splice site (Sontheimer et al., 1997). This is in direct contrast to the situation in group I introns, where the two steps of splicing are catalyzed by a single active site within which two Mg^{2+} ions directly coordinate to either side of the scissile phosphate, as revealed by metal-ion rescue experiments in which both 3′- and 5′-sulfurs were placed in short RNA substrates (Piccirilli et al., 1993; Weinstein et al., 1997).

Specific backbone modifications incorporated at positions more remote from the sites of chemical reaction have also proved to be valuable structure-function probes. For example, the introduction of individual and multiple 2′-deoxy sugars at positions −1 to −6 relative to the cleavage site of a truncated substrate for the *Tetrahymena* group I ribozyme identified contacts between the substrate backbone and bases in the ribozyme that are crucial for docking the substrate into the catalytic core (Pyle and Cech, 1991; Pyle et al., 1992). In contrast, modifications at analogous positions in a group II intron substrate had markedly less dramatic effects (Griffin et al., 1995), indicating that group I and group II introns employ different mechanisms to effect tight binding of their target sequences. Varying effects of 2′-deoxy modifications in domain V of the ai5g group II intron have been taken to indicate that this bulged hairpin domain has two important surfaces: one required for binding to the remainder of the intron, and one required to promote the splicing reactions (Abramovitz et al., 1996). It is noteworthy that site-specific incorporation of 2′-deoxy sugars in a region of U6 snRNA thought to be structurally and functionally analogous to domain V of group II introns implicated a somewhat different set of 2′-hydroxyls in substrate binding and splicing (Kim et al., 1997). This might indicate that the two regions are not as similar as previously thought.

In addition to backbone modifications, variation of individual functional groups on the bases has provided important insights into the recognition properties and functions of key nucleotide positions in intronic RNAs. For example, substitution of the branch site adenosine in nuclear pre-mRNA with purine riboside, 2,6-diaminopurine riboside, xanthosine, or guanosine revealed three sequential and distinct recognition events involving this nucleotide during the course of spliceosomal splicing (Query et al., 1996). A parallel analysis of the ai5g intron showed that the functional group requirements for branch site usage are somewhat different in group II introns and the spliceosome (Liu et al., 1997; also see Gaur et al., 1997). A conserved G·U wobble pair at the 5′ splice site of group I introns has been similarly probed by systematically varying all of the functional groups on both bases (Strobel et al., 1994; Strobel and Cech, 1995, 1996).

Whereas all of the examples previously cited involve site-specific substitutions, important structural and functional information has also emerged from studies in which the modifications were introduced randomly into the RNA. Early analysis of yeast U6 snRNA containing randomly distributed R_p phosphorothioates identified two phosphate oxygens required for the first step of splicing and a third required only for the second step (Fabrizio and Abelson, 1992). A subsequent and more extensive analysis of the effects of phosphorothioate substitution in a nematode U6 snRNA revealed that the positions of important backbone oxygens within a secondary-structure element shared by U6 snRNA and domain V of group II introns were strikingly similar (Chanfreau and Jacquier, 1994; Yu et al., 1995). These data support a common functional role for this motif in the spliceosome and in group II introns. Recently, Strobel and Shetty have developed a potentially powerful technique called nucleotide analog interference mapping to identify important functional groups at almost any position of the base or sugar moieties (Strobel and Shetty, 1997). This method uses random incorporation of nucleotide analogs containing both a functional group modification and an associated phosphorothioate tag; the latter permits identification of interfering positions by iodine cleavage of the active and inactive RNAs.

Cross-Linking

Mapping the mRNA pathway through the ribosome

An efficient cross-linking methodology for defining the pathway of mRNA through the 30S ribosomal subunit has been developed by Brimacombe, Bogdanov, and their collaborators (Stade et al., 1989; Rinke-Appel et al., 1991; Dontsova et al., 1991). Synthetic mRNAs containing one or more 4-thioU residues at defined positions were prepared by T7 transcription. In some cases, the 4-thioU residues were also derivatized with *p*-azidophenacyl bromide. After binding to ribosomes in the presence of an appropriate tRNA, the mRNAs were activated by irradiation with near UV light and the cross-linked products were isolated and characterized. All of the mRNAs became attached exclusively to the 30S subunit, and depending on the position of the photoreactive bases, these probes labeled several distinct regions of the 16S rRNA as well as a number of ribosomal proteins (Stade et al., 1989; Rinke-Appel et al., 1991; Dontosova et al., 1991).

In a comprehensive analysis of mRNA-16S rRNA contacts, a series of mRNA analogs were synthesized that contained a Shine-Dalgarno sequence, an AUG initiator codon and 4-thioU residues at every position from −8 to +16, or 6-thioG residues at every position between +4 and +14, relative to the translation initiation site (Rinke-Appel et al., 1991, 1993, 1994; Dontsova et al., 1992a; Sergiev et al., 1997). In the presence of tRNA^{fMet}, which anchors the AUG codon in the 30S-subunit P site, specific cross-links were observed between thionucleotides at positions +4, +6, +7, +8/+9, +11, and +12 of the mRNA and nucleotides 1402, 1052, 1395, 1196, 532, and 530 of the 16S rRNA, respectively (Fig. 11). Because residues +4 and +6 of the mRNA are expected to be in the 30S-subunit A site under these conditions, with residues +7 to +9 as close neighbors, the results suggest an intimate relationship between two portions of the 16S rRNA, the 1400 region, and helix 34, which are widely separated in the primary structure. Moreover, the cross-links from positions +11 and +12 suggest that a third segment of the 16S rRNA, the 530 loop, is within 15–20 Å of the site of codon-anticodon interaction. The addition of tRNA to the A site eliminates cross-linking from mRNA positions +4 to +6 and reduces that from position +7 (Rinke-Appel et al., 1993). This suggests that the A-site codon undergoes a rearrangement upon interaction with the tRNA anticodon. In contrast to the highly specific cross-links established by nucleotides on the 3′ side of the initiation codon, 4-thioU residues at positions −8 to −1, between the Shine–Dalgarno sequence and the initial AUG, almost all cross-linked to the same three sites in the 16S rRNA both in the presence and in the absence of tRNA^{fMet} (Rinke-Appel et al., 1994). These observations indicate that the mRNA chain is highly constrained relative to the 16S rRNA as it approaches the decoding site, but that after exiting the P site it acquires more flexibility as it passes on to the E site and beyond.

Topography of the A, P, and E sites on the ribosome

Early efforts to affinity label tRNA binding sites on the *E. coli* ribosome usually entailed the attachment of reactive ligands to naturally occurring modified bases in the tRNA (for a review, see Cooperman, 1988). A disadvantage of these derivatives for topographical studies was that the reactive moieties extended 10 to 20 Å from the sites into which they were incorporated, leading to uncertainties in the spatial relationship of the cross-linked ribosomal component(s) to the tRNA molecule. Nonetheless, such endeavors identified a number of ribosomal proteins that are located in the general neighborhood of the tRNA binding sites (Lin et al., 1984; Ofengand et al., 1986; Podkowinski and Gornicki, 1991). Complementary information on the relative locations of tRNA and rRNA has been obtained from cross-linking experiments in which photoreactive aryl azide or diazirine moieties were introduced into the thio and amino groups that occur naturally in certain tRNAs (summarized in Rinke-Appel et al., 1995).

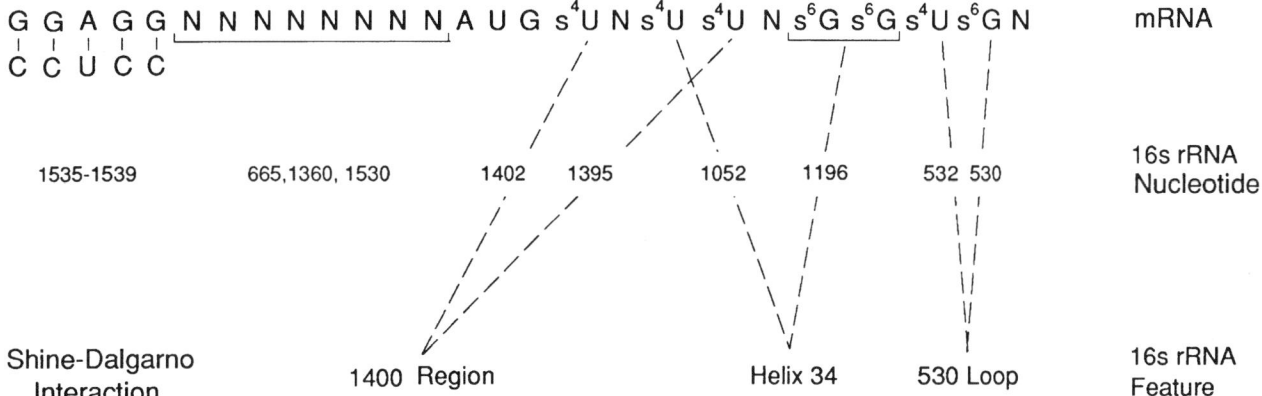

Figure 11. Summary of cross-links formed between mRNA analogs and three regions of the 16S rRNA in 30S ribosomal subunits. (Data from Rinke-Appel et al., 1991, 1993, 1994; Dontsova et al., 1992a; and Sergiev et al., 1997.)

An approach for mapping the topography of the ribosomal tRNA binding sites at higher resolution was suggested by the observation that a modified uridine residue in the anticodon of tRNA$_1^{Val}$ formed a photocross-link to nucleotide 1400 of the 16S rRNA that was only 2 to 3 Å in length (Prince et al., 1982). Thus, through the incorporation of photoreactive nucleotides into the primary structure of tRNA, either at specific sites by tRNA reconstruction or at random locations by T7 transcription, numerous short-range cross-links, on the order of 2–4 Å, have been established between tRNA and components of the ribosomal A, P, and E sites (Fig. 12). In one series of experiments, A and P site-bound tRNAs containing azidonucleotides in the 3′-terminal ACCA$_{OH}$ sequence were found to label mainly 50S-subunit protein L27 and two sequences within a region of the 23S rRNA that has been closely identified with the site of peptidyl transferase activity (Wower et al., 1989, 1995). The similar A and P site cross-linking patterns reflect the fact that the 3′ ends of the two tRNAs must be juxtaposed during peptide bond formation. In contrast, the 3′ terminus of E site-bound tRNA labels protein L33 and a different sequence of the 23S rRNA, indicating that it occupies a distinct position removed from the site of peptidyl transfer (Wower et al., 1993a). By using a similar approach, azidonucleotides placed within the anticodon loop were cross-linked to protein S7 and the 3′ domain of the 16S rRNA at the A and P sites (Sylvers et al., 1992) but labeled protein S11 and the 3′ end of the 16S rRNA from the E site (Wower et al., 1993a). In addition, the variable loop has been placed in the immediate vicinity of 30S-subunit proteins S13 and S19 (Wower et al., 1990; Rosen et al., 1993), while positions 17/17.1 and 20 of the D loop have been cross-linked to proteins L27 and S11, respectively (Rosen et al., 1993; Rosen and Zimmermann, 1997). From these experiments, it was concluded that the central fold of P site-bound tRNA is sandwiched between the ribosomal subunits, with the T-loop side facing toward the head of the small subunit, and the D-loop side positioned between the L1 ridge and the central protuberance of the large subunit.

Intermolecular interactions within the spliceosome

Much of the recent progress in elucidating the complex web of ever-changing RNA-RNA and RNA-protein interactions in the spliceosome has come from cross-linking experiments using pre-mRNA substrates containing photoreactive nucleotide derivatives at specific sites. The most widely used reagent is 4-thioU (Wyatt et al., 1992; Sontheimer and Steitz, 1993; Sontheimer, 1994; Guar et al., 1995; Teigelkamp et al., 1995a), although both 2,6-diaminopurine and a "convertible" adenosine analog coupled to benzophenone have been used to obtain RNA-protein cross-links at the pre-mRNA branch site (MacMillan et al., 1994; Query et al., 1996). Most of the RNA-RNA interactions defined by site-specific photocross-linking have been recently reviewed (Madhani and Guthrie, 1994; Nilsen, 1994). One of the themes that emerges from these studies is the remarkably dynamic nature of RNA-RNA interactions in the spliceosome. For example, the 5′ splice site is initially recognized by base pairing with the 5′ end of U1 snRNA, but this interaction is soon supplanted by new interactions between the 5′ splice site and U5 and U6 snRNAs. A concern that arises in any cross-linking experiment is whether the interaction captured in a particular cross-link is part of a normal reaction pathway. It could, for instance, reflect a competing but nonproductive side pathway or an aberrant interaction brought about by the structural peculiarities of the reactive nucleotide employed. An elegant way to address this problem is to carry out the cross-linking first, and then determine if the

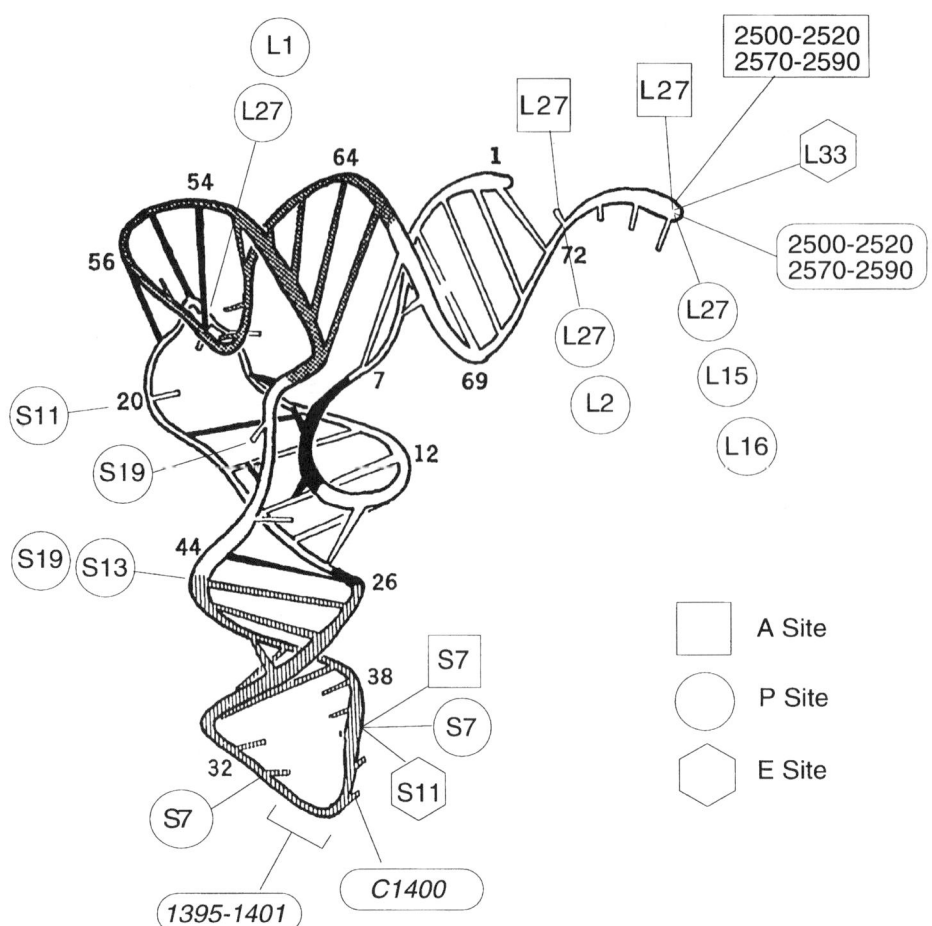

Figure 12. Short-range cross-links formed between tRNA and the A, P and E sites of the *E. coli* ribosome. SX, 30S subunit proteins; LX, 50S subunit proteins; arabic numerals, nucleotides in 23S rRNA; italic numerals, nucleotides in 16S rRNA. (Data from Prince et al., 1982; Wower et al., 1989, 1990, 1993a, and 1995; Sylvers et al., 1992; Rosen et al., 1993; and Rosen and Zimmermann, 1997 and unpublished data.)

cross-linked species is still functional. It was through the use of such an approach that Sontheimer and Steitz (1993) proved that an interaction between U5 snRNA and the 5′ exon, which could be detected by cross-linking before lariat formation, is maintained until after exon ligation.

Site-specific cross-linking has been equally important for elucidating RNA-protein interactions within the spliceosome. Cross-linking reagents at the branch site have revealed at least three distinct sets of proteins that interact with this position at different stages of spliceosome assembly (MacMillan et al., 1994; Gaur et al., 1995; Chiara et al., 1996; Query et al., 1996). One of these, PRP8p or p220, can also be crosslinked to the pre-mRNA in the vicinitiy of both the 5′ and 3′ splice sites (Wyatt et al., 1992; Teigelkamp et al., 1995a; Teigelkamp et al., 1995b; Umen and Guthrie, 1995; Chiara et al., 1997). Interestingly, experiments employing an appropriate substrate have shown that at the time of the first chemical step PRP8p is the only molecule, protein or RNA, that can be cross-linked to the highly conserved GU dinucleotide that defines the 5′ splice site (Reyes et al., 1996). This result indicates that PRP8p is likely to be an integral part of the active site for the first step of splicing, in contrast to the widely held belief that the spliceosome is essentially an RNA catalyst.

PERSPECTIVES

In this chapter, we have attempted to give the reader a sense of the wide range of modified nucleotides that can now be engineered into RNA molecules, the diverse strategies that have been employed for their incorporation, and the impact that the use of such molecules has already had in many areas of RNA research. The main advances that have occurred in recent years have been the development of better techniques for chemical and enzymatic RNA synthesis and new and innovative approaches for the introduction of site-specific modifications into functional

RNA molecules, especially those that exceed 40 to 50 nucleotides in length. In some laboratories, in fact, it has become almost routine to produce RNAs of well over 200 residues containing single-atom changes. Although engineered modifications have until now been used mainly for the study of RNAs that are naturally unmodified, we hope that the approaches described in this chapter will prove useful in the future in the many exciting areas of research involving naturally modified and edited RNAs.

REFERENCES

Aboul-ela, F., J. Karn, and G. Varani. 1995. The structure of the human immunodeficiency virus type-1 TAR RNA reveals principles of recognition by tat protein. *J. Mol. Biol.* 253:313–332.

Aboul-ela, F., J. Karn, and G. Varani. 1996. Structure of HIV-1 TAR RNA in the absence of ligands reveals a novel conformation of the trinucleotide bulge. *Nucleic Acids Res.* 24:3974–3982.

Abramovitz, D. L., R. A. Freidmen, and A. M. Pyle. 1996. Catalytic role of 2′-hydroxyl groups within a group II intron active site. *Science* 271:1410–1413.

Adams, C. J., J. B. Murray, J. R. P. Arnold, and P. G. Stockley. 1994. Incorporation of a fluorescent nucleotide into oligoribonucleotides. *Tetrahedron Lett.* 35:1597–1600.

Baidya, N., and O. C. Uhlenbeck. 1995. The role of 2′-hydroxyl groups in an RNA-protein interaction. *Biochemistry* 34:12363–12368.

Bare, L. A., and O. C. Uhlenbeck. 1986. Specific substitution into the anticodon loop of yeast tyrosine transfer RNA. *Biochemistry* 25:5825–5830.

Bare, L., A. G. Bruce, R. Gesteland, and O. C. Uhlenbeck. 1983. Uridine-33 in yeast tRNA not essential for amber suppression. *Nature* 305:554–556.

Barrio, J. R., M. C. G. Barrio, N. J. Leonard, T. E. England, and O. C. Uhlenbeck. 1978. Synthesis of modified nucleoside 3′,5′-bisphosphates and their incorporation into oligoribonucleotides with T4 RNA ligase. *Biochemistry* 17:2077–2081.

Bayer, P., M. Kraft, A. Ejchart, M. Westendorp, R. Frank, and P. Rösch. 1995. Structural studies of HIV-1 tat protein. *J. Mol. Biol.* 247:529–535.

Beauchemin, N., H. Grosjean, and R. Cedergren. 1986. Construction, aminoacylation and 80S ribosomal complex formation with yeast initiator tRNA having an arginine CCU anticodon. *FEBS Lett.* 202:8–12.

Beigelman, L., A. Karpeisky, and N. Usman. 1994. Synthesis of 1-deoxy-D-ribofuranose phosphoramidite and the incorporation of abasic nucleotides in stem-loop II of a hammerhead ribozyme. *Bioorg. Med. Chem. Lett.* 4:1715–1720.

Bevilacqua, P. C., and T. R. Cech. 1996. Minor-groove recognition of double-stranded RNA by the double-stranded RNA-binding domain from the RNA-activated protein kinase PKR. *Biochemistry* 35:9983–9994.

Bochkariov, D. E., and A. A. Kogon. 1992. Application of 3-[3-(3-trifluoromethyl-diazirin-3-yl)phenyl]-2,3-dihydroxypropionic acid, carbene generating, cleavable crosslinking reagent for photoaffinity labeling. *Anal. Biochem.* 204:90–95.

Breitschopf, K., and H. J. Gross. 1996. The discriminator bases G73 in human tRNA[Ser] and A73 in tRNA[Leu] have significantly different roles in the recognition of aminoacyl-tRNA synthetases. *Nucleic Acids Res.* 24:405–410.

Bruce, A. G., and O. C. Uhlenbeck. 1982. Enzymatic replacement of the anticodon of yeast phenylalanine transfer ribonucleic acid. *Biochemistry* 21:855–861.

Burgin, A. B., and N. R. Pace. 1990. Mapping the active site of ribonuclease P RNA using a substrate containing a photoaffinity reagent. *EMBO J.* 9:4111–4118.

Calnan, B. J., B. Tidor, S. Biancalana, D. Hudson, and A. D. Frankel. 1991. Arginine-mediated RNA recognition: the arginine fork. *Science* 252:1167–1171.

Cech, T. R. 1986. The generality of self-splicing RNA: relationship to nuclear mRNA splicing. *Cell* 44:207–210.

Cedergren, R., and H. Grosjean. 1987. RNA design by *in vitro* RNA recombination and synthesis. *Biochem. Cell Biol.* 65:677–692.

Chaix, C., A. M. Duplaa, D. Molko, and R. Teoule. 1989. Solid phase synthesis of the 5′-half of the initiator tRNA from *B. subtilis*. *Nucleic Acids Res.* 17:7381–7393.

Chamberlin, M., and T. Ryan. 1982. Bacteriophage DNA-dependent RNA polymerase, p. 87–105. *In* P. D. Boyer (ed.), *The Enzymes*, vol. 15, part B. Academic Press, New York, N.Y.

Chanfreau, G., and A. Jacquier. 1994. Catalytic site components common to both splicing steps of a group II intron. *Science* 266:1383–1387.

Chen, C. H., and D. S. Sigman. 1986. Nuclease activity of 1,10-phenanthroline-copper: sequence-specific targeting. *Proc. Natl. Acad. Sci. USA* 83:7147–7151.

Chiara, M. D., O. Gozani, M. Bennett, P. Champion-Arnaud, L. Palandjian, and R. Reed. 1996. Identification of proteins that interact with exon sequences, splice sites, and the branchpoint sequence during each stage of spliceosome assembly. *Mol. Cell. Biol.* 16:3317–3326.

Chiara, M. D., L. Palandjian, R. Feld-Kramer, and R. Reed. 1997. Evidence that U5 snRNP recognizes the 3′ splice site for catalytic step II in mammals. *EMBO J.* 16:4746–4759.

Chowrira, B. M., A. Berzal-Herranz, C. F. Keller, and J. M. Burke. 1993. Four ribose 2′-hydroxyl groups essential for catalytic function of the hairpin ribozyme. *J. Biol. Chem.* 268:19458–19462.

Churcher, M. J., C. Lamont, F. Hamy, C. Dingwall, S. M. Green, A. D. Lowe, P. J. G. Butler, M. J. Gait, and J. Karn. 1993. High affinity binding of TAR RNA by the human immunodeficiency virus type-1 *tat* protein requires base-pairs in the RNA stem and amino acid residues flanking the basic region. *J. Mol. Biol.* 230:90–110.

Cohen, S. B., and T. R. Cech. 1997. Dynamics of thermal motions within a large catalytic RNA investigated by cross-linking with thiol-disulfide interchange. *J. Am. Chem. Soc.* 119:6259–6268.

Conrad, F., A. Hanne, R. K. Gaur, and G. Krupp. 1995. Enzymatic synthesis of 2′-modified nucleic acids: identification of important phosphate and ribose moieties in RNase P substrates. *Nucleic Acids Res.* 23:1845–1853.

Cooperman, B. L. 1988. Affinity labeling of ribosomes. *Methods Enzymol.* 164:341–361.

Cunningham, P. R., and J. Ofengand. 1990. Use of inorganic pyrophosphatase to improve the yield of *in vitro* transcription reactions catalyzed by T7 RNA polymerase. *BioTechniques* 9:713–714.

Damha, M. J., and K. K. Ogilvie. 1993. Oligoribonucleotide synthesis, p. 81–114. *In* S. Agrawal (ed.), *Methods in Molecular Biology*. Humana Press, Totowa, N.J.

Davis, R. H. 1995. Large-scale oligoribonucleotide production. *Curr. Opin. Biotechnol.* 6:213–217.

Delling, U., L. S. Reid, R. W. Barnett, M. X. Y. Ma, S. Climie, M. Sumner-Smith, and N. Sonenberg. 1992. Conserved nucleotides in the TAR RNA stem of human immunodeficiency virus type 1 are critical for tat binding and *trans* activation: model for TAR RNA tertiary structure. *J. Virol.* 66:3018–3025.

Dietz, T. M., and T. H. Koch. 1987. Photochemical coupling of 5-bromouracil to tryptophan, tyrosine and histidine, peptide-like derivatives in aqueous fluid solution. *Photochem. Photobiol.* **46**: 971–978.

Donis-Keller, H. 1979. Site-specific enzymatic cleavage of RNA. *Nucleic Acids Res.* **7**:179–192.

Dontsova, O., A. Kopylov, and R. Brimacombe. 1991. The location of mRNA in the ribosomal 30S initiation complex: site-directed cross-linking of mRNA analogues carrying several photo-reactive labels simultaneously on either side of the AUG start codon. *EMBO J.* **10**:2613–2620.

Dontsova, O., S. Dokudovskaya, A. Kopylov, A. Bogdanov, J. Rinke-Appel, N. Jünke, and R. Brimacombe. 1992a. Three widely separated positions in the 16S RNA lie in or close to the ribosomal decoding region: a site-directed cross-linking study with mRNA analogues. *EMBO J.* **11**:3105–3116.

Dontsova, O. A., K. V. Rosen, S. L. Bogdanova, E. A. Skripkin, A. M. Kopylov, and A. A. Bogdanov. 1992b. Identification of the *Escherichia coli* 30S ribosomal protein neighboring mRNA during initiation of translation. *Biochimie* **74**:363–371.

Doudna, J. A., and J. H. Cate. 1997. RNA structure: crystal clear? *Curr. Opin. Struct. Biol.* **7**:310–316.

Doudna, J. A., J. W. Szostak, A. Rich, and N. Usman. 1990. Chemical synthesis of oligoribonucleotides containing 2-aminopurine: substrates for the investigation of ribozyme function. *J. Org. Chem.* **55**:5547–5549.

Dreyer, G. B., and P. B. Dervan. 1985. Sequence specific cleavage of single-stranded DNA: oligodeoxynucleotide-EDTA·Fe(II). *Proc. Natl. Acad. Sci. USA* **82**:968–972.

Eadie, J. S., L. J. McBride, J. W. Efcavitch, L. B. Hoff, and R. Cathcart. 1987. High-performance liquid chromatographic analysis of oligodeoxyribonucleotide base composition. *Anal. Biochem.* **165**:442–447.

Earnshaw, D. J., B. Masquida, S. Müller, S. T. Sigurdsson, F. Eckstein, E. Westhof, and M. J. Gait. 1997. Inter-domain cross-linking and molecular modelling of the hairpin ribozyme. *J. Mol. Biol.* **274**:197–212.

Eaton, B. E., and W. A. Pieken. 1995. Ribonucleosides and RNA. *Annu. Rev. Biochem.* **64**:837–863.

England, T. E., and O. C. Uhlenbeck. 1978. Enzymatic oligonucleotide synthesis with T4 RNA ligase. *Biochemistry* **17**: 2069–2076.

Engler, M. J., and C. C. Richardson. 1982. DNA ligases, p. 3–29. *In* P. D. Boyer (ed.), *The Enzymes*, vol. 15, part B. Academic Press, New York, N.Y.

Enright, C., and B. Sollner-Webb. 1994. Ribosomal RNA processing in vertebrates, p. 135–172. *In* S. J. Higgins and B. D. Hames (ed.), *RNA Processing: a Practical Approach*, vol. II. IRL Press, Oxford, United Kingdom.

Evans, R. K., and B. E. Haley. 1987. Synthesis and biological properties of 5-azido-2′-deoxyuridine 5′-triphosphate, a photoactive nucleotide suitable for making light-sensitive DNA. *Biochemistry* **26**:269–276.

Fabrizio, P., and J. Abelson. 1992. Thiophosphates in yeast U6 snRNA specifically affect pre-mRNA splicing in vitro. *Nucleic Acids Res.* **20**:3659–3664.

Favre, A., and J.-L. Fourrey. 1995. Structural probing of small endonucleolytic ribozymes in solution using thio-substituted nucleobases as intrinsic photolabels. *Acc. Chem. Res.* **28**:375–382.

Ferré-D'Amaré, A. R., and J. A. Doudna. 1996. Use of *cis*- and *trans*-ribozymes to remove 5′ and 3′ heterogeneities from milligrams of *in vitro* transcribed RNA. *Nucleic Acids Res.* **24**:977–978.

Fournier, M., E. Haumont, S. de Henau, J. Gangloff, and H. Grosjean. 1983. Post-transcriptional modification of the wobble nucleotide in anticodon-substituted yeast tRNA$_{II}^{Arg}$ after microinjection into *Xenopus laevis* oocytes. *Nucleic Acids Res.* **11**:707–718.

Frugier, M., C. Florentz, M. W. Hosseini, J.-M. Lehn, and R. Giegé. 1994. Synthetic polyamines stimulate *in vitro* transcription by T7 RNA polymerase. *Nucleic Acids Res.* **22**:2784–2790.

Fu, D.-J., and L. W. McLaughlin. 1992. Importance of specific adenosine N^7-nitrogens for efficient cleavage by a hammerhead ribozyme. A model for magnesium binding. *Biochemistry* **31**: 10941–10949.

Fu, D.-J., S. B. Rajur, and L. W. McLaughlin. 1993. Importance of specific N^7-nitrogens and purine amino groups for efficient cleavage by a hammerhead ribozyme. *Biochemistry* **32**: 10629–10637.

Gait, M. J., J. Grasby, J. Karn, K. Mersmann, and C. E. Pritchard. 1995. Synthetic ribonucleotide analogues for structure-function studies. *Nucleosides Nucleotides* **14**:1133–1144.

Gait, M. J., and J. Karn. 1993. RNA recognition by the human immunodeficiency virus tat and rev proteins. *Trends Biochem. Sci.* **18**:255–259.

Gait, M. J., C. E. Pritchard, and G. Slim. 1991. Oligoribonucleotide synthesis, p. 25–48. *In* M. J. Gait (ed.), *Oligonucleotides and Analogues: a Practical Approach*. IRL Press, Oxford, United Kingdom.

Gasparutto, D., T. Livache, H. Bazin, A.-M. Duplaa, A. Guy, A. Khorlin, D. Molko, A. Roget, and R. Teoule. 1992. Chemical synthesis of a biologically active natural tRNA with its minor bases. *Nucleic Acids Res.* **19**:5159–5166.

Gaur, R. K., J. Valcárcel, and M. R. Green. 1995. Sequential recognition of the pre-mRNA branch point by U2AF^{65} and a novel spliceosome-associated 28-kDa protein. *RNA* **1**:407–417.

Gaur, R. K., L. W. McLaughlin, and M. R. Green. 1997. Functional group substitutions of the branchpoint adenosine in a nuclear pre-mRNA and a group II intron. *RNA* **3**:861–869.

Goodwin, J. T., S. E. Osborne, E. J. Scholle, and G. D. Glick. 1996. Design, synthesis, and analysis of yeast tRNA^{Phe} analogs possessing intra- and interhelical disulfide cross-links. *J. Am. Chem. Soc.* **118**:5207–5215.

Gott, J. M., M. C. Willis, T. K. Koch, and O. C. Uhlenbeck. 1991. A specific, UV-induced RNA-protein cross-link using 5-bromouridine-substituted RNA. *Biochemistry* **30**:6290–6295.

Grasby, J. A., and M. J. Gait. 1994. Synthetic oligoribonucleotides carrying site-specific modifications for RNA structure-function studies. *Biochimie* **76**:1223–1234.

Grasby, J. A., P. J. G. Butler, and M. J. Gait. 1993. The synthesis of oligoribonucleotides containing O^6-methylguanosine: the role of conserved guanosine residues in hammerhead ribozyme cleavage. *Nucleic Acids Res.* **21**:4444–4450.

Grasby, J. A., M. Singh, J. Karn, and M. J. Gait. 1995. Synthesis and applications of oligoribonucleotides containing N^4-methylcytidine. *Nucleosides Nucleotides* **14**:1129–1132.

Griffin, E. A. J., Q. Zhifeng, W. J. J. Michels, and A. M. Pyle. 1995. Group II intron ribozymes that cleave DNA and RNA linkages with similar efficiency, and lack contacts with substrate 2′-hydroxyl groups. *Chem. Biol.* **2**:761–770.

Gurevich, V. V. 1996. Use of bacteriophage RNA polymerase in RNA synthesis. *Methods Enzymol.* **275**:382–397.

Haley, B. E. 1983. Development and utilization of 8-azidopurine nucleotide photoaffinity probes. *Fed. Proc.* **42**:2831–2836.

Hamm, M. L., and J. A. Piccirilli. 1997. Incorporation of 2′-deoxy-2′-mercaptocytidine into oligonucleotides via phosphoramidite chemistry. *J. Org. Chem.* **62**:3415–3420.

Hamy, F., U. Asseline, J. A. Grasby, S. Iwai, C. E. Pritchard, G. Slim, P. J. G. Butler, J. Karn, and M. J. Gait. 1993. Hydrogen-bonding contacts in the major groove are required for human immunodeficiency virus type-1 *tat* protein recognition of TAR RNA. *J. Mol. Biol.* **230**:111–123.

Hanna, M. M. 1996. Photochemical cross-linking analysis of protein-nucleic acid interactions in *Escherichia coli* transcription complexes from lambda $P_{R'}$ promoter. *Methods Enzymol.* 274: 403–418.

Haumont, E., M. Fournier, S. de Henau, and H. Grosjean. 1984. Enzymatic conversion of adenosine to inosine in the wobble position of yeast tRNA^{Asp}: the dependence on anticodon sequence. *Nucleic Acids Res.* 12:2705–2715.

Hayase, Y., H. Inoue, and E. Ohtsuka. 1990. Secondary structure in formylmethionine tRNA influences the site-directed cleavage of ribonuclease H using chimeric 2′-O-methyl oligodeoxyribonucleotides. *Biochemistry* 29:8793–8797.

Hayase, Y., M. Jahn, M. J. Rogers, L. A. Sylvers, M. Koizumi, H. Inoue, E. Ohtsuka, and D. Söll. 1992. Recognition of bases in *Escherichia coli* tRNA^{Gln} by glutaminyl-tRNA synthetase: a complete identity set. *EMBO J.* 11:4159–4165.

He, B., D. L. Riggs, and M. M. Hanna. 1995. Preparation of probe-modified RNA with 5-mercapto-UTP for analysis of protein-RNA interactions. *Nucleic Acids Res.* 23:1231–1238.

He, B., M. Rong, D. Lyakhov, H. Gartenstein, G. Diaz, R. Castagna, W. T. McAllister, and R. K. Durbin. 1997. Rapid mutagenesis and purification of phage RNA polymerases. *Protein Expr. Purif.* 9:142–151.

Higgins, N. P., and N. C. Cozzarelli. 1979. DNA-joining enzymes: a review. *Methods Enzymol.* 68:50–71.

Hixson, S. H., and S. S. Hixson. 1975. p-Azidophenacyl bromide, a versatile photolabile bifunctional reagent. Reaction with glyceraldehyde-3-phosphate dehydrogenase. *Biochemistry* 14: 4251–254.

Hou, Y.-M., and P. Schimmel. 1988. A simple structural feature is a major determinant of the identity of transfer RNA. *Nature* 333:140–145.

Inoue, H., Y. Hayase, A. Imura, S. Iwai, and E. Ohtsuka. 1987. Sequence-dependent hydrolysis of RNA using modified oligonucleotide splints and RNase H. *FEBS Lett.* 215:327–330.

Iwai, S., C. E. Pritchard, D. A. Mann, J. Karn, and M. J. Gait. 1992. Recognition of the high affinity binding site in rev-response element RNA by the human immunodeficiency virus type-1 rev protein. *Nucleic Acids Res.* 24:6465–6472.

Karn, J., M. J. Churcher, K. Rittner, N. Keen, and M. J. Gait. 1996. Control of transcriptional elongation by the human immunodeficiency virus tat protein, p. 254–286. *In* S. Goodbourn (ed.), *Eukaryotic Gene Transcription*. Oxford University Press, Oxford, United Kingdom.

Kaufmann, G., and U. Z. Littauer. 1974. Covalent joining of phenylalanine transfer ribonucleic acid half-molecules by T4 RNA ligase. *Proc. Natl. Acad. Sci. USA* 71:3741–3745.

Kim, C. H., D. E. Ryan, T. Marciniec, and J. Abelson. 1997. Site-specific deoxynucleotide substitutions in yeast U6 snRNA block splicing of pre-mRNA in vitro. *EMBO J.* 16:2119–2129.

Krug, M., P. L. de Haseth, and O. C. Uhlenbeck. 1982. Enzymatic synthesis of a 21-nucleotide coat binding fragment of R17 ribonucleic acid. *Biochemistry* 21:4713–4720.

Kuznetsova, S. A., M. G. Ivanovskaya, and Z. A. Shabarova. 1990. Chemical reactions in double-stranded nucleic acids. IX. Directed introduction of substituted pyrophosphate bonds into DNA structure. *Bioorg. Khim.* 16:219–225.

Lamond, A. I., and B. S. Sproat. 1994. Isolation and characterization of ribonucleoprotein complexes, p. 103–140. *In* S. J. Higgins and B. D. Hames (ed.), *RNA Processing: a Practical Approach*, vol. I. IRL Press, Oxford, United Kingdom.

Lin, S. Y., and A. D. Riggs. 1974. Photochemical attachment of *lac* repressor to bromodeoxyuridine-substituted *lac* operator by ultraviolet radiation. *Proc. Natl. Acad. Sci. USA* 71:947–951.

Lin, F.-L., M. Boublik, and J. Ofengand. 1984. Immunoelectron microscopic localization of the S19 site on the 30S ribosomal subunit which is crosslinked to A site bound transfer RNA. *J. Mol. Biol.* 172:41–55.

Liu, Q., J. B. Green, A. Khodadadi, P. Haeberli, L. Beigelman, and A. M. Pyle. 1997. Branch-site selection in a group II intron mediated by active recognition of the adenine amino group and steric exclusion of non-adenine functionalities. *J. Mol. Biol.* 267: 163–171.

Loria, A., and T. Pan. 1997. Recognition of the T stem-loop of a pre-tRNA substrate by the ribozyme from *Bacillus subtilis* RNase P. *Biochemistry* 36:6317–6325.

MacMillan, A. M., and G. L. Verdine. 1990. Synthesis of functionally tethered oligodeoxynucleotides by the convertible nucleoside approach. *J. Org. Chem.* 55:5931–5933.

MacMillan, A. M., C. C. Query, C. R. Allerson, S. Chen, G. L. Verdine, and P. A. Sharp. 1994. Dynamic association of proteins with the pre-mRNA branch region. *Genes Dev.* 8:3008–3020.

Madhani, H. D., and C. Guthrie. 1994. Dynamic RNA–RNA interactions in the spliceosome. *Annu. Rev. Genet.* 28:1–26.

Maschhoff, K. L., and R. A. Padgett. 1993. The stereochemical course of the first step of pre-mRNA splicing. *Nucleic Acids Res.* 21:5456–5462.

McKay, D. B. 1996. Structure and function of the hammerhead ribozyme: an unfinished story. *RNA* 2:395–403.

McLain, W. H., and K. Foss. 1988. Changing the identity of a tRNA by introducing a G·U wobble pair near the 3′ acceptor end. *Science* 240:793–796.

Meisenheimer, K. M., P. L. Meisenheimer, M. C. Willis, and T. H. Koch. 1996. High yield photocrosslinking of a 5-iodouridine (IC) substituted RNA to its associated protein. *Nucleic Acids Res.* 24:981–982.

Milligan, J. F., and O. C. Uhlenbeck. 1988. Synthesis of small RNAs using T7 RNA polymerase. *Methods Enzymol.* 180: 51–62.

Milligan, J. F., and O. C. Uhlenbeck. 1989. Determination of RNA-protein contacts using thiophosphate substitutions. *Biochemistry* 28:2849–2855.

Moore, M. J., and C. C. Query. 1998. Use of site-specifically modified RNAs constructed by RNA ligation, p. 75–108. *In* C. Smith (ed.), *RNA-Protein Interactions: a Practical Approach*. IRL Press, Oxford, United Kingdom.

Moore, M. J., and P. A. Sharp. 1992. Site-specific modification of pre-mRNA: the 2′ hydroxyl groups at the splice sites. *Science* 256:992–997.

Moore, M. J., and P. A. Sharp. 1993. Evidence for two active sites in the spliceosome provided by stereochemistry of pre-mRNA splicing. *Nature* 365:364–368.

Moroney, S. E., and J. A. Piccirilli. 1991. Abortive products as initiating nucleotides during transcription by T7 RNA polymerase. *Biochemistry* 30:10343–10349.

Musier-Forsyth, K., and P. Schimmel. 1992. Functional contacts of a transfer RNA synthetase with 2′-hydroxyl groups in the RNA minor groove. *Nature* 357:513–515.

Musier-Forsyth, K., and P. Schimmel. 1994. Acceptor helix interactions in a class II tRNA synthetase: photoaffinity cross-linking of an RNA miniduplex substrate. *Biochemistry* 33:773–779.

Musier-Forsyth, K., N. Usman, S. Scaringe, J. Doudna, R. Green, and P. Schimmel. 1991. Specificity for aminoacylation of an RNA helix: an unpaired, exocyclic amino group in the minor groove. *Science* 253:784–786.

Naryshkin, N. A., M. A. Farrow, M. G. Ivanovskaya, T. S. Orestkaya, Z. A. Shabarova, and M. J. Gait. 1997. Chemical cross-linking of the human immunodeficiency virus type 1 Tat protein to synthetic models of the RNA recognition sequence TAR containing site-specific trisubstituted pyrophosphate analogues. *Biochemistry* 36:3496–3505.

Naryshkin, N. A., M. G. Ivanovskaya, T. S. Oretskaya, E. M. Volkov, M. J. Gait, and Z.A. Shabarova. 1996. Synthesis and properties of mixed ribo- and deoxyribooligonucleotide du-

plexes containing an internucleotide trisubstituted pyrophosphate bond. *Bioorg. Khim.* 22:592–598.
Newman, A. J., S. Teigelkamp, and J. D. Beggs. 1995. snRNA interactions at 5′ and 3′ splice sites monitored by photoactivated crosslinking in yeast spliceosomes. *RNA* 1:968–980.
Nilsen, T. W. 1994. RNA–RNA interactions in the spliceosome: unraveling the ties that bind. *Cell* 78:1–4.
Nolan, J. M., D. H. Burke, and N. R. Pace. 1993. Circularly permuted tRNAs as specific photoaffinity probes of ribonuclease P RNA structure. *Science* 261:762–765.
Ofengand, J., J. Ciesiolka, R. Denman, and K. Nurse. 1986. Structural and functional interactions of the tRNA-ribosome complex, p. 473–494. *In* B. Hardesty and G. Kramer (ed.), *Structure, Function and Genetics of Ribosomes*. Springer, New York, N.Y.
Ogilvie, K. K., E. A. Thompson, M. A. Quilliam, and J. B. Westmore. 1974. Selective protection of hydroxyl groups in deoxynucleosides using alkylsilyl reagents. *Tetrahedron Lett.* 1974:2865–2868.
Ohtsuka, E., T. Doi, R. Fukumoto, J. Matsugi, and M. Ikehara. 1983. Modification of the anticodon triplet of E. coli tRNA$_f^{Met}$ by replacement with trimers complementary to non-sense codons UAG and UAA. *Nucleic Acids Res.* 11:3863–3872.
Owens, J. R., A.-Y. M. Woody, and B. E. Haley. 1987. Characterization of the guanosine-3′-diphosphate-5′-diphosphate binding site on E. coli RNA polymerase using a photoprobe, 8-azidoguanosine-3′-5′-bisphosphate. *Biochem. Biophys. Res. Commun.* 142:964–971.
Padgett, R. A., M. Podar, S. C. Boulanger, and P. S. Perlman. 1994. The stereochemical course of group II intron self-splicing. *Science* 266:1685–1688.
Paulsen, H., and W. Wintermeyer. 1984. Incorporation of 1,N^6-ethanoadenosine into the 3′ terminus of tRNA using T4 RNA ligase. *Eur. J. Biochem.* 138:117–123.
Perreault, J.-P., D. Labuda, N. Usman, J.-H. Yang, and R. Cedergren. 1991. Relationship between 2′-hydroxyls and magnesium binding in the hammerhead RNA domain: a model for ribozyme catalysis. *Biochemistry* 30:4020–4025.
Perreault, J.-P., T. Wu, B. Cousineau, K. K. Ogilvie, and R. Cedergren. 1990. Mixed deoxyribo- and ribo-oligonucleotides with catalytic activity. *Nature* 344:565–567.
Philippsen, P., R. Thiebe, W. Wintermeyer, and H. G. Zachau. 1968. Splitting of phenylalanine specific tRNA into half molecules by chemical means. *Biochem. Biophys. Res. Commun.* 33:922–928.
Piccirilli, J. A., J. S. Vyle, M. H. Caruthers, and T. R. Cech. 1993. Metal ion catalysis in the Tetrahymena ribozyme reaction. *Nature* 361:85–88.
Pieken, W. A., D. B. Olsen, F. Benseler, H. Aurup, and F. Eckstein. 1991. Kinetic characterization of ribonuclease-resistant 2′-modified hammerhead ribozymes. *Science* 253:314–317.
Podar, M., P. S. Perlman, and R. A. Padgett. 1995. Stereochemical selectivity of group II intron splicing, reverse splicing, and hydrolysis reactions. *Mol. Cell. Biol.* 15:4466–4478.
Podkowinski, J., and P. Gornicki. 1991. Neighbourhood of the central fold of the tRNA molecule bound to the E. coli ribosome-affinity labeling studies with modified tRNAs carrying photoreactive probes attached to the dihydrouridine loop. *Nucleic Acids Res.* 19:801–808.
Price, S. R., N. Ito, C. Oubridge, J. M. Avis, and K. Nagai. 1995. Crystallization of RNA-protein complexes. I. Methods for the large-scale preparation of RNA suitable for crystallographic studies. *J. Mol. Biol.* 249:398–408.
Prince, J. B., B. H. Taylor, D. L. Thurlow, J. Ofengand, and R. A. Zimmermann. 1982. Covalent cross-linking of tRNA$_1^{Val}$ at the ribosomal P site: identification of cross-linked residues. *Proc. Natl. Acad. Sci. USA* 79:5450–5454.

Pritchard, C. E., J. A. Grasby, F. Hamy, A. M. Zacharek, M. Singh, J. Karn, and M. J. Gait. 1994. Methylphosphonate mapping of phosphate contacts critical for recognition by the human immunodeficiency virus tat and rev proteins. *Nucleic Acids Res.* 22:2592–2600.
Pyle, A. M., and T. R. Cech. 1991. Ribozyme recognition of RNA by tertiary interactions with specific ribose 2′-OH groups. *Nature* 350:628–631.
Pyle, A. M., F. L. Murphy, and T. R. Cech. 1992. RNA substrate binding site in the catalytic core of the Tetrahymena ribozyme. *Nature* 358:123–128.
Query, C. C., M. J. Moore, and P. A. Sharp. 1994. Branch nucleophile selection in pre-mRNA splicing: evidence for the bulged duplex model. *Genes Dev.* 8:587–597.
Query, C. C., S. A. Strobel, and P. A. Sharp. 1996. Three recognition events at the branch-site adenine. *EMBO J.* 15:1392–1402.
Reddy, M. P., F. Farooqui, and Hanna, N. B. 1995. Methylamine deprotection provides increased yield of oligoribonucleotides. *Tetrahedron Lett.* 36:8929–8932.
Reyes, J. L., P. Kois, B. B. Konforti, and M. M. Konarska. 1996. The canonical GU dinucleotide at the 5′ splice site is recognized by p220 of the U5 snRNP within the spliceosome. *RNA* 2:213–225.
Rinke-Appel, J., N. Jünke, K. Stade, and R. Brimacombe. 1991. The path of mRNA through the Escherichia coli ribosome: site-directed cross-linking of mRNA analogues carrying a photoreactive label at various points 3′ to the decoding site. *EMBO J.* 10:2195–2202.
Rinke-Appel, J., N. Jünke, R. Brimacombe, S. Dokudovskaya, O. Dontsova, and A. Bogdanov. 1993. Site-directed cross-linking of mRNA analogues to 16S ribosomal RNA: a complete scan of cross-links from all positions between "+1" and "+16" on the mRNA, downstream from the decoding site. *Nucleic Acids Res.* 21:2853–2859.
Rinke-Appel, J., N. Jünke, R. Brimacombe, I. Lavrik, S. Dokudovskaya, O. Dontsova, and A. Bogdanov. 1994. Contacts between 16S ribosomal RNA and mRNA, within the spacer region separating the AUG initiator codon and the Shine-Dalgarno sequence: a site-directed cross-linking study. *Nucleic Acids Res.* 22:3018–3025.
Rinke-Appel, J., N. Jünke, M. Osswald, and R. Brimacombe. 1995. The ribosomal environment of tRNA: crosslinks to rRNA from positions 8 and 20:1 in the central fold of tRNA located at the A, P, or E site. *RNA* 1:1918–1028.
Romaniuk, E., L. W. McLaughlin, T. Neilson, and P. J. Romaniuk. 1982. The effect of acceptor oligonucleotide sequence in the T4 RNA ligase reaction. *Eur. J. Biochem.* 125:639–643.
Romaniuk, P. J., and O. C. Uhlenbeck. 1983. Joining of RNA molecules with RNA ligase. *Methods Enzymol.* 100:52–59.
Rosen, K. V., and R. A. Zimmermann. 1997. Phototaffinity labeling of 30S-subunit proteins S7 and S11 by 4-thiouridine-substituted tRNA^{Phe} situated at the P site of Escherichia coli ribosomes. *RNA* 3:1028–1036.
Rosen, K. V., and R. A. Zimmermann. Unpublished data.
Rosen, K. V., R. W. Alexander, J. Wower, and R. A. Zimmermann. 1993. Mapping the central fold of tRNA^{fMet} in the P site of the Escherichia coli ribosome. *Biochemistry* 32:12802–12811.
Sampson, J. R., and O. C. Uhlenbeck. 1988. Biochemical and physical characterization of an unmodified yeast phenylalanine transfer RNA transcribed in vitro. *Proc. Natl. Acad. Sci. USA* 85:1033–1037.
SantaLucia, J., Jr., R. Kierzek, and D. H. Turner. 1991. Functional group substitutions as probes of hydrogen bonding between GA mismatches in RNA internal loops. *J. Am. Chem. Soc.* 113:4313–4322.

Santoro, S. W., and G. F. Joyce. 1997. A general purpose RNA-cleaving DNA enzyme. *Proc. Natl. Acad. Sci. USA* **94**: 4262–4266.

Scaringe, S. A., C. Francklyn, and N. Usman. 1990. Chemical synthesis of biologically active oligoribonucleotides using β-cyanoethyl protected ribonucleoside phosphoramidites. *Nucleic Acids Res.* **18**:5433–5441.

Schatz, D., R. Leberman, and F. Eckstein. 1991. Interaction of *Escherichia coli* tRNA[Ser] with its cognate aminoacyl-tRNA synthetase as determined by footprinting with phosphorothioate-containing tRNA transcripts. *Proc. Natl. Acad. Sci. USA* **88**: 6132–6136.

Schenborn, E. T., and R. C. Mierendorf. 1985. A novel transcription property of SP6 and T7 RNA polymerases: dependence on template structure. *Nucleic Acids Res.* **13**: 6223–6236.

Schmidt, S., L. Beigelman, A. Karpeisky, N. Usman, U. S. Sørenson, and M. J. Gait. 1996. Base and sugar requirements for RNA cleavage of essential nucleoside residues in internal loop B of the hairpin ribozyme: implications for secondary structure. *Nucleic Acids Res.* **24**:573–581

Schulman, L. H., and H. Pelka. 1983. Anticodon loop size and sequence requirements for recognition of formylmethionine tRNA by methionyl-tRNA synthetase. *Proc. Natl. Acad. Sci. USA* **80**:6755–6789.

Scott, W. G., J. T. Finch, R. Grenfell, J. Fogg, T. Smith, M. J. Gait, and A. Klug. 1995. Rapid crystallization of chemically synthesised hammerhead RNAs using a double screening procedure. *J. Mol. Biol.* **250**:327–332.

Seela, F., and K. Mersmann. 1992. 7-Deazaguanosine: phosphoramidite and phosphonate building blocks for solid-phase oligoribonucleotide synthesis. *Heterocycles* **34**:229–236.

Seela, F., and K. Mersmann. 1993. 7-Deazaguanosine: synthesis of an oligoribonucleotide building block and disaggregation of the U-G-G-G-G-U-G$_4$ structure by the modified base. *Helv. Chim. Acta* **76**:1435–1449.

Seela, F., K. Mersmann, J. A. Grasby, and M. J. Gait. 1993. 7-Deazaadenosine: oligoribonucleotide building block synthesis and autocatalytic hydrolysis of base-modified hammerhead ribozymes. *Helv. Chim. Acta* **76**:1809–1820.

Sergiev, P. V., I. N. Lavrik, V. A. Wlassoff, S. S. Dokudovskaya, O. A. Dontsova, A. A. Bogdanov, and R. Brimacombe. 1997. The path of mRNA through the bacterial ribosome: a site-directed crosslinking study using new photoreactive derivatives of guanosine and uridine. *RNA* **3**:464–475.

Sharp, P. A. 1985. On the origins of RNA splicing and introns. *Cell* **42**:397–400.

Shibahara, S., S. Mukai, T. Hishihara, H. Inoue, E. Ohtsuka, and H. Morisawa. 1987. Site-directed cleavage of RNA. *Nucleic Acids Res.* **15**:4403–4415.

Sigurdsson, S., T. Tuschl, and F. Eckstein. 1995. Probing RNA tertiary structure: interhelical cross-linking of the hammerhead ribozyme. *RNA* **1**:575–583.

Sigurdsson, S. T., and F. Eckstein. 1996. Site specific labelling of sugar residues in oligoribonucleotides: reactions of aliphatic isocyanates with 2′-amino groups. *Nucleic Acids Res.* **24**: 3129–3133.

Sinha, N. D., P. Davis, N. Usman, J. Pérez, R. Hodga, J. Kremsky, and R. Casale. 1993. Labile exocyclic amine protection of nucleosides in DNA, RNA and oligonucleotide analog synthesis facilitating N-deacylation, minimizing depurination and chain degradation. *Biochimie* **75**:13–23.

Slim, G., and M. J. Gait. 1991. Configurationally defined phosphorothioate-containing oligoribonucleotides in the study of the mechanism of cleavage of hammerhead ribozymes. *Nucleic Acids Res.* **19**:1183–1188.

Slim, G., and M. J. Gait. 1992. The role of the exocyclic amino groups of conserved purines in hammerhead ribozyme cleavage. *Biochem. Biophys. Res. Commun.* **183**:605–609.

Sontheimer, E. J. 1994. Site-specific RNA crosslinking with 4-thiouridine. *Mol. Biol. Rep.* **20**:35–44.

Sontheimer, E. J., and J. A. Steitz. 1993. The U5 and U6 small nuclear RNAs as active site components of the spliceosome. *Science* **262**:1989–1996.

Sontheimer, E. J., S. Sun, and J. A. Piccirilli. 1997. Metal ion catalysis during splicing of premessenger RNA. *Nature* **388**: 801–805.

Sproat, B., F. Colonna, B. Mullah, D. Tsou, A. Andrus, A. Hampel, and R. Vinayak. 1995. An efficient method for the isolation and purification of oligoribonucleotides. *Nucleosides Nucleotides* **14**:255–273.

Stade, K., J. Rinke-Appel, and R. Brimacombe. 1989. Site-directed cross-linking of mRNA analogues to the ribosome: identification of 30S ribosomal components that can be cross-linked to the mRNA at various points 5′ with respect to the decoding site. *Nucleic Acids Res.* **17**:9889–9908.

Stepanova, O. B., V. G. Metelev, N. V. Chichkova, V. D. Smirnov, N. P. Rodionova, J. G. Atabekov, A. A. Bogdanov, and Z. A. Shabarova. 1979. Addressed fragmentation of RNA molecules. *FEBS Lett.* **103**:197–201.

Strobel, S. A., and T. R. Cech. 1993. Tertiary interactions with the internal guide sequence mediate docking of the P1 helix into the catalytic core of the *Tetrahymena* ribozyme. *Biochemistry* **32**:13593–13604.

Strobel, S. A., and T. R. Cech. 1995. Minor groove recognition of the conserved G·U pair at the *Tetrahymena* ribozyme reaction site. *Science* **267**:675–679.

Strobel, S. A., and T. R. Cech. 1996. Exocyclic amine of the conserved G·U pair at the cleavage site of the *Tetrahymena* ribozyme contributes to 5′-splice site selection and transition state stabilization. *Biochemistry* **35**:1201–1211.

Strobel, S. A., and K. Shetty. 1997. Defining the chemical groups essential for *Tetrahymena* group I intron function by nucleotide analog interference mapping. *Proc. Natl. Acad. Sci. USA* **94**: 2903–2908.

Strobel, S. A., T. R. Cech, N. Usman, and L. Beigelman. 1994. The 2,6-diaminopurine riboside 5-methylisocytidine wobble base pair: an isoenergetic substitution for the study of G·U pairs in RNA. *Biochemistry* **33**:13824–13835.

Sumner-Smith, M., S. Roy, R. Barnett, L. S. Reid, R. Kuperman, U. Delling, and N. Sonenberg. 1991. Critical chemical features in trans-acting-responsive RNA are required for interaction with human immunodeficiency virus type 1 Tat protein. *J. Virol.* **65**: 5196–5201.

Suntharalingam, M., E. Dulude, and M. J. Moore. 1997. Unpublished data.

Sylvers, L. A., and J. Wower. 1993. Nucleic acid-incorporated azidonucleotides: probes for studying the interaction of RNA or DNA with proteins and nucleic acids. *Bioconjug. Chem.* **4**: 411–418.

Sylvers, L. A., J. Wower, S. S. Hixson, and R. A. Zimmermann. 1989. Preparation of 2-azidoadenosine 3′,5′-[5′-^{32}P]bisphosphate for incorporation into transfer RNA: photoaffinity labeling of *Escherichia coli* ribosomes. *FEBS Lett.* **245**:9–13.

Sylvers, L. A., A. M. Kopylov, J. Wower, S. S. Hixson, and R. A. Zimmermann. 1992. Photochemical cross-linking of the anticodon loop of yeast tRNA[Phe] to 30S-subunit protein S7 at the ribosomal A and P sites. *Biochimie* **74**:381–389.

Sylvers, L. A., K. C. Rogers, M. Shimizu, E. Ohtsuka, and D. Söll. 1993. A 2-thiouridine derivative in tRNA[Glu] is a positive determinant for aminoacylation by *Escherichia coli* glutamyl-tRNA synthetase. *Biochemistry* 32:3836-3841.

Tanner, N. K., M. M. Hanna, and J. Abelson. 1988. Binding interactions between yeast tRNA ligase and a precursor transfer ribonucleic acid containing two photoreactive uridine analogues. *Biochemistry* 27:8852-8861.

Teigelkamp, S., A. J. Newman, and J. D. Beggs. 1995a. Extensive interactions of PRP8 protein with the 5′ and 3′ splice sites during splicing suggest a role in stabilization of exon alignment by U5 snRNA. *EMBO J.* 14:2602-2612.

Teigelkamp, S., E. Whittaker, and J. D. Beggs. 1995b. Interaction of the yeast splicing factor PRP8 with substrate RNA during both steps of splicing. *Nucleic Acids Res.* 23:320-326.

Tsou, D., A. Andrus, and R. Vinayak. 1995a. Improvements in large scale synthesis, isolation and purification of oligoribonucleotides. Nucleic Acids Symposium, Noordwijkerhout, The Netherlands.

Tsou, D., A. Hampel, A. Andrus, and R. Vinayak. 1995b. Large scale synthesis of oligoribonucleotides on high-loaded polystyrene (HLP) support. *Nucleosides Nucleotides* 14:1481-1492.

Tuschl, T., M. M. P. Ng, W. Pieken, F. Benseler, and F. Eckstein. 1993. Importance of exocyclic base functional groups of central core guanosines for hammerhead ribozyme activity. *Biochemistry* 32:11658-11668.

Uhlenbeck, O. C., and R. I. Gumport. 1982. T4 RNA ligase, p. 31-58. *In* P. D. Boyer (ed.), *The Enzymes*, vol. 15, part B. Academic Press, New York, N.Y.

Umen, J. G., and C. Guthrie. 1995. Prp16p, Slu7p, and Prp8p interact with the 3′ splice site in two distinct stages during the second catalytic step of pre-mRNA splicing. *RNA* 1:584-597.

Usman, N., and R. Cedergren. 1992. Exploiting the chemical synthesis of RNA. *Trends Biochem. Sci.* 17:334-339.

Wahle, E., and W. Keller. 1994. 3′ end-processing of mRNA, p. 1-34. *In* S. J. Higgins and B. D. Hames (ed.), *RNA Processing: a Practical Approach*, vol. II. IRL Press, Oxford, United Kingdom.

Wang, J.-F., W. D. Downs, and T. R. Cech. 1993. Movement of the guide sequence during RNA catalysis by a group I intron. *Science* 260:504-508.

Weinstein, L. B., B. C. Jones, R. Cosstick, and T. R. Cech. 1997. A second catalytic metal ion in group I ribozyme. *Nature* 388:805-808.

Weintraub, H. 1973. The assembly of newly replicated DNA into chromatin. *Cold Spring Harbor Symp. Quant. Biol.* 38:247-256.

Whitfeld, P. R., and R. Markham. 1953. Natural configuration of purine nucleotides in ribonucleic acids. Chemical hydrolysis of the dinucleoside phosphates. *Nature* 171:1151-1152.

Willis, M. C., B. J. Hicke, O. C. Uhlenbeck, T. R. Cech, and T. H. Koch. 1993. Photocrosslinking of 5-iodouracil-substituted RNA and DNA to proteins. *Science* 262:1255-1257.

Willis, M. C., K. A. LeCuyer, K. M. Meisenheimer, O. C. Uhlenbeck, and T. H. Koch. 1994. An RNA-protein contact determined by 5-bromouridine substitution, photocrosslinking and sequencing. *Nucleic Acids Res.* 22:4947-4952.

Wincott, F., A. DiRenzo, C. Shaffer, S. Grimm, D. Tracz, C. Workman, D. Sweedler, C. Gonzalez, S. Scaringe, and N. Usman. 1995. Synthesis, deprotection, analysis and purification of RNA and ribozymes. *Nucleic Acids Res.* 23:2677-2684.

Wittenberg, W. L., and O. C. Uhlenbeck. 1985. Specific replacement of functional groups of uridine-33 in yeast phenylalanine transfer ribonucleic acid. *Biochemistry* 24:2705-2712.

Woody, A.-Y. M., R. K. Evans, and B. E. Haley. 1988. Characterization of a photoaffinity analog of UTP, 5-azido-UTP, for analysis of the substrate binding site on *E. coli* RNA polymerase. *Biochem. Biophys. Res. Commun.* 150:917-924.

Wower, J., S. S. Hixson, and R. A. Zimmermann. 1988. Photochemical cross-linking of yeast tRNA[Phe] containing 8-azidoadenosine at positions 73 and 76 to the *Escherichia coli* ribosome. *Biochemistry* 27:8114-8121.

Wower, J., S. S. Hixson, and R. A. Zimmermann. 1989. Labeling the peptidyl transferase center of the *Escherichia coli* ribosome with photoreactive tRNA[Phe] derivatives containing azidoadenosine at the 3′ end of the acceptor arm: a model of the tRNA-ribosome complex. *Proc. Natl. Acad. Sci. USA* 86:5232-5236.

Wower, J., T. A. Malloy IV, S. S. Hixson, and R. A. Zimmermann. 1990. Probing tRNA binding sites on the *Escherichia coli* 30S ribosomal subunit with photoreactive analogs of the anticodon arm. *Biochim. Biophys. Acta* 1050:38-44.

Wower, J., P. Scheffer, L. A. Sylvers, W. Wintermeyer, and R. A. Zimmermann. 1993a. Topography of the E site on the *Escherichia coli* ribosome. *EMBO J.* 12:617-623.

Wower, J., L. A. Sylvers, K. V. Rosen, S. S. Hixson, and R. A. Zimmermann. 1993b. A model of the tRNA binding sites on the *Escherichia coli* ribosome, p. 455-464. *In* K. H. Nierhaus, F. Franceschi, A. R. Subramanian, V. A. Erdmann, and B. Wittmann-Liebold (ed.), *The Translational Apparatus*. Plenum Press, New York, N.Y.

Wower, J., K. V. Rosen, S. S. Hixson, and R. A. Zimmermann. 1994a. Recombinant photoreactive tRNA molecules as probes for cross-linking studies. *Biochimie* 76:1235-1246.

Wower, J., S. S. Hixson, L. A. Sylvers, Y. Xing, and R. A. Zimmermann. 1994b. Synthesis of 2,6-diazido-9-(β-D-ribofuranosyl)purine 3′,5′ bisphosphate: incorporation into transfer RNA and photochemical labeling of *Escherichia coli* ribosomes. *Bioconjug. Chem.* 5:158-161.

Wower, J., I. K. Wower, S. V. Kirillov, K. V. Rosen, S. S. Hixson, and R. A. Zimmermann. 1995. Peptidyl transferase and beyond. *Biochem. Cell Biol.* 73:1041-1047.

Wu, T., K. K. Ogilvie, and R. T. Pon. 1989. Prevention of chain cleavage in the chemical synthesis of 2′-silylated oligoribonucleotides. *Nucleic Acids Res.* 17:3501-3517.

Wyatt, J. R., E. J. Sontheimer, and J. A. Steitz. 1992. Site-specific crosslinking of mammalian U5 snRNP to the 5′ splice site prior to the first step of pre-messenger RNA splicing. *Genes Dev.* 6:2542-2553.

Wu, T., K. K. Ogilvie, and R. T. Pon. 1989. Prevention of chain cleavage in the chemical synthesis of 2′-silylated oligoribonucleotides. *Nucleic Acids Res.* 17:3501-3517.

Xu, Q., G. Barany, R. P. Hammer, and K. Musier-Forsyth. 1996. Efficient introduction of phosphorothioates into RNA oligonucleotides by 3-ethoxy-1,2,4-dithiazoline-5-one (EDITH). *Nucleic Acids Res.* 24:3643-3644.

Yu, Y.-T., and J. A. Steitz. 1997. A new strategy for introducing photoactivatable 4-thiouridine (^{4s}U) into specific positions in a long RNA molecule. *RNA* 3:807-810.

Yu, Y.-T., P. A. Maroney, E. Darzynkiewicz, and T. W. Nilsen. 1995. U6 snRNA function in nuclear pre-mRNA splicing: a phosphorothioate interference analysis of the U6 phosphate backbone. *RNA* 1:46-54.

Chapter 5

Biophysical and Conformational Properties of Modified Nucleosides in RNA (Nuclear Magnetic Resonance Studies)

DARRELL R. DAVIS

The location and identity of a large number of the modified nucleosides found in tRNA are highly conserved across diverse species and between tRNA isoacceptors (Sprinzl et al., 1998). Conserved modification patterns in rRNAs (Maden, 1990) and small nuclear RNAs (Gu and Reddy, 1997) also suggest that particular modified nucleosides have critical roles, although these functions are not well described compared to modifications found in tRNA (Limbach et al., 1994). In certain cases, the modified nucleoside may be functionally distinct from the parent, unmodified nucleoside due to direct interaction of unique functional groups with an RNA or protein target molecule. Examples of these direct effects are rare and the fact that the most frequent modifications are simple methylations or pseudouridylation indicates that modification generally has an indirect effect on RNA structure. The effects of nucleoside modification on intrinsic biophysical properties have been extensively investigated by nuclear magnetic resonance (NMR) spectroscopy and used to postulate structural roles for modification. For example, NMR is very useful for comparing the conformational properties of modified nucleosides with the properties of the parent, unmodified nucleosides. Changes in nucleoside conformation (Fig. 1) correlate well with global thermodynamic effects and with biochemical changes in tRNAs. The NMR determined biophysical properties have also provided an explanation for changes in important functions of tRNA during protein synthesis such as codon-anticodon recognition (Grosjean and Houssier, 1990). Similar correlations between modification and function in larger RNAs (such as rRNA) have lagged behind tRNA studies simply because of the large complex structures involved and lack of earlier recognition of the functional role of rRNA in translation (Lodmell and Dahlberg, 1997; Noller, 1993). NMR studies of modified nucleosides in the context of an RNA oligonucleotide are more relevant biochemical models than nucleosides, but have proceeded more slowly because of the difficulty of preparing site-specifically-modified RNAs. From recent studies of modified RNA oligonucleotides it has become clear that structural interactions may be different in the context of an oligonucleotide than is seen at the nucleoside or nucleotide level. NMR studies of tRNAs have also identified specific interactions involving modified nucleosides that suggest the modifications may have different structural effects in different sequence contexts. Despite the wealth of information provided by X-ray crystal structures, NMR is the current method of choice for investigating the structural and dynamic changes to RNA that result from nucleoside modification.

NMR DETERMINATION OF NUCLEOSIDE CONFORMATION

NMR Measurable Parameters

Biophysical properties of nucleosides can be determined by NMR measurement of fundamental properties that change as a function of modification, and usually by an additional environmental factor such as temperature or salt concentration. The NMR measurables are chemical shift, scalar coupling constants, and the nuclear Overhauser effect (NOE). The chemical shift is sensitive to the local diamagnetic environment of the nucleoside. In nucleoside monomers, changes in chemical shift with modification are informative with respect to the intrinsic chemical properties of the nucleoside, but provide little insight into how the modification might affect RNA struc-

Darrell R. Davis • Department of Medicinal Chemistry, University of Utah, Salt Lake City, Utah 84112-5820.

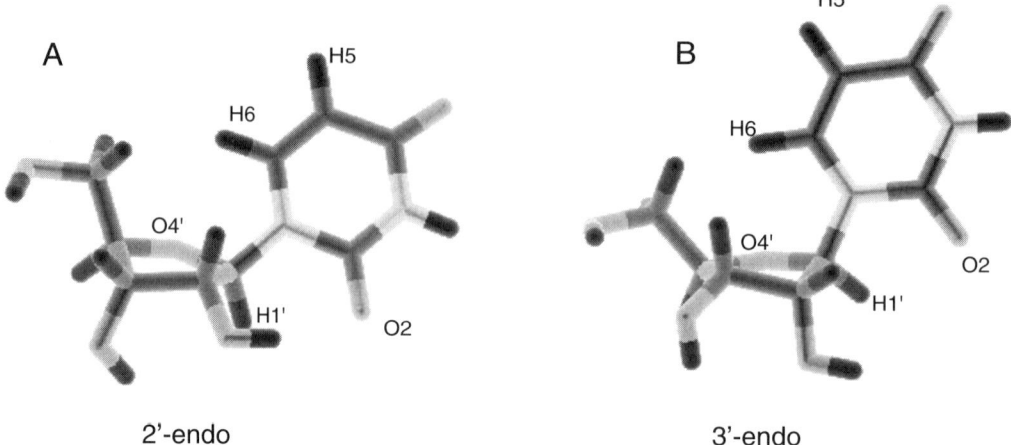

Figure 1. The two limiting conformations for ribonucleosides. (A) 2′-endo sugar with the base in the pseudoequatorial conformation; (B) 3′-endo sugar with the base pseudoaxial.

ture. Chemical shift changes for nucleosides within RNA oligonucleotides are potentially more informative because the interaction of the neighboring nucleosides can be compared for modified and unmodified forms. The aromatic base protons, in particular, demonstrate altered chemical shift positions when modification changes the extent of base stacking (Lee et al., 1976; Lee and Tinoco, 1977; Smith et al., 1992b).

The sugar protons in RNA comprise a scalar coupling network where each proton interacts with its vicinal coupled neighbors, and the magnitude of this interaction is strongly dependent on the dihedral angles. Scalar coupling measurements have been used extensively to investigate nucleoside conformation, and a sophisticated treatment has evolved that takes into account not only the respective angular dependence, but also the substitution pattern of the ribose ring (Altona and Sundaralingam, 1973; Haasnoot et al., 1980). It is well established that scalar couplings depend not only on the dihedral angle, but also on the electronegativity of the substituents around the ribose ring (Haasnoot et al., 1980). For instance, the scalar couplings will be different for the same ribose conformation when the 3′ position is a free hydroxyl compared to a nucleotide with a 3′-phosphate substituent. Altona and coworkers have developed a protocol where the sugar scalar coupling constants are iteratively fit to a model for the ribose ring that takes into account the particular substitution pattern of ribose and the functional group electronegativities (de Leeuw and Altona, 1983). This results in an accurate determination of the ribose pseudorotation and puckering amplitude, and provides for a more sophisticated description of sugar conformation than a simple 2′-endo versus 3′-endo model (Fig. 1). An exhaustive investigation of substituent and base effects on sugar conformation has been made by Chattopadhyaya and coworkers (Plavec et al., 1994a; Plavec et al., 1994b; Plavec et al., 1993; Thibaudeau et al., 1994a; Thibaudeau et al., 1996; Thibaudeau et al., 1994c; Thibaudeau et al., 1994b). These studies have further refined the approach of Altona by documenting the effects of different bases on the nucleoside conformation through base interaction with the attached sugar, resulting in a model where both the base and sugar substituents can affect the sugar conformation. An experimental limitation of the iterative fit (modified Karplus model) approach is that it requires very accurate measurement of both the H1′-H2′ and the H3′-H4′ coupling constants. The H1′-H2′ scalar couplings usually are accessible from the one-dimensional (1D) proton NMR spectrum, or from a high-resolution 2D COSY (Wuthrich, 1986) type experiment, but the corresponding H3′-H4′ couplings may be difficult to resolve, particularly in oligonucleotides. In these cases where spectral overlap precludes measurement of the H3′-H4′ values, the H1′-H2′ scalar coupling value alone can be used to determine the percent 2′-endo conformer by assuming that (H1′-H2′) + (H3′-H4′) = 10 Hz and % 2′-endo = 100 × (H1′-H2′) (Altona and Sundaralingam, 1973; Lee et al., 1976). The determination of sugar conformation from H1′-H2′ couplings alone is not sufficient for precise determination of sugar conformation thermodynamics, but provides a reasonable method for comparing modified nucleosides with chemically similar substitutions.

The ^1H-^1H NOE arises due to cross-relaxation between protons that are separated by less than 5 Å and is purely a through space interaction. This provides a method for identifying protons that are remote in terms of the covalent geometry (Neuhaus and Williamson, 1989; Noggle and Shirmer, 1971). NOE measurements of nucleosides have been used to augment the scalar coupling data that is used to determine the sugar conformation (Fig. 2); for instance, the H1'-H4' distance varies significantly for different values of sugar pseudorotation (Wuthrich, 1986). NOE experiments also provide critical data for determination of the glycosyl χ angle, using NOE connectivity between base protons and sugar protons (Rosemeyer et al., 1990). Once the sugar conformation is known from scalar couplings, the interproton distances between the purine H8 or pyrimidine H6 protons and the sugar H1', H2', and H3' protons allow the χ angle to be determined with high accuracy. By measuring the change in NOE intensity as a function of temperature, it is possible to determine the thermodynamic barrier for interconversion between *syn* and *anti* conformers (Rosemeyer et al., 1990). The relative preference for *syn* versus *anti* glycosyl conformations is one of the important properties that can be affected by nucleoside modification.

The conformational preference of a nucleoside, either isolated or in an oligonucleotide, can be determined from the relative intensities of the NOE cross-relaxation between the H6/H8 to H1' compared to the sum of the H6/H8 to H2' and H6/H8 to H3' protons (Rosemeyer et al., 1990). In the *syn* conformation the H6/H8 to H1' NOE is at its maximum while the sum of the H6/H8 to H2' and H6/H8 to H3' NOEs is almost zero. The converse is true for a nucleoside in a high *anti* conformation where the H6/H8 to H1' NOE is almost zero. This simple treatment is quite powerful, but accurate interpretation depends on knowledge of the sugar conformation. A common interpretation, typified by the approach of Seela and coworkers, is that NOE patterns between the two limiting cases are indicative of an equilibrium between *syn* and *anti* where the NOEs represent a weighted average (Rosemeyer et al., 1990). Another interpretation of medium strength NOEs to H1', H2', and H3' is that the nucleoside exists purely in an intermediate conformation, although NMR evidence in addition to simple NOE data is needed to support such a contention. The NOE is also used to determine base stacking between nucleosides in oligonucleotides. There are distinct NOE connectivities involving the base protons and sugar protons for nucleosides in a 3'-endo, A-form geometry and significantly different NOE patterns expected for 2'-endo nucleosides (Varani and Tinoco, 1991; Wuthrich, 1986). NOE experiments have been used to determine the effects of nucleoside modification on local RNA structure and these results are described in subsequent sections.

Calculation of Thermodynamic Parameters from NMR Experiments

NMR experiments have been used to measure the thermodynamic values (ΔG, ΔH, and ΔS) for the sugar conformational interconversion barrier and the thermodynamic barrier to rotation about the glycosyl bond. These are conceptually related to calorimetric heat measurements, but in practice utilize the van't Hoff relation, $\ln K = -\Delta H/RT + \Delta S/R$, where an NMR observable such as the scalar coupling constant or NOE is used to determine the change in an equi-

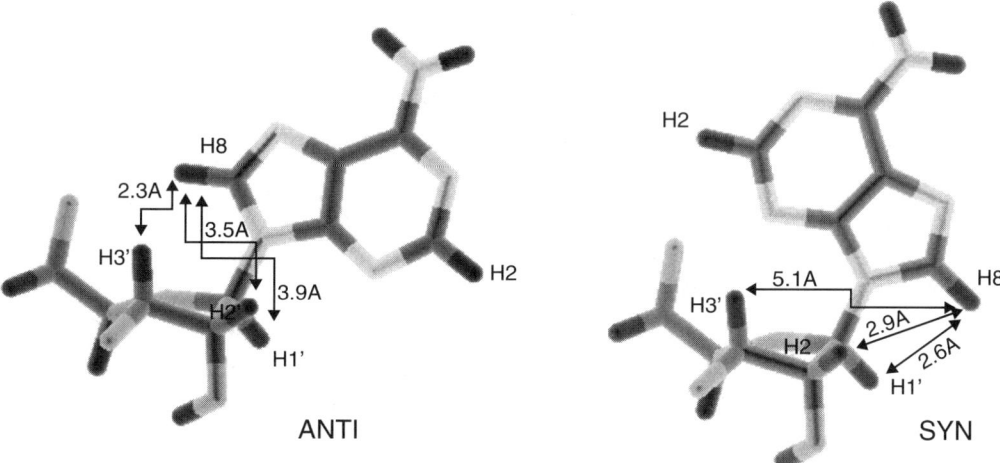

Figure 2. NMR NOE experiments are used to define the glycosyl conformation for nucleosides. The interproton distances between the base H6 (pyrimidines) or H8 (purines) and the sugar protons change as a function of glycosyl angle. The figure shows the key distances involving H1', H2', and H3' protons for adenosine and how they differ for either the *anti* or *syn* conformations.

librium constant with temperature (Rosemeyer et al., 1990; Sakamoto et al., 1996; Thibaudeau et al., 1996). For example, the K_{eq} for 2'-endo/3'-endo interconversion as a function of temperature can be measured from the scalar coupling constants as described above, and this temperature dependence can be plotted to provide the thermodynamic values. The van't Hoff enthalpy is the slope of the ln K_{eq} versus $1/T$ plot and the entropy the intercept. Once ΔH and ΔS are known, ΔG at a specified temperature can be calculated. The other assumption that is required for the van't Hoff enthalpy to accurately reflect the calorimetric enthalpy is that the measured process conform to a two-state model. This is likely to be generally true for the sugar conformational interconversion and *syn/anti* interconversion in nucleosides, but more complicated processes such as oligonucleotide melting may involve multiple states where the van't Hoff and calorimetric enthalpies do not agree (Laing and Draper, 1994).

Correlation of Sugar Conformation With Base Stacking

The use of the percentage of 3'-endo sugar conformation as a measure of base stacking originated from NMR investigations on ribonucleotide dimers and trimers (Lee et al., 1976; Lee and Tinoco, 1980; Lee and Tinoco, 1977). The chemical shifts and sugar-coupling constants were measured as a function of temperature, and dimer chemical shifts and coupling constants were compared to those of the component monomers. These early NMR studies established methods for determining structure in oligonucleotide model systems and began to catalogue NMR properties of the parent, unmodified nucleosides that are used as a baseline for understanding the effects of modification. These studies established that the percentage of 3'-endo conformer in the nucleoside monomer can be used to extrapolate from the monomer properties to make predictions about the RNA stabilization when the monomer is incorporated into an oligonucleotide. The ability to predict global structural changes from modification properties at the nucleoside level is powerful, but this may be misleading for modifications where the RNA effects require some higher-order RNA structure. Even simple functional groups, such as 5'-phosphate, may be involved in interactions with a modified functional group. Nucleoside modifications that have their effect on sugar conformation solely within a discrete nucleoside unit are the only class where base stacking effects can be accurately inferred from a change in sugar conformation.

The transition from an unstructured polynucleotide to an A-form stacked structure associated with stable, helical RNA involves the cooperative change of several structural features. The sugars change from an equilibrium mixture of 2'-endo and 3'-endo to purely 3'-endo conformations, the bases assume an axial glycosyl conformation, the bases are restricted to the *anti* conformation, and the bases interact via dipolar stacking interactions (Saenger, 1984). A nucleoside modification can therefore stabilize RNA structure by affecting any one or a combination of these equilibrium processes and the effect, then seen as a shift in equilibrium for the coupled interactions.

STEREOELECTRONIC EFFECTS AND NUCLEOSIDE CONFORMATIONAL PREFERENCES

Nucleoside modification in either the base or sugar can have significant effects on the sugar conformation, the glycosyl conformation, and base stacking. The sugar conformation and glycosyl (*syn/anti*) effects are readily quantified at the nucleoside level using NMR, and there is a rich literature for naturally occurring nucleoside modifications and many synthetic nucleosides. The 2'-O-methyl substituent is the only sugar modification that is commonly found in nature for RNA (Limbach et al., 1994), but the wealth of data on 2'-deoxy nucleosides provides information on the relative importance of electronic effects from the 2' oxygen compared to the steric effect associated with 2'-O-methyl (Kawai et al., 1992b; Plavec et al., 1994a). Modifications in the base can have significant effects on the sugar conformational preference, illustrating that there are stereoelectronic interactions between the base and sugar that affect the sugar conformation. Furthermore, the orientation of the base substituent in either an axial or equatorial conformation is coupled to the sugar conformation. The sugar conformation (2'-endo or 3'-endo), glycosyl conformation (axial or equatorial), and glycosyl conformation (*syn* or *anti*) are energetically coupled, can be modulated by modification, and ultimately relate to the structural effects of nucleoside modification.

Steric Effects

The 3'-endo conformation is stabilized in nucleosides by modification of the 2'-hydroxyl to a 2'-O-methyl (Kawai et al., 1992b; Lee and Tinoco, 1977). In 2'-O-methyl nucleosides the axial orientation is favored, which serves to minimize unfavorable interaction between the large methyl group and the base.

This modification is most effective and most common in the pyrimidines, where the equatorial conformation results in severe steric clash between the 2'-O-methyl and O2 on the base (Kawai et al., 1992b). The 2'-O-methyl guanosine modification is widely distributed in tRNA, but still rare compared to the cytosine and uridine nucleosides, while 2'-O-methyl adenosine is prevalent in thermophilic archaea, very rare in eukaryotes, and absent in bacteria (Limbach et al., 1994). Stabilization of the 3'-endo, axial sugar conformation at the nucleoside level translates to overall thermodynamic stabilization of 2'-O-methyl modified RNA (Inoue et al., 1987). The very high percentage of 2'-O-methyl modified nucleosides in tRNAs from extreme thermophiles indicates how biologically important this modification can be for stabilizing RNA because the overall stability of tRNAs from thermophiles is well above what can be obtained from increasing the G-C content at the expense of A-U base pairs (Agris et al., 1973; Kowalak et al., 1994).

The 3'-endo conformation is favored for 2'-O-methyl modified nucleosides because this minimizes steric clash with O2 on pyrimidine bases. The 2-thiouridine modification has a similar and even more dramatic effect where the 3'-endo conformation is adopted to minimize steric interactions between the sugar and the large sulfur atom (Agris et al., 1992; Sierzputowska-Gracz et al., 1987; Yamamoto et al., 1983). However, synthetic and natural RNA molecules are stabilized to such a large extent by s²U modification that additional stereoelectronic factors may play an important role (Houssier et al., 1988; Kumar and Davis, 1997a, 1997b).

The Anomeric and 3'-Gauche Effects

Modifications can affect the distribution of electrons in the base and therefore influence the sugar conformation by interactions between base and sugar. There are two molecular orbital interactions that appear to be predominantly responsible for these effects, and both involve the lone pair electrons on O4'. Base modification affects the electrostatic charge density at C1' and modulates the anomeric effect (Thibaudeau et al., 1996; Thibaudeau et al., 1994b). The source of the anomeric effect is shown in Fig. 3, where the lone pair at O4' interacts favorably with the σ^* orbital at C1' when the base is axial. Electron-withdrawing base modifications that decrease electron density at C1' increase the anomeric effect and favor the axial, 3'-endo sugar conformation; protonation of the base has the same effect (Juaristi and Cuevas, 1995; Thibaudeau et al., 1996). A second interaction has been implicated in pyrimidines where an interaction occurs between the lone pair at O4'

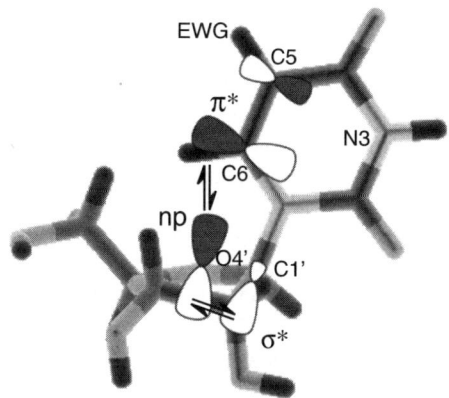

Figure 3. Molecular orbital interactions in pyrimidines that affect the thermodynamically preferred glycosyl conformation and change upon modification. Both the anomeric interaction and an interaction involving the H5-H6 bond are affected by modification. For the anomeric effect, the lone pair electrons at O4' are in an np orbital which overlaps with the σ^* orbital at C1'. In the 3'-endo conformation (shown) the C1'-N bond is axial and the σ^* orbital oriented along the glycosyl bond can bond in a π-like fashion with the np orbital. The same O4' lone pair orbitals can interact with π^* antibonding orbitals from the pyrimidine H5-H6 double bond. An electron withdrawing group (EWG) at the 5 position increases the size of the π^* orbital on H6 as shown which would increase the interaction and favor a 3'-endo conformation with a low χ angle.

and the π^* orbital of the H5-H6 double bond (Egert et al., 1980; Kawai et al., 1992a; Thibaudeau et al., 1994b; Uhl et al., 1983). Base modifications that decrease electron density in the ring will increase the strength of the interaction at C5-C6 in addition to increasing the anomeric effect, resulting in stabilization of the 3'-endo sugar conformation. Base modification effects on both the σ^* and π^* orbital energies shift the nucleoside conformation in the same direction; therefore, interaction between O4' lone pairs and either the C1' σ^*, or the H5-H6 π^* orbitals can be used to explain modification effects in pyrimidines. The C-nucleosides, (pseudouridine and related modifications) favor the 2'-endo, equatorial conformation at the nucleoside level due to a decreased anomeric effect, and consequently have a lower energy barrier to rotation about the glycosidic bond compared to uridine (Thibaudeau et al., 1994a).

The 3'-gauche effect is mechanistically related to the anomeric effect in that it arises due to a geometrically dependent overlap between σ^* and n lone pair molecular orbitals (Juaristi and Cuevas, 1995). The published NMR studies have described how the absence or presence of a 3'-phosphate affects nucleoside conformation and have shown that it is important to be aware that the 3'-phosphate will perturb the sugar conformation and affect the NMR coupling constants (Kawai et al., 1992b; Plavec et al., 1994a; Thibaudeau et al., 1994c). No specific interactions have been described between base modifications and the 3′-phosphate, but there are several examples of interaction between the 5′-phosphate and functional groups on modified bases (vide infra).

Base Stacking Effects

Nucleoside bases that are in a stacked geometry generally have aromatic NMR proton shifts that are upfield of the positions seen in the absence of significant stacking. The aromatic protons are primarily affected by the ring current of the 5′-neighboring base, with 5′-purine nucleosides having a greater shielding effect than 5′-pyrimidines. Among the major nucleosides, the purines provide the greatest stacking stabilization, but there are a significant number of nucleoside modifications that increase the stacking propensity of pyrimidines. The addition of a polarizable group to the base will result in stacking stabilization since the stacking interaction is a dipole-dipole, dipole-induced, dipole effect. Modifications that increase base stacking are potentially powerful since stacking is the most important energetic interaction for RNA stabilization (Saenger, 1984). Addition of a polarizable methyl group to the base or a sulfur at C2 of pyrimidines has been shown to directly stabilize base stacking (Plesiewicz et al., 1976; Sowers et al., 1987). Several hypermodified bases are clear candidates to act through stacking stabilization, such as ms^{2}i^{6}A (Houssier and Grosjean, 1985) and the yW (Grosjean et al., 1976; Grosjean et al., 1978) nucleoside, which are both found adjacent to the anticodon triplet at position 37 in tRNA (Björk, 1983; Grosjean and Houssier, 1990; see also Chapter 7).

CONFORMATIONAL AND THERMODYNAMIC EFFECTS OF SPECIFIC NUCLEOSIDE MODIFICATIONS

2′-O-Methylations

Modification of the 2′-OH to a 2′-O-methyl is one of the most powerful biological strategies for stabilizing RNA structure. The modified nucleoside preferentially adopts the 3′-endo, axial conformation to minimize steric interactions between the large methyl group and the base. This steric clash is more severe for pyrimidines than for purines, but 2′-O-methyl modification promotes the 3′-endo conformation for all nucleosides. The thermodynamic stabilization of the 3′-endo conformer on modification is 0.08 kcal/mol for uridine as a result of the shift in equilibrium from 55% 3′-endo for U to 60% 3′-endo for Um at 28°C (Table 1) (Kawai et al., 1992b). This relatively modest effect is increased significantly for

Table 1. Modification effects on the 2′-endo/3′-endo sugar conformational equilibrium thermodynamics^a

Nucleoside	ΔH for 3′-endo	% S conformer at 25°C (reference)
ac^{4}Cm	1.53 (a)	
ac^{4}C	1.22 (c)	
C	0.37 (a)	
Cm	0.65 (a)	
f^{5}C	1.56 (b)	
m$^2_2$Gm	−0.31 (c)	
G	−0.67 (c)	
U	0.37 (d)	88 (e)
mo^{5}U	0.58 (d)	74 (e)
cmo^{5}U	0.43 (d)	
ho^{5}U	−0.01 (d)	
mnm^{5}s^{2}U	1.32 (d)	30 (e)
m^{5}s^{2}U	0.98 (d)	
s^{2}U	1.12 (d)	41 (e)
m^{5}U	0.16 (d)	
pU	0.09 (d)	
pmo^{5}U	−0.72 (d)	
pcmo^{5}U	−0.67 (d)	
pho^{5}U	−0.28 (d)	
pmnm^{5}s^{2}U	1.10 (d)	
ps^{2}U	0.87 (d)	
pm^{5}U	−0.11 (d)	
mnm^{5}U	0.39 (f)	73 (e)
pmnm^{5}U	0.65 (f)	
Up	−0.10 (a)	
Um	0.45 (g)	
Ump	0.67 (g)	
yW		54 (h)
mcm^{5}U		81 (e)
mcm^{5}s^{2}U		35 (e)
cmnm^{5}U		77 (e)
cmnm^{5}s^{2}U		31 (e)
mo^{5}s^{2}U		21 (e)
mcmo^{5}U		76 (e)
mcmo^{5}s^{2}U		28 (e)
ψ		70 (e)
acp^{3}U		40 (i)
m^{1}acp^{3}ψ		60 (i)
Dp	−1.43 (j)	68 (j)

^{a} Positive ΔH values indicate that the 3′-endo (N) conformer is thermodynamically favored. References: a, Kawai et al., 1992a; b, Kawai et al., 1994; c, Kawai et al., 1991; d, Yokoyama et al., 1985; e, Sierzputowska-Gracz et al., 1987; f, Sakamoto et al., 1996; g, Kawai et al., 1992b; h, Sierzputowska-Gracz et al., 1991; i, Smith et al., 1992a; j, Dallüge et al., 1996.

the 3′ monophosphates; the 3′-endo conformer of Ump is stabilized by 0.8 kcal/mol compared to the 3′-endo conformer of Up since in the 3′-endo form steric interactions are minimized between the methyl group and both the 3′-phosphate and O2 on the base (Kawai et al., 1992b). Sugar methylation appears to be a general strategy for stabilizing RNA in many organisms that grow optimally at high temperature, as shown in RNAs isolated from the archaeal hyperthermophiles that contain unprecedented numbers of 2′-O-methyl nucleosides (Edmonds et al., 1991; Kowalak et al., 1994).

5-Methylated Pyrimidines (m⁵U and m⁵C)

Methyl modifications in the base are found for each of the parent nucleosides, and modification of hydrogen bond donor and acceptor sites will dramatically affect hydrogen bonding properties. Highly conserved, modified purines at position 37 in tRNA are clear examples where methylation is used to prevent undesirable interactions with mRNA and to maintain the open loop structure of the anticodon (Basti et al., 1996). The effects of 5-methyl substitution of pyrimidine nucleosides are less obvious since the introduction of a hydrophobic methyl into the major groove does not perturb base-pairing interactions. The stabilizing effect of 5-methylation generally is acknowledged in the context of nucleic acid duplexes and is attributed to better base stacking by the methyl modified nucleosides. The details of the mechanism are perhaps less appreciated, but recent studies have distinguished between the C-5 methyl and 2'-hydroxyl effects in comparing DNA and RNA stabilities (Wang and Kool, 1995). The partitioning between 2'-endo and 3'-endo conformers for m⁵U and U is nearly identical at pH 7 (Thibaudeau et al., 1996; Uhl et al., 1983), indicating that RNA stabilizing effects for 5-methylation of pyrimidines are not due to a difference in thermodynamically favored nucleoside conformations, but involve base stacking interactions. The "methyl effect" on oligonucleotide stability has been confirmed as arising from increased polarizability in the 5-methyl modified nucleosides (Sowers et al., 1987) and independent from either a hydrophobic effect or any effect on the ribose sugar conformation. Similar behavior is expected for m⁵C because the effect on base polarizability for m⁵C compared to C is even greater than the effect for T compared to U (Sowers et al., 1987). The detailed structural influence of the ubiquitous m⁵U at position 54 in the T-ψ-C loop of tRNA is still poorly understood, but studies of specifically undermodified tRNA indicate that m⁵U does stabilize tRNA (Davanloo et al., 1979). Melting experiments on native tRNAs and tRNAs lacking m⁵U-54 have shown as much as 6°C of stabilization for the methyl group, while adding a sulfur at the 2-position to give m⁵s²U-54 provides an additional 6–7°C of stabilization (Davanloo et al., 1979; Horie et al., 1985).

2-Thiouridine and 5-Methyl,2-Thiouridine (s²U and m⁵s²U)

The change from an oxygen to a sulfur at C2 in uridine has the greatest effect on nucleoside conformation and on the thermodynamic stabilization of modified RNAs of any single modification. The effect on sugar conformation is mechanistically similar to 2'-O-methylation of the ribose in that the nucleoside adopts the 3'-endo conformation to relieve steric interactions between the large sulfur atom and the 2'-OH (Sierzputowska-Gracz et al., 1987; Yokoyama et al., 1979). The thermodynamic preference for the 3'-endo conformation of uridine and 2-thiolated uridines has been reported in a series of papers. The enthalpy differences between 2'-endo and 3'-endo are as follows: U, 0.37 kcal/mol; s²U, 1.12 kcal/mol; m⁵U, 0.16 kcal/mol; m⁵s²U, 1.0 kcal/mol. For comparison, the value of 2-O-methyl uridine is 0.45 kcal/mol (Kawai et al., 1992b; Yamamoto et al., 1983; Yokoyama et al., 1985). Conformational studies of nucleoside monomers accurately predict the effect seen upon modification of RNA oligonucleotides. Compared to the parent uridine nucleoside, an s²U in RNA strongly favors the 3'-endo, axial conformation and the conformation of nucleosides adjacent to the modified uridine is also shifted toward a 3'-endo, A-form geometry compared to the unmodified oligonucleotide (Kumar and Davis, 1997a; Smith et al., 1992b).

The thermodynamic stabilizing effect of 2-thiolation has also been demonstrated for polynucleotides where poly[r(A-s²U)] is stabilized compared to the unmodified system (Mazumdar et al., 1974; Scheit and Faerber, 1975). We have shown for synthetic oligonucleotides that a single s²U nucleoside can cause as much as a 13°C increase in T_m for a short double-stranded RNA (Kumar and Davis, 1997a). In tRNA, 2-thiolation in both the T-ψ-C loop and the anticodon loop have significant effects. In thermophilic bacteria, the modified nucleoside m⁵s²U replaces the typical m⁵U at position 54 and provides significant thermostabilization as mentioned above (Davanloo et al., 1979; Horie et al., 1985). While the normal m⁵U should provide for local structural stabilization compared to an unmodified U through better stacking interactions, the addition of sulfur at C2 favors the 3'-endo sugar conformation and provides additional stacking stabilization though the highly polarizable sulfur. The thermal stabilization by m⁵s²U in this highly conserved region of tRNA structure tells us that particular structural interactions are maintained at high temperature in thermophylic bacteria and that a 3'-endo sugar conformation at position 54 plays a critical role in the thermal stability of the T-ψ-C loop. The effect of s²U modification in the anticodon is discussed below in the section on mnm⁵s²U.

4-Thiouridine (s⁴U)

There is a paucity of data on the conformational and thermodynamic effects of 4-thiouridine modification, despite the fact that it is highly conserved at position 8 of bacterial tRNAs, is found in archaeal tRNAs, also presumably at position 8, and is involved in a key tertiary interaction that helps define the L-shape of tRNA. Furthermore, this nucleoside has been used extensively as a substitute for U in RNA cross-linking experiments (Hanna, 1989; Yu and Steitz, 1997); therefore, it seems that an understanding of its effect on RNA structure and stability is of some interest. Stability measurements on polynucleotides indicate that s^4U is destabilizing in a Watson-Crick, base-pairing environment compared to uridine (Cramer et al., 1971). The NMR chemical shifts for s^4U in the s^4U8-A14 reversed Hoogsteen interaction in tRNAs indicate that this interaction is at least as stable as a Watson-Crick A-U base pair and the temperature dependence of this interaction in tRNA indicates that in fact this tertiary base pair is one of the stronger A-U type interactions in tRNA (Griffey et al., 1986). In a recent study on uridine thiolation in RNA duplexes, we observed that s^4U modification clearly destabilizes Watson-Crick A-U base pairs (Kumar and Davis, 1997a). The same RNA sequence is stabilized by s^2U, leading us to propose a general paradigm for the effects of thiolation on base pair stability. The destabilization by s^4U of Watson-Crick base pairs and potential stabilization of reversed Hoogsteen s^4U-A base pairs are simply explained by considering the effect of sulfur on hydrogen bond stability, imino proton pK_a, and base stacking interactions. As shown in Fig. 4, thiolation will stabilize base pair interactions when the sulfur is not directly involved in hydrogen bonding as occurs for s^4U in the reversed Hoogsteen geometry and s^2U-A in Watson-Crick geometry, but will be destabilizing for s^4U-A in a Watson-Crick geometry or s^2U-A in the reversed Hoogsteen geometry. The NMR assignment of s^4U8-A14 imino protons at 15.0 ppm in class I *Escherichia coli* tRNAs (Griffey et al., 1986; Hare et al., 1985; Roy et al., 1982) and *Thermus thermophilus* tRNA[Ile]

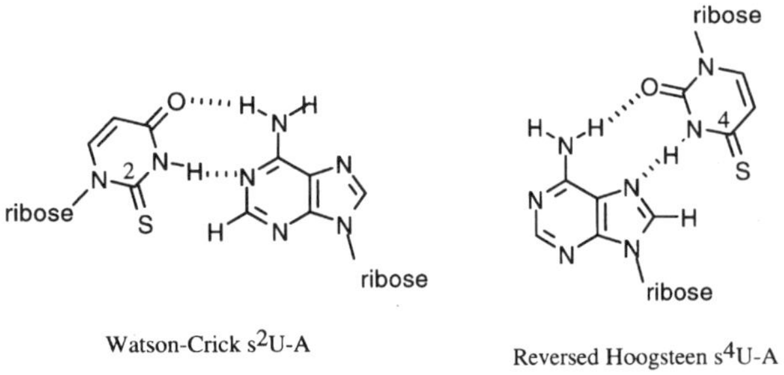

Figure 4. Watson-Crick s^2U-A base pair and reversed Hoogsteen s^4U-A base pair. The s^2 modification in a non-hydrogen bonding position has a direct effect on base stacking and strongly promotes the 3'-endo sugar conformation.

(Choi and Redfield, 1986) indicate that these reversed Hoogsteen base pairs involve hydrogen bonding interactions comparable in strength to the very stable s^2U-A Watson-Crick base pair (Kumar and Davis, 1997a). However, the s^4U8-A14 interaction in some class II *E. coli* tRNAs has an imino proton shift in the same range as a normal Watson-Crick A-U (Griffey et al., 1986), suggesting that s^4U8-A14 is slightly destabilized in these tRNAs, or that there are local structural effects that cause shielding of the imino proton.

Pseudouridine (ψ)

Conformational studies of the ψ monomer have established that the nucleoside has little preference for either the *syn* or *anti* glycosyl conformation, although the *syn* may be slightly preferred for ψ in comparison to uridine (Nanda et al., 1974; Neumann et al., 1980). The *syn* conformational preference of ψ at the nucleoside level indicates that ψ would be a destabilizing modification compared to uridine in RNA. These same properties have lead to the proposal that ψ can act as a conformational switch because it has a low energy for *syn/anti* interconversion and because the N1-H imino proton in the *syn* conformation would be available to make the same hydrogen bonding interactions as the N3-H imino proton of uridine in the *anti* conformation (Lane et al., 1995; Neumann et al., 1980). However, there are a number of studies that indicate ψ is stabilizing in the context of an RNA oligonucleotide (Davis, 1995; Davis et al., submitted; Pieles et al., 1994). We used NMR to show that ψ stabilizes the A31-ψ39 base pair in *E. coli* tRNA[Phe] and provided a structural basis for the effects of ψ on codon-anticodon recognition (Davis and Poulter, 1991). NMR has also been used to show that ψs in different locations in tRNA all have a stable interaction involving N1-H and an oxygen H-bond acceptor; and we now know that this is almost certainly a water-mediated bridge to the phosphate backbone (Fig. 5) (Davis, 1995; Griffey et al., 1985; Hall and McLaughlin, 1991; Hall and McLaughlin, 1992; Roy et al., 1984). Molecular dynamics simulations have shown that ψ provides the appropriate structural environment to coordinate a water molecule between N1-H and the phosphate backbone (Auffinger et al., 1996; Auffinger and Westhof, 1997) (see also Chapter 6).

A recent study from our laboratory used the 1D ROESY (Kessler et al., 1989) NMR experiment to investigate the conformational preference for ψ nucleoside and also to measure the conformation of ψ in the AψA trimer (Davis et al., submitted). These experiments and a related study of AAψA (Davis

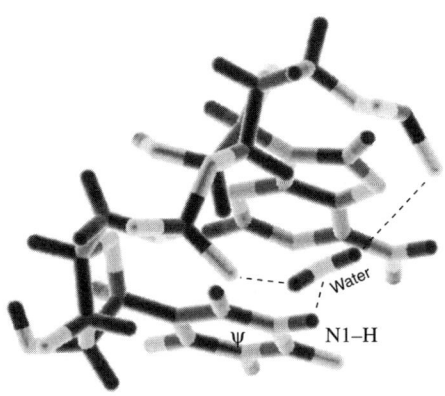

Figure 5. Three-dimensional structure of the pApψ dinucleotide step in a standard A-form geometry. The coordinated water molecule stabilizes the N1-H imino proton against facile exchange with bulk solvent and coordination to the phosphate backbone restricts the base conformation and the backbone 5' to the modification site.

for the 3'-endo sugar conformation, and dramatically increases stacking in a cooperative fashion.

NMR experiments have shown that ψ stabilizes local RNA stacking, with a structural water molecule playing a critical role in the mechanism for stabilization. The local structural stabilization leads to thermodynamic stabilization that can be measured by the standard temperature-dependent UV experiments. A somewhat surprising observation is that ψ modification in a single-stranded region significantly raises the overall T_m, which is a cooperative effect involving stacking in the base-paired region. In a model of the codon-anticodon interaction in tRNA (Fig. 6), we observed that a ψ three nucleotides away from the base-paired region increased the T_m by 2.6°C while a ψ within the base-paired region at position 35 in a model for the codon-anticodon interaction in tRNA^{Tyr} increased the T_m by 5.3°C (Davis et al., submitted). In the 17-nucleotide hairpin corresponding

1995) were designed to identify the minimal sequence of RNA that would show stabilization by ψ because the result of ψ modification in oligonucleotides and tRNA was at odds with the properties reported for ψ nucleoside. The choice of a ROESY experiment rather than the more common NOESY experiment was made so that we could compare the conformational behavior, as a function of temperature, for ψ and uridine nucleosides with the conformation of the respective nucleosides in RNA oligonucleotides. The ROESY has the advantage that the sign of the cross-relaxation is positive for all values of the molecular correlation time while the NOESY goes through a null in the intermediate correlation time regime (Bax and Davis, 1985; Ernst et al., 1987). The results for ψ were compared to the uridine nucleoside and also to AUA. The free ψ nucleoside prefers the 2'-endo, equatorial sugar conformation due to the decreased anomeric effect in C-nucleosides (Plavec et al., 1993; Thibaudeau et al., 1994a; Thibaudeau et al., 1994b). The 1D ROESY data on ψ nucleoside confirmed that there is little conformational preference, similar to the behavior of uridine, although the *syn* conformer may be slightly preferred as has been reported (Neumann et al., 1980). However, the conformational preference of ψ in the trinucleotide AψA is dramatically different than that seen for the nucleoside, with the ψ nucleoside in the trimer showing a strong conformational preference with limited interconversion between the *syn* and *anti* forms. Population of the *syn* conformer should also be accompanied by dynamic interconversion of the sugar with a preference for the 2'-endo form, but ψ substitution results in a strong preference

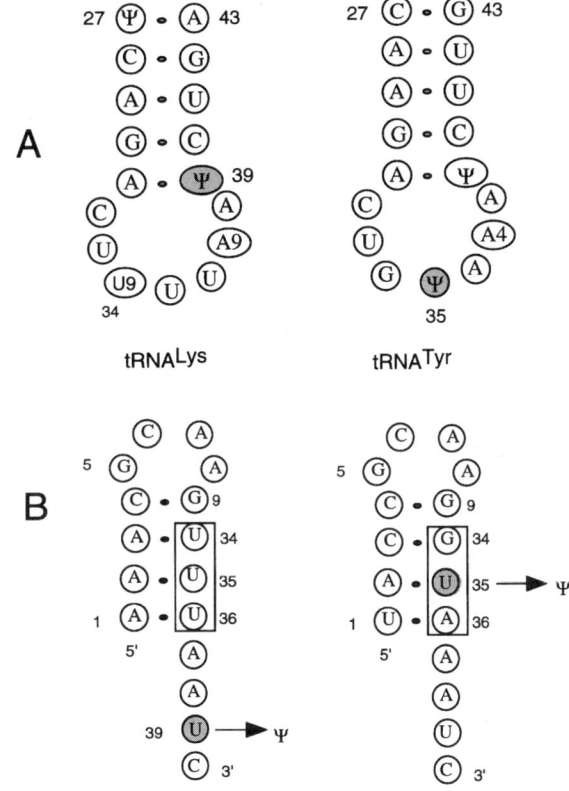

Figure 6. The anticodon domains of tRNA^{Lys} and tRNA^{Tyr} shown in panel A have pseudouridines at positions 39 and 35, respectively. As a model system for the codon-anticodon interaction where ψ would either be remote from the anticodon triplet as in tRNA^{Lys} or within the anticodon triplet as in tRNA^{Tyr}, the two RNA hairpins in panel B were used to demonstrate stabilization for ψ modification adjacent or within the double-stranded region. Pseudouridine results in an increase in the T_m of 2.6 and 5.5°C for the tRNA^{Lys} and tRNA^{Tyr} tetraloop hairpins, respectively.

to the anticodon stem-loop of tRNA^{Lys} (Fig. 6A), ψ-39 modification increases the T_m by 5°C and stabilizes stacking interactions on the 3' side of the anticodon (Durant and Davis, 1997 and submitted). The NMR NOE patterns for the tRNA^{Lys}-ψ39 hairpin define a structural environment for ψ analogous to that seen in the tRNA^{Phe} crystal structure (Westhof and Sundaralingam, 1986), and confirm that the low *anti* conformation of ψ seen in the model AψA and AAψA oligonucleotides is typical for ψ in native RNA.

There are still unanswered questions regarding the range of effects for ψ in different structural contexts as demonstrated by the modification of the minor yeast tRNA^{Ile} UAU anticodon to ψAψ (Senger et al., 1997). Pseudouridine-34 appears to be a recognition element for IleRS, a role consistent with the demonstration that inosine is a recognition element for the major tRNA$^{Ile}_{IAU}$. However, there is no definitive structural data to support the argument that ψ-34 would restrict wobble recognition of the methionine AUG codon (Keith et al., 1994; Yokoyama and Nishimura, 1995), while the role of inosine in this context is clear (Senger et al., 1997). Pseudouridine may be less likely to pair with G due to restriction to the low *anti* conformation, but a comparison of U and ψ in wobble pairs with G has not been done even though there are several known examples of G-ψ base pairs in the stems of tRNAs.

Dihydrouridine (D)

Dihydrouridine is one of the most common tRNA modifications (Limbach et al., 1994; Sprinzl et al., 1998) and has also been specifically placed within the peptidyl transferase loop in domain V of *E. coli* 23S RNA (Kowalak et al., 1995). The X-ray crystal structures of tRNA show that D adopts the 2'-endo conformation (Jack et al., 1976; Quigley and Rich, 1976) and it has been proposed that the specific role of D in RNA is to provide for extra flexibility and to extend the sugar-phosphate backbone to span loop regions (Westhof and Sundaralingam, 1986). The observation that D occurs with higher frequency in organisms that grow at low temperatures (psychrophiles) has led to the proposal that a higher percentage of D in these organisms allows for extra flexibility for RNAs that function near the freezing point of water (Dalluge et al., 1997). Dihydrouridine is structurally unique because reduction of the uridine 5–6 double bond eliminates the planar nature that is characteristic of other nucleoside bases, causing D to be a very poor stacker. NMR studies of the conformational preference of Dp and ApDpA indicate that the 2'-endo conformation of Dp is thermodynamically favored compared to Up by 1.5 kcal/mol and by a dramatic 5.3 kcal/mol for D in ApDpA compared to U in ApUpA (Dalluge et al., 1996). The magnitude of this modification effect on thermodynamic equilibrium is comparable to that of 2'-O-methylation, but in the opposite direction. The results from the trinucleotide experiments indicated that there is a direct base stacking effect on thermodynamic stability in addition to the sugar conformational effect. These two properties reinforce each other to shift the equilibrium strongly in favor of an unstacked 2'-endo conformation for both the D residue and the 5' neighboring adenosine.

The modified nucleoside acp^{3}U is often found in tRNAs with the consensus sequence D-acp^{3}U-A. Agris and coworkers investigated the conformational properties of this trinucleotide and found that in this sequence context D also strongly preferred the 2'-endo conformation (Stuart et al., 1996). A comparison of D versus U in the trimer showed that the sugar conformation of acp^{3}U was strongly affected by the adjacent D modification; the 2'-endo conformational preference for acp^{3}U was shifted from 63% with a 5'U to 72% for 5' D. 2D ROESY NMR experiments were used in both the study of ApDpA (Dalluge et al., 1996) and the study of D-acp^{3}U-A (Stuart et al., 1996) to determine solution models for D containing oligonucleotides. These two models each reflect the salient features of D modified RNA: the 2'-endo sugar conformations, and the absence of stacking interactions involving the D base.

Dihydrouridine stacks poorly compared to U and thermodynamically favors the 2'-endo conformation, but the question of whether D is involved in hydrogen bonding interactions is unresolved. The crystal structure of yeast tRNA^{Phe} (Westhof and Sundaralingam, 1986) shows D exposed to solvent and yeast tRNA^{Asp} (Westhof et al., 1985) has two Ds that are both exposed. NMR studies of *E. coli* tRNA$_f^{Met}$, tRNA^{Phe}, tRNA^{Lys}, and tRNA^{Ser} indicate that the N3-H imino protons of D residues in these tRNAs are involved in stable hydrogen bonding interactions (Davis et al., 1986; Griffey et al., 1986), and the crystal structure of *T. thermophilus* tRNA^{Ser} complexed with its cognate synthetase shows D20 participating in a G15-C48-D20 base triple (Biou et al., 1994). The NMR results and the recent crystal structure of the tRNA^{Ser}-SerRS complex indicate that D may be flexible enough to participate in a number of different structural roles in RNA.

5'-Formylcytidine (f^{5}C and f^{5}Cm)

5-Formylcytidine and the related 2'-O-methyl nucleoside have now been found at the wobble position in both mitochondrial and cytoplasmic eukaryotic tRNAs (Moriya et al., 1994; Pais de Barros et al., 1996). NMR studies have shown that the 5-formyl group results in dramatic stabilization of the 3'-endo sugar conformation with an enthalpy difference of 1.56 kcal/mol over the 2'-endo conformation (Kawai et al., 1994). Although it has yet to be measured, the additional 2'-O-methyl is expected to further increase this preference. The formyl group is extremely electron withdrawing, and this modification should stabilize the interaction between the lone pair electrons on O4' and the C5-C6 π^* antibonding orbitals, promoting the 3'-endo, axial conformation (Fig. 3). Kawai et al. favor this explanation even though the downfield NMR chemical shift of H6 expected to support this argument is confounded by shielding from the formyl carbonyl (Kawai et al., 1994). Constraining the modified cytidine to a strictly 3'-endo, axial conformation should restrict wobble pairing and eliminate interactions with C or U in the third position of the codon (Moriya et al., 1994; Pais de Barros et al., 1996), but the effect of f^{5}C on recognition of both G and A at the third codon position is still unclear (Takemoto et al., 1995).

N^4-Acetylcytidine and Its 2'-O-Methylated Derivative (ac^{4}C and ac^{4}Cm)

Acetylation of the cytidine 4-amino group shifts the sugar conformational equilibrium strongly in favor of 3'-endo. NMR measurement of sugar coupling constants as a function of temperature showed that for ac^{4}C the enthalpy difference between 2'-endo and 3'-endo conformations was 1.22 kcal/mol, compared to a difference of 0.37 kcal/mol for unmodified cytidine, an effect comparable in magnitude to that of s^{2}U compared to U (Kawai et al., 1992a; Kawai et al., 1991). Further methyl modification of the 2'-OH provides ac^{4}Cm, and an additional 0.3 kcal/mol stabilization for the 3'-endo nucleoside conformation. The doubly modified, ultra-stable ac^{4}Cm nucleoside has so far only been found in thermophilic archaea (Edmonds et al., 1991). The parent ac^{4}C nucleoside in E. coli tRNA^{Met} stabilizes the codon-anticodon interaction in ribosome binding experiments and prevents misreading of the AUA isoleucine codon (Stern and Schulman, 1978). In the tRNA anticodon loop the ac^{4}C nucleoside would preferentially exist in the 3'-endo, axial, stacked conformation, poised to pair with a complementary codon triplet. This rigid structure would also prevent the conformational flexibility needed to mispair with A. The verification that the intrinsic nucleoside conformational properties have the expected result in a bacterial elongator tRNA context suggests a general role for ac^{4}C in restricting wobble pairing.

5-Modified Uridines That Restrict Wobble Pairing (mnm^{5}s^{2}U and mnm^{5}U)

Fine tuning of the wobble interaction at position 34 appears to be critically important in tRNA as indicated by the large number of nucleoside modifications at this position (Agris, 1991). The acetylated cytidine family discussed above shows how powerful a simple modification to the base can be in affecting sugar conformation and codon-anticodon recognition. The 5-modified uridines are an unusually structurally diverse group (Limbach et al., 1994), where modifications can either restrict or expand wobble recognition depending on whether the modification promotes the 3'-endo or 2'-endo sugar conformation, respectively (Agris, 1996; Agris, 1991; Yokoyama and Nishimura, 1995). Although there are many structural questions to be resolved regarding the effect of wobble position modifications on the anticodon loop structure and function, the simple correlation between sugar conformation at position 34 and wobble pairing appears valid.

Uridine modification to promote the 3'-endo conformation and stabilize the codon-anticodon interaction is most effectively accomplished by 2-thiolation as described above for m^{5}s^{2}U. The wobble position 5-modified uridines that restrict wobble pairing usually also have a 2-thio group (Agris et al., 1973; Lustig et al., 1981; Sekiya et al., 1969; Seno et al., 1974). Because the s^{2}U modification is so effective in restricting the sugar conformation, the uridine 5-position modifications appear to serve specific functions: fine-tuning codon-anticodon recognition, participating in unique structural interactions within the anticodon (Agris et al., 1997), and serving as recognition elements for aminoacyl tRNA synthetases (Agris et al., 1973; Sylvers et al., 1993). In E. coli, the wobble position of tRNA$^{Glu}_{UUC}$, tRNA$^{Gln}_{UUG}$, tRNA$^{Lys}_{UUU}$ is modified to mnm^{5}s^{2}U, and closely related modifications are found for these tRNA isoacceptors in other bacteria and in eukaryotes (Sprinzl et al., 1998). This modified nucleoside is strongly restricted to the 3'-endo conformation and until recently this was thought to be entirely due to the 2-thione, with the sugar conformation being unaffected by the 5-substituent. However, it has recently been shown that wobble restriction is seen in the anticodon of tRNA$^{Arg}_{UCU}$ which contains only the mnm^{5}U modification without the sulfur (Sakamoto et al., 1993). The thermodynamic preferences at the nucleoside level are 1.12 kcal/mol for s^{2}U, 0.37 kcal/mol for U, and 0.39 kcal/mol for mnm^{5}U, indicating no stabilization effect for the mnm^5 substituent compared to uridine (Yokoyama et al., 1985) [Agris and coworkers have measured a similar trend, although the absolute numbers are larger (Agris et al., 1992)]. However, recent measurements on the 5'-phosphate showed that pmnm^{5}U favors the 3'-endo conformer by 0.65 kcal/mol, consistent with the wobble restriction seen in the tRNA$^{Arg}_{UCU}$ (Sakamoto et al., 1996). A mechanism involving an interaction between the positively charged amino side chain and the phosphate backbone was proposed, which would favor the low anti nucleoside conformation and restrict wobble pairing. This model is consistent with an NMR study on the anticodon of E. coli tRNA^{Glu}, which positioned the methylaminomethyl sidechain of mnm^{5}s^{2}U near the phosphate backbone (Yokoyama and Muramatsu, 1990). Whether the interaction between the 5 substituent and the phosphate backbone is a general structural motif for this family of modified nucleosides is unclear because not all are charged at physiological pH. For instance, the mammalian tRNA$^{Lys,3}_{UUU}$ isoacceptor that serves as the specific primer for HIV-1 reverse transcriptase has the modification mcm^{5}s^{2}U, which should not have the same interaction with the phosphate backbone as the bacterial tRNA^{Lys} modification (Isel et al., 1993; Litvak et al., 1994). Agris and coworkers have also proposed that other modifications within the anticodon may affect the conformation of the wobble position nucleotide and they propose that a salt bridge involving mnm^{5}s^{2}U is formed in tRNA^{Lys} that cannot be present in tRNA^{Glu} (Agris et al., 1997).

5-Modified Uridines That Expand Wobble Pairing (the xo^{5}U Family)

The xo^{5}U family of modified nucleosides found at position 34 of tRNAs expand wobble (Ishikura et al., 1971; Mitra et al., 1979; Murao et al., 1982; Samuelsson et al., 1980) by thermodynamically favoring the 2'-endo sugar conformation, which is less structurally restrictive (Yokoyama et al., 1985). tRNA isoacceptors for valine, serine, and alanine have been shown in ribosome binding experiments and in in vitro protein synthesis systems to have expanded wobble capacity when xo^{5}U nucleosides are at the wobble site. The xo^{5}U nucleotides without a sulfur at the 2 position of the uridine ring favor the 2'-endo conformation; pmo^{5}U has a ΔH of -0.72 kcal/mol for the 2'-endo/3'-endo equilibrium, and the ΔH for pcmo^{5}U is -0.67 kcal/mol compared to 0.09 kcal/mol for pU and $+1.1$ kcal/mol for pmnm^{5}s^{2}U (Yokoyama et al., 1985). NMR studies by Agris and coworkers show a similar trend for the nucleosides and clearly demonstrate that the xo^{5}U nucleosides greatly favor the 2'-endo conformation compared to the s^{2}xo^{5}U nucleosides (Agris et al., 1992; Sierzputowska-Gracz et al., 1987). The xo^{5}U nucleosides may promote the 2'-endo sugar conformation by a combination of steric repulsion with the 5'-phosphate and an inductive effect where electron donation to the uridine ring decreases the anomeric effect and the interaction between O4' and the H5-H6 double bond.

6-Threonylcarbamoyl-Adenosine (t^{6}A)

The hypermodified nucleoside t^{6}A has a number of interesting properties that have resulted in it being extensively studied. The nucleoside is widely distributed among tRNA isoacceptors and across phylogenetic kingdoms, yet is found exclusively at position 37 adjacent to the anticodon. The presence of t^{6}A strengthens the binding of tRNA to programmed ribosomes (Miller et al., 1976), and also the interaction with complementary tRNAs in anticodon-anticodon association experiments (Grosjean et al., 1976). The highly functionalized threonine side chain has the potential to interact in unique ways with other functional groups in the tRNA anticodon, serve as a specific coordination site for divalent metals, or act as a specific recognition element for other RNAs and proteins.

Schweizer and coworkers have measured the 3'-endo conformational equilibrium for t^{6}A and also the effect of metal ion binding. They observed that the nucleoside has a slight preference for the 2'-endo sugar conformation and that the nucleoside stacking self-association is comparable to 2'-deoxyadenosine (Reddy et al., 1981). NMR experiments were also used to show that the nucleoside and especially the nucleotide 5'-phosphate bind Mg^{2+} and Mn^{2+} site specifically, and with high affinity (Reddy et al., 1981; Schweizer et al., 1984). The coordination of Mg^{2+} and Mn^{2+} to t^{6}A has recently been questioned upon reinvestigation of the earlier potentiometric titrations (Varnagy et al., 1990). However, this study only investigated the nucleoside and not the 5'-phosphate, which is more biologically relevant and was shown to have higher affinity for Mg^{2+} and Mn^{2+} (Schweizer et al., 1984).

The proposed role of t^{6}A in stabilizing the codon-anticodon interaction has been expanded by Agris and coworkers, who have proposed a direct interaction between t^{6}A-37 and the mnm^{5}s^{2}U-34 nucleoside in tRNA^{Lys} (Agris et al., 1997). This interaction was proposed to explain a number of unique characteristics about the structure and chemistry in this special tRNA, but t^{6}A is found in many tRNAs that do not have a positively charged base modification at 34, suggesting that the putative interaction between t^{6}A-37 and the mnm^{5}s^{2}U-34 in tRNA^{Lys,3} represents a special case.

2′-O-Methyl-N^2,N^2-Dimethylguanosine (m$^2_2$Gm)

Modifications to purine nucleosides have modest effects on sugar conformation compared to those seen for pyrimidine modifications, such as s^{2}U, ac^{4}C, and f^{5}C. Even the ubiquititous 2′-O-methyl modification has a smaller effect because the steric interactions in the 2′-endo conformation are not as severe for purines. The archaeal thermophile nucleoside m$^2_2$Gm prefers the 3′-endo sugar conformation with a $\Delta\Delta H$ of 0.4 kcal/mol over G (Kawai et al., 1991). In the *anti* conformation, the dimethylamino group at position 2 of G sterically interferes with the 2′-O-methyl group for the 2′-endo conformation; Kawai et al. report that molecular modeling confirms this interaction. Although it is apparent that such a modification would be desirable in thermophilic organisms, and increases its response to increasing growth temperature (Kowalak et al., 1994), the specific structural interactions stabilized by m$^2_2$Gm putatively located at position 26 in tRNA have not been identified. However, McCloskey and coworkers have postulated a structural role for m$^2_2$Gm at position 26 based on stacking interactions in the crystal structure of tRNA^{Phe} (Kowalak et al., 1994).

Other Modifications

Several other modified nucleosides from tRNA are structurally intriguing, but either do not alter the NMR observable properties or have not been investigated. The prenylated nucleoside, i^{6}A, is found at position 37 of many eukaryotic and bacterial tRNAs when there is an adenosine at the third anticodon position (Sprinzl et al., 1998). The introduction of this hydrophobic group does not affect the nucleoside conformation either at the nucleoside level (Westhof et al., 1975), or in an A-i^{6}A-A trimer (Davis, unpublished data). Model studies of codon-anticodon interaction using the temperature-jump method indicate that i^{6}A has little effect on codon-anticodon recognition (Houssier and Grosjean, 1985). However, protein synthesis studies have shown a significant effect for the i^{6}A modification (Diaz and Ehrenberg, 1991; Esberg and Björk, 1995; Wilson and Roe, 1989).

The tricyclic nucleoside wyosine has been shown to stabilize stacking in the anticodon (Grosjean et al., 1978) and the structure within the tRNA^{Phe} anticodon is known from X-ray crystallography (Westhof and Sundaralingam, 1986). The solution conformations of wyosine and several isomers have been determined by NMR and generally prefer the 2′-endo conformation with little restriction about the glycosyl bond (Sierzputowska-Gracz et al., 1991). The effects of wyosine modification on RNA structure are likely to be dominated by the strong stacking propensity of the tricyclic ring.

Lysidine and queuosine are two functionally elaborate wobble position nucleosides. Lysidine is a C2 modified cytidine found at the wobble position of certain isoleucine tRNAs and changes the decoding specificity wherein the LAU anticodon does not misread the methionine AUG codon (Muramatsu et al., 1988). Lysidine and guanosine are positive recognition elements for *E. coli* IleRS (Muramatsu et al., 1992; Pallanck and Schulman, 1991) and negative determinants for MetRS (Muramatsu et al., 1988). The unique hydrogen bonding potential of lysidine is sufficient to explain both the synthetase recognition and the base-pairing effects, obviating the need to invoke conformational changes in the wobble position cytidine upon modification to lysidine. Queuosine, in contrast, has not been shown to change codon-anticodon recognition, or to serve as a protein recognition site despite the great deal of attention that this wobble position nucleoside has received (Björk, 1992, 1995). To our knowledge, no solution conformational studies have been done, but the fact that decoding is not affected by Q indicates that the elaborate functionality of Q has a neutral effect on sugar and glycosyl conformation.

CONCLUSIONS

Nucleoside modifications have distinct effects on the fundamental biophysical properties of nucleosides that in turn affect the structure, stability, and function of RNA. Sugar modification by 2′-O-methylation serves to generically stabilize RNA structure while the "non-natural" 2′-deoxy sugar modification has a general destabilizing effect (Aphasizhev et al., 1997). The mechanism of RNA stabilization by 2′-O-methylation illustrates a fundamental aspect of RNA structure; that is, the interplay between the ribose sugar conformation, the base glycosyl conformation, and RNA stacking interactions. Modifications that stabilize the 3′-endo sugar stabilize RNA stacking between bases; conversely, the sugar conformation can be affected by modifications in the base. Strongly electron withdrawing substituents in the base, 5-formylcytidine and N^4-acetylcytidine, for example, greatly influence the preferred sugar conformation and result in biochemically important restriction of the wobble interaction in tRNAs. Electron donating substituents for modified uridine nucleosides at the wobble position allow for more permissive wobble pairing by promoting the more flexible 2′-endo sugar conformation. Simple methylated nucleosides and pseudouridine can dramatically affect the local structure and global stability of RNA and are highly abundant in RNAs from high-temperature organisms. Dihydrouridine is highly conserved in loop regions where flexibility is important, is one of the few modified nucleosides that destabilize RNA, and is more abundant in low-temperature organisms. Modification of the primary nucleosides can serve to extend the capability of RNA through the addition of functional groups, but also by changing the fundamental biophysical properties that determine RNA structure and function. Modifications are found clustered in the "active regions" of RNAs: the anticodon of tRNA, the peptidyl transferase site of LSU RNA, and the conserved regions of snRNAs. These functional regions have been biochemically optimized with nucleoside modifications that provide local stability or flexibility, as needed, while maintaining critical hydrogen bonding interactions and hydrophobic environments. Our knowledge of the fundamental biophysical properties of modified nucleosides will be critical to understanding how modification has been used thoughout biology to optimize the function of biochemically critical processes involving RNA.

Acknowledgments. This work was supported by the American Cancer Society (JFRA-405), the National Science Foundation (MCB-9317196), and the National Institutes of Health (GM55508).

REFERENCES

Agris, P. F. 1991. Wobble position modified nucleosides evolved to select transfer RNA codon recognition: a modified-wobble hypothesis. *Biochimie* 73:1345–1349.

Agris, P. F. 1996. The importance of being modified: roles of modified nucleosides and Mg^{2+} in RNA structure and function, p. 74–129. *In* W. Cohn and K. Moldave (ed.), *Progress in Nucleic Acid Research and Molecular Biology*, vol. 53. Academic Press, San Diego, Calif.

Agris, P. F., R. Guenther, P. C. Ingram, M. M. Basti, J. W. Stuart, E. Sochacka, and A. Malkiewicz. 1997. Unconventional structure of tRNA^{Lys}SUU anticodon explains tRNA's role in bacterial and mammalian ribosomal frameshifting and primer selection by HIV-1. *RNA* 3:420–428.

Agris, P. F., H. Koh, and D. Söll. 1973. The effect of growth temperatures on the in vivo ribose methylation of *Bacillus stearothermophilus* transfer RNA. *Arch. Biochem. Biophys.* 154:277–282.

Agris, P. F., H. Sierzputowska-Gracz, W. Smith, A. Malkiewicz, E. Sochacka, and B. Nawrot. 1992. Thiolation of uridine carbon-2 restricts the motional dynamics of the transfer RNA wobble position nucleoside. *J. Am. Chem. Soc.* 114:2652–2656.

Agris, P. F., D. Söll, and T. Seno. 1973. Biological function of 2-thiouridine in *Escherichia coli* glutamic acid transfer ribonucleic acid. *Biochemistry* 12:4331–4337.

Altona, C., and M. Sundaralingam. 1973. Conformational analysis of the sugar ring in nucleosides and nucleotides. Improved method for the interpretation of proton magnetic resonance coupling constants. *J. Am. Chem. Soc.* 95:2333–2344.

Aphasizhev, R., A. Theobald-Dietrich, K. D. Kostyuk, S. N. Kochetkov, L. Kisselev, R. Giege, and F. Fasiolo. 1997. Structure and aminoacylation capacities of tRNA transcripts containing deoxyribonucleotides. *RNA* 3:893–904.

Auffinger, P., S. Louise-May, and E. Westhof. 1996. Hydration of C-H groups in tRNA. *Faraday Discuss.* 103:151–173.

Auffinger, P., and E. Westhof. 1997. RNA hydration: 3 ns of multiple molecular dynamics simulations of the solvated tRNA^{Asp} anticodon hairpin. *J. Mol. Biol.* 269:326–341.

Basti, M. M., J. W. Stuart, A. T. Lam, R. Guenther, and P. F. Agris. 1996. Design, biological activity and NMR-solution structure of a DNA analogue of yeast tRNA^{Phe} anticodon domain. *Nat. Struct. Biol.* 3:38–44.

Bax, A., and D. G. Davis. 1985. Practical aspects of two-dimensional transverse NOE spectroscopy. *J. Magn. Reson.* 63:207–213.

Biou, V., A. Yaremchuk, M. Tukalo, and S. Cusack. 1994. The 2.9 Å crystal structure of *T. thermophilus* seryl-tRNA synthetase complexed with tRNA^{Ser}. *Science* 263:1404–1410.

Björk, G. R. 1983. Modified nucleosides in RNA—their formation and function, p. 291–330. *In* D. Apirion (ed.), *Processing of RNA*. CRC Press Inc., Boca Raton, Fla.

Björk, G. R. 1992. The role of modified nucleosides in tRNA interactions, p. 23–85. *In* D. L. Hatfield, B. J. Lee, and R. M. Pirtle (ed.), *Transfer RNA in Protein Synthesis*. CRC Press, Ann Arbor, Mich.

Björk, G. R. 1995. Biosynthesis and function of modified nucleosides, p. 165–206. *In* D. Söll, and U. L. RajBhandary (ed.), *tRNA: Structure, Biosynthesis, and Function*. ASM Press, Washington, D.C.

Choi, B. S., and A. G. Redfield. 1986. NMR study of isoleucine transfer RNA from *Thermus thermophilus*. *Biochemistry* 25:1529–1534.

Cramer, F., E. M. Gottschalk, H. Matzura, K.-H. Scheit, and H. Sternbach. 1971. The synthesis of the alternating copolymer poly[r(A-s^{4}U)] by RNA polymerase of *Escherichia coli*. *Eur. J. Biochem.* 19:379–385.

Dallugue, J. J., T. Hamamoto, K. Horikoshi, R. Y. Morita, K. O. Stetter, and J. A. McCloskey. 1997. Posttranscriptional modification of transfer RNA in psychrophilic bacteria. *J. Bacteriol.* 179:1918–1923.

Dallugue, J. J., T. Hashizume, A. E. Sopchik, J. A. McCloskey, and D. R. Davis. 1996. Conformational flexibility in RNA: the role of dihydrouridine. *Nucleic Acids Res.* 24:1073–1079.

Davanloo, P., M. Sprinzl, K. Watanabe, M. Albani, and H. Kersten. 1979. Role of ribothymidine in the thermal stability of transfer RNA as monitored by proton magnetic resonance. *Nucleic Acids Res.* 6:1571–1581.

Davis, D. R. 1995. Stabilization of RNA stacking by pseudouridine. *Nucleic Acids Res.* 23:5020–5026.

Davis, D. R. Unpublished data.

Davis, D. R., R. H. Griffey, Z. Yamaizumi, S. Nishimura, and C. D. Poulter. 1986. ^{15}N-labeled tRNA: identification of dihydrouridine in *E. coli* tRNA^{fMet}, tRNA^{Lys}, and tRNA^{Phe} by ^{1}H-^{15}N two-dimensional NMR. *J. Biol. Chem.* 261:3584–3587.

Davis, D. R., and C. D. Poulter. 1991. ^{1}H-^{15}N NMR studies of *E. coli* tRNA^{Phe} from *hisT* mutants: a structural role for pseudouridine. *Biochemistry* 30:4223–4231.

Davis, D. R., C. A. Veltri, and L. Nielsen. An RNA model system for investigation of pseudouridine stabilization of the codon-anticodon interaction in tRNA^{Lys}, tRNA^{His} and tRNA^{Tyr}. Submitted for publication.

de Leeuw, F. A. A. M., and C. Altona. 1983. Computer-assisted pseudorotation analysis of five-membered rings by means of proton spin-spin coupling constants: program PSEUROT. *J. Comp. Chem.* 4:428–437.

Diaz, I., and M. Ehrenberg. 1991. ms^{2}i^{6}A deficiency enhances proofreading in translation. *J. Mol. Biol.* 222:1161–1171.

Durant, P. C., and D. R. Davis. 1997. The effect of pseudouridine and pH on the structure and dynamics of the anticodon stem-loop of tRNALys,3. *Nucleic Acids Symp. Ser.* **36**:56–57.

Durant, P. C., and D. R. Davis. Structure and dynamics of the anticodon stem-loop of tRNA^{Lys}. Structural stabilization of the HIV reverse transcriptase primer by an A$^+$-C base pair and by pseudouridine. Submitted for publication.

Edmonds, C. G., P. F. Crain, R. Gupta, T. Hashizume, C. H. Hocart, J. A. Kowalak, S. A. Pomerantz, K. O. Stetter, and J. A. McCloskey. 1991. Posttranscriptional modification of tRNA in thermophilic archaea (Archaebacteria). *J. Bacteriol.* **173**:3138–3148.

Egert, E., H. J. Lindner, W. Hillen, and M. C. Bohm. 1980. Influence of substituents at the 5-position on the structure of uridine. *J. Am. Chem. Soc.* **102**:3707–3713.

Ernst, R. R., G. Bodenhausen, and A. Wokaun. 1987. *Principles of Nuclear Magnetic Resonance in One and Two Dimensions.* Oxford University Press, Oxford, United Kingdom.

Esberg, B., and G. R. Björk. 1995. The methylthio group (ms^2) of N^6-(4-hydroxyisopentenyl)-2-methylthioadenosine (ms^{2}io^{6}A) present next to the anticodon contributes to the decoding efficiency of the tRNA. *J. Bacteriol.* **177**:1967–1975.

Griffey, R. H., D. R. Davis, Z. Yamaizumi, S. Nishimura, A. Bax, B. Hawkins, and C. D. Poulter. 1985. ^{15}N-labeled *E. coli* tRNA^{fMet}, tRNA^{Glu}, tRNA^{Tyr}, and tRNA^{Phe}: double resonance and two-dimensional NMR of N1-labeled pseudouridine. *J. Biol. Chem.* **260**:9734–9741.

Griffey, R. H., D. R. Davis, Z. Yamaizumi, S. Nishimura, B. L. Hawkins, and C. D. Poulter. 1986. ^{15}N-labeled tRNA: identification of 4-thiouridine in *Escherichia coli* tRNA^{Ser} and tRNA^{Tyr} by ^{1}H-^{15}N two-dimensional NMR spectroscopy. *J. Biol. Chem.* **261**:12074–12078.

Grosjean, H., and C. Houssier. 1990. Codon recognition: evaluation of the effects of modified bases in the anticodon loop of tRNA using the temperature jump relaxation method, p. A255–A297. *In* C. W. Gehrke and K. C. T. Kuo (ed.), *Chromatography and Modification of Nucleosides, Part A.* Elsevier Science Publishers, Amsterdam, The Netherlands.

Grosjean, H., D. G. Söll, and D. M. Crothers. 1976. Studies of the complex between transfer RNAs with complementary anticodons. I. Origins of enhanced affinity between complementary triplets. *J. Mol. Biol.* **103**:499–519.

Grosjean, H. J., S. de Henau, and D. M. Crothers. 1978. On the physical basis for ambiguity in genetic coding interactions. *Proc. Natl. Acad. Sci. USA* **75**:610–614.

Gu, J., and R. Reddy. 1997. Small RNA database. *Nucleic Acids Res.* **25**:98–102.

Haasnoot, C. A. G., F. A. A. M. De Leeuw, and C. Altona. 1980. The relationship between proton-proton NMR coupling constants and substituent electronegativities. I. An empirical generalization of the Karplus equation. *Tetrahedron* **36**:2783–2792.

Hall, K. B., and L. W. McLaughlin. 1991. Properties of a U1/mRNA 5' splice site duplex containing pseudouridine as measured by thermodynamic and NMR methods. *Biochemistry* **30**:1795–1801.

Hall, K. B., and L. W. McLaughlin. 1992. Properties of pseudouridine N1 imino protons located in the major groove of an A-form RNA duplex. *Nucleic Acids Res.* **20**: 1883–1889.

Hanna, M. M. 1989. Photoaffinity cross-linking methods for studying RNA-protein interactions. *Methods Enzymol.* **180**: 383–409.

Hare, D. R., S. Ribeiro, D. E. Wemmer, and B. R. Reid. 1985. Complete assignment of the imino protons of *Escherichia coli* valine transfer RNA: two-dimensional NMR studies in water. *Biochemistry* **24**:4300–4306.

Horie, N., M. Hara-Yokoyama, S. Yokoyama, K. Watanabe, Y. Kuchino, S. Nishimura, and T. Miyazawa. 1985. Two tRNA^{Ile} species from an extreme thermophile, *Thermus thermophilus* HB8: effect of 2-thiolation of ribothymidine on the thermostability of tRNA. *Biochemistry* **24**:5711–5715.

Houssier, C., P. Degee, K. Nicoghosian, and H. Grosjean. 1988. Effect of uridine dethiolation in the anticodon triplet of tRNA(Glu) on its association with tRNA(Phe). *J. Biomol. Struct. Dyn.* **5**:1259–1266.

Houssier, C., and H. Grosjean. 1985. Temperature jump relaxation studies on the interactions between transfer RNAs wih complementary anticodons. The effect of modified bases adjacent to the anticodon triplet. *J. Biomol. Struct. Dyn.* **3**:387–399.

Inoue, H., Y. Hayase, A. Imura, S. Iwai, K. Miura, and E. Ohtsuka. 1987. Synthesis and hybridization studies on two complementary nona(2'-O-methyl)ribonucleotides. *Nucleic Acids Res.* **15**:6131–6148.

Isel, C., R. Marquet, G. Keith, C. Ehresmann, and B. Ehresmann. 1993. Modified nucleotides of tRNA^{Lys,3} modulate primer/template loop-loop interaction in the initiation complex of HIV-1 reverse transcription. *J. Biol. Chem.* **34**:25269–25272.

Ishikura, H., Y. Yamada, and S. Nishimura. 1971. Structure of serine tRNA from *Escherichia coli*. 1. Purification of serine tRNAs with different codon responses. *Biochim. Biophys. Acta* **228**:471–481.

Jack, A., J. E. Ladner, and A. Klug. 1976. Crystallographic refinement of yeast phenylalanine transfer RNA at 2.5 Å resolution. *J. Mol. Biol.* **108**:619–649.

Juaristi, E., and G. Cuevas. 1995. *The Anomeric Effect.* CRC Press, Boca Raton, Fla.

Kawai, G., H. Ue, M. Yasuda, K. Sakamoto, T. Hashizume, J. A. McCloskey, T. Miyazawa, and S. Yokoyama. 1991. Relation between functions and conformational characteristics of modified nucleosides found in tRNAs. *Nucleic Acids Res. Symp. Ser.* **25**:49–50.

Kawai, G., T. Hashizume, M. Yasuda, T. Miyazawa, J. A. McCloskey, and S. Yokoyama. 1992a. Conformational rigidity of N^4-acetyl-2'-O-methylcytidine found in tRNA of extremely thermophylic archaebacteria (archaea). *Nucleosides Nucleotides* **11**: 759–771.

Kawai, G., Y. Yamamoto, T. Kamimura, T. Masegi, M. Sekine, T. Hata, T. Iimori, T. Watanabe, T. Miyazawa, and S. Yokoyama. 1992b. Conformational rigidity of specific pyrimidine residues in tRNA arises from posttranscriptional modifications that enhance steric interaction between the base and the 2'-hydroxyl group. *Biochemistry* **31**:1040–1046.

Kawai, G., T. Yokogawa, K. Nishikawa, T. Ueda, T. Hashizume, J. A. McCloskey, S. Yokoyama, and K. Watanabe. 1994. Conformational properties of a novel modified nucleoside, 5-formylcytidine, found at the first position of the anticodon of bovine mitochondrial tRNAMet. *Nucleosides Nucleotides* **13**: 1189–1199.

Keith, G., U. Englisch, F. Cramer, and F. Fasiolo. 1994. Does a ψ syn conformation of the wobble base determine the codon and amino acid specificity of a yeast isoleucine transfer RNA?, p. 9-2. *In Proceedings of the EMBO-CNRS Workshop on Nucleotide Modification and Base Conversion of RNA*, Assois, France.

Kessler, H., U. Anders, G. Gemmecker, and S. Steuernagel. 1989. Improvement of NMR experiments by employing semiselective half-gaussian-shaped pulses. *J. Magn. Reson.* **85**:1–14.

Kowalak, J. A., E. Bruenger, and J. A. McCloskey. 1995. Posttranscriptional modification of the central loop of domain V in *Escherichia coli* 23S ribosomal RNA. *J. Biol. Chem.* **270**: 17758–17764.

Kowalak, J. A., J. J. Dalluge, J. A. McCloskey, and K. O. Stetter. 1994. The role of posttranscriptional modification in stabilization of transfer RNA from hyperthermophiles. *Biochemistry* 33: 7869–7876.

Kumar, R. K., and D. R. Davis. 1997a. The effects of 2-thiouridine and 4-thiouridine on sugar conformation and base stacking in RNA oligonucleotides. *Nucleic Acids Res.* 25:1272–1280.

Kumar, R. K., and D. R. Davis. 1997b. Structural studies of 2-thiouridine in RNA. *Nucleosides Nucleotides* 16:1469–1472.

Laing, L. G., and D. E. Draper. 1994. Thermodynamics of RNA folding in a conserved ribosomal RNA domain. *J. Mol. Biol.* 237: 560–576.

Lane, B. G., J. Ofengand, and M. W. Gray. 1995. Pseudouridine and O2′-methylated nucleosides. Significance of their selective occurrence in rRNA domains that function in ribosome-catalyzed synthesis of the peptide bonds in proteins. *Biochimie* 77:7–15.

Lee, C.-H., F. S. Ezra, N. S. Kondo, R. H. Sarma, and S. S. Danyluk. 1976. Conformation properties of dinucleoside monophosphates: dipurines and dipyrimidines. *Biochemistry* 15: 3627–3638.

Lee, C. H., and I. Tinoco. 1980. Conformation studies of 13 trinucleoside phosphates by 360 MHz PMR spectroscopy. A bulged base conformation. I. Base protons and H1′ protons. *Biophys. Chem.* 11:283–294.

Lee, C. H., and I. Tinoco. 1977. Studies of the conformation of modified dinucleoside phosphates containing 1,N^6-ethenoadenosine and 2′-O-methylcytidine by 360 MHz ^{1}H nuclear magnetic resonance spectroscopy. Investigation of the solution conformations of dinucleoside phosphates. *Biochemistry* 16:5403–5414.

Limbach, P. A., P. F. Crain, and J. A. McCloskey. 1994. Summary: the modified nucleosides of RNA. *Nucleic Acids Res.* 22: 2183–2196.

Litvak, S., L. Sarih-Cottin, M. Fournier, M. Andreola, and L. Tarrago-Litvak. 1994. Priming of HIV replication by tRNA^{Lys,3}: role of reverse transcriptase. *Trends Biochem. Sci.* 19:114–118.

Lodmell, J. S., and A. E. Dahlberg. 1997. A conformational switch in *Escherichia coli* 16S ribosomal RNA during decoding of messenger RNA. *Science* 277:1262–1267.

Lustig, F., P. Elias, T. Axberg, T. Samuelsson, I. Titawella, and U. Lagerkvist. 1981. Codon reading and translational error. Reading of the glutamine and lysine codons during protein synthesis in vitro. *J. Biol. Chem.* 256:2635–2643.

Maden, B. E. H. 1990. The numerous modified nucleotides in eukaryotic ribosomal RNA. *Prog. Nucleic Acids Res. Mol. Biol.* 39:241–303.

Mazumdar, S. K., W. Saenger, and K. H. Scheit. 1974. Molecular structure of poly-2-thiouridylic acid, a double helix with nonequivalent polynucleotide chains. *J. Mol. Biol.* 85:213–229.

Miller, J. P., Z. Hussain, and M. P. Schweizer. 1976. The involvement of the anticodon adjacent modified nucleoside N-[9-(β-d-ribofuranosyl)purine-6-ylcarbamoyl]threonine in the biological function of *E. coli* tRNA^{Ile}. *Nucleic Acids Res.* 3:1185–1201.

Mitra, S. K., F. Lustig, B. Akesson, T. Axberg, P. Elias, and U. Lagerkvist. 1979. Relative efficiency of anticodons in reading the valine codons during protein synthesis in vitro. *J. Biol. Chem.* 254:6397–6401.

Moriya, J., T. Yokogawa, K. Wakita, T. Ueda, K. Nishikawa, P. F. Crain, T. Hashizume, S. C. Pomerantz, J. A. McCloskey, G. Kawai, N. Hayashi, S. Yokoyama, and K. Watanabe. 1994. A novel modified nucleoside found at the first position of the anticodon of methionine tRNA from bovine liver mitochondria. *Biochemistry* 33:2234–2239.

Muramatsu, T., T. Miyazawa, and S. Yokoyama. 1992. Recognition of the nucleoside in the first position of the anticodon of isoleucine tRNA by isoleucyl-tRNA synthetase from *Escherichia coli*. *Nucleosides Nucleotides* 11:719–730.

Muramatsu, T., K. Nishikawa, F. Nemoto, Y. Kuchino, S. Nishimura, T. Miyazawa, and S. Yokoyama. 1988. Codon and amino-acid specificities of a transfer RNA are both converted by a single post-transcriptional modification. *Nature* 336: 179–181.

Murao, K., T. Hasegawa, and H. Ishikura. 1982. Nucleotide sequence of valine tRNAmo^{5}UAC from *Bacillus subtilis*. *Nucleic Acids Res.* 10:715–718.

Nanda, R. K., R. Tewari, G. Govil, and I. C. P. Smith. 1974. The conformation of β-pseudouridine about the glycosidic bond as studied by ^{1}H homonuclear overhauser measurements and molecular orbital calculations. *Can. J. Chem.* 52:371–375.

Neuhaus, D., and M. Williamson. 1989. *The Nuclear Overhauser Effect in Structural and Conformational Analysis*. VCH Publishers, New York, N.Y.

Neumann, J. M., J. M. Bernassau, M. Gueron, and S. Tran-Dinh. 1980. Comparative conformations of uridine and pseudouridine and their derivatives. *Eur. J. Biochem.* 108:457–463.

Noggle, J. H., and R. E. Shirmer. 1971. *The Nuclear Overhauser Effect: Chemical Applications*. Academic Press, New York, N.Y.

Noller, H. F. 1993. On the origin of the ribosome: coevolution of subdomains of tRNA and rRNA, p. 137–156. *In* R. F. Gesteland and J. F. Atkins (ed.), *The RNA World*. Cold Spring Harbor Laboratory Press, Cold Spring Harbor, N.Y.

Pais de Barros, J.-P., G. Keith, C. El Adlouni, A.-L. Glasser, G. Mack, G. Dirheimer, and J. Desgres. 1996. 2′-O-methyl-5-formylcytidine (f^{5}Cm), a new modified nucleotide at the "wobble" position of two cytoplasmic tRNAs^{Leu}(NAA) from bovine liver. *Nucleic Acids Res.* 24:1489–1496.

Pallanck, L., and L. H. Schulman. 1991. Anticodon-dependent aminoacylation of a noncognate tRNA with isoleucine, valine, and phenylalanine in vivo. *Proc. Natl. Acad. Sci. USA* 88: 3872–3876.

Pieles, U. B., K. Bohmann, S. Weston, S. O'Loughlin, V. Adam, and B. S. Sproat. 1994. New and convenient protection system for pseudouridine, highly suitable for solid-phase oligoribonucleotide synthesis. *J. Chem. Soc. Perkin Trans.* 1:3423–3429.

Plavec, J., C. Thibaudeau, and J. Chattopadhyaya. 1994a. How does the 2′-hydroxy group drive the pseudorotational equilibrium in nucleoside and nucleotide by the tuning of the 3′-gauche effect? *J. Am. Chem. Soc.* 116:6558–6560.

Plavec, J., T. Thibaudeau, G. Viswanadham, C. Sund, and J. Chattopadhyaya. 1994b. How does the 3′-phosphate drive the sugar conformation in DNA? *J. Chem. Soc. Chem. Commun.*, p. 781–783.

Plavec, J., W. Tong, and J. Chattopadhyaya. 1993. How do the gauche and anomeric effects drive the pseudorotational equilibrium of the pentofuranose moiety of nucleosides? *J. Am. Chem. Soc.* 115:9734–9746.

Plesiewicz, E., E. Stepien, K. Bolewska, and K. L. Wierzchowski. 1976. Stacking self-association of pyrimidine nucleosides and cytosines: effects of methylation and thiolation. *Nucleic Acids Res.* 3:1295–1306.

Quigley, G. J., and A. Rich. 1976. Structural domains of transfer RNA molecules. *Science* 194:794–806.

Reddy, P. R., D. W. Hamill, G. B. Chheda, and M. P. Schweizer. 1981. On the function of N-[(9-β-d-ribo-furanosyl-purine-6-ylcarbamoyl)]threonine in transfer ribonucleic acid. Metal ion binding studies. *Biochemistry* 20:4979–4986.

Rosemeyer, H., G. Toth, B. Golankiewicz, Z. Kazimierczuk, W. Bourgeois, U. Kretschmer, H.-P. Muth, and F. Seela. 1990. Syn-anti conformational analysis of regular and modified nucleosides by 1D ^{1}H NOE difference spectroscopy: a simple graphical method based on conformationally rigid molecules. *J. Org. Chem.* 55:5784–5790.

Roy, S., M. Z. Papastavros, and A. G. Redfield. 1982. Nuclear Overhauser effect study of yeast aspartate transfer ribonucleic acid. *Biochemistry* 21:6081–6088.

Roy, S., M. Z. Papastavros, V. Sanchez, and A. G. Redfield. 1984. Nitrogen-15-labeled yeast tRNA[Phe]. Double and two-dimensional heteronuclear NMR of guanosine and uracil ring NH groups. *Biochemistry* 23:4395–4400.

Saenger, W. 1984. *Principles of Nucleic Acid Structure.* Springer-Verlag, New York, N.Y.

Sakamoto, K., G. Kawai, T. Niimi, T. Satoh, M. Sekine, Z. Yamaizumi, S. Nishimura, T. Miyazawa, and S. Yokoyama. 1993. A modified uridine in the first position of the anticodon of a minor species of arginine tRNA, the argU gene product, from *Escherichia coli*. *Eur. J. Biochem.* 216:369–375.

Sakamoto, K., G. Kawai, S. Watanabe, T. Niimi, N. Hayashi, Y. Muto, K. Watanabe, T. Satoh, M. Sekine, and S. Yokoyama. 1996. NMR studies of the effects of the 5′-phosphate group on conformational properties of 5-methylaminomethyluridine found in the first position of the anticodon of *Escherichia coli* tRNA[Arg,4]. *Biochemistry* 35: 6533–6538.

Samuelsson, T., P. Elias, F. Lustig, T. Axberg, G. Folsch, B. Akesson, and U. Lagerkvist. 1980. Aberrations of the classic codon reading scheme during protein synthesis in vitro. *J. Biol. Chem.* 255:4583–4588.

Scheit, K. H., and P. Faerber. 1975. The effects of thioketo substitution upon uracil-adenine interactions in polyribonucleotides. *Eur. J. Biochem.* 50:549–555.

Schweizer, M. P., N. De, M. Pulsipher, M. Brown, P. R. Reddy, C. R. Petrie, and G. B. Chheda. 1984. Quantitative aspects of metal ion binding to certain transfer RNA anticodon loop modified nucleosides. *Biochim. Biophys. Acta* 802:352–361.

Sekiya, T., K. Takeishi, and T. Ukita. 1969. Specificity of yeast glutamic acid transfer RNA for codon recognition. *Biochim. Biophys. Acta* 182:411–426.

Senger, B., S. Auxilien, U. Englisch, F. Cramer, and F. Fasiolo. 1997. The modified wobble base inosine in yeast tRNA[Ile] is a positive determinant for aminoacylation by isoleucyl-tRNA synthetase. *Biochemistry* 36:8269–8275.

Seno, T., P. F. Agris, and D. Söll. 1974. Involvement of the anticodon region of *Escherichia coli* tRNA[Gln] and tRNA[Glu] in the specific interaction with cognate aminoacyl-tRNA synthetase. *Biochim. Biophys. Acta* 349:328–338.

Sierzputowska-Gracz, H., R. H. Guenther, P. F. Agris, W. Folkman, and B. Golankiewicz. 1991. Structure and conformation of the hypermodified purine nucleoside wyosine and its isomers: a comparison of coupling constants and distance geometry solutions. *Magn. Reson. Chem.* 29:885–892.

Sierzputowska-Gracz, H., E. Sochacka, A. Malkiewicz, K. Kuo, C. W. Gehrke, and P. F. Agris. 1987. Chemistry and structure of modified uridines in the anticodon, wobble position of transfer RNA are determined by thiolation. *J. Am. Chem. Soc.* 109: 7171–7177.

Smith, W. S., B. Nawrot, A. Malkiewicz, and P. F. Agris. 1992a. RNA modified uridines. VI. Conformations of 3-[3-(S)-amino-3-carboxypropyl]uridine (acp^{3}U) from tRNA and 1-methyl-3-[3-(S)-amino-3-carboxypropyl]pseudouridine (m^{1}acp$^3\psi$) from rRNA. *Nucleosides Nucleotides* 11:1683–1694.

Smith, W. S., H. Sierzputowska-Gracz, E. Sochacka, A. Malkiewicz, and P. F. Agris. 1992b. Chemistry and structure of modified uridine dinucleosides are determined by thiolation. *J. Am. Chem. Soc.* 114:7989–7997.

Sowers, L. C., B. R. Shawk, and W. D. Sedwick. 1987. Base stacking and molecular polarizability: effect of a methyl group in the 5-position of pyrimidines. *Biochem. Biophys. Res. Commun.* 148:790–794.

Sprinzl, M., C. Horn, M. Brown, A. Ioudovitch, and S. Steinberg. 1998. Compilation of tRNA sequences and sequences of tRNA genes. *Nucleic Acids Res.* 26:148–153.

Stern, L., and L. H. Schulman. 1978. The role of the minor base N^4-acetylcytidine in the function of the *Escherichia coli* noninitiator methionine transfer RNA. *J. Biol. Chem.* 253:6132–6139.

Stuart, J. W., M. M. Basti, W. S. Smith, B. Forrest, R. Guenther, H. Sierzputowska-Gracz, B. Nawrot, A. Malkiewicz, and P. F. Agris. 1996. Structure of the trinucleotide D-acp3U-A with co-ordinated Mg^{2+} demonstrates that modified nucleosides contribute to regional conformations of RNA. *Nucleosides Nucleotides* 15:1009–1028.

Sylvers, L. A., K. C. Rogers, M. Shimizu, E. Ohtsuka, and D. Söll. 1993. A 2-thiouridine derivative in tRNAGlu is a positive determinant for aminoacylation by *E. coli* glutamyl-tRNA synthetase. *Biochemistry* 32:3836–3841.

Takemoto, C., T. Yokogawa, L. Benkowski, L. L. Spremulli, T. A. Ueda, K. Nishikawa, and K. Watanabe. 1995. The ability of bovine mitochondrial transfer RNA[Met] to decode AUG and AUA codons. *Biochimie* 77:104–108.

Thibaudeau, C., J. Plavec, and J. Chattopadhyaya. 1994a. Quantitation of the anomeric effect in adenosine and guanosine by comparison of the thermodynamics of the pseudorotational equilibrium of the pentofuranose moiety in N- and C-nucleosides. *J. Am. Chem. Soc.* 116:8033–8037.

Thibaudeau, C., J. Plavec, K. A. Watanabe, and J. Chattopadhyaya. 1994b. How do the aglycones drive the pseudorotational equilibrium of the pentofuranose moiety in C-nucleosides? *J. Chem. Soc. Chem. Commun.*, p. 537–540.

Thibaudeau, C., J. Plavec, N. Garg, A. Papchikhin, and J. Chattopadhyaya. 1994c. How does the electronegativity of the substituent dictate the strength of the Gauche effect? *J. Am. Chem. Soc.* 116:4038–4043.

Thibaudeau, C., J. Plavec, and J. Chattopadhyaya. 1996. Quantitation of the pD dependent thermodynamics of the N-S pseudorotational equilibrium of the pentofuranose moiety in nucleosides gives a direct measurement of the strength of the tunable anomeric effect and the pKa of the nucleobase. *J. Org. Chem.* 61:266–286.

Uhl, W., J. Reiner, and H. G. Gassen. 1983. On the conformation of 5-substituted uridines as studied by proton magnetic resonance. *Nucleic Acids Res.* 11:1167–1180.

Varani, G., and I. Tinoco. 1991. RNA structure and NMR spectroscopy. *Q. Rev. Biophys.* 24:479–532.

Varnagy, K., M. Jezowska-Bojczuk, J. Swiatek, H. Kozlowski, I. Sovago, and R. W. Adamiak. 1990. Metal binding ability of hypermodified nucleosides of tRNA. Potentiometric and spectroscopic studies on the metal complexes of N-[(9-β-d-ribofuranosylpurin-6-yl)carbamoyl]threonine. *J. Inorg. Biochem.* 40: 357–363.

Wang, S., and E. T. Kool. 1995. Origins of the large differences in stability of DNA and RNA helices: C-5 methyl and 2′-hydroxyl effects. *Biochemistry* 34:4125–4132.

Westhof, E., P. Dumas, and D. Moras. 1985. Crystallographic refinement of yeast aspartic acid transfer RNA. *J. Mol. Biol.* 184: 119.

Westhof, E., O. Roder, I. Croneiss, and H. D. Ludemann. 1975. Ribose conformations in the common purine (β) ribosides, in some antibiotic nucleosides, and in some isopropylidene derivatives: a comparison. *Z. Naturforsch.* 30:131–140.

Westhof, E., and M. Sundaralingam. 1986. Restrained refinement of the monoclinic form of yeast phenylalanine transfer RNA. Temperature factors and dynamics, coordinated waters, and base pair propeller twist angles. *Biochemistry* 25:4868–4878.

Wilson, R. K., and B. A. Roe. 1989. Presence of the hypermodified nucleotide N6-(Δ^2-isopentenyl)-2-methylthioadenosine prevents codon misreading by *Escherichia coli* phenylalanyl-transfer RNA. *Proc. Natl. Acad. Sci. USA* **86:**409–413.

Wuthrich, K. 1986. *NMR of Proteins and Nucleic Acids.* Wiley-Interscience, New York, N.Y.

Yamamoto, Y., S. Yokoyama, T. Miyazawa, K. Watanabe, and S. Higuchi. 1983. NMR analyses on the molecular mechanism of the conformational rigidity of 2-thioribothymidine, a modified nucleoside in extreme thermophile tRNAs. *FEBS Lett.* **157:**95–99.

Yokoyama, S., and T. Muramatsu. 1990. NMR analysis of structures and functions of modified nucleosides in transfer ribonucleic acids. *Nucleosides Nucleotides* **9:**303–310.

Yokoyama, S., and S. Nishimura. 1995. Modified nucleosides and codon recognition, p. 207–224. *In* D. Söll and U. L. Raj Bhandary (ed.), *tRNA: Structure, Biosynthesis, and Function.* ASM Press, Washington, D.C.

Yokoyama, S., T. Watanabe, K. Murao, H. Ishikura, Z. Yamaizumi, S. Nishimura, and T. Miyazawa. 1985. Molecular mechanism of codon recognition by tRNA species with modified uridine in the first position of the anticodon. *Proc. Natl. Acad. Sci. USA* **82:**4905–4909.

Yokoyama, S., Z. Yamaizumi, S. Nishimura, and T. Miyazawa. 1979. ^{1}H NMR studies on the conformational characteristics of 2-thiopyrimidine nucleotides found in transfer RNAs. *Nucleic Acids Res.* **6:**2611–2627.

Yu, Y.-T., and J. A. Steitz. 1997. A new strategy for introducing photoactivatable 4-thiouridine into specific positions in a long RNA molecule. *RNA* **3:**807–810.

Chapter 6

Effects of Pseudouridylation on tRNA Hydration and Dynamics: a Theoretical Approach

PASCAL AUFFINGER AND ERIC WESTHOF

All the essential biophysical methods used for the characterization of the various structural and functional roles of modified nucleotides are discussed in the different chapters of this book. Among theoretical methods, molecular dynamics (MD) simulations can give valuable insight, at the atomic level, not only on the intrinsic dynamics of RNA structures containing modified nucleotides, but also on the change of hydration induced by the presence of such residues.

For about 20 years, MD simulations have been widely used for studying the dynamical behavior of proteins and nucleic acids (McCammon and Harvey, 1987; Brooks et al., 1988). However, because of methodological problems essentially associated with the negative charge carried by each nucleotide, nucleic acid systems have been less studied than proteins. Given the complex tertiary folds of RNA molecules that lead to high local concentrations of charged phosphate groups, the data extracted from MD simulations on RNA molecules were found to be more simulation protocol dependent than those extracted from MD simulations on DNA duplexes (Westhof et al., 1995). Advances in the treatment of the electrostatic forces increased considerably the reliability of MD simulations applied to RNA molecules, and allowed the exploration of the dynamics and the hydration of RNA systems over time-scales reaching the nanosecond (for reviews, see Louise-May et al., 1996, and Auffinger and Westhof, 1998a, 1998b).

Recent MD simulations of solvated RNA systems include studies on duplexes (Cheatham and Kollman, 1997), hairpins (Zichi, 1995; Miller and Kollman, 1997), and large structures such as the hammerhead ribozyme (Hermann et al., 1997, 1998) and the yeast tRNA^{Asp} molecule (Auffinger et al., 1996a; Auffinger and Westhof, 1997b). However, MD simulations on RNA fragments focusing on the structural role of modified nucleotides currently are limited to investigations of the yeast tRNA^{Asp} anticodon hairpin (Auffinger et al., 1996a; Auffinger and Westhof, 1996, 1997a). This hairpin contains a pseudouridine at position 32 and a 1-methylguanine at position 37 (Fig. 1).

Pseudouridine (Ψ; modified nucleoside 1 in Appendix 1), for which a large amount of experimental data is available, is the most frequently modified nucleotide present in tRNAs. In this review, we will focus on the structural implications resulting from the occurrence of this residue in tRNA. Fig. 2 gives an overview of the tRNA pseudouridylation sites (Sprinzl et al., 1998). For further discussions on the roles of pseudouridines in tRNA, rRNA, and snRNA, see Chapters 5, 15, 26, 12, and 11. For reviews, also see Lane et al., 1995; Ofengand et al., 1995; and Agris, 1996.

MD

MD is a potential-energy-based technique that allows calculation of a possible evolution of a physical, chemical, or biochemical system over a certain period of time (van Gunsteren and Berendsen, 1990; van Gunsteren, 1993). The principle of MD resides in the numerical integration of the Newtonian equations of motion. The force acting on a specific atom is calculated iteratively by taking the derivative of the potential energy function with respect to its position. The potential energy function contains several terms that account for covalent bond stretching, bond angle bending, harmonic dihedral bending, and nonbonded

---

Pascal Auffinger and Eric Westhof • Institut de Biologie Moléculaire et Cellulaire du CNRS, 15 rue René Descartes, 67084 Strasbourg Cedex, France.

```
      5'    3'
      G - C
      G - C
      C - G
  30  G···U 40
      C - G
      Ψ·····C
      U    m¹G
       G     C
         U
         35
```

Figure 1. Secondary structure of the yeast tRNA^Asp anticodon hairpin.

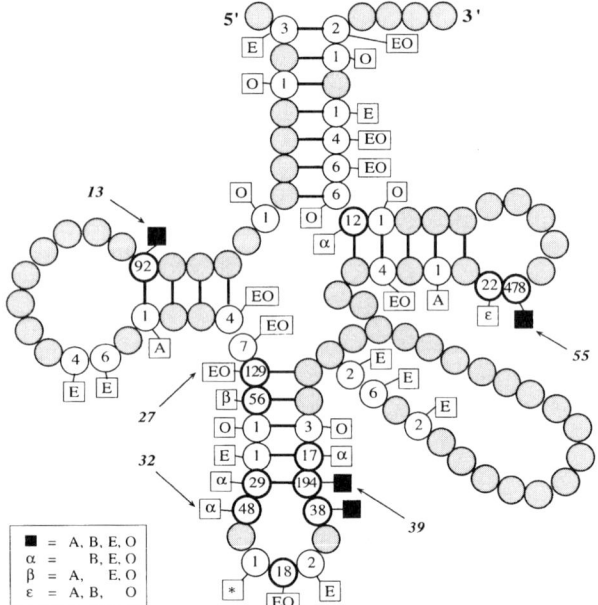

Figure 2. Locations and frequencies of tRNA pseudouridylation sites extracted from the tRNA database containing 546 tRNA sequences (Sprinzl et al., 1998). Sites for which more than 10 modifications have been counted are surrounded by a bold circle. The letters A, B and E refer to the three kingdoms (A, archaea [59 sequences]; B, eubacteria [133 sequences], and E, eukaryotes [212 sequences], while O includes the remaining 142 mitochondrial, chloroplastic and viroid tRNA sequences. The black square marks modifications occurring in all four (A, B, E and O) subdomains; the α (not A), β (not B) and ε (not E) symbols mark modifications occurring respectively in the (B, E, O), (A, E, O), and (A, B, O) subdomains. For the tRNA numbering, see Sprinzl et al. (1998) and Appendix 5. The asterisk indicates that the sequence of *Saccharomyces cerevisiae* minor tRNA^Ile containing the rare ΨAΨ anticodon (Szweykowska-Kulinska et al., 1994) has been added to the present compilation.

interactions including van der Waals and Coulombic terms. A common expression of this potential energy function, used in the AMBER 4.0 molecular dynamics package (Pearlman et al., 1994, 1995), is given below:

$$E_{total} = \sum_{bonds} K_r(r - r_{eq})^2 + \sum_{angles} K_\theta (\Theta - \Theta_{eq})^2$$

$$+ \sum_{dihedrals} \frac{1}{2} V_n[1 + \cos(n\Phi - \gamma)]$$

$$+ \sum_{i<j}^{atoms} \left(\frac{A_{ij}}{R_{ij}^{12}} - \frac{B_{ij}}{R_{ij}^{6}} \right) + \sum_{i<j}^{atoms} \left(\frac{q_i q_j}{\varepsilon R_{ij}} \right)$$

The different parameters that are used by this empirical potential energy function are obtained from experimental and quantum mechanical studies (Cornell et al., 1995). Other MD packages are available that use slightly different potential energy functions and force fields such as CHARMm (MacKerell et al., 1995) or GROMOS (van Gunsteren et al., 1996).

For a system of several thousands of atoms, such as a solvated tRNA molecule, a tremendous number of force evaluations has to be performed at each time step. Given available computational means, it is currently possible to generate accurate MD trajectories of complex nucleic acid systems including a complete representation of the environment, for example, water molecules but also neutralizing counterions and coions. These simulations reach the nanosecond time scale and allow one to estimate, among others, the stability of important tertiary interactions and the hydration characteristics of the investigated system.

THE PSEUDOURIDINE AT POSITION 32 OF THE tRNA^Asp ANTICODON HAIRPIN

Specific Hydration of Ψ_{32}

In yeast tRNA^Asp, a pseudouridine is located at position 32 of the seven-nucleotide anticodon loop (Fig. 1). MD simulations of the tRNA^Asp anticodon hairpin (Auffinger and Westhof, 1997a) have shown that Ψ_{32} stabilizes a water molecule, bridging the two adjacent $(C_{31})O_R$ and $(\Psi_{32})O_R$ anionic oxygen atoms through an N_1-H...O_w hydrogen bond (see Fig. 3; O_R stands for the pro-R_p anionic oxygen atom of the RNA backbone and O_w stands for the oxygen atom of a water molecule). The residence time of this water molecule, which is consistently observed in several MD simulations, was estimated to be significantly longer then 500 ps (time limit of the MD investigation). This water molecule forms an important structural link between the nucleotide backbone and the modified base and, thus, reduces the conforma-

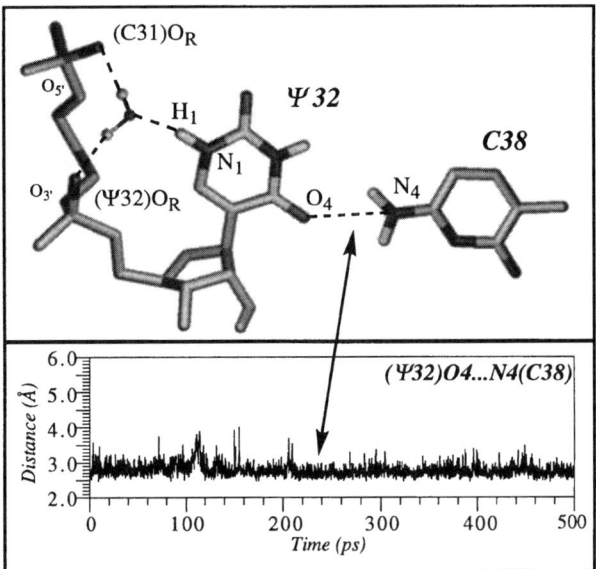

Figure 3. (Top) Snapshot extracted from a 500-ps MD simulation of the solvated yeast tRNA^{Asp} anticodon hairpin. This snapshot shows the water molecule linking the base of Ψ_{32} to its nucleotide backbone through a N_1-H...O_w and two O_w-H_w...O_R hydrogen bonds (O_R = pro-R_p). This water molecule is stable for at least 500 ps and contributes to reduce the mobility of base 32 (Auffinger and Westhof, 1997a). This reduced conformational mobility results in the stabilization of the single bifurcated (Ψ_{32})O_4...N_4(C_{38}) interaction specific of the 32-38 "pseudo-base pair". (Bottom) Typical time course of the (Ψ_{32})O_4...N_4(C_{38}) distance extracted from a 500-ps MD simulation of the solvated yeast tRNA^{Asp} anticodon hairpin (Auffinger and Westhof, 1996).

tional mobility of the RNA close to the pseudouridylation site.

Although no water molecule was observed in the vicinity of the (Ψ_{32})N_1-H group in the crystal of tRNA^{Asp}, in the orthorhombic form of yeast tRNA^{Phe} a water molecule bridges the (Ψ_{55})N_1-H group to the anionic oxygen atoms of residues T_{54} and Ψ_{55} (Westhof et al., 1988) similarly to the one bound to Ψ_{32} of tRNA^{Asp} during the MD simulations (Auffinger and Westhof, 1997a). From a difference electron density map between the complexes of modified and unmodified tRNA^{Gln} with the cognate glutaminyl-tRNA synthetase (Arnez and Steitz, 1994), identical water bridging patterns were detected near residues Ψ_{38} and Ψ_{39} of the anticodon hairpin. Furthermore, N_1-H hydrogen bonded water molecules could be modeled in hydration pockets located close to residue Ψ_{55} of tRNA^{Asp} and tRNA^{Gln}, Ψ_{13} and Ψ_{32} of tRNA^{Asp}, and Ψ_{39} of tRNA^{Phe}. Nuclear magnetic resonance (NMR) studies also indicate that the N_1-H proton has a slow exchange rate with solvent protons in agreement with the existence of a long-lived (Ψ)N_1-H...O_w hydrogen bond (Griffey et al., 1985; Davis and Poulter, 1991; Hall and McLauglin, 1991, 1992; Davis, 1995).

Thus, experimental and theoretical data confirm that, for all the previously described tRNA pseudouridylation sites, the (Ψ)N_1-H group is hydrogen bonded to a water molecule establishing a stable link between the base and the nucleotide backbone. However, water molecules bridging successive anionic O_R oxygen atoms have often been observed in A-DNA and RNA crystal structures and can be stabilized by a (C or U)C_5-H...O_w interaction (Auffinger et al., 1996a). Such C-H...O_w interactions between the C_5-H group of pyrimidines and the oxygen atom of a water molecule bridging two adjacent anionic phosphate oxygen atoms also have been characterized by MD simulations (Auffinger et al., 1996a). Although weaker, the (C or U)C_5-H...O_w interactions are isostructural to (Ψ)N_1-H...O_w hydrogen bonds. The similarity between N_1-H...O_w and C_5-H...O_w hydrogen bonds has been pointed out by Griffey et al. (1985) when they reported that, in yeast tRNA^{Phe}, the (U_{32})C_5 atom is 5–5.5 Å apart from the O_R atoms of residues 31 and 32, allowing a water molecule to link the base to the backbone through a C_5-H...O_w interaction. Therefore, the U→Ψ modification, which does not induce significant stereochemical changes as both bases adopt the same *anti* orientation in most RNA structural contexts (Davis and Poulter, 1991; Arnez and Steitz, 1994; Davis, 1995), leads essentially to a strengthening of an existing (U)C_5-H...O_w interaction through the replacement of the C_5-H group by a more polar N_1-H group. This interaction, linking Ψ_{32} to the backbone through a water molecule in yeast tRNA^{Asp}, reduces the mobility of the modified base and subsequently increases the stability of the specific Ψ_{32}-C_{38} "pseudo-base pair" (Fig. 2), which will be described next.

The Ψ_{32}-C_{38} "Pseudo-Base Pair"

In the tRNA structures that have been crystallographically determined, the first (32) and last (38) bases of the anticodon loop form an isostructural "pseudo-base pair" involving a single hydrogen bond, that is, a bifurcated (Ψ_{32})O_4...N_4(C_{38}) hydrogen bond in tRNA^{Asp} (Westhof et al., 1985) and a bifurcated (Cm_{32})O_2...N_6(A_{38}) hydrogen bond in tRNA^{Phe} (Quigley and Rich, 1976). An NMR structure determination of unmodified transcripts of initiator and elongator tRNA^{Met} anticodon hairpins (Schweisguth and Moore, 1997) suggests the occurrence of a (C_{32})O_2...N_6(A_{38}) hydrogen bond similar to the one observed in the tRNA^{Phe} crystal structure. Furthermore, an UC "pseudo-base pair," closing the seven-nucleotide loop of hepatitis delta virus (HDV) and involving a (U)O_2...N_4(C) hydrogen bond, has been characterized by NMR (Kolk et al., 1997). This

UC pseudo-base pair is isostructural to the ΨC "pseudo-base pair" found in yeast tRNA^{Asp}. It is noteworthy that an AU pseudo-base pair involving a single (U)O$_4$...N$_6$(A) hydrogen bond closes stem III of the hammerhead ribozyme (Pley et al., 1994; Scott et al., 1995). This hydrogen bond is well maintained during an MD simulation of the ribozyme (Hermann et al., 1998).

In tRNA^{Asp}, because the 32–38 pair involves a single base-to-base hydrogen bond, it is tempting to speculate that this pseudo-base pair may display a low stability. Surprisingly, MD simulations of the tRNA^{Asp} anticodon hairpin (Auffinger and Westhof, 1996) have shown that this single bifurcated (Ψ$_{32}$)O$_4$...N$_4$(C$_{38}$) hydrogen bond remains stable in several 500-ps trajectories (Fig. 3). The simulations also revealed a particularly strong hydration pattern around the (Ψ$_{32}$)O$_4$...N$_4$(C$_{38}$) base pair also involving [besides the (Ψ$_{32}$)N$_1$-H group] the (Ψ$_{32}$)N$_3$-H, (C$_{38}$)N$_4$-H and (C$_{38}$)C$_2$-O$_2$ groups (Auffinger and Westhof, 1997b). Such a strong hydration is induced by the presence of the pseudouridine and contributes to the stabilization of the 32–38 interaction.

In the crystal structure of tRNA^{Asp} complexed with its cognate synthetase, the single (Ψ$_{32}$)O$_4$...N$_4$(C$_{38}$) interaction is conserved upon complex formation despite the large conformational changes induced in the rest of the anticodon loop (Rees et al., 1996). Similarly, in the complex of unmodified transcript of tRNA^{Gln} with its synthetase, a single (U$_{32}$)N$_3$...O$_4$(U$_{38}$) interaction is observed, although, in this case, the anticodon loop structure is less altered by the protein binding (Arnez and Steitz, 1996). An identical (Um$_{32}$)N$_3$...O$_4$(Ψ$_{38}$) interaction is present in the complex formed with native tRNA^{Gln} and its cognate synthetase (Rould et al., 1991). In that complex, an additional water molecule forms a link between bases 32 and 38, stressing the importance of the hydration of these specific motifs in relation to their structural stability.

Thus, in all known tRNA structures, a 32–38 pseudo-base pair, which remains intact after complexation with their cognate synthetase, extends the RNA helix of the anticodon stem. The preservation of the structural integrity of this pseudo-base pair seems to be essential for tRNA functions. For tRNA^{Gly}, it has been shown that a mutation of the nucleotide at position 32 leads to a loss of the discriminating ability of its anticodon (Lustig et al., 1993; Claesson et al., 1995). The authors proposed that such codon-misreading events result from an alteration of the dynamical stability of the 32–38 interaction after modification of nucleotide 32. Mutations at position 32, which have been detected in malignant cells, may similarly alter the stability of the anticodon loop region (Dirheimer et al., 1995). Consequently, modified nucleotides at position 32 or 38 may be required for increasing the stability of the 32–38 pseudo-base pair. It is noteworthy that at position 32–38, no CG or GC base pairs are found, while two AU and 30 UA base pairs, which probably are not of the Watson-Crick type, are present in the tRNA database (Sprinzl et al., 1998). Thus, this 32–38 pseudo-base pair can be considered as a mandatory transition element between the helical stem and the U-turn fold of the hairpin.

PSEUDOURIDINES AT OTHER tRNA SITES

Ψ at Position 39

Modifications of nucleotides close to the 32–38 site may also contribute to the stabilization of the 32–38 pseudo-base pair. Recently, an NMR study of two anticodon hairpins of the unmodified transcript of initiator (tRNA^{iMet}) and elongator (tRNA^{mMet}) methionine tRNAs has shown that the stability of the hairpin is dependent on the closing 31–39 base pair (Schweisguth and Moore, 1997). The sequence of these tRNA hairpins differs only in their stem. The tRNA^{iMet} contains a stack of three GC base pairs at the 5′ end of the anticodon stem, while in the tRNA^{mMet} two of these GC base pairs are replaced by AU base pairs. The mobility of the tRNA^{mMet} hairpin containing the more labile AU base pair at position 31–39 was found to be larger than that of tRNA^{iMet}. The authors proposed that this dynamical difference explains the greater sensitivity of tRNA^{mMet} toward the S1 nuclease. This result correlates well with those of several thermodynamic studies on RNA hairpins, which have shown that loops closed by AU (or UA) base pairs are less stable than loops closed by GC (or CG) base pairs (Serra et al., 1993, 1997).

In the wild-type anticodon hairpin of tRNA^{mMet}, the closing AU base pair of the stem is replaced by an AΨ base pair, probably to increase the stability of the anticodon hairpin. This is supported by experimental data indicating that the replacement of U by Ψ in poly(AU) increases the helix melting temperature by 26°C (Ward and Reich, 1968), and that a U$_{39}$ to Ψ$_{39}$ modification in the anticodon hairpin of tRNA^{Lys,3} results in a 5°C increase in the melting temperature of the hairpin (Durant and Davis, 1997). Similarly, a U$_{39}$ to Ψ$_{39}$ modification in tRNA^{Phe} leads to a vanishing of the NMR signal of the uridine N$_3$-H imino proton interpreted as resulting from a destabilization of the A$_{31}$-U$_{39}$ base pair (Davis and Poulter, 1991).

To determine the occurrence of modified AU base pairs at positions 31 and 39, we performed a statistical analysis of the tRNA database (Sprinzl et al., 1998). This analysis revealed that the percentage of occurrence of AΨ base pairs at position 31-39 is close to 35%, while that of standard AU base pairs is only of 2% and that of natural GC or CG base pairs is 48% (ΨA accounts for 2%, UA accounts for 6%, and others account for 9%). Thus, while GC base pairs are only rarely subject to modifications, pseudouridylation is mandatory for stabilizing the more labile AU base pairs, at least at position 31-39 of tRNAs. As inferred from NMR and crystallographic data (Griffey et al., 1985; Davis and Poulter, 1991; Arnez and Steitz, 1994; Davis, 1995) and from MD simulations (Auffinger et al., 1996a; Auffinger and Westhof, 1997a), Ψ_{39} increases the stability of the 31-39 base pair by forming an intraresidue water mediated base-to-backbone link that reduces the conformational mobility of Ψ_{39} and, consequently, increases the stability of the AΨ base pair.

Indeed, an increased stability correlated with reduced conformational dynamics of the 31-39 base pair has a positive feedback on the stability of the interactions between bases 32 and 38. As inferred from the NMR structures of the seven-nucleotide loop of the hepatitis delta virus (Kolk et al., 1997) and of tRNA^{iMet} (Schweisguth and Moore, 1997), a CA or an UC pseudo-base pair may be stable at position 32-38 when associated with a CG, GC, or AΨ base pair at position 31-39. However, the 32-38 pseudo-base pair are less stable when associated with an A_{31}-U_{39} Watson-Crick base pair; in their unmodified form, these base pairs are almost absent from the sequences of the tRNA database (Sprinzl et al., 1998). Summarizing the preceding information suggests strongly that the main function of pseudouridylation close to the anticodon is to stabilize the structure of the loop by reducing its intrinsic dynamics, likely to avoid codon misreading. The fact that a large number of modified bases (pseudouridines, but also 2′-O-methylated bases) are clustered at the ends of the anticodon stem points to the necessity of preserving the structural stability of these regions.

Ψ at Positions 13 and 27

Two other important sites of pseudouridylation are positions 13 and 27 (Fig. 2). The base pair 27-43 forms the junction between the anticodon arm and the rest of the structure. The percentage of ΨA base pairs is close to 20%, while that of unmodified UA base pairs is only 11% (GC and CG base pairs account for 44%, others for 25%). Again, pseudouridylation may be required for stabilizing the weaker UA stem base pair. Schultz and Yarus (1994a, 1994b) noted that a substitution of the C_{27}-G_{43} base pair by AU, UA, or GC base pairs leads to a marked decrease of the ribosomal activity of tRNA su7 G_{36}. Considering the higher stability of ΨA compared to a UA base pair in helical contexts, ribosomal activity could be restored by using a ΨA base pair. Thus, at least at positions where pseudourydilation occurs frequently, ΨA or AΨ should be considered as alternative base pairs for replacing CG or GC base pairs.

The role of Ψ at position 13 is less obvious but it may reside in an increased stability for non-Watson-Crick UG (4% of UG, 7% of ΨG) and UU (0.5% of UU, 8% of ΨU) base pairs at the end of the D-stem. It is noteworthy that the 13-22 base pair of the D-loop is often involved in triple base pairs and stacks with the conserved Hoogsteen U_8-A_{14} base pair.

Ψ at Position 55

Almost 90% of the known tRNA sequences contain a Ψ at position 55. tRNAs lacking the Ψ_{55} modification essentially are found in nonelongator tRNAs. Therefore position 55 is the most prevalent tRNA pseudouridylation site (Fig. 2). It is known that the $(\Psi_{55})N_3$-H imino group interacts with an anionic oxygen atom of residue 58 and contributes to the U-turn structure of the TΨC loop. It has also been shown that the proton of the $(\Psi_{55})N_1$-H group is protected from exchange with those of water molecules (Griffey et al., 1985; Davis and Poulter, 1991) and, in the yeast tRNA^{Asp} crystal structure, the $(\Psi_{55})N_1$-H group was found hydrogen bonded to a water molecule that bridges to adjacent anionic oxygen atoms (Westhof et al., 1988). Thus, the Ψ_{55} residue is highly constrained by direct and water mediated intramolecular hydrogen bonds and contributes to the stability of the TΨC loop. This interaction scheme may be universal in the tRNA world.

FUNCTIONAL ROLES OF Ψ_{34}, Ψ_{35} AND Ψ_{36}

In the structural framework of the U-turn of the anticodon loop, the pseudouridines found at positions 34, 35, and 36 are prevented from interacting with the backbone through a water mediated interaction, as in helical regions. In the following, based on MD simulations, we propose possible structural roles for pseudouridines at positions 35, 36, and 34.

Ψ at Position 35

Pseudouridines are the only modified nucleotides that have been detected at position 35 of the anticodon. Although no MD studies were conducted on anticodon hairpins containing residue Ψ35, simulations have been performed on the yeast tRNA^{Asp} anticodon hairpin containing an uridine at position 35 (Auffinger et al., 1996b; Auffinger and Westhof, 1996). These simulations revealed the occurrence of a stable $(U_{35})C_5$-H...$O_{2'}(U_{33})$ hydrogen bond, also present in the tRNA^{Asp} crystal structure (Fig. 4) (for reviews on C-H...O hydrogen bonds, see Desiraju, 1996; Steiner, 1996; and Wahl and Sundaralingam, 1997). This interaction is similar to the $(A_{35})N_7$...H-$O_{2'}(U_{33})$ hydrogen bond inferred from the crystallographic structures of tRNA^{Phe} (Quigley and Rich, 1976). In the first case, the hydroxyl group works as a hydrogen bond acceptor and, in the second case, as a hydrogen bond donor group. However, both hydrogen bonds link base 35 to the sugar of U_{33}. Thus, at position 35, the replacement of a pyrimidine by a pseudouridine would lead to a substitution of the weak $(U_{35})C_5$-H...$O_{2'}(U_{33})$ interaction by a stronger $(\Psi_{35})N_1$-H...$O_{2'}(U_{33})$ hydrogen bond and, thus, provide additional stabilization of the anticodon loop structure. The ensuing increase in stability may be required to improve the fidelity of anticodon-codon interactions.

It has been shown experimentally that the Ψ_{35} modification increases the activity of yeast suppressor tRNA^{Tyr}(UΨA) by increasing the stability of the anticodon(Ψ)-codon(A) interaction (Johnson and Abelson, 1983). Similarly, Ψ_{35} in plant cytoplasmic tRNA^{Tyr}(GΨA) is mandatory for efficiently recognizing the UAG or UAA codons and discriminating between the UAA and UGA codons, while the unmodified GUA anticodon does not recognize the UAG or UAA codons (Zerfass and Beier, 1992). Thus, the reduced mobility of Ψ_{35}, constrained by the $(\Psi_{35})N_1$-H...$O_{2'}(U_{33})$ hydrogen bond, may constitute the determining factor leading to strengthening of the anticodon(Ψ)-codon(A) interaction and, consequently, to a better discriminating ability of GΨA compared to GUA anticodons.

This interpretation does not exclude that the pseudouridine at position 35 may be involved in the recognition process of the tRNA with its cognate synthetase or other tRNA interacting proteins. Bare and Uhlenbeck (1986) demonstrated that the $\Psi_{35} \rightarrow U_{35}$ substitution in yeast tRNA^{Tyr}(GΨA) increased twofold the K_m for aminoacylation, suggesting that specific interactions may form between the synthetase and the two imino groups of Ψ_{35}. Yet this slight change in K_m indicates that the primary function of pseudouridylation at position 35 is not to increase the efficiency of the recognition with the cognate TyrRS but rather, as previously proposed, to improve the specificity of the anticodon-codon interactions. Such complexes with cognate synthetase often involve modifications of the anticodon loop structure that are not required for anticodon-codon interactions. Thus, in some instances, a stiffening of the anticodon loop structure may be required to avoid codon misreading. For that purpose, the only modification accepted at position 35 seems to be pseudouridylation, which does not alter the Watson-Crick recognition sites.

Ψ at Position 36

MD simulations of the tRNA^{Asp} anticodon loop revealed the occurrence of a second C-H...O hydrogen bond contributing to the stability of the anticodon loop structure, that is, the $(C_{36})C_5$-H...$O_2(U_{33})$ interaction (Auffinger et al., 1996b; Auffinger and Westhof, 1996). Likewise, an equivalent C-H...O interaction may systematically occur in tRNAs with a uridine located at position 36. However, modifications at position 36 are rare (Sprinzl et al., 1998). The minor species of tRNA^{Ile} from Saccharomyces cerevisiae contain the ΨAΨ anticodon (Szweykowska-Kulinska et al., 1994). The structural role of Ψ_{36} may be to increase the stability of the anticodon loop, again by replacing the weak $(\Psi_{36})C_5$-H...$O_2(U_{33})$ interaction by a stronger $(\Psi_{36})N_1$-H...$O_2(U_{33})$ hydrogen bond (Fig. 5). Such a change does not affect the overall anticodon loop structure. The additional stabilization of the hairpin provided by Ψ_{36} may increase the efficiency of codon-anticodon interactions. Recently, it was proposed that the

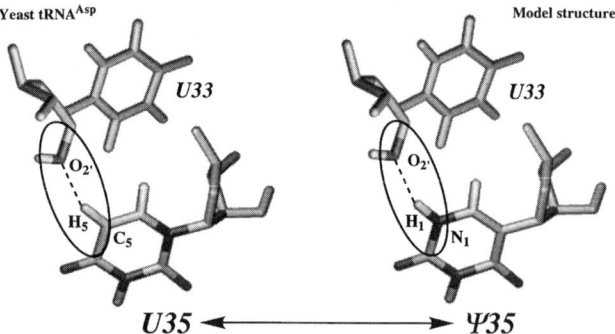

Figure 4. Substitution of a pyrimidine at position 35 of the anticodon loop by a pseudouridine. (Left) The occurrence of a $(U_{35})C_5$-H...$O_{2'}(U_{33})$ hydrogen bond is inferred from the refined crystal structure of yeast tRNA^{Asp} (Westhof et al., 1985). This interaction is analogous to the $(A_{35})N_7$...H-$O_{2'}(U_{33})$ hydrogen bond found in the crystal structure of tRNA^{Phe} (Quigley and Rich, 1976). This C-H...O interaction displays a stable dynamical behavior in several MD simulations of the tRNA^{Asp} anticodon hairpin (Auffinger et al., 1996b; Auffinger and Westhof, 1996). (Right) Substitution of a pyrimidine at position 35 by a pseudouridine increases the strength of the interaction established between base 35 and the ribose hydroxyl group of base 33, since a C-H...O contact is replaced by an N-H...O bond.

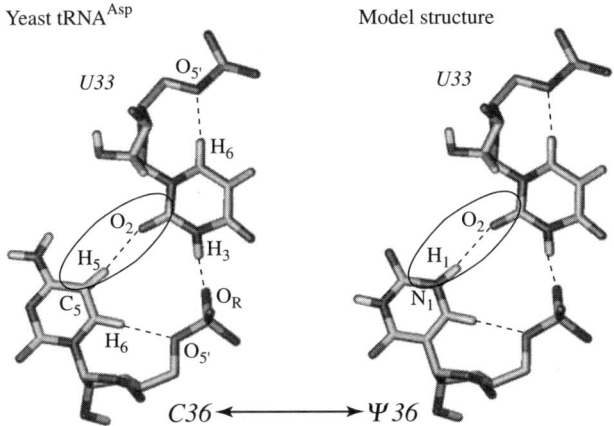

Figure 5. Substitution for a pyrimidine at position 36 of the anticodon loop by a pseudouridine. (Left) A $(C_{36})C_5$-H...$O_2(U_{33})$ interaction is present in the crystal structure of yeast tRNA^{Asp} (Westhof et al., 1985). This interaction is maintained in several 500-ps MD simulations of the anticodon hairpin (Auffinger et al., 1996b; Auffinger and Westhof, 1996). Additionally, U_{33} is stabilized by the strong $(U_{33})N_3$-H...$O_R(C_{36})$ internucleotide hydrogen bond and the weaker $(U_{33})C_6$-H...O_5' intranucleotide C-H...O interaction. The base of C_{36} is similarly linked to its backbone through a C_6-H...O_5' hydrogen bond. U_{36} is thus linked to C_{36} through an array of strong N-H...O and weaker C-H...O hydrogen bonds. (Right) A replacement of a pyrimidine at position 36 by a pseudouridine would not perturb the array of existing hydrogen bonds at positions 33 and 36. Instead, it results in a strengthening of the interaction established between the two bases through the replacement of the C_5-H...O_2 interaction by an N_1-H...O_2 hydrogen bond.

main role of the pseudouridylation in tRNA^{Ile}(ΨAΨ), which reads the AUA codon, is to prevent misreading of AUG codons (Senger et al., 1997). This may occur by stabilizing the structure of the anticodon through a (Ψ_{36})N_1-H...$O_2(U_{33})$ hydrogen bond which, consequently, reduces its dynamics. Thus, a stiffening of Ψ_{36} may favor the formation of a Watson-Crick (Ψ_{36})-A base pair, and increase the discriminating ability of the ΨAΨ anticodon.

Ψ at Position 34

The pseudouridine present at position 34 in tRNA^{Ile}(ΨAΨ) also contributes to the discriminating ability of the ΨAΨ anticodon for AUA compared to AUG codons (Grosjean et al., 1997; Senger et al., 1997). Yet, Ψ_{34} is in a specific structural context where it cannot lock a water molecule between the base and the backbone (as in helices), or interact with a hydroxyl group (such as Ψ_{35}) or with an oxygen atom of a base (as Ψ_{36}). Thus, a particular stabilization mechanism may lead to the increased discriminating ability of the ΨAΨ anticodon, which may again involve a direct or water mediated interaction of the (Ψ_{34})N_1-H group with the backbone. It would be interesting to determine if the (Ψ_{34})N_1-H proton is protected from exchange with those of water, because it is the case for the previously described pseudouridines.

Modified Pseudouridines Ψm and m$^1\Psi$

Ψm, also rare, is found at positions 32 and 38 in tRNAs (see Appendix 5) and may provide an additional strengthening of the anticodon loop structure. m$^1\Psi$ is almost entirely conserved in the 59 archaebacteria tRNA structures of the tRNA database (Sprinzl et al., 1998). Hence, m$^1\Psi$ is required for maintaining the TΨC-loop structure of the tRNAs of these organisms living in extreme conditions.

CONCLUSION

Since the time pseudouridine was described by Pochon et al. (1964) as "a natural analogue of uridine of unpredictable properties," a large number of experimental and theoretical data have been collected to define more precisely some of the structural functions of this modified base.

In tRNA, pseudouridines essentially are found at positions where they can contribute to the stabilization of specific structural motifs: Ψ_{55} is involved in the stabilization of the TΨC-loop; Ψ_{13} is involved in the stabilization of the D-stem; Ψ_{31}, Ψ_{32}, Ψ_{38}, and Ψ_{39} are involved in the stabilization of the anticodon loop by contributing to a stiffening of the 31-39 AU Watson-Crick base pairs and of the 32-38 pseudobase pair, which seems to be a specific structural feature of anticodon-like loops.

Molecular dynamics simulations and experimental data indicate that the stabilization mechanism of Ψs at the previously described locations involves a water mediated base-to-backbone interaction where the additional (Ψ)N_1-H imino group is used to establish a new (Ψ)N_1-H...O_w hydrogen bond replacing a weaker (C or U)C_5-H...O_w interaction (Fig. 3). This water mediated link reduces the conformational mobility of the Ψ nucleotide and, consequently, of the structural elements in the surroundings of the modified site. Such a reduced mobility is correlated with an increased thermodynamic stability of RNA structural motifs, as revealed by the lower melting temperatures of unmodified tRNA transcripts.

This specific stabilization mechanism, involving a water mediated base-to-backbone interaction, has as a consequence that the protected (Ψ)N_1-H group may not be generally considered as a recognition site for RNA-RNA or RNA-protein interactions. A recognition process implying the (Ψ)N_1-H imino group would probably require a local unfolding of the structure. Thus, a *syn*-to-*anti* conversion of the Ψ nucleotide seems improbable at most tRNA locations, although it may occur at other RNA sites (for a discussion see Wahl et al., 1996).

Besides, pseudouridines are found in the anticodon loop at the three positions where they could, as inferred from MD simulations, rigidify the loop structure through the $(\Psi_{36})N_1\text{-H}\ldots O_{2'}(U_{33})$ and $(\Psi_{35})N_1\text{-H}\ldots O_{2'}(U_{33})$ hydrogen bonds with the phylogenetically conserved U_{33} base. Such a strengthening of the loop structure may be required to increase the discriminating ability of the anticodon. Ψ_{34} may also be involved in direct or water mediated base-to-backbone interactions; however, more experimental data are needed to precisely determine its stereochemistry.

Pseudouridines are also important in ribosomal RNA (see Chapter 12) and small nuclear RNA (see Chapter 11). It would not be surprising that in most structural contexts pseudouridines would stabilize rRNA and snRNA in a similar manner as in tRNA. In rRNA, it was reported that pseudouridines, present in both single- and double-stranded regions, are most frequent in positions where they close a loop or a bulge (Ofengand and Bakin, 1997), in good agreement with the location of the main pseudouridylation sites in tRNA. These results, and the fact that no single deletion of pseudouridines in rRNAs was found to be lethal (see Chapter 12), support the assumption that the primary role of pseudouridines is to strengthen specific structural elements. In snRNA, the occurrence of pseudouridines in single-stranded regions may be required to improve their recognition features.

PERSPECTIVES

Given the growing potential of MD simulations to reproduce the dynamics of RNA systems, there is no doubt that this technology will be used more and more for investigating the functional and structural roles of the numerous modified bases found in tRNAs and in other RNA systems. Besides pseudouridines, the functional and structural roles of a large number of other modifications are accessible to MD simulations, such as the common methylation of the hydroxyl group of ribose sugars. Recent MD studies on the conformational behavior of 2'-OH groups in tRNA (Auffinger and Westhof, 1997b) may be considered as a first approach toward the investigation of the roles of 2'-O-methyl groups.

Acknowledgments. P.A. thanks the Fondation pour la Recherche Médicale for a fellowship. E.W. is thankful to the Institut Universitaire de France for support.

REFERENCES

Agris, P. F. 1996. The importance of being modified: roles of modified nucleosides and Mg^{2+} in RNA structure and function. *Prog. Nucleic Acid Res. Mol. Biol.* 53:79–129.

Arnez, J. G., and T. A. Steitz. 1994. Crystal structure of unmodified tRNA^{Gln} complexed with glutaminyl-tRNA synthetase and ATP suggests a possible role for pseudo-uridines in stabilization of RNA structure. *Biochemistry* 33:7560–7567.

Arnez, J. G., and T. A. Steitz. 1996. Crystal structure of three mysacylating mutants of *Escherichia coli* glutaminyl-tRNA synthetase complexed with tRNA^{Gln} and ATP. *Biochemistry* 35:14725–14733.

Auffinger, P., S. Louise-May, and E. Westhof. 1996a. Hydration of C-H groups in tRNA. *Faraday Discuss.* 103:151–174.

Auffinger, P., S. Louise-May, and E. Westhof. 1996b. Molecular dynamics simulations of the anticodon hairpin of tRNA^{Asp}: structuring effects of C-H...O hydrogen bonds and of long-range hydration forces. *J. Am. Chem. Soc.* 118:1181–1189.

Auffinger, P., and E. Westhof. 1996. H-bond stability in the tRNA^{Asp} anticodon hairpin: 3 ns of multiple molecular dynamics simulations. *Biophys. J.* 71:940–954.

Auffinger, P., and E. Westhof. 1997a. RNA hydration: three nanoseconds of multiple molecular dynamics simulations of the solvated tRNA^{Asp} anticodon hairpin. *J. Mol. Biol.* 269:326–341.

Auffinger, P., and E. Westhof. 1997b. Rules governing the orientation of the 2'-hydroxyl group in RNA. *J. Mol. Biol.* 274:54–63.

Auffinger, P., and E. Westhof. 1998a. Molecular dynamics of nucleic acids. *In* P. V. R. Schleyer (ed.), *Encyclopedia of Computational Chemistry*, in press. John Wiley & Sons, New York, N.Y.

Auffinger, P., and E. Westhof. 1998b. Simulations of the molecular dynamics of nucleic acids. *Curr. Opin. Struct. Biol.*, in press.

Bare, L. A., and O. C. Uhlenbeck. 1986. Specific substitution into the anticodon loop of yeast tyrosine transfer RNA. *Biochemistry* 25:5825–5830.

Brooks, C. L., M. Karplus, and B. M. Pettitt. 1988. Proteins: a theoretical perspective of dynamics, structure and thermodynamics, vol. LXXI. John Wiley & Sons, New York, N.Y.

Cheatham, T. E., and P. A. Kollman. 1997. Molecular dynamics simulations can reasonably represent the structural differences in DNA:DNA and RNA:RNA and DNA:RNA hybrid duplexes. *J. Am. Chem. Soc.* 119:4805–4194.

Claesson, C., F. Lustig, T. Borén, C. Simonsson, M. Barciszewska, and U. Lagerkvist. 1995. Glycine codon discrimination and the nucleotide in position 32 of the anticodon loop. *J. Mol. Biol.* 247:191–196.

Cornell, W. D., P. Cieplak, C. I. Bayly, I. R. Gould, K. M. Merz, D. M. Ferguson, D. C. Spellmeyer, T. Fox, J. W. Caldwell, and P. A. Kollman. 1995. A second generation force field for the simulation of proteins, nucleic acids, and organic molecules. *J. Am. Chem. Soc.* 117:5179–5197.

Davis, D. R. 1995. Stabilization of RNA stacking by pseudouridine. *Nucleic Acids Res.* 23:5020–5026.

Davis, D. R., and C. D. Poulter. 1991. ^{1}H-^{15}N NMR studies of *Escherichia coli* tRNA^{Phe} from *hisT* mutants: a structural role for pseudouridine. *Biochemistry* 30:4223–4231.

Desiraju, G. R. 1996. The C-H...O hydrogen bond: structural implications and supramolecular design. *Acc. Chem. Res.* 29:441–449.

Dirheimer, G., W. Baranowski, and G. Keith. 1995. Variations in tRNA modifications, particularly of their queunine content in higher eukaryotes. Its relation to malignancy grading. *Biochimie* 77:99–103.

Durant, P. C., and D. R. Davis. 1997. The effect of pseudouridine and pH on the structure and dynamics of the anticodon stem-loop of tRNA^{Lys,3}. *Nucleic Acids Symp. Ser.* 36:56–57.

Griffey, R. H., D. Davis, Z. Yamaizumi, S. Nishimura, A. Bax, B. Hawkins, and C. D. Poulter. 1985. ^{15}N-labeled *Escherichia coli* tRNA$_f^{Met}$, tRNA^{Glu}, tRNA^{Tyr}, and tRNA^{Phe}. Double resonance and two dimensional NMR of N1-labeled pseudouridine. *J. Biol. Chem.* 260:9734–9741.

Grosjean, H., Z. Szweykowska-Kulinska, Y. Motorin, F. Fasiolo, and G. Simos. 1997. Intron-dependent enzymatic formation of modified nucleosides in eukaryotic tRNAs: a review. *Biochimie* 79:293–302.

Hall, K. B., and L. W. McLaughlin. 1991. Properties of a U1/mRNA 5′ splice site duplex containing pseudouridines as measured by thermodynamics and NMR methods. *Biochemistry* 30:1795–1801.

Hall, K. B., and L. W. McLauglin. 1992. Properties of pseudouridine N1 imino protons located in the major groove of an A-form RNA duplex. *Nucleic Acids Res.* 20:1883–1889.

Hermann, T., P. Auffinger, W. G. Scott, and E. Westhof. 1997. Evidence for a hydroxide ion bridging two magnesium ions at the active site of the hammerhead ribozyme. *Nucleic Acids Res.* 25:3421–3427.

Hermann, T., P. Auffinger, and E. Westhof. 1998. Molecular dynamics investigations on the hammerhead ribozyme RNA. *Eur. J. Biophys.*, in press.

Johnson, P. F., and J. Abelson. 1983. The yeast tRNA^{Tyr} gene intron is essential for correct modification of its tRNA product. *Nature* 302:681–687.

Kolk, M. H., H. A. Heus, and C. W. Hilbers. 1997. The structure of the isolated, central hairpin of the HDV antigenomic ribozyme: novel structural features and similarity of the loop in the ribozyme and free in solution. *EMBO J.* 16:3685–3692.

Lane, B. G., J. Ofengand, and M. W. Gray. 1995. Pseudouridine and O2′-methylated nucleosides. Significance of their selective occurrence in rRNA domains that function in ribosome-catalyzed synthesis of the peptide bonds in proteins. *Biochimie* 77:7–15.

Louise-May, S., P. Auffinger, and E. Westhof. 1996. Calculation of nucleic acid conformation. *Curr. Opin. Struct. Biol.* 6:289–298.

Lustig, F., T. Borén, C. Claesson, C. Simonsson, M. Barciszewska, and U. Lagerkvist. 1993. The nucleotide at position 32 of the tRNA anticodon loop determines ability of anticodon UCC to discriminate among glycine codons. *Proc. Natl. Acad. Sci. USA* 90:3343–3347.

MacKerell, A. D., J. Wiórkiewicz-Kuczera, and M. Karplus. 1995. An all-atom empirical energy function for the simulation of nucleic acids. *J. Am. Chem. Soc.* 117:11946–11975.

McCammon, J. A., and S. C. Harvey. 1987. *Dynamics of Proteins and Nucleic Acids.* Cambridge University Press, New York, N.Y.

Miller, J., and P. A. Kollman. 1997. Theoretical studies of an exceptionally stable RNA tetraloop: observation of convergence from an incorrect NMR structure to the correct one using unrestrained molecular dynamics. *J. Mol. Biol.* 270:436–450.

Ofengand, J., and A. Bakin. 1997. Mapping to nucleotide resolution of pseudouridine residues in large subunit ribosomal RNAs from representative eukaryotes, prokaryotes, archaebacteria, mitochondria and chloroplasts. *J. Mol. Biol.* 266:246–268.

Ofengand, J., A. Bakin, J. Wrzesinski, K. Nurse, and B. G. Lane. 1995. The pseudouridine residues of ribosomal RNA. *Biochem. Cell Biol.* 73:915–924.

Pearlman, D. A., D. A. Case, J. W. Caldwell, W. S. Ross, T. E. Cheatham, S. DeBolt, D. Ferguson, G. Seibel, and P. Kollman. 1995. AMBER, a package of computer programs for applying molecular mechanics, normal mode analysis, molecular dynamics and free energy calculations to simulate the structural and energetic properties of molecules. *Comp. Phys. Commun.* 91:1–41.

Pearlman, D. A., D. A. Case, J. W. Caldwell, W. S. Ross, T. E. Cheatham, D. M. Ferguson, G. L. Seibel, U. C. Singh, P. K. Weiner, and P. A. Kollman. 1994. AMBER 4.1. University of California, San Francisco.

Pley, H. M., K. M. Flaherty, and D. B. McKay. 1994. Three-dimensional structure of a hammerhead ribozyme. *Nature* 372:68–74.

Pochon, F., A. M. Michelson, M. Grunberg-Manago, W. E. Cohn, and L. Dondon. 1964. Polynucleotide analogues. III. Polypseudouridylic acid: synthesis and some physicochemical and biochemical properties. *Biochim. Biophys. Acta* 80:441–447.

Quigley, G. J., and A. Rich. 1976. Structural domains of transfer RNA molecules. *Science* 194:796–806.

Rees, B., J. Cavarelli, and D. Moras. 1996. Conformational flexibility of tRNA: structural changes in yeast tRNA^{Asp} upon binding to aspartyl-tRNA synthetase. *Biochimie* 78:624–631.

Rould, M. A., J. J. Perona, and T. A. Steitz. 1991. Structural basis of anticodon loop recognition by glutaminyl-tRNA synthetase. *Nature* 352:213–218.

Schultz, D. W., and M. Yarus. 1994a. tRNA structure and ribosomal function. I. tRNA nucleotide 27–43 mutations enhance first position wobble. *J. Mol. Biol.* 235:1377–1380.

Schultz, D. W., and M. Yarus. 1994b. tRNA structure and ribosomal function. II. Interaction between anticodon helix and other tRNA mutations. *J. Mol. Biol.* 235:1395–1405.

Schweisguth, D. C., and P. B. Moore. 1997. On the conformation of the anticodon loops of initiator and elongator methionine tRNAs. *J. Mol. Biol.* 267:505–519.

Scott, W. G., J. T. Finch, and A. Klug. 1995. The crystal structure of an all-RNA hammerhead ribozyme: a proposed mechanism for RNA catalytic cleavage. *Cell* 81:991–1002.

Senger, B., S. Auxiliens, U. English, F. Cramer, and F. Fasiolo. 1997. The modified wobble base inosine in yeast tRNA^{Ile} is a positive determinant for aminoacylation by isoleucyl-tRNA synthetase. *Biochemistry* 36:8269–8273.

Serra, M. J., T. W. Barnes, K. Betschart, M. J. Guiterrez, K. J. Sprouse, C. K. Riley, L. Stewart, and R. E. Temel. 1997. Improved parameters for the prediction of RNA hairpin stability. *Biochemistry* 36:4844–4851.

Serra, M. J., M. H. Lyttle, T. J. Axenson, C. A. Schadt, and D. H. Turner. 1993. RNA hairpin loop stability depends on closing base pair. *Nucleic Acids Res.* 21:3845–3849.

Sprinzl, M., C. Horn, M. Brown, A. Ioudovitch, and S. Steinberg. 1998. Compilation of tRNA sequences and sequences of tRNA genes. *Nucleic Acids Res.* 26:148–153.

Steiner, T. 1996. C-H...O hydrogen bonding in crystals. *Cryst. Rev.* 6:1–57.

Szweykowska-Kulinska, Z., B. Senger, G. Keith, F. Fasiolo, and H. Grosjean. 1994. Intron-dependent formation of pseudouridines in the anticodon of *Saccharomyces cerevisiae* minor tRNA^{Ile}. *EMBO J.* 13:4636–4644.

van Gunsteren, W. F. 1993. Molecular dynamics and stochastic dynamics simulation: a primer, p. 3–36. *In* W. F. van Gunsteren, P. K. Weiner, and A. J. Wilkinson (ed.), *Computer Simulation of Biomolecular Systems*, vol. 2. ESCOM, Leiden, The Netherlands.

van Gunsteren, W. F., and H. J. C. Berendsen. 1990. Computer simulation of molecular dynamics: methodology, applications, and perspectives in chemistry. *Angew. Chem. Int. Ed. Engl.* **29**: 992–1023.

van Gunsteren, W. F., S. R. Billeter, A. A. Eising, P. H. Hünenberger, P. Krüger, A. E. Mark, W. R. P. Scott, and I. G. Tironi. 1996. Biomolecular simulation: the GROMOS96 manual and user guide. ETH Verlag, Zürich, Switzerland.

Wahl, C. M., and M. Sundaralingam. 1997. C-H...O hydrogen bonding in biology. *Trends Biochem. Sci.* **22**:97–102.

Wahl, M. C., S. T. Rao, and M. Sundaralingam. 1996. The structure of r(UUCGCG) has a 5'-UU-overhang exhibiting Hoogsteen-like trans U·U base pairs. *Nat. Struct. Biol.* **3**:24–31.

Ward, D. C., and E. Reich. 1968. Conformational properties of polyformycin: a polyribonucleotide with individual residues in the syn conformation. *Proc. Natl. Acad. Sci. USA* **61**:1494–1501.

Westhof, E., P. Dumas, and D. Moras. 1985. Crystallographic refinement of yeast aspartic acid transfer RNA. *J. Mol. Biol.* **184**: 119–145.

Westhof, E., P. Dumas, and D. Moras. 1988. Hydration of transfer RNA molecules: a crystallographic study. *Biochimie* **70**: 145–165.

Westhof, E., C. Rubin-Carrez, and V. Fritsch. 1995. The use of molecular dynamics simulations for modelling nucleic acids, p. 103–131. *In* J. M. Goodfellow (ed.), *Computer Modelling in Molecular Biology*. VCH, New York, N.Y.

Zerfass, K., and H. Beier. 1992. Pseudouridine in the anticodon GΨA of plant cytoplasmic tRNA[Tyr] is required for UAG and UAA suppression in the TMV-specific context. *Nucleic Acids Res.* **20**:5911–5918.

Zichi, D. A. 1995. Molecular dynamics of RNA with the OPLS force field. Aqueous simulation of a hairpin containing a tetranucleotide loop. *J. Am. Chem. Soc.* **117**:2957–2969.

Chapter 7

Modulatory Role of Modified Nucleotides in RNA Loop–Loop Interaction

Henri Grosjean, Claude Houssier, Pascale Romby, and Roland Marquet

Hairpin loops are the most frequent secondary motifs in RNA, and provide specific recognition sites for protein or RNA binding (reviewed in Le and Zucker, 1991; Wyatt and Tinoco, 1993; Varani, 1995; Pyle and Green, 1995; Brion and Westhof, 1997). The anticodon stem loop of tRNA is one of the most remarkable hairpins in RNAs that present several modified nucleotides arising by posttranscriptional processes. Many different modifications have been identified at position 34 (the so-called wobble base) and at position 37, 3′-adjacent to the anticodon (Fig. 1; see also Appendixes 5 and 6). These modifications are considered to be important for the specificity, fidelity, and efficiency of the tRNA functions, mainly in decoding the genetic message during translation on the ribosome (Björk, 1995a; see also Chapter 27 by Curran). The anticodon loop of tRNA is also involved in other types of RNA–RNA interaction, such as with mRNA in transcription antitermination control of the expression of a number of genes coding for aminoacyl-tRNA synthetases in *Bacillus subtilis* (for reviews, see Henkin, 1994; Putzer et al., 1995b) and with HIV-1 RNA in the initiation complex of reverse transcription (see Chapter 28 by Marquet).

Because of the remarkable binding properties of the tRNA anticodon loop with other RNA molecules, understanding of the role of modified nucleotides in these interactions is of utmost importance. The complex formed between two tRNAs harboring complementary anticodons was used as a model system for evaluating these effects. This type of loop–loop interaction was initially detected by electrophoresis on a nondenaturing polyacrylamide gel of a mixture of yeast tRNA^{Phe} (anticodon GmAA) and *Escherichia coli* tRNA$_2^{Glu}$ (anticodon mnm^{5}s^{2}UUC) (Eisinger, 1971a, 1971b; Eisinger and Blumberg, 1973). The existence of such binary complexes has been confirmed with many other pairs of naturally occurring tRNAs bearing complementary anticodons (Fig. 2) using different techniques: gel electrophoresis (Eisinger and Gross, 1974); affinity chromatography on solid supports (Grosjean et al., 1973; Vögeli et al., 1975; Buckingham 1976); temperature-jump (T-jump) relaxation technique (Grosjean et al., 1976, 1978), X-ray crystallography (Moras et al., 1980; Westhof et al., 1983); small-angle X-ray scattering (Nilson et al., 1982); and nuclear magnetic resonance (NMR) (Amano and Kyogoku, 1993).

These anticodon–anticodon interactions between tRNAs show some similarity with intermolecular loop–loop interactions between mRNA transcripts and antisense RNA involved in plasmid replication control (the so-called "kissing" loop complexes; for reviews, see Eguchi and Tomizawa, 1991; Tomizawa, 1993; Wagner and Simons, 1994). A similar type of "kissing" loop complex also exists between two genomic viral RNAs during RNA dimerization of certain retroviruses (for review see Paillart et al., 1996a).

In this chapter, we summarize the information available about the role of modified nucleotides present at positions 34 and 37 in modulating tRNA anticodon–anticodon interaction. Implications for codon–anticodon interaction and comparison with other types of RNA–RNA interacting systems are also discussed.

Henri Grosjean • Centre National de la Recherche Scientifique, Laboratoire d'Enzymologie et Biochimie Structurales, 1 avenue de la Terrasse, F-91198 Gif-sur-Yvette, France. **Claude Houssier** • Université de Liège, Laboratoire de Chimie Macromoléculaire et Chimie Physique, Sart-Tilman (B6), B-4000 Liège, Belgium. **Pascale Romby and Roland Marquet** • 3 Centre National de la Recherche Scientifique, Institut de Biologie Moléculaire et Cellulaire, 15 rue René Descartes, F-67084 Strasbourg, France.

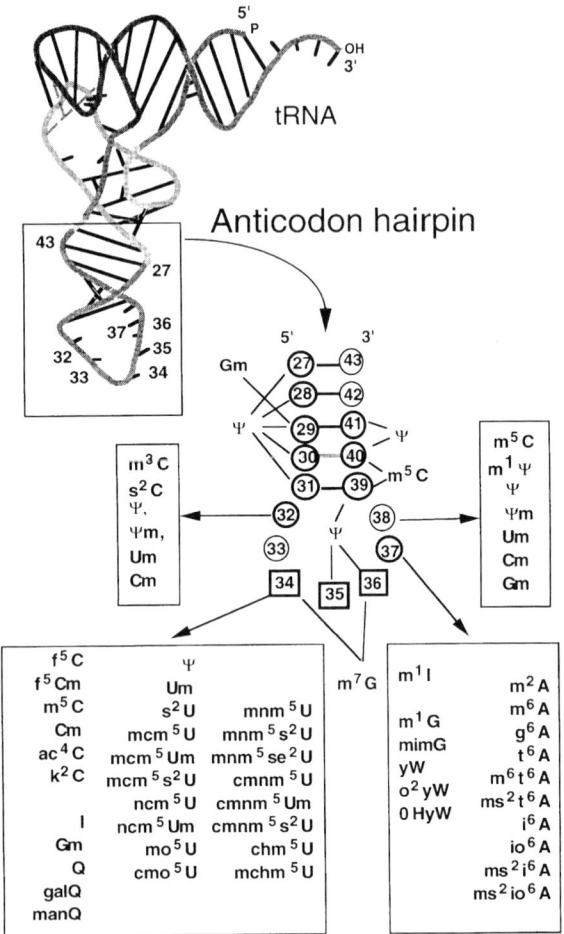

Figure 1. Type and location of modified nucleotides in the anticodon stem and loop of tRNAs. The upper part is a schematic representation of the three-dimensional architecture of tRNA. Numbering of nucleotide positions are those universally adopted (see Appendix 5 by Auffinger and Westhof). The anticodon nucleotides corresponding to positions 34, 35, and 36 are shown as square boxes in the anticodon hairpin representation (lower part of the figure). Symbols for modified nucleotides are those defined in Appendix 1 by Motorin and Grosjean. The information is derived from the tRNA data bank of Sprinzl et al. (1996). Frequency of modified nucleotides at specific positions can be obtained from the compilation of Grosjean et al. (1995) and from the data in Appendix 5 by Auffinger and Westhof. Almost all of the hypermodified nucleotides characterized so far occur exclusively in position 34 (the so-called wobble base) or in position 37 (3'-adjacent to the anticodon); see also Appendix 6 by Björk. Notice also that all positions of the anticodon hairpin (except positions 33, 42, and 43) are potential sites for modification, especially for pseudouridylation (Ψ).

THE TEMPERATURE-JUMP RELAXATION METHOD

One way to evaluate the effect of modified nucleotides on the stability and dynamic properties of anticodon–anticodon complexes is the temperature-jump relaxation technique. In this method, the temperature of a solution containing two interacting molecules at equilibrium is quickly raised by a few degrees. This is usually done by the discharge of a high-voltage capacitor into the conducting solution. The time for the solution to reach the new temperature is less than 3 microseconds. This is much faster than the time needed for most biomolecular processes to reach the new chemical equilibrium (Eigen, 1954; Eigen and DeMaeyer, 1963; Hammes, 1968). This method is effective only if an appreciable change of enthalpy (of the order of 10–25 kcal per mol) is associated with the equilibrium process as found for anticodon-anticodon interactions. Also, the important hyperchromic effect associated with dissociation of an RNA double-stranded structure facilitates detection of even a small shift in equilibrium (for details, see Crothers, 1971; Riesner and Römer, 1973).

The range of temperatures explored for anticodon–anticodon complexes by the T-jump technique is between 0 and 35°C. Also, the buffer solution always contains magnesium ions and enough salts to preserve the three-dimensional (3-D) architecture of the tRNA molecules. Therefore, only the stability of the tRNA duplex is affected by the abrupt elevation of temperature (usually between 4 and 5°C). After the T-jump, a fraction of the binary complex dissociates. This corresponds to the self-adjustment of the perturbed molecular system to its new thermal equilibrium. The time ("relaxation time") required in this process is related to the specific rates of the RNA–RNA association and dissociation reactions involved. Analysis of the relaxation signal is monitored by measuring the hyperchromic effect at 260 nm (for examples see Fig. 2 in Vacher et al., 1984; and Houssier and Grosjean, 1985). This determines both the amplitude of the relaxation signal and the relaxation time required to reach the new equilibrium. Knowing the concentration of the interacting macromolecules and the final temperature of the mixture, the corresponding rate constants and the standard thermodynamic parameters can be easily calculated (Crothers, 1971; Thusius et al., 1973; Bernasconi, 1976; Grosjean and Houssier, 1990).

FEATURES OF THE ANTICODON–ANTICODON COMPLEXES

The temperature-jump relaxation technique does not detect the binding of two complementary RNA triplets, such as UUC and GAA at concentrations up to the millimolar range, even at low temperature (K_{ass} lower than 1 at 0°C [Fig. 3a]; Yaskunas et al., 1968). Valuable thermodynamic and kinetic parameters can be obtained only for interacting polynu-

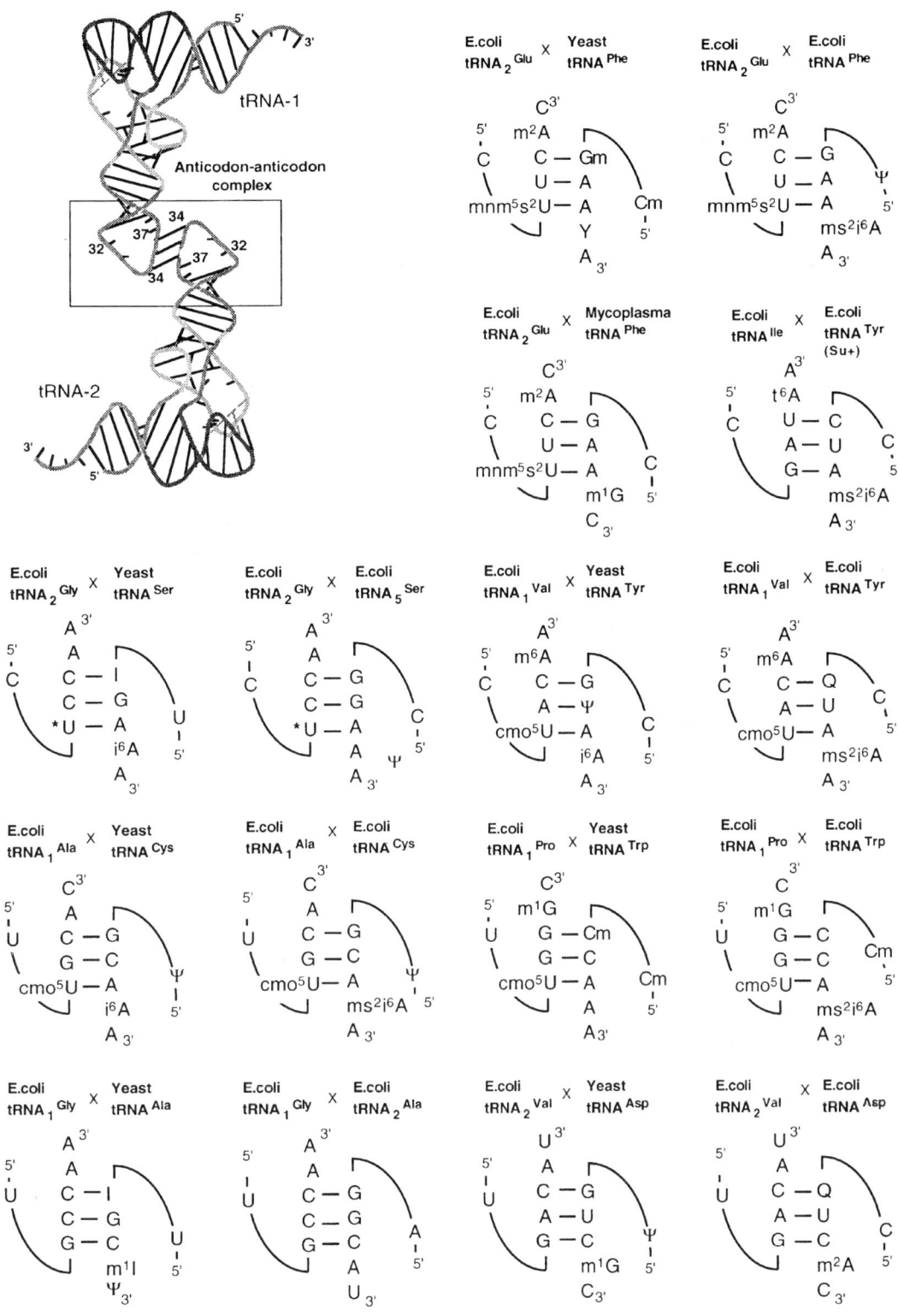

Figure 2. Schematic representation of complexes formed between two tRNAs having complementary anticodons. Only the nucleotides present in the 7-member anticodon loops are indicated. However, the invariant uridine-33, 5'-adjacent to the wobble base, is omitted for better clarity of the figure. Out of the 60 anticodon–anticodon complexes we have studied, only those having their 3 anticodon nucleotides engaged in canonical A·U, G·C or I·C Watson–Crick base pairs are represented (no mismatch base pairs). Symbols for modified nucleotides are defined in Appendix 1 by Motorin and Grosjean. U* at position 34 of *E. coli* tRNA$_2^{Gly}$ designates a yet unidentified modified uridine (possibly a derivative of nm^{5}U or mnm^{5}U). Sequence information originates from the tRNA data bank of Sprinzl et al., 1998.

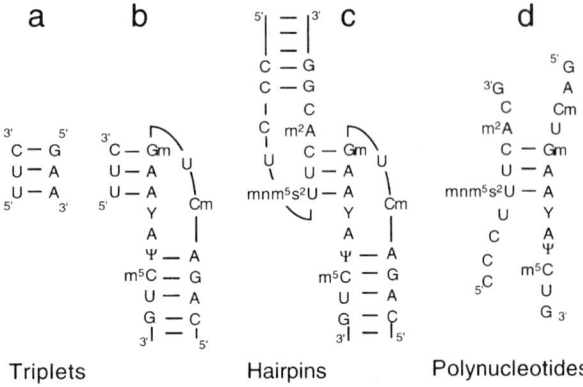

Figure 3. Comparison of complexes formed between complementary sequences in various polynucleotides. (a) Two triplets mimicking the anticodons of tRNA$_2^{Glu}$ and tRNA^{Phe}. (b) Anticodon triplet corresponding to tRNA$_2^{Glu}$ and its complementary sequence within an anticodon hairpin of yeast tRNA^{Phe} (harboring naturally occurring modified nucleotides). Only the anticodon loop and the proximal anticodon stem are represented, despite the fact that the experiments described in the text concern the entire yeast tRNA^{Phe}. (c) The complementary sequences are both part of the anticodon loop of the entire tRNAs (E. coli tRNA$_2^{Glu}$ and yeast tRNA^{Phe}) (again only the anticodon loop and proximal anticodon stem are presented). (d) The interaction between a fragment of E. coli tRNA$_2^{Glu}$ (10-mer) and fragment of yeast tRNA^{Phe} (12-mer) is shown, both fragments bearing their respective anticodon sequence. These fragments were obtained after purification of the RNase T$_1$ hydrolysate of the corresponding tRNAs. Association constants were determined at ionic strength 0.1, 10 mM Mg^{2+} at pH 6.8, by the T-jump relaxation technique. Symbols for modified nucleotides are those of Appendix 1 by Motorin and Grosjean. For more details see Grosjean et al. (1976) and Labuda et al. (1984).

cleotides longer than a triplet (for examples see Craig et al., 1971; Pörschke et al., 1973; Ravetch et al., 1974; Nelson and Tinoco, 1982). However, the UUC triplet binds to the anticodon of yeast tRNA^{Phe} (Fig. 3b) with an association constant K$_{ass}$ of about 10^3 M^{-1} at 0°C (Eisinger et al., 1971; Uhlenbeck, 1972; Labuda et al., 1984, 1985; Bujalowski et al., 1986a). When the complementary triplets are both part of the anticodon loop of tRNAs, such as the yeast tRNA^{Phe} and the E. coli tRNA$_2^{Glu}$ (Fig. 3c), the K$_{ass}$ at 0°C becomes as high as 10^7 M^{-1} (Eisinger and Gross, 1975, Grosjean et al., 1976). Interestingly, tRNA fragments containing the anticodon sequences (10 and 12 nucleotides [nt] long), thus including the modified nucleotides but unable to adopt a stable loop structure (Fig. 3d), show an association constant about two orders of magnitude lower (K$_{ass}$ = 10^5 M^{-1} at 0°C) than that obtained with intact tRNAs (Grosjean et al., 1976). Clearly, these observations indicate that complexes formed between tRNAs with a complementary anticodon are more stable by several orders of magnitude than those formed between complementary triplets. Also, both the loop structure and the presence of modified nucleotides play an important role in stabilizing such anticodon-anticodon complexes.

From a detailed analysis of the kinetic parameters of the anticodon-anticodon equilibrium, it appeared that the drastic increase in stability is almost entirely due to a remarkably low dissociation rate (k$_{-1}$ in the range of 0.1 to 6 s^{-1} at 0°C; Grosjean et al., 1976; Vacher et al., 1984; Houssier and Grosjean, 1985; Romby et al., 1985). In consequence, anticodon-anticodon complexes show unexpectedly high activation energy for the dissociation of the complexes (in the range of 20 kcal mol^{-1}). This is of the same order of magnitude as the values measured for short oligoribonucleotide duplexes having five to seven interacting base pairs (Craig et al., 1971; Pörschke et al., 1973; Nelson and Tinoco, 1982).

In contrast, the association rate constants (k$_{+1}$) of the anticodon-anticodon complex are in the range of 10^5 to 10^6 M^{-1} s^{-1} at 0°C, close to those measured for oligonucleotide duplex formation (Grosjean et al., 1976; Vacher et al., 1984; Houssier and Grosjean, 1985; Romby et al., 1985). The presence of 1.6 M ammonium sulfate (in addition to 10 mM Mg^{2+} present in all buffers used) increased this association rate constant by a factor of 2 to 3, suggesting that the anticodon loop adopts a more favorable geometry under these conditions, thereby reducing the entropy term (Romby et al., 1985). Moreover, the apparent activation energy for the formation of anticodon-anticodon duplex is close to zero (Grosjean et al., 1976; Houssier and Grosjean, 1985).

T-jump studies also revealed one striking feature of the tRNA anticodon-anticodon complexes. The observed relaxation time (or lifetime) for different pairs of tRNAs with complementary anticodons of different G or C contents varied by less than two orders of magnitude (right side of Fig. 4; Grosjean et al., 1978), whereas the results obtained for duplex formation between single-stranded oligoribonucleotides containing only G·C or A·U base pairs suggest differences in lifetimes of several orders of magnitude (left side of Fig. 4; Craig et al., 1971; Pörschke et al., 1973; Borer et al., 1974). Only duplexes at least 5 to 10 nt long (depending on the G·C content) reach a lifetime comparable with those measured for anticodon-anticodon complexes (Ravetch et al., 1974; Nelson and Tinoco, 1982).

Altogether, these results reveal three major components that contribute to the remarkable stability of the anticodon-anticodon complexes: the decreased flexibility of the anticodon loop due to constraints imposed by the base-paired stem (reduction of the unfavorable entropy term); the presence of noncomplementary nucleotides outside the anticodon triplet (the dangling effect); and the presence of character-

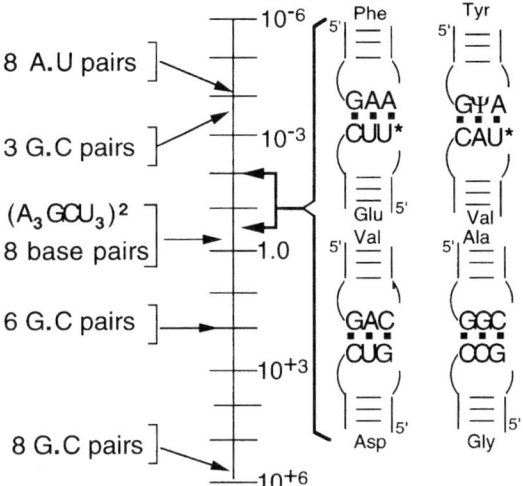

Figure 4. Comparison of relaxation time constants (lifetimes) of different types of nucleic acid associations at 20°C. The left part of the figure shows data concerning duplex formation using synthetic oligoribonucleotides. They were kindly provided by D. Pörschke (MPI of Göttingen, Germany; see also Pörschke et al., 1973; Ravetch et al., 1974). On the right part of the figure are the data concerning tRNA anticodon–anticodon complexes. Only a few selected examples are illustrated on this figure: the complete set of tRNA pairs are given in Fig. 2; see also Grosjean et al., 1978; Vacher et al., 1984; Houssier and Grosjean, 1985; and Romby et al., 1985). The time scale is logarithmic. Relaxation time constants of the bimolecular reactions are given at zero concentration and correspond to the reverse of the kinetic rate constants of the dissociation of the complexes.

istic modified nucleotides that enhance stacking interactions and impose structural constraints to the loop. These peculiarities of the anticodon loop not only allow the anticodon–anticodon complex to have a long lifetime, but also tend to minimize the difference in lifetimes of an A·U-rich and a G·C-rich anticodon–anticodon pair.

ROLE OF MODIFIED NUCLEOTIDES ON THE ANTICODON–ANTICODON INTERACTION

Stabilization/Modulation Role for Modified Bases 3′-Adjacent to the Anticodon

By using the experimental procedures previously described, the effect of the presence (or absence) of hypermodified bases at position 37 on the thermodynamic and kinetic properties of various tRNA pairs has been evaluated (Fig. 5). These values are summarized in Table 1 (see also Grosjean et al., 1976; Weissenbach and Grosjean, 1981; Vacher et al., 1984; and Houssier and Grosjean, 1985).

The wybutosine (Y-base) in yeast tRNA^{Phe} (Fig. 5a) and the 2-methylthio-N^6-isopentenyl (ms^{2}i^6) modification of adenine 37 in E. coli tRNA^{Phe} and in E. coli RNA^{Trp} (Fig. 5b and c), have a strong stabilizing effect arising mainly from a decrease of the dissociation rate constant k_{-1} by a factor of 3 to 5 at 0°C, as compared with other tRNA pairs including Mycoplasma tRNA^{Phe} having m^{1}G$_{37}$, yeast tRNA^{Phe} lacking the Y-base (Grosjean et al., 1976), or a mutant miaA E. coli RNA^{Trp} lacking the ms^{2}i^6 group (Vacher et al., 1984). In the case of the ms^{2}i^6 modification, most if not all of the effect may be attributed to the ms^2 part of this hypermodification of A$_{37}$, as tested with E. coli tRNA^{Cys} (ms^{2}i^{6}A) and yeast tRNA^{Cys}(i^{6}A), both interacting with E. coli tRNA^{Ala} (Fig. 5d; Houssier and Grosjean, 1985). Evaluation of the variation of the standard enthalpy of the reaction and inspection of the differential relaxation spectrum in the range of 300 to 350 nm (where the thio group absorbs), concludes that the stabilizing effect essentially occurs by the stacking of the thiolated purine 37 on the anticodon–anticodon complex (see Fig. 5 in Houssier and Grosjean, 1985).

A significant stabilizing effect of the N^6-threonylcarbamoyl group (t^6) on A$_{37}$ in yeast tRNA^{Arg} has also been measured. In this case, it seems that the t^6 group reinforces the intrastrand stacking of the anticodon loop, but also the interstrand stacking with nucleoside 34 of the complementary anticodon (Weissenbach and Grosjean, 1981).

One unexpected result is the remarkably high stability of the complex formed between E. coli tRNA$^{Gly}_1$ (anticodon GCC) and E. coli tRNA$^{Ala}_2$ (anticodon GGC), in which adenosine 37 is not modified in both tRNAs (Fig. 2; Grosjean et al., 1978). The same situation exists with E. coli tRNA$^{Ser}_5$ (anticodon GGA) and E. coli tRNA$^{Gly}_2$ (anticodon U*CC, where U* is an unknown modified U [Fig. 2]; Grosjean et al., 1985; Houssier and Grosjean, 1985). The most plausible explanation of these data is that the presence of a dangling adenosine 37 at the 3′ side of the G·C-rich anticodons sufficiently increases the potential for strong binding with another RNA, so that the A$_{37}$ hypermodification is unnecessary and therefore has been selectively prevented in evolution (Cedergren et al., 1981, 1986; Motorin et al., 1997).

It is noteworthy that the tRNA anticodons ending with A$_{36}$ or U$_{36}$ are almost always flanked on their 3′ side by a stabilizing hypermodified purine 37: wybutosine in eukaryotic tRNA^{Phe}, or i^{6}A, io^{6}A, ms^{2}i^{6}A, t^{6}A, mt^{6}A and ms^{2}t^{6}A in other tRNAs (Fig. 6; see also Appendix 6 by Björk). The presence of these modified nucleotides at position 37 is well correlated with the lower stacking energy of the doublets of base pairs including one or two U·A or A·U pairs. On the

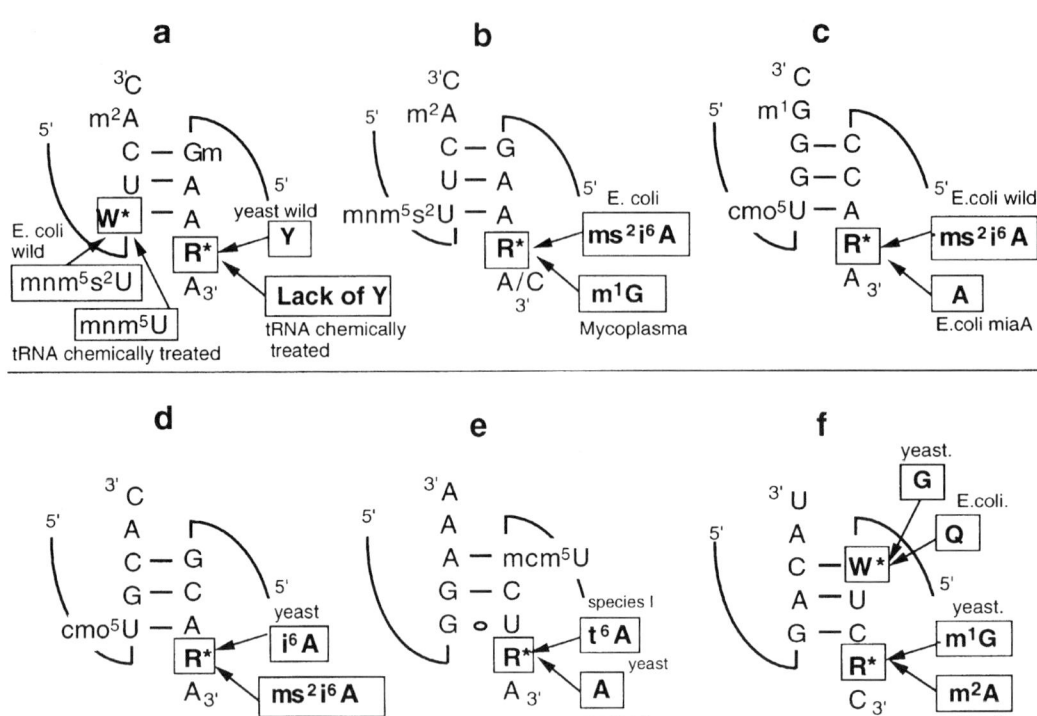

Figure 5. Schematic representation of complexes formed between two tRNAs having complementary anticodons. In boxes are the nucleotides for which the effect of alteration of the chemical structure or substitution by another nucleotide have been measured (see Table 1). Only the nucleotides present in positions 34–38 of the anticodon loops are indicated. W* stands for wobble base-34; R* stands for purine 37; all other symbols for modified nucleotides are as in Appendix 1 by Motorin and Grosjean. The different tRNA pairs are as follows. (a) *E. coli* tRNA$_2^{Glu}$ (wild type or chemically treated to oxidize the thiol group in mnm^{5}s^{2}U$_{34}$, Houssier et al., 1988) and yeast tRNA^{Phe} (wild type or chemically treated under acidic conditions to remove the Wye base; Grosjean et al., 1976). (b) *E. coli* tRNA$_2^{Glu}$ (wild type) and *E. coli* tRNA^{Phe} or *Mycoplasma* tRNA^{Phe} (both wild type; Grosjean et al., 1976). (c) *E. coli* tRNA$_1^{Pro}$ (wild type) and *E. coli* tRNA^{Trp} (wild type or the *mia* mutant lacking the ms^{2}i^6 group on A$_{37}$; Vacher et al., 1984). (d) *E. coli* tRNA$_1^{Ala}$ (wild type) and *E. coli* tRNA^{Cys} or yeast tRNA^{Cys} (both wild type; Houssier and Grosjean, 1985). (e) *E. coli* tRNA$_5^{Ser}$ (wild type) and yeast tRNA$_3^{Arg}$ (species I and II). Both species were purified from the same yeast strain; species II was shown to lack the t^6 group on A$_{37}$ (Weissenbach and Grosjean, 1981). Notice that in this example, a G·U wobble occurs within the anticodon–anticodon complex. (f) *E. coli* tRNA$_2^{Val}$ (wild type) and *E. coli* tRNA^{Asp} or yeast tRNA^{Asp} (both wild type; Romby et al., 1985). Here, two parameters vary (G$_{34}$ and m^{1}G$_{37}$ in the yeast tRNA versus Q$_{34}$ and m^{2}A$_{37}$ in the *E. coli* tRNA). Therefore, the data obtained with this tRNA pair cannot be interpreted solely on the basis of one nucleotide difference, as in all of the other cases above. The T-jump relaxation technique was used to compare the relative stability of the different tRNA pairs (wild types and mutants; see Table 1).

contrary, in the case of "G- and/or C-rich" anticodons ending with C$_{36}$ or G$_{36}$, the 3′-adjacent purine 37 is generally an unmodified A or an A or G modified by a simple methylation (m^{2}A, m^{6}A, m^{1}G or m^{1}I; Fig. 6; see also Appendix 6 by Björk). Most of these modifications are specific not for the tRNA per se but for a class of anticodons, and are therefore relevant to the codon tRNA recognition process.

Stabilization/Modulation Role for the Wobble Base

It is generally accepted that the recognition of synonymous codons occurs as a consequence of structural flexibility ("wobble") of the glycosidic bond at position 34 of the anticodon, leading to the formation of nonstandard hydrogen bonded base pairs with the third codon position (Crick, 1966). The effect of base modifications, especially uridine 34, would be to regulate the flexibility or rigidity of the anticodon so as to contribute to the correct and efficient translation of family codons by cognate tRNA species (Yokoyama and Nishimura, 1995; see also Chapters 5 and 27, by Davis and Curran, respectively).

To evaluate the possibility that certain base modifications at position 34 are related to changes in the intrinsic properties of the anticodon loop itself, thus changing its binding properties, more than 60 pairs of tRNAs with partly complementary anticodons (two correct base pairs out of three), were analyzed by the T-jump relaxation method (Grosjean et al., 1978; Weissenbach and Grosjean, 1981; Labuda et al., 1982; Romby et al., 1985). The results indicate

Table 1. Thermodynamic and kinetic parameters of selected tRNA anticodon–anticodon complexes, as determined by the temperature-jump relaxation technique[a]

tRNA:tRNA	Figure(s)	$T_{1/2}$ (°C) (±2)	$-\Delta H°$ (kcal/mol) (±2)	$-\Delta S°$ (cal/mol·degree) (±7)	K_{ass} (μM^{-1}) 0°C	K_{ass} (μM^{-1}) 20°C	$k_{-1(diss)}$ (s^{-1}) 0°C	$k_{-1(diss)}$ (s^{-1}) 20°C	$k_{+1(ass)}$ ($\mu M^{-1} s^{-1}$) 0°C	$k_{+1(ass)}$ ($\mu M^{-1} s^{-1}$) 20°C	E_{diss} (kcal/mol) (±2)	E_{ass} (kcal/mol) (±2)	Reference
Reference system: *E. coli* tRNA$_2^{Glu}$ (wild) + yeast tRNA^{Phe} (wild)	2, 5a	26	24	55	20.0	2.9	0.2	2.9	3.7	3.8	22	0	Grosjean et al., 1976
Testing removal of Wye base 37: *E. coli* tRNA$_2^{Glu}$ (wild) + yeast tRNA^{Phe} (lacking Wye base)	5a	12	18	39	1.4	0.1	1.0	12.9	1.5	1.8	20	2	Grosjean et al., 1976
Testing dethiolation of mnm^{5}s^{2}U$_{34}$: *E. coli* tRNA$_2^{Glu}$ (lacking s^2 on mnm^{5}U) + yeast tRNA^{Phe} (wild)	5a	6	22	50	1.8	0.1	0.9	15.4	1.3	1.2	22	−2	Houssier et al., 1988
ms^{2}i^{6}A$_{37}$ for Wye 37 and G$_{34}$ for Gm34: *E. coli* tRNA$_2^{Glu}$ (wild) + *E. coli* tRNA^{Phe} (wild)	2, 5b	24	24	55	23.0	1.1	0.2	4.1	3.7	4.4	26	2	Grosjean et al., 1976
Testing m^{1}G$_{37}$ in place of ms^{2}i^{6}A$_{37}$: *E. coli* tRNA$_2^{Glu}$ (wild) + *Mycoplasma* tRNA^{Phe} (wild)	2, 5b	15	22	52	3.4	0.2	0.6	13.3	2.0	2.7	25	3	Grosjean et al., 1976
Testing fragments of tRNAs: *E. coli* tRNA$_2^{Glu}$ (fragment N10) + yeast tRNA^{Phe} (fragment N12)	3d	3	23	35	1.0	0.07	3.4	58.0	3.3	4.2	23	2	Grosjean et al., 1976
Reference system: *E. coli* tRNA$_1^{Pro}$ (wild) + *E. coli* tRNA^{Trp} (wild)	2, 5c	14	23	51	5.1	0.3	0.1	2.3	0.6	0.7	25	2	Vacher et al., 1984
Testing lack of ms^{2}i^6 group on A$_{37}$: *E. coli* tRNA$_1^{Pro}$ (wild) + *E. coli* tRNA^{Trp} (*miaA* mutant)	5c	2	18	39	0.8	0.1	1.0	16.0	0.8	1.5	22	4	Vacher et al., 1984
Cm$_{34}$ for C$_{34}$ and A$_{37}$ for ms^{2}i^{6}A$_{37}$: *E. coli* tRNA$_1^{Pro}$ (wild) + yeast tRNA^{Trp} (wild)	2	~0	20	48	0.3	0.03	1.2	25.0	0.4	0.7	24	4	Houssier and Grosjean, 1985
Reference system: *E. coli* tRNA$_1^{Ala}$ (wild) + *E. coli* tRNA^{Cys} (wild)	2, 5d	16	17	32	3.9	0.5	0.6	6.5	2.5	3.0	19	2	Houssier and Grosjean, 1985
Testing i^{6}A$_{37}$ in place of ms^{2}i^{6}A$_{37}$: *E. coli* tRNA$_1^{Ala}$ (wild) + yeast tRNA^{Cys} (wild)	2, 5d	~0	14	22	0.9	0.2	5.7	32.0	4.9	5.0	14	0	Houssier and Grosjean, 1985

Continued on following page

Table 1. Continued

tRNA:tRNA	Figure(s)	$T_{1/2}$ (°C) (±2)	$-\Delta H°$ (kcal/mol) (±2)	$-\Delta S°$ (cal/mol·degree) (±7)	K_{ass} (μM^{-1}) 0°C	K_{ass} (μM^{-1}) 20°C	$k_{-1(diss)}$ (s^{-1}) 0°C	$k_{-1(diss)}$ (s^{-1}) 20°C	$k_{+1(ass)}$ (μM^{-1}s^{-1}) 0°C	$k_{+1(ass)}$ (μM^{-1}s^{-1}) 20°C	E_{diss} (kcal/mol) (±2)	E_{ass} (kcal/mol) (±2)	Reference
Reference system: *E. coli* tRNA$_5^{Ser}$ (wild) + yeast tRNA$_2^{Arg}$ (wild, species I)	5e	<0	ND	ND	0.22^b		7.7^b		1.7^b		ND	ND	Weissenbach and Grosjean, 1981
Testing lack of t^6 group on A$_{37}$: *E. coli* tRNA$_5^{Ser}$ (wild) + yeast tRNA$_3^{Arg}$ (wild, species II)	5e	<0	ND	ND	0.06^b		32.0^b		2.1^b		ND	ND	Weissenbach and Grosjean, 1981
Reference system: *E. coli* tRNA$_2^{Val}$ (wild) + yeast tRNA^{Asp} (wild)	2, 5f	10	25	60	3.9	0.2	0.2	4.1	0.9	0.7	23	−2	Romby et al., 1985
Q$_{34}$ and m^{2}A$_{37}$ instead of G$_{34}$ and m^{1}G$_{37}$: *E. coli* tRNA$_2^{Val}$ (wild) + *E. coli* tRNA^{Asp} (wild)	2, 5f	7	24	58	2.0	0.1	0.4	6.8	0.8	0.7	23	−1	Romby et al., 1985
Another Q$_{34}$-containing tRNA tested: *E. coli* tRNA$_1^{Val}$ (wild) + *E. coli* tRNA^{Tyr} (wild)	2	5	17	36	0.7	0.1	1.1	13.4	0.7	1.0	20	3	Grosjean and Houssier, unpublished results
Other systems													
No modification at position 37: *E. coli* tRNA$_1^{Gly}$ (wild) + *E. coli* tRNA$_2^{Ala}$ (wild)	2, 4	3	16	32	0.9	0.1	0.7	6.5	0.6	0.7	18	2	Grosjean and Houssier, unpublished results
No modification at position 37: *E. coli* tRNA$_2^{Gly}$ (wild) + *E. coli* tRNA$_5^{Ser}$ (wild)	2	19	20	41	7.5	0.6	0.7	9.0	5.6	5.6	20	0	Houssier and Grosjean, 1985

^{a} The temperature-jump experiments were performed with the aid of an instrument as described in Crothers, 1971; Rigler et al., 1974; or Grosjean and Houssier, 1990. Mixtures of each pair of tRNAs were prepared in 100 mM sodium sulfate, 10 mM magnesium sulfate, and 10 mM sodium cacodylate (pH 6.9), at equimolecular concentrations of each tRNA (about 1.8 μM), on the basis of amino acid acceptance measurements. The anticodon sequences are given in Fig. 2, 3, and 5. A temperature-jump of 4.2°C was applied and several signals were recorded at about 4°C intervals from 0°C to 35°C. The relaxation kinetics were followed by absorption changes at 260 nm. Assuming a simple equilibrium of the type tRNA^1 + tRNA^2 → tRNA^1:tRNA^2 complex, thermodynamic and kinetic parameters can be easily obtained from the differential melting curve profiles. They were obtained by analyzing the effect of temperature on the amplitude of the slow relaxation signal (for details see Crothers, 1971; Raveth et al., 1974; Grosjean et al., 1976; and Houssier and Grosjean, 1985). The $T_{1/2}$ corresponds to temperature of half melting of the tRNA:tRNA complex (T_{max} of the differential melting curve profile). The half-width of the differential melting curve allowed us to determine the standard enthalpy change ($\Delta H°$) and the equilibrium association constant (K_{ass}) of the bimolecular reaction at a given temperature (Van t'Hoff equation). The corresponding standard entropy change ($\Delta S°$) was then calculated from $\Delta H°$ and K_{ass} using $\Delta S° = RT \ln K_{ass} = RT \ln K_{ass} / \Delta H° \times T$. The kinetic rate constants $k_{-1(diss)}$ and $k_{+1(ass)}$ were obtained from the measuring the relaxation times and the K_{ass} at different temperatures. Analysis of the temperature dependence of the rate constants yields the activation energies E_{ass} and E_{diss}. For tRNA pairs having $T_{1/2}$ below zero, like for the complex *E. coli* tRNA$_5^{Ser}$:yeast tRNA^{Arg} bearing a G·U wobble base pair, the dependence of the relaxation time on the tRNA concentration was used to obtain the kinetic and the equilibrium constants at final temperature after the T-jump. Numbers in brackets give an idea of the accuracy of each measurements; ND means nondetermined.
^{b} At 10°C.

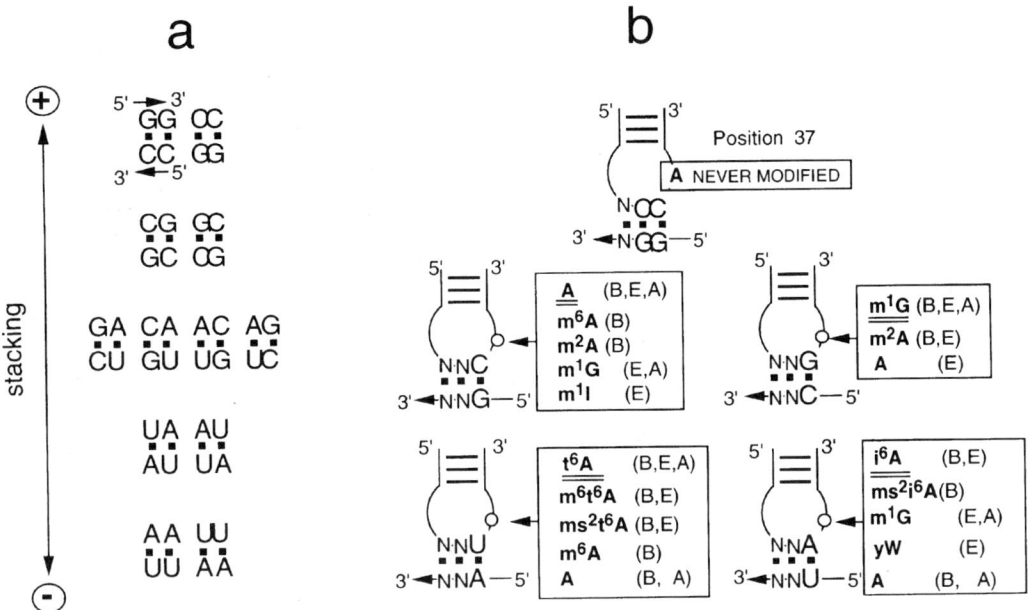

Figure 6. Correlation between the relative order of the stacking energy of base pairs (a) and the presence of modified bases located at position 37 of the anticodon loop of tRNA (b). (a) Classification of the 10 possible Watson–Crick base paired nearest-neighbor sequences (notice that GG·CC and CC·GG as well as UU·AA and AA·UU are identical) according to measured thermodynamic parameters taken from Borer et al., 1974, and largely confirmed later (reviewed in Turner and Bevilacqua, 1993). (b) Correlation between the nucleotide at position 36 of anticodon in tRNA (cardinal nucleotide; see Yarus, 1982) and the type of modified nucleotide found at position 37 of tRNA from different origins. B, E, and A in parentheses stand for eubacteria, eukaryotes, and archaea, respectively; *Mycoplasma*, mitochondria, and chloroplasts have been excluded from the compilation. Symbols for modified nucleotides are those defined in Appendix 1 by Motorin and Grosjean. The most frequent modified nucleotides are doubly underlined. N stands for any nucleotide. For more detailed information see Appendix 6 by Björk or Sprinzl et al., 1998.

that all of the "correct" base triplet interactions (as determined by the genetic coding rules) have a long lifetime (between 15 and 1500 ms at 9°C). Among those stable anticodon–anticodon complexes, the most remarkable ones involve the G·U wobble base pair between the first and the third bases of the complementary anticodons. These complexes are only 3 to 10 times less stable than those involving a G·C Watson–Crick base pair at the same positions (Weissenbach and Grosjean, 1981; Labuda et al., 1982). Interestingly, the Q·U wobble pairing increases stability by approximately 3 times compared to the identical triplet sequence containing the G·U wobble pairing, whereas the interaction involving a Q·C is slightly less stable than the identical pairing containing G·C pairing. In other words, while a G·U wobble base pair is less stable than a G·C Watson–Crick base pair, both the Q·U wobble and the Q·C Watson–Crick base pairs have about the same stability (Grosjean et al., 1978; Romby et al., 1985; see also Meier et al., 1985). Likewise, I·U wobble pairs are less stable than G·C Watson–Crick base pairs in the same sequence by a factor of 3 to 10, but form complexes of about the same relative stability as those involving I·C pairs (Grosjean et al., 1978).

Noteworthy is that RNA:RNA complexes involving I·A pairs are marginally stable (Wang and Kallenbach, 1971; Curran, 1995; Carter et al., 1997).

The complexes involving a G·U or U·G base opposition in the middle of the anticodon (position 35), as well as most "genetically incorrect" complexes with mismatches of the kind C·A, C·C, C·U, A·A, G·G, or G·A at any of the three anticodon positions 34, 35, or 36, are only marginally stable (lifetimes below 15 ms at 9°C) or even nonexistent (Grosjean et al., 1978). A remarkable exception is the yeast tRNA^{Asp} homoduplex (lifetime, 45 ms at 9°C) involving a U·U mismatch in the middle position of the pseudocomplementary GUC/GUC anticodons, as demonstrated in solution (Fig. 7a; Romby et al., 1985) and in the crystal structure (Fig. 7b: Moras et al., 1980, 1986; Westhof et al., 1985). The existence of the *E. coli* tRNA$^{Gly}_1$ homoduplex involving the hemiprotonated C$^+$·C mismatch also in the middle position of the pseudocomplementary GCC/GCC anticodons has been demonstrated in solution only under acidic pH (Romby et al., 1986; Amano and Kyogoku, 1993). The special situation of the U·U and C$^+$·C pairs "sandwiched" between two stable and coaxially stacked G·C/C·G pairs is probably the

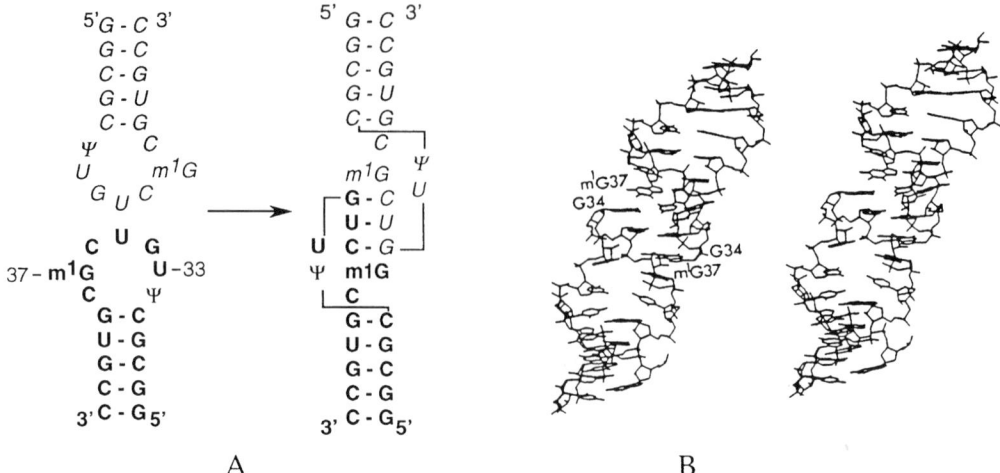

Figure 7. Structural model for the tRNA anticodon–anticodon interaction. The case of duplex formation with yeast tRNA^{Asp} is illustrated because the crystal structure of this duplex at a 3-Å resolution is known (Westhof et al., 1985; Moras et al., 1986). (A) Nucleotide sequences of the interacting anticodon hairpins; (B) stereo view of the interacting anticodon loops of yeast tRNA^{Asp} (only the segment from Ψ_{32} to A_{38} is shown). In this complex, the three bases of the anticodon stack in a helical conformation (A-form) with the purine 37 and the pyrimidine 38 on their 3' side. The two pyrimidines on the 5' side of the loop are stacked together but separated from the other stacked bases by a short bend. The two-fold rotational axis is perpendicular to this paper, through the bond joining the two middle bases of the anticodon (here a U·U mismatch). The double helix composed of the three anticodon base pairs is sandwiched (coaxially stacked) into an almost continuous double helix (A-form) which includes the anticodon stems of both tRNAs. The extensive stacking interactions generated in this structure explain the stabilizing effects of the anticodon stems and of the modified purine (often hypermodified residues) 3'-adjacent to the anticodon ending with A or U (see Fig. 6). The crystal structure (as discussed in Dirheimer et al., 1995) fully confirmed the earlier model proposed by Grosjean et al. (1976).

main determinant of this unexpected selfanticodon association (see also He et al., 1991, and Kim et al., 1996).

Mismatches involving the 2-thio derivatives of mnm^{5}U$_{34}$ with U or C (lifetimes between 50 and 100 ms at 9°C), but not with G, have also been detected. The existence of such complexes is not consistent with the genetic code rules. We have found that the thio group at position 2 of uridine 34 (Fig. 5a) is the main determinant of the stabilization of anticodon–anticodon complexes (Table 1; Houssier et al., 1988). The ^{1}H-NMR studies indicate that this type of modification makes the first position of the anticodon more rigid (Yokoyama et al., 1979, 1985; Yokoyama and Nishimura, 1995), an observation that is compatible with both our results and those of Davis (see Chapter 5).

From the previously described results, it appears that the posttranscriptional nucleotide modifications in position 34 play an important role in modulating both standard and nonstandard base pairing at the wobble position of the anticodon. This hypothesis was put forward a long time ago (Weiss, 1973; see also Cedergren et al., 1981), but until recently clearcut experimental evidence was lacking.

A Model for the Anticodon–Anticodon Interaction

Based on the complex formed between yeast tRNA^{Phe} (anticodon GmAA) and E. coli tRNA$_2^{Glu}$ (anticodon mnm^{5}s^{2}UUC) a model of the anticodon–anticodon structure has been proposed (Fig. 9 in Grosjean et al., 1976). This model was later fully confirmed by X-ray crystallography of the yeast tRNA^{Asp} (Fig. 7; Moras et al., 1980). In this model, the two anticodon stems, both purine 37 and nucleotide 38 (3' adjacent to anticodon), and the anticodon stem are coaxially stacked within an almost continuous A-form double helix (anti-parallel pairing mode). The pyrimidine at position 32 and the uracil base 33 (5'-adjacent to anticodon) are looping outside the "pseudo," quasi-perfect double helix. In this model, modification of purine 37 serves to reinforce the stability of the whole architecture by stacking interactions. Also, because modified nucleotides 37 (such as m^{1}G, m^{1}I, t^{6}A, i^{6}A, and Y base) cannot base pair in a Watson–Crick scheme, they restrict the binding within the three anticodon bases. According to this model, the base at the wobble position (position 34) is not only involved in base pairing with the complementary anticodon, but also in stacking with the neighboring base of the same anticodon (lateral stacking) and with bases of the complementary anticodon (diagonal stacking), including the purine 37 (see also Fig. 9 in Grosjean et al., 1976). Hence, multistacking effects, modulated by additional chemical groups on the purine or pyrimidine ring, are an important source of affinity enhancement in anticodon–anticodon complexes (discussed in Bubienko et al., 1983).

OTHER RNA–RNA INTERACTION INVOLVING LOOPS OF tRNA

Codon Recognition by tRNA on the Ribosome

One main function of tRNAs is the decoding of codon triplets in mRNA during translation on the ribosome. As previously discussed, built-in features of the anticodon stem loops in natural tRNA species are such that not only the binding energy of complementary nucleotides is increased by several orders of magnitude, as compared to unstructured oligonucleotides, but also, and possibly more importantly, the intrinsic difference between G·C and A·U pairs within the anticodon triplets is reduced. If this property applies to the codon–anticodon interaction on the ribosome, this would mean that during translation of an mRNA, the residence time of the various tRNAs on the decoding site would be maintained within "optimal" values ("optimal binding energy hypothesis," Grosjean et al., 1978; Grosjean and Fiers, 1982). However, these values, although characteristic of a set of codon–anticodon pairs in a given cell species, may not be necessarily the same in all types of cells. Indeed, yeast tRNAs often form slightly less stable anticodon–anticodon complexes than the corresponding *E. coli* pairs. This is probably related to the type of hypermodified bases found in the anticodons (see Appendix 6 by Björk). For example, only eukaryotic decoding systems have made systematic use of the loose I·C or I·U base pairs between the first anticodon base and the third codon base; while in prokaryotes, inosine 34 is found only in tRNA^{Arg} (anticodon ICG). Noteworthy is the absence of inosine 34 and the preferred G or C as the first base of anticodons in the major tRNA species from an organism like *Thermus thermophilus*, an extremely thermophilic bacterium that grows at temperatures as high as 75–85°C (Hara-Yokoyama et al., 1986). This observation correlated well with the paucity of U and A at the third position of the codon and the almost exclusive use of codons terminating in G or C in mRNAs from higher thermophiles, thus avoiding the less stable G·U wobble base pair (see Nakamura et al., 1996). Such a remarkable trend in the codon usage, as well as in the choice of the wobble base in tRNAs, must reflect the functional need of the cells to decode the genetic message on the ribosome at high temperature, thereby selecting the most stable codon–anticodon pairings.

Fig. 8 summarizes the various constraints, intrinsic to the anticodon stem loop of a tRNA, which can influence the efficiency and ultimately the accuracy of the translation process on the ribosome. In this cartoon, the modified nucleotides that are often present in the proximal part of the anticodon stem and in the loop (Z in Fig. 8) are supposed to play an essential role in the flexibility and hence the preferential 3′-stacked conformation adopted by the anticodon loop when it binds to the complementary codon (conformational switches). The 3′-stacked conformation schematically represented in Fig. 8b to d corresponds to the preferred anticodon conformation in solution as determined by different techniques (X-ray crystallography, T-jump relaxation, and NMR techniques) (Geerdes et al., 1980; Clore et al., 1984a, 1984b; Westhof et al., 1983; Stricker et al., 1989; reviewed in Dirheimer et al., 1995). Modulation by the immediate nucleotide context may arise from stacking by the purine, 3′-adjacent to the anticodon (R in Fig. 8c). When the first base of the codon and the third base (position 36) of the anticodon are A·U or U·A, then extra stabilization by modification or hypermodification of the R_{37} base is required (indicated by R* and an arrow in Fig. 8c). Certain modified purines at position 37, such as ms^2i^6A, t^6A and wybutosine, are particularly efficient because of their stacking potentials. Also, because most modified nucleotide 37 cannot base pair in a Watson–Crick mode, their presence 3′-adjacent to the anticodon restricts the pairing to the in-frame codon–anticodon pair. In support of this conclusion, it was shown that the absence of a methyl group on N^1 of G_{37} in tRNA^{Pro} (anticodon UGG) induces a frameshift in a run of Cs in the message (Björk et al., 1989; Hagervall et al., 1993; reviewed in Björk, 1995a, 1995b; see also Qian and Björk, 1997).

As far as the wobble base 34 is concerned (W in Fig. 8d), the binding strength (illustrated by a dashed arrow) depends on the possibility to form a stable Watson–Crick type base pair instead of a less stable mismatched pair (modulation by third codon choice; see also Weiss, 1973). The stabilizing effect by lateral stacking of the wobble base (modified or not) on the adjacent anticodon position 35 and possibly also by diagonal stacking on the middle codon base (illustrated by an arrow) can also modulate the strength of codon–anticodon binding. Except for possible steric hindrance, the type of interaction allowed at the wobble position might be a function of the overall

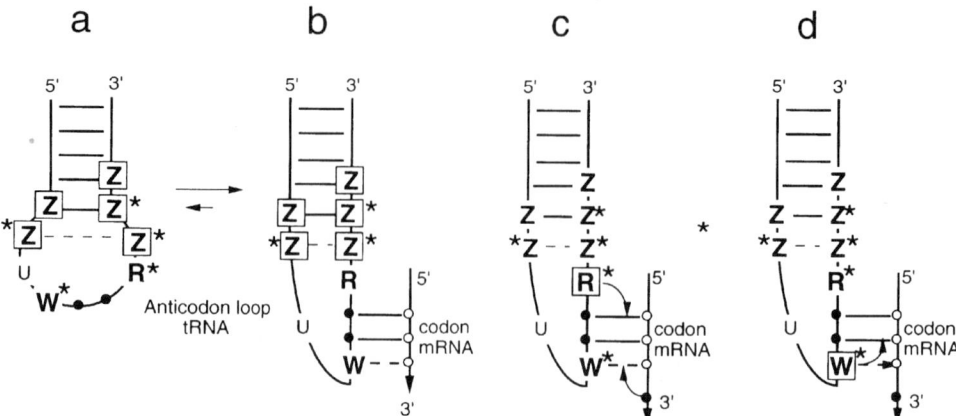

Figure 8. Modulation of codon–anticodon binding. The anticodon hairpin in tRNAs is schematically represented. Z stands for any nucleotide in the anticodon loop or in the proximal anticodon stem, R stands for purine 37 located 3′-adjacent to the anticodon, and W stands for the anticodon wobble base 34. Asterisks show positions where the bases or the ribose are most frequently modified (see Fig. 1). (a and b) Modulation of the strength of binding occurs by the loop constraint. Here the nature of the bases (Z) in the proximal anticodon stem and loop plays a role in the flexibility and hence the preferential conformation adopted by the anticodon when it binds to a complementary codon (conformational switches). The 3′-stacked conformation is schematically represented in panel b because it is the predominant conformation in solution (see Geerdes et al., 1980; Clore et al., 1984b; Bujalowski et al., 1986b; and Stricker et al., 1989) and the only conformation in the crystal (discussed in Dirheimer et al., 1995). However, major contributions of the "ribosomal milieu" as well as of divalent cations have also to be taken into account. (c) Modulation occurs by the immediate context of the codon–anticodon pairing. Here the purine (R) 3′-adjacent to the anticodon (and possibly the base 3′-adjacent to the codon) plays a role in the stabilization of the base pair between the third anticodon base and the first codon base (also discussed in Bubienko et al., 1983). Certain modified bases such as t^6A, ms^2i^6A and Y base (designated by the symbol R*) are particularly important because of their stacking potential (this chapter). (d) Modulation occurs by the so-called wobble base (W). Here the strength of the binding depends on the possibility of forming a stable Watson–Crick type base pair rather than a less stable "mismatched" pair (modulation by third codon choice). It also depends on the stabilizing effect (stacking) of the wobble base (modified or not) on the strength of pairing in the middle position of the codon–anticodon complex.

free energy of the whole tRNA-mRNA-ribosome complex. This parameter not only depends on the type of base pairing within the two other codon–anticodon pairings, but also on many other kinds of interactions that occur with the tRNA in the ribosomal environment (discussed in Buckingham and Grosjean, 1986; Yarus and Smith, 1995; reviewed in Chapter 27 by Curran). Of course, this does not exclude the possibility that the modified nucleotide has other functions in the mRNA-tRNA-ribosome complex or in the whole cellular process (discussed in Chapters 25 and 26 by Winkler and by Björk and Rasmuson, respectively).

Lastly, because the codon–anticodon pairing bears some symmetry, the bases surrounding the codon on the mRNA (codon context) are also important. However, since these nucleotides are not modified (except for inosine in certain mRNAs; see Chapter 19 by Rueter and Emeson), they are not discussed here.

Clearly, if this model is correct, the dominant stabilizing effect between codon and anticodon arises from stacking energy. Base pairing and the loop structure (at least for the tRNA counterpart) provide only the proper orientation of base rings to favor this stacking. Also, hydrogen bonds between bases hardly compensate for the energy lost by removing solvent molecules. Therefore, correct decoding may depend mainly on a short double helix formed between the "two-out-of-three" complementary anticodons (Lagerkvist, 1978), sandwiched between stacked, but not necessarily complementary nucleotides present both in the anticodon loop and in the mRNA immediate context. Evidently, modified nucleotides in the anticodon loop and proximal anticodon stem enhance and modulate the efficiency or accuracy of the translation process (Yarus, 1982; the extended anticodon theory). However, as previously stated, the enhancement of codon–anticodon interaction may not necessarily provide higher selectivity in the decoding of the genetic message. Therefore, to avoid errors, the ribosomes must amplify the small differences in the lifetime of the association between various codons and anticodons to produce a higher distinction between "genetically correct" and "genetically incorrect" pairings in protein synthesis. Mechanisms for such an amplification have been extensively discussed by Yarus and Smith (1995).

Inasmuch as anticodon–anticodon complexes can be regarded as models of codon–anticodon interactions, one can assume that a function of the ribosome is to favor the mRNA taking a complementary loop structure that can form a duplex with coaxially stacked bases. As a result, not only the lifetime for codon–anticodon interaction on the ribosome will considerably increase, but also the base pairing may become less sensitive to codon sequence and codon context (discussed in Grosjean and Chantrenne, 1980, and Buckingham and Grosjean, 1986).

Interaction Between D-Loop and TΨ-Loop in tRNAs

Since the elucidation of the crystal structure of several tRNAs, it is now well documented how the D-loop and the TΨC-loop interact "at the corner" of the universally conserved L-shape. This interaction is characterized by a $G_{19} \cdot C_{56}$ base pair and to a lesser extent by a $G_{18} \cdot \Psi_{55}$ base pair that locks the D- and TΨC-loops (T stands for m^5U) within the characteristic and ubiquitous tRNA architecture (reviewed in Dirheimer et al., 1995). The TΨC-loop itself has an internal architecture due to the synergetic effect of the $G_{53} \cdot C_{61}$ base pair stacked on a reversed-Hoogsteen $T_{54} \cdot A_{58}$ (Fig. 3 to 5 in Westhof et al., 1983; see also Romby et al., 1987; Becker et al., 1997; Yao et al., 1997).

Direct evidence that posttranscriptional modification of nucleotides at position 54 enhance the interaction between the D-loop and the TΨC-loop come from experiments with tRNAs isolated from the thermophilic bacterium *Thermus thermophilus*. When the cells were grown at high temperature (80°C), the ratio of s^2T_{54} over T_{54} in the isolated tRNA^{Ile} was significantly higher than when the same cells were grown a low temperature (50°C) (Watanabe et al., 1976). Interestingly, the simple change from an oxygen to a sulfur at C2 in T_{54}, leading to s^2T_{54}, shifts the melting temperature (T_m) of the purified tRNA^{Ile} by 3°C (Horie et al., 1985; Kawai et al., 1992). Thus, thiolation of T_{54} in *Thermus thermophilus* tRNAs and possibly other types of heat-induced nucleotide modifications in tRNAs (discussed in Yokoyama et al., 1987, and Kowalak et al., 1994) appear like a cellular strategy to overcome in part the deleterious effect of temperature on tRNA function. Noteworthy is that the stabilizing effect of thio group of s^2T_{54} on T-loop–D-loop interaction has some resemblance with the stabilizing effect of thio group of $mnm^5s^2U_{34}$ on tRNA anticodon–anticodon interaction as well as on human tRNA$_3^{Lys}$–HIV-1 RNA interaction.

Last, but not least, modified nucleotides T_{54} and Ψ_{55} are among the most universally conserved nucleotides in all known functional tRNAs. In fact, both 5-methylation of U_{54} into T_{54} and isomerization of U_{55} into Ψ_{55} have stabilizing effects on tRNA architecture (Davanloo et al., 1979), most probably by enhancing base stacking (Wang and Kool, 1995; Davis, 1995) (discussed in Chapter 5 by Davis). Also, the presence of a water-mediated bridge between Ψ_{55} and the phosphate backbone provides additional stabilizing effect (Auffinger and Westhof, 1997) (see Chapter 6 by Auffinger and Westhof).

In agreement with this conclusion, the thermal stability of naturally occurring tRNAs from bacteria and eukaryotes, bearing several modified nucleotides, has been shown to be systematically higher by 5 to 7°C as compared to the corresponding T7-transcripts lacking all the naturally occurring modified nucleotides (Derrick and Horowitz, 1993; Sampson and Uhlenbeck, 1988; Hall et al., 1989; Perret et al., 1990). With tRNAs from hyperthermophiles, this stabilization effect of modified nucleotides may be higher than 20°C (Kowalak et al., 1994). This is certainly due in large part to the stabilization of the three-dimensional structure of tRNA, for which the interaction between the D-loop and the T-loop plays a major role.

The Interaction between Human tRNA$_3^{Lys}$ and HIV-1 RNA

Retroviruses, pararetroviruses, and some retrotransposons use a specific tRNA isoacceptor from the infected cells to prime reverse transcription (for reviews see Marquet et al., 1995; Levin, 1997). The primer tRNA is usually bound to the viral RNA via its 10 to 18 3' terminal nucleotides, which are complementary to the so-called primer binding site. Beside this general tRNA and primer binding site interaction, evidence has accumulated for additional specific interactions between the primer tRNA and the genomic RNA (for details see Chapter 28 by Marquet).

In the case of the HIV-1 RNA and bovine tRNA$_3^{Lys}$ complex, extensive studies by enzymatic and chemical probing showed that formation of this "extended" complex was accompanied by numerous intra- and intermolecular rearrangements that result in a highly structured complex. In addition to the interaction between the 3' end of tRNA$_3^{Lys}$ and the primer binding site, the anticodon loop, the 3' part of the anticodon stem and part of the variable loop of tRNA$_3^{Lys}$ interact with viral sequences upstream from the primer binding site (Isel et al., 1993; Isel et al., 1995; see Fig. 4 in Chapter 28 by Marquet). This extended intermolecular interactions involve a loop–loop type of interaction between the anticodon loop of the tRNA$_3^{Lys}$ (anticodon mcm^5s^2U-U-U) and a

A-rich loop in the HIV-1 RNA. Such interaction requires the posttranscriptional modifications that are naturally present in bovine or human tRNA$_3^{Lys}$. Indeed, with a synthetic tRNA$_3^{Lys}$ devoid of posttranscriptional modifications, the intermolecular contact with the viral RNA was restricted to only the primer binding site sequence. In addition, treatment of natural tRNA$_3^{Lys}$ with hydrogen peroxide, which selectively transforms mcm^{5}s^{2}U$_{34}$ of the anticodon into thiol-depleted (oxidized) mcm^{5}U$_{34}$, strongly destabilizes the extended loop–loop type of interaction (Isel et al., 1993). Also, priming of reverse transcription was much more efficient with fully modified tRNA$_3^{Lys}$ than with its synthetic counterpart. This is partly because of the fact that the extended tRNA$_3^{Lys}$ and HIV-1 RNA interactions are involved in the transition from the initiation to the elongation phases of reverse transcription (Isel et al., 1996) (see also Chapter 28 by Marquet).

Evidently, HIV-1 has evolved to take advantage of the stabilization of the RNA–RNA interactions provided by the modified nucleotides of its replication tRNA primer. It has not yet been determined how the different modified nucleotides in the primer tRNA, in addition to the mcm^{5}s^{2}U$_{34}$, contribute to the stabilization of the complex with the viral RNA.

tRNA–mRNA Interactions in the Control of Aminoacyl–tRNA Synthetase Gene Expression

In *Bacillus subtilis*, most aminoacyl–tRNA synthetase genes are specifically induced by starvation for their cognate amino acid by a common transcription antitermination mechanism (Putzer et al., 1992; Grundy and Henkin, 1993, 1994; reviewed in Putzer et al., 1995b). This phenomenon depends on a sequence of 14 nt within an extensive conserved secondary structure located upstream from the leader Rho-independent transcription terminator. This region of the mRNA is partially complementary to the 5' stem of the terminator, allowing the formation of an alternative, but less stable, antiterminator structure. There is strong evidence that this conformational switch is responsible for the antitermination process and that the cognate uncharged tRNA stabilizes the antiterminator conformation by binding to the mRNA leader (for reviews, see Henkin, 1994; Garriti and Zahler, 1994; Putzer et al., 1995a, 1995b).

Two sites of interaction between the tRNA and the mRNA leader have been identified by genetic analysis. One region involves the tRNA anticodon interacting with a codon-like sequence located in a bulged loop of a conserved secondary structure in the leader, while the other region involves the acceptor end of the tRNA (NCCA3') interacting with a complementary sequence (UGGN) displayed in a side-bulged loop of the antiterminator structure (see Fig. 1 in Putzer et al., 1995a). The codon-like sequence in mRNA is located in a bulged loop of variable length from 5 to 10 nt. Also, the nucleotide following the codon-like sequence is always a purine residue that possibly extends base pairing with the universally conserved uridine 33 in anticodon loop of tRNA (Grundy and Henkin, 1994; Putzer et al., 1995b; Luo et al., 1997). One exception is found in the case of *argS* mRNA leader, which carries uridine instead of a purine 3'-adjacent to the codon-like CGC (Henkin, 1994). In this case the G- and C-rich codon-like sequence may provide sufficient stability to the codon–anticodon interaction so that the extra pairing with the 3'-adjacent uridine may not be essential.

The identity of the interacting tRNA is specified by a classical anticodon–codon type interaction. Indeed, it is possible to switch the specificity of the regulation by changing the nature of the codon-like sequence in the mRNA leader (Grundy and Henkin, 1993; Grundy et al., 1997). Consequently, modification of purine 37 in the anticodon loop of tRNAs may also increase the stability of the interaction, by stacking of the modified purine on the codon–anticodon complex. Therefore, one can anticipate that modified nucleotides in the anticodon loop of tRNA contribute to the efficiency of the transcription antitermination process, but this point has not yet been explored. Stabilization of the tRNA–mRNA interactions by additional protein factors has also been suggested (Grundy and Henki, 1993; Condon et al., 1996, 1997). This situation might well be analogous to the one previously described for the codon–anticodon interaction within the ribosome during translation.

OTHER TYPES OF RNA LOOP–LOOP INTERACTION

In this section, we describe two RNA loop–loop interacting systems that do not involve modified nucleotides. Their comparison with the loop–loop interacting systems involving hairpins of tRNA molecules described in the preceding sections are interesting to better understand how modified nucleotides can restrict the interaction to only a few selected nucleotides within the interacting RNA loops.

Antisense RNA Control of ColE1 Plasmid Replication

The replication of ColE1 plasmid is initiated by the transcription of a primer RNA, called RNA-II.

This RNA forms a persistent hybrid with the DNA near the replication origin and is then cleaved by RNase H at a specific site where DNA synthesis begins. The antisense RNA-I, encoded by the plasmid, controls the initiation of such DNA replication by binding to the 5′ end of RNA-II. This interaction induces a conformational change in RNA-II that prevents the formation of the persistent hybrid (Masukata and Tomizawa, 1986; reviewed in Eguchi et al., 1991; Tomizawa, 1993; Wagner and Simons, 1994).

Binding of RNA-I to the 5′ end of RNA-II is initiated by the formation of at least two loop–loop interactions (the so-called "kissing complexes"). It has been shown that the rate constant of such association is the critical step for the biological effect (Tomizawa, 1990a; Eguchi and Tomizawa, 1991). This initial "kissing" loop–loop complex formation is followed by a series of reactions that progressively lead to the formation of a more extended and very stable duplex (Tomizawa, 1984; 1990a, Eguchi and Tomizawa, 1991). An additional factor, a ColE1 encoded protein called *Rom*, binds to the transiently formed intermediary "kissing complex," thus facilitating formation of the extended duplex and therefore enhancing the inhibitory effect of RNA-I (Eguchi and Tomizawa, 1990; Tomizawa, 1990b).

Extensive physicochemical studies on the "kissing" loop–loop interaction have been performed on the isolated stem loop I (7 nucleotides in the loop) of RNA-I and its complementary sequence (Eguchi and Tomizawa, 1990, 1991). A model was proposed in which all of the seven nucleotides within the complementary loops pair with each other, leading to a coaxial stacking of the newly formed loop–loop helix with their respective stems (Fig. 9A and B; reviewed in Tomizawa, 1993). With this kind of complex, full complementarity of the two interacting loops is required to achieve enhanced affinity, while the nucleotide compositions of the stems of the two hairpins can differ (Gregorian and Crothers, 1995). The existence of such a loop–loop helix of seven base pairs, as well as of a continuous helical stacking of the loop residues on the 3′ side of their helical stems (compare Fig. 9A and B with Fig. 7A and B), has been confirmed by NMR (Marino et al., 1995). To accommodate formation of the seven base pairs, a bend of the loop–loop helix is necessary. Bending of the loop–loop helix is further enhanced by binding of the *Rom* protein (Eguchi and Tomizawa, 1990; Marino et al., 1995). Such a bended structure may have a functional role by bringing close together the distal regions of both RNAs, thereby facilitating the formation of a subsequent stable extended RNA–RNA duplex.

Interestingly, the sequences of loop I or II of the antisense RNA-I are strictly identical to the anticodon loops of *E. coli* tRNA$_1^{Thr}$ and tRNA$_3^{Thr}$, except that they lack of modified nucleotide at position equivalent to A_{37}, 3′-adjacent to the anticodon-like sequence in RNA-I (compare Fig. 9A and C). Thus, both loops may adopt an anticodon-like 3′-stacked conformation and the initial "kissing" pairing may be analogous to the one proposed for the anticodon–anticodon or codon–anticodon interactions. The nucleation step would therefore involve only a few nucleotides, most probably the GGU anticodon-like sequence. Because of the strict complementarity of the inverted loop sequences of RNA-I and RNA-II, and the absence of modified nucleotide of the kinds usually found in tRNA anticodon loop, base pairing extends rapidly to all nucleotides within the loop of RNA-I and RNA-II (Tomizawa, 1993). This is clearly at variance with the anticodon–anticodon complexes in which the nucleotide 3′-adjacent to the anticodon is often modified into a stabilizing bulky modified nucleotide that usually cannot base pair in a Watson–Crick mode (t^6A_{37} or mt^6A_{37} as in the case of the tRNA^{Thr}-anticodon GGU [Fig. 6]; see also Appendix 6 by Björk).

In ColE1, the stabilization of the loop–loop interaction is provided by hydrogen bonds between all the seven bases within the complementary loops, but also by base stacking between neighboring bases and by the binding of *Rom* protein (Eguchi and Tomizawa, 1991; Gregorian and Crothers, 1995). Only after the initial "kissing" interactions occur, the conversion to a more stable extended RNA-I:RNA-II duplex is possible, allowing efficient control of ColE1 replication.

HIV-1 RNA Dimerization

A characteristic feature of retroviruses is that they encapsidate a dimer of homologous genomic RNA molecules (for a review see Paillart et al., 1996a). In the case of HIV-1 genomic RNA, the dimerization initiation site consists of a short sequence located between the primer binding site and the major splice donor site, that adopts a stem-loop structure with nine nucleotides within the loop (Skripkin et al., 1994). Depending on the HIV-1 isolate, two main partially self-complementary sequences are found in this loop: AGGUGCACA (as in HIV-Mal) and AAGCGCGCA (as in HI-Lai) (Paillart et al., 1996a). These self-complementary sequences (*GUGCAC* or *GCGCGC*) promote RNA dimerization by forming a "kissing" loop-loop complex (Laughrea and Jetté, 1994; Skripkin et al., 1994; Muriaux et al., 1995). Mutations that destroyed the self-

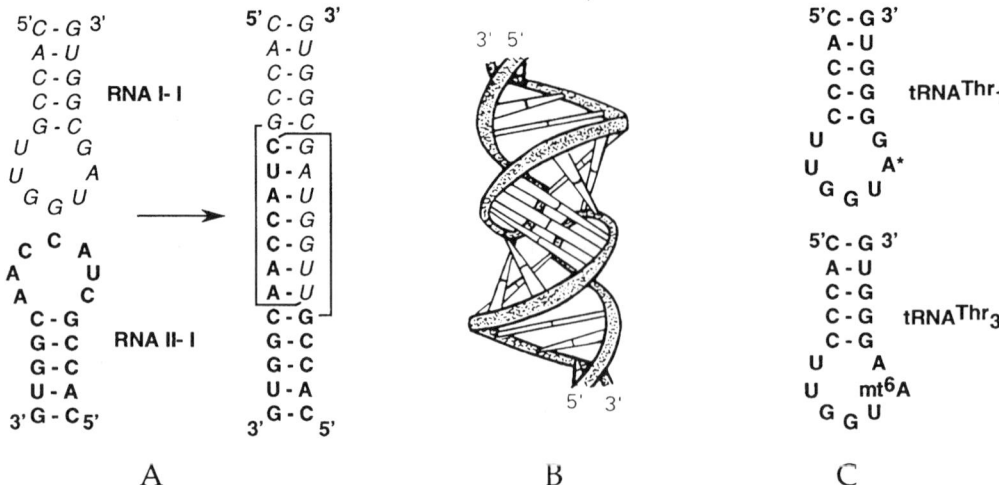

Figure 9. Structural model for the loop–loop interaction formed by two complementary hairpin loops (in fact, an "inverted" loop) found in the antisense RNA-I transcript and its target site (mRNA-II) in the ColE1 plasmid. (A) Nucleotide sequences of the interacting hairpins. (B) Corresponding schematic model as proposed by Tomizawa (1993). The phosphodiester backbone of the RNA–RNA complex is shown as a shaded ribbon (A-form RNA helix). The seven base pairs of the loop and of the proximal stems are indicated by crossed rods. This complex has also a dyad axis of symmetry located perpendicular to the paper, through the bond joining the pairing central bases of the seven-member loop (here a C·G). Determinants of the instability of such RNA hairpin loop–loop complexes have been extensively studied by Chang and Tinoco (1994) and Gregorian and Crothers (1995). More detailed model building shows that the helix bends to accommodate the full complementarity of the seven base pairs within the loop (Marino et al., 1995). Such bending is further stabilized by *Rom* protein (see text). (C) Secondary structure of the anticodon domain of *E. coli* tRNA^{Thr} (isoacceptors 1 and 3). The chemical structure of the modified nucleotide (A*) at position 37 of tRNA$_1^{Thr}$ is not known but could be a derivative of t^{6}A.

complementarity within the "kissing region" prevented dimerization, and subsequently encapsidation of the genomic RNA and reverse transcription, while complementary mutations restored dimerization (Paillart et al., 1994; Skripkin et al., 1994; Berkhout and van Wamel, 1996; Haddrick et al., 1996; McBride and Panganiban, 1996; Paillart et al., 1996b).

A tertiary structure model of the HIV-1 "kissing" loop–loop complex, based on chemical probing data, has been proposed (see Fig. 5 in Paillart et al., 1997). In this duplex, the two helices of the stems and the newly formed intermolecular helices form an almost continuous coaxially stacked helical architecture. Also, the conserved first and last adenines of the loop (AGNNGCNNA) form a *trans* base pair, while the second purine of the loop forms a base-triple at the center of the deep groove of the self-complementary intermolecular helix. The second purines of the two RNAs are stacked on each other, further stabilizing the complex. Initiation of RNA dimerization is very dependent on the two central GC bases of the self-complementary loop sequence (for details see Paillart et al., 1997).

This model is therefore different from that proposed for the "kissing" loop–loop complex formed between the ColE1 RNA-I and RNA-II described above (Fig. 9A and B). It is also distinct of the anticodon–anticodon complexes (Fig. 7). Evidently,

two loops can associate and form a complex with distinct detailed architecture. The main common feature is the formation of coaxial helices that provide a major contribution to the stability and the general architecture of these loop–loop complexes.

Because no modified nucleotides are present in the HIV-1 "kissing" loop–loop system, it can potentially "zip up" to a more extended one, a property that is clearly reminiscent of the "kissing" loop-loop complex observed with the ColE1 antisense RNA-I and mRNA-II, but not of the tRNA anticodon–anticodon complexes. However, in the absence of a protein factor, this HIV-1 RNA dimerization process occurs in vitro only at high temperature, low ionic strength or in the absence of divalent cations (Laughrea and Jetté, 1996; Muriaux et al., 1996a). No extended duplex is formed at 37°C, in the presence of magnesium (Paillart et al., 1996c). It is possible that, in vivo, the transition from the HIV-1 RNA "kissing" complex to the more extended stable complex might be promoted by the viral nucleocapsid protein (Muriaux et al., 1996b).

Other parts of the HIV-1 RNA region can also make "kissing" hairpin complexes, such as the *trans*-activation response element (TAR) at the 5′ end of the untranslated leader region. In this case, NMR studies with model systems has revealed that five base pairs are formed out of the six possible pairs within the loop–loop complex (Chang and Tinoco, 1994). Again, no modified nucleotides are involved in these complexes.

CONCLUDING REMARKS

In this chapter, we summarize our data on tRNA anticodon–anticodon complexes and stress the point that base stacking plays a major role in stabilizing such complexes. Comparison to other types of loop–loop complexes involving RNAs other than tRNAs, the so-called "kissing" complexes, is also made. The main difference between these different types of loop–loop complexes is the extent of base pairing between the nucleotides of the interacting loops. In the case of tRNA anticodon–anticodon complexes, this base pairing is restricted to the three central bases of the anticodon loops, while with the "kissing" loop complexes it involves almost all of the bases present in the loops (7 out of 7 in the case of ColE1 RNAs and 6 out of 9 in the case of HIV-1 RNA). Despite these differences in the number of base pairs formed, the anticodon–anticodon complexes appear almost as stable as the "kissing" loop–loop complexes. The major reason for this remarkable stability of anticodon–anticodon complexes is the presence of modified nucleotides at position 34 or 37 of the anticodon loop. Some of these modified nucleotides stabilize the complex and compensate for the difference of binding energy between an A·U pair and a G·C pair. Furthermore, the majority of the modified nucleotides at position 37 in the anticodon loop are unable to form a Watson–Crick base pair and, therefore, restrict the interactions to the anticodon triplet. Thus, the antipairing property of nucleotide 37 is crucially important for the correct decoding of triplets in mRNA on the ribosome, facilitating the maintenance of the translation process in the correct reading frame.

In this chapter, only a few examples of modified nucleotides in the anticodon loop are discussed. Further developments in this field will certainly aim at gaining more information on how the different modified nucleotides present in the anticodon hairpin (see Fig. 1) affect the binding capability of the anticodon (positive determinants; see also Chapters 5 and 6 by Davis and by Auffinger and Westhof, respectively) and also on how they contribute to avoid certain undesirable interactions (negative determinants). It is noteworthy that N^2,N^2-dimethylguanine at position 26 prevents the alternative misfolding of certain tRNAs, thereby facilitating the correct folding of the molecules (Steinberg and Cedergren, 1995). This methylated base is located at the junction of the D-stem and the anticodon stem where important coaxial stacking of helices, including base triples, occurs (discussed in Kim et al., 1996). N^1-methyladenosine at position 9 in human mitochondrial tRNAs appears to fulfill the same function (Helm et al., 1997). Hence, coaxial stacking of helices (see also Walter et al., 1994; Walter and Turner, 1994; Michel and Westhof, 1990; Pleij, 1994; Westhof and Jaeger, 1992; Westhof et al., 1996) acts as a bridge between secondary and tertiary structure. This property obviously also applies to RNA–hairpin interactions where the association of the two loops generates a helix portion, sandwiched between the two stems.

Another important factor that is not discussed here is the stability and rigidity of the stem bearing the interacting loops. This parameter depends on both the sequence and the G/C content of the stem (discussed in He et al., 1991; Mandal et al., 1996; Schweisguth and Moore, 1997; Motorin et al, 1997). As far as the anticodon hairpin is concerned, the stability and rigidity of the stems can also contribute to maintaining a preferred loop architecture that restricts the pairing to only the anticodon triplets (discussed in Yarus, 1982; Yarus and Smith, 1995).

Having in hand the tools for producing a large variety of RNA molecules containing or not containing modified nucleotides (see Chapter 4 by Zimmermann et al.), as a wide range of physicochemical techniques, such as gel electrophoresis, UV spectrometry, NMR, X-ray crystallography, and quenched-flow and fast relaxation kinetic techniques, it should now be possible to gain additional information on the importance of modified nucleotides in the biophysical and the biological properties of the RNA molecules.

Acknowledgments. The T-jump work on anticodon–anticodon interactions was initiated in the laboratory of Prof. D. M. Crothers and Dr. D. Söll (Yale University), pursued in the laboratory of Prof. M. Eigen (Max Planck Institute for Biophysical Chemistry, Göttingen, Germany) and further developed at the University of Liège (Liège, Belgium). H.G. was the recipient of fellowships from NIH, NATO and EMBO. We acknowledge research grants in Belgium from the Fonds National de la Recherche Scientifique (to H.G. and C.H.); in France from the CNRS, the Association pour la Recherche sur le Cancer and the Agence Nationale de Recherche sur le Sida (to H.G., R.M., and P.R.). Research concerning the HIV-1 virus was entirely performed in the laboratory of C. and B. Ehresmann in Strasbourg, France. We acknowledge critical reading of the manuscript and useful comments by Y. Motorin and J.-P. Waller (both at the CNRS, Gif-sur-Yvette, France).

REFERENCES

Amano, M., and Y. Kyogoku. 1993. Nuclear magnetic resonance study of the codon-anticodon interaction in *Bombix mori* tRNA^{Gly} (GCC). *Eur. J. Biochem.* **217:**131–136.

Auffinger, P., and E. Westhof. 1997. RNA hydration: three nonaseconds of multiple molecular dynamics simulations of the solvated tRNA^{Asp} anticodon hairpin. *J. Mol. Biol.* **269:**326–341.

Becker, H. F., Y. Motorin, M. Sissler, C. Florentz, and H. Grosjean. 1997. Major identity determinants for enzymatic formation of ribothymidine and pseudouridine in the TΨ-loop of yeast tRNAs. *J. Mol. Biol.* 274:505–518.

Berkhout, B., and J. L. B. van Wamel. 1996. Role of the DIS hairpin in replication of human immunodeficiency virus type 1. *J. Virol.* 70:6723–6732.

Bernasconi, C. F. 1976. *Relaxation Kinetics*. Academic Press, New York, N.Y.

Björk, G. R. 1995a. Biosynthesis and function of modified nucleosides, p. 165–206. *In* D. Söll and U. RajBhandary (ed.), *tRNA: Structure, Biosynthesis, and Function*. ASM Press, Washington, D.C.

Björk, G. R. 1995b. Genetic dissection of synthesis and function of modified nucleosides in bacterial tRNA. *Prog. Nucleic Acid Res. Mol. Biol.* 50:263–338.

Björk, G. R., P. Wikström, and A. S. Byström. 1989. Prevention of translational frameshifting by the modified nucleoside 1-methylguanosine. *Science* 244:986–989.

Borer, P. N., B. Dengler, Jr., I. Tinoco, and O. C. Uhlenbeck. 1974. Stability of ribonucleic acid double-stranded helices. *J. Mol. Biol.* 86:843–853.

Brion, P., and E. Westhof. 1997. Hierarchy and dynamics of RNA folding. *Annu. Rev. Biophys. Biomol. Struct.* 26:113–137.

Bubienko, E., P. Cruz, J. F. Thomason, and P. N. Borer. 1983. Nearest neighbour effects in the structure and function of nucleic acids. *Prog. Nucleic Acid Res. Mol. Biol.* 30:41–90.

Buckingham, R. H. 1976. Anticodon conformation and accessibility in wild-type and suppressor tryptophan tRNA from *E. coli*. *Nucl. Acids Res.* 3:965–975.

Buckingham, R. H., and H. Grosjean. 1986. The accuracy of mRNA-tRNA recognition, p. 83–126. *In* T. B. L. Kirkwood, R. F. Rosenberger, and D. J. Galas (ed.), *Accuracy in Molecular Processes: Its Control and Relevance to Living Systems*. Chapman and Hall, London, United Kingdom.

Bujalowski, W., M. Jung, L. W. McLaughlin, and D. Porschke. 1986a. Codon-induced association of the isolated anticodon loop of tRNA^{Phe}. *Biochemistry* 25:6372–6378.

Bujalowski, W., E. Graeser, L. W. McLaughlin, and D. Porschke. 1986b. Anticodon loop of tRNA^{Phe}: structure, dynamics and Mg2+ binding. *Biochemistry* 25:6365–6371.

Carter, R. J., K. J. Baeyens, J. SantaLucia, D. H. Turner, and S. R. Holbrook. 1997. The crystal structure of an RNA oligomer incorporating tandem adenosine-inosine mismatches. *Nucleic Acids Res.* 25:4117–4122.

Cedergren, R. J., D. Sankoff, B. LaRue, and H. Grosjean. 1981. The evolving tRNA molecule. *Crit. Rev. Biochem.* 11:35–104.

Cedergren, R. J., H. Grosjean, and B. LaRue. 1986. Primordial reading of genetic information. *BioSystems* 19:259–266.

Chang, K.-Y., and I. Tinoco, Jr. 1994. Characterization of a "kissing" hairpin complex derived from the human immunodeficiency virus genome. *Proc. Natl. Acad. Sci. USA* 91:8705–8709.

Clore, G. M., A. M. Gronenborn, and L. W. McLaughlin. 1984a. Structure of the ribonucleoside diphosphate codon UpUpC bound to tRNA^{Phe} from yeast: a time dependent transferred nuclear overhauser enhancement study. *J. Mol. Biol.* 174:163–174.

Clore, G. M., A. M. Gronenborn, E. A. Piper, L. W. McLaughlin, E. Graeser, and J. H. Van Boom. 1984b. The solution structure of a RNA pentadecamer comprising the anticodon loop and stem of yeast tRNA^{Phe}. *Biochem. J.* 221:737–751.

Condon, C., H. Putzer, and M. Grunberg-Manago. 1996. Processing of the leader mRNA plays a major role in the induction of *thrS* expression following threonine starvation in *B. subtilis*. *Proc. Natl. Acad. Sci. USA* 93:6992–6997.

Condon, C., H. Putzer, D. Luo, and M. Grunberg-Manago. 1997. Processing of the *Bacillus subtilis thrS* leader mRNA is RNase-dependent in *E. coli*. *J. Mol. Biol.* 268:235–242.

Craig, M. E., D. M. Crothers, and P. Doty. 1971. Relaxation kinetics of dimer formation by self complementary oligonucleotides. *J. Mol. Biol.* 62:383–401.

Crick, F. H. C. 1966. Codon-anticodon pairing: the wobble hypothesis. *J. Mol. Biol.* 19:548–555.

Crothers, D. M. 1971. Temperature-jump methods, p. 369–388. *In* G. L. Cantoni and D. R. Davies (ed.), *Procedures in Nucleic Acid Research*, vol. 2. Harper and Row, New York, N.Y.

Curran, J. F. 1995. Decoding with the A:I wobble pair is inefficient. *Nucleic Acids Res.* 23:683–688.

Davanloo, P., M. Sprinzl, K. Watanabe, M. Albani and H. Kersten. 1979. Role of ribothymidine in the thermal stability of tRNA as monitored by proton magnetic resonance. *Nucleic Acids Res.* 6:1571–1581.

Davis, D. R. 1995 Stabilization of RNA stacking by pseudouridine. *Nucleic Acids Res.* 23:5020–5026.

Derrick, W. B., and J. Horowitz. 1993. Probing structural differences between native and *in vitro* transcribed *E. coli* valine transfer RNA: evidence for stable base modification dependent conformers. *Nucleic Acids Res.* 21:4948–4953.

Dirheimer, G., G. Keith, P. Dumas, and E. Westhof. 1995. Primary, secondary and tertiary structures of tRNAs, p. 93–140. *In* D. Söll and U. RajBhandary (ed.), *tRNA: Structure, Biosynthesis, and Function*. ASM Press, Washington, D.C.

Eguchi, Y., and J. Tomizawa. 1990. Complex formed by complementary RNA stem-loops and its stabilization by a protein: function of ColE1 Rom protein. *Cell* 60:199–209.

Eguchi, Y., and J. Tomizawa. 1991. Complexes formed by complementary RNA stem-loops: their formations, structures, and interaction with ColE1 Rom protein. *J. Mol. Biol.* 220:831–842.

Eguchi, Y., T. Itoh, and J. Tomizawa. 1991. Antisense RNA. *Annu. Rev. Biochem.* 60:631–652.

Eigen, M., 1954. Methods for investigation of ionic reactions in aqueous solutions with half-times as short as 10^{-9} sec: application to neutralization and hydrolysis reactions. *Discuss. Faraday Soc.* 17:194–205.

Eigen, M., and L. DeMaeyer. 1963. Relaxation methods, p. 895–1054. *In* S. L. Friess, E. S. Lewis, and A. Weissberger (ed.), *Technique of Organic Chemistry*, vol. VIII, part 2. Wiley (Interscience), New York, N.Y.

Eisinger, J. 1971a. Complex formation between transfer RNA's with complementary anticodons. *Biochem. Biophys. Res. Commun.* 43:854–861.

Eisinger, J. 1971b. Visible gel electrophoresis and the determination of association constant. *Biochem. Biophys. Res. Commun.* 44:1135–1142.

Eisinger, J., B. Feuer, and T. Yamane. 1971. Codon-anticodon binding in yeast tRNA^{Phe}. *Nature* 231:126–128.

Eisinger, J., and W. E. Blumberg. 1973. Binding constants from zone transport of interacting molecules. *Biochemistry* 12:3648–3660.

Eisinger, J., and N. Gross. 1974. The anticodon-anticodon complex. *J. Mol. Biol.* 88:165–174.

Eisinger, J., and N. Gross. 1975. Conformers, dimers and anticodons complexes of tRNA$_2^{Glu}$ of *E. coli*. *Biochemistry* 14:4031–4041.

Garriti, D. B., and S. A. Zahler. 1994. Mutations in the gene for a tRNA that functions as a regulator of a transcriptional attenuator in *B. subtilis*. *Genetics* 137:627–636.

Geerdes, H. A. M., J. H. Van Boom, and C. W. Hilbers. 1980. Codon-anticodon interactions in tRNA^{Phe}: II. A NMR study of the binding of the codon UUC. *J. Mol. Biol.* 142:219–230.

Gregorian, R. S., and D. M. Crothers. 1995. Determinants of RNA hairpin loop-loop complex stability. *J. Mol. Biol.* 248:968–984.

Grosjean, H., C. Takada, and J. Petre. 1973. Complex formation between tRNAs with complementary anticodons: use of matrix bound tRNA. *Biochem. Biophys. Res. Commun.* 53:882–893.

Grosjean, H., D. G. Söll, and D. M. Crothers. 1976. Studies of the complex between tRNAs with complementary anticodons: I. Origins of enhanced affinity between complementary triplets. *J. Mol. Biol.* **103**:499–519

Grosjean, H., S. de Henau, and D. M. Crothers. 1978. On the physical basis for ambiguity in genetic coding interactions. *Proc. Natl. Acad. Sci. USA* **75**:610–614.

Grosjean, H., and H. Chantrenne. 1980. On codon-anticodon interactions, p. 347–367. In F. Chapeville and A.-L. Haenni (ed.), *Molecular Biology Biochemistry and Biophysics*, vol. 32. *Chemical Recognition in Biology*. Springer-Verlag, Berlin, Germany.

Grosjean, H., and W. Fiers. 1982. Preferential codon usage in prokaryotic genes: the optimal codon-anticodon interaction energy and the selective codon usage in efficiently expressed genes. *Gene* **18**:199–209.

Grosjean, H., K. Nicoghosian, E. Haumont, D. Söll, and R. J. Cedergren. 1985. Nucleotide sequences of two serine tRNAs with a GGA anticodon: the structure-function relationships in the serine family of E. coli tRNAs. *Nucleic Acids Res.* **13**:5697–5706.

Grosjean, H., and C. Houssier. 1990. Codon recognition: evaluation of the effects of modified bases in the anticodon loop of tRNA using temperature-jump relaxation method, p. A255–A295. In C. W. Gehrke and K. C. Kuo (ed.), *Chromatography and Modification of Nucleosides*, part A. *Analytical Methods for Major and Modified Nucleosides*. *J. Chrom. Library*, vol. 45A. Elsevier Science Publishing, Amsterdam, The Netherlands.

Grosjean, H., M. Sprinzl, and S. Steinberg. 1995. Posttranscriptionnally modified nucleotides in tRNA: their locations and frequencies. *Biochimie* **77**:139–141.

Grundy, F. J., and T. M. Henkin. 1993. tRNA as a positive regulator of transcription in B. subtilis. *Cell* **74**:475–482.

Grundy, F. J., and T. M. Henkin. 1994. Conservation of a transcription antitermination mechanism in aminoacyl-tRNA synthetase and amino acid biosynthesis genes in gram-positive bacteria. *J. Mol. Biol.* **235**:798–804.

Grundy, F. J., S. E. Hodil, S. M. Rollins, and T. M. Henkin. 1997. Specificity of tRNA-mRNA interactions in Bacillus subtilis tyrS antitermination. *J. Bacteriol.* **179**:2587–2594.

Haddrick, M., A. L. Lear, A. J. Cann, and S. Heaphy. 1996. Evidence that a kissing loop structure facilitates genomic RNA dimerisation in HIV-1. *J. Mol. Biol.* **259**:58–68.

Hagervall, T. G., T. M. F. Tuohy, J. F. Atkins, and G. R. Björk. 1993. Deficiency of 1-methylguanosine in tRNA from Salmonella typhimurium induces frameshifting by quadruplet translocation. *J. Mol. Biol.* **232**:756–765.

Hall, K. B., J. R. Sampson, O. C. Uhlenbeck, and A. G. Redfield. 1989. Structure of an unmodified tRNA molecule. *Biochemistry* **28**:5794–5801.

Hammes, G. G. 1968. Relaxation spectrometry of biological systems. *Adv. Prot. Chem.* **23**:1–57.

Hara-Yokoyama, M., S. Yokoyama, T. Watanabe, K. Watanabe, K. Kitazume, Y. Mitamura, T. Morii, S. Takahashi, Y. Kuchino, S. Nishimura, and T. Miyazawa. 1986. Characteristic anticodon sequences of major tRNA species from an extreme thermophile, T. thermophilus HB8. *FEBS Lett.* **202**:149–152.

He, L., R. Kierzek, J. SantaLucia, A. E. Walter and D. H. Turner. 1991. Nearest-neighbour parameters for G.U mismatches: GU/UG is destabilizing in the contexts CGUG/GUGC, UGUA/AUGU, and AGUU/UUGA but stabilizing in GGUC/CUGG. *Biochemistry* **30**:11124–11132.

Helm, M., H. Brulé, F. Degoul, C. Cepanec, J. P. Leroux, R. Giegé and C. Florentz. 1997. The presence of a modified nucleotide is required for the cloverleaf folding of a human mitochondrial tRNA, abstr. 3-13. 17th International tRNA Workshop, May 10-15, Kazusa Akademia Center, Chiba, Japan.

Henkin, T. M. 1994. tRNA-directed transcription antitermination. *Mol. Microbiol.* **13**:381–387.

Horie, N., M. Hara-Yokoyama, S. Yokoyama, K. Watanabe, Y. Kuchino, S. Nishimura and T. Miyazawa. 1985. Two tRNA$_1^{Ile}$ from an extreme thermophile, *Thermus thermophilus* HB8: effect of 2-thiolation of ribothymidine on the thermostability of tRNA. *Biochemistry* **24**:5711–5715.

Houssier, C., and H. Grosjean. 1985. Temperature jump relaxation studies on the interactions between tRNAs with complementary anticodons: the effect of modified bases adjacent to the anticodon triplet. *J. Biomol. Struct. Dyn.* **3**:387–408.

Houssier, C., P. Degée, K. Nicoghosian, and H. Grosjean. 1988. Effect of uridine dethiolation in the anticodon triplet of tRNA$_2^{Glu}$ on its association with tRNA^{Phe}. *J. Biomol. Struct. Dyn.* **5**:1259–1266.

Isel, C., C. Ehresmann, G. Keith, B. Ehresmann, and R. Marquet. 1995. Initiation of reverse transcription of HIV-1: secondary structure of the HIV-1 RNA/tRNA$_3^{Lys}$ (template/primer) complex. *J. Mol. Biol.* **247**:236–250.

Isel, C., J. M. Lanchy, S. F. J. Le Grice, C. Ehresmann, B. Ehresmann, and R. Marquet. 1996. Specific initiation and switch to elongation of human immunodeficiency virus type 1 reverse transcription require the post-transcriptional modifications of primer tRNA$_3^{Lys}$. *EMBO J.* **15**:917–924.

Isel, C., R. Marquet, G. Keith, C. Ehresmann, and B. Ehresmann. 1993. Modified nucleotides of transfer-RNA$_3^{Lys}$ modulate primer/template loop-loop interaction in the initiation complex of HIV-1 reverse transcription. *J. Biol. Chem.* **268**:25269–25272.

Kawai, G., Y. Yamamoto, T. Kamimura, T. Masegi, M. Sekine, T. Hata, T. Iimori, T. Watanabe, T. Miyazawa, and S. Yokoyama. 1992. Conformational rigidity of specific pyrimidine residues in tRNA arises from post-transcriptional modifications that enhance steric interaction between the base and the 2'-hydroxyl group. *Biochemistry* **31**:1040–1046.

Kowalak, J. A., J. J. Dalluge, J. A. McCloskey, and K. O. Stetter. 1994. The role of post-transcriptional modification in stabilization of tRNA from hyperthermophiles. *Biochemistry* **33**:7869–7876.

Kim, J., A. E. Walter, and D. H. Turner. 1996. Thermodynamics of coaxially stacked helixes with GA and CC mismatches. *Biochemistry* **35**:13753–13781.

Labuda, D., H. Grosjean, G. Striker, and D. Pörschke. 1982. Codon:anticodon and anticodon:anticodon interaction. Evaluation of equilibrium and kinetic parameters of complexes involving a G:U wobble. *Biochim. Biophys. Acta* **698**:230–236.

Labuda, D., G. Striker, and D. Pörschke. 1984. Mechanism of codon recognition by tRNA and codon-induced tRNA association. *J. Mol. Biol.* **174**:587–604.

Labuda, D., G. Striker, H. Grosjean, and D. Pörschke. 1985. Mechanism of codon recognition by tRNA studied with oligonucleotides larger than triplets. *Nucleic Acids Res.* **13**:3667–3683.

Lagerkvist, U. 1978. "Two-out-of-three": an alternative method of codon reading. *Proc. Natl. Acad. Sci. USA* **75**:1759–1762.

Laughrea, M., and L. Jetté. 1994. A 19-nucleotide sequence upstream of the 5' major splice donor is part of the dimerization domain of human immunodeficiency virus 1 genomic RNA. *Biochemistry* **33**:13464–13474.

Laughrea, M., and L. Jetté. 1996. Kissing-loop model of HIV-1 genome dimerization: HIV-1 RNA can assume alternative dimeric forms, and all sequences upstream or downstream of hairpin 248-271 are dispensable for dimer formation. *Biochemistry* **35**:1589–1598.

Le, S. Y., and M. Zuker. 1991. Predicting common foldings of homologous RNA. *J. Biomol. Struct. Dyn.* **8**:1027–1044.

Levin, H. L. 1997. It's prime time for reverse transcriptase. *Cell* **88:**5–8.

Luo, D., J. Leautey, M. Grunberg-Manago, and H. Putzer. 1997. Structure and regulation of expression of the *Bacillus subtilis* valyl-tRNA synthetase gene. *J. Bacteriol.* **179:**2472–2478.

Mandal, N., D. Mangroo, J. J. Dalluge, J. A. McCloskey and U. L. RajBandary. 1996. Role of the three consecutive G:C base pairs conserved in the anticodon stem of initiator tRNAs in initiation of protein synthesis in *E. coli. RNA* **2:**473–482.

Marino, J. P., R. S. Gregorian, G. Csankovvszki, and D. M. Crothers. 1995. Bent helix formation between RNA hairpins with complementary loops. *Science* **268:**1448–1454.

Marquet, R., C. Isel, C. Ehresmann, and B. Ehresmann. 1995. tRNAs as primer of reverse transcriptase. *Biochimie* **77:**113–124.

Masukata, H., and J. Tomizawa. 1986. Control of primer formation for ColE1 plasmid replication: conformational change of the primer transcript. *Cell* **44:**125–136.

McBride, M. S., and A. T. Panganiban. 1996. The human immunodeficiency virus type 1 encapsidation site is a multipartite RNA element composed of functional hairpin structures. *J. Virol.* **70:**2963–2973.

Meier, F., B. Suter, H. Grosjean, G. Keith, and E. Kubli. 1985. Queuosine modification of the wobble base in tRNA^{His} influences *in vivo* decoding properties. *EMBO J.* **4:**823–827.

Michel, F., and E. Westhof. 1990. Modelling of the three-dimensional architecture of group I catalytic introns based on comparative sequence analysis. *J. Mol. Biol.* **216:**585–610.

Moras, D., M. B. Comarmond, J. Fisher, R. Weiss, J. C. Thierry, J. P. Ebel, and R. Giegé. 1980. Crystal structure of yeast tRNA^{Asp}. *Nature* **288:**669–674.

Moras, D., A. C. Dock, P. Dumas, E. Westhof, P. Romby, J. P. Ebel, and R. Giegé. 1986. Anticodon-anticodon interaction induces conformational changes in yeast tRNA^{Asp}, a model for tRNA-mRNA recognition. *Proc. Natl. Acad. Sci. USA* **83:**932–936.

Motorin, Y., G. Bec, R. Tewari, and H. Grosjean. 1997. Transfer RNA recognition by *E. coli* isopentenyl-pyrophosphate:tRNA-isopentenyl transferase: dependence on the anticodon arm structure. *RNA* **3:**721–733.

Muriaux, D., P.-M. Girard, B. Bonnet-Mathonière, and J. Paoletti. 1995. Dimerization of HIV-1Lai RNA at low ionic strength. An autocomplementary sequence in the 5' leader region is evidenced by an antisense oligonucleotide. *J. Biol. Chem.* **270:**8209–8216.

Muriaux, D., P. Fossé, and J. Paoletti. 1996a. A kissing complex together with a stable dimer is involved in the HIV-1$_{Lai}$ RNA dimerization process *in vitro*. *Biochemistry* **35:**5075–5082.

Muriaux, D., H. DeRocquigny, B. P. Roques, and J. Paoletti. 1996b. NCp7 activates HIV-1(Lai) RNA dimerization by converting a transient loop-loop complex into a stable dimer. *J. Biol. Chem.* **271:**33686–33692.

Nakamura, Y., K.-N. Wada, Y. Wada, H. Doi, S. Kanaya, T. Gojobori, and T. Ikemura. 1996. Codon usage tabulated from the international DNA sequence databases. *Nucleic Acids Res.* **24:**214–215.

Nelson, J. W., and I. Tinoco, Jr. 1982. Comparison of the kinetics of ribooligonucleotide, deoxyribonucleotide and hybrid oligonucleotide double-strand formation by T-jump kinetics. *Biochemistry* **21:**5289–5295.

Nilson, L., R. Rigler, and P. Laggner. 1982. Structural variability of small-angle x-ray scattering of the yeast tRNA^{Phe}-*E. coli* tRNA$^{Glu}_2$ complex. *Proc. Natl. Acad. Sci. USA* **79:**5891–5895.

Paillart, J. C., R. Marquet, E. Skripkin, B. Ehresmann, and C. Ehresmann. 1994. Mutational analysis of the bipartite dimer linkage structure of HIV-1 genomic RNA. *J. Biol. Chem.* **269:**27486–27493.

Paillart, J. C., R. Marquet, E. Skripkin, C. Ehresmann, and B. Ehresmann. 1996a. Dimerization of retroviral genomic RNAs: structural and functional implications. *Biochimie* **78:**639–653.

Paillart, J.-C., L. Berthoux, M. Ottmann, J.-L. Darlix, R. Marquet, B. Ehresmann, and C. Ehresmann. 1996b. A dual role of the putative RNA dimerization initiation site of human immunodeficiency virus type 1 in genomic RNA packaging and proviral DNA synthesis. *J. Virol.* **70:**8348–8354.

Paillart, J.-C., E. Skripkin, B. Ehresmann, C. Ehresmann, and R. Marquet. 1996c. A loop-loop "kissing" complex is the essential part of the dimer linkage of genomic HIV-1 RNA. *Proc. Natl. Acad. Sci. USA* **93:**5572–5577.

Paillart, J. C., E. Westhof, C. Ehresmann, B. Ehresmann, and R. Marquet. 1997. Non-canonical interactions in a kissing loop complex: the dimerization initiation site of HIV-1 genomic RNA. *J. Mol. Biol.* **270:**36–49.

Perret, V., A. Garcia, J. Puglisi, H. Grosjean, J. P. Ebel, C. Florentz, and R. Giegé. 1990. Conformation in solution of yeast tRNA^{Asp} transcripts deprived of modified nucleotides. *Biochimie* **72:**735–743.

Pleij, C. W. A. 1994. RNA pseudoknots. *Curr. Opin. Struct. Biol.* **4:**337–344.

Pörschke, D., O. C. Uhlenbeck, and F. H. Martin. 1973. Thermodynamics and kinetics of the helix-coil transition of oligomers containing GC base pairs. *Biopolymers* **12:**1313–1335.

Putzer, H., N. Gendron, and M. Grunberg-Manago. 1992. Co-ordinate expression of the two threonyl-tRNA synthetase genes in *Bacillus subtilis*: control by transcriptional antitermination involving a conserved regulatory sequence. *EMBO J.* **11:**3117–3127.

Putzer, H., S. Laalami, A. A. Brakhage, C. Condon, and M. Grunberg-Manago. 1995a. Aminoacyl-tRNA synthetase gene regulation in *Bacillus subtilis*: induction, repression and growth-rate regulation. *Mol. Microbiol.* **16:**709–718.

Putzer, H., M. Grunberg-Manago, and M. Springer. 1995b. Bacterial aminoacyl-tRNA synthetases: genes and regulation of expression, p. 293–333. *In* D. Söll and U. RajBhandary (ed.), *tRNA: Structure, Biosynthesis, and Function*. ASM Press, Washington, D.C.

Pyle, A. M., and J. B. Green. 1995. RNA folding. *Curr. Opin. Struct. Biol.* **5:**303–310.

Qian, Q., and G. R. Björk. 1997. Structural alterations far from the anticodon of the tRNA^{Pro}(GGG) of *Salmonella typhimurium* induce +1 frameshifting at the peptidyl-site. *J. Mol. Biol.* **273:**978–992.

Ravetch, J., J. Gralla, and D. M. Crothers. 1974. Thermodynamic and kinetic properties of short RNA helices: the oligomers sequences A$_n$GCU$_n$. *Nucleic Acids Res.* **1:**109–127.

Riesner, D., and R. Römer. 1973. Thermodynamics and kinetics of conformational transitions in oligonucleotides and tRNA, p. 237–318. *In* J. Duchesne (ed.) *Physico-chemical Properties of Nucleic Acids*, vol. 2. Academic Press, New York, N.Y.

Rigler, R., C. R. Rabl, and T. M. Jovin. 1974. A T-jump apparatus for fluorescence measurements. *Rev. Sci. Instrum.* **45:**580–588.

Romby, P., P. Carbon, E. Westhof, C. Ehresmann, J.-P. Ebel, B. Ehresmann, and R. Giegé. 1987. Importance of conserved residues for the conformation of the T-loop in tRNAs. *J. Biomol. Struct. Dyn.* **5:**669–687.

Romby, P., R. Giegé, C. Houssier, and H. Grosjean. 1985. Anticodon-anticodon interactions in solution: studies of the self-association of yeast or *E. coli* tRNA^{Asp} and of their interactions with *E. coli* tRNA^{Val}. *J. Mol. Biol.* **184:**107–118.

Romby, P., E. Westhof, D. Moras, R. Giegé, C. Houssier, and H. Grosjean. 1986. Studies on anticodon-anticodon interactions.

Hemi-protonation of cytosines induces self-pairing through the GCC anticodon of *E. coli* tRNA^{Gly}. *J. Biomol. Struct. Dyn.* **4:** 193–203.

Sampson, J. R., and O. C. Uhlenbeck. 1988. Biochemical and physical characterization of an unmodified yeast phenylalanine transfer RNA transcribed *in vitro*. *Proc. Natl. Acad. Sci. USA* **85:** 1033–1037.

Schweisguth, D. C., and P. B. Moore. 1997. On the conformation of the anticodon loops of initiator and elongator methionine tRNAs. *J. Mol. Biol.* **267:**505–519.

Simons, R. W. 1988. Naturally occurring antisense RNA control- a brief review. *Gene* **72:**35–44.

Skripkin, E., J. C. Paillart, R. Marquet, B. Ehresmann, and C. Ehresmann. 1994. Identification of the primary site of human immunodeficiency virus type 1 RNA dimerization *in vitro*. *Proc. Natl. Acad. Sci. USA* **91:**4945–4949.

Sprinzl, M., C. Horn, M. Brown, A. Ioudovitch, and S. Steinberg. 1998. Compilation of tRNA sequences and sequences of tRNA genes. *Nucleic Acids Res.* **26:**148–153.

Steinberg, S., and R. Cedergren. 1995. A correlation between N^2-dimethylguanosine presence and alternate tRNA conformers. *RNA* **1:**886–891.

Stricker, G., D. Labuda and M. D. C. Vega-Carmen. 1989. The three conformations of the anticodon loop of yeast tRNA^{Phe} *J. Biomol. Struct. Dyn.* **7:**235–255.

ten Dam, E., K. Pleij, and D. Draper. 1992. Structural and functional aspects of RNA pseudoknots. *Biochemistry* **31:** 11665–11676.

Thusius, D., G. Foucault, and F. Guillain. 1973. The analysis of chemical relaxation amplitudes and some applications to reactions involving macromolecules, p. 271–284. *In* C. Sadron (ed.), *Dynamics Aspects of Conformation Changes in Biological Macromolecules*. D. Reidel Publishing Company, Dordrecht, The Netherlands.

Tomizawa, J. 1984. Control of ColE1 plasmid replication: the process of binding of RNA I to the primer transcript. *Cell* **38:** 861–870.

Tomizawa, J. 1990a. Control of ColE1 plasmid replication: intermediates in the binding of RNA I and RNA II. *J. Mol. Biol.* **212:** 683–694.

Tomizawa, J. 1990b. Control of ColE1 plasmid replication: interaction of Rom protein with an unstable complex formed by RNA I and RNA II. *J. Mol. Biol.* **212:**695–708.

Tomizawa, J. 1993. Evolution of functional structures of RNA, p. 419–445. *In* R. F. Gesteland and J. F. Atkins (ed.), *The RNA World*. Cold Spring Harbor Laboratory Press, Cold Spring Harbor, N.Y.

Turner, D. H., and P. C. Bevilacqua. 1993. Thermodynamic considerations for evolution by RNA, p. 447–464. *In* R. F. Gesteland and J. F. Atkins (ed.), *The RNA World*. Cold Spring Harbor Laboratory Press, Cold Spring Harbor, N.Y.

Uhlenbeck, O. C. 1972. Complementary oligonucleotide binding to tRNA. *J. Mol. Biol.* **65:**25–41.

Vacher, J., H. Grosjean, C. Houssier, and R. H. Buckingham. 1984. The effect of point mutations affecting *E. coli* tryptophan-tRNA on anticodon-anticodon interactions and on UGA suppression. *J. Mol. Biol.* **177:**329–342.

Varani, G. 1995. Exceptionally stable nucleic acid hairpins. *Annu. Rev. Biophys. Biomol. Struct.* **24:**379–404.

Vögeli, G., H. Grosjean, and D. Söll. 1975. A method for the isolation of specific tRNA precursor. *Proc. Natl. Acad. Sci. USA* **72:**4790–4794.

Wagner, E. G. H., and R. W. Simons. 1994. Antisense RNA control in bacteria, phages, and plasmids. *Annu. Rev. Microbiol.* **48:** 713–742.

Walter, A. E., and D. H. Turner. 1994. Sequence dependence of stability for coaxial stacking of RNA helixes with Watson-Crick base paired interfaces. *Biochemistry* **33:**12715–12719.

Walter, A. E., D. H. Turner, J. Kim, M. H. Lyttle, P. Müller, D. H. Mathews, and M. Zucker. 1994. Coaxial stacking of helixes enhances binding of oligoribonucleotides and improves predictions of RNA folding. *Proc. Natl. Acad. Sci. USA* **91:** 9218–9222.

Wang, A. C., and N. R. Kallenbach. 1971. Helical complexes of polyriboinosinic acid with copolymers of polyribocytidylic acid containing inosine, adenosine and uridine residues. *J. Mol. Biol.* **62:**591–611.

Wang, S., and E. T. Kool. 1995. Origins of the large difference in stability of DNA and RNA helices: C5-methyl and 2′-hydroxyl effects. *Biochemistry* **34:**4125–4132.

Watanabe, K., M. Shinma, and T. Oshima. 1976. Heat-induced stability of tRNA from an extreme thermophile, *Thermus thermophilus*. *Biochem. Biophys. Res. Commun.* **72:**1137–1144.

Weiss, G. B. 1973. Translational control of protein synthesis by tRNA unrelated to changes in tRNA concentration. *J. Mol. Evol.* **2:**199–204.

Weissenbach, J., and H. Grosjean. 1981. Effect of threonylcarbamoyl modification (t^{6}A) in yeast tRNA^{Arg} on codon-anticodon and anticodon-anticodon interactions: a thermodynamic and kinetic evaluation. *Eur. J. Biochem.* **116:**207–213.

Westhof, E., P. Dumas, and D. Moras. 1983. Loop stereochemistry and dynamics in tRNA. *J. Biomol. Struct. Dyn.* **1:**337–355.

Westhof, E., P. Dumas, and D. Moras. 1985. Crystallographic refinement of yeast aspartic acid tRNA. *J. Mol. Biol.* **184:**119–145.

Westhof, E., B. Masquida, and L. Jaeger. 1996. RNA tectonics: towards RNA design. *Fold Des.* **1:**R78–R88.

Westhof, E., and L. Jaeger. 1992. RNA pseudoknots. *Curr. Opin. Struct. Biol.* **2:**327–333.

Wyatt, J. R., and I. Tinoco, Jr. 1993. RNA structural elements and RNA function, p. 465–496. *In* R. F. Gesteland and J. F. Atkins (ed.), *The RNA World*. Cold Spring Harbor Laboratory Press, Cold Spring Harbor, N.Y.

Yao, L. J., T. L. James, J. T. Kealey, D. V. Santi, and U. Schmitz. 1997. The dynamic NMR structure of the TΨC-loop: implications for the specificity of tRNA methylation. *J. Biomol. NMR* **9:**229–244.

Yarus, M. 1982. Translational efficiency of tRNA's: uses of an extended anticodon. *Science* **218:**646–652.

Yarus, M., and D. Smith. 1995. tRNA on the ribosome: a waggle theory, p. 443–469. *In* D. Söll and U. RajBhandary (ed.), *tRNA: Structure, Biosynthesis, and Function*. ASM Press, Washington, D.C.

Yaskunas, S. R., C. R. Cantor, and I. Tinoco, Jr. 1968. Association of complementary oligonucleotides in aqueous solution. *Biochemistry* **7:**3164–3178.

Yokoyama, S., T. Watanabe, K. Murao, H. Ishikura, Z. Yamaizumi, S. Nishimura, and T. Miyazawa. 1985. Molecular mechanism of codon recognition by tRNA species with modified uridine in the first position of the anticodon. *Proc. Natl. Acad. Sci. USA* **82:**4905–4909.

Yokoyama, S., Z. Yamaizumi, S. Nishimura, and T. Miyazawa. 1979. 1H-NMR studies on the conformational characteristics of 2-thiopyrimidine nucleotides found in tRNA. *Nucleic Acids Res.* **6:**2611–2626.

Yokoyama, S., and S. Nishimura. 1995. Modified nucleosides and codon recognition, p. 207–223. *In* D. Söll and U. RajBhandary (ed.), *tRNA: Structure, Biosynthesis, and Function*. ASM Press, Washington, D.C.

Chapter 8

Mechanisms of RNA-Modifying and -Editing Enzymes

GEORGE A. GARCIA AND DEEANNE M. GOODENOUGH-LASHUA

While over 90 modified nucleosides have been identified to date, surprisingly few of the enzymes responsible for their syntheses and/or incorporation into nucleic acids have been well characterized. The chemical nature of the modifications is quite diverse, ranging from methylation to elaborate alterations of the heterocyclic ring system. All of the modifications appear to occur posttranscriptionally and, with only two known exceptions (queuosine and archaeosine) are carried out as chemical transformations of the genetically encoded and transcribed nucleotide.

For RNA-modifying enzymes, the issues of mechanism and recognition are inextricably linked. All enzymes must specifically recognize their substrates. All enzymes must also catalyze their respective reactions. However, the concept that "you don't get something for nothing" applies to enzymes just as well as it does in human affairs. Many enzymes utilize some energy source, often the hydrolysis of ATP, to drive catalysis. Others utilize high-energy or activated molecules and catalyze a reaction by simply bringing intrinsically reactive molecules together. Another common approach used by enzymes is taking advantage of the binding energy of the substrates (derived from direct interactions between the enzyme and the substrate, and indirect interactions such as the hydrophobic and solvation-desolvation effects) to drive catalysis. The potential for binding energy-driven catalysis is particularly great for enzymes whose substrates are macromolecules (e.g., RNAs) with large surface areas for interaction with the enzyme, and hence a potentially large amount of binding energy available. Presumably, RNA-modifying and -editing enzymes utilize this binding energy for both catalysis and recognition or discrimination.

Often the site of modification is "buried" within the RNA molecule. The modifying enzyme must gain access to this site. An example of this phenomenon occurs when the Hha I methyltransferase binds its DNA substrate. The X-ray crystal structure of this complex has been solved (Klimasauskas et al., 1994) and it reveals that the enzyme disrupts the DNA double helix and "flips out" the cytidine that will be methylated (see recent review by Roberts, 1995). Again, presumably binding energy will help to drive conformational changes of both the enzyme and substrate (DNA in this case) necessary for access to the site of modification. This chapter will focus on mechanistic issues involved in RNA-modifying enzymes. In those cases where information is available, RNA recognition, as it impacts the ability of the enzymes to catalyze their reactions, will also be discussed. It must also be noted that there are a number of examples where RNA recognition requirements for a particular modification are quite different between organisms. Different organisms have also been shown to exhibit different pathways and perhaps even different chemical mechanisms to the same modification (see section on m^1I).

A cursory examination of the structures of the known modified nucleosides (see Appendix 1) reveals that virtually every atom in the purine and pyrimidine ring systems may be a target for modification. Additionally, it is also apparent that the 2' hydroxyl of the ribose is also a target for modification. From a chemical/structural viewpoint, modified nucleosides can be divided into two groups. The first group consists of relatively "simple" modifications (e.g., methylation, thiolation, deamination, and isomerization). One criterion that could be used for classifying a modification as "simple" is that the modification arises due to the action of only one or two enzymatic activities. These relatively simple modifications will be discussed in the first section. The second group of modified nucleosides consists of more "complex" modifications (e.g., multiple modifications and hy-

---

George A. Garcia and DeeAnne M. Goodenough-Lashua • Interdepartmental Program in Medicinal Chemistry, College of Pharmacy, University of Michigan, Ann Arbor, Michigan, 48109-1065, USA.

permodifications), involving a multi-enzyme pathway or, as is the case with queuine and archaeosine, involving biosynthetic precursors synthesized by other enzymes for this purpose alone (as far as is known). These modifications will be discussed in the section describing complex modifications.

"SIMPLE" MODIFICATIONS

Methylation

The subject of RNA methylation was reviewed by Söll and Kline (1982).

Perhaps the most common form of nucleoside modification is methylation. This occurs in numerous forms from "simple" methylations such as m^5U (or ribothymidine) and m^5C to the more involved, multiple modifications such as ncm^5Um. With two exceptions (T54 in *S. faecalis* and *Bacillus subtilis*) (see below), adenosylmethionine (AdoMet) has been shown to serve as the methyl donor for all RNA-modifying methyltransferases. In at least one of these cases (see below), it has been shown that the methyltransferase binds adenosylhomocysteine (AdoHcy) as well as AdoMet. This is consistent with a proposal that AdoHcy product inhibition is a general characteristic of AdoMet-dependent methyltransferases (Eichler, 1994). It has been suggested (Eichler, 1994) that the activities of these various methyltransferases may be regulated by the relative levels of AdoMet and AdoHcy in the cell, which are in turn regulated by AdoHcy hydrolase and adenosine deaminase (Ueland, 1982). This may be especially important in those cases where methylation may be involved in regulating precursor RNA processing such as eukaryotic ribosomal RNA (see Chapter 13 by Bachellerie and Cavaillé and Chapter 14 by Mason).

The predominant difference between the various methylations has to do with the nature of the species being methylated (e.g., carbon versus nitrogen versus oxygen) (Table 1) and how that species is recognized by the enzyme. The question, from a mechanistic viewpoint, then becomes the following: how do these various methyltransferases activate the diverse species for nucleophilic attack on the methyl donor? In many cases, the species to be methylated is sufficiently nucleophilic that simple binding of both substrates and orientation of the nucleophile and the methyl donor probably suffice to catalyze the reaction. In other cases, an enzymatic general base may be required to deprotonate the potential nucleophile. From a recognition viewpoint, the question is twofold. First, how do the methyltransferases recognize the "correct" nucleotide within the RNA molecule (a problem common to all modifying enzymes); second, how do the methyltransferases recognize the correct atom within the nucleotide to methylate? It is reasonable to assume that the macromolecular structure of the RNA molecules (tRNAs, mRNAs, etc.) provides sufficient surface area for interactions that will determine the nucleotide to be modified. Examples of recognition motifs will be given as the various enzymes are discussed. The recognition of the specific atom within the "correct" nucleotide to methylate is likely to be inextricably linked with the mechanism of the specific reaction. Again, binding and orientation of the substrates will dictate which atom will be suitably positioned to attack the methyl donor.

Methylation on carbon [m^5U (rT), m^5C]

In the biosynthesis of DNA, the enzyme thymidylate synthase (TS) (EC 2.1.1.45) is solely responsible for the generation of dTMP from dUMP. This occurs through methylation of C^5, utilizing 5,10-methylenetetrahydrofolate as the methyl donor. This enzyme has proven to be an effective chemotherapeutic target for antineoplastic therapy. TS utilizes a covalent catalysis mechanism involving an enzymatic nucleophile, in this case a cysteine, to activate pyrimidine carbon 5 for nucleophilic attack. These studies have served as the foundation for extensive studies of tRNA (m^5U54)methyltransferase.

The *trmA* gene in *Escherichia coli* codes for the enzyme tRNA (m^5U54)methyltransferase (RUMT) (EC 2.1.1.35). This enzyme is responsible for the methylation of U54 at the C^5 position generating m^5U54 in essentially all tRNAs (Gustafsson and Björk, 1993). The *trmA* gene has been shown to be essential for cell viability even in the absence of RUMT enzymic activity (Persson et al., 1992). However, it has not been determined whether the protein has a second essential activity or if the gene has a second essential activity. This discussion will focus only on the RUMT enzyme and the mechanistic and tRNA recognition investigations that have been performed.

Using the well-established mechanism for thymidylate synthase, Santi and Hardy (1987) proposed a minimal chemical mechanism for RUMT that in-

Table 1. Classification of methylation modifications

Position	Example(s)
Carbon	m^5U (rT), m^5C
3° Nitrogen	m^1A, m^1G, m^7G, m^1I
1° Nitrogen	m^6A, m^2G
Oxygen	ncm^5Um, f^5Cm
Sulfur	ms^2i^6A[a]

[a] Methylation on sulfur is discussed in the section on thiolation.

volves attack at C^6 by an enzymatic cysteine leading to C^5 methylation and subsequent deprotonation at C^5 and elimination of the enzymatic cysteine (Fig. 1). The observation that the enzyme will also catalyze the exchange of H-5 with solvent supported this mechanistic proposal as depicted by the reversible formation of 2a in Fig. 1.

The stereochemistry of methyl transfer was investigated by experiments using chiral methyl AdoMet (Kealey et al., 1991). Two of the hydrogen atoms of the methyl group of AdoMet were replaced with one deuterium and one tritium atom, yielding a methyl group chiral due to the three isotopes of hydrogen present (Fig. 2). These experiments showed that the transfer of the methyl group from AdoMet to tRNA occurred with inversion of configuration about the methyl group, strongly suggesting direct displacement of the methyl group of AdoMet by C^5 of U54, consistent with the mechanism proposed in Fig. 1.

Substantial evidence for the proposed mechanism was also provided by studies of a mechanism-based inhibitor, F^5U-tRNA. tRNA molecules were generated in which the uridine bases were uniformly replaced by 5-fluorouridine (initially by isolating tRNA from cells grown on F^5U and later by in vitro transcription of the tRNAs using F^5UTP). The fluoro group at the 5 position prevents elimination of the enzymatic cysteine and causes the enzyme to remain covalently attached to the tRNA [3 (F) in Fig. 1] in a manner virtually identical to the process of mechanism-based inhibition of thymidylate synthase by 5-fluorouracil. Denaturing sodium dodecyl sulfate-polyacrylamide gel electrophoresis (SDS-PAGE) demonstrated the covalent bond between the enzyme and the tRNA (Santi and Hardy, 1987). The identity of the enzymatic nucleophile was determined by incubating RUMT with F^5U-tRNA and [^{3}H]methyl AdoMet (Kealey and Santi, 1991). Protease digestion of the now radiolabeled covalent complex followed by chromatography and sequencing revealed that cysteine 324 is the nucleophilic catalyst in RUMT. The facts that F^5U-tRNA trapped the covalent intermediate (3 in Fig. 1) and that fluorine is a convenient nuclear magnetic resonance (NMR) probe allowed Kealey et al. (1994) to determine the stereochemistry of the addition of RUMT and the methyl group from AdoMet to U54. They found that the enzyme, via cysteine 324, and the methyl group from AdoMet add to U54 in a *cis* orientation. This is possible only if the addition occurs in two steps via an enol intermediate with the double bond between C^5 and C^4 (Fig. 3). [This intermediate is shown with the phenolic oxygen protonated. It has been argued that protonation of the carbonyl oxygen is necessary to reduce the pKa of an α carbon acid, in this case at C^5, to allow the enzyme to deprotonate it at a "kinetically competent" rate (Gerlt and Gassman, 1993).] *trans* addition to the C^5-C^6 double bond has been observed for the mechanistically related enzymes thymidylate synthase (James et al., 1976) and Hha I methyltransferase (Klimasauskas et al., 1994). The observation of *cis* addition in the case of RUMT suggests that this enzyme binds U54 in such a manner that it blocks one face of the pyrimidine ring and only allows addition from the "bottom" face (as depicted in Fig. 3).

A kinetic mechanism has been proposed for the RUMT reaction (Kealey et al., 1994). This mechanism starts with a rapid association of tRNA and

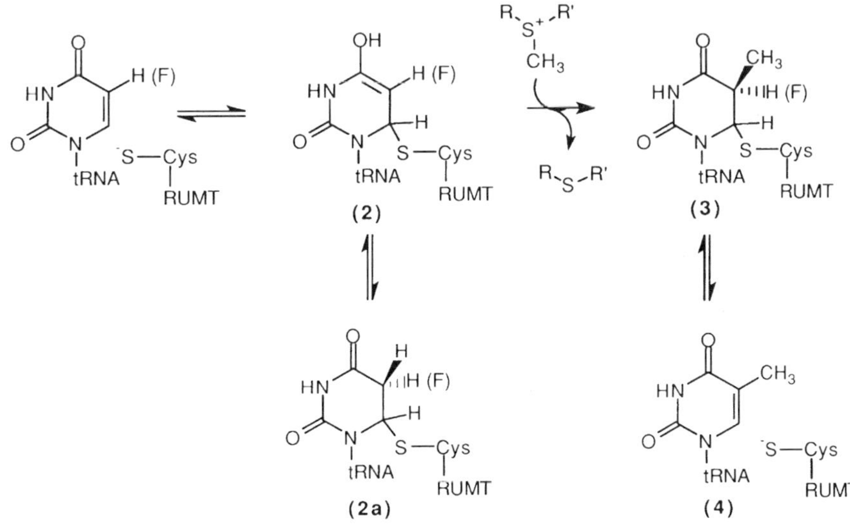

Figure 1. Proposed chemical mechanism for RUMT.

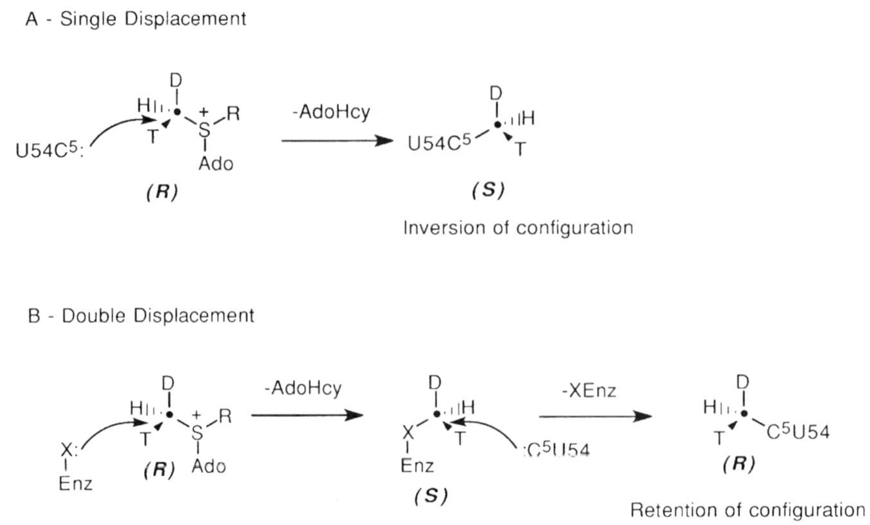

Figure 2. Chiral methyl AdoMet mechanisms. X, a nucleophile on the enzyme; Enz, enzyme.

RUMT, followed by a slow step that yields a combination of noncovalent and covalent complexes (Gu and Santi, 1992). It was suggested that this step involves conformational changes of both tRNA and RUMT. (We will return to this point.) One of the covalent species is the dihydrouridine adduct (2a [Fig. 1]). This species is an obligatory intermediate in the RUMT-catalyzed exchange of 5-H mentioned previously. The rate of this exchange, however, is less than that of the overall methylation reaction; therefore, this species is not "kinetically competent" for methylation and cannot be an intermediate on the methylation reaction pathway (Gu and Santi, 1992). The addition of AdoMet to RUMT, isolated as the dihydrouridine adduct with tRNA, resulted in complete conversion of the tRNA to the methylated form, demonstrating that the dihydrouridine adduct is "chemically competent" for methylation (Gu and Santi, 1992). The enzyme must be able to deprotonate C^5 (2 [Fig. 1]) and then continue on the methyl-

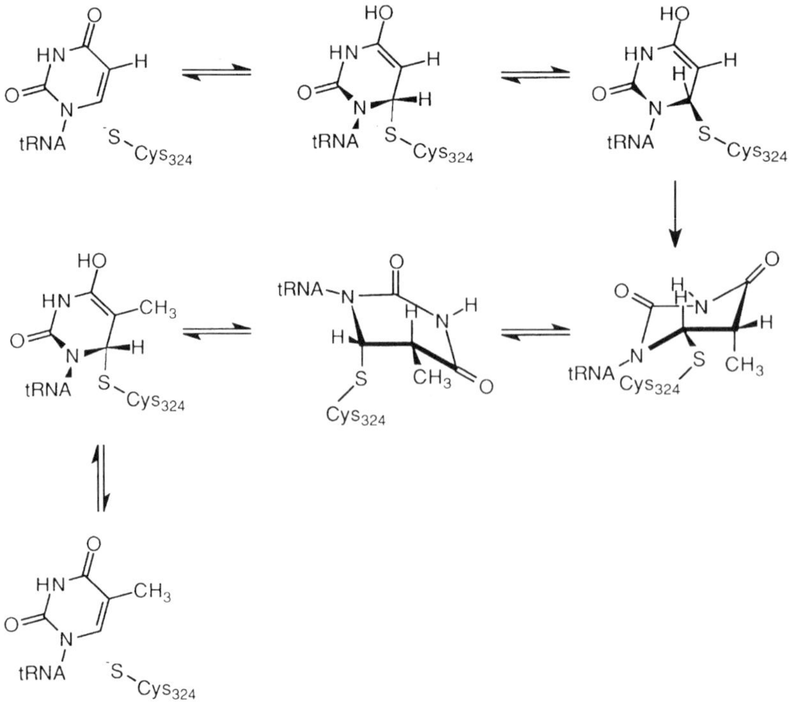

Figure 3. Stereochemistry of addition to C^5-C^6 of U54.

ation pathway. In *Escherichia coli*, RUMT is found in two forms: one free in solution, and the other covalently bound to either tRNA or 16S rRNA (Gustafsson and Björk, 1993). While it seems likely that this covalent bond is due to a dihydrouridine adduct between RUMT and the RNAs, it was shown that addition of AdoMet to the 16S rRNA·RUMT complex did not result in dissociation of the complex. Additionally, the methyltransferase activity of the 16S rRNA·RUMT complex is unaltered by removal of the RNA component by treatment with micrococcal nuclease (Gustafsson and Björk, 1993). This indicates that the covalent bond between RUMT and 16S rRNA must be distinct from the methyltransferase active site.

The chemical mechanism depicted in Fig. 3 requires that RUMT must have access to at least one face of the U54 pyrimidine ring and to the C^5 and C^6 carbons. In the tertiary structure of yeast tRNA^{Phe}, both the C^5 and C^6 of U54 are almost completely solvent-inaccessible. The crystal structure of the Hha I methyltransferase bound to (5FdCyd)DNA shows that the enzyme has "flipped out" the cytidine from the DNA double helix (Klimasauskas et al., 1994). RUMT must also disrupt the structure of the T-loop of the tRNA to gain access to U54. Experiments on D-arm mutants designed to remove specific T-loop/D-arm tertiary interactions showed that the mutant tRNAs exhibit a faster rate of complexation with RUMT (Kealey et al., 1994). The K_ms for the mutants are reduced compared to the wild type; however, the k_{cat} for methylation was essentially unaffected. These results are entirely consistent with the initial steps in RUMT-tRNA association and covalent complex formation involving a significant disruption of the tRNA structure. However, the overall rate-limiting step in the RUMT reaction must occur after the conformational change in order for k_{cat} for these mutant tRNAs to remain unaffected.

The tRNA recognition properties of RUMT have also been extensively studied. It has been found that the *E. coli* RUMT will recognize a truncated form of tRNA (17 nucleotides long) corresponding to the TΨC stem and loop of tRNA (Gu and Santi, 1991; Guenther et al., 1994). Gu and Santi found that only two base pairs in the stem of an 11-nucleotide-long analog were absolutely required for RUMT recognition (although k_{cat} was decreased by almost 2 orders of magnitude). Assays were performed at 15°C to ensure that the truncated RNA molecule was predominantly in the stem-loop conformation. Interestingly, the T_m for the 17-mer was found to be 45°C and for the 11-mer to be 38°C. These T_ms are higher than what would be expected from the stem structures alone, suggesting that interactions within the loop itself contribute significantly to the stability of the final structure. This is consistent with the proposed conformational changes that must be elicited by RUMT in order to access the C^5 and C^6 of U54 for methylation.

In addition to recognizing the truncated tRNA species, Guenther et al. showed that the *E. coli* RUMT is capable of recognizing a 2′-deoxy (DNA) analog of this 17-mer (Guenther et al., 1994). In this case, the DNA oligonucleotide corresponded exactly to the RNA 17-mer of the tRNA TΨC loop and stem with 2′-deoxyuridine substituted for thymidine in the DNA oligonucleotide. Not surprisingly, the same DNA oligonucleotide synthesized with thymidines was not a substrate for RUMT. Interestingly, it was not an inhibitor either, suggesting that the presence of thymidines in place of uridines alters the structure of the oligonucleotide such that any interaction with RUMT is abolished. It was not reported if the normal product of the RUMT reaction can bind significantly to RUMT. While it is unlikely, simple methylation of U54 may reduce the binding affinity of the normal substrate to RUMT and may account for the thymidine-containing DNA analog's inactivity.

Gu et al. have studied a large number of T-arm mutants in both full-length *E. coli* tRNA^{Phe} analogs and truncated (17-mer) T-arm analogs to investigate the sequence dependence of tRNA recognition by RUMT (Gu et al., 1996). They found that there appears to be no specific base recognition of tRNA by RUMT other than the requirement for U54. It is required that U54 be the first base in a 7-base loop at the end of a helical stem. The only base mutation that is not tolerated by RUMT is C56G. While no explanation for this observation was given, it is intriguing to speculate that perhaps G56 could form some base pairing with U54 that prevents RUMT from gaining access to U54. Alternatively, the C56G tRNA may adopt some unusual conformation that is not recognized by RUMT.

Gu and Santi proposed that the limited tRNA structural recognition motif may allow the *E. coli* RUMT to recognize and methylate other RNA species that contain this TΨC-like stem and loop structure. By homology searching an RNA sequence database, they identified a potential RUMT recognition site in 16S rRNA. Subsequently they demonstrated that RUMT will methylate, in vitro, this site in 16S rRNA (Gu et al., 1994) although this site is not normally found to be methylated in vivo. It is intriguing that this site is indistinguishable from the site in 16S rRNA where RUMT is found covalently attached.

The above studies relate only to the RUMT from *E. coli*. Grosjean and coworkers have utilized a set of 43 yeast tRNA^{Asp} mutants to study the recognition of tRNA^{Asp} by the yeast (*Saccharomyces cerevisiae*) RUMT and pseudouridine 55 synthase (Becker et al., 1997). They found that, unlike its *E. coli* counterpart, the yeast RUMT requires G53, U54, U55, C56, A58, a pyrimidine at 60, and C61 for recognition. While a stem-loop structure was recognized, a stem length longer than 4 base pairs was required for efficient T54 formation. They also permuted the tRNA by exchanging the positions of the anticodon and the TΨC arms, and found that U32 and U33 were quantitatively converted to T and Ψ, suggesting that the yeast RUMT recognizes a local structural/sequence element and that the positioning with the tRNA molecule is not important.

RUMT must make specific interactions with its RNA substrate that dictate the recognition of the T-loop (or T-loop-like structure) and U54. Based on the above studies, these interactions must be limited to the T-loop and first two base pairs of the stem. This recognition process must be a dynamic one in that the enzyme must recognize the native structure of the tRNA T-loop and it must also, presumably through a conformational change, elicit a change in conformation of the tRNA and bind the altered conformation. Further studies, ideally including the determination of the three-dimensional structure of the tRNA·RUMT complex, will be necessary to elucidate these interactions.

The lack of sequence specificity in the *E. coli* RUMT's recognition of tRNA is consistent with the enzyme's role in modifying virtually every tRNA. This enzyme has a "relaxed" sequence specificity that allows it to recognize all tRNAs and yet has some conformational specificity, in addition to the requirement for U54, to restrict the methylation to the tRNA T-arm. Gu et al. suggest that RUMT may be methylating RNA species other than tRNA and that this may be responsible for the requirement of *trmA* for cell viability. There is precedent for a dual-specificity modifying enzyme in that a pseudouridine synthase has been characterized that recognizes both tRNA and 23S RNA (Wrzesinski et al., 1995a) (see Chapter 12 by Ofengand and Fournier). The tRNA recognition determinants of the yeast RUMT are quite different and reflect a high degree of sequence specificity. This difference in recognition properties between prokaryotic and eukaryotic modifying enzymes may be a general feature as it also occurs for m^{1}G and for queuine. Further detail is provided later in this chapter.

Interestingly, the RUMTs from *S. faecalis* and *B. subtilis* have been shown to utilize 5,10-methylenetetrahydrofolate as the methyl donor (Romeo et al., 1974; Arnold and Kersten, 1975; Schmidt et al., 1975; Delk et al., 1976; Delk et al., 1980). The enzyme has also been shown to require FADH$_2$ to complete the methyl transfer (Delk et al., 1980). Radioactivity from [5-^{3}H]5-deazaFMNH$_2$ was observed to be incorporated into the methyl moiety of the ribothymidine. The requirement for FADH$_2$ suggests that the RUMT from these organisms may be a bifunctional enzyme with dihydrofolate reductase activity. The thymidylate synthase from protozoa is a bifunctional enzyme that also contains a dihydrofolate reductase activity on the same polypeptide (Garrett et al., 1984). These results suggest that, in these organisms, RUMT may have been evolutionarily recruited from thymidylate synthase (or more correctly that they had a common ancestor). These are the only reported instances of an RNA methyltransferase that does not utilize AdoMet.

The above studies also serve as a paradigm for the mechanistic issues involved in the C^5 methylation of cytidine (m^{5}C). A m^{5}C40 methyltransferase has been purified and characterized from human (HeLa) cells (Keith et al., 1980). This enzyme has a molecular mass of ca. 72 kDa and utilizes AdoMet as the methyl donor. The K_m for AdoMet was reported to be ca. 0.5 μM and a K_i for AdoHcy was 0.9 μM. These constants should be regarded as "apparent" as they were determined under only three or four different concentrations of AdoMet spanning less than a 10-fold range of concentration. Additionally, the methyl acceptor RNA was not present in saturating amount. Interestingly, the methyl transferase was found to methylate a variety of RNA substrates (five different tRNAs from *E. coli* and yeast, *E. coli* rRNA, tobacco mosaic virus [TMV] RNA, brome mosaic virus [BMV] RNA, vaccinia virus RNA, and RNA copolymers containing C and at least one other base) and was found not to methylate DNA. *E. coli* tRNA^{Phe} was the "best" acceptor. However, an RNase T$_1$ fragment of this tRNA that contains the methylation site was not a substrate, suggesting the some degree of macromolecular structure is necessary for recognition. The broad (or perhaps lack of) specificity of this enzyme may reflect a very limited sequence and/or structural requirement for RNA recognition. It is unclear why this enzyme would need such a broad specificity. It may be that the enzyme's subcellular localization may help to define RNA specificity by restricting access to many RNAs (see Chapter 24 by Maden for a discussion of this topic). The only other example of an RNA-modifying enzyme with a specificity for more than one species of RNA is the dual-specificity pseudouridine synthase (see Chapter 12 by Ofengand and Fournier). The tRNA recognition properties of the m^{5}C40 methyltransferase from yeast has recently been probed in an in vitro, cell free yeast extract system (Jiang et al., 1997). It was found that the formation of m⁵C40 in yeast tRNA^Phe absolutely required the presence of the naturally occurring intron in the anticodon loop, suggesting that this modification takes place at the level of the precursor tRNA.

Methylation on tertiary nitrogen [m¹A, m¹G, m⁷G, m¹I]

N^1 methylated purines (m¹A, m¹G, and m¹I) are found in a number of positions in tRNAs (Grosjean et al., 1995b). Chemically, N^1 can be considered to be the most nucleophilic atom in the guanine molecule. The first dissociative pK_a of the neutral form of the guanine free base is that of N^1 at a pH value of 9.3 (Pfleiderer, 1961). (This is mirrored by the 7-deazaguanine pyrrolopyrimidines [Hoops et al., 1996].) Alkylation of guanine usually occurs first at N^1. In fact, N^1 must be protected (usually by conversion of guanine to the 6-chloro derivative) to achieve selective alkylation at N9. All of this can be taken to suggest that m¹G methyltransferases have a relatively easy job of methylating a fairly nucleophilic nitrogen with an activated methyl donor, AdoMet.

The m¹G37 methyltransferase (EC 2.1.1.31) from *E. coli* appears to be the best studied of these enzymes. The gene (*trmD*) has been cloned and sequenced (Byström and Björk, 1982a; Byström and Björk, 1982b) and the protein has been overexpressed and purified (Hjalmarsson et al., 1983). The enzyme catalyzes the transfer of the methyl group of AdoMet to the N^1 position of G37 in tRNAs whose anticodons end in G. The enzyme is fairly active with a k_{cat} of ca. 10 min^{-1} and a K_m of ca. 5 μM for AdoMet (Hjalmarsson et al., 1983). A study of the tRNA structural requirements for the enzyme was carried out (Holmes et al., 1992). These studies determined that the enzyme will recognize truncated tRNAs but with significant losses in activity. The analog with the poorest activity, consisting of the anticodon arm and acceptor stem, exhibited a 200-fold loss in V_{max} but less than a doubling of K_m. (It should be noted that the kinetic parameters reported are "apparent" values, owing to the fact that they were not determined under true pseudo-first-order conditions. That is, tRNA K_ms were determined with an [AdoMet] of ca. three times its K_m rather than at a saturating concentration.) A number of anticodon arm sequence mutants were generated and their kinetic parameters were determined. The analog with the poorest activity had a diminution in V_{max}/K_m of less than 100-fold from the wild type with the majority of the analogs being within a factor of 10. The only strong correlation found was that with a G in position 36. This suggests that while the sequence of the anticodon arm may be optimized to "fine tune" recognition and catalysis by m¹G37 methyltransferase, the enzyme will recognize a number of sequences but prefers a GpG in positions 36 and 37. Unpublished observations were reported that the dinucleotide GpG and poly(G) "are both potent and specific inhibitors of the enzyme" (Holmes et al., 1995). It was also noted that AdoMet is not required for tRNA binding to the methyltransferase. In contrast to the *E. coli* enzyme, the yeast m¹G37 methyltransferase does not appear to require G36 (Droogmans and Grosjean, 1987). The sequences GAAG and GUCG (nucleotides 34 to 37) are efficiently methylated by the yeast enzyme. No studies of the chemical mechanism for this reaction have been reported; however, given the preceding discussion it seems plausible that the enzyme would act through general base catalysis by deprotonating N^1 and allowing it to nucleophilically attack the methyl group of AdoMet (Fig. 4A).

1-Methyl inosine would be considered, by the criteria set out at the beginning of this chapter, to be one of the "complex" modifications, and as such will be dealt with later in this chapter. However, the overall reaction of methylation of inosine at N^1, forming m¹I, is virtually identical to that of methylation of guanosine at N^1 and most likely occurs in a very similar, if not identical, fashion.

N^1-methyl adenosine is frequently found in position 58 of the TΨC loop. The enzyme responsible for this modification, m¹A58 methyltransferase (EC 2.1.1.36), has been purified from thermophiles (Yamazaki et al., 1992; Yamazaki et al., 1994) and from *Tetrahymena pyriformis* (Agris, personal communication). Although the chemical mechanisms of these enzymes have not been studied, the tRNA recognition properties have been probed. Both enzymes will recognize truncated tRNAs. In the case of the thermophilic enzyme, the 3′ half of the initiator tRNA from *E. coli* was a substrate, albeit with significantly lower activity. In these studies, the assays were carried out at 50°C, a suitable temperature for the enzyme; however, the stability of the 3′ half of the tRNA at this temperature was not addressed (Yamazaki et al., 1992). It may be that the diminution in activity is due to a large fraction of the tRNA fragment being "melted" and not recognizable by the enzyme (this was found to be the case for the tRNA-guanine transglycosylase from *E. coli*). Further studies determined that the K_m for the *E. coli* tRNA$_2^{Glu}$ was 100 nM and the K_m for AdoMet was 7.8 μM (Yamazaki et al., 1992). The *Tetrahymena* enzyme was found to recognize a 17-mer corresponding to the TΨC arm, but at only 10% the level of methylation of intact tRNA (Agris, personal communication). This suggests that other portions of the

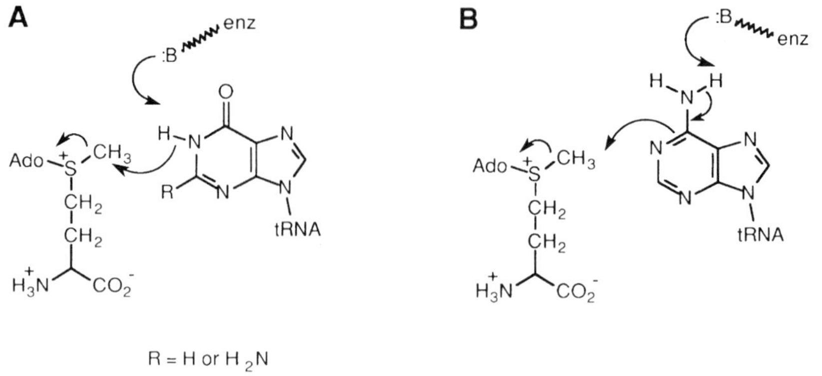

Figure 4. Possible general base mechanisms for purine N^1 methylation. (A) Deprotonation of guanine N^1; (B) deprotonation of adenine N^6. enz, enzyme.

tRNA molecule may play a role in m^1A58 methyltransferase recognition, but the primary determinants reside in the TΨC arm. This conclusion is also consistent with the observations of Grosjean and co-workers who studied the modification patterns of yeast $tRNA^{Asp}$ mutants microinjected into *Xenopus laevis* oocytes (Grosjean et al., 1996b). Their results showed that the m^1A58 modification was essentially insensitive to changes in the overall tRNA structure, and was formed in minimalist stem-loop analogs.

Chemically, methylation of the N^1 of adenosine will be more difficult than that for the 6-oxo purines (guanosine and inosine) due to the pyridine-like nature of N^1 of adenosine. However, one could envision a general base mechanism where deprotonation of N^6 would lead to an imine at the 6 position and a pyrrole-like nitrogen at N^1, which could then act as a nucleophile to attack the methyl group of AdoMet (Fig. 4B).

Methylation on primary nitrogen [m^6A, m^2G, m^2_2G]

There are essentially two forms of methylated primary amines in RNA, N^6-methyl adenosine (m^6A) and N^2-methyl (or dimethyl) guanosine (m^2G or m^2_2G). N^6-methyl adenosine is found in a variety of RNAs. The best characterized m^6A methyltransferase (EC 2.1.1.32) is that for m^6A in mRNA from eukaryotes (Bokar et al., 1994; Rottman et al., 1994; Shimba et al., 1995).

m^6A is found in eukaryotic pre-mRNA at a frequency of about one to three modifications per message. The modification site minimally involves a recognition sequence of A-<u>A</u>-C or G-<u>A</u>-C (where <u>A</u> is modified to m^6A). Based on a study of m^6A sites in mRNA, an extended recognition sequence of N_1-R-<u>A</u>-C-N_2 (where R = purine, N_1 = purine 90% of the time, and $N_2 \neq G$) has been proposed (Schibler et al., 1977). In vitro studies using partially purified methyltransferase suggest that structural and/or context effects contribute to the recognition site (Bokar et al., 1994; Rottman et al., 1994). Studies of truncated U6 snRNA methylation (Shimba et al., 1995) suggest that the adenosine must be unpaired in a bulge-like structure to be methylated. This seems reasonable as that would then allow the enzyme easy access to the exocyclic amine of the adenosine for methylation. No mechanistic details have been reported for the m^6A methyltransferase; however, simple general base catalysis, similar to that shown in Fig. 4B, seems likely.

The guanosine base found in position 26 of eukaryotic tRNAs, at the juncture between the D-stem and anticodon stem, is often found to be modified to N^2-dimethyl guanosine (m^2_2G). In yeast, this modification has been attributed to the single protein product (ca. 63 kDa) of the *trm1* gene (Ellis et al., 1986). Expression of the *trm1* gene in *E. coli* (normally devoid of m^2_2G in tRNA) converted the organism to having m^2_2G-modified tRNA and expressing the methyltransferase. This result also indicated that, in the yeast system, a single enzyme catalyzes the dimethylation of G26. Studies of tRNAs microinjected into *X. laevis* oocytes yielded both monomethylated and dimethylated G26 tRNAs with different rates of formation (Edqvist et al., 1992) and revelaed that precursor and mature length tRNAs were modified at G26 (Grosjean et al., 1990). This suggests that two different enzymes may be involved: one for monomethylation, and another for dimethylation. Alternatively, a single enzyme could be responsible but have a different active site for the second methylation or require dissociation of the monomethylated tRNA to obtain the second AdoMet (the methyl group donor). A G26 methyltransferase was purified from *T. pyriformis* and was found to have both monomethylating and dimethylating activity as well (Reinhart et al., 1986). The enzyme was also estimated to be ca. 200–250 kDa in its native form, suggesting that the G26 methyltransferase is oligomeric (possibly tetrameric). Mechanistically, the second methylation of the N^2 of G26 may be "easier" than the first due to the electron-donating property of the first methyl group; however, there may be some steric hindrance by the methyl group as well. It is likely that these effects will somewhat cancel out. The tRNA structural and sequence requirements for m^2_2G dimethylation in both *X. laevis* and yeast have also been studied (Edqvist et al., 1992; Edqvist et al., 1994; Edqvist et al., 1995). The modification of G26 requires GC base pairs at positions G11-C25 and C10-G24 as well as a five-base variable loop. The anticodon stem-loop is not required for recognition. The authors conclude that there is little base specific recognition involved and that the structural requirements seem to ensure that the G26 is presented to the methyltransferase in a favorable orientation.

Methylation on oxygen [ncm⁵Um, f⁵Cm, Cm, Gm]

2′-O-methylated nucleosides have been found in various RNAs (tRNA, mRNA, and rRNA). The methyltransferases responsible for these modifications appear to be distinct and to have quite different properties. Three tRNA 2′-O-methyl transferases (EC 2.1.1.34) have been studied (Droogmans et al., 1986; Matsumoto et al., 1990). The G34 2′-O-methyltransferase from *X. laevis* oocytes was found to be cytoplasmic. The recognition of tRNA by this enzyme was determined to not rely on the sequence surrounding the site of modification but rather on the size of the anticodon loop and structural features of the tRNA molecule beyond the anticodon loop (Droogmans et al., 1986). Interestingly, the nature of the base at position 34 was not critical either as C34, A34, and U34 mutants were also methylated. However the possibility of four different, base-specific methyltransferases, albeit unlikely, was not ruled out. The G18 2′-O-methyltransferase from *Thermus thermophilus* has also been studied using tRNA fragments to deduce the recognition requirements (Matsumoto et al., 1990). Only those fragments which contain an intact D-stem and loop structure were substrates for the enzyme. RNase T_1 footprinting experiments revealed that the methyltransferase protects only the D-loop region of the tRNA. A 2′-O-methylated nucleoside (2′-O-methyl-5-formylcytidine) has recently been discovered in position 34 of two cytoplasmic tRNA^Leu s from bovine liver (Pais de Barros et al., 1996).

In the nucleoli of mammalian cells, the 47S precursor rRNA is 2′-O-methylated at various positions before being processed to the mature forms. A 2′-O-methyltransferase has been partially purified from mouse nucleoli and characterized (Eichler, 1994). Photo-cross-linking studies revealed a native molecular mass of ca. 150 kDa for the enzyme; however, under denaturing conditions the labeled enzyme had a molecular mass of 50 kDa. This suggests that the enzyme exists in some oligomeric form, but the specific quaternary structure is not known. Like the tRNA G34 methyltransferase, this enzyme will methylate all four nucleosides. It was shown to bind S-adenosylhomocysteine (AdoHcy; $K_d = 0.17$ μM) as well as AdoMet ($K_d = 0.24$ μM) and to exhibit a 10-fold preference for the physiologically relevant S-isomer of AdoMet (Segal and Eichler, 1989). The efficient binding of AdoHcy is consistent with the proposal that AdoHcy product inhibition may regulate the activity of AdoMet-dependent methyltransferases discussed at the beginning of this section. It was determined that this enzyme was capable of methylating a unique triple methylation site in a 28S rRNA transcript (one of processed portions of the 47S precursor) in vitro (Segal and Eichler, 1991). Although the triple methylation site was methylated, other single methylation sites present in the 28S rRNA transcript were not, suggesting that structural features present in the 47S precursor are required for recognition or that other methyltransferases may be involved. Eukaryotic rRNA 2′-O-methyltransferases have been shown to utilize small nucleolar RNAs (snoRNAs) as "guide" molecules to help them find the site for ribose methylation in the rRNA substrate (Bachellerie and Cavaillé, 1997). The "guide" RNAs consist of antisense sequences (relative to the substrate rRNA) that bind to the rRNA. It is this complex that appears to be recognized by the 2′-O-methyltransferase. This subject is discussed in detail in Chapter 13 by Bachellerie and Cavaillé.

2′-O-Methyl nucleosides have also been found in mRNA. An mRNA 5′ cap-specific 2′-O-methyltransferase encoded by vaccinia virus has been cloned, sequenced, and expressed (Schnierle et al., 1994). Interestingly, this protein (VP39) has previously been shown to have mRNA polyadenylation-stimulatory activity (Gershon et al., 1991; Schnierle et al., 1992; Gershon and Moss, 1993). The protein exists in both monomeric and heterodimeric (associated with the polyadenylation enzyme) forms, both of which have methyltransferase activity. Truncated mutants of the protein have localized the RNA-binding domain to the carboxyl-terminal 112 amino acids. Approximately 20 clusters of charged amino acids appear in the sequence of VP39. Alanine-scanning mutagenesis of these clusters revealed that 11 of these mutated clusters had little to no methylation activity (Schnierle et al., 1994). Four of the eleven methylation-deficient cluster mutants were also deficient in polyadenylation stimulatory activity, suggesting a role in RNA recognition for these clusters. Further mutagenesis studies identified specific amino acid residues in VP39 that are involved in the methyltransferase activity (presumably involved in AdoMet binding) and others that are involved in RNA binding (Shi et al., 1996).

The X-ray crystal structure of a triple mutant (AS11) of VP39 with AdoMet bound (Fig. 5A) has been determined to 1.85 Å (Hodel et al., 1996). (The wild-type VP39 was unable to be crystallized.) This mutant has three positively charged amino acids replaced with alanine (R140A, K142A, and R143A). Although deficient in methyltransferase activity, this mutant is equivalent to the wild type in polyadenylation stimulation. The mutant, AS11, was found to contain a tightly bound AdoMet. It had been determined previously that this mutant has a higher binding affinity for AdoMet than the wild type (Shi et al., 1996). The crystal structure suggests that this difference may be due to the R140A mutation. In the wild type, the side chain of R140 would be close to the sulfonium of AdoMet. Electrostatic repulsion of these two positively charged groups may reduce AdoMet binding to the wild type relative to the R140A mutant. One possible explanation for the methyltransferase deficiency in AS11 is that the R140A mutation may allow the enzyme to bind AdoMet too well and that the reduction in AdoMet binding provided by R140 is necessary to guide the reaction through the transition state to the products, which would not be positively charged. This would account, in part, for the affinity of VP39 for AdoHcy. The crystal structure also reveals that there are a number of basic residues (K41, K175, and R209) in the vicinity of the methyl group of AdoMet that projects into a long cleft on the protein surface. It is possible that these residues assist in catalysis by orienting the 2′ hydroxyl of the mRNA for attack on the methyl group of AdoMet, and perhaps also assist in deprotonation of the hydroxyl via general base catalysis. Combined with the results of the mutagenesis studies discussed above, a model for RNA binding to VP39, within the long cleft on the protein surface, and for the location of the methyltransferase active site (Fig. 5B) has been proposed (Hodel et al., 1996).

The observation that many of the 2′-O-methyltransferases have relaxed base specificity is somewhat consistent with the location of the methylation on the 2′ hydroxyl of the ribose. It seems reasonable that structural features of the RNA will be required to present the 2′ hydroxyl to the methyltransferase in the proper orientation for methylation, as the model for VP39 suggests. Activation of the 2′ hydroxyl should be a relatively facile process, most likely occurring through a general base-type mechanism, similar to that suggested for some of the heterocycle methylations above. If this is the case, then it seems likely that under the right conditions these enzymes may catalyze the cleavage of the substrate

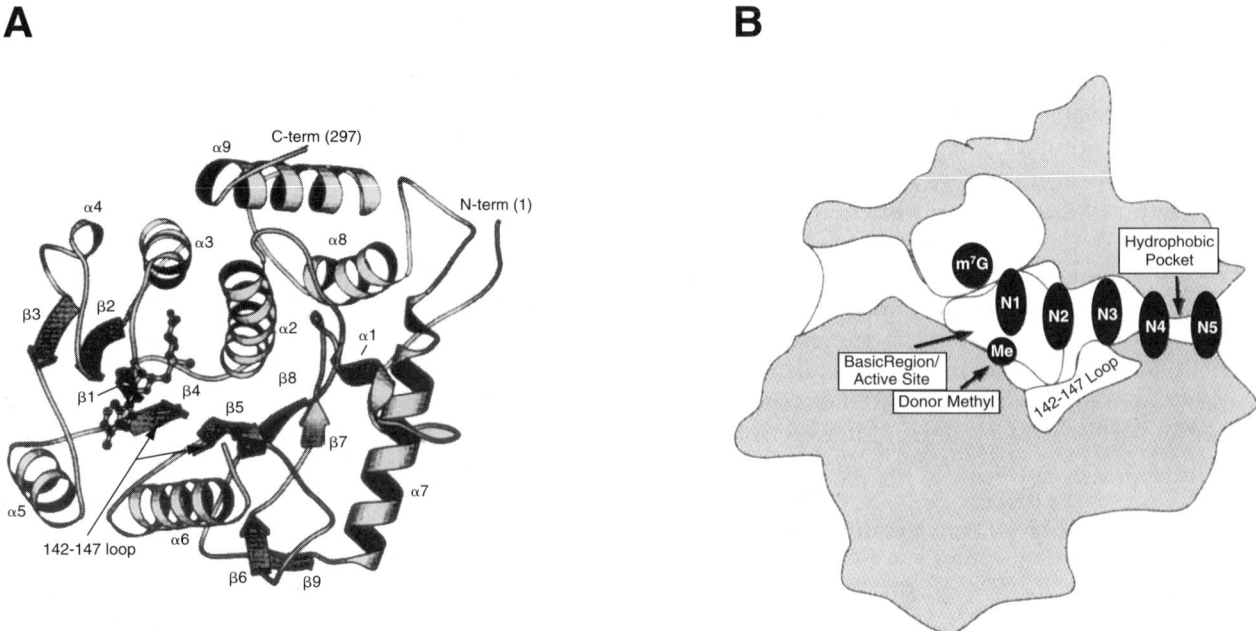

Figure 5. (A) X-ray crystal structure of VP39; (B) model for mRNA binding. term, terminus. Adapted from Hodel et al., 1996, with permission of the author and publisher.

RNA by deprotonating the 2' hydroxyl in a manner similar to RNases.

Thiolation: s⁴U, s²U, and ms²i⁶A

While thionucleosides have been found in a wide variety of species, the number and type of sulfur-containing modifications deviate significantly between organisms (Vold et al., 1981). To date, 2-thiopyrimidines and 4-thiouridine have been isolated. Thiolation of the second position of uridine is often accompanied by further modification at C^5. A variety of adenosine derivatives containing a C^2 methylthio substituent have also been observed. So far, no naturally occuring thioguanosine modifications have been reported. Each type of thiolation arises from the activity of a different enzyme (or combination of enzymes) (Ajitkumar and Cherayil, 1988).

Studies have been performed to identify the origin of the sulfur atom involved in nucleoside thiolation. In most cases, cysteine (or cystine) was identified as the source of the sulfur atom as shown through the use of [^{35}S]cysteine (Peterkofsky and Lipsett, 1965; Hayward and Weiss, 1966; Abrell et al., 1971). However, the identity of the exact molecular species involved in transfer of the sulfur to the nucleoside remains under debate. With respect to the tRNA sulfurtransferase system, β-mercaptopyruvate has been variously reported to be inactive (Peterkofsky and Lipsett, 1965; Abrell et al., 1971), a required cofactor (Lipsett et al., 1967), and the sulfur donor (Wong et al., 1970). One plausible explanation for these conflicting reports is that cysteine may be converted to β-mercaptopyruvate, which then could act as the immediate sulfur donor (Lipsett et al., 1967; Wong et al., 1970). This is consistent with the fact that the sulfur transfer reaction is pyridoxal phosphate dependent (Lipsett and Peterkofsky, 1966; Lipsett, 1972), presumably through a role in β-mercaptopyruvate formation (Lipsett et al., 1967). Thiolation of tRNA also appears to require ATP and a divalent cation, preferably Mg^{2+} (Hayward and Weiss, 1966; Lipsett and Peterkofsky, 1966; Lipsett et al., 1967; Abrell et al., 1971). Although ATP has not been shown to be involved, β-mercaptopyruvate is well known as a thiol donor (Kun, 1967).

Biosynthesis of 4-thiouridine results from the activity of two distinct enzymes (Abrell et al., 1971). Factor A, which requires ATP, Mg^{2+}, and a sulfhydryl compound, appears to "activate" the acceptor tRNA. The identity of this activated species has not been determined, but it is likely to be 4-phosphouridine (Fig. 6). The second enzyme, factor C, catalyzes sulfur transfer activity in the presence of pyridoxal phosphate, cysteine, and the product from the first reaction.

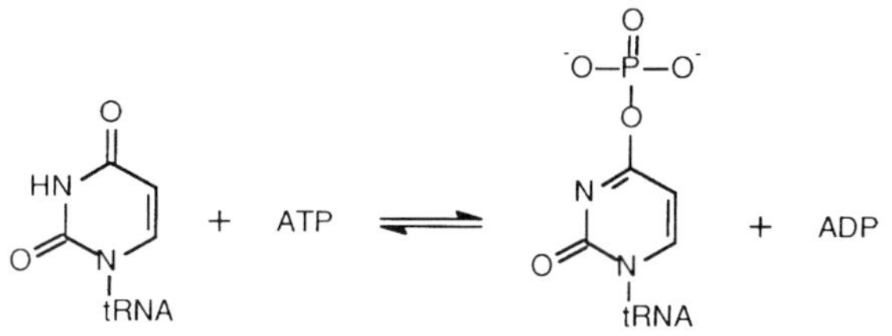

Figure 6. Proposed activation step for uridine thiolation.

Based on the current data, a mechanism for sulfur transfer can be proposed, which involves a uridine-activating step forming 4-phosphouridine from tRNA and ATP (Fig. 6). The proposed sulfur transfer mechanism utilizes cysteine and PLP (or alternatively β-mercaptopyruvate and pyridoxamine) in the form of a Schiff's base adduct in the reaction (Fig. 7). Beta-elimination of HS^- from this adduct provides the thiol nucleophile, which can attack the activated uridine and yields PLP and dehydroalanine (which may be converted to pyruvate and NH_3). It should be noted that if β-mercaptopyruvate is involved, then the resulting PLP must be recycled back to PMP after each enzymic turnover. It is possible that cysteine could act as the sulfide donor in the absence of any pyridoxal cofactor involvement, yielding dehydroalanine after beta-elimination of HS^-. It is difficult to imagine a mechanism to activate β-mercaptopyruvate without the involvement of a pyridoxal cofactor.

The 2-methylthio derivatives of adenosine also arise from multi-step processes. With respect to the C^2 substituent, thiolation occurs first, followed by S-adenosylmethionine-dependent methylation (Agris et al., 1975). It has been shown that in the case of ms²i⁶A, the isopentenyl group is added before formation of the methylthio group (Agris et al., 1975; Buck and Ames, 1984). In the presence of oxygen ms²i⁶A can be further modified to the hydroxylated form, ms²io⁶A (Buck and Ames, 1984). Based on this, it is likely that the introduction of the substituent at C^6 also occurs prior to C^2 modification for the modified nucleoside ms²t⁶A. The actual mechanism of this sulfur transfer reaction is likely to be somewhat different than that described above. In this case, sulfur is being added directly to an unsubstituted carbon atom of the purine ring. In the cases above, the sulfur is replacing an oxygen atom that is already present.

Deamination: Inosine

The modified nucleoside inosine (I) was first located at position 34 of tRNA. A collection of more

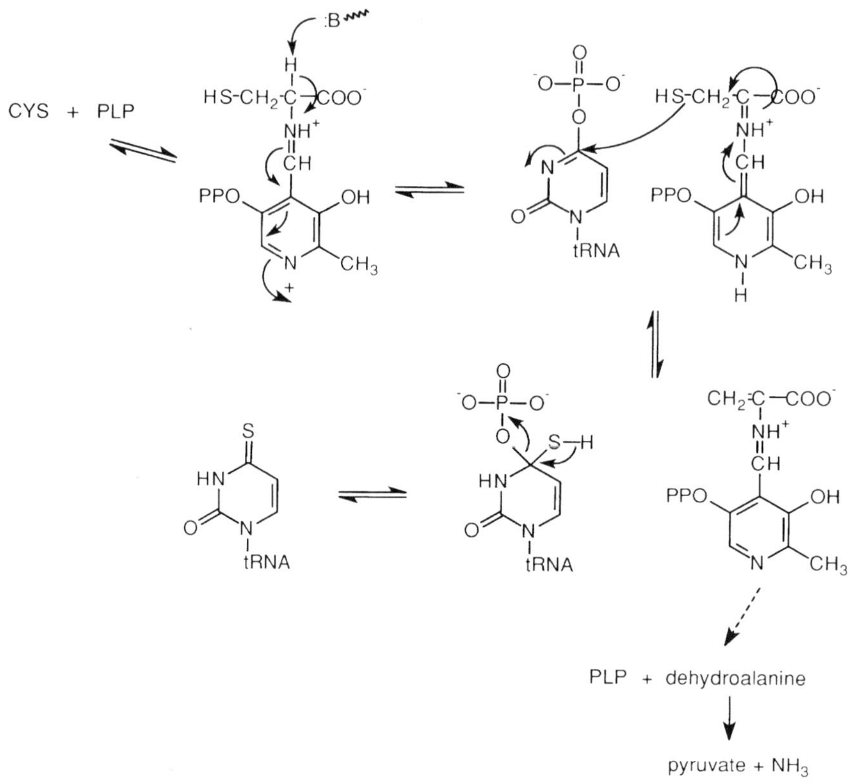

Figure 7. Proposed mechanism for sulfur transfer.

than 500 known tRNA sequences revealed that all eukaryotic cells, plant chloroplasts, and eubacteria contain at least one I34 tRNA species. While eubacteria express only a single inosine containing tRNA (tRNA^{Arg}), at least seven different yeast tRNAs exhibit this modification, illustrating a significant variance in modification between organisms (Sprinzl et al., 1996). Further comparison of the inosine-containing tRNA from yeast revealed that the only sequence requirement is a genetically encoded adenosine at position 34. A purine at position 35 is preferred, but not required (Auxilien and Grosjean, 1995). Structurally, the tRNA must maintain its natural L-shape fold and the addition of a single base in the anticodon loop (giving eight bases instead of seven), completely abolishes inosine incorporation (Auxilien and Grosjean, 1995).

Elliott and Trewyn (1984) reported the identification of an enzyme from rat liver and human (HL-60 cells) capable of incorporating hypoxanthine into tRNA. From this observation it was suggested that the inosine modification in tRNA arose from a base exchange mechanism similar to that of tRNA-guanine transglycosylase. These studies also looked at the ability of this isolated enzyme to incorporate [^{3}H]guanine. The observed lack of [^{3}H]guanine incorporation activity suggested that this enzyme was distinct from TGT (Elliott and Trewyn, 1984). More recent studies have demonstrated that although low levels of hypoxanthine-adenine exchange may occur, the biologically relevant method of inosine 34 incorporation is hydrolytic deamination of A34 (Auxilien et al., 1996). tRNA containing [^{32}P]-, [^{3}H]adenosine was used to demonstrate that the ^{32}P/^{3}H ratio did not change during the adenine to inosine conversion. Furthermore, the use of H$_2^{18}$O revealed that the oxygen in inosine originated from water (Auxilien et al., 1996). Together, these experiments eliminated the possibility of a base exchange mechanism for inosine 34 incorporation. The enzyme responsible for deamination of adenosine in tRNA has now been termed tRNA:adenosine 34 deaminase (EC 3.5.4.4).

Over the last decade, inosine has also been located in a variety of other RNAs including glutamate receptor mRNA (Sommer et al., 1991; Bass, 1995; Rueter et al., 1995; Yang et al., 1995), TAR RNA of HIV-1 (Sharmeen et al., 1991), and hepatitis delta virus antigenome RNA (Benne, 1996; Polson et al., 1996). In each of these RNAs, the inosine modification occurs in a double-stranded RNA region. Studies similar to those described above demonstrated that the inosine modification in double-stranded RNA also arises from hydrolytic deamination of adenosine (Polson et al., 1991) and a number of different double-stranded RNA deaminases have been identified in mammalian cells (dsRAD, RED1, and RED2). The exact RNA sequence and/or structural requirements of the mammalian enzymes remain to be determined, but dsRAD from *X. laevis* demonstrated a higher extent of modification in longer pieces of dsRNA with a 5′ neighbor preference (A = U > C > G), but no preference for the 3′ neighboring nucleotide (Polson and Bass, 1994). In dsRNA the adenosine to inosine modification leads to a U*I mismatch which results in a loosening or unwinding of the duplex. Although this has not been experimentally demonstrated, it is possible that the lower level of modification in shorter dsRNA fragments results from loss of a necessary duplex structure. This is in agreement with the observation that the proximity to the strand termini also influences whether the A to I conversion occurs (Polson and Bass, 1994).

The X-ray crystal structure of a nucleoside adenosine deaminase has been determined, and a zinc ion and an ordered water molecule have been located in the active site (Wilson et al., 1991; Wilson and Quiocho, 1994). The X-ray crystal structure of cytidine deaminase (EC 3.5.4.5) complexed with uridine has also recently been determined (Xiang et al., 1997). Details of the structure, including a bound water molecule in the ammonia leaving group site, led the authors to a number of conclusions regarding the mechanism. First, a glutamate residue is poised to hydrogen bond to the incoming hydroxyl and appears to transfer the proton to the leaving ammonia yielding the 4-keto group. Second, the pyrimidine ring rotates toward the zinc ion as the ammonia group moves in the opposite direction, suggesting that the enzyme "pulls" the ammonia group and the pyrimidine ring apart. Finally, a zinc-sulfur bond (to cysteine 132) lengthens in the transition state and shortens again in the product complex allowing the zinc ion to compensate for the buildup of negative charge on the attacking hydroxyl oxygen in the transition state. These conclusions are also consistent with the structure of adenosine deaminase and most likely apply as well to its mechanism. Based on this, a modification of a previously proposed mechanism for adenosine deaminase (Wilson et al., 1991) is presented in Fig. 8A. Additional support for this mechanism is provided by the demonstration that coformycin and deoxycoformycin (Fig. 8B), compounds which mimic the putative tetrahedral intermediate, are potent inhibitors of both nucleoside adenosine deaminase and AMP deaminase (Frick et al., 1986; Merkler et al., 1990).

It seems quite likely that an identical mechanism holds for tRNA:adenosine 34 deaminase and dsRAD. However, no inhibition by coformycin or deoxycoformycin has been observed for either tRNA:adenosine 34 deaminase or dsRAD (Polson et al., 1991; Auxilien et al., 1996). This absence of inhibition may be due to an enzymic requirement for some degree of RNA macromolecular structure (or architecture) resulting in a lack of affinity of these enzymes for coformycin and deoxycoformycin. To make further conclusions on the mechanism of these inosine forming enzymes, determination of the presence or absence of a catalytic zinc is essential.

A more detailed discussion of the relationship between the nucleoside deaminases and the RNA deaminases can be found in Chapter 20 by Carter. Evidence that the cytosine to uridine editing that occurs in RNA is carried out by a deaminase reaction has been reported (Yu and Schuster, 1995). This also is discussed in Chapter 20 by Carter.

Isomerization: Pseudouridine

Pseudouridine (5-β-D-ribofuranosyluracil; Ψ) is the most abundant modified nucleoside found in RNA. To date pseudouridine has been located in transfer RNA (Singer and Smith, 1972), ribosomal RNA (Erdmann et al., 1983), and nuclear RNA (Shibata et al., 1975). Naturally occurring C-nucleosides are not uncommon; however, pseudouridine and its 1- and 2′-substituted analogs are the only C-nucleosides that have been identified in RNA (Limbach et al., 1994). Despite extensive studies surrounding the pseudouridine modification, relatively little is known about the mechanism of its formation other than the fact that pseudouridine results from a posttranscriptional isomerization of the genetically encoded uridine at the appropriate positions (Kammen et al., 1988).

To date, four families of pseudouridine synthases (EC 5.4.99.12) have been isolated (Kammen et al., 1988; Nurse et al., 1995; Wrzesinski et al., 1995a; Wrzesinski et al., 1995b). Two families of enzymes appear to catalyze formation of pseudouridine in tRNA, while the others appear to be responsible for pseudouridine formation in rRNA. However, a dual-specificity pseudouridine synthase, responsible for both Ψ764 in 23S RNA and Ψ32 in tRNA[Phe], has been identified in *E. coli* (Wrzesinski et al., 1995b). Although a comparison of the protein sequences of the known pseudouridine synthases reveals little amino acid homology, three of the pseudouridine synthase families contain a conserved motif, also found in cytidine triphosphate deaminase and uridine triphosphatase (Koonin, 1996). This motif is likely to be involved in pyrimidine binding.

The *hisT* gene product of *Salmonella typhimurium* (Allaudeen et al., 1972; Singer et al., 1972) was

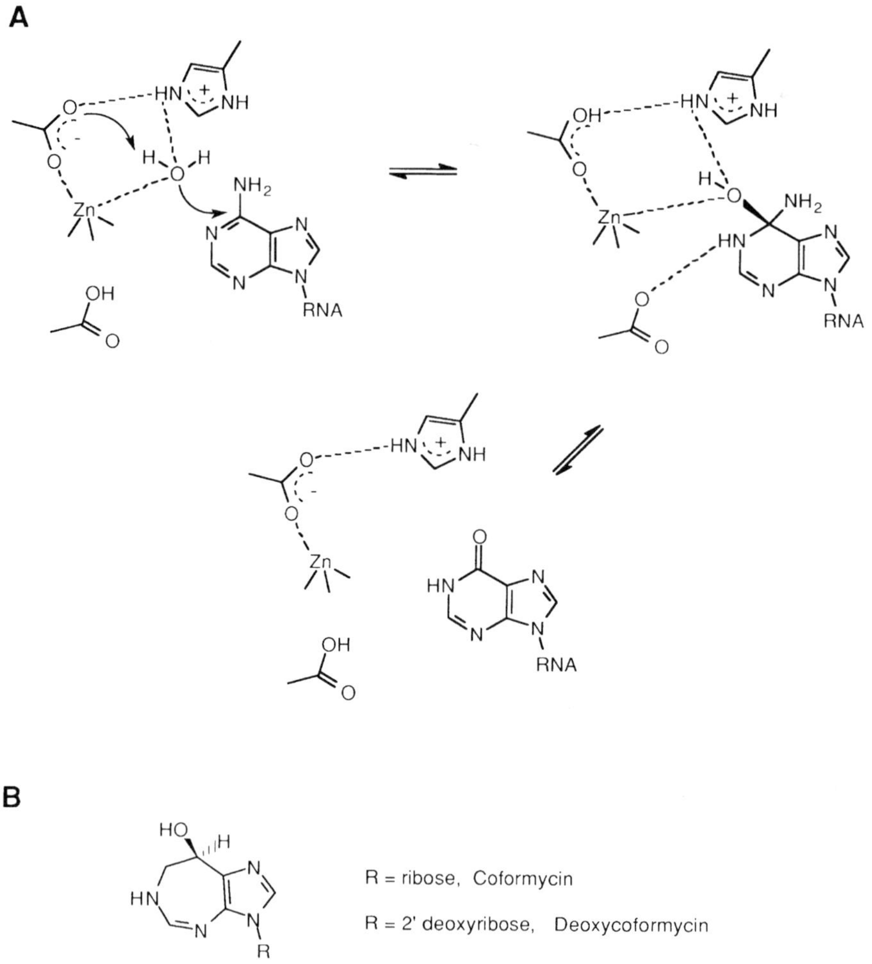

Figure 8. (A) Proposed mechanism for the nucleoside adenosine deaminase; (B) structures of its transition state analog inhibitors.

determined to have pseudouridine synthesizing properties. The resulting enzyme was found to be responsible for the modifications in positions 38–40, which lie in the anticodon loops of several tRNAs. Interestingly, *hisT* mutants still contain Ψ32 in the tRNA TΨC loops (Green et al., 1982). This was an initial indication that more than one enzyme was responsible for the pseudouridine modifications. The enzyme expressed by the *hisT* gene was termed pseudouridine synthase I (PSU-I). This is the most extensively studied of the pseudouridine synthases and therefore will be the focus of the following discussion.

An interesting issue regarding the recognition of tRNA by the various pseudouridine synthases has recently come to light. It has been known for some time that posttranscriptional modifications are introduced into precursor tRNAs and mature-length tRNAs (Nishikura and De Robertis, 1981). Pseudouridine is found at specific positions in precursor tRNAs and in additional positions in the mature tRNAs. The PUS-1 from yeast catalyzes the incorporation of Ψ in three different locations (positions 27, 34, and 36) in tRNA^{Ile}. In vitro studies with mature and intron-containing tRNA^{Ile} have revealed that the introduction of pseudouridine in position 27 is independent of the presence of the intron. However, the formation of both Ψ34 and Ψ36 is absolutely dependent upon the presence of the intron (Szweykowska-Kulinska et al., 1994; Motorin et al., 1997a), although not on the base sequence of the intron.

The *E. coli* PSU-I has been shown to bind both substrate and nonsubstrate tRNA (Kammen et al., 1988). Although no exogenous ions are necessary for binding, a monovalent cation (preferably NH_4^+ or K^+) is required for catalysis (Mullenbach et al., 1976). Furthermore, no cofactors or energy sources are required (Kammen et al., 1988). The observation that treatment of the enzyme with thiol reactive reagents (e.g., p-chloromercuribenzoate and iodoacetic acid) completely inactivates the *E. coli* PSU-I (Mullenbach et al., 1976; Kammen et al., 1988) suggests the involvement of one or more cysteine residues in the pseudouridine synthase reaction. In vitro the enzymic thiol can be protected by cysteine, β-mercaptoethanol, or dithiothreitol.

The most common method of measuring PSU-I activity involves a ^{3}H-release assay (Cortese et al., 1974). This assay stems from the fact that the pyrimidine C^5 proton must be released during rearrangement. Tritium release can be monitored because the released ^{3}H will be equilibrated with solvent H_2O, whereas the tRNA-incorporated pyrimidine will be absorbed by charcoal when the reaction is quenched. The quenched reaction mixture is vortexed and then centrifuged to separate the RNA (charcoal) and solvent H_2O (supernatant) fractions. This approach is similar to the assays performed when studying the mechanisms of thymidylate synthase and RUMT (Santi et al., 1974).

The validity of the ^{3}H-release assay is supported by the following points: (i) PSU-I preferentially catalyzes ^{3}H release from HisT$^-$ [5-^{3}H] tRNA in comparison to wild-type tRNA; (ii) catalysis is specific for C^5 protons, and no release is observed when [6-^{3}H] tRNA is used; (iii) PSU-I catalyzed reaction products appear identical to wild-type tRNA^{Phe} when analyzed by reverse-phase chromatography; and (iv) ^{3}H release occurs even in the absence of small molecule donors such as S-adenosylmethionine (Mullenbach et al., 1976) which, in crude enzyme preparations, could give false-positive results due to contaminating RUMT activity.

PSU-I was also analyzed for the ability to catalyze the exchange of exogenous uracil into wild-type and *hisT* tRNA, as well as the ability to catalyze ^{3}H release from uracil, uridine, and UTP (Kammen et al., 1988). No activity was detected in any of these studies, supporting direct isomerization of the RNA incorporated uridine and a requirement for at least some portion of the tRNA structure. Therefore, based on the known information concerning the *E. coli* PSU-I, a minimal mechanism for pseudouridine formation must involve four steps: (i) binding of substrate tRNA to the enzyme, (ii) cleavage of the glycosidic bond, (iii) a 180° rotation of the base with respect to the ribose, and (iv) re-formation of the glycosidic bond through a carbon-carbon linkage.

The apparent role of one or more cysteine residues, as well as the potent inhibitory action of 5-fluorouridine containing tRNA (Frendewey et al., 1982), led Kammen et al. (1988) to propose a conceptual model for uridine isomerization based on the mechanisms of known pyrimidine C^5 modifying enzymes such as thymidylate synthase (Pogolotti et al., 1986) and RUMT (Santi and Hardy, 1987). Kammen et al. suggest that the initiating step is attack at C^6 by a nucleophilic group, possibly one of the cysteine residues, and proceeding though a transient dihydopyrimidine-enzyme adduct (Fig. 9). The glycosidic bond is then broken in a unimolecular decomposition yielding an oxocarbonium ion intermediate. This intermediate is then trapped by the C^5 of the uridine-enzyme adduct, forming pseudouridine after elimination of the enzymatic nucleophile.

Support for this mechanism comes from the report that a 6-hydroxy-5,6-dihydrouridine intermediate is kinetically competent in the acid-catalyzed hydrolysis of uridine to uracil (Prior and Santi, 1984). While this report does indeed provide support for the proposed mechanism, the reactions being followed are different (hydrolysis versus deglycosylation/glycosylation) and the conditions for the hydrolysis are fairly extreme (1 N HCl and 100°C). The observation of a secondary kinetic isotope effect at $C^{1'}$ in the acid-catalyzed hydrolysis studies reveals some carbocation character for $C^{1'}$ in the transition state (Prior and Santi, 1984) for this reaction. This is consistent with a significant degree of concertedness involving glycosidic bond cleavage and water attack at $C^{1'}$.

Recent mutagenesis studies of the three cysteine residues found in *E. coli* PSU-I rule out the possibility of a mechanism involving a cysteine nucleophile (Zhao and Horne, 1997). Studies involving the mutants C154A and C169A, as well as a triple mutant, in which all three cysteine residues were converted to alanine, revealed that all three mutants exhibited less than 10% reduction in activity compared to the wild-type enzyme (Zhao and Horne, 1997). However, mutation of cysteines 154 and 169 to serine resulted in almost complete loss of ^{3}H release activity, whereas the C55S mutant maintained full activity. Although C55S and C154S were still able to bind tRNA, C169S could not. Based on these observations, the following conclusions regarding the mechanism of PSU-I can be made. First, uridine isomerization does not involve a covalent cysteine intermediate, nor are any cysteine residues necessary for catalysis. In fact, only a 2-fold decrease in V_{max}/K_m was observed for the alanine triple mutant (Zhao and Horne, 1997). It has been demonstrated that the interaction of thiol reactive reagents with cysteine residues located in or near an enzyme active site can inhibit catalysis, despite the lack of a direct catalytic role for the cysteine (Giorgianni et al., 1995; Garcia and Chong, 1997). It is possible that this accounts for the loss of activity observed by Mullenbach et al. for PSU-I. Second, although serine and cysteine are similar in that both can act as nucleophiles, their steric bulk and polarities differ significantly. It is possible that the introduction of the smaller and more polar serine residue perturbs the enzyme structure, leading to the observed loss in activity with the

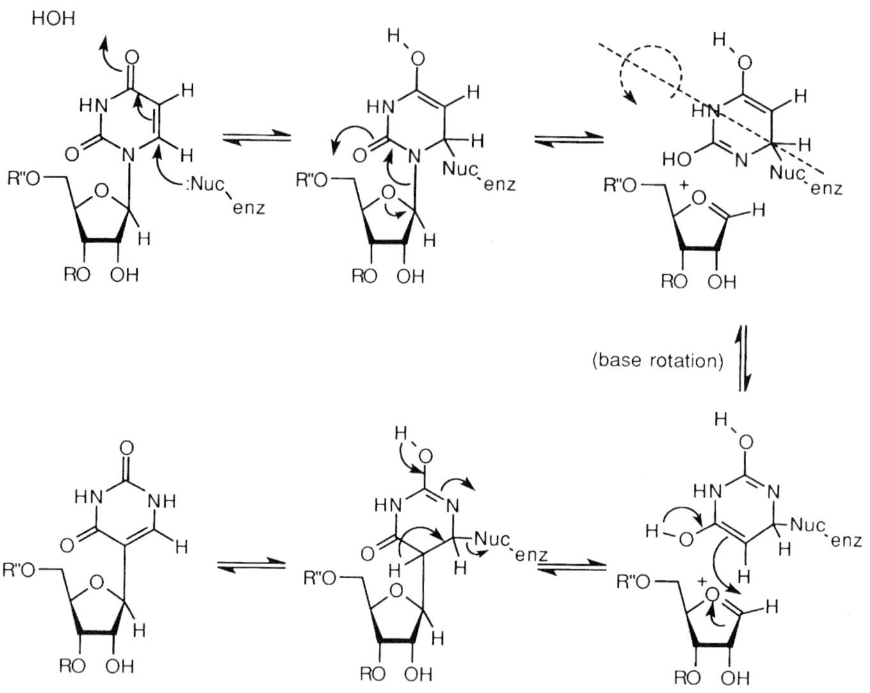

Figure 9. Hypothetical nucleophilic attack at C^5 mechanism for pseudouridine synthase. Nuc, enzymatic nucleophile; enz, enzyme.

C154S and C169S mutations. These results do not rule out a nucleophilic mechanism for the pseudouridine synthase reaction; rather, they eliminate the possibility of cysteine acting as the nucleophile. One caveat of these experiments is that the chemical mechanism for the mutant enzymes may be altered from that for the wild type, with a coincidentally similar overall activity. While unlikely, this can only be absolutely ruled out by more detailed studies.

The initial mechanistic model for pseudouridine synthase was based on the fact that it catalyzes a "C^5 modification" reaction. An alternative view is that the enzyme is catalyzing an intramolecular transglycosylation. Accordingly, a review of enzymes such as tRNA-guanine transglycosylase and uracil-DNA glycosylase suggests a potential mechanism involving enzymatic nucleophilic attack at $C^{1'}$ of the ribose rather than at C^5 of the uracil (Fig. 10). In the case of uracil-DNA glycosylase, the nucleophile has been identified as an activated water molecule (Savva et al., 1995). For tRNA-guanine transglycosylase, an aspartic acid residue is believed to be the catalytic nucleophile (Romier et al., 1996b). While this "transglycosylation" mechanism is plausible for the pseudouridine synthase reaction, it is entirely possible that the enzyme could use two nucleophiles in a combination of these mechanisms, one to attack at C^5 and the other to displace the pyrimidine at $C^{1'}$. This transglycosylation reaction may also be relevant for certain RNA editing events, although the evidence suggests that C to U and U to C editing in plants occurs via a cytidine deaminase-type activity (see Chapter 17 by Marchfelder et al.). Finally, a mechanism similar to that shown in Fig. 9 but without the intervention of any enzymatic nucleophile cannot be ruled out.

Alkylation

i⁶A (isopentenylation)

The modified base N^6-(Δ^2-isopentenyl)-adenosine (i⁶A) occurs in the anticodon loop (position 37) of many tRNAs. It is often further modified to 2-methylthio-N^6-(Δ^2-isopentenyl)-adenosine (ms²i⁶A). In some organisms, mostly plants, the Δ^2-isopentenyl group is hydroxylated at the 4 position, yielding the zeatin base (Buck and Ames, 1984). This base as well as its 2-methylthio derivative has been found in plant and some bacterial tRNAs (Agris et al., 1975). The isopentenyl group in this modified base is actually a 3-methylbut-2-enyl or dimethylallyl group (nomenclature that is used in the isoprenoid literature). For ease of recognition by the reader, the Δ^2-isopentenyl nomenclature (traditionally used in the tRNA modification literature) will be used here.

The isopentenylation of A37 in tRNA is catalyzed by the product of the *miaA* gene in *E. coli*. The genetic regulation of the *miaA* gene is complex and

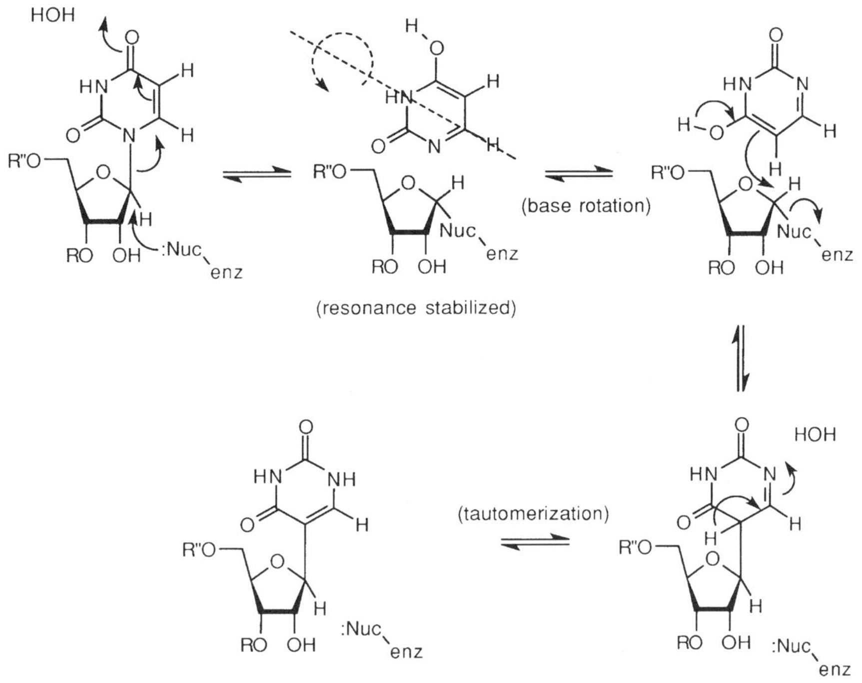

Figure 10. Hypothetical nucleophilic attack at C$^{1'}$ (or S$_N$^{2}) mechanism for pseudouridine synthase. Nuc, enzymatic nucleophile; enz, enzyme.

dealt with in Chapter 25 by Winkler. The *miaA* enzyme (EC 2.5) utilizes Δ^2-isopentenyl pyrophosphate (IPP, or dimethylallyl diphosphate) as the isopentenyl group donor. This molecule is used in a number of other biosynthetic pathways (e.g., isoprenoids, steroids, and dimethylallyltryptophan) and reviews of its biosynthesis and utilization are found in the text *Enzymatic Reaction Mechanisms* by Walsh (1979) and the chapter on isopentenyl transferases and isomerase by Poulter and Rilling in *Biosynthesis of Isoprenoid Compounds*, volume 1 (Elkins and Keller, 1974).

Δ^2-Isopentenyl pyrophosphate directly results from the isomerization of Δ^3-isopentenyl pyrophosphate, catalyzed by isopentenyl pyrophosphate isomerase (EC 5.3.3.2). Δ^2-Isopentenyl pyrophosphate is utilized by a number of isopentenyl transferases leading to various isoprenoids and ultimately to steroids. Two reaction mechanisms have been postulated for the isopentenyl transferases, an associative mechanism and a dissociative mechanism (Fig. 11A). In the associative mechanism, a nucleophile attacks C^1 of IPP, directly displacing pyrophosphate in an S$_N$^{2} fashion. In the dissociative mechanism, the C^1-oxygen bond cleaves releasing pyrophosphate and leaving a resonance-stabilized carbocation at C^1. A nucleophile can then attack the carbocation at C^1 yielding the product. Studies using fluorine-substituted, Δ^2-isopentenyl pyrophosphate (Fig. 11B) strongly support the dissociative mechanism involving the allylic carbocation intermediate (Poulter and Rilling, 1976;

Poulter et al., 1976; Poulter and Satterwhite, 1977). The electron withdrawing nature of the trifluoromethyl group has been shown to severely retard S$_N$^{1}-type solvolysis of the IPP analog while slightly enhancing S$_N$^{2}-type displacement of pyrophosphate in chemical model studies. The trifluoromethyl IPP analog is ca. 10^6-fold less active in isopentenyl transfer than IPP, consistent with the dissociative mechanism.

tRNA isopentenyl transferases have been partially purified from several sources (Fittler et al., 1968; Kline et al., 1969; Bartz et al., 1970; Bartz and Söll, 1972; Rosenbaum and Gefter, 1972; Holtz and Klambt, 1975; Holtz and Klambt, 1978), but none of these have been well characterized. The *E. coli* MiaA isopentenyl transferase has recently been cloned (Caillet and Droogmans, 1988; Connolly and Winkler, 1989), overexpressed, and purified (Leung et al., 1997; Moore and Poulter, 1997). The *E. coli* protein has a subunit molecular mass of ca. 35 kDa. Gel filtration chromatography and gel electrophoresis experiments indicate that the protein is a homodimer (Leung et al., 1997). Filter binding experiments are most consistent with the homodimer binding a single tRNA molecule with a K_d of ca. 70 nM (for in vitro-transcribed *E. coli* tRNA^{Phe}). This is in contrast to another report (Moore and Poulter, 1997) which indicates that the enzyme is monomeric and that the K_d for tRNA^{Phe} is ca. 5 nM. The gel filtration, electrophoresis, and filter binding studies were performed at a necessarily higher concentration of enzyme than the

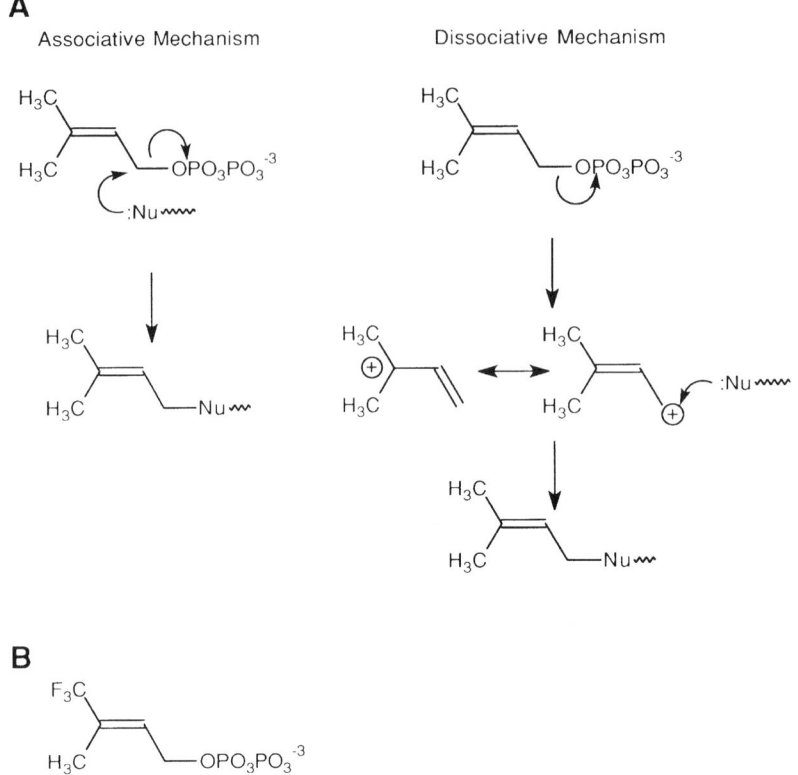

Figure 11. (A) Two proposed mechanisms for prenyl transferases; (B) trifluoromethyl analog of D^2-isopentenyl pyrophosphate. Nu, prenyl acceptor nucleophile.

activity studies of Moore and Poulter. This may account for the discrepancy in observed quaternary structure; however, more experiments are needed to sort this out.

Winkler and coworkers have found that MiaA is strongly inhibited by ATP and ADP (competitive with respect to IPP) with a K_i of ca. 60 nM (Leung et al., 1997). Nevertheless, even though inhibited in the cell, the enzyme is still present in a catalytic excess based upon the determination of MiaA protein expression levels by quantitative Western immunoblotting (660 monomers/cell) and estimations of the tRNA flux in the cell (Leung et al., 1997). Moore and Poulter found that IPP does not bind to the enzyme in the absence of tRNA, but that it does bind (K_d, ca. 3 μM) in the presence of an inhibitory tRNA analog corresponding to the anticodon stem and loop of tRNA^{Phe} with A37 replaced by inosine. These results suggest that the enzyme follows an ordered sequential kinetic mechanism for substrate binding with tRNA binding first followed by IPP. Truncated tRNAs, corresponding to the anticodon stem and loop of tRNA^{Phe}, are substrates for MiaA when assayed at 24°C with K_ms of ca. 10-fold higher and a k_{cat} of ca. 50% of those for the full-length tRNA (Leung et al., 1997). Further studies of 17 anticodon stem and loop sequence mutants of tRNA^{Ser} (GGA) and 7 other tRNAs have determined that the key feature for recognition and activity by MiaA is the A36A37A38 sequence in the 7-membered anticodon loop and the retention of helical structure and flexibility (Motorin et al., 1997b).

Much work remains to be done to characterize the MiaA enzyme and reaction to the same level as that for other isopentenyl transferases. It does seem likely that MiaA will be found to follow the same "dissociative" chemical mechanism with ordered, bi-bi kinetics. However, it has been reported that MiaA does not contain any of the conserved motifs found in the isopentenyl transferases, which is suggestive that the same chemical mechanism may not necessarily apply (Leung et al., 1997; Moore and Poulter, 1997). All of the necessary tools (substrate analog, competitive inhibitors, etc.) to firmly establish this are available.

t^{6}A (threonylation)

The modified adenosine analog N-[N-(9-β-D-ribofuranosylpurin-6-yl)carbamoyl] threonine [N^6-(N-threonylcarbonyl)adenosine; t^{6}A] is located at position 37 of several prokaryotic and eukaryotic tRNAs. Studies have identified the required characteristics of the tRNA substrate. The necessary sequence was shown to be -C-U-X-X-U-A-A- (nucleotides 32 to 38) (Elkins and Keller, 1974). Interestingly, although prokaryotic initiator tRNA^{fMet} contains the proper sequence, it fails to contain the modification. However, *E. coli* initiator tRNA^{fMet} injected into *X. laevis* oocytes becomes modified suggesting that the prokaryotic enzyme is not able to modify the initiator tRNA (Grosjean et al., 1987). Although the overall three-dimensional structure of the tRNA is important, limited perturbations are tolerated. It was observed that in *X. laevis* the anticodon loop size can be varied from 6 to 8 nucleotides without drastically affecting modification and that only U36 and A37 were absolutely required (Morin et al., 1998).

Numerous studies have identified the sources of the various components involved in the modification. First, it was demonstrated that the source of the threonyl group is free L-threonine (Chheda et al., 1972; Powers and Peterkofsky, 1972). A small amount of the glycine analog of t^{6}A has been observed in tRNA, and it has been shown that a single enzyme is responsible for both threonine and glycine incorporation (Elkins and Keller, 1974). A variety of known single-carbon donors, including biotin, *S*-adenosylmethionine, and formyltetrahydrofolate, were analyzed as the potential source of the carbamoyl carbon. However, addition of these components did not alter the level of modification (Korner and Söll, 1974), suggesting that the carbamoyl carbon does not derive from any of these carbon donors. Further studies revealed that bicarbonate is the true carbon donor for the carbamoyl carbon (Elkins and Keller, 1974). Additionally, Mg^{2+} and ATP were found to be required for modification (Elkins and Keller, 1974). Powers and Peterkofsky (1972) investigated the possibility that carbamoyl-phosphate is the active carbon-donating species. Observing that an *E. coli* mutant deficient in the biosynthesis of arginine and uracil, thought to be due to a defective carbamoyl-phosphate synthetase, was still capable of the t^{6}A modification, Powers and Peterkofsky concluded that carbamoyl-phosphate was not the source of the urea moiety of t^{6}A (Powers and Peterkofsky, 1972). However, it is possible that the t^{6}A modifying enzyme catalyzes a reaction independent of, but mechanistically similar to that of the carbamoyl-phosphate synthetases involved in uracil and arginine synthesis. Consequently, a brief review of the mechanism of carbamoyl-phosphate synthetase is pertinent.

Carbamoyl-phosphate synthetase (EC 6.3.5.5) catalyzes the reaction shown in Fig. 12. Extensive ev-

$$\text{Enz} + \text{ATP} + \text{HCO}_3^- \rightleftharpoons \text{Enz} \cdot (\text{CO}_3^-\text{PO}_3^{-2}) + \text{ADP}$$

$$\text{Enz} \cdot (\text{CO}_3^-\text{PO}_3^{-2}) + \text{Gln} \longrightarrow \text{Enz} \cdot (\text{H}_2\text{N-CO}_2^-) + \text{P}_i + \text{Glu}$$

$$\text{Enz} \cdot (\text{H}_2\text{N-CO}_2^-) + \text{ATP} \longrightarrow \text{Enz} + (\text{H}_2\text{N-CO}_2^- \cdot \text{PO}_3^{-2}) + \text{ADP}$$

Figure 12. Chemical steps for the carbamoyl phosphate synthetase reaction. Enz, enzyme.

idence demonstrates that two distinct ATP binding sites exist (Powers and Meister, 1978; Britton et al., 1979; Rubio et al., 1979). Although pulse-chase experiments revealed that both molecules of ATP could bind simultaneously (Rubio et al., 1979), the generally accepted kinetic mechanism involves the ordered binding of the reaction components (Raushel et al., 1978; Braxton et al., 1992). Evidence for this kinetic mechanism includes the following observations. Bicarbonate-dependent ATPase activity is observed (Aitken et al., 1975; Rubio et al., 1981). This reaction is reversible as shown by the exchange of oxygen atoms between HCO$_3^-$ and the γ-phosphate of ATP (Rubio et al., 1981). The addition of ammonia results in rapid release of P$_i$ and oxygen exchange is no longer observed (Rubio et al., 1981). Furthermore, the lack of ADP-ATP exchange suggests that the ADP molecule stays bound as part of the ADP-enzyme-(CO$_3^-$PO$_3^{2-}$) complex (Rubio et al., 1981; Meister, 1989). The ATP which binds in the second site acts as the phosphoryl donor for carbamoyl phosphate (Powers and Meister, 1978).

Based on the above discussion, it seems likely that the t^{6}A synthetase follows the mechanism proposed in Fig. 13. As with carbamoyl phosphate synthetase, the initiating step would be formation of an ADP-enzyme-(CO$_3^-$PO$_3^{2-}$) complex. The 6-amino group of adenosine could act as the nitrogen donor, resulting in the loss of inorganic phosphate and an N^6-carbamoyladenosine intermediate. Phosphorylation of the carbamoyl group by a second molecule of ATP leads to N^6-phosphocarbamoyladenosine. Finally attack at the carbamoyl carbon by the α-amine of threonine releases P$_i$, completing the modification.

Both carbamoyl-phosphate synthase and the t^{6}A modifying enzyme utilize ATP, Mg^{2+}, and HCO$_3^-$. Consequently the chemical mechanism proposed above is plausible. Isolation of either the N^6-carbamoyl- or the N^6-phosphocarbamoyladenosine intermediate would provide strong evidence for this proposed mechanism. Should it be determined that the chemical mechanism of t^{6}A modification is similar to that of carbamoyl phosphate synthesis, extensive studies will still be necessary to identify the kinetic mechanism, which may be distinct from that observed by carbamoyl-phosphate synthetase.

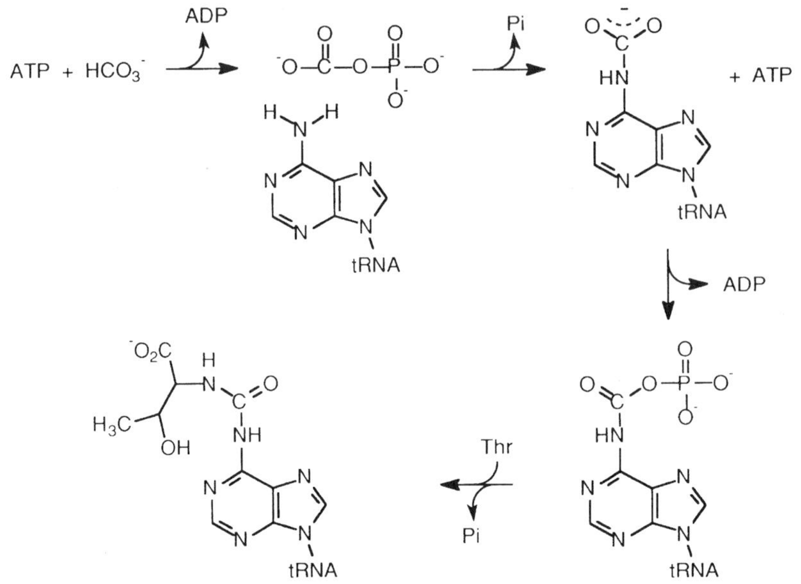

Figure 13. Proposed mechanism for threonylcarbamoylation of tRNA.

A/GrP (2'-O-ribosylphosphate)

In eukaryotic initiator tRNA^{Met}s, a unique modified base is found at position 64. This modified base contains a ribosylphosphate group attached to the 2' hydroxyl for the base through the 1' carbon of the ribosyl-5'-phosphate (Desgres et al., 1989; Glasser et al., 1991). This modification appears to be involved in preventing the initiator tRNA from interacting in the ribosomal P site and inactivating the tRNA for elongation (Desgres et al., 1989; Glasser et al., 1991; Åström and Byström, 1994). The *rit1* gene from yeast has been cloned, sequenced and expressed (Åström and Byström, 1994). The protein has a molecular mass of ca. 57 kDa and has 2'-O-ribosylphosphate transferase activity. The 2'-O-ribosylphosphate donor was found to be 5'-phosphoribosyl-1'-pyrophosphate (PRPP). A number of mutant initiator tRNA^{Met}s were studied as substrates for RIT1 (Åström and Byström, 1994). The results of these studies led to the conclusion that the primary specificity determinants for RIT1 reside in the TΨC loop and stem and that those specificity determinants for RIT1 match those previously found for initiator tRNA function. Taking into account the utilization of PRPP (which has a 1' α configuration for the pyrophosphate) and the resulting β linkage in the ribosylated ArP modification, it seems likely that a straightforward S_N2 mechanism obtains for RIT1. In this case, mechanistically similar to the 2'-O-methyltransferases, RIT1 would merely have to deprotonate the 2' hydroxyl of the base at position 64 and allow it to attack the 1' position of PRPP, displacing pyrophosphate. It would be interesting to find out if RIT1 would catalyze pyrophosphate exchange into PRPP in the same fashion that the aminoacyl-tRNA synthetases catalyze pyrophosphate exchange into ATP (a back reaction of the amino acid activation step).

Reduction: D (Dihydrouridine)

The modified base 5,6-dihydrouridine is found in the first (sequentially) loop of the cloverleaf tRNA structure so often that this loop has been trivially referred to as the D-loop. Although dihydrouridine has been known for many years, the mechanism for its incorporation into RNA remains obscure. It has been shown that 5,6-dihydro-UTP can be incorporated into RNA by the action of the *E. coli* RNA polymerase (Roy-Burman et al., 1965, 1967). Later studies with human cells revealed that the incorporation of dihydrouridine was closely coupled with tRNA synthesis (Tidwell and Howard, 1972). These studies followed the appearance of radiolabeled dihydrouridine in tRNA isolated from cells pulsed with 6-^{14}C-orotic acid. In this manner, Tidwell and Howard were able to show that both pseudouridine and dihydrouridine were formed on the same time scale as the biosynthesis of tRNA. However, it could be that the enzymes responsible for pseudouridine and dihydrouridine are present in catalytic excess such that the rate-limiting step in their formation is the production of tRNA. The results of Tidwell and Howard are also consistent with an alternative hypothesis that dihydrouridine is incorporated during transcription. The latter hypothesis is unlikely because it is difficult to imagine a mechanism for the very specific incorporation of the modified base transcriptionally that would also prevent its incorrect incorporation (presumably substituting for uridine) in other positions. This hypothesis would also require significant amounts of dihydro-UTP to be present in the cell. Studies by Bell and coworkers suggest a posttranscriptional incorporation of dihydrouridine into tRNA (Lo and Bell, 1981; Lo et al., 1982). These authors characterized a yeast mutant that accumulated isoaccepting tRNA species. In vivo pulse-chase labeling experiments showed that these unusual isoaccepting tRNAs appeared "at the expense of" the normal isoacceptors and that they could be converted over time into the normal, wild-type isoacceptors (Lo and Bell, 1981). The isoacceptors were subsequently found to lack methylated nucleosides and dihydrouridine (Lo et al., 1982). While it is possible that the conversion of these isoacceptors to wild type, manifested by a change in elution on reversed-phase chromatography, is entirely due to posttranscriptional methylation, the results are suggestive that the incorporation of dihydrouridine occurs as the authors suggest by a "simple enzymatic reduction of the 5,6 double bond of the uracil moiety."

"COMPLEX" MODIFICATIONS

Multiple Modifications

m^1I

While inosine is a fairly common modification in RNA, 1-methylinosine (m^1I) is found less frequently and is limited to tRNA. In fact 1-methylinosine occurs exclusively at position 37 of eukaryotic tRNA^{Ala} and at position 57 of several archaebacterial tRNA species (Sprinzl et al., 1996). It has been established that formation of 1-methylinosine proceeds by a two-step mechanism. In eukaryotic organisms, this two-step process begins with deamination of adenosine to give inosine 37 (see above) (Grosjean et al., 1996a). The enzyme responsible for formation of I37 is similar to, but distinct from, tRNA:adenosine 34 deaminase (Grosjean et al., 1996a). The deamination step is followed by methylation at position 1 by an S-adenosylmethionine-dependent enzyme. The order of this pathway was verified by the observation that in the absence of AdoMet, a buildup of I37-tRNA was obtained (Grosjean et al., 1996a).

However, an interesting observation was made when 1-methylinosine formation at position 57 of *Haloferax volcanii* tRNA^{Ile} was studied. In archaebacteria, such as thermophiles, halophiles, and methanogens, methylation apparently precedes deamination (Grosjean et al., 1995a). An analysis of the nucleoside content of tRNA^{Ile} demonstrated the presence of 1-methyladenosine (m^1A), a modification that does not naturally exist in this tRNA species. Consequently, it was concluded that m^1A must be an intermediate in the formation of a different modified nucleoside, possibly m^1I. Verification of this conclusion came from the following experimental observations. No inosine was detected in the archaebacteria tRNA, even in the absence of AdoMet (Grosjean et al., 1995a). Monitoring m^1A and m^1I levels over a time course demonstrated that the appearance of m^1I coincided with the depletion of m^1A. In fact, the sum of these two modifications never exceeded 1 mol/1 mol of tRNA (Grosjean et al., 1995a). Similar results were obtained with *Pyrococcus furiosus* (Constantinesco and Grosjean, personal communication), supporting the concept that this modification pathway is a general characteristic of archaea.

The exact significance of this difference in the pathway leading to the formation of m^1I remains to be determined. Both of the methylating enzymes involved require AdoMet, suggesting that the mechanisms of methylation are similar, if not identical, in each organism. On the other hand, the presence of the methyl group on the purine ring may influence the mechanism by which the deamination reaction occurs. Methylation may eliminate the requirement for zinc-catalyzed activation of the attacking water molecule (Grosjean et al., 1995a). A hypothetical mechanism for m^1A deamination is presented in Fig. 14, in which a water molecule attacks C^6 of a 6-imine or an N^1-pyridinium-type species. Alternatively, it is possible that the methyl group is simply a necessary recognition element for the m^1A deamination enzyme.

mnm^5s^2U/mnm^5se^2U

The modified thiopyrimidine, 5-methylaminomethyl-2-thiouridine (mnm^5s^2U), has been found in the wobble position of several tRNAs. Biosynthesis of mnm^5s^2U involves at least four distinct steps as shown in Fig. 15. In vivo, the first step is thiolation of uridine at C^2 by the *asuE* gene product (Sullivan et al., 1985; Hagervall et al., 1987). In vitro it has been demonstrated that the thiolation step can occur at any point along the pathway (Hagervall et al., 1987). The mechanism of this thiolation reaction is likely to be similar, if not identical to that of the enzymes involved in formation of 4-thiouridine.

A *trmE* mutant was found to contain s^2U, but none of its 5-substituted derivatives (Hagervall et al., 1987). This led to the conclusion that the *trmE* gene is responsible for the formation of 5-carboxy-

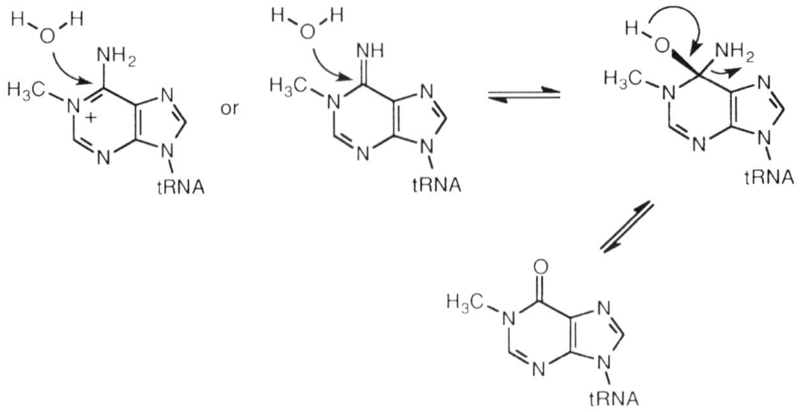

Figure 14. Proposed mechanism for m^{1}A deaminase.

methylaminomethyl-2-thiouridine (cmnm^{5}s^{2}U) (the second step in Fig. 15). The source of the 5-carboxymethylaminomethyl group has not yet been identified. Using [^{14}C-methyl]methionine, it has been determined that only one of the carbons in the methylaminomethyl side chain of the product arises from S-adenosylmethionine (Hagervall et al., 1987).

The enzyme involved in the conversion of the 5-carboxymethylaminomethyl substituent to 5-methylaminomethyl, tRNA-(5-methylaminomethyl-2-thiouridine)-methyltransferase (EC 2.1.1.61), has been found to have multiple activities (Hagervall et al., 1987). The tRNA from methionine-starved cells contains cmnm^{5}s^{2}U, as well as nm^{5}s^{2}U. In the presence of S-adenosylmethionine mnm^{5}s^{2}U is produced. This suggests that the enzyme first removes the carboxymethyl portion of the side chain, and then methylates the remaining nucleoside in an S-adenosylmethionine dependent reaction.

Studies have shown that 5-methylaminomethyl-2-selenouridine arises from a direct replacement of selenium for sulfur in mnm^{5}s^{2}U (Wittwer and Stadtman, 1986). The strongest piece of evidence for this conclusion was the observation that the level of mnm^{5}s^{2}U decreased proportionally to the increase in mnm^{5}se^{2}U. Furthermore, the use of ^{35}S-labeled tRNA demonstrated that the radiolabel was lost specifically from mnm^{5}s^{2}U. The C^5 substituent is not required for Se incorporation, but the presence of this group enhances the level of selenonucleoside formation.

The biologically active selenium donor has been identified as monoselenophosphate (Stadtman, 1994; Veres et al., 1994; Veres and Stadtman, 1994). This donor molecule is produced from selenide and ATP

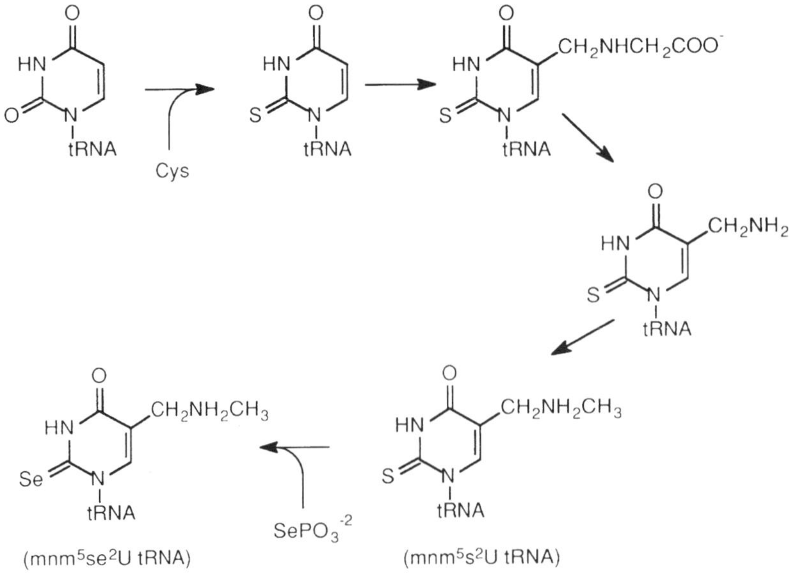

Figure 15. Biosynthetic pathway for mnm^{5}s^{2}U and mnm^{5}se^{2}U.

by selenophosphate synthetase (Veres et al., 1992; Veres et al., 1994; Veres and Stadtman, 1994), and is involved in the specific incorporation of selenium into both proteins and nucleic acids (Veres et al., 1994). Along with ATP and selenide, Mg^{2+} and a monovalent cation are also required for the formation of selenophosphate. In the absence of selenide, hydrolysis of ATP to AMP and two inorganic phosphates occurs. A multistep formation of selenophosphate is suggested by the observation that AMP is a competitive inhibitor with respect to ATP (Veres et al., 1992; Veres et al., 1994), but selenophosphate, inorganic phosphate, and sulfate are not. Although an enzyme-pyrophosphate intermediate is believed to be involved, no direct evidence for its existence has been reported.

ATP is required for the formation of selenophosphate; however, it is not required for the transfer reaction (Veres and Stadtman, 1994). This suggests that selenation occurs by a mechanism distinct from that of pyrimidine thiolation. It has been proposed that direct attack at C^2 by selenophosphate occurs, followed by the release of sulfur (Veres and Stadtman, 1994). The exact form of the released sulfur atom has not been identified. The sensitivity of the selenium incorporating enzyme to iodoacetamide (Veres and Stadtman, 1994) suggests a requirement for an enzymatic thiol group, although the role of this putative thiol is unknown.

"Hypermodifications"

Hypermodifications involve a significant alteration in the structure of the nucleoside base. Most of these hypermodifications appear to come about by alteration of the existing base. Two modifications, however, do not come about in this fashion. In the cases of queuine and archaeosine, the modified bases are biosynthesized at the heterocyclic level. The modified heterocycles are then incorporated into tRNA by an exchange mechanism. This pathway makes the queuine and archaeosine modifications the most complex of all modified bases, involving an unknown number of enzymes to generate the heterocyclic bases and additional enzymes to incorporate the bases into tRNA. [It should be noted that this definition of "hypermodification" is somewhat different from the traditional usage of the term. This chapter focuses on modified bases as chemical entities not on their locations in RNA (except as this effects enzyme-RNA recognition) or roles. The definition for hypermodification that is used in this chapter reflects this.]

Queuosine

Queuine [7-(4,5-cis-dihydroxy-1-cyclopenten-3-ylaminomethyl)-7-deazaguanine (Fig. 16)] is present only in the wobble position (position 34) of the anticodon loop of tRNAs with the anticodon sequence GUN (tRNAs Asn, Asp, His and Tyr). In higher organisms, queuine-containing tRNA is further modified by a glycosylation of the cyclopentene diol with either mannose or galactose (Kasai et al., 1976). Eukaryotes do not appear to be able to synthesize queuine. In these organisms, queuine is incorporated either from free queuine base or is salvaged from tRNA containing queuine (Kirtland et al., 1988). The queuine modification in these tRNAs arises via a posttranscriptional base exchange of queuine with the genetically encoded guanine in the intact tRNA. The enzyme responsible for this base exchange is tRNA-guanine transglycosylase (TGT) (EC 2.4.2.29). TGT has been isolated from a number of different sources and exhibits different gross structural features such as molecular weight and subunit composition (Singhal, 1983). There appear to be two functionally different classes of the enzyme. The first class, represented by the enzyme from E. coli, does not recognize queuine itself but exchanges a queuine precursor that lacks the cyclopentene diol moiety, for guanosine 34. Furthermore, the E. coli TGT does not recognize tRNA containing queuosine 34. The second class of TGT, represented by the enzyme isolated from rat liver, does recognize queuine as well as its precursors (Roe et al., 1979). In neither type of TGT has any cofactor or energy requirement been found. Evidently, TGT utilizes the binding energy of the substrates to drive catalysis, a potentially general feature of modifying enzymes discussed at the beginning of this chapter.

Studies of synthetic tRNAs microinjected into X. laevis oocytes revealed that the eukaryotic TGT requires the entire tRNA structure for recognition (Edqvist et al., 1993; Grosjean et al., 1996b). This is in sharp contrast to in vitro studies of synthetic tRNAs with the TGT from E. coli which demonstrate that this enzyme will recognize an oligoribonucleotide hairpin structure corresponding to the anticodon stem and loop (Curnow et al., 1993; Nakanishi et al., 1994; Curnow and Garcia, 1995; Mueller and Slany, 1995). Conformational studies on the recognition of this oligoribonucleotide hairpin substrate for TGT indicate that it must be in the hairpin/helical conformation for recognition; a nonhelical analog is neither a substrate nor an inhibitor of the E. coli TGT (Curnow and Garcia, 1995). Consistent with the concept of TGT recognition of the anticodon stem and loop only is the observation that the E. coli TGT will recognize an unusual dimeric form of tRNA[Tyr] in which the anticodon stems and loops are essentially unaltered (Curnow and Garcia, 1994). One aspect of the tRNA recognition that the eukaryotic and prokaryotic TGTs have in common is the sequence specific-

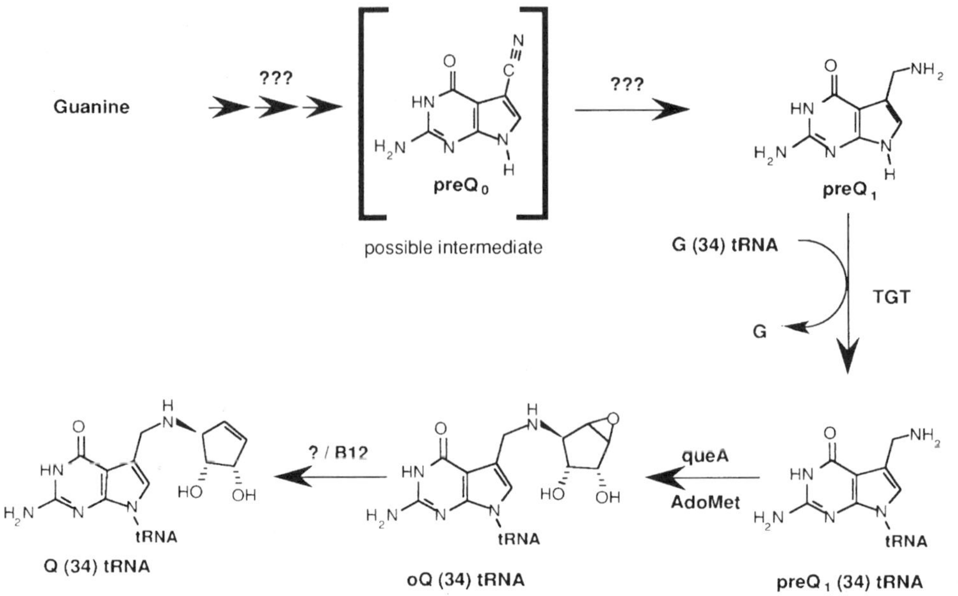

Figure 16. Queuosine 34-tRNA biosynthesis in *Escherichia coli*.

ity in the anticodon loop. Both types of TGT require the U-G-U sequence in the anticodon loop for recognition (Carbon et al., 1983; Haumont et al., 1987; Nakanishi et al., 1994; Curnow and Garcia, 1995). No other specific sequence requirements have been found.

In addition to the differences in substrate recognition between prokaryotic and eukaryotic TGTs, there also appear to be differences in quaternary structure (Singhal, 1983) and activation/regulation (Morris et al., 1995). The prokaryotic TGT appears to be a monomer (see discussion below) of ca. 42 kDa, while the eukaryotic TGT is a heterodimer of ca. 60-kDa and ca. 35-kDa subunits (Slany and Muller, 1995; Deshpande et al., 1996). The TGT from rat liver has been reported to be stimulated by phosphorylation by protein kinase C, where the 60 kDa subunit is postulated to serve a regulatory role while the 34-kDa subunit is catalytic (Morris et al., 1995). No such activation has been noted for the monomeric, prokaryotic TGT.

While it has not been fully elucidated, aspects of the biosynthesis of queuosine-tRNA in *E. coli* have been determined (Kersten and Kersten, 1990). From these data a pathway can be proposed (Fig. 16). It has been shown that queuine derives from guanine and that both C^8 and N^7 of guanine are replaced in queuine (Kuchino et al., 1976). Noguchi et al. have detected 7-cyano-7-deazaguanine (preQ$_0$) in tRNA of Q-deficient *E. coli* mutants (Noguchi et al., 1978). It is not known if preQ$_0$ is a precursor to preQ$_1$ or simply a by-product that accumulates in the mutant strain. PreQ$_1$ is the physiological substrate for the *E. coli* tRNA-guanine transglycosylase, which exchanges preQ$_1$ into tRNA replacing guanine at position 34 in substrate tRNAs. It has been shown that methionine is required for the conversion of preQ$_1$-tRNA to Q-tRNA, but that methionine itself is not incorporated into the queuine molecule (Katze et al., 1977; Okada and Nishimura, 1977). Reuter et al. (1991) have determined the DNA sequence of the queuine operon from *E. coli*. An open reading frame of 1,125 bp, coding for a 42.5-kDa protein, was found to express TGT in an in vitro expression system. Another open reading frame, ORF39, of 1,068 bp, coding for a 39-kDa protein, was identified. ORF39 was designated *queA* and was found to complement a defect in Q biosynthesis after the TGT step in a Q$^-$ *E. coli* mutant (Reuter et al., 1991). More recently, Slany et al. have reported the cloning and over expression of the *queA* gene product. They demonstrated that this enzyme utilizes S-adenosylmethionine as a cosubstrate forming epoxyqueuosine tRNA (oQ-tRNA) from preQ$_1$-tRNA (Slany et al., 1993). Frey et al. (1988) have shown that oQ-tRNA is reduced to Q-tRNA by a cobamide-dependent enzyme. None of the genes involved in the biosynthesis of preQ$_1$, nor the cobamide-dependent enzyme responsible for the reduction of oQ to queuine, have been identified. However, it seems likely that a pathway similar to that found for pterin biosynthesis, involving a GTP cyclohydrolase activity, may be responsible for the generation of preQ$_1$.

Biochemical studies of the *E. coli* TGT have revealed that it is a zinc metalloprotein (Chong et al., 1995; Garcia et al., 1996). These studies, confirmed by the X-ray crystal structure of the *Zymononas mobilis* TGT (Romier et al., 1996a) (see also Chapter 9 by Romier et al.), indicate that the zinc performs a structural role in TGT. Recombinant *E. coli* TGT appears to be a homotrimer of 42.5 kDa subunits; however, it binds tRNA in a monomer·tRNA complex (Curnow and Garcia, 1994). This trimerization, while specific and reversible in vitro, may not be significant in vivo as it has been reported that at concentrations of less than 1 mg/ml, the *E. coli* TGT appears to dissociate in the absence of tRNA (Reuter and Ficner, 1995) and that the *Z. mobilis* TGT is monomeric. Additionally, an *E. coli* TGT mutant (H317C) that does not form the homotrimer and yet has kinetic parameters similar to those of the wild type has been characterized (Garcia et al., 1996). An *E. coli* mutant described by Noguchi et al. (1982) contains tRNA lacking queuine due to an inactive TGT. The mutant *tgt* gene was cloned and sequenced; it contained a single point mutation resulting in the change of serine 90 to phenylalanine (Reuter et al., 1994). Overexpression of the mutant gene yielded TGT (S90F) that showed a reduced solubility and no detectable enzymic activity. To determine if serine 90 performs a catalytic role in the TGT reaction, alanine and cysteine mutants were generated and characterized (Reuter et al., 1994). Native PAGE of wild-type and mutant TGTs in the absence and presence of substrate tRNA exhibited band shifts indicating that both mutants retain the ability to bind tRNA. Determination of in vitro kinetic parameters showed that the S90C mutant was significantly less active than the wild type and that the S90A mutant activity was reduced by 4 orders of magnitude compared to the wild type. These results indicate that serine 90 performs a critical role in the TGT reaction.

Early studies have shown that the *E. coli* TGT will exchange free guanine for the guanine base at position 34 of substrate tRNAs (Okada and Nishimura, 1979). This property has provided a convenient assay for TGT activity by following the incorporation of radiolabeled guanine into tRNA. Although convenient and useful for studying substrate recognition, this assay has a number of drawbacks: (i) the true reaction (preQ$_1$ incorporation) is not being followed, (ii) the kinetic parameters obtained by following guanine incorporation may not reflect those that would be observed during preQ$_1$ incorporation, (iii) the endpoint of the reaction is an equilibrium state where both labeled and unlabeled guanine is being exchanged into and out of tRNA, and (iv) only the incorporation of radiolabeled guanine, or its inhibition, can be monitored. These drawbacks can be somewhat ameliorated by using TGT to preparatively exchange radiolabeled guanine into tRNA. This preexchanged tRNA is then reisolated and used in a "washout" assay in which the incorporation of preQ$_1$ or analogs is followed by monitoring the loss of the radiolabeled, preexchanged guanine from the tRNA (Okada and Nishimura, 1979). This assay has been used along with a series of synthetic 7-substituted 7-deazaguanines (more correctly named 5-substituted 2-aminopyrrolo[2,3-d]pyrimidin-4(3H)-ones; note the different numbering system for pyrrolopyrimidines versus purines) to study the substrate specificity of the *E. coli* TGT (Hoops et al., 1995b). This study reveals that the TGT will tolerate a wide variety of substituents at the 7 position, albeit with up to three orders of magnitude difference in K_m. The role of the 7 substituent appears to be entirely in binding/recognition with no apparent effects on catalysis. A correlation between N^9 pK_a and V_{max} suggests the deprotonation of N^9 during the reaction, which must occur before subsequent glycosidic bond formation, may be partially rate-determining for the natural substrate. 9-Methyl-substituted analogs were found to be competitive inhibitors of TGT. A mechanism-based inactivator of TGT has been designed and characterized (Hoops et al., 1995a). This compound (7-fluoromethyl-7-deazaguanine or 2-amino-5-fluoromethylpyrrolo[2,3-d]pyrimidin-4(3H)-one [FMPP] [Fig. 17]) was designed to take advantage of the fact that the incoming heterocyclic base (preQ$_1$ in the normal reaction) must be deprotonated at the 9 position to form the glycosidic bond. Deprotonation of FMPP leads to fluoride ion elimination and ultimately inactivation of the enzyme. The results are consistent with mechanism-based inactivation; however, the formation of a covalent bond to the enzyme has not been proven. The crystal structure of preQ$_1$ bound to the *Z. mobilis* TGT (Romier et al., 1996a) (see also Chapter 9 by Romier et al.) is consistent with the biochemical studies on the heterocyclic substrate analogs. The complex structure also reveals that cysteine 145 (*E. coli* numbering) is within 4.1 angstroms of the nitrogen of the amino methyl group of preQ$_1$. Cysteine 145 could theoretically serve as the nucleophile involved in FMPP inactivation of TGT and may be involved in the recognition of preQ$_1$ during normal TGT turnover. This is presently under investigation in our laboratory.

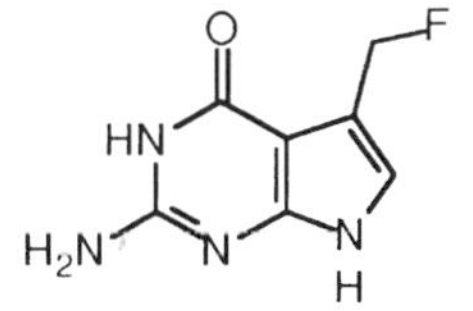

Figure 17. Structure of FMPP.

The above studies, in conjunction with literature precedents for related enzymes, suggest a nucleophilic catalysis mechanism for TGT (Fig. 18). This mechanism involves attack on the 1′ position of the G34 ribose by an enzymatic nucleophile, displacing guanine which may be protonated by an enzymatic general base in either a stepwise or concerted fashion. The incoming preQ$_1$ is then deprotonated to allow it to attack the 1′ position of the ribose, displacing the enzymatic nucleophile. While this mechanism is still speculative, evidence for a covalent TGT·tRNA intermediate has been reported (Romier et al., 1996a). It has been speculated that serine 90 may be the enzymatic nucleophile (Reuter et al., 1994); however, the crystal structure of the Z. mobilis TGT presents aspartate 89 (E. coli numbering) as a likely candidate. Mutagenesis of this aspartate to alanine has yielded inactive TGT that is incapable of forming the covalent intermediate (Romier et al., 1996b); however, the X-ray crystal structure of the mutant exhibits only subtle alterations from the wild type. Given the proximity of aspartate 89 and serine 90 (E. coli numbering), it is difficult to assign which of these residues is the true nucleophile. Mutagenesis of either of these residues could affect the reactivity of the other. More careful studies need to be done to elucidate the true nature of the covalent intermediate. A detailed discussion of the three-dimensional structure of the TGT from Z. mobilis and of the structural bases for substrate recognition and catalysis can be found in Chapter 9 by Romier et al.

Archaeosine

A novel tRNA-guanine transglycosylase has been isolated from H. volcanii that appears to be involved in the introduction of the modified base archaeosine in position 15 of the D loop of archaeal tRNAs (Watanabe et al., 1997). Archaeosine was recently discovered (Gregson et al., 1993) and the structure of the modified base was found to be a pyrrolopyrimidine derivative of guanine similar to queuine (Fig. 19). The novel TGT has a molecular mass of ca. 78 kDa and is specific for the cyano-substituted pyrrolopyrimidine, preQ$_0$ (Fig. 19). The enzyme does not incorporate the base of archaeosine nor does it incorporate the amino methyl-substituted queuine precursor, preQ$_1$ (Fig. 16). The reaction catalyzed is a base replacement, removing a guanine base and replacing it with the modified base, mirroring the reaction catalyzed by the TGTs involved in queuine biosynthesis. Two important differences are the position within the tRNA molecule where the modified base is introduced, and the base recognition properties. As mentioned above, the TGT from E. coli has been found to recognize a wide variety of 7-substituted 7-deazaguanines (Hoops et al., 1995b). The TGT from H. volcanii appears to be quite stringent in its base requirement. Presumably, subsequent to incorporation into tRNA, the cyano substituent of preQ$_0$ must be converted to the amidine to complete the biosynthesis of archaeosine. While preQ$_0$ has been found in E. coli and in E. coli tRNAs (Noguchi et al., 1978), it has not be proven to be a true intermediate in the biosynthesis of queuine. The finding that preQ$_0$ is an intermediate in the biosynthesis of archaeosine, suggests that preQ$_0$ may also be an intermediate in the queuine pathway and further suggests that a divergent evolution may have occurred taking preQ$_0$ to queuine in eubacteria and to archaeosine in archaea.

Wyosine

Position 37 of tRNA commonly contains a modified purine nucleoside. The wye nucleosides, found exclusively at this position in tRNA^{Phe}, are complex structures consisting of a tricyclic core (Fig. 20). In most species, an extensive side chain has been found at position 10 of the third ring; however, tRNA^{Phe} from Torulopsis utilis lacks a substituent at this po-

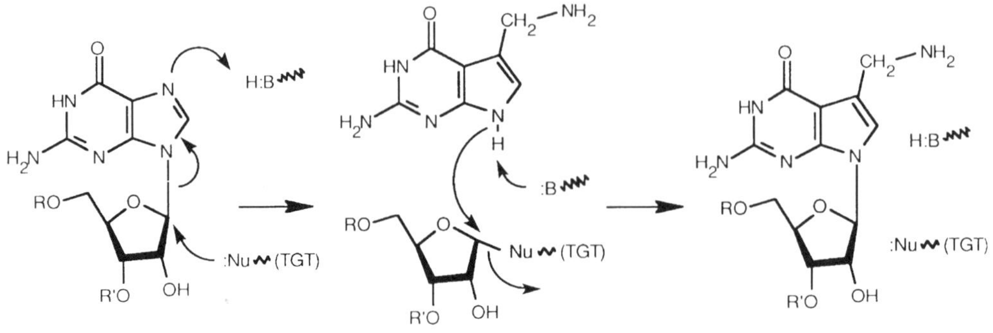

Figure 18. Postulated nucleophilic catalysis mechanism for TGT. Nu, enzymatic nucleophile.

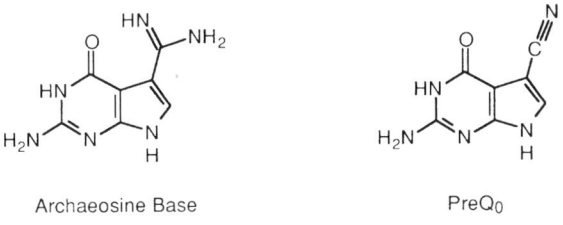

Figure 19. Structures of the base of archaeosine and preQ$_0$.

sition (Kasai et al., 1971). The studies discussed below refer to the wye nucleoside found in *S. cerevisiae*.

The biosynthesis of wyosine is a multienzymatic process (Droogmans and Grosjean, 1987). Although very little is known about the enzymes involved in wye formation, the modification requirements are slowly being elucidated. It has been demonstrated that wyosine originates from a genetically encoded guanine (Li et al., 1973; Thiebe and Poralla, 1973). The first step in wye biosynthesis appears to be methylation at N^1 by an *S*-adenosylmethionine-dependent tRNA methyl transferase (Droogmans and Grosjean, 1987). Although complete modification of guanine to wyosine requires a GAA anticodon sequence, this sequence is not necessary for methylation (Droogmans and Grosjean, 1987).

Methionine, radiolabeled at the carbonyl, was used to demonstrate that the 3-amino-3-carboxypropyl group contributes to the side chain of wyosine (Munch and Thiebe, 1975). The source of the second ester of the side chain has not yet been identified. Yeast cells grown on media containing ^{13}C-enhanced methionine produced the tRNA that was used for NMR studies. The ^{13}C-enhancements associated with both of the methyl esters (the N^3-methyl and one of the two carbons of the third ring) suggest methionine as the source of each of these carbons. Although the NMR studies could not distinguish whether C^{10} or C^{11} contained the ^{13}C-enrichment, origination of C^{10} from methionine is consistent with the observations of Droogmans and Grosjean (1987). The methyl at C^{11} clearly does not arise from methionine (Smith et al., 1985), suggesting that perhaps C^{11} and the C^{11}-methyl originate from an unidentified two-carbon donor.

Consistent with the above studies performed in yeast, methionine was shown to be required for wyosine synthesis in Vero cells and a radiolabel was incorporated into wyosine when radiolabeled methionine was used (Pergolizzi et al., 1978). Parallel studies performed using lysine demonstrated that it is also required for wye formation and becomes incorporated during modification (Pergolizzi et al., 1979). This is clearly inconsistent with the results of the studies described above. However, it is possible that the biosynthetic pathway varies from one species to another, especially because the structure of the wye nucleoside deviates.

CONCLUSIONS AND PROSPECTS FOR THE FUTURE

Although close to 100 chemically distinct modified nucleosides have been identified in nucleic acids, relatively few of the enzymes responsible for their biosynthesis and/or incorporation have been identified or characterized to any great extent. These enzymes make up a large family of enzymes that catalyze various kinds of reactions (Table 2). The best-characterized modifying enzyme is the tRNA (m^{5}U54)methyltransferase (RUMT). tRNA recognition by this enzyme has been elucidated, as have the chemical and kinetic mechanistic details of the methylation reaction.

Even at this relatively early stage in the study of RNA-modifying and -editing enzymes, we can make some conclusions and generalizations from what is known. There are enough examples of RNA recognition to say that there does not appear to be any single type of RNA recognition motif. Many examples exist of modifying enzymes that require significant three-dimensional structure in their RNA substrate, and many other enzymes exist that recognize a small structural or sequence motif. As noted previously, the macromolecular RNA substrates for the modifying enzymes have sufficiently large surface areas for interaction with the modifying enzymes to easily provide for substrate-nonsubstrate discrimination. Exactly how this is accomplished appears to vary among the various modifying enzymes. Mechanistically, RNA-modifying and -editing enzymes ap-

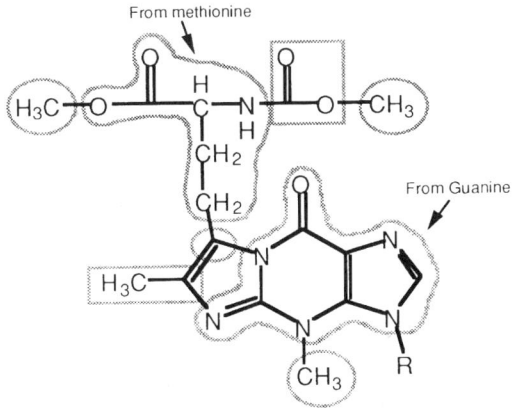

Figure 20. Wyosine from *S. cerevisiae*. Circled carbons are derived from the methyl group of methionine. Atoms in boxes are of unknown origin. R, ribose.

Table 2. Examples of RNA-modifying and -editing reactions

Type of reaction	Cofactor(s) or cosubstrate(s)	Examples
Methylation	AdoMet (in one case folate)	m^5U (rT), m^5C, m^1A, m^1G, m^7G, m^1I, m^6A, m^2G, ...
Thiolation	Cys or β-mercaptopyruvate?	s^2U, s^4U, ms^2A, s^2T, ...
Reduction	?	D, ...
Oxidation	?	ms^2io^6A, oY, ...
Isomerization	None	Ψ, ...
Alkylation	IPP, Thr, Lys, PRPP	i^6A, t^6A, L, pRibp, ...
Transglycosylation	Queuine, archaeosine base	Q, archaeosine
Deamination	H_2O	I, C to U, ...
Others	?	Y, ...

pear to follow reaction pathways similar to their mononucleotide or nucleoside counterparts (e.g., deamination of A to I) although this is not always true. There are two cases where an RNA methyltransferase utilizes 5,10-methylenetetrahydrofolate as a cofactor. However, in every other known case adenosylmethionine is used exclusively. This differentiates the formation of tRNA ribothymidine (RUMT) in most species from that of dTMP from dUMP (thymidylate synthetase).

Many questions regarding RNA-modifying and -editing enzymes remain. Where are the enzymes located in the cell? Are they free or associated to other cellular components? Do they exist in multienzyme complexes? This is especially relevant for those modifications that occur as a result of a pathway of enzymic events where substrate channeling of the various intermediate species may enhance efficiency. [For example, in the queuine pathway in *E. coli*, TGT is a very slow enzyme in vitro (k_{cat}, ca. 0.01 s^{-1}); however, in vivo TGT may associate with queA and/or other queuine pathway enzymes to form a more efficient, multienzyme complex. This is under investigation but is entirely speculative at this point.] How are modifying and editing enzymes regulated (at the transcriptional, translational, and post-translational levels)? Why are they regulated (a question that also addresses the role of the modified bases)? Some of these questions are addressed in other chapters of this book (Chapter 7 by Grosjean et al., Chapter 24 by Maden, and Chapter 25 by Winkler).

Other questions involve mechanism and RNA recognition. How do the modifying and editing enzymes work? Why are there cases of identical modifications in different species that occur via different pathways and perhaps even different chemical mechanisms for the same reaction (e.g., m^1I)? What do modifying and editing enzymes recognize in their RNA substrates? While there are many instances where sequence-specific recognition motifs have been elucidated, how are these sequences recognized by the enzymes? Many cases have also been discovered where modifying and editing enzymes recognize a particular three-dimensional structure or conformation of the substrate RNA, but how this is recognized is still unknown. What role does RNA conformational flexibility or dynamics play in modifying and editing enzyme recognition and catalysis? What is (are) the role(s) of *trans*-acting factors (e.g. guide RNAs) in RNA modifying and editing? Finally, how universal are these recognition processes? The recognition process for the same modification (same location) can be quite different across species. How does this relate to evolution, and where do modified bases become involved in evolution?

Certainly the number of characterized modifying enzymes has dramatically increased over the last 10 to 15 years. This trend toward more and better characterized modifying enzymes shows every sign of continuing and will greatly assist in our understanding of the roles of modified nucleosides. Technological advances in RNA generation (both by in vitro transcription and by chemical synthesis) along with advances in molecular biological approaches (to identify, clone and express modifying enzyme genes) have been predominantly responsible for the renewed activity in RNA modification and editing research. The combination of these advances with chemically based, mechanistic enzymology and three-dimensional structure determination holds much promise for the future understanding of the processes involved in RNA modification and editing. This understanding is key to elucidating the physiological roles of modified bases.

Acknowledgments. The research in our lab has been supported by the National Institutes of Health (GM45968), the National Science Foundation (9720139) and the Upjohn and Vahlteich Research Funds of the University of Michigan College of Pharmacy. Dee-Anne M. Goodenough-Lashua is a National Institutes of Health Pre-Doctoral Trainee (GM07767) and an American Foundation for Pharmaceutical Education Fellow. We thank Paul Agris, Henri Grosjean, Dale Poulter, and Malcolm Winkler for reprints, preprints, and personal communication of their work. We also thank

Henri Grosjean for helpful comments and suggestions on the manuscript.

REFERENCES

Abrell, J. W., E. E. Kaufman, and M. N. Lipsett. 1971. The biosynthesis of 4-thiouridylate. *J. Biol. Chem.* 246:294–301.

Agris, P. F. Personal communication.

Agris, P. F., D. J. Armstrong, K. P. Schafer, and D. Söll. 1975. Maturation of a hypermodified nucleoside in transfer RNA. *Nucleic Acids Res.* 2:691–698.

Aitken, D. M., P. F. Lue, and J. G. Kaplan. 1975. Kinetics and reaction mechanism of the carbamylphosphate synthetase of a multienzyme aggregate from yeast. *Can. J. Biochem.* 53:721–730.

Ajitkumar, P., and J. D. Cherayil. 1988. Thionucleosides in transfer ribonucleic acid: diversity, structure, biosynthesis, and function. *Microbiol. Rev.* 52:103–113.

Allaudeen, H. S., S. K. Yang, and D. Söll. 1972. Leucine tRNA from *hisT* mutant of *Salmonella typhimurium* lacks two pseudouridines. *FEBS Lett.* 28:205–208.

Arnold, H., and H. Kersten. 1975. Inhibition of the tetrahydrofolate-dependent biosynthesis of ribothymidine in tRNAs of *B. subtilis* and *M. lysodeikticus* by trimethoprim. *FEBS Lett.* 53:258–261.

Åström, S. U., and A. S. Byström. 1994. Rit1, a tRNA backbone-modifying enzyme that mediates initiator and elongator tRNA discrimination. *Cell* 79:535–546.

Auxilien, S., P. F. Crain, R. W. Trewyn, and H. Grosjean. 1996. Mechanism, specificity and general properties of the yeast enzyme catalyzing the formation of inosine 34 in the anticodon of transfer RNA. *J. Mol. Biol.* 262:437–458.

Auxilien, S., and H. Grosjean. 1995. Edition and modification of RNA from eukaryotic cells and viruses by enzymatic deamination of adenosine to inosine. *M-S (Med. Sci.)* 11:1089–1098.

Bachellerie, J.-P., and J. Cavaillé. 1997. Guiding ribose methylation of rRNA. *Trends Biochem. Sci.* 22:257–261.

Bartz, J. K., L. K. Kline, and D. Söll. 1970. N^6-(Δ^2-Isopentenyl)adenosine: biosynthesis *in vitro* in transfer RNA by an enzyme purified from *Escherichia coli*. *Biochem. Biophys. Res. Commun.* 40:1481–1487.

Bartz, J. K., and D. Söll. 1972. N^6-(Δ^2-Isopentenyl)adenosine: biosynthesis in vitro in transfer RNA by an enzyme purified from *Escherichia coli*. *Biochimie* 54:31–39.

Bass, B. 1995. An I for editing. *Curr. Biol.* 5:598–600.

Becker, H. F., Y. Motorin, M. Sissler, C. Florentz, and H. Grosjean. 1997. Major identity determinants for enzymatic formation of ribothymidine and pseudouridine in the TΨ-loop of yeast tRNAs. *J. Mol. Biol.* 274:505–518.

Benne, R. 1996. The long and short of it. *Nature* (London) 380:391–392.

Bokar, J. A., M. E. Rath-Shambaugh, R. Ludwiczak, P. Narayan, and F. Rottman. 1994. Characterization and partial purification of mRNA N^6-adenosine methyltransferase from HeLa cell nuclei. Internal mRNA methylation requires a multisubunit complex. *J. Biol. Chem.* 269:17697–704.

Braxton, B. L., L. S. Mullins, F. M. Raushel, and G. D. Reinhart. 1992. Quantifying the allosteric properties of *Escherichia coli* carbamoyl-phosphate synthetase: determination of thermodynamic linked-function parameters in an ordered kinetic mechanism. *Biochemistry* 31:2309–2316.

Britton, H. G., V. Rubio, and S. Grisolia. 1979. Mechanism of carbamoyl-phosphate synthetase. *Eur. J. Biochem.* 102:521–530.

Buck, M., and B. Ames. 1984. A modified nucleoside in tRNA as a possible regulator of aerobiosis. *Cell* 36:523–531.

Byström, A. S., and G. R. Björk. 1982a. Chromosomal location and cloning of the gene (*trmD*) responsible for the synthesis of tRNA (m^1G) methyltransferase in *Escherichia coli* K-12. *Mol. Gen. Genet.* 188:440–446.

Byström, A. S., and G. R. Björk. 1982b. The structural gene (*trmD*) for the tRNA (m^1G) methyltransferase in part of a four polypeptide operon in *Escherichia coli* K-12. *Mol. Gen. Genet.* 188:447–454.

Caillet, J., and L. Droogmans. 1988. Molecular cloning of the *Eschericia coli miaA* gene, involved in the formation of $\Delta 2$ isopentenyl adenosine in tRNA. *J. Bacteriol.* 170:4147–4152.

Carbon, P., E. Haumont, M. Fournier, S. D. Henau, and H. Grosjean. 1983. Site-directed in vitro replacement of nucleosides in the anticodon loop of tRNA: application to the study of structural requirements for queuine insertase activity. *EMBO J.* 2:1093.

Chheda, G. B., C. I. Hong, C. F. Piskorz, and G. A. Harmon. 1972. Biosynthesis of N-(purin-6-ylcarbamoyl)-L-threonine riboside. *Biochem. J.* 127:515–519.

Chong, S., A. W. Curnow, T. J. Huston, and G. A. Garcia. 1995. tRNA-guanine transglycosylase from *Escherichia coli* is a zinc metalloprotein. Site-directed mutagenesis studies to identify the zinc ligands. *Biochemistry* 34:3694–3701.

Connolly, D. M., and M. E. Winkler. 1989. Genetic and physiological relationships among the *miaA* gene, 2-methylthio-N6-(delta 2-isopentenyl)-adenosine tRNA modification, and spontaneous mutagenesis in *Escherichia coli* K-12. *J. Bacteriol.* 171:3233–3246.

Constantinesco, F., and H. Grosjean. Personal communication.

Cortese, R., H. O. Kammen, S. J. Spengler, and B. N. Ames. 1974. Biosynthesis of pseudouridine in transfer ribonucleic acid. *J. Biol. Chem.* 249:1103–1108.

Curnow, A. W., and G. A. Garcia. 1994. tRNA-guanine transglycosylase from *Escherichia coli*: recognition of dimeric, unmodified tRNA^{Tyr}. *Biochimie* 76:1183–1191.

Curnow, A. W., and G. A. Garcia. 1995. tRNA-guanine transglycosylase from *Escherichia coli*—minimal tRNA structure and sequence requirements for recognition. *J. Biol. Chem.* 270:17264–17267.

Curnow, A. W., F. L. Kung, K. A. Koch, and G. A. Garcia. 1993. tRNA-guanine transglycosylase from *Escherichia coli*: gross tRNA structural requirements for recognition. *Biochemistry* 32:5239–5246.

Delk, A. S., D. P. Nagle, and J. C. Rabinowitz. 1980. Methylenetetrahydrofolate-dependent biosynthesis of ribothymidine in transfer RNA of *Streptococcus faecalis*. Evidence for reduction of the 1-carbon unit by FADH2. *J. Biol. Chem.* 255:4387–4390.

Delk, A. S., J. M. Romeo, D. P. Nagle, and J. C. Rabinowitz. 1976. Biosynthesis of ribothymidine in the transfer RNA of *Streptococcus faecalis* and *Bacillus subtilis*. A methylation of RNA involving 5,10-methylenetetrahydrofolate. *J. Biol. Chem.* 251:7649–7656.

Desgres, J., G. Keith, K. C. Kuo, and C. W. Gehrke. 1989. Presence of phosphorylated O-ribosyl-adenosine in T-Ψ-stem of yeast methionine initiator tRNA. *Nucleic Acids Res.* 17:865–882.

Deshpande, K. L., P. H. Seubert, D. M. Tillman, W. R. Farkas, and J. R. Katze. 1996. Cloning and characterization of cDNA encoding the rabbit tRNA-guanine transglycosylase 60-kilodalton subunit. *Arch. Biochem. Biophys.* 326:1–7.

Droogmans, L., and H. Grosjean. 1987. Enzymatic conversion of guanosine 3' adjacent to the anticodon of tRNA^{Phe} to N^1-methylguanosine and the wye nucleoside: dependence on the anticodon sequence. *EMBO J.* 6:477–483.

Droogmans, L., E. Haumont, S. deHenau, and H. Grosjean. 1986. Enzymatic 2'-O-methylation of the wobble nucleoside of eukaryotic tRNA^{Phe}: specificity depends on structural elements outside the anticodon loop. *EMBO J.* **5**:1105–1109.

Edqvist, J., K. Blomqvist, and K. B. Stråby. 1994. Structural elements in yeast tRNAs required for homologous modification of guanosine-26 into dimethylguanosine-26 by the yeast *Trm1* tRNA-modifying enzyme. *Biochemistry* **33**:9546–9551.

Edqvist, J., H. Grosjean, and K. B. Stråby. 1992. Identity elements for N^2-dimethylation of guanosine-26 in yeast tRNAs. *Nucleic Acids Res.* **20**:6575–6581.

Edqvist, J., K. B. Stråby, and H. Grosjean. 1995. Enzymatic formation of N^2,N^2-dimethylguanosine in eukaryotic tRNA: importance of the tRNA architecture. *Biochimie* **77**:54–61.

Edqvist, J., K. B. Stråby, and H. Grosjean. 1993. Pleiotrophic effects of point mutation in yeast tRNA^{Asp} on the base modification pattern. *Nucleic Acids Res.* **21**:413–417.

Eichler, D. C. 1994. Characterization of a nucleolar 2'-O-methyltransferase and its involvement in the methylation of mouse precursor ribosomal RNA. *Biochimie* **76**:1115–1122.

Elkins, B. N., and E. B. Keller. 1974. The enzymatic synthesis of N-(purin-6-ylcarbamoyl)threonine, an anticodon-adjacent base in transfer ribonucleic acid. *Biochemistry* **13**:4622–4628.

Elliott, M. S., and R. W. Trewyn. 1984. Inosine biosynthesis in transfer RNA by an enzymatic insertion of hypoxanthine. *J. Biol. Chem.* **259**:2407–2410.

Ellis, S. R., M. J. Morales, J. M. Li, A. K. Hopper, and N. C. Martin. 1986. Isolation and characterization of the TRM1 locus, a gene essential for the N^2,N^2-dimethylguanosine modification of both mitochondrial and cytoplasmic tRNA in *Saccharomyces cerevisiae*. *J. Biol. Chem.* **261**:9703–9709.

Erdmann, V. A., E. Huysmans, A. Vandenbergh, and R. DeWachter. 1983. Collection of published 5S and 5.8S ribosomal RNA sequences. *Nucleic Acids Res.* **11**:r105–r133.

Fittler, F., L. K. Kline, and R. H. Hall. 1968. N^6-(Δ^2-Isopentenyl)adenosine: biosynthesis in vitro by an enzyme extract from yeast and rat liver. *Biochem. Biophys. Res. Commun.* **31**:571–576.

Frendewey, D. A., D. M. Kladianos, V. G. Moore, and I. I. Kaiser. 1982. Loss of tRNA 5-methyluridinemethyltransferase and pseudouridine synthase activities in 5-fluorouracil and 1-(tetrahydro-2-furanyl)-5-fluoouracil (FTORAFUR) treated *Escherichia coli*. *Biochim. Biophys. Acta* **697**:31–40.

Frey, B., J. McCloskey, W. Kersten, and H. Kersten. 1988. New function of vitamin B$_{12}$: cobamide-dependent reduction of epoxyqueuosine to queuosine in tRNAs of *Escherichia coli* and *Salmonella typhimurium*. *J. Bacteriol.* **170**:2078–2082.

Frick, L., R. Wolfenden, E. Smal, and D. Baker. 1986. Transition-state stabilization by adenosine deaminase: structural studies of its inhibitory complex with deoxycoformycin. *Biochemistry* **25**:1616–1621.

Garcia, G. A., and S. R. Chong. 1997. Cysteine 265 is in the active site of, but is not essential for catalysis by tRNA-guanine transglycosylase (TGT) from *Escherichia coli*. *J. Protein Chem.* **16**:11–17.

Garcia, G. A., D. L. Tierney, S. R. Chong, K. Clark, and J. E. Penner-Hahn. 1996. X-ray absorption spectroscopy of the zinc site in tRNA-guanine transglycosylase from *Escherichia coli*. *Biochemistry* **35**:3133–3139.

Garrett, C. E., J. A. Coderre, T. D. Meek, E. P. Garvey, D. M. Claman, S. M. Beverley, and D. V. Santi. 1984. A bifunctional thymidylate synthetase-dihydrofolate reductase in protozoa. *Mol. Biochem. Parasitol.* **11**:257–265.

Gerlt, J. A., and P. G. Gassman. 1993. Understanding the rates of certain enzyme-catalyzed reactions: proton abstraction from carbon acids, acyl-transfer reactions, and displacement reactions of phosphodiesters. *Biochemistry* **32**:11943–11952.

Gershon, P. D., B. Y. Ahn, M. Garfield, and B. Moss. 1991. Poly(A) polymerase and a dissociable polyadenylation stimulatory factor encoded by vaccinia virus. *Cell* **66**:1269–1278.

Gershon, P. D., and B. Moss. 1993. Stimulation of poly(A) tail elongation by the VP39 subunit of the vaccinia virus-encoded poly(A) polymerase. *J. Biol. Chem.* **268**:2203–2210.

Giorgianni, F., S. Beranova, C. Wesdimiotis, and R. E. Viola. 1995. Elimination of the sensitivity of L-aspartase to active-site-directed inactivation without alteration of the catalytic activity. *Biochemistry* **34**:3529–3535.

Glasser, A.-L., J. Desgres, J. Heitzler, C. W. Gehrke, and G. Keith. 1991. O-Ribosyl-phosphate purine as a constant modified nucleotide located at position 64 in cytoplasmic initiator tRNAs of yeasts. *Nucleic Acids Res.* **19**:5199–5203.

Green, C. J., H. O. Kammen, and E. E. Penhoet. 1982. Purification and properties of a mammalian tRNA pseudouridine synthase. *J. Biol. Chem.* **257**:3045–3052.

Gregson, J. M., P. F. Crain, C. G. Edmonds, R. Gupta, T. Hashizume, D. W. Phillipson, and J. A. McCloskey. 1993. Structure of the archaeal transfer RNA nucleoside G*-15 (2-amino-4,7-dihydro-4-oxo-β-D-ribofuranosyl-1H-pyrrolo[2,3-d]pyrimidine-5-carboximidamide (archaeosine). *J. Biol. Chem.* **268**:10076–10086.

Grosjean, H., S. Auxilien, F. Constantinesco, C. Simon, Y. Corda, H. F. Becker, D. Foiret, A. Morin, Y. X. Jin, M. Fournier, and J. L. Fourrey. 1996a. Enzymatic conversion of adenosine to inosine and to N-1–methylinosine in transfer RNAs: a review. *Biochimie* **78**:488–501.

Grosjean, H., F. Constantinesco, D. Foiret, and N. Benachenhou. 1995a. A novel enzymatic pathway leading to 1-methylinosine modification in *Haloferax volcanii* tRNA. *Nucleic Acids Res.* **23**:4312–4319.

Grosjean, H., S. De Henau, T. Doi, A. Yamane, E. Ohtsuka, M. Ikehara, N. Beauchemin, K. Nicoghosian, and R. Cedergren. 1987. The in vivo stability, maturation and aminoacylation of anticodon-substituted *Escherichia coli* initiator methionine tRNAs. *Eur. J. Biochem.* **166**:325–332.

Grosjean, H., L. Droogmans, R. Giegé, and O. C. Uhlenbeck. 1990. Guanosine modifications in runoff transcripts of synthetic transfer RNA-Phe genes microinjected into *Xenopus* oocytes. *Biochim. Biophys. Acta* **1050**:267–273.

Grosjean, H., J. Edqvist, K. B. Stråby, and R. Giegé. 1996b. Enzymatic formation of modified nucleosides in tRNA: dependence on tRNA architecture. *J. Mol. Biol.* **255**:67–85.

Grosjean, H., M. Sprinzl, and S. Steinberg. 1995b. Posttranscriptionally modified nucleosides in transfer RNA: their locations and frequencies. *Biochimie* **77**:139–141.

Gu, X., and D. V. Santi. 1991. The T-arm of tRNA is a substrate for the tRNA (m^{5}U54)-methyltransferase. *Biochemistry* **30**:2999–3002.

Gu, X., and D. V. Santi. 1992. Covalent adducts between tRNA (m^{5}U54)-methyltransferase and RNA substrates. *Biochemistry* **31**:10295–10302.

Gu, X. G., J. Ofengand, and D. V. Santi. 1994. In vitro methylation of *Escherichia coli* 16S rRNA by tRNA (m^{5}U54)-methyltransferase. *Biochemistry* **33**:2255–2261.

Gu, X. R., K. M. Ivanetich, and D. V. Santi. 1996. Recognition of the T-arm of tRNA by tRNA (m(5)U54)–methyltransferase is not sequence specific. *Biochemistry* **35**:11652–11659.

Guenther, R. H., R. S. Bakal, B. Forrest, Y. Chen, R. Sengupta, B. Nawrot, E. Sochacka, J. Jankowska, A. Kraszewski, A. Malkiewicz, and P. F. Agris. 1994. Aminoacyl-tRNA synthetase and U-54 methyltransferase recognize conformations of the yeast tRNA(Phe) anticodon and T stem/loop domain. *Biochimie* **76**:1143–1151.

Gustafsson, C., and G. R. Björk. 1993. The tRNA-(m⁵U54)-methyltransferase of *Escherichia coli* is present in two forms in vivo, one of which is present as bound to tRNA and to a 3'-end fragment of 16S rRNA. *J. Biol. Chem.* 268:1326–1331.

Hagervall, T. G., C. G. Edmonds, J. A. McCloskey, and G. R. Björk. 1987. Transfer RNA (5-methylaminomethyl-2-thiouridine)-methyltransferase from *Escherichia coli* K-12 has two enzymatic activities. *J. Biol. Chem.* 262:8488–8495.

Haumont, E., L. Droogmans, and H. Grosjean. 1987. Enzymatic formation of queuosine and of glycosyl queuosine in yeast tRNAs microinjected into *Xenopus laevis* oocytes. *Eur. J. Biochem.* 168:219.

Hayward, R. S., and S. B. Weiss. 1966. RNA thiolase: the enzymatic transfer of sulfur from cysteine to sRNA in *Escherichia coli* extracts. *Biochemistry* 55:1161–1168.

Hjalmarsson, K. J., A. S. Byström, and G. R. Björk. 1983. Purification and characterization of transfer RNA (guanine-1)methyltransferase from *Escherichia coli*. *J. Biol. Chem.* 258:1343–1351.

Hodel, A. E., P. D. Gershon, X. Shi, and F. A. Quiocho. 1996. The 1.85 Å structure of vaccinia protein VP39: a bifunctional enzyme that participates in the modification of both mRNA ends. *Cell* 85:247–256.

Holmes, M. W., C. Adraos-Selim, and M. Redlak. 1995. tRNA-m¹G methyltransferase interactions: touching bases with structure. *Biochimie* 77:62–65.

Holmes, M. W., C. Adraos-Selim, I. Roberts, and S. Z. Wahab. 1992. Structural requirements for tRNA methylation. *J. Biol. Chem.* 267:13440–13445.

Holtz, J., and D. Klambt. 1975. tRNA isopentenyltransferase from *Lactobacillus acidophilus* ATCC 4963. *Hoppe Seylers Z. Physiol. Chem.* 356:1459–1464.

Holtz, J., and D. Klambt. 1978. tRNA isopentenyltransferase from *Zea mays* L. Characterization of the isopentenylation reaction of tRNA, oligo (A) and other nucleic acids. *Hoppe Seylers Z. Physiol. Chem.* 359:89–101.

Hoops, G. C., J. Park, G. A. Garcia, and L. B. Townsend. 1996. The synthesis and determination of acidic ionization constants of certain 5-substituted 2-aminopyrrolo[2,3-d]pyrimidin-4-ones and methylated analogs. *J. Heterocycl. Chem.* 33:767–781.

Hoops, G. C., L. B. Townsend, and G. A. Garcia. 1995a. Mechanism-based inactivation of tRNA-guanine transglycosylase from *Escherichia coli* by 2-amino-5-(fluoromethyl)pyrrolo[2,3-d] pyrimidin-4(3H)-one. *Biochemistry* 34:15539–15544.

Hoops, G. C., L. B. Townsend, and G. A. Garcia. 1995b. tRNA-guanine transglycosylase from *Escherichia coli*:structure-activity studies investigating the role of the aminomethyl substituent of the heterocyclic substrate preQ(1). *Biochemistry* 34:15381–15387.

James, T. L., A. L. Pogolotti, K. M. Ivanetich, Y. Wataya, S. M. Lam, and D. V. Santi. 1976. Thymidylate synthase: fluorine-19 NMR characterization of the active site peptide covalently bound to 5-fluoro-2'-deoxyuridylate and 5,10-methylenetetrahydrofolate. *Biochem. Biophys. Res. Commun.* 72:404–410.

Jiang, H.-Q., Y. Motorin, Y.-X. Jin, and H. Grosjean. 1997. Pleiotropic effects of intron removal on base modification pattern of yeast tRNA^Phe: an in vitro study. *Nucleic Acids Res.* 25:2694–2701.

Kammen, H. O., C. C. Marvel, L. Hardy, and E. E. Penhoet. 1988. Purification, structure, and properties of *Escherichia coli* tRNA pseudouridine synthase I. *J. Biol. Chem.* 263:2255–2263.

Kasai, H., M. Goto, S. Takemura, T. Goto, and S. Matsurra. 1971. Structure and synthesis of a fluorescent Y-like base from *Torulopsis utilis* tRNA. *Tetrahedron Lett.* 29:2725–2728.

Kasai, H., K. Nakanishi, R. D. Macfarlane, D. F. Torgerson, Z. Ohashi, J. A. McCloskey, H. J. Gross, and S. Nishimura. 1976. The structure of Q nucleoside isolated from rabbit liver transfer ribonucleic acid. *J. Am. Chem. Soc.* 98:5044.

Katze, J. R., M. H. Simonian, and R. B. Mosteller. 1977. Role of methionine in the synthesis of nucleoside Q in *Escherichia coli* transfer ribonucleic acid. *J. Bacteriol.* 132:174–179.

Kealey, J. T., X. Gu, and D. V. Santi. 1994. Enzymatic mechanism of tRNA (m⁵U54)methyltransferase. *Biochimie* 76:1133–1142.

Kealey, J. T., S. Lee, H. G. Floss, and D. V. Santi. 1991. Stereochemistry of methyl transfer catalyzed by tRNA (m⁵U54)-methyltransferase-evidence for a single displacement mechanism. *Nucleic Acids Res.* 19:6465–6468.

Kealey, J. T., and D. V. Santi. 1991. Identification of the catalytic nucleophile of tRNA (m⁵U54)methyltransferase. *Biochemistry* 30:9724–9728.

Keith, J. M., E. M. Winters, and B. Moss. 1980. Purification and characterization of a HeLa cell transfer RNA (cytosine-5-)-methytransferase. *J. Biol. Chem.* 255:4636–4644.

Kersten, H., and W. Kersten. 1990. Biosynthesis and function of queuine and queuosine tRNAs, p. B69–B108. In C. Gehrke and K. Kuo (ed.), *Chromatography and Modification of Nucleosides*, part B. *Biological Roles and Function of Modification*. Elsevier, Amsterdam, The Netherlands.

Kirtland, G. M., T. D. Morris, P. H. Moore, J. J. O'Brian, C. G. Edmonds, J. A. McCloskey, and J. R. Katze. 1988. Novel salvage of queuine from queuosine and absence of queuine synthesis in *Chlorella pyrenoidosa* and *Chlamydomonas reinhardtii*. *J. Bacteriol.* 170:5633–5641.

Klimasauskas, S., S. Kumar, R. J. Roberts, and X. D. Cheng. 1994. Hha I methyltransferase flips its target base out of the DNA helix. *Cell* 76:357–369.

Kline, L. K., F. Fittler, and R. H. Hall. 1969. N^6-(Δ^2-Isopentenyl) adenosine. Biosynthesis in transfer ribonucleic acid in vitro. *Biochemistry* 8:4361–4371.

Koonin, E. V. 1996. Pseudouridine synthases: four families of enzymes containing a putative uridine-binding motif also conserved in dUTPases and dCTP deaminases. *Nucleic Acids Res.* 24:2411–2415.

Körner, A., and D. Söll. 1974. N-(Purin-6-ylcarbamoyl)threonine: biosynthesis *in vitro* in transfer RNA by an enzyme purified from *Escherichia coli*. *FEBS Lett.* 39:301–306.

Kuchino, Y., H. Kasai, K. Nihei, and S. Nishimura. 1976. Biosynthesis of the modified nucleoside Q in transfer RNA. *Nucleic Acids Res.* 3:393–398.

Kun, E. 1967. Pages 375–401. In D. M. Greenberg (ed.), *Metabolic Pathways*. Academic Press, New York, N.Y.

Leung, H.-C. E., Y. Chen, and M. E. Winkler. 1997. Regulation of substrate recognition by the MiaA tRNA prenyl transferase modification enzyme of *Escherichia coli* K-12. *J. Biol. Chem.* 272:13073–13083.

Li, H. J., K. Nakanishi, D. Grunberger, and I. B. Weinstein. 1973. Biosynthestic studies of the Y base in yeast phenylalanine tRNA. Incorporation of guanine. *Biochem. Biophys. Res. Commun.* 55:818–823.

Limbach, P. A., P. F. Crain, and J. A. McCloskey. 1994. Summary: the modified nucleosides of RNA. *Nucleic Acids Res.* 22:2183–2196.

Lipsett, M. N. 1972. Biosynthesis of 4-thiouridylate. *J. Biol. Chem.* 247:1458–1461.

Lipsett, M. N., J. S. Norton, and A. Peterkofsky. 1967. A requirement for β-mercaptopyruvate in the in vitro thiolation of transfer ribonucleic acid. *Biochemistry* 6:855–860.

Lipsett, M. N., and A. Peterkofsky. 1966. Enzymatic thiolation of *E. coli* sRNA. *Biochemistry* 5:1169–1174.

Lo, R. Y., and J. B. Bell. 1981. Characterization of a mutation in *Saccharomyces cerevisiae* that produces mutant isoaccepting tRNAs for several of its tRNA species. *Curr. Genet.* 3:73–82.

Lo, R. Y., J. B. Bell, and K. L. Roy. 1982. Dihydrouridine-deficient tRNAs in *Saccharomyces cerevisiae*. *Nucleic Acids Res.* 10:889–902.

Matsumoto, T., K. Nishikura, H. Hori, T. Ohta, K. Miura, and K. Watanabe. 1990. Recognition sites of tRNA by a thermostable tRNA (guanosine-2′-)methyltransferase from *Thermus thermophilus* HB27. *J. Biochem.* 107:331–338.

Meister, A. 1989. Mechanism and regulation of the glutamine-dependent carbamyl phosphate synthetase of *Escherichia coli*. *Adv. Enzymol.* 62:315–374.

Merkler, D., M. Brenowitz, and V. Schramm. 1990. The rate constant describing slow-onset inhibition of yeast AMP deaminase by coformycin analogues is independent of inhibitor structure. *Biochemistry* 29:8358–8364.

Moore, J. A., and C. D. Poulter. 1997. *Escherichia coli* dimethylallyl diphosphate:tRNA dimethylallyl transferase: a binding mechanism for recombinant enzyme. *Biochemistry* 36:604–614.

Morin, A., S. Auxilien, B. Senger, R. Tewari, and H. Grosjean. 1998. Structural requirements for enzymatic formation of threonylcarbamoyl adenosine (t^{6}A) in tRNA: an in vivo study with *Xenopus laevis* oocytes. *RNA* 4:24–37.

Morris, R. C., B. J. Brooks, P. Eriotou, D. F. Kelly, S. Sagar, K. L. Hart, and M. S. Elliott. 1995. Activation of transfer RNA-guanine ribosyltransferase by protein kinase C. *Nucleic Acids Res.* 23:2492–2498.

Motorin, Y., V. Arluison, H. Becker, G. Simos, E. Hurt, and H. Grosjean. 1997a. Pseudouridine formation in yeast tRNAs: cloning and characterization of the corresponding enzymes. In *Proceedings of 17th International tRNA Workshop, Chiba, Japan*.

Motorin, Y., G. Bec, R. Tewari, and H. Grosjean. 1997b. Transfer RNA recognition by the *Escherichia coli* Δ^2-isopentenyl-pyrophosphate:tRNA Δ^2-isopentenyl transferase: dependence on the anticodon arm structure. *RNA* 3:721–733.

Mueller, S. O., and R. K. Slany. 1995. Structural analysis of the interaction of the tRNA modifying enzymes Tgt and QueA with a substrate tRNA. *FEBS Lett.* 361:259–264.

Mullenbach, G. T., H. O. Kammen, and E. E. Penhoet. 1976. A heterologous system for detecting eukaryotic enzymes which synthesize pseudouridine in transfer ribonucleic acids. *J. Biol. Chem.* 251:4570–4578.

Munch, H.-J., and R. Thiebe. 1975. Biosynthesis of the nucleoside Y in yeast tRNA^{Phe}: incorporation of the 3-amino-3-carboxy-propyl-group from methionine. *FEBS Lett.* 51:257–258.

Nakanishi, S., T. Ueda, H. Hori, N. Yamazaki, N. Okada, and K. Watanabe. 1994. A UGU sequence in the anticodon loop is a minimum requirement for recognition by *Escherichia coli* tRNA-guanine transglycosylase. *J. Biol. Chem.* 269:32221–32225.

Nishikura, K., and E. M. De Robertis. 1981. RNA processing in microinjected *Xenopus* oocytes: sequential addition of base modifications in a spliced transfer RNA. *J. Mol. Biol.* 145:405–420.

Noguchi, S., Y. Nishimura, Y. Hirota, and S. Nishimura. 1982. Isolation and characterization of an *Escherichia coli* mutant lacking tRNA-guanine transglycosylase. *J. Biol. Chem.* 257:6544–6550.

Noguchi, S., Z. Yamaizumi, T. Ohgi, T. Goto, Y. Nishimura, Y. Hirota, and S. Nishimura. 1978. Isolation of Q nucleoside precursor present in tRNA of an *E. coli* mutant and its characterization as 7-(cyano)-7-deazaguanine. *Nucleic Acids Res.* 5:4215–4223.

Nurse, K., J. Wrzesinski, A. Bakin, B. G. Lane, and J. Ofengand. 1995. Purification, cloning, and properties of the tRNA Ψ55 synthase from *Escherichia coli*. *RNA* 1:102–112.

Okada, N., and S. Nishimura. 1977. Enzymatic synthesis of Q* nucleoside containing mannose in the anticodon of tRNA: isolation of a novel mannosyltransferase from a cell-free extract of rat liver. *Nucleic Acids Res.* 4:2931–2937.

Okada, N., and S. Nishimura. 1979. Isolation and characterization of a guanine insertion enzyme, a specific tRNA transglycosylase, from *Escherichia coli*. *J. Biol. Chem.* 254:3061–3066.

Pais de Barros, J. P., G. Keith, C. El Adlouni, A. L. Glasser, G. Mack, G. Dirheimer, and J. Desgres. 1996. 2′-O-methyl-5-formylcytidine (f^{5}Cm), a new modified nucleotide at the "wobble" position of two cytoplasmic tRNAs^{Leu} (NAA) from bovine liver. *Nucleic Acids Res.* 24:1489–1496.

Pergolizzi, R. G., D. L. Engelhardt, and D. Grunberger. 1978. Formation of phenylalanine transfer RNA lacking the wye base in vero cells during methionine starvation. *J. Biol. Chem.* 253:6341–6343.

Pergolizzi, R. G., D. L. Engelhardt, and D. Grunberger. 1979. Incorporation of lysine into Y base of phenylalanine tRNA in vero cells. *Nucleic Acids Res.* 6:2209–2216.

Persson, B., C. Gustafsson, D. Berg, and G. Björk. 1992. The gene for a tRNA modifying enzyme, m^{5}U54-methyltransferase, is essential for viability in *Escherichia coli*. *Proc. Natl. Acad. Sci. USA* 89:3995–3998.

Peterkofsky, A., and M. N. Lipsett. 1965. The origin of the sulfur in s-RNA. *Biochem. Biophys. Res. Commun.* 20:780–786.

Pfleiderer, W. 1961. Über die Methylierung des 9-Methylguanins und die Struktur des Herbipolins. *Liebigs Ann. Chem.* 647:167–173.

Pogolotti, A. L., K. M. Ivanetich, H. Sommer, and D. V. Santi. 1986. Thymidylate synthase: studies on the peptide containing covalently bound 5-fluoro-2′-deoxyuridylate and 5,10-methylenetetrahydrofolate. *Biochem. Biophys. Res. Commun.* 70:972–978.

Polson, A., B. Bass, and J. Casey. 1996. RNA editing of hepatitis delta virus antigenome by dsRNA-adenosine deaminase. *Nature (London)* 380:454–456.

Polson, A., P. Crain, S. Pomerantz, J. McCloskey, and B. Bass. 1991. The mechanism of adenosine to inosine conversion by the double-stranded RNA unwinding/modifying activity: a high-performance liquid chromatography-mass spectrometry analysis. *Biochemistry* 30:11507–11514.

Polson, A. G., and B. L. Bass. 1994. Preferential selection of adenosines for modification by double-stranded RNA adenosine deaminase. *EMBO J.* 13:5701–5711.

Poulter, C. D., and H. C. Rilling. 1976. Prenyltransferase: the mechanism of the reaction. *Biochemistry* 15:1079–1083.

Poulter, C. D., and D. M. Satterwhite. 1977. Mechanism of the prenyl-transfer reaction. Studies with (E)- and (Z)-3-trifluoromethyl-2-buten-1-yl pyrophosphate. *Biochemistry* 16:5470–5478.

Poulter, C. D., D. M. Satterwhite, and H. C. Rilling. 1976. Prenyltransferase. The mechanism of the reaction. *J. Am. Chem. Soc.* 98:3376–3377. (Letter).

Powers, D. M., and A. Peterkofsky. 1972. Biosynthesis and specific labeling of N-(purin-6-ylcarbamoyl)threonine of *Escherichia coli* transfer RNA. *Biochem. Biophys. Res. Commun.* 46:831–838.

Powers, S. G., and A. Meister. 1978. Mechanism of the reaction catalyzed by carbamyl phosphate synthetase. *J. Biol. Chem.* 253:800–803.

Prior, J. J., and D. V. Santi. 1984. On the mechanism of the acid-catalyzed hydrolysis of uridine to uracil. *J. Biol. Chem.* 259:2429–2434.

Raushel, F. M., P. M. Anderson, and J. J. Villafranca. 1978. Kinetic mechanism of *Escherichia coli* carbamoyl-phosphate synthetase. *Biochemistry* 17:5587–5591.

Reinhart, M. P., J. M. Lewis, and P. S. Leboy. 1986. A single tRNA (guanine)-methyltransferase for *Tetrahymena pyriformis* with both mono- and di-methylating activity. *Nucleic Acids Res.* 14:1131–1148.

Reuter, K., S. Chong, F. Ullrich, H. Kersten, and G. A. Garcia. 1994. Serine-90 is required for enzymic activity by tRNA-guanine transglycosylase from *Escherichia coli*. *Biochemistry* 33:7041–7046.

Reuter, K., and R. Ficner. 1995. Sequence analysis and overexpression of the *Zymomonas mobilis* tgt gene encoding tRNA-guanine transglycosylase: purification and biochemical characterization of the enzyme. *J. Bacteriol.* 177:5284–5288.

Reuter, K., R. Slany, F. Ullrich, and H. Kersten. 1991. Structure and organization of *Escherichia coli* genes involved in biosynthesis of the deazaguanine derivative queuine, a nutrient factor for eukaryotes. *J. Bacteriol.* 173:2256–2264.

Roberts, R. J. 1995. On base flipping. *Cell* 82:9–12.

Roe, B. A., A. F. Stankiewicz, H. L. Rizi, C. Weisz, M. DiLauro, D. Pike, C. Y. Chen, and E. Y. Chen. 1979. Comparison of rat liver and Walker 256 carcinosarcoma tRNAs. *Nucleic Acids Res.* 6:673.

Romeo, J. M., A. S. Delk, and J. C. Rabinowitz. 1974. The occurrence of a transmethylation reaction not involving S-adenosylmethionine in the formation of ribothymidine in *Bacillus subtilis* transfer-RNA. *Biochem. Biophys. Res. Commun.* 61:1256–1261.

Romier, C., K. Reuter, D. Suck, and R. Ficner. 1996a. Crystal structure of tRNA-guanine transglycosylase: RNA modification by base exchange. *EMBO J.* 15:2850–2857.

Romier, C., K. Reuter, D. Suck, and R. Ficner. 1996b. Mutagenesis and crystallographic studies of *Zymomonas mobilis* tRNA-guanine transglycosylase reveal aspartate 102 as the active site nucleophile. *Biochemistry* 35:15734–15739.

Rosenbaum, N., and M. L. Gefter. 1972. Δ^2-Isopentenylpyrophosphate: transfer ribonucleic acid Δ^2-isopentenyltransferase from *Escherichia coli*. Purification and properties of the enzyme. *J. Biol. Chem.* 247:5675–5680.

Rottman, F. M., J. A. Bokar, P. Narayan, M. E. Shambaugh, and R. Ludwiczak. 1994. N^6-adenosine methylation in mRNA: substrate specificity and enzyme complexity. *Biochimie* 76:1109–1114.

Roy-Burman, P., S. Roy-Burman, and D. W. Visser. 1965. Incorporation of 5,6-dihydrouridine triphosphate into ribonucleic acid by DNA-dependent RNA polymerase. *Biochem. Biophys. Res. Commun.* 20:291–297.

Roy-Burman, P., S. Roy-Burman, and D. W. Visser. 1967. Utilization of 5,6-dihydrouridine 5'-triphosphate in the reaction catalyzed by *Escherichia coli* RNA polymerase. *Biochim. Biophys. Acta* 142:355–367.

Rubio, V., H. G. Britton, and S. Grisolia. 1979. Mechanism of carbamoyl phosphate synthetase. *Eur. J. Biochem.* 93:245–256.

Rubio, V., H. G. Britton, S. Grisolia, B. S. Sproat, and G. Lowe. 1981. Mechanism of activation of bicarbonate ion by mitochondrial carbamoyl-phosphate synthetase: formation of enzyme-bound adenosine diphosphate from the adenosine triphosphate that yields inorganic phosphate. *Biochemistry* 20:1969–1974.

Rueter, S. M., C. M. Burns, S. A. Coode, P. Mookherjee, and R. B. Emeson. 1995. Glutamate receptor RNA editing in vitro by enzymatic conversion of adenosine to inosine. *Science* (Washington, D.C.) 267:1491–1494.

Santi, D. V., and L. W. Hardy. 1987. Catalytic mechanism and inhibition of tRNA-(uracil-5-)methyltransferase: evidence for covalent catalysis. *Biochemistry* 26:8599–8606.

Santi, D. V., C. S. McHenry, and E. R. Perriard. 1974. A filter assay for thymidylate synthetase using 5-fluoro-2'deoxyuridylate as an active site titrant. *Biochemistry* 13:467–470.

Savva, R., K. McAuley-Hecht, T. Brown, and L. Pearl. 1995. The structural basis of specific base excision repair by uracil-DNA glycosylase. *Nature* (London) 373:487–493.

Schibler, U., D. E. Kelley, and R. P. Perry. 1977. Comparison of methylated sequences in messenger RNA and heterogeneous nuclear RNA from mouse L cells. *J. Mol. Biol.* 115:695–714.

Schmidt, W., H. Arnold, and H. Kersten. 1975. Biosynthetic pathway of ribothymidine in *B. subtilis* and *M. lysodeikticus* involving different coenzymes for transfer RNA and ribosomal RNA. *Nucleic Acids Res.* 2:1043–1051.

Schnierle, B. S., P. D. Gershon, and B. Moss. 1992. Cap-specific mRNA (nucleoside-O2'-)-methyltransferase and poly(A) polymerase stimulatory activities of vaccinia virus are mediated by a single protein. *Proc. Natl. Acad. Sci. USA* 89:2897–2901.

Schnierle, B. S., P. D. Gershon, and B. Moss. 1994. Mutational analysis of a multifunctional protein, with mRNA 5' cap-specific (nucleoside-2'-O-)-methyltransferase and 3'-adenylyltransferase stimulatory activities, encoded by vaccinia virus. *J. Biol. Chem.* 269:20700–20706.

Segal, D. M., and D. C. Eichler. 1989. The specificity of interaction between S-adenosyl-L-methionine and a nucleolar 2'-O-methyltransferase. *Arch. Biochem. Biophys.* 275:334–343.

Segal, D. M., and D. C. Eichler. 1991. A nucleolar 2'-O-methyltransferase: specificity and evidence for its role in the methylation of 28S precursor ribosomal RNA. *J. Biol. Chem.* 266:24385–24389.

Sharmeen, L., B. Bass, N. Sonenberg, H. Weintraub, and M. Groudine. 1991. Tat-dependent adenosine-to-inosine modification of wild-type transactivation response RNA. *Proc. Natl. Acad. Sci. USA* 88:8096–8100.

Shi, X., P. Yao, T. Jose, and P. D. Gershon. 1996. Methyltransferase-specific domains within VP39, a bifunctional protein that participates in the modification of both mRNA ends. *RNA* 2:88–101.

Shibata, H., T. S. Ro-Choi, P. Reddy, Y. C. Choi, D. Henning, and H. Busch. 1975. The primary nucleotide sequence of nuclear U-2 ribonucleic acid. *J. Biol. Chem.* 250:3909–3920.

Shimba, S., J. A. Bokar, F. Rottman, and R. Reddy. 1995. Accurate and efficient N-6-adenosine methylation in spliceosomal U6 small nuclear RNA by HeLa cell extract in vitro. *Nucleic Acids Res.* 23:2421–2426.

Singer, C. E., and G. R. Smith. 1972. Histidine regulation in *Salmonella typhimurium*. *J. Biol. Chem.* 247:289-300.

Singer, C. E., G. R. Smith, R. Cortese, and B. N. Ames. 1972. Mutant tRNA[His] ineffective in repression and lacking two pseudouridine modifications. *Nat. New Biol.* 238:72–74.

Singhal, R. P. 1983. Queuine: an addendum. *Prog. Nucleic Acids Res. Mol. Biol.* 28:75–80.

Slany, R. K., M. Bösl, P. F. Crain, and H. Kersten. 1993. A new function of S-adenosylmethionine: the ribosyl moiety of AdoMet is the precursor of the cyclopentenediol moiety of the tRNA wobble base queuine. *Biochemistry* 32:7811–7817.

Slany, R. K., and S. O. Muller. 1995. tRNA-guanine transglycosylase from bovine liver-purification of the enzyme to homogeneity and biochemical characterization. *Eur. J. Biochem.* 230:221–228.

Smith, C., P. G. Schmidt, J. Petsch, and P. F. Agris. 1985. Nuclear magnetic resonance signal assignments of purified [^{13}C]methyl-enriched yeast phenylalanine transfer ribonucleic acid. *Biochemistry* 24:1434–1440.

Söll, D., and L. Kline. 1982. RNA methylation, p. 557–566. *In* P. D. Boyer (ed.), *The Enzymes*. Academic Press, New York, N.Y.

Sommer, B., M. Kohler, R. Sprengel, and P. H. Seeburg. 1991. RNA editing in brain controls a determinant of ion flow in glutamate-gated channels. *Cell* 67:11–19.

Sprinzl, M., C. Steegborn, F. Hubel, and S. Steinberg. 1996. Compilation of tRNA sequences and sequences of tRNA genes. *Nucleic Acids Res.* 24:68–72.

Stadtman, T. C. 1994. Emerging awareness of the critical roles of S-phosphocysteine and selenophosphate in biological systems. *Biofactors* 4:181–185.

Sullivan, M. A., J. F. Cannon, F. H. Webb, and R. M. Bock. 1985. Antisuppressor mutation in *Escherichia coli* defective in biosynthesis of 5-methylaminomethyl-2-thiouridine. *J. Bacteriol.* 161:368–376.

Szweykowska-Kulinska, Z., B. Senger, G. Keith, F. Fasiolo, and H. Grosjean. 1994. Intron-dependent formation of pseudouridines in the anticodon of *Saccharomyces cerevisiae* minor tRNA(Ile). *EMBO J.* 13:4636–4644.

Thiebe, R., and K. Poralla. 1973. Origin of the nucleoside Y in yeast tRNA^{Phe}. *FEBS Lett.* 38:27–28.

Tidwell, T., and E. Howard. 1972. The thiolation, methylation, and formation of pseudouridine and dihydrouridine in tRNA of regenerating rat liver, human phytohemagglutinin stimulated lymphocytes, and Novikoff ascites cells. *Cell Differ.* 1:199–207.

Ueland, P. M. 1982. Pharmacological and biochemical aspects of S-adenosylhomocysteine and S-adenosylhomocysteine hydrolase. *Pharmacol. Rev.* 34:223–246.

Veres, Z., I. Y. Kim, T. D. Scholz, and T. C. Stadtman. 1994. Selenophosphate synthetase. *J. Biol. Chem.* 269:10597–10603.

Veres, Z., and T. C. Stadtman. 1994. A purified selenophosphate-dependent enzyme from *Salmonella typhimurium* catalyzes the replacement of sulfur in 2-thiouridine residues in tRNAs with selenium. *Proc. Natl. Acad. Sci. USA* 91:8092–8096.

Veres, Z., L. Tsai, T. D. Scholz, M. Politino, and R. S. Balaban. 1992. Synthesis of 5-methylaminomethyl-2-selenouridine in tRNAs: ^{31}P NMR studies show the labile selenium donor synthesized by the selD gene product contains selenium bonded to phosphorus. *Proc. Natl. Acad. Sci. USA* 89:2975–2979.

Vold, B. S., M. E. Longmire, and D. E. Keith. 1981. Thiolation and 2-methylthio- modification of *Bacillus subtilis* transfer ribonucleic acids. *J. Bacteriol.* 148:869–876.

Walsh, C. T. 1979. *Enzymatic Reaction Mechanisms*. W. H. Freeman and Co., San Francisco, Calif.

Watanabe, M., M. Matsuo, S. Tanaka, H. Akimoto, S. Asahi, S. Nishimura, J. Katze, T. Hashizume, P. F. Crain, J. A. McCloskey, and N. Okada. 1997. Biosynthesis of archaeosine, a novel derivative of 7-deazaguanosine specific to archaeal tRNA, proceeds via a pathway involving base replacement on the tRNA polynucleotide chain. *J. Biol. Chem.* 272:20146–20151.

Wilson, D., and F. Quiocho. 1994. Crystallographic observation of a trapped tetrahedral intermediate in a metalloenzyme. *Nat. Struct. Biol.* 1:691–694.

Wilson, D., F. Rudolph, and F. Quiocho. 1991. Atomic structure of adenosine deaminase complexed with a transition-state analog: understanding catalysis and immunodeficiency mutations. *Science* (Washington, D.C.) 252:1278–1284.

Wittwer, A. J., and T. C. Stadtman. 1986. Biosynthesis of 5-methylaminomethyl-2-selenouridine, a naturally occurring nucleoside in *Escherichia coli* tRNA. *Arch. Biochem. Biophys.* 248:540–550.

Wong, T. W., S. B. Weiss, G. L. Eliceiri, and J. Bryant. 1970. Ribonucleic acid sulfurtransferase from *Bacillus subtilis* W168. Sulfuration with β-mercaptopyruvate and properties of the system. *Biochemistry* 9:2376–2386.

Wrzesinski, J., K. Nurse, A. Bakin, B. G. Lane, and J. Ofengand. 1995a. A dual-specificity pseudouridine synthase: an *Escherichia coli* synthase purified and cloned on the basis of its specificity for Ψ746 in 23S RNA is also specific for Ψ32 in tRNA^{Phe}. *RNA* 1:437–448.

Wrzesinski, J., K. Nurse, A. Bakin, B. G. Lane, and J. Ofengand. 1995b. Purification, cloning and properties of the 16S RNA pseudouridine 516 synthase from *Escherichia coli*. *Biochemistry* 34:8904–8913.

Xiang, S., S. Short, R. Wolfenden, and C. Carter. 1997. The structure of cytidine deaminase-product complex provides evidence for efficient proton transfer and ground-state stabilization. *Biochemistry* 36:4768–4774.

Yamazaki, N., H. Hori, K. Ozawa, S. Nakanishi, T. Ueda, I. Kumagai, K. Watanabe, and K. Nishikawa. 1992. Purification and characterization of tRNA (adenosine-1-)-methyltransferase from *Thermus thermophilus* HB27. *Nucleic Acids Symp. Ser.* 27:141–142.

Yamazaki, N., H. Hori, K. Ozawa, S. Nakanishi, T. Ueda, I. Kumagai, K. Watanabe, and K. Nishikawa. 1994. Substrate specificity of tRNA (adenine-1-)-methyltransferase from *Thermus thermophilus* HB27. *Biosci. Biotechnol. Biochem.* 58:1128–1133.

Yang, J., P. Sklar, R. Axel, and T. Maniatis. 1995. Editing of glutamate receptor subunit B pre-mRNA by site-specific deamination of adenosine. *Nature* (London) 374:77–81.

Yu, W., and W. Schuster. 1995. Evidence for a site-specific cytidine deamination reaction involved in C to U RNA editing of plant mitochondria. *J. Biol. Chem.* 270:18227–18233.

Zhao, X. M., and D. A. Horne. 1997. The role of cysteine residues in the rearrangement of uridine to pseudouridine catalyzed by pseudouridine synthase I. *J. Biol. Chem.* 272:1950–1955.

Chapter 9

Structural Basis of Base Exchange by tRNA-Guanine Transglycosylases

CHRISTOPHE ROMIER, RALF FICNER, AND DIETRICH SUCK

INTRODUCTION AND BACKGROUND

Location and Role of Queuine and Archaeosine

Base modification in RNA ranges from mere methylations and thiolations to complex "hyper"-modifications (Limbach et al., 1994). With a few exceptions, all hypermodified bases are located in the anticodon loop of tRNAs—in positions 34 (wobble position) and 37 (position following the anticodon)—where they generally are involved in the regulation of codon–anticodon interactions (Björk, 1995; Yokoyama and Nishimura, 1995) (also see Chapter 27 and Appendix 1).

The 7-deazaguanine derivative queuine (Fig. 1) belongs to this class of hypermodified bases. It is found at the wobble position of tRNAs specific for Asn, Asp, His and Tyr (Goodman et al., 1968, 1970; Harada and Nishimura, 1972), in most organisms with the exception of yeast and archaebacteria (Kasai et al., 1975; Katze et al., 1982; Grosjean et al., 1995). The exact biological role of queuine is not yet fully understood and may vary between prokaryotes and eukaryotes. However, several studies have indicated an involvement of queuine at the translational level (reviewed by Curran in Chapter 27) but also in the cellular metabolism (reviewed by Björk and Rasmuson in Chapter 26).

Recently, another 7-deazaguanine derivative, archaeosine (Fig. 1), has been found exclusively at position 15 in the dihydrouridine loop of most archaebacterial tRNAs (Gregson et al., 1993). Due to its location at a position that is important for the maintenance of the tertiary structure of tRNAs, it has been proposed that archaeosine might also play a structural role by enhancing tertiary interactions to adapt to the harsh living conditions of the archaebacteria (Gregson et al., 1993).

Biosynthesis of Queuine and Archaeosine

Contrary to most other modified bases, queuine and archaeosine biosyntheses start outside the tRNA and require a base exchange at the tRNA level, a reaction catalyzed by enzymes known as tRNA-guanine transglycosylases (TGTs).

In prokaryotes where queuine is synthesized de novo in a complex biosynthetic pathway (reviewed by Slany and Kersten [1994] and by Garcia and Goodenough-Lashua [Chapter 8]), the encoded guanine at the wobble position is replaced with the queuine precursor preQ$_1$ (Fig. 1) (Okada et al., 1978). Two other enzymes, QueA (Reuter et al., 1991; Slany et al., 1993) and an unknown enzyme requiring vitamin B$_{12}$ (Frey et al., 1988), are responsible for the modification of preQ$_1$ into queuine.

In eukaryotes, queuine is a nutrient and the replacement of the encoded wobble guanine is carried out in a single enzymatic step (Shindo-Okada et al., 1980; Slany and Mueller, 1995). Queuine in eukaryotic tRNA^{Asp} and tRNA^{Tyr} is further modified to mannosyl- and galactosylqueuine (Fig. 1), respectively, by unknown enzymes specifically recognizing the anticodon loop of these tRNAs (Kasai et al., 1975, 1976; Haumont et al., 1987).

Finally, even though the archaeosine biosynthetic pathway is not yet fully characterized, very recent results (Watanabe et al., 1997) show that the base replacing the encoded guanine at position 15 of archaebacterial tRNAs is the 7-deazaguanine derivative preQ$_0$ (Fig. 1), another queuine precursor (Noguchi et al., 1978).

Christophe Romier and Dietrich Suck • European Molecular Biology Laboratory, Structural Biology Programme, Meyerhofstr. 1, 69117 Heidelberg, Germany.   **Ralf Ficner** • Institut für Molekularbiologie und Tumorforschung, Philipps-Universität Marburg, Emil-Mannkopff-Str. 2, 35037 Marburg, Germany.

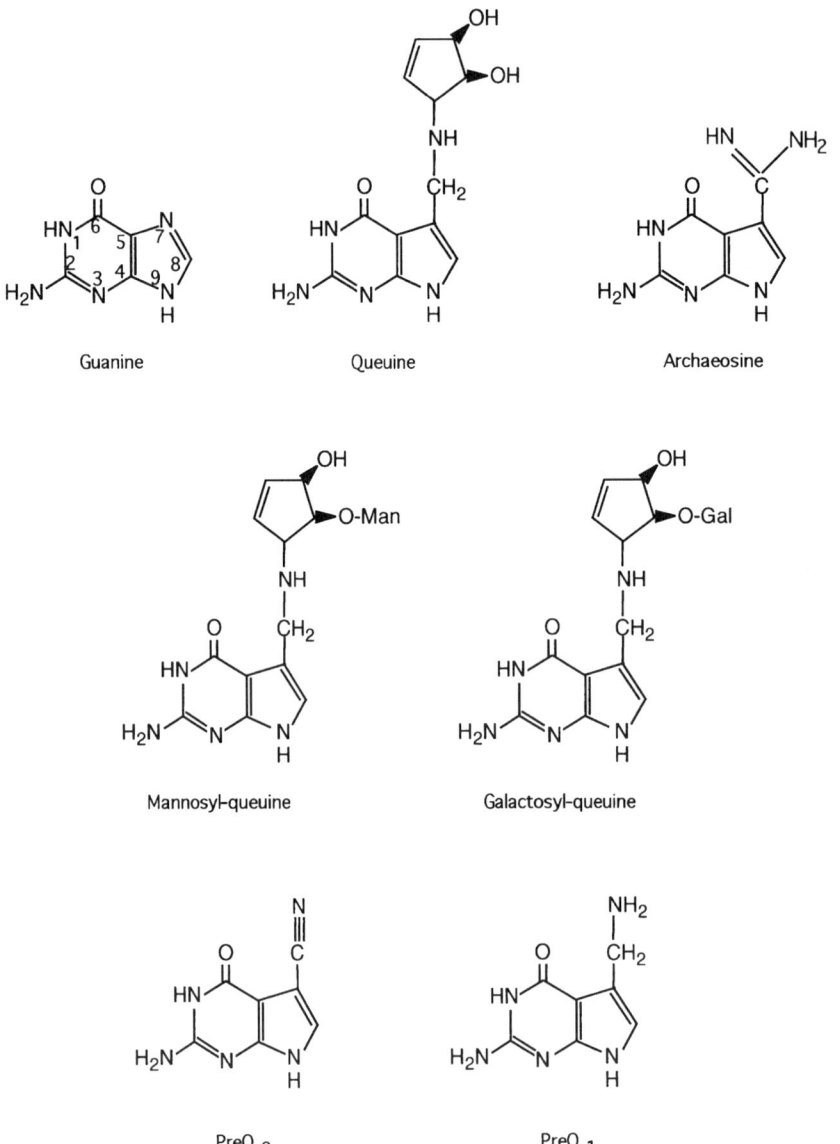

Figure 1. Chemical structures of guanine and the different 7-deazaguanine derivatives. The numbering scheme displayed on guanine is used throughout this chapter.

tRNA-Guanine Transglycosylases from the Three Kingdoms

The prokaryotic TGT is an ~43-kDa enzyme (Okada and Nishimura, 1979; Reuter et al., 1991; Garcia et al., 1993; Reuter and Ficner, 1995) containing one zinc ion coordinated by three cysteines and one histidine (Chong et al., 1995). Even though preQ$_1$ is the natural substrate for this enzyme, guanine and preQ$_0$ (but not queuine) have also been shown to be substrates in vitro (Okada et al., 1979). This latter study also revealed that no additional energy source is required for the reaction which proceeds through a cleavage of the glycosidic bond without breakage of the phosphodiester backbone.

Additionally, the prokaryotic TGT can use a folded anticodon stem-loop as a substrate (Curnow et al., 1993; Mueller and Slany, 1995), with a U$_{33}$G$_{34}$U$_{35}$ sequence being a specific requirement for the reaction to proceed (Nakanishi et al., 1994; Curnow and Garcia, 1995). Two highly conserved residues, Ser103 and Cys281 (*Zymomonas mobilis* numbering), have been proposed to be involved in the enzymatic reaction (Reuter et al., 1994; Chong et al., 1995). Finally, replacement of the 7-aminomethyl group of preQ$_1$ by various substituents showed that the role of this group is entirely in binding/recognition with no apparent effect on catalysis (Hoops et al., 1995a). Replacement of the additional amino group of preQ$_1$ by fluor leads to a suicide inhibitor of TGT (Hoops et al., 1995b). A correlation was observed in these studies between the pK_a of atom N9 and V_{max}, suggesting that deprotonation of nitrogen N9 during the reaction is partially rate-determining for the natural substrate.

Several groups have reported the isolation of the eukaryotic TGT, but these reports do not agree on its oligomeric state and the size of its subunits (Howes and Farkas, 1978; Shindo-Okada et al., 1980; Walden et al., 1982; Slany and Mueller, 1995; Deshpande et al., 1996). However, these studies essentially confirm the use of queuine as a natural substrate, with guanine, preQ$_0$, and preQ$_1$ also being substrates in vitro. Interestingly, the eukaryotic TGT also requires specifically a U$_{33}$G$_{34}$U$_{35}$ sequence for activity (Carbon et al., 1983).

An archaebacterial TGT has been isolated very recently (Watanabe et al., 1997). This enzyme is a 78-kDa protein that has been shown through partial sequencing to be sequence-related to the prokaryotic TGT. Even though preQ$_0$ seems to be the natural substrate for archaebacterial TGT, guanine, but not preQ$_1$ or archaeosine, is also a substrate in vitro (Watanabe et al., 1997).

THREE-DIMENSIONAL STRUCTURE OF PROKARYOTIC TGT

TGT is certainly one of the most interesting enzymes of the queuine biosynthesis pathway because it catalyzes a reaction (a base exchange) which is also observed in many other cellular processes involving DNA and RNA. This enzyme was therefore chosen first for the structural analysis of the different enzymes involved in this pathway. Since its gene had been cloned previously (Reuter et al., 1991), the *Escherichia coli* enzyme was initially used for crystallization attempts. However, possibly due to its heterogeneous oligomeric state at high concentrations (Garcia et al., 1993; Reuter and Ficner, 1995), no suitable crystals could be obtained. Consequently another *tgt* gene, from *Zymomonas mobilis*, was cloned (Reuter and Ficner, 1995). Although both proteins are highly homologous in sequence, *Z. mobilis* TGT is clearly monomeric in solution. Furthermore, the protein could be easily purified by precipitation as microcrystalline material and larger crystals suitable for X-ray diffraction experiments were obtained by macroseeding techniques (Romier et al., 1996a).

The structure of *Z. mobilis* TGT was solved by the well-known technique of Multiple Isomorphous Replacement (MIR). The starting model was refined initially at a resolution of 1.85 Å (Romier et al., 1996b), and more recently at 1.5 Å resolution (Romier, unpublished data). One of the most amazing aspects of the prokaryotic TGT structure is the fact that the enzyme belongs to the large family of $(\beta/\alpha)_8$-barrel enzymes whose characteristic is a barrel formed by eight parallel β-strands (Fig. 2A) (Farber and Petsko, 1990; Reardon and Farber, 1995). Indeed, this enzyme is the first nucleic acid-binding protein shown to adopt such a fold. It does not, however, form a canonical $(\beta/\alpha)_8$-barrel with a direct succession of eight β/α units. While the characteristic eight strands are present, the enzyme contains a number of different secondary structure elements partly included within the connections between the strands (Fig. 3).

The structure starts at the N terminus with a three-stranded antiparallel β-sheet that caps the N-terminal face of the barrel (Fig. 2B and 3). The central part of the protein forms the barrel itself. The C-terminal domain is termed the zinc-binding subdomain as it contains the three cysteines and a histidine (residues 318, 320, 323 and 349, respectively) liganding the zinc ion. It is interesting to note that the helix following the eighth strand of the barrel (α12 in Fig. 3 and highlighted in white in Fig. 2) does not pack against the barrel as would be expected. Rather, it is located above the barrel forming a link to the zinc-binding subdomain (Fig. 2B). The role of the "eighth helix of the barrel" seems in fact to be played by a long α-helix located almost at the C terminus (α15 in Fig. 3 and shown in black in Fig. 2) which is part of the zinc-binding subdomain and packs tightly against and closes the barrel (Fig. 2B). The histidine ligand of the zinc ion is located in the middle of this long helix, whereas the three cysteine ligands are found in loop regions at the very tip of the zinc-binding subdomain. Therefore, by its coordination, the zinc ion keeps the subdomain tightly packed against the barrel and plays an essential role for the structural integrity of the protein.

PreQ$_1$ BINDING BY PROKARYOTIC TGT

A careful structural analysis of the three-dimensional structure of prokaryotic TGT did not reveal how this enzyme recognizes specifically its substrates—both the tRNA with its specific U$_{33}$G$_{34}$U$_{35}$ sequence and preQ$_1$—and how it performs its catalysis. The small size of the preQ$_1$ molecule allowed however its diffusion into the TGT crystals through the large solvent channels they contain and, from data collected from a preQ$_1$-soaked crystal, a preQ$_1$ molecule could easily be fitted in the additional electron density (Romier et al., 1996b). The preQ$_1$-binding pocket is located at the C-terminal face of

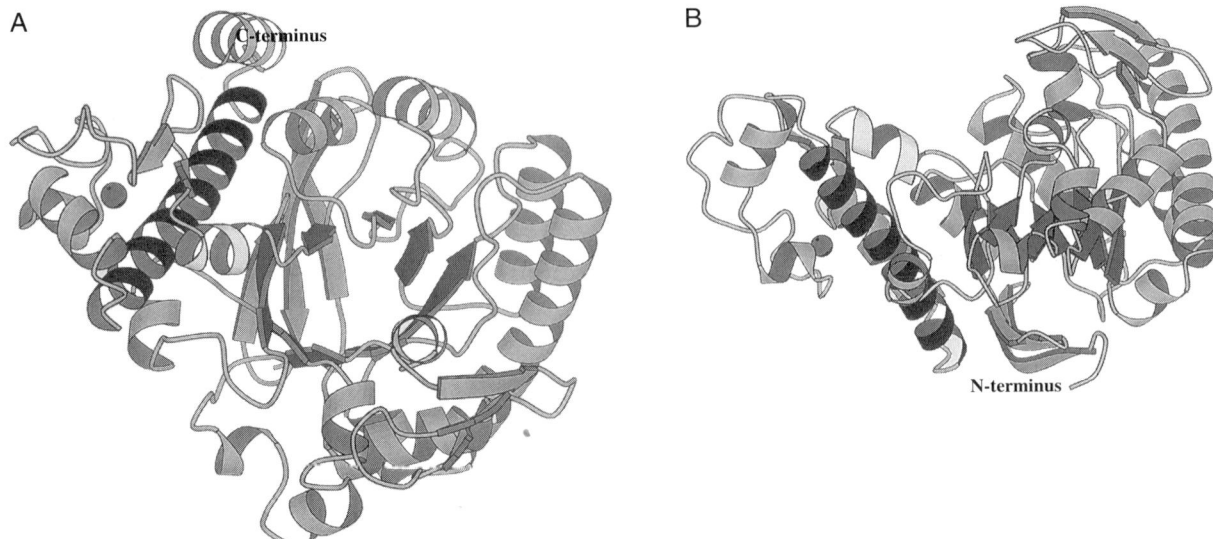

Figure 2. Ribbon representation of TGT. The eight strands forming the barrel are colored dark gray. The zinc ion is represented as a sphere. The helix following the eighth strand of the barrel (colored white) is assumed to interact with the phosphate backbone of the anticodon stem-loop of the tRNA. Highlighted in black is the helix which plays the role of the "eighth helix of the barrel." (A) View parallel to the barrel axis; (B) view perpendicular to the barrel axis.

the barrel, which, contrary to the N-terminal face, is accessible to the solvent. The pocket could not be identified during the initial inspection of the TGT structure since it was blocked by the side chain of Tyr106, which formed a hydrogen bond with the carboxylate of Asp156.

Upon preQ$_1$ binding, this hydrogen bond is broken. The base itself is sandwiched between the side chains of Met260 on one side, and Tyr106 and Cys158 on the other side (Fig. 4). Specific recognition occurs through hydrogen bonding between the side chain of Asp156 and the 1-NH and 2-NH$_2$ groups, the amide of Gly230 and oxygen O6, and the carbonyl oxygen of Leu231 and the additional amino group of preQ$_1$ (Fig. 4 and 5). All these residues are conserved in *E. coli* TGT with the exception of Tyr106, which is replaced by a phenylalanine.

Interestingly, a preQ$_0$ molecule could also be recognized in this pocket. Indeed, the partial negative charge of the cyano nitrogen would be in hydrogen bonding distance to the amide nitrogen of Leu231 due to the linearity of the CCN group (data not shown). Such a recognition would require a slight tilting of the base, which has also been observed with preQ$_1$ when bound to TGT. On the other hand, a queuine molecule would not fit in the pocket due to Val233, which prevents the binding of the cyclopentenediol moiety of this molecule (Fig. 4). This is in agreement with the specificity of *E. coli* TGT (Okada et al., 1979).

It should be pointed out that the preQ$_1$-binding pocket also contains a guanine recognition motif. Indeed, recognition of the 1-NH and 2-NH$_2$ groups by a carboxylate and of oxygen O6 by an amide has frequently been observed in enzymes recognizing guanine, GTP and pterins (Kjeldgaard et al., 1996). However, soaking experiments with guanine did not lead to clear results, probably due to the extremely low solubility of this base around neutral pH. Nevertheless, it is clear that in the guanine-soaked crystals the hydrogen bond between Tyr106 and Asp156 is disrupted and additional electron density is observed

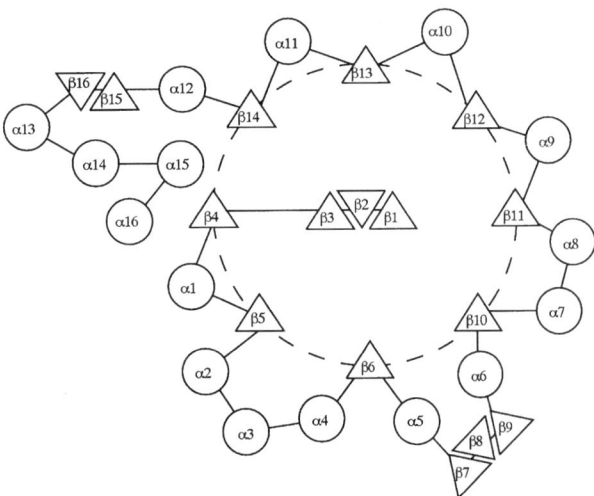

Figure 3. Topology scheme of the structure of TGT. β-strands are represented as triangles, and α-helices are shown as circles. The eight strands forming the barrel are linked by a dashed circle. Reprinted from *The EMBO Journal* (Romier et al., 1996b) with permission of the publisher.

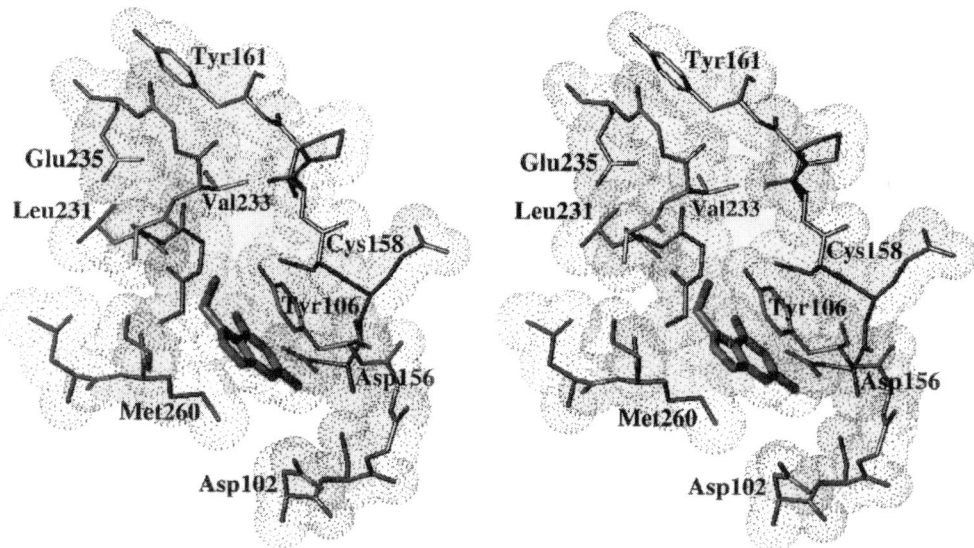

Figure 4. Recognition of preQ$_1$ by *Z. mobilis* TGT. Van der Waals surfaces are represented by dots. Asp102 at the bottom of the figure is the active site nucleophile of *Z. mobilis* TGT. Asp156, the amide group of G230 and the carbonyl oxygen of L231 are involved in specific recognition of preQ$_1$. Reprinted from *FEBS Letters* (Romier et al., 1997) with permission of the publisher.

in the preQ$_1$ binding pocket, suggesting that guanine may indeed bind in this pocket (Romier, unpublished results). Hoops et al. (1995a) have shown by replacing the aminomethyl moiety of preQ$_1$ by various chemical groups that a change in the K_m value depends on the charge and the bulkiness of the new group. These results are easily explained by the structure of the complex showing that a proper interaction with the enzyme requires a positively charged terminal group, small enough to fit into the pocket.

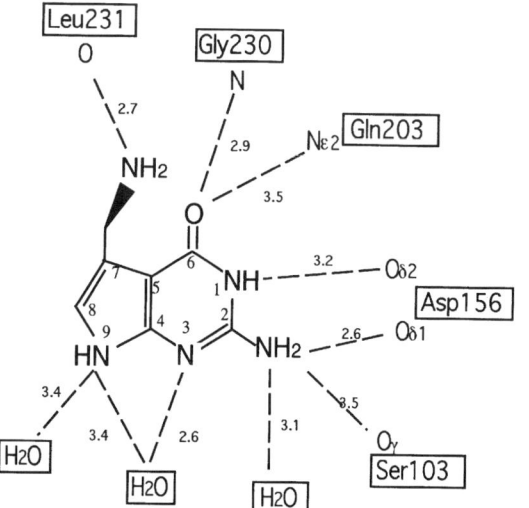

Figure 5. TGT-preQ$_1$ hydrogen bonding contacts. Distances are indicated in angstroms. Reprinted from *The EMBO Journal* (Romier et al., 1996b) with permission of the publisher.

tRNA BINDING AND RECOGNITION BY PROKARYOTIC TGT

The soaking approach employed with the small preQ$_1$ molecule cannot be used with a tRNA macromolecule and the only way to get the structure of a prokaryotic TGT-tRNA complex is by cocrystallization of the components. While this structure determination is clearly the ultimate goal, the structural knowledge previously gained allowed the modeling of the interaction between both macromolecules. Indeed, since preQ$_1$ is specifically recognized at the C-terminal face of the barrel, it is clear that the tRNA should also be recognized on this side of the protein, which shows a strong electrostatic bipolarity with negatively charged pockets in the center of the barrel and a positively charged zinc-binding subdomain (see back cover).

Guided by the three-dimensional structure of the *Z. mobilis* TGT-preQ$_1$ complex, a model of the interaction of a tRNA molecule with prokaryotic TGT based on electrostatic complementarity was established. For this purpose the coordinates of a tRNA^{Asp} molecule derived from the crystal structure of a tRNA-tRNA-synthetase complex (Ruff et al., 1991) were used. In this complex, the bases of the anticodon are flipped outward and interact with the synthetase, a phenomenon that should also be expected upon TGT binding. The final model, which maximizes both the buried area and the electrostatic complementarity between the macromolecules, shows a strong interaction between the positively charged zinc-binding subdomain and the phosphate backbone of the anticodon stem-loop (see back cover). The rest of the tRNA does not seem to interact with the enzyme. This result is not unexpected since it had already been shown that the anticodon stem-loop interacts primarily with E. coli TGT (Curnow et al., 1993; Mueller and Slany, 1995). Interestingly, the helix following the eighth strand of the barrel (shown in white in Fig. 2) contains two highly conserved arginines (residues 286 and 289) that could interact with the phosphate backbone. This helix being rather mobile could play the role of a structural "reading-head" which would move slightly upon tRNA binding to cope with the different tRNAs recognized by prokaryotic TGT.

As for the specific recognition of the UGU sequence, our model suggests that the bases are recognized within the negatively charged pockets (see back cover). Modeling of the specific binding of the UGU sequence by TGT was not attempted because of the inherent flexibility observed for the anticodon loop (Rould et al., 1989; Ruff et al., 1991).

PROPOSED MECHANISM OF BASE EXCHANGE FOR PROKARYOTIC TGT

The three-dimensional structure of queuosine monophosphate shows that the β-configuration of the ribose is preserved (Yokoyama et al., 1979). A direct attack of the tRNA-G_{34} glycosidic bond by preQ$_1$ is therefore excluded because it would lead to an inversion of configuration at C1'. Consequently, either an S_N1 or two consecutive S_N2 reactions could be catalyzed by the enzyme. The former mechanism is unlikely because of the presence of many water molecules in the center of the barrel and the absence of a site in the vicinity of the preQ$_1$-binding pocket where the ribose could be hidden from the solvent. Because the prokaryotic TGT reaction does not require any additional source of energy provided by GTP or ATP, the latter mechanism—i.e., two consecutive S_N2 reactions—most likely involves the formation of a covalent intermediate. Proving that such an intermediate is formed has been complicated by the fact that prokaryotic TGT is able to replace the wobble guanine with free guanine (Okada et al., 1979). Therefore, when the wobble guanine is released upon cleavage of the glycosidic bond, it can attack the covalent intermediate in a reverse reaction. The strategy used was to incubate unmodified tRNA with Z. mobilis TGT, then to unfold the protein by sodium dodecyl sulfate (SDS), and finally to analyze this sample by SDS-polyacrylamide gel electrophoresis (PAGE) (Romier et al., 1996c). Inspection of the resulting gel shows the presence of shifted protein bands, which clearly indicate the formation of a covalent intermediate during catalysis (Fig. 6, lanes b and c). The presence of multiple bands has not yet been clarified, in view especially of the presence of a single band in native gels (data not shown).

The formation of a covalent intermediate implies that an amino acid of prokaryotic TGT plays the role of the active site nucleophile. This residue, besides the fact that it should be a potential nucleophile, has also to be located close enough to the preQ$_1$-binding pocket. Inspection of the TGT structure revealed that only one residue, Asp102 (Z. mobilis numbering), meets these requirements (Fig. 4). An aspartate residue as nucleophile may seem unusual because carboxylates are often considered poor nucleophiles compared to histidines and cysteines. However, 2-deoxyribosyltransferase, which catalyzes the same kind of reaction, also has a catalytic mechanism involving a covalent intermediate and makes use of a carboxylate as nucleophile (Porter et al., 1995; Porter and Short, 1995; Short et al., 1996). It should also be mentioned that many retaining glycoside hydrolases, which catalyze analogous reactions, are enzymes that also utilize the carboxyl group from either a glutamate or an aspartate as active site nucleophiles (McCarter and Withers, 1994).

To test this hypothesis, Asp102 was mutated into an alanine TGT(D102A). As expected, the mutant did not show any remaining activity and its structure did not reveal any significant structural change,

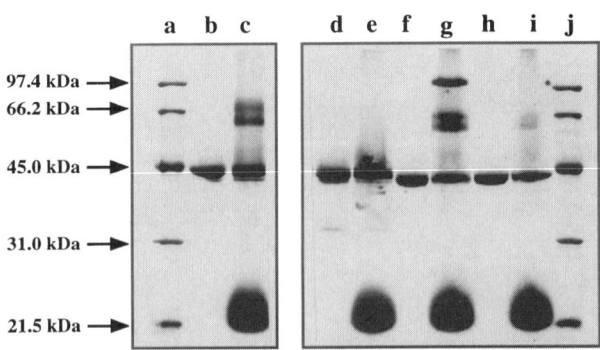

Figure 6. Silver-stained SDS-PAGE of wild-type TGT [TGT(wt)] and TGT mutants D102A, D156A and D156Y, in absence or presence of tRNA^{Tyr}(G$_{34}$). Shifted bands indicating the formation of a covalent intermediate are seen only with wild-type TGT and the D156A and D156Y mutants. The lack of a shifted band with the D102A mutant identifies D102 as the active site nucleophile of TGT. Lane a, molecular mass standards; lane b, TGT(wt); lane c, TGT(wt) plus tRNA; lane d, TGT(D102A); lane e, TGT(D102A) plus tRNA; lane f, TGT(D156A); lane g, TGT(D156A) plus tRNA; lane h, TGT(D156Y); lane i, TGT(D156Y) plus tRNA; lane j, molecular mass standards. Reprinted from Biochemistry (Romier et al., 1996c) with permission of the publisher.

besides the mutation, which could explain this loss of activity. Analysis of a TGT(D102A)-tRNA mixture by SDS-PAGE did not show, contrary to the wild-type TGT-tRNA sample, any shifted protein bands that could reveal the presence of a covalent intermediate (Fig. 6, lanes d and e). These results clearly indicate that Asp102 is the active site nucleophile of Z. mobilis TGT.

Mutagenesis was also used to check the assumption that the specific recognition of the wobble guanine takes place in the preQ$_1$-binding pocket since, as discussed earlier, this pocket also contains a guanine recognition motif and no other guanine-binding pocket could be found in the vicinity of the active site. Asp156, a major recognition element of preQ$_1$ (Fig. 4 and 5), was mutated both into an alanine TGT(D156A) to reduce the large negative charge of the pocket and into a tyrosine TGT(D156Y) to prevent proper binding of a base in the pocket. Both mutants are totally inactive and no structural changes besides the mutations were observed, which could explain the loss of activity of the mutants. However, contrary to the TGT(D102A) mutant, both mutants TGT(D156A) and TGT(D156Y) are able to form a covalent intermediate with tRNA (Fig. 6, lanes f and g and lanes h and i, respectively). The amounts of covalent intermediate formed are vastly different for these mutants: whereas with the TGT(D156A) mutant it is equivalent to what is observed with the wild-type protein, only an extremely small amount, hardly detectable by silver staining, is observed with the TGT(D156Y) mutant (Fig. 6). This difference strongly suggests that the preQ$_1$-binding pocket is involved in the binding of the UGU sequence, and most probably of the wobble guanine. The presence of some shifted material in the case of the TGT(D156Y) mutant suggests that alternative conformations of the anticodon loop enable Asp102 to come close enough to the C1′ position for nucleophilic attack. These results also imply that, although space is needed to accommodate the wobble guanine in the preQ$_1$-binding pocket, Asp156 is not absolutely necessary for its recognition. Actually, the hydrogen bond formed between oxygen 6 of guanine and the amide of Gly230 might be sufficient for recognition. On the other hand, the presence of Asp156 at the bottom of the pocket is clearly essential for the second step of the reaction, i.e., the substitution of Asp102 by preQ$_1$, to proceed.

So far, pockets recognizing specifically U$_{33}$ and U$_{35}$ have not been identified. Soaking experiments with guanosine, 5′-GMP, uracil, uridine, and even a UGU sequence did not give interpretable difference densities contrary to what was observed with preQ$_1$ (Romier, unpublished data). Although it is not clear why no binding is observed in these crystals, it should be remembered that only a folded—and not a nonfolded—anticodon stem-loop is a substrate for TGT (Curnow et al., 1993). The steric and electrostatic perturbations created upon binding of a properly folded substrate could therefore affect specific recognition of the UGU sequence and/or the catalytic mechanism itself. This would help prevent arbitrary modification of UGU sequences present in the RNA of the cell.

The role of both residues Ser103 and Cys281 (Z. mobilis numbering), which have been shown to affect catalysis in E. coli TGT (Reuter et al., 1994; Chong et al., 1995), remain unclear. The former forms a weak hydrogen bond to the 2-NH$_2$ group of preQ$_1$ (Fig. 4) and could be involved in preQ$_1$ orientation. Its hydroxyl also forms hydrogen bonds with atoms of the protein main chain and this residue could play a mere structural role. However, the distance of ~7 Å observed between this hydroxyl and nitrogen N9 of preQ$_1$ clearly shows that this residue cannot be the active site nucleophile of prokaryotic TGT. Cys281, found rather far away from both the preQ$_1$-binding pocket and Asp102, could help stabilize the "reading-head" α-helix thought to interact with the tRNA backbone (see above) and has been shown to have no direct influence on the catalytic mechanism (Garcia and Chong, 1997).

Taken together, the experiments and results presented lead to the proposal of a catalytic mechanism for prokaryotic TGT involving the following steps (Fig. 7): (i) orienting of the tRNA on the surface of the enzyme by electrostatic contacts between the anticodon stem-loop backbone and the zinc-binding subdomain, including the "reading-head" helix; (ii) conformational adjustment of the anticodon loop to enable recognition of the UGU sequence, especially the binding of the wobble guanine in the preQ$_1$-binding pocket; (iii) nucleophilic attack by the carboxylate of Asp102 at C1′, resulting in the cleavage of the glycosidic bond and formation of a covalent TGT-tRNA intermediate; (iv) replacement of the guanine by preQ$_1$; and (v) nucleophilic attack at C1′ by the N9 atom of preQ$_1$ with formation of the new glycosidic bond.

This proposed catalytic mechanism is certainly incomplete. For instance, it is unclear whether any protonation or deprotonation occurs to facilitate the attack at C1′ in both nucleophilic replacement steps. A deprotonation of nitrogen N9 of preQ$_1$ prior to its attack at C1′ has however been hypothesized (Hoops et al., 1995a). Surprisingly, no side chains are found in the immediate vicinity of the preQ$_1$-binding pocket and Asp102 that could play such roles. The cocrystallization of a TGT-tRNA complex, which is under

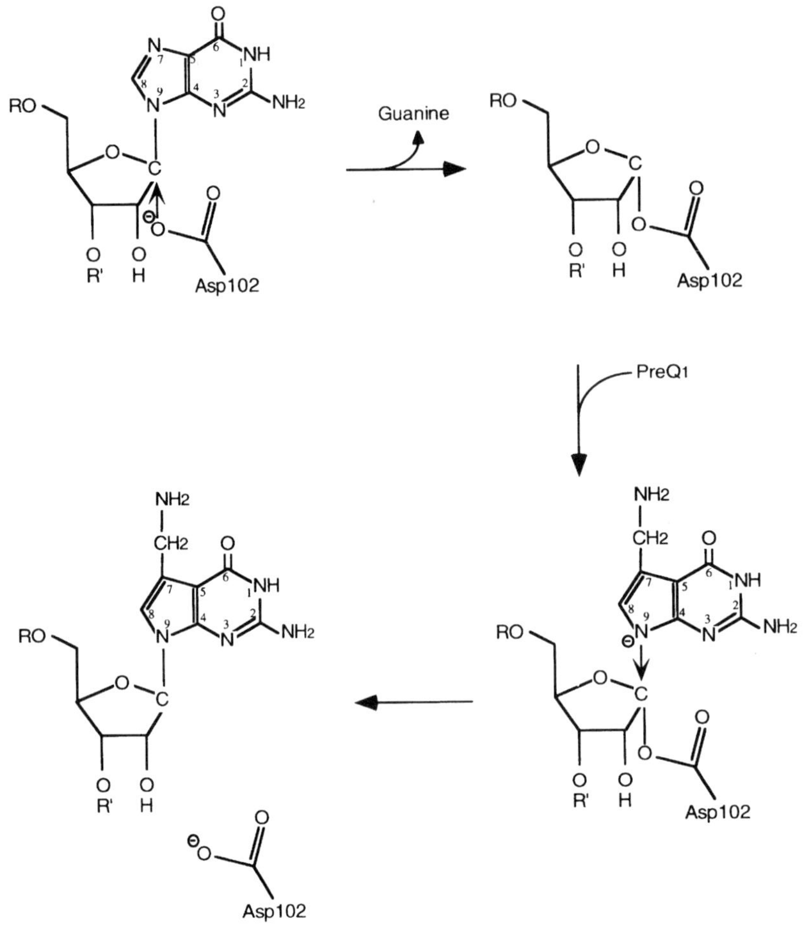

Figure 7. Proposed catalytic mechanism of TGT. A covalent intermediate is formed following the nucleopholic attack of aspartate 102 at C1'. Subsequently the deprotonated preQ$_1$ molecule attacks the C1' atom, restoring the β-configuration and leading to the modified preQ$_1$-tRNA. Reprinted from *Biochemistry* (Romier et al., 1996c) with permission of the publisher.

way in our laboratory, will help to answer the remaining open questions concerning tRNA binding, UGU specific recognition and catalysis, and the design of further biochemical experiments.

EUKARYOTIC AND ARCHAEBACTERIAL TGTs

A Tale of Databases

During the last few years and even months, several genome sequences of prokaryotes, eukaryotes and archaebacteria have been partially or fully released. Scanning of the sequence databases for protein sequences homologous to *Z. mobilis* TGT surprisingly gives several clear hints not only for other prokaryotic TGTs but also for full and partial eukaryotic (*Caenorhabditis elegans*, rice, *Drosophila melanogaster*, mouse, and human) and archaebacterial (*Methanococcus jannaschii*, *Archaeoglobus fulgidus* and *Methanobacterium thermoautotrophicum*) proteins (Fig. 8) (Romier et al., 1997).

Eukaryotic TGTs

The eukaryotic proteins are about 40% identical in sequence to the prokaryotic TGTs and are phylogenetically very close to them. Especially, the nucleophilic aspartate and the zinc ligands are strictly conserved (Fig. 8). The high homology between the eukaryotic and prokaryotic enzymes allowed the modeling of the three-dimensional structure of the *C. elegans* protein. Only a few amino acid substitutions are observed within or in the vicinity of the prokaryotic preQ$_1$-binding pocket (Fig. 9). Most importantly, Val233, which is preventing the binding of queuine in the prokaryotic TGT, is here replaced by a glycine. This replacement, together with the mutation of the bulky Cys158 into a valine, enlarges the binding pocket and allows the recognition of a queuine molecule (Fig. 9). While the interactions between

Figure 8. See following page for legend.

Figure 8. Alignment of the *Z. mobilis* (Z.mobi), *E. coli* (E.coli), *Shigella flexneri* (S.flex), *Haemophilus influenzae* (H.infl), *Helicobacter pylori* (H.pylo), *Synechocystis* sp. (S.sp.), *Thermotoga maritima* (T.mari), *Caenorhabditis elegans* (C.eleg), mouse, human, *Methanococcus jannaschii* (M.jann), *Archaeoglobus fulgidus* (A.fulg) and *Methanobacterium thermoautotrophicum* (M.ther) TGT sequences. Highly conserved regions together with other important regions are shaded. The human and mouse sequences are incomplete and the gaps they contain may indicate missing data. Important residues have been labeled with *Z. mobilis* numbers. Asp102, marked with an asterisk, is the active site nucleophile of *Z. mobilis* TGT. The four zinc ligands are marked. The amino acid marked "Additional Proline" is the proline residue found exclusively in the archaebacterial sequences. The additional C-terminal residues of these later sequences are not shown (marked as "...").

the protein and the preQ$_1$ moiety are strongly conserved, the cyclopentene ring also forms a hydrophobic interaction with the side chain of Val158, and the two additional hydroxyls are in hydrogen bonding distance to several main chain carbonyl oxygens. Because the rest of the binding pocket is relatively unchanged, it is clear that guanine, preQ$_0$ and preQ$_1$ should also be recognized by the eukaryotic protein, in agreement with the specificity of eukaryotic TGT (Shindo-Okada et al., 1980; Slany and Mueller, 1995). The high sequence homology could also explain the identical specificity toward the U$_{33}$G$_{34}$U$_{35}$ sequence by both eukaryotic and prokaryotic TGTs (Carbon et al., 1983; Nakanishi et al., 1994; Curnow and Garcia, 1995).

Interestingly, the molecular mass of the eukaryotic proteins (~45 kDa) is identical to that of the smallest subunit of the eukaryotic TGT isolated by Howes and Farkas (1978). However, Deshpande et al. (1996) recently reported that this 45-kDa subunit is not important for eukaryotic TGT activity and have only sequenced the other 60-kDa subunit. The close resemblance of this latter subunit to an ubiquitin carboxyl-terminal hydrolase and the presence of homologs in the genomes of *Saccharomyces cerevisiae* and *Schizosaccharomyces pombe*, organisms devoid of any 7-deazaguanine derivatives, strongly suggest that it is not the eukaryotic TGT.

Archaebacterial TGTs

The archaebacterial proteins are rather different because they have about 300 additional residues at their C terminus (data not shown). These proteins therefore have a higher molecular mass, close to the 78 kDa observed for the *Haloferax volcanii* TGT (Watanabe et al., 1997). This C-terminal tail is not homologous to any other protein within the sequence databases. Its central part is not very well conserved, whereas its extremities are highly conserved with successions of charged and hydrophobic residues. On the other hand, the N-terminal part of the archaebacterial proteins is clearly homologous to the prokaryotic sequences, even though it is not as close phylogenetically to them as are the eukaryotic proteins (Fig. 8). In particular, the nucleophilic aspartate and the zinc ligands are strictly conserved.

Once again, the strong homology between these archaebacterial proteins and the prokaryotic TGTs allowed the modeling of their three-dimensional struc-

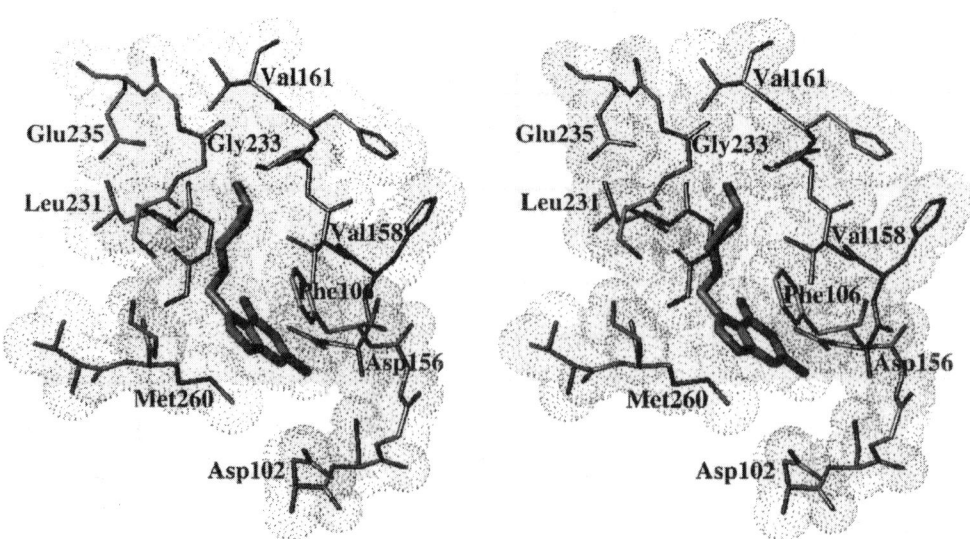

Figure 9. Recognition of queuine by *C. elegans* TGT. Van der Waals surfaces are represented with dots. Reprinted from *FEBS Letters* (Romier et al., 1997) with permission of the publisher.

ture, especially in the active site vicinity. More amino acid substitutions are observed in this case. Notably, Val233 is here replaced by a leucine. This bulkier residue cannot be accommodated where the valine previously was located due to the presence opposite to it of a proline (numbered 161)—totally conserved in all the archaebacterial sequences—which bulges above the binding pocket (Fig. 10). Consequently, the whole loop formed by residues 231 to 235 is slightly pushed toward the center of the pocket. Additionally, a proline is inserted within this loop whose role might be either to rigidify or change the conformation of the loop. Altogether, these modifications prevent the recognition of $preQ_1$—and consequently of archaeosine—by restricting the volume of the pocket in the region where its additional amino group binds. However, there is still enough space for the binding of a $preQ_0$ (Fig. 10) or a guanine molecule, in agreement with the observed substrate specificity of archaebacterial TGT (Watanabe et al., 1997).

It is interesting to consider that the archaebacterial TGT does not incorporate $preQ_0$ at the wobble position but rather at position 15, a totally different location in the tRNA. The lower homology observed between these proteins and the eukaryotic and prokaryotic proteins and/or their 300 additional residues are probably related to their difference in tRNA recognition.

A Common Catalytic Mechanism?

Altogether, these results strongly favor the hypothesis that the eukaryotic and archaebacterial proteins found by the database search represent the tRNA-guanine transglycosylases of these two kingdoms. Prokaryotic TGTs catalyze their reaction via the formation of a protein-tRNA covalent intermediate, Asp102 being the active site nucleophile of *Z. mobilis* TGT (Romier et al., 1996c). Based on mutagenesis and structural data, it has also been postulated that the wobble guanine is recognized within the $preQ_1$-binding site of prokaryotic TGTs. The total conservation of the nucleophilic aspartate and other active site residues, and the fact that all enzymes are able to recognize guanine in their 7-deazaguanine-binding pockets, is in agreement with a universal base exchange mechanism for the synthesis of the various 7-deazaguanine-modified tRNAs based upon a common protein fold.

EVOLUTIONARY RELATIONSHIPS AND PROSPECTS FOR THE FUTURE

The finding that the eukaryotic and archaebacterial TGTs most certainly share a common fold and a common catalytic mechanism with the prokaryotic TGTs will certainly affect the biochemical and structural research on these proteins and the other enzymes responsible for the synthesis of the various 7-deazaguanine derivatives in the three kingdoms. It also clarifies the debate concerning the eukaryotic enzyme whose transglycosylase activity was the first to be described.

Clearly, one of the immediate questions raised concerns the evolutionary relationship between these different enzymes. In this respect, the case of *Ther-*

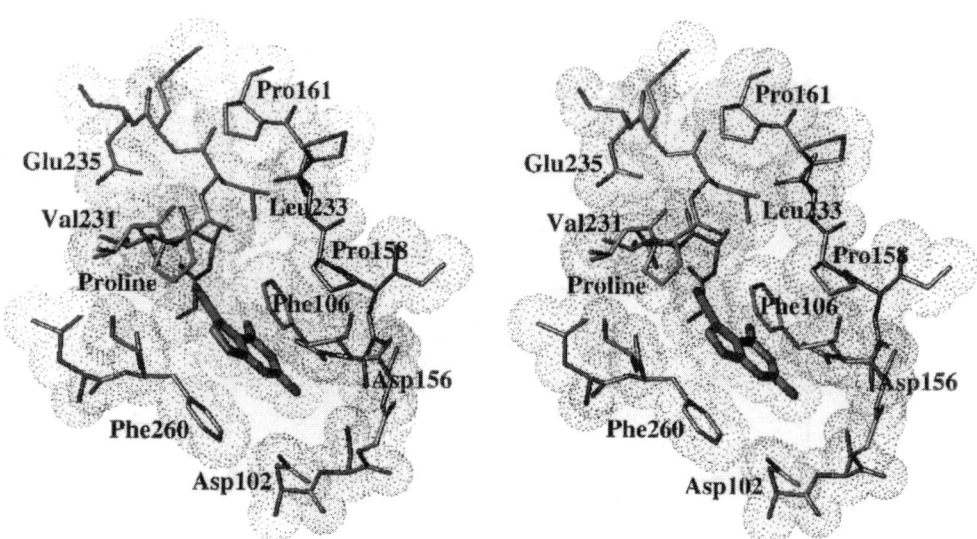

Figure 10. Recognition of $preQ_0$ by *M. jannaschii* TGT. Van der Waals surfaces are represented with dots. Reprinted from *FEBS Letters* (Romier et al., 1997) with permission of the publisher.

motoga maritima, an organism that diverged quite early from the other prokaryotes, is interesting. Indeed, even though it is highly homologous to the other prokaryotic TGTs, its own enzyme displays some features characteristic of the archaebacterial enzymes. Particularly, a proline is found at position 161 and an isoleucine is found at position 233, but no additional proline is found in the loop formed by residues 231 to 235 (Fig. 8). It is therefore unclear whether this enzyme uses preQ$_0$ or preQ$_1$ as a substrate, with the former also being found in prokaryotes (Noguchi et al., 1978). This could indicate two possible evolutionary pathways for these enzymes. In the first, the archaebacterial proteins could have evolved by protein fusion from an ancestor, involved in the synthesis of a 7-deazaguanine derivative at the wobble position and possibly using preQ$_0$, to synthesize archaeosine to adapt to their challenging environments. Alternatively, the prokaryotic and eukaryotic proteins could have evolved from a common ancestor involved in archaeosine synthesis and used it to synthesize queuine to cope with more complicated translational events. In either cases, the eukaryotes would have further evolved to use queuine as a mere nutrient. It should be noted that the generally accepted representation of the universal phylogenetic tree favors the first hypothesis.

More generally, the surprising fact that TGT is a $(\beta/\alpha)_8$-barrel protein raises the question of its evolutionary relationship to other members of this large family of proteins, adding still another facet to the ongoing debate about the convergent and/or divergent evolution of $(\beta/\alpha)_8$-proteins (Farber and Petsko, 1990; Reardon and Farber, 1995). In the discussion about the mechanism of TGT we mentioned the class of pertaining glycosidases that catalyze a similar reaction apparently using a related mechanism involving the nucleophilic attack by a carboxylate group at the C1 position (McCarter and Withers, 1994). The majority of these enzymes also have the $(\beta/\alpha)_8$-barrel fold, and it is therefore tempting to speculate that an evolutionary link exists between these two families of enzymes. One could imagine that a $(\beta/\alpha)_8$-barrel enzyme with glycosidase activity has evolved into a tRNA transglycosylase through attachment of a C-terminal zinc-binding domain which enabled it to bind to tRNA.

In this context it would be interesting to know whether the other enzymes of the queuine, mannosyl- and galactosyl-queuine, and archaeosine biosynthetic pathways are also $(\beta/\alpha)_8$-barrel proteins, analogous to the tryptophane biosynthetic pathway where all enzymes involved belong to this group. So far only the prokaryotic QueA has been isolated and biochemically characterized, and crystallization experiments have been started in our laboratory. The last enzyme of this pathway, which reduces epoxy-queuine to queuine, is still unknown, but has been shown to require vitamin B$_{12}$ (Frey et al., 1988). Recently the first structure of a vitamin B$_{12}$-requiring enzyme has been determined, revealing an $(\beta/\alpha)_8$-barrel fold with the B$_{12}$ moiety embedded in the barrel center (Mancia et al., 1996).

Finally, it should be noticed that the biosyntheses of queuine- and archaeosine-modified tRNAs are the only pathways in base modification and editing where a base exchange mechanism has clearly been identified (Garcia and Goodenough-Lashua, Chapter 8; Price and Gray, Chapter 16; Marchfelder et al., Chapter 17). Many modification or editing pathways have not been fully characterized and the occurrence of other similar mechanisms could well be discovered in the future. Whether or not such reactions would be catalyzed by enzymes belonging to the large family of the $(\beta/\alpha)_8$-barrel proteins remains an open question.

Prokaryotic, eukaryotic, and archaebacterial tRNA-guanine transglycosylases are enzymes catalyzing a base exchange reaction in the biosynthesis of the queuine- and archaeosine-modified tRNAs, with preQ$_1$, queuine, and preQ$_0$, respectively, being their natural substrates. We have solved the structure of *Zymomonas mobilis* TGT, showing that prokaryotic TGT adopts a $(\beta/\alpha)_8$-barrel fold. The structure of *Z. mobilis* TGT in complex with preQ$_1$ revealed a specific recognition pocket for this base at the C-terminal face of the barrel and allowed the modeling of the interaction between prokaryotic TGT and a tRNA macromolecule. Using band-shift assays and mutagenesis, we have shown that a covalent intermediate is formed between the protein and the tRNA during catalysis and we have identified Asp102 as the active site nucleophile of *Z. mobilis* TGT. The recent occurrence in sequence databases of eukaryotic and archaebacterial proteins highly homologous to the prokaryotic TGTs defines a clear phylogenetic link and suggests that TGTs from all three kingdoms have a common fold and a common catalytic mechanism.

Acknowledgments. This work was partly supported by a grant from the Deutsche Forschungsgemeinschaft. We gratefully acknowledge the contribution of Klaus Reuter, whose groundbreaking work in cloning and characterizing the prokaryotic TGTs made this work possible. We thank our colleague Joachim Meyer, who was involved in the identification and the alignment of eukaryotic and archaebacterial TGT sequences.

ADDENDUM IN PROOF

The recently published complete genome sequence of the archaeon *A. fulgidus* contains a second TGT gene of unknown function which is more closely related to prokaryotic TGTs and does not have the additional 300 residues found at the C terminus of the other archaeal enzymes (H.-P. Klenk et al., *Nature* 390: 364–370, 1997).

REFERENCES

Björk, G. R. 1995. Biosynthesis and function of modified nucleosides, p. 165–205. *In* D. Söll and U. RajBhandary (ed.), *tRNA: Structure, Biosynthesis, and Function*. American Society for Microbiology, Washington, D.C.

Carbon, P., E. Haumont, M. Fournier, S. de Henau, and H. Grosjean. 1983. Site-directed *in vitro* replacement of nucleosides in the anticodon loop of tRNA: application to the study of structural requirements for queuine insertase activity. *EMBO J.* 2: 1093–1097.

Chong, S., A. Curnow, T. J. Huston, and G. A. Garcia. 1995. tRNA-guanine transglycosylase from *E. coli* is a zinc metalloprotein. Site-directed mutagenesis studies to identify the zinc ligands. *Biochemistry* 34:3694–3701.

Curnow, A. W., F. L. Kung, K. A. Koch, and G. A. Garcia. 1993. tRNA-guanine transglycosylase from *Escherichia coli*: gross tRNA structural requirements for recognition. *Biochemistry* 32: 5239–5246.

Curnow, A. W., and G. A. Garcia. 1995. tRNA-guanine transglycosylase from *E. coli*. Minimal tRNA structure and sequence requirements for recognition. *J. Biol. Chem.* 270: 17264–17267.

Deshpande, K. L., P. H. Seubert, D. M. Tillman, W. R. Farkas, and J. R. Katze. 1996. Cloning and characterization of cDNA encoding the rabbit tRNA-guanine transglycosylase 60-kilodalton subunit. *Arch. Biochem. Biophys.* 326:1–7.

Farber, G. K., and G. A. Petsko. 1990. The evolution of α/β barrel enzymes. *Trends Biochem. Sci.* 15:228–234.

Frey, B., J. McCloskey, W. Kersten, and H. Kersten. 1988. New function of vitamin B_{12}: cobamide-dependent reduction of epoxyqueuosine in tRNAs of *Escherichia coli* and *Salmonella typhimurium*. *J. Bacteriol.* 170:2078–2082.

Garcia, G. A., K. A. Koch, and S. Chong. 1993. tRNA-guanine transglycosylase from *Escherichia coli*. Overexpression, purification and quaternary structure. *J. Mol. Biol.* 231:489–497.

Garcia, G. A., and S. Chong. 1997. Cysteine 265 is in the active site of, but is not essential for catalysis by tRNA-guanine transglycosylase (TGT) from *Escherichia coli*. *J. Protein Chem.* 16: 11–18.

Goodman, H. M., J. Abelson, A. Landy, S. Brenner, and J. D. Smith. 1968. Amber suppression: a nucleotide change in the anticodon of a tyrosine transfer RNA. *Nature* 217:1019–1024.

Goodman, H. M., J. N. Abelson, A. Landy, S. Zadrazil, and J. D. Smith. 1970. The nucleotide sequences of tyrosine transfer RNAs of *Escherichia coli*. *Eur. J. Biochem.* 13:461–483.

Gregson, J. M., P. F. Crain, C. G. Edmonds, R. Gupta, T. Hashizume, D. W. Phillipson, and J. A. McCloskey. 1993. Structure of the archaeal tRNA nucleoside G*-15. *J. Biol. Chem.* 268: 10076–10086.

Grosjean, H., M. Sprinzl, and S. Steinberg. 1995. Posttranscriptionally modified nucleosides in tRNA: their locations and frequencies. *Biochimie* 77:139–141.

Harada, F., and S. Nishimura. 1972. Possible anticodon sequences of tRNA^{His}, tRNA^{Asn}, and tRNA^{Asp} from *Escherichia coli* B. Universal presence of nucleoside Q in the first position of the anticodons of these tRNAs. *Biochemistry* 11:301–308.

Haumont, E., L. Droogmans, and H. Grosjean. 1987. Enzymatic formation of queuosine and of glycosyl queuosine in yeast tRNAs microinjected into *Xenopus laevis* oocytes: the effect of the anticodon loop sequence. *Eur J. Biochem.* 168:219–225.

Hoops, G. C., L. B. Townsend, and G. A. Garcia. 1995a. tRNA-guanine transglycosylase from *E. coli*: structure-activity studies investigating the role of the aminomethyl substituent of the heterocyclic substrate preQ$_1$. *Biochemistry* 34:15381–15387.

Hoops, G. C., L. B. Townsend, and G. A. Garcia. 1995b. Mechanism-based inactivation of tRNA-guanine transglycosylase from *E. coli* by 2-amino-5-(fluoromethyl)pyrrolo[2,3-d]pyrimidin-4(3H)-one. *Biochemistry* 34:15539–15544.

Howes, N. K., and W. R. Farkas. 1978. Studies with a homogeneous enzyme from rabbit erhthrocytes catalyzing the insertion of guanine into tRNA. *J. Biol. Chem.* 253:9082–9087.

Kasai, H., Y. Kuchino, K. Nihei, and S. Nishimura. 1975. Distribution of the modified nucleoside Q and its derivatives in animal and plant tRNAs. *Nucleic Acids Res.* 2:1931–1939.

Katze, J. R., B. Basile, and J. A. McCloskey. 1982. Queuine, a modified base incorporated posttranscriptionally into eukaryotic tRNA: wide distribution in nature. *Science* 216:55–56.

Kjeldgaard, M., J. Nyborg, and B. F. C. Clark. 1996. The GTP binding motif: variations on a theme. *FASEB J.* 10:1347–1368.

Limbach, P. A., P. F. Crain, and A. McCloskey. 1994. Summary: the modified nucleosides of RNA. *Nucleic Acids Res.* 22: 2183–2196.

Mancia, F., N. H. Keep, A. Nakagawa, P. F. Leadlay, S. McSweeney, B. Rasmussen, P. Bosecke, O. Diat, and P. R. Evans. 1996. How coenzyme B_{12} radicals are generated: the crystal structure of methylmalonyl-coenzyme A mutase at 2 Å resolution. *Structure* 4:339–350.

McCarter, J. D., and S. G. Withers. 1994. Mechanisms of enzymatic glycoside hydrolysis. *Curr. Opin. Struct. Biol.* 4:885–892.

Mueller, S. O., and R. K. Slany. 1995. Structural analysis of the interaction of the tRNA modifying enzymes Tgt and QueA with a substrate tRNA. *FEBS Lett.* 361:259–264.

Nakanishi, S., T. Ueda, H. Hori, N. Yamazaki, N. Okada, and K. Watanabe. 1994. A UGU sequence in the anticodon loop is a minimum requirement for recognition by *E. coli* tRNA-guanine transglycosylase. *J. Biol. Chem.* 269:32221–32225.

Noguchi, S., Z. Yamaizumi, T. Ohgi, T. Goto, Y. Nishimura, Y. Hirota, and S. Nishimura. 1978. Isolation of Q nucleoside precursor present in tRNA of an *E. coli* mutant and its characterization as 7-(cyano)-7-deazaguanosine. *Nucleic Acids Res.* 5: 4215–4223.

Okada, N., S. Noguchi, S. Nishimura, T. Ohgi, T. Goto, P. F. Crain, and J. A. McCloskey. 1978. Structure determination of a nucleoside Q precursor isolated from *E. coli* tRNA: 7-(aminomethyl)-7-deazaguanosine. *Nucleic Acids Res.* 5: 2289–2297.

Okada, N., and S. Nishimura. 1979. Isolation and characterization of a guanine insertion enzyme, a specific tRNA transglycosylase, from *Escherichia coli*. *J. Biol. Chem.* 254:3061–3066.

Okada, N., S. Noguchi, H. Kasai, N. Shindo-Okada, T. Ohgi, T. Goto, and S. Nishimura. 1979. Novel mechanism of posttranscriptional modification of tRNA. *J. Biol. Chem.* 254: 3067–3073.

Porter, D. J. T., B. M. Merrill, and S. A. Short. 1995. Identification of the active site nucleophile in nucleoside 2-deoxyribosyltransferase as glutamic acid 98. *J. Biol. Chem.* 270: 15551–15556.

Porter, D. J. T., and S. A. Short. 1995. Nucleoside 2-deoxyribosyltransferase. Pre-steady-state kinetics analysis of native enzyme and mutant enzyme with an alanyl residue replacing Glu98. *J. Biol. Chem.* 270:15557–15562.

Reardon, D., and G. K. Farber. 1995. The structure and evolution of α/β barrel proteins. *FASEB J.* 9:497–502.

Reuter, K., R. Slany, F. Ullrich, and H. Kersten. 1991. Structure and organization of *Escherichia coli* genes involved in biosynthesis of the deazaguanine derivative queuine, a nutrient factor for eukaryotes. *J. Bacteriol.* 173:2256–2264.

Reuter, K., S. Chong, F. Ullrich, H. Kersten, and G. A. Garcia. 1994. Serine 90 is required for enzymic activity by tRNA-guanine transglycosylase from *E. coli*. *Biochemistry* 33:7041–7046.

Reuter, K., and R. Ficner. 1995. Sequence analysis and overexpression of the *Zymomonas mobilis tgt* gene encoding tRNA-guanine transglycosylase: purification and biochemical characterization of the enzyme. *J. Bacteriol.* 177:5284–5288.

Romier, C. Unpublished data.

Romier, C., R. Ficner, K. Reuter, and D. Suck. 1996a. Purification, crystallization, and preliminary X-ray diffraction studies of tRNA-guanine transglycosylase from *Zymomonas mobilis*. *Proteins Struct. Funct. Genet.* 24:516–519.

Romier, C., K. Reuter, D. Suck, and R. Ficner. 1996b. Crystal structure of tRNA-guanine transglycosylase: RNA modification by base exchange. *EMBO J.* 15:2850–2857.

Romier, C., K. Reuter, D. Suck, and R. Ficner. 1996c. Mutagenesis and crystallographic studies of *Zymomonas mobilis* tRNA-guanine transglycosylase reveal aspartate 102 as active site nucleophile. *Biochemistry* 35:15734–15739.

Romier, C., J. E. W. Meyer, and D. Suck. 1997. Slight sequence variations of a common fold explain the substrate specificities of tRNA-guanine transglycosylases from the three kingdoms. *FEBS Lett.* 416:93–98.

Rould, M. A., J. J. Perona, D. Söll, and T. A. Steitz. 1989. Structure of *E. coli* glutaminyl-tRNA synthetase complexed with tRNA[Gln] and ATP at 2.8 Å resolution. *Science* 246:1135–1142.

Ruff, M., S. Krishnaswamy, M. Boeglin, A. Poterszman, A. Mitschler, A. Podjarny, B. Rees, J. C. Thierry, and D. Moras. 1991. Class II aminoacyl transfer RNA synthetases: crystal structure of yeast aspartyl-tRNA synthetase complexed with tRNA[Asp]. *Science* 252:1682–1689.

Shindo-Okada, N., N. Okada, T. Ohgi, T. Goto, and S. Nishimura. 1980. Transfer ribonucleic acid guanine transglycosylase isolated from rat liver. *Biochemistry* 19:395–400.

Short, S. A., S. R. Armstrong, S. E. Ealick, and D. J. T. Porter. 1996. Active site amino acids that participate in the catalytic mechanism of nucleoside 2'-deoxyribosyltransferase. *J. Biol. Chem.* 271:4978–4987.

Slany, R. K., M. Bösl, P. F. Crain, and H. Kersten. 1993. A new function of S-adenosylmethionine: the ribosyl moiety of AdoMet is the precursor of the cyclopentenediol moiety of the tRNA wobble base queuine. *Biochemistry* 32:7811–7817.

Slany, R. K., and H. Kersten. 1994. Genes, enzymes and coenzymes of queuosine biosynthesis in procaryotes. *Biochimie* 76:1178–1182.

Slany, R. K., and S. O. Mueller. 1995. tRNA-guanine transglycosylase from bovine liver: purification of the enzyme to homogeneity and biochemical characterization. *Eur. J. Biochem.* 230:221–228.

Walden, T. L., N. Howes, and W. R. Farkas. 1982. Purification and properties of guanine, queuine-tRNA transglycosylase from wheat germ. *J. Biol. Chem.* 257:13218–13222.

Watanabe, M., M. Matsuo, S. Tanaka, H. Akimoto, S. Asahi, S. Nishimura, J. R. Katze, T. Hashizume, P. F. Crain, J. A. McCloskey, and N. Okada. 1997. Biosynthesis of archaeosine, a novel derivative of 7-deazaguanosine specific to archaeal tRNA, proceeds via a pathway involving base replacement on the tRNA polynucleotide chain. *J. Biol. Chem.* 272:20146–20151.

Yokoyama, S., T. Miyazawa, Y. Iitaka, Z. Yamaizumi, H. Kasai, and S. Nishimura. 1979. Three-dimensional structure of hypermodified nucleoside Q located in the wobbling position of tRNA. *Nature* 282:107–109.

Yokoyama, S., and S. Nishimura. 1995. Modified nucleosides and codon recognition, p. 207–223. *In* D. Söll and U. RajBhandary (ed.), *tRNA: Structure, Biosynthesis, and Function*. American Society for Microbiology, Washington, D.C.

Chapter 10

Biosynthesis and Functions of Modified Nucleosides in Eukaryotic mRNA

JOSEPH A. BOKAR AND FRITZ M. ROTTMAN

mRNA in eukaryotic organisms contains a limited and well-defined spectrum of modified nucleosides. With the exception of the recent finding of deamination of specific adenosine and cytosine residues in several mRNAs to form inosine and uracil residues, respectively (discussed in detail in Chapter 18 by Chang et al. and in Chapter 19 by Rueter and Emeson), all of the nucleoside modifications in mRNA involve methylation events. The terminal guanine residue, first and second transcribed nucleosides, and specific internal residues can be methylated at positions either on the base or on the ribose moiety. As is the case for most nucleoside modifications in rRNA, tRNA, and snRNA, the functional significance of these methylated nucleosides is not well understood. However, studies using methylation inhibitors suggest that mRNA methylation affects the efficiency of pre-mRNA splicing, and possibly transport (Bachellerie et al., 1978; Stoltzfus and Dane, 1982; Finkel and Groner, 1983; Camper et al., 1984; Carroll et al., 1990). This review focuses on the sequence-specific methylation of internal adenosine residues to form N^6-methyladenosine. Recent advances in the characterization and cDNA cloning of a critical subunit of the mRNA N^6-adenosine methyltransferase are highlighted. Interesting features of the enzyme include its multicomponent nature, the presence of methyltransferase consensus domains within the S-adenosylmethionine (SAM)-binding subunit, and colocalization with splicing factors in nuclear speckles.

The production of a mature mRNA in eukaryotic cells involves transcription of a gene by RNA polymerase II, the synthesis of the 5'-terminal cap, the addition of the 3'-polyadenylic acid terminus, removal of intron sequences, and in some instances, the specific alteration of nucleotide sequences by RNA editing (as described in Chapters 16 to 19 and 22). With the possible exception of mRNA editing, these steps in mRNA biogenesis appear to occur in a concerted fashion, linked both temporally and spatially with transcription (for a review, see Steinmetz, 1997). Modification of nucleosides, both within the cap structure and at positions within the body of the pre-mRNA, occurs after transcription. With the exception of the deamination of adenosine and cytosine residues at specific sites, all other nucleoside modifications in mRNA involve methylation of either the ribose or base moieties. Unlike the wide array of nucleoside modifications described in rRNA (Maden, 1990, and Appendix 1) and tRNA (Björk et al., 1987, and Appendix 1), there is a more limited set of modified nucleosides in mRNA (Limbach et al., 1994). These can be separated into two groups—those found within the 5'-terminal cap, and those within internal regions of the mRNA. This review will focus on those modifications present in cellular mRNA; the nucleoside modifications present in virion RNA and viral mRNA will also be discussed because of the nearly complete overlap of the spectrum of modifications present. Much of what is known about the location and potential functions of these modifications comes from studies of viral systems. It is also important to note that unlike mRNA from higher organisms, prokaryotic mRNA does not appear to contain nucleoside modifications.

MODIFIED NUCLEOSIDES PRESENT IN THE 5'-TERMINAL CAP STRUCTURE

The general structure of the mRNA cap is $m^7GpppN1pN2pN\ldots$ (Fig. 1), where the terminal

Joseph A. Bokar and Fritz M. Rottman • Department of Molecular Biology and Microbiology, and Department of Medicine, Case Western Reserve University, Cleveland, Ohio 44106.

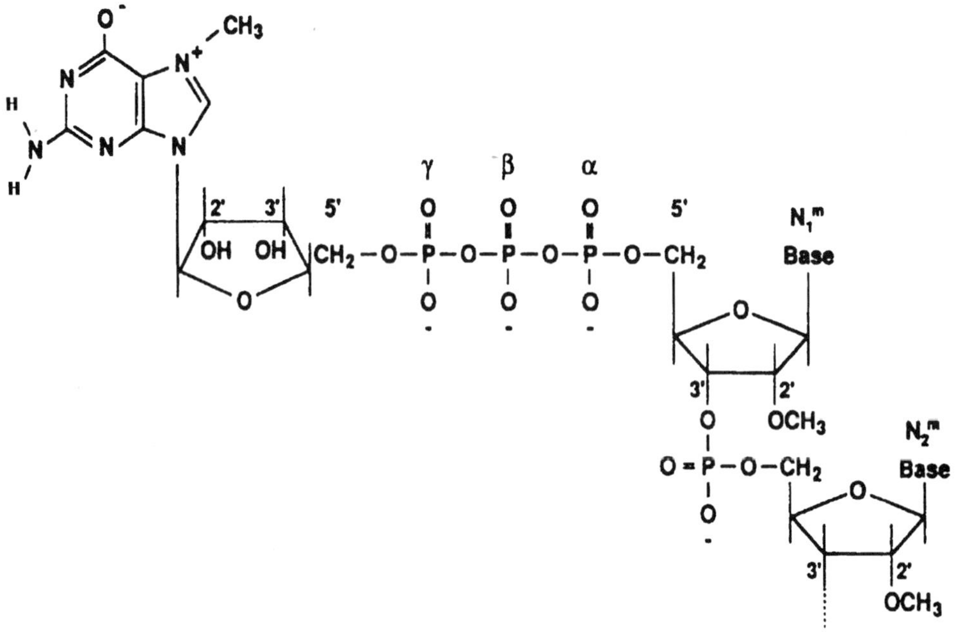

Figure 1. Chemical structure of eukaryotic mRNA cap. The general structure of a typical mRNA cap is shown. The terminal N^7-methylguanosine is present linked to the first transcribed nucleoside (N_1) via a 5′-5′-triphosphate bridge. If neither N_1 nor N_2 is methylated at the 2′-O position, the structure is designated cap 0. If N_1 is methylated at the 2′-O position, the structure is designated cap 1. If N_1 and N_2 are methylated at the 2′-O position, the structure is designated cap 2.

guanine is invariably methylated at the N-7 position on the purine ring. N_1, which can be any of the four nucleosides, is generally methylated on the 2′-O position of the ribose ring, and N_2 can also be methylated in the same fashion (Rottman et al., 1974; Wei et al., 1975; Shatkin, 1976; Banerjee, 1980; and reviewed by Reddy et al., 1992). Cap structures containing only 7-methylguanine are designated cap 0, those containing a 2′-O-methylnucleoside at position N_1 are designated cap 1, and if both N_1 and N_2 are 2′-O-methylnucleosides the structure is designated cap 2. Furthermore, if N_1 is 2′-O-methyladenosine, it can also be further methylated at the N-6 position of the purine ring. Both cap 1 and cap 2 structures predominate in higher eukaryotic mRNAs, with cap 1 being found with five times the frequency of cap 2 in mammalian mRNA (Perry and Kelley, 1976). Most animal virus mRNAs also contain caps 1 and 2 (Banerjee, 1980), as does Drosophila mRNA (Levis and Penman, 1978). Lower eukaryotes such as yeast (Sripati et al., 1976) and Neurospora (Seidel and Somberg, 1978) contain only cap 0. Collectively, these cap methylation events account for the presence of most of the nucleoside modifications in eukaryotic mRNA, namely, 7-methylguanine (m^7G), all four 2′-O-methylnucleosides (Nm), and N^6,2′-O-dimethyladenosine (m^6Am).

In addition to the methylated nucleosides previously discussed, other more extensively methylated nucleosides in trypanosomatid and Caenorhabditis elegans mRNA caps have been described. C. elegans mRNAs that result from the trans-splicing of an SL RNA to the pre-mRNA contain a terminal 2,2,7-trimethylguanosine cap that is originally synthesized on the separately transcribed SL RNA (Liou and Blumenthal, 1990; Van Doren and Hirsh, 1990). In the case of Trypanosoma brucei and Crithidia fasciculata (both kinetoplastid protozoa), a more extensively methylated Cap 4 structure has been described with the structure $m^7GpppNm^6_2AmpAmpCmpm^3Ump\ldots$ where m^6_2Am is N^6,N^6,2′-O-trimethyladenosine, and m^3Um is 3,2′-O-dimethyluridine (Bangs et al., 1992). This novel mRNA cap structure is also the result of the trans-splicing of a preformed 5′ terminus from an snRNA (termed the medRNA) to the pre-mRNA.

BIOLOGICAL FUNCTION OF METHYLATED NUCLEOSIDES WITHIN THE CAP STRUCTURE

The 7-methylguanosine cap is important for a number of biochemical processes involving mRNA. The most studied of these is the effect of the cap on translation initiation. The m^7G cap structure enhances mRNA translation in Xenopus oocytes, in in vitro translation systems, and in cells directly transfected with mRNAs (for a review, see Merrick and

Hershey, 1996). The rate limiting step for initiation is the binding of the mRNA to the 40S ribosomal subunit. This appears to be facilitated by the binding of cap-binding protein (CBP, also called eIF4E) to the m^7G cap. eIF4E then can be coupled to the 40S subunit via a bridging interaction with eIF4G and eIF3 (for a review, see Hentze, 1997). Translation can occur in the absence of a cap; however, the efficiency is markedly diminished. It has been shown in in vitro systems that an unmethylated guanine cap is not sufficient for the translational effect (Muthukrishnan et al., 1975). N-7-ethyl-substituted guanosine cap was equivalent in activity to m^7G, suggesting that alkylation, and not methylation per se, at the N-7 position is the important determinant for increasing translation efficiency (Furuichi et al., 1979). 2,7-Dimethylguanosine was more active than m^7G at stimulating translation, while 2,2,7-trimethylguanosine was less active (Darzynkiewicz et al., 1988). A mechanism has been proposed for the effect of the N-7-substitution based on studies using substituted cap analogs as inhibitors of translation. The positive charge on the imidazole of the N-7-alkylated purine ring may interact with the phosphate oxygens, resulting in a conformer that can then efficiently interact with the cap-binding protein (Adams et al., 1978; Rhoads et al., 1983).

More recently the cap ribose methylations have also been shown to be an important regulator of translation efficiency in a *Xenopus* oocyte system. Kuge and Richter (1995) demonstrated that the increase in translation efficiency of specific mRNAs, mediated by progesterone-induced polyadenylation of the mRNAs, involves stimulation of cap ribose methylation. They demonstrated that upon progesterone stimulation of oocytes injected with a cap 0-containing reporter mRNA, an increase in poly(A) tail length and conversion of cap 0 to cap 1 and 2 structures occur in a coordinated fashion. They also observed a simultaneous increase in translation of the reporter mRNA. Treatment of the oocytes with cordycepin, which blocks polyadenylation, led to inhibition of cap methylation and a lack of translation of the mRNA. When a methylation inhibitor was used, polyadenylation was not affected, while ribose methylation was eliminated. In this setting translation was also inhibited, suggesting that the increased translation efficiency seen with progesterone induction was mediated at least in part by conversion of cap 0 to cap 1 and 2 structures (that is, via ribose methylation) in concert with polyadenylation. The stimulatory effect of polyadenylation on translation efficiency has been described in a number of other systems, including models of development in mice (Huarte et al., 1987) and *Drosophila* (Salles et al., 1994). It is important to note that the poly(A) tail itself has also been shown to increase the efficiency of translation initiation (Hentze, 1997). This likely occurs via a bridging interaction between the preinitiation complex and the poly(A) tail mediated by eIF4G and poly(A) binding protein. Whether an additional mechanism involving cap ribose methylation occurs generally in these systems is not known. However, it is interesting that in vaccinia virus, the poly(A) polymerase activity and 2'-O-ribose methyltransferase activities are both contained on a single heterodimeric enzyme (see the following paragraphs), suggesting that the interplay of polyadenylation and cap ribose methylation may be functionally linked in other systems.

The m^7G cap also functions to decrease the susceptibility of mRNA molecules to 5' exoribonuclease activity, and is involved in both splicing and transport. Izaurralde et al. (1994) have characterized a nuclear cap-binding complex that is required for efficient splicing of adenoviral pre-mRNA in vitro. Free m^7G and 7-ethylguanosine cap analogs both efficiently inhibited splicing in vitro. In contrast, $m_2^{2,7}GpppG$, $m_2^{2,2,7}GpppG$, and ApppG were not effective inhibitors, suggesting that the methylation status of the guanosine is important for the affinity of the cap binding complex. The effect of unmethylated cap analogs could not be definitively determined because the nuclear extracts contained cap-methylating activity. The monomethylated cap has also been shown to be important for transport of mRNA from the nucleus to the cytoplasm (Hamm and Mattaj, 1990). In contrast, di- and trimethylguanosine caps prevent transport of RNA from the nucleus.

ENZYMES INVOLVED IN CAP METHYLATION

The cellular enzymes that catalyze methylation of the nucleosides comprising the cap structure have been partially purified and characterized, mainly by B. Moss and coworkers. Four separate activities were identified in HeLa cell extracts: RNA (guanine-7-)-methyltransferase (m^7G methyltransferase), cap 1- and cap 2-specific 2'-O-methyltransferases (Nm methyltransferases), and 2'-O-methyladenosine-N^6-methyltransferase (m^6Am methyltransferase) (for a review, see Mizumoto and Kaziro, 1987). RNA (guanine-7)methyltransferase was purified 165-fold and had an apparent molecular mass of 56 kDa (Ensinger and Moss, 1976). The enzyme was specific for the guanosine involved in the terminal 5'-dinucleoside triphosphate in capped RNA substrates, and could also methylate G(5′)pppG, but not GTP, GDP, or G(5′)pppG. To date, further characterization and cDNA cloning of the enzyme have not been reported. Two separate activities were partially purified that catalyze the 2′-O-methylation of the first (cap 1) and second (cap 2) transcribed nucleosides (Langberg and Moss, 1981). These were not extensively characterized, but the data suggest that cap 1 methyltransferase can utilize m^{7}G(5′)pppN or G(5′)pppN as a substrate with equivalent efficiency in vitro. The enzyme appears to fractionate as a nuclear protein during the early steps of purification. The cap 2 methyltransferase activity, which appears to fractionate as a cytoplasmic enzyme, methylates N$_2$ only when N$_1$ is already methylated. The fourth cap methyltransferase, also described by Moss and colleagues, is (2′-O-methyladenosine-N^6)-methyltransferase (Keith et al., 1978). This enzyme was purified >340-fold from HeLa cytoplasmic extracts and had a molecular mass of 65 kDa. This partially purified enzyme fraction had a marked preference for m^{7}G(5′)pppA$_m$, and was much less efficient at methylating m^{7}G(5′)pppApN or G(5′)pppA$_m$pN. This enzyme preparation was not able to methylate internal adenosine residues, suggesting that it is distinct from the internal mRNA N^6-adenosine methyltransferase (m^{6}A-MT).

Compared to the cellular cap methyltransferases, the vaccinia virus capping enzyme and cap methyltransferases have been studied in greater detail (Martin et al., 1975; Higman et al., 1992, 1994a, 1994b; Cong and Shuman, 1992; Myette and Niles, 1996). The vaccinia virus capping enzyme is a heterodimeric protein composed of a 97-kDa and a 33-kDa subunit. The RNA triphosphatase, nucleoside triphosphate phosphohydrolase, and guanylyltransferase activities have been localized to a 60-kDa peptide from the amino-terminal portion of the large subunit, while the (N^7-guanine-)methyltransferase activity is contained within the 33-kDa subunit and the carboxyl one third of the large subunit. The 2′-O′methyltransferase that converts the vaccinia virus cap 0 structure to cap 1 is contained within another bifunctional protein, VP39 (Shi et al., 1996). In addition to methyltransferase activity, VP39 with VP55 comprises the heterodimeric vaccinia virus poly(A) polymerase. The biochemical linkage of polyadenylation at the 3′-end with ribose methylation within the 5′-cap is intriguing in the context of the functional linkage of these two events as previously described for oocyte mRNA.

MODIFIED NUCLEOSIDES PRESENT WITHIN INTERNAL REGIONS OF EUKARYOTIC mRNA

Three modified nucleosides have been reported within internal regions of eukaryotic mRNA: N^6-methyladenosine (m^{6}A), 5-methylcytidine (m^{5}C), and inosine. In addition, deamination of cytosine to yield uridine residues must also be considered a nucleoside modification, although the resulting nucleoside is not a "minor" nucleoside. The most studied and the most abundant of these is m^{6}A, which is the major focus of this review, following brief discussions of inosine, uridine, and m^{5}C.

Inosine

The most recently described nucleoside modification in eukaryotic cellular mRNA involves the conversion of specific adenosine residues to inosine residues, catalyzed by a double-stranded RNA adenosine deaminase (dsRAD) (see Chapter 19 by Rueter and Emeson; Hough and Bass, 1997; Kim et al., 1994). The specificity of this modification is directed by intron encoded "editing site complementary sequences" that form intramolecular dsRNA regions with the target exonic sequences to be edited. Several groups recently have reported evidence that mRNAs encoding subunits of glutamate receptor channels in mammalian brain and the hepatitis delta virus antigenome RNA, undergo posttranscriptional editing in this way (Melcher et al., 1995; Polson et al., 1995). In the case of the glutamate receptor channels, these edited sites result in altered coding sequences and altered ion permeability of the resulting receptor activated channels. The conversion of internal adenosine residues to inosine may turn out to be a more widespread mechanism of posttranscriptional regulation of gene expression. This topic is presented in detail in Chapter 19 by Rueter and Emeson.

In a somewhat analogous fashion, specific cytidine residues can also undergo deamination to yield uridine residues. This has been studied mainly for the apolipoprotein B mRNA, and appears to control the switch between expression of the ApoB-100 and ApoB-48 isoforms. The details of this modification and characterization of the responsible enzyme (Apobec-1) are presented in Chapter 18 by Chang et al.

5-Methylcytidine

The presence of small amounts of m^{5}C in Sindbis viral mRNA, adenovirus 2 mRNA, and cellular mRNA from cultured hamster (BHK-21) cells was reported in early studies (Dubin and Stollar, 1975; Dubin and Taylor, 1975; Dubin et al., 1977). Interestingly, m^{5}C was not detected in mRNA from Novikoff hepatoma cells (Desrosiers et al., 1974, 1975), HeLa (Wei et al., 1976; Wei and Moss, 1977), or mouse L cells (Perry et al., 1975). The significance of this discrepancy is unclear, and further studies of m⁵C content in mRNA from various sources have not been reported. The possibility that the m⁵C detected represented contamination of the mRNA fraction with other RNA species cannot be ruled out, because the RNA studied was not double-selected with oligo(dT)-cellulose, as was the case for the studies in the other cell lines. Studies addressing neither the specific sequences methylated nor the functional significance of m⁵C in cellular mRNA have been reported. It also remains possible that low levels of m⁵C are present in mRNA as a result of the activity of tRNA m⁵C methyltransferase. Grosjean and coworkers have shown that enzymes capable of catalyzing m^5C_{40} and m^5C_{49} formation in yeast tRNA^Phe are independent of the three-dimensional structure of the substrate and can act on fragments of the tRNA (Jiang et al., 1997; Grosjean et al., 1997; Grosjean et al., 1996). It might well be that the same methyltransferase acts on mRNA.

N⁶-Methyladenosine

In contrast to the apparent infrequent occurrence of m⁵C, inosine, and deaminated cytidine in cellular mRNA, m⁶A is present in easily detectable amounts in mRNA isolated from all higher eukaryotes tested, including plants (Nichols, 1979), mammals (Desrosiers et al., 1975; Perry and Scherrar, 1975; Perry et al., 1975; Furuichi et al., 1975; Wei et al., 1976; Adams and Cory, 1975), and *Drosophila* (Levis and Penman, 1978). mRNAs isolated from lower eukaryotes such as *S. cerevisiae* (Sripati et al., 1976), *Dictyostelium* (Dottin et al., 1976), and *Neurospora* (Seidel and Somberg, 1978) do not contain detectable amounts of m⁶A. This modification is present in viral RNAs that replicate in the nucleus (adenovirus, simian virus 40 [SV40], herpesviruses, Rous sarcoma virus, and influenza virus) (Somner et al., 1976; Aloni et al., 1979; Canaani et al., 1979; Bartuski and Roizman, 1978; Beemon and Keith, 1977; Narayan et al., 1987), but not in the RNA from viruses that do not have a nuclear phase in their life cycle (Sindbis virus, vaccinia virus, reovirus, and vesicular stomatitis virus) (for a review, see Narayan and Rottman, 1992). Analysis of the methylated constituents of mouse L cell, Novikoff hepatoma cell, and HeLa cell mRNA has revealed that approximately 50% of incorporated methyl label is in m⁶A, making it the most abundant modified nucleoside in mRNA. The average content of m⁶A has been estimated to be 3–5 residues per mRNA in Novikoff, HeLa, mouse L cell, and Chinese hamster ovary (CHO) cell mRNA. The number of m⁶A residues in viral RNAs ranges from 1–15 per RNA, depending on the virus. Although m⁶A is present at an average of 3–5 residues per mRNA in mammalian cells when total polyadenylated RNA is analyzed, its distribution among individual mRNA is not uniform. Examples of m⁶A abundance in specific cellular mRNAs are approximately one residue in bovine prolactin mRNA (Horowitz et al., 1984) and three residues in mouse dihydrofolate reductase mRNA (Rana and Tuck, 1990). Some cellular mRNAs, including bovine growth hormone (Horowitz et al., 1984) and mouse globin (Perry and Scherrer, 1975), do not contain detectable amounts of m⁶A, suggesting that this modification may not be an absolute requirement for mature mRNA function.

Sequence Specificity of m⁶A In Vivo

The distribution of m⁶A within an RNA molecule is nonrandom, and it is not found within the poly(A) tract (Desrosiers et al., 1975; Perry et al., 1975). m⁶A occurs only within sequences matching PuAC (where A is the methylated residue) in all organisms tested including plants and mammals (Schibler et al., 1977; Wei and Moss, 1977; Nichols, 1979; Bartkuski and Roizman, 1978; Dimock and Stoltzfus, 1977). Using a variety of specific ribonucleases in combination with column and thin-layer chromatography (TLC), Schibler et al. (1977) analyzed *methyl*-³H-labeled mRNA from mouse L cells. They demonstrated that there is a preferred five nucleotide consensus sequence, PuPuACH (where H = A, C, or U) for methylation. Based on the frequency with which this degenerate sequence occurs and the estimated m⁶A content of mRNA, it is clear that only a minority of sequences matching the consensus actually are methylated. This suggests that there are other constraints in addition to the primary sequence that affect m⁶A formation.

Specific m⁶A sites have been mapped in only two individual mRNAs. Rous sarcoma virus (RSV) virion RNA contains 12 m⁶A residues per RNA molecule of 9,500 nucleotides (Beemon and Keith, 1977). A transformation deficient mutant of RSV, lacking the *src* sequence, contains only 7 m⁶A residues, suggesting that 5 m⁶A residues are contained within the *src* sequence. In contrast, the *gag* and *pol* sequences did not contain m⁶A. The individual sites were further localized by hybrid selection of RNA fragments, followed by RNase T₁ digestion and two-dimensional fingerprinting. Thirteen m⁶A sites were precisely mapped (Kane and Beemon, 1985; Csepany et al., 1990). All of these sites matched the sequence PuGACU, in agreement with the consensus sequence suggested by the analysis of heterogenous mRNA. An important feature of m⁶A occurrence in viral and cellular mRNA was also documented in these studies; the modification occurs nonstoichiometrically, that is, individual sites are methylated on only a portion of the mRNA molecules, ranging from 20–90% for these RSV sites.

The only cellular mRNA for which individual m^6A sites have been mapped is that encoding bovine prolactin (bPRL) (Horowitz et al., 1984; Narayan and Rottman, 1992). Unlike viral mRNA, labeling steady-state mRNA populations in eukaryotic cells with ^{32}P-orthophosphate or [^3H-*methyl*]methionine results in the incorporation of low amounts of radioactivity into individual mRNAs. Therefore, for these studies, unlabeled bPRL mRNA from pituitary was first purified by hybrid selection, followed by T_1 nuclease digestion. Individual T_1 fragments were then isolated and digested to 3′-mononucleotides, and the resulting nucleotides were then 5′-labeled in vitro with [γ-^{32}P]ATP and T4 polynucleotide kinase. The 3′ phosphates were removed using P1 nuclease, and the resulting nucleotides were analyzed for the presence of m^6A by TLC. Hybrid selection of fragments of the bPRL mRNA localized the predominant m^6A site to the 3′-terminal 130 nucleotide fragment. By digestion of this hybrid-selected fragment with RNase T_1 and analysis of each T_1 fragment for the presence of m^6A, the m^6A site was localized to an AGACU consensus sequence in the 3′ untranslated region. The extent of methylation at this site was estimated to be approximately 20%, also illustrating the lack of stoichiometry of m^6A formation at individual sites.

In Vitro System for Sequence Specific Formation of m^6A

The development of a cell-free system capable of accurately catalyzing N^6-adenosine methylation led to the ability to further study the methyltransferase activity and the RNA features that were required for an efficient methylation substrate (Narayan and Rottman, 1988). Briefly, nuclear extracts prepared from cultured HeLa cells were used as an enzyme source. The substrates were synthetic RNA fragments derived from the bPRL mRNA sequence and [^3H-*methyl*]S-adenosylmethionine ([^3H-*methyl*]SAM). Using this system, the same adenosine (within the AGACU consensus) was methylated on the synthetic transcript as was shown to be methylated in vivo. Harper et al. (1990) examined short (20 nucleotide) synthetic RNA substrates containing single or multiple consensus methylation sequences, or single nucleotide mutations derived from the consensus sequence. This study confirmed that the in vitro specificity closely paralleled the frequency with which sequences are methylated in vivo as originally described by Schibler (1977).

From the localization studies of total cellular mRNA, and individual mRNA and the results of the in vitro studies previously described, it appears that the five nucleotide consensus sequence is the primary determinant for m^6A formation. However, because the expected frequency of the degenerate five nucleotide consensus methylation sequence far exceeds the m^6A content of mRNA, and because identical sequences are methylated to varying degrees in vivo (for example, in RSV), the effects of RNA context were further explored. Two regions of PRL mRNA were examined in vitro (Narayan et al., 1994): the major methylation site (AGACU) found in the 3′ untranslated region and the 5′-terminal 192 nucleotides of bPRL mRNA, which normally is unmethylated. Mutation of the normally strong AGACU sequence in the 3′ untranslated region of PRL mRNA to the generally weaker AAACU sequence surprisingly resulted in enhanced methylation at this site, whereas mutation of the wild-type AGACU to several nonconsensus sequences resulted in methylation of a nearby cryptic AAACA site, located adjacent to the normal site of methylation, to levels similar to the wild-type sequence. The same series of substrate sequences was placed in a different context, namely, the 5′ 192 nucleotide bPRL fragment. Interestingly, a different relative order of efficiency with which the individual sequences were methylated was observed when in the context of the 5′ fragment versus the normal 3′ site. Finally, a synthetic 20 nucleotide fragment of RNA complementary (antisense) to the 3′-terminal PRL methylation site was used in the in vitro system to demonstrate that a consensus m^6A site present in duplex structure was incapable of being methylated.

This series of experiments demonstrates a number of interesting points regarding the RNA substrate. The nucleotide sequence of the substrate is degenerate, as was expected. The efficiency with which a given sequence is methylated appears to depend on both the primary sequence and the context within which the sequence occurs. A sequence that normally is methylated efficiently cannot be methylated when the consensus region is in a stable RNA duplex. The abolishment of a methylation site by mutation of the potentially modified adenosine residue can lead to the enhanced methylation of adjacent consensus sites that are otherwise not methylated. These data suggest that methylation of a strong site somehow inhibits additional methylation of the adjacent weaker sites. The mechanism by which this occurs is unknown, but may explain how the loss of a methylation site by site-directed mutagenesis may lead to compensatory use of an adjacent cryptic site. This phenomenon has been observed for bPRL mRNA sequences both in vitro (as previously discussed) and in vivo, as discussed later in this chapter.

There are no published reports of studies directly testing the hypothesis that RNA secondary structure is a determinant for an efficient m⁶A site. However, examination of the sequence surrounding the major m⁶A site in bPRL mRNA (Fig. 2A) and two sites in RSV RNA (Fig. 2B and C) has suggested that each of these methylation sites may be displayed on the loop of a stem-loop structure. Regions of RNA sequence surrounding these sites were analyzed for potential secondary structures by computer using the method of Zuker (Jaeger et al., 1989a, 1989b). In each instance evaluated, the m⁶A site is predicted to lie within the loop of a stem-loop structure. Although it is important to note that this prediction has not been experimentally verified, such structural features could contribute to the context effects previously described.

FUNCTION OF m⁶A IN mRNA

Despite two decades of study of m⁶A in mRNA, the biological function of this ubiquitous posttranscriptional modification remains unclear. Two general approaches have been used by several groups in an effort to define the role of m⁶A in mRNA biogenesis and function. First, mutations of the major methylation sites within RSV and bPRL derived expression constructs have been studied to determine the effects on mRNA biogenesis. Second, cells grown in culture have been treated with methylation inhibitors to explore the effects of mRNA undermethylation. These experiments suggest that m⁶A affects the efficiency of pre-mRNA splicing or transport of mRNA from the nucleus to the cytoplasm. However, for reasons to be discussed, inherent limitations of both lines of experimentation have led to somewhat inconclusive results.

Mutation of m⁶A Sites

Studies on the location of m⁶A in RSV RNA had shown that there was a cluster of seven specific m⁶A sites in the region of the *src* and *env* genes (Beemon and Keith, 1977). Kane and Beemon (1987) generated a mutant virus containing point mutations at two of these sites located just downstream of the *src* splice acceptor site, and just upstream of a potential or "cryptic" splice acceptor site. Mutation of two GAC sites to GAU sequences rendered them incapable of being methylated. These mutations did not lead to

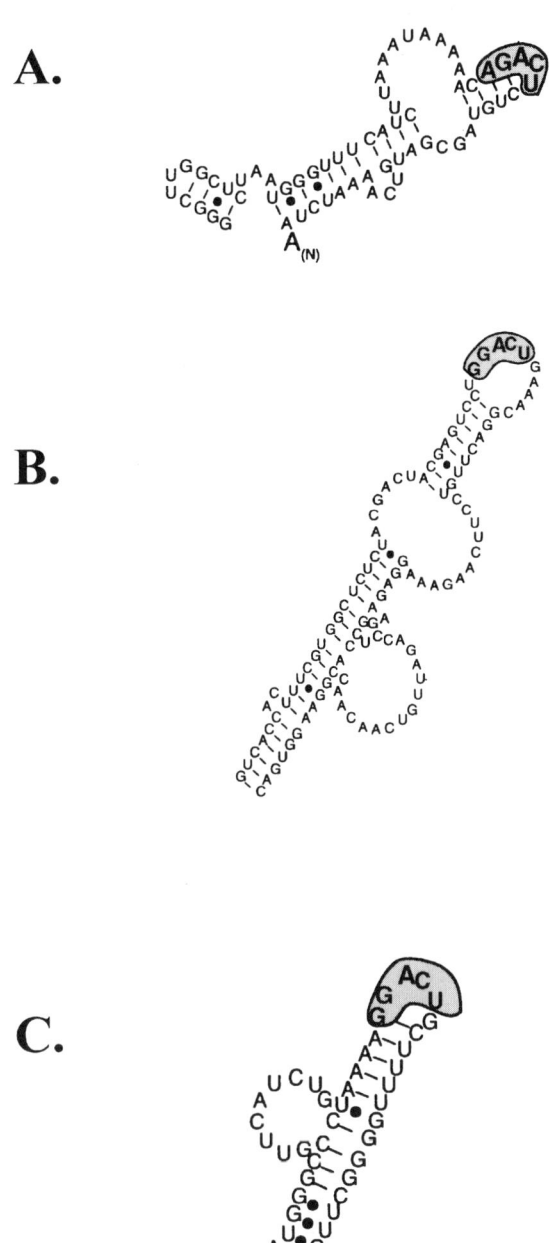

Figure 2. Potential secondary structure of mRNAs containing m⁶A sites. Partial sequences of mRNA from bPRL and RSV surrounding selected m⁶A sites were analyzed for potential secondary structures using the method of Zuker (1989). The methylation consensus sequences are highlighted in gray; the methylated adenosyl residue is shown in larger type. (A) 62 nucleotides of sequence corresponding to the 3′-terminus of the bPRL mRNA. This is also the sequence of the RNA used in in vitro assays for the purification of m⁶A-MT (Bokar et al. 1994, 1997). (B) 105 nucleotides of sequence corresponding to the first methylation site in the RSV *src* gene at position 7414 (Kane and Beemon, 1985). (C) 42 nucleotides of sequence corresponding the methylation site at position 6718 in the RSV *env* gene (Kane and Beemon, 1985).

any differences in steady-state levels of *src* mRNA in mutant versus wild-type virus in infected CEF cells. No differences in the levels of *src* mRNA in the nuclear fraction were detected either, and no unusual spliced products were observed. Furthermore, there was no difference in the translation of the *src* protein, packaging of RNA into virions or infectivity of mutant virus compared to wild type. Csepany et al. (1990) extended this study by evaluating an additional two m^{6}A sites in a similar fashion. Mutation of the four major m^{6}A sites in the *src* gene did not have an appreciable effect on steady-state levels of viral RNA nor on viral infectivity. Several explanations for the lack of an effect of these methylation mutations have been proposed. If methylation affects the kinetics of processing or transport, it may be necessary to study only newly synthesized mRNA by pulse-labeling studies to observe the effect. Alternatively, the multiple methylation sites may be redundant for function; therefore, all would need to be removed for an effect to be observed. This level of mutation was not studied. Finally, compensatory increases in methylation at nonmutated sites may have occurred in the mutant viral mRNA, but were not detected.

Carroll et al. (1990) documented the presence of m^{6}A within an intron-encoded region of an RNA transcribed from a bPRL minigene. This minigene contained the last two exons and intervening intron and was expressed in stably transfected CHO cells. Following hybrid selection of this RNA, analysis of RNase T$_1$ generated oligonucleotides from the intron-encoded region revealed the presence of m^{6}A in three oligonucleotides containing sites matching the five nucleotide consensus sequence. In subsequent studies, these methylation sites were mutated to determine if any alteration in splicing pattern or efficiency would result from hypomethylation of the intron. Unfortunately, these mutations led to an apparently compensatory increase in methylation of nearby consensus sequences that were not normally methylated (Kienzle and Rottman, unpublished results). Therefore, it was impossible to study the effect of hypomethylation of intron D using this approach. This line of experimentation was not pursued, in favor of developing more specific and complete methods for inhibiting m^{6}A formation.

Studies Utilizing Methylation Inhibitors

A series of studies evaluating the effects of methylation inhibitors on mRNA synthesis, translation, and stability is described in the literature. These studies collectively suggest that undermethylation affects the efficiency of pre-mRNA splicing or transport. However, these studies uniformly suffer from the limitation that the observed effects may be due to inhibition of methylation events other than m^{6}A in mRNA. In addition, they do not allow the discrimination of effects mediated by inhibition of mRNA cap versus internal methylation, even though partial selectivity for m^{6}A inhibition over cap methylation is reported with some inhibitors.

Bachellerie et al. (1978) studied the effect of cycloleucine treatment on CHO cell mRNA formation. Cycloleucine is a competitive inhibitor in vitro of methionine adenosyl methyltransferase, and rapidly blocks the synthesis of SAM in vivo. It acts as a potent and reversible inhibitor of nucleic acid methylation. In this study, high concentrations of cycloleucine were used during a 20-minute pulse-labeling with [^{3}H-*methyl*]methionine or [^{3}H]uridine. This treatment resulted in a 96% decrease in ^{3}H-methyl incorporation into polyadenylated RNA, and the decay of this hypomethylated polyadenylated RNA from the nucleus was dramatically prolonged during a subsequent chase period in the absence of the inhibitor. This suggested that methylation of mRNA affects the efficiency of processing or transport, but does not discriminate between the effect of inhibiting cap methylation versus internal methylation. The possibility that inhibition of methylation of molecules other than the polyadenylated RNA (that is, proteins and snRNAs) may be important for mRNA processing also could not be ruled out. The short window of exposure to the methylation inhibitor, however, makes it unlikely that the effects are mediated via inhibition of DNA, rRNA, or snRNA because of the presumed relatively long half-lives of these molecules.

Stoltzfus and Dane (1982) studied the effect of cycloleucine treatment on B77 avian sarcoma virus RNA. The concentration of cycloleucine used in these studies significantly inhibited m^{6}A and cap-associated 2'-O-methylation, but not m^{7}G formation. A striking difference was observed in the size distribution of the virus-specific RNA when compared with RNA from the untreated control cells. Subgenomic *env* RNA did not accumulate, whereas genome-length RNA accumulated in larger than normal amounts. An increase in synthesis of a protein encoded by the genome-length RNA (Pr76^{gag}) and a decrease in synthesis of a protein encoded by a spliced subgenomic RNA (gPr92^{env}) were also observed, consistent with the notion that alterations in RNA methylation ultimately can result in altered gene expression, by altering the efficiency of pre-mRNA splicing.

To dissect the effect of internal methylation from cap methylation, Finkel and Groner (1983) utilized lower concentrations of cycloleucine that decreased m^{6}A content by >90% with minimal effect on the extent of cap methylation. In infected BSC-1 cells, hypomethylated SV40 late mRNA accumulates in the cytoplasm at much lower levels. There was no measurable difference in the stability of the cytoplasmic mRNA; therefore, this study supported an effect of m⁶A on processing or transport.

Camper et al. (1984) utilized another inhibitor, S-tubericidinylhomocysteine (STH), which is a structural analog of SAM and a potent inhibitor of SAM-dependent methyltransferases. In this study, the effects of the inhibitor on cap methylation and internal methylation were closely monitored. At an inhibitor concentration at which internal m⁶A was inhibited by 80% and cap methylation was inhibited by 50–60%, there was no measurable change in the half-life of HeLa mRNA, but STH caused a significant lag in the time of cytoplasmic appearance of newly synthesized polyadenylated RNA (Fig. 3A). Carroll et al. (1990) extended this result by studying the effect of another methylation inhibitor, neplanocin, on the processing of a specific mRNA transcribed from a transfected minigene. A stable cell line was generated that expressed a transfected bPRL minigene, which contains a single intron. Treatment of these cells with the inhibitor resulted in increased levels of unspliced minigene pre-mRNA in the nucleus relative to untreated cells (Fig. 3B). The effects of the inhibitor on cap methylation relative to internal adenosine methylation were also determined, and found to be nearly identical to those reported by Camper et al. (1984) using STH.

In summary, these studies utilizing different methylation inhibitors in a variety of cell lines yielded fairly consistent results; that is, inhibition of methylation leads to an apparent accumulation of unspliced pre-mRNA in the nucleus. The studies do not differentiate between a block in splicing only and a block in both splicing and transport, and they do not definitively differentiate between effects at the level

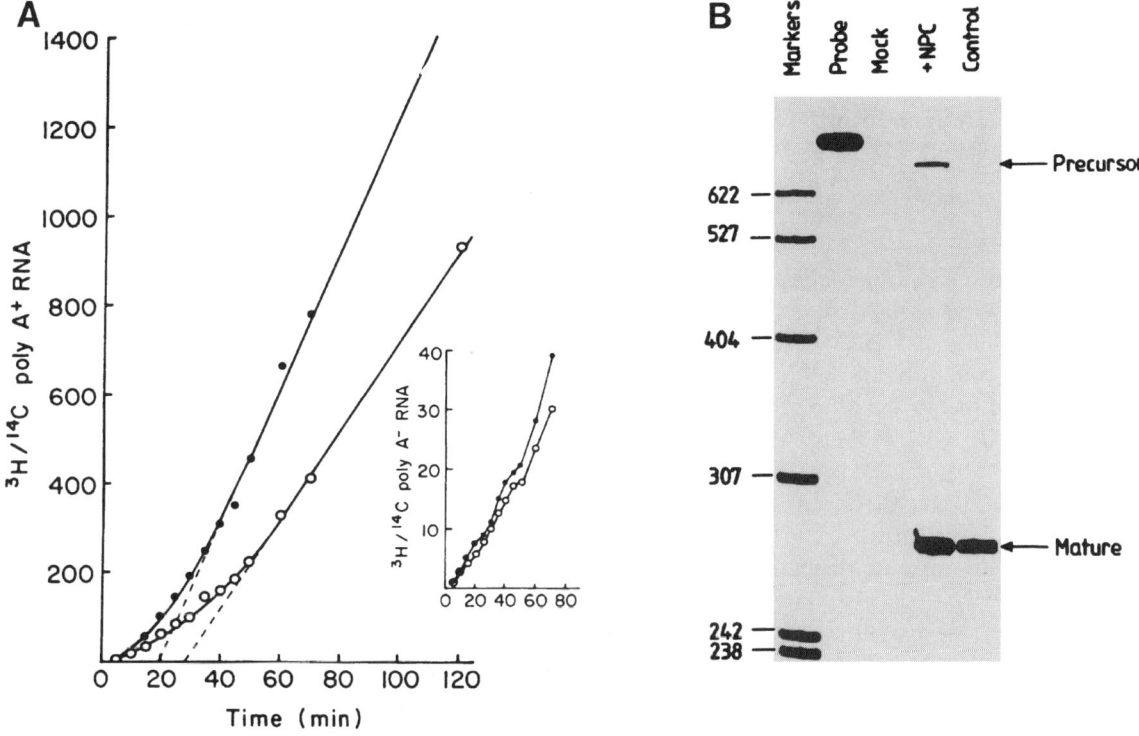

Figure 3. Effects of methylation inhibitors on the cytoplasmic appearance of newly synthesized RNA and on nuclear splicing of a bPRL derived pre-mRNA. (A) Effect of a methylation inhibitor on the cytoplasmic appearance of newly synthesized RNA. HeLa cells were prelabeled with a low level of [¹⁴C]uridine (0.48 mCi/mmol) for 12 h, treated with 500 μM STH (open circles) or no STH (closed circles) for 90 min, and then labeled with [³H]uridine (40 mCi/mmol). At various times after the addition of [³H]uridine, cells were placed on ice, and cytoplasmic poly(A)⁺ RNA and nonpolyadenylated RNA were prepared subsequently. The ratio of [³H]uridine to [¹⁴C]uridine in cytoplasmic poly(A)⁺ RNA and nonpolyadenylated RNA is shown. Appearance of newly synthesized polyadenylated RNA was delayed in STH treated cells as compared to control cells. No difference was seen for nonpolyadenylated RNA (inset). (B) Quantitative S1 nuclease mapping of nuclear bPRL precursor and mature-form RNA in stably transfected cells treated with the methylation inhibitor neplanocin (NPC). An autoradiogram of a gel showing the DNA fragments protected by 5 μg of total nuclear RNA isolated from cells treated with 10 μM NPC for 8 h (+NPC) and untreated cells (control). The unspliced (precursor) and spliced (mature) bPRL minigene transcripts are shown. The lane labeled mock contains probe that was hybridized in the absence of RNA followed by treatment with S1 nuclease.

of internal methylation versus cap methylation. Furthermore, each of these studies is also subject to the criticism that the observed effects could be mediated by hypomethylation of factors (protein or nucleic acid) other than mRNA, although the short time of exposure to the inhibitor in several of the studies (20–90 minutes) partially decreases this concern. It has been apparent for a number of years that further elucidation of the role of m⁶A in mRNA depended on developing new tools or techniques that would allow targeted disruption of m⁶A formation in mRNA. For this reason the focus has shifted toward characterization, purification, and cDNA cloning of a critical components of the mRNA m⁶A-methyltransferase.

CHARACTERIZATION OF mRNA N⁶-ADENOSINE METHYLTRANSFERASE FROM HeLa CELLS

A rapid assay for m⁶A formation and the development of an in vitro system for studying m⁶A formation in nuclear extracts facilitated the study of this methyltransferase (Narayan and Rottman, 1988; Narayan et al., 1994; Bokar et al., 1994). The assay is performed as follows: a 60 nucleotide segment from the 3'-end of the bPRL cDNA was subcloned into a T7 expression vector, containing a poly(A) tract, for in vitro production of ^{35}S-uridine-labeled and polyadenylated RNA substrate. This RNA substrate is incubated along with [^{3}H-*methyl*]SAM and a source of enzyme, and after incubation the RNA is separated from unreacted [^{3}H-*methyl*]SAM by binding to oligo(dT)-cellulose. The RNA is then eluted, the ^{3}H/^{35}S ratio is determined by scintillation counting, and the extent of methylation is determined for the recovered RNA. It has been shown by HPLC analysis that all of the counts per minute of incorporated ^{3}H are present in m⁶A.

An unanticipated result upon initial attempts at purification of the enzyme was the finding that multiple separable components are required for m⁶A-MT activity in vitro (Bokar et al., 1994). Fractionation of HeLa nuclear extract on a DEAE Sepharose column led to a flowthrough and a bound fraction, neither of which contained significant m⁶A-MT activity when assayed individually. Full activity was restored when an aliquot of each was included in the reaction mixture. The factor or component that is initially in the flowthrough fraction has been termed MT-A, while the factor that binds to the DEAE column has been named MT-B. MT-A and MT-B have both been further purified using a variety of chromatography columns. It initially appeared that the fraction containing MT-A could also be further fractionated into two separate and necessary factors. However, in subsequent experiments this was not always the case. Therefore, it is currently believed that there are two components to the m⁶A-MT enzyme that are separable using nondenaturing techniques.

Purification and cDNA Cloning of MT-A

The series of steps used for further purification and characterization of MT-A included anion exchange, cation exchange, gel filtration, and heparin affinity chromatography (Fig. 4A). Based on the sedimentation in glycerol gradients and its mobility on a size exclusion column, MT-A has an apparent molecular mass of 200 kDa (Bokar et al., 1994). Aliquots of column fractions that contained MT-A activity were treated by UV-cross-linking with [^{3}H-*methyl*]SAM and then analyzed by sodium dodecyl sulfate-polyacrylamide gel electrophoresis (SDS-PAGE) and fluorography. A labeled protein of 70 kDa was observed only in the fractions that contained MT-A activity, and its elution paralleled that of the enzymatic activity on multiple columns (Fig. 4B). This protein, termed MT-A70, was therefore a likely candidate for the SAM-binding subunit of the m⁶A-MT. MT-A70 was purified on a preparative scale using FPLC by following enzymatic activity and cross-linking activity, and then purified to homogeneity by SDS-PAGE. Tryptic peptide microsequence analysis was performed, yielding three peptide sequences. One of these was used to design a degenerate oligodeoxynucleotide probe, which in turn was used to isolate a cDNA clone. Recombinant protein expressed from this cDNA was used to generate specific antisera. These antisera can deplete m⁶A-MT activity from HeLa nuclear extract (Bokar et al. 1997), and recognize an immunoreactive protein on Western blots of column fractions that coelutes precisely with MT-A activity (Fig. 5). This series of findings has led to a model in which MT-A70 is a 70-kDa protein that contains a SAM-binding site, which in turn is a subunit of the 200-kDa factor MT-A. MT-A is necessary, but is not sufficient, for m⁶A-MT activity, as the presence of the MT-B component is also always required.

GenBank and Expressed Sequence Tag (EST) Database Homology Searches

The sequence of pMT-A70 was used to search for homologous genes in the nonredundant (nr) GenBank database using the basic local alignment search tool (BLAST) (Altschul et al., 1990). A search

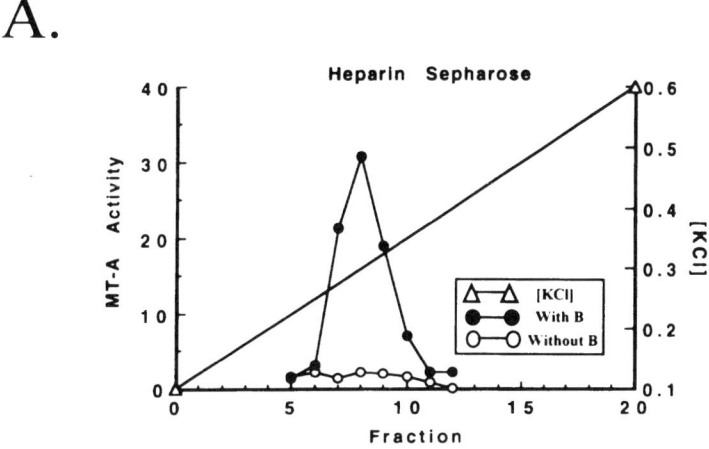

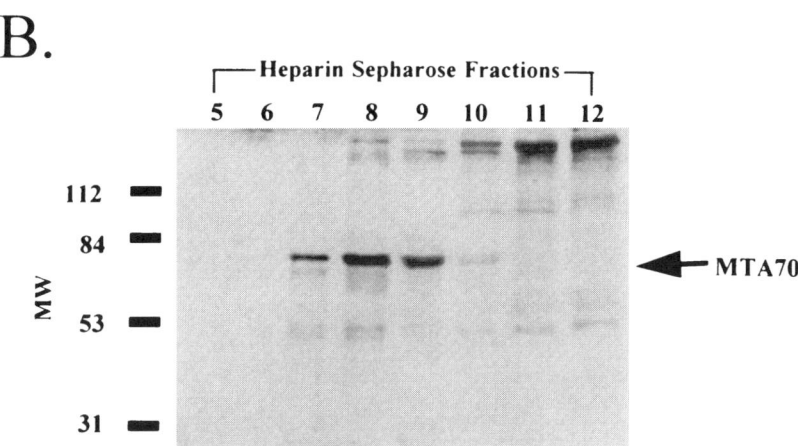

Figure 4. Purification of the MT-A70 subunit of mRNA m⁶A-MT. MT-A was purified as previously described (Bokar et al., 1994) by cation exchange, anion exchange, and size exclusion chromatography. The active fractions were pooled and further fractionated on a heparin-Sepharose column. (A) A representative activity profile of the heparin-Sepharose column fractions is shown. Closed circles represent the activity of aliquots of the fractions when assayed with supplemented partially purified fractions of MT-B; open circles are without MT-B. The KCl concentration units are molar. (B) [³H-*methyl*]SAM-UV-cross-linking. Aliquots of the same fractions assayed in panel A were cross-linked to [³H-*methyl*]SAM using UV light, and were then separated by SDS–8% PAGE, transferred to a polyvinylidene difluoride (PVDF) membrane, and fluorographed. The 70-kDa band that coelutes with MT-A activity is shown (MT-A70). For details of experimental methodology see Bokar et al., 1994 and 1997.

using the predicted amino acid sequence of MT-A70 resulted in the identification of two genes encoding proteins with significant amino acid sequence homology to MT-A70. The first, M. Mun1 methyltransferase ($P = 0.005$), is a DNA N⁶-adenine-specific methyltransferase isolated from a *Mycoplasma* species. Three homologous regions of 23–30 amino acids were identified containing from 47–61% conservation and 37–43% amino acid identity (Fig. 6A). The homologous regions encompass methyltransferase consensus motifs I and II that have been shown to be highly conserved among 69 prokaryotic DNA methyltransferases (Timinskas et al., 1995). Both the relative order and spacing of these sequences is preserved between M. Mun1 methyltransferase and MT-A70. This observation is consistent with the identification of MT-A70 as the SAM-binding subunit of an N⁶-adenosine methyltransferase, and suggests a functional conservation of this region in genes from widely divergent species. Double-stranded adenosine deaminase (discussed previously; see Chapter 19 by Rueter and Emeson) has also been reported to contain domains that are homologous to these same consensus methylation motifs (Hough and Bass, 1997). Although the significance of this finding is not yet known, it is interesting that two separate mRNA nu-

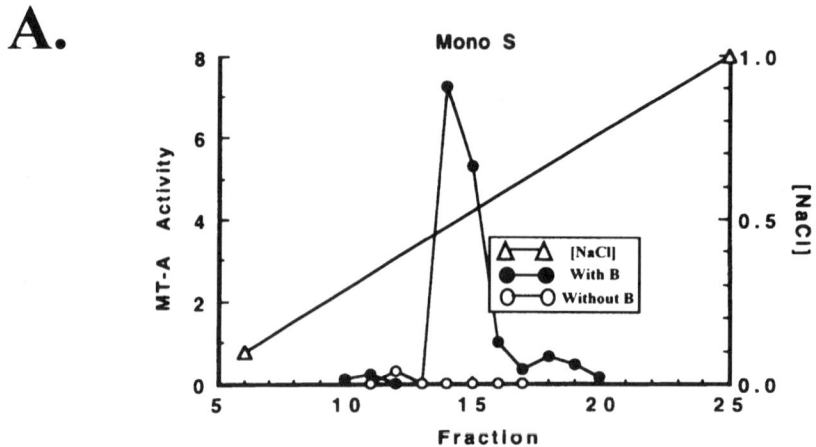

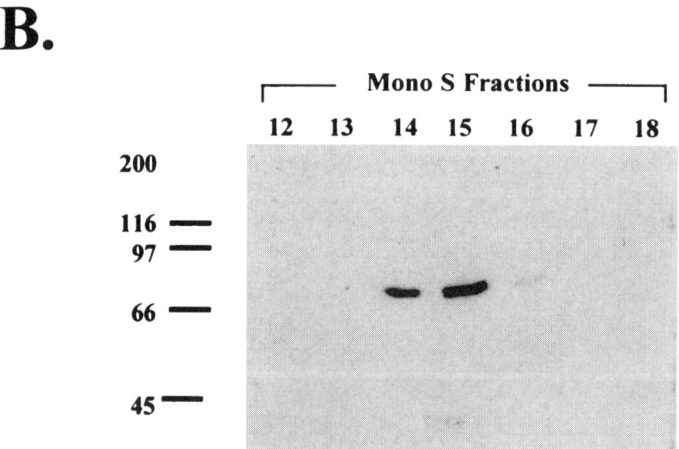

Figure 5. An anti-MT-A70 immunoreactive protein copurifies with MT-A activity. (A) A representative activity profile from a Mono S (cation exchange) column is shown. Closed circles represent the activity of aliquots of the fractions when assayed with supplemented partially purified fractions of MT-B; open circles are without MT-B. The NaCl concentration units are molar. MT-A activity elutes mainly in fractions 14 and 15. (B) Western blot analysis of Mono S column fractions. Aliquots of the fractions were separated by SDS-PAGE and transferred to a PVDF membrane. The membrane was incubated with anti-MT-A70 antisera and then a horseradish peroxidase conjugated secondary antibody, and bands were visualized by chemiluminescence and fluorography. A 70-kDa immunoreactive protein coelutes with methyltransferase activity. (For details of experimental methodology, see Bokar et al., 1997.)

cleoside modification enzymes, dsRAD and MT-A70, contain similar functionally conserved motifs.

The second gene identified as homologous to MT-A70 is from *S. cerevisiae* ($P = 1.4 \times 10^{-88}$). It encodes SPO8, which was previously shown to be an activator of genes involved in the sporulation pathway (Smith et al., 1988; Esposito and Klapholz, 1981). MT-A70 and SPO8 share 72% amino acid similarity and 54% identitiy over a region of 270 amino acids near their carboxyl termini (Fig. 6B). The consensus methylation motifs both fall within the region of MT-A70 that is most highly conserved with SPO8, suggesting that SPO8 may also function as a methyltransferase, although no evidence for this has been reported because the molecular function of SPO8 is unknown. A search for protein motifs within the pMT-A70 encoded sequence was also performed. A sequence that matches the nuclear localization signal (NLS) consensus identified by Robbins and Dingwall (Robbins et al., 1991) was present near the amino terminus (amino acid 11), and a second near-match is present approximately 150 amino acids downstream.

A third protein with characteristics that are similar to MT-A70 is the yeast tRNA (N^2,N^2-dimethylguanosine$_{26}$)-methyltransferase, encoded by

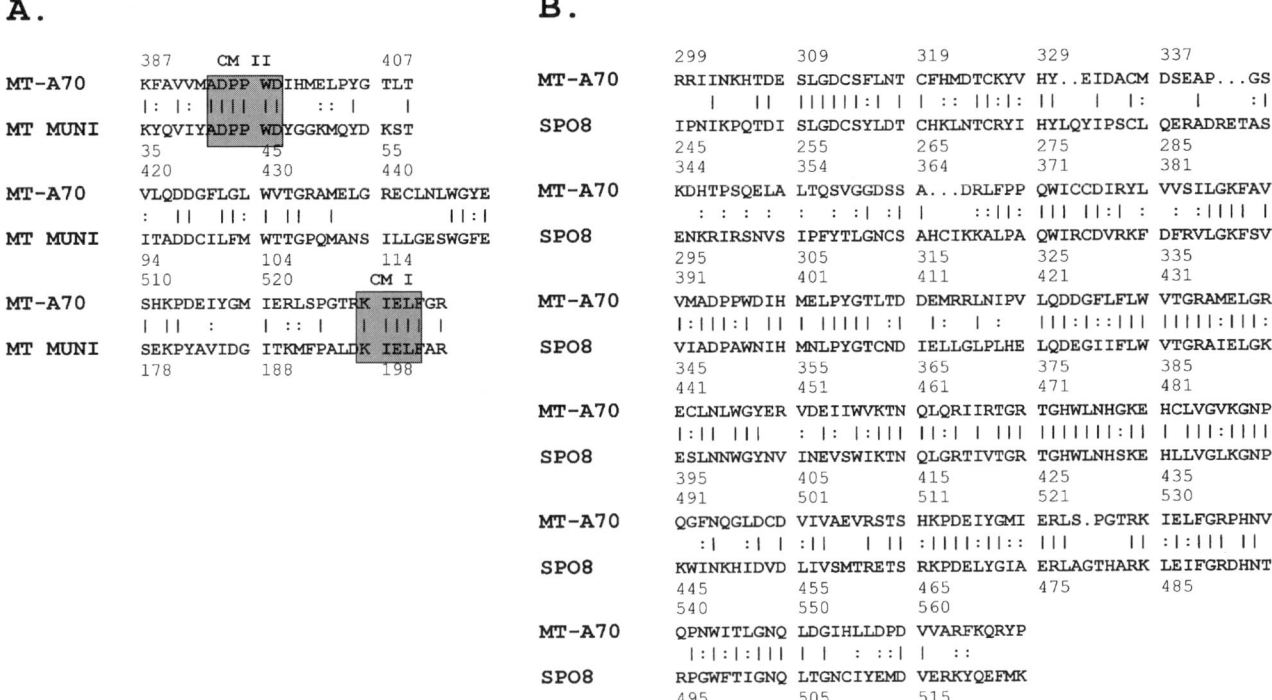
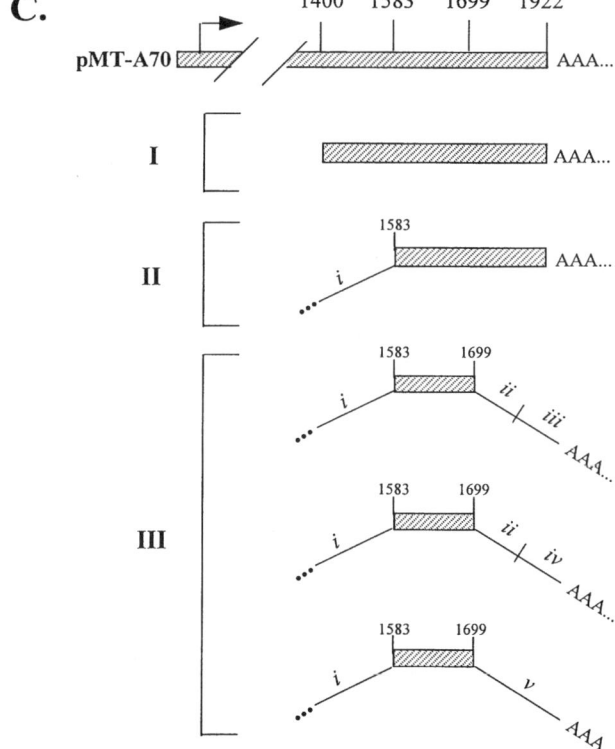

Figure 6. MT-A70 sequence homology to bacterial N^6-adenine DNA methyltransferase motifs and to SPO8. (A) Alignment of similar regions of MT-A70 and M. Mun1 methyltransferase. A vertical line denotes identical amino acids, and a colon denotes similar amino acids. The regions that correspond to consensus DNA methyltransferase motifs are highlighted (see Timinskas et al., 1995). (B) Alignment of similar regions of MT-A70 and SPO8. (C) Schematic representation of three groups of EST sequences that contain regions of identity to the 3′-portion pMT-A70. The stippled box represents pMT-A70 sequence. The thin lines represent EST sequence not homologous to pMT-A70. Blocks of sequence that appear to be alternatively included exons that can be deduced from alignment of the EST sequences are denoted by lowercase roman numerals. The GenBank accession numbers for the EST sequences are as follows: type I, N39589 and W93679; type II, N55548, W04670, R42072, and W95745; and type III, N94880, N95688, N66219, W58127, F09834, N66255, and C00061.

the *S. cerevisiae TRM1* gene (Ellis et al., 1986). Although alignment of MT-A70 and TRM1 does not reveal obvious sequence similarity, the sizes and multi-subunit nature of these two enzymes are similar. *MT-A70* and *TRM1* encode proteins with predicted molecular masses of 65 kDa and similar lengths (579 and 570 amino acids, respectively). Also, both MT-A70 and TRM1 appear to be subunits of larger heteromeric enzymes, each with a native molecular mass of 200 kDa (Edqvist et al., 1995). This may be purely coincidental, or may suggest that these two RNA methyltransferases share structural properties.

A BLAST search of the Database of Expressed Sequence Tags (dbEST) identified 36 human cDNA entries with long regions of sequence identity to pMT-A70. These EST clones are derived from libraries prepared from human fetal brain, heart, kidney, liver, spleen, and parathyroid tumor, confirming that this gene is expressed in a wide variety of tissues. The sequences of many of these cDNA clones indicate that the pMT-A70 pre-mRNA undergoes alternative splicing. A subset of 13 of the EST clones that contain regions of sequence identity to pMT-A70 at the 3′-end were aligned with the pMT-A70 sequence (Fig. 6C). Three classes of EST sequences are schematically represented, with each type appearing to result from different patterns of alternative exon inclusion. Two EST sequences (Fig. 6C, type I) were identified that are identical to pMT-A70 from positions 1337 and 1388 to the 3′-terminus at position 1851 (for GenBank accession numbers, see Fig. 6 legend). At least four EST sequences (type II) were identified that are identical to pMT-A70 sequence from position 1514 to the 3′-terminus at 1851, but contain varying lengths of sequence upstream of postion 1514. This upstream sequence is similar or identical among these ESTs, but bears no similarity to any region of pMT-A70. This finding is consistent with the inclusion of an exon that is not present in pMT-A70 (designated exon i). The full sequence of this putative exon (~174 nucleotides [not shown]) is present in only a single EST entry and it includes unidentified nucleotides at several postions. Therefore, it is not possible to determine whether exon i contains an open reading frame. Nine EST sequences (type III) were identified that begin within the putative exon i, followed by 116 nucleotides identical to pMT-A70 sequence (1514–1629), and finally entirely different 3′-termini. Comparison of these novel 3′-termini to each other reveals that they are also composed of varying blocks of sequence most easily explained as alternatively spliced exons (putative alternative exons ii, iii, iv, and v). These cDNAs predict protein products that are different at their carboxyl termini by virtue of their divergence from pMT-A70 within the predicted coding region. Also of note, all 36 EST clones identified contain long stretches of sequence that are identical to the corresponding regions of pMT-A70, suggesting that they are all derived from expression of the same gene, and not different closely related genes.

Northern Blot Analysis of MT-A70 Expression

Northern blot analysis of polyadenylated RNA from a panel of human tissues was performed (Fig. 7). Two RNA species with approximate lengths of 2 and 3 kb were detected in polyadenylated RNA isolated from eight human tissues upon hybridization with an MT-A70-specific probe. The relative intensities of these two bands varied both within each tissue and between tissues. The exact mobility of the individual bands also varied somewhat between tis-

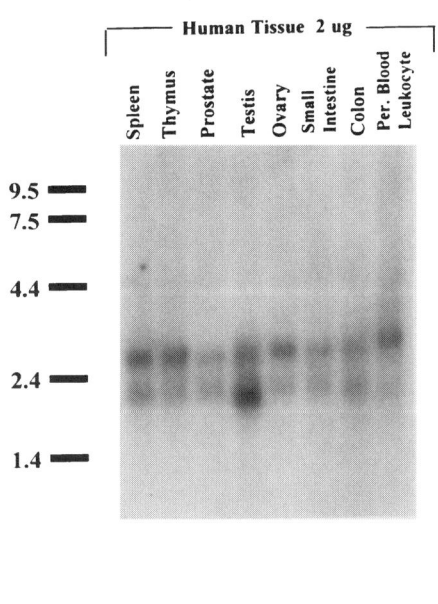

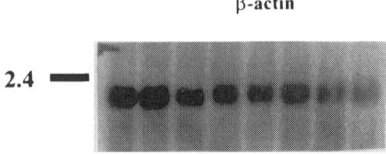

Figure 7. Northern blot analysis of MT-A70 mRNA. ^{32}P-labelled probe was prepared by random primer labelling of a pMT-A70 cDNA insert and was hybridized to blots containing 2 μg of human tissue poly(A)+ RNA. The blots were washed using high-stringency conditions and exposed to X-ray film overnight with an intensifying screen. Two RNA species of approximately 2 and 3 kb were seen in each tissue sample. The blots were stripped and rehybridized with a probe specific for β-actin mRNA.

sues, over a fairly narrow range of 100–200 bp. As a control, the blot was rehybridized to a probe specific for β-actin. Although there are differences in the intensity of the β-actin mRNA signal, these differences do not parallel the distribution seen with the pMT-A70 probe. The presence of two discrete MT-A70 mRNA bands of very different size can result from two possibilities. There may be another gene with sequence homology to pMT-A70 that is expressed in all of the tissues tested, which cross-hybridizes with the MT-A70 derived probe. Alternatively, two or more different transcripts may arise from alternative splicing of a single pre-mRNA transcribed from the gene that codes for MT-A70. The heterogeneity of the sizes within the large and small transcripts between tissues may also arise due to alternative splicing, or may be due to differences in poly(A) tail length, or to both. These data, in conjunction with the EST database analysis previously described, are highly suggestive that the MT-A70 pre-mRNA undergoes alternative splicing.

Subnuclear Localization of MT-A70 in HeLa Cells

With the use of antibodies and specific nucleic acid probes, it has become increasingly apparent that many nucleoplasmic components are nonrandomly located throughout the nucleus (Fakan, 1994; Misteli and Spector, 1996). Of recent interest is the observation that apparently disparate components of the mRNA transcription and processing machinery can be colocalized and specifically interact. For example, the large subunit of RNA polymerase II has been shown to colocalize with the splicing protein SC35 in nuclear "speckles" (interchromatin granule clusters) and the physical interaction of these components has been documented by immunoprecipitation (McCracken et al., 1997). The TRM1 gene product [yeast tRNA (N^2,N^2-dimethylguanosine$_{26}$)-methyltransferase] localizes to the yeast nuclear periphery with a ring-like appearance by immunofluorescence (Rose et al., 1995).

In an experiment to explore the possible distribution of MT-A70 among discrete structural domains of HeLa cell nuclei, the anti-MT-A70 antiserum was used to probe paraformaldehyde-fixed HeLa cells. The results shown (Fig. 8A) demonstrate a striking localization of MT-A70 to speckles in these human interphase nuclei. Colocalization of U2B″ protein (Fig. 8B), which is an snRNP protein and is used as a marker for speckles, clearly demonstrates that these two proteins are found in identical regions of the nucleus. This distribution is similar to that found for the large subunit of RNA polymerase II and splicing factor SC35, and is in contrast to the pattern seen with the TRM1 gene product. This is an exciting result, that suggests that mRNA m^6A-MT may indeed be associated with nuclear pre-mRNA splicing components.

Further Characterization of MT-B

MT-B has not been as extensively characterized as MT-A. By glycerol gradient sedimentation and size

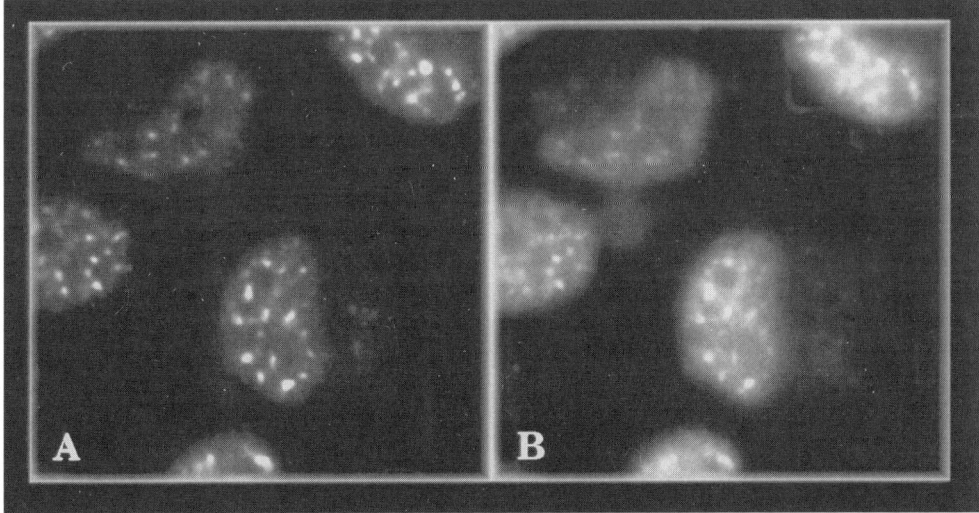

Figure 8. MT-A70 localizes to speckled domains in human interphase nuclei. Double labeling of anti-MT-A70 and anti-U2B″ (a U2-specific snRNP protein) was performed on paraformaldehyde-fixed HeLa cell nuclei. (A) MT-A70 was detected with polyclonal rabbit antiserum 12622 and fluorescein isothiocyanate-conjugated anti-rabbit IgG. (B) U2B″ protein was detected with monoclonal antibody 4G3 and Texas Red-conjugated anti-mouse IgG. Rabbit preimmune serum showed only background fluorescence (not shown).

exclusion chromatography it has an apparent molecular mass of 875 kDa (Bokar, 1994). Unlike MT-A, MT-B binds to single-stranded DNA cellulose, suggesting that it may contain the nucleic acid binding site of the enzyme. The large size of MT-B and its chromatographic properties led to the suspicion that it may contain an snRNA. However, treatment of nuclear extract with micrococcal nuclease and immunoprecipitation with an antitrimethylguanosine antibody did not lead to any change in m^{6}A-MT activity, or in the apparent size of MT-B as judged by its elution on the size exclusion column (unpublished result). However, the involvement of snRNAs and snoRNAs in splicing, rRNA methylation (Chapter 13 by Bachellerie and Cavaillé), pseudouridine formation (Chapter 12 by Ofengand and Fournier), and most recently in snRNA nucleoside modification (Chapter 11 by Massenet et al.), continue to make the involvement of a small RNA cofactor an intriguing possibility. If an RNA component is required, it must be inaccessible to micrococcal nuclease and either does not have a trimethylguanosine cap or the cap is inaccessible to antibody. Clearly there are precedents for these possibilities. More direct experiments aimed at detecting an RNA cofactor will need to be performed to test this hypothesis.

CONCLUSION

Posttranscriptional modification of nucleosides within mRNA molecules is an integral aspect of pre-mRNA processing. These modifications are characteristic of eukaryotic mRNA only; no equivalent nucleotide modifications have been found in any of the prokaryotic mRNAs characterized so far. The individual modifications are fairly simple in nature, that is, methylation of specific groups on the sugar or base, and deamination. However, in reality, these modifications are not simple in that they occur only at precise positions, require complex enzymes, and are likely to be regulated. The biological functions of the individual modifications are either poorly understood, or not understood at all. Recent advances in characterization of the modifying enzymes (for example, vaccinia virus capping enzymes, dsRAD, Apobec-1, and m^{6}A-MT) and rapid developments in understanding the formation and functions of analogous modifications in other types of RNA are sure to lead to a better understanding of the importance of these modifications in the near future.

Acknowledgments. This work was supported in part by National Institutes of Health grant CA31810 to F.M.R. J.A.B. was supported by a Howard Hughes Medical Institute Physician Research Fellowship and an American Association for Cancer Research Fellowship.

REFERENCES

Adams, J. M. and S. Cory. 1975. Modified nucleosides and bizarre 5′-termini in mouse myeloma mRNA. *Nature* 255:28–33.

Adams, B. L., M. Morgan, S. Muthukrishnan, S. M. Hecht, and A. J Shatkin. 1978. The effect of "cap" analogs on reovirus mRNA binding to wheat germ ribosomes. *J. Biol. Chem.* 253: 2589–2595.

Aloni, Y., R. Dhar, and G. Khoury. 1979. Methylation of nuclear simian virus 40 RNAs. *J. Virol.* 32:52–60.

Altschul, S. F., W. Gish, W. Miller, E. W. Myers, and D. J. Lipman. 1990. Basic local alignment search tool. *J. Mol. Biol.* 215: 403–410.

Bachellerie, J. P., F. Amalric, and M. Caboche. 1978. Biosynthesis and utilization of extensively undermethylated poly (A)+ RNA in CHO cells during a cycloleucine treatment. *Nucleic Acids Res.* 5:2927–2943.

Banerjee, A. K. 1980. 5′-terminal cap structure in eucaryotic messenger ribonucleic acids. *Microbiol. Rev.* 44:175–205.

Bangs, J. D., P. F. Crain, T. Hashizume, J. A. McCloskey, and J. C. Boothroyd. 1992. Mass spectrometry of mRNA cap 4 from trypanosomatids reveals two novel nucleosides. *J. Biol. Chem.* 267:9805–9815.

Bartkoski, M. J., and B. Roizman. 1978. Regulation of herpesvirus macromolecular synthesis. VII. Inhibition of internal methylation of mRNA late in infection. *Virology* 85:146–156.

Beemon, K., and J. Keith. 1977. Localization of N^6-methyladenosine in the Rous sarcoma virus genome. *J. Mol. Biol.* 113: 165–179.

Björk, G. R., J. U. Ericson, C. E. D. Gustafsson, T. G. Hagervall, Y. H. Jonsson, and P. M. Wikström. 1987. Transfer RNA modification. *Annu. Rev. Biochem.* 56:263–287.

Bokar, J. A., M. E. Rath-Shambaugh, R. L. Ludwiczak, P. Narayan, and F. M. Rottman. 1994. Characterization and partial purification of mRNA N^6-adenosine methyltransferase from HeLa cell nuclei. *J. Biol. Chem.* 269:17697–17704.

Bokar, J. A., M. E. Shambaugh, D. Polayes, A. G. Matera, and F. M. Rottman. 1997. Purification and cDNA cloning of the AdoMet-binding subunit of the human mRNA (N6-adenosine)-methyltransferase. *RNA* 3:1233–1247.

Camper, S. A., R. J. Albers, J. K. Coward, and F. M. Rottman. 1984. Effect of undermethylation on mRNA cytoplasmic appearance and half-life. *Mol. Cell. Biol.* 4:538–543.

Canaani, D., C. Kahana, S. Lavi, and Y. Groner. 1979. Identification and mapping of N6-methyladenosine containing sequences in simian virus 40 RNA. *Nucleic Acids Res.* 6: 2879–2899.

Carroll, S. M., P. Narayan, and F. M. Rottman. 1990. N^6-methyladenosine residues in an intron-specific region of prolactin pre-mRNA. *Mol. Cell. Biol.* 10:4456–4465.

Cong, P., and S. Shuman. 1992. Methyltransferase and subunit association domains of vaccinia virus mRNA capping enzyme. *J. Biol. Chem.* 267:16424–16429.

Csepany, T., A. Lin, C. J. Baldick, and K. Beemon. 1990. Sequence specificity of mRNA N^6-adenosine methyltransferse. *J. Biol. Chem.* 265:20117–20122.

Darzynkiewicz, E., J. Stepinski, I. Ekiel, Y. Jin, D. Haber, T. Sijuwade, and S. M. Tahara. 1988. β-globin mRNAs capped with m^{7}G, m$_2^{2,7}$G or m$_3^{2,2,7}$G differ in intrinsic translation efficiency. *Nucleic Acids Res.* 16:8953–8962.

Desrosiers, R. C., K. H. Friderici, and F. M. Rottman. 1974. Identification of methylated nucleosides in messenger RNA from

Novikoff hepatoma cells. *Proc. Natl. Acad. Sci. USA* **71**: 3971–3975.

Desrosiers, R. C., K. H. Friderici, and F. M. Rottman. 1975. Characterization of Novikoff hepatoma mRNA methylation and heterogeneity in the methylated 5' terminus. *Biochemistry* **14**: 4367–4374.

Dimock, K., and C. M. Stoltzfus. 1977. Sequence specificity of internal methylation in B77 avian sarcoma virus RNA subunits. *Biochemistry* **16**:471–478.

Dottin, R. P., A. M. Weiner, and H. F. Lodish. 1976. 5'-terminal nucleotide sequences of the messenger RNAs of *Dictyostelium discoideum*. *Cell* **8**:233–244.

Dubin, D. T., and V. Stollar. 1975. Methylation of Sindbis virus "26s" messenger RNA. *Biochem. Biophys. Res. Commun.* **66**: 1373–1379.

Dubin, D. T. and R. H. Taylor. 1975. The methylation state of poly A-containing-messenger RNA from cultured hamster cells. *Nucleic Acids Res.* **2**:1653–1668.

Dubin, D. T., V. Stollar, C. C. Hsuchen, K. Timko, and G. M. Guild. 1977. Sindbis virus messenger RNA: the 5' termini and methylated residues of 26 and 42 S RNA. *J. Virol.* **77**:457–470.

Edqvist, J., K. B. Stråby, and H. Grosjean. 1995. Enzymatic formation of N^2, N^2-dimethylguanosine in eukaryotic tRNA: importance of the tRNA architecture. *Biochimie* **77**:54–61.

Ellis, S. R., M. J. Morales, J. M. Li, A. K. Hopper, and N. C. Martin. 1986. Isolation and characterization of the TRM1 locus, a gene essential for the N2,N2-dimethylguanosine modification of both mitochondrial and cytoplasmic tRNA in *Saccharomyces cerevisiae*. *J. Biol. Chem.* **261**:9703–9709.

Ensinger, M. J., and B. Moss. 1976. Modification of the 5' terminus of mRNA by an RNA (Guanine-7-)-methyltransferase from HeLa cells. *J. Biol. Chem.* **251**:5283–5291.

Esposito, R. E., and S. Klapholz. 1981. *The Molecular Biology of the Yeast Saccharomyces. Life Cycle and Inheritance.* Cold Spring Harbor Laboratory Press, Cold Spring Harbor, N.Y.

Fakan, S. 1994. Perichromatin fibrils are *in situ* forms of nascent transcripts. *Trends Cell Biol.* **4**:86–90.

Finkel, D., and Y. Groner. 1983. Methylations of adenosine residues (m^6A) in pre-mRNA are important for formation of late simian virus 40 mRNAs. *J. Virol.* **131**:409–425.

Furuichi, Y., M. A. Morgan, and A. J. Shatkin. 1979. Synthesis and translation of mRNA containing 5'-terminal 7-ethylguanosine cap. *J. Biol. Chem.* **254**:6732–6738.

Grosjean, H., J. Edqvist, K. B. Stråby, and R. Giege. 1996. Enzymatic formation of modified nucleosides in tRNA: dependence on tRNA architecture. *J. Mol. Biol.* **255**:67–85.

Grosjean, H., Z. Szweykowska-Kulinska, Y. Motorin, F. Fasiolo, and G. Simos. 1997. Intron-dependent enzymatic formation of modified nucleosides in eukaryotic tRNAs: a review. *Biochimie.* **79**:293–302.

Hamm, J., and I. W. Mattaj. 1990. Monomethylated cap structures facilitate RNA export from the nucleus. *Cell* **63**:109–118.

Harper, J. E., S. M. Miceli, R. J. Roberts, and J. L. Manley. 1990. Sequence specificity of the human mRNA N^6-adenosine methylase *in vitro*. *Nucleic Acids Res.* **18**:5735–5741.

Hentze, M. W. 1997. eIF4G: a multipurpose ribosome adapter? *Science* **275**:500–501.

Higman, M. A., N. Bourgeois, and E. G. Niles. 1992. The vaccinia virus mRNA (guanine-N^7-)-methyltransferase requires both subunits of the mRNA capping enzyme for activity. *J. Biol. Chem.* **267**:16430–16437.

Higman, M. A., and E. G. Niles. 1994a. Location of the S-adenosyl-L-methionine binding region of the vaccinia virus mRNA (guanine-7-)methyltransferase. *J. Biol. Chem.* **269**: 14982–14987.

Higman, M. A., L. A. Christen, and E. G. Niles. 1994b. The mRNA (guanine-7-)methyltransferase domain of the vaccinia virus mRNA capping enzyme: expression in *Escherichia coli* and structural and kinetic comparison to the intact capping enzyme. *J. Biol. Chem.* **269**:14974–14981.

Horowitz, S., A. Horowitz, T. W. Nilsen, T. W. Munns, and F. M. Rottman. 1984. Mapping of N^6-methyladenosine residues in bovine prolactin mRNA. *Proc. Natl. Acad. Sci. USA* **81**: 5667–5671.

Hough, R. F., and B. L. Bass. 1997. Analysis of *Xenopus* dsRNA adenosine deaminase cDNAs reveals similarities to DNA methyltransferases. *RNA* **3**:356–370.

Huarte, J., D. Belin, A. Vassalli, S. Strickland, and J. D. Vassalli. 1987. Meiotic maturation of mouse oocytes triggers the translation and polyadenylation of dormant tissue-type plasminogen activator mRNA. *Genes Dev.* **1**:1201–1211.

Izaurralde, E., J. Lewis, M. McGuigan, E. Jankowska, E. Darzynkiewicz, and I. W. Mattaj. 1994. A nuclear cap binding protein complex involved in pre-mRNA splicing. *Cell* **78**:657–668.

Jaeger, J. A., D. H. Turner, and M. Zuker. 1989a. Improved predictions of secondary structure for RNA. *Proc. Natl. Acad. Sci. USA.* **86**:7706–7710.

Jaeger, J. A., D. H. Turner, and M. Zuker. 1989b. Predicting optimal and suboptimal secondary structure for RNA. *Methods Enzymol.* **183**:281–306.

Jiang, H. Q., Y. Motorin, Y. X. Jin, and H. Grosjean. 1997. Pleiotropic effects of intron removal on base modification pattern of yeast $tRNA^{Phe}$: an *in vitro* study. *Nucleic Acids Res.* **25**: 2694–2701.

Kane, S. E., and K. Beemon. 1985. Precise localization of m^6A in Rous sarcoma virus RNA reveals clustering of methylation sites: implications for RNA processing. *Mol. Cell. Biol.* **5**:2298–2306.

Kane, S. E., and K. Beemon. 1987. Inhibition of methylation at two internal N^6-methyladenosine sites caused by GAC to GAU mutations. *J. Biol. Chem.* **262**:3422–3427.

Keith, J. M., M. J. Ensinger, and B. Moss. 1978. HeLa cell RNA (2'-O-methyladenosine-N^6-)-methyltransferase specific for the capped 5'-end of messenger RNA. *J. Biol. Chem.* **253**: 5033–5041.

Kienzle, T., and F. Rottman. Unpublished results.

Kim, U., T. L. Garner, T. Sanford, D. Speicher, J. V. K. Murray, and K. Nishikura. 1994. Purification and characterization of double-stranded RNA adenosine deaminase from bovine nuclear extracts. *J. Biol. Chem.* **269**:13480–13489.

Kuge, H., and J. D. Richter. 1995. Cytoplasmic 3' poly (A) addition induces 5' cap ribose methylation: implications for translational control of maternal RNA. *EMBO J.* **14**:6301–6310.

Langberg, S. R., and B. Moss. 1981. Posttranscriptional modifications of mRNA: purification and characterization of cap 1 and cap 2 RNA (nucleoside-2'-)-methyltransferases from HeLa cells. *J. Biol. Chem.* **256**:10054–10060.

Levis, R., and S. Penman. 1978. 5'-terminal structures of poly (A)+ cytoplasmic messenger RNA and of poly (A)+ and poly (A)- heterogeneous nuclear RNA of cells of the dipteran *Drosophila melanogaster*. *J. Mol. Biol.* **120**:487–515.

Limbach, P. A., P. F. Crain, and J. A. McCloskey. 1994. Summary: the modified nucleosides of RNA. *Nucleic Acids Res.* **22**: 2183–2196.

Liou, R. F., and T. Blumenthal. 1990. *trans*-spliced *Caenorhabditis elegans* mRNAs retain trimethylguanosine caps. *Mol. Cell. Biol.* **10**:1764–1768.

Martin, S. A., E. Paoletti, and B. Moss. 1975. Purification of mRNA guanylyltransferase and mRNA (guanine-7-)methyltransferase from vaccinia virions. *J. Biol. Chem.* **250**:9322–9329.

McCracken, S., N. Fong, K. Yankulov, S. Ballantyne, G. Pan, J. Greenblatt, S. D. Patterson, M. Wickens, and D. L. Bentley. 1997. The C-terminal domain of RNA polymerase II couples mRNA processing to transcription. *Nature* 385:357–361.

Melcher, T., S. Maas, M. Higuchi, W. Keller, and P. H. Seeburg. 1995. Editing of a-amino-3-hydroxy-5-methylisoxazole-4-propionic acid receptor GluR-B pre-mRNA *in vitro* reveals site-selective adenosine to inosine conversion. *J. Biol. Chem.* 270:8566–8570.

Merrick, W. C., and J. W. B. Hershey. 1996. The pathway and mechanism of eukaryotic protein synthesis, p. 31–69. In J. W. B. Hershey, M. B. Mathews, and N. Sonenberg (ed.), *Translational Control*. Cold Spring Harbor Laboratory Press, Cold Spring Harbor, N.Y.

Misteli, T., and D. L. Spector. 1996. Protein phosphorylation and the nuclear organization of pre-mRNA splicing. *Trends Cell Biol.* 135–138.

Mizumoto, K., and Y. Kaziro. 1987. Messenger RNA capping enzymes from eukaryotic cells. *Prog. Nucleic Acid Res.* 34:1–28.

Muthukrishnan, S., G. W. Both, Y. Furuichi, and A. J. Shatkin. 1975. 5′-terminal 7-methylguanosine in eukaryotic mRNA is required for translation. *Nature* 255:33–37.

Myette, J. R., and E. G. Niles. 1996. Characterization of the vaccinia virus RNA 5′-triphosphatase and nucleoside triphosphate phosphohydrolase activities: demonstration that both activities are carried out at the same active site. *J. Biol. Chem.* 271:11945–11952.

Narayan, P., D. F. Ayers, F. M. Rottman, P. A. Maroney, and T. W. Nilsen. 1987. Unequal distribution of N^6-methyladenosine in influenza virus mRNAs. *Mol. Cell. Biol.* 7:1572–1575.

Narayan, P., and F. M. Rottman. 1988. An *in vitro* system for accurate methylation of internal adenosine residues in messenger RNA. *Science* 242:1159–1162.

Narayan, P., and F. M. Rottman. 1992. Methylation of mRNA, p. 225–285. In A. Meister (ed.), *Advances in Enzymology and Related Areas of Molecular Biology*. John Wiley and Sons, Inc., New York, N.Y.

Narayan, P., R. L. Ludwiczak, E. Goodwin, and F. M. Rottman. 1994. Context effects of N^6-adenosine methylation sites in prolactin mRNA. *Nucleic Acids Res.* 22:419–426.

Nichols, J. L. 1979. "Cap" structures in maize poly(A)-containing RNA. *Biochim. Biophys. Acta* 563:490–495.

Perry, R. P., and K. Scherrer. 1975. Methylated constituents of globin mRNA. *FEBS Lett.* 57:73–78.

Perry, R. P., D. E. Kelley, K. Fridirici, and F. M. Rottman. 1975. The methylated constituents of L cell messenger RNA: evidence for an unusual cluster at the 5′-terminus. *Cell* 4:387–394.

Perry, R. P., and D. E. Kelley. 1976. Kinetics of formation of 5′ terminal caps in mRNA. *Cell* 8:433–442.

Polson, A. G., B. L. Bass, and J. L. Casey. 1995. RNA editing of hepatitis delta virus antigenome by dsRNA-adenosine deaminase. *Nature* 380:454–456.

Rana, A. P., and M. T. Tuck. 1990. Analysis and *in vitro* localization of internal methylated adenine residues in dihydrofolate reductase mRNA. *Nucleic Acids Res.* 18:4803–4807.

Reddy, R., R. Singh, and S. Shimba. 1992. Methylated cap structures in eukaryotic RNAs: structure, synthesis and functions. *Pharmacol. Ther.* 54:249–267.

Rhoads, R. E., G. M. Hellmann, P. Remy, and J. P. Ebel. 1983. Translational recognition of messenger ribonucleic acid caps as a function of pH. *Biochemistry* 22:6084–6088.

Robbins, J., S. M. Dilworth, R. A. Laskey, and C. Dingwall. 1991. Two interdependent basic domains in nucleoplasmin nuclear targeting sequence: identification of a class of bipartite nuclear targeting sequence. *Cell* 64:615–623.

Rose, A. M., H. G. Belford, W. C. Shen, C. L. Greer, A. K. Hopper, and N. C. Martin. 1995. Location of N^2,N^2-dimethylguanosine-specific tRNA methyltransferase. *Biochimie* 77:45–53.

Rottman, F. M., A. J. Shatkin, and R. P. Perry. 1974. Sequences containing methylated nucleotides at the 5′-termini of messenger RNAs: possible implications for processing. *Cell* 3:197–199.

Salles, F. J., M. E. Lieberfarb, C. Wreden, J. P. Gergen, and S. Strickland. 1994. Coordinate initiation of *Drosophila* development by regulated polyadenylation of maternal messenger RNAs. *Science* 266:1996–1999.

Schibler, U., D. E. Kelley, and R. P. Perry. 1977. Comparison of methylated sequences in messenger RNA and heterogenous nuclear RNA from mouse L cells. *J. Mol. Biol.* 115:695–714.

Seidel, B. L., and E. W. Somberg. 1978. Characterization of Neurospora crassa polyadenylated messenger ribonucleic acid structure of the 5′ terminus. *Biochem. Biophys. Res. Commun.* 187:108–112.

Shatkin, A. J. 1976. Capping of eukaryotic mRNAs. *Cell* 9:645–653.

Shi, X., P. Yao, T. Hose, and P. D. Gershon. 1996. Methyltransferase-specific domains within VP39, a bifunctional protein that participates in the modification of both mRNA ends. *RNA* 2:88–101.

Smith, L. M., L. G. Robbins, A. Kennedy, and P. T. Magee. 1988. Identification and characterization of mutations affecting sporulation in *Saccharomyces cerevisiae*. *Genetics* 120:899–907.

Sommer, S., M. Salditt-Georgieff, S. Bachenheimer, J. E. Darnell, Y. Furuichi, M. Morgan, and A. J. Shatkin. 1976. The methylation of adenovirus-specific nuclear and cytoplasmic RNA. *Nucleic Acids Res.* 3:749–765.

Sripati, C. E., Y. Groner, and J. R. Warner. 1976. Methylated, blocked 5′ termini of yeast mRNA. *J. Biol. Chem.* 251:2898–2904.

Steinmetz, E. J. 1997. Pre-mRNA processing and the CTD of RNA polymerase II: the tail that wags the dog? *Cell* 89:491–494.

Stoltzfus, C. M., and R. W. Dane. 1982. Accumulation of spliced avian retrovirus mRNA is inhibited in S-adenosylmethionine-depleted chicken embryo fibroblasts. *J. Virol.* 42:918–931.

Timinskas, A., V. Butkus, and A. Janulaitis. 1995. Sequence motifs characteristic for DNA [cytosine-N4] and DNA [adenine-N6] methyltransferases. Classification of all DNA methyltransferases. *Gene* 157:3–11.

Van Doren, K., and D. Hirsh. 1990. mRNAs that mature through trans-splicing in *Caenorhabditis elegans* have a trimethylguanosine cap at their 5′ termini. *Mol. Cell. Biol.* 10:1769–1772.

Wei, C. M., A. Gershowitz, and B. Moss. 1975. Methylated nucleotides block 5′ terminus of HeLa cell messenger RNA. *Cell* 4:379–386.

Wei, C. M., A. Gershowitz, and B. Moss. 1976. 5′-terminal and internal methylated nucleotide sequences in HeLa cell mRNA. *Biochemistry* 15:397–401.

Wei, C. M., and B. Moss. 1977. Nucleotide sequences at the N^6-methyladenosine sites of HeLa cell messenger ribonucleic acid. *Biochemistry* 16:1672–1676.

Zuker, M. 1989. On finding all suboptimal foldings of an RNA molecule. *Science* 244:48–52.

Chapter 11

Posttranscriptional Modifications in the U Small Nuclear RNAs

SÉVERINE MASSENET, ANNIE MOUGIN, AND CHRISTIANE BRANLANT

In eukaryotes, the nucleus is the site for RNA transcription and RNA maturation and most of the RNA maturation processes involve small nuclear ribonucleoprotein particles (snRNPs) containing metabolically stable small nuclear RNAs (snRNAs). This chapter deals with the posttranscriptional modifications of these snRNAs and the possible link to the function of these molecules.

In the nucleolus, the 18S, 5.8S, and 28S rRNAs are transcribed by RNA polymerase I as a long precursor, and this primary transcript is processed by an ordered series of RNase cleavages and posttranscriptional modifications (for reviews, see Maden, 1990; Henry et al., 1994; Eichler and Craig, 1994; Raué and Planta, 1995; Ofengand et al., 1995; Bachellerie et al., 1995) (see also Chapters 12 to 15). This complex maturation process involves the participation of a large series of small nucleolar ribonucleoparticles (snoRNPs) containing small nucleolar RNAs (snoRNAs) (for reviews, see Maxwell and Fournier, 1995; Tollervey and Kiss, 1997; Smith and Steitz, 1997). The snoRNAs can be divided into two classes: the snoRNAs required for the ordered cascade of RNase cleavage (U3, U8, U14, U22, and the MRP RNA) and the snoRNAs that serve as guides for the 2'-O-methylation and pseudouridine formation of rRNAs. U14 snoRNA is a special case because it seems to be required for both processes. The snoRNAs are either transcribed from their own genes by RNA polymerase II or RNA polymerase III or derived from the maturation of protein gene introns (for reviews, see Maxwell and Fournier, 1995; Bachellerie et al., 1995; Tollervey and Kiss, 1997; Smith and Steitz, 1997) (also see Chapter 13). The latter situation is more frequent for the snoRNAs that serve as a guide for posttranscriptional modifications. Depending on how they are produced, snoRNAs have different 5'-terminal structures: a trimethylated cap structure for RNA polymerase II transcripts (for reviews, see Busch et al., 1982; Reddy et al., 1992); a γ-methyl triphosphate for RNA polymerase III transcripts (Singh and Reddy, 1989; for a review, see Reddy et al., 1992); and a phosphate when they are produced from introns (for review, Maxwell and Fournier, 1995). Because most of the U snoRNA primary structures have been established from DNA sequence analysis, except for the terminal structure, the information on snoRNA posttranscriptional modifications is very limited. Only the posttranscriptional modifications of the rat hepatoma (Reddy et al., 1985; Reddy et al., 1979) and the *Saccharomyces cerevisiae* (our unpublished results) U3 snoRNAs and of the mouse U8 snoRNA (Kato and Harada, 1984) have been investigated. The results obtained will be presented in this chapter, because they may be meaningful for understanding the relationships between U snoRNA function and posttranscriptional modifications.

In addition to rRNAs, the second main class of transcripts produced in the nucleus is pre-mRNAs. They are transcribed in the nucleoplasm by RNA polymerase II and are maturated into messenger RNAs before their transport to the cytoplasm. Indeed, most of the nuclear genes coding for proteins contain introns that interrupt the coding sequences. They are eliminated in the pre-mRNAs by assembly onto the intron of a large macrocomplex, called the spliceosome, that catalyzes the two transesterification steps of the splicing reaction (Fig. 1) (for a review, see Moore et al., 1993). The first step is the nucleophilic attack of the phosphate at the 5' end of the intron by the 2'-OH of an adenosine residue from the intron branch-site sequence. For vertebrate introns, the branch-site sequence is located between positions 18 and 40 upstream of the 3'-splice site (Keller and Noon, 1984; Ruskin et al., 1984; Shapiro and Sen-

---

Séverine Massenet, Annie Mougin, and Christiane Branlant • Laboratoire d'Enzymologie et de Génie Génétique, URA-CNRS 457, Université de Nancy I, Faculté des Sciences, Bld des Aiguillettes, 54506 Vandoeuvre-lès-Nancy Cedex, France.

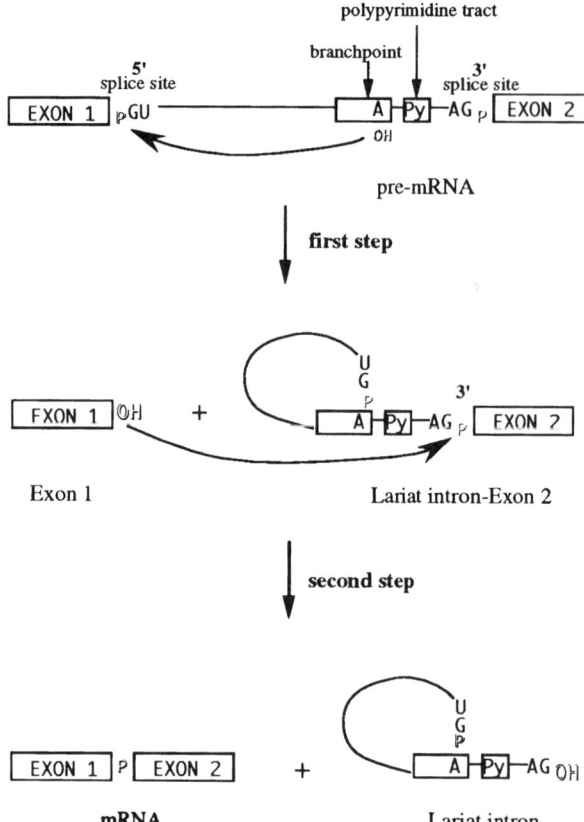

Figure 1. The two chemical steps of nuclear pre-mRNA splicing. The highly conserved nucleotides and structures of the intermediates and products are shown (for a review, see Moore et al., 1993).

ner et al., 1980; Rogers and Wall, 1980; Zhuang and Weiner, 1986; Séraphin et al., 1988; Siliciano and Guthrie, 1988; Séraphin and Rosbash, 1990) and is recognized by the U1 snRNP in an early step of spliceosome assembly (Mount et al., 1983; Black et al., 1985) (Fig. 2). The U1 snRNA interaction is later disrupted and replaced by an interaction with U6 snRNA (Sawa and Shimura, 1992; Lesser and Guthrie, 1993; Kandels-Lewis and Séraphin, 1993). The intron branch-site sequence, which displays a complementarity with U2 snRNA, is also essential. It is recognized by factor U2AF (Kohtz et al., 1994; Valcarcel et al., 1996) and then by U2 snRNA (Parker et al., 1987; Zhuang and Weiner, 1989; Wu and Manley, 1989) (Fig. 2). Finally, the third intron element required for spliceosome assembly and function is the AG 3'-terminal dinucleotide (Krainer et al., 1984; Parker and Siliciano, 1993). Hence, U1 and U2 snRNAs and a certain number of splicing factors play an essential role in the first steps of spliceosome assembly. First, the U1 snRNP is associated, then the U2 snRNP, and finally the U4–U6 and U5 snRNPs as a preformed (U4/U6,U5) tri-snRNP complex (Behrens and Lührmann, 1991) (Fig. 2). At this stage, important conformational changes occur: the U4–U6 snRNA interaction that is preserved in the tri-snRNP

apathy, 1987). In yeast, the situation is more complex: two classes of introns that differ by the distance between the branch-site sequence and the 3'-splice site were characterized (Parker and Patterson, 1987). As a result of the first nucleophilic attack, a lariat intermediate is formed and the first exon is released. The second step consists of the nucleophilic attack of the phosphate at the 3' end of the intron by the 3' hydroxyl of the free exon 1. The intron is liberated in a lariat form and the mature mRNA is produced (Ruskin et al., 1984; Padgett et al., 1984; Konarska et al., 1985). Catalysis requires a very precise assembly of the spliceosomal components: U1 and U2 snRNAs play a crucial role in intron recognition, U5 snRNA aligns the exon sequences for correct ligation, and U2 and U6 snRNAs probably play a direct role in catalysis (for reviews, Moore et al., 1993; Madhani and Guthrie, 1994; Umen and Guthrie, 1995).

Introns are defined by a limited number of conserved sequences. One essential sequence consists of the 5' extremity of the intron and the few nucleotides at the 3' end of the first exon; this segment displays complementarity with the 5' end of U1 snRNA (Ler-

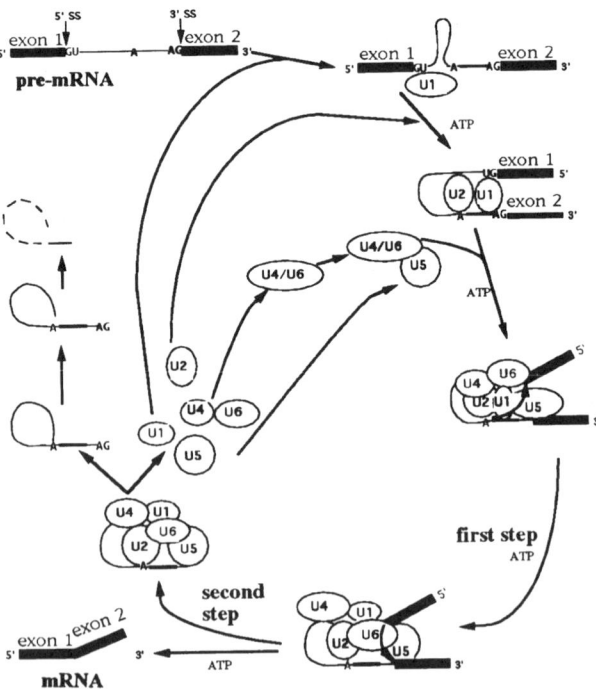

Figure 2. Steps of spliceosome assembly in the vertebrates. Substrate (pre-mRNA), intermediate (lariat-intron-exon2), and products (exon1, mRNA and lariat intron) of the reaction are indicated as well as the order of assembly of the snRNP components. First and second steps are the same as first and second steps in Fig. 1. (For references, see text.)

complex (our unpublished results) is dissociated and a U6-U2 interaction is formed (Hausner et al., 1990; Wu and Manley, 1991; Datta and Weiner, 1991; Madhani and Guthrie, 1992; Sun and Manley, 1995); U6 snRNA displaces the U1-intron interaction (Sawa and Abelson, 1992; Kandels-Lewis and Séraphin, 1993; Lesser and Guthrie, 1993; Sun and Manley, 1995); and U5 snRNA may participate to this transition and interacts with the two extremities of the exons (Newman and Norman, 1991 and 1992; Wyatt et al., 1992; Cortes et al., 1993; Sontheimer and Steitz, 1993; Newman et al., 1995). After splicing completion, U snRNAs are dissociated from the introns, the U2-U6 snRNA interaction is probably disrupted and the U4-U6 snRNA interaction reformed. This complete process implies the successive formation and dissociation of several intermolecular RNA-RNA interactions and as it is described later, the U snRNA sequences involved in these interactions are highly posttranscriptionally modified.

Indeed, posttranscriptional modifications in the spliceosomal U snRNAs have been studied long ago for vertebrate species, plants, and *Drosophila* (for reviews, see Jacob et al., 1984; Reddy, 1988; Myslinski et al., 1984; Solymosy and Pollàk, 1993) and more recently for a few lower eukaryotes, *Physarum polycephalum* (Szkukàlek et al., 1995), *Schizosaccharomyces pombe* (Gu et al., 1996) and *Saccharomyces cerevisiae* (our unpublished data). As mentioned above, U1, U2, U4, and U5 snRNAs that are transcribed by RNA polymerase II (Hellung-Larsen et al., 1980; Chadrasekharappa et al., 1983; for a review, see Dahlberg and Lund, 1988) have a trimethylated cap structure (Reddy et al., 1992), and this cap structure plays a very important role for the spliceosomal U snRNP biogenesis. U6 snRNA, which is transcribed by RNA polymerase III (Kunkel et al., 1986; Reddy et al., 1987; Krol et al., 1987), has a γ-methyl phosphate at the 5′ end (Singh and Reddy, 1989; for a review, see Reddy et al., 1992), except in *Physarum*, where it bears a trimethyl G cap (Adams et al., 1987).

In addition to this major spliceosomal machinery that is universal in eukaryotes, a minor spliceosomal machinery was recently found in metazoan cells. It is required for the splicing of a minor class of introns, of which 17 members have been detected to date (Wu and Krainer, 1997). They are characterized by the replacement of the 5′-GU and 3′-AG dinucleotides, which are universally conserved in the major form of introns, by 5′-AU and 3′-AC dinucleotides, respectively (Jackson, 1991; Hall and Padgett, 1994). In addition, their 5′-terminal sequence is highly conserved and different from that of the classical introns. It is complementary to the 5′-terminal sequence of a minor nucleoplasmic snRNA denoted U11 (Kolossova and Padgett, 1977). Their branch-site sequence is also different from that of classical introns, and it was found to be complementary to another minor snRNA called U12 (Tarn et al., 1995; Hall and Padgett, 1996). By depletion experiments, U11 and U12 snRNAs and U5 snRNA (which is used for assembly of the major form of spliceosome) were found to be required for excision of the AT-AC introns (AT-AC I introns) (Tarn and Steitz, 1996a). Finally, minor snRNAs U4atac and U6atac, which are the counterparts of the U4 and U6 snRNAs of the major form of spliceosome, have been identified (Tarn and Steitz, 1996b). Hence, this minor form of intron is spliced in a minor form of spliceosome that involves four U snRNAs distinct from those involved in the major form of spliceosome. However, the minor form of spliceosome seems to use the same U5 snRNA as the major form of spliceosome. Recently, a functionally distinct subclass of AT-AC introns (AT-AC II introns) was shown to be spliced by the major spliceosomal snRNAs (Wu and Krainer, 1997). Therefore, it is also possible that there are introns spliced by hybrid pathways, involving combinations of snRNAs or protein factors from the GT-AG and AT-AC I pathways. In both splicing machineries, the modified nucleotides may play a role. We started a study of the posttranscriptional modifications in the AT-AC specific spliceosomal snRNAs, and the results obtained will be described in this chapter. By several lines of evidence, there are similarities between the minor splicing machinery in vertebrates and the yeast splicing machinery, and this also will be discussed.

TECHNIQUES USED FOR DETECTION OF POSTTRANSCRIPTIONAL MODIFICATIONS IN U snRNAs AND U snoRNAs

The first identification of posttranscriptional modifications in U snRNAs were made long ago when RNA nucleotides sequences were determined directly using the chemical sequencing method of Peattie (1979). This method and two-dimensional, thin-layer chromatography were used for the identification of modified nucleotides. The 2′-O-methylations were easily detected after alkaline hydrolysis of end-labeled RNA because the phosphodiester bond in which a 2′-O-methyl nucleotide is implicated is resistant to hydrolysis.

One way to identify pseudouridine residues was the observation of an absence of reactivity of these residues to hydrazine in the chemical sequencing conditions described by Peattie. The presence of pseudouridine residues was verified by thin-layer chromatography (Harada et al., 1980; Reddy et al., 1981;

Krol et al., 1981a and 1981b, as examples). A rapid method for the detection of pseudouridine residues was recently described (Bakin and Ofengand, 1993) (also see Chapter 12) and used for the detection in the *S. pombe* (Gu et al., 1996) and *S. cerevisiae* snRNAs and in the human AT–AC snRNAs (our unpublished results).

Other base modifications in U6 snRNA (Gu et al., 1996; for reviews, see Reddy, 1988; Solymosy and Pollak, 1993) and U2 snRNA (Reddy et al., 1981; Westin et al., 1984; Gu et al., 1996) were identified by thin-layer chromatography.

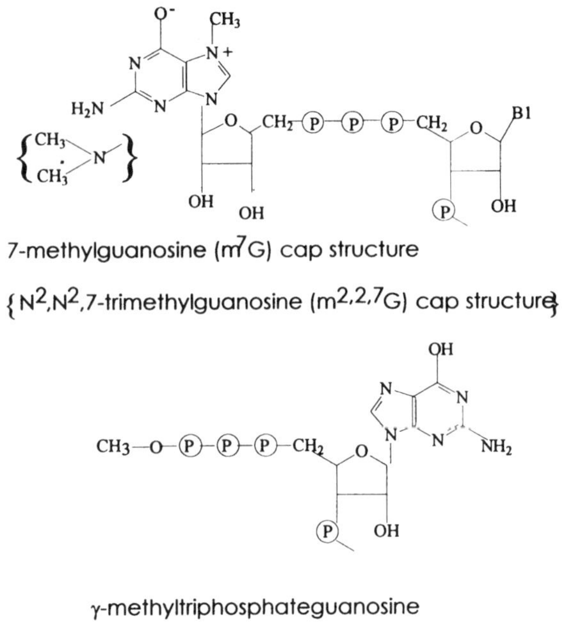

POSTTRANSCRIPTIONAL MODIFICATIONS OF NUCLEOLAR snoRNAS

5′-Terminal Posttranscriptional Modifications

All the nucleolar snoRNAs that were characterized in vertebrates, insects, and yeast were found to be transcribed by RNA polymerase II, either from a classical gene or from an intron sequence (for reviews, see Maxwell and Fournier, 1995; Bachellerie et al., 1995; Tollervey and Kiss, 1997; Smith and Steitz, 1997); to date, the only exceptions to this rule are the plant and the *Chlamydomonas reinhardtii* U3 snoRNAs, which are transcribed by RNA polymerase III (Kiss et al., 1985; Kiss and Solymosy, 1990; Marshallsay et al., 1990; Leader et al., 1994; Antal et al., personal communication). Hence, depending on the phylum, a gene coding for a given RNA can be transcribed by different RNA polymerases and, as a consequence, the RNAs produced have different 5′-terminal structures.

All RNA polymerase II transcripts (as well the premessenger RNAs and the U snRNAs) undergo addition of a monomethylated guanosine residue to the first encoded nucleotide via a 5′–5′ linkage to the 5′-terminal triphosphate (Fig. 3) (also see Chapter 10). This addition takes place shortly after the start of transcription (Shatkin, 1976; Salditt-Georgieff et al., 1980; Coppola et al., 1983). In addition, for all the characterized U snRNAs (the nucleolar and nucleoplasmic U snRNAs), this cap structure has the peculiarity to undergo two further methylations on position 2 of the terminal guanosine (Fig. 3) (Busch et al., 1982; Mattaj, 1986; for a review, see Reddy et al., 1992). In vertebrates, this is the case for the nucleolar U3, U8, and U13 RNAs (Tyc and Steitz, 1989; Reddy et al., 1979), the other snoRNAs being produced from introns. In yeast, a larger number of snoRNAs have a trimethylated cap structure, U3, snR3, R4, R5, R8, R9, R10, R11, R13, R30, R31, R32, R33, R34, R35, R37, and R189 (Tollervey, 1987; Riedel et al., 1986; Bally et al., 1988; Thompson et al., 1988; Balakin et al., 1993 and 1996; Samarsky et al., 1995).

Figure 3. The various 5′-cap structures of U snoRNAs and spliceosomal U snRNAs. The m^{7}G cap of the RNA polymerase II transcripts (Shatkin, 1976); N^2,N^2,7-trimethylguanosine cap structure (in square brackets) of the spliceosomal U1, U2, U4, U5, U11, U12, and U4atac snRNAs (Busch et al., 1982; Mattaj, 1986) and of the vertebrate U3, U8 and U13 snoRNAs (Tyc and Steitz, 1989; Reddy et al., 1979); and γ-methyl phosphate guanosine cap structure of U6 snRNA, U6atac snRNA (Singh and Reddy, 1989; Tarn and Steitz, 1996b), and plant U3 snoRNA (Marshallsay et al., 1990) are shown.

According to the analysis of Zieve and Penman (1976), the idea was that the precursor for U3 snoRNA, like the precursors for the spliceosomal U1, U2, U4, and U5 snRNAs, is exported to the cytoplasm where it is maturated (3′-terminal trimming and cap trimethylation), before being reimported into the nucleus. Based on *Xenopus laevis* oocyte microinjection assays, Baserga et al. (1992) also concluded that the biogenesis of the U3 snoRNP includes a cytoplasmic step. However, based on the same approach, a common maturation pathway was recently proposed for U3 and U8 snoRNAs that does not include a cytoplasmic step (Terns et al., 1995). Indeed, precursors of these snoRNAs were found to be retained and maturated in the nucleus. Box D (see Fig. 4), a conserved element found in these and several of the snoRNAs (Hughes et al., 1987), plays a key role in their nuclear retention, probably by binding to specific proteins. Retention of U3 and U8 snoRNAs in the nucleus is saturable and relies on one or more common factors. This saturation property may explain the observation of U3 snoRNA in the cytoplasm after microinjection of a large quantity of U3

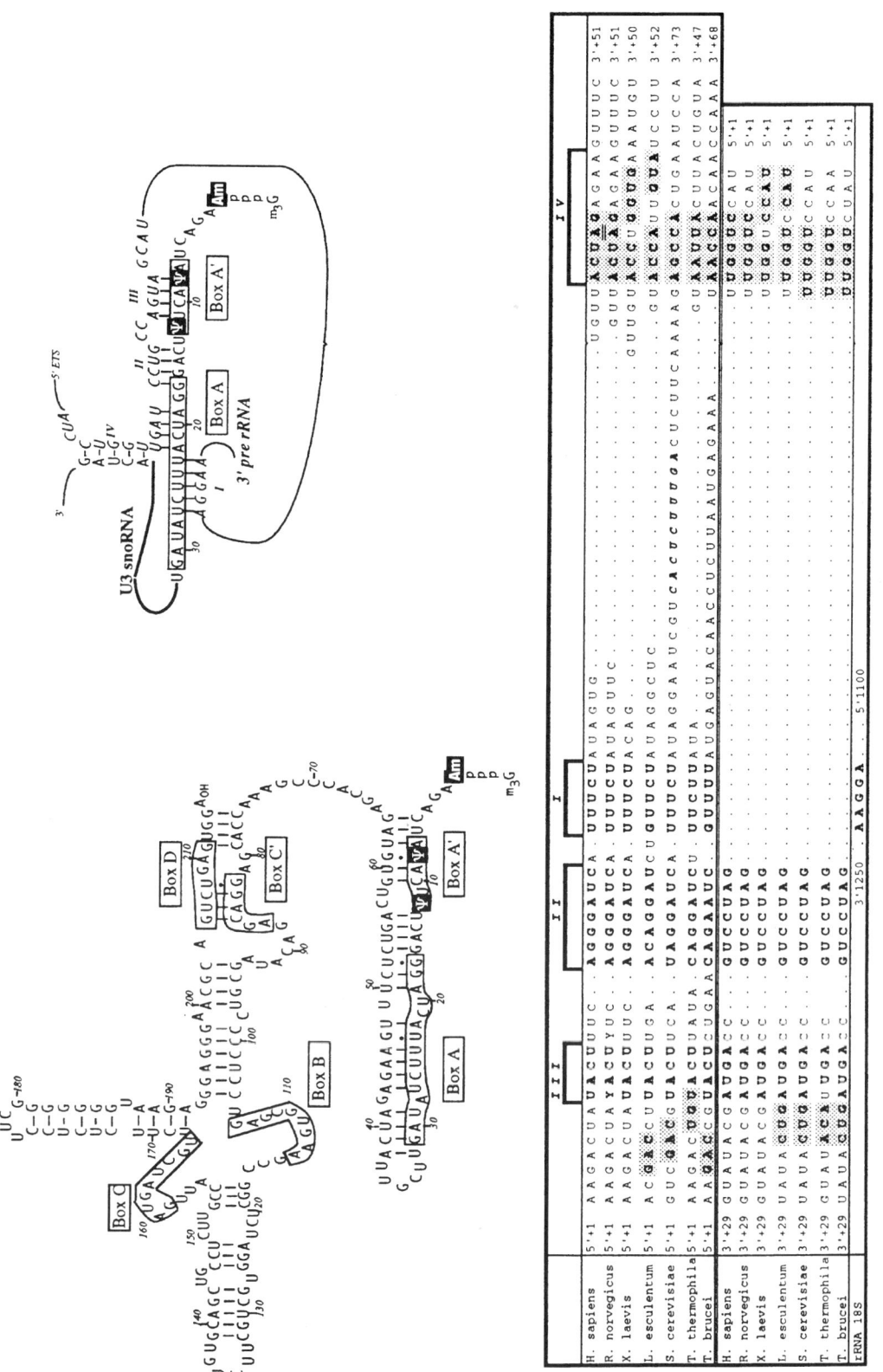

Figure 4. Rat U3B snoRNA. On the left, the possible secondary structure of the rat U3B snoRNA, based on the secondary structure experimentally established for the HeLa cell U3 snoRNA (Stroke and Weiner, 1985; Parker and Steitz, 1987). The primary sequence was from Reddy et al. (1985). The two pseudouridines identified in the rat hepatoma U3 snoRNA are shown (squared in black). The evolutionarily conserved sequences denoted Box A (Wise and Weiner, 1980), A' (Myslinski et al. 1990), B (Wise and Weiner, 1980), C (Wise and Weiner, 1980), C' (Tycowski et al., 1993) and D (Hughes et al., 1987) are indicated. One the right, interaction of U3 snoRNA with 18S pre-rRNA, as proposed by Hughes (1996) (helices I, II, and III) and Méreau et al. (1997) (helix IV). In the table at the bottom, phylogenetic conservation of the heterologous helices, I, II, III and IV between U3 snoRNA and 18S rRNA sequences according to Hughes (1996), and Méreau et al. (1997) is shown. The 5' domains of U3 snoRNAs from a wide range of organisms (top) are aligned to the complementary sequences in the respective SSU (small subunit) rRNA (bottom). The sequences are derived from the following GenEMBL database files, in order of listing in the table: U3 snoRNAs, M14061, K00780, X07318, X14411, M26649, X63788, and S74818; SSU rRNAs, X03205, V01270, XLRN01, X51576, J01353, M10932, and 152676. The putative extension of helix III in lower eukaryotes and the putative helix IV are in grey boxes.

snoRNA in the nucleus of X. laevis oocytes (Baserga et al., 1992). In agreement with the observation of a nuclear retention of U3 and U8 snoRNAs, Terns and Dahlberg (1994) observed the presence of a nuclear methylating activity that specifically converts the m^7G cap of U3 snoRNA into an $m^{2,2,7}G$ cap. Trimethylation of U3 snoRNA can also occur in the cytoplasm, but is much slower than in the nucleus (Baserga et al., 1992), suggesting that in vivo U3 snoRNA is mainly hypermethylated in the nucleus. The methylase activity is probably located within the nucleoplasm because hypermethylation can occur in nuclear extracts lacking nucleoli (Terns and Dahlberg, 1994). Hence, in vertebrates, after transcription by polymerase II, U3 snoRNA is retained within the nucleoplasm by a mechanism to be identified, and after maturation and assembly into a U3 snoRNP is transported to the nucleolus by a mechanism that involves RCC1, the abundant nuclear guanine nucleotide exchange factor for the guanosine triphosphatase Ran/TC4 (Cheng et al. 1995).

Internal Posttranscriptional Modifications

Studies on the internal posttranscriptional modifications of snoRNAs have been very limited. They were restricted to the rat hepatoma (Reddy et al., 1985) and *Saccharomyces cerevisiae* (our unpublished results) U3 snoRNAs and to the mouse U8 snoRNA (Kato and Harada, 1984) (Table 1).

In rat U3 snoRNA, three modified nucleotides were detected (Fig. 4): a 2′-O-methyladenosine residue on the 3′ side of the cap triphosphate, and two pseudouridines that are located in the highly phylogenetically conserved box A′ (Reddy et al., 1985; Myslinski et al., 1990) (see Fig. 4). U3 snoRNA is known to be required for the first nucleolytic cleavage at the 5′ extremity of the 45S pre-rRNA precursor (Kass et al., 1990; Hughes and Ares, 1991; Mougey et al., 1993) and it was also shown to be required for further endonucleolytic steps that generate the mature 18S rRNA (Li et al., 1990; Hughes and Ares, 1991; Morrisey and Tollervey, 1993). In yeast, U3 snoRNA was first shown to base pair with a 10-nt segment of the 5′-external spacer (5′ETS) of the 45S pre-rRNA (Beltrame et al., 1994; Brulé et al., 1996) (Fig. 4). More recently, it was proposed to interact with 18S rRNA sequences from the precursor rRNA that are involved in the formation of a pseudoknot structure in mature 18S rRNA (Hughes, 1996; Méreau et al., 1997) (see Fig. 4). Using DMS as the probe, we got very strong arguments in favor of the formation of this base pair interaction in vivo (Méreau et al., 1997). Interestingly, Box A′, which contains the two pseudouridines, is involved in one of the proposed base pair interactions with 18S rRNA sequences (helix III [Fig. 4]). Based on our recent analysis, there is no pseudouridine in *Saccharomyces cerevisiae* U3 snoRNA. In this species, as well as in the other lower eukaryotes studied, the possible interaction between the 5′-terminal region of U3 snoRNA including box A′ and the 18S rRNA sequence (helix III) is three base pairs longer than in vertebrates (Fig. 4) (Hughes, 1996). Substitution of uridines by pseudouridines is known to stabilize RNA–RNA interactions (for a review, see Agris, 1996) (also see Chapters 5 and 6). Hence, it might be that the presence of pseudouridines in the box A′ of the vertebrate U3 snoRNAs compensates for the lower complementarity between box A′ and the 18S rRNA.

The mouse U8 snoRNA was found to contain a 2′-O-methylated adenine and a 2′-O-methylated uridine on the 3′ side of the triphosphate cap, and a pseudouridine residue at position 20 (Fig. 5) (Kato and Harada, 1984). U8 snoRNA was shown to be required for the formation of 5.8S and 28S rRNAs (Peculis and Steitz, 1993). The sequence of the 5′ end of this snoRNA is able to base pair with the 5′ end of 28S rRNA (Peculis, 1997). The role of this base pairing is to prevent the formation of premature 5.8S–28S or inappropriate 28S interactions during the pre-rRNA maturation process. It facilitates pre-rRNA processing and U8 snoRNA works as a "guide" to initiate the 5.8S–28S interaction essential for the completion of ribosome biogenesis (Peculis, 1997). None of the modified nucleotides is implicated in this U8-28S interaction (see Fig. 5). Hence, at the moment it is difficult to know whether there is a link between the presence of a pseudouridine at position 20 in U8 snoRNA and the mechanism of action of this molecule.

POSTTRANSCRIPTIONAL MODIFICATIONS OF THE SPLICEOSOMAL U snRNAs INVOLVED IN THE SPLICING OF THE GT–AG INTRONS

5′-Terminal Structure and Nuclear Export

Except for U6 snRNA, which is transcribed by RNA polymerase III (Kunkel et al., 1986; Reddy et al., 1987; Krol et al., 1987) (but not for the slime mold *P. polycephalum* [see the introduction]), the other spliceosomal RNAs are transcribed by RNA polymerase II (for reviews, see Dahlberg and Lund, 1988; Reddy et al., 1992). U6 snRNA is retained in the nucleus probably because of the γ methylation of its 5′-terminal tri-phosphate (Fig. 3) (Singh and Reddy, 1989). Like the precursors of the vertebrate

Table 1. Distribution of intramolecular nucleotide modifications in analyzed spliceosomal U snRNAs and U snoRNAs

RNA[n]	Modifications[n]
U1	
Hs[a]	Am$_1$ Um$_2$ Ψ$_5$ Ψ$_6$ ... Am$_{70}$
Rn[a]	Am$_1$ Um$_2$ Ψ$_5$ Ψ$_6$ ... Am$_{70}$
Gg[a]	Am$_1$ Um$_2$ Ψ$_5$ Ψ$_6$ ... Am$_{70}$
Mm[a]	Am$_1$ Um$_2$ Ψ$_5$ Ψ$_6$ ... Am$_{70}$
Dm[a]	Am$_1$ Um$_2$ Ψ$_5$ Ψ$_6$ ... Am$_{70}$
Cs[b]	Ψm$_5$ Um$_6$
Sp[c]	Ψ$_3$
Sc[g]	Ψ$_5$ Ψ$_6$
U2	
Hs[d]	Am$_1$ Um$_2$ Gm$_{11}$Gm$_{12}$ Gm$_{19}$ Gm$_{25}$ Am$_{30}$ Cm$_{40}$ Ψ$_{41}$ Ψ$_{43}$ Ψ$_{44}$ Um$_{47}$ Ψ$_{54}$ Ψ$_{58}$ (Ψ$_{60}$) Cm$_{61}$ (Ψ$_{69}$)(Ψ$_{72}$)(Ψ$_{89}$) Ψ$_{91}$
Rn[e]	Am$_1$ Um$_2$ Ψ$_6$ Ψ$_7$ Gm$_{11}$Gm$_{12}$ Ψ$_{15}$ Gm$_{10}$ Gm$_{25}$ m^{5}Am$_{30}$ Ψ$_{34}$ Ψ$_{37}$ Ψ$_{39}$ Cm$_{40}$ Ψ$_{41}$ Ψ$_{43}$ Ψ$_{44}$ Um$_{47}$ Ψ$_{54}$ Ψ$_{58}$(Ψ$_{60}$) Cm$_{61}$ Ψ$_{58}$ Ψ$_{60}$
Vf[a]	Ψ$_6$ Ψm$_7$ Ψ$_9$ Ψ$_{15}$ Ψ$_{17}$ Gm$_{19}$ Gm$_{25}$ Cm$_{28}$ Am$_{30}$ Ψ$_{34}$ Ψ$_{37}$ Am$_{38}$ Ψ$_{39}$ Cm$_{40}$ Ψ$_{41}$ Ψ$_{43}$ Ψ$_{44}$ Ψ$_{58}$
Sp[c,f]	Um$_9$ Um$_{11}$ Gm$_{25}$ m^{6}A$_{30}$ Ψ$_{34}$ Ψ$_{37}$ Ψ$_{39}$ Cm$_{40}$ Ψ$_{41}$ Ψ$_{43}$ Ψ$_{44}$ Ψ$_{42}$ Ψ$_{44}$
Sc[g]	Ψ$_{35}$
U12	
Hs[g]	Ψ$_{19}$ Ψ$_{28}$
U4	
Hs[b]	Am$_1$ Gm$_2$ Ψ$_4$ Cm$_8$
Rn[b]	Am$_1$ Gm$_2$ Ψ$_4$ Cm$_8$
Gg[b]	Am$_1$ Gm$_2$ Ψ$_4$ Cm$_8$
Mm[a]	Am$_1$ Gm$_2$ Ψ$_4$ Cm$_8$
Dm[a]	Am$_1$
Vf[i]	Ψm$_5$ Ψm$_{11}$ Ψ$_{38}$
U4atac	
Hs[g]	Ψ$_{12}$
U5	
Hs[a]	Am$_1$ Um$_2$ Gm$_{37}$ Um$_{41}$ Ψ$_{43}$ Cm$_{45}$ Ψ$_{46}$ Ψ$_{53}$ Am$_{65}$ Ψ$_{72}$ Ψ$_{79}$
Rn[a]	Am$_1$ Um$_2$ Gm$_{37}$ Um$_{41}$ Ψ$_{43}$ Cm$_{45}$ Ψ$_{46}$ Ψ$_{53}$ Am$_{65}$ Ψ$_{72}$ Ψ$_{79}$
Mm[a]	Am$_1$ Um$_2$ Gm$_{37}$ Um$_{41}$ Gm$_{35}$ Um$_{39}$ Ψ$_{41}$ Cm$_{43}$ Ψ$_{44}$ Am$_{65}$ Ψ$_{72}$ Ψ$_{79}$
Cc[a]	m^{6}A$_1$ Um$_2$ Cm$_3$ Ψ$_{10}$ Ψ$_{29}$ Ψ$_{32}$ Ψ$_{33}$ Gm$_{36}$ Ψm$_{40}$ Ψ$_{42}$ Cm$_{44}$ Ψ$_{45}$ Ψ$_{52}$ Am$_{65}$ Ψ$_{72}$ Ψ$_{79}$
Tt[a]	Am$_1$ Um$_2$ Um$_{30}$ Ψ$_{33}$ Gm$_{36}$ Ψm$_{40}$ Ψ$_{42}$ Cm$_{44}$ Ψ$_{45}$ Ψ$_{55}$ Am$_{65}$ Ψ$_{131}$ Ψ$_{132}$
Ps[a]	Ψ$_{34}$ Um$_{35}$ Gm$_{37}$ Um$_{41}$ Ψ$_{43}$ Cm$_{45}$ Ψ$_{46}$ Ψ$_{57}$ Am$_{67}$ Ψ$_{74}$
Pp[i]	Am$_1$ Cm$_2$ Ψ$_{12}$ Ψ$_{14}$ Ψ$_{30}$
Dm[j]	Am$_1$ Um$_2$ Ψ$_{32}$ Gm$_{38}$ Um$_{42}$ Ψ$_{44}$ Cm$_{46}$ Ψ$_{47}$ Ψ$_{54}$
Cr[i]	Am$_1$ Um$_2$ Gm$_{24}$ Ψ$_{31}$ Ψ$_{36}$ Gm$_{39}$ Um$_{43}$ Ψ$_{45}$ Cm$_{47}$ Ψ$_{48}$
Tp[i]	Am$_1$ Um$_2$ Ψ$_{33}$ Gm$_{36}$ Um$_{42}$ Ψ$_{44}$ Cm$_{44}$ Ψ$_{45}$ Ψ$_{52}$
Sp[c]	Ψ$_{38}$ Gm$_{46}$ Um$_{50}$ Ψ$_{52}$ Cm$_{54}$
Sc[g]	Ψ$_{99}$
U6	
Rn[a]	Ψ$_{31}$ Ψ$_{40}$ m^{6}A$_{43}$ Am$_{47}$ Am$_{53}$ Gm$_{54}$ Cm$_{60}$ Cm$_{62}$ Cm$_{63}$ Am$_{70}$ mG$_{72}$ Cm$_{77}$ Ψ$_{86}$

Continued on following page

Table 1. Continued

RNA	Modifications[n]
Mm[a]	Ψ_{31} Ψ_{40} m^{6}A$_{43}$ Am$_{47}$ Am$_{53}$ Gm$_{54}$ Cm$_{60}$ Cm$_{62}$ Cm$_{63}$ Am$_{70}$ mG$_{72}$ Cm$_{77}$ Ψ_{86}
Vf[k]	Ψ_{26} m^{6}A$_{38}$ Am$_{42}$ Am$_{48}$ Cm$_{57}$ Am$_{65}$ Gm$_{75}$ Cm$_{84}$ Ψ_{90}
Sp[c]	Ψ_{26} Gm$_{25}$ Ψ_{40} m^{6}A$_{37}$ Am$_{41}$ Cm$_{56}$ Cm$_{57}$ Cm$_{63}$
U3	
Rn[l]	Am$_1$ Ψ_8 Ψ_{12}
U8	
Mm[m]	Am$_1$ Um$_2$ Ψ_{20}

[a] Reddy, 1988 (review).
[b] Kiss et al., 1988a.
[c] Gu et al., 1996.
[d] Westin et al., 1984.
[e] Reddy et al., 1981. The pseudouridines in parentheses are additional pseudouridines found in a study of posttranscriptional modifications in the 5' half of rat liver and brain snRNA (Branlant et al., 1982).
[f] For the S. pombe U2 snRNA, the presence of Ψ59 is uncertain, as it was shown in a figure but not cited in the text in a paper by Gu et al. (1996). No Ψ or 2'-O methylation was detected in S. pombe U4 snRNA.
[g] Massenet et al., unpublished results. Only the presence of pseudouridines was studied in the S. cerevisiae U1, U2, U4, U5, and U6 snRNAs and in the atac-specific U snRNAs from HeLa cells (U11, U12, U4atac, and U6atac). No pseudouridine was detected in the S. cerevisiae U4 or U6 snRNA or in the HeLa cell U11 or U6atac snRNA.
[h] Branlant et al., 1982.
[i] Kiss et al., 1988b.
[j] Szkukálek et al., 1995.
[k] Solymosy and Pollák, 1993 (review).
[l] Reddy et al., 1979.
[m] Kato and Harada, 1984.
[n] Abbreviations: Hs, Homo sapiens; Rn, Rattus norvegicus; Gg, Gallus gallus; Mm, Mus musculus; Vf, Vicia faba; Ps, Pisum sativum; Dm, Drosophila melanogaster; Cs, Chlorella saccharophila; Cr, Chlamydomonas reinhardtii; Cc, Crypthecodinium cohnii; Tp, Tetrahymena pyriformis; Tt, Tetrahymena thermophila; Sp, Schizosaccharomyces pombe; Sc, Saccharomyces cerevisice; Am, Gm, Um and Cm, 2'-O-methyl adenosine, 2'-O-methyl guanosine, 2'-O-methyl uridine, and 2'-O-methyl cytidine, respectively; m^{6}A, N^6-methyladenosine; Ψ, pseudouridine; mG, 2-methylguanosine. All these U snRNAs are capped by a 5' m$_3$G (N^{2,2,7}-trimethylguanosine), except the U6 and U6atac snRNAs, which are capped by a 5' γmG (γ-monomethyltriphosphate guanosine).

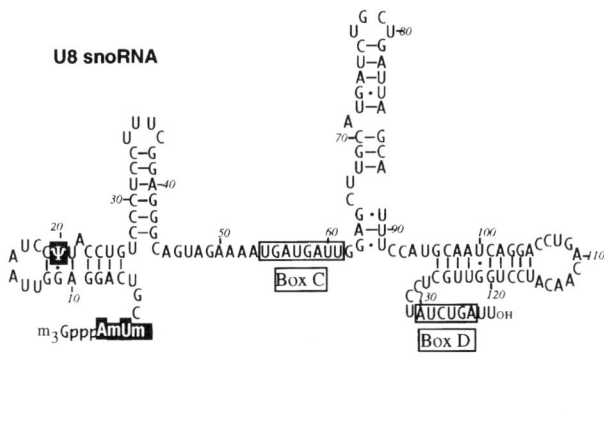

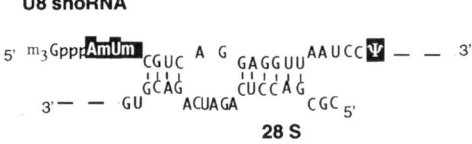

Figure 5. Secondary structure proposed for mouse U8 snoRNA. The sequence was from Kato and Harada (1984), and the structure was proposed by Tyc and Steitz (1989). The evolutionarily conserved sequences denoted Box C and D are indicated, as well as the identified posttranscriptional modifications. The interaction of U8 snoRNA with 28S rRNA, as proposed by Peculis (1997), is shown. The posttranscriptional modifications are squared in black.

U3 and U8 snoRNAs, after transcription by RNA polymerase II, the precursors of the U1, U2, U4, and U5 snRNAs carry a monomethyl-G-cap structure (for reviews, see Busch et al., 1982; Dahlberg and Lund, 1988). However, in contrast to the U snoRNA precursors, the U1, U2, U4, and U5 snRNA precursors are transported to the cytoplasm where they are maturated (hypermethylation of the cap structure, 3' trimming and binding of Sm proteins) (see Fig. 6).

As a first step in nuclear export of U1, U2, U4, and U5 snRNA precursors, the monomethyl-G-cap structure is recognized by the nuclear cap-binding-protein complex (CBC), which was first identified by Ohno et al. (1990) (Fig. 6). This complex is composed of two proteins, CBP80 and CBP20 (Izaurralde et al., 1994) (Fig. 6), and it is also involved in the recognition of the 5'-cap structure of premessenger RNAs (Jarmolowski et al., 1994). Protein CBP20 belongs to the large family of proteins containing a conserved (RNP) motif, including the highly conserved RNP1 and RNP2 sequences. However, it does not exhibit specific cap-binding activity in the absence of protein CBP80 (Izaurralde et al., 1995). Demonstration of the role of protein CBP20 in pre-U1 and pre-U5 snRNA export was shown by microinjection of CBP20 antibodies in *X. laevis* oocytes (Izaurralde et al., 1995). The effect of cap structure elimination was found to be much more dramatic for U snRNA export than for mRNA export, probably because other proteins like hnRNP A1 are also involved in mRNA export (Piñol-Roma and Dreyfuss, 1992; Michael et al., 1995; Lee et al., 1996; Visa et al., 1996). However, based on microinjection of various U1 snRNA mutants, the m7G cap structure was proposed to be required but not sufficient for nuclear export of U1 snRNA. Both the 3' terminal stemloop structure and sequence in the 5' terminal 124-nt region were proposed to contribute to efficient export of this RNA (Terns et al., 1993). Nevertheless, in this kind of experiment, it is somehow difficult to make a distinction between the effect on export and the effect on RNA stability. By using U1 snRNA derivatives carrying different cap structures as competitors of export, it was demonstrated unequivocally that the export of U snRNAs requires saturable factors that bind to their 5' caps, CBP20 and CBP80 proteins (Jarmolowski et al., 1994). Their distribution in the nucleoplasm was shown to be relatively uniform (Izaurralde et al., 1994; Visa et al., 1996). The yeast counterpart of the vertebrate CBP20, the protein Mud13p, was recently isolated (Colot et al., 1996) and it was shown to be involved in the formation of splicing commitment complexes. Protein CBP80 contains leucine-rich sequences that somehow resemble the hydrophobic nuclear export signal (NES). However, there is no evidence that these sequences are required for CBC function in export. CBC acts rather by the recruitment of a NES-containing polypeptide to the RNP (for a review, see Görlich and Mattaj, 1996). Recent results demonstrate that importin-α, a protein first characterized as a protein nuclear localization signal (NLS) receptor (Görlich et al., 1995a and 1995b), forms a trimeric complex with proteins CBP20 and CBP80 and this trimeric complex binds RNA in a cap-dependent fashion, which is consistent with the idea that the CBC RNA complex is exported from the nucleus as a complex with importin-α (Görlich et al., 1996a) (Fig. 6).

In addition to proteins CBP20, CBP80 and importin-α, the GTPase RAN/TC4 and its nucleotide exchange factor RCC1 were found to be required for nuclear export of spliceosomal RNAs (Kadowaki et al., 1992 and 1993; Belhumeur et al., 1993; Amberg et al., 1993; Schlenstedt et al., 1995; Cheng et al., 1995). The abundant RCC1 protein was found to stimulate GDP to GTP exchange at the catalytic center of the Ran/TC4 GTPase (Moore and Blobel, 1994; Bischoff et al., 1995). Ran is essentially present in a form associated with GDP in the cytoplasm and it plays a function in protein import (Görlich et al., 1996b). In contrast, due to the activity of RCC1 in the nucleus, Ran is essentially present in a form associated with GTP in this compartment (Ohtsubo et

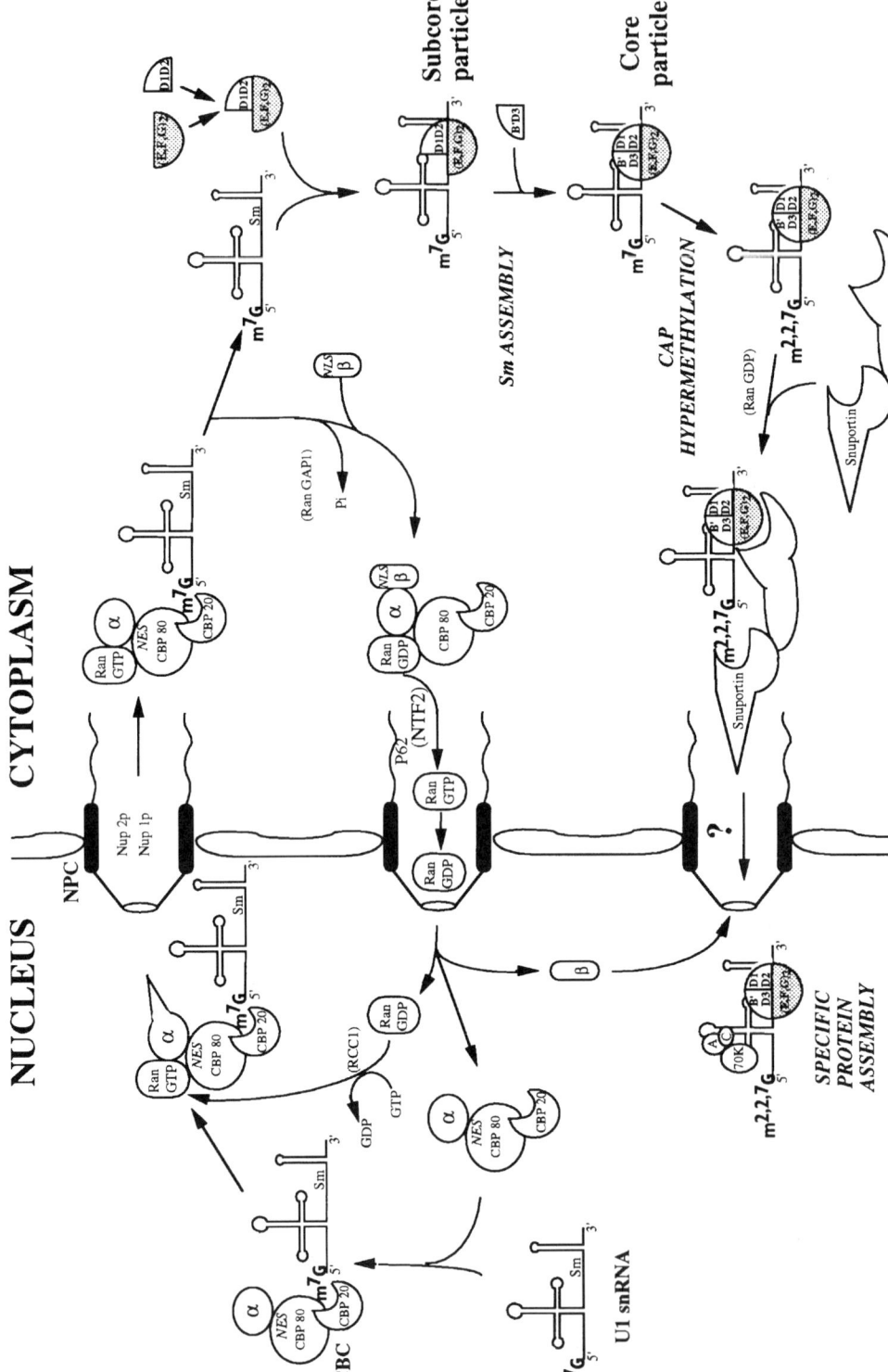

Figure 6. A model for the nuclear transport cycle of the spliceosomal U snRNPs. In the nucleus, newly transcribed snRNA (U1 as an example) binds the CBC-importin-α complex, and the Ran/TC4 guanosine triphosphatase in its GTP-associated form (Ran GTP). This complex is exported to the cytoplasm through the nuclear pore complex (NPC). Importin-α (α) interacts with the pore proteins Nup1p and Nup2p. In the cytoplasm, GTP is hydrolyzed in the presence of Ran GAP1, association of importin-β promotes the liberation of the m^7G U snRNA, and the protein complex (Ran; importins α, β; CBC) is allowed to move back to the nucleus. The m^7G U snRNAs associate with the Sm proteins in a sequential pathway, and hypermethylation of the cap structure occurs. Both events provide the bipartite nuclear localization signal (NLS), recognized by a cytosolic factor (snuportin) which drives the core snRNP back to the nucleus, where specific proteins can associate. (For references, see the text.)

al., 1989; Bischoff and Ponstingl, 1991; Cheng et al. 1995). It associates to the CBC-importin-α-U snRNA complex to allow its translocation through the NPC where importin-α interacts with the NPC proteins Nup1p and Nup2p (Belanger et al., 1994) (Fig. 6).

Once in the cytoplasm, importin-α forms a heterodimer with importin-β (Görlich et al., 1996a) (Fig. 6), which results in the dissociation of the RNA from the RNA-CBC-importin-α complex delivered to the cytoplasm (Görlich et al., 1996c). Based on these data, it was proposed that the ability of importin-β to dissociate the export complex is blocked by Ran GTP in the nucleus (Görlich et al., 1996c). In the cytoplasm, the Ran-GAP1-GTPase activating protein activates the hydrolysis of GTP by the Ran protein, allowing the binding of importin-β. The importin-β has an NLS sequence that drives back the CBC-importin-α complex to the nucleus, to enter a new transport cycle (Görlich et al., 1996a). This implicates that the binding of Ntf2 (linked to the nucleoporin P62) to Ran-GDP and a GDP to GTP exchange, followed by GTP hydrolysis, allow translocation of the import complex across the central channel of the NPC (Wong et al., 1997; Panté and Aebi, 1996).

Hence, in the present stage of the knowledge, the presence of the m^7G cap structure on U1, U2, U4, and U5 snRNA precursors allows recognition by the cap-binding protein complex (CBP) and associated proteins and this mediates nuclear export. Although bearing the same m^7G cap structure, the nucleolar U3, U8, and U13 snoRNAs escape from this process by a retention mechanism involving the phylogenetically conserved Box D.

5'-Terminal Structure and Nuclear Import

After their liberation in the cytoplasm, the precursors of U1, U2, U4, and U5 snRNAs associate with a common set of eight proteins, called the Sm proteins (Lerner and Steitz, 1979; Liautard et al., 1982; Kastner et al., 1990) (Fig. 6). The Sm binding sites of the 4 snRNAs is a single-stranded segment with the consensus sequence $PuA(U)_{3-6}GPu$, which is flanked by two stable helical structures (Branlant et al., 1982; Jarmolowski and Mattaj, 1993). Correct assembly of the eight Sm proteins is needed to get the conversion of the m^7G-cap structure into a $m^{2,2,7}G$-cap structure (Mattaj, 1986; Plessel et al., 1994). Trimming also occurs at the 3' end of the precursor molecules, but the information on this step is very limited (Jacobson et al., 1993; Huang et al., 1997; Chanfreau et al., 1997). Formation of the U snRNP core complexes, upon association of U snRNAs with the Sm proteins, involves both protein–protein and RNA–protein interactions and the assembly pathway of the U1 snRNP core complex has been established in vitro (Raker et al., 1996) (Fig. 6). The Sm proteins E, F, and G first form an hexamer containing two copies of each protein. In association with proteins D1 and D2, the E.F.G. hexamer can form a stable complex with U1 snRNA (U1 snRNP-subcore particle), which is converted into a U1 snRNP-core particle by association with a preformed B/B'.D3 complex. When microinjected into X. laevis oocyte cytoplasm, this reconstituted U1-core snRNP particle is the substrate of the trans-active-snRNA-(guanosine-N2)-methyl transferase leading to the trimethylated cap structure and is transported to the nucleus (Plessel et al., 1994).

Both binding of Sm proteins and hypermethylation of the cap provide the two parts of the bipartite NLS of the core-U snRNP particles (Mattaj and De Robertis, 1985; Fisher and Lührmann, 1990; Hamm et al., 1990). Only the presence of the core Sm protein complex seems to be completely indispensable for nuclear import of UsnRNAs (Fisher et al., 1993). The importance of the m_3G cap-structure for nuclear import is variable from one U snRNA to the others. It is more marked for U1 and U2 snRNAs than for U4 and U5 snRNAs. U snRNPs containing only the Sm-core RNP domain can enter the nucleus (Fisher et al., 1993), which suggests that the m_3G cap and the Sm-core-RNP domain may be recognized independently by the nuclear receptor(s) involved in the nuclear import of UsnRNPs, with the binding of the second element certainly strengthening the signal of the first one. In agreement with this hypothesis, a specific m_3G-cap-binding protein, called snuportin, was cross-linked to an RNA oligonucleotide corresponding to the 5' terminus of U1 snRNA (Marshallsay and Lührmann, 1994; Huber and Lührmann, personal communication). Once in the nucleus, completion of the snRNP is performed by the assembly of the specific proteins. Until now, nothing is known about the step(s) of U snRNP assembly where internal posttranscriptional modifications of U snRNA take place. Hence U1, U2, U4, and U5 snRNAs' biogenesis involves a bidirectional traffic of the U snRNAs throughout the NPC, and the posttranscriptional modifications at the 5' end of these RNAs plays an essential role in this traffic.

The Internal Posttranscriptional Modifications and Their Biogenesis

Internal posttranscriptional modifications in spliceosomal snRNAs were studied for four vertebrate species (chicken, mouse, and human [Harada et al., 1980; Kato and Harada, 1981; Krol et al., 1981a,

1981b; Branlant et al., 1982] and rat [for a review, see Reddy, 1988]), for the insect *Drosophila melanogaster* (Myslinski et al., 1984), for plants (Kiss et al., 1988a, 1988b; Solymosy and Pollàk, 1993), for the fission yeast *S. pombe* (Gu et al., 1996) and for the budding yeast *S. cerevisiae* (our unpublished data). The study was extended to a larger number of species in the case of U5 snRNA (Fig. 7).

The identity and location of all the detected posttranscriptional modifications are shown in Table 1. In addition to 2'-O-methylations, they essentially consist of pseudouridines and a few methylations at position 6 of adenosines (Table 2). For most of the U snRNAs, modifications are clustered in specific areas of the molecules and they are highly conserved throughout evolution: the 5' terminal region of U1 snRNA, the 5' third of U2 snRNA, the 5'-terminal and central region of U4 snRNA, the terminal loop of stem-loop structure 1 of U5 snRNA, and the central region of U6 snRNA (Table 1; Fig. 7, 8, and 9A).

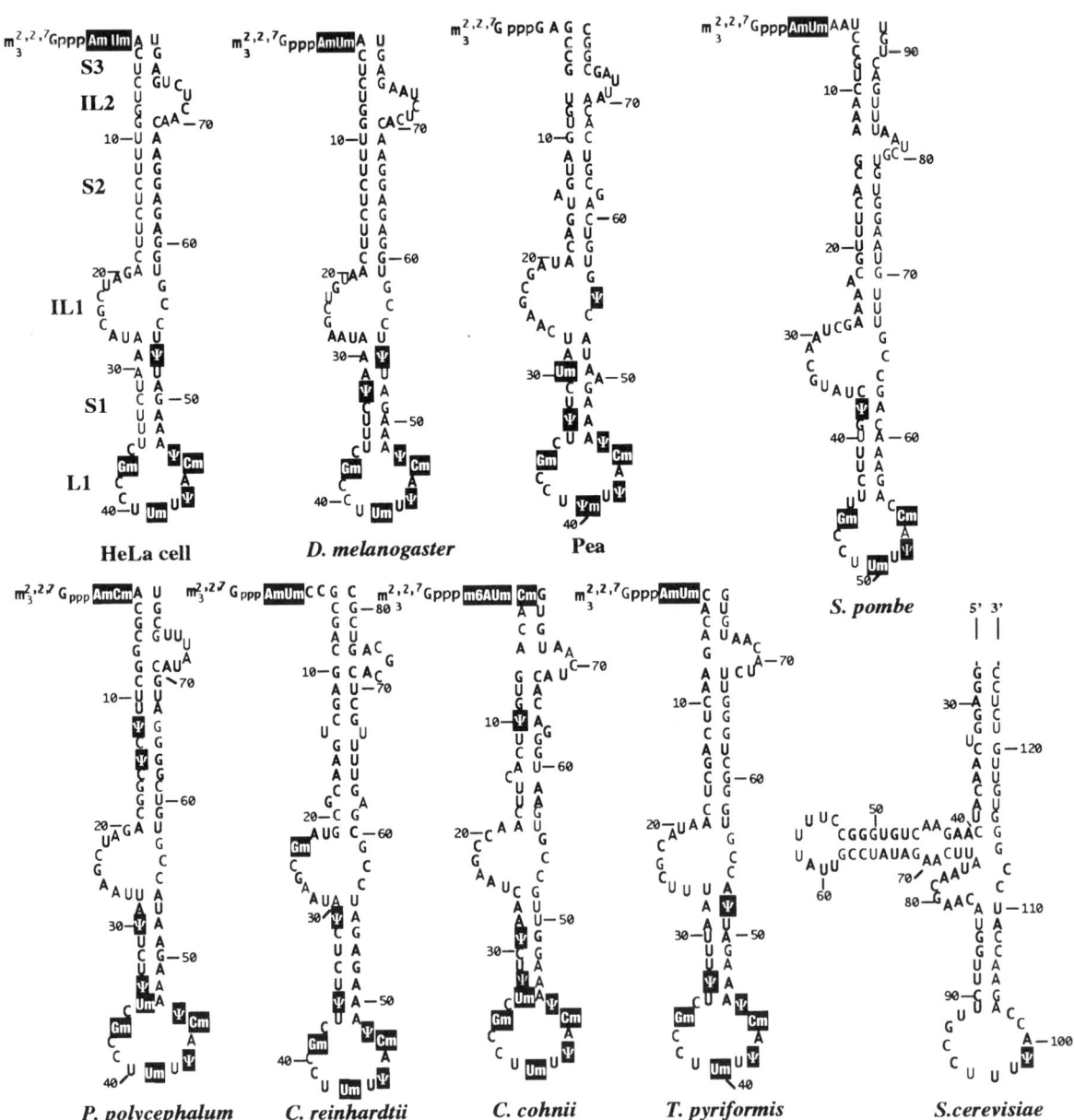

Figure 7. A comparison of modified nucleotides in U5 snRNAs from different species. Only the stem-loop structure I that contains modified nucleotides is shown. The nucleotide sequences are from HeLa cells (Reddy, 1988), *Drosophila melanogaster* (Myslinski et al., 1984), Pea (Krol et al., 1983), *Saccharomyces pombe* (Gu et al., 1996), *Physarum polycephalum* (Szkukàlek et al., 1995), *Chlamydomonas reinhardtii* (Jakab et al., 1992), *Crypthecodinium cohnii* (Reddy et al., 1983), *Tetrahymena pyriformis* (Branlant et al., 1983) and *Saccharomyces cerevisiae* (Massenet et al., unpublished results). The posttranscriptional modifications are squared in black.

Table 2. Internal posttranscriptional modifications in analyzed U snRNAs and spliceosomal U snRNAs[a]

Common name	Symbol	*Chemical Abstracts Index* name
N^6-Methyladenosine	m^6A	Adenosine, N-methyl-
2'-O-Methyladenosine	Am	Adenosine, 2'-O-methyl-
N^6,2'-O-Methyladenosine	m^6Am	Adenosine, N-methyl-2'-O-methyl-
2'-O-Methylcytidine	Cm	Cytidine, 2'-O-methyl-
N^2-Methylguanosine	m^2G	Guanosine, N-methyl-
2'-O-Methylguanosine	Gm	Guanosine, 2'-O-methyl-
Pseudouridine	Ψ	2,4(1H,3H)-Pyrimidinedione, 5-β-D-ribofuranosyl-
2'-O-Methyluridine	Um	Uridine, 2'-O-methyl-
2'-O-Methylpseudouridine	Ψm	2,4(1H,3H)-Pyrimidinedione, 5-(2-O-methyl-β-D-ribofuranosyl)-

[a] For a review, see Solymosy and Pollàk (1993). The chemical modification structures are shown in Appendix 1.

With respect to the established secondary structures of U1, U2, U4, U5, and U6 snRNAs (Branlant et al., 1981; Mount and Steitz, 1981; Krol et al. 1981a, 1981b; Branlant et al. 1982; Myslinski and Branlant, 1991; Zavanelli et al., 1994), a large number of the posttranscriptionally modified nucleotides are located in RNA segments predicted to be single stranded.

Both 2'-O-methylations and pseudouridines in *S. pombe* U snRNAs as well as pseudouridines in *S. cerevisiae* U snRNAs were studied (Table 1). Surprisingly, U snRNA posttranscriptional modifications are considerably less numerous in these two species than in all the other species studied, and this difference is more marked for *S. cerevisiae* than for *S. pombe*. We detected only six pseudouridines in the five spliceosomal U snRNAs of *S. cerevisiae*: two in U1 snRNA, three in U2 snRNA (against 12 in rat hepatoma) and one in U5 snRNA.

Little is known on the biogenesis of the U snRNA internal modifications. Only the in vitro conversion of uridines into pseudouridines of U snRNAs produced by in vitro transcription was studied in HeLa cell nuclear extract or S100 extract (for a review, see Patton, 1994a). The strategy used to determine whether synthesis of pseudouridine moieties in U1, U2, and U5 snRNAs was catalyzed by the same or by different pseudouridine synthases was to test for the pseudouridine synthase activity of HeLa cell extracts, in the presence of U1, U2, or U5 snRNA containing 5-fluorouridine (5-FU). Indeed, RNAs containing 5-FU at positions where uridines are converted into pseudouridines were previously shown to

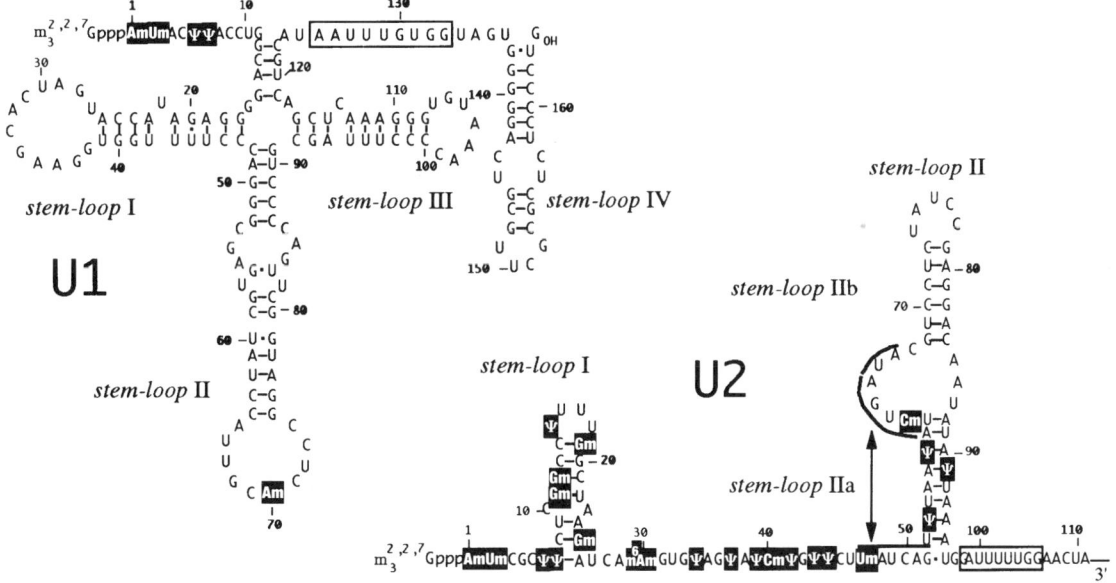

Figure 8. Secondary structures of rat spliceosomal U1 and U2 snRNAs. The secondary structures are from Branlant et al. (1981) and Mount and Steitz (1981) (U1) and Branlant et al. (1982) and Zavanelli et al. (1994) (U2). For the U2 snRNA structure, the sequences implicated in the proposed pseudoknot (Zavanelli et al., 1994) are overlined and joined by a double arrow. The Sm binding sites are boxed. The 3' extremity is not shown. The posttranscriptional modifications are squared in black.

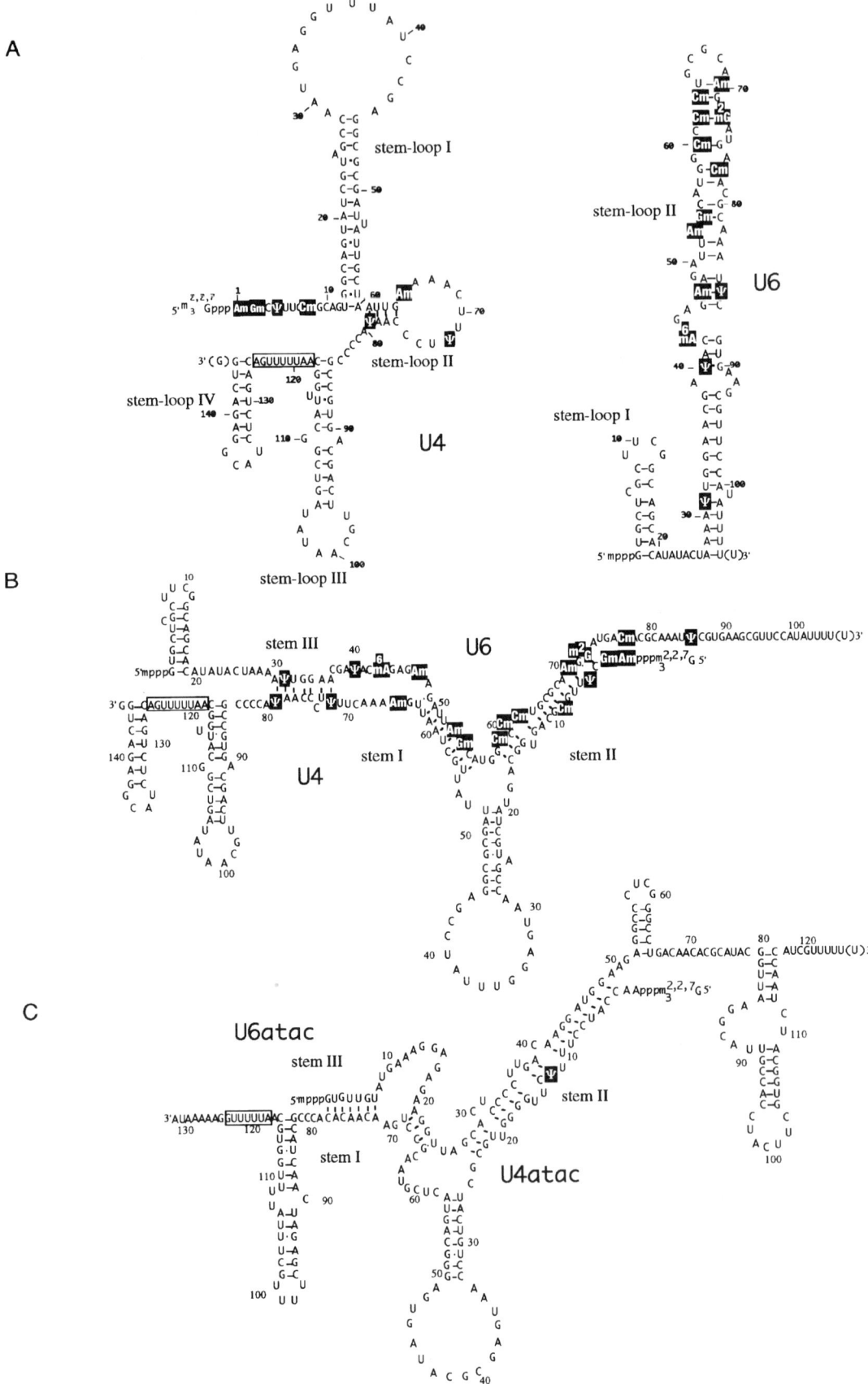

block pseudouridine-synthase activity by formation of an irreversible complex (Samuelson, 1991). Using this strategy, U1 snRNA containing 5-FU was found to inhibit pseudouridine formation in U1 snRNA transcripts, but not in U2 or U5 snRNA transcripts (Patton, 1993). The same situation was found for U2 and U5 snRNAs. Hence, distinct pseudouridine synthases catalyze the pseudouridine formation in each of these three U snRNAs. A more detailed analysis was performed on pseudouridine formation in U2 and U5 snRNAs. For U2 snRNA, the inhibition experiments with 5-FU containing RNAs were extended to a large series of U2 snRNAs containing deletions and point mutations, and the data supported the hypothesis that there is no obligatory pathway for pseudouridine formation in U2 snRNA because formation at one site in U2 snRNA was not dependent on the formation of pseudouridine at any other site. As a whole, the data also suggest that there are probably multiple pseudouridine synthase activities that modify U2 snRNA (Patton et al., 1994).

Studies of pseudouridine formation in U5 snRNA found additional evidence for the existence of multiple pseudouridine synthase activities that modify a single U snRNA. Indeed, among the three pseudouridines that are present in the HeLa cell U5 snRNA (Kato and Harada, 1981; Branlant et al., 1983), formation of pseudouridine at positions 43 and 46 in the highly phylogenetically conserved terminal loop of the stem-loop structure I (Fig. 7) is obtained in vitro using an S100 extract or a nuclear extract (Patton, 1991, 1994b). Formation of pseudouridine 53 in the stem of this stem-loop structure is not obtained with an S100 extract. However, it can be obtained with a nuclear extract (Patton, 1991; Patton, 1994b). Hence, formation of pseudouridine 53, on the one hand, and formation of pseudouridines 43 and 46, on the other hand, are catalyzed by at least two distinct pseudouridine synthases. A combination of nuclear and S100 extracts gave the most efficient total formation of pseudouridines in the human U5 snRNA (Patton, 1994b).

Formation of pseudouridines in U5 snRNA at all three positions was found to be dependent on the integrity of the Sm binding site (Patton, 1991; Patton, 1994b). Such dependence on the Sm binding site was not observed for U2 snRNA, although it cannot be ruled out that some of the pseudouridines are not formed in U2 snRNA upon deletion of the Sm binding site. Indeed, a detailed analysis of the pseudouridines formed in these conditions was not performed (Kleinschmidt et al. 1989). In fact, no systematic analysis of pseudouridines in the human U2 snRNA has ever been performed, and to interpret their results, Patton and coworkers made the hypothesis that the formation of pseudouridines in the human U2 snRNAs takes place at the same positions as in rat hepatoma cells.

The requirements for pseudouridine formation in U4 and U6 snRNAs in HeLa cell nuclear extract and S100 extract were also investigated in vitro. For U4 snRNA, optimal pseudouridine formation was observed when U4 RNA transcripts were first incubated in S100 extract, followed by addition of nuclear extract. The presence of an intact Sm binding site was also found to strongly increase the efficiency of pseudouridine formation (Zerby and Patton, 1996). In vitro formation of pseudouridines in U6 snRNA transcripts is optimal with incubation in nuclear extract, followed by the addition of S100 extract. This is surprising because U6 snRNA does not exit the nucleus in vivo. Interestingly, pseudouridine formation in both U4 and U6 snRNA transcripts was found to be highly increased on formation of a U4-U6 snRNA duplex (Fig. 9B) (Zerby and Patton, 1996, 1997). Two distinct U4 pseudouridine synthase activities seem to be required for U4 full modification (Zerby and Patton, 1997), and U4 snRNA with a truncated Sm binding site was found to be less efficient in increasing pseudouridine formation in U6 snRNA (Zerby and Patton, 1996). Among the three pseudouridines of the human U6 snRNA (positions 31, 40, and 86), only pseudouridine residues at positions 40 and 86 could be formed in vitro (Zerby and Patton, 1996). Interestingly, in addition to the heterologous stems I and II, proposed by Brow and Guthrie (1988), a third heterologous stem can be formed between U4 and U6 snRNAs, and this possibility is phylogenetically conserved (Fig. 9B) (Jakab et al., 1997). Residue 31 belongs to the U6 snRNA segment involved in the formation of this third putative heterologous stem, whereas residue 40 is located at one of its extremities.

Figure 9. Secondary structures of rat spliceosomal U4 and U6 snRNAs and U4-U6 and U4atac-U6atac snRNA interactions in vertebrates. (A) The secondary structures are from Krol et al. (1981a) and Myslinski and Branlant (1991) (U4) and Mougin et al. (unpublished results) (U6). (B) U4-U6 snRNA duplex. The nucleotide sequences shown are from rat (Reddy, 1988). The heterologous stems I and II are as proposed by Brow and Guthrie (1988), and the additional stem III is as proposed by Jakab et al. (1997). (C) U4atac-U6atac snRNA duplex. The nucleotide sequences are from HeLa cells as determined by Tarn and Steitz (1996b). The heterologous stems I and II were proposed by Tarn and Steitz (1996b) based on the model proposed by Brow and Guthrie (1988). An additional potential stem III is shown as proposed by Jakab et al. (1997). Pseudouridines were investigated in both RNAs; only one residue (pseudouridine 14 in U4atac) was detected (Massenet et al., unpublished results). The posttranscriptional modifications are squared in black.

Residue 86 is located in a single-stranded region at the 3' extremity of the molecule (Fig. 9B). Hence, difference of conformation of the substrate-RNA segment may explain the different levels of modifications observed in vitro at positions 31, 40, and 86 of U6 snRNA. U4 snRNA is considered to be a modulator of U6 snRNA activity by formation of the U4–U6 RNA duplex that masks U6 snRNA activity. U4 snRNA may also play a positive role for U6 snRNA catalytic activity, by allowing pseudouridine formation in this molecule (Zerby and Patton, 1996). Hence, according to these data, after transcription, U4 snRNA exits the nucleus and is assembled into a core ribonucleoprotein particle with its 5' cap hypermethylated before returning to the nucleus (Fig. 6). Because base pair interaction between U4 and U6 snRNAs allows pseudouridine formation in both U4 and U6 snRNAs, it suggests that modifications in both snRNAs might occur simultaneously, probably in the nuclear compartment.

Nevertheless, based on the present knowledge, there is no clear idea of the compartment where internal posttranscriptional modifications of the spliceosomal U snRNAs take place. As suggested by the efficiency of S100 extracts for producing some of the peudouridines, they are probably initiated in the cytoplasm after core-U snRNP formation and completed in the nucleus after the nuclear import of the core-U snRNP particles, U6 and U4 snRNAs being modified after their association.

Possible Implication of Posttranscriptional Modifications in Spliceosome Assembly and Function

Interestingly, the posttranscriptional modifications identified in spliceosomal RNAs are mainly concentrated in the U snRNA segments that were shown to be involved in intermolecular RNA–RNA interactions or in U snRNA segments showing two alternative secondary structures.

In U1 snRNA, except for the 2'-O-methyladenosine at position 70, all the other posttranscriptional modifications are located at the 5' end of the molecule (Fig. 8). The two pseudouridines at positions 5 and 6, which are conserved in vertebrates, Drosophila, and S. cerevisiae, are involved in the base pair interaction with the 5' extremity of the intron (Fig. 10A). Nucleotides at positions 5 and 6 are also modified in plants (Ψm$_5$Um$_6$ in the alga Chlorella saccharophila) (Kiss et al., 1988a). In S. pombe, only one pseudouridine is detected at position 3 (Table 1). The 2'-O methylated residue 70 belongs to the binding site of protein U1A (Scherly et al., 1989; Bach et al., 1990; Oubridge et al., 1994) and it may play a role

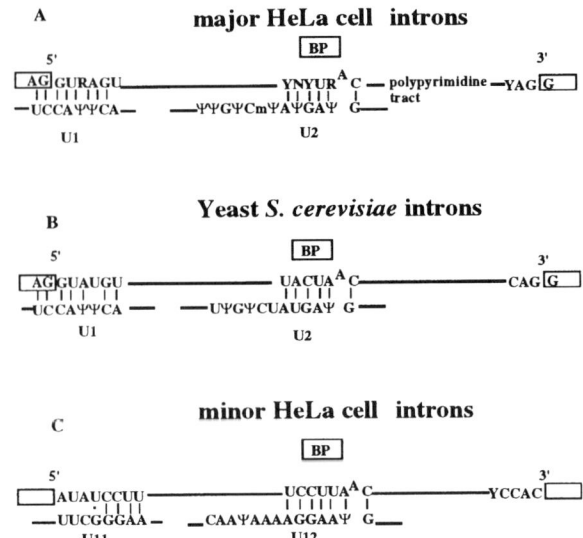

Figure 10. The 5' and 3' splice sites and branch-site consensus sequences of HeLa cells and S. cerevisiae introns and their interactions with the spliceosomal U snRNAs. (A) Consensus sequences of the major introns from vertebrates; (B) consensus sequences of the yeast introns; (C) consensus sequences of the minor introns from vertebrates. The complementary sequences of U1 and U11 for 5' splice sites and U2 and U12 for the branch-point sequences (BP) are shown.

in RNA–protein interaction. U1 snRNA transcripts produced in vitro, could complement a HeLa nuclear extract depleted in U1 snRNAs (Will et al., 1996). This means that 2'-O methylations are not essential for U1 snRNA function. No conclusion can be drawn on the role of pseudouridines in the human U1 snRNA function in splicing because no investigation on the in vitro formation of pseudouridines in U1 snRNA was performed.

A huge number of posttranscriptional modifications are gathered in the 5'-terminal third of U2 snRNA. This region was found to contain a pseudoknot structure (Zavanelli et al., 1994) (Fig. 8). Interestingly, the sequence involved in where this pseudoknot structure is located contains posttranscriptional modifications: two 2'-O methylations and three pseudouridines (Fig. 8). Here, posttranscriptional modifications may be required for stabilization of the pseudoknot structure.

Upon association with the intron, U2 snRNA interacts with the intron branch-site sequence. As found for the 5' segment of U1 snRNA, in vertebrates the segment of U2 snRNA that interacts with the branch-site sequence contains two pseudouridines (Fig. 10A). They are conserved in plants and in S. pombe (Table 1). However, only one of them is present in S. cerevisiae (Fig. 10B). Within the base pair interaction formed by U2 snRNA and the branch-site sequence, the adenosine residue that makes the first nucleophilic attack at the 5' end of the intron is a bulged nucleotide. Interestingly, the unique pseudouridine conserved in *S. cerevisiae* is involved in the formation of the base pair located on the 5' side of the bulged adenosine residue. It may be required for the stabilization of the interaction or for its recognition by proteins. The branch-site sequence is highly conserved in *S. cerevisiae*, in contrast to the situation in vertebrates (for reviews, see Jackson, 1991, and Rymond and Rosbash, 1992). Hence, the presence of a second pseudouridine residue in the vertebrate U2 snRNA sequence that base pairs with the branch-site sequence may be required to compensate for the lower level of complementarity with the introns.

After association of the (U4/U6, U5) tri-snRNP particle with prespliceosomal complexes, U2 snRNA undergoes a strong conformational change and forms several base pair interactions with U6 snRNA (Fig. 11). The ability of U2 and U6 snRNAs to form helices Ia, II, and III was found to be required for splicing in vertebrates (Fig. 11A). However, it is not known whether all of these intermolecular interactions take place together. Interestingly, the entire region involved in the formation of the four intermolecular interactions with U6 snRNA is highly posttranscriptionally modified in the vertebrate and plant U2 snRNAs (Fig. 11A; Table 1). We only detected three pseudouridines in the *S. cerevisiae* U2 snRNA: the one in the segment complementary to the branch-site sequence that we mentioned above and two others in the segment that is involved in helix III formation (Fig. 11B). No strict requirement for the formation of helix III was found in *S. cerevisiae* (Yan and Ares, 1996). However, the U2 snRNA region downstream from the branch-site complementary sequence was found to be the binding site of several splicing factors both in yeast (Ruby et al., 1993; Hodges and Beggs, 1994; Wells and Ares, 1994; Wells et al., 1996; Yan and Ares, 1996) and in vertebrates (Behrens et al., 1993). Hence, the two highly conserved pseudouridines in this area may either facilitate the formation of helix III or be involved in RNA–protein interaction. In yeast, mutation at position 35, corresponding to the pseudouridine residue in the sequence complementary to the branch site, strongly reduced splicing efficiency (McPheeters and Abelson, 1992). However, because this destroys the base pair, one cannot conclude whether the presence of a pseudouridine is absolutely required at this position. Mutation at position 42 had no phenotype (Yan and Ares, 1996), which indicates that this pseudouridine is dispensable in yeast. In contrast, mutation at position 44 strongly affects growth, and it was proposed that the presence of a uridine at this position (in fact, a pseudouridine according to our unpublished results) is required to avoid the formation of an artifactual nonfunctional U2 snRNA secondary structure (Yan and Ares, 1996). This hypothesis is not exclusive of the involvement of pseudouridine 44 in RNA–protein interactions or helix III formation.

The importance of the posttranscriptional modifications of the human U2 snRNA for spliceosome activity was clearly demonstrated by reconstitution of the spliceosome with HeLa cell nuclear extracts depleted of U2 snRNA and complemented with in vitro transcribed U2 snRNA. Whereas active spliceosomes were obtained when the same kinds of experiments were performed with U5 snRNA transcripts, no active spliceosome could be obtained with U2 snRNA transcripts (Ségault et al., 1995). Active spliceosomes were obtained when the same kinds of experiments were performed in yeast (McPheeters et al., 1989). This confirmed the lower importance of the U snRNA posttranscriptional modifications for the yeast splicing machinery and this shows that some of the posttrancriptional modifications, which are not formed upon incubation of an in vitro transcribed human U2 snRNA in a HeLa nuclear extract, are required for the formation of efficient spliceosomes.

In vertebrates, the U6 snRNA regions involved in base pair interactions with U2 snRNA are also highly modified (Fig. 11A). One highly phylogenetically conserved U6 snRNA sequence (the ACAGAG sequence) is essential for the function of U6 snRNA in splicing (Fabrizio and Abelson, 1990 and 1992; Madhani et al., 1990; Madhani and Guthrie, 1992). Interestingly, this sequence contains an m^6 residue that was found to be conserved in all the species studied: vertebrates, plants, and *S. pombe* (Table 1). It will be interesting to see whether it is also conserved in *S. cerevisiae*. As U6 snRNA is expected to play a direct role in catalysis with a possible implication of the ACAGAG sequence, the m^6 residue of the ACAGAG sequence may be important for the splicing reaction. However, this residue is certainly not essential, because active spliceosomes could be obtained upon substitution for the authentic U6 snRNA by an in vitro produced transcript in the HeLa, *Xenopus*, and yeast systems (Fabrizio et al., 1989; Bindereif et al., 1990; Vankan et al., 1990; Wolff and Bindereif, 1995). In vitro, in the absence of S-adenosyl methionine (the expected methyl donor), no m^6A formation is expected. It may be noticed that methylation at position 6 of adenine in U6 snRNA occurs in a sequence that fits the consensus sequence of messenger RNA m^6A methylation sites (see Chapter 10). Another part of U6 snRNA, which was found to be very important for splicing, is the segment 57 to 78 (Wolff and Bindereif, 1993 and 1995; Fortner et al., 1994; Sun and Manley, 1995 and 1997; McPheeters,

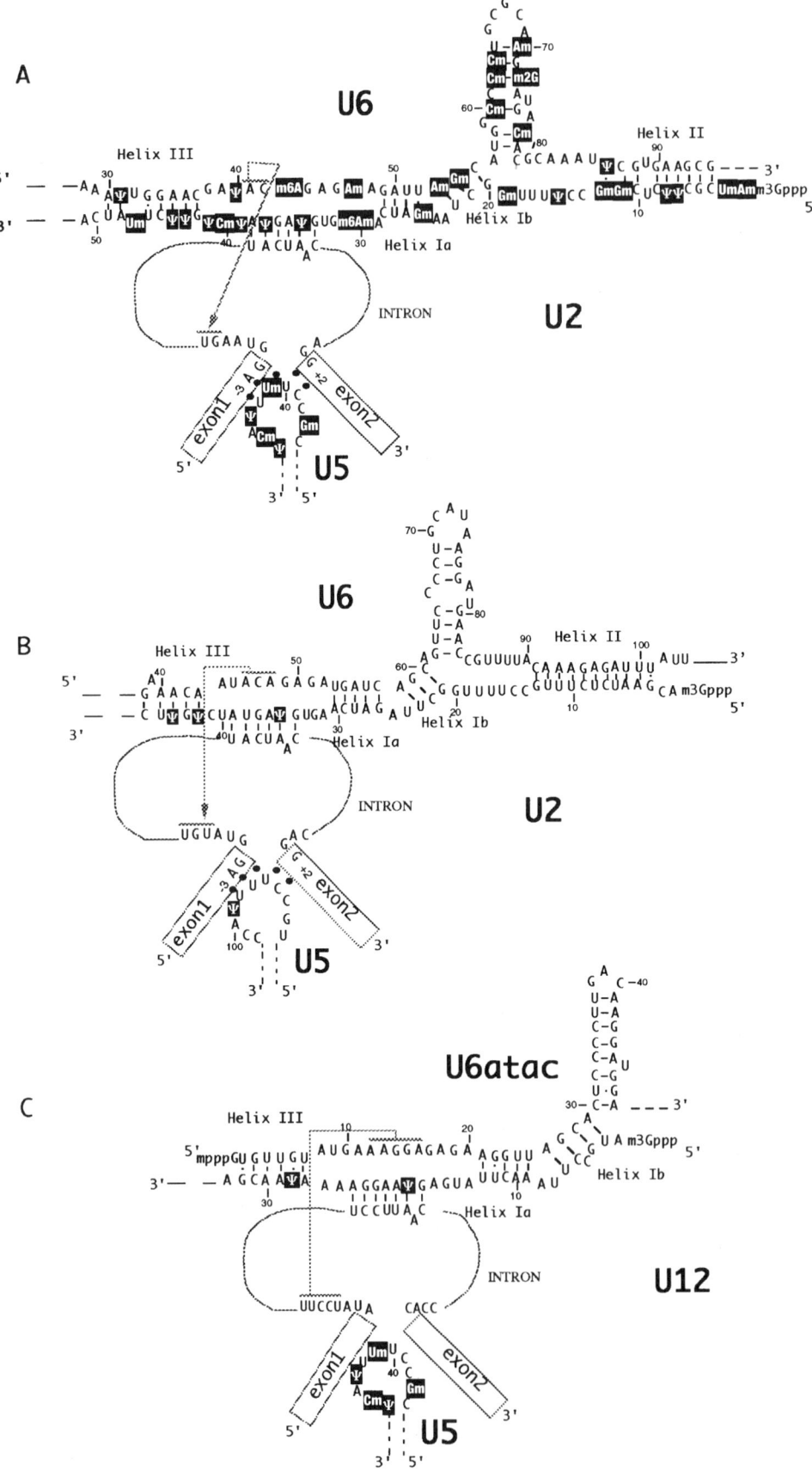

1996), that forms a stem-loop structure interrupting the U2–U6 interactions. Interestingly, this segment is also hypermodified in vertebrates (Fig. 11A), and mutation of some of the modified residues was found to impair splicing (Wolff and Bindereif, 1995; Sun and Manley, 1997). However, in the present stage of the study, there is no direct evidence that the effect is linked to the absence of posttranscriptional modifications. Before the interaction with U2 snRNA, the U6 snRNA segment 59 to 72 interacts with U4 snRNA and forms the heterologous stem II of the U4–U6 snRNA duplex (Brow and Guthrie, 1988) (Fig. 9B). As previously mentioned, in addition to the previously proposed heterologous stems I and II, a third intermolecular stem III can be formed (Jakab et al., 1997). U4 snRNA contains a low level of posttranscriptional modifications, even in vertebrates. Only three pseudouridines were identified. All three are contained in segments that can form intermolecular interactions with U6: stem II and stem III.

U5 snRNA carries two types of posttranscriptional modifications: species-specific pseudouridines in the stem of stem-loop structure I (Fig. 7), and highly phylogenetically conserved pseudouridines and 2′-O methylated residues in the terminal loop of stem-loop structure I (Fig. 7). In S. pombe and S. cerevisiae, this terminal loop I shows one difference as compared to other species: the pseudouridine at the 3′ end of the terminal loop is replaced by a C residue (Fig. 7). In addition, no 2′-O methylation was detected in the terminal loop I of the S. cerevisiae U5 snRNA (our unpublished results); therefore, this loop contains a single modified residue. Hence, except for S. cerevisiae, the terminal loop of the U5 snRNA stem-loop structure I, which aligns the exon extremities for correct ligation (Newman and Norman, 1991 and 1992; Wyatt et al., 1992; Sontheimer and Steitz, 1993; Cortes et al., 1993; Newman et al., 1995) is highly posttranscriptionally modified, and modifications are highly conserved throughout evolution, probably because they ensure a precise conformation of this loop. Such peculiar structure may help U5 snRNA to bridge the exon extremities together with proteins. In yeast, both U5 snRNA and the internal structure of the intron may be involved in bridging the exon extremities. Indeed, constraints on the intron secondary structure were specifically observed in S. cerevisiae, and it was proposed that the observed base pair interactions are required to limit the distance between the 5′ and 3′ splice sites (Goguel and Rosbash, 1993; Libri et al., 1995; Mougin et al., 1996; Charpentier and Rosbash, 1996). This may explain a lower pressure for the conservation of posttranscriptional modifications in the S. cerevisiae U5 snRNA terminal loop I. However, yeast spliceosome reconstitution with a synthetic U5 snRNA bearing a deleted or a mutated terminal loop I was found to block the second step of splicing (O'Keefe et al., 1996), whereas splicing could be obtained when the same kind of experiment was performed in the HeLa cell system (Ségault et al., unpublished data). Based on these two observations, a conclusion in contradiction with the preceding one can be drawn: the yeast splicing reaction seems to be more dependent on the integrity of the U5 snRNA terminal loop I than on the vertebrate spliceosomes. However, it should be pointed out that the premessenger RNA used for the assay in the HeLa cell nuclear extract is very efficiently spliced in vitro and it might be that splicing dependence upon the structure of the U5 snRNA terminal loop I is greater for other pre-mRNAs. On the other hand, the terminal loop of U5 snRNA may have additional functions in the formation of active spliceosomes. Indeed, psoralen was recently found to produce a U5–U1 snRNA cross-link before the first step of splicing and, interestingly, the cross-linked residue of U5 snRNA was pseudouridine 43 (Ast and Weiner, 1997). Based on these data, U5 snRNA was proposed to be involved in the process of replacement of U1 snRNA by U6 snRNA in the interaction with the intron 5′-extremity. The facts that pseudouridines located in the stem I of the human U5 snRNA are species specific, whereas the pseudouridines located in the terminal loop I are phylogenetically conserved, are in accord with the observation that these two kinds of pseudouridines are formed by different pseudouridine synthases in human cells (Patton, 1991 and 1994b). Furthermore, the possible formation of these pseudouridines in nuclear cell extract may explain the functionality of in

Figure 11. U snRNA–U snRNA and U snRNA–pre-mRNA interactions at the catalytic center of the spliceosomes. (A) Major spliceosome of vertebrates; (B) yeast spliceosome. In panel A, the nucleotide sequences are from rat. The heterologous helices I, II, and III between U2 and U6 snRNAs are shown, as well as the base-pairing interaction between U2 snRNA and the branch-point sequence, and the interaction between U6 snRNA and the sequence close to the 5′-splice site. Interactions between the terminal loop of U5 snRNA and the exon extremities are also represented (for a review, see Moore et al., 1993, and text for references). (C) Minor spliceosome of vertebrates. The nucleotide sequences are from HeLa cells (Montzka and Steitz, 1988; Tarn and Steitz, 1996b). The predicted interactions (Tarn and Steitz, 1996b) between U12 and U6atac snRNAs are shown. Potential interactions deduced from comparisons with the major spliceosome are also shown: the potential interaction between U6atac snRNA and the sequence at the 5′ end of the intron and the potential interaction of U5 snRNA with the exon extremities are shown. The posttranscriptional modifications are shown in black squares.

vitro transcribed U5 snRNA in an in vitro splicing reaction (Ségault et al., 1995).

In conclusion, as shown in Fig. 11, all the U snRNA segments that are expected to be at or near the spliceosome catalytic center are highly modified, and most of the modifications are conserved throughout evolution. These observations are strongly in favor of an important role of the internal posttranscriptional modifications of U snRNAs in splicing. In contrast, the 5'-terminal cap structure is probably not involved in the splicing reaction because in vitro transcribed U1 snRNA was found to be active in splicing when it carries an ApppAm cap structure instead of the m$_3$GpppAm cap structure (Will et al., 1996). A m^{7}GpppAm cap structure was found to impair U1 snRNA function probably because it binds the CBP protein complex (Will et al., 1996).

POSTTRANSCRIPTIONAL MODIFICATIONS OF THE SPLICEOSOMAL U snRNAs INVOLVED IN THE SPLICING OF THE AT–AC INTRONS

Terminal and Internal Posttranscriptional Modifications

As previously mentioned, a new class of introns recently have been characterized, and they were found to be spliced in a spliceosome structure gathering around the intron, the U11, U12, U4atac, U6atac, and U5 snRNAs. U11 and U12 (Fig. 10) have been characterized previously (Montzka and Steitz, 1988) as minor nucleoplasmic U snRNAs. They were shown to bear a m^{2,2,7}G cap structure, together with an Sm-binding site (Montzka and Steitz, 1988), which suggests that they are processed in the cytoplasm as U1 and U2 snRNAs, their counterparts of the main splicing machinery.

U4atac and U6atac snRNAs (Fig. 9C) were recently characterized and, as was found for the U4 and U6 snRNAs of the main splicing machinery, U4atac snRNA has a m^{2,2,7}G cap structure and an Sm-binding site, whereas U6atac snRNA, which is transcribed by RNA polymerase III, has a γ-methyl triphosphate at its 5'-extremity. Hence, with respect to the 5'-terminal modifications, the situation is completely identical for the two classes of spliceosomal snRNAs.

We looked for the presence of pseudouridines in the AT–AC spliceosomal RNAs. The results obtained show some similarity with the S. cerevisiae GT–AG spliceosomal RNAs. U11 does not contain any pseudouridine; U12 contains only two pseudouridines (Fig. 10; Table 1). As found for the S. cerevisiae U2 snRNA, one pseudouridine residue is involved in the interaction with the branch-site sequence and is located in the base pair on the 5' side of the bulged A residue. The second one is located in the segment upstream of the branch-site complementary sequence, which is involved in formation of the putative heterologous helix III of the U12–U6atac interaction (Tarn and Steitz, 1996b) (Fig. 11C). No pseudouridine residue was detected in the U6atac snRNA, and only one was found in the U4atac snRNA (Fig. 9C). It belongs to the helix II of the U4–U6atac RNA duplex. Analysis of the 2'-O methylations is underway in our laboratory.

Possible Implications of the AT–AC U snRNAs Posttranscriptional Modifications for Splicing of AT–AC Introns

The 5'-terminal modifications of the AT–AC U snRNAs probably play the same role in AT–AC U snRNP biogenesis as those of the GT–AG U snRNPs.

Concerning the internal posttranscriptional modifications, the low level of pseudouridine residues observed may be explained by the strong conservation of the 5'-terminal segment of the intron that base pairs with U11 snRNA and of the branch-site sequence that base-pairs with U12 snRNAs (Jackson, 1991; Hall and Padgett, 1994; Wu and Krainer, 1997). An optimal complementarity with U12 snRNA is found for all the introns detected in metazoan cells (Tarn et al., 1995). Hence, as it is the case for the S. cerevisiae GT–AG machinery, there is no need to reinforce the U snRNAs–pre-mRNA interactions by several posttranscriptional modifications. Only the pseudouridine residue that forms the base pair on the 5' side of the bulged branched site is conserved. This strongly suggests that this pseudouridine residue is essential in both splicing machineries, as is the case for the pseudouridine residue located a few nucleotides downstream from the branch-site complementary sequence. The ACm^{6}AGAG sequence, that is essential for the GT–AG U6 snRNA function, is replaced by GGAGAG sequence in the U6atac snRNA. It will be interesting to test whether the first adenosine residue bears a methylation on position 6 in the U6atac snRNA. Several of the residues that carry a 2'-O methylation in the GT–AG U6 snRNA are conserved in the AT–AC U6 snRNA; it will be interesting to see whether they are also 2'-O methylated. Experiments are under way in our laboratory to test this possibility.

CONCLUSIONS

It has now been clearly demonstrated that 5'-terminal posttranscriptional modifications play an essential role for the biogenesis of the spliceosomal U snRNPs in eukaryotes, and this is probably also the case for the U snRNAs involved in the recently discovered minor splicing machinery of vertebrates. The role of the internal modifications of the spliceosomal U snRNAs from the major splicing machinery is not yet clear. A trivial explanation could be that the U snRNA sequences involved in intermolecular RNA–RNA base pair interactions are free of protein interactions and freely accessible for the numerous posttranscriptional modifying enzymes present in the cytoplasm and the nucleus. If this was the only explanation, the strongly observed conservation of several of the modifications, throughout evolution, would not have been expected. Furthermore, the atac U snRNAs would carry a similar number of posttranscriptional modifications. We showed that this is not the case.

Another trivial explanation could be that posttranscriptional modifications of U snRNA segments involved in intermolecular base pair interactions are needed to confer a resistance of these RNA segments to the nucleases of the nuclear compartment. Although this explanation cannot be completely ruled out, on one hand, the nucleus is not known to contain high levels of nucleases; on the other hand, atac U snRNAs would also need to be protected by posttranscriptional modifications. Finally, the various recent experiments that we previously summarized are strongly in favor of a direct functional role of U snRNA posttranscriptional modifications in spliceosome assembly and splicing.

Although clear proof is still lacking, the essential role of human U2 snRNA internal posttranscriptional modifications in splicing was already shown (Ségault et al., 1995). It should be pointed out that, taken individually, the posttranscriptional modifications of vertebrate U snRNAs are perhaps not essential for splicing, but together they may have a positive effect on splicing, especially for introns that deviate from the consensus sequences.

The current view is that the pre-mRNA splicing machinery has evolved from a common ancestor to the autocatalytic introns of group II (Sharp, 1985; Cech, 1986). According to this theory, some of the DNA segments coding for autocatalytic intron sequences would have dissociated to generate the spliceosomal U snRNA genes (Sharp, 1991). Arguments in favor of this hypothesis are the following. First, splicing of the autocatalytic group II introns and of the pre-mRNA introns proceeds through the same two chemical steps (for a review, see Moore et al., 1993). Second, some of the secondary structure elements formed by U snRNAs and the pre-mRNA at the catalytic center of the spliceosome share strong homology with secondary structure elements of introns of group II (Wu and Manley, 1989; Jacquier, 1990; Newman and Norman, 1992; for a review, see Madhani and Guthrie, 1994). Based on its high dependence on the pre-mRNA secondary structure, the nuclear pre-mRNA splicing machinery of S. cerevisiae seems to be closer to the autocatalytic intron ancestor of the spliceosome than the nuclear pre-mRNA splicing machinery of vertebrates. Introns have probably played a very important role in the evolution of the genomes of pluricellular eukaryotes that developed complex cell differentiation processes. It seems likely that for an easier evolution of the genome, the dependence on intron primary and secondary structure, which probably existed in the ancestral pre-mRNA introns and whose vestige may be the yeast introns, has been decreased in the course of evolution. One may imagine that the multiplication of internal posttranscriptional modifications in some of the U snRNAs participated in this evolution of introns in pluricellular eukaryotes. The participation of proteins, in particular the SR proteins (for a review, see Manley and Tacke, 1996), may also have contributed. Introns are rare in yeast. They probably did not contribute strongly in the evolution of the yeast genomes, and this may explain the presence of premessenger introns that may be more similar to the ancestral ones than those in vertebrates, with as corollary a more limited number of U snRNA internal modifications in yeasts.

The recent discovery of a new category of intron in vertebrates, spliced by a minor form of spliceosome, opens new questions concerning the phylogeny of pre-mRNA introns. Interestingly, the pattern of pseudouridine formations in the atac U snRNAs has similarity to that of S. cerevisiae. However, the functional sequences of these introns that are recognized by the atac splicing machinery are very different from the ones recognized by the main splicing machinery of yeast and vertebrates. Hence, the similarity of posttranscriptional modifications in the yeast S. cerevisiae U snRNAs and in the atac U snRNAs may result from convergent evolution rather than from the existence of a common ancestor.

Little is known about the internal posttranscriptional modifications in the U snoRNAs. However, clearly for U3 snoRNA, as is the case for spliceosomal U snRNAs, posttranscriptional modifications are found in a segment involved in an intermolecular RNA–RNA interactions.

Acknowledgments. Work in the authors' laboratory is supported by the Centre National de la Recherche Scientifique and the French Ministère de la Recherche et de l'Enseignement Supérieur. S. Massenet is a fellow of the French Ministère de la Recherche et de l'Enseignement Supérieur. We thank H. Grosjean, R. Benne, J.-P.

Bachellerie, and S. Jacquenet for critical reading of the manuscript and useful comments.

REFERENCES

Adams, D. S., D. Noonan, T. C. Burn, and H. B. Skinner. 1987. A library of trimethylguanosine-capped small RNAs in *Physarum polycephalum*. *Gene* 54:93–103.

Agris, P. F. 1996. The importance of being modified: roles of modified nucleosides and Mg^{2+} in RNA structure and function. *Prog. Nucleic Acid Res. Mol. Biol.* 53:79–129.

Amberg, D. C., M. Fleischmann, I. Stagljar, C. N. Cole, and M. Aebi. 1993. Nuclear PRP20 protein is required for mRNA export. *EMBO J.* 12:233–241.

Antal et al. Personal communication.

Ast, G., and A. M. Weiner. 1997. A novel U1/U5 interaction indicates proximity between U1 and U5 snRNAs during an early step on mRNA splicing. *RNA* 3:371–381.

Bach, M., A. Krol, and R. Lührmann. 1990. Structure-probing of U1 snRNPs gradually depleted of the U1-specific proteins A, C, and 70k. Evidence that A interacts differentially with developmentally regulated mouse U1 snRNA variants. *Nucleic Acids Res.* 18:449–457.

Bachellerie, J. P., B. Michot, M. Nicoloso, A. Balakin, J. Ni, and M. J. Fournier. 1995. Antisense snoRNAs: a family of nucleolar RNAs with long complementarities to rRNA. *Trends Biochem. Sci.* 20:261–264.

Bakin, A., and J. Ofengand. 1993. Four newly located pseudouridylate residues in *Escherichia coli* 23S ribosomal RNA are all at the peptidyltransferase center: analysis by the application of a new sequencing technique. *Biochemistry* 32:9754–9762.

Balakin, A. G., G. S. Schneider, M. S. Corbett, J. Ni, and M. J. Fournier. 1993. SnR31, snR32, and snR33: three novel, nonessential snRNAs from *Saccharomyces cerevisiae*. *Nucleic Acids Res.* 21:5391–5397.

Balakin, A. G., L. Smith, and M. J. Fournier. 1996. The RNA world of the nucleolus: two major families of small RNAs defined by different elements with related functions. *Cell* 86:823–834.

Bally, M., J. Hughes, and G. Cesareni. 1988. SnR30: a new, essential small nuclear RNA from *Saccharomyces cerevisiae*. *Nucleic Acids Res.* 16:5291–5303.

Baserga, S. J., M. Gilmore-Hebert, and X. W. Yang. 1992. Distinct molecular signals for nuclear import of the nucleolar snRNA U3. *Genes Dev.* 6:1120–1130.

Behrens, S. E., and R. Lührmann. 1991. Immunoaffinity purification of a [U4/U6.U5] tri-snRNP from human cells. *Genes Dev.* 5:1439–1452.

Behrens, S. E., K. Tyc, B. Kastner, J. Reichelt, and R. Lührmann. 1993. Small nuclear ribonucleoprotein (RNP) U2 contains numerous additional proteins and has a bipartite RNP structure under splicing conditions. *Mol. Cell. Biol.* 13:307–319.

Belanger, K. D., M. A. Kenna, S. Wei, and L. I. Davis. 1994. Genetic and physical interactions between Srp1p and nuclear pore complex proteins Nup1p and Nup2p. *J. Cell Biol.* 126:619–630.

Belhumeur, P., A. Lee, R. Tam, T. Di Paolo, N. Fortin, and M. W. Clark. 1993. GSP1 and GSP2, genetic suppressors of the prp20-1 mutant in *Saccharomyces cerevisiae*: GTP-binding proteins involved in the maintenance of nuclear organization. *Mol. Cell. Biol.* 13:2152–2161.

Beltrame, M., Y. Henry, and D. Tollervey. 1994. Mutational analysis of an essential binding site for the U3 snoRNA in the 5' external transcribed spacer of yeast pre-rRNA. *Nucleic Acids Res.* 22:4057–4065. (Corrected and republished in *Nucleic Acids Res.*, 22:5139–5147.)

Bindereif, A., T. Wolff, and M. R. Green. 1990. Discrete domains of human U6 snRNA required for the assembly of U4/U6 snRNP and splicing complexes. *EMBO J.* 9:251–255.

Bischoff, F. R., and H. Ponstingl. 1991. Mitotic regulator protein RCC1 is complexed with a nuclear Ras-related polypeptide. *Proc. Natl. Acad. Sci. USA* 88:10830–10834.

Bischoff, F. R., H. Krebber, T. Kempf, I. Hermes, and H. Ponstingl. 1995. Human RanGTPase-activating protein RanGAP1 is a homolog of yeast RNA1p involved in messenger RNA processing and transport. *Proc. Natl. Acad. Sci. USA* 92:1749–1753.

Black, D. L., B. Chabot, and J. A. Steitz. 1985. U2 as well as U1 small nuclear ribonucleoproteins are involved in premessenger RNA splicing. *Cell* 42:737–750.

Branlant, C., A. Krol, J. P. Ebel, H. Gallinaro, E. Lazar, and M. Jacob. 1981. The conformation of chicken, rat and human U1A RNAs in solution. *Nucleic Acids Res.* 9:841–858.

Branlant, C., A. Krol, J. P. Ebel, E. Lazar, B. Haendler, and M. Jacob. 1982. U2 RNA shares a structural domain with U1, U4, and U5 RNAs. *EMBO J.* 1:1259–1265.

Branlant, C., A. Krol, E. Lazar, B. Haendler, M. Jacob, L. Galego-Dias, and C. Pousada. 1983. High evolutionary conservation of the secondary structure and of certain nucleotide sequences of U5 RNA. *Nucleic Acids Res.* 11:8359–8367.

Brow, D. A., and C. Guthrie. 1988. Spliceosomal RNA U6 is remarkably conserved from yeast to mammals. *Nature* 334:213–218.

Brulé, F., J. Venema, V. Ségault, D. Tollervey, and C. Branlant. 1996. The yeast *Hansenula wingei* U3 snoRNA gene contains an intron and its coding sequence co-evolved with the 5'ETS region of the pre-ribosomal RNA. *RNA* 2:183–197.

Busch, H., R. Reddy, L. Rothblum, and Y. C. Choi. 1982. SnRNAs, SnRNPs, and RNA processing. *Annu. Rev. Biochem.* 51:617–654.

Cech, T. R. 1986. The generality of self-splicing RNA: relationship to nuclear mRNA splicing. *Cell* 44:207–210.

Chandrasekharappa, S. C., J. H. Smith, and G. L. Eliceiri. 1983. Biosynthesis of small nuclear RNAs in human cells. *J. Cell. Physiol.* 117:169–174.

Chanfreau, G., S. A. Elela, M. Ares, Jr., and C. Guthrie. 1997. Alternative 3'-end processing of U5 snRNA by RNase III. *Genes Dev.* 11:2741–2751.

Charpentier, B., and M. Rosbash. 1996. Intramolecular structure in yeast introns aids the early steps in vitro spliceosome assembly. *RNA* 2:509–522.

Cheng, Y., J. E. Dahlberg, and E. Lund. 1995. Diverse effects of the guanine nucleotide exchange factor RCC1 on RNA transport. *Science* 267:1807–1810.

Colot, H. V., F. Stutz, and M. Rosbash. 1996. The yeast splicing factor Mud13 is a commitment complex component and corresponds to CBP20, the small subunit of the nuclear cap-binding complex. *Genes Dev.* 10:1699–1708.

Coppola, J. A., A. S. Feild, and D. S. Luse. 1983. Promoter-proximal pausing by RNA polymerase 2 in vitro: transcripts shorter than 20 nucleotides are not capped. *Proc. Natl. Acad. Sci. USA* 80:1251–1255.

Cortes, J. J., E. J. Sontheimer, S. D. Seiwert, and J. A. Steitz. 1993. Mutations in the conserved loop of human U5 snRNA generate use of novel cryptic 5' splice sites in vivo. *EMBO J.* 12:5181–5189.

Dahlberg, J. E., and E. Lund. 1988. The genes and transcription of the major small nuclear RNAs, p. 100–114. *In* M. L. Birnstiel (ed.), *Structure and Function of the Major and Minor Small Nuclear Ribonucleoprotein Particles*. Springer Verlag, Berlin, Germany.

Datta, B., and A. M. Weiner. 1991. Genetic evidence for base-pairing between U2 and U6 snRNA in mammalian mRNA splicing. *Nature* 352:821–824.

Eichler, D. C., and N. Craig. 1994. Processing of eukaryotic ribosomal RNA. *Prog. Nucleic Acid Res. Mol. Biol.* 49:197–239.

Fabrizio, P., D. S. McPheeters, and J. Abelson. 1989. In vitro assembly of yeast U6 snRNP: a functional assay. *Genes Dev.* 3:2137–2150.

Fabrizio, P., and J. Abelson. 1990. Two domains of yeast U6 small nuclear RNA required for both steps of nuclear precursor messenger RNA splicing. *Science* 250:404–409.

Fabrizio, P., and J. Abelson. 1992. Thiophosphates in yeast U6 snRNA specifically affect pre-mRNA splicing *in vitro*. *Nucleic Acids Res.* 20:3659–3664.

Fisher, U., and R. Lührmann. 1990. An essential signaling role for the m_3G cap in the transport of U1 snRNP to the nucleus. *Science* 249:786–790.

Fisher, U., V. Sumpter, M. Sekine, T. Satoh, and R. Lührmann. 1993. Nucleocytoplasmic transport of UsnRNPs: definition of a nuclear location signal in the Sm core domain that binds a transport receptor independently of the m_3G cap. *EMBO J.* 12:573–583.

Fortner, D. M., R. G. Troy, and D. A. Brow. 1994. A stem/loop in U6 RNA defines a conformational switch required for pre-mRNA splicing. *Genes Dev.* 8:221–223.

Goguel, V., and M. Rosbash. 1993. Splice site choice and splicing efficiency are positively influenced by pre-mRNA intramolecular base pairing in yeast. *Cell* 72:893–901.

Görlich, D., S. Kostka, R. Kraft, C. Dingwall, R. A. Laskey, E. Hartmann, and S. Prehn. 1995a. Two different subunits of importin cooperate to recognize nuclear localization signals and bind them to the nuclear envelope. *Curr. Opin. Cell Biol.* 5:383–392.

Görlich, D., F. Vogel, A. D. Mills, E. Hartmann, and R. A. Laskey. 1995b. Distinct functions for the two importin subunits in nuclear protein import. *Nature* 377:246–248.

Görlich, D., and I. W. Mattaj. 1996. Nucleocytoplasmic transport. *Science* 271:1513–1518.

Görlich, D., P. Henklein, E. Hartmann, and R. A. Laskey. 1996a. A 41 amino acid motif in importin-α confers binding to importin-β and hence transit into the nucleus. *EMBO J.* 15:1810–1817.

Görlich, D., N. Panté, U. Kutay, U. Aebi, and F. R. Bischoff. 1996b. Identification of different roles for RanGDP and RanGTP in nuclear protein import. *EMBO J.* 15:5584–5594.

Görlich, D., R. Kraft, S. Kostka, F. Vogel, E. Hartmann, R. A. Laskey, I. W. Mattaj, and E. Izaurralde. 1996c. Importin provides a link between nuclear protein import and UsnRNA export. *Cell* 87:21–32.

Gu, J., J. R. Patton, S. Shimba, and R. Reddy. 1996. Localisation of modified nucleotides in *Saccharomyces pombe* spliceosomal small nuclear RNAs: modified nucleotides are clustered in functionally important regions. *RNA* 2:909–918.

Hall, S. L., and R. A. Padgett. 1994. Conserved sequences in a class of rare eukaryotic nuclear introns with non-consensus splice sites. *J. Mol. Biol.* 239:357–365.

Hall, S. L., and R. A. Padgett. 1996. Requirement of U12 snRNA for the *in vivo* splicing of a minor class of eukaryotic nuclear pre-mRNA introns. *Science* 271:1716–1718.

Hamm, J., E. Darzynkiewicz, S. M. Tahara, and I. W. Mattaj. 1990. The trimethylguanosine cap structure of U1 snRNA is a component of a bipartite nuclear targeting signal. *Cell* 62:569–577.

Harada, F., N. Kato, and S. Nishimura. 1980. The nucleotide sequence of nuclear 4.8S RNA of mouse cells. *Biochem. Biophys. Res. Comm.* 95:1332–1340.

Hausner, T. P., L. M. Giglio, and A. M. Weiner. 1990. Evidence for base-pairing between mammalian U2 and U6 small nuclear ribonucleotprotein particles. *Genes Dev.* 4:2146–2156.

Hellung-Larsen, P., I. Kulamowicz, and S. Frederiksen. 1980. Synthesis of low molecular weight RNA components in cells with a temperature-sensitive polymerase II. *Biochem. Biophys. Acta* 609:201–204.

Henry, Y., H. Wood, J. P. Morrissey, E. Petfalski, S. Kearsey, and D. Tollervey. 1994. The 5' end of yeast 5.8S rRNA is generated by exonucleases from an upstream cleavage site. *EMBO J.* 13:2452–2463.

Hodges, P. E., and J. D. Beggs. 1994. RNA splicing. U2 fulfils a commitment. *Curr. Oping Cell Biol.* 4:264–267.

Huang, Q., M. R. Jacobson, and T. Pederson. 1997. 3' processing of human pre-U2 small nuclear RNA: a base-pairing interaction between the 3' extension of the precursor and an internal region. *Mol. Cell. Biol.* 17:7178–7185.

Huber. Personal communication.

Hughes, J. M. X., D. A. Konings, and G. Cesareni. 1987. The yeast homologue of U3 snRNA. *EMBO J.* 6:2145–2155.

Hughes, J. M. X., and M. Ares. 1991. Depletion of U3 small nucleolar RNA inhibits cleavage in the 5' external transcribed spacer of yeast pre-ribosomal RNA and impairs formation of 18S ribosomal RNA. *EMBO J.* 10:4231–4239.

Hughes, J. M. X. 1996. Functional base-pairing interaction between highly conserved elements of U3 small nucleolar RNA and the small ribosomal subunit. *J. Mol. Biol.* 259:645–654.

Izaurralde, E., J. Lewis, C. McGuigan, M. Jankowska, E. Darzynkiewicz, and I. W. Mattaj. 1994. A nuclear cap binding protein complex involved in pre-mRNA splicing. *Cell* 78:657–668.

Izaurralde, E., J. Lewis, C. Gamberi, A. Jarmolowski, C. McGuigan, and I. W. Mattaj. 1995. A cap-binding protein complex mediating UsnRNA export. *Nature* 376:709–712.

Jackson, I. J. 1991. A reappraisal of non-consensus mRNA splice sites. *Nucleic Acids Res.* 19:3794–3798.

Jacob, M., E. Lazar, B. Haendler, H. Gallinaro, A. Krol, and C. Branlant. 1984. A family of small nucleoplasmic RNAs with common structural features. *Biol. Cell.* 51:1–9.

Jacobson, M. R., M. Rhoadhouse, and T. Pederson. 1993. U2 small nuclear RNA 3' end formation is directed by a critical internal structure distinct from the processing site. *Mol. Cell. Biol.* 13:1119–1129.

Jacquier, A. 1990. Self-splicing group II and nuclear pre-mRNA introns: how similar are they? *Trends Biochem. Sci.* 15:351–354.

Jakab, G., A. Mougin, M. Kis, T. Pollàk, M. Antal, C. Branlant, and F. Solymosy. 1997. *Chlamydomonas* U2, U4 and U6 snRNAs. An evolutionarily conserved putative third interaction between U4 and U6 snRNAs which has a counterpart in the U4atac-U6atac snRNA duplex. *Biochimie* 79:387–395.

Jakab, G., M. Kis, and F. Solymosy. 1992. Nucleotide sequence of U5 RNA from a green alga, *Chlamydomonas reinhardtii*. *Nucleic Acids Res.* 20:5224.

Jarmolowski, A., and I. W. Mattaj. 1993. The determinants for Sm protein binding to *Xenopus* U1 and U5 snRNAs are complex and non-identical. *EMBO J.* 12:223–232.

Jarmolowski, A., W. C. Boelens, E. Izaurralde, and I. W. Mattaj. 1994. Nuclear export of different classes of RNA is mediated by specific factors. *J. Cell Biol.* 124:627–635.

Kadowaki, T., Y. Zhao, and A. Tartakoff. 1992. Mtr1, a conditional yeast mutant deficient in mRNA transport from nucleus to cytoplasm. *Proc. Natl. Acad. Sci. USA* 89:2312–2316.

Kadowaki, T., D. Goldfarb, L. M. Spitz, A. M. Tartakoff, and M. Ohno. 1993. Regulation of RNA processing and transport by a nuclear guanine nucleotide release protein and members of the Ras superfamily. *EMBO J.* 12:2929–2937.

Kandels-Lewis, S., and B. Séraphin. 1993. Role of U6 snRNA in 5' splice site selection. *Science* 262:2035–2039.

Kass, S., K. Tyc, J. A. Steitz, and B. Sollner-Webb. 1990. The U3 small nucleolar ribonucleoprotein functions in the first step of pre-ribosomal RNA processing. *Cell* 60:897–908.

Kastner, B., M. Bach, and R. Lührmann. 1990. Electron microscopy of small nuclear ribonucleoprotein (snRNP) particles U2 and U5: evidence for a common structure-determining principle in the major U snRNP family. *Proc. Natl. Acad. Sci. USA* 87: 1710–1714.

Kato, N., and F. Harada. 1981. Nucleotide sequence of nuclear 5S of mouse cells. *Biochem. Biophys. Res. Commun.* 99: 1468–1476.

Kato, N., and F. Harada. 1984. Nucleotide sequence of nuclear 5.4S RNA of mouse cells. *Biochem. Biophys. Acta* 782:127–131.

Keller, E. B., and W. A. Noon. 1984. Intron splicing: a conserved internal signal in introns of animal pre-mRNAs. *Proc. Natl. Acad. Sci. USA* 81:7417–7420.

Kiss, T., M. Toth, and F. Solymosy. 1985. Plant small nuclear RNAs. Nucleolar U3 snRNA is present in plants: partial characterization. *Eur. J. Biochem.* 152:259–266.

Kiss, T., M. Antal, H. Hegyi, and F. Solymosy. 1988a. Plant small nuclear RNAs. IV. The structure of U1 RNA from *Chlorella saccharophila*: a phylogenetic support, in terms of RNA structure, for the probable interaction between U1 and U2 RNPs during the splicing of pre-mRNA. *Nucleic Acids Res.* 16:2734.

Kiss, T., G. Jakab, M. Antal, Z. Palfi, H. Hegyi, M. Kis, and F. Solymosy. 1988b. Plant small nuclear RNAs. V. U4 RNA is present in broad bean plants in the form of sequence variants and is base-paired with U6 RNA. *Nucleic Acids Res.* 16:5407–5426.

Kiss, T., and F. Solymosy. 1990. Molecular analysis of a U3 RNA gene locus in tomato: tanscription signals, the coding region, expression in transgenic tobacco plants and tandemly repeated pseudogenes. *Nucleic Acids Res.* 18:1941–1949.

Kleinschmidt, A. M., J. R. Patton, and T. Pederson. 1989. U2 small nuclear RNP assembly *in vitro*. *Nucleic Acids Res.* 17: 4817–4828.

Kohtz, J. D., S. F. Jamison, C. L. Will, P. Zuo, R. Lührmann, M. A. Garcia-Blanco, and J. L. Manley. 1994. Protein-protein interactions and 5′ splice site recognition in mammalian mRNA precursors. *Nature* 368:119–124.

Kolossova, I., and R. A. Padgett. 1997. U11 snRNA interacts *in vivo* with the 5′ splice site of U12-dependent (AU-AC) pre-mRNA introns. *RNA* 3:227–233.

Konarska, M. M., R. A. Padgett, and P. A. Sharp. 1985. Trans-splicing of mRNA precursors *in vitro*. *Cell* 42:165–171.

Krainer, A. R., T. Maniatis, B. Ruskin, and M. R. Green. 1984. Normal and mutant human beta-globin pre-mRNAs are faithfully and efficiently spliced *in vitro*. *Cell* 36:993–1005.

Krol, A., C. Branlant, E. Lazar, H. Gallinaro, and M. Jacob. 1981a. Primary and secondary structures of chicken, rat and man nuclear U4 RNAs. Homologies with U1 and U5 RNAs. *Nucleic Acids Res.* 9:2699–2716.

Krol, A., H. Gallinaro, E. Lazar, M. Jacob, and C. Branlant. 1981b. The nuclear 5S RNAs from chicken, rat and man. U5 RNAs are encoded by multiple genes. *Nucleic Acids Res.* 9: 769–787.

Krol, A., J. P. Ebel, J. Rinke, and R. Lührmann. 1983. U1, U2 and U5 small nuclear RNAs are found in plant cells. Complete nucleotide sequence of the U5 RNA family from pea nuclei. *Nucleic Acids Res.* 11:8583–8594.

Krol, A., P. Carbon, J. P. Ebel, and B. Appel. 1987. *Xenopus tropicalis* U6 snRNA genes transcribed by Pol III contain the upstream promoter elements used by Pol II dependent UsnRNA genes. *Nucleic Acids Res.* 15:2463–2478.

Kunkel, G. R., R. L. Maser, J. P. Calvet, and T. Pederson. 1986. U6 small nuclear RNA is transcribed by RNA polymerase III. *Proc. Natl. Acad. Sci. USA* 83:8575–8579.

Leader, D. J., S. Connely, W. Filipowicz, and J. W. Brown. 1994. Characterization and expression of a maize U3 snRNA gene. *Biochim. Biophys. Acta* 1219:145–147.

Lee, M. S., M. Henry, and S. A. Silver. 1996. A protein that shuttles between the nucleus and the cytoplasm is an important mediator of RNA export. *Genes Dev.* 10:1233–1246.

Lerner, M. R., and J. A. Steitz. 1979. Antibodies to small nuclear RNAs complexed with proteins are produced by patients with systemic lupus erythematosus. *Proc. Natl. Acad. Sci. USA* 76: 5495–5499.

Lerner, M. R., J. A. Boyle, S. M. Mount, S. L. Wolin, and J. A. Steitz. 1980. Are snRNPs involved in splicing? *Nature* 283: 220–224.

Lesser, C. F., and C. Guthrie. 1993. Mutations in U6 snRNA that alter splice site specificity: implications for the active site. *Science* 262:1982–1988.

Li, H. V., J. Zagorski, and M. J. Fournier. 1990. Depletion of U14 small nuclear RNA (snR128) disrupts production of 18S rRNA in *Saccharomyces cerevisiae*. *Mol. Cell. Biol.* 10:1145–1152.

Liautard, J. P., J. Sri-Widada, C. Brunel, and P. Jeanteur. 1982. Structural organization of ribonucleoproteins containing small nuclear RNAs from HeLa cells. Proteins interact closely with a similar structural domain of U1, U2, U4 and U5 small nuclear RNAs. *J. Mol. Biol.* 162:623–643.

Libri, D., F. Stutz, T. McCarthy, and M. Rosbash. 1995. RNA structural patterns and splicing: molecular basis for an RNA-based enhancer. *RNA* 1:425–436.

Maden, B. E. 1990. The numerous modified nucleotides in eukaryotic ribosomal RNA. *Prog. Nucleic Acid Res. Mol. Biol.* 39: 241–303.

Madhani, H. D., R. Bordonne, and C. Guthrie. 1990. Multiple roles for U6 snRNA in the splicing pathway. *Genes Dev.* 4: 2264–2277.

Madhani, H. D., and C. Guthrie. 1992. A novel base-pairing interaction between U2 and U6 snRNAs suggests a mechanism for the catalytic activation of the spliceosome. *Cell* 71:803–817.

Madhani, H. D., and C. Guthrie. 1994. Dynamic RNA–RNA interactions in the spliceosome. *Annu. Rev. Genet.* 28:1–26.

Manley, J. L., and R. Tacke. 1996. SR proteins and splicing control. *Genes Dev.* 10:1569–1579.

Marshallsay, C., T. Kiss, and W. Filipowicz. 1990. Amplification of plant U3 and U6 snRNA gene sequences using primers specific for an upstream promoter element and conserved intragenic regions. *Nucleic Acids Res.* 18:3459–3466.

Marshallsay, C., and R. Lührmann. 1994. *In vitro* nuclear import of snRNPs: cytosolic factors mediate m$_3$G-cap dependence of U1 and U2 snRNPs transport. *EMBO J* 13:222–231.

Massenet S., et al. Unpublished results.

Mattaj, I. W., and E. M. De Robertis. 1985. Nuclear segregation of U2 snRNA requires binding of specific snRNP proteins. *Cell* 40:111–118.

Mattaj, I. W. 1986. Cap trimethylation of UsnRNA is cytoplasmic and dependent on UsnRNP protein binding. *Cell* 46:905–911.

Maxwell, E. S., and M. J. Fournier. 1995. The small nucleolar RNAs. *Annu. Rev. Biochem.* 67:6523–6527.

McPheeters, D. S., P. Fabrizio, and J. Abelson. 1989. *In vitro* reconstitution of functional yeast U2 snRNPs. *Genes Dev.* 3: 2124–2136.

McPheeters, D. S., and J. Abelson. 1992. Mutational analysis of the yeast U2 snRNA suggests a structural similarity to the catalytic core of group I introns. *Cell* 71:819–831.

McPheeters, D. S. 1996. Interactions of the yeast U6 RNA with the pre-mRNA branch site. *RNA* 2:1110–1123.

Méreau, A., R. Fournier, A. Grégoire, A. Mougin, P. Fabrizio, R. Lührmann, and C. Branlant. 1997. An *in vivo* and *in vitro* structure-function analysis of the *Saccharomyces cerevisiae* U3A snoRNP: protein-RNA contacts and base-pair interaction with the pre-ribosomal RNA. *J. Mol. Biol.* 273:552–571.

Michael, W. M., M. Choi, and G. Dreyfuss. 1995. A nuclear export signal in hnRNP A1: a signal-mediated, temperature-dependent nuclear protein export pathway. *Cell* **83:**415–422.

Montzka, K. A., and J. A. Steitz. 1988. Additional low-abundance human small nuclear ribonucleoproteins: U11, U12, etc. *Proc. Natl. Acad. Sci. USA* **85:**8885–8889.

Moore, M. J., C. C. Query, and P. A. Sharp. 1993. Splicing of precursors to messenger RNAs by the spliceosome, p. 303–358. *In* R. Gesteland and J. Atkins (eds.), *The RNA World*. Cold Spring Harbor Laboratory Press, Cold Spring Harbor, N.Y.

Moore, M. S., and G. Blobel. 1994. A G protein involved in nucleocytoplasmic transport: the role of Ran. *Trends Biochem. Sci.* **19:**211–216.

Morrissey, J. P., and D. Tollervey. 1993. Yeast snR30 is a small nucleolar RNA required for 18S rRNA synthesis. *Mol. Cell. Biol.* **13:**2469–2477.

Mougey, E. B., L. K. Pape, and B. Sollner-Webb. 1993. A U3 small nuclear ribonucleoprotein-requiring processing event in the 5' external transcribed spacer of *Xenopus* precursor rRNA. *Mol. Cell. Biol.* **13:**5990–5998.

Mougin, A., et al. Unpublished results.

Mougin, A., A. Grégoire, J. Banroques, V. Ségault, R. Fournier, F. Brulé, M. Chevrier-Miller, and C. Branlant. 1996. Secondary structure of the yeast *Saccharomyces cerevisiae* pre-U3A snoRNA and its implication for splicing efficiency. *RNA* **2:**1079–1093.

Mount, S. M., and J. A. Steitz. 1981. Sequence of U1 RNA from *Drosophila melanogaster*: implications for U1 secondary structure and possible involvement in splicing. *Nucleic Acids Res.* **9:**6351–6368.

Mount, S. M., I. Petterson, M. Hinterberger, A. Karmas, and J. A. Steitz. 1983. The U1 small nuclear RNA-protein complex selectively binds a 5' splice site *in vitro*. *Cell* **33:**509–518.

Myslinski, E., C. Branlant, E. D. Wieben, and T. Pederson. 1984. The small nuclear RNAs of Drosophila. *J. Mol. Biol.* **180:**927–945.

Myslinski, E., V. Ségault and C. Branlant. 1990. An intron in the genes for U3 small nucleolar RNAs of the yeast *Saccharomyces cerevisiae*. *Science* **247:**1213–1216.

Myslinski, E., and C. Branlant. 1991. A phylogenetic study of U4 snRNA reveals the existence of an evolutionarily conserved secondary structure corresponding to "free" U4 snRNA. *Biochimie*. **73:**17–28.

Newman, A. J., and C. Norman. 1991. Mutations in yeast U5 snRNA alter the specificity of 5' splice site cleavage. *Cell* **65:**115–123.

Newman, A. J., and C. Norman. 1992. U5 snRNA interacts with exon sequences at 5' and 3' splice sites. *Cell* **68:**743–754.

Newman, A. J., S. Teigelkamp, and J. D. Beggs. 1995. SnRNA interactions at 5' and 3' splice sites monitored by photoactivated crosslinking in yeast spliceosomes. *RNA* **1:**968–980.

Ofengand, J., A. Bakin, J. Wrzesinski, K. Nurse, and B. G. Lane. 1995. The pseudouridine residues of ribosomal RNA. *Biochem. Cell Biol.* **73:**915–924.

Ohno, M., N. Kataoka, and Y. Shimura. 1990. A nuclear cap binding protein from Hela cells. *Nucleic Acids Res.* **18:**6989–6995.

Ohtsubo, M., H. Okasaki, and T. Nishimoto. 1989. The RCC1 protein, a regulator for the onset of chromosome condensation, locates in the nucleus and binds to DNA. *J. Cell. Biol.* **109:**1389–1397.

O'Keefe, R. T., C. Norman, and A. J. Newman. 1996. The invariant U5 snRNA loop I sequence is dispensable for the first catalytic step of pre-mRNA splicing in yeast. *Cell* **86:**679–689.

Oubridge, C., N. Ito, P. R. Evans, C. H. Teo, and K. Nagai. 1994. Crystal structure at 1.92 A resolution of the RNA-binding domain of the U1A spliceosomal protein complexed with an RNA hairpin. *Nature* **372:**432–438.

Padgett, R. A., M. M. Konarska, P. J. Grabowski, S. F. Hardy, and P. A. Sharp. 1984. Lariat RNAs as intermediates and products in the splicing of messenger RNA precursors. *Science* **225:**898–903.

Panté, N., and U. Aebi. 1996. Toward the molecular dissection of protein import into nuclei. *Curr. Opi. Cell Biol.* **8:**397–406.

Parker, R., P. G. Siliciano, and C. Guthrie. 1987. Recognition of the TACTAAC box during mRNA splicing in yeast involves base pairing to the U2-like snRNA. *Cell* **49:**229–239.

Parker, R., and B. Patterson. 1987. Architecture of fungal introns: implications for spliceosome assembly, p. 133–148. *In* M. Inouye and B. Dubock (eds.), *Molecular Biology of RNA: New Perspective*. Academic Press, New York, N.Y.

Parker, K. A., and J. A. Steitz. 1987. Structural analysis of the human U3 ribonucleoprotein particle reveal a conserved sequence available for base pairing with pre-rRNA. *Mol. Cell. Biol.* **7:**2899–2913.

Parker, R., and P. G. Siliciano. 1993. Evidence for an essential non-Watson–Crick interaction between the first and last nucleotides of a nuclear pre-mRNA intron. *Nature* **361:**660–662.

Patton, J. R. 1991. Pseudouridine modification of U5 RNA in ribonucleoprotein particles assembled *in vitro* *Mol. Cell. Biol.* **11:**5998–6006. (Erratum, **12:**904, 1992.)

Patton, J. R. 1993. Multiple pseudouridine synthase activities for small nuclear RNAs. *Biochem. J.* **290:**595–600.

Patton, J. R. 1994a. Pseudouridine formation in small nuclear RNAs. *Biochimie* **76:**1129–1132.

Patton, J. R., M. R. Jacobson, and T. Pederson. 1994. Pseudouridine formation in U2 small nuclear RNA. *Proc. Natl. Acad. Sci. USA* **91:**3324–3328.

Patton, J. R. 1994b. Formation of pseudouridine in U5 small nuclear RNA. *Biochemistry* **33:**10423–10427.

Peattie, D. A. 1979. Direct chemical method for sequencing RNA. *Proc. Natl. Acad. Sci. USA* **76:**1760–1764.

Peculis, B. A., and J. A. Steitz. 1993. Disruption of U8 nucleolar snRNA inhibits 5.8S and 28S rRNA processing in the *Xenopus* oocyte. *Cell* **73:**1233–1245.

Peculis, B. A. 1997. The sequence of the 5' end of the U8 small nucleolar RNA is critical for 5.8S and 28S rRNA maturation. *Mol. Cell. Biol.* **17:**3702–3713.

Piñol-Roma, S., and G. Dreyfuss. 1992. Shuttling of pre-mRNA binding proteins between nucleus and cytoplasm. *Nature* **355:**730–732.

Plessel, G., U. Fisher, and R. Lührmann. 1994. m_3G cap hypermethylation of U1 small nuclear ribonucleoprotein (snRNP) in vitro: evidence that the U1 small nuclear RNA-(guanosine-N2)-methyl-transferase is a non-snRNP cytoplasmic protein that requires a binding site on the Sm core domain. *Mol. Cell. Biol.* **14:**4160–4172.

Raker, V. A., G. Plessel, and R. Lührmann. 1996. The snRNP core assembly pathway: identification of stable core protein heterodimeric complexes and an snRNP subcore particle *in vitro*. *EMBO J.* **15:**2256–2269.

Raué, H. A., and R. J. Planta. 1995. The pathway to maturity: processing of ribosomal RNA in *Saccharomyces cerevisiae*. *Gene Expression* **5:**71–77.

Reddy, R., D. Henning, and H. Busch. 1979 Nucleotide sequence of nucleolar U3B RNA. *J. Biol. Chem.* **254:**11097–11105.

Reddy, R., D. Henning, P. Epstein, and H. Busch. 1981. Primary and secondary structure of U2 snRNA. *Nucleic Acids Res.* **9:**5645–5658.

Reddy, R., D. Spector, D. Henning, M. H. Liu, and H. Busch. 1983. Isolation and partial characterization of dinoflagellate U1-U6 small RNAs homologous to rat U small nuclear RNAs. *J. Biol. Chem.* **258:**13965–13969.

Reddy, R., D. Henning, S. Chirala, L. Rothblum, D. Wright, and H. Busch. 1985. Isolation and characterization of three rat U3 RNA pseudogenes colinear with U3 RNA. *J. Biol. Chem.* 260: 5715–5719.

Reddy, R., D. Henning, G. Das, M. Harless, and D. Wright. 1987. The capped U6 small nuclear RNA is transcribed by RNA polymerase III. *J. Biol. Chem.* 262:75–81.

Reddy, R. 1988. Compilation of small RNA sequences. *Nucleic Acids Res.* 16:71–85.

Reddy, R., and H. Busch. 1988. Small nuclear RNAs: RNA sequences, structures and modifications. *In* M. L. Birnstiel (ed.), *Structure and Function of Major Minor Small Nuclear Ribonucleoprotein Particles.* Springer-Verlag, Berlin, Germany.

Reddy, R., R. Singh, and S. Shimba. 1992. Methylated cap structures in eukaryotic RNAs: structure, synthesis and functions. *Pharmacol. Ther.* 54:249–267.

Riedel, N., J. A. Wise, H. Swerdlow, A. Mak, and C. Guthrie. 1986. Small nuclear RNAs from *Saccharomyces cerevisiae*: unexpected diversity in abundance, size and molecular complexity. *Proc. Natl. Acad. Sci. USA* 83:8097–8101.

Rogers, J., and R. Wall. 1980. A mechanism for RNA splicing. *Proc. Natl. Acad. Sci. USA* 77:1877–1879.

Ruby, S. W., T. H. Chang, and J. Abelson. 1993. Four yeast spliceosomal proteins (PRP5, PRP9, PRP11, and PRP21) interact to promote U2 snRNP binding to pre-mRNA. *Genes Dev.* 7: 1909–1925.

Ruskin, B., A. R. Krainer, T. Maniatis, and M. R. Green. 1984. Excision of an intact intron as a novel lariat structure during pre-mRNA splicing *in vitro*. *Cell* 38:317–331.

Rymond, B. C., and M. Rosbash. 1992. Yeast pre-mRNA splicing, p. 143–192. *In* E. W. Jones, J. R. Pringle, and J. R. Broach (ed.), *Yeast Pre-mRNA Splicing.* Cold Spring Harbor Laboratory Press, Cold Spring Harbor, N.Y.

Salditt-Georgieff, M., M. Harpold, S. Chen-Kiang, and J. J. Darnell. 1980. The addition of 5' cap structures occurs early in hnRNA synthesis and prematurely terminated molecules are capped. *Cell* 19:69–78.

Samarsky, D. A., A. G. Balakin, and M. J. Fournier. 1995. Characterization of three new snRNAs from *Saccharomyces cerevisiae*: snR34, snR35, and snR36. *Nucleic Acids Res.* 23: 2548–2554.

Samuelson, T. 1991. Interactions of tRNA pseudouridine synthases with RNAs substituted with fluorouracil. *Nucleic Acids Res.* 19: 6139–6144.

Sawa, H., and Y. Shimura. 1992. Association of U6 snRNA with the 5' splice site region of pre-mRNA in the spliceosome. *Genes Dev.* 6:244–254.

Sawa, H., and J. Abelson. 1992. Evidence for a base-pairing interaction between U6 small nuclear RNA and 5' splice site during the splicing reaction in yeast. *Proc. Natl. Acad. Sci. USA* 89: 11269–11273.

Scherly, D., W. Boelens, W. J. Van Venrooij, N. A. Dathan, J. Hamm, and I. W. Mattaj. 1989. Identification of the RNA binding segment of human U1 A protein and definition of its binding site on U1 snRNA. *EMBO J.* 8:4163–4170.

Schlenstedt, G., D. H. Wong, D. M. Koepp, and P. A. Silver. 1995. Mutants in a yeast Ran binding protein are defective in nuclear transport. *EMBO J.* 14:5367–5378.

Ségault, V., et al. Unpublished data.

Ségault, V., C. L. Will, B. S. Sproat, and R. Lührmann. 1995. In vitro reconstitution of mammalian U2 and U5 snRNPs active in splicing: Sm proteins are functionally interchangeable and are essential for the formation of functional U2 and U5 snRNPs. *EMBO J.* 14:4010–4021.

Séraphin, B., L. Kretzner, and M. Rosbash. 1988. A U1 snRNA: pre-mRNA base pairing interaction is required early in yeast spliceosome assembly but does not uniquely define the 5' cleavage site. *EMBO J.* 7:2533–2538.

Séraphin, B., and M. Rosbash. 1990. Exon mutations uncouple 5' splice site selection from U1 snRNA pairing. *Cell* 63:619–629.

Shapiro, M. B., and P. Senapathy. 1987. RNA splice junctions of different classes of eukaryotes: sequence statistics and functional implications in gene expression. *Nucleic Acids Res.* 15: 7155–7174.

Sharp, P. A. 1985. On the origin of RNA splicing and introns. *Cell* 42:397–400.

Sharp, P. A. 1991. Five easy pieces. *Science* 254:663.

Shatkin, A. 1976. Capping of eukaryotic mRNAs. *Cell* 9:645–653.

Siliciano, P. G., and C. Guthrie. 1988. 5' splice site selection in yeast: genetic alterations in base-pairing with U1 reveal additional requirements. *Genes Dev.* 2:1258–1267.

Singh, R., and R. Reddy. 1989. Gamma-monomethyl phosphate: a cap structure in spliceosomal U6 small nuclear RNA. *Proc. Natl. Acad. Sci. USA* 86:8280–8283.

Smith, C., and J. A. Steitz. 1997. Sno storm in the nucleolus: new roles for myriad small RNPs. *Cell* 89:669–672.

Solymosy, F., and T. Pollàk. 1993. Uridylate-rich small nuclear RNAs (UsnRNAs), their genes and pseudognes, and UsnRNPs in plants: structure and function. A comparative approach. *Crit. Rev. Plant Sci.* 12:275–369.

Sontheimer, E. J., and J. A. Steitz. 1993. The U5 nd U6 small nuclear RNAs as active site components of the spliceosome. *Science* 262:1989–1996.

Stroke, I. L., and A. M. Weiner. 1985. Genes and pseudogenes for rat U3A and U3B small nuclear RNAs. *J. Mol. Biol.* 184: 183–193.

Sun, J. S., and J. L. Manley. 1995. A novel U2-U6 snRNA structure is necessary for mammalian mRNA splicing. *Genes Dev.* 9: 843–854.

Sun, J. S., and J. M. Manley. 1997. The human U6 snRNA intramolecular helix: structural constraints and lack of sequence specificity. *RNA* 3:514–526.

Szkukàlek, A., E. Myslinski, A. Mougin, R. Lührmann, and C. Branlant. 1995. Phylogenetic conservation of modified nucleotides in the terminal loop 1 of the spliceosomal U5 snRNA. *Biochimie* 77:16–21.

Tarn, W., T. R. Yario, and J. A. Steitz. 1995. U12 snRNA in vertebrates: Evolutionary conservation of 5' sequences implicated in splicing of pre-mRNAs containing a minor class of introns. *RNA* 1:644–656.

Tarn, W. Y., and J. A. Steitz. 1996a. A novel spliceosome containing U11, U12 and U5 snRNPs excises a minor class (AT-AC) intron *in vitro*. *Cell* 84:1–20.

Tarn, W. Y., and J. A. Steitz. 1996b. Highly diverged U4 and U6 small nuclear RNAs required for splicing rare AT-AC introns. *Science* 273:1824–1832.

Terns, M. P., J. E. Dahlberg, and E. Lund. 1993. Multiple cis-acting signals for export of pre-U1 snRNA from the nucleus. *Genes Dev.* 7:1898–1908.

Terns, M. P., and J. E. Dahlberg. 1994. Retention and 5' cap trimethylation of U3 snRNA in the nucleus. *Science* 2641: 959–961.

Terns, M. P., C. Grimm, E. Lund, and J. E. Dahlberg. 1995. A common maturation pathway for small nucleolar RNAs. *EMBO J.* 14:4860–4871.

Thompson, J. R., J. Zagorski, J. L. Woolford, and M. J. Fournier. 1988. Sequence and genetic analysis of a dispensible 189 nucleotides snRNA from *Saccharomyces cerevisiae*. *Nucleic Acids Res.* 16:5587–5601.

Tollervey, D. 1987. A yeast small nuclear RNA is required for normal processing of preribosomal RNA. *EMBO J.* 6: 4169–6175.

Tollervey, D., and T. Kiss. 1997. Function and synthesis of small nucleolar RNAs. *Curr. Opin. Cell Biol.* 9337–342.

Tyc, K., and J. A. Steitz. 1989. U3, U8 and U13 comprise a new class of mammalian snRNPs localized in the cell nucleolus. *EMBO J.* 8:3113–3119.

Tycowski, K. T., M. D. Shu, and J. A. Steitz. 1993. A small nucleolar RNA is processed from an intron of the human gene encoding ribosomal protein S3. *Genes Dev.* 7:1176–1190.

Umen, J. G., and C. Guthrie. 1995. The second catalytic step of pre-mRNA splicing. *RNA* 1:869–885.

Valcarcel, J., R. K. Gaur, R. Singh, and M. R. Green. 1996. Interaction of U2AF^{65}RS region with pre-mRNA of branch point and promotion base-pairing with U2 snRNA. *Science* 273: 1706–1709.

Vankan, P., C. McGuigan, and I. W. Mattaj. 1990. Domains of U4 and U6 snRNAs required for snRNP assembly and splicing complementation in *Xenopus* oocytes. *EMBO J.* 9:3397–3404.

Visa, N., A. T. Alzhanova-Ericsson, X. Sun, E. Kiselava, B. Björkroth, T. Wurtz, and B. Daneholt. 1996. A pre-mRNA binding protein accompanies the RNA from the gene through the nuclear pores and into polysomes. *Cell* 84:253–264.

Wells, S. E., and M. Ares. 1994. Interactions between highly conserved U2 small nuclear RNA structures and Prp5p, Prp9p, Prp11p, and Prp21p proteins are required to ensure integrity of the U2 small nuclear ribonucleoprotein in *Saccharomyces cerevisiae*. *Mol. Cell. Biol.* 14:6337–6349.

Wells, S. E., M. Neville, M. Haynes, J. Wang, H. Igel, and M. Ares. 1996. CUS1, a suppressor of cold-sensitive U2 snRNA mutations, is a novel yeast splicing factor homologous to human SAP 145. *Genes Dev.* 10:220–232.

Westin, G., E. Lund, J. T. Murphy, U. Petterson, and J. E. Dahlberg. 1984. Human U2 and U1 RNA genes use similar transcription signals. *EMBO J.* 3:3295–3301.

Will, C. L., S. Rümpler, J. K. Gunnewiek, W. J. VanVenrooij, and R. Lührmann. 1996. *In vitro* reconstitution of mammalian U1 snRNPs active in splicing: the U1-C protein enhances the formation of early (E) spliceosomal complexes. *Nucleic Acids Res.* 24:4614–4623.

Wise, J. A., and A. M. Weiner. 1980. *Dictyostelium* small nuclear RNA D2 is homologous to rat nucleolar RNA U3 and is encoded by a dispersed multigene family. *Cell* 22:109–118.

Wolff, T., and A. Bindereif. 1993. Conformational changes of U6 RNA during the spliceosome cycle: an intramolecular helix is essential both for initiating the U4–U6 interaction and for the first step of splicing. *Genes Dev.* 7:1377–1389.

Wolff, T., and A. Bindereif. 1995. Mutational analysis of human U6 RNA: stabilizing the intramolecular helix blocks the spliceosomal assembly pathway. *Biochim. Biophys. Acta* 1263:39–44.

Wong, N., A. H. Corbett, H. M. Kent, M. Steward, and P. A. Silver. 1997. Interaction between the small GTPase Ran/Gsp1p and Ntf2p is required for nuclear transport. *Mol. Cell. Biol.* 17: 3755–3767.

Wu, J., and J. L. Manley. 1989. Mammalian pre-mRNA branch site selection by U2 snRNP involves base-pairing. *Genes Dev.* 3: 1553–1561.

Wu, J. A., and J. L. Manley. 1991. Base-pairing between U2 and U6 snRNAs is necessary for splicing of a mammalian pre-mRNA. *Nature* 352:818–821.

Wu, Q., and A. R. Krainer. 1997. Splicing of a divergent subclass of AT–AC introns requires the major spliceosomal snRNAs. *RNA* 3:586–601.

Wyatt, J. R., E. J. Sontheimer, and J. A. Steitz. 1992. Site-specific cross-linking of mammalian U5 snRNP to the 5′ splice site before the first step of pre-mRNA splicing. *Genes Dev.* 6: 2542–2553.

Yan, D., and M. J. Ares. 1996. Invariant U2 RNA sequences bordering the branchpoint recognition region are essential for interaction with yeast SF3A and SF3B subunits. *Mol. Cell. Biol.* 16:818–828.

Zavanelli, M. I., J. S. Britton, A. H. Igel, and M. Ares. 1994. Mutations in an essential U2 small nuclear RNA structure cause cold-sensitive U2 small nuclear ribonucleoprotein function by favoring competing alternative U2 RNA structures. *Mol. Cell. Biol.* 14:1689–1697.

Zerby, D. B., and J. R. Patton. 1997. The modification of U4 RNA requires U6 RNA and multiple pseudouridine synthases. *Nucleic Acids Res.* 25:4808–4815.

Zerby, D. B., and J. R. Patton. 1996. Metabolism of pre-messenger RNA splicing cofactors: modification of U6 RNA is dependent on its interaction with U4 RNA. *Nucleic Acids Res.* 24: 3583–3589.

Zhuang, Y., and A. M. Weiner. 1986. A compensatory base change in U1 snRNA suppresses a 5′ splice site mutation. *Cell* 46: 827–835.

Zhuang, Y., and A. M. Weiner. 1989. A compensatory base change in human U2 snRNA can suppress a branch site mutation. *Genes Dev.* 3:1545–1552.

Zieve, G., and S. Penman. 1976. Small RNA species of the HeLa cell: metabolism and subcellular localization. *Cell* 8:19–31.

Chapter 12

The Pseudouridine Residues of rRNA: Number, Location, Biosynthesis, and Function

JAMES OFENGAND AND MAURILLE J. FOURNIER

Pseudouridine (5-β-D-ribofuranosyluracil; Ψ), the 5-ribosyl isomer of uridine, was first detected as a new nucleoside by Davis and Allen (1957). Its structure (Appendix 1) was proposed shortly thereafter (Yu and Allen, 1959; Scannell et al., 1959) and proven by Cohn (1959, 1960). Ψ is found in tRNA (Sprinzl et al., 1996), rRNA (Maden, 1990), and sn(o)RNA (Gu and Reddy, 1997) (see Chapter 11 by Massenet et al.) but not so far in mRNA or viral RNAs. In other words, Ψ appears to be confined to those RNAs whose function is dependent on a specific tertiary structure. Ψ is formed at the polynucleotide level (Johnson and Söll, 1970; Ciampi, et al., 1977) by isomerization of a specific uridine residue in a reaction that requires neither energy source nor cofactors (Arena et al., 1978; Green et al., 1982; Kammen et al., 1988; Samuelsson and Olsson, 1990; Nurse et al., 1995; Wrzesinski et al., 1995a, 1995b).

rRNA plays a key role in the functioning of the ribosome (Green and Noller, 1997). Two features of rRNA that are noteworthy in this context are (i) sequences that have been strongly conserved across species lines, most notably residues 518 to 533, 1394 to 1408, and 1492 to 1505 (*Escherichia coli* numbering) in the small-subunit (SSU) RNA, which have been almost totally conserved among all sequenced rRNAs (Van de Peer et al., 1997) and which are intimately involved in the decoding of mRNA (Ofengand et al., 1993; Brimacombe, 1995), and (ii) the modified nucleoside content of rRNA. These modified nucleosides consist of Ψ, the most common single modified nucleoside in bulk rRNA, methylated nucleosides (both on the base and on the 2'-OH of the ribose) and much less frequently, other more esoteric nucleosides such as 2'-O-methyl inosine (Gray, 1976), dihydrouridine (Kowalak et al., 1995), 1-methyl-3-(3-amino-3-carboxypropyl)pseudouridine (m^{1}acp^{3}Ψ) (Saponara and Enger, 1974; Maden et al. 1975), N^4-acetylcytidine (Thomas et al., 1978), and N^3-methyl Ψ (Kowalak, et al., 1996). See Crain and McCloskey (1997) for a current summary of the modified nucleosides in rRNA.

This review will focus on the Ψ residues in rRNA. This subject has been reviewed previously both in a detailed analysis of the work on eukaryotic rRNA up to circa 1990, which also includes methylated nucleosides (Maden, 1990), and in a review of more recent work (Ofengand et al., 1995).

NUMBER AND LOCATION OF Ψs IN SSU AND LSU rRNAs

Methodology

The earlier "classical" methods for identifying and locating Ψ residues in an RNA sequence have been ably described by Maden (1990). Briefly, alkaline hydrolysis of ^{32}P-labelled RNA followed by chromatographic separation of Ψp from other nucleosides and quantitation gave the amount of Ψ present. Alternatively, all of the RNase T$_1$ oligonucleotides were separated and each was analyzed for the amount of Ψ. For sequence localization, the RNase T$_1$ oligonucleotides, oriented on the basis of the known complete RNA sequence, were further analyzed by conventional enzymatic digestion and chromatography. The arduous nature of these analyses inhibited development of the study of the role of Ψ, especially in large RNA molecules like rRNA, until the development of a more facile reverse transcriptase approach

---

James Ofengand • Department of Biochemistry and Molecular Biology, University of Miami School of Medicine, Miami, Florida 33101.   **Maurille J. Fournier** • Department of Biochemistry and Molecular Biology, University of Massachusetts, Amherst, Massachusetts 01003.

(Bakin and Ofengand, 1993). This method takes advantage of the alkaline stability of the N^3-1-cyclohexyl-3-(2-morpholinoethyl)carbodiimide metho-p-toluenesulfonate (CMC)–Ψ adduct compared to N^3-CMC-U, N^3-CMC-rT(m^5U), and N^1-CMC-G. Using this approach, under suitable conditions a statistical proportion of the Ψ residues can be modified with CMC without concomitant modification of other nucleosides. The CMC-Ψ adduct after alkaline treatment is derivatized at the N_3 position (Ho and Gilham, 1971). Because the N_3 of Ψ is used for base-pairing during reverse transcription, CMC-Ψ formation inhibits reverse transcription. This is revealed by gel electrophoresis on a sequencing gel as a band 1 residue 3′ to the Ψ residue. For reasons unknown, some sites are prone to stuttering (Denman et al., 1988), resulting in a doublet band both one residue 3′ and at the Ψ residue itself (Bakin and Ofengand, 1993). When this phenomenon occurs in runs of U residues, it is not possible to distinguish between stuttering and two adjacent Ψ residues unless the Ψ in question is at the 5′ end of the run. In these cases (Bakin et al., 1994b), a hydrazine-aniline procedure (Bakin and Ofengand, 1993) usually suffices to resolve the ambiguity.

Although the reaction is not completely specific for Ψ, for example s^4U forms an adduct with CMC which blocks reverse transcription, several other modified U residues frequently found in tRNA do not yield reverse transcriptase stops (Bakin and Ofengand, 1993). Unlike Ψ, s^4U reacts with hydrazine/aniline and therefore can be distinguished from Ψ by this reaction. The U-derived nucleoside specificity for reaction with CMC/OH and hydrazine/aniline is summarized in Table 1.

The advantages of this approach are several-fold. First, it allows the placement of Ψ residues in large RNAs to nucleotide resolution with relative ease. Second, a virtually complete scan of such an RNA can be done simply by the choice of primer for reverse transcription. Third, the RNA to be sequenced need not be purified, since the reverse transcription primer selects the RNA species to be transcribed. For example, analysis of the *S. cerevisiae* large-subunit (LSU) mitochondrial rRNA was carried out on a preparation which not only contained the SSU RNA but also was only 60–70% mitochondrial in origin (Bakin et al., 1994b).

Number of Ψ Residues in rRNA

SSU RNA

By both the "classical" approach and the newer reverse transcriptase sequencing procedure, the number of Ψ in SSU RNA from a number of species has been determined (Table 2). In some species many or all were exactly localized. *Escherichia coli* and *Saccharomyces cerevisiae* have been completely screened, and about three-fourths of the mammalian Ψ have been mapped. Although only one Ψ was detected in *Bacillus subtilis*, at the same site as in *E. coli*, other sites may exist because only 15% of the RNA was examined (Niu and Ofengand, unpublished results). No Ψ was found in *Halobacter halobium* even after screening all of the corresponding regions known to contain Ψ in yeast or mammals. However, because 30% of the SSU RNA was not examined (Bakin and Ofengand, 1995), it is not possible to be sure that there is no Ψ present in this molecule. A striking feature of the number of Ψs in SSU RNA is the marked percent increase in Ψ content in the following order: prokaryotes, unicellular eukaryotes, and multicellular eukaryotes. A similar trend is noted below for LSU RNA.

LSU RNA

The total number of Ψs and number exactly positioned in LSU RNA is shown in Table 3. Previous work, not listed in Table 3, includes a determination of 8 Ψs (Gehrke and Kuo, 1989) and the identification

Table 1. Reactivity of U-derived modified nucleosides with CMC/OH and hydrazine-aniline

Base	CMC/OH[a]	Hydrazine/aniline[b]
Uridine (U)	−	+
Pseudouridine (Ψ)	+	−
5-Methyluridine (m^5U)	−	−
4-Thiouridine (s^4U)	+	+
Uridine-5-oxyacetic acid (cmo^5U)	−	+
2-Thio-5-methylaminomethyluridine (mam^5s^2U)	−	+

[a] Formation of an alkali-stable adduct which blocks reverse transcription and produces a gel band is scored as + (Bakin and Ofengand, 1993).
[b] Opening of the pyrimnidine ring and chain cleavage which blocks reverse transcription and produces a gel band is scored as + (Bakin and Ofengand, 1993; Lankat-Buttgereit et al., 1987).

Table 2. Number of pseudouridine and modified pseudouridine residues in small subunit rRNAs and number positioned in the RNA sequence

Organism	Length of RNA[a] (nucleotides)	Total no. of Ψ	Ψ as % of total	No. of Ψs located to nucleotide resolution
Escherichia coli	1,542	1[b]	0.06	1[b]
Bacillus subtilis	1,550	≥1[c]	≥0.06	≥1[c]
Saccharomyces carlsbergensis	1,798[e]	11[d]	0.6	–
Saccharomyces cerevisiae	1,798	14[e]	0.7	14[e]
Xenopus laevis	1,825	44[f]	2.4	–
Mus musculus	1,869	~36[f]	1.9	–
Rattus norvegicus	1,870, 1,874	~37[f]	2.0	30[f,g]
Homo sapiens	1,868–1,870	~36[f]	1.9	32[f,g]

[a] Van de Peer et al. (1997).
[b] Bakin et al. (1994a).
[c] Wrzesinski et al. (1995a).
[d] Brand et al. (1979).
[e] Bakin and Ofengand (1995).
[f] Located to within 1 to 4 residues (Maden, 1990).
[g] Ganot et al. (1997a).

of Ψ746, Ψ1911, and Ψ1917 (Branlant et al., 1981) in *E. coli* LSU RNA, the localization of 4 of the Ψs in *S. carlsbergensis* LSU RNA (Veldman et al., 1981), the positioning of 3 Ψs in *X. laevis* and *H. sapiens* LSU RNAs (Maden, 1988), the finding of not more than one Ψ in *S. cerevisiae* mitochondrial LSU RNA (Klootwijk et al., 1975), and the localization of 2 Ψ residues in vertebrate 5.8S RNA (Nazar et al, 1976; Khan and Maden, 1977). The larger base of data in this table compared to Table 2 reinforces the fact that the percent Ψ increases from prokaryotes and organelles to yeast to multicellular eukaryotes. The difference is quite marked but so far there is no explanation for the phenomenon.

Locations of the Ψ Residues

SSU RNA

A summary of the known Ψ sites in SSU RNA is shown in Fig. 1. In addition to the *S. cerevisiae* and *H. sapiens* sites shown, the figure also indicates the location of the single Ψ in *E. coli* SSU RNA and the only known site in *B. subtilis* SSU RNA. The triangles mark the sites of base and ribose methylation, and will be considered below. There are several points to note about the Ψ distribution among the different organisms. First, the discrepancy in number between *E. coli* and *B. subtilis* on the one hand, and yeast and human on the other, is marked. Note also that the single Ψ found in *E. coli* and *B. subtilis* does not have a counterpart in the eukaryotes despite the 30- to 40-fold increase in number. Second, there is considerable correspondence between yeast and human, yet both species also have their specific sites. Third, there is no apparent grouping of the Ψ residues in any particular locale, either in the secondary structure shown or when the tertiary structure of the RNA is examined (not shown). In fact, they are notably absent from the three highly conserved sequence elements of SSU RNA, the 564-581, 1628-1643, and 1753-1766 (yeast numbering) segments. This result stands in sharp contrast to what is found for LSU RNA (see the next section) and supports the view that Ψ residues may fulfill more than one type of function in rRNA.

LSU RNA

Figure 2 shows the positions of the 9 Ψ residues in *E. coli* LSU RNA. The locations of these sites are in marked contrast to those in SSU RNA. They are clustered in three distinct areas, which we have termed the 5′, central, and 3′ regions. The 3′ region sites are in or adjacent to the peptidyl transferase center (PTC). Cross-linking results show that the central region is also close to the PTC in the LSU. Thus, A1913 has been cross-linked to G1964, which in turn can be cross-linked to U1940, which can be cross-linked to U2554 at the PTC (Brimacombe, 1995). As these are all zero-length or short cross-linkers, these results, albeit circuitously, place the central region stem-loop within a short distance of the PTC. Ψ746 and Ψ955 in the 5′ region have also been placed near the PTC. Ψ746 was directly cross-linked to U2613A2614 (Mitchell et al., 1990), and Ψ955 was placed close to C2475 via zero-length cross-links from A960 to U89 of 5S RNA and from U89 to C2475 (Dokudovskaya et al., 1996). Additional evidence suppporting the juxtaposition of all 3 cluster regions when in the 50S subunit is cited in Bakin and Ofengand (1993) and Bakin et al. (1994b).

This surprising clustering behavior at the most important functional site of the ribosome, the PTC,

Table 3. Number of pseudouridine and modified pseudouridine residues in large subunit rRNAs and number positioned in the RNA sequence

Organism (GenBank accession no.)	Length of RNA[a] (nucleotides)	Total no. of Ψs	Ψ as % of total	No. of Ψs located to nucleotide resolution
Cytoplasm				
E. coli (J01695)	2,904	9[b,c]	0.31	9[b,c]
B. subtilis (K00637)	2,927	5[c,d]	0.17	5[c,d]
H. halobium (X03407)	2,905	4[d]	0.14	4[d]
S. carlsbergensis (X00468)	3,393	23–24[e], ~32[f]	–	4[g]
S. cerevisiae (J01355)	3,392	30[h]	0.88	30[h]
D. melanogaster (M21017)	3,900[i]	–	1.44	57[d]
X. laevis (X59734)	4,082	~52[f]	1.27	–
M. musculus (X00525)	4,712	57[d]	1.21	57[d]
H. sapiens (M11167)	5,025	55[d]	1.09	55[d]
Mitochondria				
S. cerevisiae (J01527)	3,273	1[j]	0.03	1[j]
M. musculus (J01420)	1,582	≥1[d]	0.06	≥1[d]
H. sapiens (J01415, V00710)	1,559	≥1[d]	0.06	≥1[d]
T. brucei (X02547)	1,149	6[d]	0.35	6[d]
Chloroplasts (Z. mays [Z00028, X01365])[c]	2,888	4[d,d]	0.14	4[c,d]

[a] According to GenBank version 88.0.
[b] Bakin and Ofengand (1993).
[c] Includes the N^1-methyl Ψ in E. coli (Kowalak et al., 1996) and the unidentified modified U found at the identical position in B. subtilis and Z. mays (Ofengand and Bakin, 1997).
[d] Ofengand and Bakin, (1997).
[e] Brand et al. (1979).
[f] Maden (1990).
[g] Veldman et al. (1981).
[h] Bakin et al. (1994b).
[i] Gene sequence of 3,945 residues.
[j] Exactly 1 (Bakin et al., 1994b); 0–1 (Klootwijk et al., 1975).

inspired a more extensive series of analyses encompassing a phylogenetically broad range of organisms, including organelles (Bakin et al., 1994b; Ofengand and Bakin, 1997). The results are summarized in Fig. 3 through 5. In the prokaryote and organelle LSU RNAs examined, the Ψ residues cluster tightly into two of the three regions found in E. coli RNA. The close evolutionary relationship between prokaryotes and organelles appears to also hold true at the level of Ψ number and placement. In view of the same core secondary structure for these RNAs and the fact that all ribosomes carry out the same function by the same mechanism (as far as one knows), we assume that the juxtaposition of the two regions in E. coli also occurs in these other organisms. Therefore, it appears that all of the Ψ residues in the LSU RNA of a gram-positive and gram-negative prokaryote, a representative of the Archaea, and examples of both chloroplast and mitochondrial organelles are concentrated at or near the PTC. This geography must have meaning. Possibilities are discussed later in this chapter.

Extension of the analyses to eukaryotes showed that the clustering phenomenon persisted, although with more dispersion as the total number of Ψs increased. Note the unusually large increase in Ψ in the 5′ region as compared to E. coli. Comparative analyses of a number of organisms are shown in Fig. 3 through 5. Figure 4 shows the nucleotide present at the equivalent of the Ψ site in those organisms that lack Ψ. Thus it can be immediately ascertained whether Ψ is missing because the precursor U is absent or because of a failure to modify this particular U. Perusal of the figure shows that both situations occur. Similar results are obtained when the other regions are examined (see Table 3 in Ofengand and Bakin, 1997). The main results from this data set are that the clustering phenomenon persists, and that certain sites appear as universal or semiuniversal sites for Ψ. In the central region, the two boxed Ψ residues are present in all the organisms tested, although note that one is a modified Ψ in E. coli and probably also in B. subtilis and Zea mays chloroplasts. These Ψs should have some special function. In the 3′ region (Fig. 5), there are two sites denoted by two pair of dashed ovals where most of the organisms tested, except for mitochondrial ribosomes, have a Ψ residue at or adjacent to the same site. These may also be sites of a more general importance. A more complete discussion of these results can be found in the original publication (Ofengand and Bakin, 1997).

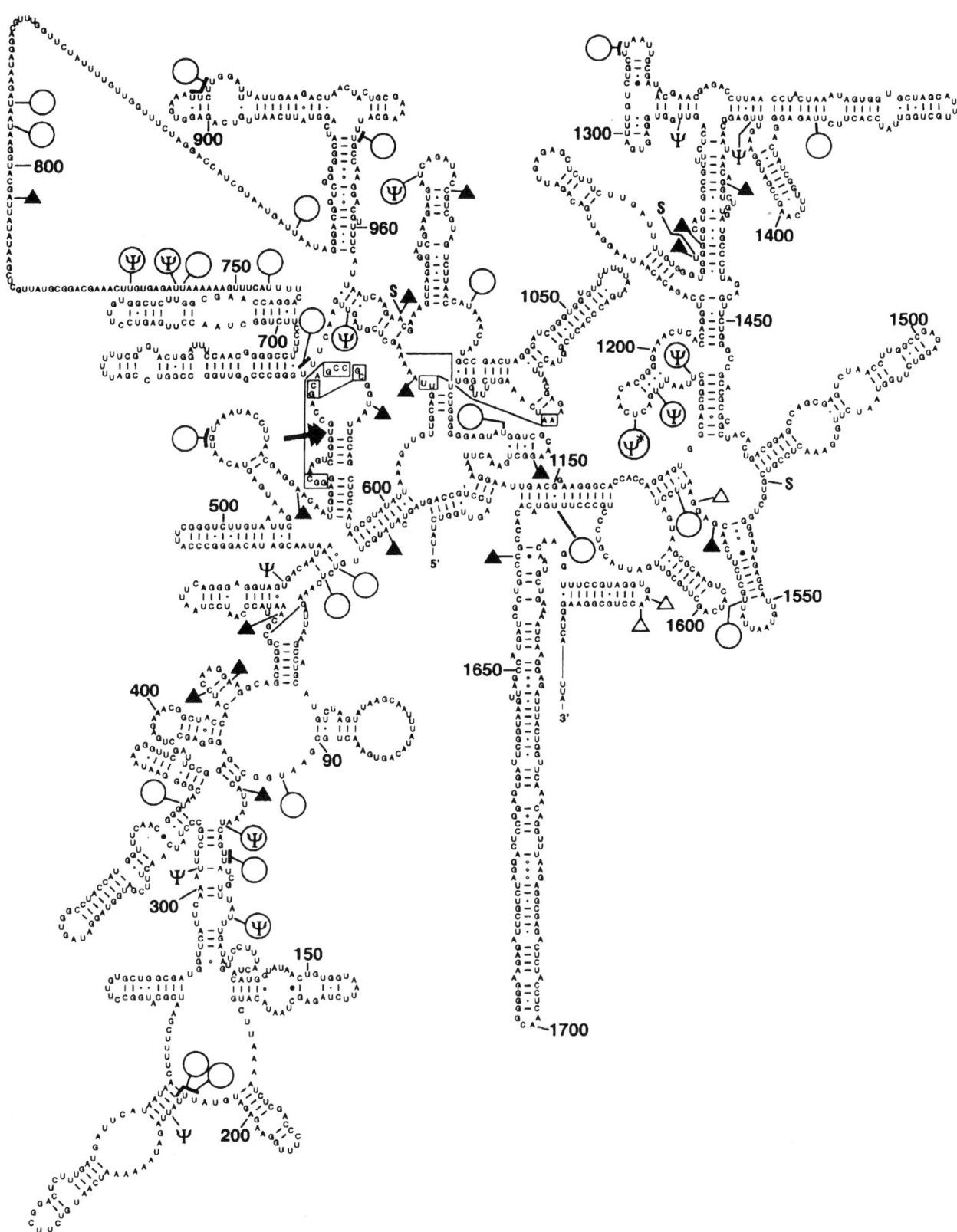

Figure 1. Location of Ψ and other modified residues in SSU RNA. The secondary structure is that of *S. cerevisiae*. Ψ and Ⓨ, pseudouridines in *S. cerevisiae* (Bakin and Ofengand, 1995); ○ and Ⓨ, mammalian pseudouridine positions (Maden, 1990; Ganot et al., 1997a); Ⓨ, pseudouridines at the same site in both *S. cerevisiae* and mammals; Ψ*, m¹acp³Ψ (Saponara and Enger, 1974; Maden et al., 1975); △, base-methylated, and ▲, 2'-O-methyl, nucleosides in *S. cerevisiae* (Maden, 1990); arrow, site of Ψ in *E. coli* (Bakin et al., 1994a) and *B. subtilis* (Wrzesinski et al., 1995a). Adapted from Bakin and Ofengand (1995).

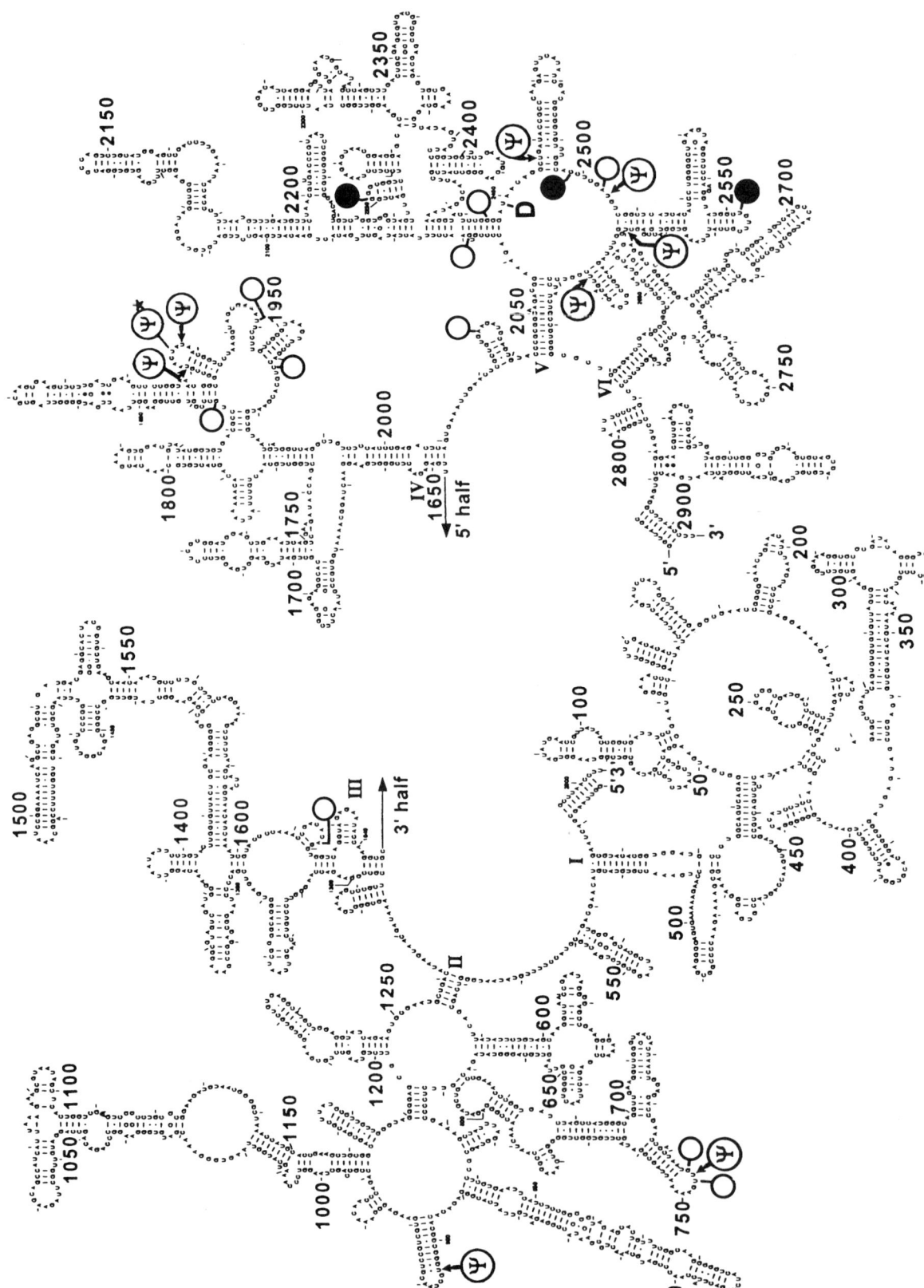

Figure 2. Location of Ψ and other modified residues in *E. coli* LSU RNA. The secondary structure is from Gutell et al. (1993). Location of the Ψ residues is from Bakin and Ofengand (1993) and Bakin et al. (1994b). Open circles, base-methyl, and filled circles, 2'-O-methyl nucleosides (see Table II of Bakin and Ofengand, 1993); Ψ*, N^3-methyl Ψ (Kowalak et al., 1996).

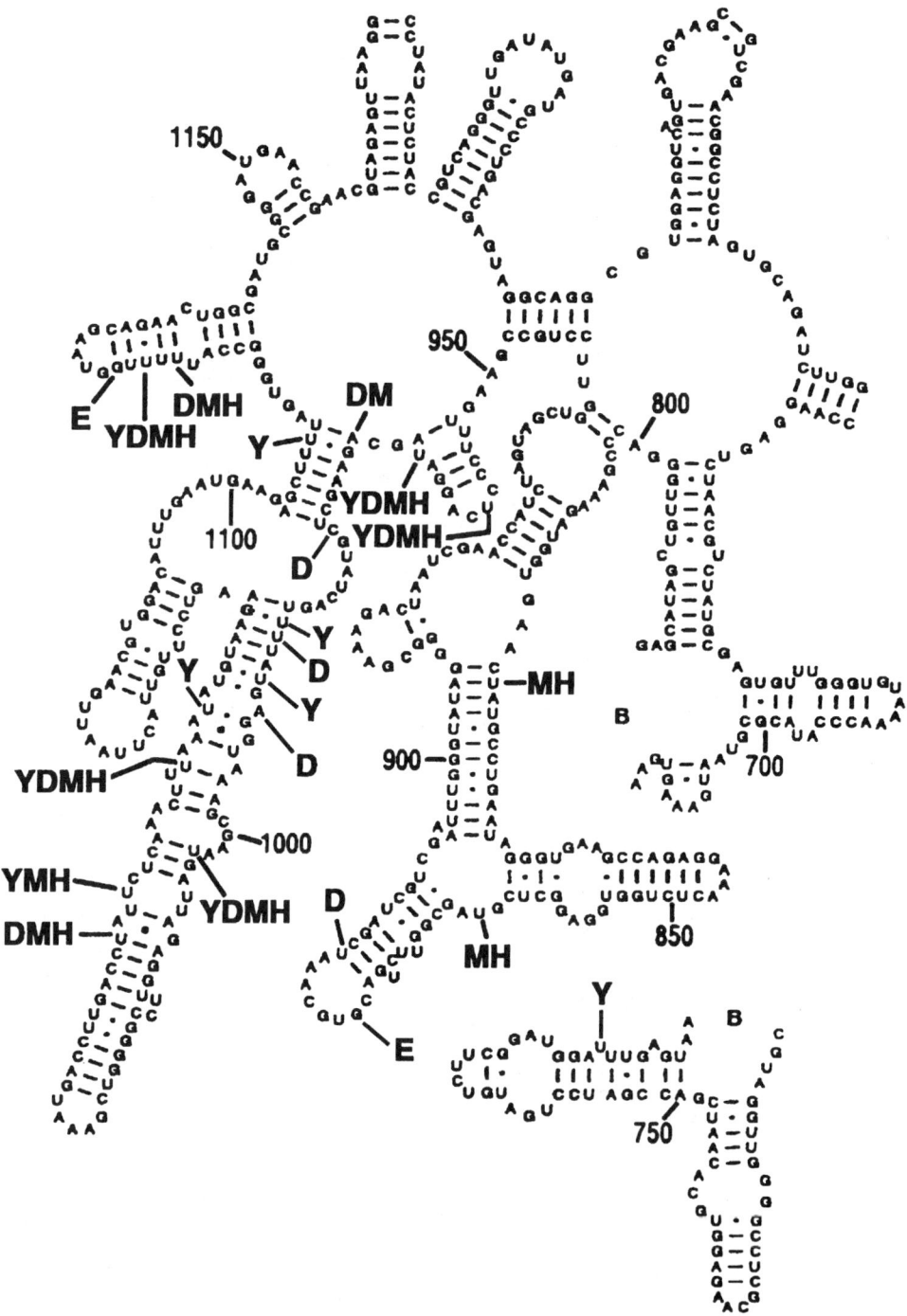

Figure 3. Comparative positions of Ψ residues in the 5' region of LSU RNAs. The sequence is that of *S. cerevisiae*. E, *E. coli*; Y, *S. cerevisiae*; D, *D. melanogaster*; M, *M. musculus*; H, *H. sapiens*. Reprinted from Ofengand et al. (1995) with permission.

Correlation of Ψ Sites with Other Modified Nucleosides

In SSU RNA, there does not appear to be any spatial relationship between the sites of Ψ occcurrence and those for the base-methylated and 2'-O-methyl nucleosides (Fig. 1). This holds true for yeast (Bakin and Ofengand, 1995) and also for mammals (Maden, 1990). On the other hand, the reverse is true for LSU RNA. Both in *E. coli* (Fig. 2) and in yeast (Bakin et al., 1994b), there is a strong correlation between the locations of the two sets of modified nucleosides. This is also true for human LSU RNA

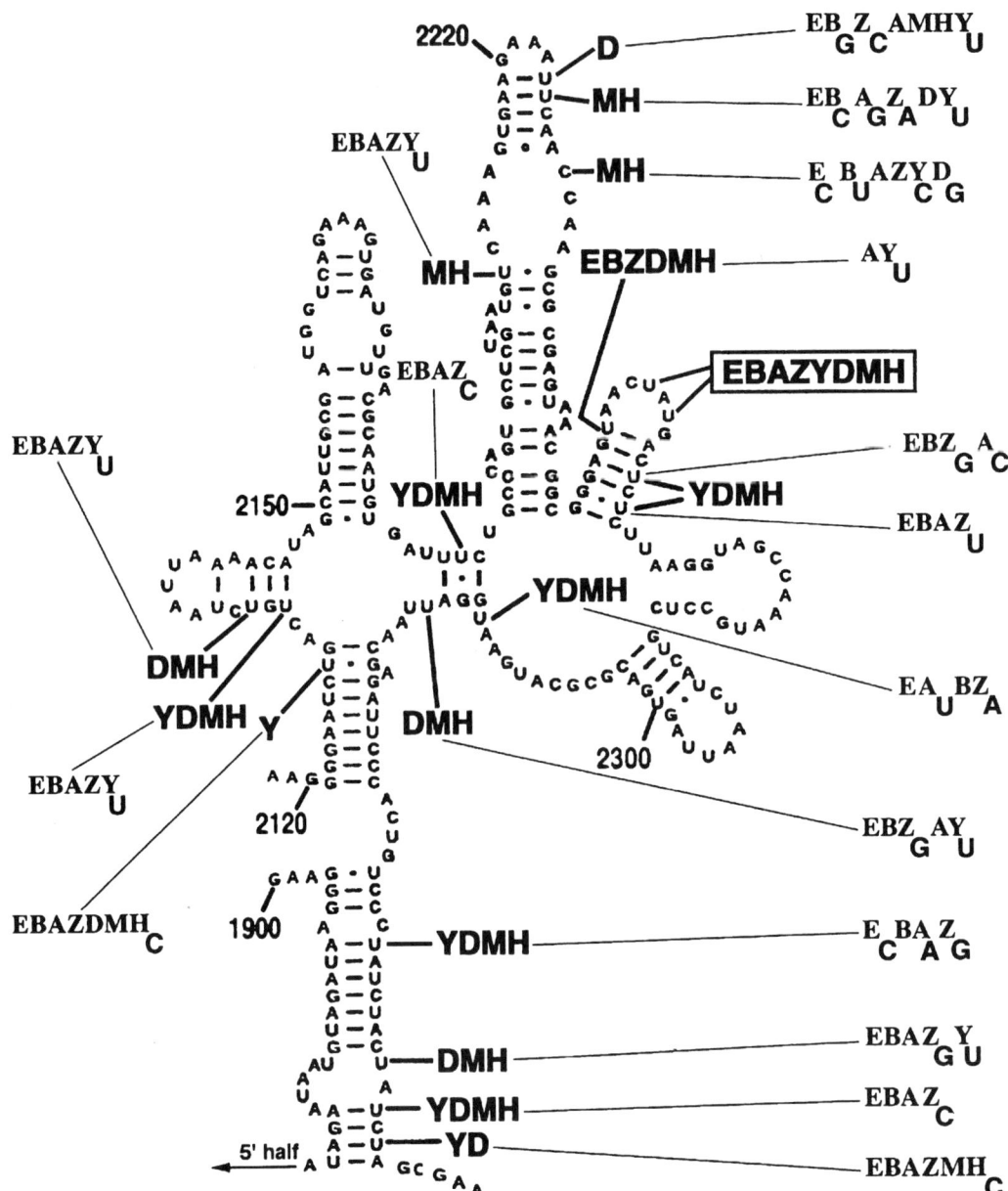

Figure 4. Comparative positions of Ψ residues in the central region of LSU RNAs. The sequence is that of *S. cerevisiae*. E, *E. coli*; B, *B. subtilis*; A, *H. halobium*; Z, *Z. mays* chloroplasts; Y, *S. cerevisiae*; D, *D. melanogaster*; M, *M. musculus*; H, *H. sapiens*. E*, B*, and Z*, see legend to Fig. 5. The boxed residues are invariant among the tested cytoplasmic and chloroplast species. Also shown are all of the sequence variations at the sites for Ψ in each of the examined species with the uppercase letter(s) indicating the organism(s) and the subscript letter indicating the nucleoside in those species. Reprinted from Ofengand and Bakin (1997) with permission.

(Ofengand and Bakin, 1997). It was previously suggested (Bakin et al., 1994b) that by mixing methylation sites that are hydrophobic with Ψ sites that are hydrophilic, nature may have found a way to create a complex pattern of noncovalent forces for some specific interactions on the surface of the ribosome. Possible reaction partners could include mRNA, tRNA, and initiation, elongation, and termination factors.

Structural Environment of Ψ Residues in rRNA

The SSU and LSU RNAs in which Ψ occurs are structural RNA molecules, and thus one of the functions of Ψ may be as a structural stabilizer (see below). Enough sites for Ψ have now been determined (281 total and 255 in known secondary structure) that their distribution in different secondary structural environments can be analyzed. These have been

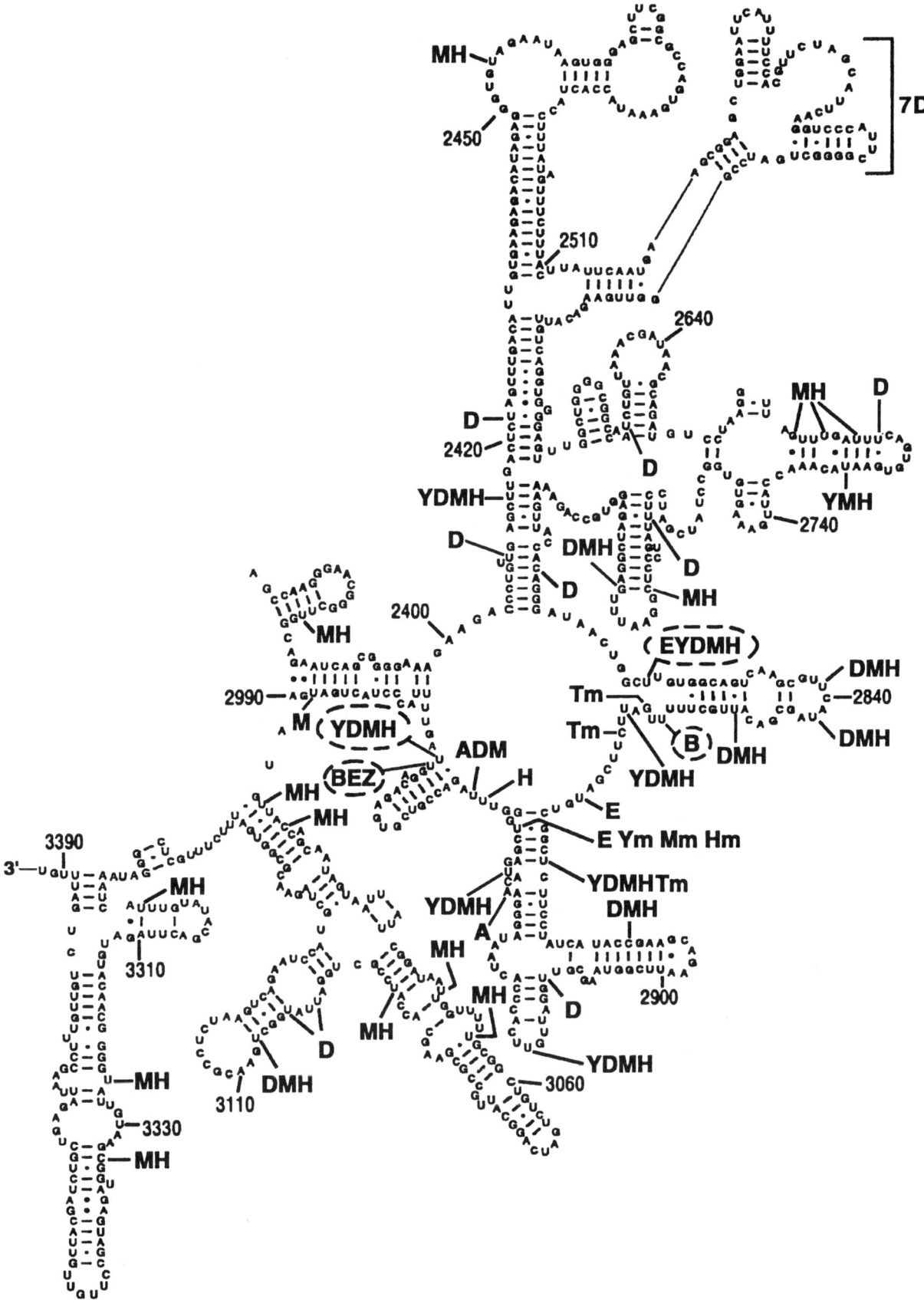

Figure 5. Comparative positions of Ψ residues in the 3' region, including the PTC, of LSU RNAs. The sequence is that of *S. cerevisiae*. E, *E. coli*; B, *B. subtilis*; A, *H. halobium*; Z, *Z. mays* chloroplasts; Y, *S. cerevisiae*; D, *D. melanogaster*; M, *M. musculus*; H, *H. sapiens*; Ym, *S. cerevisiae* mitochondria; Mm, *M. musculus* mitochondria; Hm, *H. sapiens* mitochondria; T, *T. brucei* mitochondria. The two sets of paired dashed ovals denote two semi-invariant sites. Reprinted from Ofengand and Bakin (1997) with permission.

classified into 5 groups (Table 4). The distribution of Ψ within these groups was previously reported for the LSU RNAs (Ofengand and Bakin, 1997). Here we summarize the data for both SSU and LSU RNA. All of the defined sites for Ψ listed in Tables 2 and 3 were used except for the rat SSU data which, being identical to human, were not included. Surprisingly, the percent in each category did not change whether all of the sites or only the "distinct" sites were counted. Whereas one might have expected that most Ψs would be found in single-stranded regions where they could most readily contribute to tertiary structure, in fact close to 60% were found in base-paired contexts. However, most of the sites were in base pairs closing a loop, and the percent decreased steadily as the Ψ-Pu base pair became further and further embedded in a base-paired stem.

BIOSYNTHESIS

Prokaryotes

Cloning of pseudouridine synthases

Previous work on Ψ synthases for tRNA, where the ubiquitous Ψ55 has long been a distinctive feature, had shown the existence of separate enzymes for different sites in the tRNA molecule (Schaefer et al., 1973; Ciampi et al., 1977; Samuelsson and Olsson, 1990; Szweykowska-Kulinska et al., 1994) and one such enzyme has been cloned (Arps et al., 1985; Kammen et al., 1988). It seemed reasonable, therefore, that the small number of Ψ found in prokaryotic and organelle rRNAs would also utilize a small set of site-specific enzymes for their synthesis. Whether the large number of Ψ found in eukaryotes would be made by a similarly large number of specific synthases was not clear, because this would require a major commitment of resources by the cell. We now know from recent work that a different system is used in eukaryotes (see below).

The Ψ synthases in *E. coli* were found by the time-honored procedure of fractionating a cell-free extract and looking for an in vitro activity. Since the conversion of U to Ψ is an isomerization that requires no cofactors or energy, the usual labelling methods are inadequate. The only useful rapid quantitative assay is that devised by Cortese et al. (1974) in which a 5-[^{3}H]uracil-labelled substrate RNA upon conversion to Ψ releases the 5-^{3}H atom as water. With the advent of in vitro RNA transcription technology using either T7 or SP6 RNA polymerase, such RNA substrates are readily obtained from commercially available 5-[^{3}H]UTP. By using this assay and standard

Table 4. Structural environment of Ψ and modified Ψ residues in SSU and LSU RNAs[f]

Organism	RNA	No. of Ψs in structural type[a]					
		A	B	C	D	E	Total
E. coli	SSU	–	–	–	–	1	1
	LSU	4	1	–	4	–	9
B. subtilis	SSU					1	1
	LSU	1	1	–	2	1	5
H. halobium	LSU	–	–	–	3	1	4
S. cerevisiae	SSU	2	3	1	4	2	12[b]
	LSU	9	8	2	7	4	30
X. laevis	LSU	–	1	–	2	–	3
D. melanogaster	LSU	19	8	4	10	9	50[c]
M. musculus	LSU	18	11	4	12	12	57
H. sapiens	SSU	3	3	2	6	4	18[d]
	LSU	17	11	4	12	11	55
S. cerevisiae mitochondria	LSU	–	–	1	–	–	1
M. musculus mitochondria	LSU	–	–	1	–	–	1
H. sapiens mitochondria	LSU	–	–	1	–	–	1
T. brucei mitochondria	LSU	1	–	1	1	–	3[e]
Z. mays chloroplasts	LSU	1	1	–	2	–	4
Sum (%)		75 (29)	48 (19)	21 (8)	65 (25)	46 (18)	255 (100)
Sum of distinct sites (%)		35 (30)	23 (20)	11 (9)	25 (21)	23 (20)	117 (100)

[a] Structural types are as follows: A, Pu-Ψ base pair closing a loop or bulge; B, Pu-Ψ base pair in a helix one base pair removed from a loop or bulge; C, Pu-Ψ base pair in a helix surrounded by at least two base pairs on either side; D, Ψ in loops or single-stranded regions but not adjacent to a base pair; E, Ψ in loops or single-stranded regions adjacent to a base pair.
[b] Two Ψ residues were in a region of undetermined secondary structure.
[c] Seven Ψ residues were in a region of undetermined secondary structure.
[d] Fourteen Ψ residues were in a region of undetermined secondary structure.
[e] Three Ψ residues were in a region of undetermined secondary structure.
[f] The LSU RNA data are summarized from Table 3 of Ofengand and Bakin (1997) except for *X. laevis*, which is from Maden (1990). The SSU RNA data are from Fig. 3 and sources cited therein. For the sum of distinct sites, Ψ residues occurring in identical structural contexts have been counted only once.

chromatographic methods, it was possible to purify three Ψ synthases with distinct and different specificities.

One was a tRNA-specific synthase responsible only for formation of Ψ55. Another was specific for formation of Ψ516 in *E. coli* SSU RNA, and a third was specific for a single site, Ψ746, in LSU RNA and a single site, Ψ32, in certain tRNA molecules. This "dual specificity" will be discussed below. In standard fashion, the N-terminal amino acid sequence of these three purified proteins was determined, appropriate open reading frames (ORFs) were located in GenBank, and the genes were cloned. In each case, overexpression with a His tag, affinity purification, and in vitro testing of activity and specificity confirmed that each gene product corresponded to the expected enzyme. These three ORFs are listed in Table 5. In addition, the previously cloned tRNA Ψ synthase gene formerly known as *hisT* and now designated *truA*, another rRNA synthase gene (*rluB*) identified by sequence homology (see below), and three recently cloned tRNA synthase genes from yeast are listed. These eight gene products are so far the only known cloned Ψ synthases from any organism.

RNA recognition by synthases

The truBp enzyme only reacted at one site and only with tRNA as substrate (Nurse et al., 1995). As a tRNA synthase, however, further discussion is beyond the scope of this chapter. The rsuAp makes the only Ψ in *E. coli* SSU RNA (Wrzesinski et al., 1995a). How does it distinguish U516 from the other 313 U residues? What is known so far is that free full-length (1-1542) or truncated (1-526 or 1-678) SSU RNA transcripts are not substrates, but an RNP particle assembled from the 1-678 transcript and ribosomal proteins is a substrate. However, 30S ribosomes assembled in vitro from full-length synthetic transcripts and ribosomal proteins (Krzyzosiak et al., 1987) are at best only a poor substrate, while a miniribosome (Weitzmann et al., 1993) made from the 1-526 fragment and ribosomal proteins is not active at all. This result implies that the synthase requires a defined RNA structure for recognition. Presumably, the active RNP substrate mimics an intermediate state of ribosome assembly which is the true substrate for Ψ516 formation. If that is the case, then only a small subset of U residues may be available to the synthase. In that context, it may be significant that there are only three other U residues in the 30-mer comprising the stem-loop where Ψ516 is found.

The rluAp synthase (Wrzesinski et al., 1995b), on the other hand, needs little if any structure in its substrate RNA. A fragment of LSU RNA from resi-

Table 5. Cloned pseudouridine synthases

Name[a]	Gene location (min or °)	RNA substrate	Position modified	Specificity class	Monomer molecular mass (kDa) of ORF	No. of amino acids	Accession no.	Reference(s)
rluA (E.c.)	1.30	23S rRNA	746	IV	24.9	219	P39219	Wrzesinski et al., 1995b
		tRNA	32					
rsuA (E.c.)	49.12	16S rRNA	516	I	25.9	231	P33918	Wrzesinski et al., 1995a
rluA (B.s.) (YpuL)	206.80°	23S rRNA	2605	I	26.0	229	P35159	Niu and Ofengand, unpublished
truB (E.c.)	71.36	tRNA	55	I	35.1	314	P09171	Nurse et al., 1995
truA (E.c.)	52.46	tRNA	38–40	II	30.4	270	P07649	Kammen et al., 1988
pus1 (S.c.)	—	tRNA	34–36 27–28	III	62.1	544	Q12211	Simos et al., 1996; Grosjean et al., 1997
pus3 (S.c.)	—	tRNA	38, 39	II	50.9	442	P31115	Lecointe et al., 1997
pus4 (S.c.)	—	tRNA	55	I	45.3	403	P48567	Becker et al., 1997

[a] E.c., *Escherichia coli*; B.s., *Bacillus subtilis*; S.c., *Saccharomyces cerevisiae*.

dues 1-847, the Ψ being at position 746, is more active than the full-length (1-2904) LSU RNA, and both RNAs are more active in EDTA than in Mg^{2+}, implying that tertiary, and possibly also secondary, structure is not an important determinant for this enzyme. The synthase was unreactive with SSU RNA transcripts whether as fragments or full-length, and unmodified 30S subunits were also inactive. An important observation was that this cloned and affinity-purified protein that was highly specific for a single site, 746, in LSU RNA was also able, with good efficiency, to convert U32 to Ψ in $tRNA^{Phe}$ transcripts, but not in transcripts of $tRNA^{Val}$. Only U32 was modified. All other U residues, including U55 which is Ψ55 in natural tRNA, were not affected. This specificity is the same as the natural state because normally $tRNA^{Phe}$, but not $tRNA^{Val}$, bears a Ψ32.

Comparison of the sequence and structure surrounding Ψ746 in LSU RNA, Ψ32 in $tRNA^{Phe}$, and the corresponding region in $tRNA^{Val}$ explained the "dual specificity" of this synthase (Ofengand et al., 1995). Both RNAs have a similar stem-loop structure, although the 23S RNA loop is 8-membered whereas the tRNA loops are 7 in size and while the Ψ in 23S RNA is one residue removed from the stem, in $tRNA^{Phe}$ it is adjacent to the stem. However, the most important feature is the sequence 3' to the Ψ which is the same in both cases, UGAAAA. $tRNA^{Val}$, with the same secondary structure as $tRNA^{Phe}$, is unreactive because it has a different sequence. All three of the other tRNAs in E. coli that normally have Ψ32—$tRNA^{Cys}$, $tRNA^{Leu4}$, and $tRNA^{Leu6}$—have an almost identical sequence with only the single base change from G to U or C ($RNA^{Leu4,6}$) or a change of the 5'-most A to C ($tRNA^{Cys}$). Moreover, $tRNA^{Cys}$ and $tRNA^{Leu4}$ transcripts react with rluAp to form Ψ with kinetics and yield similar to those of $tRNA^{Phe}$ (Nurse and Ofengand, unpublished results). It appears, therefore, that the primary recognition determinant of rluAp is the consensus sequence, UUNNAAA, where the 5' U is modified to Ψ. It is not yet clear whether the stem-loop configuration is helpful or required or whether the entire sequence need be single-stranded, but it is noteworthy that the sequence UUNNAAA only occurs at one other site in E. coli 23S RNA, at residues 1781 to 1787. While U1781, the putative Ψ site, is in a loop, the 3' AAA sequence is base-paired. Moreover, no Ψ is found at this site naturally (Bakin and Ofengand, 1993) nor in the in vitro reaction (Wrzesinski et al., 1995b). In the latter, 1 mol of Ψ was found per mole of RNA both with full-length RNA and with the 1-847 fragment. Because sequencing showed Ψ at position 746 in both cases, it is highly unlikely that any reaction could have occurred at position 1781 in the full-length RNA.

The RNA specificity of rluBp also appears to be due more to surrounding sequence than secondary structure. While the main in vitro reaction product was Ψ2605, three other sites were also modified although at a much slower rate. From the four sites, a consensus site was derived (Niu and Ofengand, unpublished results).

The concept of multiple specificity

The finding that a single purified protein was capable of modification at a single specific site in more than one class of RNA was a surprise. Previous work with RNA-modifying enzymes had shown either reaction at a specific site in a specific RNA or, like the truA synthase, reaction at one of several adjacent sites in different RNAs of the same class (in this case different tRNAs). This new result, which we termed "dual specificity," if applicable generally, means that the known substrate for a modifying enzyme may not be the only substrate, and indeed may not even be the meaningful substrate. Although the rluAp synthase was identified using 23S RNA as a substrate, it may be that the real function of this enzyme is to form Ψ32 in certain tRNAs, and reaction with 23S RNA is tolerated as a benign artifact of accidental sequence homology. The true situation could also be the reverse, and it is also possible that both reactions are important to the cell. In any case, the existence of this first example of a high specificity for a single site in more than one class of RNA should be a warning signal that this situation could occur elsewhere. Unfortunately, this line of reasoning also means that even gene disruption and restoration experiments cannot with certainty link a specific metabolic effect with a specific modification. It could always be the modification in some unknown species of RNA that is responsible.

Other types of multiple specificity are now known. We propose, therefore, that Ψ synthase specificity be classified into four subtypes. In class I, only a single specific site in one type of RNA is modified. rsuAp and truBp would be examples of this class. In class II, nearby residues are also modified, where nearby is defined as up to n (say 5-6) residues of separation. truAp would be an example of this class, and one in rRNA is also known (Conrad et al., unpublished results). Class III would encompass reaction at more distant sites but in the same molecule or class of molecules. An example of class III specificity in a tRNA Ψ synthase (pus1p) has recently been described (Table 5), and one in rRNA is now known (Sun et al., unpublished results). Class IV would be reserved for those cases, so far only rluAp, where the sites are even more distant, for example, in different classes of RNA molecules.

Are there guide RNAs in prokaryotes?

As will be described in the following section, the locations of the many Ψs in eukaryotic SSU and LSU RNAs are determined by specific snoRNAs called guide RNAs. These base pair in a specific manner with nucleotides on either side of the U destined for conversion to Ψ, and signal the (probably) single protein synthase to act. Although the prokaryotic cloned Ψ synthases each show specificity for their individual sites, in light of the eukaryotic mechanism, one may ask whether these synthases contain or use guide RNAs as well. If not, eukaryotes would appear to have evolved a completely different system for recognition of the correct sites for Ψ synthesis.

Two kinds of guide RNAs in prokaryotes are possible. Separate RNA molecules could complex with the different protein synthases to select the correct site, as is done in eukaryotes, or RNA sequences embedded in the rRNA substrate itself could serve as the guide. In the former case, one needs to postulate that the amount of preexisting guide RNA is normally more or less equimolar with the protein and thus forms a complex with only a small percentage of the overexpressed protein synthase since the proteins are overexpressed at least fifty-fold (Wrzesinski et al., 1995a, 1995b). The complexes need also to survive affinity purification and dialysis. Sodium dodecyl sulfate gel analysis would not reveal them even if they were stable because the percent protein in complexed form would probably be too low. In this scenario, the purified "enzyme" would be a mixture of a few percent of active RNP but mostly nonfunctional protein apoenzyme. This mechanism is unlikely to explain the structural requirements of rsuAp but perhaps could account for the behavior of rluAp and rluBp. If the putative guide RNAs were similar in size to those in eukaryotes (see below), complexation with the synthase protein should produce a substantial shift in size in view of the modest molecular weights of the synthases (Table 5). Sizing column chromatography of the purified synthases with activity assay of the various fractions should then reveal whether the activity migrated like the protein peak or was faster. Such experiments are in progress but no information is yet available. The presence of the postulated few percent of active RNP should also be revealed by comparing the specific activity of the overexpressed protein versus that of the enzyme purified from cells. If a guide RNA were required, there should be a large decrease in the specific activity, measured as units/mg of protein, of the recombinant preparation. In fact, the specific activities of the purified and recombinant forms of rluAp did not differ by more than 2-fold, and the overexpressed protein was the more active one (Wrzesinski et al., 1995b)

The question of a guide RNA sequence in the substrate RNA itself is more complex. If one assumes that the mechanism of guide RNA function is the same as in eukaryotes, then it is possible to search the sequence of 23S or 16S RNA for any potential guide RNA sequences. Moreover, it is possible to test putative sequences against the known reactivity of rRNA fragments to see if they even contain the putative guide sequence, and to test new RNA fragments with the potential guide sequence removed. In a preliminary test of this approach, a search was performed for a guide sequence for rluAp, the enzyme that forms Ψ746 in 23S RNA (Rudd and Ofengand, unpublished results). The guide RNA structure proposed by Ganot et al. (1997a) and shown in Fig. 6 was used with G-U base pairing or one mismatch allowed in the guide-substrate base pair interaction and a 20 to 70 nucleotide link between the two base-pairing segments. These boundary conditions encompass the majority of guide RNA structures described by these workers. A total of 11 sites were obtained, but 4 were eliminated because they were not contained in the 1-847 active fragment that is readily recognized by the enzyme. Seven sites remained but all failed the requirement for a minimum of three base-pairs, either A-U, G-C, or G-U, in the stem immediately adjacent to the base-pairing wings of the guide RNA (Fig. 6). Less-stringent conditions have not been tested, and other sites have not yet been

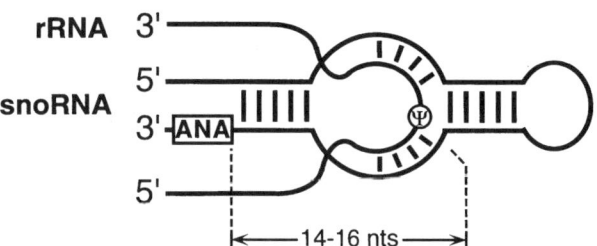

Figure 6. Hypothetical pairing of a guide snoRNA with rRNA sites of Ψ formation. Targeting involves base pairing of the guide RNA with complementary rRNA sequences which flank the uridine to be modified. The lengths of the guide sequences vary, but the target uridine sits in an unpaired pocket that is quite constant in size. Pairing on the 5' side of the uridine involves 4 to 10 base pairs, and that on the 3' side involves 3 to 10. The distance between the target uridine and the helix on the 5' side is mostly 0 but occasionally 1 residue and that on the 3' side is mostly one and rarely 2. In addition to the guide sequences, the distance to the H or ACA box is also a determinant in site selection. This spacing is a nearly constant 14 to 16 nucleotides. One or both domains shown in Fig. 7 can function in site selection. Adapted from Ganot et al. (1997a).

examined, but the success of this early attempt suggests that it may be feasible to lay the internal guide RNA hypothesis to rest by computer analysis alone.

Amino acid sequence relationships among Ψ synthases and their use in identifying new synthase genes

An alternate approach to locating genes for desired enzymes once a few members of the class are known is amino acid sequence homology. Therefore, once the sequence of the 4th Ψ synthase gene to be cloned was determined, an attempt was made to locate homologous regions by amino acid sequence comparison (Wrzesinski et al., 1995b). It was expected that at least the catalytic center of all four synthases would share a similar pattern of amino acids that would be evident in the primary sequence. However, no such pattern emerged. Subsequently, Koonin (1996) and Gustafsson et al. (1996) used this same general approach in a more sophisticated way to search genome data banks using the sequences of the same four genes: *truA*, *truB*, *rsuA*, and *rluA*. ORFs in both prokaryotes and eukaryotes were found, and in addition, 4 distinct subclasses could be detected, corresponding to each of the original genes used for screening (Koonin, 1996). In Table 6, the putative Ψ synthases in *E. coli* are collated. This table includes the addition of YmfC, detected by K. Rudd (personal communication) upon screening the entire *E. coli* genome for each of the 4 subclasses (only ca. 75% of the *E. coli* genome was available to Koonin). The validity of this approach is shown by the fact that before the publication by Koonin (1996), one of the ORFs, SfhB, was independently identified as a Ψ synthase by sequencing of a purified enzyme (Table 6) and by the demonstration in one of our laboratories (J. Ofengand) that the gene products of most of the putative synthases identified by Koonin show activity.

In *E. coli* at least, these may be all of the Ψ synthases because a reiterative search by K. Rudd (personal communication) using the identified genes did not reveal any additional ORFs. It is always possible, of course, that a Ψ synthase exists with a sufficiently different homology pattern that it has so far escaped detection. Similar analyses could be carried out in other organisms whose complete genome sequence is now known.

Eukaryotes

Guide RNAs as determinants of specificity: overview

Synthesis of cytoplasmic rRNAs in eukaryotic cells involves the action of a large population of small nucleolar RNAs (snoRNAs) (reviewed in Maxwell and Fournier, 1995; Sollner-Webb et al., 1995; Venema and Tollervey, 1995; Maden, 1997; Peculis, 1997; Smith and Steitz, 1997; Tollervey and Kiss, 1997) (see also Chapter 11 by Massenet et al. and Chapter 13 by Bachellerie and Cavaillé). A small number of these RNAs are required for processing the precursor which encodes the 18S, 5.8S, and 25/28S rRNAs. Nearly all of the others are now known to target ribose methylation and Ψ formation by acting as guide RNAs. These modifications are believed to occur in the primary transcript, based on results from modified nucleoside analyses of precursor rRNA (reviewed in Maden, 1990). Site selection in each case involves base pairing of a guide snoRNA with the rRNA segment to be modified, and selection of a nucleotide located at a constant distance from an additional determinant(s) in the snoRNA. The two types of guide function are provided by snoRNAs in separate families known as the box C/D and H/ACA box families, respectively. Each family contains snoRNAs required for rRNA processing, but the main function of these RNAs is the modification of rRNA nucleotides. Some snoRNAs influence both processing and modification. Guide snoRNAs are expected to be involved in the formation of most, and possibly

Table 6. Pseudouridine synthases in *E. coli* identified by sequence homology

Name	Gene location (min)	Predicted molecular mass (kDa)	No. of amino acids	Accession no.	Cloned	Ψ synthase activity
SfhB[a,b,c]	58.93	37.1	326	P33643	Yes	Yes
YceC[b,c]	24.66	36.0	319	P23851	Yes	Yes
YciL[b,c]	28.56	32.7	291	P37765	Yes	Yes
YjbC[b,c]	91.14	32.5	290	P32684	Yes	Yes
YqcB[b,c]	62.98	29.7	260	Q46918	Yes	Yes
YmfC[d]	25.74	24.9 (23.7)	217 (207)	P75966	In progress	?

[a] N-terminal sequence obtained on a purified Ψ synthase and ORF identified (Wrzesinski and Ofengand, unpublished results).
[b] Koonin, 1996. In this report, SfhB was YfiI and YqcB was ECU29581_5.
[c] Gustafsson et al., 1996. In this report, SfhB was YfiI and YqcB was f260.
[d] Rudd (personal communication). Values in parentheses were obtained by using the next downstream initiator AUG. There is uncertainty regarding the true start site.

all, 2'-O-methylated nucleotides and Ψ residues in cytoplasmic rRNA. Thus, all such sites may be served by a single methylase or pseudouridylate synthase. Also to be determined is whether the enzyme activity is a stable component of the snoRNP complex, or rather associates with it transiently at the time of reaction.

The box H/ACA snoRNAs

This newly defined family of snoRNAs is characterized by two conserved box elements—the ACA box, which is located three nucleotides from the 3' end of the mature snoRNA and is nearly always the triplet ACA (Balakin et al., 1996), and the H box, which is defined as AgA in yeast and ANANNA for all species analyzed, and is located in a hinge region of the molecule (Balakin et al., 1996; Ganot et al., 1997b). RNAs in the H/ACA box family do not contain statistically long sequences complementary to rRNA, but rather contain short complementarities that provide the guide function. The snoRNAs in the H/ACA family can be represented by a common secondary structure, which consists of 5' and 3' hairpin domains, connected by a hinge segment (Fig. 7). The conserved H (hinge) and ACA boxes are located at the base of the 5' and 3' hairpin domains respectively, on the 3' side. The lower portions of the hairpins have similar folding properties, but vary in overall length among individual RNAs. The length of the hinge segments also vary. The H/ACA RNAs in yeast are considerably larger than those in mammals, with a capped 5' leader and additional stem-loop domains accounting for most of the difference. While all of the snoRNAs analyzed conform to this consensus structure, some have additional secondary domains (Ganot et al., 1997b; Ni and Fournier, unpublished results). This generally is true for the yeast RNAs and this property complicates efforts to model RNA folding. Mutational studies with yeast and human showed that both box elements are necessary for snoRNA accumulation (Balakin et al., 1996; Ganot et al., 1997b), and the ACA box (and neighboring helix) is involved in 3' end formation of the snoRNA (Balakin et al., 1996). The need for these box elements is presumed to relate to binding of proteins required for maturation and stability. The two major snoRNA families are also distinguished by the existence of family-specific proteins that are diagnostic for each class. In the case of the box C/D snoRNAs the protein is fibrillarin, which is known as Nop1p in yeast. snoRNAs in the H/ACA family are associated with a glycine- and arginine-rich protein known as Gar1p (Girard et al., 1992). Both proteins are essential for growth and each also affects ribose methylation or Ψ synthesis (see Chapter 13 by Bachellerie and Cavaillé).

The discovery of the H/ACA family yielded a simple classification scheme for the snoRNAs (Balakin et al., 1996). All known snoRNAs fall into two major families—the box C/D and box H/ACA families. A third subset containing a single RNA also exists. This exception to the rule is the ribozyme-like MRP snoRNA, which lacks the defining box elements (Maxwell and Fournier, 1995; Tollervey and Kiss, 1997).

H/ACA box snoRNAs direct Ψ site selection

The involvement of the H/ACA snoRNAs in Ψ formation was first demonstrated in yeast, by analyzing sites of modification in cells genetically depleted of individual snoRNAs (Ni et al., 1997; Ganot et al., 1997a). At this writing, 19 yeast snoRNAs have been shown or predicted to be required for Ψ synthesis, demonstrating that this function is widespread and not limited to a few snoRNAs (Table 7). The guide function became clear from correlating the snoRNA secondary structures and sequences flanking the site of modification. A consensus snoRNA:rRNA pairing arrangement was defined, and this general motif can occur in either or both hairpin domains (Fig. 7). Individual sites of modification correlate with one domain in the snoRNA. Thus, some snoRNAs specify modifications at two rRNA sites, using each guide domain. This feature suggests that the two guide domains act independently, but this has not been established. The pairing motif in Fig. 6 appears to accommodate all snoRNAs known or predicted to target modification (Ganot et al., 1997a; Ni et al., 1997; Ni and Fournier, unpublished results). Key features of the model include base pairing of the snoRNA with the region of modification, through sites that flank the uridine to be modified, and loca-

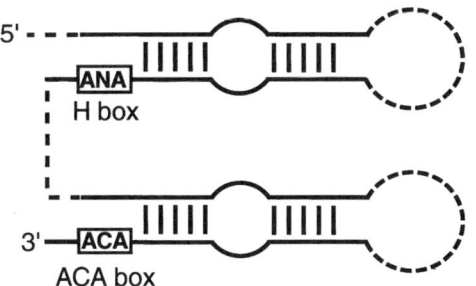

Figure 7. Consensus secondary structure of the H/ACA box snoRNAs. snoRNAs in this family share common secondary structure domains, which can be represented schematically in a simple 'hairpin-hinge-hairpin' arrangement. Adapted from Balakin et al. (1996) and Ganot et al. (1997b).

Table 7. ACA snoRNAs in *Saccharomyces cerevisiae*

RNA	Chain length (nucleotides)	TMG Cap[a]	Disruption phenotype	Function[g]	Evidence
snR3	194	+	None	2263 in LSU rRNA	Experimental[a]
snR5	197	+	None	1003 in LSU rRNA	Experimental[c,d]
				1123 in LSU rRNA	Experimental[c,d]
snR8	189	+	None	959 in LSU rRNA	Experimental[b]
				985 in LSU rRNA	Experimental[b]
snR9	187	+	None	2339 in LSU rRNA	Experimental[d]
snR10	245	+	cs slow growth	rRNA processing	Experimental[f]
				2919 in LSU rRNA	Experimental[b]
snR11	258	+	None	2415 in LSU rRNA	Experimental[d]
snR30	609	+	Lethal	rRNA processing	Experimental[e]
snR31	222	+	None	1000 in SSU rRNA	Experimental[b]
snR32	188	+	None	2190 in LSU rRNA	Experimental[b]
snR33	183	+	None	1041 in LSU rRNA	Experimental[b]
snR34	203	+	None	2822 in LSU rRNA	Experimental[d]
				2876 in LSU rRNA	Experimental[b]
snR35	204	+	None	1189 in SSU rRNA	Motif analysis[c,d]
snR36	182	−	None	1185 in SSU rRNA	Experimental[c]
snR37	386	+	None	2940 in LSU rRNA	Experimental[b]
snR42	352	+	None	2971 in LSU rRNA	Experimental[b]
snR43	209	−	n.d.	965 in LSU rRNA?	Motif analysis[d]
snR44	211	−	n.d.	106 in SSU rRNA	Motif analysis[c,d]
				1055 in LSU rRNA	Motif anlaysis[c,d]
snR46	196	−	None	2861 in LSU rRNA	Experimental[b]
snR49	~170	?	n.d.	989 in LSU rRNA	Motif analysis[c,d]
snR189	192	+	None	466 in SSU rRNA	Motif analysis[c,d]
				2730 in LSU rRNA	Experimental[d]

[a] TMG, 2,2,7-trimethyl guanosine.
[b] Ni et al. (1997).
[c] Ganot et al. (1997a).
[d] Ni et al. (unpublished results).
[e] Morrissey and Tollervey (1993).
[f] Tollervey (1987).
[g] The number indicates the Ψ formation site.

tion of the target uridine at an apparently fixed distance of 14–16 nucleotides from the H or ACA box (14 in yeast, 14–16 in humans). Evidence that these features are indeed determinants in site selection was provided by mutational analyses. Complementarity was demonstrated to be essential with snoRNA substitution mutations that altered the base pairing potential on the 3′ side of the target uridine or changed both regions flanking the site of modification (Ni et al., 1997; Ganot et al., 1997a). This pairing arrangement is also supported by rDNA deletion results showing that the sequences flanking the target uridine are essential for Ψ formation (Ganot et al., 1997a). As predicted, site selection is also influenced by altering the distance to the ACA box (Ni et al., 1997). An initial analysis showed that small insertions or deletions led either to a reduced level of Ψ at the canonical position or to reduced modification at that site and simultaneous modification at a neighboring uridine that normally is not modified.

It remains to be learned if site selection involves other determinants, and whether an alternate pairing model is relevant. The alternate model also consists of two guide sequences that flank the site of modification (Ni et al., 1997). The sequence on the 3′ side of the target uridine is the same as that identified in Fig. 6. The other involves different conserved complementarities in both RNAs. If both pairing schemes are involved, these might function at different stages of modification—for example, in recognition and in catalysis. If only one pairing scheme is relevant, the model in Fig. 6 provides the best fit. A key attraction of this model is that the complex has a consistent structure at the modification position. In addition to the sites of base pairing, the model includes nearly constant spacings between the snoRNA:rRNA helices and the subject uridine, and an open "pocket" in the duplex that can be imagined to accommodate the isomerization reaction (Ganot et al., 1997a; Maden, 1997; Peculis, 1997; Smith and Steitz, 1997).

Are all Ψs in rRNA formed with guide RNAs?

The large size of the H/ACA snoRNA family and large number of Ψs in rRNA suggest that most Ψ modifications are mediated by guide RNAs. The total number of H/ACA snoRNAs is not known for any organism. However, 20 H/ACA snoRNAs have been identified thus far in *S. cerevisiae* (Table 7), 15 of which have been experimentally tied to pseudouridylation. Together with the 4 identified by motif analysis, they acount for the placement of 24 Ψs out of the known 44 sites (Tables 2 and 3). In human, putative guide snoRNAs have been described (Ganot et al., 1997a) that account for 17 Ψ out of the 93–95 sites (Maden, 1990; Ofengand and Bakin, 1997). All of the yeast species known were identified by direct characterization of individually purified nuclear small RNAs (Balakin et al., 1996) (see Table 7 and citations therein). The human RNAs were identified by sequencing as well, most by analysis of cDNAs prepared from noncapped, presumably intron-encoded snRNAs (Ganot et al., 1997a, and citations therein).

Computer search strategies will be useful for identification of new candidate guide RNAs. This can be done by predicting secondary structures of the snoRNAs and identifying conserved guide sequences. A reverse search strategy can also be used, in which guide sequences can be predicted from known sites of Ψ modification. The yeast snoRNA set will likely be defined first because the complete genome sequence is known and modification maps are available for both the SSU and LSU rRNAs. Other RNA substrates also need to be considered in light of the recent discovery showing that box C/D guide snoRNAs target ribose methylation in vertebrate splicing snRNAs (Tycowski et al., personal communication). The fact that some guide RNAs have two targeting motifs indicates that the total number of guide RNAs can be smaller than the number of Ψ modifications. Thus, the maximum number of guide snoRNAs required for the yeast SSU and LSU rRNAs will be less than 44 and that in humans less than 93–95 (Maden, 1990; Tables 2, 3).

It remains to be seen if all Ψs in eukaryotic rRNA are produced by a guide RNA mechanism. A reasonable hypothesis is that Ψ residues that occur at the same site in both prokaryotic and eukaryotic rRNAs might be candidates for guide-independent synthesis. In yeast, these are the positions 2257, 2259, and 2822, corresponding to *E. coli* positions 1915, 1917, and 2457. There is also 2971, which is adjacent to *E. coli* 2605. The same positions as in yeast are also found in human rRNA (Fig. 4 and 5). Unfortunately for this hypothesis, snR34 acts as a guide RNA for Ψ2822 in yeast, and snR42 does the same for Ψ2971 (Table 7). In human, U19 has the motifs for guiding the equivalent of *E. coli* Ψ1915 and Ψ1917 synthesis, and U65 has the same for the *E. coli* 2457 equivalent. Therefore, while it may be the case that guide-independent sites exist in eukaryotes, just as it is still formally possible that guide RNAs function in prokaryotes, there is no evidence so far to support either of these premises. Rather, the weight of the evidence supports the view that the mechanisms of synthesis in eukaryotes and prokaryotes are fundamentally different (see also below).

Are all H/ACA box snoRNAs guide RNAs?

Two yeast H/ACA snoRNAs—snR10 and snR30—are known to influence growth (Table 7). Loss of the snR30 species disrupts processing of 18S rRNA and is lethal (Bally et al., 1988; Morrissey and Tollervey, 1993). It is not known if this RNA affects Ψ formation. Depletion of snR10 causes a more general defect in processing and a cold sensitive phenotype (Tollervey and Guthrie, 1985; Tollervey, 1987). This snoRNA is known to direct synthesis of a Ψ in the core of the peptidyltransferase center (Ni et al., 1997). The growth effects associated with depletion of this RNA could be related to Ψ modification or alternatively to other functions. A precedence for multiple functions has already been decribed for one of the box C/D snoRNAs, U14 (Li et al., 1990; Liang and Fournier, 1995; Kiss-Laszlo et al., 1996).

Proteins implicated in snoRNA-directed Ψ synthesis

The Ψ modification reaction is presumed to be catalyzed by a protein enzyme, although until this has been demonstrated, a ribozyme catalyst cannot be formally excluded. The prediction that guide RNAs are involved in the formation of most, and perhaps all, Ψ residues in eukaryotic rRNA suggests that the number of pseudouridine synthases involved is small, perhaps only one. This view is consistent with the lower yield of candidate synthases in the rluA/rsuA subclasses identified in a search of the complete yeast and *E. coli* genomes. Whereas 4 potential Ψ synthases were found for yeast (Rudd, personal communication), 8 were found for *E. coli* (Tables 5 and 6) even though the number of Ψ residues is only one-fourth that of yeast (Tables 2 and 3). However, as already noted by Koonin (1996), there may well be synthases whose sequences are not detectably related to the four known classes.

One yeast protein deserves special notice. This is the yeast protein Cbf5p, which, along with its rat equivalent (Nap57p), belongs to a sequence family defined by the *E. coli* Ψ synthase truBp, the enzyme responsible for synthesis of Ψ55 in tRNA. Both eukaryotic proteins are known to be nucleolar and on that basis are candidates for a role in modification of rRNA (Jiang et al., 1993; Meier and Blobel, 1994). Cbf5p has been shown to associate with H/ACA box snoRNAs, and its depletion results in a general loss of Ψ in rRNA (Lafontaine et al., 1998). The sequence homology with other known Ψ synthases suggests that it is the catalytic protein but this needs to be established directly. Gar1p is also associated with the H/ACA box snoRNAs, and its depletion also blocks global rRNA Ψ formation (Bousquet-Antonelli et al., 1997). However, the Gar1p sequence bears no resemblance to any of the 4 subclasses of Ψ synthases and is most likely a snoRNP structural protein. Disruption of modification could simply reflect loss of functional snoRNP complexes. Other yeast ORFs with synthase homology are known, and these must be considered candidate Ψ synthases for rRNA as well. It will be interesting to learn if the eukaryotic modifying enzyme(s) is an integral component of the snoRNP complexes that contain the guide snoRNAs or only associate(s) with them at the time of modification.

Organelles

The ribosomes of mitochondria and chloroplasts also contain Ψ, although to a much lesser extent than their cytoplasmic colleagues. The most reliable values are those obtained by sequencing (Table 3) because contamination of isolated organelle rRNA with either tRNA or cytoplasmic rRNA can result in artificially high values. The enzymes and/or guide RNAs responsible for Ψ synthesis in organelles have not yet been determined. The main question is whether they are synthesized by individual specific enzymes as in prokaryotes, i.e., *E. coli*, or use a guide RNA mechanism like their cytoplasmic counterparts. The single Ψ in yeast mitochondrial LSU RNA does not correspond in location to any of the 30 Ψ in cytoplasmic LSU RNA but rather corresponds to Ψ2580 in *E. coli*. The same is true for mouse and human RNAs, although these mitochondrial LSU RNAs have not been completely sequenced as was done for yeast. This might suggest the existence of a nuclear gene coding for a mitochondrion-specific enzyme that was likely inherited from the mitochondrion's prokaryotic ancestor. This seems more likely than mitochondrial synthesis of a specific guide RNA because the rest of the snoRNP complex, including the catalytic protein, would need to be imported from the cytoplasm. This situation differs from Ψ synthesis in yeast tRNA, where recent work has shown that the same enzyme directs Ψ synthesis at the same specific sites in both mitochondrial and cytoplasmic tRNAs (Becker et al., 1997; Lecointe et al., 1997).

Less information is available on rRNA Ψ synthesis in chloroplasts. In the one case examined, *Z. mays*, all four Ψ residues were in the LSU RNA and corresponded to sites common to prokaryotes, but the sites were also largely shared by eukaryotes (Fig. 4 and 5). In view of the small number of Ψ residues, which resemble prokaryotes, it seems likely that a small set of distinct protein enzymes are involved rather than a guide RNA system, but there is no evidence so far available to support either view.

Archaea

These organisms, formerly referred to as archaebacteria, would be most interesting to examine in terms of their biosynthetic mechanism because many previous studies have shown that they often possess an unusual mosaic of prokaryotic and eukaryotic features (Olsen and Woese, 1997). In the one example studied so far, *H. halobium*, the distribution (Fig. 4 and 5) of the four Ψ residues is more like that in eukaryotes than prokaryotes, although the small total number is more like prokaryotes (Table 3). With complete genomes available now for three species (Pennisi, 1997), it should be possible to search for Ψ synthases and for putative guide RNAs.

FUNCTION

Ribosome Biosynthesis

Processing of the RNA

In both prokaryotes and eukaryotes, the initial transcript of an rRNA operon is a large pre-rRNA molecule that must be processed by a series of nucleolytic steps to yield ultimately the mature sized RNAs found in ribosomes. Ψ is found in the 30S precursor of *E. coli* (Dahlberg et al., 1975), in the 45S precursor of mammals (Jeanteur et al., 1968), and in the 35S precursor of yeast (Brand et al., 1979), indicating that Ψ formation is a very early process. Although it is not known if this is true for all of the Ψ sites, Bousquet-Antonelli et al. (1997) found 30–35 of the expected 44 Ψ in yeast 35S RNA. In *E. coli* it seems unlikely that Ψ516 is made on the 30S precursor because the RNA substrate specificity of rsuAp suggests that it is made only after the RNA is partially folded and complexed with some ribosomal proteins. The synthesis of most Ψ on precursor rRNA suggests a possible role in rRNA processing. Ψ could signal "cut here" or "don't cut here." The former role seems unlikely as many Ψ residues are found internally and localized to the "universal core" (Maden, 1990) of rRNA. The latter role, in which Ψ residues are placed so as to block undesirable nuclease cuts that would otherwise occur, is plausible, although it does not provide any simple explanation for the wide range in number of Ψ per RNA, especially in SSU RNA. Moreover, recent experiments by Bousquet-Antonelli et al. (1997) show that for LSU RNA of yeast this is not the case. These authors were able to inhibit virtually all Ψ synthesis by depletion of an essential protein, gar1p, yet mature sized LSU RNA was made. If Ψ were acting as a blocking agent, the LSU RNA should have been cut into small pieces. SSU RNA could not be examined this way since it failed to appear at its mature size. This result might suggest, among other explanations, that the above hypothesis holds for SSU RNA, but it clearly is not a general explanation for the presence of Ψ in rRNA.

Folding of the RNA

Ψ residues could be involved in several ways in the process of folding the RNA into its secondary and tertiary structure and assembly into a correctly structured ribosome. First, the presence of a Ψ residue might help to direct folding along a productive, as opposed to a dead-end, path. It could do this either by lowering the activation energy for the "correct" path at a branch point or by raising the activation energy for the "incorrect" path. The variation in number and location of the Ψ residues in both SSU and LSU RNAs would not be in opposition to this notion. In a variant of this hypothesis applicable to eukaryotes in particular, the guide RNAs that direct Ψ formation could actually function as chaperones in the folding process, and Ψ formation would merely be a benign byproduct designed as a signal to tell the guide RNA to dissociate from the rRNA once its folding job is done. (Independently, D. A. Samarsky also conceived of this hypothesis.) So far, the ability of guide RNAs to stably base pair with unmodified versus modified rRNA has not been tested.

Assembly with ribosomal proteins and formation of the correct tertiary structure

Ψ residues may help the interaction with ribosomal proteins during the assembly process and/or participate in the establishment of the correct tertiary structure. Three facets of Ψ structure could contribute to this process. First, as pointed out by Maden (1990), in either the *syn* or *anti* base-pairing configuration the atoms lining the major groove side of Ψ differ from those of U, creating the potential for differential recognition by proteins or other ligands. Second, the N-1 proton of Ψ could act as a hydrogen-bond donor cross-strut to stabilize secondary or tertiary structure and to interact with proteins. Stabilization could occur locally by H-bonding via a water molecule to its own or 5'-adjacent 5'-phosphate (Arnez and Steitz, 1994) or at a longer range by triple-strand formation (Trapane et al. 1994; Bandaru et al. 1995). Third, stabilization could also occur as a result of the increase in the 3'-endo conformation of the ribose moiety in Ψ (Davis, 1995) (see also Chapter 5 by Davis). Stabilization mechanisms are also discussed in Chapter 6 by Auffinger and Westhof. The occurrence of variable numbers of Ψ at sometimes different positions in the secondary structure of different rRNAs would fit such a structural role for Ψ because in order to achieve the same three-dimensional result each rRNA could require a somewhat different placement of stabilizers as a result of its particular sequence.

Ribosome Activity

SSU RNA

It is difficult to discern any unitary functional role for Ψ in SSU RNA. Initially, the finding of Ψ516 in *E. coli* and its analog in *B. subtilis* suggested a functional role in the "530" stem-loop, a structure known to be involved in codon recognition (references cited in Wrzesinski et al., 1995a). However, the absence of an analogous site in yeast SSU RNA, despite the presence of 14 other Ψs, and its absence in *H. halobium* indicate that the role is not universal. Overall, the facts that the number of Ψs varies from possibly 0 to over 40 and that their location in eukaryotes seems to have no relation to other known structural or functional features (Fig. 1) do not support any hypothesis that would be connected to a general ribosomal function.

LSU RNA

The situation is different in LSU RNA. All organisms examined had at least one Ψ, and even when the overall number expanded drastically, all were clustered into three distinct areas, all three of which either make up or are near to the peptidyl transferase region. These facts alone suggest a role for Ψ in the LSU that is distinct from its role in the SSU and that should involve some aspect of the peptide bond formation process. The most obvious role is to use the available N-1 position as an intermediate acceptor of the carboxyl end of the growing peptide chain before it reacts with the amino group of the incoming amino acid (Lane et al., 1995). However, it is hard to support this hypothesis, given the large variation in number and exact location of the Ψ residues. Even though there are several semi-invariant sites at the PTC (Ofengand and Bakin, 1997), the universal mechanism of peptide bond formation would appear to require at least one universal site for the occurrence of Ψ if it were directly involved as an intermediate in forming the peptide bond.

The only universal site found so far is in a small loop in the central region, residues 1915 and 1917 (*E. coli* numbering), which is near, but not at, the PTC. Ψ1915 is sometimes additionally modified. The nearby site, 1911, is also frequently Ψ but not in yeast or *H. halobium*, despite the presence of the precursor U. These residues, because of their location, are unlikely to be the peptide acceptor, but instead are close to the decoding region of 16S RNA in the 30S subunit based on the following facts. Association with the 30S subunit protects A1916, A1918, and U1926 (unpublished results of Merryman et al. cited in Joseph and Noller, 1996), and residues 1912–1920 can be cross-linked to residues 1408–1411 in the 16S RNA (Mitchell et al., 1992). Residues 1408–1411 are at the decoding site of the 30S subunit (Ofengand et al., 1993). Moreover, mutation at C1914 or A1916 affected the fidelity of codon-anticodon recognition (O'Connor and Dahlberg, 1995). These results suggest that rather than being part of the peptide bond formation step, these 2–3 Ψ residues may play a role in subunit–subunit interaction and/or in the decoding process.

The coclustering of the methylated nucleosides and Ψ (Fig. 2) suggests another functional role. Except for m^7G, which is charged, the methyl groups contribute a hydrophobic patch to the ribosomal surface. On the other hand, Ψ is hydrophilic by virtue of its extra N_1-H group. Together, these residues create an area on the interface side of the LSU that is a mosaic of hydrophilic and hydrophobic centers. The subunit interface is where mRNA, tRNA, and the protein factors are bound and do their work. Thus, it is possible that this mosaic pattern of noncovalent forces is what creates the specific binding sites for the essential ligands of the ribosome. In this model, each modified residue might contribute toward the total binding energy, but no one need be essential. Deletion of multiple residues might be necessary to detect an effect.

Are Any Ψ Residues Essential?

SSU RNA

It is possible to assemble in vitro an active *E. coli* SSU from an SSU RNA transcript devoid of all modified nucleosides and ribosomal proteins. However, the assembly of the subunit has abnormal features and its functional activity in protein synthesis is only about half that of a modified control SSU (Krzyzosiak et al., 1987; Denman et al., 1989a, 1989b). These defects were attributed to the absence of all of the modified nucleosides in the SSU RNA (Cunningham et al., 1991). Therefore, while Ψ516 is clearly not essential for SSU function, it may be necessary for optimum assembly and/or function. In agreement with this conclusion, *E. coli* mutants in which the *rsuA* gene is deleted are viable and, in preliminary studies, show no major metabolic defects (Niu and Ofengand, unpublished results). Moreover, in yeast SSU RNA, Ψ466, Ψ1000, Ψ1185, and Ψ1189 are individually not likely to be essential since depletion of the guide RNAs known or predicted to be responsible for their formation was not lethal. In the case of Ψ1000 and Ψ1185, their absence was directly shown (Table 7).

LSU RNA

The situation is less clear for LSU RNA. *E. coli* LSU reconstitution using fragments of natural, modified RNA with the complementary part made by in vitro transcription showed that a segment of the RNA from ca. 2445 to 2523, containing m^2G2445, D2449, Ψ2457, Cm2498, m^2A2503, and Ψ2504 appears to be required in its modified form to retain in vitro reconstitution and peptidyl transferase activity (Green and Noller, 1996). Presumably, one or more of these modified residues should be essential. Other functions, such as subunit–subunit interaction, were not tested and possible enhancing effects of the other modified residues were not examined. Attempts to specifically remove Ψ residues from *E. coli* LSU RNA by inactivation of specific synthases have so far been confined to Ψ746 formed by the *rluA* gene product. Deletion of this synthase was not lethal, and the mutant was not metabolically altered in any major way (Niu and Ofengand, unpublished results). Deletion of other *E. coli* synthases is being pursued. In eukaryotes, 16 of the 30 Ψ residues in yeast LSU RNA either have experimentally been singly removed by deletion of the gene for their guide RNA or are supposed to be absent because the snoRNA believed to act as their specific guide RNA was deleted (Table 7). So far, the only multiple Ψ deletions are in yeast, where depletion of four (snR33 to -36) or five (snR3, -5, -8, -9, and -10) H/ACA snoRNAs was not lethal (Parker et al., 1988; Samarsky et al., 1995). These combinations presumably blocked formation of five and seven Ψs, respectively (Table 7). Moreover, all 8 of the Ψs in yeast which surround the PTC (Fig. 5) have been deleted singly, with no effect on viability or growth (Ni et al., 1997 and unpublished results). The fact that each Ψ is dispensable individually indicates that none plays an essential role in protein synthesis or any other critical process. It will now be interesting to learn if protein synthesis is affected by multiple deletions at the PTC or by the loss of any specific Ψ or combination of Ψ in other rRNA domains.

In summary, there is no firm evidence so far for an essential role for any Ψ in the cell, and a number of cases are known where deletion of a single Ψ has no obvious effect. It remains to be seen if there is some redundancy in the function of Ψ so that deletion of a number will be needed to show a significant metabolic effect.

MODIFIED PSEUDOURIDINES

Simple Modifications: $m^1\Psi$, $m^3\Psi$, Ψm

$m^1\Psi$ has been found in eukaryotic SSU RNA, where it is a precursor to $m^1acp^3\Psi$ (Brand et al., 1978). $m^3\Psi$ (Kowalak et al., 1996) is the unidentified methylated U derivative described previously (Smith et al., 1992, and references therein). It is found in *E. coli* and possibly also in *B. subtilis* and *Z. mays* at position 1915 (*E. coli* numbering), which may be a universal site for Ψ or its derivatives and which has a potentially important function. Note that methylation covered the N_3-H but not the N_1-H, which is the one made available by the U to Ψ conversion. Could there be some special feature of the N_1-H that needs to be conserved at that position? Ψm has been found in HeLa cell LSU RNA (Maden and Salim, 1974) and in wheat embryo total rRNA (Gray, 1974). No specific function is known.

Complex Modifications: $m^1acp^3\Psi$

This baroquely modified Ψ has both ring N atoms derivatized. It was first described in CHO cells by Saponara and Enger (1974) and in yeast and HeLa cells by Maden et al. (1975). The methyl group is added first in the nucleolus, and then the acp^3 group is put on outside of the nucleolus (Brand et al., 1978). It occurs in SSU RNA at the apex of a small stem-loop (residue 1189, yeast numbering) which in *E. coli* is known to be part of the P binding site for tRNA (Moazed and Noller, 1990). Curiously, in *E. coli*, the position of $m^1acp^3\Psi$, which is not found so far in prokaryotes, is taken by two methylated residues—m^2G966 at the equivalent site and m^5C967 at the adjacent position. The specific methylase responsible for forming m^5C967 has been cloned (Tscherne et al., unpublished results). Deletion of this methylase gene, and hence specific deletion of m^5C967, may reveal its role in ribosome structure and/or function and by extension the role played in eukaryotes by $m^1acp^3\Psi$. It should be noted, however, that the guide RNA for $\Psi1189$ is supposed to be snR35, and this snoRNA has already been deleted without affecting cell viability (Samarsky et al., 1995; Ganot et al., 1997a).

CONCLUSIONS AND PROSPECTS

Despite the fact that no clear function for Ψ in rRNA has yet emerged, it seems inescapable that they are there for some important purpose. The clustering in the LSU in 3 distinct areas that are part of or near to the PTC, the functional raison d'être of the ribosome, strongly suggests a functional role. Moreover, the care with which nature has ensured the precise placement of the Ψ residues, using a series of specific enzymes in prokaryotes and an elaborate sytem of guide RNAs in eukaryotes, is further evidence for the importance of Ψ. The fact that no specific function for Ψ has emerged so far should not be viewed as a discouragement but rather as a challenge.

Important progress can be expected on several fronts. It is likely that additional pseudouridine synthases will be identified and characterized and that they will come from new and familiar sources. Excellent progress is being made with the bacterial enzymes and identifying all of the *E. coli* synthases now seems within reach. Here and in other organisms computer screening of genomic sequences is sure to play a major role in identifying new enzymes. The discovery that guide snoRNAs provide targeting specificity in eukaryotes suggests that the number of synthases in each organism is small, making the goal of cataloging and describing these more attainable as well. The novelty of the snoRNA-directed mechanism itself has led to new efforts to study Ψ synthesis and function among eukaryotes, specifically in yeast and mammalian cells. Characterization of Ψ formation in eukaryotic organelles, such as the mitochondrion and chloroplast, and in archaebacteria will be another exciting area of study.

Major advances can be anticipated in defining the structure and function of the guide RNPs in eukaryotes and precisely how the modification process(es) that utilizes guide RNAs compares with those that do not. As more enzymes and guide RNAs are characterized, phylogenetic analyses will become more fruitful. The results from these studies should provide insight into the origin of the synthase proteins and guide RNAs and into the origin of the isomerization reaction itself.

The mechanism of isomerization, an aspect that has been dormant for many years, is now amenable to attack (see also Chapter 8 by Garcia and Goodenough-Lashua). Recent work from the Santi laboratory, both in identifying an aspartic acid residue in truAp which may be involved in forming an enzyme-substrate intermediate (Huang et al., 1998) and in obtaining the first crystals of a Ψ synthase, again truAp (Foster et al., 1997), are harbingers of many studies to come on other Ψ synthases.

The structural effects of converting uridine to Ψ in different sequence and folding contexts are still a major goal. Efforts in this area will benefit from both the development of new in vitro synthesis systems and the ability to control modification in vivo in a site-specific way. Synthesis of specific Ψ residues can now be blocked in vivo and, by introduction of mutant guide RNAs, it should be possible to target modifications to entirely new sites, as can be done now for 2'-O methylation of rRNA (see Chapter 13 by Bachellerie and Cavaillé). Determining the structural effects of modification currently involves sample-intensive methodologies, such as NMR and X-ray analysis of crystals. It should be possible to eventually meet these needs with chemically synthesized modified RNA fragments, or by in vitro enzymatic modification of RNAs generated in vivo or in vitro.

Defining the function of Ψ in rRNA will require both in vivo and in vitro studies. Modifications that cause a notable in vivo phenotype, such as temperature sensitivity, slow growth, or death, will of course be most useful. Thus far, no single Ψ in rRNA has been directly tied to such an effect, and it remains to be seen if any are, and what effects might occur with losses of Ψ at multiple sites. However, where no phenotype is observed or the phenotype is weak, it should be possible in principle to develop useful phenotypes by synthetic lethal strategies, in which nonlethal mutations in Ψ biosynthesis components are combined with nonlethal mutations in different, functionally related genes. This strategy can be used in yeast to identify molecular interactions involved in synthesis of Ψ, and also should be useful in identifying functional effects. Genetic approaches, including two- and three-hybrid screens, should also be useful for identifying interacting molecules.

A direct in vitro approach of studying the partial reactions of protein synthesis and ribosome assembly in mutant strains lacking one or more Ψ in their rRNA may also be informative, even when in vivo parameters such as growth rate are not detectably changed. While developing full descriptions of Ψ synthesis and function in rRNA will be a long-term affair, many important discoveries are sure to occur in the near future. In particular, the new capabilities for manipulating the synthase and guide RNA genes ensure that the period just ahead will be very fruitful and exciting.

Acknowledgments. The authors thank Jingwei Ni (University of Massachusetts) for preparing Table 7 and Fig. 6 and 7. This work was supported in part by a Markey Foundation grant to the Department of Biochemistry and Molecular Biology, University of Miami (J.O.), and NIH grant GM19351 (M.J.F.).

REFERENCES

Arena, F., G. Ciliberto, S. Ciampi, and R. Cortese. 1978. Purification of pseudouridylate synthetase I from *Salmonella typhimurium*. *Nucleic Acids Res.* 5:4523–4536.

Arnez, J. G., and T. A. Steitz. 1994. Crystal structure of unmodified tRNA^{Gln} complexed with glutaminyl-tRNA synthetase and ATP suggests a possible role for pseudo-uridines in stabilization of RNA structure. *Biochemistry* 33:7560–7567.

Arps, P. J., C. C. Marvel, B.C. Rubin, D. A. Tolan, E. E. Penhoet, and M. E. Winkler. 1985. Structural features of the hisT operon of *Escherichia coli* K-12. *Nucleic Acids Res.* 13:5297–5315.

Bakin, A., and J. Ofengand. 1993. Four newly located pseudouridylate residues in *Escherichia coli* 23S ribosomal RNA are all at the peptidyl transferase center: analysis by the application of a new sequencing technique. *Biochemistry* 32:9754–9762.

Bakin, A, and J. Ofengand. 1995. Mapping of the thirteen pseudouridine residues in *Saccharomyces cerevisiae* small subunit ribosomal RNA to nucleotide resolution. *Nucleic Acids Res.* 23:3290–3294.

Bakin, A., J. A. Kowalak, J. A. McCloskey, and J. Ofengand. 1994a. The single pseudouridine residue in *Escherichia coli* 16S RNA is located at position 516. *Nucleic Acids Res.* 22:3681–3684.

Bakin, A., B. G. Lane, and J. Ofengand. 1994b. Clustering of pseudouridine residues around the peptidyl transferase center of yeast cytoplasmic and mitochondrial ribosomes. *Biochemistry* 33:13475–13483.

Balakin, A. G., L. Smith, and M. J. Fournier. 1996. The RNA world of the nucleolus: two major families of small RNAs defined by different box elements with related functions. *Cell* 86:823–834.

Bally, M., J. Hughes, and G. Cesareni. 1988. snR30: a new essential small nuclear RNA from Saccharomyces cerevisiae. *Nucleic Acids Res.* 16:5291–5303.

Bandaru, R., H. Hashimoto, and C. Switzer. 1995. An inverted motif for oligonucleotide triplexes: adenosine-pseudouridine-adenosine (A-Ψ-A). *J. Org. Chem.* 60:786–787.

Becker, H. F., Y. Motorin, R. J. Planta, and H. Grosjean. 1997. The yeast gene YNL292w encodes a pseudouridine synthase (Pus4) catalyzing the formation of Ψ55 in both mitochondrial and cytoplasmic tRNAs. *Nucleic Acids Res.* 25:4493–4499.

Bousquet-Antonelli, C., Y. Henry, J.-P. Gélugne, M. Caizergues-Ferrer, and T. Kiss. 1997. A small nucleolar RNP protein is required for pseudouridylation of eukaryotic ribosomal RNAs. *EMBO J.* 16:4770–4776.

Brand, R. C., J. Klootwijk, R. J. Planta, and B. E. Maden. 1978. Biosynthesis of a hypermodified nucleotide in *Saccharomyces carlsbergensis* 17S and HeLa-cell 18S ribosomal ribonucleic acid. *Biochem. J.* 169:71–77.

Brand, R. C., J. Klootwijk, C. P. Sibum, and R. J. Planta. 1979. Pseudouridylation of yeast ribosomal precursor RNA. *Nucleic Acids Res.* 7:121–134.

Branlant, C., A. Krol, M. A. Machatt, J. Pouyet, and J. P. Ebel. 1981. Primary and secondary structures of *Escherichia coli* MRE 600 23S ribosomal RNA. Comparison with models of secondary structure for maize chloroplast 23S rRNA and for large portions of mouse and human 16S mitochondrial rRNAs. *Nucleic Acids Res.* 9:4303–4324.

Brimacombe, R. 1995. The structure of ribosomal RNA: a three-dimensional jigsaw puzzle. *Eur. J. Biochem.* 230:365–383.

Ciampi, M. S., F. Arena, R. Cortese, and V. Daniel. 1977. Biosynthesis of pseudouridine in the in vitro transcribed tRNA^{Tyr} precursor. *FEBS Lett.* 77:75–82.

Cohn, W. E. 1959. 5-Ribosyl uracil, a carbon-carbon ribofuranosyl nucleoside in ribonucleic acids. *Biochim. Biophys. Acta* 32: 569–571.

Cohn, W. E. 1960. Pseudouridine, a carbon-carbon linked ribonucleoside in ribonucleic acids: isolation, structure, and chemical characteristics. *J. Biol. Chem.* 235:1488–1498.

Conrad, J., S. Raychaudhuri, B. Hall, and J. Ofengand. Unpublished results.

Cortese, R., H. O. Kammen, S. J. Spengler, and B. N. Ames. 1974. Biosynthesis of pseudouridine in transfer ribonucleic acid. *J. Biol. Chem.* 249:1103–1108.

Crain, P. F., and J. A. McCloskey. 1997. The RNA modification database. *Nucleic Acids Res.* 25:126–127.

Cunningham, P. R., R. B. Richard, C. J. Weitzmann, K. Nurse, and J. Ofengand. 1991. The absence of modified nucleotides affects both in vitro assembly and in vitro function of the 30S ribosomal subunit of *Escherichia coli*. *Biochimie* 73:789–796.

Dahlberg, J. E., N. Nikolaev, and D. Schlessinger. 1975. Posttranscriptional modification of nucleotides in *E. coli* ribosomal RNAs. *Brookhaven Symp. Biol.* 26:194–200.

Davis, D. R. 1995. Stabilization of RNA stacking by pseudouridine. *Nucleic Acids Res.* 23:5020–5026.

Davis, F. F., and F. W. Allen. 1957. Ribonucleic acids from yeast which contain a fifth nucleotide. *J. Biol. Chem.* 227:907–915.

Denman, R., J. Colgan, K. Nurse, and J. Ofengand. 1988. Crosslinking of the anticodon of P site bound tRNA to C-1400 of *E. coli* 16S RNA does not require the participation of the 50S subunit. *Nucleic Acids Res.* 16:165–178.

Denman, R., C. Weitzmann, P. R. Cunningham, D. Negre, K. Nurse, J. Colgan, Y. C. Pan, M. Miedel, and J. Ofengand. 1989a. In vitro assembly of 30S and 70S bacterial ribosomes from 16S RNA containing single base substitutions, insertions, and deletions around the decoding site (C1400). *Biochemistry* 28:1002–1011.

Denman, R., D. Negre, P. R. Cunningham, K. Nurse, J. Colgan, C. Weitzmann, and J. Ofengand. 1989b. Effect of point mutations in the decoding site (C1400) region of 16S ribosomal RNA on the ability of ribosomes to carry out individual steps of protein synthesis. *Biochemistry* 28:1012–1019.

Dokudovskaya, S., O. Dontsova, O. Shpanchenko, A. Bogdanov, and R. Brimacombe. 1996. Loop IV of 5S ribosomal RNA has contacts both to domain II and to domain V of the 23S RNA. *RNA* 2:146–152.

Foster, P. G., L. Huang, D. V. Santi, and R. M. Stroud. 1997. The crystal structure of *E. coli* tRNA pseudouridine synthase I. *FASEB J.* 11:A862.

Ganot, P., M.-L. Bortolin, and T. Kiss. 1997a. Site-specific pseudouridine formation in preribosomal RNA is guided by small nucleolar RNAs. *Cell* 89:799–809.

Ganot, P., M. Caizergues-Ferrer, and T. Kiss. 1997b. The family of box ACA small nucleolar RNAs is defined by an evolutionarily conserved secondary structure and ubiquitous sequence elements essential for RNA accumulation. *Genes Dev.* 11:941–956.

Gehrke, C. W., and K. C. Kuo. 1989. Ribonucleoside analysis by reversed-phase high-performance liquid chromatography. *J. Chromatogr.* 47:3–36.

Girard, J.-P., H. Lehtonen, M. Caizergues-Ferrer, F. Amalric, D. Tollervey, and B. Lapeyre. 1992. GAR1 is an essential small nucleolar RNP protein required for pre-rRNA processing in yeast. *EMBO J.* 11:673–682.

Gray, M. W. 1976. O2'-Methylinosine, a constituent of the ribosomal RNA of *Crithidia fasciculata*. *Nucleic Acids Res.* 3: 977–988.

Gray, M. W. 1974. The presence of O-2'-methylpseudouridine in the 18S + 26S ribosomal ribonucleates of wheat embryo. *Biochemistry* 13:5453–5463.

Green, R., and H. F. Noller. 1996. In vitro complementation analysis localizes 23S rRNA posttranscriptional modifications that are required for *Escherichia coli* 50S ribosomal subunit assembly and function. *RNA* 2:1011–1021.

Green, R., and H. F. Noller. 1997. Ribosomes and translation. *Annu. Rev. Biochem.* 66:679–716.

Green, C. J., H. O. Kammen, and E. E. Penhoet. 1982. Purification and properties of a mammalian tRNA pseudouridine synthase. *J. Biol. Chem.* 257:3045–3052.

Grosjean, H., Z. Szweykowska-Kulinska, Y. Motorin, F. Fasiolo, and G. Simos. 1997. Intron-dependent enzymatic formation of modified nucleosides in eukaryotic tRNAs: a review. *Biochimie* 79:293–302.

Gu, J., and R. Reddy. 1997. Small RNA database. *Nucleic Acids Res.* 25:98–101.

Gustafsson, C., R. Reid, P. J. Greene, and D. V. Santi. 1996. Identification of new RNA modifying enzymes by iterative genome search using known modifying enzymes as probes. *Nucleic Acids Res.* 24:3756–3762.

Gutell, R. R., M. W. Gray, and M. N. Schnare. 1993. A compilation of large subunit (23S- and 23S-like) ribosomal RNA structures. *Nucleic Acids Res.* 21:3055–3074.

Ho, N. W. Y., and P. T. Gilham. 1971. Reaction of pseudouridine and inosine with N-cyclohexyl-N'-β-(4-methylmorpholinium)ethylcarbodiimide. *Biochemistry* 10:3651–3657.

Huang, L., M. Pookanjanatavip, X. Gu, and D. V. Santi. 1998. A conserved aspartate of tRNA pseudouridine synthase is essential for activity and a probable nucleophilic catalyst. *Biochemistry* 37:344–351.

Jeanteur, P., F. Amaldi, and G. Attardi. 1968. Partial sequence analysis of ribosomal RNA from HeLa cells. II. Evidence for sequences of non-ribosomal type in 45S and 32S ribosomal RNA precursors. *J. Mol. Biol.* 33:757–775.

Jiang, W., K. Middleton, H.-J. Yoon, C. Fouquet, and J. Carbon. 1993. An essential yeast protein, CBF5, binds in vitro to centromeres and microtubules. *Mol. Cell. Biol.* 13:4884–4893.

Johnson, L., and D. Söll. 1970. In vitro biosynthesis of pseudouridine at the polynucleotide level by an enzyme extract from *Escherichia coli*. *Proc. Natl. Acad. Sci. USA* 67:943–950.

Joseph, S., and H. F. Noller. 1996. Mapping the rRNA neighborhood of the acceptor end of tRNA in the ribosome. *EMBO J.* 15:910–916.

Kammen, H. O., C. C. Marvel, L. Hardy, and E. E. Penhoet. 1988. Purification, structure, and properties of *Escherichia coli* tRNA pseudouridine synthase I. *J. Biol. Chem.* 263:2255–2263.

Khan, M. S. N., and B. E. H. Maden. 1977. Nucleotide sequence relationships between vertebrate 5.8 S ribosomal RNAs. *Nucleic Acids Res.* 4:2495–2505.

Kiss-Laszlo, Z., Y. Henry, J. P. Bachellerie, M. Caizergues-Ferrer, and T. Kiss. 1996. Site-specific ribose methylation of preribosomal RNA: a novel function for small nucleolar RNAs. *Cell* 85: 1077–1088.

Klootwijk, J., I. Klein, and L. A. Grivell. 1975. Minimal posttranscriptional modification of yeast mitochondrial ribosomal RNA. *J. Mol. Biol.* 97:337–350.

Koonin, E. V. 1996. Pseudouridine synthases: four families of enzymes containing a putative uridine-binding motif also conserved in dUTPases and dCTP deaminases. *Nucleic Acids Res.* 24:2411–2415.

Kowalak, J. A., E. Bruenger, and J. A. McCloskey. 1995. Posttranscriptional modification of the central loop of domain V in *Escherichia coli* 23S ribosomal RNA. *J. Biol. Chem.* 270: 17758–17764.

Kowalak, J. A., E. Bruenger, T. Hashizume, J. M. Peltier, J. Ofengand, and J. A. McCloskey. 1996. Structural characterization of 3-methylpseudouridine in domain IV from *E. coli* 23S ribosomal RNA. *Nucleic Acids Res.* **24**:688–693.

Krzyzosiak, W., R. Denman, K. Nurse, M. Hellmann, M. Boublik, C. W. Gehrke, P. F. Agris, and J. Ofengand. 1987. In vitro synthesis of 16S ribosomal RNA containing single base changes and assembly into a functional 30S ribosome. *Biochemistry* **26**:2353–2364.

Lafontaine, D. L. J., C. Bousquet-Antonelli, Y. Henry, M. Caizergues-Ferrer, and D. Tollervey. 1998. The box H+ACA snoRNAs carry Cbf5p, the putative rRNA pseudouridine synthase. *Genes Dev.* **12**:527–537.

Lane, B. G., J. Ofengand, and M. W. Gray. 1995. Pseudouridine and $O^{2'}$-methylated nucleosides. Significance of their selective occurrence in rRNA domains that function in ribosome-catalyzed synthesis of the peptide bonds in proteins. *Biochimie* **77**:7–15.

Lankat-Buttgereit, B., H. J. Gross, and G. Krupp. 1987. Detection of modified nucleosides by rapid RNA sequencing methods. *Nucleic Acids Res.* **15**:7649.

Lecointe, F., G. Simos, A. Sauer, E. C. Hurt, Y. Motorin, and H. Grosjean. 1997. Characterization of yeast protein Deg1 as pseudouridine synthase (Pus3) catalyzing the formation of Ψ_{38} and Ψ_{39} in tRNA anticodon loop. *J. Biol. Chem.* **273**:1316–1323.

Li, H. V., J. Zagorski, and M. J. Fournier. 1990. Depletion of U14 small nuclear RNA (snR128) disrupts production of 18S rRNA in *Saccharomyces cerevisiae*. *Mol. Cell. Biol.* **10**:1145–1152.

Liang, W.-Q., and M. J. Fournier. 1995. U14 base pairs with 18S rRNA: a novel snoRNA interaction required for rRNA processing. *Genes Dev.* **9**:2433–2443.

Maden, B. E. H. 1988. Locations of methyl groups in 28S rRNA of *Xenopus laevis* and man. Clustering in the conserved core of the molecule. *J. Mol. Biol.* **201**:289–314.

Maden, B. E. H. 1990. The numerous modified nucleotides in eukaryotic ribosomal RNA. *Prog. Nucleic Acid Res. Mol. Biol.* **39**:241–300.

Maden, B. E. H. 1997. Guides to 95 new angles. *Nature* **389**:129–131.

Maden, B. E., J. Forbes, P. de Jong, and J. Klootwijk. 1975. Presence of a hypermodified nucleotide in HeLa cell 18S and *Saccharomyces carlsbergensis* 17S ribosomal RNAs. *FEBS Lett.* **59**:60–63.

Maden, B. E. H., and M. Salim. 1974. The methylated nucleotide sequences in HeLa cell ribosomal RNA and its precursors. *J. Mol. Biol.* **88**:133–164.

Maxwell, E. S., and M. J. Fournier. 1995. The small nucleolar RNAs. *Annu. Rev. Biochem.* **35**:897–934.

Meier, U. T., and G. Blobel. 1994. NAP57, a mammalian nucleolar protein with a putative homolog in yeast and bacteria. *J. Cell Biol.* **127**:1505–1514.

Mitchell, P., M. Osswald, D. Schüler, and R. Brimacombe. 1990. Selective isolation and detailed analysis of intra-RNA cross-links induced in the large ribosomal subunit of *E. coli*; a model for the tertiary structure of the tRNA binding domain in 23S RNA. *Nucleic Acids Res.* **18**:4325–4333.

Mitchell, P., M. Osswald, and R. Brimacombe. 1992. Identification of intermolecular RNA cross-links at the subunit interface of the *Escherichia coli* ribosome. *Biochemistry* **31**:3004–3011.

Moazed, D., and H. F. Noller. 1990. Binding of tRNA to the ribosomal A and P sites protects two distinct sets of nucleotides in 16 S rRNA. *J. Mol. Biol.* **211**:135–145.

Morrissey, J. P., and D. Tollervey. 1993. Yeast snR30 is a small nucleolar RNA required for 18S rRNA synthesis. *Mol. Cell. Biol.* **13**:2469–2477.

Nazar, R. N., T. O. Sitz, and H. Busch. 1976. Sequence homologies in mammalian 5.8S ribosomal RNA. *Biochemistry* **15**:505–508.

Ni, J., A. L. Tien, and M. J. Fournier. 1997. Small nucleolar RNAs direct site-specific synthesis of pseudouridine in ribosomal RNA. *Cell* **89**:565–573.

Ni, J., A. L. Tien, and M. Fournier. Unpublished results.

Niu, L., and J. Ofengand. Unpublished results.

Nurse, K., and J. Ofengand. Unpublished results.

Nurse, K., J. Wrzesinski, A. Bakin, B. G. Lane, and J. Ofengand. 1995. Purification, cloning, and properties of the tRNA $\Psi 55$ synthase from *Escherichia coli*. *RNA* **1**:102–112.

O'Connor, M., and A. E. Dahlberg. 1995. The involvement of two distinct regions of 23 S ribosomal RNA in tRNA selection. *J. Mol. Biol.* **254**:838–847.

Ofengand, J., and A. Bakin. 1997. Mapping to nucleotide resolution of pseudouridine residues in large subunit ribosomal RNAs from representative eukaryotes, prokaryotes, archaebacteria, mitochondria, and chloroplasts. *J. Mol. Biol.* **266**:246–268.

Ofengand, J., A. Bakin, and K. Nurse. 1993. The functional role of conserved sequences of 16S ribosomal RNA in protein synthesis. p. 489–500. *In* K. H. Nierhaus, A. R. Subramanian, V. A. Erdmann, F. Franceschi, and B. Wittman-Liebold (ed.), *The Translational Apparatus*. Plenum Press, New York, N.Y.

Ofengand, J., A. Bakin, J. Wrzesinski, K. Nurse, and B. G. Lane. 1995. The pseudouridine residues of ribosomal RNA. *Biochem. Cell Biol.* **73**:915–924.

Olsen, G. J., and C. R. Woese. 1997. Archaeal genomics: an overview. *Cell* **89**:991–994.

Parker, R., T. Simmons, E. O. Shuster, P. G. Siliciano, and C. Guthrie. 1988. Genetic analysis of small nuclear RNAs in *Saccharomyces cerevisiae*: viable sextuplet mutant. *Mol. Cell. Biol.* **8**:3150–3159.

Peculis, B. 1997. RNA processing: pocket guides to ribosomal RNA. *Curr. Biol.* **7**:R480–R482.

Pennisi, E. 1997. Microbial genomes come tumbling in. *Science* **277**:1433.

Rudd, K. Personal communication.

Rudd, K., and J. Ofengand. Unpublished results.

Samarsky, D. A., A. G. Balakin, and M. J. Fournier. 1995. Characterization of three new snRNAs from *Saccharomyces cerevisiae*: snR34, snR35 and snR36. *Nucleic Acids Res.* **23**:2548–2554.

Samuelsson, T., and M. Olsson. 1990. Transfer RNA pseudouridine synthases in *Saccharomyces cerevisiae*. *J. Biol. Chem.* **265**:8782–8787.

Saponara, A. G., and M. D. Enger. 1974. The isolation from ribonucleic acid of substituted uridines containing α-aminobutyrate moieties derived from methionine. *Biochim. Biophys. Acta* **349**:61–77.

Scannell, J. P., A. M. Crestfield, and F. W. Allen. 1959. Methylation studies on various uracil derivatives and on an isomer of uridine isolated from ribonucleic acids. *Biochim. Biophys. Acta* **32**:406–412.

Schaefer, K. P., S. Altman, and D. Söll. 1973. Nucleotide modification in vitro of the precursor of transfer RNA^{Tyr} of *Escherichia coli*. *Proc. Natl. Acad. Sci. USA* **70**:3626–3630.

Simos, G., H. Tekotte, H. Grosjean, A. Segref, K. Sharma, D. Tollervey, and E. C. Hurt. 1996. Nuclear pore proteins are involved in the biogenesis of functional tRNA. *EMBO J.* **15**:2270–2284.

Smith, C. M., and J. A. Steitz. 1997. Sno storm in the nucleolus: new roles for myriad small RNPs. *Cell* **89**:669–672.

Smith, J. E., B. S. Cooperman, and P. Mitchell. 1992. Methylation sites in *Escherichia coli*. Ribosomal RNA: localization and identification of four new sites of methylation in 23S RNA. *Biochemistry* **31:**10825–10834.

Sollner-Webb, B., K. Tyc, and J. A. Steitz. 1995. Ribosomal RNA processing in eukaryotes. *In* R. A. Zimmermann and A. E. Dahlberg (ed.), *Ribosomal RNA Structure, Evolution, Processing, and Function in Protein Biosynthesis*. Telford, Cadwell, N.J.

Sprinzl, M., C. Steegborn, F. Hübel, and S. Steinberg. 1996. Compilation of tRNA sequences and sequences of tRNA genes. *Nucleic Acids Res.* **24:**68–72.

Sun, D., F. Nallaseth, and J. Ofengand. Unpublished results.

Szweykowska-Kulinska, Z., B. Senger, G. Keith, F. Fasiolo, and H. Grosjean. 1994. Intron-dependent formation of pseudouridines in the anticodon of *Saccharomyces cerevisiae* minor tRNA(Ile). *EMBO J.* **13:**4636–4644.

Thomas, G., J. Gordon, and H. Rogg. 1978. N^4-Acetylcytidine. A previously unidentified labile component of the small subunit of eukaryotic ribosomes. *J. Biol. Chem.* **253:**1101–1105.

Tollervey, D. 1987. A yeast small nuclear RNA is required for normal processing of pre-ribosomal RNA. *EMBO J.* **13:**4169–4175.

Tollervey, D., and C. Guthrie. 1985. Deletion of a yeast small nuclear RNA gene impairs growth. *EMBO J.* **6:**4169–4175.

Tollervey, D., and T. Kiss. 1997. Function and synthesis of small nucleolar RNAs. *Curr. Opin. Cell Biol.* **9:**337–342.

Trapane, T. L., M. S. Christopherson, C. D. Roby, P. O. P. Ts'o, and D. Wang. 1994. DNA triple helices with C-nucleosides (deoxypseudouridine) in the second strand. *J. Am. Chem. Soc.* **116:**8412–8413.

Tscherne, J., P. Popieniek, K. Nurse, H. Michel, M. Sochacki, and J. Ofengand. Unpublished results.

Tycowski, K. T., Z. You, and J. A. Steitz. Personal communication.

Van de Peer, Y., J. Jansen, P. De Rijk, and R. De Wachter. 1997. Database on the structure of small ribosomal subunit RNA. *Nucleic Acids Res.* **25:**111–116.

Veldman, G. M., J. Klootwijk, V. C. H. F. de Regt, R. J. Planta, C. Branlant, A. Krol, and J. P. Ebel, 1981. The primary and secondary structure of yeast 26S rRNA. *Nucleic Acids Res.* **9:**6935–6952.

Venema, J., and D. Tollervey. 1995. Processing of pre-ribosomal RNA in *Saccharomyces cerevisiae*. *Yeast* **11:**1629–1650.

Weitzmann, C., P. R. Cunningham, K. Nurse, and J. Ofengand. 1993. Chemical evidence for domain assembly of the *E. coli* 30S ribosome. *FASEB J.* **7:**177–180.

Wrzesinski, J., A. Bakin, K. Nurse, B. G. Lane, and J. Ofengand. 1995a. Purification, cloning, and properties of the 16S RNA Ψ516 synthase from *Escherichia coli*. *Biochemistry* **34:**8904–8913.

Wrzesinski, J., K. Nurse, A. Bakin, B. G. Lane, and J. Ofengand. 1995b. A dual-specificity pseudouridine synthase: purification and cloning of a synthase from *Escherichia coli* which is specific for both Ψ746 in 23S RNA and for Ψ32 in tRNA^{Phe}. *RNA* **1:**437–448.

Yu, C. T., and F. W. Allen. 1959. Studies on an isomer of uridine isolated from ribonucleic acids. *Biochim. Biophys. Acta* **32:**393–405.

Chapter 13

Small Nucleolar RNAs Guide the Ribose Methylations of Eukaryotic rRNAs

JEAN-PIERRE BACHELLERIE AND JÉRÔME CAVAILLÉ

The biogenesis of eukaryotic ribosomes in the nucleolus involves an intricate superimposition of complex processes including transcription of rRNA genes, maturation of the primary transcript, and its stepwise assembly with ribosomal proteins. The long pre-rRNA is processed through a series of endo- and exonucleolytic cleavages, which removes the extra-spacer sequences and produces stoichiometric amounts of the small and large subunit mature rRNAs (Eichler and Craig, 1994; Sollner-Webb et al., 1995; Venema and Tollervey, 1995). Maturation of pre-rRNA also includes an elaborate pattern of nucleoside modifications occurring mostly on the nascent primary transcript, which has been identified in detail in a few vertebrates (Maden, 1986, 1988, 1990) and in the yeast *Saccharomyces carlsbergensis* (Veldman et al., 1981), a close relative of *S. cerevisiae*. Nucleotide modifications in eukaryotic rRNAs are essentially of two types—methylations (the vast majority of which are added on the ribose) and pseudouridylations. Vertebrate rRNAs contain about one hundred ribose methylated nucleotides and also about one hundred pseudouridines, i.e., about twice as many as yeast rRNAs (Table 1). Although the numbers of the two major types of rRNA modified nucleotides vary substantially among distant eukaryotes, their pattern along the rRNA sequence is largely conserved (Maden, 1990). So far, the role of nucleotide modifications in the assembly or function of the eukaryotic ribosome remains speculative (Maden, 1990; Lane et al., 1995). For a long time, another baffling question has been how the rRNA nucleotides to be modified are selected within the long pre-rRNA sequence, given that the different sites of ribose methylation in eukaryotic rRNAs do not exhibit any obvious common feature, either in sequence or in secondary structure, which could mediate their recognition by a common modifying enzyme. Very recently, major advances in the characterization of the snoRNA population of eukaryotic cells have illuminated the process by which most (if not all) sites of rRNA ribose methylation are specified. Throughout its synthesis and processing, pre-rRNA transiently associates with a large number of snoRNAs, much larger than anticipated 2–3 years ago (Maxwell and Fournier, 1995). All of these newly discovered snoRNAs fall into two major families, based on structural features (Balakin et al., 1996) (see also Chapter 12 by Ofengand and Fournier). One of the two families, termed box C/D antisense snoRNAs because these all contain two common short sequence motifs and long complementarities to rRNA (Bachellerie et al., 1995a), has turned out to play a crucial role in rRNA ribose methylation. At each site of ribose methylation in rRNA a specific snoRNA of this family is associated which targets precisely the nucleotide to be methylated, through transient formation of a canonical duplex with pre-rRNA at the modification site, as briefly reported in recent review articles (Tollervey, 1996a; Maden, 1996; Bachellerie and Cavaillé, 1997; Smith and Steitz, 1997).

DISCOVERY OF MODIFICATION GUIDE snoRNAs

The landmarks in the work on box C/D antisense snoRNAs can be demarcated as follows: (1) U14, identified in 1986–1988, remained for long an oddity, the only snoRNA with long phylogenetically conserved complementarities to rRNA (Trinh-Rohlik and Maxwell, 1988); (2) in the wake of the finding

---

Jean-Pierre Bachellerie and Jérôme Cavaillé • Laboratoire de Biologie Moléculaire Eucaryote du CNRS, Université Paul-Sabatier, 118, route de Narbonne, 31062 Toulouse Cédex, France.

Table 1. Numbers of modified nucleotides detected in rRNAs of representative species

rRNA	2'-O-ribose methylations				Base methylations				Pseudouridines[a,d]			
	Human[a]	X. laevis[a]	S. carlsbergensis[a]	E. coli[b,c]	Human[a]	X. laevis[a]	S. carlsbergensis	E. coli[b,c]	Human	X. laevis	S. carlsbergensis	E. coli
18S-16S	40	33	18	1	5	5	4	10	36	44	13	1
28S-23S	65	64	37	3	5	5	6	14	~57	~52	30	9
5.8S	2	2	–						2	2	1	

[a] Maden, 1990.
[b] Cunningham et al., 1990.
[c] Smith et al., 1992.
[d] Ofengand et al., 1995.

that U14 was intron-encoded in vertebrates (Liu and Maxwell, 1990; Leverette et al., 1992), other intronic snoRNAs were soon discovered in 1993–1994, among them a few with extended rRNA complementarities (Prislei et al., 1993; Nicoloso et al., 1994; Qu et al., 1994); (3) the serendipitous way these additional intronic antisense snoRNAs were detected tantalizingly suggested their representing only the tip of the iceberg (clearly, hints as to their function were to come from an enlarged snoRNA repertoire); (4) as their number steadily increased, it became clear that their rRNA complementarities systematically matched sites of ribose methylation in rRNA (Qu et al., 1995; Bachellerie et al., 1995a); (5) comparison of a large set of snoRNA:rRNA complementarities then plainly revealed that the 2'-O-ribose methylation site always had an invariant position in the duplex, unambiguously pointing to a role of guide for the antisense snoRNA in the modification (Nicoloso et al., 1996; Kiss-Laszlo et al., 1996); (6) finally, the guide notion was directly confirmed by different experimental approaches in yeast and vertebrates (Kiss-Laszlo et al., 1996; Cavaillé et al., 1996), showing that the canonical snoRNA/rRNA duplex is not only required but sufficient for site-specific ribose methylation.

This article provides a progress report on developments that should set the stage for dissecting the rRNA-ribose methylation machinery and for deciphering the role of these modifications in ribosome assembly and function. Remarkably, following the finding that box C/D antisense snoRNAs guide ribose methylation of rRNA, ACA snoRNAs, the other major family of snoRNAs (Balakin et al., 1996; Ganot et al., 1997a), have been shown to guide the other major type of nucleotide modification in rRNA, pseudouridylation, again through base-paired interactions at the modification site (Ni et al., 1997; Ganot et al., 1997b; Smith and Steitz, 1997) (see also Chapter 12 by Ofengand and Fournier). Selected aspects of the study of rRNA nucleoside modifications are also reviewed in this volume by Maden (Chapter 24). Because this article is focused on the guide role of snoRNAs in the ribose methylation of rRNA, the readers are encouraged to consult reviews on rRNA processing (Eichler and Craig, 1994; Sollner-Webb et al., 1995; Tollervey, 1996b) and posttranscriptional modifications (Maden, 1990), ribosome biogenesis (Warner, 1989), and the nucleolus (Hadjiolov, 1985).

IDENTIFYING THE PATTERNS OF rRNA RIBOSE METHYLATION: AN EXPERIMENTAL CHALLENGE

Our present knowledge on the location of the 2'-O-ribose methylations in eukaryotic rRNAs comes essentially from studies performed in the years 1973–1988 by Maden and coworkers on several vertebrates and by Veldman et al. (1981) on the yeast *Saccharomyces carlsbergensis*. While data more particularly relevant to the topic of this article are given below, additional information on the patterns of eukaryotic rRNA methylation can be found in the comprehensive review by Edward Maden (1990). In that article the author provides, beyond a thorough depiction of the patterns of eukaryotic rRNA ribose methylated sites known at that time, an insightful discussion as to their enigmatic roles, based on a detailed analysis of their location and occasional clustering with reference to local features of rRNA structure and to the position of pseudouridylation sites.

The paucity of information as to the function of rRNA ribose methylations largely stems from the technical difficulties associated with their precise mapping in large RNA molecules. Most data on rRNA ribose methylation sites come from oligonucleotide fingerprint analyses on methyl-labeled rRNA purified from cells grown for at least two generations in medium containing radioactive methionine. Very early, such analyses provided an accurate determination of the number of ribose methylated nucleotides in rRNA, as well as the major information that the modification was stoichiometric for all sites (but one) in mature eukaryotic rRNAs (Maden, 1986, 1988). Combination of RNA oligonucleotide data with rRNA sequence data generally proved sufficient to map precisely most sites of ribose methylation. However, while all the 40 ribose methylated sites of human 18S rRNA have been unambiguously located, about 11–13 of the 63–65 ribose methylated sites of human 28S rRNA still remain to be placed (Maden, 1990). The advent of direct chemical or enzymatic RNA sequencing methods has done little to obviate these limitations. Indeed, ribose methylated nucleotides appear as gaps in sequencing gel ladders of primer extension products generated on partially alkali-hydrolyzed RNAs, due to the resistance conferred by the 2′-O-methyl group (Kiss-Laszlo et al., 1996). However, even when a nonmethylated RNA is partially hydrolyzed, band intensities may vary dramatically between adjacent positions of the ladder, thus making problematic the assignment of ribose methylation sites in a modified RNA, particularly for sites that are not fully modified. Reverse transcription of intact (nonhydrolyzed) RNAs can also provide useful diagnostic information when the reaction is performed at low dNTP concentrations, taking advantage of concentration-dependent pauses at 2′-O-ribose methylated sites in the template (Maden et al., 1995). However, because individual 2′-O-methyls widely differ in intensities of their effect on reverse transcription and because the progression of the enzyme may be also hampered by a variety of other structural features in the substrate, these experiments cannot suffice to identify, reliably and in a straightforward way, the complete methylation pattern of an rRNA molecule. The development of a new, efficient mapping method, possibly based on prior chemical modification of the RNA substrate, such as has been recently achieved for locating pseudouridines in large RNA molecules (Bakin and Ofengand, 1993), would definitely represent a major breakthrough in the field, allowing the derivation of rRNA ribose methylation patterns from a much larger repertoire of species. Comparative analyses of an extended database will undoubtedly provide further insight into the precise contribution of these modifications to rRNA folding and thereby to ribosomal function.

PATTERNS OF RIBOSE METHYLATION WITHIN EUKARYOTIC rRNA STRUCTURE

In mammals, the RNA molecules of a mature ribosome contain slightly more than 100 sites of methylation on the ribose, while only 10 rRNA nucleotides are methylated on the base. As reviewed in detail by Maden (1990), human and rodents share an almost identical pattern of ribose methylation. Sites of rRNA ribose methylation are also conserved among distant vertebrates (Khan et al., 1978), with amphibian *Xenopus laevis* rRNAs showing a pattern identical to that of mammals, except for the absence of 7–8 of the mammalian modifications. The yeast *S. carlsbergensis* exhibits a much simpler pattern with only 18 sites of ribose methylation in 18S rRNA and 37 sites in 25S rRNA, but most of the sites ribose methylated in yeast are also ribose methylated in vertebrates. As shown in Table 1, the numbers of ribose methylations and of pseudouridines seem roughly correlated in the rRNAs of the different organisms analyzed to date (see also Chapter 12 by Ofengand and Fournier). In all these species, sites of ribose methylations are always located within the universally conserved core of rRNA secondary structure despite their large variation in number among very distant organisms. Strikingly, the expansion segments of eukaryotic rRNAs, in which most of the sequence and size variations of eukaryotic rRNAs are restricted, do not contain ribose methylated nucleotides, although they amount collectively to a very large fraction of the mature rRNA sequence in some groups of organisms (e.g., about 40% of mammalian 28S rRNA length). Ribose methylations generally occur on nucleotides invariant among eukaryotic rRNAs, even when the modification is not conserved between yeast and vertebrates (Maden, 1990). Ribose methylated sites in eukaryotic rRNAs display a large diversity, in both sequence and secondary structure, around the modified nucleotide. No common oligonucleotide motif can be recognized in the immediate neighborhood of the different ribose methylation sites, which are found both in single-stranded regions and in stems in the secondary structure of mature rRNAs (Maden, 1990). Interestingly, ribose methylations are far from being uniformly distributed within the universally conserved domains of eukaryotic rRNAs. For large subunit rRNA for which a three-dimensional model is not known, a significant clustering is apparent at the secondary structure level, with most ribose methylations regrouped in three major areas: near the boundaries of domains II and III in the 5′ half; in central domain IV (human positions 3600–3800); and in parts of 3′ domains V and VI. The clustering in 28S rRNA is even more apparent when the sites are not considered collectively but as subsets with a similar degree of phylogenetic conservation, such as on one hand the sites that are conserved between yeast and vertebrates (Fig. 1A) and on the other hand those that are specific to vertebrates (Fig. 1B). Re-

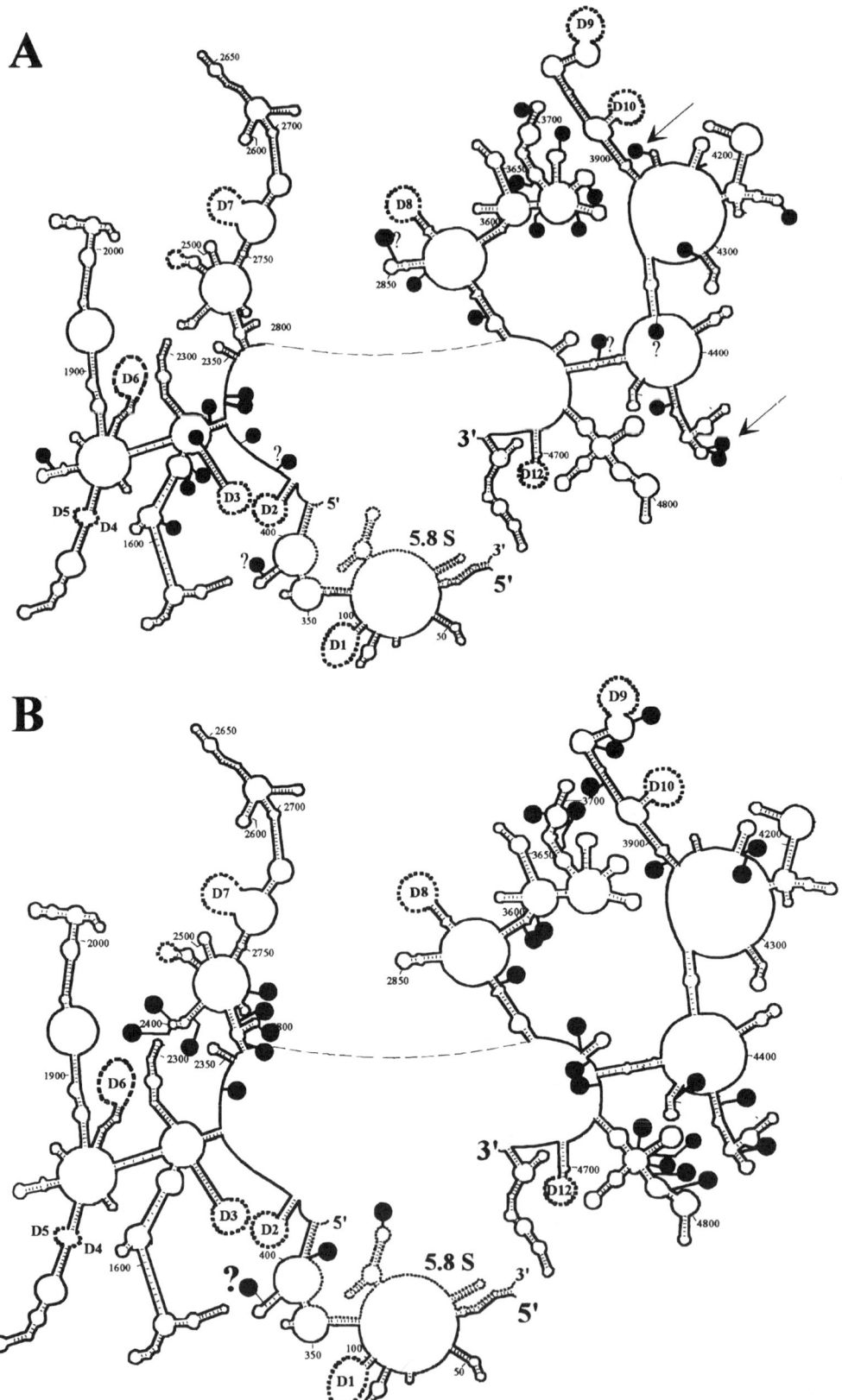

Figure 1. Clustering of ribose methylation sites in eukaryotic 28S rRNA secondary structure. Only the universally conserved domains of the rRNA molecule are represented (the evolutionarily variable D domains are indicated by dots; the junction between the 5' and 3' halves of the molecule is schematized by a broken line). Ribose methylated sites are shown by filled circles (sites that are still presumptive are denoted by question marks). (A) Sites conserved between the yeast *S. cerevisiae* and vertebrates (the two arrows point to sites conserved in *E. coli* 23 S rRNA). (B) Ribose methylated sites specific to vertebrates.

markably, several ribose methylations occur on nucleotides directly involved in the few tertiary interactions identified so far within the large subunit rRNA (Fig. 2), in line with the notion that the modification may modulate the three-dimensional folding of rRNA and its interaction with ligands. In eukaryotic large subunit rRNA a major fraction of ribose methylated nucleosides and pseudouridines co-cluster in the secondary structure, essentially in the central and 3' domains, i.e., over regions that contain elements of the peptidyl transferase center of the ribosome (see Chapter 12 by Ofengand and Fournier). For eukaryotic small subunit rRNA, ribose methylations do not show any obvious singular distribution in the primary or secondary structures but they definitely appear clustered in the three-dimensional model, mostly found in the body and head and absent from the platform (Maden, 1990). Taken together with the substantial conservation of their pattern during evolution, these observations point to a role of ribose methylated nucleotides in key steps of protein synthesis. Finally, the individual ribose methylations in eukaryotic rRNAs range from universal to species-specific and there is no reason why a unitary role should be assigned to all of them. The point is best illustrated by a prokaryotic example. The 23S rRNA of a thiostrepton-producing *Streptomyces* sp. contains an additional, species-specific site of ribose methylation for which the function has been unambiguously identified. Catalyzed by a specific methylase, the ribose methylation of this particular rRNA nucleotide confers to the organism resistance against its toxic peptide antibiotic product by altering thiostrepton binding to 23S rRNA (Cundliffe, 1989). By contrast, other site-specific methylations of rRNA leading to antibiotic resistance in antibiotic-producing organisms occur on the bases, not on the ribose.

RIBOSE METHYLATION OF PROKARYOTIC rRNAs

In prokaryotes, the pattern of rRNA ribose methylation is dramatically less complex than in eukaryotes and most methylations are found on the bases. *E. coli* rRNAs contain only four ribose methylated nucleotides, one in 16S rRNA, at C1402 (Cunningham et al., 1990), and three in 23S rRNA, at G2251, C2498 and U2552 (Smith et al., 1992). Remarkably, the 16S rRNA site and two of the three 23S rRNA sites (Gm2251 and Um2552) are at positions also ribose methylated in eukaryotic rRNAs. The pseudouridine content of prokaryotic rRNAs is also dramatically lower than that of eukaryotic rRNAs (see Chapter 12 by Ofengand and Fournier). Unmethylated 16S rRNA can reconstitute into 30S ribosomal subunits in vitro but the functional activity of the synthetic particles is reduced to about half that of subunits containing a fully methylated 16S rRNA (Cunningham et al., 1990). While this suggests that methylation may optimize function, the contribution to this effect of the single ribose methylation of *E. coli* 16S rRNA remains unknown. Most modified nucleotides in *E. coli* rRNAs, not only the few ribose methylated ones, are clustered at the functional center of the ribosomal RNA molecules, suggesting their being functionally important (Brimacombe et al., 1993). It is noteworthy that one of the 23S rRNA nucleotide positions which is ribose methylated in *E. coli* like in eukaryotic rRNAs, Gm2251, has been linked to tRNA binding by chemical footprinting experiments (Moazed and Noller, 1989). As for Cm1402, the single *E. coli* 16S rRNA ribose methylated nucleotide, also methylated in eukaryotic 18S rRNA, it is adjacent to a string of nucleotides (positions 1399–1401) protected from chemical probes by P-site bound tRNA (Noller et al., 1996). Remarkably, the peptidyl transferase activity of reconstituted *E. coli* ribosomal subunits does not seem to require at least two of the three ribose methylations of 23S rRNA, the universally conserved Gm2251 and Um2552, while a few other nucleoside modifications seem important, as indicated by in vitro complementation analyses (Green and Noller, 1996).

RIBOSE METHYLATION OF rRNAs AND RIBOSOME BIOGENESIS

The complete pattern of rRNA ribose methylations can be detected on the rRNA primary transcript (Maden and Salim, 1974). Remarkably, the modifications exclusively occur on the mature rRNA portions of the precursor sequence. The lack of ribose methylation in the transcribed spacer segments must be stressed, given that these evolutionarily variable regions processed out of the precursor amount to a large fraction of the pre-rRNA length in some eukaryotic groups (e.g., about 50% in mammals). Several lines of evidence (see Chapter 24 by Maden) indicate that most, if not all, rRNA ribose methyl groups are added on the nascent, elongating transcript before transcription is completed, at least in vertebrates. This underlines the remarkable rapidity of the entire methylation process and points to a possible relationship with transcription of rRNA genes. Editing of the mitochondrial RNAs of *Physarum polycephalum* by nucleotide insertion in isolated mitochondria is also closely linked to RNA synthesis; it proceeds very

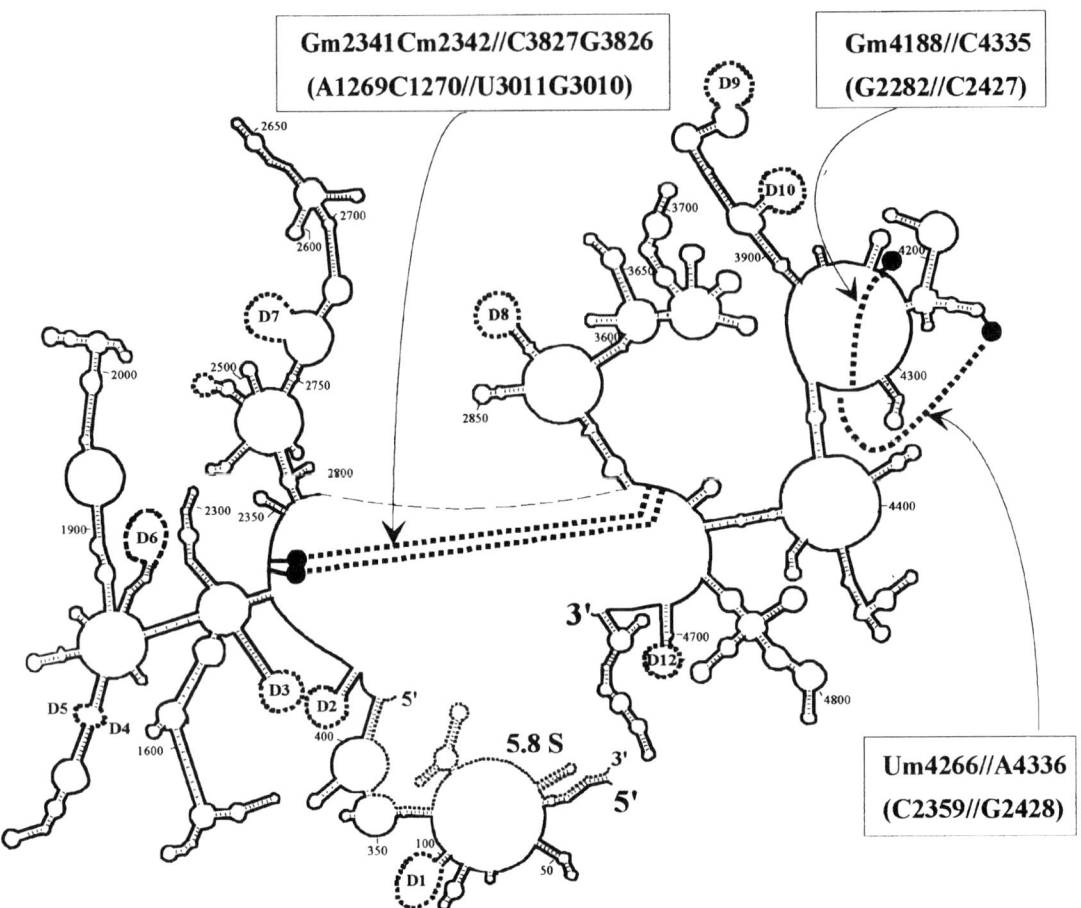

Figure 2. Ribose methylated nucleotides involved in tertiary interactions in 28S rRNA structure. Among the few tertiary interactions by canonical base pairing which are supported by comparative sequence analysis (Schnare et al., 1996), only the ones involving ribose methylated nucleotides are shown here (dotted lines). In each case the precise nucleotide positions involved are indicated for human 28S rRNA (the corresponding *E. coli* 23S rRNA coordinates are also given in parentheses).

close to the 3' end of the nascent transcript, possibly in association with the transcription complex (Visomirski-Robic and Gott, 1997) (see Chapter 22 by Gott and Visomirski-Robic). Indeed, early observations on cultured mammalian cells seemed to support the notion that methylations might be essential for rRNA processing and ribosome assembly (Vaughan et al., 1967). However, subsequent studies have shown that blockade of pre-rRNA methylation does not obligatorily result in a complete inhibition of pre-rRNA maturation and ribosome assembly, indicating that methyl groups are not essential elements of rRNA processing signals. Thus, unmethylated ribosomes are still formed in mammalian cells treated with cycloleucine, an inhibitor of S-adenosylmethionine production (Caboche and Bachellerie, 1977), or in a yeast temperature-sensitive fibrillarin mutant (Tollervey et al., 1993). Obviously, this is not to say that inhibition of methylation is not accompanied by substantial alterations in ribosome biogenesis: the overall efficiency of rRNA processing is substantially decreased and its kinetics are delayed during the cycloleucine treatment. Methylations, although not essential, could play a significant part in facilitating ribosomal subunit assembly and accelerating the whole process. In that regard some individual methylations could be more particularly important. Remarkably, one of the universally conserved ribose methylations of large subunit rRNA, corresponding to Gm2251 in *E. coli*, seems to be required for the formation of functional large subunits of the mitochondrial ribosome, as shown by the disruption of the *S. cerevisiae PET56* nuclear gene, the product of which catalyzes that particular modification (Sirum-Connolly and Mason, 1993; Sirum-Connolly et al., 1995). However, analysis of an extragenic mutation that suppresses *pet56* loss-of-function mutations (Mason et al., 1996) suggests that neither the cognate rRNA ribose methylation nor Pet56 protein is absolutely essential for the formation of functional ribosomes. Taken together, the available evidence leaves open the possibility that ribose methylations are added early because of an early, nonessential role in ribosome assembly, possibly through a fine-tuning of pre-rRNA folding and ribosomal protein assembly, as further discussed in the following paragraphs. Alternatively, they could function much later in the mature ribosome, but still have to be added early because target sites in the substrate are accessible only at this stage, when they are not yet buried into the complex preribosomal RNP structure. Determining how close to the growing end of the nascent transcript ribose methylations are added and identifying precisely their individual timings, by reference to definite steps of pre-rRNA folding and ribosomal protein binding, could illuminate their precise role in ribosome assembly. This definitely represents a daunting experimental challenge for the years to come.

AN AMAZING DIVERSITY OF INTRON-ENCODED SMALL NUCLEOLAR RNAs

By analogy to snRNAs involved in pre-mRNA splicing, snoRNAs have long been suspected to play a role in pre-rRNA processing. However, over the last 2 or 3 years the number of newly identified snoRNAs has become so high that the spliceosomal snRNA paradigm has lost much of its allure. Of the more than 80 snoRNAs characterized so far, only a very small fraction appears to be required at specific steps of pre-rRNA cleavage, as shown for U3, U8, U22 or MRP-7.2 RNAs (reviewed by Maxwell and Fournier, 1995; Tollervey, 1996b). Another major surprise has been that all the novel snoRNA species in vertebrates, and several of them in yeast, have very peculiar features in terms of gene organization and biosynthetic pathway. They are encoded in introns and are not transcribed from their own promoter but formed by posttranscriptional processing of the pre-mRNA intron. The fact that a majority of host-genes for intronic snoRNAs encode proteins involved in ribosome assembly or function has initially suggested (reviewed by Sollner-Webb, 1993) that their peculiar mode of formation could provide the basis for a coordinated production of functionally linked gene products, assuming that intronic snoRNAs were involved in rRNA processing. However, a few host-genes for intronic snoRNAs do not code for proteins directly associated with ribosome function (Renalier et al., 1996), and some of them do not even seem to code for proteins (Tycowski et al., 1996a). Moreover, intronic snoRNAs may move between different host-genes during evolution, and several yeast homologs of vertebrate intronic snoRNAs are not encoded in introns. While the biological significance of this bizarre mode of production so widespread in vertebrates remains elusive, the critical feature might be that snoRNA host-genes must be very actively transcribed, ubiquitous genes. A very large number of intronic and nonintronic snoRNAs belong to the so-called box C/D snoRNA family, because they all contain the short sequence motifs box C (5'-RUGAUGA-3') and box D (5'-CUGA-3'). These boxes correspond to two of the five short sequence elements conserved in U3 snoRNA of all eukaryotes (Maxwell and Fournier, 1995). Boxes C and D, together with a short adjoining helical stem involving the 5'- and 3'-terminal nucleotides of the mature snoRNA, are essential for accumulation of nonintronic yeast snoRNA U14, which is processed from a longer transcript (Huang et al., 1992). The same structural elements are also found at both ends of a large fraction of intron-encoded snoRNAs, in which they are required for faithful intronic RNA processing (Cavaillé and Bachellerie, 1996; Caffarelli et al., 1996; Watkins et al., 1996). All box C/D snoRNAs associate with fibrillarin, an abundant nucleolar protein, which mostly localizes to the dense fibrillar component of the nucleolus, the site of early stages of preribosome formation (Tollervey et al., 1991; Jansen et al., 1991). While box C and D elements are required for fibrillarin association (Baserga et al., 1991; Maxwell and Fournier, 1995), their role in the binding might be indirect. Most, but not all, box C/D snoRNAs contain a long (at least 10-nucleotide [nt]) complementarity to a conserved mature rRNA sequence—hence their name of box C/D antisense snoRNAs (Bachellerie et al., 1995a). Besides U3, the only other box C/D snoRNAs devoid of an uninterrupted complementarity to mature rRNA longer than 10 nt are U8 (Tyc and Steitz, 1989), which is like U3 encoded by an autonomous gene, and intron-encoded U22 (Tycowski et al., 1994). All of these seem to have a peculiar role at specific steps of rRNA processing, unlike the antisense snoRNAs which have turned out to guide the ribose methylations of rRNA. Very recently, a growing number of snoRNAs which do not contain boxes C and D have been identified. Remarkably, they are all structurally related, first by the presence of a conserved ACA trinucleotide motif invariably found 3 nt upstream from their 3' terminus (Balakin et al., 1996; Ganot et al., 1997a). snoRNAs of the ACA family have a mode of formation clearly analogous to that of the box C/D antisense snoRNA family. They do not correspond to primary transcripts, but they are produced, generally from intronic RNA in vertebrates, by an exonucleolytic process exclusively directed by sequences in the mature snoRNA coding region. ACA snoRNAs are associated with the GAR1 nucleolar protein, required for pseudouridylation of *S. cerevisiae* rRNAs (Bousquet-Antonelli et al., 1997). As detailed in Chapter 12 by Ofengand and Fournier, the tantalizing possibility that ACA snoRNAs direct the selection of pseudouridylation sites in rRNA has recently turned out to be true (Ni et al., 1997; Ganot et al., 1997b).

BOX C/D ANTISENSE snoRNAs WITH rRNA COMPLEMENTARITIES MATCHING SITES OF RIBOSE METHYLATION

U14 has been the first, and for a long time the only, box C/D snoRNA with extended, phylogenetically conserved complementarities to mature rRNAs. Unlike most antisense snoRNAs later identified, U14 contains two separate antisense elements (13 and 14 nt long), both matching mature 18S rRNA sequences. Yeast U14 is an essential snoRNA required for the production of 18S rRNA through a mechanism that is still obscure (Liang and Fournier, 1995). This function stringently requires only one of the two antisense elements of U14, domain A, which unlike antisense domain B is not directly involved in guiding an individual 18S rRNA ribose methylation, for reasons detailed below. Another first from the U14 studies was the landmark finding that vertebrate U14 snoRNA was not only intron-encoded but was produced by processing of intronic RNA, the first reported case of this kind (Leverette et al., 1992). This outstanding observation stimulated a search for additional intronic snoRNAs, which rapidly proved fruitful and paved the way for the ultimate elucidation of box C/D antisense snoRNA function. A growing set of box C/D antisense snoRNAs were soon identified after intronic sequence databases were searched for statistically significant tracts of complementarity to rRNA. In 1995, it became clear that antisense elements of box C/D snoRNAs generally encompassed sites of ribose methylation in rRNA, suggesting for the first time their potential guide function in the posttranscriptional modification of rRNA nucleotides (Qu et al., 1995; Bachellerie et al., 1995a). As for the few rRNA complementarities that did not follow the rule, they had not been phylogenetically tested at that time and they later turned out to be probably fortuitous; another antisense, bona fide methylation guide sequence was eventually detected in each "deviant" snoRNA. All doubts as to the significance of the link between rRNA methylation sites and snoRNA complementarities were dissipated in 1996, after the characterization of scores of new box C/D antisense snoRNAs, a large proportion of which had phylogenetically tested rRNA complementarities (Nicoloso et al., 1996). In addition to sequence database searches, a direct method also proved highly valuable for the identification of new family members. Specifically devised for the characterization of intronic nucleolar RNAs, it relied on the fact that these are not primary transcripts but processing products with unusual 5' monophosphate and 3' hydroxyl ends (Kiss-Laszlo et al., 1996). When a low molecular weight fraction of purified nucleolar RNAs is incubated with a 5' phosphorylated oligoribonucleotide in the presence of T4 RNA ligase, only snoRNAs with such unusual termini can be tagged at both ends. Through a reverse transcription/amplification protocol, a specific cDNA library was constructed, which allowed the identification of 20 new box C/D antisense snoRNAs in humans. Their rRNA complementarities were 10–17 nt long and all again matched sites of ribose methylation. Taken together, more than 40 members of this snoRNA family were eventually known in early 1996, and the comparison of their potential duplexes with rRNA ribose methylation sites provided an unambiguous clue as to their function (see Table 2 for a compilation of the methylation guide snoRNAs reported to date).

A CANONICAL STRUCTURE FOR snoRNA/ rRNA DUPLEXES AT RIBOSE METHYLATION SITES

A first striking observation was that rRNA complementarities were far from being uniformly distributed along the 70–100 nt long snoRNA sequences (Bachellerie et al., 1995a). About half of them were located immediately upstream from the box D (CUGA) motif in the 3'-terminal region of the snoRNA, while the others were all found in the 5' half of the snoRNA sequences, at a variable distance downstream from box C (Fig. 3). This bimodal distribution was still observed with a much larger repertoire of antisense snoRNAs. Remarkably, all upstream antisense elements were found to be immediately followed by another CUGA motif, termed box D' (Tycowski et al., 1996a). The importance of the antisense element/CUGA systematic linkage was illuminated by detailed examination of the large collection of snoRNA/rRNA duplexes. In all duplexes the methylated site is found at the same location, always paired to the fifth nucleotide upstream from the CUGA motif (Fig. 4), suggesting that the snoRNA pairing guides the methylation (Nicoloso et al., 1996; Kiss-Laszlo et al., 1996). Reinforcing the notion that each site of ribose methylation in rRNA is associated with a specific snoRNA guide, a few pairs of snoRNAs with overlapping antisense ele-

Table 2. Ribose-methylated sites in eukaryotic rRNAs associated with a known snoRNA guide

Site[a]	snoRNA complementarity[b]	Guide snoRNA(s)[c]
18S rRNA		
Am27	CAU<u>A</u>UGCUUGUC	U27
Am99	GCUC<u>A</u>UUAAAUCAGUU	U57
UM116	UGG<u>U</u>UCCUUUG	U42
Am159	GUA<u>A</u>UUCUAGAGCUAA	U45
AM166	UAG<u>A</u>GCUAAU	U44
Um172	CUA<u>A</u>UACAUGCCGAC	U45
Cm462	ACA<u>U</u>CCAAGGAAGG	U14
Am484	CGC<u>A</u>AAUUAC	U16
Cm517	UGA<u>C</u>GAAAAAUA	U56
Am590	GAGG<u>A</u>UCCAUUGGA	U62
Gm644	AAU<u>A</u>GCGUAUA	U54
Am668	GUU<u>A</u>AAAAGCUCGU	U36
Am1031	AAG<u>A</u>ACGAAAGUCG	U59
Gm1123(y)	CAAGG<u>C</u>UGAAACUU	snR41
Um1326	UGG<u>U</u>GGUGCAUG	U33
Gm1328	GUG<u>G</u>UGCAUGG	U32, snR40
Cm1391	ACU<u>C</u>UGGCAUGCUA	U28
Um1442	AAC<u>U</u>UCUUAGAGGG	U61
Gm1490	ACA<u>G</u>GUCUGUGA	U25
Cm1703	CCG<u>C</u>CCGUCG	U43
Um1804	ACU<u>U</u>GACUAUCUAGAGGAA	U20
28S rRNA		
? Am389	AAG<u>A</u>GAGAGUUCA	U26
? Gm1296	CCC<u>G</u>UCUUGAAAC	U21
Am1306	AAA<u>C</u>ACGGACCA	U18
Gm1501	CCC<u>G</u>AAAGAUGGUG	snR39B
Am1503	CGA<u>A</u>AGAUGGUGA	snR39, U32, U51
Am1849	AGC<u>A</u>GAACUGG	U38
Cm2328	GAU<u>C</u>UUGGUGGU	U24
Am2340	AGU<u>A</u>GCAAAUAU	S. cerevisiae U24
Cm2342	UAG<u>C</u>AAAUAUU	Vertebrate U24
Cm2781	UC<u>U</u>CCAAGGU	U39
Um2814	ACA<u>A</u>UGUAGGUA	U34
?Gm2853	UUC<u>G</u>GGAUAAGG	U50
Am3687	GUC<u>A</u>AAGUGAAGAA	U37
Am3693	GUG<u>A</u>AGAAAUUC	U36
Am3729	GUA<u>A</u>CUAUGAC	U40
Am3754	CAA<u>A</u>UGCCUC	U15
Am2794	UGA<u>A</u>CGAGAUUC	U30
Cm3838	ACA<u>G</u>CCAAGGGAA	U53
Cm3856	UUG<u>G</u>CGGAAUCA	U47
? Gm3868	GCG<u>G</u>GGAAAGAAGA	snR190
Um3894	GAC<u>U</u>CUAGUCUG	U52
*Gm4156	ACU<u>G</u>GGGCGGUA	U31
Gm4188	AGG<u>U</u>GUCCUAAGG	U58
Um4266	UCAG<u>U</u>ACGAAUAC	U41
Gm4330	AgG<u>A</u>GGUGUCAGAA	U60
? Gm2805(y)	ACAG<u>G</u>GAUAACUG	snR38
Cm4416	CGG<u>C</u>UCUUCCUAU	U49
Am4483	GUG<u>A</u>GCUGGGUUU	U29
Cm4496	AGA<u>C</u>CGUCGUGAGA	U35
Am4531	GAU<u>A</u>UGUGUUG	U63

[a] Coordinates are for human rRNAs, except for the sites recognized by snR41 and snR38, for which the yeast *Saccharomyces cerevisiae* position is given. Methylated nucleotides conserved between vertebrates and yeast are in boldface characters. Sites denoted by a question mark are presumptive but probably correspond to some of the few unprecisely mapped ribose-methylated nucleotides in 28S rRNA (Maden, 1990). The site denoted by a star is also ribose methylated in *E. coli* 23S rRNA.
[b] RNA nucleotides conserved between yeast and vertebrates are in boldface characters, and the site of ribose methylation is underlined.
[c] References to the corresponding snoRNA complementarities are listed in Table 1 of the review by Bachellerie and Cavaillé (1997).

ments were detected (Nicoloso et al., 1996). In such cases the rRNA complementarities in the two snoRNAs are shifted from each other by the precise number of nucleotides separating the two vicinal methylations in rRNA, to the effect that each methylation site is paired to the fifth position upstream from box D, in either one of the two mutually exclusive snoRNA/rRNA duplexes (Fig. 5). To date, approximately half of the 100 ribose methylation sites present in human rRNAs are associated to an identified member of the box C/D antisense snoRNA family, with in all cases the ribose methylation sites at the same location relative to box D (or D') in the canonical 10-21 bp duplex with the snoRNA (Table 2). Remarkably, within this large repertoire of box C/D snoRNA antisense elements longer than 10 nt, domain A of U14, which has an essential but still obscure role, remains unique in that it does not pair to a rRNA ribose methylation site and is not imme-

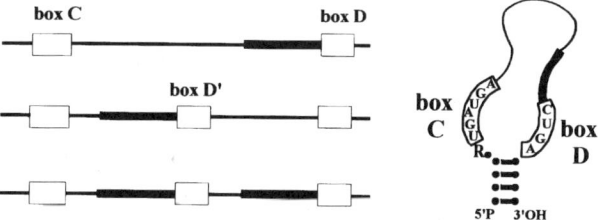

Figure 3. Generic structure of snoRNAs belonging to the box C/D antisense family. Most snoRNAs contain a single antisense element (filled bar), located either immediately upstream from box D (top) or in the 5' half, immediately upstream from another copy of box D, box D' (middle). A few snoRNAs (U24, U32, U36, U45 and U50) contain two different antisense elements (bottom). The 5'-3' terminal stem-box structure shared by most members of the family is schematized on the right.

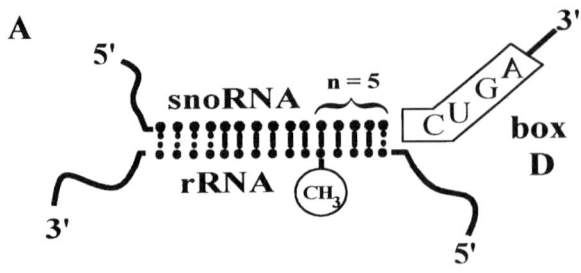

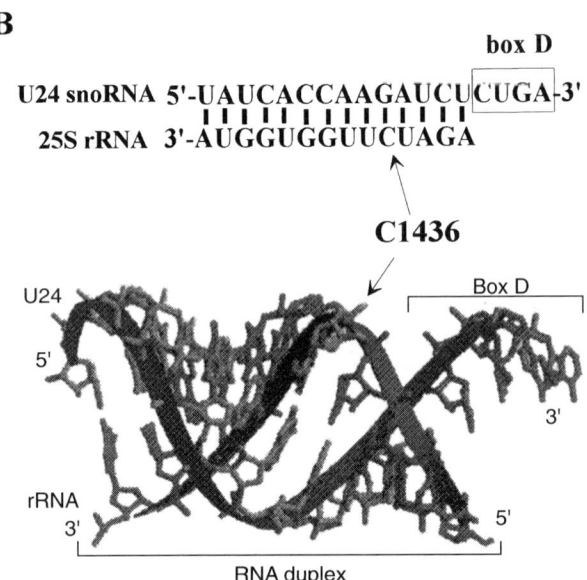

Figure 4. The snoRNA/rRNA guide duplex at ribose methylation sites. (A) Structure of the canonical base pairing (dots denote base pairs not present in all duplexes). (B) Representation of the RNA duplex as an A-form helix, in the case of the U24-25S rRNA interaction. The nucleotides shown are written in the upper part and the site of ribose methylation is denoted by an arrow. (Panel B is reproduced with permission from Tollervey, 1996a.)

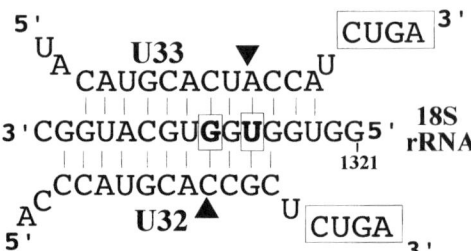

Figure 5. Mutually exclusive snoRNA/rRNA duplexes at adjacent ribose methylation sites. The two sites of ribose methylation in the 18S rRNA sequence (middle line) are boxed. Arrowheads point to the fifth nucleotide upstream from box D in each cognate guide snoRNAs, U32 and U33 (Nicoloso et al., 1996).

diately followed by a CUGA motif. U14, the prototypic box C/D antisense snoRNA, seems definitely to have a more complex role than that of a mere methylation guide, unlike most snoRNAs of this family. Interestingly, in addition to U24 a few other box C/D snoRNAs exhibit a sort of dimeric structure, with the presence of two antisense element/CUGA blocks, each one forming the above-mentioned canonical duplex around a rRNA ribose methylation site (Fig. 3, bottom). Structural comparisons among vertebrates and yeast suggest that recombinations between elementary RNA modules may have played a significant part in the evolution of this snoRNA family, which implies that the identification of faithful homologs among distant species may be problematic in some instances (Nicoloso et al., 1996). For several antisense snoRNA species, homologs have been characterized in different vertebrates, in *Drosophila* and in the yeast *S. cerevisiae*. The antisense element always stands out as the best conserved portion of the snoRNA sequence, in addition to boxes C and D, likely reflecting its major role in snoRNA function (Bachellerie et al., 1995b). The notion inferred from comparative structural analysis that these snoRNAs guide the ribose methylation of rRNA has been unambiguously confirmed by different experimental approaches in yeast and vertebrate cells, as summarized in the following section.

SITE-SPECIFIC rRNA RIBOSE METHYLATION IS DEPENDENT ON THE COGNATE ANTISENSE snoRNA

U24, one of the few box C/D snoRNAs with two separate antisense elements, each one matching ribose methylation sites in 25S rRNA, has been characterized both in vertebrates and in yeast, and found to be intron-encoded in both groups of organisms (Qu et al., 1995). Disruption of the yeast U24 snoRNA coding sequence (Kiss-Laszlo et al., 1996) resulted in the selective disappearance of ribose methylations at the predicted 25S rRNA sites, that is, the positions paired to the fifth nucleotide upstream from box D (or D') in the duplex with each antisense element of U24, C1436 and A1448, and unexpectedly also at G1449. Conversely, ribose methylations at other positions in the large subunit rRNA were not affected. The lack of the three 2'-O-methyl groups in 25S rRNA had no significant effect on the growth rate of yeast cells and on the accumulation of precursor and mature rRNAs. The three ribose methylations of 25S rRNA were fully restored by ectopic expression of U24, following insertion of the snoRNA coding region into the intron of an actin gene in an appropriate expression construct, confirming that the specific methylation defect was directly related to the lack of U24, not to the disruption of its host-gene. As for the unexpected role of U24 in the ribose methylation of G1449, in addition to the predicted vicinal A1448, it suggests that the action of a so-far unidentified snoRNA guiding the methylation of G1449 is tightly dependent on that of U24. The same study also showed that the nucleotides to be methylated in the U24/25S rRNA duplexes were determined by the position of box D, because deletion of one nucleotide adjacent to the CUGA shifted the methylation by one nucleotide, as expected. These conclusions were confirmed by independent analyses of other box C/D antisense snoRNAs, U18, snR38, snR39a/b, snR40, snR41, and snR47 (Ni et al., personal communication), again by gene disruption in yeast, suggesting the possibility that most rRNA ribose methylations could be dispensable for growth. Finally, disruption of box C/D antisense snoRNA U25 in the *Xenopus* oocyte specifically inhibits ribose methylation of the predicted 18S rRNA nucleotide, and the methylation defect can be reversed by microinjecting either the *Xenopus* or human U25 transcript into U25-depleted oocytes (Tycowski et al., 1996b).

THE SITE SPECIFICITY OF RIBOSE METHYLATION IS EXCLUSIVELY DETERMINED BY THE snoRNA/rRNA CANONICAL DUPLEX

In another set of experiments, the rRNA/snoRNA duplex was shown to be the sole determinant of the site of methylation, by directing a novel ribose methylation to an arbitrarily selected nucleotide in endogenous ribosomal RNA, through the expression of an artificial box C/D snoRNA with a tailored antisense element (Cavaillé et al., 1996). An intron-encoded, box C/D antisense snoRNA can be ectopically expressed in mouse cells transfected with a construct in which the snoRNA-containing intron and the two flanking exons of the host-gene are transcribed under the control of the cytomegalovirus promoter. Faithful processing of the intronic snoRNA and its stable cellular accumulation are exclusively directed by the snoRNA terminal stem-box structure and is not dependent on its complementarity region (Cavaillé and Bachellerie, 1996). This allowed for the replacement of the natural antisense element by a 16-nt complementarity to a rRNA sequence naturally devoid of sugar-methylated site in the ribosome, selected so as to provide for an easy and sensitive assay of a novel methylation. Not only ribose methylation was observed at the predicted nucleotide in rRNA, but the reaction was quite extensive. Most rRNA molecules synthesized during the expression of the transfected snoRNA guide were methylated at this site, showing that the rRNA complementarity in the snoRNA was sufficient to specify a methylation site. The snoRNA nucleotides essential for the modification was then further analyzed by mutagenesis and the regularity of a duplex of a minimal length (in this case more than 12 bp) was found to be essential. The fundamental role of the box D motif downstream from the antisense element in the site-specific reaction was also established. Any single mutation in the CUGA dramatically affected the degree of ribose methylation at the target site and deletions modifying the spacing between box D and the antisense element shifted the site of ribose methylation as predicted—to the nucleotide paired to the fifth nucleotide upstream from box D. In the next step, the analysis was switched to the substrate to identify its structural elements essential for the reaction. The rRNA nucleotides that participate in the duplex with the snoRNA were found to be sufficient in the substrate to determine a site-specific ribose methylation. Thus, a 20-nt-long RNA segment of 28S rRNA sequence corresponding to a natural site of ribose methylation in intact endogenous rRNA, i.e., exclusively encompassing the rRNA nucleotides complementary to the cognate snoRNA guide, was faithfully methylated at the expected position, at a high level (more than 0.9 mol of methyl added per mole of RNA substrate), when ectopically expressed from a ribosomal minigene by RNA polymerase I transcription (Cavaillé et al., 1996). As expected, the reaction was completely abolished by mutations in the minisubstrate which prevent formation of the canonical duplex with the endogenous snoRNA guide, and efficiently restored by coexpression of a tailored snoRNA guide bearing the compensatory mutations. Clearly, the snoRNA-guided reaction can be directed to nonribosomal RNA sequences and is not dependent on specific features of the pre-rRNP architecture. Remarkably, when the same minisubstrate was transcribed by RNA polymerase II instead of polymerase I, the degree of site-directed methylation was dramatically decreased, suggesting that RNA polymerase I transcription and/or nucleolar localization of the substrate might be critical for the reaction (Cavaillé et al., 1996). However, ectopically expressed RNA polymerase II transcripts engineered to be directed to the nucleolus can be efficiently ribose methylated, indicating that the snoRNA-guided reaction is not obligatorily linked to RNA polymerase I transcription (Cavaillé, 1997). Given that some cellular pre-mRNAs localize to the nucleolus (Bond and Wold, 1993), this result suggests that targeted ribose methylation could serve to specifically alter gene expression, at the RNA processing level, by means of appropriate tailored snoRNA guides.

BIOSYNTHESIS OF rRNA RIBOSE-METHYLATED NUCLEOTIDES IN PROKARYOTES

The finding that small RNA guides specify the sites of ribose methylation in eukaryotic rRNAs raises the issue that RNA cofactors also could be involved in prokaryotes. Prokaryotic rRNAs have considerably fewer ribose methylations than eukaryotic rRNAs, with only 4 sites in *E. coli* (Table 1). Two prokaryotic rRNA ribose methylases have been characterized so far that are each specific for a single site in rRNA (Cundliffe, 1989; Sirum-Connolly and Mason, 1993). Moreover, using these two enzymes as probes in an iterative search of previously uncharacterized open reading frames in prokaryotic genomes, Gustafsson et al. (1996) have recently identified additional, presumptive rRNA 2'-O-methyltransferases, suggesting that each site of ribose methylation in prokaryotic rRNA has its own associated, site-specific methylase and that guide RNAs have therefore no "raison d'être" in prokaryotes. Consistent with this notion, site-specific methylation can be reproduced in vitro on truncated, naked prokaryotic rRNA substrates by incubation with each of the two purified prokaryotic enzymes without any other added cellular cofactor (Sirum-Connolly and Mason, 1993; Cundliffe, 1989). Obviously, unidentified small RNAs could have copurified with the enzyme or the prokaryotic guide RNA structure could be internal, i.e., provided in *cis* in the rRNA substrate. However, it is noteworthy that in eukaryotic cells an ectopically expressed rRNA substrate did not undergo site-specific ribose methylation when the cognate snoRNA guide sequence was inserted in *cis* into the substrate (Cavaillé, 1997). Computer searches of eubacterial genomes coupled with Northern analyses have failed so far to detect potential homologs of antisense box C/D eukaryotic snoRNAs (Chetouani et al., unpublished results), but negative results are inconclusive at this stage because a hypothetic prokaryotic double-stranded RNA guide structure could substantially differ from the eukaryotic canonical duplex previously described. Nevertheless, any evidence of an RNA-guided process in prokaryotes is still lacking and this type of site selection could well be inherent to the high complexity of the rRNA methylation pattern in eukaryotic cells. Similar conclusions seem to apply for snoRNAs of the ACA family guiding pseudouridylations of eukaryotic rRNA, the pattern of which is also far more complex than in prokaryotic rRNA (see Chapter 12 by Ofengand and Fournier).

CATALYSIS OF THE 2'-O-RIBOSE METHYLATION

The enzymology of the snoRNA-guided reaction is completely unknown, and the protein(s) likely to catalyze the addition of methyl groups onto eukaryotic rRNAs is not identified. Paradoxically, despite their dramatically enlarged number of methylations as compared to prokaryotes, eukaryotes could have a smaller set of rRNA ribose methylases. The presence of a complex repertoire of site-specific, *trans*-acting snoRNA guides sufficient to precisely target the modification through the formation of a common canonical duplex structure with the rRNA substrate strongly suggests that the enzymology of the reaction is quite simple, at least as to the variety of protein enzymes involved. Possibly a single methylase could catalyze the formation of the entire pattern of eukaryotic rRNA ribose methylations. As suggested by a study of 2'-O-methyltransferase modifying position 34 in tRNA (Droogmans et al., 1986), the nature of the base should not have a major impact on the methylation on the 2' hydroxyl of the ribose (see also Chapter 8 by Garcia and Goodenough-Lashua). This is all the more likely in the present case because the base of the rRNA target nucleotide is buried within the RNA double helix of the guide structure. An rRNA 2'-O-ribose methylase activity that methylates all four nucleosides has been partially purified from nucleolar extracts of mammalian cells (Segal and Eichler, 1991) (see also Chapter 8 and Appendix 3). In vitro this activity can selectively ribose methylate a unique cluster of three adjacent nucleotides in a truncated 28S rRNA, whereas other natural sites of ribose methylation in the same substrate are not modified. Remarkably, for one of these three adjacent nucleotides in 28S rRNA, the cognate snoRNA guide U24 has been identified and its requirement for the site-specific reaction in vivo has been experimentally demonstrated (Kiss-Laszlo et al., 1996). Accordingly, the selectivity observed in the above-mentioned in vitro assays might be ascribed to the presence of copurifying snoRNA guides in the active fraction or might merely reflect the preferential accessibility of these sites in a truncated substrate. Undoubtedly, further study of the activity characterized by Segal and Eichler (1991) could help illuminate many questions now raised as to the structure and functioning of the rRNA methylation apparatus and its evolution among the major groups of organisms. Is the methylase a snoRNP-associated protein or is it a diffusible factor which recognizes and transiently binds the preformed canonical double-stranded RNA structure? Does the methylase binding involve direct recognition of the RNA duplex or is it mediated by protein-protein interactions? What are the snoRNP-associated proteins and auxiliary factors in the reaction? One of the few characterized nucleolar proteins that seems to be involved in the process is fibrillarin. Certain mutations in NOP1, the yeast fibrillarin homolog, specifically inhibit the ribose methylations of rRNA (Tollervey et al., 1993). However, the role of fibrillarin in the modification reaction might be indirect, given that the protein also binds box C/D snoRNAs devoid of the hallmark, bipartite, methylating signal—an antisense element located immediately upstream from box D—such as U3, U8 or U22, which are required for pre-rRNA cleavages. Moreover, it does not obviously exhibit the structural signatures of known 2'-O-ribose methylases (Koonin and Rudd, 1993; Gustafsson et al., 1996). Indeed, analysis of temperature-sensitive mutations points to a more complex role of fibrillarin in pre-rRNA processing and ribosome assembly (Tollervey et al., 1993). Intriguingly, fibrillarin-like proteins are present in organisms devoid of a nucleus, belonging to the domain *Archaea* (Amiri, 1994). In this context it is worth mentioning that the intriguing report of a U3-like RNA involved in the processing of 16S rRNA in an archaeal species (Potter et al., 1995), which had raised the tantalizing possibility that homologs of box C/D antisense snoRNAs could exist in non-eukaryotic organisms, has proved to be unsubstantiated (Dennis et al., 1997). What could be the role in the methylation reaction of box D, and also possibly of box C, which are both required for fibrillarin binding? How is the nucleotide paired to the fifth nucleotide upstream from box D selected for ribose methylation? Does the fixed distance reflect a rather direct role of the CUGA motif, which is remarkably close to the target ribose in the 3D structure of the canonical snoRNA/rRNA duplex (see Fig. 4b)? Interestingly, a CUGA is involved in a sharp turn of the hammerhead ribozyme (Pley et al., 1994) and the motif is also strikingly related to the U-turn found in tRNAs (Cech and Uhlenbeck, 1994), suggesting that it might play a direct part in the formation or dissociation of the double-stranded RNA structure. Alternatively, the role of box D in the fixed spacing might be indirect, mediated by a key role of the motif in the snoRNP architecture which could itself, by mere steric hindrance toward the methylase, largely contribute to define the target ribose within the RNA duplex. The amino acid domain structure of the methylase could provide clues on these questions, particularly by reference to the detailed structural information now available on an mRNA 5' cap-specific 2'-O-ribose methyltransferase, the VP39 protein encoded by vaccinia virus (Shi et al., 1996; Hodel et al., 1996) (see also Chapter 8). Searches on the complete yeast *S. cerevisiae* genome sequence may help in identifying the methylase(s), given that the prokaryotic RNA ribose methylases known so far exhibit, in addition to an amino acid motif resembling a binding site for AdoMet (the donor of methyl groups for rRNA ribose methylations), two conserved motifs of unknown role, providing the basis for homology searches in distant organisms (Koonin and Rudd, 1993; Gustafsson et al., 1996).

Box C/D antisense snoRNAs specific for a majority of the vertebrate rRNA ribose methylations have already been identified (Bachellerie and Cavaillé, 1997) and many more are likely to await identification. Nevertheless, the possibility that a few sites in eukaryotic rRNAs are specified by a different mechanism cannot be formally ruled out for the moment. In this regard, the three ribose-methylated sites conserved among prokaryotic and eukaryotic rRNAs obviously deserve particular attention. Remarkably, a vertebrate snoRNA belonging to the antisense box C/D family, U31 (Tycowski et al., 1996a), displays all the structural characteristics expected for guiding one of these, corresponding to Gm2251 in *E. coli* 23S rRNA (Nicoloso et al., 1996). Whether U31 snoRNA is actually required for this universal ribose methylation remains to be checked.

MODIFICATION GUIDE snoRNAs AS RNA CHAPERONES?

snoRNA-guided ribose methylation of rRNA involves the formation of scores of 10–21 bp long intermolecular RNA duplexes along the elongating pre-rRNA transcript in eukaryotes. Similarly, pseudouridylation of the rRNA precursor requires the pairing to the nascent precursor of tens of ACA snoRNAs, through two shorter regions of complementarity to rRNA around each pseudouridylation site (Ganot et al., 1997b; Ni et al., 1997). As a result, the secondary structure of a very large portion of the mature rRNA sequences in the pre-rRNA molecule must be dramatically constrained. The two families of modification guide snoRNAs are therefore likely also to have a major role in directing and/or accelerating pre-rRNA folding, and their binding to the nascent precursor could dramatically affect early stages of ribosome assembly in the nucleolus (Bachellerie et al., 1995a; Steitz and Tycowski, 1995). In line with this notion the thorough inhibition of rRNA methylations, either in mammalian cells treated with cycloleucine (Caboche and Bachellerie, 1977) or in a yeast temperature-sensitive fibrillarin mutant (Tollervey et al., 1993), has been correlated with delays in the synthesis of mature ribosomal subunits. The additional, intrinsic RNA chaperone function of modification guide snoRNAs could be similar to that of the few box C/D snoRNAs that have a role not directly related to rRNA methylation. Thus, for U3 the experimental evidence points to its facilitating the correct folding of pre-rRNA (Abou-Elela et al., 1996; Hughes, 1996) and the same could hold true for domain A of U14. Are the two functions—guide for rRNA methylation and chaperone for rRNA folding—inextricably linked for each snoRNA species in all eukaryotes? Alternatively, in organisms such as those with a minimal number of rRNA methylations, could some snoRNAs of this family have solely the RNA chaperone function, not the associated methylation guide function as in other eukaryotes? While no such case has been reported so far, the identification of antisense snoRNA repertoires and rRNA modification patterns in a larger panel of eukaryotic organisms could help answer this point. Along this line, searches of the complete *S. cerevisiae* genome sequence for hypothetic RNAs carrying antisense elements homologous to that of snoRNA guides for vertebrate-specific ribose methylations of rRNA might be worthwhile.

It would also be particularly interesting to know the relative timing of the complex sequence of snoRNA/rRNA duplex formation and dissociation, relative to key steps of pre-rRNA folding and ribosomal protein assembly. Is there a preferential order for the addition of the different ribose methylations along the nascent pre-rRNA molecule or do they occur stochastically, even those involving the formation of mutually exclusive guide duplexes (such as depicted in Fig. 5)? What about the functioning of the few guide snoRNAs that contain two antisense sequences (Fig. 3, bottom)? For this subset of guides can the same snoRNA molecule, through a single round of pre-rRNA binding, direct two separate methylations located sometimes far apart in the pre-rRNA sequence (Nicoloso et al., 1996)? Alternatively, are the two modification events mechanistically unlinked, with the formation of the two cognate guide duplex structures involving two different snoRNA molecules?

In exponentially growing mammalian cells cultured in vitro, methylation guide snoRNAs are stable RNA species (with a lifetime very similar to that of mature rRNAs), only a minor fraction of which is not bound to nascent pre-rRNA (Cavaillé et al., 1996). Moreover, their copy numbers are all in the same range, at about 10^4 molecules per cell (Nicoloso et al., 1994; Qu et al., 1995; Nicoloso et al., 1996), a value very close to the number of elongating pre-rRNA transcripts. Taken together, these data are consistent with an antisense snoRNA molecule remaining bound to nascent pre-rRNA throughout transcription and with its rapid recycling for a new round of action on another nascent transcript after its release from a methylated rRNA precursor. Dissociation of the long, thermodynamically stable snoRNA/rRNA duplexes is likely to involve RNA helicases (Liang et al., 1997), the characterization of which could illuminate the biological significance of the entire process. Could the RNA helicase action be itself triggered by the addition of the methyl group? Even more far-fetched, might the sole role of some methyl additions be to signal that the snoRNA has properly completed its chaperone role and that the RNA duplex has to be disrupted?

AN rRNA NUCLEOTIDE MODIFICATION FACTORY?

Presumably, many of the elementary modification reactions taking place in rapid succession on the nascent transcript are precisely concerted, possibly through the higher-order organization of a complex modification apparatus incorporating multiple guide snoRNAs, which remains to be characterized. In vertebrates, an actively transcribed ribosomal RNA gene bears about 100 densely packed elongating transcripts, as seen in the well-known "Christmas-tree" structures of Miller spreads (Miller, 1981). For topological reasons it seems likely that the template—and not the polymerase—moves, pulled through fixed polymerization sites, as nascent transcripts are elongated and assembled into closely packed RNP particles (Hozak et al., 1994). Integration of different guide snoRNPs into a highly organized rRNA nucleotide modifying complex, possibly associated with the fixed RNA polymerases, could ensure the efficient coordination of scores of elementary reactions and the constant recycling of active modifying factors onto new transcripts, in addition to the ordered packaging of preribosomal subunits. Along this line it is noteworthy that internal base methylation of mRNA might involve an association of the methylase with nuclear pre-mRNA splicing components (see Chapter 10 by Bokar and Rottman). Further studies of modification guide snoRNPs, focused on their ultrastructural organization in the nucleolar architecture, might elucidate their role(s) in ribosome biogenesis (see also Chapter 24 by Maden).

RNA-GUIDED SITE SELECTION: A COMMON THEME FOR RNA MODIFICATION AND EDITING REACTIONS

The involvement of a double-stranded RNA structure in specifying the target nucleotide makes the two major processes of rRNA modification, ribose methylation and pseudouridylation (see Chapter 12 by Ofengand and Fournier), clearly related to a wide range of RNA editing and modification mechanisms. RNA editing through the insertion or deletion of specific nucleotides in trypanosome mitochondria involves the pairing of a small *trans*-acting guide RNA to the site of RNA alteration (Benne, 1992; Seiwert and Stuart, 1994) (see also Chapter 21 by Hajduk and Sabatini). As for the conversion of adenosine to inosine, which takes place in various cellular and viral RNAs, the modification sites are also generally located in double-stranded RNA structures (Bass, 1997; Polson et al., 1996; Grosjean et al., 1996). In this category, the editing of RNA transcripts encoding glutamate-gated ion channel subunits (GluR) in the brain appears even more particularly related to snoRNA-guided rRNA modifications in that the guide RNA duplex for adenosine/inosine conversion also involves an intronic RNA, acting in *cis* in this case (Cattaneo, 1994) (see also Chapter 19 by Rueter and Emeson). Remarkably, *cis*-acting intronic RNA sequences in the primary transcript also cooperate in the recognition of sites of pseudouridylation in yeast tRNA (Szweykowka-Kulinska et al., 1994; Grosjean et al., 1997) (see also Chapter 16 by Price and Gray), through the formation of an RNA secondary structure reminiscent of that proposed to specify pseudouridylation sites in eukaryotic rRNAs (Ganot et al., 1997b). Finally, very recent results suggest that the conserved ribose methylations of spliceosomal RNAs might also be site-specified by small RNA guides, at least in the case of U6 (Tycowski et al., personal communication).

CONCLUSIONS AND PERSPECTIVES

Elucidation of how the nucleosides to be 2'-O-methylated in eukaryotic rRNAs are specifically recognized should pave the way for rapid progress as to the enzymology of the reaction and the still-elusive role of these modifications in ribosome assembly and function. Remarkably, for the two prevalent types of nucleoside modifications in eukaryotic rRNA, 2'-O-ribose methylations and pseudouridylations, the complex pattern of modifications turns out to be specified in a rather simple way, by base pairing to the rRNA modification site of a specific snoRNA belonging either to the box C/D antisense or the box ACA snoRNA family. The discovery of a complex repertoire of *trans*-acting snoRNA guides, each one specific for a particular nucleotide modification site in rRNA, solves a long-standing riddle and points to a rather simple, common enzymology for the elementary reactions. The new findings also bring about a flurry of new questions and offer many exciting prospects. First of all, what are the mechanisms of the reaction and how is the precise selection achieved, within the double-stranded RNA structure, of the nucleotide to be ribose methylated? What is the actual molecular basis of the "box D + 5 nt" targeting rule? Could a few eukaryotic rRNA ribose methylations be specified by an entirely distinct process? How different is the site selection process for prokaryotic rRNAs? A major issue is now to isolate the protein enzyme(s) catalyzing the eukaryotic rRNA ribose methylation. Analyzing the structure of the methylase(s) and pseudouridine synthase(s) involved in the modification of eukaryotic rRNAs, with particular reference to RNA editing enzymes, also recognizing double-stranded RNA structures (Smith and Sowden, 1996; Hough and Bass, 1997) (see also Chapter 20 by Carter), could be illuminating. What is the biological significance of the complex pattern of rRNA ribose methylations, for which eukaryotic cells have evolved an elaborate system of guide RNAs to ensure a stringent site specificity? The great conservation of this pattern during evolution and the fact that it is restricted to the functionally important regions of the mature rRNA sequences in the nascent pre-rRNA obviously point to a very significant role. However, the synthesis of functional ribosomes does not require a full complement of rRNA 2'-O-ribose methyl groups and the methylation guide snoRNAs tested so far in yeast are dispensable for growth, even those specifying highly conserved ribose methylated sites, consistent with a mere role of rRNA ribose methylation in the fine-tuning of rRNA structure and rRNA-ribosomal protein interactions. Could the fine-tuning be important at the level of ribosome assembly, possibly by accelerating the whole process, or only later, at specific stages of the ribosome lifecycle in the cytoplasm? Remarkably, the way the site selection is achieved not only for ribose methylation but also for pseudouridylation suggests that the two prevalent types of nucleoside modifications in eukaryotic rRNAs could be closely linked to an RNA chaperone action of the snoRNA guides. What is the significance of the intronic location and unusual biosynthetic pathway of the snoRNAs that guide the modification of rRNA nucleotides? Different box C/D snoRNAs (belonging or not to the antisense family) can be found in multiple introns of a single host-gene (Nicoloso et al., 1996; Tycowski et al., 1996a), and the same gene (e.g., nucleolin gene) may host both box C/D and box ACA intronic snoRNAs. As mobile genetic elements (Bachellerie et al., 1995b; Nicoloso et al., 1996), have intronic snoRNAs had a palpable impact on the evolution of eukaryotic genomes, particularly in vertebrates?

These studies also reveal unanticipated relationships between mechanisms of rRNA ribose methylation and other types of posttranscriptional modification and editing processes occurring on different cellular RNA species, for which site selection also involves double-stranded RNA guide structures. Fresh insights into rRNA ribose methylation undoubtedly promise to shed new light on the links between nucleoside modification and RNA editing. Finally, the demonstration that RNA ribose methylation can be targeted at novel sites by expressing tailored snoRNAs with artificial antisense elements (Cavaillé et al., 1996) offers the exciting prospect of using this knowledge for applied purposes. Site-directed ribose methylation of cellular RNAs other than rRNAs could represent a new means for altering gene expression in vivo in a highly selective way, at the posttranscriptional level, by targeting the modification to critically important, reactive 2'-hydroxyl groups along the RNA sequence, such as those that play a key role in pre-mRNA splicing and polyadenylation (Moore and Sharp, 1992; Bardwell et al., 1991). Clearly, the story of intron-encoded snoRNPs is far from being over and we can expect further exciting developments over the years to come, not only as to the mechanisms of rRNA nucleotide modification and ribosome biogenesis, but also in relation to the role of intron-encoded genetic information on the expression and evolution of eukaryotic genomes.

Acknowledgments. This research was supported by the C.N.R.S., by the Université Paul-Sabatier and by a grant (ACC-SV1) from the Ministère de l'Education Nationale, de l'Enseignement Supérieur et de la Recherche (MENESR) to J.-P.B.; J.C. was supported by Ph.D. fellowships from MENESR and the Association pour la Recherche sur le Cancer (ARC).

REFERENCES

Abou-Elela, S., H. Igel, and M. Ares, Jr. 1996. RNase III cleaves eukaryotic preribosomal RNA at a U3 snoRNP-dependent site. *Cell* 85:115-124.

Amiri, K. A. 1994. Fibrillarin-like proteins occur in the domain *Archaea. J. Bacteriol.* 176:2124-2127.

Bachellerie, J.-P., and J. Cavaillé. 1997. Guiding ribose methylation of rRNA. *Trends Biochem. Sci.* 22:257-261.

Bachellerie, J.-P., B. Michot, M. Nicoloso, A. Balakin, J. Ni, and M. J. Fournier. 1995a. Antisense snoRNAs: a family of nucleolar RNAs with long complementarities to rRNA. *Trends Biochem. Sci.* 20:261-264.

Bachellerie, J.-P., M. Nicoloso, L. H. Qu, B. Michot, M. Caizergues-Ferrer, J. Cavaillé, and M. H. Renalier. 1995. Novel intron-encoded small nucleolar RNAs with long sequence complementarities to mature rRNAs involved in ribosome biogenesis. *Biochem. Cell Biol.* 73:835-843.

Bakin, A., and J. Ofengand. 1993. Four newly located pseudouridine residues in *Escherichia coli* 23S ribosomal RNA are all at the peptidyltransferase center: analysis by the application of a new sequencing technique. *Biochemistry* 32:9754-9762.

Balakin, A. G., L. Smith, and M. J. Fournier. 1996. The RNA world of the nucleolus: two major families of small nucleolar RNAs defined by different box elements with related functions. *Cell* 86:823-834.

Balakin, A. G., R. A. Lempicki, G. M. Huang, and M. J. Fournier. 1994. *Saccharomyces cerevisiae* U14 small nuclear RNA has little secondary structure and appears to be produced by post-transcriptional processing. *J. Biol. Chem.* 269:739-746.

Bardwell, V. J., M. Wickens, S. Bienroth, W. Keller, B. S. Sproat, and A. I. Lamond. 1991. Site-directed ribose methylation identifies 2'-OH groups in polyadenylation substrate critical for AAUAAA recognition and poly(A) addition. *Cell* 65:125-133.

Baserga, S. J., X. W. Yang, and J. A. Steitz. 1991. An intact box C sequence in the U3 snRNA is required for binding of fibrillarin, the protein common to the major family of nucleolar snRNPs. *EMBO J.* 10:2645-2651.

Bass, B. L. 1997. RNA editing and hypermutation by adenosine deamination. *Trends Biochem. Sci.* 22:157-162.

Benne, R. 1992. RNA editing in trypanosomes. *Mol. Biol. Rep.* 16:217-227.

Bond, V. C., and B. Wold. 1993. Nucleolar localization of *myc* transcripts. *Mol. Cell. Biol.* 13:3221-3230.

Bousquet-Antonelli, C., Y. Henry, J.-P. Gélugne, M. Caizergues-Ferrer, and T. Kiss. 1997. A small nucleolar RNP protein is required for pseudouridylation of eukaryotic ribosomal RNAs. *EMBO J.* 15:4770-4776.

Brimacombe, R., P. Mitchell, M. Osswald, K. Stade, and D. Bochkariov. 1993. Clustering of modified nucleotides at the functional center of bacterial ribosomal RNA. *FASEB J.* 7:161-167.

Caboche, M., and J.-P. Bachellerie. 1977. RNA methylation and control of eukaryotic RNA biosynthesis. Effects of cycloleucine, a specific inhibitor of methylation, on ribosomal RNA maturation. *Eur. J. Biochem.* 74:19-29.

Caffarelli, E., A. Fatica, S. Prislei, E. De Gregorio, P. Fragapane, and I. Bozzoni. 1996. Processing of the intron-incoded U16 and U18 snoRNAs: the conserved C and D boxes control both the processing and the stability of the mature snoRNA. *EMBO J.* 15:1121-1131.

Cattaneo, R. 1994. RNA duplexes guide base conversions. *Curr. Biol.* 4:134-136.

Cavaillé, J. 1997. Biosynthèse et fonction dans la méthylation de l'ARNr du petit ARN nucléolaire d'origine intronique U20. Ph.D. thesis. Université Paul-Sabatier, Toulouse, France.

Cavaillé, J., and J.-P. Bachellerie. 1996. Processing of fibrillarin-associated snoRNAs from pre-mRNA introns: an exonucleolytic process exclusively directed by the common stem-box terminal structure. *Biochimie* 78:443-456.

Cavaillé, J., M. Nicoloso, and J.-P. Bachellerie. 1996. Targeted ribose methylation of RNA in vivo directed by tailored antisense RNA guides. *Nature* 383:732-735.

Cech, T. R., and O. C. Uhlenbeck. 1994. Hammerhead nailed down. *Nature* 372:39-40.

Chetouani, F., M. Nicoloso, and J.-P. Bachellerie. Unpublished results.

Cundliffe, E. 1989. How antibiotic-producing organisms avoid suicide. *Annu. Rev. Microbiol.* 43:207-233.

Cunningham, P. R., C. J. Weitzmann, D. Nègre, J. G. Sinning, V. Frick, K. Nurse, and J. Ofengand. 1990. In vitro analysis of the role of rRNA in protein synthesis: site-specific mutation and methylation, p. 243-252. *In* W. E. Hill, A. Dahlberg, R. A. Garrett, P. B. Moore, D. Schlessinger, and J. R. Warner (ed.), *The Ribosome: Structure, Function, and Evolution.* American Society for Microbiology, Washington, D.C.

Dennis, P. P., A. G. Russell, and M. M. de Sà. 1997. Formation of the 5' end pseudoknot in small subunit ribosomal RNA: involvement of U3-like sequences. *RNA* 3:337–343.

Droogmans, L., E. Hautmont, S. de Henau, and H. Grosjean. 1986. Enzymatic 2'-O-methylation of the wobble nucleoside of eukaryotic tRNA[Phe]: specificity depends on structural elements outside the anticodon loop. *EMBO J.* 5:1105–1109.

Eichler, D. C., and N. Craig. 1994. Processing of eukaryotic ribosomal RNA. *Prog. Nucleic Acid Res. Mol. Biol.* 49:197–239.

Ganot, P., M. Caizergues-Ferrer, and T. Kiss. 1997a. The family of box ACA snoRNAs is defined by an evolutionarily conserved secondary structure and ubiquitous sequence elements essential for RNA accumulation. *Genes Dev.* 11:946–972.

Ganot, P., M. L. Bortolin, and T. Kiss. 1997b. Site-specific pseudouridine formation in eukaryotic pre-rRNAs is guided by small nucleolar RNAs. *Cell* 89:799–809.

Girard, J.-P., H. Lehtonen, M. Caizergues-Ferrer, F. Amalric, D. Tollervey, and B. Lapeyre. 1992. GAR1 is an essential small nucleolar RNP protein required for pre-rRNA processing in yeast. *EMBO J.* 11:673–682.

Green, R., and H. F. Noller. 1996. In vitro complementation analysis localizes 23S rRNA posttranscriptional modifications that are required for *Escherichia coli* 50S ribosomal subunit assembly and function. *RNA* 2:1011–1021.

Grosjean, H., S. Auxilien, F. Constantinesco, C. Simon, Y. Corda, H. F. Becker, D. Foiret, A. Morin, Y. X. Jin, M. Fournier, and J. L. Fourrey. 1996. Enzymatic conversion of adenosine to inosine and to N1-methylinosine in transfer RNAs: a review. *Biochimie* 78:488–501.

Grosjean, H., Z. Szweykowska-Kulinska, Y. Motorin, F. Fasiolo, and G. Simos. 1997. Intron-dependent enzymatic formation of modified nucleosides in eukaryotic tRNAs: a review. *Biochimie* 79:293–302.

Gustafsson, C., R. Reid, P. J. Greene, and D. V. Santi. 1996. Identification of new RNA modifying enzymes by iterative genome search using known modifying enzymes as probes. *Nucleic Acids Res.* 24:3756–3762.

Hadjiolov, A. A. 1985. *The Nucleolus and Ribosome Biogenesis.* Springer-Verlag, New York, N.Y.

Hodel, A. E., P. D. Gershon, X. Shi, and F. A. Quiocho. 1996. The 1.85 A° structure of vaccinia protein VP39, a bifunctional protein that participates in the modification of both mRNA ends. *Cell* 85:247–256.

Hough, R. F., and B. L. Bass. 1997. Analysis of *Xenopus* dsRNA adenosine deaminase cDNAs reveals similarities to DNA methyltransferases. *RNA* 3:356–370.

Hozak, P., P. R. Cook, C. Schöfer, W. Mosgöller, and F. Wachtler. 1994. Sites of transcription of ribosomal RNA and intranucleolar structure in HeLa cells. *J. Cell Sci.* 107:639–648.

Huang, G. M., A. Jarmolowski, J. C. R. Struck, and M. J. Fournier. 1992. Accumulation of U14 small nuclear RNA in *Saccharomyces cerevisiae* requires box C, box D and a 5', 3' terminal stem. *Mol. Cell. Biol.* 12:4456–4463.

Hughes, J. M. X. 1996. Functional base-pairing interaction between highly conserved elements of U3 small nucleolar RNA and the small ribosomal subunit RNA. *J. Mol. Biol.* 259:645–654.

Jansen, R. P., E. C. Hurt, H. Kern, H. Lehtonen, M. Carmo-Fonseca, B. Lapeyre, and D. Tollervey. 1991. Evolutionary conservation of the human nucleolar protein fibrillarin and its functional expression in yeast. *J. Cell Biol.* 113:715–729.

Khan, M. S. N., M. Salim, and B. E. H. Maden. 1978. Extensive homologies between the methylated nucleotide sequences in several vertebrate ribosomal ribonucleic acids. *Biochem. J.* 169:531–542.

Kiss-Laszlo, Z., Y. Henry, J.-P. Bachellerie, M. Caizergues-Ferrer, and T. Kiss. 1996. Site-specific ribose methylation of preribosomal RNA: a novel function for small nucleolar RNAs. *Cell* 85:1077–1088.

Koonin, E. V., and K. E. Rudd. 1993. SpoU protein of *Escherichia coli* belongs to a new family of putative rRNA methylases. *Nucleic Acids Res.* 21:5519.

Lane, B. G., J. Ofengand, and M. W. Gray. 1995. Pseudouridine and $O^{2'}$-methylated nucleosides. Significance of their selective occurrence in rRNA domains that function in ribosome-catalyzed synthesis of the peptide bonds in proteins. *Biochimie* 77:7–15.

Leverette, R. D., M. T. Andrews, and E. S. Maxwell. 1992. Mouse U14 snRNA is a processed intron of the cognate hsc70 heat-shock pre-messenger RNA. *Cell* 71:1215–1221.

Liang, W.-Q., and M. J. Fournier. 1995. U14 base-pairs with 18S rRNA: a novel snoRNA interaction required for rRNA processing. *Genes Dev.* 9:2433–2443.

Liang, W.-Q., J. A. Clark, and M. J. Fournier. 1997. The rRNA-processing function of the yeast U14 small nucleolar RNA can be rescued by a conserved RNA-helicase-like protein. *Mol. Cell. Biol.* 17:4124–4132.

Liu, J., and E. S. Maxwell. 1990. Mouse U14 snRNA is encoded in an intron of the mouse cognate hsc70 heat shock gene. *Nucleic Acids Res.* 18:6565–6571.

Maden, B. E. H., and M. Salim. 1974. The methylated nucleotide sequences in HeLa cell ribosomal RNA and its precursors. *J. Mol. Biol.* 88:133–164.

Maden, B. E. H. 1986. Identification of the location of the methyl groups in 18S ribosomal RNA from *Xenopus laevis* and man. *J. Mol. Biol.* 189:681–699.

Maden, B. E. H. 1988. Location of methyl groups in 28S rRNA of *Xenopus laevis* and man: clustering in the conserved core of the molecule. *J. Mol. Biol.* 201:289–314.

Maden, B. E. H. 1990. The numerous modified nucleotides in eukaryotic ribosomal RNA. *Prog. Nucleic Acid Res. Mol. Biol.* 39:241–301.

Maden, B. E. H. 1996. Click here for methylation. *Nature* 383:675–676.

Maden, B. E. H., M. E. Corbett, P. A. Heeney, K. Pugh, and P. M. Ajuh. 1995. Classical and novel approaches to the detection and localization of the numerous modified nucleotides in eukaryotic ribosomal RNA. *Biochimie* 77:22–29.

Mason, T. L., C. Pan, M. E. Sanchirico, and K. Sirum-Connolly. 1996. Molecular genetics of the peptidyl transferase center and the unusual Var1 protein in yeast mitochondrial ribosomes. *Experientia* 52:1148–1157.

Maxwell, E. S., and M. J. Fournier. 1995. The small nucleolar RNAs. *Annu. Rev. Biochem.* 35:897–934.

McCallum, F. S., and B. E. H. Maden. 1985. Human 18S ribosomal RNA sequence inferred from DNA sequence. Variations in 18S sequences and secondary modification patterns between vertebrates. *Biochem. J.* 232:725–733.

Miller, O. L. 1981. The nucleolus, chromosomes, and visualization of genetic activity. *J. Cell Biol.* 91:15s–27s.

Moazed, D., and H. F. Noller. 1989. Interaction of tRNA with 23S rRNA in the ribosomal A, P and E sites. *Cell* 57:589–597.

Moore, M. J., and P. A. Sharp. 1992. Site-specific methylation of pre-mRNA: the 2'-hydroxyl groups at the splice sites. *Science* 256:992–997.

Nègre, D., C. Weitzmann, and J. Ofengand. 1989. In vitro methylation of *Escherichia coli* 16S ribosomal RNA and 30S ribosomes. *Proc. Natl. Acad. Sci. USA* 86:4902–4906.

Ni, J., A. G. Balakin, and M. J. Fournier. Personal communication.

Ni, J., A. L. Tien, and M. J. Fournier. 1997. Small nucleolar RNAs direct site-specific synthesis of pseudouridine in ribosomal RNA. *Cell* 89:565–573.

Nicoloso, M., M. Caizergues-Ferrer, B. Michot, M. C. Azum, and J.-P. Bachellerie. 1994. U20, a novel small nucleolar RNA, is encoded in an intron of the nucleolin gene in mammals. *Mol. Cell. Biol.* 14:5766–5776.

Nicoloso, M., L.-H. Qu, B. Michot, and J.-P. Bachellerie. 1996. Intron-encoded, antisense small nucleolar RNAs: the characterization of nine novel species points to their direct role as guides for the 2′-O-ribose methylation of rRNAs. *J. Mol. Biol.* 260: 178–195.

Noller, H. F., T. Powers, P. N. Allen, D. Moazed, and S. Stern. 1996. Ribosomal RNA and translation: tRNA selection and movement in the ribosome, p. 239–258. *In* R. A. Zimmermann and A. E. Dahlberg (ed.), *Ribosomal RNA. Structure, Evolution, Processing, and Function in Protein Biosynthesis.* Telford, Caldwell, N.J.

Ofengand, J., A. Bakin, J. Wrzesinski, K. Nurse, and B. G. Lane. 1995. The pseudouridine residues of ribosomal RNA. *Biochem. Cell Biol.* 73:915–924.

Pley, H. W., K. M. Flaherty, and D. B. McKay. 1994. Three dimensional structure of a hammerhead ribozyme. *Nature* 372: 68–74.

Polson, A. G., B. L. Bass, and J. L. Casey. 1996. RNA editing of hepatitis delta antigenome by dsRNA-adenosine deaminase. *Nature* 380:454–456.

Potter, S., P. Durovic, and P. P. Dennis. 1995. Ribosomal RNA precursor processing by a eukaryotic U3 small nucleolar RNA-like molecule in an archaeon. *Science* 268:1056–1060.

Prislei, S., A. Michienzi, C. Presutti, P. Fragapane, and I. Bozzoni. 1993. Two different snoRNAs are encoded in introns of amphibian and human L1 ribosomal protein genes. *Nucleic Acids Res.* 21:5824–5830.

Qu, L. H., M. Nicoloso, B. Michot, M. C. Azum, M. Caizergues-Ferrer, M. H. Renalier, and J.-P. Bachellerie. 1994. U21, a novel small nucleolar RNA with a 13 nt. complementarity to 28S rRNA, is encoded in an intron of ribosomal protein L5 gene in chicken and mammals. *Nucleic Acids Res.* 22:4073–4081.

Qu, L. H., Y. Henry, M. Nicoloso, B. Michot, M. C. Azum, M. H. Renalier, M. Caizergues-Ferrer, and J.-P. Bachellerie. 1995. U24, a novel intron-encoded small nucleolar RNA with two 12 nt long, phylogenetically conserved complementarities to 28S rRNA. *Nucleic Acids Res.* 23:2669–2676.

Renalier, M. H., M. Nicoloso, L. H. Qu, and J.-P. Bachellerie. 1996. SnoRNA U21 is also intron-encoded in *Drosophila melanogaster* but in a different host-gene as compared to warm-blooded vertebrates. *FEBS Lett.* 379:212–216.

Schnare, M. N., S. H. Damberger, M. Gray, and R. R. Gutell. 1996. Comprehensive comparison of structural characteristics in eukaryotic cytoplasmic large subunit (23S-like) ribosomal RNA. *J. Mol. Biol.* 256:701–719.

Segal, D. M., and D. C. Eichler. 1991. A nucleolar 2′-O-methyltransferase. Specificity and evidence for its role in the methylation of mouse 28S precursor ribosomal RNA. *J. Biol. Chem.* 266: 24385–24389.

Seiwert, S. D., and K. Stuart. 1994. RNA editing: transfer of genetic information from gRNA to precursor mRNA in vitro. *Science* 266:114–117.

Shi, X., P. Yao, T. Jose, and P. D. Gershon. 1996. Methyltransferases specific domains within VP39, a bifunctional protein that participates in the modification of both mRNA ends. *RNA* 2: 88–101.

Simpson, L., and D. A. Maslov. 1994. RNA editing and the evolution of parasites. *Science* 264:1870–1871.

Sirum-Connolly, K., and T. L. Mason. 1993. Functional requirement of a site-specific ribose methylation in ribosomal RNA. *Science* 262:1886–1889.

Sirum-Conolly, K., J. M. Peltier, P. F. Crain, J. A. McCloskey, and T. L. Mason. 1995. Implications of a functional large ribosomal RNA with only three modified nucleotides. *Biochimie* 77:30–39.

Smith, H. C., and M. P. Sowden. 1996. Base-modification mRNA editing through deamination—the good, the bad and the unregulated. *Trends Genet.* 12:418–424.

Smith, J. E., B. S. Cooperman, and P. Mitchell. 1992. Methylation sites in *E. coli* ribosomal RNA: localization and identification of four new sites of methylation in 23S rRNA. *Biochemistry* 31: 10825–10834.

Smith, C. M., and J. A. Steitz. 1997. Sno storm in the nucleolus: new roles for myriad small RNPs. *Cell* 89:669–672.

Sollner-Webb, B. 1993. Novel intron-encoded small nucleolar RNAs. *Cell* 75:403–405.

Sollner-Webb, B., K. Tyc, and J. A. Steitz. 1995. Ribosomal RNA processing in eukaryotes, p. 469–490. *In* R. A. Zimmermann and A. E. Dahlberg (ed.), *Ribosomal RNA. Structure, Evolution, Processing, and Function in Protein Biosynthesis.* Telford, Caldwell, N.J.

Steitz, J. A., and K. T. Tycowski. 1995. Small RNA chaperones for ribosome biogenesis. *Science* 270:1626–1627.

Szweykowska-Kulinska, Z., B. Senger, G. Keith, F. Fasiolo, and H. Grosjean. 1994. Intron-dependent formation of pseudouridines in the anticodon of *Saccharomyces cerevisiae* minor tRNA^{Ile}. *EMBO J.* 13:4636–4644.

Tollervey, D. 1996a. Small nucleolar RNAs guide ribosomal RNA methylation. *Science* 273:1056–1057.

Tollervey, D. 1996b. Trans-acting factors in ribosome biogenesis. *Exp. Cell Res.* 229:226–232.

Tollervey, D., H. Lehtonen, M. Carmo-Fonseca, and E. C. Hurt. 1991. The small nucleolar RNP protein NOP1 (fibrillarin) is required for pre-rRNA processing in yeast. *EMBO J.* 10:573–583.

Tollervey, D., H. Lehtonen, R. Jansen, H. Kern, and E. C. Hurt. 1993. Temperature-sensitive mutations demonstrate roles for yeast fibrillarin in pre-rRNA processing, pre-rRNA methylation, and ribosome assembly. *Cell* 72:443–457.

Trinh-Rohlik, Q., and E. S. Maxwell. 1988. Homologous genes for mouse 4.5S hybRNA are found in all eukaryotes and their low molecular weight transcripts intermolecularly hybridize with eukaryotic 18S ribosomal RNAs. *Nucleic Acids Res.* 16: 6041–6056.

Tyc, K., and J. A. Steitz. 1989. U3, U8 and U13 comprise a new class of mammalian snRNPs localized in the cell nucleolus. *EMBO J.* 8:3113–3119.

Tycowski, K. T., M.-D. Shu, and J. A. Steitz. 1994. Requirement for intron-encoded U22 small nucleolar RNA in 18S rRNA maturation. *Science* 266:1558–1561.

Tycowski, K. T., M.-D. Shu, and J. A. Steitz. 1996a. A mammalian gene with introns instead of exons generating stable RNA products. *Nature* 379:464–466.

Tycowski, K. T., C. M. Smith, M.-D. Shu, and J. A. Steitz. 1996b. A small nucleolar RNA requirement for site-specific ribose methylation of rRNA in *Xenopus*. *Proc. Natl. Acad. Sci. USA* 93: 14480–14485.

Tycowski, K. T., Z. You, and J. A. Steitz. Personal communication.

Vaughan, M. H., R. Soeiro, J. R. Warner, and J. E. Darnell. 1967. The effects of methionine deprivation on ribosome synthesis in HeLa cells. *Proc. Natl. Acad. Sci. USA* 58:1527–1534.

Veldman, G. M., J. Klootwijk, V. C. H. F. De Regt, R. J. Planta, C. Branlant, A. Krol, and J. P. Ebel. 1981. The primary and secondary structure of yeast 26S rRNA. *Nucleic Acids Res.* 9: 6935–6952.

Venema, J., and D. Tollervey. 1995. Processing of pre-ribosomal RNA in *Saccharomyces cerevisiae*. *Yeast* 11:1629–1650.

Visomirski-Robic, L. M., and J. M. Gott. 1997. Insertional editing in isolated *Physarum* mitochondria is linked to RNA synthesis. *RNA* 3:821–837.

Warner, J. R. 1989. Synthesis of ribosomes in *Saccharomyces cerevisiae*. *Microbiol. Rev.* 53:256–271.

Watkins, N. J., R. D. Leverette, L. Xia, M. T. Andrews, and E. S. Maxwell. 1996. Elements essential for processing intronic U14 snoRNA are located at the termini of the mature snoRNA sequence and include conserved nucleotide boxes C and D. *RNA* 2:118–133.

Chapter 14

Functional Aspects of the Three Modified Nucleotides in Yeast Mitochondrial Large-Subunit rRNA

Thomas L. Mason

The ribosome is the universal organelle of biological protein synthesis, and its fundamental structural and functional properties have been highly conserved in evolution. The ribosomes in prokaryotes and in the cytoplasm of eukaryotes are required for the synthesis of thousands of different cellular proteins; hence, cell growth is highly dependent on efficient ribosome formation and function. In contrast, the ribosomes in eukaryotic organelles, mitochondrial ribosomes in particular, typically have a much more limited role. For example, the mitochondrial ribosomes in the budding yeast *Saccharomyces cerevisiae* produce only eight major polypeptides that are required for aerobic energy metabolism but not for cell viability. Mitochondrial ribosomes therefore have evolved under markedly different constraints than those imposed on cytoplasmic ribosomes, and the unique features of mitochondrial ribosomes are probably evolutionary adaptations to their limited and specialized role in the cell. On the other hand, many highly conserved features of mitochondrial ribosomes highlight structure-function relationships that are potentially important in all ribosomes.

One of the most striking variables among mitochondrial, eubacterial, and eukaryotic cytoplasmic ribosomes is the extent of posttranscriptional modification of the rRNAs. Yeast mitochondrial rRNAs are considered to be minimally modified because the small-subunit rRNA (SSU rRNA) apparently is not modified at all, although the presence of possibly one pseudouridine (Ψ) has not been rigorously excluded, and the large-subunit rRNA (LSU rRNA) contains only two 2'-O-ribose methylated nucleosides and one Ψ (Klootwijk et al., 1975; Sirum-Connolly et al., 1995). In *Escherichia coli*, the rRNAs contain approximately 35 modified nucleosides (see Sirum-Connolly et al., 1995), and the cytoplasmic rRNAs of higher eukaryotes have over 3% modified nucleotides (see Chapter 24 by Maden). The clustering of modified nucleotides in the conserved core regions of the rRNAs suggests a role in the formation or activity of important functional centers of the ribosome. In the context of the evidence that rRNA is intimately involved in the catalysis of peptide bond formation (for recent reviews, see Lieberman and Dahlberg, 1995; Garrett and Rodriguez-Fonseca, 1996; and Green and Noller, 1997), it is significant that the peptidyl transferase center (PTC) of domain V in LSU rRNAs is modification rich, and the three modified nucleotides in yeast mitochondrial LSU rRNA are all located at highly conserved positions in the PTC.

This chapter focuses on the possible functional significance of the three highly conserved modified nucleotides in the otherwise minimally modified yeast mitochondrial LSU rRNA. The modification of mitochondrial rRNAs in other organisms has been reviewed recently (Sirum-Connolly et al., 1995) and is also covered in other chapters in this volume (see Chapter 1 by Lane, Chapter 12 by Ofengand and Fournier, and Chapter 24 by Maden).

THE THREE MODIFIED NUCLEOTIDES IN YEAST LSU rRNA ARE LOCATED AT STRATEGIC POSITIONS IN THE PTC

The three modified nucleotides in yeast mitochondrial LSU rRNA, Gm2270, Um2791, and Ψ2819, are equivalent to the Gm2251, Um2552, and Ψ2580, respectively, in PTC of *E. coli* LSU rRNA. Because the sequence and secondary structure of the PTC are very highly conserved, it seems valid to extrapolate from what is known about the PTC in the prototype *E. coli* ribosome to the eubacterial-like ri-

---

Thomas L. Mason • Department of Biochemistry and Molecular Biology, University of Massachusetts, Amherst, Massachusetts 01003.

bosomes in yeast mitochondria, and even to ribosomes in general. Significantly, as shown in Fig. 1, each position in *E. coli* LSU rRNA that is equivalent to the three modified nucleotides in yeast mitochondrial LSU rRNA has been implicated in the interaction between the ribosome and the 3′-end of either P-site- or A-site-bound tRNA (for a recent review, see Green and Noller, 1997).

Gm2251 (*E. coli* Numbering)

The G nucleotides at positions 2251, 2252, and 2253 in the 2250 loop in the *E. coli* LSU rRNA are universally conserved, and these nucleotides are protected from chemical probing by the 3′-terminal CCA of tRNA bound in the P site (Moazed and Noller, 1989; Green et al., 1997). Mutations at Gm2251 (Green et al., 1997; Gregory and Dahlberg, personal communication) and G2252 cause a dominant lethal phenotype in cells expressing the mutant rRNAs from a plasmid, whereas mutations at G2253 cause less severe phenotypes (Gregory et al., 1994; Samaha et al., 1995; Porse et al., 1996). Most importantly, a 3′-end fragment of tRNA is unable to bind to ribosomes with mutations at either Gm2251 or G2252, and binding is restored in G2252 mutant ribosomes by compensatory Watson–Crick substitutions at C74 in the tRNA fragment (Samaha et al., 1995; Porse et al., 1996). Substitutions at either C74 or C75 in the tRNA fragment do not offset the effects of the rRNA mutations at Gm2251 (Green et al., 1997). Thus, the CCA end of P-site-bound tRNA interacts strongly with the three universally conserved G nucleotides in the 2250 loop, including Gm2251. This interaction is stabilized by Watson–Crick base pairing between C74 in the tRNA and G2252 in the rRNA.

Although Gm2251 apparently does not interact with the 3′-end of tRNA through Watson–Crick base pairing, it is nevertheless essential for the formation of a functional P site. Substitutions at G2251 do not interfere with the in vivo assembly of 50S subunits or the formation of tight-couple 70S ribosomes (Green et al., 1997; Gregory and Dahlberg, personal communication), and the dominant lethal phenotype caused by these mutations has been attributed to a deficiency in peptidyl transferase activity (Green et al., 1997). Interestingly, the substitutions at G2251 enhance the chemical reactivities of U2584 and U2586 in the mutant ribosomes (Green et al., 1997), suggesting that the 2250 loop interacts directly or indirectly with nucleotides in close proximity to Ψ2580.

Base substitutions at G2251 could affect 2′-O-methylation at this position, and conceivably, loss of methylation might be a determining factor in the phenotypes of G2251 mutants. To examine this possibility, the extent of modification at 2251 was compared in wild-type and mutant rRNAs using a primer extension assay (Maden et al., 1995; Sirum-Connolly et al., 1995). The results showed that in vivo modification of the A2251 and U2251 variants is roughly equivalent to wild type whereas modification of C2251 is reduced by approximately 50% (Gregory et al., unpublished results). Clearly, the lack of 2′-O-methylation cannot account for the strong phenotypes associated with base substitutions at 2251. These results also show that the *E. coli* G2251 2′-O-methyltransferase does not have strong specificity for G at position 2251.

Um2552 (*E. coli* Numbering)

Um2552 and the adjacent nucleotide G2553 are universally conserved, and this dinucleotide has been hypothesized to base pair with the CA at the 3′-end of aminoacyl-tRNA bound in the A site (Moazed and Noller, 1989). G2553 is protected from chemical modification by tRNA bound in the A site and this protection is dependent on the presence of the A at the 3′-end of the tRNA (Moazed and Noller, 1989). Protection of Um2552 could not be examined because the modified nucleotide itself blocks reverse transcription. Puromycin also weakly protects G2553 (cited in Green and Noller, 1997). These footprinting studies strongly implicate the Um2552–U2555 region in the interaction with the 3′-end of the A-site-bound tRNA. However, Um2552A and Um2552C mutants do not have dominant lethal phenotypes, and the mutant ribosomes retain significant levels of peptidyl transferase activity in vitro (Porse and Garret, 1995). Therefore, if Watson–Crick base pairing occurs between Um2552 and A76 in the aminoacyl-tRNA, it does not appear to be essential for a functional PTC.

Ψ2580 (*E. coli* Numbering)

Uridine is universally conserved at position 2580, but pseudouridine formation is highly variable at this position although all LSU rRNAs examined have one or more pseudouridines within or very near the central loop in domain V (Ofengand and Bakin, 1997) (see Chapter 12 by Ofengand and Fournier). A variety of evidence implicates the region from Ψ/U2580 to U2585 in the binding of the 3′-end of tRNA bound in the P site. The phylogenetically invariant residues U2584 and U2585 are protected from chemical modification by P-site-bound tRNA, and this protection requires the presence of the 3′-terminal A in the tRNA (Moazed and Noller, 1989). Mutations at U2585 (Green et al., 1997; Porse et al.,

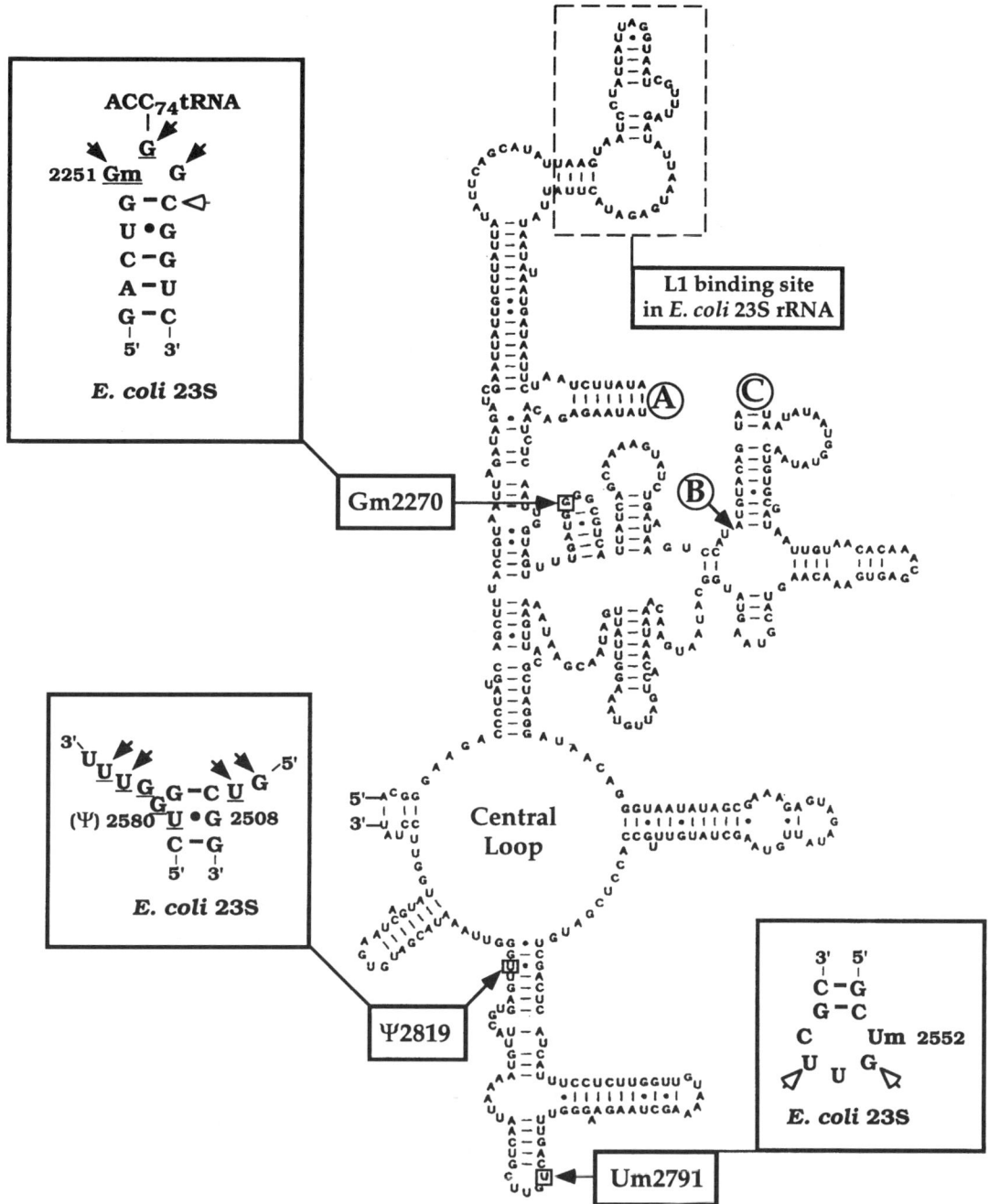

Figure 1. The three modified nucleotides in yeast mitochondrial LSU rRNA are located at positions implicated in binding the 3′-terminal end of either A-site- or P-site-bound tRNA. The secondary structure is that of domain V from *S. cerevisiae* mitochondrial LSU rRNA. The circled A, B, and C indicate the positions of variable sequences that are not shown. The locations of three modified nucleotides are indicated by arrows, and the sequences corresponding to the equivalent modifications in *E. coli* 23S rRNA are boxed. The closed arrowheads indicate sites that are protected from chemical modification by P-site-bound tRNA; the open arrowheads indicate protections by A-site-bound tRNA; positions where base substitutions cause strong dominant negative growth phenotypes are underlined. The region corresponding to the binding site for ribosomal protein L1 in *E. coli* 23S rRNA is enclosed by the dashed line. See the text for references.

1996), U2584 (Porse and Garrett, 1995; Porse et al., 1996), G2583 (Saarma and Remme, 1992; Porse et al., 1996), G2582 (Porse et al., 1996; cited in Spahn et al., 1996b), and G2581 (Spahn et al., 1996a, 1996b; Porse et al., 1996) cause strong growth phe-

notypes and affect in vitro peptidyl transferase activity to variable extents. The phenotypes associated with substitution at U2580 indicate that Ψ2580 is not essential for PTC activity (Porse and Garrett, 1995; Spahn et al., 1996a, 1996b). Compensatory

Watson–Crick substitutions at A76 and C75 of tRNA do not suppress the effects of substitutions at U2585 (Porse et al., 1996; Green et al., 1997) and U2581 (Porse et al., 1996; Spahn et al., 1996b), respectively. Therefore, while the U/Ψ2580–U2585 region is clearly implicated in binding the CCA 3′-end of tRNA, this rRNA–tRNA interaction apparently does not involve canonical base pairing or have an absolute requirement for Ψ2580.

MODIFIED NUCLEOTIDES ARE REQUIRED FOR THE RECONSTITUTION OF FUNCTIONAL E. coli 50S RIBOSOMAL SUBUNITS

Functional E. coli 30S and 50S ribosomal subunits can be reconstituted in vitro from their respective component natural rRNAs and ribosomal proteins (Traub and Nomura, 1968; Nierhaus and Dohme, 1974), and it has also been possible to reconstitute functional small ribosomal subunits, although at reduced efficiency, using unmodified in vitro transcripts of E. coli SSU rRNA (Krzyzosiak et al. 1987). In contrast, there has been no success in the reconstitution of functional 50S subunits from in vitro transcripts of E. coli LSU rRNA, although in vitro transcripts can be reconstituted into particles that bear some physical resemblance to native 50S subunits (Green and Noller, 1996). This suggests that the assembly of catalytically active 50S subunits has a strict requirement for posttranscriptionally modified nucleotides.

To localize modifications in E. coli LSU rRNA that are required for the in vitro assembly of 50S subunits, Green and Noller (1996) developed a novel chimeric rRNA reconstitution system in which partial fragments of natural rRNA are combined with partial in vitro transcripts. Surprisingly, the requirement for natural rRNA could be narrowed down to a short stretch of approximately 80 nucleotides in domain V (from position 2445 to 2523), which contains six posttranscriptional modifications (Kowalak et al., 1995), none of which is universally conserved. Thus, the three modifications that are conserved in yeast mitochondrial LSU rRNA are not essential for either in vitro reconstitution or peptidyl transferase activity. The fact that peptidyl transferase activity can be reconstituted with rRNAs devoid of universally conserved modifications means that the catalysis of peptide bond formation is not directly dependent on any specific nucleotide modification. It remains to determined, however, whether these in vitro studies are relevant to the possible requirement for modified nucleotides in ribosome assembly in vivo.

PET56 ENCODES AN ESSENTIAL rRNA RIBOSE METHYLTRANSFERASE

The PET56 nuclear gene encodes an rRNA ribose methyltransferase (Pet56p) required for the formation of 2′-O-methylguanosine at G2270 in yeast mitochondrial LSU rRNA (G2251 in E. coli numbering), and PET56 function is essential for the formation of functional mitochondrial ribosomes (Sirum-Connolly and Mason, 1993). Sucrose gradient analysis of mitochondrial ribosomes isolated from conditional pet56 mutants showed that loss of PET56 function causes a dramatic decrease in the recovery of mitochondrial large ribosomal subunits and a concomitant appearance of slowly sedimenting particles containing large-subunit ribosomal proteins (Sirum-Connolly et al., 1995). From these results, it appears that PET56 function could be required either at an early step in the assembly pathway or possibly for the stability of large-subunit particles during isolation. The loss of PET56 function does not affect the accumulation of full-length LSU rRNA (Sirum-Connolly and Mason, unpublished data), indicating that the formation of Gm2770 is not required for processing of the primary rRNA transcript or for protecting the LSU rRNA from nucleolytic cleavage.

Base substitutions at Gm2251 in E. coli do not affect LSU assembly (Green et al., 1997), so it is surprising that inactivation of the PET56 methyltransferase has such a dramatic effect on the assembly or stability of the yeast mitochondrial LSU particle. It is important to note, however, that base substitutions at Gm2251 in E. coli do not block 2′-O-methylation at position 2251. Therefore, the mutational studies in E. coli do not exclude the possibility that a 2′-O-methyl group at 2251 is a critical determinant for subunit assembly in vivo.

How could a single CH_3 group have a major impact on the structure of a $>1.5 \times 10^6$ Da macromolecular complex? Conceivably, the local conformation of the 2250 loop region (E. coli numbering) could be affected by a number of factors, including the hydrophobicity of the methyl group, the loss of hydrogen bonding at the 2′-OH, or the stabilization of the C3′-endo form of the ribose. Such local changes could in turn enhance or restrict specific intramolecular and intermolecular interactions. For example, structural changes at position 2251 could perturb the intramolecular interaction between the 2250 loop and the U2584 to U2586 region in the central loop of domain V (Samaha et al., 1995; Green et al.,

1997). The presence of a 2'-O-methyl at 2251 could also prevent forbidden interactions that disrupt the normal folding pathway.

Intermolecular interaction between rRNA and a ligand can be profoundly affected by a single modification, as is exemplified by the 2'-O-methylation at A1067 in LSU rRNA that blocks the binding of the modified peptide antibiotic thiostrepton (for example, see Thompson and Cundliffe, 1991). The Gm2251-specific rRNA methyltransferases and the 3'-end of tRNA are the only identified candidates for intermolecular interactions with Gm2551 and the 2250 loop, and this region of LSU rRNA is not known to be a binding site for any ribosomal protein, although contacts with a ribosomal protein cannot be ruled out. The presence or absence of methylation at 2251 could also influence interactions with an auxiliary protein factor similar to DbpA, which is believed to facilitate the in vivo folding of *E. coli* LSU rRNA (Nicol and Fuller-Pace, 1995). At present, there is no evidence to support any of these purely hypothetical consequences of the presence or absence of 2'-O-methylation at 2251.

Is defective LSU assembly in *pet56* mutants caused by the lack of methylation or by inactivation of the Pet56 protein? If Pet56p is bifunctional, serving both as a methyltransferase and as an rRNA chaperone, then defective LSU assembly could be because of the loss of the chaperone function, not the absence of the 2'-O-methyl group. The importance of distinguishing between the role(s) of the modification enzyme and that of the modification itself is exemplified by the function of *DIM1* in the posttranscriptional modification and processing of yeast cytoplasmic rRNA (Lafontaine et al., 1995) (see also Chapter 1 by Lane). Dim1p catalyzes the $m^6_2Am^6_2A$ dimethylation at the 3'-end of SSU rRNA, and depletion of Dim1p causes the concomitant loss of dimethylation and a block in endonucleolytic processing of pre-rRNA. This suggests that dimethylation might be a prerequisite to processing. However, a *cis*-acting mutation that replaces both dimethylated adenosines with guanosines prevents methylation at this site but has no apparent effect on in vivo rRNA processing. Therefore, the presence of Dim1p, not the actual $m^6_2Am^6_2A$ modification, is the critical factor for pre-rRNA processing. The interaction between Dim1p and pre-rRNA is apparently a checkpoint in the processing pathway that serves to prevent the incorporation of nondimethylated SSU rRNA into ribosomal subunits (see Chapter 15 by Lafontaine and Tollervey). This type of monitoring mechanism could be very important if the lack of a particular modification compromises ribosomal function. It remains to be determined whether the methyltransferase activity of Pet56p can be uncoupled from a possible involvement in LSU assembly.

AN EXTRAGENIC MUTATION SUPPRESSES THE REQUIREMENT FOR *PET56* FUNCTION

While *PET56* is normally essential for the formation of functional yeast mitochondrial ribosomes, a dominant extragenic mutation has been isolated (*SRM1-1*) that suppresses, albeit very weakly, *pet56* loss-of-function mutations without restoring methylation at G2270 in yeast mitochondrial LSU rRNA (Sirum-Connolly and Mason, unpublished data). Although the identity of the suppressor gene is not yet known, a genetic interaction was detected between the suppressor allele and the gene for the mitochondrial homolog of *E. coli* L1 ribosomal protein (Mason et al., 1996), that is, the gene for mitochondrial L1 is a multicopy suppressor of an independent phenotype associated with the *SRM1-1* allele. *E. coli* L1 binds to LSU rRNA in the peptidyl transferase region of domain V (Fig. 1), less than 60 nucleotides upstream of Gm2251, and is involved in tRNA binding to the P site (see Draper, 1996). L1 also cross-links to L33 (Walleczek et al., 1989), suggesting that the suppression activity of *SRM1-1* might be because of a mutation in the *MRPL39* gene for mitochondrial L33. However, the nucleotide sequence of *MRPL39* and its flanking DNA in the *SRM1-1* mutant is identical to the published sequence of wild-type *MRPL39* (Sirum-Connolly and Mason, unpublished data). Even though the *pet56* suppressor analysis is not complete, the available data support the important conclusion that neither 2'-O-methylation at G2270 nor Pet56p itself is absolutely essential for the in vivo formation and function of the PTC. This is in general agreement with the conclusions reached in the analysis of *E. coli* large subunits reconstituted from partial in vitro transcripts (Green and Noller, 1996).

CONSERVATION OF ENZYMATIC ACTIVITIES THAT CATALYZE THE THREE NUCLEOTIDE MODIFICATIONS IN YEAST MITOCHONDRIAL LSU rRNA

The G at position 2270 in yeast mitochondrial rRNA is universally conserved and the 2'-O-methylation of this nucleotide also appears to be widespread (Lane et al., 1995). The *E. coli yjfH* gene product has been identified as a probable homolog of the Pet56p methyltransferase (Gustafsson et al., 1996), and genes related to *yjfH* are present in the genomes of several other bacteria. This opens the way to genetic and biochemical approaches to confirm whether the YjfH protein is the Gm2251 methyltransferase, and, if so, to ascertain the functional significance of Pet56-like enzymes in *E. coli* and other bacteria.

Site-specific 2'-O-methylation of eukaryotic cytoplasmic rRNAs is guided by small nucleolar RNAs (snoRNAs) that are complementary to the rRNA at the site of modification (see Chapter 13 by Bachellerie and Cavaillé). The mouse and human U31 snoRNAs fit the criteria for directing 2-O'-methylation at the position equivalent to *E. coli* Gm2251 (Nicoloso et al., 1996; Tycowski et al., 1996). Although this modification has not been unambiguously mapped in yeast or vertebrate LSU rRNAs, it has been detected in a eukaryotic cytoplasmic LSU rRNA in *Euglena gracilis* (cited in Nicoloso et al., 1996). Surprisingly, a snoRNA comparable to U31 has not been identified in yeast, which is unfortunate given the facility of the yeast system for testing the essentiality of specific 2'-O-methyl groups and pseudouridines through disruption of the genes encoding the corresponding guide snoRNAs.

The *E. coli* yibK gene is predicted to encode the Um2552 2'-O-methyltransferase (Gustafsson et al., 1996), but the yeast genome does not contain an obvious candidate gene for a YibK homolog that might catalyze formation of Um2791 in yeast mitochondrial LSU rRNA. This is unexpected in light of the clearcut relatedness between Pet56p and YjfH. It seems imperative therefore to confirm the identity of *yibK*. The position equivalent to Um2552 is also modified in eukaryotic cytoplasmic LSU rRNA, but there is no published report of a complementary snoRNA for this site (Bachellerie and Cavaillé, 1997).

Several pseudouridine synthase genes have been identified in *E. coli* (Koonin, 1996; Gustafsson et al., 1996) (see Chapter 12 by Ofengand and Fournier) and deletion studies in progress will undoubtedly lead to the identification of the gene(s) responsible for the formation of Ψ2580. On the basis of sequence comparisons, YceC and YfiI (SfhB) appear to be good candidates for the Ψ2580 synthase (Gustafsson et al., 1996). Searches of the yeast genomic database with the YceC and YfiI sequences as probes revealed good alignment with the YDL036c open reading frame (Mason, unpublished data). The predicted Ydl036 protein is a likely candidate for a mitochondrial pseudouridine synthase because it has a putative mitochondrial targeting presequence and the pattern of codon usage is similar to that of poorly expressed mitochondrial proteins such as Pet56p. Gene-disruption experiments are in progress to test whether Ydl036p is responsible for the formation of Ψ2819 in mitochondrial LSU rRNA (Mason, unpublished data).

The high level of sequence conservation in the PTC of domain V in LSU rRNA suggests that the yeast mitochondrial rRNA modification enzymes might be able to use the *E. coli* LSU rRNA as a substrate. Indeed, partially purified Pet56p catalyzes the site-specific formation of Gm2251 on in vitro transcripts of *E. coli* LSU rRNA (Sirum-Connolly and Mason, 1993). To test whether the mitochondrial enzymes responsible for Um and Ψ formation can also correctly modify *E. coli* LSU rRNA, the *E. coli* LSU rRNA was expressed in yeast mitochondria using an in vivo T7 RNA polymerase system in which the polymerase is expressed from a nuclear plasmid and imported into mitochondria (Pinkham et al., 1994). Plasmid DNA containing the *E. coli* LSU rRNA gene under control of the T7 promoter was transformed into [rho^0] mitochondria, which lack all endogenous transcripts. Analysis of modification in the heterologous rRNA transcripts in the transformed mitochondria confirmed the presence of Um2552 and Ψ2580; however, surprisingly, the results were equivocal for the presence of Gm2251 (Sirum-Connolly and Mason, unpublished data). Thus, the sequences recognized by each of the three mitochondrial modification activities are conserved in *E. coli* LSU rRNA, although for unknown reasons the heterologous rRNA appears to be a better substrate for Pet56p in vitro than in vivo.

Although it seems unlikely that eubacterial and mitochondrial pseudouridine synthases and 2'-O-methyltransferases function by a small RNA guide mechanism, this possibility has not been rigorously excluded. A number of prokaryotic-like enzymes, including the Pet56p methyltransferase, have been cloned and overexpressed, and site-specific enzyme activity was observed to increase in proportion to the level of protein overexpression (see Chapter 12 by Ofengand and Fournier) (Sirum-Connolly and Mason, unpublished data). If guide RNAs were required, such increases in activity would not be expected without concomitant overexpression of the requisite guide RNAs. Nonetheless, definitive biochemical studies, rather than circumstantial arguments, will be needed before conclusions can be drawn about the possible involvement of guide RNAs in the modification of bacterial and mitochondrial rRNAs.

CONCLUSION AND PROSPECTS

The presence of only three modified nucleotides at highly conserved positions in the PTC of yeast mitochondrial LSU rRNA has fueled speculation that these particular modification might have special functional significance. It hardly seems fortuitous that these modified nucleotides are located at sites in *E. coli* LSU rRNA strongly implicated in the interaction with the 3'-terminal end of tRNA bound in either the A site or the P site. However, the notion that one or more of the three conserved modifications might be essential for the structure and function of the PTC was quashed when active 50S subunits were reconstituted with rRNA lacking these modifications (Green and Noller, 1996). On the other hand, the strong phenotype associated with inactivation of the Pet56p rRNA methyltransferase suggests that formation of Gm2251 might be required for more than "fine-tuning" the translational apparatus, at least in yeast mitochondria.

The picture emerging from the genetic analysis of rRNA modification in *E. coli* and yeast cytoplasmic ribosomes is that several Ψ and sugar methylations can be eliminated individually with no apparent effect on cell growth (see Chapters 12 and 13), and thus far there is no case where the loss of a single Ψ or 2'-O-methyl group has a strong detrimental effect. This fits with the idea that modifications might be important collectively but not individually. In this context, elimination of one of the three nucleotide modifications in the minimally modified yeast mitochondrial LSU rRNA might be expected to have more impact than the elimination of a single modification in the much more highly modified rRNAs in bacterial and eukaryotic cytoplasmic rRNAs. The recent progress in the identification of genes responsible for specific modifications in *E. coli*, yeast mitochondrial, and yeast cytoplasmic rRNAs will accelerate the genetic analysis of rRNA modification in these systems. It will be particularly informative to compare the in vivo effects of depleting Ψ2580 in *E. coli* and yeast mitochondria and Gm2251 and Um2552 in all three ribosomal systems.

Acknowledgment. This work was supported in part by NSF grant MCB-9419340.

REFERENCES

Bachellerie, J.-P., and J. Cavaillé. 1997. Guiding ribose methylation of rRNA. *Trends Biochem. Sci.* 22:257–261.

Draper, D. E. 1996. Ribosomal-protein interactions, p. 171–197. *In* R. A. Zimmermann and A. E. Dahlberg (ed.), *Ribosomal RNA: Structure, Evolution, Processing, and Function in Protein Biosynthesis.* CRC Press, Boca Raton, Fla.

Garrett, R. A., and C. Rodriguez-Fonseca. 1996. The peptidyl transferase center, p. 327–355. *In* R. A. Zimmermann and A. E. Dahlberg (ed.), *Ribosomal RNA: Structure, Evolution, Processing, and Function in Protein Biosynthesis.* CRC Press, Boca Raton, Fla.

Green, R., and H. F. Noller. 1996. In vitro complementation analysis localizes 23S rRNA posttranscriptional modifications that are required for *Escherichia coli* 50S ribosomal subunit assembly and function. *RNA* 2:1011–1021.

Green, R., and H. F. Noller. 1997. Ribosomes and translation. *Annu. Rev. Biochem.* 66:679–716.

Green, R., R. S. Samaha, and H. F. Noller. 1997. Mutations at nucleotides G2251 and U2585 of 23S rRNA perturb the peptidyl transferase center of the ribosome. *J. Mol. Biol.* 266:40–50.

Gregory, S., and A. Dahlberg. Personal communication.

Gregory, S., K. Sirum-Connolly, T. Mason, and A. Dahlberg. Unpublished results.

Gregory, S. T., K. R. Lieberman, and A. E. Dahlberg. 1994. Mutations in the peptidyl transferase region of *E. coli* 23S rRNA affecting translational accuracy. *Nucleic Acids Res.* 22:279–284.

Gustafsson, C., R. Reid, P. J. Greene, and D. Santi. 1996. Identification of new modifying enzymes by iterative genome search using known modifying enzymes as probes. *Nucleic Acids Res.* 24:3756–3762.

Klootwijk, J., I. Klein, and L. A. Grivell. 1975. Minimal posttranscriptional modification of yeast mitochondrial ribosomal RNA. *J. Mol. Biol.* 97:337–350.

Koonin, E. V. 1996. Pseudouridine synthases: four families of enzymes containing a putative uridine-binding motif also conserved in dUTPases and dCTP deaminases. *Nucleic Acids Res.* 24:2411–2415.

Kowalak, J. A., E. Bruenger, and J. A. McCloskey. 1995. Posttranscriptional modification of the central loop of domain V in *Escherichia coli* 23S ribosomal RNA. *J. Biol. Chem.* 270:17758–17764.

Krzyzosiak, W. R., R. Denman, K. Nurse, W. Hellman, M. Boublik, C. W. Gehrke, P. F. Agris, and J. Ofengand. 1987. In vitro synthesis of 16S ribosomal RNA containing single base changes and assembly into a functional 30S ribosome. *Biochemistry* 26:2353–2364.

Lafontaine, D., J. Vandenhaute, and D. Tollervey. 1995. The 18S rRNA dimethylase Dim1p is required for pre-ribosomal RNA processing in yeast. *Genes Dev.* 9:2470–2481.

Lane, B. G., J. Ofengand, and M. W. Gray. 1995. Pseudouridine and $O^{2'}$-methylated nucleosides. Significance of their selective occurrence in rRNA domains that function in ribosome-catalyzed synthesis of the peptide bonds in proteins. *Biochimie* 77:7–15.

Lieberman, K. R., and A. E. Dahlberg. 1995. Ribosome-catalyzed peptide-bond formation. *Prog. Nucleic Acid Res. Mol. Biol.* 50:1–23.

Maden, B. E. H., M. E. Corbett, P. A. Heeney, K. Pugh, and P. M. Ajuh. 1995. Classical and novel approaches to the detection and localization of the numerous modified nucleotides in eukaryotic ribosomal RNA. *Biochimie* 77:22–29.

Mason, T. Unpublished data.

Mason, T. L., C. Pan, M. E. Sanchirico, and K. Sirum-Connolly. 1996. Molecular genetics of the peptidyl transferase center and the unusual Var1 protein in yeast mitochondrial ribosomes. *Experientia* 52:1148–1157.

Moazed, D., and H. F. Noller. 1989. Interaction of tRNA with 23S rRNA in the ribosomal A, P, and E sites. *Cell* 57:585–597.

Nicol, S. M., and F. V. Fuller-Pace. 1995. The "DEAD box" protein Dbp interacts specifically with the peptidyltransferase center in 23S rRNA. *Proc. Natl. Acad. Sci. USA* 92:11681–11685.

Nicoloso, M., L.-H. Qu, B. Michot, and J.-P. Bachellerie. 1996. Intron-encoded, antisense small nucleolar RNAs: the characterization of nine novel species points to their direct role as guides for the 2'-O-ribose methylation of rRNAs. *J. Mol. Biol.* 260:178–195.

Nierhaus, K., and F. Dohme. 1974. Total reconstitution of functionally active 50S ribosomal subunits from *E. coli*. *Proc. Natl. Acad. Sci. USA* 71:4713–4717.

Ofengand, J., and A. Bakin. 1997. Mapping to nucleotide resolution of pseudouridine residues in large subunit ribosomal RNAs from representative eukaryotes, prokaryotes, archaebacteria, mitochondria and chloroplasts. *J. Mol. Biol.* 266:246–268.

Pinkham, J. L., A. M. Dudley, and T. L. Mason. 1994. T7 RNA polymerase-dependent expression of COXII in yeast mitochondria. *Mol. Cell. Biol.* 14:4643–4652.

Porse, B. T., and R. A. Garrett. 1995. Mapping important nucleotides in the peptidyl transferase center of 23S rRNA using a random mutagenesis approach. *J. Mol. Biol.* 249:1–10.

Porse, B. T., H. P. Thi-Ngoc, and R. A. Garrett. 1996. The donor substrate site within the peptidyl transferase loop of 23S rRNA and its putative interactions with the CCA-end of N-blocked aminoacyl-tRNA(Phe). *J. Mol. Biol.* 264:472–483.

Saarma, U., and J. Remme. 1992. Novel mutants of 23S RNA: characterization of functional properties. *Nucleic Acids Res.* 20:3147–3152.

Samaha, R. R., R. R. Green, and H. F. Noller. 1995. A base pair between tRNA and 23S rRNA in the peptidyl transferase centre of the ribosome. *Nature* 377:309–314.

Sirum-Connolly, K., and T. L. Mason. Unpublished data.

Sirum-Connolly, K., and T. L. Mason. 1993. Functional requirement of a site-specific ribose methylation in ribosomal RNA. *Science* 262:1886–1889.

Sirum-Connolly, K., J. M. Peltier, P. F. Crain, J. McCloskey and T. L. Mason. 1995. Implications of a functional large ribosomal RNA with only three modified nucleotides. *Biochimie* 77:30–39.

Spahn, C. M. T., J. Remme, M. A. Schafer, and K. H. Nierhaus. 1996a. Mutational analysis of two highly conserved UGG sequences of 23S rRNA from *Escherichia coli*. *J. Biol. Chem.* 271:32849–32856.

Spahn, C. M. T., M. A. Schafer, A. A. Krayevsky, and K. H. Nierhaus. 1996b. Conserved nucleotides of 23S rRNA located at the ribosomal peptidyltransferase center. *J. Biol. Chem.* 271:32857–32862.

Thompson, J., and E. Cundliffe, 1991. The binding of thiostrepton to 23S RNA. *Biochimie* 73:1131–1135.

Traub, P., and M. Nomura. 1968. Structure and function of *E. coli* ribosomes. V. Reconstitution of functionally active 30S ribosomal particles from RNA and protein. *Proc. Natl. Acad. Sci. USA* 59:777 784.

Tycowski, K. T., M. D. Shu, and J. A. Steitz. 1996. A mammalian gene with introns instead of exons generating stable RNA products. *Nature* 379:464–466.

Walleczek, J., B. Redl, M. Stöffler-Meilicke, and G. Stöffler. 1989. Protein-protein cross-linking of the 50S ribosomal subunit of *Escherichia coli* using 2-iminothiolane. Identification of cross-links by immunoblotting techniques. *J. Biol. Chem.* 264:4231–4237.

Chapter 15

Regulatory Aspects of rRNA Modification and Pre-rRNA Processing

DENIS L. J. LAFONTAINE AND DAVID TOLLERVEY

In most eukaryotes, three out of the four mature rRNA species are produced from a single large RNA polymerase I transcript (designated the 35S pre-rRNA in yeast), which is processed in a complex pathway involving both endonuclease cleavage and exonuclease digestion (Fig. 1) (reviewed in Lafontaine and Tollervey, 1995b; Venema and Tollervey, 1995; Tollervey, 1996). The fourth mature species, 5S rRNA, is transcribed and processed independently. During pre-rRNA processing, many specific nucleotides within the rRNAs are covalently modified (Maden, 1990; Maden and Hughes, 1997). The major types of posttranscriptional modification are the isomerization of uracil to pseudouridine, 2'-O-methylation of the ribose moieties and base methylation (see accompanying Chapter 12 by Ofengand and Fournier and Chapter 13 by Bachellerie and Cavaillé). Concomitantly with the pre-rRNA processing and modification reactions, the ribosomal proteins assemble with the pre-rRNAs to form preribosomal particles; mechanisms are likely to exist that coordinate all of these activities.

Recent analyses have identified three rRNA-modifying enzymes in yeast. Dim1p carries out the conserved base dimethylation that converts two adenosine residues at the 3'-end of the small subunit rRNA (SSU-rRNA) to $m_2^6A_{1779}m_2^6A_{1780}$ (yeast numbering); Cbf5p is the presumed rRNA pseudouridine synthase, which potentially modifies 13 sites in 18S rRNA and 30 sites in 25S rRNA (see Chapter 12 by Ofengand and Fournier for their locations); and Pet56p is a 2'-O-methylase specific for G_{2270} in the 21S yeast mitochondrial rRNA (see Chapter 14 by Mason). In addition to their roles in pre-rRNA modification, both Dim1p and Cbf5p are required for pre-rRNA processing, and Pet56p is required for synthesis of mitochondrial large ribosomal subunits.

The data on Dim1p and Cbf5p have given us insights on the type of systems that may act to coordinate the many steps in ribosome synthesis and the origins of the snoRNA-directed rRNA modification systems present in eukaryotic cells. These will be discussed in this chapter.

Dim1p, A CASE OF QUALITY CONTROL

$m_2^6Am_2^6A$ is one of the few rRNA base modifications that have been conserved from bacteria to eukaryotes. The modification is almost universally conserved; the only known exceptions are the yeast mitochondrial ribosomes that are not dimethylated and the chloroplast ribosomes from *Euglena gracilis* that are partially dimethylated (Klootwijk et al., 1975; van Buul et al., 1984; reviewed in van Knippenberg, 1986). The site of modification lies at the 3'-end of the SSU-rRNA in the loop of a hairpin structure, which is highly conserved in sequence (Fig. 2). In the ribosome, the $m_2^6Am_2^6A$ residues are located at the interface between the subunits at a site where crucial recognition reactions occur during translation (Thamana and Cantor, 1978; Maden, 1990; Brimacombe et al., 1993).

Yeast Dim1p was identified by complementation of an *Escherichia coli* mutant defective in the rRNA dimethylase ksgAp (Lafontaine et al., 1994). Dim1p and ksgAp show substantial sequence homology (27% identity and 50% similarity) and Dim1p can dimethylate the *E. coli* SSU-rRNA in vivo, showing them to be orthologs (that is, the "same" gene from

---

Denis L. J. Lafontaine and David Tollervey • Institute of Cell and Molecular Biology, The University of Edinburgh, Mayfield Road, Edinburgh EH9 3JR, Scotland.

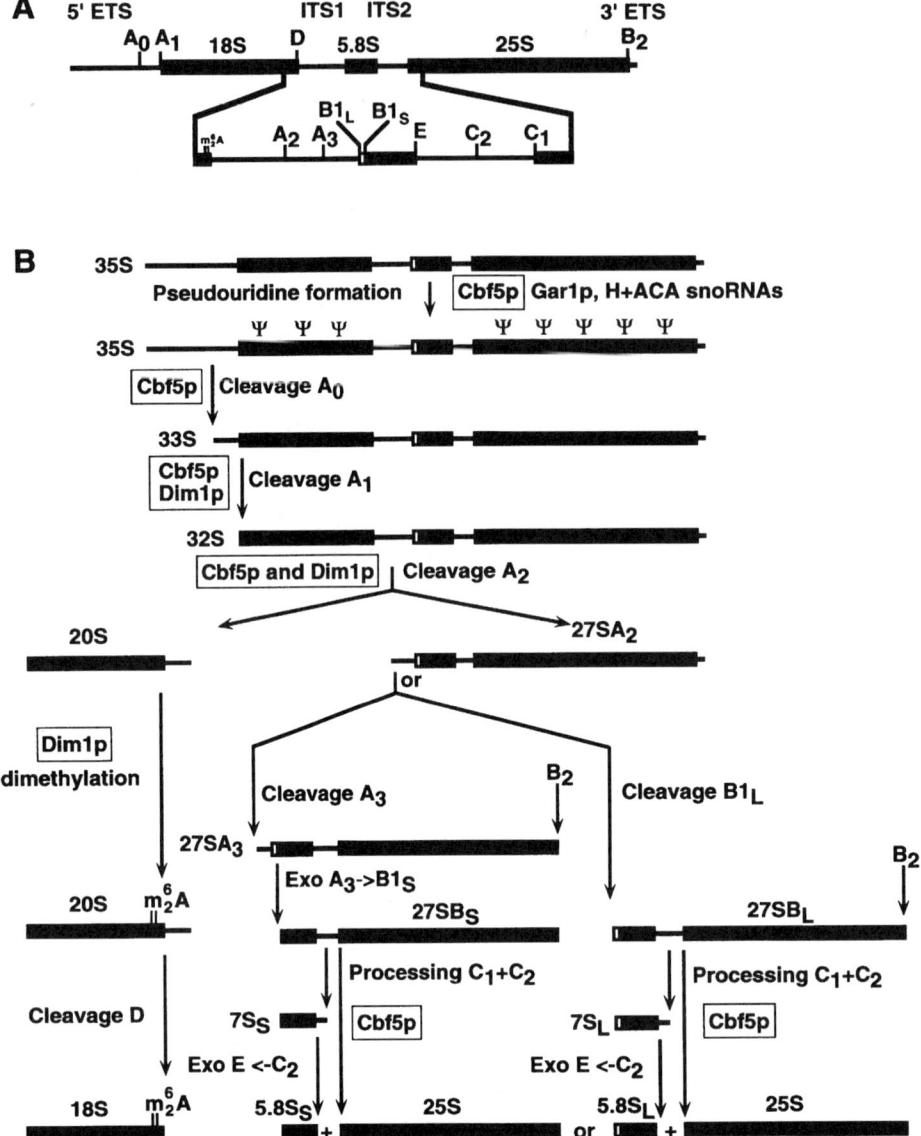

Figure 1. Structure of the yeast pre-rRNA and its processing pathway. (A) The 35S pre-rRNA. The sequences encoding the mature 18S, 5.8S and 25S rRNAs (thick lines) are flanked by the 5' and 3' external transcribed spacers (5' ETS and 3' ETS) and separated by internal transcribed spacers 1 and 2 (ITS1 and ITS2). Sites of pre-rRNA processing are indicated with uppercase letters (A_0 to E). The site of dimethylation is represented by m_2^6A. (B) The pre-rRNA processing pathway. Pseudouridine formation occurs shortly after or during transcription. Pseudouridine synthesis is targeted by the H + ACA snoRNAs and requires Cbf5p assisted by Gar1p. Processing of the primary 35S precursor starts at site A_0, yielding the 33S pre-rRNA. This molecule is subsequently processed at sites A_1 and A_2, giving rise successively to the 32S pre-rRNA and to the 20S and 27SsA$_2$ precursors. Cleavage at A_2 separates the pre-rRNAs destined for the small and large ribosomal subunit. The 20S precursor is dimethylated by Dim1p and then cleaved endonucleolytically at site D to yield the mature 18S rRNA. The 27SA$_2$ precursor is processed by two alternative pathways to form the mature 5.8S and 25S rRNAs. The major pathway involves cleavage at a second site in ITS1, A_3, rapidly followed by exonucleolytic digestion to site $B1_S$, generating the $27SB_S$ precursor. Approximately 15% of the 27SA$_2$ molecules are processed by an alternative pathway at site $B1_L$, producing the $27SB_L$ pre-rRNA. At the same time as processing at B1 is completed, the 3'-end of mature 25S rRNA is generated by processing at site B2. The subsequent processing of both 27SB species appears to follow a similar pathway. Cleavage at sites C_1 and C_2 releases the mature 25S rRNA and the 7S pre-rRNAs, which undergo rapid 3'→5' exonuclease digestion to site E, generating the mature 3'-end of 5.8S rRNA. Cbf5p and Dim1p are required for the early cleavages at sites A_1 and A_2; loss of these cleavages inhibits formation of the 20S and 27SA$_2$ pre-rRNAs, preventing synthesis of 18S rRNA. In addition, Cbf5p is required for efficient processing at site A_0 and efficient processing of the 27SB and 7S pre-rRNAs in ITS2.

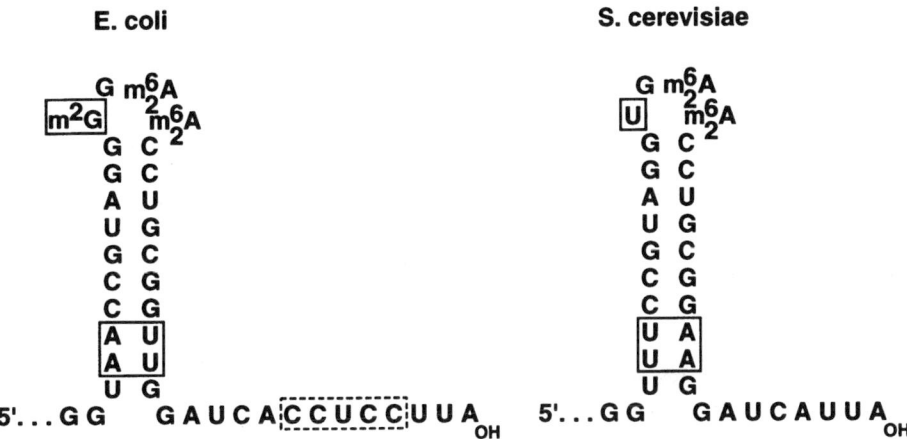

Figure 2. 3'-end of the SSU-rRNA in *E. coli* and yeast cytoplasmic ribosomes. Divergent nucleotides are boxed. A precise deletion of the anti-Shine-Dalgarno box (CCUCC) has occurred in the eukaryotic SSU-rRNA. The twin adenosine substrates (1518–1519 in *E. coli* and 1779–1780 in yeast) of Dim1p are universally conserved.

different organisms). Genetic depletion of Dim1p under the control of a regulated *GAL* promoter led to the inhibition of dimethylation, confirming that it is the only dimethylase in yeast. Deletion of the *DIM1* gene in yeast was found to be lethal, in contrast to *E. coli* in which *ksgA* mutants are viable although mildly impaired in growth. Surprisingly, however, the lack of the dimethylation was not the cause of this lethality. Genetic depletion of Dim1p inhibited pre-rRNA processing, preventing the synthesis of the mature SSU-rRNA, and it is the loss of this rRNA that is responsible for the lethality of *dim1* mutants (Lafontaine et al., 1995a).

Depletion of Dim1p inhibited the cleavage of sites A_1 and A_2 (see Fig. 1 and 3). These cleavages generate the 20S pre-rRNA, which is the substrate for the dimethylation reaction in wild-type cells. Because the processing reactions normally occur before dimethylation of the pre-rRNA (Fig. 1), it seemed unlikely that processing was dependent on dimethylation. This conclusion was supported by the observation that replacement of the A residues at the site of dimethylation with G residues that cannot be modified did not interfere with pre-rRNA processing (Lafontaine et al., 1995a). This was subsequently confirmed by the finding that point mutations in Dim1p could uncouple the pre-rRNA processing defect from the dimethylation defect (Lafontaine et al., 1998b).

In *dim1-2* mutant strains, dimethylation of the pre-rRNA is strongly inhibited with little effect on pre-rRNA processing at permissive temperature (Lafontaine et al., 1998b). In consequence, the *dim1-2* strains accumulated normal levels of small ribosomal subunits, but these lacked the $m_2^6Am_2^6A$ dimethylation. The strain grew normally, demonstrating that the dimethylation is not required for ribosome function in vivo. However, extracts prepared from the *dim1-2* mutant strain did not support in vitro translation. This suggested that the $m_2^6Am_2^6A$ dimethylation "fine-tunes" the function of the ribosome in vivo, but becomes much more important for function in the suboptimal in vitro conditions. Related observations have been made for pre-mRNA splicing; mutations that drastically inhibit in vitro splicing in cell extracts frequently have little effect on splicing activity in vivo (Jacquier et al., 1985; Séraphin et al., 1988).

The uncoupling of the dimethylation and pre-rRNA processing defects showed that Dim1p rather than the dimethylation activity is required for pre-rRNA processing. This observation did not, however, determine whether Dim1p is itself directly required for pre-rRNA processing. This question was resolved in an unexpected manner by the observation that when the transcription of an rDNA unit was driven by an RNA polymerase II (pol II) *PGK* promoter, processing of the pre-rRNAs became insensitive to temperature-sensitive mutations in *DIM1* or depletion of Dim1p (Lafontaine et al., 1998b). Dim1p is not, therefore, directly required for pre-rRNA processing.

These observations led to the conclusion that an active repression system blocks pre-rRNA processing in the absence of the binding of Dim1p to the pre-rRNA. According to this model, Dim1p binds to the pre-rRNA in the nucleolus at an early stage in ribosome synthesis. This step is monitored by a component of the processing machinery such that processing at sites A_1 and A_2 occurs only if Dim1p has bound to the pre-rRNA. In mutant strains that lack Dim1p this leads to the synthesis of a dead-end intermediate, the 22S pre-rRNA, and prevents synthesis of unmodified 18S rRNA (Fig. 3). The pre-rRNAs that are

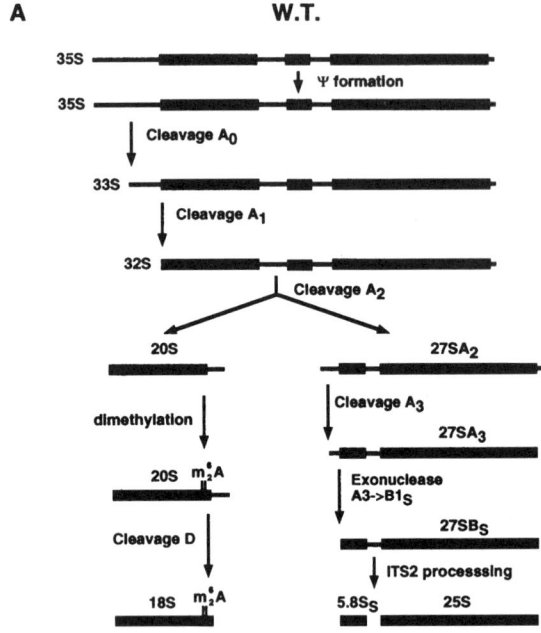

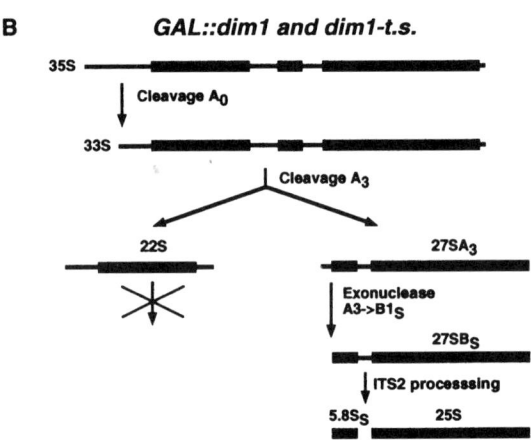

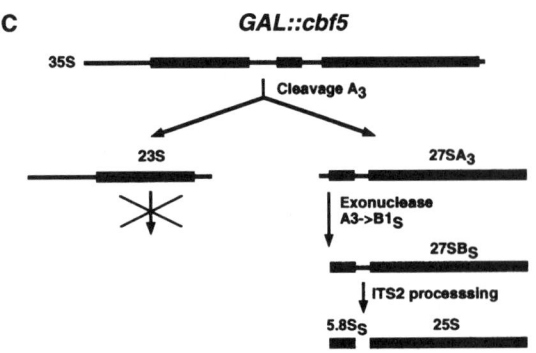

Figure 3. Pre-rRNA processing in the wild type (A) and *dim1* (B) and *cbf5* (C) mutants. (A) Simplified version of the wild-type pre-rRNA processing pathway, as described in Fig. 1. (B) GAL::dim1

transcribed from the pol II *PGK* promoter are presumably associated with a different set of hnRNP proteins as compared to the pol I transcripts. In HeLa cells the fate of β-globin transcripts can be altered by the identity of the pol II promoters from which they are transcribed (Enssle et al., 1993). We speculate that in the *PGK*-driven pre-rRNAs an hnRNP protein occupies the Dim1p binding site and is detected by the pre-rRNA processing machinery, thus alleviating the need for authentic Dim1p.

In *E. coli*, mutations in ksgAp block the $m_2^6Am_2^6A$ dimethylation of the 16S rRNA, but do not interfere with processing of the rRNA. In consequence, the mutants synthesize small ribosomal subunits that lack the modification. These allow the growth of the mutant strains, although at a reduced rate, and the unmodified ribosomes are defective in several aspects of translation in vitro (reviewed in van Knippenberg, 1986). This contrasts with the situation in yeast, where mutations in Dim1p are lethal because pre-rRNA processing is blocked. On first inspection the situation in *E. coli* appears preferable because the mutants are at least alive. However, wild-type cells are unlikely to be faced with the situation in which the dimethylase is actually absent or defective. The regulatory system in yeast presumably evolved to deal with those preribosomal particles to which Dim1p binds late in the process. In this situation, pre-rRNA processing would simply be delayed until Dim1p binding occurs and processing would then resume. This is desirable because the unmodified ribosomal subunits are impaired in function.

An obvious question that arises from these observations is why the dimethylation is not directly monitored. One possible explanation is that a system that detected the binding of the Dim1p protein to the preribosome developed more readily than a system that could detect the presence or absence of the methyl groups themselves. An alternative explanation was offered by the observation that all *dim1* mutants tested were hypersensitive to the aminoglycoside antibiotics paromomycin and neomycin B, even under conditions where neither dimethylation nor processing was clearly affected (Lafontaine et al., 1998b). This suggested that Dim1p plays an additional role in ribosome assembly. The binding of Dim1p to the

strains and *dim1-t.s.* strains are inhibited in cleavage at sites A_1 and A_2. Consequently, the 22S pre-rRNA accumulates and no 18S rRNA is made. The 27SA$_3$ pre-rRNA is normally processed to 25S and 5.8S rRNAs. (C) GAL::cbf5 strains are inhibited in cleavage at sites A_0, A_1 and A_2. In this case the 23S pre-rRNA is accumulated and no 18S rRNA is made. The 27SA$_3$ pre-rRNA can be processed to 25S and 5.8S rRNAs, although processing of the 27SB and 7S pre-rRNAs is delayed.

pre-rRNA may therefore be monitored to ensure that it fulfills its functions in both modification and assembly.

Ribosome synthesis is a highly dynamic process during which a vast number of processing, modification, and assembly reactions occur simultaneously. How can all of these reactions be coordinated? In any given preribosomal subunit there must be a significant chance that some components will bind late. This is a particular problem in eukaryotes because the assembly, modification, and processing reactions proceed very quickly with the preribosomal particles being rapidly exported to the cytoplasm, where they are likely to be inaccessible to nucleolar assembly and processing factors. We anticipate that many other quality control systems of the type proposed for Dim1p exist to coordinate assembly, modification, processing, and export to ensure that no irrevocable steps occur before all major components are assembled. It is notable that the depletion or mutation of any of several different ribosomal proteins has been found to inhibit pre-rRNA processing (Moritz et al., 1990; Moritz et al., 1991; Deshmukh et al., 1993; Vilardell and Warner, 1997). It is very likely that such inhibition also reflects the activity of quality control systems, which would normally simply delay processing to allow time for the assembly of the missing protein.

Pet56p, A MITOCHONDRIAL rRNA-MODIFYING ENZYME

Pet56p is a mitochondrial rRNA 2'-O-methylase specific to position G_{2270} in the peptidyl transferase center of the 21S yeast mitochondrial rRNA. The equivalent position in the 23S rRNA of *E. coli*, position G_{2251}, is also ribose methylated (see Chapter 14 by Mason).

Like Dim1p, Pet56p is required for ribosome synthesis. Strains depleted of Pet56p or carrying a deletion of *PET56* strongly underaccumulate large ribosomal subunits (Sirum-Connolly and Mason, 1993). Extragenic mutations were isolated that weakly suppress the *pet56* loss-of-function mutations, and this was interpreted as showing that neither the Gm_{2270} modification nor the Pet56p protein itself is absolutely required for the synthesis of functional ribosomes (Sirum-Connolly and Mason, 1995) (also see Chapter 14 by Mason). Pet56p may therefore be the target of a quality control system that functions in a manner analogous to that postulated for Dim1p.

Cbf5p, A COMPONENT OF MULTIPLE snoRNPs

Yeast Cbf5p is a nucleolar protein that shows strong sequence homology to the prokaryotic tRNA: $\Psi 55$ synthase truBp (Nurse et al., 1995) and to the rat nucleolar protein Nap57p (Meier and Blobel, 1994). The latter is tightly associated with Nop140p, a protein that is reported to shuttle between the nucleolus and the nuclear envelope (Meier and Blobel, 1992). These observations suggested that Cbf5p might be a nucleolar pseudouridine synthase. Genetic depletion of Cbf5p impaired total rRNA Ψ formation with no clear effect on formation of Ψ in tRNA (Lafontaine et al., 1998a). As with depletion of Dim1p, pre-rRNA processing was also strongly impaired, leading to the loss of the SSU-rRNA (Fig. 1 and 3).

Formation of Ψ in the rRNA is guided by the H + ACA class of snoRNAs (Ni et al., 1997; Ganot et al., 1997; see Chapter 12 by Ofengand and Fournier and recent reviews by Maden, 1997; Peculis, 1997; and Smith and Steitz, 1997) and Cbf5p is associated with this class of snoRNA. Epitope tagged Cbf5p was able to efficiently coprecipitate all tested H + ACA snoRNAs, while genetic depletion of Cbf5p led to the loss of these snoRNAs (Lafontaine et al., 1998a). Gar1p is also associated with the box H + ACA snoRNAs (Girard et al., 1992; Lübben et al., 1995) and depletion of Cbf5p led to the loss of Gar1p. Cbf5p is therefore an integral component of the box H + ACA class of small nucleolar ribonucleoproteins (snoRNPs). Depletion of Gar1p also leads to a global defect in rRNA Ψ formation (Bousquet-Antonelli et al., 1997), but it does not show any of the known pseudouridine synthase motifs and is unlikely to act enzymatically. Gar1p is not required for the stability of the box H + ACA snoRNAs tested, snR10, snR30, and snR36 (Girard et al., 1992; Bousquet-Antonelli et al., 1997), but it may play an important role in their association with the pre-rRNA. In Gar1p-depleted cells, snR36 snoRNPs are unable to associate with higher order nucleolar particles (Bousquet-Antonelli et al., 1997). The function of Gar1p in rRNA pseudouridine formation is therefore likely to be the stabilization of Cbf5p-containing snoRNP complexes at the sites of Ψ formation.

A conserved central domain of Gar1p is sufficient to allow nucleolar localization and fulfill the essential function of the protein (Girard et al. 1994), and is both necessary and sufficient for in vitro binding to the H + ACA snoRNAs snR30 and snR10 (Bagni and Lapeyre, personal communication). We speculate that the central domain of Gar1p specifically binds to the H + ACA snoRNAs, while the two external glycine and arginine rich (GAR) domains

(Girard et al., 1992) stabilize the snoRNA and pre-rRNA interactions. GAR domains have been reported both to confer RNA-binding activity (Kiledjian and Dreyfuss, 1992) and to destabilize RNA structures (Ghisolfi et al., 1992). An interesting possibility is that the GAR domains could both open up the complex structure of the pre-rRNAs and stabilize the interaction between the H + ACA snoRNAs and the pre-rRNA at the sites of modification. Such an activity may be crucial because the sequence complementarity between the H + ACA snoRNAs and the rRNA consists only of two short motifs of 3–10 nucleotides (see Chapter 12 by Ofengand and Fournier; Ganot et al., 1997). Rok1p, a putative ATP-dependent RNA helicase that functionally interacts both with Gar1p and the H + ACA snoRNA snR10 (Venema et al., 1997) may also be involved in this process.

The pre-rRNA processing defects observed on depletion of Cbf5p, inhibition of cleavage at sites A_1 and A_2 and delay of cleavage at site A_0 (Fig. 1 and 3), closely resemble those observed on depletion of the box H + ACA snoRNA, snR30 (Morrissey and Tollervey, 1993). snR30 is lost on depletion of Cbf5p, suggesting that the pre-rRNA processing defect is due to the lack of snR30. Depletion of Gar1p leads to a similar pre-rRNA processing defect (Girard et al., 1992), but the effects of Gar1p depletion on the association of snR30 and the pre-rRNA have not been reported.

The requirement for Cbf5p in pre-rRNA processing appears to be quite different from the requirement for Dim1p. In no case has the pseudouridine guide activity of a box H + ACA snoRNA been found to be required for pre-rRNA processing. In the absence of Cbf5p or Gar1p, 25S rRNA synthesis continues, leading to the synthesis of highly undermodified 60S subunits. Similarly, no 2′-O-methylation is known to be required for pre-rRNA processing and *nop1-3* mutants that are globally inhibited for 2′-O-methylation show little inhibition of processing (Tollervey et al., 1993). This raises an obvious question: why do quality control systems not exist to ensure that pseudouridine formation and 2′-O-methylation also occur? In this case we speculate that the sheer number of modifications (43 Ψ and 55 2′-O-methyl groups are present in the yeast rRNAs) has made the development of a regulatory system for each site of modification impractical. A predicted consequence of the absence of such regulatory systems is that some proportion of ribosomal subunits will lack one or more pseudouridine or 2′-O-methyl modifications. This in turn would be expected to apply selective pressure for the absence of individual modifications, not to result in any major impairment in ribosome function. This appears to be the case because deletion of individual Ψ guide or methylation guide snoRNAs has no detectable effect on growth (see Chapter 12 by Ofengand and Fournier and Chapter 13 by Bachellerie and Cavaillé and references therein).

During pseudouridine formation in the rRNA, snoRNAs base paired to the rRNA provide a common signal, allowing a single pseudouridine synthase to recognize multiple sites. This system shows some resemblance to the pseudouridylation of positions 34 and 36 in the anticodon loop of minor-tRNA^{Ile}. In yeast, modification of these sites by Pus1p is dependent on the presence of the tRNA intron; the mature tRNA is not a substrate for modification (Szweykowska-Kulinska et al., 1994; Simos et al., 1996). The intron therefore acts in *cis* as an internal guide sequence, allowing recognition of sites in the mature tRNA region (Grosjean et al., 1997).

MODEL FOR THE EVOLUTIONARY ORIGIN OF MODIFICATION GUIDE snoRNAs

Comparison of the number and distribution of Ψ residues in the rRNA of bacteria and eukaryotes raises two questions. First, why is there such a large (~10-fold-higher) number of Ψ residues in eukaryotic rRNA compared to bacterial rRNA? Second, why is there such a poor correspondence between the actual sites that are modified in each kingdom? The work on Cbf5p suggests possible explanations for these puzzling observations.

Modern eukaryotes are equipped with a huge array of pseudouridine guide H + ACA snoRNAs (see Chapter 12), but this system clearly did not spring into its existence fully formed, and it is reasonable to assume that a single snoRNA originally acquired the ability to select a site of pseudouridine formation. The yeast tRNA:Ψ55 pseudouridine synthase Pus4p (Becker et al., 1997) and Cbf5p both show high homology to the *E. coli* tRNA:Ψ55 pseudouridine synthase truBp (Koonin, 1996). This suggests that an early eukaryote (or archaea) had a single "truBp-like" tRNA pseudouridine synthase which, like truBp, recognized its substrate via structural features in the RNA. This enzyme then acquired the ability to recognize an RNA structure comprised of an snoRNA base paired in *trans* to the rRNA (Lafontaine et al., 1998a). Gene duplication and divergence of function would then lead to two forms of the pseudouridine synthase: one specialized for Ψ formation in the pre-rRNA (Cbf5p), and one for the tRNA (Pus4p). Over time new box H + ACA snoRNAs would arise, allowing the system to modify new sites in the rRNA.

This would slowly replace the preexisting system of "bacterial-like" rRNA pseudouridine synthases.

According to this model, new sites of Ψ formation in the eukaryotic rRNA arose when complementarity to new rRNA sequences was generated by mutations in the H + ACA snoRNAs. Only two short stretches of 3–10 nucleotides in the snoRNAs are base paired to the rRNA, and it is evident that new snoRNA and rRNA interactions will arise more often than will protein enzymes with specificity for a new site in the rRNA. The snoRNA-directed system, therefore, offered much greater flexibility in generating new sites. Moreover, the requirement that each site be recognized both as a binding site for the snoRNA and as a substrate for a structure-dependent pseudouridine synthase allowed greater selectivity, reducing the problems of misrecognition that would be associated with the presence of a large number of distinct pseudouridine synthases. These observations may explain why the eukaryotic rRNAs have many more modified sites than their bacterial counterparts.

This model also predicts that the new sites would have been generated by mutations in the guide snoRNAs independently of the preexisting proteinaceous system. This leads to the conclusion that the Ψ sites in eukaryotic rRNA are related to the sites in the bacterial rRNA by convergent, not divergent, evolution. This potentially explains why there is so little correspondence between the sites of Ψ in bacteria and eukaryotes. A more complete description of this model can be found in Lafontaine and Tollervey (1998c).

CONCLUSIONS AND PROSPECTS

Eukaryotic ribosome synthesis involves very complex pre-rRNA processing and assembly pathways. These include numerous steps occurring in different cellular compartments and requiring a plethora of RNA and proteins, many of them in the form of snoRNPs that only transiently associate with the pre-ribosomal particles. Such a complex process would be expected to be highly regulated, and there is increasing evidence that this is the case. Some processing steps appear to occur in an obligatory order, and there are several examples of coupling between processing reactions that occur at sites that are distant in the primary sequence. Presumably these interactions help to ensure the coordinated processing of the pre-rRNA. Moreover, the pre-rRNAs are extensively modified with the bulk of modification (2'-O-methylation and pseudouridylation) being snoRNP dependent and occurring at an early stage in the ribosome synthesis, but with some positions, mostly bases and a few sugars, being specifically methylated at later stages. $m_2^6Am_2^6A$ belongs to this latter class and it will be interesting to determine whether the other late modifications are also subject to quality control systems of the type proposed here for Dim1p.

All tested Ψ and 2'-O-methyl residues are dispensable for ribosome function in vivo. This is not evidence that these modifications do not make important contributions to ribosome function. The $m_2^6Am_2^6A$ dimethylation is also dispensable in vivo but is required for translation in vitro, showing that it does play an important role in the normal function of the ribosome. It also remains possible that a small number of individual Ψ or 2'-O-methyl modifications are essential for ribosome function, and it seems likely that the absence of multiple modifications will not be tolerated.

The archaea may hold the final clues to understanding the origins of the snoRNA-directed systems of rRNA modification. The pattern of Ψ formation and of 2'-O-methylation in archaeal rRNAs would be a usefull indication of the time at which the guide snoRNA system developed. The archaea have a homolog of the box C + D snoRNA-associated protein Nop1p (fibrillarin) (Amiri, 1994); it remains to be determined whether they also have homologs of Cbf5p, Gar1p, and, particularly, the modification guide snoRNAs.

Acknowledgments. We thank C. Bagni and B. Lapeyre for communicating results prior to publication. This work was partially supported by the Wellcome Trust and by postdoctoral fellowships from the EMBO and the European Commission.

REFERENCES

Amiri, K. A. 1994. Fibrillarin-like proteins occur in the domain of Archaea. *J. Bacteriol.* **176:**2124–2127.

Bagni, C., and B. Lapeyre. Personal communication.

Becker, H. F., Y. Motorin, R. J. Planta, and H. Grosjean. 1997. The yeast gene YNL292w encodes a pseudouridine synthase (Pus4) catalyzing the formation of Ψ55 in both mitochondrial and cytoplasmic tRNAs. *Nucleic Acids Res.* **25:**4493–4499.

Bousquet-Antonelli, C., Y. Henry, J.-P. Gélugne, M. Caizergues-Ferrer, and T. Kiss. 1997. A small nucleolar RNP protein is required for pseudouridylation of eukaryotic ribosomal RNAs. *EMBO J.* **16:**4770–4776.

Brimacombe, R., P. Mitchell, M. Osswald, K. Stade, and D. Bochkarriov. 1993. Clustering of modified nucleotides at the functional center of bacterial ribosomal RNA. *FASEB J.* **7:**161–167.

Deshmukh, M., Y.-F. Tsay, A. G. Paulovitch and J. L. Woolford. 1993. Yeast ribosomal protein L1 is required for the stability of newly synthesized 5S rRNA and the assembly of 60S ribosomal subunits. *Mol. Cell. Biol.* **13:**2835–2845.

Enssle, J., W. Kugler, M. W. Hentze, and A. E. Kulozik. 1993. Determination of mRNA fate by different RNA polymerase II promoters. *Proc. Natl. Acad. Sci. USA* **90:**10091–10095.

Ganot, P., M.-L. Bortolin, and T. Kiss. 1997. Site-specific pseudouridine formation in preribosomal RNA is guided by small nucleolar RNAs. *Cell* **89:**799–809.

Ghisolfi, L., G. Joseph, F. Amalric, and M. Erard. 1992. The glycine-rich domain of nucleolin has an unusual supersecondary structure responsible for its RNA-helix-destabilizing properties. *J. Biol. Chem.* **267:**2955-2959.

Girard, J. P., H. Lehtonen, M. Caizergues-Ferrer, F. Amalric, D. Tollervey, and B. Lapeyre. 1992. GAR1 is an essential small nucleolar RNP protein required for pre-rRNA processing in yeast. *EMBO J.* **11:**673-682.

Girard, J. P., C. Bagni, M. Caizergues-Ferrer, F. Amalric, and B. Lapeyre. 1994. Identification of a segment of the small nucleolar ribonucleoprotein-associated protein Gar1 that is sufficient for nucleolar accumulation. *J. Biol. Chem.* **269:**18499-18506.

Grosjean, H., Z. Szweykowska-Kulinska, Y. Motorin, F. Fasiolo, and G. Simos. 1997. Intron-dependent enzymatic formation of modified nucleosides in eukaryotic tRNAs: a review. *Biochimie* **79:**293-302.

Jacquier, A., J. R. Rodriguez, and M. Rosbash. 1985. A quantitative analysis of the effects of 5′ junction and TACTAAC box mutants and mutant combinations on yeast mRNA splicing. *Cell* **43:**423-430.

Kiledjian, M., and G. Dreyfuss. 1992. Primary structure and binding activity of the hnRNP U protein: binding through RGG box. *EMBO J.* **11:**2655-2664.

Klootwijk, H. A., I. Klein, and L. A. Grivell. 1975. Minimal posttranscriptional modification of yeast mitochondrial ribosomal RNA. *J. Mol. Biol.* **97:**337-350.

Koonin, E. V. 1996. Pseudouridine synthases: four families of enzymes containing a putative uridine-binding motif also conserved in dUTPases and dCTP deaminases. *Nucleic Acids Res.* **24:**2411-2415.

Lafontaine, D., J. Delcour, A.-L. Glasser, J. Desgrès, and J. Vandenhaute. 1994. The *DIM1* gene responsible for the conserved $m_2^6Am_2^6A$ dimethylation in the 3′ terminal loop of 18S rRNA is essential in yeast. *J. Mol. Biol.* **241:**492-497.

Lafontaine, D., J. Vandenhaute, and D. Tollervey. 1995a. The 18S rRNA dimethylase Dim1p is required for pre-ribosomal RNA processing in yeast. *Genes Dev.* **9:**2470-2481.

Lafontaine, D., and D. Tollervey. 1995b. *Trans*-acting factors in yeast pre-rRNA and pre-snoRNA processing. *Biochem. Cell. Biol.* **73:**803-812.

Lafontaine, D. L. J., C. Bousquet-Antonelli, Y. Henry, M. Caizergues-Ferrer, and D. Tollervey. 1998a. The box H + ACA snoRNAs carry Cbf5p, the putative rRNA pseudouridine synthase. *Genes Dev.* **12:**527-537.

Lafontaine, D. L. J., T. Preiss, and D. Tollervey. 1998b. Yeast 18S rRNA dimethylase Dim1p: a quality control mechanism in ribosome synthesis? *Mol. Cell. Biol.* **18:**2360-2370.

Lafontaine, D. L. J., and D. Tollervey. 1998c. Birth of the snoRNPs: the origin of the eukaryotic rRNA modification system. Submitted for publication.

Lübben, B., P. Fabrizio, B. Kastner, and R. Lührmann. 1995. Isolation and characterization of the small nucleolar ribonucleoprotein particle snR30 from *Saccharomyces cerevisiae*. *J. Biol. Chem.* **270:**11549-11554.

Maden, B. E. H. 1990. The numerous modified nucleotides in eukaryotic ribosomal RNA. *Prog. Nucleic Acid Res. Mol. Biol.* **39:**241-303.

Maden, B. E. H. 1997. Guides to 95 new angles. *Nature* **389:**129-131.

Maden, B. E. H., and J. M. X. Hughes. 1997. Eukaryotic ribosomal RNA: the recent excitement in the nucleotide modification problem. *Chromosoma* **105:**391-400.

Meier, U. T., and G. Blobel. 1992. Nopp140 shuttles on tracks between nucleolus and cytoplasm. *Cell* **70:**127-138.

Meier, U. T., and G. Blobel. 1994. NAP57, a mammalian nucleolar protein with a putative homolog in yeast and bacteria. *J. Cell Biol.* **127:**1505-1514.

Moritz, M., A. G. Paulovitch, Y. Tsay and J. L. Woolford. 1990. Disruption of yeast ribosomal proteins L16 or rp59 disrupts ribosome assembly. *J. Cell Biol.* **111:**2261-2274.

Moritz, M., B. A. Pulaski, and J. L. Woolford. 1991. Assembly of 60S ribosomal subunits is perturbed in temperature-sensitive yeast mutants defective in ribosomal protein L16. *Mol. Cell. Biol.* **11:**5681-5692.

Morrissey, J. P., and D. Tollervey. 1993. Yeast snR30 is a small nucleolar RNA required for 18S rRNA synthesis. *Mol. Cell. Biol.* **13:**2469-2477.

Ni, J., A. L. Tien, and M. J. Fournier. 1997. Small nucleolar RNAs direct site-specific synthesis of pseudouridine in ribosomal RNA. *Cell* **89:**565-573.

Nurse, K., J. Wrzesinski, A. Bakin, B. Lane, and J. Ofengand. 1995. Purification, cloning, and properties of the tRNA $\Psi 55$ synthase from *Escherichia coli*. *RNA* **1:**102-112.

Peculis, B. 1997. RNA processing: pocket guides to ribosomal RNA. *Curr. Biol.* **7:**480-482.

Séraphin, B., L. Kretzner, and M. Rosbach. 1988. A U1 snRNA: pre-mRNA base pairing interaction is required early in yeast spliceosome assembly but does not uniquely define the 5′ cleavage site. *EMBO J.* **7:**2533-2538.

Simos, G., H. Tekotte, H. Grosjean, A. Segref, K. Sharma, D. Tollervey, and E. C. Hurt. 1996. Nuclear pore proteins are involved in the biogenesis of functional tRNAs. *EMBO J.* **15:**2270-2284.

Sirum-Connolly, K., and T. L. Mason. 1993. Functional requirement of a site-specific ribose methylation in ribosomal RNA. *Science* **262:**1886-1889.

Sirum-Connolly, K., and T. L. Mason. 1995. The role of nucleotide modifications in the yeast mitochondrial ribosome. *Nucleic Acids Symp.* **33:**73-75.

Smith, C. M., and J. A. Steitz. 1997. Sno storm in the nucleolus: new roles for myriad small RNPs. *Cell* **89:**669-672.

Szweykowska-Kulinska, Z., B. Senger, G. Keith, F. Fasiolo, and H. Grosjean. 1994. Intron-dependent formation of pseudouridine in the anticodon of *Saccharomyces cerevisiae* minor tRNA[Ile]. *EMBO J.* **13:**4636-4644.

Thamana, P., and C. R. Cantor. 1978. Studies on ribosome structure and interactions near the $m_2^6Am_2^6A$ sequence. *Nucleic Acids Res.* **5:**805-823.

Tollervey, D. 1996. *Trans*-acting factors in ribosome synthesis. *Exp. Cell Res.* **229:**226-232.

Tollervey, D., H. Lehtonen, R. Jansen, H. Kern, and E. C. Hurt. 1993. Temperature-sensitive mutations demonstrate roles for yeast fibrillarin in pre-rRNA processing, pre-rRNA methylation, and ribosome assembly. *Cell* **72:**443-457.

van Buul, C. P., J., J., M. Hamersma, W. Visser, and P. H. van Knippenberg. 1984. Partial methylation of two adjacent adenosines in ribosomes from Euglena gracilis chloroplasts suggests evolutionary loss of an intermediate stage in the methyl-transfer reaction. *Nucleic Acids Res.* **12:**9205-9208.

van Knippenberg, P. H. 1986. Structural and functional aspects of the N^6,N^6 dimethyladenosines in 16S ribosomal RNA, p. 412-424. *In* B. Hardesty and G. Kramer (ed.), *Structure, Function and Genetics of Ribosomes*. Springer-Verlag, New York, N.Y.

Venema, J., C. Bousquet-Antonelli, J.-P. Gelugne, M. Caizergues-Ferrer, and D. Tollervey. 1997. Rok1p is a putative helicase required for rRNA processing. *Mol. Cell. Biol.* **17:**3398-3407.

Venema, J., and D. Tollervey. 1995. Processing of pre-ribosomal RNA in *Saccharomyces cerevisiae*. *Yeast* **11:**1629-1650.

Vilardell, J., and J. Warner. 1997. Ribosomal protein L32 of *Saccharomyces cerevisiae* influences both the splicing of its own transcript and the processing of rRNA. *Mol. Cell. Biol.* **17:**1959-1965.

Chapter 16

Editing of tRNA

DAVID H. PRICE AND MICHAEL W. GRAY

The term "RNA editing" was first coined more than a decade ago to describe the phenomenon of uridine insertion into trypanosomatid mitochondrial transcripts (Benne et al., 1986; also see Chapter 21). Since then, RNA editing has become something of a catch-all for a wide variety of unusual RNA processing events that are related neither by biochemistry nor by evolution (see Appendix 2). One definition of RNA editing (Gray and Covello, 1993) describes it as "any process (co- or posttranscriptional) that changes the primary nucleotide sequence of an RNA molecule from that encoded by the corresponding gene." These "processes" include changes made "by nucleotide addition or deletion, or by base exchange" (Adler and Hajduk, 1994), but also encompass base conversion reactions (e.g., C to U and A to I). Whereas this definition and these processes may suffice to describe the editing of mRNAs, whose function is to encode protein sequence, they do not span the full range of posttranscriptional alterations sustained by tRNAs (or rRNAs) during their maturation.

Unlike mRNAs, transcripts of tRNA and rRNA genes are themselves converted into the functional entities they encode. To generate this functionality, it is often necessary to modify, sometimes radically, the chemistry of particular residues within these RNA molecules. These programmed alterations (of which over 93 have been described to date; see Appendix 1) greatly expand the range of possible interactions in which these nucleic acids can engage. For example, in most tRNAs, the motif TΨC appears at the same position in the so-called T loop of the secondary structure (see Appendix 1). When present, the Ψ (pseudouridine) residue in this motif participates in the maintenance of the three-dimensional structure of the tRNA (Dirheimer et al., 1995). As this book demonstrates, such modified nucleosides are likely to be critical for the proper functioning of tRNA and rRNA molecules. Because these modifications represent specific changes to the primary structure of RNA molecules, and because certain modifications have the same functional consequences as certain editing events, it would seem reasonable to include examples of RNA modification under the banner of RNA editing (Grosjean, 1996). A case in point is lysidine (L), a derivative of C that pairs with A, not G (Muramatsu et al., 1988b). Both C-to-L modification and C-to-U editing generate the same phenotype, that is, pairing with A rather than G (Muramatsu et al., 1988a).

This concept is further emphasized by the recent identification of a yeast protein, HRA400, that appears to be responsible for site-specific formation of inosine, via adenosine deamination, in certain tRNAs (Gerber et al., 1997). Within its deaminase domains, HRA400 shows a clear evolutionary relationship with ADAR1 and ADAR2, the enzymes responsible for the editing of a number of vertebrate pre-mRNAs, including those encoding glutamate receptor subunits (see Chapter 19). These observations argue strongly for a fundamental unity of "editing" and "nucleoside modification" in tRNA. Nevertheless, for the sake of brevity, and because they are dealt with extensively elsewhere in this volume (Chapters 1 to 15 and 24 to 29), we will not discuss modified nucleosides in this article. Rather, we will distinguish between those sequence alterations that introduce modified nucleosides into RNA (and therefore must be performed posttranscriptionally) and those that introduce one of the four canonical nucleosides (and therefore could, in theory, have been accomplished by an encoded sequence change at the DNA level). Accordingly, for the purposes of this article, we define RNA editing as a programmed alteration of RNA primary structure such that the resulting sequence could have been directly encoded in the corresponding gene.

As is evident from the foregoing discussion, our definitions do not unite the various examples of RNA editing according to their biochemistry, but rather ac-

---

David H. Price and Michael W. Gray • Department of Biochemistry, Dalhousie University, Halifax, Nova Scotia, Canada B3H 4H7.

cording to their substrates and their consequences. RNA editing systems are also united by the types of questions we ask about them (Benne, 1996). How does RNA editing occur? When does it occur? Why (in an evolutionary sense) does it persist? What directs the editing activity to particular sites? How is editing integrated with the other RNA processing events? Is anything gained by making these "corrections" at the RNA level? We shall examine these questions as they apply to the editing of tRNAs in various systems.

Transfer RNAs, the adapter molecules that link the two languages of nucleic acid (nucleotide) and protein (amino acid), are usually between 70 and 85 nucleotides in length. Following conventional nomenclature, the nucleotide at the 5'-end of the mature tRNA is designated N_1, whereas the position preceding the -CCA_{OH} extension at the 3'-end of the tRNA is numbered N_{73} (Appendix 1). Positions 34–36 encompass the anticodon. N_{73}, the discriminator nucleotide, is unpaired in all tRNAs except tRNA^{His}; in that tRNA, the nucleotide opposite N_{73} is designated N_{-1}.

Although tRNAs share common sequence motifs and the same general structure, they differ significantly from one another in nucleotide sequence, and by the presence or absence of particular modified nucleosides. In their various permutations, these differences provide tRNAs with their individual identities. The tRNA structure must therefore strike a balance between the general and the particular, enabling tRNAs to interact very specifically with some enzymes (for examples, see Farruggio et al., 1996; Freist et al., 1996; Gabriel et al., 1996; Meinnel et al., 1995; Sherman et al., 1995) but nonspecifically with others (e.g., Clark et al., 1995; Deutscher, 1995; Lee et al., 1997; Levinger et al., 1995; Tallsjö et al., 1996). These interactions are particularly important at various stages in the process by which mature tRNAs are generated from their precursor transcripts.

When tRNA genes are transcribed, often as multimers, they are flanked by extra nucleotides at both the 3'- and 5'-ends of the coding sequence. The reactions that remove these flanking sequences may occur by a number of different pathways. In eukaryotic systems (both nuclear and organellar), processing of pre-tRNA (Fig. 1) appears to proceed via two endonucleolytic cleavages, one at the mature 5'-end of the tRNA, the other just 3' to the discriminator nucleotide (Martin, 1995). Following 5' and 3' trimming, the sequence -CCA_{OH} is added to the 3'-end. At specific points along the processing pathway, the various nucleoside modifications previously mentioned are generated. In addition, for certain tRNAs, maturation has been shown to involve the editing of their RNA sequence (Fig. 2). These editing events represent additional steps in posttranscriptional processing and, like nucleoside modifications, they may occur at different stages in the pathway (Fig. 1).

DETECTION AND QUANTIFICATION OF RNA EDITING

Novel RNA editing events usually are discovered as discrepancies between gene and (corresponding) mature RNA sequences. In many cases, the existence of an RNA editing event is suggested by anomalies in the predicted structure of the primary transcript. These anomalies might include a frameshift within the coding region of an mRNA (Benne et al., 1986), or an unstable helix within a conserved region of a structural RNA (Lonergan and Gray, 1993a). However, in all cases, conclusive demonstration of RNA editing is dependent on both sequencing the mature transcript and demonstrating that no gene copies exist that encode the "edited" sequence. There are a number of methods for determining the sequence of the putatively edited RNA, including reverse transcriptase sequencing (RT-sequencing) (Lonergan and Gray, 1993a), sequencing of cDNA clones (Tomita et al., 1996a), mobility shift analysis of RNase T_1 fragments (Beier et al., 1992), and chromatography of nuclease P1 digestion products (Mörl et al., 1995). Evidence for insertional editing can also come from primer extension assays that reveal discrepancies between the actual length of an RNA and that predicted from its gene sequence.

To quantify RNA editing (that is, to determine the proportion of an RNA population that has been edited) one usually compares the intensity of the bands on a sequencing (or primer extension) ladder that correspond to the edited and unedited versions of the RNA. In a variant of this technique, the minisequencing assay, a radiolabeled deoxynucleotide corresponding to either the encoded or introduced nucleotide (at the site of editing) is added to the end of an oligonucleotide that pairs with the sequence immediately downstream of the site of editing (Mörl et al., 1995). The relative intensity of the bands that result from the incorporation of each dNTP gives the desired ratio. An alternative to the direct sequencing of cDNAs is to screen cDNA clones and calculate the percentage of clones that possess the edited sequence. It is also possible to analyze the edited RNA directly by chromatography of nuclease P1 digestion products (Mörl et al., 1995). As mentioned above, the relative intensity of the spots corresponding to the edited and unedited nucleotides indicates the frequency of editing. A less-common approach to

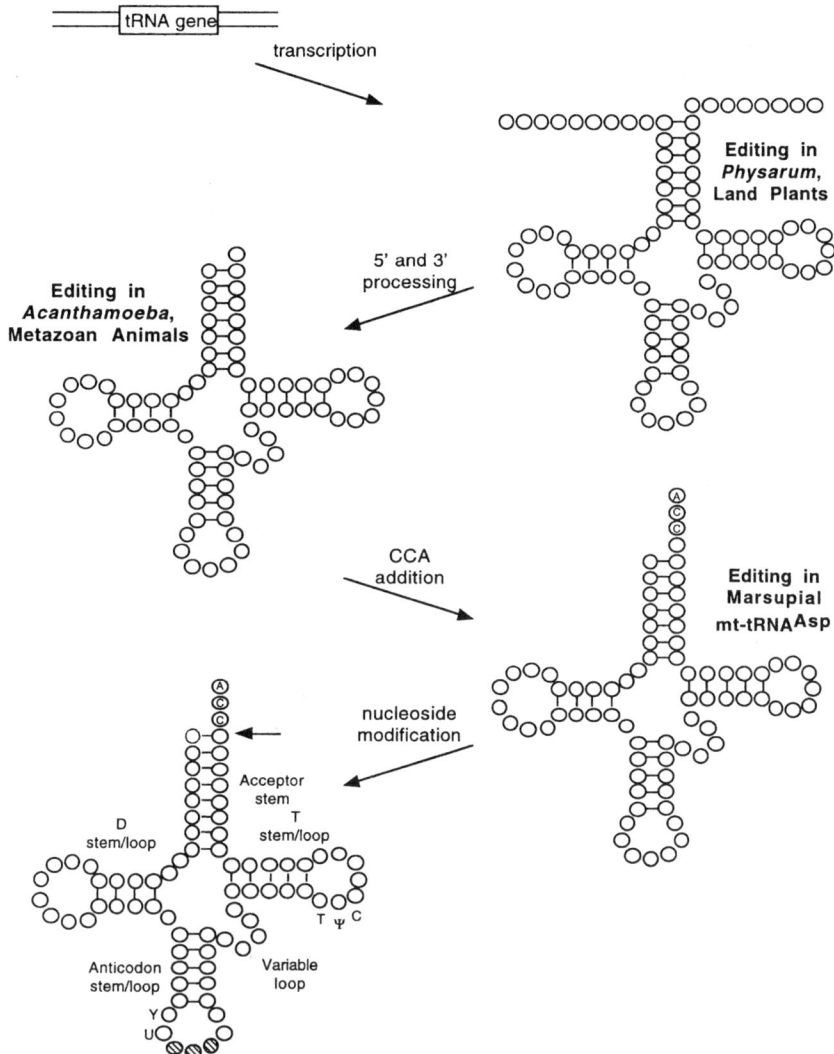

Figure 1. Generalized tRNA processing pathway, showing the major steps in the maturation of a typical pre-tRNA. The various known tRNA editing systems (bold captions) act at different stages in the maturation of tRNA, thereby constituting additional steps in the processing pathway. The final diagram is of a generalized tRNA secondary structure, showing the various helical and loop regions and conserved nucleosides that are discussed in the text [Y, pyrimidine; T, 5-methyluridine ('ribothymidine'); Ψ, pseudouridine (5-ribosyluracil)]. Positions encompassing the anticodon sequence (residues 34–36) are shaded. The extra 5'-nucleotide that is present only in tRNA^{His} is drawn lightly.

quantifying RNA editing is the use of restriction enzymes to digest reverse transcriptase-polymerase chain reaction (RT-PCR) products (see Chapter 22). If editing introduces (or removes) a particular restriction site, the proportion of the PCR product that is digested (or undigested) will correspond to the proportion of the RNA population that is edited.

FALSE STARTS?

The first case purporting to provide evidence of tRNA editing (Diamond et al., 1990) concerned a selenocysteine-inserting tRNA from bovine liver. It was reported that the newly determined gene sequence for this tRNA did not match that of either of two previously sequenced isoacceptors. The discrepancies were rationalized by invoking an RNA editing system in which pyrimidines were interconverted by nucleotide amination and deamination. This would be similar to the RNA editing system that has been described in plant mitochondria (see Chapter 17). However, subsequent work (Amberg et al., 1993) demonstrated that this selenocysteine tRNA is not, in fact, edited. Nevertheless, this report was the first to suggest the possibility that structural RNA might be edited.

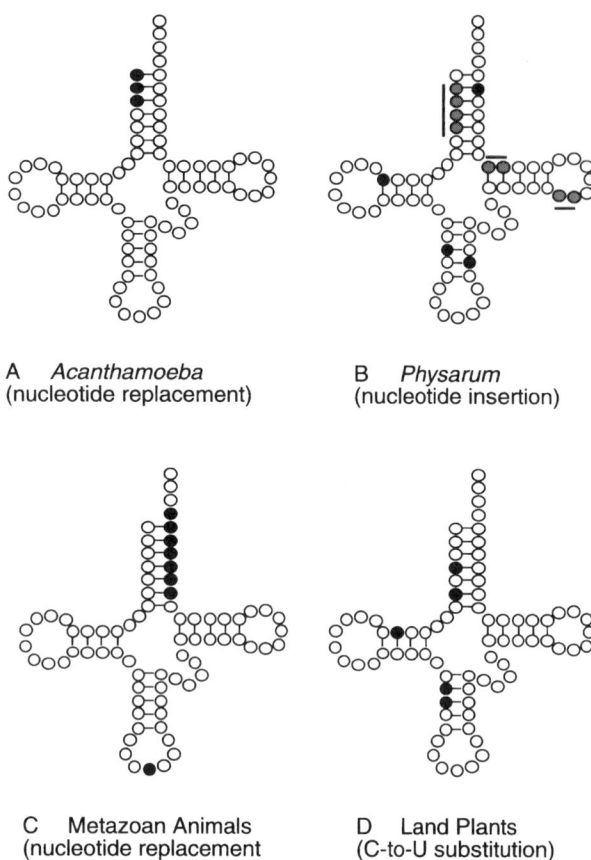

Figure 2. Diagrammatic representations of tRNA secondary structure indicating the positions (filled circles) at which editing is known to occur in different mitochondrial tRNA editing systems. (A) *Acanthamoeba* spp., nucleotide replacements at one or more of the first three positions at the 5′-end; (B) *Physarum polycephalum*, nucleotide insertions at the indicated positions (solid lines denote that a single-nucleotide insertion occurs within the delineated region [shaded circles], but cannot be localized precisely); (C) nucleotide replacements within the 3′-half of the acceptor stem in metazoan animals; C-to-U substitution in the central position of the anticodon sequence of marsupial mitochondrial tRNA^{Asp}; (D) land plants (angiosperms and gymnosperms), C-to-U substitutions at the indicated positions.

Another putative example of tRNA editing concerns a mammalian cytoplasmic tRNA, tRNA^{Asp}(GUC), from rat liver. In this case, two isoforms had been identified by mobility shift analysis of RNase T$_1$ fragments: a major form containing the sequence U$_{32}$C$_{33}$ in the anticodon loop, and a minor form having the sequence C$_{32}$U$_{33}$ at the same position (see Appendix 1). When the corresponding gene was examined both by sequencing of PCR amplification products and by single-strand conformation polymorphism (Beier et al., 1992), it was shown to encode these particular nucleotides as C$_{32}$T$_{33}$. Curiously, position 33 is occupied by a uridine in almost all known tRNAs (Dirheimer et al., 1995). Such an editing event would therefore be decidedly atypical in that it would alter the genomically encoded sequence to one less conserved among tRNA species. In fact, given the absence of the conserved U$_{33}$ residue, it is questionable whether the edited isoform would be functional in vivo. Moreover, in this tRNA G$_{34}$, the nucleoside in the first position of the anticodon is partially converted to the modified nucleoside queuosine (Q) or one of its derivatives (Q*) through a mechanism involving an exchange of guanine for queuine (the base moiety of Q) (Chapter 9). In *Escherichia coli* and *Xenopus* oocytes, the queuine-incorporating enzyme recognizes the motif UGU, the modified G being flanked by two uridine residues (Mörl et al., 1995). This corresponds to the unedited sequence of tRNA^{Asp}(GUC). It is therefore unclear what purpose would be served by the alteration of nucleosides C$_{32}$ and U$_{33}$, and what selective pressure might be acting to maintain such an editing system.

To confound the situation, a recent reexamination of this proposed editing event found no evidence of nucleoside alteration at the proposed sites (Tomita et al., 1996a). In this study, the sequence of tRNA^{Asp}(GUC) was determined by both direct enzymatic sequencing and the sequencing of clones of the anticodon stem-loop that had been generated by RT-PCR. Both of these methods are more direct than the method used in the study of Beier et al. (1992), and are less susceptible to artifacts resulting from the presence of hypermodified bases such as Q. Such modifications are known to introduce migration anomalies during electrophoretic analysis, which can result in sequence ambiguities. Tomita et al. (1996a) suggested that the electrophoretic pattern observed by Beier et al. (1992), which had been interpreted as editing, was instead a consequence of the partial conversion of G$_{34}$ to Q*$_{34}$ in rat cytoplasmic tRNA^{Asp}. In the absence of any other data supporting the editing of positions 32 and 33 in this tRNA, it seems most likely that these nucleosides are not edited, and that the mature sequence is, in fact, C$_{32}$U$_{33}$Q*$_{34}$.

TRANSFER RNA EDITING IN EUBACTERIA?

A provocative case in eubacteria that fits the formal definition of editing is that of tRNA^{Ser}(GGA) in *E. coli* K-12. Two chromatographically distinguishable forms, I and V, were found to differ at a single position (N$_{20}$), which was either C (in V, minor species) or D (dihydrouridine; in I, major species) (Grosjean et al., 1985). Although the initial expectation was that two separate tRNA^{Ser}(GGA) genes existed, one (V) encoding C$_{20}$ and the other (I) encoding U$_{20}$ within the D loop, only the C$_{20}$ version was found during characterization of the complete set of *E. coli* tRNA genes (Komine et al., 1990). This suggests that (an) enzymatic posttranscriptional editing/modification event(s) may convert C_{20} to D_{20}, possibly via U_{20} (Motorin et al., 1996).

To date, efforts to demonstrate such conversion in vitro, using *E. coli* extracts and synthetic tRNA substrates, have not been successful. Although a control substrate containing U_{20} is efficiently converted to a D_{20} product, a C_{20}-containing substrate is not (Grosjean, 1997). This raises the possibility that the presumed conversion of C_{20} to U_{20} occurs cotranscriptionally rather than posttranscriptionally (Grosjean, 1997). If this example is eventually confirmed as a case of C-to-U editing, it would be the first such case documented for a eubacterium; moreover, it would represent an example of obligatory coupling of editing and modification.

EDITING OF MITOCHONDRIAL tRNAs

Acanthamoeba spp. and Chytridiomycete Fungi

The first example of tRNA editing in a mitochondrial (mt) system is that reported to occur in the ameboid protozoon *Acanthamoeba castellanii* (Lonergan and Gray, 1993a, 1993b; Gray and Lonergan, 1993). In the course of sequencing *A. castellanii* mtDNA, five tRNA genes were found in the region between the genes for the large and small subunit mt-rRNAs. The predicted tRNA secondary structures structures had all of the features expected of conventional tRNAs, except for one striking anomaly: the consistent presence of one or more mismatches within the first three base pairs (1:72, 2:71, 3:70) of the aminoacyl acceptor stem (Fig. 2A and 3). Because this region is important in defining the tertiary structure of tRNA molecules (Dirheimer et al., 1995; Martin, 1995), and in some cases provides major identity elements for tRNA recognition by aminoacyl-tRNA synthetases (Ibba et al., 1996), these predicted acceptor stem mismatches were judged to be incompatible with tRNA function. In fact, analysis of the mature tRNAs revealed that the sequences of the 5'-portions of all of the anomalous acceptor stems were altered in a way that corrected the gene-predicted mismatches. The observed nucleotide substitutions consisted of both purine-to-purine and pyrimidine-to-purine changes, suggesting a mechanism involving base or nucleotide replacement.

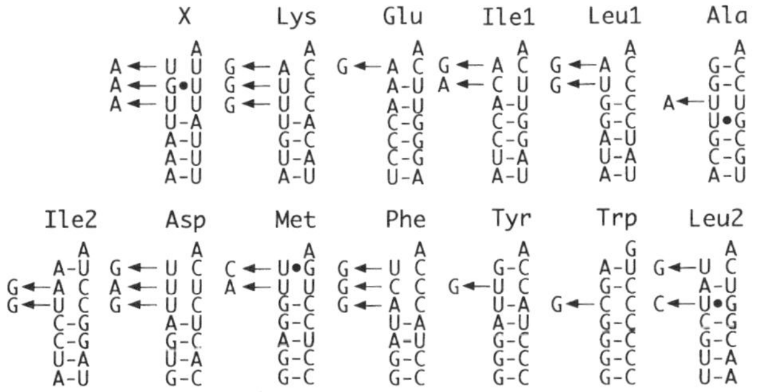

Since this discovery, sequencing of the entire mitochondrial genome of *A. castellanii* has revealed that 13 of 16 encoded tRNA genes predict this pattern of mismatches in their tRNA transcripts (Burger et al., 1995) (Fig. 3). The RNA sequence of each of these tRNAs was determined either by RT sequencing or by sequencing of PCR products that had been generated from circularized tRNAs (Price and Gray, 1997). (Because tRNAs have 3'-hydroxyl and 5'-monophosphate termini, the two ends can be joined by T_4 RNA ligase. RT-PCR amplification primed off the resulting circular RNA can then be used to generate PCR products that span the 5'- and 3'-halves of the acceptor stem [Yokobori and Pääbo, 1995a].) Analysis of these PCR products confirmed that sequence alterations are confined to the 5'-side of the acceptor stem. These results also have demonstrated that editing restores standard (G:C or A:U) base pairing in all cases (Fig. 3).

These studies have also shown that in addition to purine-to-purine and pyrimidine-to-purine conversions, two pyrimidine-to-pyrimidine conversions occur. In both of these cases, editing changes a U·G base pair to a C:G base pair (Price and Gray, 1997). Notably, in mt-tRNA^{Ala}(UGC), a $U_3·U_{70}$ mismatch is converted to an $A_3:U_{70}$ base pair, rather than the usual $G_3·U_{70}$ pair, which is a critical identity determinant in tRNAs^{Ala} (Schimmel and Ripmaster, 1995). Significantly, a $U_4·G_{69}$ base pair in the same tRNA remains unchanged, suggesting that the editing activity is only able to alter the first three residues at the 5'-end of a tRNA (see the following paragraph), possibly due to structural constraints at the active site(s) in one or more components of the editing machinery.

These observations have been extended by a study of mitochondrial sequence variation among 16 genetically divergent species of *Acanthamoeba* within the region encompassed by the two rRNA genes (Ledee and Byers, 1997). In these species, similar patterns of mismatches are observed in at least four of the five tRNA genes in the rRNA intergene region. Sequencing of some of the mature tRNAs has con-

Figure 3. Structures of the acceptor stems of the 13 *Acanthamoeba castellanii* mtDNA-encoded tRNAs that are edited, with arrows indicating the 26 edits that generate G:C or A:U base pairs at initially mispaired, or U·G/G·U-paired, positions. All of the predicted edits have been verified experimentally (Lonergan and Gray, 1993a; Price and Gray, 1997).

firmed that these base pairs are restored at the RNA level.

Curiously, this particular editing pattern has also been observed in the mitochondrial tRNAs of *Spizellomyces punctatus*, a chytridiomycete fungus (Laforest et al., 1997; Lang, 1997). Because these fungal species are not known to be specific evolutionary relatives of *Acanthamoeba* spp., it must be presumed that the editing activity was either (i) present in the last common ancestor of *Acanthamoeba* and the chytridiomycete fungi, and selectively retained in only a few lineages upon subsequent divergence; (ii) transferred laterally between phyla; or (iii) acquired independently. The first scenario seems unlikely in view of the fact that in many species that are more closely related to either *Acanthamoeba* or *Spizellomyces* than these two genera are to each other, there is currently no evidence of this mode of tRNA editing. Even among chytridiomycete fungi, some species, such as a *Harpochytrium* sp. and a *Monoblepharella* sp., seem to possess the editing activity (Lang, 1997), whereas others, such as *Allomyces macrogynus*, do not (Laforest et al., 1997). Postulating a common origin for this mode of tRNA editing would consequently imply its independent loss from many lineages, an occurrence we consider unlikely. Invoking one of the other two explanations has significant implications for how we view the activity that constitutes the editing machinery. That the same machinery could arise more than once, or be transferred between distantly related species, suggests that this particular type of tRNA editing is mediated by a small number of proteins, perhaps even a single polypeptide.

Among the 16 tRNA genes encoded in the mitochondrial genome of *A. castellanii* is an mt-tRNA^X gene, whose product would possess an eight-nucleotide anticodon loop (Burger et al., 1995). This unusual feature, coupled with the similarity of the primary structure of the mt-tRNA^X to that of the immediately upstream mt-tRNA^{Phe}(GAA) gene (Burger et al., 1995), suggests that *trnX* may not express a functional product. This inference is supported by analysis of PCR products generated from circularized tRNA^X transcripts: in none of 21 clones did the gene product contain the 3'-CCA$_{OH}$ tail that is present in all mature, functional tRNAs (Price and Gray, 1997). Surprisingly, though, seven of these clones were fully edited. It thus appears that the tRNA^X gene is expressed, and that a tRNA-like product is generated. This RNA is a substrate for the editing activity; however, the edited product does not appear to be functional in protein synthesis. These observations suggest that editing may occur prior to addition of the 3'-CCA$_{OH}$ tail, which is clearly not a required recognition element for the editing activity (see Fig. 1).

In *A. castellanii*, recent work has focused on the development of an in vitro editing system to facilitate the analysis of the mechanism and substrate specificity of the editing activity, as well as to facilitate the temporal relationship of editing to the other steps in tRNA processing. Our proposed two-step working model is one in which the first three nucleotides at the 5'-end of the acceptor stem are removed and then replaced, using the 3'-half of the acceptor stem as an internal guide sequence (Lonergan and Gray, 1993a). This process would require at least two different enzymatic activities: an endo- or exonuclease (5'-to-3') to remove the nucleotides, and a 3'-to-5' nucleotidyltransferase activity to replace them (Fig. 4A). This second activity, like 5'-to-3' RNA polymerases, would rely for specificity on the formation of standard base pairs between the newly synthesized strand and its template. However, unlike standard RNA polymerases, it would have to act in a 3'-to-5' direction; moreover, unlike some other nucleotidyltransferases, it would not extend the new strand beyond the mature 5'-position in the acceptor stem of the mature tRNA (Lonergan and Gray, 1993a).

Work to date (Price and Gray, 1997) has demonstrated the presence, within intact mitochondria, of an activity capable of incorporating [α-^{32}P]rNTPs into the 5'-ends of tRNA-sized RNA molecules. In experiments in vitro, using a crude mitochondrial ex-

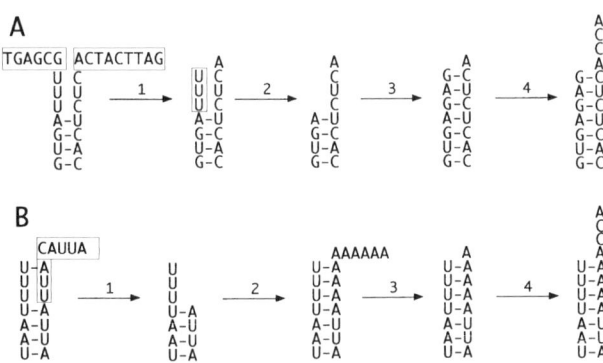

Figure 4. Working models of mitochondrial tRNA editing. (A) *Acanthamoeba castellanii* (mt-tRNA^{Asp}): (1) 5' and 3' processing by specific endo- and exonucleases to remove flanking sequences (boxed); (2) removal of the first three 5'-nucleotides (boxed) by a specific endonuclease or 5'-to-3' exonuclease; (3) replacement of the first three 5'-nucleotides by a template-dependent 3'-to-5' nucleotidyltransferase; (4) addition of the -CCA$_{OH}$ extension by a template-independent 5'-to-3' tRNA nucleotidyltransferase. (B) Metazoan animals (the example provided is mt-tRNA^{Gly} of *Euhadra herklotsi*): (1) 5' processing of a downstream, overlapping tRNA sequence results in loss of several 3'-nucleotides (boxed nucleotides); (2) template-independent 5'-to-3' addition of A residues (oligoadenylation); (3) 3' processing immediately after the discriminator nucleotide (position 73 [Fig. 1]); (4) addition of the -CCA$_{OH}$ extension by a template-independent 5'-to-3' tRNA nucleotidyltransferase.

tract depleted of endogenous tRNAs, this activity incorporates nucleotide residues exclusively into the first three positions of a number of substrates, including an *E. coli* tRNA^{Tyr} and a full-length runoff transcript of the *A. castellanii* mt-tRNA^{Leu1}(UAG) gene. The labeling of these two substrates would require the removal of nucleotides at the 5'-end of the tRNAs before nucleotide incorporation, and this is what is observed. It also appears that the nucleotides are, in fact, incorporated in a 3'-to-5' direction, because if one omits from the in vitro system the rNTP corresponding to the nucleotide to be incorporated at positions N_1 and N_2, the resulting product is two nucleotides shorter than the product obtained in the presence of all of the required nucleotides. As well, in the absence of a 5'-triphosphate, nucleotide incorporation does not occur when ATP or GTP is omitted from the reaction. These (unincorporated) nucleotides may be required to activate the 5'-terminus, converting the 5'-monophosphate to a 5'-triphosphate before the RNA can be extended in the 3'-to-5' direction. Significantly, cytoplasmic tRNAs from *A. castellanii* are not labeled by the mitochondrial extract. Together, these results provide strong evidence that the activity present in an *A. castellanii* mitochondrial extract is the same as that responsible for editing of the mt-tRNAs in vivo. Current work is directed toward further defining the mechanism and substrate specificity of this novel activity.

Physarum polycephalum

Another quite distinct example of tRNA editing has been found in the mitochondria of the plasmodial slime mold *Physarum polycephalum* (Chapter 22). Of the five *Physarum* mitochondrial tRNA (mt-tRNA) genes so far identified, the transcripts of four of them (mt-tRNA^{Glu}, mt-tRNA^{Met}, mt-tRNA^{Lys} and mt-tRNA^{Pro}) appear to be edited by cytidine insertion (Miller et al., 1993) (see Fig. 2B). Editing consists of a single C insertion (at position 29) in mt-tRNA^{Lys}, but insertion of two C residues in both mt-tRNA^{Pro} (positions 13 and 71) and mt-tRNA^{Met} (positions 40 and either 64 or 65; the precise location of the 64/65 insertion has not been determined because it falls next to an encoded C). The mt-tRNA^{Glu} also requires two insertions to complete its maturation, one C (at position 2/3/4/5) and one U (at position 54/55) (Miller, 1997). The six C insertions all create C:G or G:C base pairs in double-stranded regions. The U insertion generates the potential for subsequent formation of the almost universally conserved TΨC motif in the T loop.

A number of studies have provided insights into the mechanism of editing in this system. In one case, partially edited sequences were reported in tRNA precursors containing two adjacent tRNAs (Miller et al., 1993). Moreover, it was found that different products contained insertions at different sites, suggesting that each site is edited independently and that there is no overall polarity to editing. These observations also indicate that 5' and 3' processing of the tRNA precursor is not a prerequisite for editing. More recently, it has been shown that nascent mRNAs present in stalled RNA polymerase complexes are substrates for editing by both cytidine and dinucleotide insertion (Visomirski-Robic and Gott, 1997a). In this study, the isolated RNAs were found to be edited to within 14 to 22 nucleotides of the stalled polymerase, suggesting that the insertional editing activity in *Physarum* is able to act quite close to the site of RNA synthesis. Thus, editing in this in vitro system may occur very soon after transcription, although whether it does so in vivo remains to be determined. Nevertheless, it seems that sequences located more than 22 nucleotides downstream of the insertion sites are not required for editing. These results, as well as those of other studies (Gott et al., 1993; Rundquist and Gott, 1995), seemingly are at odds with reports of partially edited transcripts. It may be that there is a "window of opportunity," closely following the transcription complex, in which nucleotide addition can occur, and any (insertional) edits that are missed at this stage are not able to be corrected later (Visomirski-Robic and Gott, 1997b). Questions regarding the mechanism of *Physarum* mitochondrial RNA editing await further study (Chapter 22).

Didelphis virginiana (Marsupial)

In the preceding examples, although tRNA editing is clearly important for correcting base pair mismatches, in no instance is the identity (that is, the decoding properties or aminoacylation specificity) of the tRNA altered (a possible exception is the mt-tRNA^{Ala} from *A. castellanii* previously noted [Lonergan and Gray, 1993a]). By contrast, for the marsupial mt-tRNA^{Asp}, a change in identity is the sole consequence of editing. During the sequencing of the mitochondrial genome of the North American opossum, *Didelphis virginiana*, it was noted that the mt-tRNA^{Asp} gene of this species encodes a GCC (Gly) anticodon (Janke and Pääbo, 1993) rather than the expected GUC (Asp) anticodon found in the homologous gene of both monotreme and placental mammals (Janke et al., 1994) (Fig. 2C). When examined at the RNA level, the nucleoside at the second position of the anticodon (N_{35}, a C in the genomic sequence) was found to be an equimolar mixture of C and U. Subsequent work (Mörl et al., 1995) demonstrated that the nucleoside produced by the editing activity is in fact a uridine, as opposed to a hypermodified nucleoside (such as lysidine) that could mimic uridine in the decoding process (Muramatsu et al., 1988a). Interestingly, the anticodon sequence GCC is present in the mt-tRNA^{Asp} genes of both American and Australian marsupials. This suggests that this editing event was present in the last common ancestor of all marsupials, and has since persisted in at least a half dozen species (Janke and Pääbo, 1993; Janke et al., 1994).

Analysis of mt-tRNA^{Asp} isolated from *Didelphis virginiana* revealed three different isoforms (Mörl et al., 1995), all of which contained the 3'-CCA$_{OH}$ extension and a number of modified residues (three pseudouridines and two methylated nucleosides). However, each of the isoforms had a different anticodon, either GCC, GUC, or QUC. As noted earlier, Q formation requires the target sequence UGU. It was therefore proposed that, in this case, editing follows methylation and pseudouridylylation but precedes Q formation. Thus, for this tRNA, editing follows the 5' and 3' trimming reactions, and is integral to the process of nucleoside modification.

In light of this, the presence within the steady-state population of a high proportion (50%) of the unedited tRNA was initially problematic. It suggested that the unedited mt-tRNA^{Asp}(GCC) is not just a processing intermediate, but a functional tRNA in its own right. Further support for this hypothesis came from the observation that the unedited mt-tRNA^{Asp}(GCC) is aminoacylated in vivo, and in fact is equivalent in abundance to the edited mt-tRNA^{Asp}(GUC) within the pool of aminoacylated tRNAs. Subsequent study revealed that, in vitro, mt-tRNA^{Asp}(GCC) is specifically charged with glycine, whereas mt-tRNA^{Asp}(GUC) is aminoacylated exclusively with aspartate (Börner and Pääbo, 1996) (Fig. 5). In addition, mt-tRNA^{Asp}(GU/CC) (a mixture of the two isoforms) produced in vivo can be aminoacylated in vitro with glycine and aspartate, but not with any other amino acid. It therefore appears that editing provides the major identity elements that enable the glycyl- and aspartyl-tRNA synthetases to distinguish between the two mt-tRNA^{Asp} isoforms.

That the unedited mt-tRNA^{Asp} functions as a mt-tRNA^{Gly} was quite unexpected, because a functional mt-tRNA^{Gly}, with anticodon UCC, was already known to exist in marsupial mitochondria (Börner and Pääbo, 1996). Usually, an mt-tRNA^{Gly}(UCC) is able to recognize all four glycine codons (GGN). By contrast, the unedited mt-tRNA^{Asp}(GCC) would only be able to recognize two of the four glycine codons (GGU and GGC). Furthermore, considering that animal mitochondria encode a limited set of tRNAs, and even resort to extended wobble pairing in decoding their mRNAs, it is surprising that two tRNAs with seemingly overlapping specificities would persist in this genetic system. A possible explanation (Börner et al., 1996) is that the mt-tRNA^{Gly}(UCC) is, for some reason, only able to decode the two glycine codons, GGA and GGG, which are not recognized by the unedited mt-tRNA^{Asp}(GCC). In this regard, it has been observed (Börner et al., 1996) that the nucleotide at position 32 of marsupial mt-tRNA^{Gly}(UCC) is uridine, rather than the cytidine found at that position in the mt-tRNA^{Gly}(UCC) of other animal species. As well, in *E. coli*, it has been shown that a C-to-U substitution at this position limits the tRNA to decoding GGA and GGG codons. It therefore seems likely that the two functional mt-tRNAs^{Gly} do, in fact, have complementary, rather than overlapping, specificities. How such an arrangement evolved and why it persists are unknown, but a possible model is described below.

Land Plants

Unlike in animal mitochondria, where C-to-U editing is a rare and phylogenetically restricted event, C-to-U editing in plant mitochondria (Covello and Gray, 1989; Gualberto et al., 1989; Hiesel et al., 1989) is rampant, affecting nearly all mRNAs (Chapter 17). To a much lesser extent, the same type of RNA editing is also found in plant mitochondrial intron sequences and tRNAs, where editing has the general effect of enhancing the stability of secondary structures by correcting base pair mismatches (Gray and Covello, 1993).

To date, C-to-U editing has been reported for three plant mitochondrial tRNAs (Fig. 2D and Table 1), all being of the "native" type [inherited from the genome of the eubacteria-like endosymbiont ancestor (Joyce and Gray, 1989; Dietrich et al., 1996)]. The first such documented example was that of mt-tRNA^{Phe}(GAA) from bean and potato, in which a $C_4 \cdot A_{69}$ mismatch was found to be converted to a $U_4:A_{69}$ base pair in the mature tRNA (Maréchal-Drouard et al., 1993). Unlike the marsupial mt-tRNA^{Asp} case previously discussed, editing in the plant case appears to be complete because the unedited nucleoside was not detected in the population of mature tRNA species (Maréchal-Drouard et al., 1993). More recently, it has been shown that the homologous tRNA is also edited in *Oenothera berteriana* (evening primrose) (Binder et al., 1994).

A second edited plant mt-tRNA is the mt-tRNA^{Cys}(GCA) of *O. berteriana*, where a $C_{28} \cdot U_{42}$ mismatch (in the anticodon stem) is converted to a

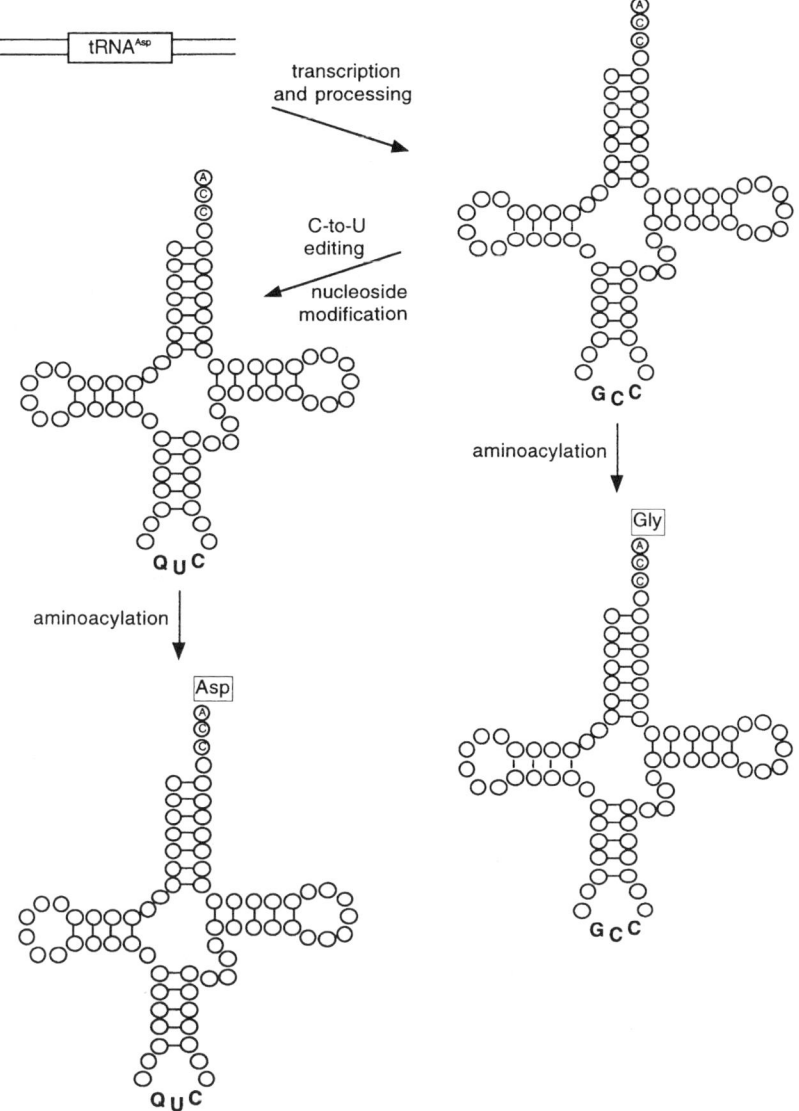

Figure 5. Aminoacylation of edited and unedited mt-tRNA^Asp in *Didelphis virginiana*. Following transcription from a single gene, the 5′- and 3′-processed mt-tRNA^Asp transcript may undergo C-to-U editing at position 35. The unedited isoform, mt-tRNA^Asp(GCC), is aminoacylated with glycine, whereas the edited isoform, mt-tRNA^Asp(GUC), is aminoacylated with aspartate.

$U_{28} \cdot U_{42}$ mismatch (Binder et al., 1994). This may be an example of "serendipitous" editing, in which an RNA is altered because it meets the substrate requirements of the editing activity, even though the editing event does not correct a structural defect. On the other hand, this U·U mismatch is encoded in the homologous mitochondrial tRNA gene in liverwort (*Marchantia polymorpha*). Such U·U mismatches have been noted in several mt-tRNAs, and may be less structurally disruptive than C·U mismatches (Maréchal-Drouard et al., 1993).

The third example of tRNA editing is that of larch mt-tRNA^His(GUG), the first such report for a gymnosperm. Here, three C·A mismatches, one in each of the acceptor ($C_6 \cdot A_{67}$), D ($C_{12} \cdot A_{23}$) and anti-codon ($A_{29} \cdot C_{41}$) stems, are converted to U:A or A:U base pairs (Maréchal-Drouard et al., 1996a).

Studies of each of the three edited tRNAs have shown that, as in *Physarum*, editing occurs early in tRNA processing (Table 1). In *O. berteriana*, one-third of the clones generated from 5′- and 3′-unprocessed mt-tRNA^Cys precursors were found to be edited at position 28. For mt-tRNA^Phe, almost half of the cloned precursors showed editing at position 4. When edited and unedited runoff transcripts of *O. berteriana* mt-tRNA^Phe were incubated in a pea mitochondrial extract, editing had a significant effect on the efficiency of both 5′- and 3′-end maturation (Binder et al., 1996; Marchfelder et al., 1996). Whereas mature tRNAs were rapidly generated in vi-

Table 1. Editing and processing of plant mt-tRNAs

tRNA	Species	Editing event	Processing of edited precursor	5' Processing of unedited precursor	3' Processing of unedited precursor	CCA added to unedited tRNA
mt-tRNA^{Phe}(GAA)	Bean[a]	$C_4 \cdot A_{69} \rightarrow U_4 : A_{69}$	Accurate and efficient	Accurate, but slow	Accurate and efficient	Yes[e]
mt-tRNA^{Phe}(GAA)	Potato[a]	$C_4 \cdot A_{69} \rightarrow U_4 : A_{69}$	Accurate and efficient	Accurate, but slow	Almost no genuine processing	No[d]
mt-tRNA^{Phe}(GAA)	Oenothera berteriana[b]	$C_4 \cdot A_{69} \rightarrow U_4 : A_{69}$				
mt-tRNA^{Cys}(CGA)	Oenothera berteriana[b]	$C_{28} \cdot A_{42} \rightarrow U_{28} : U_{42}$				
mt-tRNA^{His}(GUG)	Larch[c]	$C_6 \cdot A_{67} \rightarrow U_6 : A_{67}$ $C_{12} \cdot A_{23} \rightarrow U_{12} : A_{23}$ $A_{29} \cdot C_{41} \rightarrow U_{29} : A_{41}$	Accurate and efficient	No processing	No processing	No[c]

[a] Maréchal-Drouard et al., 1993.
[b] Binder et al., 1994.
[c] Maréchal-Drouard et al., 1996a.
[d] Binder et al., 1996; Marchfelder et al., 1996.
[e] Maréchal-Drouard et al., 1996b.

tro from the edited precursor, unedited precursors were quickly degraded. Primer extension analysis of processing products generated in vitro revealed that as well as being efficient, 5' and 3' processing of the edited precursor was highly accurate. The unedited precursor was correctly (albeit very slowly) processed only at the 5'-end of the tRNA; almost no genuine 3' processing (that is, immediately after the discriminator nucleotide) of the unedited precursor was observed. As well, lead cleavage experiments revealed significant differences in the higher-order structures of the two substrates, particularly around the D loop and at the base of the acceptor stem. Because secondary and tertiary structural features are crucial for the recognition of tRNA precursors by their processing enzymes (Martin, 1995; Yuan and Altman, 1995), these differences in higher-order structure may help to explain the different efficiencies with which the edited and unedited precursors are processed. It is also likely that the "looser" structure of the unedited precursor rendered it susceptible to attack by other RNases that were present in the pea mitochondrial extract, contributing to its rapid degradation.

Similar results (Maréchal-Drouard et al., 1996b) were observed for the potato mt-tRNA^{Phe}. Although editing did not affect the efficiency of aminoacylation in vitro, there was a 10-fold increase in the rate of processing of edited versus unedited precursor when in vitro transcripts were incubated in a potato mitochondrial extract. Another precursor, which would produce an mt-tRNA^{Phe} with a $C_4:G_{69}$ base pair, was processed four times as quickly as the unedited (genomic sequence) version. For both edited and unedited precursors, 5' and 3' processing was accurate, and the 3'-CCA$_{OH}$ sequence was added to the excised tRNA. Analysis of partially processed intermediates revealed that, while the efficiency of 5' processing was strongly affected by editing, 3' processing appeared to be unaffected. Thus, when clones of 3'-processed intermediates were isolated, only 50% were found to be edited. It was not, however, determined whether these unedited intermediates were products of accurate or aberrant 3' processing. In this case also, it was proposed that reduced processing efficiency of the unedited precursor was the result of an alteration in secondary and tertiary structure resulting from the absence of the $U_4:A_{69}$ base pair.

In larch, editing of the mt-tRNA^{His} also precedes 5' and 3' processing (Maréchal-Drouard et al., 1996a). Of 11 clones derived from mt-tRNA^{His} precursors, 5 contained the genomic sequence and 6 showed full editing. No partially edited precursors (altered at only one or two of the three sites) were detected. This suggests that editing, once initiated, is rapid and efficient. Other results have shown that editing of mt-tRNA^{His} is an absolute requirement for both 5′ and 3′ processing. When incubated in a potato mitochondrial extract, an in vitro-transcribed mt-tRNA^{His} precursor was processed properly only if it contained the edited sequence. As well, in contrast to the results with the potato mt-tRNA^{Phe}, the 3′-CCA$_{OH}$ sequence was not added to the processing products of the unedited precursor. This suggests that, for this tRNA, 5′ and 3′ processing of the unedited precursor does not occur. It may well be that a greater disruption of higher-order structure is present in the unedited mt-tRNA^{His} compared to the unedited mt-tRNA^{Phe} and mt-tRNA^{Cys}, as a consequence of the absence in the unedited mt-tRNA^{His} of three base pairs, compared to a single base pair in the other two unedited tRNAs. If so, this could account for the stricter requirement for editing prior to 5′ and 3′ processing of the mt-tRNA^{His} precursor.

Although results from studies of these three RNAs differ in detail, in all cases it is clear that editing precedes 5′ and 3′ processing, and that the edited precursor is more efficiently (and more accurately) processed than the unedited one (Table 1). This has led to the suggestion that editing may be used as a means of regulating the processing of these tRNAs, and hence the steady-state level of the mature tRNAs in mitochondria. At this point, we can only say that this is a possibility, and that further work is needed to determine how tRNA levels are regulated in plant mitochondria and other systems.

Also unclear is how editing of plant mt-tRNAs (and plant mt-RNAs in general) occurs. The observed mode of tRNA editing, from cytidine to uridine only, suggests that the responsible activity is some sort of cytidine deaminase. As well, editing of tRNAs has been found only in base-paired regions, an observation suggesting that the presence of a C·A/A·C or C·U/U·C mismatch is an important recognition element. The currently favored model for RNA editing in plant mitochondria (Schuster and Brennicke, 1994) (see Chapter 17) hypothesizes the involvement of *trans*-acting guide sequences analogous to those observed in trypanosome mitochondrial RNA editing (Benne, 1996) (Chapter 21). For the edited tRNAs, the base-paired region opposite the edited nucleotide may act as such a guide sequence (this arrangement has also been proposed for tRNA editing in the mitochondria of *A. castellanii* [Gray and Lonergan, 1993] and *P. polycephalum* [Chapter 22]). If this is the case, then, in the absence of a *trans*-acting guide, editing of plant mt-tRNAs should only occur in stem regions of a tRNA, as is observed. In fact, it could be that editing of mitochondrial tRNA and intron sequences, which is relatively rare, is a consequence of the resemblance of the secondary structures of these RNAs to those formed between mRNAs and their putative guide RNAs. Although (as previously noted) it has been proposed that editing occurs in tRNA stem regions to restore base pairing, and so enhances the stability of the tRNA secondary structure, these two interpretations are not mutually exclusive. Transfer RNA editing may have been selected to function in restoring base pairing in stem regions, and it may occur only in base-paired regions because of the need for a guide-type sequence. Unfortunately, attempts to develop in vitro RNA editing systems from plant mitochondrial extracts have not met with much success (Chapter 17). Furthermore, analyses of the sequences around editing sites in plant mtRNAs have revealed no obvious general motifs that might serve to identify such sites in the absence of guide RNAs (Gray and Covello, 1993). If guide sequences are necessary components of the editing machinery, then tRNAs, with their presumably internal guide sequences, should be particularly attractive model substrates with which to search for the protein(s) responsible for editing.

Metazoan Animals

The most recently described mode of tRNA editing is that found in the mitochondria of metazoa. First identified in the platypus (*Ornithorhynchus anatinus*), the existence of a tRNA editing system was hinted at by the presence of mispairings in the predicted acceptor stem of the mt-tRNA^{Ser}(GCU) (Yokobori and Pääbo, 1995b). Because no other tRNA^{Ser}(GCU) gene is present in the platypus mitochondrial genome and because mitochondrial tRNA import is believed not to occur in mammals, it was presumed that the very unstable stem structure in this tRNA would be "repaired" at the RNA level (Fig. 2C). When clones resulting from RT-PCR amplification of the acceptor stem were examined, all of them showed a change in the sequence in the 3′-half of the stem from the predicted $A_{68}C_{69}U_{70}U_{71}$ to $A_{68}C_{69}C_{70}C_{71}C_{72}A_{73}$CCA. It was noted that, at the DNA level, residues 70 and 71 overlap the downstream mt-tRNA^{Leu} gene. These two tRNAs are therefore likely to be produced from the same primary transcript. If the initial tRNA processing event were to occur at the 5′-end of mt-tRNA^{Leu}, then editing, by adding the sequence $C_{70}C_{71}C_{72}A_{73}$, would serve to generate a complete mt-tRNA^{Ser}(GCU) from one that lacked the four nucleotides preceding the -CCA$_{OH}$ extension.

Transfer RNA editing in the 3′-half of the acceptor stem has also been found in the mitochondria of land snails, squid, and chicken (Yokobori and Pääbo, 1995a, 1997; Tomita et al., 1996b). However, in these cases the added residues are all adenosines. For example, in the mitochondria of *Euhadra herklotsi* (the Japanese land snail) editing was shown to alter 3 mt-tRNAs—mt-tRNA^{Gly}(UCC), mt-tRNA^{Tyr}(GUA), and mt-tRNA^{Lys}(UUU)—at a total of 13 sites (Yokobori and Pääbo, 1995a). Of 10 mismatches within the acceptor stem, 9 are restored to standard base pairs and 1 is converted from a C·A to an A·A mismatch. Mitochondrial DNA sequences from other pulmonate gastropods predict a similar pattern of mismatches in the mt-tRNAs from *Cepaea nemoralis* (a British land snail; Yamazaki et al., 1997) and *Albinaria coerulea* (a Greek island snail; Hatzoglou et al., 1995). As in the platypus, genes for the edited tRNAs almost all overlap downstream tRNA genes over the edited sites, the one exception being the mt-tRNA^{Tyr}(GUA) gene of *E. herklotsi*. In *C. nemoralis*, however, this gene does overlap a downstream mt-tRNA^{Trp}(UCA) gene, and it has been suggested (Yamazaki et al., 1997) that this is the ancestral gene arrangement.

In the squid *Loligo bleekeri*, the mt-tRNA^{Tyr}(GUA) gene overlaps at its 3'-end the mt-tRNA^{Cys}(GCA) gene. Here, editing replaces G_{72} and G_{73} with A_{72} and A_{73}. In the chicken (*Gallus gallus*), the mt-tRNA^{Tyr} also overlaps the downstream mt-tRNA^{Cys} gene by one nucleotide; here, too, editing restores the phylogenetically conserved adenosine at position 73 (the discriminator nucleotide). However, in this instance, editing does not correct a mismatched base pair because the discriminator base is unpaired. The discriminator residue is also edited in three tRNAs in *E. herklotsi* and one in *L. bleekeri*. These multiple examples of discriminator nucleotide replacement suggest that this type of editing does not occur through a template-dependent mechanism. This hypothesis is further supported by the observation of an $A_2 \cdot C_{71}$ mismatch in the mature platypus mt-tRNA^{Ser}(GCU) and an $A_1 \cdot A_{72}$ mismatch in the mature mt-tRNA^{Tyr}(GUA) of *E. herklotsi*.

One possibility is that editing occurs by polyadenylation of mt-tRNAs that have been truncated within the 3'-portion of the acceptor stem by the 5' processing of overlapping downstream tRNA sequences (Yokobori and Pääbo, 1995a). Polyadenylation of mRNAs occurs in animal mitochondria in the absence of any apparent signal (Clayton, 1992), and has been shown to generate translation stop codons at the 3'-ends of mRNAs that terminate in -U$_{OH}$ or -UA$_{OH}$ (Attardi, 1985). Oligoadenylation of rRNA also has been observed in animal mitochondria (Dubin et al., 1981). Support for this proposed mechanism has come from an analysis of editing intermediates in squid mitochondria. When 45 clones of PCR products generated from circularized squid mt-tRNA^{Tyr} were sequenced (Tomita et al., 1996a), 32 had the sequence expected of the mature tRNA, that is, $U_{71}A_{72}A_{73}$CCA. Seven other clones ended at U_{71}, two ended at $U_{71}A_{72}$, and four had the sequence $U_{71}G_{72}G_{73}$CCA. The first three sets of clones suggest a pathway in which a cleavage after U_{71} is followed by the addition of two adenosine residues and then the -CCA$_{OH}$ tail. The last set of clones may represent an aberrant processing pathway.

Further support for this mechanism has come from a similar analysis of the chicken mt-tRNA^{Tyr} (Yokobori and Pääbo, 1997). In this case, only 4 of 28 clones had the mature sequence, that is, $C_{67}U_{68}U_{69}A_{70}U_{71}C_{72}A_{73}$CCA. However, another 15 clones (of types A_{73}CC and A_{73}C) lacked only a complete CCA tail. Another set of three clones had the edited A_{73} residue, but also had a tail of at least five additional adenosine residues. A fourth set was truncated at U_{69}, with a varying number of adenosine residues added to the 3'-end of the RNA. These observations provide strong support for editing by oligoadenylation of partially processed tRNAs. Furthermore, they suggest a two-step process in which a poly(A) polymerase first adds adenosine residues 3' of C_{72}, following which a 3'-nuclease cleaves the oligo(A) sequence on the 3'-side of the discriminator nucleotide in the pre-tRNA. The presence of aberrantly processed tRNAs (fourth set of clones previously mentioned) is consistent with the nonspecific nature of polyadenylation in animal mitochondria. A pathway for this mode of editing (Fig. 4B) can therefore be described as follows: (i) 5' processing of the two overlapping tRNAs (presumably by an RNase P-like activity) leaves an unpaired acceptor stem in the upstream tRNA (step 1); (ii) a tail of adenosine residues is then added to the 3'-end of the latter tRNA by a poly(A) polymerase (step 2); (iii) processing of the tRNA then occurs by way of the usual pathway, with 3' processing (step 3) being followed by -CCA$_{OH}$ addition (step 4) and nucleoside modification. Because the *E. herklotsi* mt-tRNA^{Tyr} gene does not overlap a downstream gene, editing of its transcript would necessarily require a separate 3' processing event, which in this case would presumably occur via an endonucleolytic cleavage immediately upstream of the first added nucleotide (N_{69}). Editing of the platypus mt-tRNA^{Ser}(GCU) can be accommodated within this scheme by proposing that, in this organism, the poly(A) polymerase is also able to add cytidine residues. Alternatively, editing of this tRNA may occur in a two-stage process in which three cytidine residues are added by one enzyme, following which the adenosine at the discriminator position is added by the poly(A) polymerase (step 2). This issue awaits clarification.

EVOLUTION OF tRNA EDITING

Presumably, each of the different modes of RNA editing has arisen according to the same evolutionary principles that apply to the origin of all other biochemical pathways. While it may seem counterintuitive that such baroque and potentially avoidable biochemistry should arise and persist at all, the evolution of RNA editing systems can be accommodated in a rather simple three-step model (Covello and Gray, 1993) (Fig. 6). In the first step, the RNA editing activity appears as an alteration in a preexisting enzyme activity, for example, through a process of gene duplication and divergence. In the next phase, a mutation occurs in a gene, with the nascent RNA editing activity able to interact with and act on the transcript of the mutated gene in such a way as to restore functional sequence in the mature RNA. In this way, what normally might be a deleterious (or even lethal) mutation is rendered neutral by the presence of the editing activity, which effectively acts as a "repair system." This neutral mutation is then fixed by genetic drift, that is, the mutation is propagated by chance within a population until all members of the population possess it. At the same time, mutation and genetic drift can occur at other loci, generating a number of sites that require editing at the RNA level to generate functional gene products. Finally, natural selection comes into play to maintain the editing activity: as the number of sites requiring editing increases, the system becomes increasingly dependent on the editing activity. Furthermore, because of the very low likelihood at this point of simultaneous reversion of all editable sites to a state that does not require editing, it ultimately becomes very difficult, if not impossible, to lose the editing system, which is now an indispensable part of the general RNA processing pathway. Hence, the activity is retained over time.

In *A. castellanii*, the observed pattern of secondary structure anomalies prompts the questions of why tRNA editing in this system is restricted to one or more of the first three positions in the 5'-half of the acceptor stem and why most *A. castellanii* mtDNA-encoded tRNAs (13 of 16) show this editing pattern. Our working model, in the context of the general evolutionary model previously outlined, envisages the emergence of a system that is capable of replacing the first three nucleotides at the 5'-end of a tRNA by using information in the 3'-half of the acceptor stem as a template (guide), with incorporation of a nucleotide at any mispaired positions to generate a standard G:C or A:U pair. The presence of such an activity would effectively lead to a relaxation of functional constraints (such as the necessity to maintain base pairing) at the first three positions at the 5'-end of the acceptor stem. This would allow these positions (which we define as editable positions) to undergo genetic drift according to the evolutionary pattern of the genome in which they reside. In the case of *A. castellanii* mtDNA, which is 70.6% A+T (Burger et al., 1995), A-T drift appears to be a major component of evolutionary change. Considering the 15 bona fide (excluding tRNA^X) *A. castellanii* mtDNA-encoded tRNAs (12 of which require editing), overall A+T composition at editable sites is 35% in the mature mt-tRNAs, but 67% in the corresponding gene sequences. If only the 12 bona fide tRNAs that require editing are considered, this value rises to 74% at editable sites, but to 85% if the calculation is further restricted to those editable sites at which a nucleotide change actually results from editing. These last two values are in excess of the overall A+T content (70%) of coding regions in *A. castellanii* mtDNA. In contrast, in the three bona fide tRNAs that do not require editing (because their acceptor stem regions are already fully paired), the

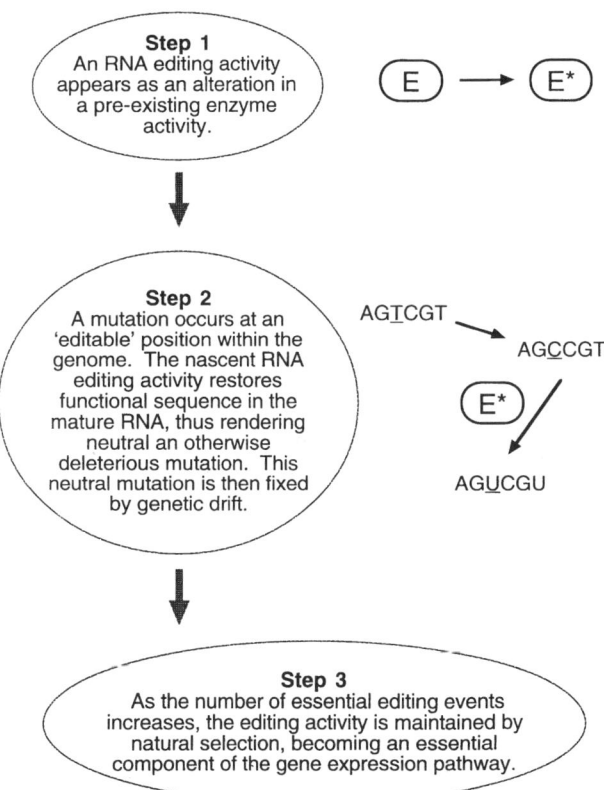

Figure 6. Generalized evolutionary scheme for the emergence of an RNA editing activity.

overall A+T content is only 45% at editable sites. Clearly, the presence of the editing activity has freed the *A. castellanii* mt-tRNA genes to become far more AT-rich at editable positions than would be possible without a relaxation of functional constraints (i.e., the need to encode acceptor stem base pairing). The continuing requirement to produce functional tRNAs from these genes provides a selective pressure for retention of the editing activity.

In the case of marsupial mt-tRNA^{Asp}(GUC), the editing activity might have arisen from a cytidine deaminase activity that acquired the ability to act on RNA substrates, as suggested in the case of C-to-U editing in plant mitochondria (Covello and Gray, 1993), and postulated (Covello and Gray, 1993) and affirmed (Navaratnam et al., 1995) in the case of the C-to-U editing of apolipoprotein B mRNA. In the presence of this activity, the nucleotide corresponding to tRNA position 35 might then have mutated from T to C in the mt-tRNA^{Asp}(GUC) gene. The presence of the new mt-tRNA^{Asp} derivative, mt-tRNA'^{Gly}'(GCC), now aminoacylatable with glycine, could then have allowed the mt-tRNA^{Gly}(CCU) to undergo a C-to-T mutation at position 32. This new mutation would have limited the decoding capacity of mt-tRNA^{Gly}(CCU) to GGG and GGA. However, because the mt-tRNA'^{Gly}'(GCC) was available to read the other two glycine codons, GGC and GGU, this mutation would not be deleterious. At this point, two simultaneous back mutations would be required to obviate the need for the editing activity. Because this is rather unlikely, the editing activity has been maintained in this system (Börner and Pääbo, 1996).

In metazoan mitochondria, strong selective pressure to minimize genome size has resulted in the creation of a number of overlapping genes. All edited tRNAs in metazoan mitochondria currently overlap, or ancestrally overlapped, downstream genes (Yokobori and Pääbo, 1995a, 1995b; Yokobori et al., 1996). How are these overlapping genes related to tRNA editing? In the presence of a polyadenylation activity of relaxed specificity, which may have arisen in response to the selective pressure to minimize genome size (because it obviated the need to encode polyadenylation signals in the DNA), certain tRNA genes might lose a number of nucleotides at their 3'-ends as a result of gene overlap. This would not render the encoded tRNAs nonfunctional, because the lost nucleotides could be reintroduced at the RNA level by poly(A) polymerase. Which genes could be truncated in this way would be determined by the characteristics of the editing system and the sequence requirements governing tRNA identity. Only those genes whose encoded tRNAs could accommodate a run of adenosines at the 3'-half of the acceptor stem or an adenosine in the discriminator position would be candidates for such truncation. An alternative model proposes that RNA editing in animal mitochondria serves to counteract the accumulation of mutations that occurs in asexual genetic systems (Müller's ratchet) (Börner et al., 1997). Metazoan mitochondria therefore provide a clear example of a genetic system in which a novel activity has been used to satisfy or counter the selective pressures that govern that system.

Similarly, in plant mitochondria, editing of mRNAs almost certainly preceded the appearance of tRNA editing. Nevertheless, the presence of this activity has had important consequences for the expression of certain mt-tRNA genes. For example, the larch mt-tRNA^{His}(GUG) gene encodes cytidines at positions 6, 12, and 41. Were it not for the presence of a C-to-U editing activity, this particular gene would not be able to produce a functional product. The editing activity has effectively neutralized three otherwise deleterious mutations. The presence of the editing activity is not, however, able to overcome the evolutionary tendency in plants to replace "native" tRNA with either chloroplast-derived or nucleus-encoded ones (Maréchal-Drouard et al., 1996a). Thus, in the case of the mtDNA-encoded tRNA^{His} gene, whose transcript is edited in a gymnosperm, editing has not prevented loss of this gene in the lineage leading to angiosperms. Similarly, in the cases of the mt-tRNA^{Phe} that is edited in bean, potato, and *Oenothera* and the mt-tRNA^{Cys} that is edited in *Oenothera*, the genes for these two tRNAs have been have been replaced in maize, wheat, and larch (Dietrich et al., 1996). As one might reasonably expect, whereas the presence of many edited sites provides a strong selective pressure for the maintenance of an editing activity per se, the presence of an editing activity does not appear to provide a selective pressure for the retention of any particular site of editing. A mutation that has become fixed within a population due to neutral drift can just as easily be lost due to neutral drift. This likely accounts for the limited phylogenetic distribution that is characteristic of all editing activities.

CONCLUSION

From the previous discussion, it seems that tRNA editing, rather than simply being a subcategory of RNA editing, is really a microcosm of this process. Generalized RNA editing and tRNA editing have similarly wide phylogenetic distributions, similar diversities of mechanism acting on a single class of substrate, and similar degrees of evolutionary innovation in the adoption of novel enzymatic activities in response to the various selective pressures that govern genome evolution. Moreover, they prompt the same questions about their potential advantages to organisms and genetic systems. Is RNA editing used as another level in the regulation of the activity of certain genes? Does it permit the production of polymorphic gene products (see, Lu et al., 1996), and if so, what advantage has been gained from such polymorphism? Is it useful to think of some cases of mt-tRNA editing (such as polyadenylation editing in metazoan mitochondria) as examples of the transfer of mitochondrially encoded information to the nucleus? Why are so many of the RNA editing systems, especially the ones involving nucleotide replacement, found in mitochondria?

Beyond these questions, there is still a pressing need to develop in vitro systems for studying the various modes of RNA editing. For this purpose, tRNAs may prove to be ideal substrates. They are easily produced in vitro and are stable in solution. Furthermore, for cases where a guide RNA has been proposed as a component of the editing machinery, the guide may be present in the tRNA substrate itself. Considering how much remains to be done on tRNA editing, and how much progress has been made over the past few years, surely the best is yet to come.

Acknowledgments. We thank T. J. Byers, H. Grosjean, B. F. Lang, D. Miller, and T. Ueda for providing unpublished data. Work on RNA editing in the authors' laboratory is supported by a grant (MT-4124) from the Medical Research Council of Canada to M.W.G. Salary support in the form of an NSERC 1967 Science and Engineering Scholarship (D.H.P.) and a fellowship from the Canadian Institute for Advanced Research, Program in Evolutionary Biology (M.W.G.), is gratefully acknowledged.

REFERENCES

Adler, B. K., and S. L. Hajduk. 1994. Mechanisms and origins of RNA editing. *Curr. Opin. Genet. Dev.* 4:316–322.

Amberg, R., C. Urban, B. Reuner, P. Scharff, S. C. Pomerantz, J. A. McCloskey, and H. J. Gross. 1993. Editing does not exist for mammalian selenocysteine tRNAs. *Nucleic Acids Res.* 21:5583–5588.

Attardi, G. 1985. Animal mitochondrial DNA: an extreme example of genetic economy. *Int. Rev. Cytol.* 93:93–145.

Beier, H., M. C. Lee, T. Sekiya, Y. Kuchino, and S. Nishimura. 1992. Two nucleotides next to the anticodon of cytoplasmic rat tRNA^{Asp} are likely generated by RNA editing. *Nucleic Acids Res.* 20:2679–2683.

Benne, R. 1996. RNA editing: how a message is changed. *Curr. Opin. Genet. Dev.* 6:221–231.

Benne, R., J. Van Den Burg, J. P. J. Brakenhoff, P. Sloof, J. H. Van Boom, and M. C. Tromp. 1986. Major transcript of the frameshifted *coxII* gene from trypanosome mitochondria contains four nucleotides that are not encoded in the DNA. *Cell* 46:819–826.

Binder, S., A. Marchfelder, and A. Brennicke. 1994. RNA editing of tRNA^{Phe} and tRNA^{Cys} in mitochondria of *Oenothera berteriana* is initiated in precursor molecules. *Mol. Gen. Genet.* 244:67–74.

Binder, S., A. Brennicke, and A. Marchfelder. 1996. RNA editing is required for efficient 5' and 3' end processing of tRNA^{Phe} in mitochondria of higher plants, p. 56. *In Proceedings of the EMBO Workshop on RNA Editing.*

Börner, G. V., and S. Pääbo. 1996. Evolutionary fixation of RNA editing. *Nature* (London) 383:225.

Börner, G. V., M. Mörl, A. Janke, and S. Pääbo. 1996. RNA editing changes the identity of a mitochondrial tRNA in marsupials. *EMBO J.* 15:5949–5957.

Börner, G. V., S.-I. Yokobori, M. Mörl, M. Dörner, and S. Pääbo. 1997. RNA editing in metazoan mitochondria: staying fit without sex. *FEBS Lett.* 409:320–324.

Burger, G., I. Plante, K. M. Lonergan, and M. W. Gray. 1995. The mitochondrial DNA of the amoeboid protozoon, *Acanthamoeba castellanii*: complete sequence, gene content and genome organization. *J. Mol. Biol.* 245:522–537.

Clark, B. F. C., M. Kjeldgaard, J. Barciszewski, and M. Sprinzl. 1995. Recognition of aminoacyl-tRNAs by protein elongation factors, p. 423–442. *In* D. Söll and U. L. RajBhandary (ed.), *tRNA: Structure, Biosynthesis, and Function.* American Society for Microbiology, Washington, D.C.

Clayton, D. A. 1992. Transcription and replication of animal mitochondrial DNAs. *Int. Rev. Cytol.* 141:217–232.

Covello, P. S., and M. W. Gray. 1989. RNA editing in plant mitochondria. *Nature* (London) 341:662–666.

Covello, P. S., and M. W. Gray. 1993. On the evolution of RNA editing. *Trends Genet.* 9:265–268.

Deutscher, M. P. 1995. tRNA processing nucleases, p. 51–65. *In* D. Söll and U. L. RajBhandary (ed.), *tRNA: Structure, Biosynthesis, and Function.* American Society for Microbiology, Washington, D.C.

Diamond, A. M., Y. Montero-Puerner, B. J. Lee, and D. Hatfield. 1990. Selenocysteine inserting tRNAs are likely generated by tRNA editing. *Nucleic Acids Res.* 18:6727.

Dietrich, A., I. Small, A. Cosset, J. H. Weil, and L. Maréchal-Drouard. 1996. Editing and import: strategies for providing plant mitochondria with a complete set of functional transfer RNAs. *Biochimie* 78:518–529.

Dirheimer, G., G. Keith, P. Dumas, and E. Westhof. 1995. Primary, secondary, and tertiary structures of tRNAs, p. 93–126. *In* D. Söll and U. L. RajBhandary (ed.), *tRNA: Structure, Biosynthesis, and Function.* American Society for Microbiology, Washington, D.C.

Dubin, D. T., K. D. Timko, and R. J. Baer. 1981. The 3' terminus of the large ribosomal subunit ("17S") RNA from hamster mitochondria is ragged and oligoadenylated. *Cell* 23:271–278.

Farruggio, D., J. Chaudhuri, U. Maitra, and U. L. RajBhandary. 1996. The A1·U72 base pair conserved in eukaryotic initiator tRNAs is important specifically for binding to the eukaryotic translation initiator factor eIF2. *Mol. Cell. Biol.* 16:4248–4256.

Freist, W., D. T. Logan, and D. H. Gauss. 1996. Glycyl-tRNA synthetase. *Biol. Chem. Hoppe-Seyler* 377:343–356.

Gabriel, K., J. Schneider, and W. H. McClain. 1996. Functional evidence for indirect recognition of G·U in tRNA^{Ala} by alanyl-tRNA synthetase. *Science* 271:195–197.

Gerber, A., M. A. O'Connell, L. P. Keegan, S. Krause, P. H. Seeburg, and W. Keller. 1997. Cloning, expression and characterization of yeast and *Drosophila* proteins which are homologous to the RNA editing enzymes DRADA and RED1, p. 247. *In Proceedings of the Second Annual Meeting of the RNA Society.*

Gott, J. M., L. M. Visomirski, and J. L. Hunter. 1993. Substitutional and insertional RNA editing of the cytochrome *c* oxidase subunit I mRNA of *Physarum polycephalum. J. Biol. Chem.* **268:** 25483–25486.

Gray, M. W., and P. S. Covello. 1993. RNA editing in plant mitochondria and chloroplasts. *FASEB J.* **7:**64–71.

Gray, M. W., and K. M. Lonergan. 1993. Transfer RNA editing in *Acanthamoeba castellanii* mitochondria, p. 15–22. *In* A. Brennicke and U. Kück (ed.), *Plant Mitochondria.* VCH Publishers, Inc., New York, N.Y.

Grosjean, H. 1997. Personal communication.

Grosjean, H. 1996. RNA editing or RNA modification: that is the question, p. 58–59. *In Proceedings of the EMBO Workshop on RNA Editing.*

Grosjean, H., K. Nicoghosian, E. Haumont, D. Soll, and R. Cedergren. 1985. Nucleotide sequences of two serine tRNAs with a GGA anticodon: the structure-function relationships in the serine family of *E. coli* tRNAs. *Nucleic Acids Res.* **13:**5697–5706.

Gualberto, J. M., L. Lamattina, G. Bonnard, J. H. Weil, and J. M. Grienenberger. 1989. RNA editing in wheat mitochondria results in the conservation of protein sequences. *Nature* (London) **341:**660–662.

Hatzoglou, E., G. Rodakis, and R. Lecanidou. 1995. Complete sequence and gene organization of the mitochondrial genome of the land snail *Albinaria coerulea. Genetics* **140:**1353–1366.

Hiesel, R., B. Wissinger, W. Schuster, and A. Brennicke. 1989. RNA editing in plant mitochondria. *Science* **246:**1632–1634.

Ibba, M., K.-W. Hong, and D. Söll. 1996. Glutaminyl-tRNA synthetase: from genetics to molecular recognition. *Genes to Cells* **1:**421–427.

Janke, A., and S. Pääbo. 1993. Editing of a tRNA anticodon in marsupial mitochondria changes its codon recognition. *Nucleic Acids Res.* **21:**1523–1525.

Janke, A., G. Feldmaier-Fuchs, W. K. Thomas, A. von Haeseler, and S. Pääbo. 1994. The marsupial mitochondrial genome and the evolution of placental mammals. *Genetics* **137:**243–256.

Joyce, P. B. M., and Gray, M. W. 1989. Chloroplast-like transfer RNA genes expressed in wheat mitochondria. *Nucleic Acids Res.* **17:**5461–5476.

Komine, Y., T. Adachi, H. Inokuchi, and H. Ozeki. 1990. Genomic organization and physical mapping of the transfer RNA genes in *Escherichia coli* K12. *J. Mol. Biol.* **212:**579–598.

Laforest, M.-J., I. Roewer, and B. F. Lang. 1997. Mitochondrial tRNAs in the lower fungus *Spizellomyces punctatus*: tRNA editing and UAG "stop" codons recognized as leucine. *Nucleic Acids Res.* **25:**626–632.

Lang, B. F. 1997. Personal communication.

Ledee, D. R., and T. J. Byers. 1997. Personal communication.

Lee, Y., D. W. Kindelberger, J.-Y. Lee, S. McClennen, J. Chamberlain, and D. R. Engelke. 1997. Nuclear pre-tRNA terminal structure and RNase P recognition. *RNA* **3:**175–185.

Levinger, L., V. Vasisht, V. Greene, R. Bourne, A. Birk, and S. Kolla. 1995. Sequence and structure requirements for *Drosophila* tRNA 5′- and 3′-end processing. *J. Biol. Chem.* **270:**18903–18909.

Lonergan, K. M., and M. W. Gray. 1993a. Editing of transfer RNAs in *Acanthamoeba castellanii* mitochondria. *Science* **259:**812–816.

Lonergan, K. M., and M. W. Gray. 1993b. Predicted editing of additional transfer RNAs in *Acanthamoeba castellanii* mitochondria. *Nucleic Acids Res.* **21:**4402.

Lu, B., R. K. Wilson, C. G. Phreaner, R. M. Mulligan, and M. R. Hanson. 1996. Protein polymorphism generated by differential RNA editing of a plant mitochondrial *rps12* gene. *Mol. Cell. Biol.* **16:**1543–1549.

Marchfelder, A., A. Brennicke, and S. Binder. 1996. RNA editing is required for efficient excision of tRNA^{Phe} from precursors in plant mitochondria. *J. Biol. Chem.* **271:**1898–1903.

Maréchal-Drouard, L., D. Ramamonjisoa, A. Cosset, J. H. Weil, and A. Dietrich. 1993. Editing corrects mispairing in the acceptor stem of bean and potato mitochondrial phenylalanine transfer RNAs. *Nucleic Acids Res.* **21:**4909–4914.

Maréchal-Drouard, L., R. Kumar, C. Remacle, and I. Small. 1996a. RNA editing of larch mitochondrial tRNA^{His} precursors is a prerequisite for processing. *Nucleic Acids Res.* **24:**3229–3234.

Maréchal-Drouard, L., A. Cosset, C. Remacle, D. Ramamonjisoa, and A. Dietrich. 1996b. A single editing event is a prerequisite for efficient processing of potato mitochondrial phenylalanine tRNA. *Mol. Cell. Biol.* **16:**3504–3510.

Martin, N. C. 1995. Organellar tRNAs: biosynthesis and function, p. 127–140. *In* D. Söll and U. L. RajBhandary (ed.), *tRNA: Structure, Biosynthesis, and Function.* American Society for Microbiology, Washington, D.C.

Meinnel, T., Y. Mechulam, and S. Blanquet. 1995. Aminoacyl-tRNA synthetases: occurrence, structure, and function, p. 251–292. *In* D. Söll and U. L. RajBhandary (ed.), *tRNA: Structure, Biosynthesis, and Function.* American Society for Microbiology, Washington, D.C.

Miller, D. 1997. Personal communication.

Miller, D., R. Mahendran, M. Spottswood, H. Costandy, S. Wang, M.-L. Ling, and N. Yang. 1993. Insertional editing in mitochondria of *Physarum. Semin. Cell Biol.* **4:**261–266.

Mörl, M., M. Dörner, and S. Pääbo. 1995. C to U editing and modifications during the maturation of the mitochondrial tRNA^{Asp} in marsupials. *Nucleic Acids Res.* **23:**3380–3384.

Motorin, Y. A., S. Auxilien, and H. Grosjean. 1996. Potential editing of cytidine to dihydrouridine in *E. coli* tRNA^{Ser}(GGA), p. 75. *In Proceedings of the EMBO Workshop on RNA Editing.*

Muramatsu, T., K. Nishikawa, F. Nemoto, Y. Kuchino, S. Nishimura, T. Miyazawa, and S. Yokoyama. 1988a. Codon and amino-acid specificities of a transfer RNA are both converted by a single post-transcriptional modification. *Nature* (London) **336:**179–181.

Muramatsu, T., S. Yokoyama, N. Horie, A. Matsuda, T. Ueda, Z. Yamaizumi, Y. Kuchino, S. Nishimura, and T. Miyazawa. 1988b. A novel lysine-substituted nucleoside in the first position of the anticodon of minor isoleucine tRNA from *Escherichia coli. J. Biol. Chem.* **263:**9261–9267.

Navaratnam, N., S. Bhattacharya, T. Fujino, D. Patel, A. L. Jarmuz, and J. Scott. 1995. Evolutionary origin of apoB mRNA editing: catalysis by a cytidine deaminase that has acquired a novel RNA-binding motif at its active site. *Cell* **81:**187–195.

Price, D. H., and M. W. Gray. 1997. Unpublished results.

Rundquist, B. A., and J. M. Gott. 1995. RNA editing of the *col* mRNA throughout the life cycle of *Physarum polycephalum. Mol. Gen. Genet.* **247:**306–311.

Schimmel, P., and T. Ripmaster. 1995. Modular design of components of the operational RNA code for alanine in evolution. *Trends Biochem. Sci.* **20:**333–334.

Schuster, W., and A. Brennicke. 1994. The plant mitochondrial genome: physical structure, information content, RNA editing, and gene migration to the nucleus. *Annu. Rev. Plant Physiol. Plant Mol. Biol* **45:**61–78.

Sherman, J. M., M. J. Rogers, and D. Söll. 1995. Recognition in the glutamine tRNA system: from structure to function, p. 395–409. *In* D. Söll and U. L. RajBhandary (ed.), *tRNA: Structure, Biosynthesis, and Function.* American Society for Microbiology, Washington, D.C.

Tallsjö, A., J. Kufel, and L. A. Kirsebom. 1996. Interaction between *Escherichia coli* RNase P RNA and the discriminator base results in slow product release. *RNA* **2:**299–307.

Tomita, K., T. Ueda, and K. Watanabe. 1996a. Two nucleotides 5'-adjacent to the anticodon of rat cytoplasmic tRNA^{Asp} are not edited. *Biochimie* 78:1001–1006.

Tomita, K., T. Ueda, and K. Watanabe. 1996b. RNA editing in the acceptor stem of squid mitochondrial tRNA^{Tyr}. *Nucleic Acids Res.* 24:4987–4991.

Visomirski-Robic, L. M., and J. M. Gott. 1997a. Insertional editing of nascent mitochondrial RNAs in *Physarum*. *Proc. Natl. Acad. Sci. USA* 94:4324–4329.

Visomirski-Robic, L. M., and J. M. Gott. 1997b. Insertional editing in isolated *Physarum* mitochondria is linked to RNA synthesis. *RNA* 3:821–837.

Yamazaki, N., R. Ueshima, J. A. Terrett, S. Yokobori, M. Kaifu, R. Segawa, T. Kobayashi, K. Numachi, T. Ueda, K. Nishikawa, K. Watanabe, and R. H. Thomas. 1997. Evolution of pulmonate gastropod mitochondrial genomes: comparisons of gene organizations of Euhadra, Cepaea and Albinaria and implications of unusual tRNA secondary structures. *Genetics* 145:749–758.

Yokobori, S.-I., and S. Pääbo. 1995a. Transfer RNA editing in land snail mitochondria. *Proc. Natl. Acad. Sci. USA* 92:10432–10435.

Yokobori, S.-I., and S. Pääbo. 1995b. tRNA editing in metazoans. *Nature* (London) 377:490.

Yokobori, S., and S. Pääbo. 1997. Polyadenylation creates the discriminator nucleotide of chicken mitochondrial tRNA^{Tyr}. *J. Mol. Biol.* 265:95–99.

Yokobori, S.-I., M. Mörl, and S. Pääbo. 1996. Transfer RNA editing in metazoan mitochondria, p. 74. *In Proceedings of the EMBO Workshop on RNA Editing*.

Yuan, Y., and S. Altman. 1995. Substrate recognition by human RNase P: identification of small, model substrates for the enzyme. *EMBO J.* 14:159–168.

Chapter 17

RNA Editing by Base Conversion in Plant Organellar RNAs

ANITA MARCHFELDER, STEFAN BINDER, AXEL BRENNICKE, AND VOLKER KNOOP

DISCOVERY OF RNA EDITING

Awareness of the RNA editing process in plants grew gradually. The first discrepancies in nucleotide identities between mRNA and DNA were noted as the appearance of thymidylates in the sequence ladders of cDNAs at positions of cytidylates in the genomic sequence. These were reported without any further comment, just stating the misfit in sequence as potential reverse transcriptase or cloning errors (Hiesel et al., 1987). Finally, three laboratories simultaneously reported their recognitions of the underlying process and identified C to U RNA editing in plant mitochondria, in analogy to the term coined for the trypanosome mitochondrial RNA sequence alteration (Covello and Gray, 1989; Gualberto et al., 1989; Hiesel et al., 1989).

The existence of RNA editing in plant mitochondria resolved initial problems with nonconserved amino acid codons found in the plant gene sequences. One of the striking differences in protein coding genes in mitochondria of plants was the numerous CGG arginine codons in the position of highly conserved TGG tryptophan codons in the respective animal and yeast genes (Gualberto et al., 1990a). The discrepancies between otherwise evolutionary highly conserved amino acid residues and the genomic codons used had nourished the suspicion of a divergent genetic code in plant mitochondria (Fox and Leaver, 1981; Hiesel and Brennicke, 1983). This was clarified and resolved with the nucleotide changes introduced at the RNA level by editing, which result generally in sequences encoding conserved amino acids without deviation from the standard genetic code.

Soon after the recognition of the frequent C to U transitions, the occurrence of reverse reactions changing U to C was realized (Schuster et al., 1990a; Gualberto et al., 1990b). These turned out to be quite infrequent in flowering plants, but much more common in ferns (Hiesel et al., 1994; Malek et al., 1997) and the hornwort *Anthoceros crispulus* (Malek et al., 1996).

Several years later, Hans Kössel and his coworkers realized that in plastids unusual translation initiation codons, ACG instead of ATG, had been reported for some genes. With several complete plastid genome sequences available, they noticed that in other plant species genomic ATG codons are found at homologous positions. Comparison of cDNA and genomic sequences at these loci subsequently confirmed that C to U editing also occurred in plastids (Hoch et al., 1991).

Thus far, U to C editing has not been seen in plastids of flowering plant species, but it occurs frequently in plastids of the hornwort *Anthoceros* (Yoshinaga et al., 1996; Freyer et al., 1997). A schematic view of RNA editing in plant organelles as discussed in this chapter is shown in Fig. 1.

DISTRIBUTION OF C TO U AND U TO C RNA EDITING IN PLANT ORGANELLES

In mitochondria of flowering plants, up to 500 RNA editing sites have already been reported in species such as *Oenothera berteriana* and wheat. The total number of the C-to-U-type editing sites per mitochondrial genome can be extrapolated from these numbers and from the complete sequence of the *Arabidopsis thaliana* mitochondrial genome to be between 500 and 1,000. These numbers suggest that about 2–5% of the C's in the transcribed coding nucleotides in the RNA will be altered to U in *Arabidopsis*, with similar numbers for other plant species.

Anita Marchfelder, Stefan Binder, Axel Brennicke, and Volker Knoop • Allgemeine Botanik Universität Ulm, Albert Einstein Allee, D-89069 Ulm, Germany.

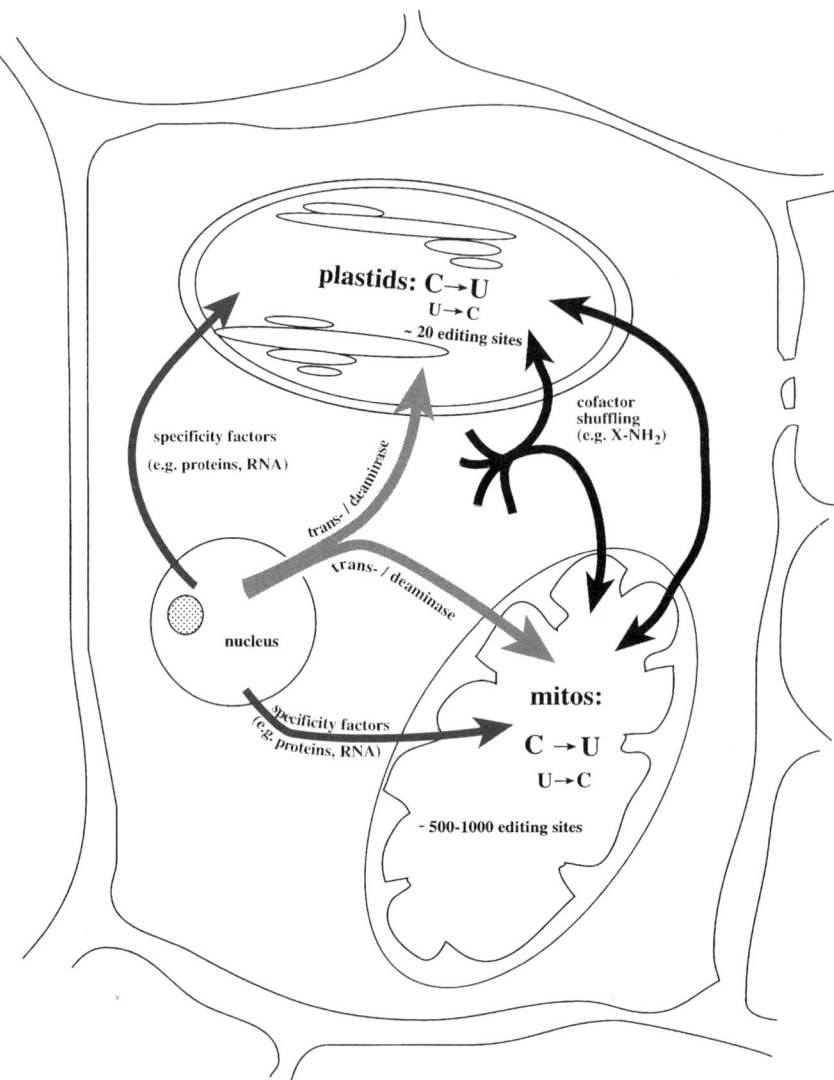

Figure 1. RNA editing in plastids and mitochondria requires specificity and/or enzymatic factors from the cytoplasm. The enzymatic reaction from C to U has all the characteristics of a deamination, but the reverse reaction U to C requires the amino group at an accessible redox potential. This could possibly be achieved by a transamination rather than a deamination, parking the amino group at an unidentified cofactor X, which presumably can be moved between the different compartments.

In wheat mitochondria, for example, 1,200 editing sites have been estimated (Bonen, 1991; Bonnard et al., 1992; Maier et al., 1996; Pring et al., 1993; Schuster and Brennicke, 1994; Schuster et al., 1993b). In flowering plants, these numbers contrast with the 1 or 2 identified reverse U to C editing sites per genome, showing the C to U reaction to dominate the editing process in these species by a ratio of more than 200:1.

Comparing editing sites in a given gene between various flowering plants shows that many sites are conserved between different species (for example, see Zanlungo et al., 1993). This conservation in editing patterns is closely linked to highly conserved amino acid sequences, such as those of the respiratory chain subunits. While many editing sites are thus similar between different plants, there are always additional unique sites and no evolutionary trend toward more or fewer sites can be discerned.

RNA editing in plant mitochondria and also in plastids generally increases conservation of amino acid sequences, thus correcting for genomic drift. About 90% of all RNA editing events in mitochondria and 100% in chloroplasts occur in the reading frames of mRNAs and alter the genomically predicted codons. Some of the mRNA editing sites appear to be neutral, changing the codon specificity from one nonconserved amino acid to another without any obvious advantage for protein function. Likewise without apparent effect are the editing events at silent codon positions, which do not alter the amino acid specified.

To date, RNA editing in plant chloroplasts has been studied quite comprehensively in numerous detailed investigations of unusual codons and variations between codons in different plant species. In a given plant, about 20 C's are edited in mRNAs of chloroplasts and in other developmental stages of the plastid (Maier et al., 1996). Some small quantitative differences in relative editing efficiency have been observed between different tissues, but no real tissue specificity has been reported (Bock et al., 1993).

In ferns and fern allies, and particularly in hornworts, the U to C reaction appears to be much more frequent both in mitochondria and in plastids (Malek et al., 1996, 1997; Yoshinaga et al., 1996; Steinhauser et al., unpublished results). UAA, UAG, and UGA stop codons are found at least once in most of the mitochondrial genes that have been analyzed in these plants. These coding region-internal translational termination codons are edited by U to C conversion, creating arginine or glutamine codons to restore the conserved reading frame. Extrapolating from the still rather scattered data set, some 500–1,000 U to C editing events are expected to occur in the mitochondrial compartments of species like the hornwort *Anthoceros*.

RNA Editing in Noncoding Regions of mRNAs

The few editing sites in noncoding RNA regions, and similarly the editing sites in silent positions of reading frames, are often found to be edited less efficiently, as deduced from the percentage of edited sequences in a population of cDNA clones. At least some of these apparently "useless" editing sites may reflect side effects of a relaxed site specificity of the organellar editing machinery.

The number of such "useless" sites may actually be smaller than assumed, since several of these sites could have some regulatory function, particularly in those instances where an editing event is located close to the first codon of an open reading frame. In the *rps14* gene in *Oenothera*, a C to U transition occurs in a position where ribosomal binding sites are located in prokaryotic genes (Schuster et al., 1991). Although in plant mitochondria ribosomal binding sites have not been clearly defined (Carillo and Bonen, 1997), such a function is a distinct possibility considering that complementarity to the 3'-terminal nucleotides of the 18S rRNA is improved by the edited U and gains one more potential base pairing.

RNA Editing in Introns

In plant mitochondria, about half of the editing sites outside open reading frames are located in intron sequences, while none have yet been observed in plastids. Although intron editing is exceptional, it may improve the canonical features of the introns in which it occurs. Nearly all of the vascular plant mitochondrial introns are classified as group II introns, usually potentially capable of folding into the corresponding secondary structure. Some mismatched nucleotides in conserved and base-pairing regions such as the stems of domains V, VI, and I are cured by restoring U–A pairings from C:A mismatches (Wissinger et al., 1991b; Binder et al., 1992; Knoop et al., 1991; Carillo and Bonen, 1997).

Because these editing events are necessary to allow helix formation in some base-paired regions, several instances of editing in introns have been considered to be important, if not essential, for efficient splicing (Wissinger et al., 1991b; Binder et al., 1992; Knoop et al., 1991; Zanlungo et al., 1995; Carillo and Bonen, 1997). Indirect evidence for this theory has been obtained from experiments with chimeric constructs between plant mitochondrial (mt) intron segments and autocatalytic yeast introns, which showed that in this context only the edited version of a plant domain V will support in vitro splicing (Börner et al., 1995). However, in another recent investigation (Carillo and Bonen, 1997) it was found that in vivo excised intron RNAs contained unedited nucleotides in some positions where a C to U transition should have occurred, as dictated by the structures modelled on paper. This observation suggests that individual sites differ in importance for the splicing competence of the intron in which they are located. In addition, cofactors that probably are essential for splicing in vivo may have made some of the otherwise essential features of the group II introns in plant mitochondria superfluous. The latter contention is supported by the observation that several introns in these organelles lack one or the other secondary or tertiary interaction and structural element found necessary for splicing competence, for example, in autocatalytic reactions such as those of some yeast mitochondrial introns (Chapdelaine and Bonen, 1991). Thus, only limited predictions about potential editing sites can be made from the secondary structure models of plant mitochondrial introns and experimental evidence is required to confirm the editing at each site in an intron. One experimental approach has used the transplastomic technique to insert the intron of the mitochondrial gene for the small-subunit ribosomal protein 10 (rps10) into the plastid genome in its unedited and edited forms. Apparently, mitochondrial specific factors are required to facilitate splicing, because neither the edited intron nor the unedited intron was processed in the plastid (Bock and Knoop, unpublished results).

Editing of Mitochondrial tRNAs

Functionally important RNA editing in structural RNAs has been reported in plant mitochondrial tRNAs (described in detail in Chapter 16). These tRNA editing events have been confirmed in several different tRNA molecules and various plant species, including potato, pea, *Oenothera*, and larch. Thus far the observed edited nucleotides are exclusively located in base-paired regions (Marechal-Drouard et al., 1993a, 1993b, 1996; Binder et al., 1994; Marchfelder et al., 1996), which could imply *cis*-guiding of the RNA to determine the editing sites. In addition, editing of tRNAs is important for their maturation (Binder et al., 1994; Marechal-Drouard et al., 1996; Marchfelder et al., 1996). Only edited tRNA sequences are efficiently excised from the longer precursor transcripts, raising the possibility that RNA editing is a limiting step in tRNA maturation and could be used to regulate tRNA pools in the mitochondrion.

Translational Consequences of mRNA Editing

Many editing events in mitochondrial mRNAs induce changes of amino acid identity that drastically alter the predicted secondary and tertiary protein structure (for example, the elimination of proline codons). Only the polypeptide produced from RNA with edited codons will be fully active. In plastids, only some of the transcripts require editing, but editing is nevertheless also essential to synthesize the correct proteins. The necessity of RNA editing for protein function has been demonstrated in both organelles. In tobacco mitochondria, the translation product of unedited atp9 RNA severely disturbs mitochondrial peak performance and prevents pollen maturation (Hernould et al., 1993). In transplastomic tobacco, the protein made from unedited spinach psbF RNA does not integrate correctly into photosystem II, resulting in a lowered chlorophyll content and an impaired photosynthetic capacity, with the plant remaining light green in color (Bock et al., 1994).

The silent RNA editing sites are part of that minority of editing events for which a clear use has not yet been identified. A potential importance for these editing sites may be sought in the codon usage and the resulting ease of codon translatability, as determined, for example, by the differential abundances of the corresponding tRNA species. However, when the occurrence of unedited codons and their edited counterparts in plant mitochondrial RNAs is compared, no coherent trend in a meaningful direction can be statistically verified, and unedited and edited codons are used with equal frequencies (Marienfeld et al., unpublished results).

Possibly, direct determination of individual tRNA abundance in plant mitochondria may provide a reason for the third codon position editing. In plastids, the overall number of editing sites is too small to have any effect on codon usage. Therefore, the role (if any) of silent editing in these organelles remains obscure.

Do Partially Edited mRNAs Specify a Family of Protein Isoforms?

When the sequences of individual cDNA clones covering a given editing site in plant mitochondria are considered, the editing status will often vary between the different clones (Schuster et al., 1990b). Some editing sites will be virtually completely edited in all of the cDNA clones, whereas others may be edited in only half or even fewer of the cDNAs. Theoretically, the existence of a single RNA editing site in a genetic system results in the possibility of producing two types of RNA molecules from one DNA region: the unedited and the edited transcripts. With more than just one editing site, there are many more possibilities: for n editing sites per RNA, the number of differentially edited RNAs is 2^n.

As an example, the complexity of the ccl577 mRNA population, which codes for a 577-amino-acid protein homologous to bacterial polypeptides involved in cytochrome *c* biogenesis (Schuster et al., 1993a), assuming that each of the 46 editing sites in *Oenothera* mitochondria is independently edited, could be 2^{46} different mRNAs—easily exceding the total number of transcripts per mitochondrion. In vivo, the relative abundances of the differently edited RNA species vary; in many tissues editing is very efficient and not all theoretically possible RNA species will be represented in detectable amounts (Schuster et al., 1990b; Wissinger et al., 1990; Grosskopf and Mulligan, 1996). Furthermore, different mRNA species covering a given coding region may vary in the percentage of unedited and edited nucleotides. For example, shorter transcripts covering the *nad3-rps12* locus (coding for subunit 3 of the NADH dehydrogenase and small-subunit ribosomal protein 12) in wheat mitochondria generally are more edited than the longer mRNAs, which presumably represent precursor molecules (Gualberto et al., 1991). Similarly, in potato mitochondria, shorter nad9 mRNAs (coding for subunit 9 of the NADH dehydrogenase) are edited at more sites than the longer (putative) precursor transcripts (Lu and Hanson, 1996).

Particularly in nongreen tissues, the mRNAs of some genes are low in abundance, in spite of the fact that these tissues strongly depend on mitochondrial activity for ATP production. In these cells, most of the transcripts represent various forms of partially edited RNAs, potentially specifying numerous protein isoforms (Fig. 2). These RNAs could specify a group of similar polypeptides reminiscent of the products of alternative splicing in nuclear genomes, which code for specialized derivatives of a sequence motif with an array of slightly altered functions presumably optimized for different tasks in different cells, such as the fibronectin RNAs (Sharp, 1994).

When mitochondrial subfractions enriched for ribosomes are analyzed for the distribution of the mRNA variants, the entire array of intermediates is found to associate to the translational machinery (Gualberto et al., 1991; Phreaner et al., 1996). There is no apparent reason to suspect diminished ribosome entry for nonedited sequences for most of the transcripts, because except for the 5' UTR editing previously described, editing does not affect mRNA regions that could be important for translation initiation. Preferential selection for translation of edited transcripts would thus have to be achieved by compartmentalization and structural differentiation, for example, by spatially ordered editing and translation complexes. A similar problem is the translatability of unspliced precursor mRNAs with introns in fungal and plant organelles, which a priori cannot be ex-

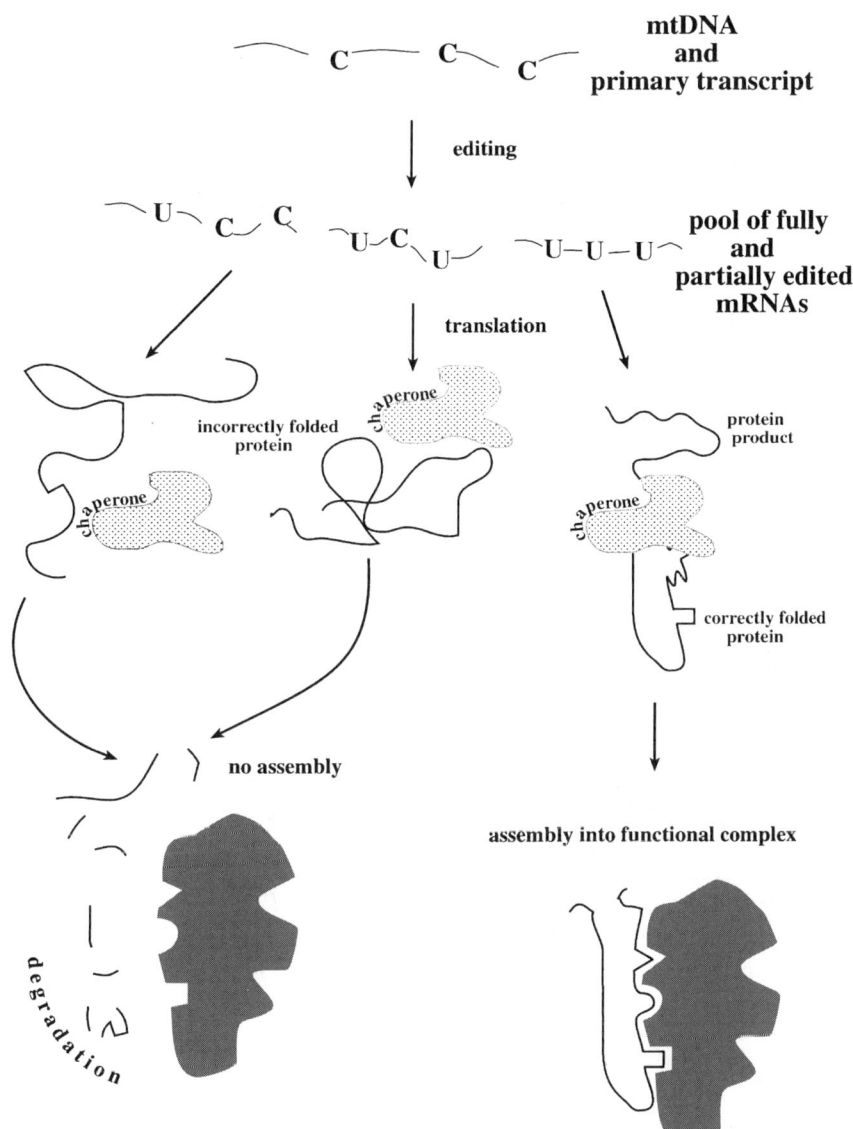

Figure 2. Protein products of incompletely edited transcripts probably are indiscriminately synthesized and degraded only at the level of folding or complex assembly, both processes being partially organized by chaperones. A large number of slightly variant proteins probably are made from the population of partially edited mRNAs in plant organelles. However, in a mature enzymatic complex only the single functional and "fully edited" protein sequence appears to be stably present.

cluded, and which will result in chimeric proteins, some of which actually may have a function in facilitating splicing or intron mobility. Indeed, intron-containing RNAs are also observed to indiscriminately associate with ribosome fractions (Yang and Mulligan, 1993).

When cellular fractions enriched for cytoplasmic ribosomes, and therefore presumed to be also enriched for mitochondrial ribosomes, are probed with antibodies specific for unedited, partially edited, and completely edited proteins, partially edited RNA-encoded polypeptides have been found for ribosomal protein S12. Therefore, it was assumed that RNAs were translated irrespective of their editing status (Phreaner et al., 1996; Lu et al., 1996). However, the same investigations gave different results on the integration of the "unedited" protein into the mature ribosomes; in maize this variant is not detectable in the ribosome (Phreaner et al., 1996), but it is in *Petunia hybrida* (Lu et al., 1996).

On the other hand, subunits of respiratory chain complexes appear to contain only the edited mRNA-derived protein sequence when isolated from the purified membrane complex, for example, subunit 9 of complex I of the NADH dehydrogenase from potato (Grohmann et al., 1994), subunit 9 of the mitochondrial ATPase (ATP9) of wheat mitochondria (Begu et al., 1990), and subunit 6 of the mitochondrial ATPase (ATP6) from *Petunia* mitochondria (Lu and Hanson, 1994; Hanson et al., 1996). Connecting these observations into a coherent picture suggests that protein variants are made in mitochondria and presumably also in plastids from the pool of partially and fully edited mRNA molecules. However, for some of these proteins the in vivo percentage of partially edited transcripts may be too low for the detection of the corresponding protein variants (Begu et al., 1990; Lu and Hanson, 1996).

Selection of the correct mitochondrial proteins occurs potentially with the aid of chaperones at the protein level and during complex assembly, with only the correctly folded and functionally competent proteins being integrated into mitochondrial or plastid protein complexes (Fig. 2). All the other polypeptide variants presumably are rapidly degraded. This conclusion confirms the initial observations that the plant cell is not limited with respect to its energy supply and that it will synthesize apparently useless proteins. As a bonus, potentially improved polypeptide variants eventually could arise, but this remains to be proven.

However, the concept of proteins synthesized from partially edited mRNAs in plant mitochondria is at odds with results obtained from transgenic plants in which ATP9 proteins were made from unedited transcripts in the cytoplasm (Hernould et al., 1993; Zabaleta et al., 1996). Upon import of these variant proteins into mitochondria, organellar function is disturbed and plants are rendered male sterile. Thus, the synthesis of proteins from partially edited and unedited mRNAs is not always without functional consequences in the mitochondrion.

The effect of in vivo detected RPS12 variant proteins may differ from the effect of the ATP9-like protein(s) in the transgenic tobacco plants because of the very different functions in the mitochondrion. The deleterious effect of transgenic unedited ATP9 on male fertility suggests that in vivo, at most very little of this protein is made from the unedited or partially edited atp9 mRNAs. Indeed, very little unedited RNA (and thus unedited protein) is made in the normal situation. The disturbance of vital mitochondrial functions by unedited ATP9 protein may be precisely the reason why atp9 RNAs are almost completely edited.

RNA Editing in Pseudogenes Defines Some Specificity Requirements

The plastid genomes are apparently too tightly organized to tolerate propagation of much extra sequence; accordingly, no pseudogenes have been identified. In these organelles, all editing events are within comparatively closely packed protein coding genes. However, in the mitochondrial genomes of plants we find numerous pseudogenes (Wissinger et al., 1991a; Andre et al., 1992), that is, sequences similar or identical to parts of coding regions. In mitochondria, the pseudogenes are created and amplified by genomic recombinations, sometimes becoming parts of larger mosaic frames (Marienfeld et al., 1997), which generally are not used for functional translation. In some instances, expression of such mosaic genes disturbs mitochondrial function during pollen maturation and renders the plant male sterile (Levings and Brown, 1989). In plant mitochondria we can discern two types of pseudogene sequences, distinguished by their evolutionary history.

One class arises by duplication of entire genes, or fragments thereof, and the integration of these sequences in dispersed regions of the genome. In these fragments, editing patterns are often identical to those in the genuine gene. This is, of course, not really surprising, considering that the editing in the real gene sequence is essential and must be efficiently executed. The size of the pseudogene fragment and the position of the editing site therein will tell what sequence context will be sufficient to specify a given site. The 193-nucleotide-long duplicated fragment of the *cox2* gene in wheat, for example, is sufficient to confer correct editing at three sites (Maier et al., 1996; further detail is given below).

The second type of mitochondrial pseudogenes is remnants of genes, an active copy of which has moved to the nucleus. In the mitochondrial genome, mutations slowly accumulate in the other copy, resulting in the creation of a pseudogene. In *Oenothera*, the leftover mitochondrial copy of the *rps12* gene has lost part of the coding region, but has retained several RNA editing sites that improve conservation of the encoded amino acid (Grohmann et al., 1994). Some of the originally required editing sites are still apparent, but new sites have arisen.

The transcript of the *rps19* pseudogene in the same plant is edited on either side of a genomic reading frame-disrupting stop codon. The editing does not improve the similarity of the protein sequence with homologs from other plants. Such erratic editing can also be observed at very low frequency in coding regions of real genes, introducing, for example, stop codons (Schuster and Brennicke, 1991).

Similarly, in the mitochondrial genome of *Arabidopsis*, the *rps14* gene (coding for small-subunit ribosomal protein 14) is a leftover gene, which is transcribed and edited in various positions, but which contains stop codons that render the sequence nonfunctional (Brandt et al., 1993). Apparently, such superfluous RNA editing sites are not eliminated immediately to save energy, but upon released evolutionary pressure editing specificity is slowly loosened. Supposedly, editing of such sites will eventually disappear with the loss or complete degeneration of the gene remnant.

THE MECHANISM OF RNA EDITING IN PLANT ORGANELLES

In analogy to the analysis of RNA editing in other systems (as discussed in Chapters 16 and 18–23), we will also present a separate discussion of the process in plant organelles, that is, the biochemistry of the modification of nucleotide identity and the specificity determinants, which decide where editing should occur.

Base Conversions Proceed without Order in the mRNA

Whereas RNA editing in trypanosome mitochondria proceeds stepwise from 3' to 5' along the RNA strand (Abraham et al. 1988, Blum et al., 1990; Maslov et al., 1992; see Chapter 21), the conversion of each editing site in plant organelles, particularly in mitochondria, appears to occur independently of the others (Schuster et al., 1990b). Even closely spaced editing sites do not always have the same editing status. Different stages of RNA editing can be observed in partially processed RNA molecules. For some transcripts, all permutations of edited and unedited nucleotides can be found in steady-state RNA. It is attractive to assume, therefore, that the hypothetical editosome does not scan the RNA chain but instead directly homes in on nucleotides to be edited.

Biochemistry of C to U and U to C RNA Editing

The most straightforward biochemical reaction to change a C to a U would be a deamination (Fig. 3A), similar to the editing of apolipoprotein B mRNA (see Navaratnam et al., 1995, and Chapter 18), rather than a more complex reaction cascade of deletion/insertion such as in trypanosomes. The in vitro experiments reported to date support such a chemical deamination reaction, because independent approaches (Araya et al., 1992; Blanc et al., 1996; Rajasekhar and Mulligan, 1993; Yu and Schuster, 1995) have consistently led to the observation that the sugar-phosphate backbone of the RNA remains intact during the editing process. Furthermore, differential isotope labeling showed that while the amino group is removed, the labeled base remains with the RNA chain (Blanc et al., 1995, 1996).

This result makes transglycosylation (see Chapter 9) as a mechanism for the C to U conversion unlikely. In addition, zinc ion requirements of the in vitro C to U reaction (Yu and Schuster, 1996) are similar to those seen for typical C deaminases and the ApoB mRNA editing enzyme APOBEC-1 (see Chapters 18 and 20 and references therein), although a Zn requirement has also been reported for tRNA-guanine transglycosylase.

The mechanism for the reverse U to C reaction is even less clear. Thus far, no cytosine deaminase that is capable of adding the amino group back onto the uridine has been described. If a transaminase rather than a cytosine deaminase is involved, this enzyme could utilize an amino group parked at some acceptor. These acceptor molecules vary between different transaminase enzymes, being glutamate for some and aspartate or the uracil base of UMP, UDP, and UTP for others (Scott, 1995) (Fig. 3B). Such transaminases often require cofactors such as GTP and a pyridoxal phosphate to transiently bind the amino group. Alternatively, an independent biochemical activity such as a CTP synthase (see, e.g., Zalkin, 1985; Scott, 1995) could be required to catalyze the U to C reaction. Last but not least, transglycosylation remains a possible mechanism for the U to C reaction, at least in theory.

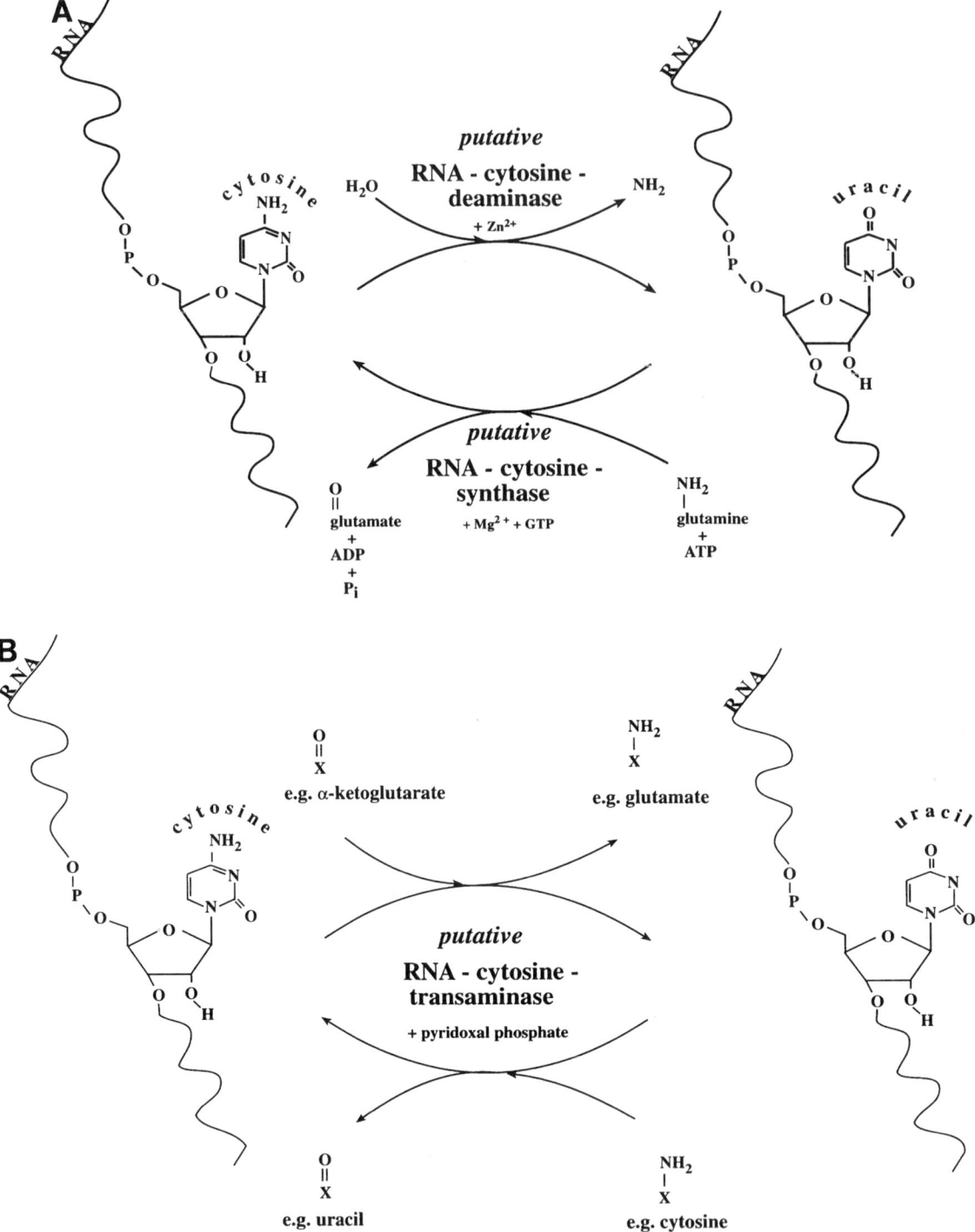

Figure 3. The biochemical process of RNA editing in plant organelles is most likely a deamination or transamination reaction. (A) A deamination reaction catalyzed by an RNA cytosine deaminase would be characterized by the in vitro ion and metal dependencies for the C to U conversion predominant in vascular plants. The reverse reaction, which is much more frequent in hornworts, would require a different enzyme, an RNA cytosine synthase possibly of the CTP synthase type, which requires magnesium ions and GTP as cofactors. (B) Alternatively, a transamination reaction by a putative RNA cytosine transaminase would store the amino group at a fairly high energy level to facilitate the reverse reaction that may be catalyzed by the same enzyme. Such transaminases often require pyridoxal phosphate as a cofactor. Experimental approaches must determine potential dependencies on amino group carriers.

Experimental Difficulties

The different in vitro systems for plant mitochondrial editing were developed along the design of the mammalian in vitro editing assays (see Chapter 18), involving incubation of synthetic, unedited RNAs in mitochondrial lysates followed by analysis of the editing frequency by primer extension and/or restriction analysis of PCR-amplified cDNAs (Araya et al., 1992; Blanc et al., 1995, 1996; Rajasekhar and Mulligan, 1993; Yu and Schuster, 1995). Unfortunately, these systems have turned out to be notoriously difficult and unreliable, varying from lysate preparation to preparation, and at their best they detect alterations by RNA editing just above background levels. The reason for this lack of success remains speculative at present. Possibly, the editing factors become organized only when anchored at a specific membrane site which is neither maintained well nor accessibly reconstituted in vitro once mitochondrial substructures have been disrupted. In addition, the biochemical reactions may require additional cofactors which are lacking or rapidly depleted without the context of a functional mitochondrion.

Nevertheless, in the absence of mitochondrial transformation systems and considering the very laborious plastid transformation assays, efficient and reproducible in vitro systems will be required in order to elucidate the mechanisms of plant organellar RNA editing and to identify the enzymes involved. As a possible solution, the selection of the starting material (e.g., the choice of the species and tissue from which to prepare the mitochondrial extract) may have to be reconsidered; it must be kept in mind that for the development of (for example) in vitro transcription assays, only selected species and cell types have turned out to provide an active system (Binder et al., 1995, 1996).

Selection of the Nucleotide To Be Edited

A second obstacle to efficient in vitro systems may be missing (complex) specificity determinants. As has been outlined and argued before (Bonnard et al., 1992; Gualberto et al., 1990b; Schuster and Brennicke, 1994), the more than 500 different editing sites in RNA sequences show no obvious common denominator that could act as an anchor for the editing activity, making it unlikely that sufficient protein determinants are available to recognize all the different sites. Instead, we consider it very likely that RNA determinants are involved, acting either in *trans*, such as the guide RNAs (gRNAs) in kinetoplastid editing (Benne, 1996) (see Chapter 21), or in *cis*, similar to the secondary RNA structures directing A to I conversions in animals (see Chapter 19).

Comparisons between the transcripts of duplicated mitochondrial gene fragments have shown that upstream sequences are more likely involved in determining editing sites than downstream regions. For some of the mitochondrial sites, 30 nucleotides appeared sufficient, while for others longer stretches seemed required. One of the most intriguing examples is a duplication of 39 nucleotides in the C-cytoplasm of maize, in which the central C nucleotide is edited to a U in both copies of the duplication (Kumar and Levings, 1993). The upstream and downstream 19 nucleotides therefore are sufficient to determine this editing event. In *Oenothera* mitochondria, a stretch of 48 nucleotides is identical in two related group II introns, the third nucleotide from the 5' end being edited in only one of the copies. Thus, 2 nucleotides upstream and 45 nucleotides downstream are not sufficient to specify this site (Lippok et al., 1994).

The base pairing stem of domain IV, of which the mismatching nucleotide needs to be edited, may act as a guide to create the U–A pair in the instance of the edited copy. In the other, nonedited copy of the 48-nucleotide duplication, a C–G pair is maintained without editing.

Editing specificity determined by the neighboring nucleotides is also indicated by a mosaic duplication of 380 nucleotides from the gene *orfB* in sugar beet (Kubo and Mikami, 1996), in which editing of two C's that are edited in *orfB* has been lost, presumably as a result of nucleotide alterations 2 and 5 nucleotides (nt) upstream of these C's (*orfB* is a conserved open reading frame in plant and algal mitochondria; its function is unclear). Intriguingly, one of these sites is edited in a mosaic *orf522* in a line of sunflower, in which 26 nucleotides upstream are identical to the *orfB* sequence and presumably sufficient to determine this site (Moneger et al., 1994). The other editing event of *orfB* is not observed in this fragment, with an altered nucleotide 9 bases upstream of this site possibly being responsible, because 36 nucleotides downstream of this site are identical to those of the original *orfB*.

The examples of editing sites compared in different contexts in vivo suggest that upstream sequences define at least some of the editing sites, but they do not rigorously exclude an involvement of downstream sequences. These observations are similar to those obtained with transgenic chloroplasts. When an editing site and surrounding sequences were introduced into a plastid genome, different extensions were required for correct recognition. At least for some of the sites, downstream sequences appear not to be crucially important and can be altered up to one or very few nucleotides from the editing site without any negative influence. At one site, upstream sequences down to 16 nt from the edited nucleotide were altered without any effect, while for another site changing nucleotides more than 150 nt upstream effectively eliminated editing (Chaudhuri and Maliga, 1996, 1997; Bock et al., 1996).

Recent evidence implies nucleus-encoded information in the specificity determination in transgenic chloroplasts (Bock and Koop, 1997). A spinach plastid editing site is not recognized in transplastomic tobacco cells, but becomes activated in cybrids containing tobacco plastids in a spinach nuclear background. If RNA molecules are indeed involved as *trans*-acting specificity determinants, this observation raises the interesting possibility that at least some guide RNAs are imported from the nucleocytoplasmic compartment. Clear precedence for RNA import into plant mitochondria exists, because several tRNAs somehow cross the membrane barriers to be used in translation (Marechal-Drouard et al., 1993b; Dietrich et al., 1996) (also see Chapter 11).

However, such guide RNA import will be as compartment specific as protein import, considering that sequences edited in the plastid are not edited at all in transcripts derived from a copy of this plastid DNA region following its integration in the mitochondrial genome of the same plant (Zeltz et al., 1996). In an analogous experimental assay, a mitochondrial sequence is not edited in plastids of a transplastomic line (Sutton et al., 1995).

Guide RNA sequences have been sought by different approaches, most of them using specific regions with editing sites as probes. In *Oenothera* mitochondria, RNA molecules with limited antisense pairing capabilities were identified by Schuster and coworkers. One of these possible guide RNAs is a fragment of the 26S rRNA, which base pairs to a highly edited region of a transcript from one of the genes involved in cytochrome *c* biogenesis (Yu and Schuster, 1996; Schuster, personal communication). To evaluate the significance of these findings it will be necessary to check their general relevance in other plants.

The flowering plant mitochondrial genome has ample room to code for even complex gRNA molecules; for example, *Arabidopsis* mt DNA has more than 200,000 nucleotides without any obvious function (Unseld et al., 1997). Interestingly, in *Marchantia*, which has no mt RNA editing, the sequence surplus in the mt genome is substantially smaller, with about 70,000 nucleotides (one third of that in the flowering plant), but still quite large (Ohyama et al., 1993). However, the screening of the mitochondrial genome of *Arabidopsis* for potentially base pairing guide RNAs does not reveal obvious antisense sequences beyond the statistically random similarities (Fig. 4).

In plastid genomes, octanucleotides with the potential for base pairing immediately upstream of an editing site have been found for some of the sites, but not for others. Similarities in sequence contexts around different sites could indicate that a common gRNA is used to direct the editing of these sites (Maier et al., 1996). The longer stretches of complementarity required in these cases have not been found in the completely sequenced plastid genomes and, provided such gRNAs indeed exist, they must be encoded in the nucleus.

In transplastomic experiments, the frequency of editing of some transcripts, but not all, is lowered by the introduction of a transgene (segment) encoding additional RNA copies. This suggests that for some editing sites the recognition factors are present in limiting amounts, and provides indirect evidence for the existence of *trans*-acting specificity guides (Chaudhuri et al., 1995; Bock et al., 1996).

DOES RNA EDITING PROVIDE A FUNCTIONAL ADVANTAGE?

Evolutionary selection and streamlining should have eliminated RNA editing, unless there is some advantage to be gained for the genetic system. One positive effect that has been speculated could be a regulatory use in fine-tuning the quantities of functionally translatable mRNAs (Chapdelaine and Bonen, 1991). Generating a correct, translatable message by RNA editing, for example, at ACG-initiation codons in the wheat *nad1* gene (Chapdelaine and Bonen, 1991) may be more rapid than de novo RNA synthesis or other ways of processing. On the other hand, physiological situations requiring such a rapid response in mitochodrial gene expression in plants have not yet been documented (Topping and Leaver, 1990) and, as previously discussed, partially edited transcripts appear to be translated (Gualberto et al., 1991; Phreaner et al., 1996). RNA editing in plant organelles most likely costs energy and thus should be selected against. This seems reflected in the distribution of editing sites between coding and noncoding regions, with only very few editing sites located in noncoding regions in mitochondria and none in plastids.

From an economic view, the negative features of RNA editing seem to far outweigh the potential advantages. The advantage of a faster regulatory response by editing a preexisting pool of mRNAs ap-

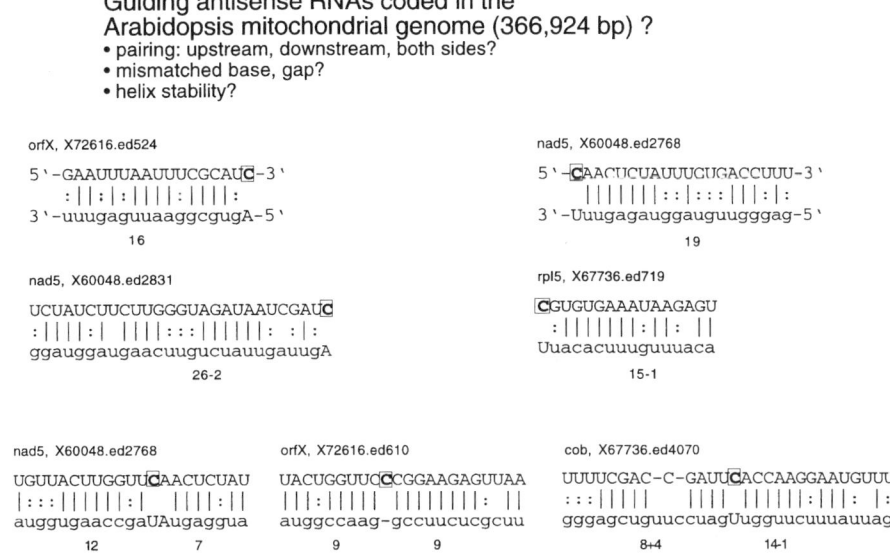

Figure 4. There could be ample space for guide RNAs in the mitochondrial genomes of plants. Similarity searches for sequences encoding potential guide RNAs for some selected editing sites in *Arabidopsis* identify candidates that, however, could also be attributed to a chance coincidence in the large chondriome. Shown are examples of potential gRNAs from searches of sequence trains upstream, straddling, and downstream of several identified editing sites. The edited nucleotide position is given for the corresponding database entry, and the edited C nucleotides are highlighted in bold and framed. Analogous guide RNAs (small nucleolar RNAs [snoRNAs]) have been found to specify ribose methylation sites in ribosomal RNA molecules by base pairing, which is not restricted to the rRNAs, but can also target this modification to other RNA molecules (Bachellerie and Cavaille, 1997, and references therein).

pears to be rather limited, considering how rapidly genetic systems can react by other mechanisms, such as transcriptional regulation or translational control. Furthermore, there currently is no example of a physiological or environmental requirement for a rapid response of the mitochondrial genetic system. In plastids, transcriptional and processing control of transcript abundance appears perfectly adequate to adjust the photosystems to changing conditions (such as light). Therefore, RNA editing appears to be costly without any obvious advantage.

The Evolutionary Origin of RNA Editing in Plants

A biochemical process increases in fashionable importance with its assumed evolutionary age; particularly, RNA processes become more interesting if there is any suspicion of connections to the primeval RNA world. Unfortunately, organelle RNA editing in plant mitochondria is most certainly a derived trait that arose eons after any possible RNA world was superseded by DNA and proteins. In an article from M. W. Gray's laboratory, the rise of plant organelle editing as a latecomer was formulated and detailed; the article also outlined the theoretical framework of establishing RNA editing by converting specificities of de-(trans-)aminases to bind and act on polynucleotides, and by creating a need for RNA editing in these organelles through genomic mutations that must be compensated for (Covello and Gray, 1993; see also Chapter 16).

Around the time plants rose from the water, RNA editing must have taken a foothold in the organelles (Fig. 5). The indirect evidence for this timing of the editing origin is the general presence of editing in organelles of all land plants, including mosses, liverworts, and hornworts, and the absence of RNA editing from organelles of green algae (Malek et al.,

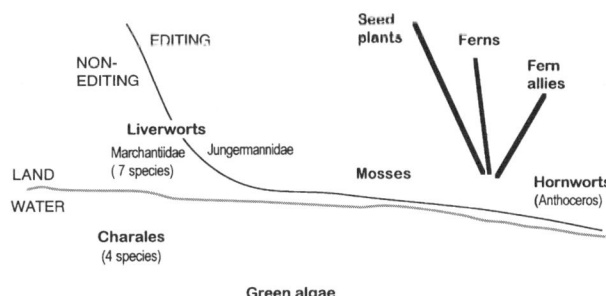

Figure 5. RNA editing is seen in all extant land plants except the Marchantiidae liverworts. The border between plants with editing and plants without (continuous line) is not congruent with the dividing line between land plants and algae (dotted line). Interpretation of independent loss or absence of gain of RNA editing in the Marchantiidae depends on the actual phylogeny, which is not clear at present.

1996). The apparent absence of RNA editing from mitochondria of *Marchantia* and six other liverworts of the Marchantiidae (Oda et al., 1992; Ohyama et al., 1993; Steinhauser et al., unpublished results) is intriguing, particularly because liverworts of the Jungermaniidae do show plant-typical RNA editing patterns in mitochondria and plastids.

An incentive or even a necessity to evolve and establish RNA editing could be a pressure to increase the G+C content in a genome with C to U RNA editing compensating for the increase in C residues. Comparing editing frequencies has indeed shown a direct correlation between the number of editing sites in a given gene and the G+C content (Malek et al., 1996), although it remains difficult to rule out the possibility that the increased editing is not compensating anything but simply reflects the increase in the number of C's.

Can a genetic system ever get rid of RNA editing after it has been established? The pertinent arguments for the establishment of this process and the continued requirement that keeps it around in a genetic compartment (Covello and Gray, 1993) would all argue against a potential loss. Such a loss of editing should theoretically be initiated by a traumatic change, such as an inactivation of the gene for the editing enzyme. Following this event, enough time must be allowed to compensate at the gene level for the now-missing C to U changes in the RNA.

It is likely that RNA editing can be lost at individual sites and, particularly after the transfer of genes from the mitochondrial genome to the nucleus, the requirements for editing of the mt remnant gene transcript are eliminated. In mitochondria of several plant species of the Brassicaceae, such as *Arabidopsis*, there is very little editing or even none required (such as in the atp9 mRNAs), which in other plants are altered by a number of such modification events. One theory would be that the requirement for editing has been lost by alterations of the genomic sequence, but these investigations have thus far given us little hard data on how this could happen. Nevertheless, it should theoretically be possible to lose the requirement for editing altogether, possibly proceeding via retrotransposition of edited cDNAs into the genome, similar to that observed during the transfer of an organellar gene to the nucleus (Nugent and Palmer, 1991, 1993; Grohmann et al., 1992). In these examples, the need for RNA editing and for splicing of the mitochondrial group II introns has been eliminated from the nuclear gene.

Is the lack of editing in *Marchantia polymorpha* mt RNAs (Oda et al., 1992; Ohyama et al., 1993) really a "loss of editing" also to be explained along the lines of retrotransposition into the mitochondrial genome of edited RNAs? The identification of reverse transcriptase activity in plant mitochondrial lysates would support such a scenario and shows that it could be a distinct possibility (Moenne et al., 1996). However, as is clear from the genes transferred to the nucleus, reverse sequence correction from RNA to DNA should also eliminate all of the intron sequences if it starts from the mature RNA. Comparison of the intron distribution between *Marchantia* and other plants should thus give some indication of the probability of such a secondary loss of editing. In the meantime, the alternative, that RNA editing in *Marchantia* organelles has never existed, cannot be ruled out as long as the phylogenetic relationships have not firmly been established.

Some kind of biochemical or functional difference may have been established between the flowering plant lineages on one hand and the hornworts and ferns on the other. The high incidence of U to C editing in the latter group suggests evolutionarily altered, different, or additional enzymes involved, and different biochemical parameters such as amino group cofactor concentrations.

Consequences of RNA Editing for the Analysis of Gene Expression

RNA editing significantly increases the efforts required to analyze gene expression in plant organelles. For the informational analysis of any given gene in plastids or mitochondria, cDNA analysis is mandatory. This includes sequencing multiple cDNA clones to ensure that all editing sites are identified.

RNA editing in plastids and mitochondria also complicates in vitro manipulation of organellar genes and expression. A selectable marker gene transformed into the organellar genome may be accidentally subject to alterations by RNA editing, leading to an inactive enzyme and a failing selection system. In plastids, the probability for inhibitory editing is much lower considering the limited number of editing sites, and sucessful transformation with expressed selection markers has shown that these techniques are not compromised by spurious editing.

Tissue-Specific Differences in RNA Editing

A priori, there appears to be no connection between the function of a gene and the number of editing sites in its transcript. Despite coding for the same function in different plant species, a gene such as atp6 or atp9 may require very little editing in its mRNAs in one plant and much editing in another one.

Considering a potential regulatory role of RNA editing, we would expect differences in editing patterns in different environmental situations. With this question in mind, RNA editing patterns of individual transcripts were investigated in different tissues and organs of several plant species. In potato mitochondria, different percentages of edited versus unedited or partially edited transcripts were indeed found when individual positions of a given message in different tissues were examined (Grohmann et al., 1994). The highest percentage of editing is generally found in flowers, correlated with the quantity of the respective mRNAs. It remains to be investigated whether this observation is an effect of prolonged half-lives of the RNAs or indeed is an accelerated editing processs. The specific requirements of mitochondrial activities by flower tissues are explained by a little-understood physiological connection, which particularly in developing pollen tissues demands peak performance of the respiratory organelles.

In another analysis, the atp9 transcript population was found to be virtually completely edited in all tissues of maize investigated, whereas the nad3 mRNA population showed some difference between different developmental stages with respect to the percentages of editing (Grosskopf and Mulligan, 1996). This difference is interpreted to be apparently without any consequence for mitochondrial function and regulation. It should be explicitly noted that the observed tissue-specific differences in editing status are indeed comparatively small, not only in mitochondria but also in plastids (further detailed above).

In plastids, RNA editing efficiency appears to be downregulated in some nonphotosynthetic tissues such as seeds and roots (Bock et al., 1993). RNA editing efficiency generally is not correlated to the photosynthetic activity, because in etiolated seedlings and in chromoplasts editing is virtually complete, as it is in green chloroplasts (Kuntz et al., 1992). If there is any connection between the regulation of gene expression and RNA editing in plastids, it can only be very subtle and accordingly hard to detect.

Thus, a regulatory role of RNA editing in preparing mRNA for producing the correct, mature protein required for organelle function cannot be excluded at present. However, considering the apparently very variable efficiency in a given tissue of editing at different sites of one coding region, it may be more reasonable and realistic to interpret the tissue-specific differences in editing as reflecting different turnover rates of the respective mRNA species.

Similar Editing Processes in Mitochondria and Plastids

The similarity of the RNA editing processes in plastids and mitochondria of a given plant cell (discussed above in the section entitled "Biochemistry of C to U and U to C RNA Editing") suggests a common evolution of the respective editing processes and very similar mechanistic features (Fig. 1). The same or a very similar enzyme(s) (for example, carrying out a transamination reaction) could be used in both organelles (Fig. 1, dividing arrow labeled trans-/deaminase). The enzymes may be encoded either by a single gene, the translation product being targeted to each of the compartments, or by evolutionarily closely related genes. An analogous example has been found for the phage-like organellar RNA polymerases, which are encoded by two very similar genes in plants (Hedke et al., 1997). Targeting a unique protein product of a single nuclear gene to different compartments has been well documented for tRNA processing and modification enzymes, which in yeast are partially kept in the cytoplasm and partially imported into the mitochondria (Martin and Hopper, 1994).

Another possible connection between editing in plastids and in mitochondria could be found in the amino group receptors, which if indeed required could be the same compound in both organelles (Fig. 1, cofactor collecting and cycling between the organelles).

The cell-fusion experiments between spinach and tobacco discussed above suggest that some kind of specificity determinant for plastid RNA editing is provided by the nucleus (Bock and Koop, 1997). If such factors are indeed encoded in the nuclear genome, it can be envisaged that they also get partitioned between plastids and mitochondria. However, specificity determinants generally are not shared by both organelles because the transcript of a plastid gene fragment that has been duplicated and integrated into mitochondria of the same plant is edited in plastids but not in mitochondria (Zeltz et al., 1996). If indeed RNA antisense base-pairing molecules guide the editing specificity, it is more likely that they will be encoded in the target compartment than that they will all be imported from the cytoplasm. Particularly, the 500–1,000 editing sites that need to be addressed in mitochondria would cause a massive traffic of polyribonucleotides through the mitochondrial membranes that is hard to imagine.

An additional problem faced by a plant cell in which all the editing components come through the cytoplasm would be to avoid the editing of cytoplasmic RNAs. If specificity determinants and the modifying enzymes are all synthesized from nuclear genes (Fig. 1), assembly of active complexes on cytoplasmic RNAs will have to be rigorously suppressed. Given the much higher complexity of the nucleus-encoded RNA pool, chance editings should be occurring, but have thus far not been reported in plant cells.

Potential Applications of RNA Editing

RNA editing is required for the production of a fully functional protein. Therefore, nonfunctional genomic sequences could be used to create cytoplasmically induced male sterile lines of transgenic plants as a method of "soft" manipulation strategically employing "native" components to introduce the controlled male sterility for plant breeding in crop plants. In these experiments (Araya et al., 1993; Hernould et al., 1993; Zabaleta et al., 1996), an unedited mitochondrial gene (*atp9*) was used to transform the nuclear genome of tobacco. The encoded variant of the ATP9 protein was efficiently imported into the organelle and rendered the mitochondria insufficiently productive to support pollen formation and maturation. Presumably this nucleus-induced male sterility is based on the slightly "wrong" ATP9 protein interfering with the normal ATPase peak function required during pollen maturation. Expression of the fully edited ATP9 coding sequence in the nuclear genome did not interfere with pollen growth, showing that the observed effect is indeed caused by the unedited status of the transcript and the aberrant protein produced. Antisense inhibition of the unedited *atp9* copy restored fertility (Zabaleta et al., 1996). In this approach, the existence of RNA editing in plant mitochondria has been exploited for the development of a controllable male sterility system.

CONCLUSIONS

RNA editing in plant mitochondria and chloroplasts interconverts C's and U's in mRNAs and tRNAs. In mitochondria of seed plants, nearly all mRNAs and some tRNAs are altered predominantly in the C to U direction, while in plastids only a few RNAs are edited. Transgenic plastids show that the length of flanking regions required to recognize individual sites varies, although in some cases relatively short stretches are sufficient. Biochemically, the cytidine appears to be deaminated; consequently, deaminases or transaminases are being considered as participating activities. The increasing number of U to C editing events being identified in more primitive plants, however, suggests a transaminating mechanism to deposit the amino group from the deamination of the cytidine onto a cofactor such as glutamate or uracil, which in the reverse U to C reaction can provide the reservoir for the amino group. RNA editing in plant organelles cosegregates with the oldest phylogenetic lineages among liverworts and has not yet been found in an alga. The simultaneous presence of editing in both organelles of most land plants suggests at least some factors common to this process in mitochondria and plastids.

Acknowledgments. Work in the authors' laboratory is supported by grants from the Deutsche Forschungsgemeinschaft; the Bundesministerium für Bildung, Wissenschaft, Forschung und Technologie; the Human Frontier Science Program; the European Community; the Fonds der Chemischen Industrie; a Landesforschungsschwerpunkt des Landes Baden-Württemberg; and the University of Ulm.

REFERENCES

Abraham, J. M., J. E. Feagin, and K. Stuart. 1988. Characterization of cytochrome *c* oxidase III transcripts that are edited only in the 3' region. *Cell* 55:267–272.

Andre, C., A. Levy, and V. Walbot. 1992. Small repeated sequences and the structure of plant mitochondrial genomes. *Trends Genet.* 8:128–131.

Araya, A., D. Begu, P. V. Graves, M. Hernould, S. Litvak, A. Mouras, and S. Suharsono. 1993. Of RNA editing and cytoplasmic male sterility in plants, p. 83–91. *In* A. Brennicke, and U. Kück (ed.), *Plant Mitochondria*. VCH Chemie, Weinheim, Germany.

Araya, A., C. Domec, D. Begu, and S. Litvak. 1992. An in vitro system for the editing of ATP synthase subunit 9 mRNA using wheat mitochondrial extracts. *Proc. Natl. Acad. Sci USA* 89:1040–1044.

Bachellerie, J.-P., and J. Cavaille. 1997. Guiding ribose methylation of rRNA. *Trends Biol. Sci.* 22:257–261.

Begu, D., P. V. Graves, C. Domec, G. Arselin, S. Litvak, and A. Araya. 1990. RNA editing of wheat mitochondrial ATP synthase subunit 9: direct protein and cDNA sequencing. *Plant Cell* 2:1283–1290.

Benne, R. 1996. RNA editing: how a message is changed. *Curr. Opin. Genet. Dev.* 6:221–231.

Binder, S., F. Hatzack, and A. Brennicke. 1995. A novel pea mitochondrial in vitro transcription system recognizes homologous and heterologous mRNA and tRNA promoters. *J. Biol. Chem.* 270:22182–22189.

Binder, S., A. Marchfelder, A. Brennicke, and B. Wissinger. 1992. RNA editing in intron sequences may be required for trans-splicing of nad2 transcripts in Oenothera mitochondria. *J. Biol. Chem.* 267:7615–7623.

Binder, S., A. Marchfelder, and A. Brennicke. 1994. RNA editing of tRNAPhe and tRNACys in mitochondria of Oenothera berteriana is initiated in precursor molecules. *Mol. Gen. Genet.* 244:67–74.

Binder, S., A. Marchfelder, and A. Brennicke. 1996. Regulation of gene expression in plant mitochondria. *Plant Mol. Biol.* 32:303–314.

Blanc, V., S. Litvak, and A. Araya. 1995. RNA editing in wheat mitochondria proceeds by a deamination mechanism. *FEBS Lett.* 373:56–60.

Blanc, V., X. Jordana, S. Litvak, and A. Araya. 1996. Control of gene expression by base deamination: the case of RNA editing in wheat mitochondria. *Biochimie* 78:511–517.

Blum, B., N. Bakalara, and M. Simpson. 1990. A model for RNA editing in kinetoplast mitochondria: "guide" RNA molecules transcribed from maxicircle DNA provide the edited information. *Cell* 57:355–366.

Bock, R., R. Hagemann, H. Kössel, and J. Kudla. 1993. Tissue- and stage-specific modulation of RNA editing of the psbF and psbL transcript from spinach plastids—a new regulatory mechanism? *Mol. Gen. Genet.* 240:238–244.

Bock, R., and V. Knoop. Unpublished results.

Bock, R., M. Herrmann, and H. Kössel. 1996. In vivo dissection of *cis*-acting determinants for plastid RNA editing. *EMBO J.* 15: 5052–5059.

Bock, R., H. Kössel, and P. Maliga. 1994. Introduction of a heterologous editing site into the tobacco plastid genome: the lack of RNA editing leads to a mutant phenotype. *EMBO J.* 13: 4623–4628.

Bock, R., and H. U. Koop. 1997. Extraplastidic site-specific factors mediate RNA editing in chloroplasts. *EMBO J.* 16:3282–3288.

Börner, G. V., M. Mörl, B. Wissinger, A. Brennicke, and C. Schmelzer. 1995. RNA editing of a group II intron in *Oenothera* as a prerequisite for splicing. *Mol. Gen. Genet.* 246:739–744.

Bonen, L. 1991. The mitochondrial genome: so simple yet so complex. *Curr. Opinion Genet. Dev.* 1:515–522.

Bonnard, G., J. M. Gualberto, L. Lamattina, and J. M. Grienenberger. 1992. RNA editing in plant mitochondria. *Crit. Rev. Plant Sci.* 10:503–524.

Brandt, P., M. Unseld, U. Eckert-Ossenkopp, and A. Brennicke. 1993. An rps14 pseudogene is transcribed and edited in *Arabidopsis* mitochondria. *Curr. Genet.* 24:330–336.

Carillo, C., and L. Bonen. 1997. RNA editing status of nad7 intron domains in wheat mitochondria. *Nucleic Acids Res.* 25:403–409.

Chapdelaine, Y., and L. Bonen. 1991. The wheat mitochondrial gene for subunit I of the NADH dehydrogenase complex: a trans-splicing model for this gene-in-pieces. *Cell* 65:465–472.

Chaudhuri, S., H. Carrer, and P. Maliga. 1995. Site specific factor involved in the editing of the psbL mRNA in tobacco plastids. *EMBO J.* 14:2951–2957.

Chaudhuri, S., and P. Maliga. 1996. Sequences directing C to U editing of the plastid psbL mRNA are located within a 22 nucleotide segment spanning the editing site. *EMBO J.* 15: 5958–5964.

Chaudhuri, S., and P. Maliga. 1997. New insights into plastid RNA editing. *Trends Plant Sci.* 2:5–6.

Covello, P. S., and M. W. Gray. 1989. RNA editing in plant mitochondria. *Nature* 341:662–666.

Covello, P. S., and M. W. Gray. 1993. On the evolution of RNA editing. *Trends Genet.* 9:265–268.

Dietrich, A., I. Small, A. Cosset, J.-H. Weil, and L. Marechal-Drouard. 1996. Editing and import: strategies for providing plant mitochondria with a complete set of functional transfer RNAs. *Biochimie* 78:518–529.

Fox, T. D., and C. J. Leaver. 1981. The Zea mays mitochondrial gene coding cytochrome oxidase subunit II has an intervening sequence and does not contain TGA codons. *Cell* 26:315–323.

Freyer, R., M.-C. Kiefer-Meyer, and H. Kössel. 1997. Occurrence of plastid RNA editing in all major lineages of land plants. *Proc. Natl. Acad. Sci. USA* 94:6285–6290.

Grohmann, L., A. Brennicke, and W. Schuster. 1992. The gene for mitochondrial ribosomal protein S12 has been transferred to the nuclear genome in *Oenothera*. *Nucleic Acids Res.* 20: 5641–5646.

Grohmann, L., O. Thieck, U. Herz, W. Schröder, and A. Brennicke. 1994. Translation of nad9 mRNAs in mitochondria from *Solanum tuberosum* is restricted to completely edited transcripts. *Nucleic Acids Res.* 22:3304–3311.

Grosskopf, D., and R. M. Mulligan. 1996. Developmental- and tissue-specificity of RNA editing in mitochondria of suspension-cultured maize cells and seedlings. *Curr. Genet.* 29:556–563.

Gualberto, J. M., G. Bonnard, L. Lamattina, and J.-M. Grienenberger. 1991. Expression of the wheat mitochondrial nad3-rps12 transcription unit: correlation between editing and mRNA maturation. *Plant Cell* 3:1109–1120.

Gualberto, J. M., C. Domon, J.-H. Weil, and J.-M. Grienenberger. 1990a. Structure and transcription of the gene coding for subunit 3 of cytochrome oxidase in wheat mitochondria. *Curr. Genet.* 17:41–47.

Gualberto, J. M., L. Lamattina, G. Bonnard, J.-H. Weil, and J.-M. Grienenberger. 1989. RNA editing in wheat mitochondria results in the conservation of protein sequences. *Nature* 341:660–662.

Gualberto, J. M., J.-H. Weil, and J.-M. Grienenberger. 1990b. Editing of the wheat coxIII transcript: evidence for twelve C to U and one U to C conversions and for sequence similarities around editing sites. *Nucleic Acids Res.* 18:3771–3776.

Hanson, M. R., C. A. Sutton, and B. Lu. 1996. Plant organellar gene expression: altered by RNA editing. *Trends Plant Sci.* 1: 57–64.

Hedke, T., T. Börner, and A. Weihe. 1997. Mitochondrial and chloroplast phage-type RNA polymerases in *Arabidopsis*. *Science* 277:809–811.

Hernould, M., S. Suharsono, S. Litvak, A. Araya, and A. Mouras. 1993. Male-sterility induction in transgenic tobacco plants with an unedited atp9 mitochondrial gene from wheat. *Proc. Natl. Acad. Sci. USA* 90:2370–2374.

Hiesel, R., and A. Brennicke. 1983. Cytochrome oxidase subunit II gene in mitochondria of *Oenothera* has no intron. *EMBO J.* 2:2173–2178.

Hiesel, R., W. Schobel, W. Schuster, and A. Brennicke. 1987. The cytochrome oxidase subunit I and subunit III genes in *Oenothera* mitochondria are transcribed from identical promoter sequences. *EMBO J.* 6:29–34.

Hiesel, R., B. Combettes, and A. Brennicke. 1994. Evidence for RNA editing in mitochondria of all major groups of land plants except the bryophyta. *Proc. Natl. Acad. Sci. USA* 91:629–633.

Hiesel, R., B. Wissinger, W. Schuster, and A. Brennicke. 1989. RNA editing in plant mitochondria. *Science* 246:1632–1634.

Hoch, B., R. M. Maier, K. Appel, G. L. Igloi, and H. Kössel. 1991. Editing of a chloroplast mRNA by creation of an initiation codon. *Nature* 353:178–180.

Knoop, V., W. Schuster, B. Wissinger, and A. Brennicke. 1991. *Trans*-splicing integrates an exon of 22 nucleotides into the nad5 mRNA in higher plant mitochondria. *EMBO J.* 10:3483–3493.

Kubo, T., and T. Mikami. 1996. A duplicated sequence in sugar-beet mitochondrial transcripts is differentially edited: analysis of orfB and its derivative orf324 mRNAs. *Biochim. Biophys. Acta* 1307:259–262.

Kumar, R., and C. S. Levings III. 1993. RNA editing of a chimeric maize mitochondrial gene transcript is sequence specific. *Curr. Genet.* 23:154–159.

Kuntz, M., B. Camara, J.-H. Weil, and R. Schantz. 1992. The psbL gene from bell pepper (*Capsicum annuum*): plastid RNA editing occurs in non-photosynthetic chromoplasts. *Plant Mol. Biol.* 20: 1185–1188.

Levings, C. S., III, and G. G. Brown. 1989. Molecular biology of plant mitochondria. *Cell* 56:171–179.

Lippok, B., A. Brennicke, and B. Wissinger. 1994. Differential RNA editing in closely related introns in *Oenothera* mitochondria. *Mol. Gen. Genet.* 243:39–46.

Lu, B., and M. R. Hanson. 1994. A single homogeneous form of ATP6 protein accumulates in petunia mitochondria despite the presence of differentially edited atp6 transcripts. *Plant Cell* 6: 1955–1968.

Lu, B., and M. R. Hanson. 1996. Fully edited and partially edited nad9 transcripts differ in size and both are associated with polysomes in potato mitochondria. *Nucleic Acids Res.* 24: 1369–1374.

Lu, B., R. K. Wilson, C. G. Phreaner, R. M. Mulligan, M. R. Hanson. 1996. Protein polymorphism generated by differential RNA editing of a plant mitochondrial *rps12* gene. *Mol. Cell. Biol.* 16:1543–1549.

Maier, R. M., P. Zeltz, H. Kössel, G. Bonnard, G. Gualberto, and J.-M. Grienenberger. 1996. RNA editing in plant mitochondria and chloroplasts, p. 303-314. In W. Filipowicz and T. Hohn (ed.), *Post-transcriptional Control of Gene Expression in Plants*. Kluwer Academic Publishers, Dordrecht, The Netherlands.

Malek, O., K. Lättig, R. Hiesel, A. Brennicke, and V. Knoop. 1996. Mitochondrial RNA editing in bryophytes and molecular phylogeny of land plants. *EMBO J.* 15:1403-1411.

Malek, O., A. Brennicke, and V. Knoop. 1997. Evolution of trans-splicing plant mitochondrial introns in pre-Permian times. *Proc. Natl. Acad. Sci. USA* 94:553-558.

Marchfelder, A., A. Brennicke, and S. Binder. 1996. RNA editing is required for efficient excision of tRNAPhe from precursors in plant mitochondria. *J. Biol. Chem.* 271:1898-1903.

Marechal-Drouard, L., D. Ramamonjisoa, A. Cosset, J.-H. Weil, and A. Dietrich. 1993a. Editing corrects mispairing in the acceptor stem of bean and potato mitochondrial phenylalanine transfer RNAs. *Nucleic Acids Res.* 21:4909-4914.

Marechal-Drouard, L., J.-H. Weil, and A. Dietrich. 1993b. Transfer RNAs and transfer RNA genes in plants. *Annu. Rev. Plant Mol. Biol.* 44:13-32.

Marechal-Drouard, L., R. Kumar, C. Remacle, and I. Small. 1996. RNA editing of larch mitochondrial tRNA-His precursors is a prerequisite for processing. *Nucleic Acids Res.* 24:3229-3234.

Marienfeld, J., M. Unseld, P. Brandt, and A. Brennicke. Unpublished results.

Marienfeld, J., M. Unseld, P. Brandt, and A. Brennicke. 1997. Mosaic open reading frames in the *Arabidopsis thaliana* mitochondrial genome. *Biol. Chem.* 378:859-862.

Martin, N. C., and A. K. Hopper. 1994. How single genes provide tRNA processing enzymes to mitochondria, nuclei and the cytosol. *Biochemie* 76:1161-1167.

Maslov, D. A., N. R. Sturm, B. M. Niner, E. S. Gruszynski, M. Peris, and L. Simpson. 1992. An intergenic G-rich region in *Leishmania tarentolae* kinetoplast maxicircle DNA is a pan-edited cryptogene encoding ribosomal protein S12. *Mol. Cell. Biol.* 12:56-67.

Moenne, A., D. Begu, and X. Jordana. 1996. A reverse transcriptase activity in potato mitochondria. *Plant Mol. Biol.* 31:365-372.

Moneger, F., C. J. Smart, and C. J. Leaver. 1994. Nuclear restoration of cytoplasmic male sterility in sunflower is associated with the tissue-specific regulation of a novel mitochondrial gene. *EMBO J.* 13:8-17.

Navaratnam, N., S. Bhattacharya, T. Fujino, D. Patel, A. L. Jarmuz, and J. Scott. 1995. Evolutionary origins of apoB mRNA editing: catalysis by a cytidine deaminase that has acquired a novel RNA-binding motif at its active site. *Cell* 81:187-195.

Nugent, J. M., and J. D. Palmer. 1991. RNA-mediated transfer of the gene coxII from the mitochondrion to the nucleus during flowering plant evolution. *Cell* 66:473-481.

Nugent, J. M., and J. D. Palmer. 1993. Evolution of gene content and gene organization in flowering plant mitochondrial DNA: a general survey and further studies on coxII gene transfer to the nucleus, p. 163-170. In A. Brennicke and U. Kück (ed.), *Plant Mitochondria*. VCH Chemie, Weinheim, Germany.

Oda, K., K. Yamato, E. Ohta, Y. Nakamura, M. Takemura, N. Nozato, K. Akashi, T. Kanegae, Y. Ogura, T. Kohchi, and K. Ohyama. 1992. Gene organization deduced from the complete sequence of liverwort *Marchantia polymorpha* mitochondrial DNA: a primitive form of plant mitochondrial genome. *J. Mol. Biol.* 223:1-7.

Ohyama, K., K. Oda, E. Ohta, and M. Takemura, M. 1993. Gene organization and evolution of introns of a liverwort, *Marchantia polymorpha*, mitochondrial genome, p. 115-129. In A. Brennicke, and U. Kück (ed.), *Plant Mitochondria*. VCH Chemie, Weinheim, Germany.

Phreaner, C. G., M. A. Williams, and R. M. Mulligan. 1996. Incomplete editing of rps12 transcripts results in the synthesis of polymorphic polypeptides in plant mitochondria. *Plant Cell* 8:107-117.

Pring, D., A. Brennicke, and W. Schuster. 1993. RNA editing gives a new meaning to the genetic information in mitochondria and chloroplasts. *Plant Mol. Biol.* 21:1163-1170.

Rajasekhar, V. K., and R. M. Mulligan. 1993. RNA editing in plant mitochondria: a phosphate is retained during C to U conversion in mRNAs. *Plant Cell* 5:1843-1852.

Schuster, W., and A. Brennicke. 1994. The plant mitochondrial genome: structure, information content, RNA editing and gene transfer. *Annu. Rev. Plant Physiol. Plant Mol. Biol.* 45:61-78.

Schuster, W., B. Combettes, K. Flieger, and A. Brennicke. 1993a. A plant mitochondrial gene encodes a protein involved in cytochrome c biogenesis. *Mol. Gen. Genet.* 239:49-57.

Schuster, W., and A. Brennicke. 1991. RNA editing makes mistakes in plant mitochondria: editing loses sense in transcripts of a rps19 pseudogene and in creating stop codons in coxI and rps3 mRNAs of Oenothera. *Nucleic Acids Res.* 24:6923-6928.

Schuster, W., R. Hiesel, and A. Brennicke. 1993a. RNA editing in plant mitochondria. *Semin. Cell Biol.* 4:279-284.

Schuster, W., R. Hiesel, B. Wissinger, and A. Brennicke. 1990a. RNA editing in the cytochrome *b* locus of the higher plant *Oenothera berteriana* includes a U-to-C transition. *Mol. Cell. Biol.* 10:2428-2431.

Schuster, W., V. Knoop, R. Hiesel, L. Grohmann, and A. Brennicke. 1993b. The mitochondrial genome on its way to the nucleus: different stages of gene transfer in plants. *FEBS Lett.* 325:140-144.

Schuster, W., M. Unseld, B. Wissinger, and A. Brennicke. 1991. Ribosomal protein S14 transcripts are edited in Oenothera mitochondria. *Nucleic Acids Res.* 18:229-233.

Schuster, W., B. Wissinger, M. Unseld, and A. Brennicke. 1990b. Transcripts of the NADH-dehydrogenase subunit 3 gene are differentially edited. *EMBO J.* 9:263-269.

Scott, J. 1995. A place in the world for RNA editing. *Cell* 81:833-836.

Sharp, P. 1994. Split genes and RNA splicing. *Cell* 77:805-816.

Steinhauser, S., S. Beckert, I. Capesius, A. Brennicke, and V. Knoop. Unpublished results.

Sutton, C. A., O. V. Zoubenko, M. R. Hanson, and P. Maliga. 1995. A plant mitochondrial sequence transcribed in transgenic tobacco chloroplasts is not edited. *Mol. Cell. Biol.* 15:1377-1381.

Topping, J. F., and C. J. Leaver. 1990. Mitochondrial gene expression during wheat leaf development. *Planta* 182:399-407.

Unseld, M., J. Marienfeld, P., Brandt, and A. Brennicke. 1997. The mitochondrial genome in *Arabidopsis thaliana* contains 57 genes in 366,924 nucleotides. *Nat. Genet.* 15:57-61.

Wissinger, B., R. Hiesel, W. Schobel, M. Unseld, A. Brennicke, and W. Schuster. 1991a. Duplicated sequence elements and their function in plant mitochondria. *Z. Naturforsch.* 46:709-716.

Wissinger, B., W. Schuster, and A. Brennicke. 1990. Species-specific RNA editing patterns in the mitochondrial rps13 transcripts of *Oenothera* and *Daucus*. *Mol. Gen. Genet.* 224:389-395.

Wissinger, B., W. Schuster, and A. Brennicke. 1991b. Trans splicing in Oenothera mitochondria: nad1 mRNAs are edited in exon and trans splicing group II intron sequences. *Cell* 65:473-482.

Yang, A. J., and R. M. Mulligan. 1993. Distribution of maize mitochondrial transcripts in polysomal RNA: evidence for non-selectivity in recruitment of mRNAs. *Curr. Genet.* 23:532-536.

Yoshinaga, K., H. Iinuma, T. Masuzawa, and K. Ueda. 1996. Extensive RNA editing of U to C in addition to C to U substitution in the *rbcL* transcripts of hornwort chloroplasts and the origin of RNA editing in green plants. *Nucleic Acids Res.* 24:1008–1014.

Yu, W., and W. Schuster. 1995. Evidence for a site-specific cytidine deamination reaction involved in C-to-U RNA editing of plant mitochondria. *J. Biol. Chem.* 270:18227–18233.

Yu, W., and W. Schuster. 1996. Requirements of a plant mitochondrial in vitro RNA editing system, p. 30. *In Proceedings of the EMBO Workshop on RNA Editing*, Maastricht, The Netherlands.

Zabaleta, E., A. Mouras, M. Hernould, M. Suharsano, and A. Araya. 1996. Transgenic male-sterile plant induced by an unedited atp9 gene is restored to fertility by inhibiting its expression with antisense RNA. *Proc. Natl. Acad. Sci. USA* 93:11259–11263.

Zalkin, H. 1985. CTP synthase. *Methods Enzymol.* 113:282–287.

Zanlungo, S., D. Begu, V. Quinones, A. Araya, and X. Jordana. 1993. RNA editing of apocytochrome b (cob) transcripts in mitochondria from two genera of plants. *Curr. Genet.* 24:344–348.

Zanlungo, S., V. Quinones, A. Moenne, L. Holuigue, and X. Jordana. 1995. Splicing and editing of rps10 transcripts in potato mitochondria. *Curr. Genet.* 27:565–571.

Zeltz, P., K. Kadowaki, N. Kubo, R. M. Maier, A. Hirai, and H. Kössel. 1996. A promiscuous chloroplast DNA fragment is transcribed in plant mitochondria but the encoded RNA is not edited. *Plant Mol. Biol.* 31:647–656.

Chapter 18

Apolipoprotein B mRNA Editing

BENNY HUNG-JUNN CHANG, PAUL P. LAU, AND LAWRENCE CHAN

Apolipoprotein B (apoB) mRNA editing was the first instance of RNA editing described in vertebrates. It consists of the conversion of the first base of the codon CAA, encoding glutamine-2153, from a C to a U, and forming UAA, an in-frame stop codon (Chen et al., 1987; Powell et al., 1987).

Since its description about a decade ago, substantial progress has been made on the specificity and regulation of the process, and on the characterization of the enzyme complex involved in the C → U conversion, a sequence-specific hydrolytic deamination reaction.

PHYSIOLOGY OF THE APOLIPOPROTEIN B-CONTAINING LIPOPROTEINS

Plasma lipoproteins are macromolecular complexes of proteins and lipids (Havel and Kane, 1995). They are classified by their flotation densities into chylomicrons, very-low-density lipoprotein (VLDL), intermediate-density lipoprotein (IDL), low-density lipoproteins (LDL), and high-density lipoproteins (HDL). The lipids, in the forms of phospholipids, cholesterol, cholesteryl esters, and triglycerides, are held in plasma in association with amphipathic proteins known as apolipoproteins (Li et al., 1988). There are nine major plasma apolipoproteins, apoA-I, A-II, A-IV, B-100, B-48, C-I, C-II, C-III, and E, and also several minor ones.

The structure and function of apoB-100 and apoB-48 have been reviewed recently (Kane and Havel, 1995; Chan, 1992). apoB-100 is synthesized in the liver and is secreted into the circulation as VLDL. In the vascular compartment, VLDL is metabolized to IDL by the action of lipoprotein lipase, and subsequently to LDL by the action of lipoprotein lipase, and hepatic lipase. Both IDL and LDL are recognized by the LDL receptor and are taken up by the liver by receptor-mediated endocytosis (Brown and Goldstein, 1986). The ligand for this process is apoB-100 in LDL. apoB-48 is synthesized only in the small intestine in humans; in some other mammals it is also synthesized in the liver (see below under "Species and Tissue Specificity of apoB mRNA Editing"). It is an essential component of intestinal chylomicrons, which are formed in response to a fatty meal.

STRUCTURE OF apoB-100 AND apoB-48: PHYSIOLOGICAL CONSEQUENCE OF apoB mRNA EDITING

Structurally, apoB-48 corresponds to the N-terminal 48% of apoB-100. apoB-100 contains 25 cysteine residues, at least 16 of which are involved in intramolecular disulfide bridge formation. There is a cluster of 18 cysteine residues in the N-terminal 40% of the molecule; these residues are thus present in both apoB-100 and apoB-48. The seven cysteine residues in the C-terminal half of apoB-100 are missing in apoB-48. apoB-100 on LDL can be divided into five subdomains, which are defined by their relative sensitivity to trypsin digestion and subsequent release of the tryptic peptides from the lipoprotein particle (Yang et al., 1989). Based on the linear alignment of apoB-48 and apoB-100, the apoB-48 sequence covers domains I and II and part of domain III of apoB-100. It would miss domains IV and V. Domain IV (spanning residues 3071–4100) is an important functional domain in apoB-100. It contains a preponderance of trypsin-releasable peptides. Most importantly, it also encompasses the putative LDL receptor-binding domain. The last domain, domain V (residues 4101–4536), is highly hydrophobic and contains mostly non-trypsin-releasable peptides. Domain V

Benny Hung-Junn Chang, Paul P. Lau, and Lawrence Chan • Departments of Cell Biology and Medicine, Baylor College of Medicine, Houston, Texas 77030.

contains two cysteine residues, one of which (C-4326) has been shown to mediate the covalent linkage of apoB-100 to apo(a) (Callow and Rubin, 1995) in lipoprotein (a), a unique LDL-like lipoprotein that is strongly correlated with atherosclerosis development (Utermann, 1995). Because domains IV and V are absent from apoB-48, this apoprotein does not have the capacity to be a ligand for the LDL receptor, or to contribute to lipoprotein (a) formation.

DISCOVERY OF apoB mRNA EDITING

The plasma apoBs are central molecules in lipoprotein metabolism, and their structure has been a subject of intense research. In 1986, the structure of apoB100 deduced from overlapping cDNA sequences (Yang et al., 1986; Knott et al., 1986) together with extensive direct peptide sequence information (Yang et al., 1989; Yang et al., 1986) showed that apoB-100 contains 4,536 amino acid residues. The following year, direct sequencing of proteolytic peptides from apoB-48 established that this protein corresponds to the N-terminal region of apoB-100 (Chen et al., 1987). Sequence analysis of multiple cDNA clones covering the C-terminal region of this protein revealed that Ile-2152 is the C-terminus of apoB-48 (Chen et al., 1987; Powell et al., 1987). apoB-100 sequences downstream from Ile-2152 are missing in apoB-48 because the CAA codon for Gln-2153 in apoB-100 mRNA has been changed to UAA, a stop codon, in apoB-48 mRNA. Thus, translation of apoB-48 mRNA stops prematurely in the middle of the mRNA molecule, and the translation product apoB-48 misses the C-terminal 2,384 residues (i.e., from Gln-2153 to Leu-4536), or over 52%, of apoB-100. Subsequent investigations indicated that the C → U change that converts apoB-100 mRNA to apoB-48 mRNA is mediated by a novel mechanism known as RNA editing (Fig. 1).

As reviewed in Chapter 21 of this volume, RNA editing was first described to occur in kinetoplastids in 1986. The description of apoB mRNA editing in 1987 represents the first example of RNA editing in vertebrates.

SPECIES AND TISSUE SPECIFICITY OF apoB mRNA EDITING

A simple sensitive assay to determine the percentage of edited apoB was developed by Driscoll et al. (1989) and greatly facilitated reseach on apoB mRNA editing (Fig. 2). Using this assay, Greeve et al. (1993) studied the tissue specificity of apoB mRNA in 12 mammalian species and detected significant editing of apoB mRNA in the small intestine of all 12 mammals (Table 1). The percentage of editing was about 40% in sheep and 73% in horse, but well over 80% in the other mammalian species. In liver, they detected apoB mRNA editing at 18% in dog, 43% in horse, 62% in rat, and 70% in mouse. Low levels (<1%) of editing were also detected in the liver of rabbit and guinea pig; it was undetectable in the other mammals studied. An interesting correlation exists between the presence of significant hepatic apoB mRNA editing and low concentrations of apoB-100-containing plasma lipoproteins (VLDL + LDL) (Table 1). The reason behind the rate-limiting role for apoB-100 in LDL production is the fact that only apoB-100-containing VLDL, but not apoB-48-containing VLDL, is metabolized to LDL (Van't Hooft et al., 1982). Thus, hepatic editing appears to be a determinant for the plasma concentrations of the atherogenic plasma lipoproteins VLDL, IDL, and LDL.

SUBCELLULAR LOCATION OF apoB mRNA EDITING

The subcellular location for apoB mRNA editing has been examined in rat liver. Lau et al. (1991) separated the cellular mRNA into cytoplasmic and nuclear pools, and fractionated the nuclear RNA into spliced and unspliced fractions (with respect to the intron immediately upstream of exon 26, which contains the edited site) by the polymerase chain reaction and polyadenylated and unpolyadenylated species by oligo(dT) binding. The percentage of editing was determined in each pool of apoB mRNA and pre-mRNA. They found the following proportions of edited mRNA: nuclear unspliced unpolyadenylated, 1.56%; nuclear spliced unpolyadenylated, 7.8% ± 0.6%; nuclear unspliced polyadenylated, 25.4% ± 0.05%; nuclear spliced polyadenylated, 53% ± 0.6%; and polysomal, 62.15% ± 6.2%. From these values, it is clear that apoB mRNA editing is a predominantly intranuclear reaction, being essentially complete before the fully mature nuclear apoB mRNA is exported into the cytoplasm, where little additional editing occurs. Furthermore, the data indicate that editing is a posttranscriptional event, since the nascent unspliced unpolyadenylated apoB pre-mRNA is essentially unedited. The apparent stepwise increase in the amount of edited apoB mRNA with individual maturation events (splicing and polyadenylation) led Lau et al. (1991) to speculate that editing might be coupled with one or both of these events. The edited codon is in a 7.5-kb exon (no. 26) (Fig.

CHAPTER 18 • apoB mRNA EDITING

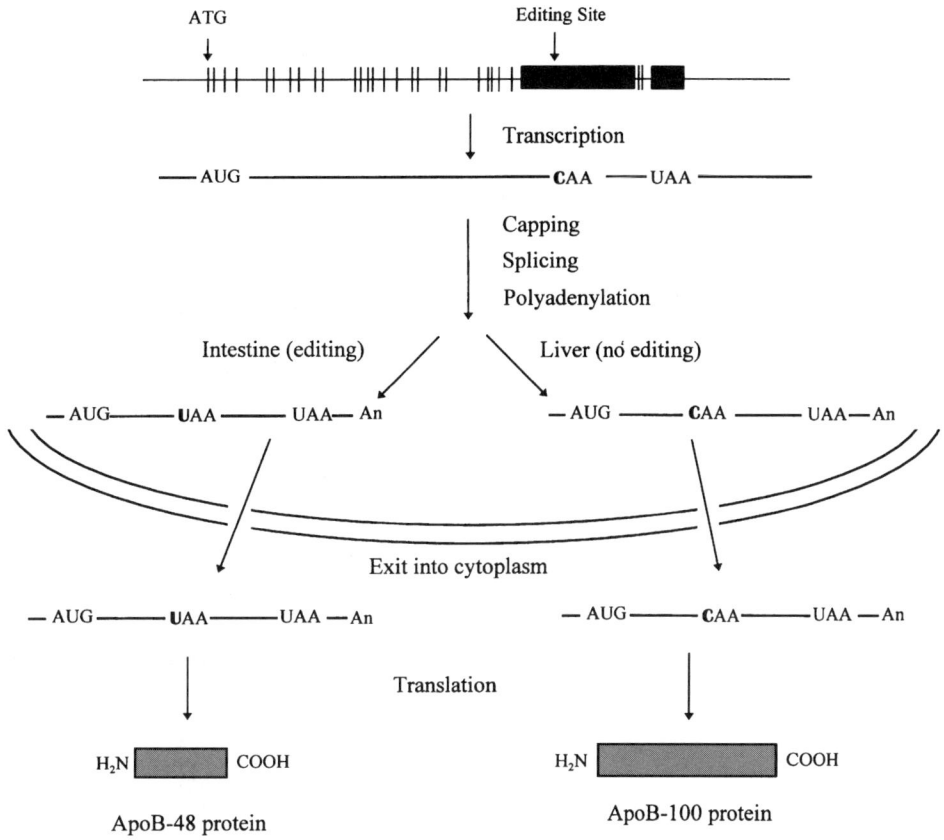

Figure 1. Schematic diagram of apoB mRNA editing. The genomic structure of human apoB is at the top. The 29 exons are depicted by vertical bars, and the 28 introns are depicted by lines between the exons. The translation initiation codon and editing site are as indicated. In humans, apoB mRNA editing occurs only in the intestine. The nuclear compartment is separated from the cytoplasmic compartment by nuclear membranes. Mature apoB mRNAs are shown exported through nuclear pores into the cytoplasm, where they are translated into proteins.

1). In in vitro experiments, splicing efficiency falls precipitously when exon size is artificially increased (Robberson et al., 1990). The splicing of exon 26 in vivo may involve unique *trans*-acting factors, which in turn interact with proteins involved in editing. The possible coupling of editing to these RNA maturation events is an interesting hypothesis. Chen and Chan reevaluated the possible mechanisms behind the apparent stepwise increase in editing with apoB mRNA maturation (Chen and Chan, 1996). They found that kinetic parameters, such as apoB mRNA transcription rate and degradation rates at individual stages of mRNA maturation, are important in determining the degree of apoB mRNA editing at each stage. When they computed the expected editing using published kinetic parameters, the degree of editing at each maturation stage was not dissimilar to the experimentally derived values of Lau et al. (1991). In other words, it is unnecessary to invoke some coupling mechanism to explain the stepwise increase in editing during apoB mRNA maturation in the nucleus (Chen and Chan, 1996).

SEQUENCE SPECIFICITY OF apoB mRNA EDITING

The primary transcript of the apoB gene is >43 kb long, and the mature mRNA is ~14 kb long (Fig. 1). There must be a high degree of sequence specificity of editing; otherwise, numerous mutations would be introduced at the posttranscriptional level. The sequence specificity of apoB mRNA editing was characterized by a number of laboratories (Backus and Smith, 1992; Backus and Smith, 1991; Shah et al., 1991; Chen et al., 1990). Our current knowledge of the specificity has been summarized in two recent review articles (Hodges and Scott, 1992; Smith, 1993). The edited C is at nucleotide position 6666 in human apoB mRNA and has been referred to as C-6666 for all species. The sequence alignment of mammalian and chicken apoB mRNAs around the editing site is shown in Fig. 3. Other sequences that show partial similarity to this region are included in this figure because they also shed light on the sequence specificity of editing. The target substrate sequence can be

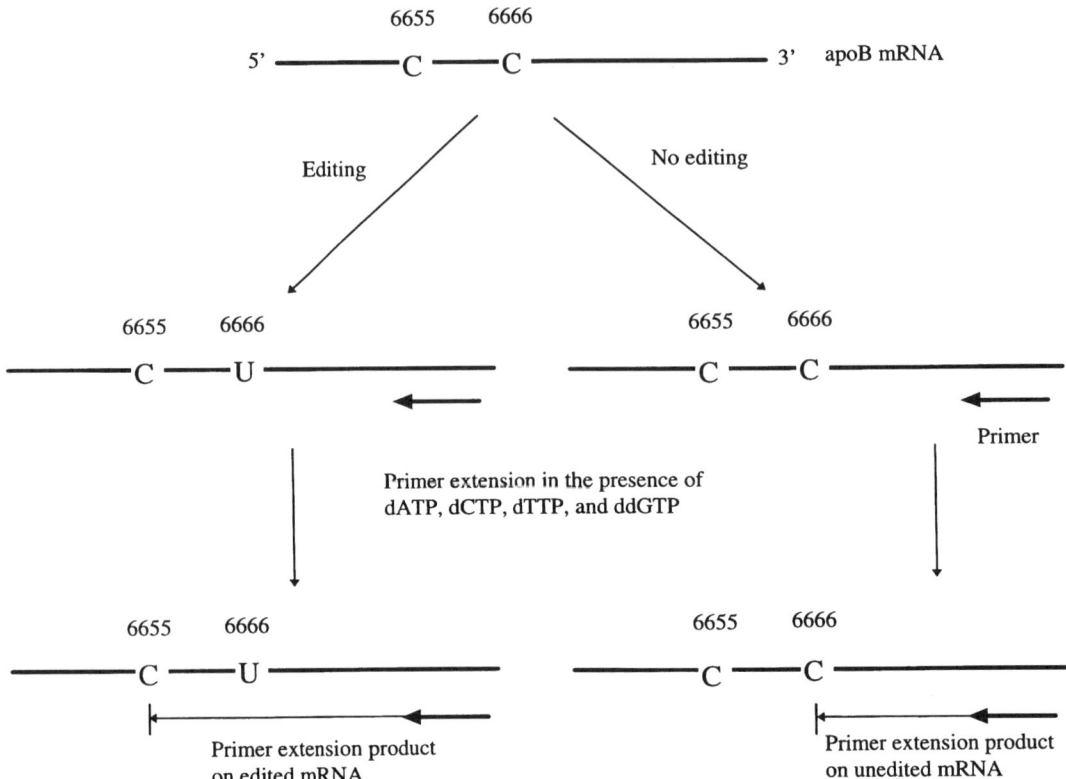

Figure 2. Primer extension assay for the degree of apoB mRNA editing. A radiolabeled antisense oligonucleotide, shown by an arrow, downstream of the apoB mRNA (human apoB mRNA is used as an example) editing site serves as a primer for reverse transcription of the apoB mRNA mixture to be assayed. In the reverse transcriptase (RT) reaction, dATP, dCTP, dTTP, and ddGTP are added into the mix so that the primer extension reaction will stop at the first cytidine of the RNA template because of the incorporation of the chain-terminating ddGTP. When the canonical C-6666 is changed to a U by RNA editing, the reverse transcription product will be extended to the next upstream C, resulting in a longer primer extension product. The relative size of the two primer extension products can be resolved by polyacrylamide gel electrophoresis and autoradiography.

Table 1. apoB mRNA editing in the liver and intestine and the (VLDL + LDL)/HDL ratio of 12 mammalian species [a]

Species	% of apoB mRNA edited in:		(VLDL + LDL)/HDL
	Liver	Intestine	
Human	0	98	1.92
Monkey	0	97	0.91
Pig	0	82	1.40
Cow	0	95	1.04
Sheep	0	40	0.65
Cat	0	84	0.47
Rabbit	<1	90	0.32
Guinea pig	<1	87	12.3
Dog	18	84	0.26
Horse	43	73	0.44
Rat	62	88	0.41
Mouse	70	89	0.25

[a] Adapted from Greeve et al. (1993).

	Regulator	(Editing site) 6666 ↓	Spacer	Mooring sequence	
human	UAUCUCAACUGCAGACAUAUA	UGAUA	C	AAUU U GAU CAGUAUA	UUAAAGAUAGUUAUGAUUUACA
dog	-G-------AA-A------G	-----	-	---- - --- --------	---------A-----------G
cat	-------G--U---------G	-----	-	---- - --- --------	---------A------------
horse	-------------A------G	-----	-	---- - --- --------	---------A------------
pig	-------G--A--A------G	-----	-	---- - --- --------	---------A------------
cow	-G-----G-----A------G	-----	-	---- - --- --------	---------A--------G--
sheep	-G-----G------------G	-----	-	---- - --- --------	---G--------------G--
guinea pig	-C-----G------------G	----G	-	---- - --- --U----	----U-A--A-------C---G
chicken	-GCUC--G--U--AGUG---C	--G-G	-	--A- A --G -----C-	-C-----G-A---C---CAGUU
human 2nd site		G	-	A-A- - ---Cca- ----	-UAC--U -GUUUA
mouse NAT1*		C		UUGC- - --AAc - ---t	----U-G AG-G
human NF1†		U		-AU-- -g----g - ----	--CAUCC -CUG
rat NF1†		U		--U-- -g---Ca - U--	--CAUCC- CCG

Figure 3. Sequence alignment of apoB mRNA from various vertebrate species, neurofibromatosis type 1 (NF1) mRNA, and NAT1 mRNA around the edited site. C-6666 in human apoB mRNA is indicated by an arrow. The chicken apoB mRNA sequence is not edited and is included for comparison. In human apoB mRNA there is a second site that is edited at about 10% efficiency compared with the canonical C-6666. It displays substantial variation in the mooring sequence region. Putative regulatory elements (regulator, spacer, and mooring sequence) are indicated by overlines. Residues that are identical to the human sequence are indicated by dashes, substitutions are indicated by uppercase letters, and insertions are indicated by lowercase letters. *, NAT1 mRNA has five mooring-like consensus sequences, but only the major one associated with an edited C is shown here. †, NF1 was originally identified by the presence of mooring-like consensus sequence, but the enzyme responsible for editing is probably not apobec-1, so the significance of the presence of the mooring-like sequence in this mRNA is unclear. Even more perplexing is the fact that in contrast to the human NF1 mRNA, the rat NF1 mRNA is not edited (Skuse et al., 1996).

divided into three functional motifs from the 3' end of the molecule going in a 5' direction: they are a mooring sequence; a spacer sequence immediately 3' of C-6666; and a regulator region immediately 5' of the edited C. In in vitro editing experiments, the mooring sequence was found to be both necessary and sufficient for conferring editing susceptibility on a C located 4–6 bases 5' of the mooring sequence (Fig. 3) (Backus and Smith, 1991). Sequences further removed from the editing site also contribute to a poorly defined bulk RNA "context" that enhances editing in vitro. The enhancing activity of such bulk RNA is related to the length and AU content of the sequence. The length of the spacer sequence is not absolute; C's that are 4–6 bases upstream of the mooring sequence are all susceptible to editing in vitro (Chen et al., 1990). The situation in vivo, however, is much more complex, because the presence of the mooring sequence is not necessarily associated with editing of its upstream C. When Yamanaka et al. (1997) overexpressed apobec-1 (the enzyme responsible for apoB mRNA editing [see below]) in mice, a novel mRNA, NAT1, was found to be edited (a discussion on the physiology of apobec-1 overexpression on Nat1 mRNA is found below in "Pathophysiology of Hepatic Overexpression of apobec-1"). Analysis of the NAT1 cDNA clones from these transgenic mice revealed that over 100 sites in NAT1 are edited to various degrees. However, only 5' mooring-like consensus sequences are found, including some that have an upstream C that is not edited. Indeed, in this mRNA, there are more mooring sequence-independent than -dependent sites that are edited. In another study, Skuse et al. (1996) used the apoB mRNA mooring sequence as a query sequence and found the mRNA for NF1 (neurofibromatosis type 1 tumor suppressor) which was edited in the correct sequence context (Fig. 3). It would have been an in vivo validation of the "necessary and sufficient" hypothesis of the role of the mooring sequence, except that apobec-1 appears not to be the responsible enzyme because the degree of editing remained unchanged when apobec-1 was markedly up-regulated by transfection in cultured cells. Interestingly, the rat NF1 mRNA, which has only one residue difference in the mooring sequence region compared with the human sequence (Fig. 3), is not edited at all. Although Skuse et al. (1996) offered a tenuous rationalization as to why the human but not the rat NF1 mRNA is edited, these observations indicate that we are still a long way from fully understanding the sequence specificity of apobec-1-mediated editing.

The sequence specificity summarized above was determined by using in vitro tissue extracts from either mammalian small intestinal cells (Driscoll et al., 1989) or rodent liver cells (Backus and Smith, 1992; Backus and Smith, 1991; Shah et al., 1991; Chen et al., 1990), which are competent in editing synthetic apoB mRNA fragments in vitro. The catalytic protein in these extracts was cloned initially from rat small intestine (Teng et al., 1993) and subsequently from human (Lau et al., 1994; Hadjiagapiou et al., 1994), rabbit (Yamanaka et al., 1994), and mouse (Nakamuta et al., 1995) small intestines as well. The cloned protein has been called apobec-1 (for apoB mRNA editing enzyme, catalytic polypeptide 1) (Davidson et al., 1995).

apobec-1: THE CATALYTIC SUBUNIT OF THE EDITOSOME COMPLEX

apobec-1 is a 229-amino-acid [rat (Teng et al., 1993) and mouse (Nakamuta et al., 1995)] or 236-amino acid [human (Lau et al., 1994; Hadjiagapiou et al., 1994) and rabbit (Yamanaka et al., 1994)] protein that is the catalytic subunit of the apoB mRNA editing protein complex (also called editosome). apobec-1 appears to largely determine the tissue specificity of apoB mRNA editing. In humans and rabbits, editing occurs in the small intestine, and apobec-1 mRNA occurs exclusively in that tissue. In rats and mice, editing occurs in both the liver and small intestine, and apobec-1 mRNA is expressed in both tissues. Moreover, in rodents apobec-1 is expressed in many other tissues that do not express apoB. The wide distribution of apobec-1 mRNA has led to speculations about its possible function in other tissues. If indeed there are other mRNAs that are modified by apobec-1-mediated editing, they have not been identified to date. It is unlikely that apoB mRNA editing subserves any unique vital functions because inactivation of apobec-1 in mice seems to be compatible with relatively good health.

In the mouse, apobec-1 mRNA exists in two different sizes, an ~2.2-kb form in the small intestine and an ~2.4-kb form in liver, spleen, kidney, lung, muscle, and heart (Fig. 4A). The tissue-specific expression of apobec-1 mRNA is determined by differential promoter utilization, and the mRNAs expressed have different sizes as a result of alternative splicing (Nakamuta et al., 1995). The mouse apobec-1 gene is located on chromosome 6. It contains eight exons and spans ~25 kbp. The major hepatic mRNA contains all eight exons, whereas the major small intestinal mRNA misses the first three exons and its transcription is initiated in exon 4, where the translation initiation codon is located (Fig. 4B). There are also two alternatively spliced hepatic apobec-1 mRNAs with different acceptor sites in exon 4 (Nakamuta et al., 1995) (Fig. 4B). Chloramphenicol acetyltransferase reporter assays revealed that liver expression of apobec-1 in mouse is driven by a promoter at the 5' flank region, while intestinal expression of apobec-1 is predominantly driven by a promoter within intron 3 (Nakamuta et al., 1995). A similar observation was found in the rat, in that the major apobec-1 mRNA transcripts are smaller in intestine than in liver (Teng et al., 1993). Recently, Hirano et al. (Hirano et al., 1997b) found that the rat apobec-1 gene contains only 6 exons; the nonintestinal mRNA is transcribed from all 6 exons, whereas the intestine-specific mRNA is transcribed from exons 2–6, which are the equivalents of exons 4–8 in mouse. The biological relevance of the tissue-specific promoter preference and of the alternative splicing is not known. However, because all the sequence variations occur in the noncoding regions, the end products are not changed. In contrast to rodents, the human gene appears to contain only 5 exons, with the translation initiation codon located in exon 1 (Hirano et al., 1997a). Interestingly, a differentially spliced apobec-1 mRNA that misses exon 2 appears to be a major transcript in the human small intestine. Immunoreactive material cross-reacting with its translation product, a nonfunctional nonapeptide, has been detected in villus cells of the adult small intestine.

The amino acid sequence of human apobec-1 is shown in Fig. 5. apobec-1 has cytidine deaminase activity (Navaratnam et al., 1993a) and shows sequence similarity to the cytidine/cytidylate deaminase housekeeping enzymes from prokaryotes and eukaryotes (Nakamuta et al., 1995; Navaratnam et al., 1993a). apobec-1 expression in the rat and mouse is very low, but can be detected by Western blotting as a single band of 27 kDa by using monospecific antisera against the protein (Yeung and Chan, unpublished observation). Furthermore, by subcellular fractionation, immunoreactive mouse apobec-1 is found to be present mainly in the nucleus, although small amounts of the protein can also be detected in the cytoplasm.

apobec-1 occurs as a spontaneous homodimer (Lau et al., 1994). Dimerization does not require disulfide bond formation and is dependent on the pres-

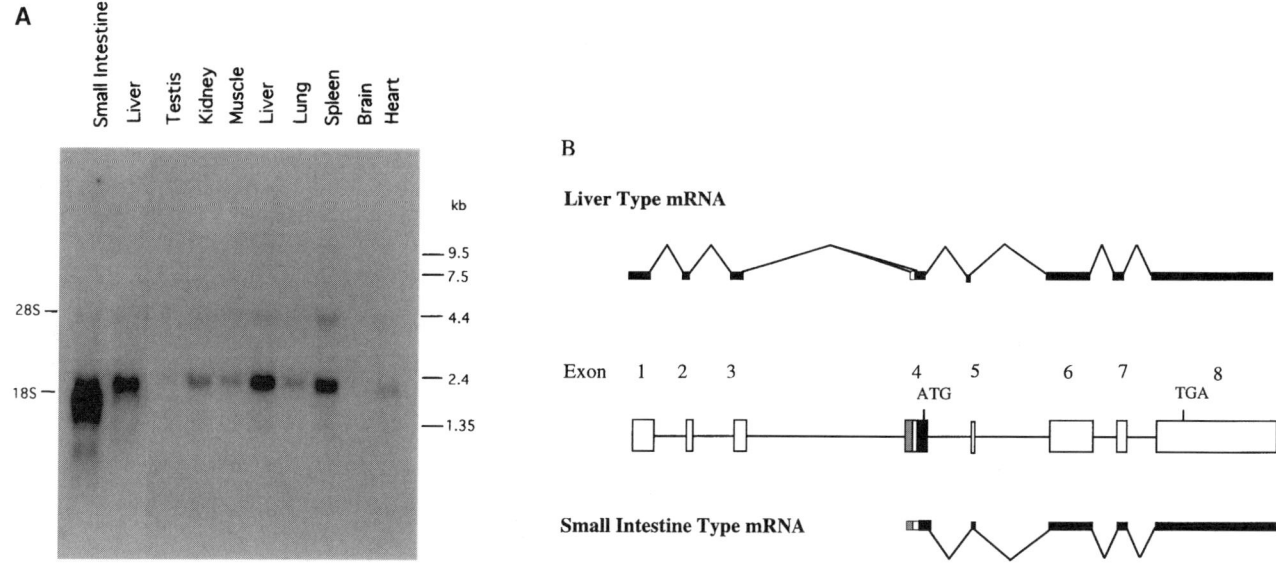

Figure 4. Tissue-specific mRNA expression of apobec-1 in mice. (A) Northern blot analysis of apobec-1 mRNA in mouse tissues. The mRNA exists in two different sizes, an ~2.2-kb form in small intestine and an ~2.4-kb form in liver, spleen, kidney, lung, muscle, and heart. The 4.4-kb band found in liver, lung, and spleen may represent yet another alternatively spliced apobec-1 transcript which has not been characterized. (B) Structure of the mouse apobec-1 gene and its major mRNA transcripts. The gene comprises eight exons (indicated by boxes). In the liver (and presumably other tissues except small intestine), there are two alternatively spliced transcripts, resulting from the presence of two different splice acceptor sites in exon 4. Transcription of the major small intestinal form, in contrast, is initiated within intron 3. The intestinal mRNA thus contains exons 4 (with a unique 102-nucleotide fragment, shown by a shaded box, at its 5' end derived from the intron 3 region of the liver transcript) through 8. Panel A is reproduced with permission from Nakamuta et al. (1995).

ence of the C-terminal region of apobec-1, which contains overlapping heptad leucine-rich motifs (Fig. 5A). The capacity to dimerize is abolished if this region of the molecule is deleted, or if 2 of the 10 leucine residues (Leu182 and Leu187 in mouse apobec-1) making up the leucine-rich motifs are replaced by arginine residues (Oka et al., 1997). Loss of dimerization potential is invariably associated with loss of editing activity, suggesting that dimer formation may be required for activity. Further support for the dimeric enzyme being the active form is the demonstration that apoB mRNA editing activity can be inhibited both in vitro and in vivo by the presence of excess amounts of a catalytically inactive mutant apobec-1 only when the mutant form remains dimerization-competent (a dominant negative mutant). Dimerization-incompetent mutant apobec-1s do not inhibit the activity of the wild-type enzyme (Oka et al., 1997).

Apart from the C-terminal leucine-rich region, apobec-1 contains some other potential functional motifs (Fig. 5B). In the N-terminal region is a sequence, $RRR(X_{11})PR/QXXRK$, which is a putative nuclear localization signal. This sequence may determine the predominantly intranuclear nature of apobec-1 and of apoB mRNA editing (see above).

Enzyme catalysis of many cytidine deaminases requires zinc for activity (Navaratnam et al., 1995). Two stretches of sequence in apobec-1, residues 61–66 and 84–93, can be aligned with the active zinc-chelating domains of both prokaryotic and eukaryotic cytidine deaminases. In these regions, H61 and C93 are potential zinc-chelating sites, and E63 is the proton-shuttling residue essential for catalysis. Encompassed within these regions are two phenylalanine residues (F66 and F87) that may be involved in binding to apoB mRNA (Navaratnam et al., 1995). These residues are unique to apobec-1 and not found in other members of the cytidine deaminase gene family (see below). A third region (residues 154–182), which overlaps the leucine-rich region, shows similarity to a leucine-rich sequence stretch in the C-terminal part of Escherichia coli cytidine deaminase (Fig. 5C). Most of the cytidine/cytidylate deaminases from other organisms are shorter proteins than the E. coli enzyme and do not have this region. Protein crystallography of E. coli cytidine deaminase indicates that this part of the molecule is not directly involved in enzyme catalysis (Betts et al., 1994). The similarity can be attributed entirely to the presence of 3 proline and 4 leucine residues in apobec-1[169–182] and E. coli cytidine deaminase[235–249] which are com-

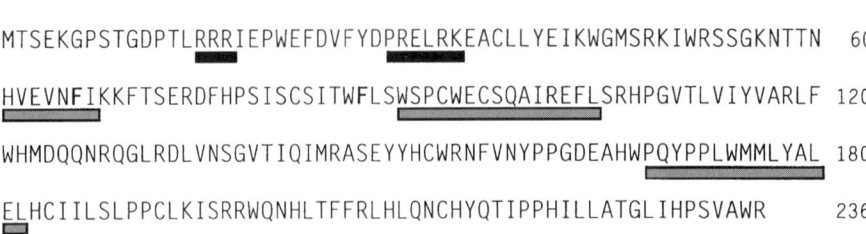

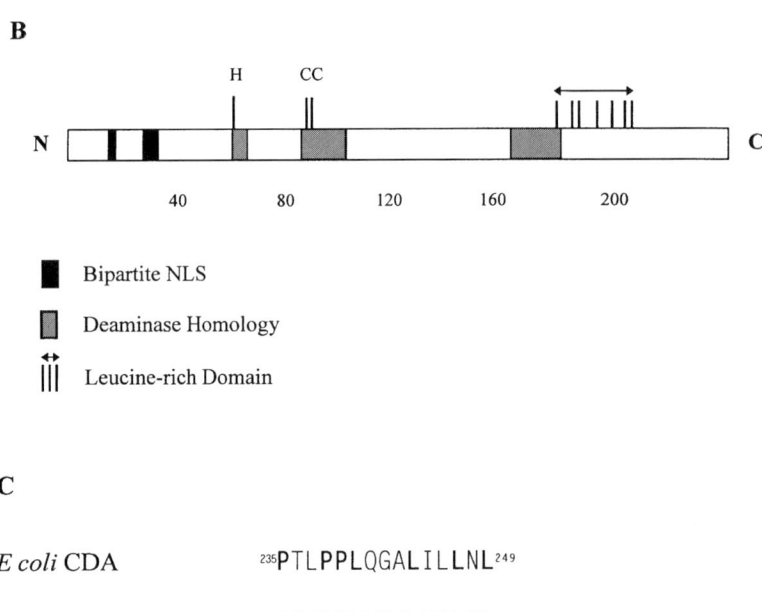

Figure 5. (A) Human apobec-1 amino acid sequence. The bipartite nuclear localization signal is highlighted by two solid boxes separated by 11 residues. The sequences homologous to highly conserved sequence blocks in *E. coli* cytidine deaminase are underlined by three shaded boxes. The first two shaded boxes are the cytidine deaminase homology motifs. Two phenylalanine residues (F66 and F87, in bold) encompassed within these two regions were found to be important for binding to apoB mRNA in vitro. The third box shows similarity to a region of *E. coli* cytidine deaminase only. The leucine-rich domain is indicated by a double-arrowheaded underline. (B) Schematic drawing of the sequence motifs identified in panel A. The zinc-coordinating residues H61, C93 and C96 are shown above the first two cytidine deaminase motifs. The NLS (nuclear localization signal) and deaminase homology domains correspond to those in panel A. The individual leucine residues in the leucine-rich domain are marked by vertical lines. (C) Alignment of the last cytidine deaminase homology domain in apobec-1 with a region in the C-terminal half of *E. coli* cytidine deaminase. The cytidine deaminase isolated from other organisms is generally shorter than the *E. coli* enzyme and misses this region shown. As is evident from the alignment, the apparently high homology is entirely the result of the matching Pro and Leu residues in this regions. The significance of this short homologous stretch of sequences, which is missing in cytidine deaminase from most other organisms, is unknown.

pletely conserved between the two enzymes (marked by bold letters in Fig. 5C). The significance of this match between the two sequences is not clear.

EVOLUTION AND MECHANISM OF ACTION OF apobec-1

The evolutionary relationship of apobec-1 with the cytidine/cytidylate deaminases has been noted by a number of investigators (Nakamuta et al., 1995; Navaratnam et al., 1993a), and has been discussed in a recent review article (Chan et al., 1997). Here we outline a few salient points concerning the evolution of the apobec-1-related deaminase gene family, taking into consideration some novel homologous sequences published since these earlier reviews.

The apobec-1-related group of deaminases appears to have similar action in that it requires zinc at the catalytic center for coordinating deaminase activity. The putative zinc-chelating motifs CHAE and PCG are conserved across most of these proteins. By aligning the conserved sequences in the catalytic domain of all the prokaryotic and eukaryotic deaminases, we constructed a phylogenetic tree of the apobec-1-related sequences (Fig. 6). Phylogenetic analysis suggests that these deaminases can be subdivided into four clusters: cytidine deaminase, cytidylate deaminase, double-stranded RNA adenosine deaminase (ADAR, for adenosine deaminase acting on RNA), and apobec-1; the former two use mononucleotides as substrates, while the latter two use RNA trascripts as substrates.

Covello and Gray (1993) outlined a three-step model for the emergence of a new editing enzyme (also see Chapter 16) that involved the duplication of the original gene, the divergence of the duplicated gene, the and its fixation toward new function. Based on such an evolutionary scenario, one can postulate that the zinc-requiring deaminases were derived from an ancestral cytidine deaminase gene through multiple independent duplication events. One of these

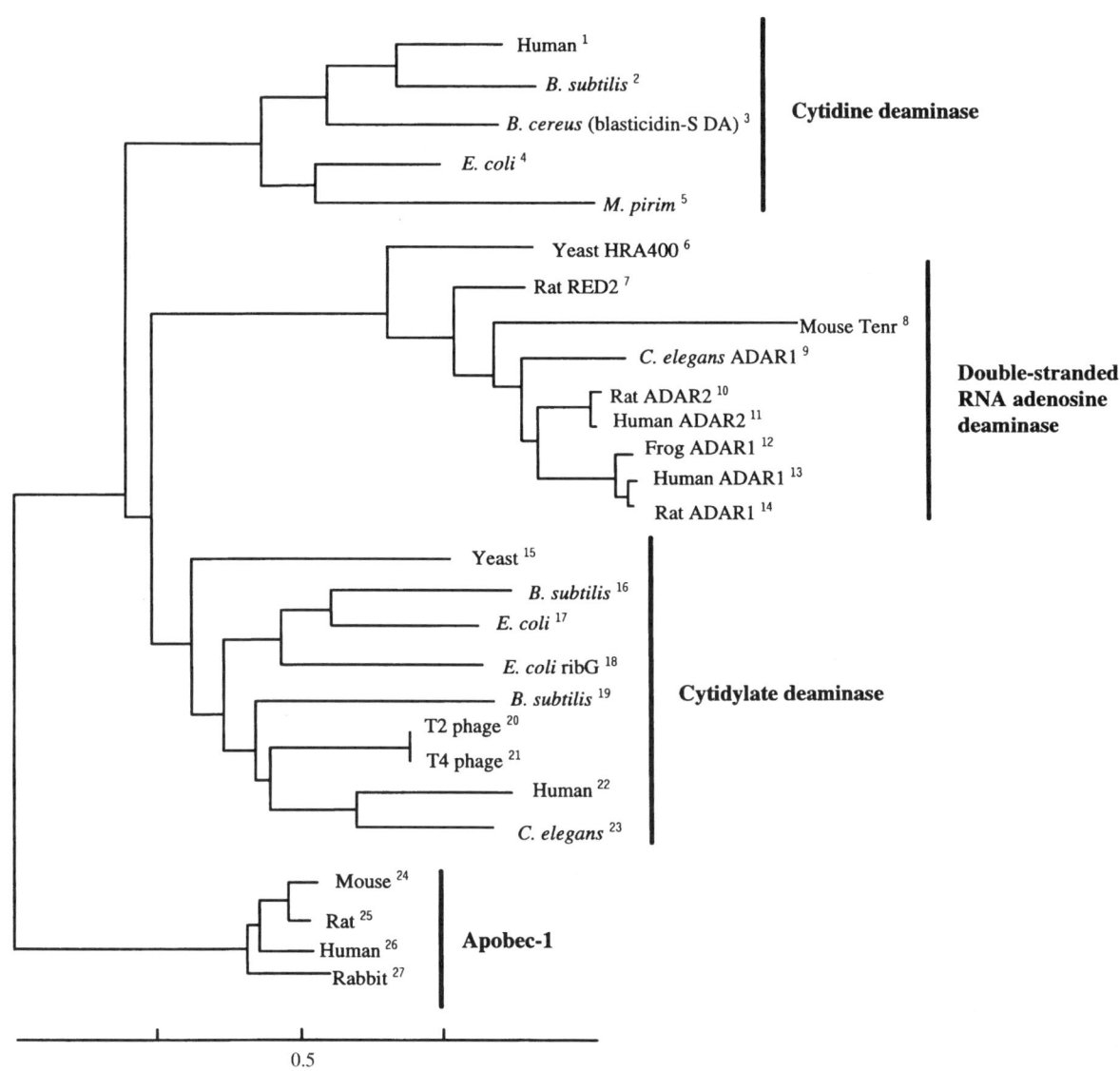

Figure 6. Phylogenetic tree of the apobec-1-related sequences reconstructed by the neighbor-joining method (Saitou and Nei, 1987). Only the putative catalytic domains of the various proteins consisting of 34 amino acid residues (some with fewer residues because of deletions) were aligned and analyzed. The GenBank accession numbers of these sequences are as follows: 1, L27943; 2, P19079; 3, JS0609; 4, P13652; 5, L13289; 6, Z49149; 7, U74586; 8, X84693; 9, U00037; 10, U43534; 11, U82120; 12, U88006; 13, U10439; 14, U18942; 15, P06773; 16, P21335; 17, P30314; 18, P25539; 19, P32393; 20, P00814; 21, P16006; 22, 12136; 23, P30648; 24, U22262; 25, L07114; 26, L26234; 27, U10695. The horizontal scale at the bottom of the figure indicates the numbers of amino acid substitutions per position along the tree branches.

events gave rise to cytidine deaminase, another gave rise to the common ancestor of cytidylate deaminase and ADAR, and a third gave rise to apobec-1.

To date, nine sequences in the ADAR family have been reported and are included in this analysis. All these enzymes appear to be involved in A → I editing in different eukaryotic systems (see other chapters on A → I editing in this volume). Two new genes were cloned recently by sequence homology to ADAR (Gerber et al., 1997). One was from *Drosophila,* and the other was from yeast. The yeast enzyme (HRA 400) was overexpressed in the yeast *Pichia pastoris,* and the purified recombinant protein was found to deaminate N37 adenosine adjacent to the anticodon loop in yeast tRNA-Ala under in vitro conditions in the absence of any cofactors. The sequence of the *Drosophila* gene is not available, but the region of HRA400 around the catalytic domain shows sequence similarity to other members of the ADAR family, which suggests that this tRNA-editing enzyme and ADAR have similar mechanisms of action and a common ancestor. With the exception of HRA400, all members of the ADAR family with known function contain double-stranded RNA-binding motifs. These motifs presumably enable these deaminases to seek out the double-stranded regions of the RNA substrate for the deamination reaction. The fact that HRA400 lacks such a motif suggests that editing of N37 in tRNA does not require a double-stranded structure for recognition. These dsRNA-binding motifs may be acquired by other ADAR enzymes later on in evolution to allow them to work on double-stranded RNA as substrates. It is also possible that HRA400 has lost these motifs during evolution. It should be pointed out that both intron-dependent and intron-independent base modifications of tRNAs have been described (Grosjean et al., 1997), but the specific enzymes involved and whether some of them have dsRNA-binding motifs currently are not known.

The structure of the editing enzymes involved in the editing of plant RNA in mitochondria or chloroplast has not been reported. It is likely that these enzymes represent yet another offshoot from the cytidine deaminase superfamily. Like other members of the superfamily, they seem to require zinc for activity (see Chapter 17), and the predominant editing reaction is a C to U transition. However, they display significant differences in how they recognize their substrates, because a consensus recognition sequence around the edited sites has not been identified. For many plant mRNAs, multiple sites are edited within the same RNA. It is highly unlikely that a large number of deaminase enzymes are involved in the editing, each effecting the modification of a unique site or a unique set of sites. It is much more likely that one enzyme or a small number of enzymes that recognize some as-yet-unidentified secondary structure in the RNA, perhaps involving guide RNAs, are involved (see Chapter 17).

apobec-1 diverged from the other members of the cytidine deaminases superfamily quite early in evolution (Fig. 6). In contrast to ADAR, which does not require complementation factors and whose sequence specificity appears to depend solely on ADAR interacting with some secondary structure of the substrate RNA, apobec-1 requires complementation factors for activity and for optimal interaction with the substrate. Thus, the evolution of apobec-1 has to be interpreted in the context of the coevolution of the auxiliary proteins that modulate its editing function. Another interesting caveat is the absence of apoB mRNA editing in nonmammalian species, suggesting that apobec-1 has acquired its editing function with the emergence of mammals. In contrast, ADARs have been identified in diverse organisms such as yeast, worm, and fly. Perhaps for the same reason, apobec-1 is the only member identified in the apobec-1 family shown in Fig. 6, whereas many ADAR homologs have been identified in the ADAR family.

Except for the ADARs, which seem to be monomeric enzymes, all members of the superfamily (cytidine/cytidylate deaminases with known subunit structures as well as apobec-1) exist as homodimers or homomultimers. The crystal structure of *E. coli* cytidine deaminase has been determined (Betts et al., 1994). The impressive sequence conservation at the active site between apobec-1 and this prokaryotic enzyme strongly indicates a parallelism in the mechanisms of action of the two deaminases. The possible mechanism of action of apobec-1 has been reviewed recently (Chan et al., 1997) and is dealt with in depth in Chapter 20, which should be consulted for details.

STRUCTURE OF THE apoB mRNA EDITING ENZYME COMPLEX

apobec-1 by itself has little editing activity on synthetic apoB mRNA substrates in vitro. Efficient editing requires the participation of complementation or auxiliary factors that appear to be almost ubiquitous in various mammalian organs, including many tissues that do not synthesize apoB. apobec-1 has some RNA-binding activity (Anant et al., 1995; Navaratnam et al., 1995), the specifics of which are discussed in Chapter 20, but the high apoB mRNA sequence specificity of editing appears to be conferred by an auxiliary protein(s) that also interacts with apobec-1.

Components in the editosome complex must recognize apoB mRNA or apobec-1 for them to confer specificity and efficiency to the latter enzyme. apoB mRNA-binding proteins have been identified by UV cross-linking of radiolabelled apoB mRNA to specific proteins in various tissue extracts. An ~40-kDa protein (Lau et al., 1990) and later both an ~44-kDa and an ~66-kDa protein (Navaratnam et al., 1993b; Harris et al., 1993) that interact with apoB mRNA have been identified. The role of these proteins in apoB mRNA editing, if any, is unknown.

Using classical biochemical separation methods and complementation editing assays, Mehta et al. (1996) enriched a protein fraction from baboon kidney extract that complements editing. Although the major protein bands had a molecular mass of ~65 kDa under nondenaturing conditions, subsequent direct apobec-1-blotting experiments suggest that an apobec-1-interacting protein band(s) might be present in a higher molecular weight region of the gel away from the major protein bands.

Smith et al. (1991; Harris et al., 1993) used a different approach toward the structure of the editosome complex. Using glycerol gradients, they isolated crude 27S editosome complexes and produced monoclonal antibodies against these preparations (Schock et al., 1996). One of the antibodies recognized a 240-kDa protein complex (on a native gel). Immunoadsorption of tissue extracts with this antibody inhibited editing activity and editosome assembly. The 240-kDa protein did not bind to apobec-1 or apoB mRNA. Schock et al. (1996) postulated that it might be an auxiliary protein complex required for the assembly of an active editosome.

Another approach to isolating editosome components that might interact with apobec-1 was by the yeast two-hybrid system. Using this technique, Lau et al. (1997) identified several proteins that specifically bind to apobec-1. One of these proteins was fully characterized and has been named apobec-1-binding protein 1 (ABBP-1). ABBP-1 contains 331 amino acid residues and is identical to a previously published human type A/B hnRNP protein (Khan et al., 1991), except for a 47-residue insertion at its C-terminal region that was produced as a result of alternate RNA splicing. ABBP-1 mRNA is widely distributed in all examined human tissues. ABBP-1 contains two ~80-amino-acid-long RNA-binding domains, each containing two short sequences, RNP-1 (RNP octamer) and RNP-2 (RNP hexamer), that are typical of many RNA-binding proteins. The C-terminal part of the protein is very glycine-rich and contains a stretch of eight glycine residues immediately preceding the extra exon sequence (produced by alternate splicing). The eight-glycine motif appears to be required for interaction with apobec-1. Interestingly, ABBP-1 also binds to apoB mRNA and can be UV-cross-linked to it in vitro (Lau et al., 1997). Immunodepletion of ABBP-1 from an active apoB mRNA editing tissue extract inhibits its editing activity. Down-regulation of ABBP-1 in an apobec-1-expressing HepG2 cell line by transfection with an antisense ABBP-1 cDNA construct results in the inhibition of endogenous apoB mRNA editing. Thus, ABBP-1 appears to be a component of the apoB mRNA editing enzyme complex. It recognizes both apobec-1 and apoB mRNA. It may be involved in recruiting and possibly disrupting the secondary structure of apoB mRNA and bringing it to the vicinity of apobec-1 for editing. It is the only auxiliary factor that has been demonstrated to play a role in apoB mRNA editing in cultured cells in vivo.

Based on our current knowledge of the area, we present a model of the apoB mRNA editing enzyme complex (Fig. 7). The central protein in this complex is apobec-1, whose only currently known function is the specific hydrolytic deamination of apoB mRNA. All the noncatalytic auxiliary proteins in the complex appear to be proteins normally involved in other structural or metabolic functions in the cell. The best characterized of these is ABBP-1, which is a type A/B hnRNP that may be involved in pre-mRNA processing. Antibodies against type A/B hnRNP proteins have been found to inhibit RNA splicing in vitro (Sierakowska et al., 1986), possibly by interfering with reactions at the 5' splice site (Mayeda and Krainer, 1992). In binding to pre-mRNA, the hnRNP causes the hnRNA to become extended and accessible to the other factors involved in RNA splicing (Dreyfuss et al., 1993). As ABBP-1 binds to the vicinity of the editing site, it also unfolds the RNA, making it accessible to apobec-1. ABBP-1-mediated facilitation of editing occurs mainly in the nucleus via apoB pre-mRNA binding. Its interaction with apoB pre-mRNA on one hand and its affinity for apobec-1 on the other bring the latter protein into a proper alignment with the editing site, where it effects the hydrolytic deamination. We do not know, however, whether the same hnRNPs involved in apoB pre-mRNA splicing act as auxiliary proteins in editing. Multiple hnRNPs are involved in splicing (Dreyfuss et al., 1993), and more than one hnRNP may have the capacity to participate in the editing reaction. There is unpublished evidence that specific hnRNPs other than ABBP-1 may also bind to apobec-1 in vitro and modulate the editing reaction. In addition to ABBP-1 (and other hnRNPs), other putative auxiliary proteins seem to play a role (Fig. 7). The other auxiliary proteins may represent factors involved in the assembly/disassem-

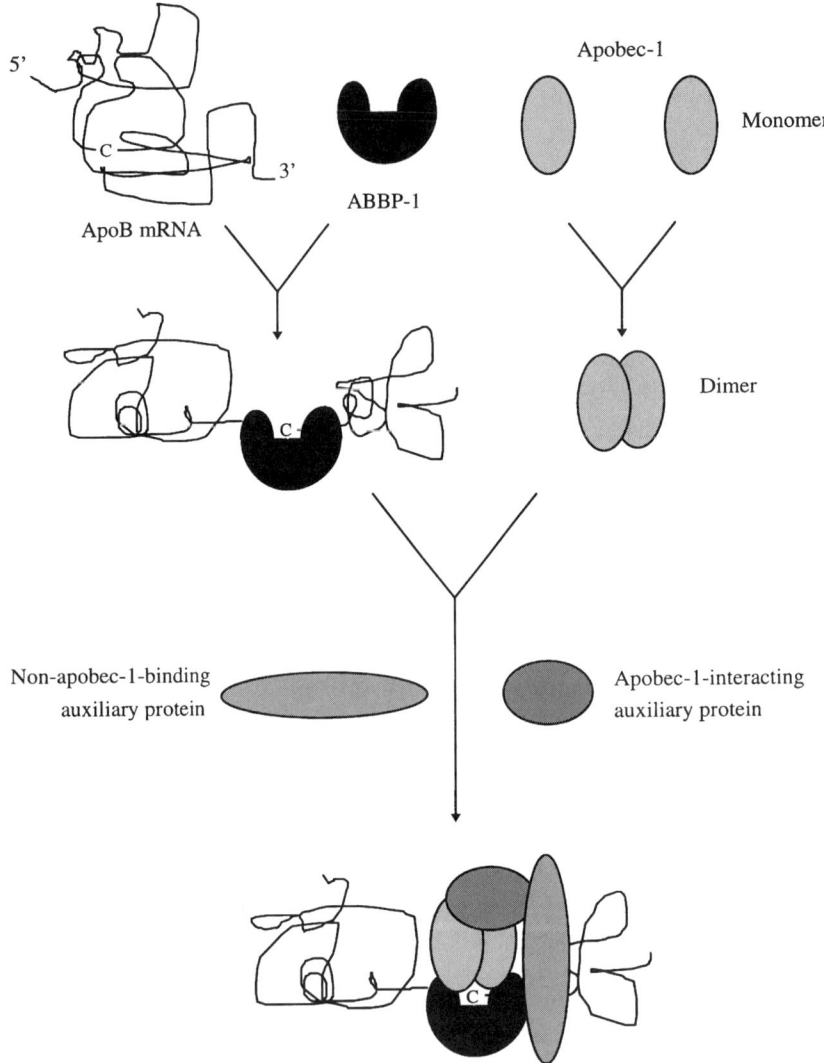

Figure 7. Hypothetical model for the structure of the apoB mRNA editing enzyme complex. ABBP-1, an hnRNP A/B-type protein, binds to apoB mRNA, causing it to become extended and exposing the canonical C-6666 editing site. Other hnRNP proteins may also serve a similar function. It is unclear if these hnRNPs are involved in apoB pre-mRNA splicing. apobec-1 undergoes spontaneous dimerization. It has affinity for apoB mRNA at the editing site, but its interaction with the latter is probably enhanced by ABBP-1 and/or other auxiliary proteins. Additional auxiliary proteins that function as assembly-disassembly proteins and editing-modulating proteins are also depicted. Some of these may also bind to apobec-1, whereas others do not directly interact with it. See the text for further details.

bly of the editosome complex, as well as those that modulate editing activity or, less likely, specificity. Some of these factors do not bind to apobec-1 and others, like ABBP-1, may be apobec-1-binding proteins. Down-regulation of some of the other apobec-1-binding proteins by immunodepletion inhibits editing (Lau and Chan, unpublished observation). The 240-kDa protein complex of Schock et al. (1996) has been postulated to be one of the auxiliary factors that do not directly interact with apobec-1, although its functional/structural role in the editosome complex has not been rigorously tested.

DEVELOPMENTAL, HORMONAL, AND NUTRITIONAL CONTROL OF apobec-1 mRNA EXPRESSION

apoB mRNA editing is regulated in a tissue-specific manner, and develops relatively late during fetal development. In humans, edited apoB mRNA is detected in the small intestine at around 11 weeks of gestation, in parallel with the appearance of apobec-1 mRNA. The amount of apoB mRNA in the small intestine increases with time of gestation (Giannoni et al., 1995). There is a strong but nonlinear correlation between apobec-1 mRNA levels and the proportion of edited apoB mRNA in human fetal and adult small intestinal RNA, suggesting an important role for apobec-1 mRNA expression in RNA editing in the human small intestine.

The developmental regulation of apoB mRNA editing in rodents differs from that in humans. In the rat, the competence to edit apoB mRNA in the small intestine is acquired prenatally between days 17 and 19 after conception and increases rapidly at birth and thereafter (Wu et al., 1990). In the liver, edited mRNA is barely detectable until about 2 weeks after birth. In both tissues, there is a nonparallel relationship between the proportion of edited apoB mRNA and apobec-1 mRNA levels during development. Therefore, factors other than apobec-1 are also important in modulating apoB mRNA editing in the small intestine and liver in rat (Funahashi et al., 1995).

Essentially all the published experiments on the control of apoB mRNA editing have examined the process in the rat liver. Hormonal, metabolic, and nutritional factors seem to play important roles in determining editing in this organ. The genetic type II diabetic GK (Goto-Kokiziki) rat has an elevated plasma apoB-48 level compared to controls. This is accompanied by an increased proportion of edited apoB-48 mRNA in the liver (Yamane et al., 1995). However, the pathogenesis underlying this condition is unclear. It could be genetic or metabolic in origin. The increased editing, however, could not be attributed to hyperinsulinemia because GK animals that developed hypoinsulinemia as a result of streptozotocin treatment, or animals with pancreatic islet β-cell exhaustion secondary to induced ventromedial hypothalamus lesions, also exhibited increased hepatic apoB-48/apoB-100 mRNA ratios. Other factors, genetic or otherwise, probably induce the increased apoB mRNA editing in the GK rats, a genetic model of non-insulin-dependent diabetes mellitus. In another study, long-term high-dose insulin treatment of isolated rat hepatocytes in vitro stimulates apoB mRNA editing and apoB-48 production from the cells (Thorngate et al., 1994). It is unknown if this pharmacological effect of insulin in vitro plays any role in the observations in GK rats.

Thyroid hormone treatment of rats (Inui et al., 1994; Davidson et al., 1988) leads to a marked increase in the proportion of edited apoB mRNA in the liver, in the absence of a change in the relative abundance of apobec-1 mRNA. When a rat is subjected to fasting and then refed a high-carbohydrate diet, a maneuver that produces a 30-fold increase in hepatic triglyceride content, the proportion of apoB-48 mRNA decreases from 60–70% to about 30–40% during fasting and increases to 80–90% in 2–3 days of refeeding (Leighton et al., 1990; Baum et al., 1990). The changes in apoB-48 mRNA are reflected by changes in the capacity of liver extracts for editing synthetic apoB mRNA substrates in vitro (Harris and Smith, 1992) and an increase in apobec-1 mRNA level (Funahashi et al., 1995).

Ethanol is another dietary factor that modulates apoB mRNA editing. Experimentally, ethanol ingestion in rats stimulates VLDL synthesis and secretion (Baraona and Leiber, 1979). Lau et al. (1995) fed rats one of three different diets: (i) a regular chow, (ii) an isocaloric liquid diet, or (iii) an isocaloric ethanol-liquid diet in which ethanol accounts for 35.5% of the total calories. They found that the ethanol diet resulted in a time-dependent increase in the proportion of hepatic apoB-48 mRNA, which reached 100% by day 40 of the ethanol diet without any detectable change in the level of apobec-1 mRNA. There was no change in apoB-48 mRNA in animals on the regular chow or the isocaloric liquid diet. The change in the amount of edited apoB mRNA was accompanied by an increase in the relative amount of newly synthesized apoB-48 protein, from 30–50% to >99% of the total plasma apoB. The ethanol diet concomitantly induced hypertriglyceridemia as a result of a marked elevation of the plasma VLDL. There is a positive correlation between plasma triglyceride concentration and the proportion of apoB-48 mRNA in the liver, but no correlation of the latter with the intrahepatic triglyceride content.

PATHOPHYSIOLOGY OF HEPATIC OVEREXPRESSION OF apobec-1

Adenovirus-mediated gene transfer (Wilson, 1996; Chan, 1995) was used to induce transient hepatic overexpression of apobec-1 in mice and rabbits. Teng et al. (1994) found that hepatic transduction of the apobec-1 gene led to marked overexpression of apobec-1 mRNA and protein. apoB mRNA editing activity assayed in vitro was found to be increased at day 3, peaked at day 7 and remained high at day 12, and returned toward basal activity at day 39. When plasma apoBs were analyzed, there was a gradual reduction in the amount of apoB-100, which reached a nadir at day 12, when it was essentially undetectable. Interestingly, early in the course of adenovirus treatment (days 1–3), there was a reciprocal increase in the amount of apoB-48. Later on, however, at days 7 and 12, there was an actual decrease in plasma apoB-48, which may be related to either decreased production or increased removal of this protein. The proportion of plasma apoB-100 in adenovirus-transduced animals decreased from ~50% to <10% of total plasma apoB concentration. Concomitant with the marked reduction of plasma apoB-100 on day 12 of treatment, there was essentially complete elimination of the plasma LDL in these animals.

The apobec-1 gene transfer experiments have been repeated in LDL receptor-knockout mice (Teng et al., 1997), an animal model for familial hypercholesterolemia, an autosomal dominant disorder in humans characterized by LDL receptor deficiency, marked hypercholesterolemia and premature coronary artery disease (Ishibashi et al., 1993). Induced hepatic overexpression of apobec-1 in these animals was found to result in marked lowering of plasma apoB-100 and significant reductions in atherogenic apoB-100-containing lipoproteins (VLDL and LDL). Similar observations were made with rabbits, in which apoB mRNA is not normally edited in the liver (Greeve et al., 1993). In these animals, apobec-1 gene transduction in vivo led to the appearance of editing activity in the liver (Hughes et al., 1996). Furthermore, in LDL receptor-deficient rabbits (Watanabe, 1980), the reduction in plasma apoB-100 was accompanied by a transient reduction of the apoB-100-containing lipoproteins VLDL and LDL (Kozarsky et al., 1996).

The gene transfer experiments indicate the efficacy of induced apobec-1 hepatic overexpression in ameliorating the hypercholesterolemia involving VLDL and LDL. Furthermore, another unique and highly atherogenic lipoprotein, lipoprotein (a), requires apoB-100 for its assembly. Hughes et al. (1996) showed that apobec-1 gene transfer to the liver in vivo also lowers lipoprotein (a) levels in transgenic mice that produce apo(a). Thus, *apobec-1* seems to be a promising therapeutic gene for the treatment of elevated VLDL and LDL associated with LDL receptor deficiency and hyperlipoproteinemia (a), a highly atherogenic condition.

The enthusiasm for the use of *apobec-1* as a therapeutic gene was dampened considerably when Yamanaka et al. (1995) showed that marked hepatic overexpression of apobec-1 was associated with hepatocellular dysplasia and carcinoma in transgenic mice and rabbits. This was probably caused by the aberrant editing of growth-related or differentiation-related genes that normally are not edited. One novel mRNA that was edited has been called NAT1 (for novel apobec-1 target 1). Mouse NAT1 mRNA contains five mooring sequence-like elements. One of these (shown in Fig. 3) is located immediately downstream of an edited C (see discussion above in "Sequence Specificity of apoB mRNA Editing"). NAT1 shows homology to the C-terminal portion of the eukaryotic translation initiation factor eIF4G. NAT1 inhibits cap-dependent and cap-independent translation in vitro. Yamanaka et al. (1997) suggested that NAT1 is a translational repressor; the abnormal editing of NAT1 mRNA causing its C-terminal truncation might interfere with its repressor function and contribute to the tumorigenicity resulting from marked overexpression of apobec-1. It is important to point out that low to moderate expression of apobec-1 (up to about 10 times its endogenous level in mice) is not associated with any liver cell dysplasia or tumorigenesis (Lau and Chan, unpublished observation; Yamanaka et al., 1995). At these levels, it is still highly effective in lowering plasma IDL/LDL in transgenic rabbits (Yamanaka et al., 1995). Therefore, to use apobec-1 as a therapeutic gene for the treatment of hypercholesterolemia, either by its induction by pharmaceutical agents or by gene therapy, one should aim at low-level regulatable expression of the gene in the liver (Miller and Whelan, 1997).

PATHOPHYSIOLOGY OF PARTIAL OR COMPLETE ANNULMENT OF apobec-1 EXPRESSION

apobec-1 is essential for apoB mRNA editing. This was clearly demonstrated by the fact that the genetic inactivation of *apobec-1* by gene targeting completely abolishes apoB mRNA editing in mice. *apobec-1*$^{-/-}$ mice are fertile and in relatively good health. Tissue extracts from these animals have no detectable editing activity in vitro. The liver and small intestine apoB mRNA is entirely in the unedited form and these animals have circulating apoB-100 but no apoB-48 (Hirano et al., 1996; Morrison et al., 1996; Nakamuta et al., 1996).

A most unexpected finding in the *apobec-1*$^{-/-}$ mice is the relative paucity of lipoprotein changes in these animals. Different laboratories reported either a total absence of lipoprotein phenotype (Hirano et al., 1996), a reduction in HDL only (Nakamuta et al., 1996), or a minor alteration of LDL (Morrison et al., 1996). This absence of a significant phenotypic effect occurs in the presence of a 2- to 3-fold increase in plasma apoB-100 concentration (Nakamuta et al., 1996). It must be pointed out, however, that the impressive increase in plasma apoB-100 occurs in a species that normally has an extremely low circulating apoB-100 concentration. Even with a 200–300% increase in apoB-100 in the *apobec-1*$^{-/-}$ animals, the plasma apoB-100 level is still only about 5–10% of that in humans (Nakamuta et al., 1996), a level that seems to be too low to effect a detectable increase in VLDL and IDL/LDL, the lipoproteins that are associated with apoB-100. Nakamuta et al. (1996) tested the effect of further increasing the plasma apoB-100 by cross-breeding these editing-incompetent mice with transgenic mice overexpressing human apoB in the liver. The presence of the human *apoB* transgene in the *apobec-1*$^{-/-}$ mice further boosted their plasma apoB-100 level another two- to threefold. This additional increase in apoB-100 was sufficient to cause a marked elevation of the plasma apoB-100-containing lipoproteins, VLDL and IDL/LDL, an observation consistent with an important role for apoB-100 in the biogenesis of these lipoproteins.

In apobec-1 knockout mice, apoB mRNA editing is absent in all tissues, including the small intestine in which editing occurs in all mammals examined. A humanized mouse with respect to apoB mRNA editing can be created either by liver-specific knockout, or by crossbreeding of complete knockout mice with transgenic mice that express apobec-1 in an intestine-specific manner. Genetic manipulations of this sort are being pursued in various laboratories. In the meantime, Oka et al. (1997) engineered a dominant negative mutant apobec-1 that they delivered to the liver of wild-type mice by adenovirus-mediated gene transfer. Overexpression of the mutant apobec-1 partially inhibited apobec-1 activity. Interestingly, they observed an acute increase in plasma IDL/LDL concentration in these animals. The presence of a lipoprotein phenotypic effect in animals with only partial inhibition of editing activity in the liver contrasts sharply with the relative lack of an effect in apobec-1 knockout mice that do not edit apoB mRNA either in the liver or the small intestine. The explanation for the difference between the two models probably lies in the acute nature of the gene transfer experiments, which did not allow sufficient time for the animals to marshal a compensatory response that may play a role in minimizing the lipoprotein changes in the *apobec-1*$^{-/-}$ animals.

CONCLUDING REMARKS

As discussed in other chapters of this volume, RNA editing encompasses many mechanistically different and unrelated instances of transcriptional or posttranscriptional modification of RNA transcripts. apoB mRNA editing was the first example of the single-base substitution type of editing described in vertebrates. It is an important genetic phenomenon, not only because of the different physiological functions of apoB-100 and apoB-48, but also because it betrays the plasticity of genetic information. By taking advantage of an existing housekeeping function, which they have duplicated and modified with time, organisms have evolved a mechanism to produce two functionally diverse proteins from a single gene. The large number of examples that are discussed in the other chapters of this book suggest that RNA editing is not simply a quirk of nature but may be an important and perhaps common method by which living organisms adapt what they are endowed with to cope with the ever-changing environment.

Acknowledgments. The experiments described in this chapter that were done in the authors' laboratory were supported by grant HL-56668 from the U.S. National Institutes of Health. B. Chang was supported by training grant 5-T32-DK07664-06 and fellowship 1-F32-HL09738-01 from the National Institutes of Health.

REFERENCES

Anant, S., A. J. MacGinnitie, and N. O. Davidson. 1995. Apobec-1, the catalytic subunit of the mammalian apolipoprotein B mRNA editing enzyme, is a novel RNA-binding protein. *J. Biol. Chem.* 270:14762–14767.

Backus, J. W., and H. C. Smith. 1991. Apolipoprotein B mRNA sequences 3' of the editing site are necessary and sufficient for editing and editosome assembly. *Nucleic Acids Res.* 19: 6781–6786.

Backus, J. W., and H. C. Smith. 1992. Three distinct RNA sequence elements are required for efficient apolipoprotein B (apoB) RNA editing *in vitro*. *Nucleic Acids Res.* 20:6007–6014.

Baraona, E., and C. S. Leiber. 1979. Effects of ethanol on lipid metabolism. *J. Lipid Res.* 20:289–315.

Baum, C. L., B.-B. Teng, and N. O. Davidson. 1990. Apolipoprotein B messenger RNA editing in the rat liver. Modulation by fasting and refeeding a high carbohydrate diet. *J. Biol. Chem.* 265:19263–19270.

Betts, L., S. Xiang, S. A. Short, R. Wolfenden, and C. W. Carter. 1994. Cytidine deaminase. The 2.3 Å crystal structure of an enzyme: transition-state analog complex. *J. Mol. Biol.* 235: 635–656.

Brown, M. S., and J. L. Goldstein. 1986. A receptor-mediated pathway for cholesterol homeostasis. *Science* 232:34–47.

Callow, M. J., and E. M. Rubin. 1995. Site-specific mutagenesis demonstrates that cysteine 4326 of apolipoprotein B is required for covalent linkage with apolipoprotein(a) *in vivo*. *J. Biol. Chem.* 270:23914–23917.

Chan, L. 1992. Apolipoprotein B, the major protein component of triglyceride-rich and low density lipoproteins. *J. Biol. Chem.* 267:25621–25624.

Chan, L. 1995. Use of somatic gene transfer to study lipoprotein metabolism in experimental animals *in vivo*. *Curr. Opin. Lipidol.* 6:335–340.

Chan, L., B. H.-J. Chang, M. Nakamuta, W.-H. Li, and L. C. Smith. 1997. Apobec-1 and apolipoprotein B mRNA editing. *Biochim. Biophys. Acta* 1345:11–26.

Chen, L., and L. Chan. 1996. Control of apolipoprotein B mRNA editing: implications of mRNA dynamics at various maturation stages. *J. Theor. Biol.* 183:391–407.

Chen, S.-H., G. Habib, C.-Y. Yang, Z.-W. Gu, B. R. Lee, S.-A. Weng, S. R. Silberman, S.-J. Cai, J. P. Deslypere, M. Rosseneu, A. M. Gotto, Jr., W.-H. Li, and L. Chan. 1987. Apolipoprotein B-48 is the product of a messenger RNA with an organ-specific in-frame stop codon. *Science* 238:363–366.

Chen, S.-H., X. Li, W. S. L. Liao, J. H. Wu, and L. Chan. 1990. RNA editing of apolipoprotein B mRNA: sequence specificity determined by *in vitro* coupled transcription-editing. *J. Biol. Chem.* 265:6811–6816.

Covello, P. S., and M. W. Gray. 1993. On the evolution of RNA editing. *Trends Genet.* 9:265–268.

Davidson, N. O., L. M. Powell, S. C. Wallis, and J. Scott. 1988. Thyroid hormone modulates the introduction of a stop codon in rat liver apolipoprotein B messenger RNA. *J. Biol. Chem.* 263:13482–13485.

Davidson, N. O., T. L. Innerarity, J. Scott, H. C. Smith, D. M. Driscoll, B.-B. Teng, and L. Chan. 1995. Proposed nomenclature for the catalytic subunit of the mammalian apolipoprotein B mRNA editing enzyme: apobec-1. *RNA* 1:3.

Dreyfuss, G., M. J. Matunis, S. Pinol-Roma, and C. G. Burd. 1993. hnRNP proteins and the biogenesis of mRNA. *Annu. Rev. Biochem.* 62:289–321.

Driscoll, D. M., J. K. Wynne, S. C. Wallis, and J. Scott. 1989. An *in vitro* system for the editing of apolipoprotein B mRNA. *Cell* 58:519–525.

Funahashi, T., F. Giannoni, A. M. DePaoli, S. F. Skarosi, and N. O. Davidson. 1995. Tissue-specific, developmental and nutritional regulation of the gene encoding the catalytic subunit of the rat apolipoprotein B mRNA editing enzyme: functional role in the modulation of apoB mRNA editing. *J. Lipid Res.* 36:414–428.

Gerber, A., M. A. O'Connell, T. Melcher, L. P. Keegan, P. H. Seeburg, and W. Keller. 1997. Cloning, expression and characterization of yeast and drosophila proteins which are homologous to the RNA editing enzymes drada and Red1. *RNA Meeting*, abstract 247.

Giannoni, F., S.-C. Chou, S. F. Skarosi, M. S. Verp, F. J. Field, R. A. Coleman, and N. O. Davidson. 1995. Developmental regulation of the catalytic subunit of the apolipoprotein B mRNA editing enzyme (APOBEC-1) in human small intestine. *J. Lipid Res.* 36:1664–1675.

Greeve, J. C., I. Altkemper, J.-H. Dieterich, H. Greten, and E. Windler. 1993. Apolipoprotein B mRNA editing in 12 different mammalian species: hepatic expression is reflected in low concentrations of apoB-containing plasma lipoproteins. *J. Lipid Res.* 34:1367–1383.

Grosjean, H., Z. Szweykowska-Kulinska, Y. Motorin, F. Fasiolo, and G. Simons. 1997. Intron-dependent enzymatic formation of modified nucleosides in eukaryotic tRNAs: a review. *Biochimie* 79:293–302.

Hadjiagapiou, C., F. Giannoni, T. Funahashi, S. F. Skarosi, and N. O. Davidson. 1994. Molecular cloning of a human small intestinal apolipoprotein B mRNA editing protein. *Nucleic Acids Res.* 22:1874–1879.

Harris, S. G., I. Sabio, E. Mayer, M. F. Steinburg, J. W. Backus, C. E. Sparks, and H. C. Smith. 1993. Extract-specific heterogeneity in high-order complexes containing apolipoprotein B mRNA editing activity and RNA-binding proteins. *J. Biol. Chem.* 268:7382–7392.

Harris, S. G., and H. C. Smith. 1992. *In vitro* apolipoprotein B mRNA editing activity can be modulated by fasting and refeeding rats with a high carbohydrate diet. *Biochem. Biophys. Res. Commun.* 183:899–903.

Havel, R. J., and J. P. Kane. 1995. Introduction: structure and metabolism of plasma lipoproteins, p. 1841–1851. In C. R. Scriver, A. L. Beaudet, W. S. Sly, and D. Valle (ed.), *The Metabolic and Molecular Bases of Inherited Disease*. McGraw-Hill, Inc., New York, N.Y.

Hirano, K., J. Min, T. Funahashi, D. A. Baunoch, and N. O. Davidson. 1997a. Characterization of the human *apobec-1* gene: expression in gastrointestinal tissues determined by alternative splicing with production of a novel truncated peptide. *J. Lipid Res.* 38:847–859.

Hirano, K., J. Min, T. Funahashi, and N. O. Davidson. 1997b. Cloning and characterization of the rat *apobec-1* gene: a comparative analysis of gene structure and promoter usage in rat and mouse. *J. Lipid Res.* 38:1103–1119.

Hirano, K.-I., S. G. Young, R. V. Farese, Jr., J. Ng, E. Sande, C. Warburton, L. M. Powell-Braxton, and N. O. Davidson. 1996. Targeted disruption of the mouse *apobec-1* gene abolishes apolipoprotein B mRNA editing and eliminates apolipoprotein B48. *J. Biol. Chem.* 271:9887–9890.

Hodges, P. E., and J. Scott. 1992. Apolipoprotein B mRNA editing: a new tier for the control of gene expression. *Trends Biochem. Sci.* 17:77–81.

Hughes, S. D., D. Rouy, N. Navaratnam, J. Scott, and E. M. Rubin. 1996. Gene transfer of cytidine deaminase apoBEC-1 lowers lipoprotein(a) in transgenic mice and induces apolipoprotein B editing in rabbits. *Hum. Gene Ther.* 7:39–49.

Inui, Y., F. Giannoni, T. Funahashi, and N. O. Davidson. 1994. REPR and complementation factor(s) interact to modulate rat apolipoprotein B mRNA editing in response to alterations in cellular cholesterol flux. *J. Lipid Res.* 35:1477–1489.

Ishibashi, S., M. S. Brown, J. L. Goldstein, R. D. Gerard, R. E. Hammer, and J. Herz. 1993. Hypercholesterolemia in LDL receptor knockout mice and its reversal by adenovirus-mediated gene delivery. *J. Clin. Invest.* 92:883–893.

Kane, J. P., and R. J. Havel. 1995. Disorders of the biogenesis and secretion of lipoproteins containing the B apolipoproteins, p. 1853–1885. In C. R. Scriver, A. L. Beaudet, W. S. Sly, and D. Valle (ed.), *The Metabolic and Molecular Bases of Inherited Disease*. McGraw-Hill, Inc. New York, N.Y.

Khan, F. A., A. K. Jaiswal, and W. Szer. 1991. Cloning and sequence analysis of a human type A/B hnRNP protein. *FEBS Lett.* 290:159–161.

Knott, T. J., R. J. Pease, L. M. Powell, S. C. Wallis, S. C. Rall, Jr., T. L. Innerarity, B. Blackhart, W. H. Taylor, Y. L. Marcel, R. Milne, D. F. Johnson, M. Fuller, A. J. Lusis, B. J. McCarthy, R. W. Mahley, B. Levy-Wilson, and J. Scott. 1986. Complete protein sequence and identification of structural domains of human apolipoprotein B. *Nature* 323:734–738.

Kozarsky, K. F., D. K. Bonen, F. Giannoni, T. Funahashi, J. M. Wilson, and N. O. Davidson. 1996. Hepatic expression of the catalytic subunit of the apolipoprotein B mRNA editing enzyme (*apobec*-1) ameliorates hypercholesterolemia in LDL receptor-deficient rabbits. *Hum. Gene Ther.* 7:943–957.

Lau, P., and L. Chan. Unpublished observation.

Lau, P. P., S.-H. Chen, J. C. Wang, and L. Chan. 1990. A 40 kilodalton rat liver nuclear protein binds specifically to apolipoprotein B mRNA around the RNA editing site. *Nucleic Acids Res.* 18:5817–5821.

Lau, P. P., W. Xiong, H.-J. Zhu, S.-H. Chen, and L. Chan. 1991. Apolipoprotein B mRNA editing is an intranuclear event that occurs posttranscriptionally coincident with splicing and polyadenylation. *J. Biol. Chem.* 266:20550–20554.

Lau, P. P., H.-J. Zhu, A. Baldini, C. Charnsangavej, and L. Chan. 1994. Dimeric structure of a human apolipoprotein B mRNA editing protein and cloning and chromosomal localization of its gene. *Proc. Natl. Acad. Sci. USA* 91:8522–8526.

Lau, P. P., D. J. Cahill, H.-J. Zhu, and L. Chan. 1995. Ethanol modulates apolipoprotein B mRNA editing in the rat. *J. Lipid Res.* 36:2069–2078.

Lau, P. P., H.-J. Zhu, M. Nakamuta, and L. Chan. 1997. Cloning of an apobec-1-binding protein that also interacts with apolipoprotein B mRNA and evidence for its involvement in RNA editing. *J. Biol. Chem.* 272:1452–1455.

Leighton, J. K., J. Joyner, J. Zamarripa, M. Deines, and R. A. Davis. 1990. Fasting decreases apolipoprotein B mRNA editing and the secretion of small molecular weight apoB by rat hepatocytes: evidence that the total amount of apoB secreted is regulated post-transcriptionally. *J. Lipid Res.* 31:1663–1668.

Li, W.-H., M. Tanimura, C. C. Luo, S. Datta, and L. Chan. 1988. The apolipoprotein multigene family: biosynthesis, structure, structure-function relationships and evolution. *J. Lipid Res.* **29**: 245–271.

Mayeda, A., and A. R. Krainer. 1992. Regulation of alternative pre-mRNA splicing by hnRNP A1 and splicing factor SF2. *Cell* **68**:365–375.

Mehta, A., S. Banerjee, and D. M. Driscoll. 1996. Apobec-1 interacts with a 65-kDa complementing protein to edit apolipoprotein-B mRNA *in vitro*. *J. Biol. Chem.* **271**:28294–28299.

Miller, N., and J. Whelan. 1997. Progress in transcriptionally targeted and regulatable vectors for genetic therapy. *Hum. Gene Ther.* **8**:803–815.

Morrison, J. R., C. Paszty, M. E. Stevens, S. D. Hughes, T. Forte, J. Scott, and E. M. Rubin. 1996. Apolipoprotein B RNA editing enzyme-deficient mice are viable despite alterations in lipoprotein metabolism. *Proc. Natl. Acad. Sci. USA* **93**:7154–7159.

Nakamuta, M., K. Oka, J. Krushkal, K. Kobayashi, M. Yamamoto, W.-H. Li, and L. Chan. 1995. Alternative mRNA splicing and differential promoter utilization determine tissue-specific expression of the apolipoprotein B mRNA-editing protein (Apobec1) gene in mice. *J. Biol. Chem.* **270**:13042–13056.

Nakamuta, M., B. H.-J. Chang, E. Zsigmond, K. Kobayashi, H. Lei, B. Y. Ishida, K. Oka, E. Li, and L. Chan. 1996. Complete phenotypic characterization of *apobec-1* knockout mice with a wild-type genetic background and a human apolipoprotein B transgenic background, and restoration of apolipoprotein B mRNA editing by somatic gene transfer of apobec-1. *J. Biol. Chem.* **271**:25981–25988.

Navaratnam, N., J. R. Morrison, S. Bhattacharya, D. Patel, T. Funahashi, F. Giannoni, B.-B. Teng, N. O. Davidson, and J. Scott. 1993a. The p27 catalytic subunit of the apolipoprotein B mRNA editing enzyme is a cytidine deaminase. *J. Biol. Chem.* **268**:20709–20712.

Navaratnam, N., R. R. Shah, D. Patel, V. Fay, and J. Scott. 1993b. Apolipoprotein B mRNA editing is associated with UV crosslinking of proteins to the editing site. *Proc. Natl. Acad. Sci. USA* **90**: 222–226.

Navaratnam, N., S. Bhattacharya, T. Fujino, D. Patel, A. L. Jarmuz, and J. Scott. 1995. Evolutionary origins of apoB mRNA editing: catalysis by a cytidine deaminase that has acquired a novel RNA-binding motif at its active site. *Cell* **81**:187–195.

Oka, K., K. Kobayashi, M. Sullivan, J. Martinez, B.-B. Teng, K. Ishimura-Oka, and L. Chan. 1997. Tissue-specific inhibition of apolipoprotein B mRNA editing in the liver by adenovirus-mediated transfer of a dominant negative mutant APOBEC-1 leads to increased low density lipoprotein in mice. *J. Biol. Chem.* **272**:1456–1460.

Powell, L. M., S. C. Wallis, R. J. Pease, Y. H. Edwards, T. J. Knott, and J. Scott. 1987. A novel form of tissue-specific RNA processing produces apolipoprotein B-48 in intestine. *Cell* **50**: 831–840.

Robberson, B. L., G. J. Cote, and S. M. Berget. 1990. Exon definition may facilitate splice site selection in RNAs with multiple exons. *Mol. Cell. Biol.* **10**:84–94.

Saitou, N., and M. Nei. 1987. The number of nucleotides required to determine the branching order of three species with special reference to the human-chimpanzee-gorilla divergence. *J. Mol. Evol.* **4**:406–425.

Schock, D., S.-R. Kuo, M. F. Steinburg, M. Bolognino, J. D. Sparks, C. E. Sparks, and H. C. Smith. 1996. An auxiliary factor containing a 240-kDa protein complex is involved in apolipoprotein B RNA editing. *Proc. Natl. Acad. Sci. USA* **93**: 1097–1102.

Shah, R. R., T. J. Knott, J. E. Legros, N. Navaratnam, J. C. Greeve, and J. Scott. 1991. Sequence requirements for the editing of apolipoprotein B mRNA. *J. Biol. Chem.* **266**: 16301–16304.

Sierakowska, A., W. Szer, P. J. Furodon, and R. Kole. 1986. Antibodies to hnRNP core proteins inhibit in vitro splicing of human beta-globin pre-mRNA. *Nucleic Acids Res.* **14**:5241–5254.

Skuse, G. R., A. J. Cappione, M. Sowden, L. J. Metheny, and H. C. Smith. 1996. The neurofibromatosis type 1 messenger RNA undergoes base-modification RNA editing. *Nucleic Acids Res.* **24**:478–486.

Smith, H. C., S.-R. Kuo, J. W. Backus, S. G. Harris, C. E. Sparks, and J. D. Sparks. 1991. *In vitro* apolipoprotein B mRNA editing: Identification of a 27S editing complex. *Proc. Natl. Acad. Sci. USA* **88**:1489–1493.

Smith, H. C. 1993. Apolipoprotein B mRNA editing: the sequence to the event. *Semin. Cell Biol.* **4**:267–278.

Teng, B.-B., C. F. Burant, and N. O. Davidson. 1993. Molecular cloning of an apoB mRNA editing protein. *Science* **260**: 1816–1819.

Teng, B.-B., S. Blumenthal, T. Forte, N. Navaratnam, J. Scott, A. M. Gotto, Jr., and L. Chan. 1994. Adenovirus-mediated gene transfer of rat apolipoprotein B mRNA editing protein in mice virtually eliminates apolipoprotein B-100 and normal low density lipoprotein production. *J. Biol. Chem.* **269**:29395–29404.

Teng, B.-B., B. Y. Ishida, T. M. Forte, S. Blumenthal, L.-Z. Song, A. M. Gotto, Jr., and L. Chan. 1997. Effective lowering of plasma, LDL, and esterified cholesterol in LDL receptor-knockout mice by adenovirus-mediated gene delivery of apoB mRNA editing enzyme (apobec-1). *Arterioscler. Thromb. Vasc. Biol.* **17**:889–897.

Thorngate, F. E., R. Raghow, H. G. Wilcox, C. S. Werner, M. Heimberg, and M. B. Elam. 1994. Insulin promotes the biosynthesis and secretion of apolipoprotein B-48 by altering apolipoprotein B mRNA editing. *Proc. Natl. Acad. Sci. USA* **91**: 5392–5396.

Utermann, G. 1995. Lipoprotein(a), p. 1887–1912. *In* C. R. Scriver, A. L. Beaudet, W. S. Sly, and D. Valle (ed.), *The Metabolic and Molecular Bases of Inherited Disease.* McGraw-Hill, Inc., New York, N.Y.

Van't Hooft, F. M., D. A. Hardman, J. P. Kane, and R. J. Havel. 1982. Apolipoprotein B (B-48) of rat chylomicrons is not a precursor of the apolipoprotein of low density lipoproteins. *Proc. Natl. Acad. Sci. USA* **79**:179–182.

Watanabe, Y. 1980. Serial inbreeding of rabbits with hereditary hyperlipidemia (WHHL rabbit). *Atherosclerosis* **36**:261–268.

Wilson, J. M. 1996. Adenoviruses as gene-delivery vehicles. *N. Engl. J. Med.* **334**:1185–1187.

Wu, J. H., C. F. Semenkovich, S.-H. Chen, W.-H. Li, and L. Chan. 1990. Apolipoprotein B mRNA editing: validation of a sensitive assay and developmental biology of RNA editing in the rat. *J. Biol. Chem.* **265**:12312–12316.

Yamanaka, S., K. S. Poksay, M. E. Balestra, G.-Q. Zeng, and T. L. Innerarity. 1994. Cloning and mutagenesis of the rabbit ApoB mRNA editing protein. A zinc motif is essential for catalytic activity, and noncatalytic auxiliary factor(s) of the editing complex are widely distributed. *J. Biol. Chem.* **269**: 21725–21734.

Yamanaka, S., M. E. Balestra, L. D. Ferrell, J. Fan, K. S. Arnold, S. Taylor, J. M. Tayor, and T. L. Innerarity. 1995. Apolipoprotein B mRNA-editing protein induces hepatocellular carcinoma and dysplasia in transgenic animals. *Proc. Natl. Acad. Sci. USA* **92**:8483–8487.

Yamanaka, S., K. S. Poksay, K. S. Arnold, and T. L. Innerarity. 1997. A novel translational repressor mRNA is edited extensively in livers containing tumors caused by the transgene expression of the apoB mRNA-editing enzyme. *Genes Dev.* 11: 321–333.

Yamane, M., S. Jiao, S. Kihara, I. Shimomura, K. Yanagi, K. Tokunaga, S. Kawata, H. Odaka, H. Ideda, S. Yamashita, K. Kameda-Takemura, and Y. Matsuzawa. 1995. Increased proportion of plasma apoB-48 to apoB-100 in non-insulin-dependent diabetic rats: contribution of enhanced apoB mRNA editing in the liver. *J. Lipid Res.* 36:1676–1685.

Yang, C.-Y., S.-H. Chen, S. H. Gianturco, W. A. Bradley, J. T. Sparrow, M. Tanimura, W.-H. Li, D. A. Sparrow, H. DeLoof, M. Rosseneu, F.-S. Lee, Z.-W. Gu, A. M. Gotto, Jr., and L. Chan. 1986. Sequence, structure, receptor-binding domains and internal repeats of human apolipoprotein B-100. *Nature* 323: 738–742.

Yang, C.-Y., Z.-W. Gu, S.-A. Weng, T. W. Kim, S.-H. Chen, H. J. Pownall, P. M. Sharp, S.-W. Liu, W.-H. Li, A. M. Gotto, Jr., and L. Chan. 1989. Structure of apolipoprotein B-100 of human low density lipoproteins. *Arteriosclerosis* 9:96–108.

Yeung, S. J., and L. Chan. Unpublished observation.

Chapter 19

Adenosine-to-Inosine Conversion in mRNA

SUSAN M. RUETER AND RONALD B. EMESON

The conversion of adenosine to inosine by RNA editing, at specific positions within RNAs, represents an increasingly common posttranscriptional modification for generating diversity and flexibility in eukaryotic gene expression. Initially identified as an RNA modification in yeast transfer RNA^{Ala} (Holley et al., 1965; Grosjean et al., 1996), the occurrence of inosine nucleosides has more recently been observed in viral RNA transcripts and nuclear-encoded mRNAs (Simpson and Emeson, 1996; Bass, 1997). As a consequence of such editing events, the primary sequence of mature mRNAs can be subtly altered, resulting in specific changes in amino acid coding potential that affect the biological activity of the resultant protein.

The first example of editing for a nucleus-encoded mRNA was the conversion of a glutamine codon (CAA) to a stop codon (UAA) in transcripts encoding apolipoprotein B (apoB) via the actions of a specific cytidine deaminase (see Chapter 18). This single nucleotide modification within apoB RNAs results in the production of a novel apoB protein isoform, referred to as apoB-48, which has physiological properties distinct from the cholesterol-transporting functions of the larger apoB-100 protein (Davidson, 1993). Although this cytidine-to-uridine editing event can readily be observed as a nucleotide sequence discrepancy between apoB genomic DNA and cDNA sequences generated from rat small intestine, the identification of adenosine-to-inosine (A-to-I) editing events in mRNA transcripts is somewhat more problematic.

Initial comparisons between the genomic and cDNA sequences for transcripts encoding subunits of certain glutamate-responsive ion channels (GluR) identified apparent adenosine-to-guanosine discrepancies (Sommer et al., 1991). It was not until the characterization of regulatory elements within GluR pre-mRNAs and the development of in vitro editing systems that it was determined that such A-to-G discrepancies actually resulted from the enzymatic modification of specific adenosine residues to a nucleotide with guanosine-like base pairing properties, inosine in this case (Higuchi et al., 1993; Rueter et al., 1995; Yang et al., 1995; Melcher et al., 1995). The conversion of adenosine to inosine can readily be accomplished by hydrolytic deamination at the C-6 position of the purine ring (see Appendix 1). Studies showing that such A-to-I editing events required the presence of an RNA duplex (Higuchi et al., 1993; Egebjerg et al., 1994) suggested that adenosine deaminases directed toward double-stranded RNA (dsRNA) could represent the enzymatic activities responsible for catalyzing such posttranscriptional modifications. The recent purification, cloning, and characterization of at least two of these dsRNA-specific adenosine deaminases (Hough and Bass, 1994; Kim et al., 1994a, 1994b; O'Connell and Keller, 1994; Melcher et al., 1996a) has indicated that such proteins can selectively modify GluR transcripts.

Additional substrates for RNA editing have recently been identified, based largely on A-to-G nucleotide discrepancies between genomic and cDNA sequences. These edited RNAs include viral transcripts (Cattaneo et al., 1988; Murphy et al., 1991; Rueda et al., 1994; Wu et al., 1994; Kumar and Carmichael, 1997) and RNAs encoding neuronal signaling molecules (Burns et al., 1997; Patton et al., 1997), a sialyltransferase (Ma et al., 1997), and a putative *Drosophila* RNA-binding protein of unknown function (Petschek et al., 1996). In several instances, the enzymatic activities responsible for GluR modification have also been shown to edit these newly identified transcripts in a site-selective manner (Polson et al., 1996; Burns et al., 1997). More recently, an assay to detect inosine-containing transcripts has been devel-

Susan M. Rueter and Ronald B. Emeson • Department of Pharmacology, Vanderbilt University, 460 Medical Research Building 2, Nashville, Tennessee 37232-6600.

oped (Morse and Bass, 1997) and should dramatically increase our knowledge of the repertoire of RNAs that undergo this type of posttranscriptional modification. This chapter will describe a number of mRNA substrates that undergo A-to-I editing events, the effects that such alterations in coding potential have upon protein function, and the candidate enzymes responsible for such posttranscriptional modifications.

RNA SUBSTRATES

Glutamate Receptor Subunits

L-Glutamate is the principal mediator of excitatory synaptic transmission in the vertebrate central nervous system. Pharmacological and physiological studies have identified multiple ion-transporting glutamate receptors that have been divided into three subtypes based on binding of the selective agonists N-methyl-D-aspartate (NMDA), α-amino-3-hydroxy-5-methyl-isoxazole-4-propionate (AMPA), and kainic acid. In addition to their roles in fast excitatory neurotransmission, these ionotropic channels also appear to be critical in processes of synaptic plasticity (Collingridge and Singer, 1990; Bliss and Collingridge, 1993) and in chronic and acute neurodegenerative disorders, including stroke, epilepsy, amyotrophic lateral sclerosis, and Parkinson's disease (Choi and Rothman, 1990; Olney, 1990). Many of the physiological and pathological functions of glutamate receptors have been attributed to the Ca^{2+} permeability of these integral ion channels (Dubinsky and Rothman, 1991; Pizzi et al., 1991; Randall and Thayer, 1992), yet until recently only NMDA receptor channels were shown to demonstrate significant Ca^{2+} permeability. Several types of AMPA and kainate receptors are now known to efficiently flux Ca^{2+} ions (Ogura et al., 1990; Lino et al., 1990; Gilbertson et al., 1991; Pruss et al., 1991; Burnashev et al., 1992a) depending on their subunit composition (Verdoorn et al., 1991; Hollmann et al., 1991).

The AMPA subtype of glutamate receptor consists of homomeric and heteromeric combinations of four subunits, GluR-A, -B, -C, and -D (Boulter et al., 1989; Hollmann et al., 1989; Keinanen et al., 1990; Nakanishi et al., 1990; Sakimura et al., 1990), although the exact number and stoichiometry of subunits that form functional receptor channels have not yet been determined. Each subunit is approximately 900 amino acids in length and is predicted to contain four hydrophobic domains representing the major determinants of channel topology and architecture (Fig. 1A); three of these hydrophobic domains are predicted to traverse the plasma membrane, while the

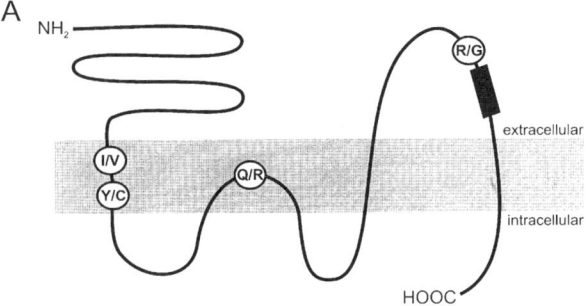

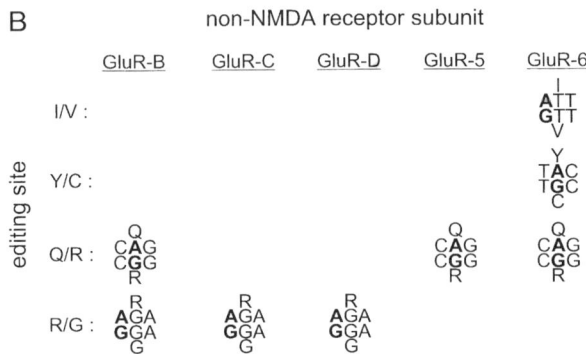

Figure 1. A-to-I editing events in transcripts encoding non-NMDA glutamate receptor subunits. A schematic representation of the predicted topology for GluR-A is presented indicating the relative positions of amino acid alterations produced as a result of A-to-I editing events and the location of the flip/flop domain (black rectangle); the topology of subunits other than GluR-A (Hollman et al., 1994) has not been determined. The posttranscriptional conversion of adenosine to inosine can be seen as an A-to-G discrepancy between the nucleotide sequences of genomic and complementary DNA (cDNA). The genomic (upper), cDNA (lower) and predicted amino acid alterations resulting from these editing events are presented.

second hydrophobic region (TM2) is thought to loop into the membrane without traversing it (Hollmann et al., 1994). Overall, there is a 70% amino acid sequence identity among the four subunits, although significantly greater sequence conservation is exhibited within the four hydrophobic domains (Keinanen et al., 1990). A 38 amino acid region immediately preceding the fourth hydrophobic domain, termed the flip/flop domain, results from alternative splicing of the GluR-A, -B, -C, and -D transcripts (Sommer et al., 1990), giving rise to two distinct protein isoforms for each AMPA receptor subunit. Expression of these alternatively spliced isoforms is regulated in both a cell- and development-specific fashion within the brain (Sommer et al., 1990, Monyer et al., 1991), imparting distinct functional properties differing in both the amplitude and kinetics of glutamate-induced responses (Sommer et al., 1990).

The divalent cation permeability of AMPA receptors is dictated by the presence or absence of the

GluR-B subunit (Hollman et al., 1991; Verdoorn et al., 1991). AMPA receptors that contain the GluR-B subunit are impermeant to calcium ions, whereas those that lack the GluR-B subunit demonstrate a dramatic increase in their relative permeability to divalent cations (Verdoorn et al., 1991; Hollman et al., 1991). This dominant property of GluR-B, in regulating both the electrophysiological and ion transport properties of heteromeric AMPA receptors, depends on a specific arginine residue located in the TM2 region of the subunit (Fig. 1A) (Hume et al., 1991; Verdoorn et al., 1991; Dingledine et al., 1992). This positively charged residue is unique to GluR-B; the GluR-A, -C, and -D subunits contain a neutral glutamine residue at the analogous position. The low calcium permeability exhibited by AMPA receptor channels in cultured hippocampal neurons (Lino et al., 1990) and in hippocampal slices (Jonas and Sakmann, 1992) has indicated that the GluR-B subunit is present in most naturally occurring AMPA receptors. However, higher calcium permeability through AMPA receptors is observed in cerebellar Bergmann glial cells (Burnashev et al., 1992a), hippocampal basket cells (Koh et al., 1995), and neocortical nonpyramidal cells (Jonas et al., 1994) where expression of GluR-B is low or absent.

While nucleotide sequence analyses have revealed the presence of a CGG codon encoding the critical regulatory arginine residue within the TM2 region of GluR-B cDNAs, a glutamine (CAG) codon was found in GluR-B genomic DNA at this position (Q/R site [Fig. 1]) (Sommer et al., 1991). This A-to-G nucleotide discrepancy gave rise to the hypothesis that RNA transcripts encoding the GluR-B subunit were modified co- or posttranscriptionally such that the adenosine residue of the glutamine codon was converted to the guanosine residue present in the arginine triplet (Sommer et al., 1991). Quantitative PCR analyses of adult rat and mouse brain RNAs revealed that the efficiency of editing at the Q/R site was quite high with >99% of all GluR-B mRNA transcripts containing the arginine codon (Sommer et al., 1991; Burnashev et al., 1992b); this correlates well with the generally low calcium permeability observed in native AMPA receptor channels. By contrast, GluR-A, -C, and -D transcripts encoded only the glutamine-containing form (Sommer et al., 1991).

RNA editing has also been shown to be responsible for an apparent adenosine-to-guanosine conversion resulting in the alteration of an arginine (AGA) to a glycine (GGA) codon in cDNAs encoding the GluR-B, -C, and -D AMPA receptor subunits. This second editing site (R/G site) is located immediately upstream of the alternatively spliced flip/flop cassette (Fig. 1) just preceding the fourth hydrophobic domain (TM4), and this single amino acid alteration increases the rate of recovery from receptor desensitization (Lomeli et al., 1994). The extent of editing at the R/G site increases with the development of the brain in a subunit-dependent and flip/flop splice variant-dependent manner (Lomeli et al., 1994).

In addition to RNA editing events mediating the electrophysiological and ion-permeation properties of AMPA receptors, subunits of heteromeric kainate receptors (GluR-5 and GluR-6) are also regulated by RNA editing processes similar to those observed for AMPA receptor subunits (Sommer et al., 1991; Egebjerg and Heinemann, 1993; Köhler et al., 1993). In addition to Q/R editing sites in GluR-5 and GluR-6 transcripts, analogous to the Q/R site in GluR-B, two further adenosine residues are selectively targeted for modification (Fig. 1). Both of these adenosine moieties are located in the region encoding the first hydrophobic domain (TM1) of GluR-6 and their apparent conversion to guanosine leads to the substitution of a valine (GTT) for a genomically specified isoleucine (ATT) codon (I/V site) and the substitution of a cysteine (TGC) for a tyrosine (TAC) codon (Y/C site). Whereas GluR-B mRNAs are efficiently edited at the Q/R site in the brain at all developmental stages examined (>99%), GluR-5 and GluR-6 transcripts were found to be differentially edited at the Q/R site at different developmental stages, and the extent of editing seen in the adult brain was only 40% and 80%, respectively (Sommer et al., 1991). Editing at the GluR-6 Q/R site can also regulate calcium permeability through kainate receptor channels; however, the amino acid residue present at this position determines calcium permeability in a manner opposite to that observed for AMPA receptors (Köhler et al., 1993). A kainate receptor with a glutamine-containing GluR-6 subunit exhibits low calcium permeability, whereas one with the arginine-containing isoform exhibits a high calcium conductance (Köhler et al., 1993). Editing at the I/V and Y/C sites in TM1 is also involved in regulating calcium permeability (Köhler et al., 1993), indicating that the determinants of ion permeation in kainate channels reside in both the TM1 and TM2 regions of GluR-6.

Although the nucleotide sequence of GluR-B mRNAs surrounding the Q/R site (exon 11) is virtually identical to the analogous region in other AMPA receptor subunit RNAs, editing at the Q/R site is specific for transcripts encoding GluR-B. The nature of this subunit specificity was revealed when sequence analyses of the proximal portion of intron 11, immediately downstream from the GluR-B Q/R site, identified at 10-nucleotide (nt) sequence having exact complementarity to the exonic sequence immediately surrounding the Q/R site (Higuchi et al., 1993; Egebjerg et al., 1994). This region of complementarity, referred to as the editing site complementary sequence (ECS), is located within an even larger inverted repeat sequence and is hypothesized to form a duplex RNA structure interrupted by single-stranded bulges and loops that is critical for the efficient and accurate editing of GluR-B pre-mRNAs (Fig. 2) (Higuchi et al., 1993; Egebjerg et al., 1994). Mutations disrupting the proposed base pairing within the inverted repeat, or between the ECS and the Q/R site, lead to a decrease or complete loss of editing, while compensatory mutations designed to restore base pairing interactions generally lead to a restoration of editing (Higuchi et al., 1993; Rueter et al., 1995). In addition to editing at the Q/R site, these studies identified additional guanosine residues in place of genomically encoded adenosine moieties at multiple sites within the intron; however, the extent of editing at these intronic positions or "hot spots" was significantly less than that observed for the Q/R site (Higuchi et al., 1993; Rueter et al., 1995). These results have indicated that the secondary structure, rather than the primary nucleotide sequence, is critical for the editing process. This dsRNA structure is unique to GluR-B transcripts and does not exist in the GluR-A, -C, and -D pre-mRNAs, thus explaining the subunit-specific modification of GluR-B RNAs at the Q/R site. The requirement for intronic sequence information further demonstrates that editing must be a nuclear event because this type of RNA modification must precede, or be coincident with, the process of splicing.

Similar RNA duplex structures were also identified in the region surrounding the R/G site for pre-mRNAs encoding the GluR-B, -C, and -D subunits consisting of base pairing interactions between the 3' end of exon 13 (the exon containing the R/G site) and the proximal portion of intron 13 (Fig. 2) (Lomeli et al., 1994). Although this predicted region of dsRNA was not as extensive as that for the Q/R site, mutational analyses to disrupt and reform the RNA duplex confirmed the absolute requirement for this secondary structure. In addition to the proposed region of dsRNA, distal intronic elements were necessary for maximal editing efficiency, although not absolutely required for R/G site modification (Lomeli et al., 1994). Characterization of Q/R site editing for RNAs encoding the GluR-5 and GluR-6 kainate receptor subunits also revealed a putative RNA duplex region (Fig. 2); however, in contrast to the dsRNA structure surrounding the GluR-B Q/R site, the editing site complementary sequence was located approximately 1,900 nt downstream from the modified nucleoside rather than in the more proximal portion of the intron (Herb et al., 1996). The cis-active regulatory sequences required for modification of GluR-6 transcripts at the I/V and Y/C sites have not yet been identified.

The editing of RNAs encoding glutamate receptor subunits results from the conversion of genomically encoded adenosine residues to the guanosine-like nucleotides found in cDNAs generated from mature GluR mRNA transcripts. These edited nucleotides direct the incorporation of cytosine by reverse transcriptase during the synthesis of cDNA, thereby suggesting that guanosine (or a nucleotide with similar base pairing properties) is the product of the editing reaction. Alterations in base pairing potential from adenosine to guanosine could occur by three general biochemical mechanisms, including nucleotide excision and replacement, base exchange through transglycosylation (see Chapter 9), or direct enzymatic modification of the base (Bass, 1993; Simpson and Emeson, 1996). Based on precedents found in the purine biosynthetic pathway, the conversion from adenosine to guanosine would require multiple enzymatic steps (Bass, 1993). Previous studies of apolipoprotein B editing have demonstrated that this posttranscriptional modification occurs by hydrolytic deamination through the actions of a specific cytidine deaminase (see Chapter 18). Similarly, the editing of GluR transcripts could proceed through the actions of an adenosine deaminase converting adenosine to inosine by deamination at the C-6 position of the purine ring. Inosine, like guanosine, preferentially base pairs with cytosine, suggesting that the actual RNA modification involved in the editing of GluR transcripts at the Q/R site could result from a CIG (arginine) codon in the mature mRNA rather than the CGG (arginine) codon inferred from the isolated cDNA sequence. Similar deamination events could also be responsible for the observed changes in coding potential seen for other edited sites within GluR transcripts.

The development of an in vitro editing system utilizing nuclear extracts derived from cultured human epithelial (HeLa) cells and synthetic GluR RNA transcripts led to the discovery that the adenosine moieties at the GluR-B Q/R and R/G sites were actually converted to inosine residues, rather than guanosines, by enzymatic base modification (deamination) (Rueter et al., 1995; Yang et al., 1995; Melcher et al., 1995). These studies took advantage of synthetic GluR RNA substrates transcribed in the presence of either [α-^{32}P]ATP or [2,8-^{3}H]ATP. The radiolabeled RNAs were incubated with HeLa cell nuclear extract and the in vitro reaction products were digested with nuclease P1, thereby yielding 5'-

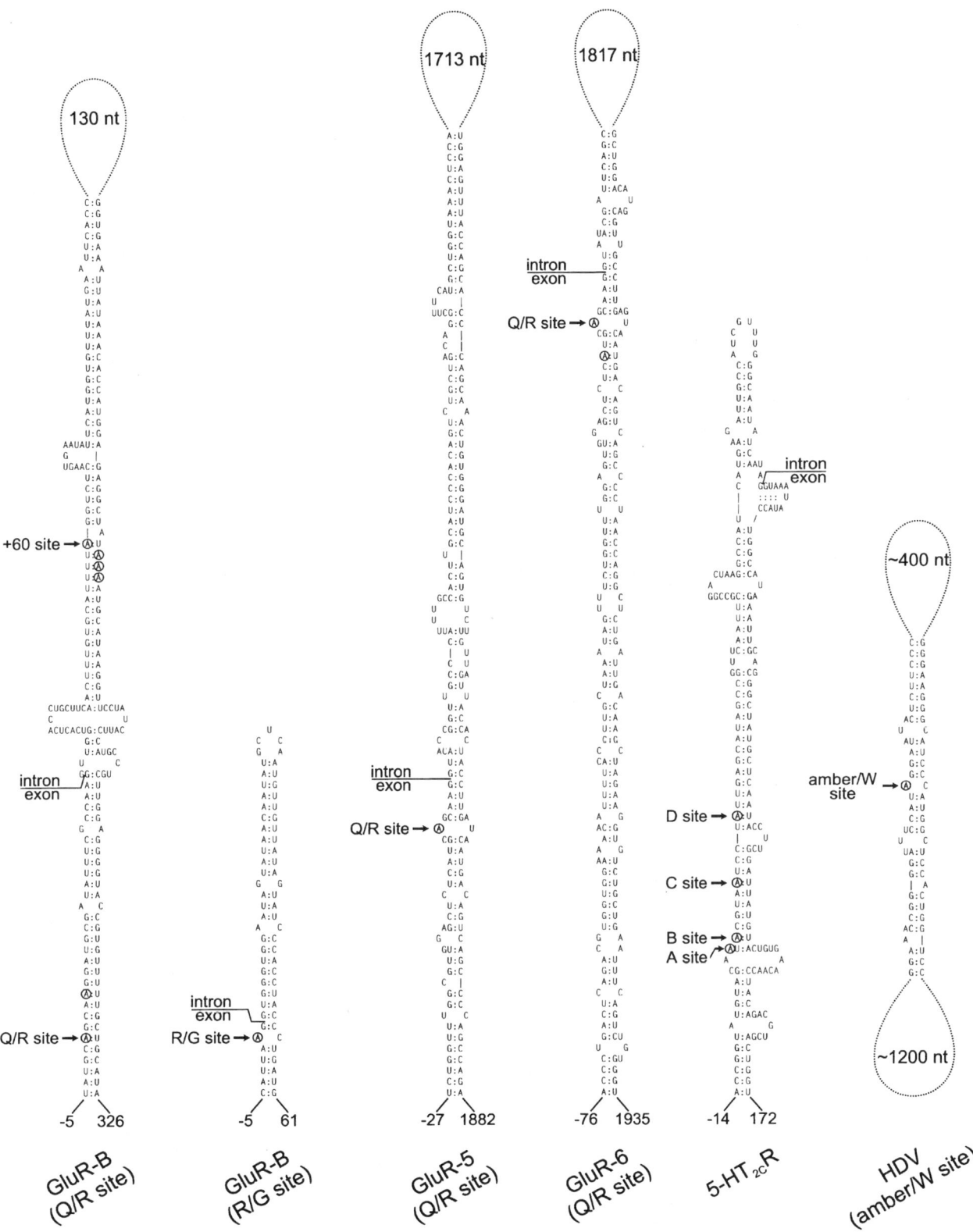

Figure 2. Predicted secondary structure of pre-mRNA transcripts encoding non-NMDA receptor subunits (GluR-B, GluR-5, and GluR-6), the 2C subtype of serotonin receptor (5-HT$_{2C}$R) and the hepatitis delta virus (HDV) antigenome in the regions of major editing modifications using an RNA folding algorithm (RNAFOLD; Scientific & Educational Software). The positions of edited nucleotides are indicated by open circles, intron-exon boundaries are designated, and nucleotides omitted from the figure are indicated in the loops. Nucleotide coordinates are relative to the Q/R, R/G or A editing sites.

nucleoside monophosphates that could be subsequently resolved by thin-layer chromatography (Fig. 3). Additional analytical strategies for the identification of modified nucleotides can be seen in Chapter 2. Results from these studies have demonstrated that radiolabeled inosine 5′-monophosphate (IMP), rather than guanosine 5′-monophosphate (GMP), was produced in the in vitro editing reaction. Since there are no known polymerases that incorporate nucleotides into a polynucleotide chain without a 5′-phosphate group, these results suggested that the editing of GluR-B transcripts was not mediated by a mechanism of nucleoside excision and replacement in which the phosphodiester backbone of the RNA was cleaved and then esterified. Furthermore, the generation of purine ring-labeled [2,8-^{3}H]IMP in this assay system demonstrated that the processing of GluR transcripts must not involve transglycosylation (base exchange), as seen for the introduction of hypoxanthine and queuosine into several tRNAs (see Chapter 9), but rather an enzymatic base modification converting adenosine to inosine.

Analyses of mutant GluR-B transcripts further demonstrated that this enzymatic activity required an RNA duplex, suggesting that the GluR editing reaction was mediated by a double-stranded, RNA-specific adenosine deaminase. Biochemical fractionation of HeLa cell (Yang et al., 1995; Maas et al., 1996) or rat brain (Burns et al., 1997) nuclear extracts identified two distinct dsRNA-directed adenosine deaminase activities that could modify specific A residues within GluR transcripts. Two enzymes with catalytic activities consistent with these biochemical properties are dsRNA-specific adenosine deaminase (Hough and Bass, 1994; Kim et al., 1994a, 1994b; O'Connell and Keller, 1994; Herbert et al., 1995) (referred to as dsRAD, DRADA or ADAR1), and dsRNA-specific editase 1 (Melcher et al., 1996a) (referred to as RED1, DRADA2 or ADAR2). Further details regarding the isolation, cloning, structure and activities of these enzymes will be presented later in this chapter.

Hepatitis Delta Virus

The hepatitis delta virus (HDV) is a subviral human pathogen whose packaging and propagation are dependent on concurrent infection with the hepatitis B virus (Ponzetto et al., 1984; Bonino et al., 1986; Wu et al., 1991; Ryu et al., 1992). The HDV genome consists of a 1.7-kb single-stranded circular RNA molecule that can form an intramolecular RNA duplex to produce a rod-like structure with about 70% of its bases paired (Chen et al., 1986; Wang et al., 1986). Within an infected cell, the HDV genome replicates independently of the hepatitis B virus by using a rolling-circle mechanism of replication involving RNA-directed RNA synthesis of genomic RNA. This mode of genome replication utilizes an RNA inter-

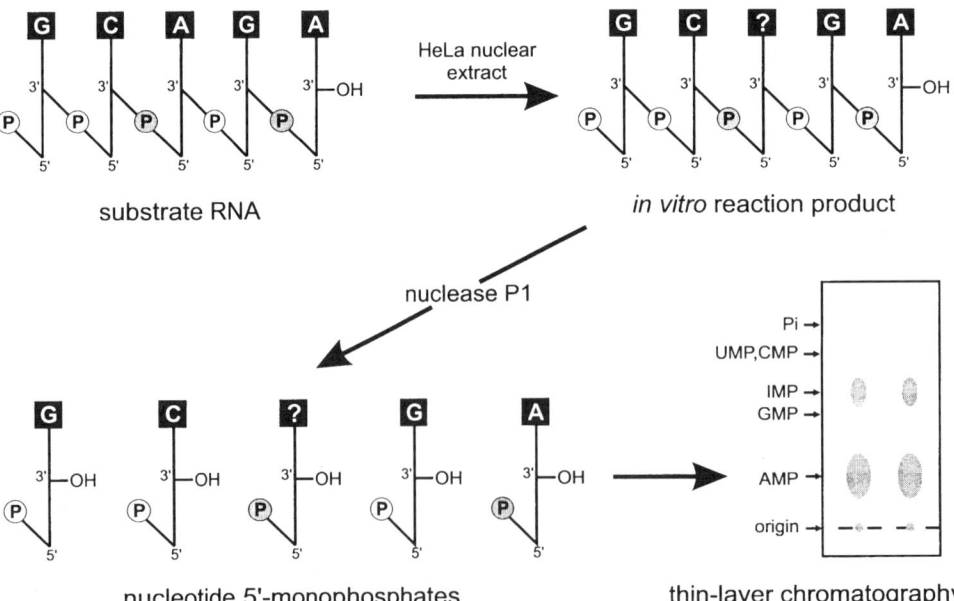

Figure 3. Experimental strategy for determining the identity of a modified nucleotide(s) subsequent to RNA editing. A biochemical strategy is presented in which an RNA substrate uniformly labeled with [α-^{32}P]-ATP is subjected to in vitro editing using HeLa cell nuclear extract as a source of enzymatic activity. The resulting in vitro reaction product is digested to completion with nuclease P1, yielding nucleotide 5′-monophosphates that can be resolved by thin-layer chromatography.

mediate, referred to as the HDV antigenome, which serves as a template for synthesis of HDV genomic RNA (Chen et al., 1986). It is the antigenomic RNA that encodes the delta antigen (HDAg), the only known protein product of HDV. The HDAg exists in two forms, a short form (HDAg-p24) and a long form (HDAg-p27) that are 195 and 214 amino acids in length, respectively. HDAg-p27 is identical to its shorter counterpart except for a 19-amino-acid extension at the carboxyl terminus of the protein (Kuo et al., 1989). These two proteins play essential but opposing roles in the life cycle of the virus; HDAg-p24 is required for replication of the HDV genome, whereas HDAg-p27 inhibits replication of the genome and is required for viral packaging (Kuo et al., 1989; Chao et al., 1990; Wu et al., 1991; Ryu et al., 1992).

The production of HDAg-p27 occurs as a result of an RNA editing event in which a particular adenosine moiety within HDV antigenomic RNA (Luo et al., 1990; Casey and Gerin, 1995) is converted to inosine (Polson et al., 1996), thus altering the coding potential of a single codon within the antigenome (amber/W site) from an amber stop (UAG) to a tryptophan (UIG) codon and resulting in the extension of the open reading frame by 19 amino acids. Preliminary studies suggested that the editing of HDAg transcripts occurred within the genomic RNA by converting a uridine to a cytidine (Zheng et al., 1992; Casey et al., 1992); however, more recent studies have demonstrated that this editing event actually occurs in the antigenomic RNA, converting an adenosine to an inosine residue (Casey and Gerin, 1995; Polson et al., 1996). Analyses of RNA from replicating viral populations have revealed a high level of specificity for HDV editing; 20–50% of the steady-state antigenome transcripts are modified at the amber/W site, while less than 1% of the adenosine residues within a 350-nt region surrounding this site show similar modifications (Bass, 1997).

The editing of HDV antigenomic RNA, similar to the editing of the AMPA and kainate subunit mRNAs, requires the formation of a dsRNA structure surrounding the amber/W editing site (Fig. 2) (Casey et al., 1992; Polson et al., 1996). The introduction of mutations near the amber/W adenosine moiety dramatically decreases the efficiency of editing both in vivo and in vitro (Polson et al., 1996). Although contained within a region of duplex RNA, the amber/W adenosine residue is predicted to be unpaired with the opposing cytosine nucleoside on the opposite side of the duplex. Surprisingly, mutation of this A–C mismatch to an A–U base pair also reduced the extent of editing (Polson et al., 1996). A similar structure, in which the modified nucleoside occurs opposite an unpaired cytosine moiety, has also been predicted for the R/G sites of GluR-B (Fig. 2), -C, and -D RNAs (Lomeli et al., 1994). The frequent occurrence of an A–C mismatch at targeted adenosine residues has suggested that this structural motif may represent an important feature by which specific A residues are targeted for modification.

Serotonin 2C Receptor

Serotonin (5-hydroxytryptamine [5-HT]) is a monoaminergic neurotransmitter that modulates numerous sensory and motor processes and a wide variety of behaviors, including sleep, appetite, pain perception, locomotion, thermoregulation, hallucinations, and sexual behavior. The multiple actions of 5-HT are mediated by specific interactions with multiple receptor subtypes. Pharmacological, physiological, and molecular cloning studies have provided evidence for fourteen distinct 5-HT receptor subtypes that have been subdivided into seven families (5-HT_1–5-HT_7) based on relative ligand binding affinities, genomic structure, amino acid sequence similarities, and coupling to specific signal transduction pathways (Hoyer et al., 1994). The 5-HT_2 family of receptors includes three receptor subtypes, 5-HT_{2A} (formerly 5-HT_2), 5-HT_{2B} (formerly 5-HT_{2F}) and 5-HT_{2C} (formerly 5-HT_{1C}), which are linked to phospholipase C, promoting the hydrolysis of membrane phosphoinositides and a subsequent rise in intracellular levels of inositol phosphates and diacylglycerol. The 5-HT_{2A} and 5-HT_{2C} receptors have been implicated in a number of human psychiatric and behavioral disorders, including psychotic depression, dysthymia, obsessive-compulsive disease, anxiety, and schizophrenia (Teitler and Herrick-Davis, 1994; Julius, 1991; Pandey et al., 1995; Dubovsky and Thomas, 1995). Mice lacking 5-HT_{2C} receptor expression are overweight, establishing a role for this receptor in appetite control, and are prone to spontaneous death from seizures, suggesting that 5-HT_{2C} receptors mediate tonic inhibition of neuronal excitability (Tecott et al., 1995).

Sequence comparisons between rat 5-HT_{2C} receptor (5-HT_{2C}R) genomic DNA and cDNAs generated from rat striatum identified four A-to-G discrepancies (Fig. 4) predicted to alter three amino acids within the second intracellular loop of the receptor (Burns et al., 1997). Therefore, it was proposed that the mRNA species encoding the 5-HT_{2C} receptor was modified such that four specific adenosine residues within 5-HT_{2C}R RNA (termed sites A, B, C, and D) were converted to inosines in the mature mRNA by a process analogous to that seen for RNA editing

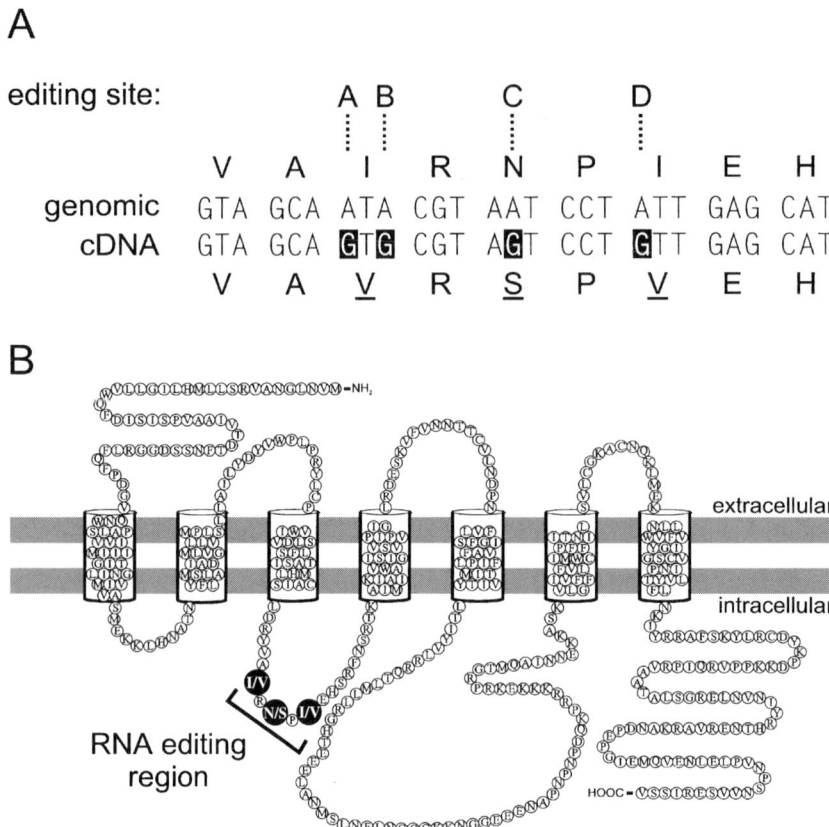

Figure 4. RNA editing of 5-HT$_{2C}$R transcripts. (A) Nucleotide and predicted amino acid sequence alignments between 5-HT$_{2C}$R genomic and cDNA sequences; A-to-G nucleotide discrepancies and predicted alterations in amino acid sequence are indicated in inverse and underlined lettering, respectively. (B) A schematic representation of the predicted topology for the 5-HT$_{2C}$ receptor is presented, indicating the sites of amino acid alteration within the second intracellular loop resulting from RNA editing events. The amino acid sequence of the 5-HT$_{2C}$R is indicated with the one-letter amino acid code (Julius et al., 1988).

events in transcripts encoding the HDV antigenome and subunits of the AMPA and kainate subtypes of ionotropic glutamate receptors (Polson et al., 1996; Sommer et al., 1991). Sequence analysis of cDNAs isolated from dissected rat brain indicated the region-specific expression of seven 5-HT$_{2C}$R isoforms encoded by eleven distinct RNA species, suggesting that differentially edited 5-HT$_{2C}$ receptors may serve distinct biological functions in those regions in which they are expressed. Further DNA sequence analyses of the rat 5-HT$_{2C}$R gene, in the region surrounding the proposed editing sites, suggested the existence of an imperfect inverted repeat forming a putative RNA duplex between the 3' end of exon 3 and the proximal region of intron 3 (Fig. 2). Mutations introduced within this region, disrupting proposed RNA base pairing interactions, abolished editing at all four sites, while the introduction of compensatory sequence alterations restored editing to wild-type levels (Burns et al., 1997).

In vitro studies of 5-HT$_{2C}$R RNA editing, using rat brain nuclear extracts fractionated by cation exchange chromatography and a 288-nt 5-HT$_{2C}$R RNA substrate, revealed two peaks of inosine (rather than guanosine) production from a radiolabeled 5-HT$_{2C}$R substrate; these two separable activities co-eluted with two distinct peaks of dsRNA adenosine deaminase activity obtained by incubating fractionated extracts with a synthetic dsRNA substrate. These two peaks also contained two distinct 5-HT$_{2C}$R editing activities. The first peak of activity was responsible for editing 5-HT$_{2C}$R transcripts at the A, B, and C sites and coeluted with the activity mediating the modification of GluR-B at intronic position +60. The second peak of activity specifically modified the D site of 5-HT$_{2C}$R transcripts and the Q/R site of GluR-B. These results suggested that the editing of 5-HT$_{2C}$R transcripts in vitro was mediated by the actions of two separable double-stranded RNA adenosine deaminases, similar to that described for

GluR-B RNAs (Yang et al., 1995; Melcher et al., 1996a; O'Connell et al., 1997), and further suggested that the editing of both 5-HT$_{2C}$R and GluR-B transcripts is mediated by common cellular machineries.

The 5-HT$_{2C}$R belongs to a large family of seven transmembrane-spanning, G-protein-coupled receptors. G-proteins are heterotrimeric complexes composed of three subunits (α, β, and γ). Binding a receptor to the heterotrimer causes dissociation of the α subunit (which regulates signaling molecules such as adenylate cyclase and phospholipase C) from the $\beta\gamma$ subunit, which also plays a role in intracellular signaling (Birnbaumer, 1992). Overwhelming evidence has shown that the third intracellular loop of receptors is an important site of regulation for receptor:G-protein coupling, although the second intracellular loop has also been shown to play a role in the coupling to G-proteins, especially for receptors linked to phospholipase C (Wong et al., 1990; Pin et al., 1994). Since RNA editing alters the primary amino acid sequence in the second intracellular loop of the 5-HT$_{2C}$ receptor, initial analyses of receptor function focused on alterations in the ability of edited 5-HT$_{2C}$R isoforms to activate phospholipase C and the subsequent accumulation of inositol phosphates. Results from such studies indicated that both fully edited and nonedited receptors were coupled to the activation of phosphoinositide hydrolysis; however, a comparison of the 5-HT dose response demonstrated that serotonergic agonists were 10-fold to 15-fold less potent when acting through the fully edited, rather than the nonedited, receptor isoform. These results suggest that edited receptor isoforms couple less efficiently to the intracellular signaling machinery. Region-specific differences in 5-HT$_{2C}$R editing could represent a mechanism by which receptor activation could be subtly modulated, thereby altering cellular responses to ambient levels of neurotransmitter.

RNA Substrates Containing Multiple Editing (Hypermutation) Sites

Despite the extensive length of predicted RNA duplexes for transcripts encoding non-NMDA glutamate receptor subunits, the 5-HT$_{2C}$ receptor and the HDV antigenome (Fig. 2), and the number of potential adenosine targets contained within these structures, the editing of such RNA substrates is extremely specific, limited to a single or a few modified nucleotides (Rueter et al., 1995; Herb et al., 1996; Polson et al., 1996; Burns et al., 1997). The identification of additional edited substrates, based largely on A-to-G discrepancies between genomic and cDNA sequences, has revealed a second category of modified substrates in which many of the adenosine residues within a defined region are converted to inosine. This type of extensive editing has been referred to as hypermutation (Cattaneo et al., 1988; Cattaneo, 1994; Bass, 1997) and was first observed in the cDNA encoding the measles virus matrix (M) protein in which 132 of 266 genomically encoded uridine residues appeared as cytidines in the matrix protein cDNA; these U-to-C discrepancies were proposed to result from A-to-G (I) conversions in the negative-strand viral RNA genome (Cattaneo et al., 1988; Bass et al., 1989). In the case of the measles virus M transcript, RNA editing appears to result in matrix protein inactivation, allowing the persistence of measles virus infection rather than a lytic state for the virus (Billeter and Cattaneo, 1991; Billeter et al., 1994). Similar hypermutations have also been observed for transcripts encoded by other viruses, including parainfluenza virus 3 (Murphy et al., 1991), vesicular stomatitis virus (O'Hara et al., 1984), avian leukosis virus (Hajjar and Linial, 1995), and polyomavirus (Kumar and Carmichael, 1997).

Recent studies of RNA editing in polyomavirus have suggested that this posttranscriptional modification represents a mechanism by which RNA transcripts expressed early after viral infection are inactivated subsequent to viral DNA replication (the late phase). Nucleotide sequence analyses of early-strand transcripts isolated 30 h postinfection have revealed that approximately one-half of the adenosine residues specified by the viral genome were replaced by guanosines on the early sense strand-derived cDNAs (Kumar and Carmichael, 1997). Reverse transcription-polymerase chain reaction (RT-PCR) analysis of viral RNAs, isolated at different times postinfection, indicated that unmodified transcripts were apparent throughout infection and in the absence of DNA replication, while modified RNAs could only be detected during late periods and were not present in the absence of viral replication (Kumar and Carmichael, 1997). Furthermore, these modified transcripts could only be identified in nuclear fractions, while unmodified species were found in both the nuclear and cytoplasmic compartments. These results suggest that the polyomavirus switch from early- to late-strand expression is mediated by the posttranscriptional modification of early-strand transcripts, resulting in the retention of such early-strand mRNAs within the nucleus. Alternatively, modified RNAs could be transported to the cytoplasm but would be preferentially degraded in that compartment. The cellular mechanisms by which such inosine-containing transcripts are preferentially retained or degraded within the nuclear or cytoplasmic compartments, respectively, have not yet been elucidated. However, recent studies have identified a novel ribonuclease activity that preferentially degrades inosine-containing RNAs (Scadden and Smith, 1997).

The first example of an extensive editing pattern for a nonviral RNA transcript was observed in the cDNA encoding a putative RNA-binding protein from *Drosophila*, referred to as *4f-rnp* (Petschek et al., 1996). Analysis of *4f-rnp* cDNAs has revealed that the primary RNA transcript is alternatively spliced to produce two protein isoforms (class 1 and 2), each containing an RNA recognition motif (RRM), a domain common to many RNA-binding proteins (Burd and Dreyfuss, 1994). Nucleotide sequence comparisons between class 1-derived cDNAs and the *4f-rnp* gene have revealed multiple A-to-G discrepancies dispersed throughout the gene that are predicted to introduce numerous and significant amino acid substitutions in the resultant protein. At the present time, it is unclear whether editing is confined to class 1 *4f-rnp* transcripts or if mRNAs from both classes are subjected to the presumed A-to-I modifications. Although the precise biological functions of the *4f-rnp* gene in the fruitfly are currently unknown, preliminary genetic analyses have demonstrated that deletion of the *4f-rnp* locus results in lethality during larval development, suggesting that *4f-rnp* is an essential gene (Petschek et al., 1996). It is anticipated that the powerful genetic approaches that are possible with *Drosophila* will enable rapid progress in elucidating the molecular mechanisms underlying these A-to-G discrepancies and the functional significance of *4f-rnp* RNA editing.

The most recent example of multiple A-to-G discrepancies between genomic and cDNA sequences was identified in cDNA clones encoding a voltage-dependent potassium channel (sqKv2) isolated from the optic lobe of the squid *Loligo peali* (Patton et al., 1997). Voltage-gated potassium channels contribute to the excitability of neurons and signaling in the nervous system, and exist as a tetrameric complex in which each subunit contains six-membrane-spanning segments (S1–S6), with both the amino and carboxyl termini located on the cytoplasmic face (Shih and Goldin, 1997). The functional properties of these ion channels (potassium permeation, channel gating, and subunit interaction) are regulated by specific protein domains within the channel, including the region between S5 and S6 (the pore region), which is thought to be involved in ion conduction (Yellen et al., 1991; Slesinger et al., 1993), and the S4 segment (voltage sensor), which mediates the channel response to alterations in membrane potential (Bezanilla and Stefani, 1994; Jan and Jan, 1997). Nucleotide sequence analysis of multiple cDNA clones identified seventeen A-to-G transitions in the region between the S4 and S6 channel segments, resulting in 12 alterations in coding potential from the genomically encoded protein sequence. The relative proportion of cDNA sequences that differed from the genomic sequence at specific positions varied from 3–100% with a total of 28 unique editing pattern combinations in cDNAs isolated from the two animals examined (Patton et al., 1997). Preliminary analysis of channel function for two potassium channel isoforms revealed that single amino acid alterations in the pore region (Y576C) and the S6 segment (I597V), produced as a result of editing, had significant effects on both the rates of slow inactivation and channel closure subsequent to repolarization. While the analysis of sqKv2 was limited to the small region between the S4 and S6 channel segments, it has been suggested that the total number of editing sites within sqKv2 transcripts may be much higher, dramatically increasing the diversity of invertebrate potassium channel expression (Patton et al., 1997).

ADENOSINE DEAMINASES THAT ACT ON RNA (ADAR)

For many types of RNA editing, the identification of cellular factors that mediate such posttranscriptional events followed the initial discovery of an edited substrate. In the case of A-to-I modifications, however, studies demonstrating the conversion of adenosine to inosine and the requirements for a dsRNA structure suggested that the characterization of an enzymatic activity responsible for these editing events may have preceded the identification of A-to-I modified RNAs.

dsRNA-Specific Adenosine Deaminase (dsRAD/DRADA/ADAR1)

Several years before the discovery of GluR-B RNA editing (Sommer et al., 1991), an activity that unwinds dsRNA was identified in *Xenopus laevis* upon microinjection of a synthetic antisense RNA complementary to a cellular RNA into *Xenopus* embryos (Bass and Weintraub, 1987; Rebagliati and Melton, 1987). This novel activity was referred to as a dsRNA unwinding activity because of its apparent ability to unwind RNA:RNA hybrids formed between sense and antisense RNA transcripts. As in *Xenopus* embryos, a similar dsRNA unwinding activity was identified in the nuclei of a wide variety of mammalian somatic cells (Wagner and Nishikura, 1988; Wagner et al., 1990) and this activity was referred to as an unwindase. Subsequent studies revealed that rather than unwinding duplex RNA, this activity catalyzed the conversion of adenosine to inosine by hydrolytic deamination (see Appendix 1) (Polson et al., 1991). As a result of such enzymatic base modification, the RNA became more single-stranded in character, because of the conversion of stable A–U base pairs to less stable I–U pairs, thereby leading to destabilization of the RNA duplex (Bass and Weintraub, 1988; Wagner et al., 1989). This enzyme became referred to as dsRNA-specific adenosine deaminase (dsRAD or DRADA), and for several years following its initial discovery researchers searched for potential roles that this enzymatic activity might play in vivo. The recent identification of additional members of this dsRNA-specific adenosine deaminase family (Melcher et al., 1996a, 1996b) has resulted in a common nomenclature for these proteins that are currently referred to as adenosine deaminases that act on RNA (ADAR); dsRAD/DRADA is known as ADAR1 (Bass et al., 1997).

ADAR1 has been biochemically purified from several sources, including *Xenopus* eggs, bovine liver, calf thymus, and chicken lung (Hough and Bass, 1994; Kim et al., 1994a; O'Connell and Keller, 1994; Herbert et al., 1995), and the human, rat, and *Xenopus* cDNAs have been cloned (Kim et al., 1994b; O'Connell et al., 1995; Hough and Bass, 1997; Liu et al., 1997). The human and rat ADAR1 proteins are approximately 136 and 130 kDa in size, respectively. The predicted protein sequence of ADAR1 reveals that it contains a putative bipartite nuclear localization signal, consistent with previous immunolocalization and biochemical studies (Fig. 5) (Wagner et al., 1990; O'Connell and Keller, 1994). ADAR1 contains a region mediating high affinity binding to Z-DNA, a DNA configuration thought to occur behind a moving RNA polymerase during transcription. The presence of such a Z-DNA binding domain (Zα) represents a mechanism by which ADAR1 can be targeted to nascent RNA transcripts, allowing editing to occur before splicing (Herbert et al., 1995, 1997). In addition, three copies of a dsRNA-binding domain (dsRBD), a motif shared among a number of dsRNA-binding proteins, including Staufen and DAI kinase (Burd and Dreyfuss, 1994), and a region homologous to the catalytic domain of other known adenosine and cytidine deaminases were also identified (Fig. 5). X-ray crystallographic studies of *Escherichia coli* cytidine deaminase have indicated that this enzyme requires the presence of an active-site zinc atom for catalytic activity (Betts et al., 1994). Although the presence of such a zinc atom within ADAR1 has not been demonstrated directly, mutagenesis of critical cysteine and histidine residues, thought to be involved in zinc coordination (Chapters 8 and 20), resulted in enzyme inactivation (Lai et al., 1995). Mutation of an essential glutamate moiety, predicted to play a role in the proton transfer functions required for deamination, also abolished catalytic activity, further elucidating the biochemical mechanisms by which ADAR1 modifies adenosine residues (Lai et al., 1995). Although catalysis by ADAR1 involves nucleophilic attack at the C-6 position of adenine, the C-6 atom is presumed to be buried deep within the major groove of the A-form RNA duplex, making it unclear how the enzyme gains access to its nucleotide target. Nucleotide sequence similarities between the carboxyl-terminal region of ADAR1 and type II DNA methyltransferases have suggested that ADAR1 might employ a base-flipping strategy similar to that of the methyltransferases to sequester target nucleotides into its active site by extracting them from the nucleotide duplex, leaving the helical structure relatively unperturbed (Hough and Bass, 1997).

In vitro analyses of ADAR1 substrate specificity have indicated that it will recognize dsRNA, generated by inter- or intramolecular duplex formation, and not ssRNA, ssDNA, or dsDNA (Bass and Weintraub, 1987; Wagner and Nishikura, 1988). ADAR1 is capable of modifying dsRNAs made from various in vitro synthesized sense and antisense transcripts (Bass and Weintraub, 1987; Rebagliati and Melton, 1987; Wagner and Nishikura, 1988) and can convert up to 50% of the adenosine residues in both the sense and antisense strands of the RNA substrate to inosine (Bass and Weintraub, 1988; Wagner et al., 1989). The efficiency of adenosine modification by ADAR1 is thought to be determined by both a nearest-neighbor preference (Chapter 8) surrounding the targeted adenosines (Polson and Bass, 1994) and the overall length of the double-stranded region (Nishikura et al., 1991; Polson and Bass, 1994).

In vitro editing analyses using recombinant ADAR1 have indicated that this enzymatic activity is not responsible for modifying the Q/R site of GluR-B pre-mRNAs (Hurst et al., 1995; Maas et al., 1996) and this conclusion was further supported by biochemical fractionation studies demonstrating that the enzymatic activity responsible for modifying the Q/R adenosine moiety did not cofractionate with ADAR1 (Yang et al., 1995; Maas et al., 1996). Although ADAR1 does not effectively modify the Q/R site of GluR-B transcripts, it can efficiently edit the intronic "hot spot" adenosine at the +60 site of GluR-B pre-mRNAs (Hurst et al., 1995; Maas et al., 1996), the R/G site of GluR-B pre-mRNA, the Q/R site of GluR-5 RNA (Maas et al., 1996; Herb et al., 1996), and the A, B, and C sites of 5-HT$_{2C}$R transcripts (Burns et al., 1997) (Fig. 2). Similar in vitro strategies using either purified or recombinant

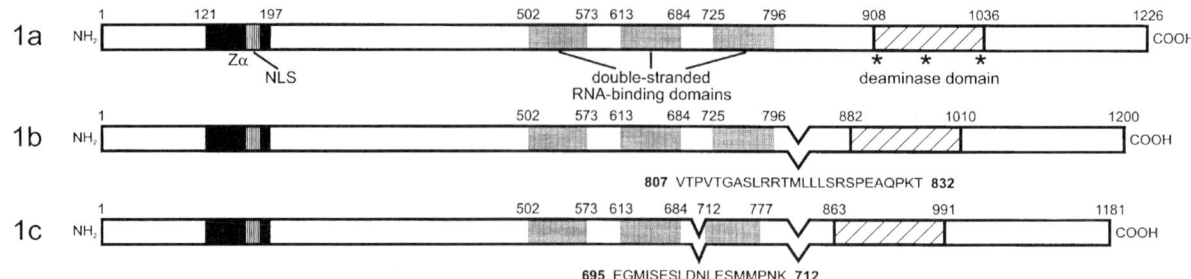

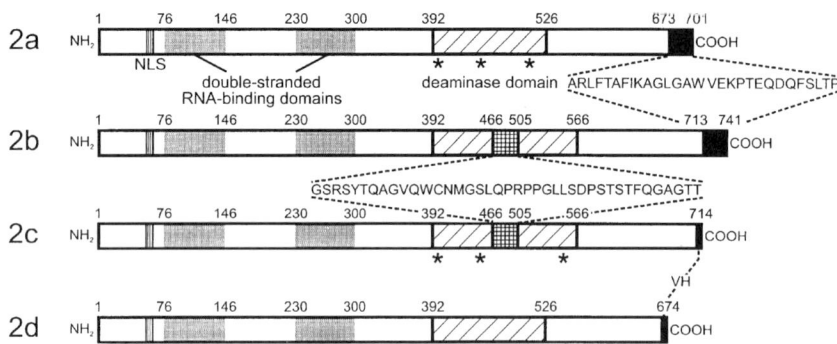

Figure 5. Alternatively spliced variants of human ADAR1 and ADAR2 mRNA produce multiple protein isoforms. Schematic representations of ADAR1 (A) and ADAR2 (B) protein isoforms are presented, indicating the locations of the putative nuclear localization signal (NLS; vertical stripe), Z-DNA binding domain (Zα; black), dsRNA-binding domains (gray) and the adenosine deaminase domain (angled stripe). The location of zinc-coordination residues within the deaminase domain are designated by asterisks and the amino acid coordinates for each domain, relative to the amino terminus, are indicated. The specific amino acid residues deleted in the ADAR1b and ADAR1c isoforms or inserted in the deaminase domain of the ADAR2b and ADAR2c isoforms (crosshatch) are indicated, as well as the amino acid residues present in unique ADAR2 carboxyl termini (black) with the one-letter amino acid code.

ADAR1 with HDV or GluR-6 RNA substrates have revealed that this dsRNA adenosine deaminase can selectively modify the amber/W and Q/R sites, respectively, although modifications at additional A residues were also observed that are not seen in vivo (Polson et al., 1996, Herb et al., 1996). It should be noted, however, that in vitro assays with purified or recombinant ADAR1, as well as transfection studies in model cell lines, do not directly mirror editing patterns observed for in vivo tissues (Hurst et al., 1995; Rueter et al., 1995; Herb et al., 1996), suggesting that additional studies will be required before one can assign ADAR1 as the enzyme responsible for deaminating a particular adenosine moiety in vivo.

The fact that ADAR1 can demonstrate apparent adenosine target specificity for certain RNA substrates (GluR, HDV, and 5-HT$_{2C}$R), while demonstrating relative nonspecificity for other transcripts, has raised a number of questions regarding the molecular mechanisms by which such specificity of ADAR1 adenosine selection is achieved. It has been proposed that the primary sequence (nearest-neighbor preference), and both the secondary (duplex length) and tertiary structures of the RNA may be required for ADAR1 to achieve its substrate specificity (Nishikura et al., 1991; Polson and Bass, 1994; Hurst et al., 1995; Bass, 1997), although extensive experimental analysis of the secondary and tertiary structures for duplex regions in edited RNA transcripts has not yet been performed. Alternatively, accessory proteins similar to those identified for the editing of apoB transcripts (see Chapter 18) could also play an important role by activating or suppressing the modification of individual adenosine residues (Saccomanno and Bass, 1994). However, in vitro studies of ADAR1 activity have demonstrated both selective and nonselective substrate modification in the absence of any additional protein factors, depending on the RNA substrate (Hough and Bass, 1994; Maas et al., 1996; Herb et al., 1996; Burns et al., 1997).

Another strategy by which to achieve the selective modification of specific adenosine residues for various RNA substrates is through the expression of multiple ADAR protein isoforms. At least two isoforms of ADAR1 have been identified in human amnion and neuroblastoma cell lines, including a constitutively expressed 110-kDa form that was located exclusively in the nucleus and a 150-kDa interferon-inducible form present in both the nucleus and cytoplasm (Patterson and Samuel, 1995). Antisera generated against an amino-terminal epitope of the full-length ADAR1 protein recognized only the 150-kDa form, whereas antisera directed against either internal or carboxyl-terminal portions of the protein recognized both forms, indicating that the 110-kDa form of ADAR1 is truncated at the amino-terminus (Patterson and Samuel, 1995).

More recent studies involving the isolation and characterization of human genomic and cDNA clones encoding ADAR1 have indicated the existence of two additional deaminase variants produced as a result of alternative splicing within exons 6 and 7 of the ADAR1 pre-mRNA (Liu et al., 1997). The initial human ADAR1 cDNA clone (Kim et al., 1994b; O'Connell et al., 1995; Patterson and Samuel, 1995) is now referred to as ADAR1a (Fig. 5), while the ADAR1b cDNA predicts a protein containing a 26-amino-acid deletion within the linker region located between the second and third dsRBDs (Liu et al., 1997). The ADAR1c protein isoform is predicted to contain the 26-amino-acid deletion observed for ADAR1b and a 19-amino-acid deletion containing a portion of the third dsRBD and the linker region located between the third dsRBD and the deaminase domain (Fig. 5) (Liu et al., 1997). All three of these ADAR1 isoforms exhibit similar dsRNA adenosine deaminase activities for a synthetic RNA duplex; however, mutagenic analysis of each of the three dsRBDs in the ADAR1a, -b, and -c isoforms has indicated differential effects on the extent of editing, suggesting that the role of each dsRBD in the activity or specificity of spliced ADAR1 isoforms may not be identical (Liu et al., 1997).

dsRNA-Specific Editase 1 (ADAR2/RED1/DRADA2)

The observation that ADAR1 was ineffective in modifying the Q/R site of GluR-B RNA substrates suggested that additional dsRNA adenosine deaminases may exist with distinct or overlapping substrate specificities. Biochemical fractionation studies had further revealed that there were at least two separable ADAR-like activities in HeLa cell nuclear extracts that could effectively modify a synthetic RNA duplex, yet only one of these activities could deaminate the Q/R site adenosine of GluR-B pre-mRNAs (Yang et al., 1995; Maas et al., 1996) or react with an antiserum directed against ADAR1 (Maas et al., 1996). In search of additional members within this family of enzymes, researchers screened a rat hippocampal cDNA library under low stringency hybridization conditions with a probe complementary to the deaminase domain of ADAR1 (Melcher et al., 1996a). Complementary DNA clones isolated during this screening procedure were assessed for their ability to selectively modify the Q/R site adenosine of GluR-B upon transfection into HEK293 cells. Using this experimental strategy, a cDNA clone was identified that could efficiently catalyze deamination of the Q/R site, and the protein encoded by this cDNA was referred to as dsRNA-specific editase 1 (RED1; Melcher et al., 1996a); this enzyme has also been referred to as DRADA2 (Lai et al., 1995, 1997), but is currently known as ADAR2 (Bass et al., 1997). Biochemical purification of the enzymatic activity responsible for editing GluR-B pre-mRNA at the Q/R site provided additional evidence that this editing event is indeed mediated by ADAR2 (O'Connell et al., 1997; Yang et al., 1997).

ADAR2 is an 80-kDa protein with structural features quite similar to those previously observed for ADAR1 (Fig. 5) (Melcher et al., 1996a). ADAR2 contains a putative bipartite nuclear localization signal and two dsRNA-binding domains sharing approximately 25% amino acid sequence similarity with the three dsRBDs of ADAR1. ADAR2 also contains an adenosine deaminase domain sharing 70% amino acid similarity with that of ADAR1, as well as three putative zinc-chelating residues conserved in the deaminase domain of both enzymes. Overall, the two proteins share only 31% amino acid identity, largely owing to conserved residues in the deaminase domain. ADAR2, like ADAR1, exhibits adenosine deaminase activity for a synthetic RNA duplex and for the R/G site of GluR-B, -C, and -D pre-mRNAs (Melcher et al., 1996a; O'Connell et al., 1997; Yang et al., 1997). However, ADAR2 can also efficiently modify the Q/R site of GluR-B pre-mRNAs and the D site of 5-HT_{2C}R transcripts (Melcher et al., 1996a; Burns et al., 1997). Both enzymes appear to be expressed in most tissues and tissue culture cell lines (Wagner et al., 1990), suggesting that they may be responsible for the modification of other, as yet unidentified, RNA substrates.

A novel cDNA isoform of human ADAR2 (hADAR2b), produced by alternative splicing, has recently been identified and contains an insertion of 120 nt in the region between the second and third zinc-coordination residues of the deaminase domain (Fig. 5) (Lai et al., 1997; Gerber et al., 1997); this 120-nt cassette is 88% identical to a segment of the human Alu-J subfamily consensus sequence (Gerber et al., 1997). Comparisons between the activities of recombinant hADAR2a and hADAR2b protein isoforms have indicated that both forms of hADAR2 are capable of adenosine deamination using a synthetic dsRNA substrate and selective modification of the Q/R and R/G sites of GluR-B pre-mRNA. Although there was some variation between the assays, hADAR2a was approximately twice as active as hADAR2b for all RNA substrates, indicating that insertion of the Alu-J cassette may result in a slight reduction in the catalytic activity of hADAR2b, without effecting the specificity of the editing reaction (Gerber et al., 1997). It is interesting to note that an alternatively spliced rat ADAR2 (rADAR2) cDNA clone was identified that lacks a 30-nt cassette within the deaminase domain at the same position as the insertion point for the Alu-J cassette in hADAR2b (Melcher et al., 1996b; Maas et al., 1997; Bass et al., 1997). Despite the fact that alternative splicing events occur for both the rat and human ADAR2 genes at identical positions within the pre-mRNAs, there are no nucleotide or predicted amino acid sequence similarities between these species-specific insertions.

Generation of hADAR2 diversity by alternative splicing has also been observed from multiple cDNA isolates differing at the 3'-end of their coding regions (Gerber et al., 1997; Lai et al., 1997). These cDNAs predict the synthesis of four hADAR2 isoforms that can contain an insertion in the deaminase domain and terminate with a valine-histidine dipeptide (hADAR2c and hADAR2d) or can contain a 29-amino-acid extension at the carboxyl terminus (hADAR2a and hADAR2b; Fig. 5). The protein isoforms of hADAR2 that contain the truncated carboxyl terminus exhibit negligible activity at the Q/R site, +60, and R/G sites of GluR-B pre-mRNA and a 5- to 10-fold reduction in deaminase activity for a synthetic dsRNA (Lai et al., 1997). It has been suggested that the hADAR2c and hADAR2d proteins could serve as competitive inhibitors of the other two editing-competent enzymes or that the altered C-terminal structure may be specialized to modify adenosine residues within currently unidentified RNA targets (Lai et al., 1997).

A Growing Family of ADAR Enzymes?

Alternative splicing of ADAR1 and ADAR2 primary RNA transcripts provides a cellular strategy for the generation of multiple ADAR protein isoforms that could differ in their catalytic activity or substrate specificity. A second mechanism by which to attain such diversity in dsRNA adenosine deaminase expression is the occurrence of multiple ADAR-encoding genes. Recent attempts to identify additional ADAR-like proteins have utilized RT-PCR amplification of rat brain RNA using degenerate oligonucleotide primers to conserved amino acid residues in the deaminase domains of ADAR1 and ADAR2. Using such a strategy, a novel cDNA with 60% amino acid identity to ADAR2 was isolated and is referred to as RED2 (Melcher et al., 1996b). The structural features of RED2 are identical to those of ADAR2 except for a 54-residue, arginine-rich extension at the amino terminus. In situ hybridization analysis of RED2 expression has revealed that unlike the wide expression of ADAR1 and ADAR2, RED2 is localized specifically to the brain (Melcher et al., 1996b). Functional studies of RED2 activity have demonstrated that it does not exhibit any adenosine deaminase activity using a synthetic dsRNA substrate, nor does it modify any of the major editing sites within GluR-B (Q/R, R/G and +60). Chimeric molecules of RED2, in which the deaminase domain was replaced by the corresponding region from ADAR2, demonstrated moderate editing activity for the Q/R and R/G sites. However, the converse chimera, in which the deaminase domain of ADAR2 was replaced with that of RED2, was devoid of detectable editing activity (Melcher et al., 1996b). These results suggest that despite its similarity to ADAR2, RED2 may not be a member ADAR protein family or, alternatively, RED2 could be specific for adenosine residues within an RNA substrate that has not yet been identified.

Two additional ADAR-like activities, and their corresponding cDNAs, have recently been identified from the yeast *Saccharomyces cerevisiae* and *Drosophila melanogaster*, referred to as HRA400 and Tad1p, respectively (Gerber, personal communication). As previously described for RED2, cDNA clones encoding these enzymes were isolated based on their sequence similarities to the deaminase domains of ADAR1 and ADAR2. The deaminase domain of the fruitfly enzyme shows 32% sequence identity to ADAR2, while the yeast protein lacks any discernible dsRNA binding motifs, but shares approximately 23% sequence identity to the mammalian enzymes. The *HRA400* gene is not essential in yeast and may play a role in the generation of inosine residues within the anticodon loop of tRNAs (Gerber, personal communication).

Recently, two proteins have been suggested to represent additional members of the ADAR family: Tenr, a nuclear RNA-binding protein expressed exclusively in the testes (Schumacher et al., 1995), and T20H4.4, a predicted open reading frame from *Caenorhabditis elegans* (Hough and Bass, 1997; Bass, 1997). The Tenr protein, predicted to contain both a dsRNA-binding motif and deaminase domain, is localized to round and early elongating spermatid cells. Confocal microscopy has revealed a lattice-like nuclear distribution, suggesting the association of Tenr with the nuclear scaffold. It has been suggested that the Tenr protein may be involved in testis-specific nuclear posttranscriptional processes such as heterogeneous nuclear RNA (hnRNA) packaging, alternative splicing, or nuclear and cytoplasmic transport of mRNAs (Schumacher et al., 1995). The amino acid sequence of T20H4.4 is predicted from the *C. elegans* genome project (Wilson et al., 1994) to encode a 40-kDa protein containing both a dsRBD and a deaminase domain. To date, no studies have been performed to show either expression of T20H4.4 mRNA or protein, and studies to examine potential deaminase activity for this protein are currently under way. A phylogenetic tree demonstrating the relationship between members of the ADAR family, ADAR-like proteins, cytidine deaminases, and apobec-1, the cytidine deaminase involved in the editing of apoB transcripts, is presented in Fig. 6 of Chapter 18.

CONCLUDING REMARKS

Initially identified as an RNA modification in the anticodon loop of tRNAs from animal, plant and eubacterial origins (Björk, 1995), the deamination of adenosine to inosine in mRNA transcripts has become an increasingly common posttranscriptional modification for the generation of proteins with altered biological functions. Although a great deal of progress has been achieved concerning the identification of mRNA substrates containing inosine residues and the enzymatic activities responsible for such editing events, it has been suggested that the original observations of inosine occurrence in tRNAs are not merely "RNA modifications," but a natural extension of the editing events seen for nucleus-encoded messenger RNAs. Preliminary studies suggesting that ADAR2 may be able to selectively modify several yeast tRNA^{Asp} variants (Grosjean, personal communication) provide additional data in support of this hypothesis.

To date, all known substrates for RNA editing have been identified based on A-to-G nucleotide discrepancies between genomic and cDNA sequences, reflecting conversion of a genomically encoded adenosine to an inosine in the mature mRNA. While the identification of such substrates has provided invaluable information concerning the sequence and structural requirements necessary for A-to-I conversion, the serendipitous nature of such substrate identification has raised numerous questions regarding the number of RNAs regulated by this posttranscriptional modification. Toward the goal of finding a systematic strategy for the identification of novel deaminase substrates, Morse and Bass (1997) have developed an assay to identify additional inosine-containing RNAs based on the properties of RNase T_1 and its ability to specifically cleave RNA after both guanosine and inosine residues. Despite their structural similarity, RNAs treated with glyoxal form stable adducts with guanosine but not inosine, resulting in resistance to RNase T_1 cleavage after the guanosine:glyoxal adduct. The selective cleavage of mRNAs after inosine residues provides a powerful strategy for identifying novel transcripts that undergo A-to-I conversion. It is anticipated that the identification of additional inosine-containing RNAs and the characterization of regulatory sequences required for A-to-I conversion will dramatically improve our ability to identify both the RNA sequences and structural features leading to the modification of specific adenosine residues.

While some of the information required for adenosine target specificity is thought to be inherent to the RNA, the identification of multiple dsRNA adenosine deaminases with differing substrate specificities has demonstrated that characterization of *trans*-acting factors is as important as the definition of *cis*-active regulatory sequences. The posttranscriptional generation of multiple ADAR protein isoforms by alternative splicing and the occurrence of multiple ADAR-encoding genes provide diverse strategies for the generation of substrate specificity and tissue-specific regulation of ADAR expression. Other than the interferon-dependent induction of ADAR1 expression (Patterson and Samuel, 1995; Patterson et al., 1995), little is known of the cellular regulation of ADAR expression or activity. Posttranslational modifications of ADAR protein isoforms also represent a plausible strategy by which to affect the extent of deaminase activity or specificity, as does specific interactions with unidentified regulatory proteins.

The last 10 years of investigative inquiry into the mechanisms of A-to-I conversion have provided dramatic advances in our understanding of the RNA substrates, enzymatic activities, and biochemical mechanisms underlying this type of RNA editing. While future work will further define the cellular and molecular basis for dsRNA adenosine deamination, the functional consequences of subtle A-to-I changes in mRNAs and the alterations in amino acid sequence for resultant proteins are bound to be profound.

Acknowledgments. Work in the authors' laboratory on RNA editing is supported by grants (NS33323 and NS35891) from the National Institutes of Health.

REFERENCES

Bass, B. L., H. Weintraub, R. Cattaneo, and M. A. Billeter. 1989. Biased hypermutation of viral RNA genomes could be due to unwinding/modification of double-stranded RNA. *Cell* 56:331.

Bass, B. L., and H. Weintraub. 1987. A developmentally regulated activity that unwinds RNA duplexes. *Cell* 48:607–613.

Bass, B. L., and H. Weintraub. 1988. An unwinding activity that covalently modifies its double-stranded RNA substrate. *Cell* 55:1089–1098.

Bass, B. L. 1993. RNA editing, p. 383–418. *In* R. Gesteland and J. Atkins (ed.), *The RNA World*. Cold Spring Harbor Laboratory Press, Cold Spring Harbor, N.Y.

Bass, B. L. 1997. RNA editing and hypermutation by adenosine deamination. *Trends Biochem. Sci.* 22:157–162.

Bass, B. L, K., Nishikura, W. Keller, P. H. Seeburg, R. B. Emeson, M. A. O'Connell, C. E. Samuel, and A. Herbert. 1997. A standardized nomenclature for adenosine deaminases that act on RNA. *RNA* 3:947–949.

Betts, L., S. Xiang, S. A. Short, R. Wolfenden, and C. W. Carter, Jr. 1994. Cytidine deaminase. The 2.3 Å crystal structure of an enzyme: transition-state analog complex. *J. Mol. Biol.* 235:635–656.

Bezanilla, F., and E. Stefani. 1994. Voltage-dependent gating of ionic channels. *Annu. Rev. Biophys. Biomol. Struct.* 23:819–846.

Billeter, M. A., and R. Cattaneo. 1991. Molecular biology of defective measles virus in the human central nervous system, p. 323–345. *In* D. W. Kingsbury (ed.), *The Paramyxoviruses*. Plenum Press, New York, N.Y.

Billeter, M. A., R. Cattaneo, P. Spielhofer, K. Kaelin, M. Huber, A. Schmid, K. Baczko, and V. ter Meulen. 1994. Generation and properties of measles virus mutations typically associated with subacute sclerosing panencephalitis. *Ann. N. Y. Acad. Sci.* 724:367–377.

Birnbaumer, L. 1992. Receptor-to-effector signaling through G proteins: roles for beta gamma dimers as well as alpha subunits. *Cell* 71:1069–1072.

Björk, G. R. 1995. Biosynthesis and function of modified nucleosides, p. 165–205. *In* D. Söll and U. L. RajBhandary (ed.), *tRNA: Structure, Biosynthesis, and Function*. American Society for Microbiology, Washington, D.C.

Bliss, T. V. P., and G. L. Collingridge. 1993. A synaptic model of memory: long-term potentiation in the hippocampus. *Nature* 361:31–39.

Bonino, F., K. H. Heermann, M. Rizzetto, and W. H. Gerlich. 1986. Hepatitis delta virus: protein composition of delta antigen and its hepatitis B virus-derived envelope. *J. Virol.* 58:945–950.

Boulter, J., M. Hollmann, A. O'Shea-Greenfield, M. Hartley, E. Deneris, C. Maron, and S. Heinemann. 1989. Molecular cloning and functional expression of glutamate receptor subunit genes. *Science* 249:1033–1037.

Burd, C. G., and G. Dreyfuss. 1994. Conserved structures and diversity of functions of RNA-binding proteins. *Science* 265:615–621.

Burnashev, N., A. Khodorova, P. Jonas, P. J. Helm, W. Wisden, H. Monyer, P. H. Seeburg, and B. Sakmann. 1992a. Calcium-permeable AMPA-kainate receptors in fusiform cerebellar glial cells. *Science* 256:1566–1570.

Burnashev, N., H. Monyer, P. H. Seeburg, and B. Sakmann. 1992b. Divalent ion permeability of AMPA receptor channels is dominated by the edited form of a single subunit. *Neuron* 8:89–98.

Burns, C. M., H. Chu, S. M. Rueter, L. K. Hutchinson, H. Canton, E. Sanders-Bush, and R. B. Emeson. 1997. Regulation of serotonin-2C receptor G-protein coupling by RNA editing. *Nature* 387:303–308.

Casey, J. L., K. F. Bergmann, T. L. Brown, and J. L. Gerin. 1992. Structural requirements for RNA editing in hepatitis δ virus: evidence for a uridine-to-cytidine editing mechanism. *Proc. Natl. Acad. Sci. USA* 89:7149–7153.

Casey, J. L., and J. L. Gerin. 1995. Hepatitis D virus RNA editing: specific modification of adenosine in the antigenomic RNA. *J. Virol.* 69:7593–7600.

Cattaneo, R., A. Schmid, D. Eschle, K. Baczko, V. ter Meulen, and M. A. Billeter. 1988. Biased hypermutation and other genetic changes in defective measles viruses in human brain infections. *Cell* 55:255–265.

Cattaneo, R. 1994. Biased (A→I) hypermutation of animal RNA virus genomes. *Curr. Opin. Genet. Dev.* 4:895–900.

Chao, M., S.-Y. Hsieh, and J. Taylor. 1990. Role of two forms of hepatitis delta virus antigen: evidence for a mechanism of self-limited genome replication. *J. Virol.* 64:5066–5069.

Chen, P.-J., G. Kalpana, J. Goldberg, W. Mason, B. Werner, J. Gerin, and J. Taylor. 1986. The structure and replication of the genome of the hepatitis delta virus. *Proc. Natl. Acad. Sci. USA* 83:8774–8778.

Choi, D. W., and S. M. Rothman. 1990. The role of glutamate neurotoxicity in hypoxic-ischemic neuronal death. *Annu. Rev. Neurosci.* 13:171–182.

Collingridge, G. L., and W. Singer. 1990. Excitatory amino acid receptors and synaptic plasticity. *Trends Pharmacol. Sci.* 11:290–296.

Davidson, N. O. 1993. Apolipoprotein B mRNA editing: a key controlling element targeting fats to proper tissue. *Ann. Med.* 25:539–543.

Dingledine, R., R. I. Hume, and S. F. Heinemann. 1992. Structural determinants of barium permeation and rectification in non-NMDA glutamate receptor channels. *J. Neurosci.* 12:4080–4087.

Dubinsky, J. M., and S. M. Rothman. 1991. Intracellular calcium concentrations during "chemical hypoxia" and excitotoxic neuronal injury. *J. Neurosci.* 11:2545–2551.

Dubovsky, S. L., and M. Thomas. 1995. Serotonergic mechanisms and current and future psychiatric practice. *J. Clin. Psychiatry* 2:38–48.

Egebjerg, J., and S. F. Heinemann. 1993. Ca^{2+} permeability of unedited and edited versions of the kainate selective glutamate receptor GluR6. *Proc. Natl. Acad. Sci. USA* 90:755–759.

Egebjerg, J., V. Kukekov, and S. F. Heinemann. 1994. Intron sequence directs RNA editing of the glutamate receptor subunit GluR2 coding sequence. *Proc. Natl. Acad. Sci. USA* 91:10270–10274.

Gerber, A. Personal communication.

Gerber, A., M. A. O'Connell, and W. Keller. 1997. Two forms of human double-stranded RNA-specific editase 1 (hRED1) generated by the insertion of an Alu cassette. *RNA* 3:453–463.

Gilbertson, T. A., R. Scobey, and M. Wison. 1991. Permeation of calcium ions through non-NMDA glutamate channels in retinal bipolar cells. *Science* 251:1613–1615.

Grosjean, H. Personal communication.

Grosjean, H., S. Auxilien, F. Constantinesco, C. Simon, Y. Corda, H. F. Becker, D. Foiret, A. Morin, Y. X. Jin, M. Fournier, and J. L. Fourrey. 1996. Enzymatic conversion of adenosine to inosine and to N1-methylinosine in transfer RNAs: a review. *Biochimie* 78:488–501.

Hajjar, A. M., and M. L. Linial. 1995. Modification of retroviral RNA by double-stranded RNA adenosine deaminase. *J. Virol.* 69:5878–5882.

Herb, A., M. Higuchi, R. Sprengel, and P. H. Seeburg. 1996. Q/R site editing in kainate receptor GluR5 and GluR6 pre-mRNAs requires distant intronic sequences. *Proc. Natl. Acad. Sci. USA* 93:1875–1880.

Herbert, A., K. Lowenhaupt, J. Spitzner, and A. Rich. 1995. Chicken double-stranded RNA adenosine deaminase has apparent specificity for Z-DNA. *Proc. Natl. Acad. Sci. USA* **92:** 7550–7554.

Herbert, A., J. Alfken, Y. G. Kim, I. S. Mian, K. Nishikura, and A. Rich. 1997. A Z-DNA binding domain present in the human editing enzyme, double-stranded RNA adenosine deaminase. *Proc. Natl. Acad. Sci. USA* **94:**8421–8426.

Higuchi, M., F. N. Single, M. Köhler, B. Sommer, R. Sprengel, and P. H. Seeburg. 1993. RNA editing of AMPA receptor subunit GluR-B: a base-paired intron-exon structure determines position and efficiency. *Cell* **75:**1361–1370.

Holley, R. W., G. A. Everett, J. T. Madison, and A. Zamir. 1965. Nucleotide sequences in the yeast alanine transfer RNA. *J. Biol. Chem.* **240:**2122–2127.

Hollmann, M., A. O'Shea-Greenfield, S. W. Rogers, and S. Heinemann. 1989. Cloning by functional expression of a member of the glutamate receptor family. *Nature* **342:**643–648.

Hollmann, M., M. Hartley, and S. Heinemann. 1991. Calcium permeability of kainate-AMPA-gated glutamate channels depends on subunit composition. *Science* **252:**851–854.

Hollmann, M., C. Maron, and S. Heinemann. 1994. N-glycosylation site tagging suggests a three transmembrane domain topology for the glutamate receptor GluR1. *Neuron* **13:** 1331–1343.

Hough, R. F., and B. L. Bass. 1994. Purification of the *Xenopus laevis* double-stranded RNA adenosine deaminase. *J. Biol. Chem.* **269:**9933–9939.

Hough, R. F., and B. L. Bass. 1997. Analysis of Xenopus dsRNA adenosine deaminase cDNAs reveals similarities to DNA methyltransferases. *RNA* **3:**356–370.

Hoyer, D., D. E. Clarke, J. R. Fozard, P. R. Hartig, G. R. Martin, E. J. Mylecharane, P. R. Saxena, and P. P. Humphrey. 1994. International Union of Pharmacology classification of receptors for 5-hydroxytryptamine (serotonin). *Pharmacol. Rev.* **46:** 157–203.

Hume, I. R., R. Dingledine, and S. F. Heinemann. 1991. Identification of a site in glutamate receptor subunits that controls calcium permeability. *Science* **253:**1028–1031.

Hurst, S. R., R. F. Hough, P. J. Aruscavage, and B. L. Bass. 1995. Deamination of mammalian glutamate receptor RNA by *Xenopus* dsRNA adenosine deaminase: similarities to in vivo RNA editing. *RNA* **1:**1051–1060.

Jan, L. Y., and Y. N. Jan. 1997. Cloned potassium channels from eukaryotes and prokaryotes. *Annu. Rev. Neurosci.* **20:**91–123.

Jonas, P., and B. Sakmann. 1992. Glutamate receptor channels in isolated patches from CA1 and CA3 pyramidal cells of rat hippocampal slices. *J. Physiol.* **455:**143–147.

Jonas, P., C. Racca, B. Sakmann, P. H. Seeburg, and H. Monyer. 1994. Differences in Ca^{2+} permeability of AMPA-type glutamate receptor channels in neocortical neurons caused by differential GluR-B subunit expression. *Neuron* **12:**1281–1289.

Julius, D. 1991. Molecular biology of serotonin receptors. *Annu. Rev. Neurosci.* **14:**335–360.

Julius, D., A. B. MacDermott, R. Axel, and T. M. Jessell. 1988. Molecular characterization of a functional cDNA encoding the serotonin 1c receptor. *Science* **241:**558–564.

Keinanen, K., W. Wisden, B. Sommer, P. Werner, A. Herb, T. A. Verdoorn, B. Sakmann, and P. H. Seeburg. 1990. A family of AMPA-selective glutamate receptors. *Science* **249:**556–560.

Kim, U., T. L. Garner, T. Sanford, D. Speicher, J. M. Murray, and K. Nishikura. 1994a. Purification and characterization of double-stranded RNA adenosine deaminase from bovine nuclear extracts. *J. Biol. Chem.* **269:**13480–13489.

Kim, U., Y. Wang, T. Sanford, Y. Zeng, and K. Nishikura. 1994b. Molecular cloning of cDNA for double-stranded RNA adenosine deaminase, a candidate enzyme for nuclear RNA editing. *Proc. Natl. Acad. Sci. USA* **91:**11457–11461.

Koh, D. S., J. R. P. Geiger, P. Jonas, and B. Sakmann. 1995. Ca^{2+} permeable AMPA and NMDA receptor channels in basket cells of rat hippocampal dentate gyrus. *J. Physiol.* **485:**383–402.

Köhler, M., N. Burnashev, B. Sakmann, and P. H. Seeburg. 1993. Determinants of Ca^{2+} permeability in both TM1 and TM2 of high affinity kainate receptor channels: diversity by RNA editing. *Neuron* **10:**491–500.

Kumar, M., and G. G. Carmichael. 1997. Nuclear antisense RNA induces extensive adenosine modifications and nuclear retention of target transcripts. *Proc. Natl. Acad. Sci. USA* **94:**3542–3547.

Kuo, M. Y.-P., M. Chao, and J. Taylor. 1989. Initiation of replication of the human hepatitis delta virus genome from clone DNA: role of delta antigen. *J. Virol.* **63:**1945–1950.

Lai, F., R. Drakas, and K. Nishikura. 1995. Mutagenic analysis of double-stranded RNA adenosine deaminase, a candidate enzyme for RNA editing of glutamate-gated ion channel transcripts. *J. Biol. Chem.* **270:**17098–17105.

Lai, F., C. X. Chen, K. C. Carter, and K. Nishikura. 1997. Editing of glutamate receptor B subunit ion channel RNAs by four alternatively spliced DRADA2 double-stranded RNA adenosine deaminases. *Mol. Cell. Biol.* **17:**2413–2424.

Lino, M., S. Ozawa, and K. Tsuzuki. 1990. Permeation of calcium through excitatory amino acid receptor channels in cultured rat hippocampal neurons. *J. Physiol.* **424:**151–165.

Liu, Y., C. X. George, J. B. Patterson, and C. E. Samuel. 1997. Functionally distinct double-stranded RNA-binding domains associated with alternative splice site variants of the interferon-inducible double-stranded RNA-specific adenosine deaminase. *J. Biol. Chem.* **272:**4419–4428.

Lomeli, H., J. Mosbacher, T. Melcher, T. Hoger, J. R. P. Geiger, T. Kuner, H. Monyer, M. Higuchi, A. Bach, and P. H. Seeburg. 1994. Control of kinetic properties of AMPA receptor channels by nuclear RNA editing. *Science* **266:**1709–1713.

Luo, G., M. Chao, S.-Y. Hsieh, C. Sureau, K. Nishikura, and J. Taylor. 1990. A specific base transition occurs on replicating hepatitis delta virus RNA. *J. Virol.* **64:**1021–1027.

Ma, J., R. Qian, F. M. Rausa, and K. J. Colley. 1997. Two naturally occurring alpha2,6-sialyltransferase forms with a single amino acid change in the catalytic domain differ in their catalytic activity and proteolytic processing. *J. Biol. Chem.* **272:**672–679.

Maas, S., T. Melcher, A. Herb, P. H. Seeburg, W. Keller, S. Krause, M. Higuchi, and M. A. O'Connell. 1996. Structural requirements for RNA editing in glutamate receptor pre-mRNAs by recombinant double-stranded RNA adenosine deaminase. *J. Biol. Chem.* **271:**12221–12226.

Maas, S., T. Melcher, and P. H. Seeburg. 1997. Mammalian RNA-dependent deaminases and edited mRNAs. *Curr. Opin. Cell. Biol.* **9:**343–349.

Melcher, T., S. Maas, M. Higuchi, W. Keller, and P. H. Seeburg. 1995. Editing of α-amino-3-hydroxy-5-methylisoxazole-4-propionic acid receptor GluR-B pre-mRNA *in vitro* reveals site-selective adenosine to inosine conversion. *J. Biol. Chem.* **270:** 8566–8570.

Melcher, T., S. Maas, A. Herb, R. Sprengel, P. H. Seeburg, and M. Higuchi. 1996a. A mammalian RNA editing enzyme. *Nature* **379:**460–464.

Melcher, T., S. Maas, A. Herb, R. Sprengel, M. Higuchi, and P. H. Seeburg. 1996b. RED2, a brain-specific member of the RNA-specific adenosine deaminase family. *J. Biol. Chem.* **271:** 31795–31798.

Monyer, H., P. H. Seeburg, and W. Wisden. 1991. Glutamate-operated channels: developmentally early and mature forms arise by alternative splicing. *Neuron* **6:**799–810.

Morse, D. P., and B. L. Bass. 1997. Detection of inosine in messenger RNA by inosine-specific cleavage. *Biochemistry* 36:8429–8434.

Murphy, D. G., K. Dimock, and C. Y. Kang. 1991. Numerous transitions in human parainfluenza virus 3 RNA recovered from persistently infected cells. *Virology* 181:760–763.

Nakanishi, N., N. A. Shneider, and R. Axel. 1990. A family of glutamate receptor genes: evidence for the formation of heteromultimeric receptors with distinct channel properties. *Toxicon* 28:1333–1346.

Nishikura, K., C. Yoo, U. Kim, J. M. Murray, P. A. Estes, F. E. Cash, and S. A. Liebhaber. 1991. Substrate specificity of the dsRNA unwinding/modifying activity. *EMBO J.* 10:3523–3532.

O'Connell, M. A., and W. Keller. 1994. Purification and properties of double-stranded RNA-specific adenosine deaminase from calf thymus. *Proc. Natl. Acad. Sci. USA* 91:10596–10600.

O'Connell, M. A., S. Krause, M. Higuchi, J. J. Hsuan, N. F. Totty, A. Jenny, and W. Keller. 1995. Cloning of cDNAs encoding mammalian double-stranded RNA-specific adenosine deaminase. *Mol. Cell. Biol.* 15:1389–1397.

O'Connell, M. A., A. Gerber, and W. Keller. 1997. Purification of human double-stranded RNA-specific editase 1 (hRED1) involved in editing of brain glutamate receptor B pre-mRNA. *J. Biol. Chem.* 272:473–478.

Ogura, A., K. Akita, and Y. Kudo. 1990. Non-NMDA receptor mediates cytoplasmic calcium elevation in cultured hippocampal neurones. *Neurosci. Res.* 9:103–113.

O'Hara, P. J., S. T. Nichol, F. M. Horodyski, and J. J. Holland. 1984. Vesicular stomatitis virus defective interfering particles can contain extensive genomic sequence rearrangements and base substitutions. *Cell* 36:915–924.

Olney, J. W. 1990. Excitotoxic amino acids and neruopsychiatric disorders. *Annu. Rev. Pharmacol. Toxicol.* 30:47–71.

Pandey, S. C., J. M. Davis, and G. N. Pandey. 1995. Phosphoinositide system-linked serotonin receptor subtypes and their pharmacological properties and clinical correlates. *J. Psychiatry Neurosci.* 20:215–225.

Patterson, J. B., and C. E. Samuel. 1995. Expression and regulation by interferon of a double-stranded-RNA-specific adenosine deaminase from human cells: evidence for two forms of the deaminase. *Mol. Cell. Biol.* 15:5376–5388.

Patterson, J. B., D. C. Thomis, S. L. Hans, and C. E. Samuel. 1995. Mechanism of interferon action: double-stranded RNA-specific adenosine deaminase from human cells is inducible by alpha and gamma interferons. *Virology* 210:508–511.

Patton, D. E., T. Silva, and F. Bezanilla. 1997. RNA editing generates a diverse array of transcripts encoding squid Kv2 K+ channels with altered functional properties. *Neuron* 19:711–722.

Petschek, J. P., M. J. Mermer, M. R. Scheckelhoff, A. A. Simone, and J. C. Vaughn. 1996. RNA editing in Drosophila 4f-rnp gene nuclear transcripts by multiple A-to-G conversions. *J. Mol. Biol.* 259:885–890.

Pin, J. P., C. Joly, S. F. Heinemann, and J. Bockaert. 1994. Domains involved in the specificity of G protein activation in phospholipase C-coupled metabotropic glutamate receptors. *EMBO J.* 13:342–348.

Pizzi, M., M. Ribola, A. Valerio, M. Memo, and P. Spano. 1991. Various Ca2+ entry blockers prevent glutamate-induced neurotoxicity. *Eur. J. Pharmacol.* 209:169–173.

Polson, A. G., P. F. Crain, S. C. Pomerantz, J. A. McCloskey, and B. L. Bass. 1991. The mechanism of adenosine to inosine conversion by the double-stranded RNA unwinding/modifying activity: a high-performance liquid chromatography-mass spectrometry analysis. *Biochemistry* 30:11507–11514.

Polson, A. G., and B. L. Bass. 1994. Preferential selection of adenosines for modification by double-stranded RNA adenosine deaminase. *EMBO J.* 13:5701–5711.

Polson, A. G., B. L. Bass, and J. L. Casey. 1996. RNA editing of hepatitis delta virus antigenome by dsRNA-adenosine deaminase. *Nature* 380:454–456.

Ponzetto, A., P. J. Cote, H. Popper, B. H. Hoyer, W. T. London, E. C. Ford, F. Bonino, R. H. Purcell, and J. L. Gerin. 1984. Transmission of the hepatitis B virus-associated δ agent to the eastern woodchuck. *Proc. Natl. Acad. Sci. USA* 81:2208–2212.

Pruss, R. M., R. L. Akeson, M. M. Racke, and J. L. Wilburn. 1991. Agonist-activated cobalt uptake identifies divalent cation-permeable kainate receptors on neurones and glial cells. *Neuron* 7:509–518.

Randall, R. D., and S. A. Thayer. 1992. Glutamate-induced calcium transient triggers delayed calcium overload and neurotoxicity in rat hippocampal neurons. *J. Neurosci.* 12:1882–1895.

Rebagliati, M. R., and D. A. Melton. 1987. Antisense RNA injections in fertilized frog eggs reveal an RNA duplex unwinding activity. *Cell* 48:599–605.

Rueda, P., B. García-Barreno, and J. A. Melero. 1994. Loss of conserved cysteine residues in the attachment (G) glycoprotein of two human respiratory syncytial virus escape mutants that contain multiple A-G substitutions (hypermutations). *Virology* 198:653–662.

Rueter, S. M., C. M. Burns, S. A. Coode, P. Mookherjee, and R. B. Emeson. 1995. Glutamate receptor RNA editing in vitro by enzymatic conversion of adenosine to inosine. *Science* 267:1491–1494.

Ryu, W.-S., M. Bayer, and J. Taylor. 1992. Assembly of hepatitis delta virus particles. *J. Virol.* 66:2310–2315.

Saccomanno, L., and B. L Bass. 1994. The cytoplasm of *Xenopus* oocytes contains a factor that protects double-stranded RNA from adenosine-to-inosine modification. *Mol. Cell. Biol.* 14:5425–5432.

Sakimura, K., H. Bujo, E. Kushiya, K. Araki, M. Yamazaki, H. Meguro, A. Warashina, S. Numa, and M. Mishina. 1990. Functional expression from cloned cDNAs of glutamate receptor species responsive to kainate and quisqualate. *FEBS Lett.* 272:73–80.

Scadden, A. D., and C. W. Smith. 1997. A ribonuclease specific for inosine-containing RNA: a potential role in antiviral defence? *EMBO J.* 16:2140–2149.

Schumacher, J. M., K. Lee, S. Edelhoff, and R. E. Braun. 1995. Distribution of Tenr, an RNA-binding protein, in a lattice-like network within the spermatid nucleus in the mouse. *Biol. Reprod.* 52:1274–1283.

Shih, T. M., and A. L. Goldin. 1997. Topology of the Shaker potassium channel probed with hydrophilic epitope insertions. *J. Cell Biol.* 136:1037–1045.

Simpson, L., and R. B. Emeson. 1996. RNA editing. *Annu. Rev. Neurosci.* 19:27–52.

Slesinger, P. A., Y. N. Jan, and L. Y. Jan. 1993. The S4-S5 loop contributes to the ion-selective pore of potassium channels. *Neuron* 11:739–749.

Sommer, B., K. Keinanen, T. A. Verdoorn, W. Wisden, N. Burnashev, A. Herb, M. Köhler, T. Takagi, B. Sakmann, and P. H. Seeburg. 1990. Flip and flop: a cell-specific functional switch in glutamate-operated channels of the CNS. *Science* 249:1580–1585.

Sommer, B., M. Köhler, R. Sprengel, and P. H. Seeburg. 1991. RNA editing in brain controls a determinant of ion flow in glutamate-gated channels. *Cell* 67:11–19.

Tecott, L. H., L. M. Sun, S. F. Akana, A. M. Strack, D. H. Lowenstein, M. F. Dallman, and D. Julius. 1995. Eating disorder and epilepsy in mice lacking 5-HT2c serotonin receptors. *Nature* 374:542–546.

Teitler, M., and K. Herrick-Davis. 1994. Multiple serotonin receptor subtypes: molecular cloning and functional expression. *Crit. Rev. Neurobiol.* 8:175–188.

Verdoorn, T. A., N. Burnashev, H. Monyer, P. H. Seeburg, and B. Sakmann. 1991. Structural determinants of ion flow through recombinant glutamate receptor channels. *Science* **252:**1715–718.

Wagner, R. W., and K. Nishikura. 1988. Cell cycle expression of RNA duplex unwindase activity in mammalian cells. *Mol. Cell. Biol.* **8:**770–777.

Wagner, R. W., J. E. Smith, B. S. Cooperman, and K. Nishikura. 1989. A double-stranded RNA unwinding activity introduces structural alterations by means of adenosine to inosine conversions in mammalian cells and *Xenopus* eggs. *Proc. Natl. Acad. Sci. USA* **86:**2647–2651.

Wagner, R. W., C. Yoo, L. Wrabetz, J. Kamholz, J. Buchhalter, N. F. Hassan, K. Khalili, S. U. Kim, B. Perussia, and F. A. McMorris. 1990. Double-stranded RNA unwinding and modifying activity is detected ubiquitously in primary tissues and cell lines. *Mol. Cell. Biol.* **10:**5586–5590.

Wang, K.-S., Q.-L. Choo, A. J. Weiner, J.-H. Ou, R. C. Najarian, R. M. Thayer, G. T. Mullenbach, K. J. Denniston, J. L. Gerin, and M. Houghton. 1986. Structure, sequence and expression of the hepatitis delta viral genome. *Nature* **323:**508–513.

Wilson, R., R. Ainscough, K. Anderson, C. Baynes, M. Berks, J. Bonfield, J. Burton, M. Connell, T. Copsey, and J. Cooper. 1994. 2.2 Mb of contiguous nucleotide sequence from chromosome III of *C. elegans*. *Nature* **368:**32–38.

Wong, S. K., E. M. Parker, and E. M. Ross. 1990. Chimeric muscarinic cholinergic: beta-adrenergic receptors that activate Gs in response to muscarinic agonists. *J. Biol. Chem.* **265:**6219–6224.

Wu, J.-C., P.-J. Chen, M. Y.-P. Kuo, S.-D. Lee, D.-S. Chen, and L.-P. Ting. 1991. Production of hepatitis delta virus and suppression of helper hepatitis B virus in a human hepatoma cell line. *J. Virol.* **65:**1099–1104.

Wu, T. T., H. J. Netter, V. Bichko, D. Lazinski, and J. Taylor. 1994. RNA editing in the replication cycle of human hepatitis delta virus. *Biochimie* **76:**1205–1208.

Yang, J.-H., P. Sklar, R. Axel, and T. Maniatis. 1995. Editing of glutamate receptor subunit B pre-mRNA in vitro by site-specific deamination of adenosine. *Nature* **374:**77–81.

Yang, J. H., P. Sklar, R. Axel, and T. Maniatis. 1997. Purification and characterization of a human RNA adenosine deaminase for glutamate receptor B pre-mRNA editing. *Proc. Natl. Acad. Sci. USA* **94:**4354–4359.

Yellen, G., M. E. Jurman, T. Abramson, and R. MacKinnon. 1991. Mutations affecting internal TEA blockade identify the probable pore-forming region of a K+ channel. *Science* **251:**939–942.

Zheng, H., T.-B. Fu, D. Lazinski, and J. Taylor. 1992. Editing of the genomic RNA of human hepatitis delta virus. *J. Virol.* **66:**4693–4697.

Chapter 20

Nucleoside Deaminases for Cytidine and Adenosine: Comparison with Deaminases Acting on RNA

CHARLES W. CARTER, JR.

NUCLEOSIDE DEAMINATION AND BASE CONVERSION RNA EDITING

Hydrolytic deamination of nucleosides is of considerable interest both from the standpoint of mechanistic enzymology and from that of biology. Fortunately, it represents a substantial barrier to the interconversion of nucleotide bases in aqueous solution. The activation free energies for uncatalyzed deamination of cytosine or adenine in water are so high that these bases remain stable for centuries at 37°C with respect to the more stable deamination products, uracyl and hypoxanthine. To facilitate the chemical engineering of the purine and pyrimidine nucleosides, nature evolved a broad spectrum of deaminases to catalyze these interconversions. Most, if not all, of these use zinc to activate water to form a hydroxide nucleophile. The existence of these enzymes raises an intriguing question. How are A and C bases in nucleic acids protected from wholesale deamination? The answer, from the X-ray crystal structures and mutational analysis of adenosine and cytidine deaminases (Betts et al., 1994; Carlow et al., in press; Wilson et al., 1991) appears to be that the nucleoside deaminases have incorporated into their very mechanisms a requirement for binding both 3' and 5' hydroxyl groups of the ribose, thereby enclosing the nucleoside entirely within the active site.

The phenomenon of base-conversion editing (Bass, 1993; Scott, 1995) (also see Chapters 16–19 in this volume), in which A and C residues in RNA molecules are converted to I and U by deamination, has recently turned this mechanistic question around. How, after pyrimidine and purine deaminases had evolved to avoid acting on RNA, did they learn to transcend the carefully constructed barrier to oligomeric substrates and gain access to a very discrete set of substrates, without reopening the Pandora's box so successfully avoided? Although firm answers to these questions must await structural studies on the editing deaminases themselves, general understanding of the range of possible answers has accumulated from models based on sequence homologies for both C-to-U (Navaratnam et al., 1997) and A-to-I editing (Hough and Bass, 1997).

The editing deaminases have apparently exploited at least two different strategies to ensure the selection of only the intended bases in target RNAs. C-to-U editing of the apolipoprotein B mRNA apparently involves a dimeric core enzyme with extensive structural homology to the *Escherichia coli* cytidine deaminase (ECCDA). Models suggest that specificity involves simultaneous recognition of the targeted C and a downstream U by the two active sites of the dimer (Navaratnam et al., 1995; Navaratnam et al., 1998; Scott, 1995). The likely RNA conformation includes a stem-loop exposing the C and presumably a similar presentation of the U. In contrast, deamination of adenine in mRNA apparently requires double-stranded RNA substrates, and may involve transient "flipping" of the targeted base from the double helix (Hough and Bass, 1997).

A substantial fraction of base-conversion editing of both cytosine and adenosine in loop regions of mitochondrial tRNA is probably due to deamination (see Chapter 16 by Price and Gray). Enzymatic pathways for the conversion of adenosine to inosine (Auxilien et al., 1996) and N^1-methylinosine (Grosjean et al., 1995) have been well characterized. The latter process involves methylation from S-adenosylmethionine (SAM) and deamination. Interestingly, methylation precedes deamination in archaea,

---

Charles W. Carter, Jr. • Department of Biochemistry and Biophysics, CB 7260, University of North Carolina at Chapel Hill, Chapel Hill, North Carolina 27599-7260.

whereas in eukarya the two steps are reversed (Grosjean et al., 1996, 1995). At least one of the enzymes responsible for deamination of tRNA bases has been partially purified (Auxilien et al., 1996). The yeast protein HRA400 appears to be homologous to enzymes that carry out adenosine deamination in double-stranded regions of mRNAs (Gerber et al., 1997). Aside from these suggestive leads, little is known of these deaminase structures.

I will review here the structure of *E. coli* cytidine deaminase and implications for evolutionary relationships between members of the cytidine deaminase superfamily. I will then summarize details of the ECCDA reaction mechanism, with particular reference to the role of conserved residues and aspects of substrate specificity relating to the nucleoside modifying and editing enzymes. I will conclude with a discussion of models for RNA editing deaminases derived from sequence homologies and structural studies.

THE CYTIDINE DEAMINASE (SUPER)FAMILY

Deaminases acting on the cytosine ring form a family whose members are apparently related by genetic ancestry (Betts et al., 1994; Navaratnam et al., 1993a) (also see Chapter 18 by Chang et al.). They all have the distinctive sequence signature shown in Fig. 1, containing all critical residues involved in catalysis. Active-site contributors occur in two clusters, separated by about 30 residues. The first cluster contains a single zinc ligand and a crucial glutamic acid residue that participates in transition-state stabilization and mediates the two proton transfer steps. The second contains the other two zinc ligands, which evidently are always cysteine residues. The first of these

is preceeded by an invariant proline residue, whose role appears to be to orient the amino group to be removed by hydration (Betts et al., 1994; Xiang et al., 1996). APOBEC-1, the sole representative of what will likely be a family of editing cytidine deaminases acting on RNA substrates, shares this signature and, as outlined in greater detail in the following section, it likely has significant tertiary and quaternary homology to the *E. coli* cytidine deaminase, ECCDA.

ECCDA, the only family member whose structure is known, is a dimer whose 31,540-kDa subunits have very strong internal twofold symmetry (Fig. 2) relating nearly identical domains. They are connected by an extended chain running the full length of the molecule from the outside of the first core domain to the outside of the second. Only the N-terminal core domain of ECCDA contains the two stretches of "signature" sequences (Fig. 1). Although these residues provide all important chemical functions for catalysis, the intact active site is actually formed from contributions from the C-terminal domain of the opposing subunit. This composite construction of the active site is apparently conserved in APOBEC-1, the editing deaminase subunit (Navaratnam et al., 1995, 1998).

The dimeric ECCDA structure suggests a homotetramer, owing to the conservation of tertiary structure in the N- and C-terminal core domains (Fig. 2). In fact, another branch of this family, including the human and *Bacillus subtilis* CDAs, are homotetramers whose monomers have approximately the same molecular weight as one core domain in ECCDA. The subunits of the tetrameric CDAs have all the catalytic residues in the signature shown in Fig. 2, and therefore probably also have nearly the same tertiary structure as the ECCDA core domains. Remarkably,

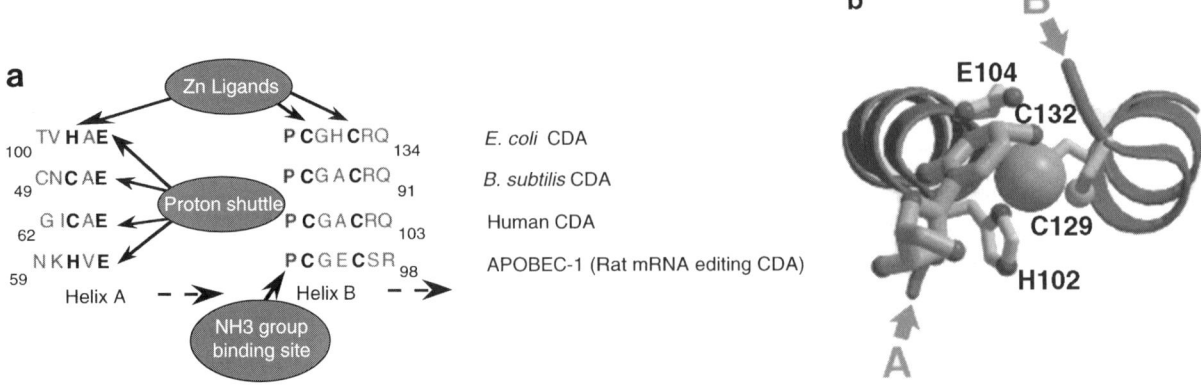

Figure 1. The signature of cytidine deaminases. (a) Conserved active-site residues and their catalytic roles as determined from studies of ECCDA. (b) ECCDA active site consisting of the segments shown in panel a. Residues in boldface in panel a are indicated, as are helices A and B.

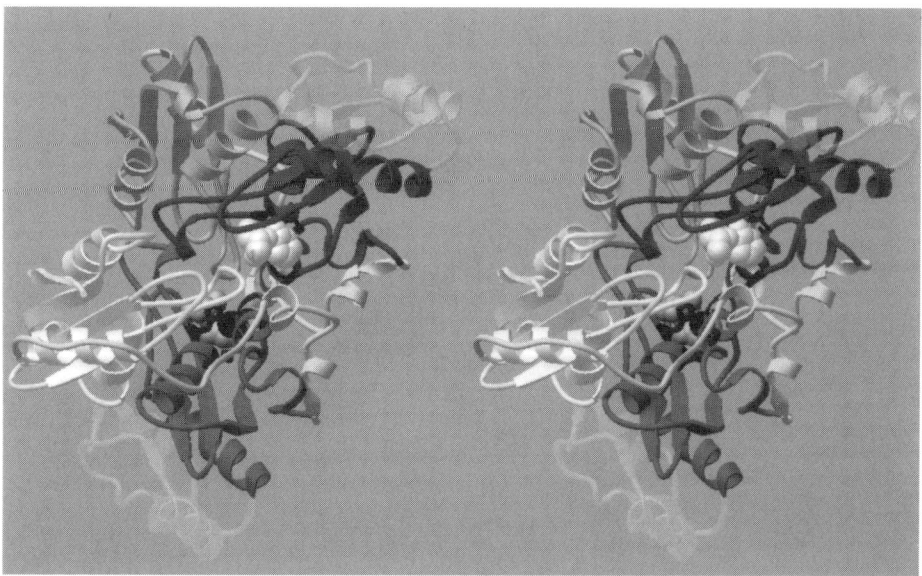

Figure 2. Internal twofold rotation symmetry in the ECCDA monomer lends approximate 222 symmetry to the dimer (a stereo view). The plane separating the monomers is approximately horizontal. The amino-terminal core domain of each monomer is darker gray and contains the active-site residues shown in Fig. 1. Residues 1–49 of each subunit preceeding the first core domain are half-saturated colors, to emphasize relationships between the four core domains in the dimer. The view illustrates the composite active-site construction, drawing on loops from both subunits.

they also have 7–10 extra residues, relative to ECCDA, in a loop immediately preceding the second α-helix, B, containing the PCGXCRQ catalytic signature. Because this loop surrounds the active site and because the *B. subtilis* enzyme has four independent and fully competent active sites (Carlow et al., unpublished data), it is likely that these extra residues comprise a "flap" that closes over the active sites of the tetrameric enzymes, in place of the second monomer in the dimeric ECCDA.

The tetrameric human and *B. subtilis* deaminases use cysteine, rather than histidine, as the first zinc ligand. This substitution chemically is rather significant, but it has essentially no detectable effect on the ligand-binding or catalytic properties of the tetrameric enzymes (Carlow et al., unpublished data). Moreover, the four subunits have equivalent activities and show no detectable interaction. This full independence suggests that the four monomers do not share the composite active-site juxtaposition observed in the dimeric ECCDA and that the tetrameric quaternary structure is likely to differ from that observed in ECCDA.

A scheme for the possible evolution of the cytidine deaminase family is shown in Fig. 3. Two configurations of the same original 12–14-kDa domain appear to have diverged, one branch giving rise to the tetrameric enzymes, with four independent active sites, the other to the dimeric species, with two composite active sites. Homology between ECCDA and APOBEC-1, the catalytic subunit of the complex that edits rat apolipoprotein B mRNA, extends throughout the second core domain of ECCDA. It, too, has a dimeric quaternary structure, apparently with two composite active sites. A more detailed model for APOBEC-1 is described in a subsequent section.

Sequences of the adenosine deaminases involved in RNA editing appear to possess variations of the CDA family signature, and have been postulated to derive from the same ancestral deaminase (see Chapter 18 by Chang et al.) (Backus et al., 1994). Structural evidence for this hypothesis is intriguing, but incomplete. Since such a possible superfamily cannot be easily discounted, we should comment on a problem of nomenclature. An attempt has been made to rationalize the naming of editing adenosine deaminases, calling them ADARs (adenosine deaminases acting on RNA) (Bass et al., 1997). A natural extension of this scheme, referring to cytidine deaminases with RNA substrates as CDARs, may be useful until more extensive structural data are available to form a basis for a more rational naming convention. However, the possibility that all known editing deaminases for both A-to-I and C-to-U editing arose from a common CDA ancestor may eventually complicate this naming scheme.

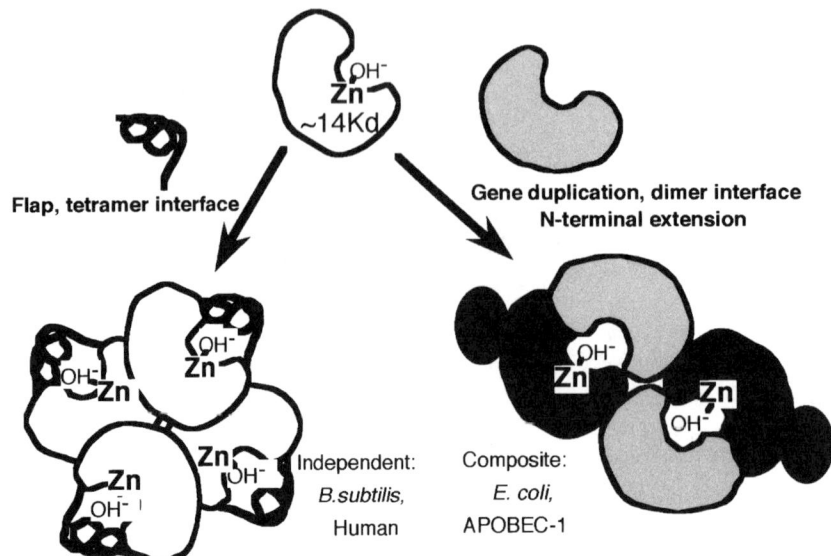

Figure 3. Schematic diagram showing probable evolutionary relationships among cytidine deaminases. The amino-terminal domain of ECCDA has sequence homology to the monomers of the tetrameric CDAs, all of which are likely to have a common ancestor that bound zinc as shown in Fig. 1a.

STRUCTURAL REACTION PROFILE OF *E. coli* CYTIDINE DEAMINASE

X-ray crystal structures have been solved for *E. coli* cytidine deaminase complexed to a series of stable ligands resembling successive stages of the presumed reaction pathway, thereby trapping the enzyme in flagrante delicto. This series constitutes a "structural reaction profile" by which the enzyme is able to accomplish the events summarized in the schematic reaction mechanism in Fig. 4 (Betts et al., 1994; Xiang et al., 1996, 1995, 1997).

Hydrolytic Deamination

Cytidine inserts head-on, deep into the enzyme interior, projecting the 4-NH$_2$ group into a pocket providing a hydrogen bond to the backbone carbonyl oxygen immediately preceding the highly conserved residue, Pro 128 (Xiang et al., 1996). Water is activated to an otherwise rarely observed nucleophilic hydroxide (Lewis et al., in press) by its interaction with zinc bound in an unusual (Cys)$_2$His ligand field of the α-β-α motif fold at the base of the active site (Fig. 1b). Conversion of cytidine to uridine involves two successive steps: hydration of the 3–4 double bond, and subsequent elimination of the leaving ammonia molecule with formation of the keto tautomer of the pyrimidine. A delicate balance of effects lowers the free energy of the chemical transition state by nearly simultaneous protonation of N3 and stabilization of the O4 hydroxyl group. The Glu 104 side chain assists both processes by donating a proton to N3 and sharing an electron with the proton on the hydroxyl group in the first step and by protonating the leaving ammonia group in the second, giving the appropriate tautomer of the product, uridine. Negative charge built up on O4 is accommodated progressively by a "valence buffer" provided by lengthening the bond between the zinc and one of its ligands, Sγ132 (Xiang et al., 1996). The 4-one carbonyl oxygen remains bound to the zinc, and constraints on the ribose bend and distort the glycosidic linkage, forcing the bound product to assume a strained conformation (Xiang et al., 1997). This mechanism attributes functional roles to all of the highly conserved residues in the sequences of CDA family enzymes (see Fig. 1 and Chapter 18 by Chang et al.).

The Inaccessible ECCDA Active Site and Editing Specificity

With respect to the question of C-to-U editing, the most striking aspect of this profile is that the ECCDA active site is inaccessible to solvent throughout, even including the ligand-free enzyme. The active-site domain itself wraps around approximately 60% of pyrimidine ligand bound at the active site. The second monomer in the dimer covers the remaining 40% of the surface area.

Access to the ribose is especially well protected in ECCDA. In all of the bound states, β-structure in residues 78–101 curls around the ribose, making spe-

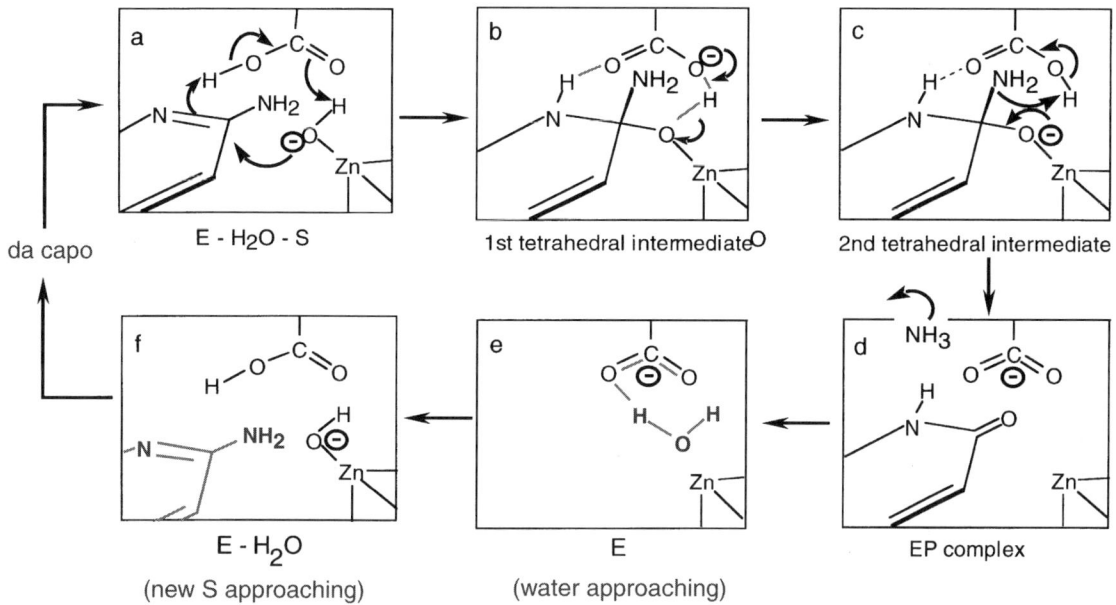

Figure 4. Schematic diagram of the hydrolytic deamination catalyzed by ECCDA. (Adapted from Carter, 1995, with permission.)

cific hydrogen bonds to the 3′ and 5′ hydroxyl groups (Fig. 5). These interactions are crucial to binding and catalysis. Cytosine itself is not a substrate for ECCDA, and mutation of Glu 91, which makes a key hydrogen bond to the ribose 3′ hydroxyl group, to alanine eliminates roughly four orders of magnitude in catalytic activity, k_{cat}/K_m, including a decrease in k_{cat} by more than an order of magnitude. This ribose-binding loop is therefore crucial to catalytic activity (Carlow et al., in press). The interaction of residues 78–101 with the ribose is reinforced by the fact that this loop lies beneath an "extra layer" formed by the interaction across the dimer interface between symmetry-related α-helices containing residues 272–284.

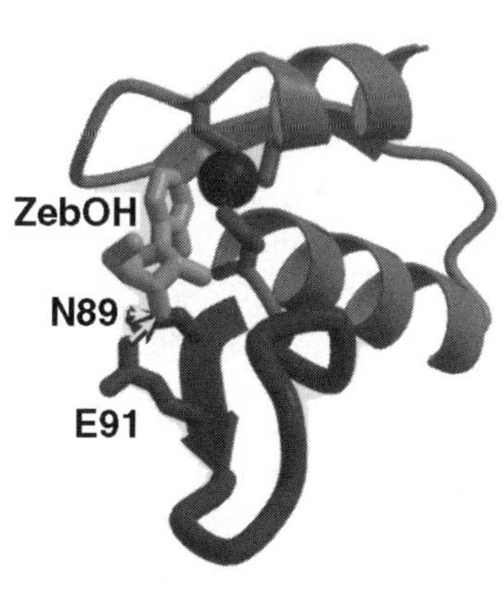

Seclusion of the active site in this manner and the important catalytic contribution of Glu 91 raise intriguing questions about turnover (Kuyper and Carter, 1996). It seems also reasonable that they are important in preventing ECCDA from deaminating cytosines in nucleic acids. This conjecture is supported by homology modeling (Navaratnam et al., 1998).

Configurational Differences in Substrate and Product Complexes of ECCDA

The NH_2 group thus binds early and remains essentially fixed during the reaction (Xiang et al., 1996), while the pyrimidine ring moves toward the activated nucleophilic substrate water molecule, changing its conformation substantially, relative to the active site (Xiang et al., 1997). Solution of the product complex of ECCDA with uridine (Xiang et al., 1997) revealed that the uridine remains bound to the active-site zinc ion through the 2-one oxygen, and that this interaction distorted the glycosidic linkage to the ribose. When superimposed on the structure of a substrate analog, this strained uridine configuration involved a rotation of the pyrimidine ring by approximately 20 degrees toward the zinc. In turn, this implies that the symmetry of the enzyme dimer may break down between two ground-state configurations during catalysis, with one site conforming to the substrate and the other conforming to the product (Fig. 6).

Figure 5. Ribose-binding interactions critical for catalysis by ECCDA. E91 and N89 form hydrogen bonds to the 3′ OH group of the nucleoside inhibitor, zebularine hydrate (ZebOH). The darkly shaded loop providing these residues corresponds exactly to the GAP-1 peptide in the apolipoprotein mRNA editing deaminase subunit (Navaratnam et al., 1998).

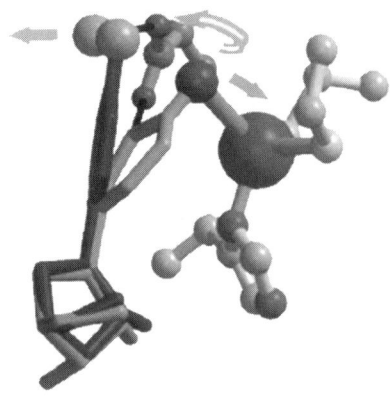

Figure 6. Comparison of initial and final ground states on the ECCDA reaction pathway. The darkly-shaded nucleoside is 3-deazacytidine, a stable substrate analog. The lightly shaded nucleoside is the product, uridine. As the reaction proceeds, the 4-NH_2 group is pulled into a binding site to the left, while the pyrimidine is pulled toward the zinc by the attacking nucleophile. After the first tetrahedral transition state (Fig. 4), the carboxylate sidechain of Glu 104 rotates (curved arrow), transferring a proton from the nucleophilic hydroxyl group to the leaving ammonia. (Adapted with permission from Xiang et al., 1997.)

C-TO-U EDITING

The Composite, Dimeric Configuration of ECCDA Provides a Model for APOBEC-1

C-to-U editing of the apolipoprotein B message involves a 27S "editosome complex" consisting of several polypeptides (Backus and Smith, 1991; MacGinnitie et al., 1995; Yamanaka et al., 1995, 1994), of which only the catalytic subunit, APOBEC-1, has been characterized (see Chapter 18 by Chang et al.). The homology between the APOBEC-1 subunit and ECCDA had been noticed simultaneously by several groups (Betts et al., 1994; Navaratnam et al., 1993a, 1993b). The implications of the homology were pursued by mutagenesis of both enzymes and a model RNA substrate and by chemical cross-linking (MacGinnitie et al., 1995; Navaratnam et al., 1995). These studies also showed that the APOBEC system recognized the targeted C6666 only when accompanied by a downstream, uridine-rich mooring sequence. These studies strengthened the possibility that the two enzymes shared important features, including the composite active-site construction of the dimeric ECCDA.

The substrate specificities of ECCDA and APOBEC-1 differ markedly. ECCDA is specific for nucleoside substrates; APOBEC-1 has acquired the capacity to deaminate a single C in a specific RNA context. Determinants for RNA editing are contained in a highly conserved, 26–30 nucleotide sequence, which can confer editing on other genes (Bostrum et al., 1989; Chen et al., 1990; Davies et al., 1989; Hodges and Scott, 1992) (also see Chapter 18 by Chang et al.). This sequence consists of six nucleotides upstream of the edited C, in which most mutations enhance editing, and a mooring sequence at a fixed distance downstream, in which most alterations reduce or abolish editing, which has been proposed to bind auxilliary editing factors (Backus et al., 1994; Backus and Smith, 1991; Hodges and Scott, 1992; Shah et al., 1991; Smith, 1993; Smith and Sowden, 1996).

Previous mutational analysis of the rat apoB RNA editing enzyme suggested a provocative model for RNA substrate recognition by the APOBEC-1 subunit (Navaratnam et al., 1995). That model hypothesized that APOBEC-1, also a dimer, recognized two specific bases. Binding to a downstream U was necessary, in addition to the targeted C. Because downstream binding could be eliminated by mutation of the zinc-binding and other active-site residues, a downstream U appeared to bind specifically to the other active site across the dimer interface. Scott (1995) pointed out that a composite, dimeric form of APOBEC-1 could achieve specificity by configuring the RNA to present a product, U, to one active site to bind the targeted C productively to the other (Scott, 1995). That proposal suggested the appealing notion that the catalytic moiety in apolipoprotein B mRNA editing might itself be capable of some measure of discrimination apart from the "auxiliary factors."

The model provided little insight, however, about the important structural relationships between APOBEC-1 and either ECCDA or the RNA substrate. Homology modeling of APOBEC-1, together with extensive mutational analysis of APOBEC-1, revealed detailed and substantive new descriptions of both relationships (Navaratnam et al., 1998). Remodeling of ECCDA was suggested by sequence alignment of ECCDA with APOBEC-1 (Fig. 7) in the light of the ECCDA tertiary and quaternary structures (Betts et al., 1994).

Three significant gaps (GAP-0, -1, and -2) (Fig. 7) occur in the APOBEC-1 sequence, relative to that of ECCDA. Their combined mass (10 kDa) matches that of a minimal RNA substrate. Moreover, GAP-1 coincides with the region in the first core domain where ECCDA closes over the ribose 3' and 5' hydroxyl groups (Fig. 5) and must in any case be removed for an RNA substrate to access the active site. Removing the GAP-1 peptide from the ECCDA structure provides access to the active sites for an RNA substrate, by freeing the region where ECCDA binds to nucleoside ribose 3' and 5' hydroxyl groups and leaving ample room for a single RNA strand to enter the dimeric enzyme through the tunnel vacated by GAP-1 on one subunit and room to leave it

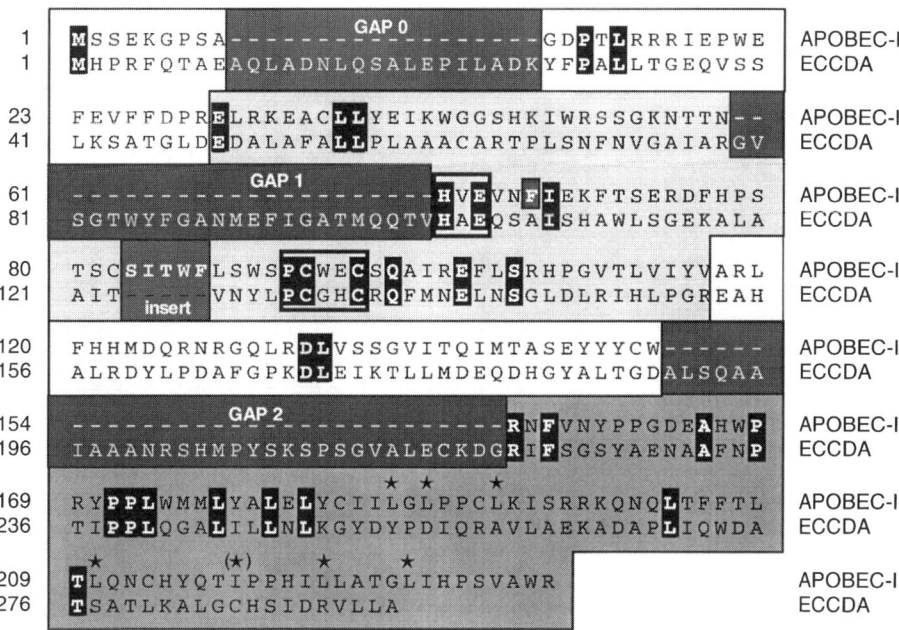

Figure 7. Sequence alignment of a consensus APOBEC-1 sequence with that of ECCDA. The two large, mid-gray blocks correspond to the two core domains in ECCDA, and the lightest gray box between them is the linker. Residue identities are shown in reverse contrast. Deletions (GAP-0, -1, and -2) and an insertion (SITWF) are shown in dark gray and in reverse contrast. (Adapted from Navaratnam et al., 1998, with permission.)

through the corresponding tunnel on the other subunit.

GAP-2 contains the middle β-strand from the second core domain and an associated α-helix that lies at the molecular surface. Deleting GAP-2 splits the second core domain into two parts, which must be resealed by closing the two remaining β-strands. There is precedent for this type of structural modification in the serpin rearrangement (Stein and Chothia, 1992). Moreover, none of the crucial nonpolar ECCDA hydrophobic packing interactions are compromised if the resealing is done in the obvious way, which is by moving the two separated β-strands together (Navaratnam et al., 1998). However, doing so does reposition the two α-helices (residues 272–284 from each subunit; shaded darkest gray in Fig. 8) outward, away from the bound ligands. The remodeling opens a large crevice directly into both active sites that is continuous from one side of the enzyme to the other (Fig. 8). The major functional differences between ECCDA and APOBEC-1 can thus be specifically related to the creation in APOBEC-1 of a large cavity capable of binding an RNA tertiary structure of sufficient complexity to endow it with the requisite binding specificity for RNA editing.

This new model is supported by extensive mutational analysis (Navaratnam et al., 1997). A broad, representative sample of APOBEC-1 mutants was examined by biochemical assays for homodimerization, RNA binding, and RNA editing. The mutagenesis strategy was designed to encompass the established features of the ECCDA structure, namely, catalytic residues and the structure of the active site, the domain organization of the monomer, the configuration of the dimer, and the evident differences between APOBEC-1 and ECCDA (that is, the gaps and the leucine-rich region). The mutant enzymes behaved consistently with the model in all but a small fraction of cases for which there were rational alternative explanations.

A PEPTIDE MIMIC OF THE RNA SUBSTRATE

The tunnel created by removing GAP-1 (Fig. 9a) is lined by amino acid residues homologous to the two phenylalanines previously cross-linked to an RNA substrate (Navaratnam et al., 1998). Moreover, the extensive network of crevices created by the remodeling of ECCDA (Fig. 8) is complementary to a macromolecular substrate with a central, globular structure flanked by single RNA strands at both ends and with two exposed bases separated by ~21 Å (Fig. 9c). Rearranging the peptide segments from GAP-1 and GAP-2 (and perhaps GAP-0) to form a single, globular entity (Fig. 9b) that fits into that network in the APOBEC-1 model suggests a crude peptide mimic of the RNA substrate when combined with the two

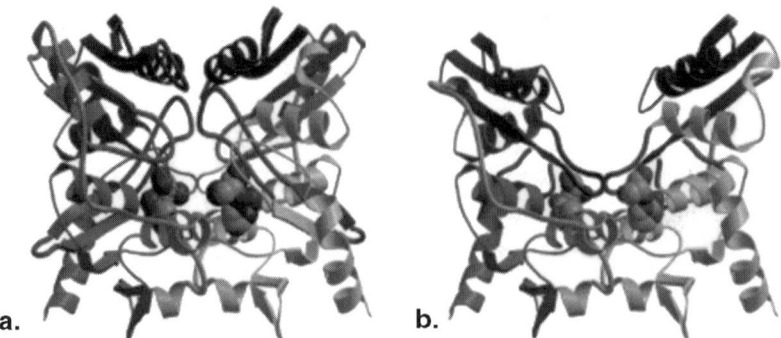

Figure 8. Comparison of the active-site access in ECCDA (a) and the APOBEC-1 homology model (b). CPK models at the center are the two bound transition-state analog molecules, ZebOH. The dark gray β-α-β crossover at the top is formed by residues 272–284, which undergo the largest rearrangement upon removal of GAP-2. The medium gray extended loop in the center of panel a is the GAP-1 loop. GAP-2 can be identified from the third β-strand from the top in either ECCDA subunit. The dimers are considerably elongated along this view, which foreshortens the large cavity that results from structural remodeling in APOBEC-1. (Adapted from Navaratnam et al., 1998, with permission.)

active-site ligands in the ECCDA transition-state analog complexes. The central, helical regions of this mimic would represent a helical or triple-helical tertiary structure, with leading and trailing segments of single-stranded RNA. Once assembled, this structure could create a loop at each end of the central region, exposing the two bases that bind to the two active sites.

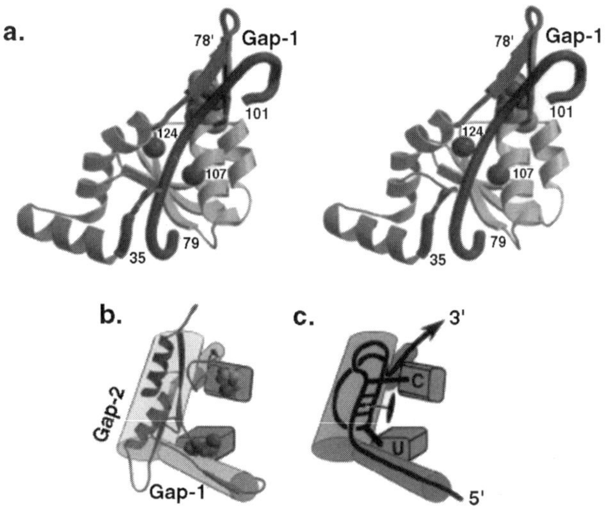

Use of a stem, pseudoknot, or even triple helical region to anchor an RNA substrate might be adapted by enzymes acting on loop regions of tRNA (see Chapter 16 by Price and Gray).

DOES ASYMMETRY IN SUBSTRATE BINDING IMPLY ALLOSTERIC ACTIVATION?

There is a fundamental asymmetry to the proposed RNA recognition, which depends on binding a product to one of the two active sites while the other site catalyzes deamination of the substrate. Using a bound active-site product complex as a second, confirmatory recognition site makes sense because the deamination product, uridine, closely resembles the substrate. Furthermore, under ambient conditions uridine is thermodynamically stable with respect to the reverse reaction (Cohen and Wolfenden, 1971a, 1971b), and would therefore be unchanged by interaction with the catalytic residues. The asymmetry of the proposed RNA recognition recalls significant differences between substrate and product binding to ECCDA (see Fig. 6) (Betts et al., 1994; Xiang et al., 1996, 1997). These differences provide, in turn, a possible structural rationale for the use of the downstream U as an allosteric effector to verify selection of the targeted C by APOBEC-1. Recent studies of ECCDA suggest possibilities inherent in that asymmetry. The two ECCDA monomers are related by crystallographic symmetry in most of the crystals examined (Betts et al., 1994) and are therefore, presum-

Figure 9. Evidence that GAP-1 and GAP-2 peptides actually represent the mass of an RNA substrate. (a) Phenylalanines crosslinked to substrate RNA occur where the spheres (residues 107 and 124) are located in ECCDA. The GAP-1 peptide (residues 79–101) is a thick, dark gray coil. Residue 78′ in the APOBEC-1 model actually precedes residue 79 of GAP-1 in ECCDA. Residue 35 shows the location of the "putative nuclear localization signal," containing the sequence RRR in APOBEC-1, which could also therefore be involved in RNA binding. (b) Two GAP-1 and two GAP-2 peptides are assembled with the two active-site pyrimidine ligands to form a mimic of the RNA substrate. The GAP-1 peptides are in identical locations to those in ECCDA. GAP-2 peptides have essentially been interchanged with the relocated helical segments 272–284. (c) Model of the RNA substrate with two exposed bases representing the targeted C6666 and the downstream U. (Adapted from Navaratnam et al., 1998, with permission.)

ably, functionally equivalent. However, because the product and a substrate analog bind quite differently to the ECCDA active site, two ECCDA active sites may alternate between asymmetric dimer conformations during catalysis. This alternation would occur whenever one active site bound a product (uridine) while the other bound a substrate (cytidine) as in the model for RNA substrate recognition by APOBEC-1. A new ECCDA crystal structure prepared at close-to physiological temperatures revealed an asymmetric dimer in the crystallographic asymmetric unit, consistent with this hypothesis (Kuyper and Carter, 1996).

Presumably, deamination of nucleosides by ECCDA preserves twofold symmetry only in time, cycling the two active sites through substrate and product binding. The detailed similarity documented in our homology model suggests that this broken symmetry of the ECCDA dimer may also be conserved in APOBEC-1. If so, then the downstream U might also serve as an allosteric effector, complementing its role in substrate recognition by signaling to the opposite monomer and activating it to catalyze deamination.

Despite extensive mutagenesis, a specific U has not been identified as an essential requirement for RNA editing and UV cross-linking (Anant et al., 1995; Backus et al., 1994; Backus and Smith, 1991; Davies et al., 1989; MacGinnitie et al., 1995; Navaratnam et al., 1995; Shah et al., 1991; Smith and Sowden, 1996). These and previous observations might, therefore, indicate that APOBEC-1 is tuned to search for a product among one of several downstream U residues (Bostrom et al., 1989; Davies et al., 1989). Such promiscuity would be surprising from the standpoint of most enzyme:substrate and other protein:RNA mechanisms, which are quite stereospecific. Nonetheless, a tendency to search for a product might also explain the editing of multiple C's in certain in vitro conditions and the hyperediting of multiple C's in transgenic animals that overexpress APOBEC-1, with mass action being the driving force (Sowden et al., 1996; Yamanaka et al., 1996).

SIGNATURES OF C-TO-U EDITING

The model of Navaratnam et al. (1998) suggests structural features likely to be charateristic of enzymes involved in other instances of C-to-U editing. These include the conservation of tertiary structure in the ECCDA core domain (Fig. 1) and conservation of a dimeric quaternary structure and exploitation of the composite dimeric active-site organization for specific substrate recognition. These features (Fig. 7) are likely, in turn, to permit additional C-to-U editing cytidine deaminase (CDARs) to be identified from among sequences that show, in addition to the catalytic signatures, significant gaps in the APOBEC-1 sequence, compared to ECCDA, immediately preceding the HAE signature (~23 residues) and at the amino terminus of the second core domain (~31 residues); an insert of the five residues with consensus sequence, SITWF, just before the PCXXC sequence containing the second group of zinc ligands; and two phenylalanine residues, one just after the first cluster of catalytic residues, and the other in the insert preceding the second group, which cross-link to the substrate RNA (Navaratnam et al., 1995). Two further sequence motifs where APOBEC-1 differs from ECCDA have been identified by other authors. A putative bipartite basic nuclear localization signal occurs near the amino terminus (Smith and Sowden, 1996; Teng et al., 1993). In addition, a distinctive pattern of leucine residues occurs toward the carboxy terminus, some of which correspond to leucine residues in ECCDA (Davidson et al., 1995; Navaratnam et al., 1993a; Scott et al., 1994; Teng et al., 1993; Yamanaka et al., 1994) (also see Chapter 18 by Chang et al.). The model (Navaratnam et al., 1998) suggests that apparently neither of these is relevant to the specific differentiation of nucleoside and editing cytidine deaminases. The leucine-rich region is discussed in the next section.

The "Leucine-Rich" C-Terminal Region?

APOBEC-1 differs from ECCDA in three major respects. In addition to the gaps absent from APOBEC-1 and the leucine-rich region, the catalytic process is absolutely dependent on as-yet uncharacterized protein cofactors. There are several ways in which these three distinctive features might be related. The leucine-rich sequences have been discussed as possible dimerization interfaces, either between APOBEC-1 monomers or with auxiliary editing factors in the editosomal complex. While such roles cannot be ruled out by currently available data, the modeling described by Navaratnam et al. (1998) makes it likely that many, if not all of these residues participate in the hydrophobic core in the second homologous ECCDA domain. The sequence motif **PPLWMMLYALEL** (see Chapter 18 by Chang et al.) aligns directly with the corresponding sequence in ECCDA (Fig. 7) where these residues form a loop from the second core domain that interacts with the active site on the other monomer. That sequence probably is actually a dimeric deaminase signature characteristic of the second core domain, and not an editing signature.

The proposed homodimer does not provide obvious interaction surfaces for macromolecular cofactors without substantive rearrangement. The most likely candidate for such a rearrangement is the carboxy-terminal crossover connection (represented by dark gray in Fig. 8). The leucines in this region are themselves evenly distributed between the inner surface of this module and the complementary face from the N terminus of the second core domain. One possibility is that the binding sites for accessory factors become available only if the carboxy-terminal crossover connection unfolds from the rest of the APOBEC-1, exposing the nonpolar surface populated by the leucines. Thus, if the numerous other leucines in this region help to mediate intermolecular intractions with auxiliary factors, their role in heterologous interactions is likely to be cryptic (Drugan et al., 1996).

A-TO-I EDITING

Adenosine Deaminase Is Apparently Not Involved in RNA Editing

The X-ray structure has been determined for murine adenosine deaminase, which catalyzes deamination of adenosine to inosine. Although adenosine deaminase is also a zinc enzyme and shares many mechanistic features with ECCDA, it has an entirely different architecture. The secondary and tertiary structure motifs of the two nucleoside deaminases are completely unrelated. In contrast to the dimeric or tetrameric architecture of ECCDA (Betts et al., 1994), ADA is monomeric and assumes a typical α-β TIM barrel tertiary structure (Wilson et al., 1991) that is unrelated to that of ECCDA. The distribution of zinc ligands in the amino acid sequence bears only a superficial resemblance to that in the CDA family. Indeed, the sole parallel between the two enzymes is that a catalytic glutamic acid follows closely on one of the zinc ligands. The sequences, H/CAE in the CDA family and HAGE in ADA, form comparable structures in the respective active sites, but they are mirror images and their active configurations are achieved by completely different folds (Carter, 1995).

The zinc ligands are also different in the two enzymes, with CDA using (His, Cys2) and ADA using (His3, Asp). The chemical differences between the two types of ligands may be correlated with the structural differences between the two types of topological switch points, as well as with subtle differences in mechanism. Cysteine thiolate groups are much stronger ligands, and those in ECCDA experience significant positive electrostatic potential from the amino termini of the two active-site helices (Hol et al., 1978; Lockhart and Kim, 1992). This environment may modulate the strong zinc-sulfhydryl affinity of the thiolate ligands Cys 129 and Cys 132 by repelling the zinc, relative to the effects of the imidazole and carboxyl groups favored for "catalytic" zinc atoms in ADA and other enzymes. The active sites of the two enzymes thus provide an interesting example of convergent evolution.

There is no evidence that enzymes related to ADA are involved in RNA editing. In fact, as previously noted, evidence from sequence comparisons points to a relationship with the cytidine deaminase family. ADAR enzymes contain the HAE and PC parts of the CDA signature, but the third zinc ligand does not follow the PC as it does in the CDA PCXXC sequences. Whether or not the ADAR family enzymes belong to a cytidine deaminase superfamily, there is also some evidence that they are related to the DNA methyltransferases (modifying enzymes for the restriction systems). Alignment of ADAR sequences suggested a global distribution of motifs (Hough and Bass, 1997) which could be aligned with corresponding motifs among the methyltransferases (Malone et al., 1995). Without presenting details, Hough and Bass (1997) argued that the conservation of motifs placed the HAE sequence in close proximity to the binding site for the methyl donor, SAM. Juxtaposition of the CDA-like motifs with the SAM binding site suggested, in turn, that ADARs operate on double-stranded RNA by extruding the selected base from the double helix in the manner shown for the methyltransferases.

There is significant support for this "base-flipping" hypothesis in tertiary structural alignments of the methyltransferases with the ECCDA active site (Fig. 10). Specifically, the suggested sequence alignments place the HAE sequences near the amino termini of α-helices, similar to helix A from ECCDA (Fig. 1b). An especially close correspondence is found with the hha1 cytosine transferase, in which the adenine of the SAM cofactor is oriented such that the 6-NH_2 group is within several angstroms of the residues homologous to the HAE sequence (Cheng et al., 1993) (Fig. 10a and b). A schematic diagram of this transferase (Fig. 11) confirms that the suggestion of Hough and Bass (1997) is quite plausible. However, the comparison falls short of providing convincing evidence for descent from a common evolutionary ancestor, as the authors point out.

ADARs may have evolved by combining a cytidine deaminase-like active site with the SAM binding site of the DNA methyltransferases. If so, however, the process involved a substantially smaller piece of

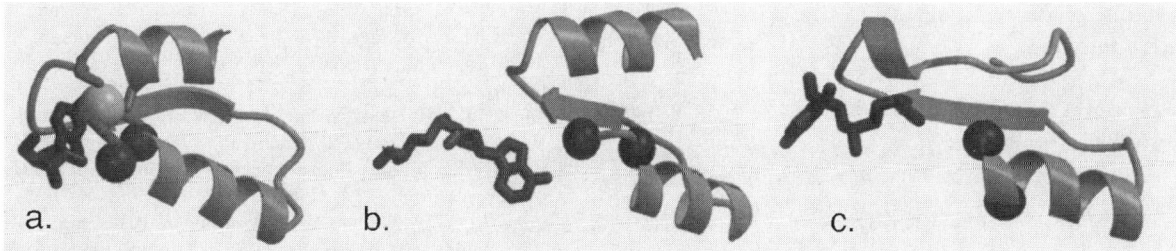

Figure 10. Comparison of the ECCDA zinc-binding motif (a) with structural fragments of the hha1 (b) and taq1 (c) DNA methyltransferases. The latter family has been proposed as a model for the ADAR family of adenosine deaminases (Hough & Bass, 1994). The spheres show, in each case, the **HAE** sequences in ECCDA and in alignments described for the corresponding HAE peptides by these authors. Dark gray rod representations are, respectively, the ZebOH ligand for ECCDA and the S-adenosylmethionine substrates of the two methylases.

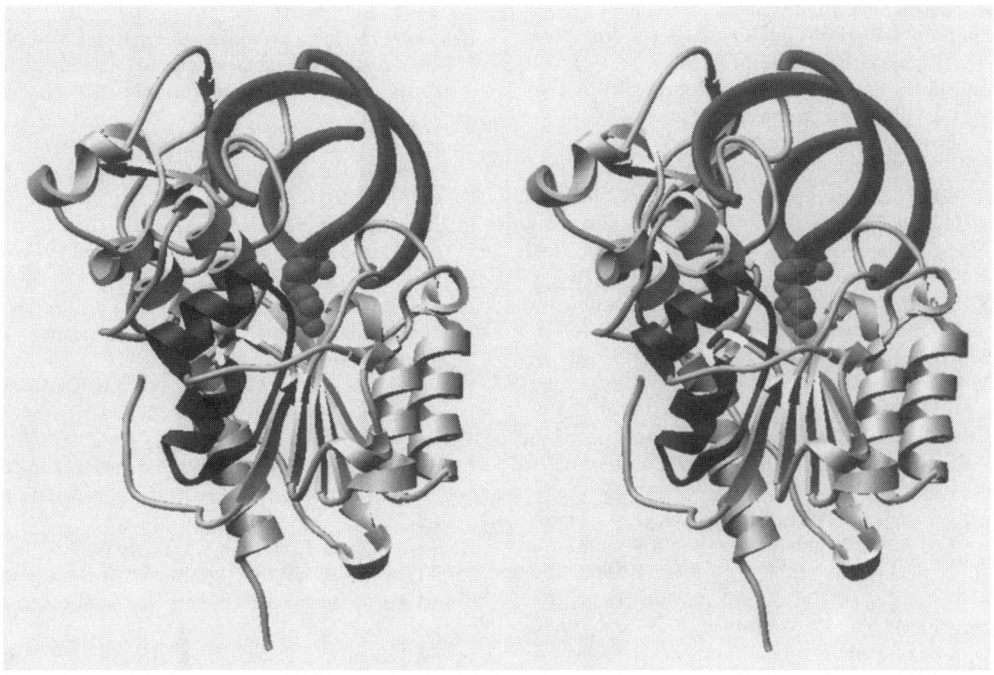

Figure 11. Schematic of the hha1 methylase, complexed to a DNA duplex substrate (medium gray), showing how the extruded base is drawn into proximity to the SAM methyl donor (Klimasauskas et al., 1994). The location of the α-β-α fagments shown in Fig. 9 are indicated in dark gray, and the location of putative catalytic residues **HAE** are emphasized in white.

genetic information than coding for the CDA core domain structure. The surrounding structure in the methyltransferases has little obvious homology with the CDA domain fold. Nevertheless, the most attractive working hypothesis for how ADARs work is that they flip out a specific base from the double-stranded substrate.

CONCLUSIONS AND PROSPECTS

The outlines described in this chapter of how C-to-U and A-to-I editing occurs provide coherent hypotheses based on known structures of related enzymes. The former process is apparently identified with specialized RNA structures that expose specific cytosine bases via tertiary structures. The latter involves quite different recognition mechanisms. Many important questions are only dimly perceived on the basis of these models, however. The specific recognition of editing sites by the ADARs, for example, is problematic because the methyltransferases are by contrast quite promiscuous. Moreover, little can be said about structures or mechanisms of enzymes involved in deamination of bases in tRNA (see Chapter 16 by Price and Gray). It is imperative, therefore, that structures of the catalytic and other components of all systems be determined before the models evolve lives of their own.

Acknowledgment. Support from American Cancer Society grant BE54 is gratefully acknowledged.

REFERENCES

Anant, S., A. J. MacGinnitie, and N. O. Davidson. 1995. Apobec-1, the catalytic subunit of the mammalian apolipoprotein B mRNA editing enzyme, is a novel RNA-binding protein. *J. Biol. Chem.* 270:14762–14767.

Auxilien, S., P. F. Crain, R. W. Trewyn, and H. Grosjean. 1996. Mechanism, specificity, and general properties of the yeast enzyme catalysing the formation of inosine 34 in the anticodon of transfer RNA. *J. Mol. Biol* 262:437–458.

Backus, J. W., D. Schock, and H. C. Smith. 1994. Only cytidines 5′ of the apolipoprotein B mRNA mooring sequence are edited. *Biochim. Biophys. Acta* 1219:1–14.

Backus, J. W., and H. C. Smith. 1991. Apolipoprotein B mRNA sequences 3′ to the editing site are necessary and sufficient for editing and editosome assembly. *Nucleic Acids Res.* 19:6781–6786.

Bass, B. 1993. RNA editing: new uses for old players in the RNA world, p. 219–237. *In* R. F. Gesteland and J. F. Atkins (ed.), *The RNA World.* Cold Spring Harbor Laboratory Press, Cold Spring Harbor, N.Y.

Bass, B. L., K. Nishikura, W. Keller, P. H. Seeburg, R. B. Emeson, M. A. O'Connell, C. E. Samuel, and A. Herbert. 1997. A standardized nomenclature for adenosine deaminases that act on RNA. *RNA* 3:947–949.

Betts, L., S. Xiang, S. A. Short, R. Wolfenden, and C. W. J. Carter. 1994. Cytidine deaminase. The 2.3 Å crystal structure of an enzyme: transition-state analog complex. *J. Mol. Biol.* 235:635–656.

Bostrum, K., S. J. Lauer, K. S. Poksay, Z. Garcia, J. M. Taylor, and T. L. Innerarity. 1989. Apolipoprotein B48 RNA editing in chimeric apolipoprotein EB mRNA. *J. Biol. Chem.* 264:15701–15708.

Carlow, D., S. A. Short, and R. Wolfenden. Site- and ligand-directed truncations of a ribose-binding function in transition-state stabilization by cytidine deaminase. *Biochemistry,* in press.

Carlow, D. A., J. Neuhard, N. Meljhede, C. W. Carter, Jr., and R. Wolfenden. Unpublished data.

Carter, C. W., Jr. 1995. The nucleoside deaminases for cytidine and adenosine structure transition state stabilization mechanism and evolution. *Biochimie* 77:92–98.

Chen, S.-H., X. X. Li, W. S. Liao, J. H. Wu, and L. Chan. 1990. RNA editing of apolipoprotein B mRNA. *J. Biol. Chem.* 265:6811–6816.

Cheng, X., S. Kumar, J. Posfai, J. W. Pflugrath, and R. J. Roberts. 1993. Crystal structure of the HhaI DNA methyltransferase complexed with S-adenosyl-L-methionine. *Cell* 74:299–307.

Cohen, R. M., and R. Wolfenden. 1971a. Cytidine deaminase from *Escherichia coli:* purification, properties, and inhibition by the potential transition state analog inhibitor, 3,4,5,6-tetrahydrouridine. *J. Biol. Chem.* 246:7561–7565.

Cohen, R. M., and R. Wolfenden. 1971b. The equilibrium of hydrolytic deamination of cytidine and N4-methyl cytidine. *J. Biol. Chem.* 246:7566–7568.

Davidson, N. O., S. Anant, and A. J. MacGinnitie. 1995. Apolipoprotein B messenger RNA editing: insights into the molecular regulation of post-transcriptional cytidine deamination. *Curr. Opin. Lipidol.* 6:70–74.

Davies, M. S., S. C. Wallis, D. M. Driscoll, J. K. Wynne, G. W. Williams, L. M. Powell, and J. Scott. 1989. Sequence requirements for apolipoprotein B RNA editing in transfected rat hepatoma cells. *J. Biol. Chem.* 264:13395–13398.

Drugan, J. K., R. Khosravi-Far, M. A. White, C. J. Der, Y.-J. Sung, Y.-W. Hwang, and S. L. Campbell. 1996. Ras interaction with two distinct binding domains in Raf-1 may be required for Ras transformation. *J. Biol. Chem.* 271:233–237.

Gerber, A., M. A. O'Connell, T. Melcher, L. P. Keegan, S. Krause, P. Seeburg, and W. Keller. 1997. Cloning, expression and characterization of yeast and drosophila proteins which are homologous to the RNA editing enzymes DRADA and RED1, p. 247. *In Proceedings of the Second Annual Meeting of the RNA Society.*

Grosjean, H., S. Auxilien, F. Constantinesco, C. Simon, Y. Corda, H. F. Vecker, D. Foiret, A. Morin, Y. X. Jin, M. Fournier, and J.-L. Fourrey. 1996. Enzymatic conversion of adenosine to inosine and to N^1-methylinosine in transfer RNAs: a review. *Biochimie* 78:488–501.

Grosjean, H., F. Constantinesco, D. Foiret, and N. Benachenhou. 1995. A novel enzymatic pathway leading to 1-methylinosine modification in *Haloferax volcanii* tRNA. *Nucleic Acids Res* 23:4312–4319.

Hodges, P., and J. Scott. 1992. Apolipoprotein B mRNA editing: a new tier for the control of gene expression. *Trends Biochem. Sci.* 17:77–81.

Hol, W. G. J., P. T. van Duijnen, and H. J. C. Berendsen. 1978. The α-helix dipole and the properties of proteins. *Nature* 273:443–446.

Hough, R. F., and B. L. Bass. 1994. Purification of the *Xenopus laevis* double stranded TNR adenosine deaminase. *J. Biol. Chem.* 269:9933–9939.

Hough, R. F., and B. L. Bass. 1997. Analysis of *Xenopus* dsRNA deaminase cDNAs reveals similarities to DNA methyltransferases. *RNA* 3:356–370.

Klimasauskas, S., S. Kumar, R. J. Roberts, and X. Cheng. 1994. HhaI methyltransferase flips its target base out of the DNA helix. *Cell* 76:357–369.

Kuyper, L., and C. W. Carter, Jr. 1996. Resolving crystal polymorphisms by finding "stationary points" from quantitative analysis of crystal growth response surfaces. *J. Crystal Growth* 168:155–169.

Lewis, J. P., C. W. Carter, Jr., J. Hermans, W. Pan, T.-S. Lee, and W. Yang. Active species for the ground-state complex of cytidine deaminase: a linear scaling quantum mechanical investigation. *J. Am. Chem. Soc.,* in press.

Lockhart, D. J., and P. S. Kim. 1992. Internal stark effect measurement of the electric field at the amino terminus of an α helix. *Science* 257:947–951.

MacGinnitie, A. J., S. Anant, and N. O. Davidson. 1995. Mutagenesis of apobec-1, the catalytic subunit of the mammalian apolipoprotein B mRNA editing enzyme, reveals distinct domains that mediate cytosine nucleoside deaminase, RNA binding, and RNA editing activity. *J. Biol. Chem.* 270:14768–14775.

Malone, T., R. M. Blumenthal, and X. Cheng. 1995. Structure-guided analysis reveals nine sequence motifs among DNA aminomethyl-transferases, and suggests a catalytic mechanism for these enzymes. *J. Mol. Biol* 253:618–632.

Navaratnam, N., S. Bhattacharya, T. Fujino, D. Patel, A. L. Jarmuz, and J. Scott. 1995. Evolutionary origins of apoB mRNA editing: catalysis by a cytidine deaminase that has acquired a novel RNA-binding motif at its active site. *Cell* 81:187–195.

Navaratnam, N., T. Fujino, J. Bayliss, A. Jarmuz, A. How, N. Richardson, S. Angelika, S. Battacharya, C. W. Carter, Jr., and J. Scott. 1998. *E. coli* cytidine deaminase provides molecular model for ApoB RNA editing: implications for asymmetric RNA substrate presentation and evolution of C to U editing. *J. Mol. Biol.* 275:695–714.

Navaratnam, N., J. R. Morrison, S. Bhattacharya, D. Patel, T. Funahashi, F. Giannoni, B. B. Teng, N. O. Davidson, and J. Scott. 1993a. The p27 catalytic subunit of the apolipoprotein B mRNA editing enzyme is a cytidine deaminase. *J. Biol. Chem.* 268:20709–20712.

Navaratnam, N., R. Shah, D. Patel, V. Fay, and J. Scott. 1993b. Apolipoprotein B mRNA editing is associated with UV crosslinking of proteins to the editing site. *Proc. Natl. Acad. Sci. USA* 90: 222–226.

Scott, J. 1995. A place in the world for RNA editing. *Cell* 81: 833–836.

Scott, J., N. Navaratnam, S. Bhattacharya, and J. R. Morrison. 1994. The apolipoprotein B messenger RNA editing enzyme. *Curr. Opin. Lipidol.* 5:87–93.

Shah, R. R., T. J. Knott, J. E. Legros, N. Navaratnam, J. C. Greeve, and J. Scott. 1991. Sequence requirements for the editing of apolipoportien B mRNA. *J. Biol. Chem.* 266: 16301–16304.

Smith, H. C. 1993. Apolipoprotein B mRNA editing: the sequence to the event. *Semin. Cell Biol.* 4:267–278.

Smith, H. C., and M. P. Sowden. 1996. Base-modification mRNA editing through deamination—the good, the bad, and the unregulated. *Trends Genet.* 12:418–424.

Sowden, M., J. K. Hamm, and H. C. Smith. 1996. Overexpression of APOBEC-1 results in mooring sequence-dependent promiscuous RNA editing. *J. Biol. Chem.* 271:3011–3017.

Stein, P., and C. Chothia. 1992. Serpin tertiary structure transformation. *J. Mol. Biol.* 221:615–621.

Teng, B., C. F. Burant, and N. O. Davidson. 1993. Molecular cloning of an apolipoprotein B messenger RNA editing protein. *Science* 260:1816–1819.

Wilson, D. K., F. B. Rudolph, and F. A. Quiocho. 1991. Atomic structure of adenosine deaminase complexed with a transition-state analog: understanding catalysis and immunodeficiency mutations. *Science* 252:1278–1284.

Xiang, S., S. A. Short, R. Wolfenden, and C. W. Carter, Jr. 1996. Cytidine deaminase complexed to 3-deazacytidine a "valence-buffer" in zinc enzyme catalysis. *Biochemistry* 35:1335–1341.

Xiang, S., S. A. Short, R. Wolfenden, and C. W. J. Carter. 1995. Transition state selectivity for a single hydroxyl group during catalysis by cytidine deaminase. *Biochemistry* 34:4516–4523.

Xiang, S., S. A. Short, R. Wolfenden, and C. W. J. Carter. 1997. Structure of the cytidine deaminase: product complex provides evidence for efficient proton transfer and ground-state destabilization. *Biochemistry* 36:4768–4774.

Yamanaka, S., M. E. Balestra, L. D. Ferrell, J. Fan, K. S. Arnold, S. Taylor, J. M. Taylor, and T. L. Innerarity. 1995. Apolipoprotein B mRNA-editing protein induces hepatocellular carcinoma and dysplasia in transgenic animals. *Proc. Natl. Acad. Sci. USA* 92:8483–8487.

Yamanaka, S., K. S. Poksay, M. E. Balestra, G. Q. Zeng, and T. L. Innerarity. 1994. Cloning and mutagenesis of the rabbit ApoB mRNA editing protein. A zinc motif is essential for catalytic activity, and noncatalytic auxiliary factor(s) of the editing complex are widely distributed. *J. Biol. Chem.* 269: 21725–21734.

Yamanaka, S., K. S. Poksay, D. M. Driscoll, and T. L. Innerarity. 1996. Hyperediting of multiple cytidines of apolipoprotein B mRNA by APOBEC-1 requires auxiliary protein(s) but not a mooring sequence motif. *J. Biol. Chem.* 271:11506–11510.

Chapter 21

Mitochondrial mRNA Editing in Kinetoplastid Protozoa

STEPHEN L. HAJDUK AND ROBERT S. SABATINI

The mRNAs in the mitochondrion of kinetoplastid protozoa are often modified by the precise addition or deletion of uridylates (U's) by a process termed RNA editing (Benne et al., 1986). Both the number of U's and the sites of insertion and deletion are specified by small guide RNAs (gRNAs), which form binary RNA complexes with the pre-mRNAs and orchestrate the editing reactions. RNA editing can be very limited, such as the case for cytochrome oxidase II (COII) mRNA where only 4 U's are added. However, more extensive editing is seen in other mRNAs; for example, in cytochrome oxidase III (COIII) mRNA of *Trypanosoma brucei* hundreds of U's are added and dozens of U's are deleted. (For other recent reviews on kinetoplastid RNA editing, see Hajduk et al., 1993; Benne, 1994; Seiwert, 1995; Simpson and Thiemann, 1995; Arts and Benne, 1996; Stuart et al., 1997; Kable et al., 1997).

The organization of the mitochondrial genome of trypanosomes is highly unusual and the genes for the pre-mRNAs and gRNAs are often encoded on separate circular DNAs that are topologically connected by DNA catenation. The gRNAs are usually encoded by small circular DNAs called minicircles. The pre-mRNAs are encoded by less abundant maxicircles that are topologically interlocked with each other and with the minicircles to form the kinetoplast DNA network. Thus, RNA editing provides a mechanism by which the incomplete genetic information of the pre-mRNA genes is compensated for by the information carried by the gRNAs encoded by separate genes. This chapter will focus on the molecular basis of RNA editing, and the integration of biochemical mechanism with the structure of the editing machinery. Finally, the more complex issues of biological function, regulation, and the evolutionary origin of editing will be considered.

KINETOPLASTID PROTOZOA

Phylogenic Distribution of RNA Editing

The order Kinetoplastida consists of flagellated protozoa and includes both free-living and parasitic organisms. Several members of the suborder Trypanosomatina cause important diseases in humans and domesticated animals (Vickerman, 1976; Vickerman, 1994). Other Trypanosomatina and several members of the suborder Bodonina infect either vertebrate or invertebrate hosts without apparent pathology. In addition, some members of the suborder Bodonina are nonparasitic and inhabit both fresh and marine environments (Vickerman, 1994).

All kinetoplastids are characterized by the presence of a single mitochondrion and a high concentration of mitochondrial DNA within a portion of the mitochondrion called the kinetoplast. Eight different genera of kinetoplastids have been shown to edit mitochondrial mRNAs. In each case deletion or insertion of U's has been described, suggesting a common mechanism of editing.

Thus far, U insertion and deletion editing has only been described for the mitochondrion of kinetoplastid protozoa. While insertional editing has been described for *Physarum* mitochondria, the basic mechanism may be very different and is likely to have evolved independently of kinetoplastid RNA editing (Simpson and Thiemann, 1995; Mahendran et al., 1991; Gott et al., 1993) (see Chapter 22 by Gott and Visomirski-Robic).

The origin of RNA editing in kinetoplastids remains unknown. Most of the studies on RNA editing have been on the members of the suborder Trypanosomatina and have focused on only three species, *Leishmania tarentolae*, *Crithidia fasciculata*, and *T. brucei* (Benne et al., 1986; van der Spek et al., 1988;

Stephen L. Hajduk and Robert S. Sabatini • Department of Biochemistry and Molecular Genetics, University of Alabama at Birmingham School of Medicine, Birmingham, Alabama 35294.

1990; Shaw et al., 1988). Recent studies have shown that U insertion and deletion editing occurs in the mitochondrion of *Trypanoplasma borreli* (Maslov and Simpson, 1994; Lukes et al., 1994; Maslov et al., 1994). This indicates that RNA editing is a trait that may have been acquired very early in kinetoplastid evolution. However, evidence that RNA editing is truly an ancient trait is lacking. The kinetoplastids and euglenoid flagellates are among the earliest eukaryotes containing mitochondria. Studies on the mitochondrial genome of *Euglena gracilis* have failed to reveal the presence of potentially edited mRNAs (Yasuhira and Simpson, 1997), suggesting that U insertion and deletion editing may be found only in kinetoplastid flagellates (Simpson and Maslov, 1994).

The Mitochondrial Genomes of Kinetoplastids

Mitochondrial biogenesis in eukaryotes involves the expression of protein-coding genes in both the nucleus and the mitochondrion. Despite great differences in genome size, each mitochondrial genome encodes approximately 15 mRNAs that must be transcribed and processed for the production of a functional mitochondrion. In addition to protein-coding genes, mitochondria typically encode rRNAs and a complete set of mitochondrial tRNAs. The mitochondrial mRNAs are translated on mitochondrial ribosomes and produce functional proteins. The remainder of the mitochondrial proteins, numbering in the hundreds, are synthesized on cytoplasmic ribosomes and posttranslationally targeted to the mitochondrion (Neupert, 1997).

The arrangement of the mitochondrial genome in kinetoplastids is unique (Simpson, 1986; Shapiro and Englund, 1995). All of the protein-coding genes and the mitochondrial ribosomal RNA genes are found on large circular DNA molecules called maxicircles (Fig. 1). Each mitochondrion contains 25–50 identical maxicircles, depending on the species. tRNA genes have not been identified on the maxicircle or minicircles, and it appears that all of the mitochondrial tRNAs of trypanosomes are encoded by nuclear genes, exported from the nucleus, targeted to the mitochondrion, and imported (Simpson et al., 1989; Hancock and Hajduk, 1990; Hancock et al., 1992; Mottram et al., 1991, Schneider, 1994; Schneider et al., 1994; Mahapatra et al., 1994). The complete sequence of the *T. brucei* maxicircle is known (Sloof et al., 1992; Myler et al., 1993) and a number of genes for known mitochondrial proteins and the 9S and 12S rRNA genes have been identified (Fig. 1). In addition, there are several unassigned open reading frames. Many of the protein-coding genes are transcribed into extensively edited mRNAs, which are unrecognizable in their unedited form.

RNA editing was discovered by Benne and co-workers (1986) by comparative analysis of maxicircle genes and their corresponding cDNAs. In *T. brucei*, all but four of the protein coding genes are incomplete and encode mRNAs that are modified to varying extents by U insertion and deletion (Table 1). In addition to mRNAs and rRNAs, the maxicircles of *L. tarentolae* and *C. fasciculata* also encode gRNAs. A total of 13 gRNA genes have been identified on the maxicircle of one *Leishmania* strain (Blum et al., 1990). In contrast, the maxicircle of *T. brucei* does not contain any gRNA genes, with the possible exception of two MURF2 gRNA genes (van der Spek et al., 1991) and a potential intramolecular guide sequence in the 3' untranslated region of cytochrome oxidase II mRNA (Blum et al., 1990; Kim et al., 1994).

There are 5,000–10,000 minicircles, ranging in size from approximately 0.8 kb in *Leishmania* to 2.5 kb in *Crithidia*. The minicircles of *T. brucei* are about 1 kb (Fig. 1). Depending on the species, there may be up to 250 different minicircle sequence classes within a single kinetoplast DNA (kDNA) network. The genetic function of minicircles was elucidated with the discovery of gRNA genes (Pollard et al., 1990; Sturm and Simpson, 1990a; Pollard and Hajduk, 1991; van der Spek et al., 1991). This finding eliminated decades of speculation that denied minicircles a genetic role and instead proposed that they were merely structural elements, perhaps providing a scaffolding for maxicircle replication or segregation. It is clear that while minicircles do not encode proteins directly, they play an important role in RNA maturation by encoding most of the gRNAs necessary for the editing of mitochondrial mRNAs.

The minicircle gRNA genes of *T. brucei* are flanked by an imperfect 18 bp inverted repeat sequence with transcription beginning 32 bp from the upstream repeat potentially extending through the downstream gRNA gene (Fig. 2) (Pollard et al., 1990; Grams and Hajduk, unpublished data). In vitro "capping" reactions with the enzyme guanylyl transferase demonstrated that each of the gRNA genes contained transcription start sites (Pollard and Hajduk, 1991). Primary gRNA transcripts beginning with the sequence 5'-RYAYA-3' were identified. While the positioning of the inverted repeat sequences suggests that these sequences might function as promoters for gRNA transcription or as processing signals for post-transcriptional events, experimental evidence supporting either possibility is lacking. Moreover, the lack of comparable inverted repeats in the minicircles of *L. tarentolae*, *C. fasciculata* and some *T. brucei*

Trypanosoma brucei Maxicircle

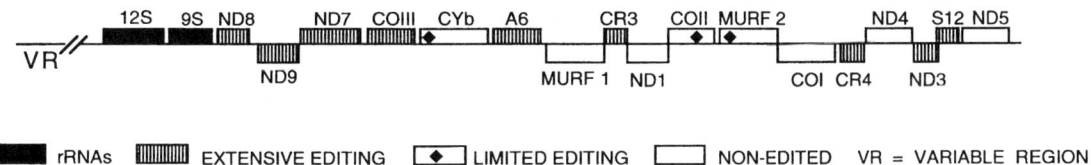

Trypanosoma brucei Minicircle

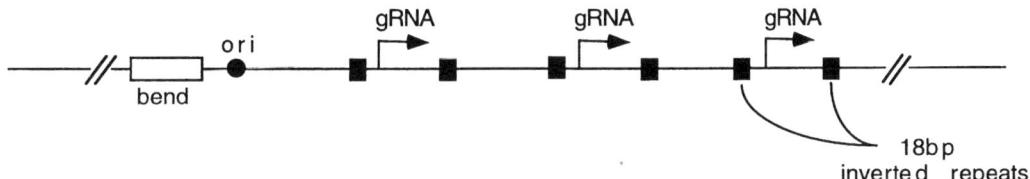

Figure 1. Transcription maps of the maxicircle and minicircle of *T. brucei*. The upper part depicts the linear map of the 23,019 bp maxicircle (Sloof et al., 1992) of *T. brucei*. The genes above the line are transcribed from left to right, while the genes beneath the line are transcribed from right to left. The ribosomal RNAs (12S and 9S) have added U's at their 3' ends (Adler et al., 1991). Transcripts from cytochrome *b* (CYb), cytochrome oxidase II (COII) and maxicircle unidentified reading frame (MURF) 2 have limited amounts of internal editing (black diamonds). The transcripts from the genes encoding NADH dehydrogenase 3, 7, 8, and 9 (ND3, ND7, ND8, and ND9), cytochrome oxidase III (COIII), ATPase subunit 6 (A6), small subunit ribosomal protein S12, and GC-rich regions 3 and 4 (CR3 and CR4) are all extensively edited (shaded boxes). The variable region of the maxicircle is indicated (VR). In the lower part, a linear map of a 1 kb minicircle of *T. brucei* is given. The bent helical region of this minicircle (open box) and the origin of replication (ori) are within the conserved region of the minicircle. The gRNA genes (arrows) are flanked by 18 bp imperfect inverted repeat sequences (dark boxes). (Adapted from Hajduk and Sabatini, 1996.)

Table 1. mRNAs in *T. brucei*

Gene	No. of uridines Added	No. of uridines Deleted	Edited size (nt)	Lifestage edited
CYb	34	0	1,151	Pro[a]
A6	447	28	811	Pro/Bs[b]
COI	0	0		Unedited[c]
COII	4	0	663	Pro
COIII	547	41	969	Pro/BS
ND1	0	0		Unedited
ND3	210	13	452	Unknown
ND4	0	0		Unedited
ND5	0	0		Unedited
ND7	553	89	1,238	5' Pro/BS, 3' BS[d]
ND8	259	46	574	BS
ND9	345	20	649	BS
S12	132	28	325	BS
MURF 1	0	0		Unedited
MURF 2	26	4	1,111	Pro/BS
CR 3	148	13	299	BS
CR 4	325	40	567	BS

[a] Pro, transcript is edited only in the procyclic developmental stage.
[b] Pro/BS, transcript is edited in both bloodstream and procyclic stages.
[c] Editing of these transcripts has not been reported.
[d] The ND7 transcript is differentially edited in the procyclic and bloodstream developmental stages.

gRNA genes suggests that other sequences might have similar functions in these minicircles (van der Spek et al., 1991; Riley et al., 1994).

T. brucei contains at least 250 different minicircle sequence classes and each minicircle contains 3 gRNA genes; thus, a minimum of 750 different gRNAs are potentially encoded by these molecules. Strains of *L. tarentolae* (UC strain), *C. fasciculata* or *T. cruzi* that are maintained in the laboratory for extended periods of cultivation have a much lower minicircle sequence heterogeneity. In *L. tarentolae* (UC strain) there are only 17 different minicircle encoded gRNA genes, consistent with the reduced amount of editing seen in these organisms (Thiemann et al., 1994; Yasuhira and Simpson, 1995; Avila and Simpson, 1995; Sturm and Simpson, 1991). The reduction in RNA editing and minicircle sequence diversity is likely to be a consequence of laboratory growth because more recently established isolates of *L. tarentolae* (LEM 125) show more extensive heterogeneity of minicircle sequences and a corresponding increase in the extent of RNA editing. For example, editing of maxicircle encoded transcripts for components of NADH dehydrogenase (complex I) is

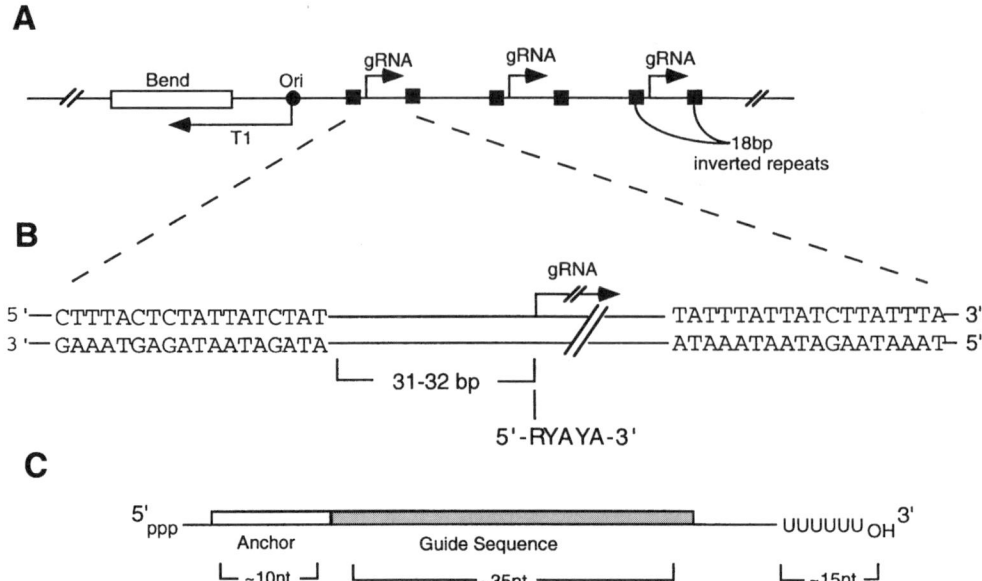

Figure 2. Genomic organization of minicircle gRNA genes. (A) Linear map of the minicircle of *T. brucei* (see Fig. 1). The transcript T1 is probably a primer for DNA replication. (B) A single gRNA transcription unit, showing the 18 bp repeats, start site for transcription and spacing between upstream repeat and transcription start site. (C) General features of gRNAs. $5'_{ppp}$ indicates that gRNAs are primary transcripts, the anchor base pairs with the preedited mRNA, the guide sequence directs U addition or deletion, and the 3' oligo(U) tail is added posttranscriptionally. (Adapted from Hajduk and Sabatini, 1996.)

reduced in the UC strain of *L. tarentolae*. These transcripts are extensively edited in the LEM 125 strain, correlating with additional minicircle and maxicircle encoded gRNAs (Thiemann et al., 1994).

The intriguing structure of the kDNA network was the first of many features of trypanosomatids that attracted the interest of molecular biologists (Borst, 1991). With the newfound knowledge that minicircles do indeed have a coding function, less attention has been paid to the function of the minicircle as an element of the kDNA network. It is worth considering that this unique DNA structure might be involved in the assembly or segregation of the editing machinery within the mitochondria, reminiscent of the sequestration of spliceosomes within the nucleus of mammals.

Recent studies on the kDNA of *T. borreli* suggest that the minicircle may have evolved from large, 180-kb, circular DNAs that contain tandemly arranged gRNA genes (Maslov and Simpson, 1994; Yasuhira and Simpson, 1996). This raises the interesting possibility that the minicircles of trypanosomatids may be the product of recombination events resulting in the generation of multiple small gRNA-containing molecules. The catenated structure of the kDNA network may provide a mechanism to ensure equal segregation of the different minicircle sequence classes (Borst, 1991).

Developmental Regulation of Mitochondrial Biogenesis

Many trypanosomatids have complicated life cycles, alternating between a vertebrate host and an insect vector (Vickerman, 1985). This is particularly true of the African trypanosomes that include *T. brucei*. A number of morphological and biochemical changes accompany the transition from mammalian to insect environment. The most striking of these changes involve mitochondrial ATP production (Priest and Hajduk, 1994; Opperdoes, 1987). In the bloodstream of the mammal mitochondrial activities are suppressed and ATP production is restricted to substrate level phosphorylation by glycolysis (Fairlamb, 1989). Following ingestion by the insect vector, the bloodstream trypanosomes differentiate to the procyclic developmental stage in the midgut of the tsetse fly. Procyclics have a complete cytochrome-mediated electron transport system and ATP is produced by oxidative phosphorylation (Bienen et al., 1991; Clarkson et al., 1989).

Regulation of mitochondrial biogenesis is poorly understood in trypanosomes. Many of the developmentally regulated proteins are nucleus encoded, and differences in the stability of both mRNAs and newly synthesized proteins contribute to the regulation of mitochondrial assembly (Torri et al., 1993; Priest and Hajduk, 1994). In addition, RNA editing is regulated during the developmental cycle of *T. brucei* and may play a critical role in the control of mitochondrial biogenesis in these organisms.

RNA EDITING IS GUIDE DIRECTED

Primary Structure of Guide RNAs

The discovery of gRNAs established an information base for the editing of mRNAs (Blum et al., 1990). With the possible exception of the internal guide sequence within the cytochrome oxidase subunit II 3' untranslated region, all gRNAs appear to be transcribed from their own genes and associate with preedited mRNAs within mitochondrial RNPs.

Each gRNA is complementary to the sequence of a portion of an edited mRNA (Fig. 3). Typically, the region of complementarity is short (about 45 nucleotides [nt]) and is defined both by conventional Watson–Crick base pairing (A–U and G–C) and also G–U base pairing. Thus, while gRNAs are complementary to edited sequences, they cannot serve as conventional templates for the polymerization of edited sequences.

Near the 5' terminus of each gRNA is a sequence of approximately 10–15 nt that is complementary to unedited mRNA sequences immediately 3' to an editing region. The base pairing of this "anchor" sequence with the mRNA at the editing region is an important step in editing (Blum et al., 1990). This short duplex region could be necessary for the assembly of the editing complex or could be directly involved in the formation of the catalytic site for editing.

Another universal feature of gRNAs is the presence of a 5–15 nt oligo(U) tail at the 3' terminus (Fig. 3) (Blum and Simpson, 1990; Arts et al., 1995; Arts et al., 1993; Thiemann and Simpson 1996). The oligo(U) tail is added posttranscriptionally by the terminal uridylyl transferase (TUTase) present in kinetoplastid mitochondria (Bakalara et al., 1989). The presence of an oligo(U) tail led to the proposal that gRNAs might serve as the donor of added U's or the acceptor of deleted U's during editing (Cech, 1991; Blum et al., 1991). Direct evidence that the oligo(U) is involved in donating or accepting U's is lacking and, as will be discussed later in this chapter, there is substantial evidence that free UTP is the source of added U's in *Leishmania* and trypanosomes.

Pre-mRNA Editing Proceeds 3' to 5'

Some of the earliest studies on kinetoplastid mRNA editing focused on the sequencing of partially edited cDNAs (Decker and Sollner-Webb, 1990; Koslowsky et al., 1990; 1991; Maslov et al., 1992; Read et al., 1992; Shaw et al., 1988; Souza et al., 1992; 1993; Sturm and Simpson, 1990b; Sturm et al.,

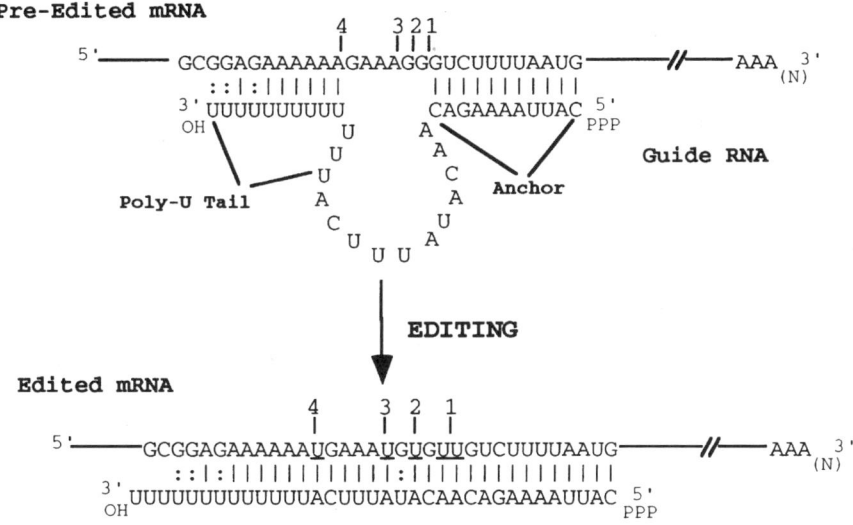

Figure 3. Proposed structure of a gRNA hybridized to preedited and edited mRNA. Two distinct duplexes are formed between preedited mRNA and its cognate gRNA. An "anchor" duplex exists where the 5' end of the gRNA hybridizes to the preedited RNA immediately 3' to the first editing site. The first base pair mismatch of this duplex may direct the editing site-specific endonuclease reaction. A second duplex exists between the oligo(U) tail of the gRNA and the purine rich region 5' to the editing site. This duplex may serve to hold the 5' cleavage fragment after endonuclease cleavage occurs. The sequence of the gRNA between these two duplexes directs the editing at sites 1 through 4. Editing is complete when the gRNA is able to hybridize throughout its entire region with the mRNA.

1992; Abraham et al., 1988). A general pattern of cDNAs edited in 3' portions of the mRNA but containing unedited domains that were more 5' along the mRNA suggested that editing of an mRNA might initiate at the 3' most editing domain and proceed in a 3' to 5' direction along the mRNA. The mechanism for this apparent directionality was suggested by the analysis of gRNA sequences. While a large number of gRNAs have been identified that are complementary to the edited mRNAs, only a minor subset of the gRNAs showed sequence complementary to unedited mRNA sequences (Koslowsky et al., 1992; Maslov and Simpson, 1992). This complementarity is between sequences near the 5' end of the gRNA and pre-mRNA sequences immediately adjacent to the 3' most editing site. Base pairing leads to the formation of the short 10–15 nt anchor duplex. New gRNA–mRNA anchors are formed by the editing of the mRNA with the initial gRNA. Following editing of the first sequence block, a new gRNA is able to base pair, via its anchor, with the partially edited mRNA and direct the editing of an extended portion of the pre-mRNA (Maslov and Simpson, 1992; Corell et al., 1993).

Because editing of a region of the mRNA results in the formation of a duplex between the mRNA and gRNA, it is necessary that the 3' part of the first gRNA dissociates from the mRNA so that the next gRNA can form an anchor duplex with the newly edited sequence. Little is known about the mechanism for extending the editing domain of an mRNA, but it is tempting to speculate that the dissociation of the initial gRNA is mediated by an RNA helicase activity associated with the editing complex. Recently, a novel mitochondrial DEAD-box protein, mHEL61p, has been identified in *T. brucei* (Missel and Göringer, 1994; Missel et al., 1995) and its gene has been cloned (Missel et al., 1997). When the gene for this protein is deleted, the abundance of edited mRNA is reduced in vivo, while the in vitro editing at a single editing site is unaffected (Missel et al., 1997). These results are consistent with mHEL61p being involved in the formation of a fully edited mRNA but not as a catalytic component of the editing complex.

BIOCHEMICAL MECHANISMS FOR RNA EDITING

Transesterification and Enzymatic Models

Several different pathways for RNA editing have been proposed. All propose a role for the gRNA in directing editing and allow for an overall 3' to 5' directionality of the process. However, the models differ significantly in the proposed chemical reactions that result in internal U addition or deletion. One model proposes a mechanism analogous to the transesterification events that occur in RNA splicing. The other model proposes enzymatic cleavage and ligation in a mechanism that is more analogous to tRNA splicing.

The transesterification model was based on the direct involvement of the oligo(U) tail of the gRNA in addition or deletion reactions. Cech (1991) and Blum et al. (1991) proposed that the 3'OH of the gRNA could serve as a nucleophile attacking the phosphodiester bond at the editing site. This would result in the formation of an intermediate in editing that would be formed by the covalent joining of the 3' end of the gRNA with the 3' half of the mRNA at an editing site. Such chimeric molecules have been detected by polymerase chain reaction analysis of mitochondrial RNA (Blum et al., 1991; Arts et al., 1993; Read et al., 1992) and are formed in vitro when complementary gRNAs and mRNAs are incubated with mitochondrial extracts (Koslowsky et al., 1992; Harris and Hajduk, 1992; Blum and Simpson, 1992; Arts et al., 1995). As described in the next section, an alternative mechanism called enzymatic cleavage and ligation is the more likely mechanism of both insertion and deletion editing in kinetoplastids.

Cleavage-Ligation Mechanism of RNA Editing

Currently, the most favored model for RNA editing involves the addition of U residues directly to internal sites in the preedited mRNA by the action of the mitochondrial TUTase (Bakalara et al., 1989). Interestingly, this model is very similar to the first model proposed for RNA editing by Blum et al. (1990). Support for this model comes from studies that revealed the presence of editing site endonuclease, RNA ligase and TUTase activities in *Leishmania* and trypanosome mitochondrial extracts (White and Borst, 1987; Bakalara et al., 1989; Pollard et al., 1992; Sabatini and Hajduk, 1995; Rusché et al., 1995) and, more recently, from in vitro editing assay systems (Seiwert and Stuart, 1994; Seiwert et al., 1996; Kable et al., 1996; Cruz-Reyes and Sollner-Webb, 1996; Corell et al., 1996; Byrne et al., 1996).

In this mechanism, a binary complex between the gRNA and the preedited mRNA forms by base pairing of the anchor and the oligo(U) tail of the gRNA 3' and 5' to the editing sites, respectively (Fig. 4). This complex is recognized and cleaved by an editing site-specific endoribonuclease at the phosphodiester bond immediately 5' to the anchor duplex. U's are added to the 5' mRNA cleavage product by the mitochondrial TUTase. Following U addition, the

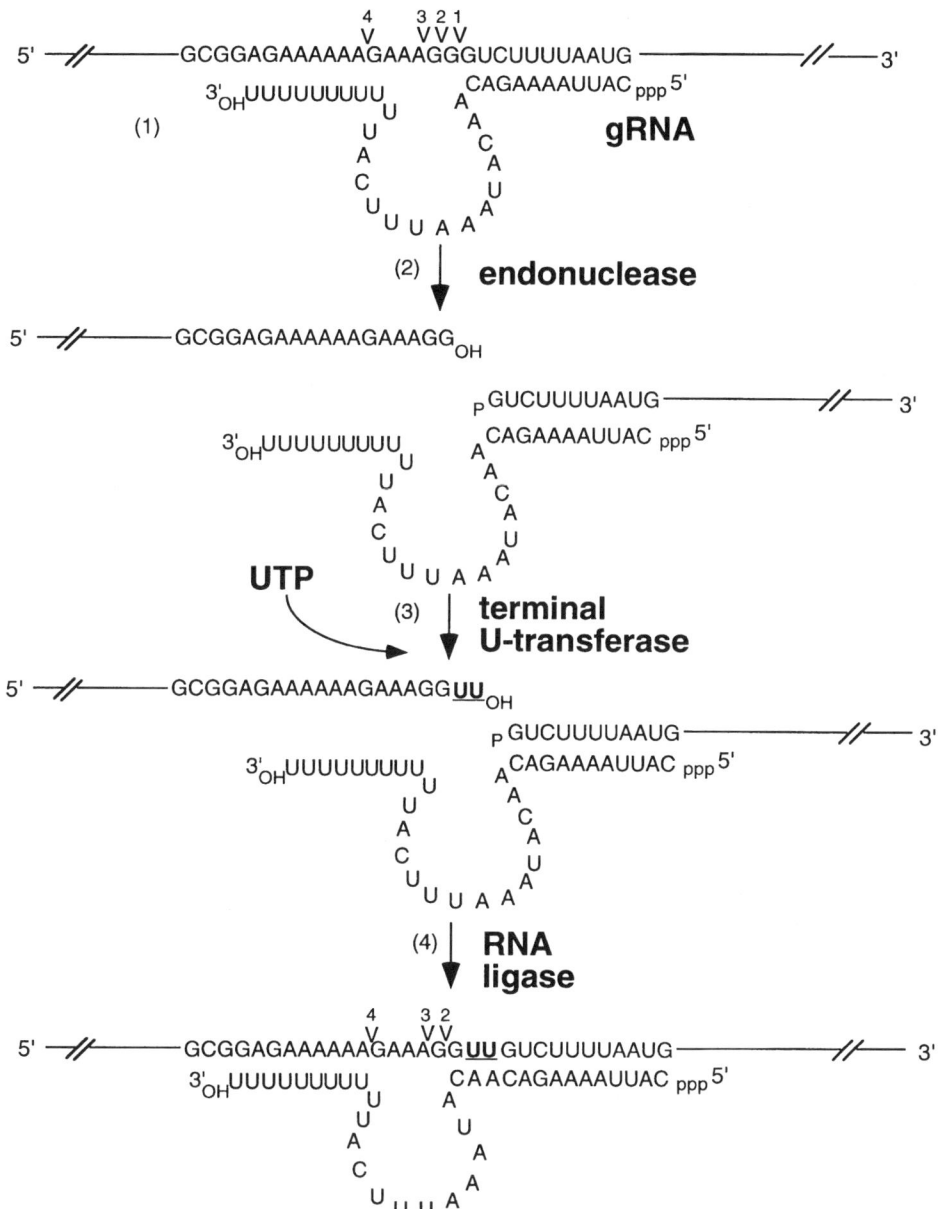

Figure 4. Schematic drawing of the cleavage-ligation mechanism for insertional RNA editing. An edited cytochrome *b* mRNA is formed by the sequential formation of the binary gRNA-mRNA structure (step 1). This is followed by endonuclease cleavage of the pre-mRNA at the editing site immediately adjacent to the duplex gRNA-mRNA anchor (step 2). The 5' fragment of the pre-mRNA contains a 3' terminal hydroxyl which is the substrate for the addition of U's by the mitochondrial TUTase (step 3). Following U addition the cleavage fragments are joined by the mitochondrial RNA ligase to form the edited mRNA product (step 4).

mRNA fragments are rejoined by RNA ligase. Deletion editing is likely to be mediated by a similar sequence of cleavage, U removal, and ligation. This reaction leads to the release of UMP and may be mediated either by the reverse action of TUTase or by a specific exoribonuclease activity. Evidence for the latter has recently been provided (Cruz-Reyes and Sollner-Webb, 1996). The overall pathway for the cleavage ligation mechanism of RNA editing is depicted in Fig. 4. The editing reaction can be broken down into four distinct steps. Step 1 involves the formation of a binary RNA complex between the gRNA and its cognate preedited mRNA. This is likely to require conformation changes in the structure of the gRNA and preedited mRNA to facilitate formation of stable anchor and oligo(U) duplexes between the RNAs (Schmid et al., 1995). The incoming gRNA and pre-mRNAs may bring together a complement of editing complex proteins, or the formation of the binary RNA structure may initiate editing complex assembly by the formation of unique RNA structures contributed by the heterodimeric RNA structure. After the formation of the gRNA and preedited mRNA duplex, in step 2 the preedited mRNA is cleaved by an editing site-specific endoribonuclease. This enzyme is part of a mitochondrial RNP and cleaves the preedited mRNA immediately adjacent to the anchor duplex formed with the 5' sequences of the gRNA (Adler and Hajduk, 1997; Seiwert et al., 1996; Piller et al., 1997). Endonuclease cleavage is protease sensitive and ATP independent. In vitro, the site of cleavage can be altered by changes to the gRNA sequences that lead to the creation of a new anchor duplex with the preedited mRNA substrate (Adler and Hajduk, 1997; Seiwert et al., 1996). In step 3, cleavage of the preedited mRNA produces a 5' fragment with a 3' terminal OH and a 3' fragment with a 5' terminal phosphate. U's are now either added to or deleted from the 3' end of the 5' cleavage product by TUTase or deleted by an exoribonuclease activity. U's added during in vitro editing reaction are derived from the free pool of mitochondrial UTP (Kable et al., 1996). It is likely that deleted U's are returned directly to the free pool of nucleotides, but this has not been demonstrated. In step 4, formation of the final mRNA product is mediated by the action of a mitochondrial RNA ligase (Sabatini and Hajduk, 1995; Rusché et al., 1995). Formation of the final edited product is protein dependent and requires the hydrolysis of ATP at the α-β position, as predicted if an RNA ligase were to catalyze the reaction. Two proteins of 50 and 57 kDa have been identified in *T. brucei* mitochondria (Sabatini and Hajduk, 1995; Rusché et al., 1995) that are adenylated in the presence of ATP and have been shown to be the adenylated forms of RNA ligase.

Several lines of evidence argue against the transesterification model and support the cleavage-ligation model. First, stereochemical analysis of in vitro U insertions into mRNAs by using *L. tarentolae* mitochondrial extracts is not consistent with U's being transferred directly from the oligo(U) tail of the gRNAs to the editing site (Frech and Simpson, 1996). Second, the abundance of chimeric molecules within the mitochondrion of *T. brucei* is extremely low, numbering less than a molecule per cell (Riley et al., 1995). Third, 3' cleavage products appear in time course experiments before the formation of edited RNAs, suggesting that they are intermediates in the reactions (Seiwert et al., 1996). Fourth, the initial pre-mRNA 5' cleavage product formed in the ATPase 6 deletion assay has two U's at its 3' terminus that are removed following cleavage (Seiwert et al., 1996). Finally, in time course experiments chimeras fail to appear until after the detection of edited products, suggesting that they are not intermediates in the formation of edited mRNAs but may be aberrant products of the in vitro reactions (Seiwert et al., 1996; Kable et al., 1997).

INVOLVEMENT OF MITOCHONDRIAL RNPs IN RNA EDITING

Identification of Mitochondrial RNPs

The proposed mechanism of RNA editing in kinetoplastids suggests that multiple proteins and RNAs are involved in a series of enzymatic reactions. It was natural to predict that the preedited mRNA substrate and cognate gRNAs might be assembled with the predicted editing site-specific endoribonuclease, exoribonuclease, TUTase, RNA ligase, and RNA helicase into a ribonucleoprotein complex (RNP) analogous to the spliceosome. Recent studies have shown that both the predicted components of the editing machinery and RNA editing activity are RNP associated.

Mitochondrial editing complexes were first identified in mitochondrial extracts of *T. brucei* (Pollard et al., 1992). Glycerol gradient ultracentrifugation resolved two gRNA-containing mitochondrial RNPs: one of 19S, and a second more heterogeneous complex of 35–40S. Under physiological conditions, preedited and edited mRNAs were found associated exclusively with the 35–40S complexes. Several enzymatic activities that might be associated with editing (including RNA ligase and TUTase) were found to cosediment with both complexes on glycerol gradients. The RNA ligase was evenly distributed in the 19S and 35–40S complexes, while the TUTase activity was mainly associated with the 19S complexes. The editing site-specific endonuclease was not found associated with either of the mitochondrial RNPs (Pollard, et al., 1992). However, more recently an editing site-specific endonuclease has been found associated both the 19S and 35–40S complexes (Adler and Hajduk, 1997; Piller et al., 1997). The failure to initially detect this activity in the mitochondrial RNPs was because of the presence of competing RNA ligase activity in the same gradient fractions. By depletion of the extract for ATP, ligase activity diminished while endonuclease cleavage was unaffected, allowing the accumulation of cleavage products. Other studies with *T. brucei* mitochondrial extracts (Göringer et al., 1994, 1995; Corell et al., 1996) are largely consistent with the presence of two classes of gRNA associated mitochondrial RNPs in trypanosomes.

Mitochondrial extracts from *L. tarentolae* also contain two gRNA-containing RNPs, which appear to be more heterogeneous than the mitochondrial

RNPs of *T. brucei*. The "T" class of *L. tarentolae* mitochondrial RNPs is composed of multiple RNP complexes that sediment at about 10S (Peris et al., 1994, 1997). The "T" complexes superficially resemble the 19S RNPs of *T. brucei* by containing TUTase activity, some RNA ligase activity and gRNAs. These complexes have been proposed to be involved in the addition of U's to the 3' ends of gRNAs. This is consistent with the inability to detect either editing site-specific endonuclease activity or in vitro RNA editing activity with these complexes (Peris et al., 1994, 1997). The "G" class of mitochondrial RNPs is large particles of 170–300 nm (approximately 20S) containing gRNAs and a gRNA independent RNA editing activity. The exact nature of this U addition activity is not known; however, it appears to be dependent on the structure of the mRNA (Frech et al., 1995; Frech and Simpson, 1996; Connell et al., 1997).

It is tempting to speculate that the 19S complexes of *T. brucei* and the T complexes of *L. tarentolae* might be functionally related. Despite several unresolved differences in the size and enzymatic activities associated with these complexes, it seems likely that they function in the maturation of the gRNAs by the addition of the 3' oligo(U) tail. In addition, these preassembled complexes might play a role in the selection of the cognate preedited mRNAs. Again, despite obvious differences, it is also possible that 35–40S complexes of *T. brucei* and the G complexes of *L. tarentolae* also share a common function. While the *Leishmania* complexes lack the gRNA dependent editing activity, this might be a consequence of the overall low levels of this activity in *Leishmania* mitochondria.

In Vitro Assembly of Editing Complexes

To determine the function of the individual components of the editing complexes, it will first be necessary to develop in vitro reconstitution assays with purified components of the mitochondrial RNPs. While detailed analysis awaits the cloning and purification of the editing complex components, preliminary analysis of RNP assembly using *T. brucei* mitochondrial extracts with synthetic gRNA (Read et al., 1994a; Göringer et al., 1994) and pre-mRNAs (Koslowsky et al., 1996) has been reported. Formation of these RNPs appears to be specific for gRNAs and pre-mRNAs because addition of nonspecific RNAs failed to block assembly. These studies suggest that as the components of the editing RNPs are identified and purified it may be possible to assemble functional RNPs. It is important to note that the relationship of these in vitro assembled RNPs to native mitochondrial RNPs has not yet been firmly established.

COMPONENTS OF THE EDITING COMPLEX

Structure of gRNAs, Pre-mRNAs, and gRNA–mRNA Binary Complexes

Guide RNAs play a central role in the editing process. The primary sequence of the gRNAs both directs the initial assembly of the editing complex by the formation of the anchor duplex with pre-mRNA sequences immediately 3' to an editing region and functions to direct the correct editing of the mRNA by serving as a template for the editing machinery. The conservation of functions in RNA editing, despite differences in both the length and the primary sequence of gRNAs, suggests that gRNAs might adopt similar higher-order structures. While a large number of gRNA sequences have been reported from *T. brucei*, *L. tarentolae*, *C. fasciculata*, and *T. borreli*, very little is known about the secondary structure of these molecules.

The secondary structures of four gRNAs from *T. brucei* have been determined by enzymatic and chemical probing (Schmid et al., 1995). These studies show that the gRNAs adopt a structure in solution that differs from the predicted lowest energy conformations. Despite a lack of primary sequence similarity, the gRNAs fold into similar secondary structures containing a short, unpaired 5' end (5–10 nt) followed by two hairpin elements that are separated by a short (3–12 nt) single-stranded region. The 3' oligo(U) tail of the gRNAs is most likely single stranded, but enzymatic analysis suggests that it may be in a helical single-strand conformation. Changes in the length or deletion of the oligo(U) tail do not influence the folding of the gRNAs, which is consistent with the length variation in oligo(U) tails seen in native gRNA populations.

The secondary structures determined for the four gRNAs, while similar, have Gibbs free energies well below those predicted for the most stable structures for these RNAs. This is consistent with the low thermodynamic stability of these RNA structures. The four gRNAs melt at 33–39°C and with enthalpies of -32 to -38 kcal/mol (Schmid et al., 1995). This suggests that gRNAs may have conserved secondary structures, perhaps functioning as protein binding sites (Köller et al., 1997), but can be unfolded with a minimal amount of energy. This is appealing since both recognition of the editing site on the pre-mRNA and the templating of editing would require conformation changes in the gRNA.

The secondary structures of cytochrome *b* and cytochrome oxidase II pre-mRNAs from *T. brucei* have been examined by enzymatic probing (Piller et al., 1995a, 1995b). These studies suggest that the editing site of the pre-mRNA may form a hairpin structure that is recognized by the editing site-specific endonuclease of the editing complex. There is support for this model in the cleavage of the mRNA at the editing site by single strand-specific mung bean nuclease (Piller et al., 1995b). However, in vitro editing (Seiwert et al., 1996; Kable et al., 1996) and pre-mRNA cleavage reactions (Adler and Hajduk, 1997) indicate that the efficient editing and cleavage of pre-mRNA by editing complex associated endoribonuclease requires the formation of a duplex anchor between the pre-mRNA and its cognate gRNA. An interesting possibility is that the secondary structures seen in the pre-mRNAs might be protein binding sites and might be important either in the assembly of the editing complex or in regulation of editing.

The structure of the gRNA–mRNA binary complex is completely unknown. Several structures have been proposed (Blum et al., 1991). Based on the current model for RNA editing, it is likely that both 5' and 3' gRNA sequences participate in the formation of duplex regions with the pre-mRNA (Blum and Simpson, 1990; Seiwert et al., 1996; Kable et al., 1996). The functional significance of the anchor duplex, formed by base pairing of the 5' sequences of gRNAs with complementary pre-mRNA sequences 3' to the editing site, is well documented. Less clear is the potential interaction of the 3' oligo(U) tail of the gRNA with the pre-mRNA. It has been suggested that the oligo(U) tail might base pair with the purine-rich region of the pre-mRNA 5' to the editing site (Blum and Simpson, 1990). This duplex RNA of 10–15 nt might serve to further stabilize the initial interactions of the cognate gRNA and pre-mRNA, and might facilitate editing complex assembly.

Protein Components of RNA Editing Complexes

As previously described, specific protein–RNA interactions have been studied by the in vitro assembly of gRNA–RNPs (Göringer et al., 1994; Read et al., 1994; Frech et al., 1995). Two *L. tarentolae* mitochondrial proteins of 18 kDa and 51 kDa associate with gRNAs (Bringaud et al., 1995). The genes for the 18- and 51-kDa proteins have been cloned and sequenced. The 51-kDa protein shows significant homology to mitochondrial aldehyde dehydrogenase (Bringaud et al., 1995), while the 18-kDa protein may represent the F_0F_1 ATPase subunit b homolog (Speijer et al., 1997). A third *L. tarentolae* gRNA binding protein was identified in UV cross-linking experiments. This 110-kDa protein is abundant in *L. tarentolae* mitochondria and was shown to associate with the 10S T complexes, which are rich in TUTase activity (Bringaud et al., 1997). The sequence of the gene encoding this protein was determined and found to have approximately 37% identity and 57% similarity to the mitochondrial glutamate dehydrogenase of *Neurospora crassa* (Kapoor et al., 1993). In addition, a highly conserved NAD binding domain was present in the *Leishmania* sequence. This protein binds specifically to the oligo(U) tail of gRNA, since heterologous RNAs with 3' oligo(U) tails also bind the 110-kDa protein. These studies indicate that glutamate dehydrogenase is an oligo(U)-binding protein and capable of binding gRNAs in vitro with oligo(U) tails. However, experiments to examine the direct involvement of this protein in RNA editing have not been reported.

The identification of abundant metabolic enzymes as editing complex proteins should be viewed with some caution. However, other dinucleotide binding proteins have been proposed to play important roles as RNA-binding proteins in other systems (Hentze, 1994; Singh and Green, 1993; Preiss et al., 1993).

In a series of UV cross-linking studies, gRNAs binding proteins have also been identified from *T. brucei* mitochondrial extracts. Proteins of 90, 25, and 9 kDa bind gRNA in the presence of 100 mM KCl, which suggests high affinities for gRNAs (Read et al., 1994; Köller et al., 1994). An additional five proteins ranging in size from 124 to 9 kDa bind to gRNAs at a lower affinity. When different gRNAs specific to different mRNA editing sites are incubated with mitochondrial extracts, the same eight proteins are able to bind. This suggests that there may be a very limited number of proteins in trypanosome mitochondria that recognize and bind gRNAs (Köller et al., 1994).

Binding of the 90-kDa protein to gRNAs is dependent of the presence of an oligo(U) tail (Köller et al., 1994) The sequence of this protein and a detailed analysis of its RNA binding properties await production of recombinant protein.

A nuclear gene for the 21-kDa protein, which corresponds to the 25-kDa protein identified by UV cross-linking experiments, has been cloned from *T. brucei*. The protein does not share significant homology with known proteins, is rich in arginine, and specifically binds gRNAs (Köller et al., 1997). The protein, designated gRNA-binding protein 21 (gBP21), binds gRNAs with different primary sequences with high affinity, nanomolar dissociation constants. Furthermore, binding does not require the presence of the 3' oligo(U) tail on the gRNA, suggesting that gBP21 recognizes a higher-order structure conserved in divergent gRNAs.

In mitochondrial extracts from *C. fasciculata*, three polypeptides of 88, 65, and 30 kDa cross-link to gRNAs (Leegwater et al., 1995). Binding was dependent on the presence of the 3' oligo(U) tail on the gRNA, suggesting that these proteins may represent a class of oligo(U) associating proteins. Evidence that these proteins are directly involved in RNA editing has not been provided.

Recently, purification of *T. brucei* editing activity from mitochondrial extract has identified a 20S complex composed of eight polypeptides that is competent for a single round of deletional editing in vitro (Rusché et al., 1997). Two of these polypeptides include the 50- and 57-kDa adenylatable RNA ligases. This purified fraction also contains RNA ligase, TUTase and endonuclease activities that are believed to be required for editing, but not gBP21 or RNA helicase activity. gRNA and preedited RNA are also absent from this complex. While it is interesting that a purified fraction containing eight visible polypeptides is able to catalyze in vitro editing, further work is needed to identify the proteins and determine their role (if any) in RNA editing.

Using an alternative approach to identify the individual components of editing complexes, a battery of monoclonal antibodies was prepared against the 30–45S editing complexes from *T. brucei* (Pollard et al. 1992; Antenucci et al., submitted for publication). One of these monoclonal antibodies reacted with of protein of 45 kDa in the mitochondrial 30–45S RNPs. Immunofluorescence microscopy localized the protein to the trypanosome mitochondria. The presence of p45 within the mitochondrial matrix was confirmed by immunoelectron microscopy (Russell, personal communication). The gene for the p45 protein was cloned from a *T. brucei* cDNA expression library and sequenced. The predicted protein contains a series of internal 21 amino acid repeats rich in positively charged amino acids, and a putative mitochondrial localization signal at the N terminus of the protein (Fig. 5). The role of p45 in RNA editing was demonstrated by the inhibition of in vitro RNA editing assays by antibody to p45. This RNA editing-associated protein, REAP-1, is the first protein shown physically and functionally to be an essential component of an editing complex. While the function of REAP-1 in editing is unknown, the presence of positively charged domains suggests that it may be an RNA-binding protein. Experiments are under way to address the function of this protein in RNA editing.

Enzymatic Activities Associated with Kinetoplastid Mitochondrial RNPs

Based on the cleavage-ligation model for RNA editing previously proposed, there are a number of

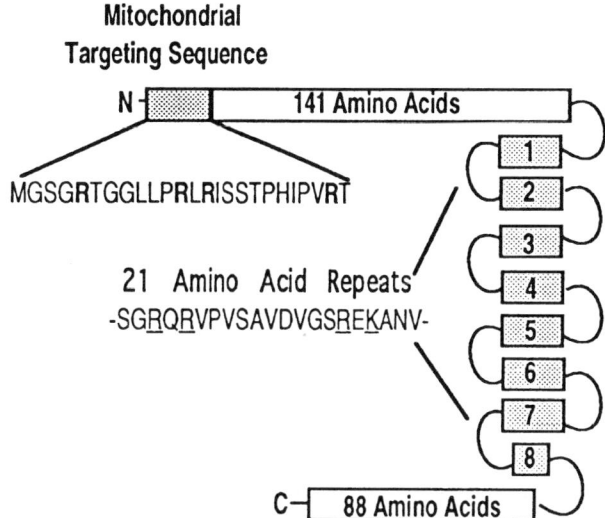

Figure 5. Mitochondrial editing complex-associated protein 1 (REAP-1). Schematic drawing showing the general features of REAP-1. The sequence and relative position of the positively charged 21 amino acid repeats (positively charged residues are underlined) and a putative mitochondrial targeting sequence are shown. The role of the targeting sequence in mitochondrial localization has not been established.

enzymatic activities predicted to be associated with RNA editing complexes. If one examines the pathway for U insertion, the initiating enzymatic reaction is the cleavage of the pre-mRNA by an editing site-specific endoribonuclease. As previously mentioned, an endoribonuclease activity with specificity for cleavage of preedited, but not edited, mRNAs was first described for mitochondrial extracts from *T. brucei* (Harris et al., 1992). A similar editing site-specific activity has also been identified in *L. tarentolae* mitochondrial extracts (Simpson et al., 1992).

Several studies have shown that the editing site-specific cleavage activity is dependent on the presence of the cognate gRNA (Adler and Hajduk, 1997; Piller et al., 1997). Additional gRNA-independent endoribonuclease activities have also been detected at varying positions in glycerol gradients (Piller et al., 1997). The relationship of these endoribonuclease activities to one another and their role in RNA editing are not well established. None of the mitochondrial endoribonucleases has been purified, nor have their genes been cloned.

According to the model, U insertion is likely to be mediated by the activity of a mitochondrial TUTase. The mitochondrial TUTase may have more than one function, however, because U's are added not only to internal sites within mRNAs but also to the 3' ends of gRNAs and of the 9S and 12S rRNAs (Adler et al., 1991). While it has not been experimentally demonstrated that RNA tailing and insertional editing are mediated by the same enzyme, it seems likely. If so, then the enzyme may react differently with, for example, the gRNA and mRNA substrates because the product of gRNA tailing is a extension of 10–15 U's, suggesting that the TUTase may display some degree of processivity while the internal addition of U's into mRNA may proceed one nucleotide at a time (McManus and Hajduk, unpublished data).

The rejoining of the two halves of the mRNA following addition of U's to the 5' cleavage fragment by TUTase requires the presence of RNA ligase. The presence of RNA ligase in L. tarentolae mitochondria was demonstrated shortly after the discovery of RNA editing (Bakalara et al., 1989) in L. tarentolae. The enzyme was shown to be associated with the 19S and 34–40S T. brucei mitochondrial RNPs, and adenylation of complexes revealed the presence of 50- and 57-kDa adenylated RNA ligase intermediates (Sabatini and Hajduk, 1995; Rusché et al., 1995). Similar adenylated proteins, of 45 and 50 kDa, are associated with a 20S mitochondrial RNP in L. tarentolae (Peris et al., 1997). While the 50- and 57-kDa adenylated proteins of T. brucei have been shown to be ATP activated forms of RNA ligase and necessary for RNA editing, the genes for these proteins have not been cloned. It is also unclear what function two distinct RNA ligases might have in RNA editing.

A surprising omission from the list of enzymes associated with RNA editing RNPs is an RNA helicase. While an RNA helicase has been identified within T. brucei mitochondria (Missel and Göringer, 1994) and its role in RNA editing has been established by gene disruption experiments (Missel et al., 1997), this protein is not stably associated with a mitochondrial RNP (Corell et al., 1996).

A speculative model for the involvement of mitochondrial RNPs in the editing of T. brucei mRNAs is schematically shown in Fig. 6. In 19S complexes, gRNAs are associated with a set of specific proteins, including mitochondrial RNA ligase, editing site endonuclease, gBP21, TUTase, p90, and 9-kDa gRNA-binding proteins (Pollard et al., 1992; Adler and Hajduk, 1997; Koller et al., 1994). The major function of this complex may be the maturation of gRNA by the addition of a 3' oligo(U) tail. A similar proposal has been made for the T complexes of L. tarentolae (Peris et al., 1997). Association of preedited mRNAs with the 19S complexes results in the formation of the 35–40S RNPs, which may most closely resemble the native editing complexes. It is likely that a subclass of editing complex proteins is recruited along with pre-mRNAs in the assembly to the 35–40S complexes. Among these is the 45-kDa protein that we have termed REAP-1 (Antenucci et al., submitted for publication). This pathway of RNP assembly and RNA editing complex formation is supported by an analysis of the composition of the T. brucei complexes. In addition, pulse-labeling studies indicate that UTP is initially incorporated into the 3' oligo(U) tail of the gRNAs in 19S RNPs. It then rapidly chases to a gRNA population in the 35–40S complexes. This precedes the incorporation of UTP into internal sites in the mRNA (Pollard and Hajduk, unpublished data). Finally, edited mRNAs leave the 30–45S complexes and associate with mitochondrial ribosomes. While highly speculative, this model is supported by the presence of several shared proteins in the 19S and 30–45S RNPs and pulse-chase experiments with isolated mitochondria (Pollard and Hajduk, unpublished data).

THE FUNCTION OF RNA EDITING

Role of RNA Editing in Regulating Gene Expression

A small but indispensable fraction of the mitochondrial proteins required for electron transport and ATP production is encoded by the maxicircle. Based entirely on analysis of genomic and cDNA sequences, we can predict with some certainty that RNA editing is obligatory for the formation of mitochondrial mRNAs with correct initiation and termination codons and continuous open reading frames to encode these mitochondrial proteins. However, the direct demonstration that edited mRNAs are translated into mitochondrial proteins has not been accomplished (Priest and Hajduk, 1992; Speijer et al., 1996a, 1996b). A universal feature of mitochondrial encoded proteins is their extreme hydrophobicity. This has frustrated all attempts at purification and subsequent sequencing of proteins formed by the translation of edited mRNAs. An indirect approach to demonstrate proteins specified by edited mRNAs is the use of antibodies specific for peptide epitopes specified by edited mRNA sequences. There is a single account of the successful use of this approach. Antibodies prepared against a synthetic peptide corresponding to the carboxy terminal domain of cytochrome oxidase subunit II were used to identify a protein by Western blot analysis (Shaw et al., 1989). Only the edited form of the cytochrome oxidase II mRNA could be translated to form this portion of the protein because the unedited mRNA contains a termination codon that would result in a truncated protein lacking the carboxy terminal domain. The unequivocal demonstration that edited mRNAs actually specify proteins remains an important future experiment.

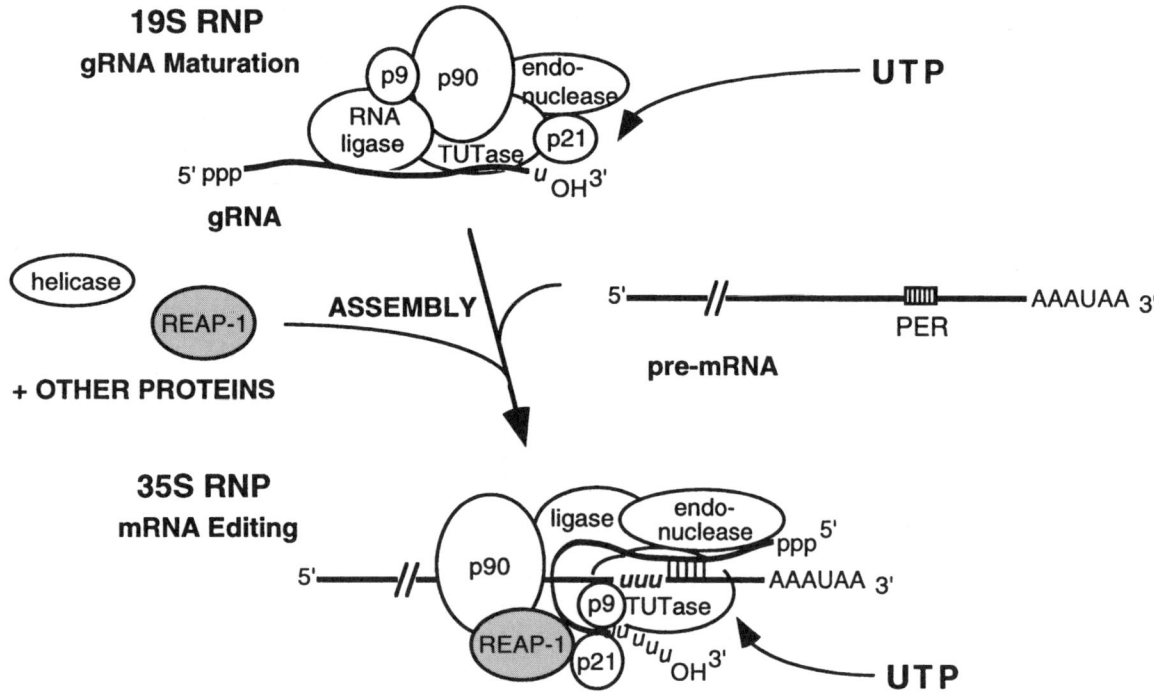

Figure 6. Proposed model for assembly and maturation of editing complexes. gRNAs are associated with specific proteins in a 19S RNP which is the initial site of posttranscriptional U addition to the gRNAs. The 19S RNP can bind pre-mRNAs and associated proteins to form a 35–40S RNP. The assembly of 35–40S RNP probably requires base pairing of the anchor region of the gRNA and mRNA immediately 3' to the preedited region (PER) of the mRNA. The complex may function to hold the 5' mRNA fragment in position for the second step in editing. Specific proteins of 9, 21, and 90 kDa have been identified by UV cross-linking (Goringer et al., 1994), while RNA ligase, TUTase, and endonuclease activities have been assayed directly (Pollard et al., 1992; Sabatini and Hajduk, 1995).

Recent results indicate that RNA editing may also influence the relative abundance of mitochondrial mRNAs. Much of the maxicircle is composed of tightly packed genes in which the 5' terminus of a downstream gene and the 3' terminus of an upstream gene are separated by only a few nucleotides. These genes are immediately juxtaposed to one another or actually overlap (Simpson, 1986). Several studies suggest that the maxicircle is initially transcribed as one or potentially several polycistronic transcripts (Read ct al., 1992, 1994b). Editing of a polycistronic precursor RNA spanning the sequence that overlaps the 3' end of S12 and the 5' end of ND5 produces a transcript with edited S12 mRNA, while effectively destroying the 5' portion of the ND5 mRNA (Read et al., 1992). Conversely, if processing of the polycistronic transcript to form the 5' terminus of the ND5 precedes editing, then the S12 sequence may be truncated. Thus, the rate of editing versus 5' end formation may be important in modulating the relative amounts of individual mRNAs derived from the same transcription unit.

In a number of eukaryotic organisms, selective or alternative mRNA processing (either by splice site selection or differential polyadenylation of mRNAs) leads to the production of different mRNAs encoded by the same gene. It is appealing to speculate that editing of trypanosome mRNAs might also allow alternative RNAs to be produced. While the production of two functionally distinct molecules has not been demonstrated at the protein level, because of the problems previously discussed, the possibility clearly exists. There are a number of examples of either selective or differential editing that could produce multiple mRNAs. The edited and unedited forms of the cytochrome *b* mRNA contain in-frame methionine codons and long opening reading frames (Feagin and Stuart, 1988; Feagin et al., 1988). Editing of the cytochrome *b* mRNA forms an AUG codon in-frame with a protein coding sequence, which shows significant homology to cytochrome *b* from other organisms. Translation of the unedited cytochrome *b* mRNA would produce a protein that differs at the N terminus from the protein produced from the edited mRNA.

RNA Editing Is Developmentally Regulated

Developmental changes in mitochondrial activities correlate with changes in the editing of specific mRNA. For example, in both *T. brucei* and *T. congolense*, the abundance of edited cytochrome *b* and cytochrome oxidase II mRNA is low in bloodstream trypanosomes but high in the procyclic trypanosomes (Feagin and Stuart, 1988; Read et al., 1994b). On the other hand, the amounts of edited mRNA from NADH dehydrogenase subunits 8 and 9 are low in the procyclic trypanosomes but increase in the bloodstream stages (Koslowsky et al., 1990; Souza et al., 1992). In the case of NADH dehydrogenase subunit 7, the developmental regulation of editing is not only transcript specific but also domain specific. In bloodstream trypanosomes, a fully edited NADH dehydrogenase subunit 7 transcript is produced, while in procyclics, editing is restricted to the 5' domain (Koslowsky et al., 1990). Differential editing of the 3' domain is predicted to yield two forms of the mRNA for NADH dehydrogenase subunit 7.

It is now clear that the editing of specific mRNAs, and even domains within an mRNA, can be developmentally modulated. Since the transcription of mitochondrial genes in trypanosomes is constitutive during the life cycle (Michelotti et al., 1992), it is reasonable to speculate that editing may play a central role in controlling mitochondrial biogenesis during the developmental cycle of the African trypanosome. The obvious question is how trypanosomes accomplish this feat. Perhaps the most appealing possibility is that the extent of editing of a given mRNA is modulated by the abundance of its cognate gRNAs. This possibility has now been examined for a number of gRNAs. Currently, no correlation has been established between the abundance of a gRNA and its edited mRNA (Koslowsky et al., 1992; Riley et al., 1995). If the availability of gRNAs does not regulate editing, then what is the mechanism? Without a detailed knowledge of the mechanism of editing it may be premature to speculate. However, one is tempted to predict that developmentally regulated protein factors, specific for particular mRNAs, might either suppress or activate the editing machinery. It will be interesting to explore the factors involved in this process as we gain a better understanding of the mechanism of editing and the components of the editosome.

REFERENCES

Abraham, J. M., J. E. Feagin, and K. Stuart. 1988. Characterization of cytochrome c oxidase III transcripts that are edited only in the 3' region. *Cell* **55**:267–272.

Adler, B. K., and S. L. Hajduk. 1997. Guide RNA requirement for editing-site-specific endonucleolytic cleavage of preedited mRNA by mitochondrial ribonucleoprotein particles in *Trypanosoma brucei*. *Mol. Cell. Biol.* **17**:5377–5385.

Adler, B. K., M. E. Harris, K. I. Bertrand, and S. L. Hajduk. 1991. Modification of *Trypanosoma brucei* mitochondrial rRNA by posttranscriptional 3' polyuridine tail formation. *Mol. Cell. Biol.* **11**:5878–5884.

Antenucci, S. M., R. S. Sabatini, V. W. Pollard, and S. L. Hajduk. Submitted for publication.

Arts, G. J., and R. Benne. 1996. Mechanism and evolution of RNA editing in kinetoplastida. *Biochim. Biophys. Acta* **1307**:39–54.

Arts, G. J., P. Sloof, and R. Benne. 1995. A possible role for the guide RNA U-tail as a specificity determinant in formation of guide RNA-messenger RNA chimeras in mitochondrial extracts of *Crithidia fasciculata*. *Mol. Biochem. Parasitol.* **73**:211–222.

Arts, G. J., H. van der Spek, D. Speijer, J. van den Burg, H. van Steeg, P. Sloof, and R. Benne. 1993. Implications of novel guide RNA features for the mechanism of RNA editing in *Crithidia fasciculata*. *EMBO J.* **12**:1523–1532.

Avila, H. A., and L. Simpson. 1995. Organization and complexity of minicircle-encoded guide RNAs in *Trypanosoma cruzi*. *RNA* **1**:939–947.

Bakalara, N., A. M. Simpson, and L. Simpson. 1989. The *Leishmania* kinetoplast-mitochondrion contains terminal uridylyltransferase and RNA ligase activities. *J. Biol. Chem.* **264**:18679–18686.

Benne, R. 1994. RNA editing in trypanosomes. *Eur. J. Biochem.* **221**:9–23.

Benne, R., J. van den Burg, J. P. J. Brakenhoff, P. Sloof, J. H. van Boom, and M. C. Tromp. 1986. Major transcript of the frameshifted coxII gene from trypanosome mitochondria contains four nucleotides that are not encoded in the DNA. *Cell* **46**:819–816.

Bienen, E. J., M. Saric, G. Pollakis, R. W. Grady, and A. B. Clarkson, Jr. 1991. Mitochondrial development in *Trypanosoma brucei brucei* transitional bloodstream forms. *Mol. Biochem. Parasitol.* **45**:185–192.

Blum, B., N. Bakalara, and L. Simpson. 1990. A model for RNA editing in kinetoplast mitochondria: "guide" RNA molecules transcribed from maxicircle DNA provide the edited information. *Cell* **60**:189–198.

Blum, B., and L. Simpson. 1990. Guide RNAs in kinetoplastid mitochondria have a nonencoded 3' oligo(U) tail involved in recognition of the preedited region. *Cell* **62**:391–397.

Blum, B., and L. Simpson. 1992. Formation of guide RNA/messenger RNA chimeric molecules *in vitro*, the initial step of RNA editing, is dependent on an anchor sequence. *Proc. Natl. Acad. Sci. USA* **89**:11944–11948.

Blum, B., N. R. Sturm, A. M. Simpson, and L. Simpson. 1991. Chimeric gRNA-mRNA molecules with oligo(U) tails covalently linked at sites of RNA editing suggest that U addition occurs by transesterification. *Cell* **65**:543–550.

Borst, P. 1991. Why kinetoplast DNA networks? *Trends Genet.* **7**:139–141.

Bringaud, F., M. Peris, K. H. Zen, and L. Simpson. 1995. Characterization of two nuclear-encoded protein components of mitochondrial ribonucleo-protein complexes from *Leishmania tarentolae*. *Mol. Biochem. Parasitol.* **71**:65–79.

Bringaud, F., R. Stripeche, G. C. Frech, S. Freedland, C. Turck, E. M. Byrne and, L. Simpson. 1997. Mitochondrial glutamate dehydrogenase from *Leishmania tarentolae* is a guide RNA-binding protein. *Mol. Cell. Biol.* **17**:3915–2913.

Byrne, E. M., G. J. Connell, and L. Simpson. 1996. Guide RNA directed uridine insertion RNA editing in vitro. *EMBO J.* **15**:6758–6765.

Cech, T. R. 1991. RNA editing: world's smallest introns? *Cell* **64**:667–669.

Clarkson, A. B., Jr., E. J. Bienen, G. Pollakis, and R. W. Grady. 1989. Respiration of bloodstream forms of the parasite *Trypanosoma brucei brucei* is dependent on a plant-like alternative oxidase. *J. Biol. Chem.* **264**:17770–17776.

Connell, G. J., E. M. Byrne, and L. Simpson. 1997. Guide RNA-independent and guide RNA-dependent uridine insertion into cytochrome b mRNA in a mitochondrial extract from *Leishmania tarentolae*. *J. Biol. Chem.* 272:4212–4218.

Corell, R. A., L. K. Read, G. R. Riley, J. K. Nellissery, T. E. Allen, M. L. Kable, M. D. Wachal, S. D. Seiwert, P. J. Myler, and K. Stuart. 1996. Complexes from *Trypanosoma brucei* that exhibit deletion editing and other editing-associated properties. *Mol. Cell. Biol.* 16:1410–1418.

Corell, R. A., J. E. Feagin, G. R. Riley, T. Strickland, J. A. Guderian, P. J. Myler, and K. Stuart. 1993. *Trypanosoma brucei* minicircles encode multiple guide RNAs which can direct editing of extensively overlapping sequences. *Nucleic Acids Res.* 21:4313–4320.

Cruz-Reyes, J., and B. Sollner-Webb. 1996. Trypanosome U-deletional RNA editing involves gRNA-directed endonuclease cleavage, terminal U exonuclease, and RNA ligase activities. *Proc. Natl. Acad. Sci. USA* 93:8901–8906.

Decker, C. J., and B. Sollner-Webb. 1990. RNA editing involves indiscriminate U changes throughout precisely defined editing domains. *Cell* 61:1001–1011.

Fairlamb, A. H. 1989. Novel biochemical pathways in parasitic protozoa. *Parasitology* 99S:S93–112.

Feagin, J. E., J. M. Abraham, and K. Stuart. 1988. Extensive editing of the cytochrome *c* oxidase III transcript in *Trypanosoma brucei*. *Cell* 53:413–422.

Feagin, J. E., D. P. Jasmer, and K. Stuart. 1987. Developmentally regulated addition of nucleotides within apocytochrome *b* transcripts in *Trypanosoma brucei*. *Cell* 49:337–345.

Feagin, J. E., J. M. Shaw, L. Simpson, and K. Stuart. 1988. Creation of AUG initiation codons by addition of uridines within cytochrome *b* transcripts of kinetoplasts. *Proc. Natl. Acad. Sci. USA* 85:539–543.

Feagin, J. E., and K. Stuart. 1988. Developmental aspects of uridine addition with mitochondrial transcripts of *Trypanosoma brucei*. *Mol. Cell. Biol.* 8:1259–1265.

Frech, G. C., N. Bakalara, L. Simpson, and A. M. Simpson. 1995. In vitro RNA editing-like activity in a mitochondrial extract from *Leishmania tarentolae*. *EMBO J.* 14:178–187.

Frech, G. C., and L. Simpson. 1996. Uridine insertion into pre-edited mRNA by a mitochondrial extract from *Leishmania tarentolae*: stereochemical evidence for the enzyme cascade model. *Mol. Cell. Biol.* 16:4584–4589.

Göringer, H. U., D. J. Koslowsky, T. H. Morales, and K. Stuart. 1994. The formation of mitochondrial ribonucleoprotein complexes involving guide RNA molecules in *Trypanosoma brucei*. *Proc. Natl. Acad. Sci. USA* 91:1776–1780.

Göringer, H. U., J. Köller, and H. H. Shu. 1995. Multicomponent complexes involved in kinetoplastid RNA editing. *Parasitol. Today* 11:265–267.

Gott, J. M., L. M. Visomirski, and J. L. Hunter. 1993. Substitutional and insertional RNA editing of the cytochrome *c* oxidase subunit 1 messenger RNA of *Physarum polycephalum*. *J. Biol. Chem.* 268:25483–25486.

Grams, J. M., and S. Hajduk. Unpublished data.

Hajduk, S. L., M. E. Harris, and V. W. Pollard. 1993. RNA editing in kinetoplastid mitochondria. *FASEB J.* 7:54–63.

Hajduk, S. L., and R. S. Sabatini. 1996. RNA editing: posttranscriptional restructuring of genetic information, p. 134–158. In D. F. Smith and M. Parsons (ed.), *Molecular Biology of Parasitic Protozoa*. Oxford University Press, New York, N.Y.

Hancock, K., and S. L. Hajduk. 1990. The mitochondrial tRNAs of *T. brucei* are nuclear encoded. *J. Biol. Chem.* 265:19208–19215.

Hancock, K., A. J. LeBlanc, D. Donze, and S. L. Hajduk. 1992. Identification of nuclear encoded precursor tRNAs within the mitochondrion of *Trypanosoma brucei*. *J. Biol. Chem.* 267:23963–23971.

Harris, M., C. Decker, B. Sollner-Webb, and S. Hajduk. 1992. Specific cleavage of pre-edited mRNAs in trypanosome mitochondrial extracts. *Mol. Cell. Biol.* 12:2591–2598.

Harris, M. E., and S. L. Hajduk. 1992. Kinetoplastid RNA editing: in vitro formation of cytochrome *b* gRNA-mRNA chimeras from synthetic substrate RNAs. *Cell* 68:1091–1099.

Hentze, M. W. 1994. Enzymes as RNA-binding proteins: role for (di)nucleotide-binding domains. *Trends Biochem. Sci.* 19:101–103.

Kable, M. L., S. D. Seiwert, S. Heidmann, and K. Stuart. 1996. RNA editing: a mechanism for gRNA-specified uridylate insertion into precursor mRNA. *Science* 273:1189–1195.

Kable, M. L., S. Heidmann, and K. D. Stuart. 1997. RNA editing: getting U into RNA. *Trends Biochem. Sci.* 22:162–166.

Kapoor, M., Y. Vijayaraghavan, R. Kadonaga, and K. E. LaRue. 1993. NAD(+)-specific glutamate dehydrogenase of *Neurospora crassa*: cloning, complete nucleotide sequence, and gene mapping. *Biochem. Cell Biol.* 71:205–219.

Kim, K. S., S. M. R. Teixeira, L. V. Kirchhoff, and J. E. Donelson. 1994. Transcription and editing of cytochrome oxidase II RNAs in *Trypanosoma cruzi*. *J. Biol. Chem.* 269:1206–1211.

Köller, J., U. F. Müller, B. Schmid, A. Missel, K. Stuart, and H. U. Göringer. 1997. *Trypanosoma brucei* gBP21: an arginine-rich mitochondrial protein that binds to guide RNA with high affinity. *J. Biol. Chem.* 272:3749–3757.

Köller, J., G. Nörskau, A. S. Paul, K. Stuart, and H. U. Göringer. 1994. Different *Trypanosoma brucei* guide RNA molecules associate with an identical complement of mitochondrial proteins in vitro. *Nucleic Acids Res.* 22:1988–1995.

Koslowsky, D. J., G. J. Bhat, A. L. Perrolaz, J. E. Feagin, and K. Stuart. 1990. The MURF3 gene of *T. brucei* contains multiple domains of extensive editing and is homologous to a subunit of NADH dehydrogenase. *Cell* 62:901–911.

Koslowsky, D. J., G. J. Bhat, L. K. Read, and K. Stuart. 1991. Cycles of progressive realignment of gRNA with mRNA in RNA editing. *Cell* 67:537–546.

Koslowsky, D. J., H. U. Göringer, T. H. Morales, and K. Stuart. 1992. In vitro guide RNA/mRNA chimaera formation in *Trypanosoma brucei* RNA editing. *Nature* 356:807–809.

Koslowsky, D. J., S. M. Katus, and K. Stuart. 1996. Distinct differences in the requirements for ribonucleoprotein complex formation on differentially regulated pre-edited mRNA in *Trypanosoma brucei*. *Mol. Biochem. Parasitol.* 80:1–14.

Leegwater, P., D. Speijer, and R. Benne. 1995. Identification by UV cross-linking of oligo(U)-binding proteins in mitochondria of the insect trypanosomatid *Crithidia fasciculata*. *Eur. J. Biochem.* 227:780–786.

Lukes, J., G. J. Arts, J. Van den Burg, A. de Haan, F. Opperdoes, P. Sloof, and R. Benne. 1994. Novel pattern of editing regions in mitochondrial transcripts of the cryptobiid *Trypanoplasma borreli*. *EMBO J.* 13:5086–5098.

Mahapatra, S., T. Ghosh, and S. Adhya. 1994. Import of small RNAs into *Leishmania* mitochondria in vitro. *Nucleic Acids Res.* 22:3381–3386.

Mahendran, R., M. R. Spottswood, and D. L. Miller. 1991. RNA editing by cytidine insertion in mitochondria of *Physarum polycephalum*. *Nature* 349:434–438.

Maslov, D. A., H. A. Avila, J. A. Lake, and L. Simpson. 1994. Evolution of RNA editing in kinetoplastid protozoa. *Nature* 368:345–348.

Maslov, D. A., and L. Simpson. 1994. RNA editing and mitochondrial genomic organization in the cryptobiid kinetoplastid protozoan *Trypanoplasma borreli*. *Mol. Cell. Biol.* 14:8174–8182.

Maslov, D. A., and L. Simpson. 1992. The polarity of editing within a multiple gRNA-mediated domain is due to formation of anchors for upstream gRNAs by downstream editing. *Cell* **70:** 459–467.

Maslov, D. A., N. R. Sturm, B. M. Niner, E. S. Gruszynski, M. Peris, and L. Simpson. 1992. An intergenic G-rich region in *Leishmania tarentolae* kinetoplast maxicircle DNA is a pan-edited cryptogene encoding ribosomal protein S12. *Mol. Cell. Biol.* **12:**56–67.

McManus, M., and S. Hajduk. Unpublished data.

Michelotti, E. F., M. E. Harris, B. K. Adler, A. F. Torri, and S. L. Hajduk. 1992. Mitochondrial ribosomal RNA synthesis, processing and developmentally regulated expression in *Trypanosoma brucei*. *Mol. Biochem. Parasitol.* **54:**31–42.

Missel, A., and H. U. Göringer. 1994. *Trypanosoma brucei* mitochondria contain RNA helicase activity. *Nucleic Acids Res.* **22:** 4050–4056.

Missel, A., G. Nörskau, H. H. Shu, and H. U. Göringer. 1995. A putative RBA helicase of the DEAD-box family from *Trypanosoma brucei*. *Mol. Biochem. Parasitol.* **75:**123–126.

Missel, A., A. E. Souza, G. Nörskau, and H. U. Göringer. 1997. Disruption of a gene encoding a novel mitochondrial DEAD-box protein in *Trypanosoma brucei* affects edited mRNAs. *Mol. Cell. Biol.* **17:**4895–4903.

Mottram, J. C., S. D. Bell, R. G. Nelson, and J. D. Barry. 1991. tRNAs of *Trypanosoma brucei*. Unusual gene organization and mitochondrial importation. *J. Biol. Chem.* **266:**18313–18317.

Myler, P. J., D. L. Glick, J. E. Feagin, T. H. Morales, and K. Stuart. 1993. Structural organisation of the maxicircle variable region in *Trypanosoma brucei*: identification of potential replication origins and topoisomerase II binding sites. *Nucleic Acids Res.* **21:**687–694.

Neupert, W. 1997. Protein import into mitochondria. *Annu. Rev. Biochem.* **66:**863–917.

Opperdoes, F. R. 1987. Compartmentation of carbohydrate metabolism in trypanosomes. *Annu. Rev. Microbiol.* **4:**127–151.

Peris, M., G. C. Frech, A. M. Simpson, F. Bringuad, E. Byrne, A. Bakker, and L. Simpson. 1994. Characterization of two classes of ribonucleoprotein complexes possibly involved in RNA editing from *Leishmania tarentolae* mitochondria. *EMBO J.* **13:** 1664–1672.

Peris, M., A. M. Simpson, J. Grunstein, J. E. Lilietal, G. C. Frech, and L. Simpson. 1997. Native gel analysis of ribonucleoprotein complexes from a *Leishmania tarentolae* mitochondrial extract. *Mol. Biochem. Parasitol.* **85:**9–24.

Piller, K. J., C. J. Decker, L. N. Rusché, M. E. Harris, S. L. Hajduk, and B. Sollner-Webb. 1995a. Editing domains of *Trypanosoma brucei* mitochondrial RNAs identified by secondary structure. *Mol. Cell. Biol.* **15:**2916–2924.

Piller, K. J., C. J. Decker, C. J. Rusché, and B. Sollner-Webb. 1995b. *Trypanosoma brucei* mitochondrial guide RNA-mRNA chimera-forming activity cofractionates with an editing-domain-specific endonuclease and RNA ligase and is mimicked by heterologous nuclease and RNA ligase. *Mol. Cell. Biol.* **15:** 2925–2932.

Piller, K. J., C. J. Rusché, J. Cruz-Reyes, and B. Sollner-Webb. 1997. Resolution of the RNA editing gRNA-directed endonuclease from two other endonucleases of *Trypanosoma brucei* mitochondria. *RNA* **3:**279–290.

Pollard, V. W., and S. L. Hajduk. Unpublished data.

Pollard, V. W., and S. L. Hajduk. 1991. *Trypanosoma equiperdum* minicircles encode three distinct primary transcripts which exhibit guide RNA characteristics. *Mol. Cell. Biol.* **11:**1668–1675.

Pollard, V. W., M. E. Harris, and S. L. Hajduk. 1992. Native messenger RNA editing complexes from *Trypanosoma brucei* mitochondria. *EMBO J.* **11:**4429–4438.

Pollard, V. W., S. P. Rohrer, E. F. Michelotti, K. Hancock, and S. L. Hajduk. 1990. Organization of minicircle genes for guide RNAs in *Trypanosoma brucei*. *Cell* **63:**783–790.

Preiss, T., A. G. Hall, and R. N. Lightowlers. 1993. Identification of bovine glutamate dehydrogenase as an RNA-binding protein. *J. Biol. Chem.* **268:**24523–24526.

Priest, J. W., and S. L. Hajduk. 1992. Cytochrome *c* reductase purified from *Crithidia fasciculata* contains an atypical cytochrome c_1. *J. Biol. Chem.* **267:**20186–20195.

Priest, J. W., and S. L. Hajduk. 1994. Developmental regulation of mitochondrial biogenesis in *Trypanosoma brucei*. *Bioenerg. Biomembr.* **26:**179–191.

Read, L. K., R. A. Corell, and K. Stuart. 1992. Chimeric and truncated RNAs in *Trypanosoma brucei* suggest transesterifications at non-consecutive sites during RNA editing. *Nucleic Acids Res.* **20:**2341–2347.

Read, L. K., H. U. Göringer, and K. Stuart. 1994a. Assembly of mitochondrial ribonucleoprotein complexes involves specific guide RNA (gRNA)-binding proteins and gRNA domains but does not require preedited mRNA. *Mol. Cell. Biol.* **14:** 2629–2639.

Read, L. K., K. A. Stankey, W. R. Fish, A. M. Muthiani, and K. Stuart. 1994b. Developmental regulation of RNA editing and polyadenylation in four life cycle stages of *Trypanosoma congolense*. *Mol. Biochem. Parasitol.* **68:**297–306.

Read, L. K., P. J. Myler, and K. Stuart. 1992. Extensive editing of both processed and preprocessed maxicircle CR6 transcripts in *Trypanosoma brucei*. *J. Biol. Chem.* **267:**1123–1128.

Read, L. K., K. D. Wilson, P. J. Myler, and K. Stuart. 1994. Editing of *Trypanosoma brucei* maxicircle CR5 mRNA generates variable carboxy terminal predicted protein sequences. *Nucleic Acids Res.* **22:**1489–1495.

Riley, G. R., R. A. Corell, and K. Stuart. 1994. Multiple guide RNAs for identical editing of *Trypanosoma brucei* apocytochrome *b* mRNA have an unusual minicircle location and are developmentally regulated. *J. Biol. Chem.* **269:**6101–6108.

Riley, G. R., P. J. Myler, and K. Stuart. 1995. Quantitation of RNA editing substrates, products and potential intermediates: implications for developmental regulation. *Nucleic Acids Res.* **23:** 708–712.

Rusché, L. N., J. Cruz-Reyes, K. J. Piller, and B. Sollner-Webb. 1997. Purification of a functional enzymatic editing complex from *Trypanosoma brucei* mitochondria. *EMBO J.* **16:** 4069–4081.

Rusché, L. N., K. J. Piller, and B. Sollner-Webb. 1995. Guide RNA-mRNA chimeras, which are potential RNA editing intermediates, are formed by endonuclease and RNA ligase in a trypanosome mitochondrial extract. *Mol. Cell. Biol.* **15:** 2933–2941.

Russell, D., Personal communication.

Sabatini, R., and S. L. Hajduk. 1995. RNA ligase and its involvement in guide RNA/mRNA chimera formation: evidence for a cleavage-ligation mechanism of *Trypanosoma brucei* mRNA editing. *J. Biol. Chem.* **270:**7233–7240.

Schmid, B., G. R. Riley, K. Stuart, and H. U. Göringer. 1995. The secondary structure of guide RNA molecules from *Trypanosoma brucei*. *Nucleic Acids Res.* **23:**3093–3102.

Schneider, A. 1994. Import of RNA into mitochondria. *Trends Cell. Biol.* **4:**282–286.

Schneider, A., J. Martin, and N. Agabian. 1994. A nuclear encoded tRNA of *Trypanosoma brucei* is imported into mitochondria. *Mol. Cell. Biol.* **14:**2317–2322.

Seiwert, D. S., S. Heidmann, and K. Stuart. 1996. Direct visualization of uridylate deletion *in vitro* suggests a mechanism for kinetoplastid RNA editing. *Cell* **84:**831–841.

Seiwert, S. D., and K. Stuart. 1994. RNA editing: transfer of genetic information from gRNA to precursor mRNA *in vitro*. *Science* 266:114–117.

Seiwert, S. D. 1995. The ins and outs of editing RNA in kinetoplastids. *Parasitol. Today* 11:362–368.

Shapiro, T. A., and P. T. Englund. 1995. The structure and replication of kinetoplast DNA. *Annu. Rev. Microbiol.* 49:117–143.

Shaw, J. M., D. Campbell, and L. Simpson. 1989. Internal frameshifts within the mitochondrial genes for cytochrome oxidase subunit II and maxicircle unidentified reading frame 3 of *Leishmania tarentolae* are corrected by RNA editing: evidence for translation of the edited cytochrome oxidase II mRNA. *Proc. Natl. Acad. Sci. USA* 86:6220–6224.

Shaw, J. M., J. E. Feagin, K. Stuart, and L. Simpson. 1988. Editing of kinetoplastid mitochondrial mRNAs by uridine addition and deletion generates conserved amino acid sequences and AUG initiation codons. *Cell* 52:401–411.

Simpson, A. M., Y. Suyama, H. Dewes, D. A. Campbell, and S. Simpson. 1989. Kinetoplastid mitochondria contain functional tRNAs which are encoded in nuclear DNA and also contain small minicircle and maxicircle transcripts of unknown function. *Nucleic Acids Res.* 17:5427–5445.

Simpson, A. M., N. Bakalara, and L. Simpson. 1992. A ribonuclease activity is activated by heparin or by digestion with proteinase K in mitochondrial extracts of *Leishmania tarentolae*. *J. Biol. Chem.* 267:6782–6788.

Simpson, L. 1986. Kinetoplast DNA in trypanosomatid flagellates. *Int. Rev. Cytol.* 99:119.

Simpson, L., and D. A. Maslov. 1994. Ancient origin of RNA editing in kinetoplastid protozoa. *Curr. Opin. Genet. Dev.* 4:887–894.

Simpson, L., and O. H. Thiemann. 1995. Sense from nonsense: RNA editing in mitochondria of kinetoplastic protozoa and slime molds. *Cell* 81:837–840.

Singh, R., and M. R. Green. 1993. Sequence-specific binding of transfer RNA by glyceraldehyde-3-phosphate dehydrogenase. *Science* 259:365–368.

Sloof, P., A. de Haan, W. Eier, M. van Iersel, E. Boel, H. van Steeg, and R. Benne. 1992. The nucleotide sequence of the variable region in *Trypanosoma brucei* completes the sequence analysis of the maxicircle component of mitochondrial kinetoplast DNA. *Mol. Biochem. Parasitol.* 56:289–300.

Souza, A. E., P. J. Myler, and K. Stuart. 1992. Maxicircle CR1 transcripts of *Trypanosoma brucei* are edited and developmentally regulated and encode a putative iron-sulfur protein homologous to an NADH dehydrogenase subunit. *Mol. Cell. Biol.* 12:2100–2107.

Souza, A. E., H. H. Shu, L. K. Read, P. J. Myler, and K. D. Stuart. 1993. Extensive editing CR2 maxicircle transcripts of *Trypanosoma brucei* predicts a protein with homology to a subunit of NADH dehydrogenase. *Mol. Cell. Biol.* 13:6832–6840.

Speijer, D., A. O. Muijsers, H. Dekker, A. de Haan, C. K. D. Breek, S. P. J. Albracht, and R. Benne. 1996a. Purification and characterization of cytochrome c oxidase from the insect trypanosomatid *Crithidia fasciculata*. *Mol. Biochem. Parasitol.* 79:47–59.

Speijer, D., C. K. D. Breek, A. O. Muijsers, P. X. Groenevelt, H. Dekker, A. de Haan, and R. Benne. 1996b. The sequence of a small subunit of cytochrome c oxidase from *Crithidia fasciculata* which is homologous to mammalian subunit IV. *FEBS Lett.* 381:123–126.

Speijer, D., C. K. D. Breek, A. O. Muijsers, A. F. Hartog, J. A. Berden, S. P. J. Albracht, B. Samyn, J. Van Beeumen, and R. Benne. 1997. Characterization of the respiratory chain from cultured *Crithidia fasciculata*. *Mol. Biochem. Parasitol.* 85:171–186.

Stuart, K., T. E. Allen, S. Heidmann, and S. D. Seiwert. 1997. RNA editing in kinetoplastid protozoa. *Microbiol. Mol. Biol. Rev.* 61:105–120

Sturm, N. R., D. A. Maslov, B. Blum, and L. Simpson. 1992. Generation of unexpected editing patterns in *Leishmania tarentolae* mitochondrial mRNAs: misediting produced by misguiding. *Cell* 70:469–476.

Sturm, N. R., and L. Simpson. 1990a. Kinetoplast DNA minicircles encode guide RNAs for editing of cytochrome oxidase subunit-III messenger RNA. *Cell* 61:879–884.

Sturm, N. R., and L. Simpson. 1990b. Partially edited mRNAs for cytochrome *b* and subunit-III of cytochrome oxidase from *Leishmania tarentolae* mitochondria: RNA editing intermediates. *Cell* 61:871–878.

Sturm, N. R., and L. Simpson. 1991. *Leishmania tarentolae* minicircles of different sequence classes encode single guide RNAs located in the variable region approximately 150 bp from the conserved region. *Nucleic Acids Res.* 19:6277–6281.

Thiemann, O. H., D. A. Maslov, and L. Simpson. 1994. Disruption of RNA editing in *Leishmania tarentolae* by the loss of minicircle-encoded guide RNA genes. *EMBO J.* 13:5689–5700.

Thiemann, O. H., and L. Simpson. 1996. Analysis of the 3′ uridylation sites of guide RNAs from *Leishmania tarentolae*. *Mol. Biochem. Parasitol.* 79:229–234.

Torri, A. F., K. I. Bertrand, and S. L. Hajduk. 1993. Protein stability regulates the expression of cytochrome c during the developmental cycle of *Trypanosoma brucei*. *Mol. Biochem. Parasitol.* 57:305–316.

van der Spek, H., G. J. Arts, R. R. Zwaal, J. van den Burg, P. Sloof, and R. Benne. 1991. Conserved genes encode guide RNAs in mitochondria of *Crithidia fasciculata*. *EMBO J.* 10:1217–1224.

van der Spek, H., D. Speijer, G. J. Arts, J. van den Burg, H. van Steeg, P. Sloof, and R. Benne. 1990. RNA editing in transcripts of the mitochondrial genes of the insect trypanosome *Crithidia fasciculata*. *EMBO J.* 9:257–262.

van der Spek, H., J. van den Burg, A. Croiset, M. van den Broek, P. Sloof, and R. Benne. 1988. Transcripts from the frameshifted MURF3 gene from *Crithidia fasciculata* are edited by U insertion at multiple sites. *EMBO J.* 7:2509–2514.

Vickerman, K. 1976. The diversity of the kinetoplastid flagellates, p. 1–34. *In* W. H. R. Lumsden and D. A. Evans (ed.), *Biology of the Kinetoplastida*. Academic Press, Inc., London, United Kingdom.

Vickerman, K. 1994. The evolutionary expansion of the trypanosomatid flagellates. *Int. J. Parasitol.* 24:1317–1331.

Vickerman, K. 1985. Developmental cycle and biology of pathogenic trypanosomes. *Br. Med. Bull.* 41:105.

White, T. C., and P. Borst. 1987. RNA end-labeling and RNA ligase activities can produce a circular rRNA in whole cell extracts from trypanosomes. *Nucleic Acids Res.* 15:3275–3290.

Yasuhira, S., and L. Simpson. 1995. Minicircle-encoded guide RNAs from *Crithidia fasciculata*. *RNA* 1:634–643.

Yasuhira, S., and L. Simpson. 1996. Guide RNAs and guide RNA genes in the cryptobiid kinetoplastid protozoan, *Trypanoplasma borreli*. *RNA* 2:1153–1160.

Yasuhira, S. and L. Simpson. 1997. Phylogenetic affinity of mitochondria of *Euglena gracilis* and kinetoplastids using cytochrome oxidase I and hsp60. *J. Mol. Evol.* 44:341–347.

Modification and Editing of RNA
Edited by Henri Grosjean and Rob Benne
© 1998 ASM Press, Washington, D.C.

Chapter 22

RNA Editing in *Physarum* Mitochondria

JONATHA M. GOTT AND LINDA M. VISOMIRSKI-ROBIC

INTRODUCTION

Editing in the Mitochondria of the Acellular Slime Mold *Physarum polycephalum*

RNA editing in the mitochondria of the myxomycete *Physarum polycephalum* was first described by Dennis Miller and colleagues, who reported the presence of 54 nonencoded cytidines in the α-ATPase mRNA (Mahendran et al., 1991). These extra residues were present as widely spaced (26 ± 10 nucleotides [nt]) single nucleotide insertions. Since that time, a large number of additional editing sites have been identified (Table 1), including the insertion of each of the four nucleotides. Thus far, insertions of single C and U residues and the dinucleotides CU, CG, GU, UA, and AA have been observed, while no deletions have yet been reported (Mahendran et al., 1991; Gott et al., 1993; Mahendran et al., 1994; Visomirski-Robic and Gott, 1995). The *coI* mRNA is also edited by four apparent C to U changes, in which C residues are present in the mitochondrial DNA, but U residues (or at least bases that pair with A) are found in mitochondrial RNAs and in cDNAs (Gott et al., 1993). Thus, *Physarum* editing is unusual in two regards: it involves mixed nucleotide addition, and it encompasses both nucleotide insertion and base substitution.

Although the mitochondrial genome has not been entirely sequenced, it has been estimated that there are approximately 1,000 different editing sites in *Physarum* mitochondrial RNAs (see reviews by Miller et al., 1993a, 1993b). The RNAs transcribed from most (~90%), but not all, of the characterized genes encoded in the 60-kb *Physarum* mitochondrial genome require editing to produce fully mature RNAs (Miller et al., 1993a, 1993b). Editing is not limited to mRNAs, as both the large and small rRNA subunits and many tRNAs are also subject to nucleotide insertion (Table 1). Editing sites are found only within coding regions of mRNAs and in regions of tRNAs and rRNAs that are present in mature molecules, falling within both highly conserved and variable regions of these structured RNAs (Miller et al., 1993a, 1993b).

Biological Function

It is clear from the widespread nature of editing in *Physarum* mitochondria that, in its absence, few functional gene products could be made. Long open reading frames are rare in *Physarum* mitochondrial DNA, and frequent frameshifting events are required to create open reading frames encoding proteins homologous to those found in the mitochondria of other organisms (Mahendran et al., 1991; Gott et al., 1993). These frameshifts are accomplished by site-specific nucleotide insertions (Fig. 1). The four known instances of C to U changes in the *coI* mRNA also result in codon changes, and in each case, the amino acid encoded by the edited mRNA more closely resembles the residue present in CoI proteins from other organisms (Gott et al., 1993). Likewise, insertion of nonencoded nucleotides into rRNAs and tRNAs results in the formation of conserved primary, secondary, and presumably tertiary structures required for their function (Miller et al., 1993b; Mahendran et al., 1994) (see Chapter 16). RNAs that are fully edited by both nucleotide insertions and base substitutions are present at all stages of the *Physarum* life cycle (Rundquist and Gott, 1995), reflecting the critical role that editing plays in these mitochondria. How and why this unusual mode of gene expression initially arose remain to be determined.

Jonatha M. Gott • Center for RNA Molecular Biology, Department of Molecular Biology and Microbiology, Case Western Reserve University, Cleveland, Ohio 44106.   **Linda M. Visomirski-Robic** • Department of Biological Sciences, Carnegie Mellon University, Pittsburgh, Pennsylvania 15213.

Table 1. Characterized editing events in *Physarum* mitochondria

RNA	Nucleotide insertions and base substitutions	Reference
α-ATPase mRNA	54 C	Mahendran et al., 1991
Cytochrome *c* oxidase subunit 1 mRNA	59 C	Gott et al., 1993
	1 U	Visomirski-Robic and Gott, 1995
	1 UA	
	1 CU	
	1 GU	
	4 C-to-U changes	
Cytochrome *c* oxidase subunit 2 mRNA	31 C	Visomirski-Robic, Webb, and Gott, unpublished data
Small subunit rRNA	40 C	Mahendran et al., 1994
	2 U	
	1 CU	
	2 AA	
Large subunit rRNA	52 C	Unpublished; see Miller et al., 1993
	2 CU	
	1 GU	
	2 AA	
Proteolipid subunit ATPase mRNA	9 C	Unpublished; see Miller et al., 1993
Cytochrome *b* mRNA	31 C	Unpublished; see Miller et al., 1993
	6 U	
	2 CU	
	1 GC	
NADH dehydrogenase subunit 1 mRNA	38 C	Unpublished; see Miller et al., 1993
	2 U	
tRNA^K	1 C	Unpublished; see Miller et al., 1993
tRNA^E	1 C	Unpublished; see Miller et al., 1993
	1 U	
tRNA^M	2 C	Unpublished; see Miller et al., 1993
tRNA^P	2 C	Unpublished; see Miller et al., 1993

Characteristics of *Physarum* Editing

The sequences surrounding editing sites in *Physarum* do not appear to be entirely random because approximately 70% of the unambiguous sites of C insertion are preceded by a purine-pyrimidine (Gott et al., 1993; Miller et al., 1993a). However, almost all possible combinations of nucleotides are allowed in these positions (Gott et al., 1993; Miller et al., 1993a), and no consensus elements that might direct editing have been identified thus far. These analyses have been complicated by the functional constraints imposed by the need to maintain highly conserved amino acid sequences in protein coding RNAs and secondary and tertiary structures in the case of rRNAs and tRNAs. Indeed, two thirds of the insertional events in mRNAs occur at the third position of a codon (Miller et al., 1993a), most likely due to the lower degree of evolutionary pressure at this position. Despite these constraints, the overall pattern of editing sites is very similar in all *Physarum* mitochondrial RNAs that have been characterized to date (Miller et al., 1993a, 1993b), suggesting that these features may be relevant to editing site selection and that the same machinery is likely to catalyze nucleotide addition into mRNAs, tRNAs, and rRNAs.

Attempts to identify structural elements that might be used to specify editing sites have also not yielded convincing motifs. Although the U-rich mitochondrial mRNAs can be folded into many possible configurations, no consistent structural context has been found. It is somewhat easier to predict the secondary structure of the rRNAs and tRNAs, at least in their mature forms, but as discussed in the following sections, no single "rule" can be used to correlate structural features and editing sites. Sites of nucleotide insertion fall in both double-stranded stems and single-stranded regions within both conserved and variable regions of the phylogenetically conserved secondary structures characteristic of rRNAs (Miller et al., 1993b; Mahendran et al., 1994) and within base-paired regions of the acceptor, anticodon, D and T stems and the TΨC loop of tRNAs (see Chapter 16).

One of the more striking features of editing in *Physarum* is the spacing between inserted nucleotides. On average, nucleotide insertions occur once every 25 nt in mRNAs and every 40–45 nt in rRNAs

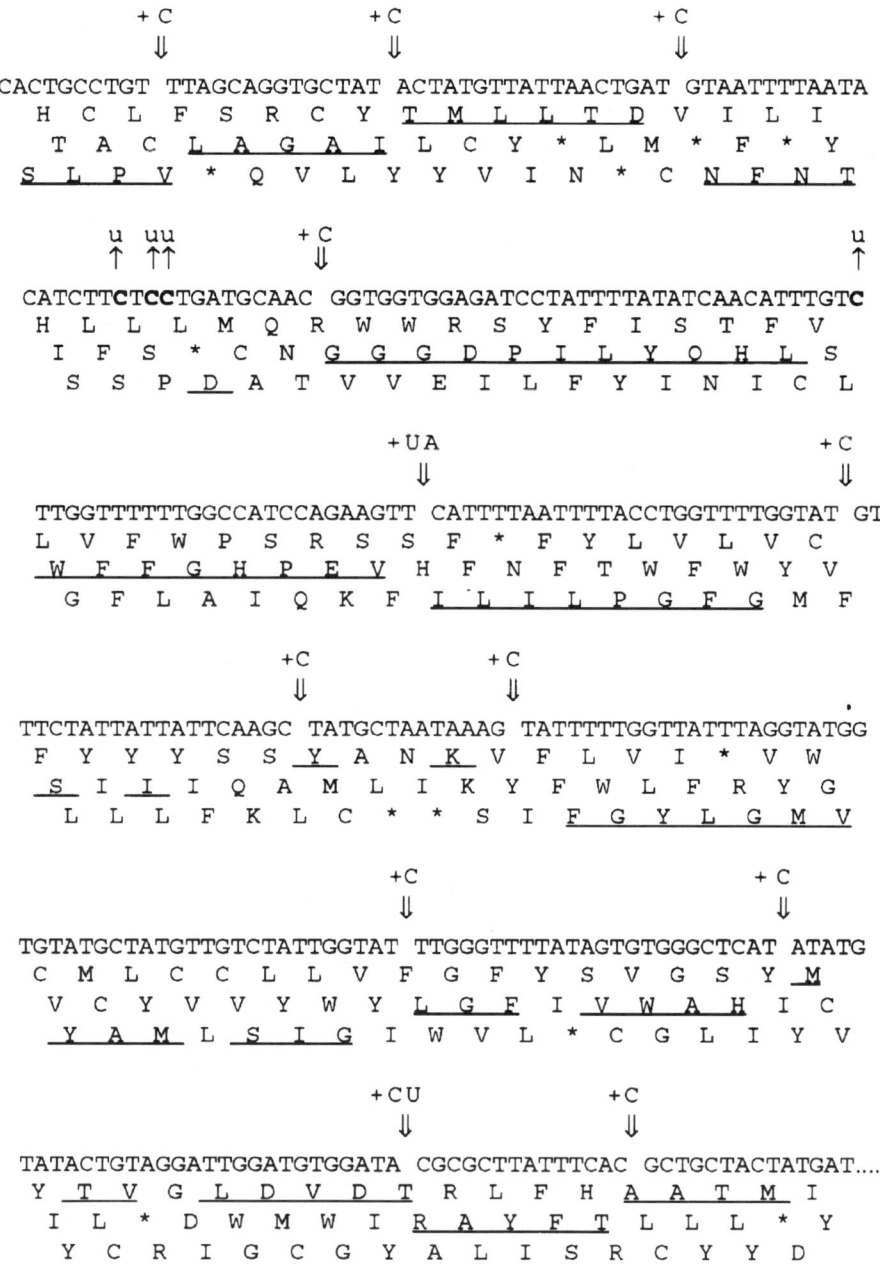

Figure 1. Functional outcome of editing in *Physarum* mitochondria. A region of the *coI* DNA sequence is shown, with gaps at sites of nucleotide insertion at the RNA level. Nucleotides added at each site are indicated (+N), as are the four sites of apparent C to U changes (C → u). Predicted translation products in all three reading frames (single-letter symbols) are given below the mitochondrial DNA sequence. Underlined amino acids are those that match the consensus sequence for CoI proteins from a wide range of organisms (Gott et al., 1993; Visomirski-Robic and Gott, 1995).

(Miller et al., 1993a, 1993b). These latter values are somewhat skewed, however, as the rRNAs have long stretches that do not contain editing sites. Except in the case of the dinucleotide additions, *Physarum* insertional editing sites are at least 9 nt apart and fairly evenly spaced, generally extending almost the entire length of the RNA (see Fig. 1 and Miller et al., 1993a). While there may be a mechanistic basis for this spacing, it remains to be elucidated.

Unedited and partially edited RNAs are extremely rare in the steady-state mitochondrial RNA pool in *Physarum*. Essentially all of the cloned cDNAs derived from the *coI* gene were completely edited (Gott et al., 1993) and similar results have been observed for other genes (Miller et al., 1993b, and our unpublished results). Reverse transcription (RT)-PCR assays used to determine the extent of editing within total RNA demonstrated that greater than 98% of the *coI* mRNA was edited by nucleotide insertion at all sites examined (Gott et al., 1993; Rundquist and Gott, 1995). Similarly, primer extension sequencing of bulk mitochondrial RNAs indicated that the vast majority (>95%) of steady-state *coI* mRNA contained base substitutions at each of the four sites of apparent C to U changes and all nucleotide insertions (Gott et al., 1993; Rundquist, 1995; Rundquist and Gott, 1995). The few partially edited cDNAs that have been generated are not suggestive of any particular mechanism or polarity of editing (see Miller et al., 1993b). For instance, unedited regions within α-ATPase cDNAs have been observed upstream of edited regions in some cDNAs and downstream in others (Miller et al., 1993b). Perhaps significantly, however, in none of these clones were edited and unedited sites interspersed; nevertheless, given the scarcity of these RNAs and the in vitro data described below, it is unlikely that these partially edited molecules represent true editing intermediates.

tRNA precursors that were unedited or partially edited have also been detected in *Physarum* mitochondrial RNA by RT-PCR. In one case involving 4 potential editing sites in two adjacent tRNA molecules, 8 of 16 possible editing patterns were found, while in a second precursor 6 of 8 possible combinations of edited and unedited sites were observed within these unprocessed molecules (see Miller et al., 1993b). Given that the folding of these *Physarum* tRNAs is likely to be affected by the absence of editing, and that in other systems the lack of editing affects the processing of tRNAs (see Chapter 16), the unedited and partially edited molecules within the pool of incompletely processed tRNAs most likely represent dead-end side products.

Editing in *Physarum* mitochondria is also highly accurate (Visomirski-Robic and Gott, 1995, 1997a, 1997b). "Misediting," in which either a nucleotide is inserted at an inappropriate position or the wrong nucleotide is added at an editing site, has not been observed in cloned cDNAs or in an in vitro mitochondrial system (see below). This is an important feature, as the site of addition, the number of residues inserted, and the identity of the inserted nucleotide(s) all influence the amino acid sequence of the resulting proteins and the final conformation of rRNAs and tRNAs. It is currently difficult to envision the basis of this specificity; ultimately, any models for editing will have to account for these characteristics.

METHODS USED IN THE STUDY OF *PHYSARUM* EDITING

Two complementary approaches have proven useful in the study of *Physarum* editing: analyses of total mitochondrial RNA, and characterization of labeled RNAs synthesized in isolated mitochondria. Examination of the steady-state population of mitochondrial RNAs has allowed the identification of editing sites, the determination of the frequency of editing at individual positions, and the isolation (via RT-PCR) of potential editing intermediates (Mahendran et al., 1991; Gott et al., 1993; Miller et al., 1993b; Mahendran et al., 1994; Rundquist and Gott, 1995). In vitro approaches have facilitated mechanistic studies designed to look at the timing, accuracy and efficiency of insertional editing of newly synthesized RNAs (Visomirski-Robic and Gott, 1995, 1997a, 1997b). Because different assays are required for these distinct RNA populations, the two approaches are discussed separately. Similar methodologies are also being used in combination with other means of investigating *Physarum* editing mechanisms currently under development, including biolistic transformation, soluble editing systems, and genetic approaches.

Insertional and Substitutional Editing in Steady-State RNA

Editing in *Physarum* was first discovered by comparing the sequence of cDNA clones with that of the mitochondrial genome (Mahendran et al., 1991), and this method is still the first level of analysis when new regions of the genome are characterized. As illustrated in Fig. 1, sites of nucleotide insertion can often be predicted by translating the mitochondrial DNA sequence in all three reading frames and looking for amino acid homologies based on aligned protein sequences. Primer extension sequencing using end-labeled oligonucleotides and reverse transcriptase can also be carried out on bulk mitochondrial RNA to confirm the presence of both nucleotide insertions and base substitutions in the steady-state RNA pool (Gott et al., 1993; Rundquist and Gott, 1995).

The frequency of editing in total *Physarum* RNA has been estimated by two assays employed in other editing systems, "poisoned" primer extension and restriction analyses of PCR products. The poisoned primer extension assay, which involves the use of a chain-terminating nucleotide in primer extension reactions, has been used previously to assay both C to U changes in apolipoprotein B (apoB) mRNAs (Driscoll et al., 1989) and uridine insertion and deletion in trypanosomatid protozoa (Seiwert and Stuart, 1994; Byrne et al., 1996). This assay has proven useful for examining the presence of both nucleotide insertions and base substitutions in *Physarum* (Rundquist, 1995). A more commonly used method for analyzing steady-state RNA is provided by a restriction enzyme assay similar to that used to quantify base substitutions in hepatitis delta virus RNA (Casey et al., 1992; Zheng et al., 1992). Insertion of a nucleotide into *Physarum* RNA followed by cDNA synthesis and PCR yields a product that differs from the mitochondrial DNA template. At some sites, editing results in a cDNA in which the recognition sequence for a given restriction enzyme is either created or destroyed. For example, editing of the first C insertion site in the *coI* mRNA destroys a *Hin*fI site (GAGTC) and creates a *Sac*I site (GAGCTC) in its cDNA product. Thus, editing at this site can be quantified for an entire population of RNAs by using total RNA in the analysis. Quantification can be simplified by using an end-labeled PCR primer in the assay (Gott et al., 1993; Rundquist and Gott, 1995). Variations on this technique have also been used to confirm the lack of DNA templates containing edited sequence in *Physarum* (Gott et al., 1993) and to select for partially edited molecules (see Miller et al., 1993b).

Phylogenetic comparisons are also being pursued in hopes of identifying *cis*-acting sequences or structural elements that specify editing sites. Miller and colleagues have begun a comparative sequence analysis involving organisms that are closely related to *Physarum*. Interestingly, at least two other myxomycetes, *Didymium* and *Stemonitis*, have been found to insert nonencoded nucleotides into their mitochondrial rRNAs, usually at sites identical to those observed in *Physarum* rRNAs (Miller, personal communication). A comparison of the contexts of these sites, including both highly conserved and variable nucleotides, may yield important information regarding the location of editing signals.

In Vitro Systems: Isolated Mitochondria

Isolated *Physarum* mitochondria are capable of taking up exogenously supplied nucleotides and incorporating them into previously initiated transcripts via run-on transcription (Visomirski-Robic and Gott, 1995). The development of this in vitro system has permitted the analysis of RNAs produced under a wide variety of defined experimental conditions and has allowed specific mechanistic questions to be addressed. However, since the RNAs produced in this in vitro system are present amid a background of fully edited, endogenous RNAs, the assays described above could not be used to determine the frequency of editing of these newly synthesized RNAs. Instead, RNAs made in vitro are labeled by the incorporation of [^{32}P]nucleoside triphosphates ([^{32}P]NTP) during synthesis and are examined by the assays depicted in Fig. 2. These assays are also proving useful in characterizing RNAs synthesized in more purified in vitro systems.

The population of labeled RNAs synthesized in isolated mitochondria includes transcripts from many genes. To analyze an individual RNA species, single-stranded DNA (ssDNA) complementary to the transcript of interest is annealed to total RNA and unhybridized molecules are degraded with S1 nuclease. Hybrid-protected RNA, which generally constitutes less than 1% of the total labeled RNA, is then purified for further analysis. To minimize cleavage at single nucleotide bulges that would be present in hybrids between edited and unedited molecules, the conditions used for hybrid protection differ slightly from those that are generally used in S1 nuclease protection experiments. However, cleavage at dinucleotide bulges and dinucleotide mismatches is still efficient under these conditions, making this technique a convenient assay for dinucleotide insertions and for substitutional editing at adjacent sites (Visomirski-Robic and Gott, 1995, 1997a).

One of the most powerful assays used to determine whether newly synthesized RNAs are edited has been RNA fingerprinting (Visomirski-Robic and Gott, 1995, 1997a, 1997b). In this assay, hybrid-protected RNAs are digested with ribonuclease T$_1$ (RNase T$_1$), which cuts after G residues, and the resulting oligonucleotides are separated in two dimensions. Because separation depends in the first dimension on base composition and in the second dimension on charge (size), the addition of even a single nucleotide or the substitution of one base for another is easily detected. In addition, shifts in the first dimension yield information on the identity of the added residue, as each nucleotide affects migration differently. A further advantage of this technique is that levels of misediting and effects on transcription can also be assessed from the fingerprint pattern. Alternatively, RNase T$_1$ fragments can be separated in a single dimension on denaturing polyacrylamide gels (Visomirski-Robic and Gott, 1997a, 1997b). This latter assay yields information on only the size of individual oligonucleotides, and therefore does not indicate which nucleotide has been added. However, this is a simpler means of comparing the frequency of editing under multiple conditions (Visomirski-Robic and Gott, 1997b), and can often be combined with nearest neighbor analyses of isolated bands to determine the identity of the added nucleotide (Visomirski-Robic and Gott, 1997a). For example, if RNA were synthesized in the presence of [α-^{32}P]ATP, digestion of the oligonucleotide CUAU(c)ACUAUG with RNase T$_2$ would yield only labeled Up* if unedited, but would give Up* and Cp* in a 2:1 ratio if edited (Fig. 2). Similarly, if [α-

^{32}P]CTP were used in the experiment, the unedited oligonucleotide would yield only Ap*, while the edited oligonucleotide would give Ap* and Up* in a 1:1 ratio.

MECHANISTIC FEATURES OF *PHYSARUM* EDITING

Different Mechanisms for Insertional and Substitutional Editing

Physarum is the only organism known to alter its RNAs by both substitution of one base for another and insertion of nonencoded nucleotides (Gott et al., 1993). Thus, it is possible that the C to U base changes are due to the deletion of a C and insertion of a U at these positions, similar to the proposed mechanism of base substitution in *Acanthamoeba castellanii* mitochondrial tRNAs (see Chapter 16). However, three lines of evidence argue in favor of separate mechanisms for nucleotide insertion and base substitution in *Physarum*. First, there is currently no evidence for nucleotide deletion in *Physarum* mitochondria. Second, labeled RNA synthesized in isolated mitochondria is edited by nucleotide insertion, but not by base substitution (Visomirski-Robic and Gott, 1995). Third, nascent *coI* mRNA synthesized in vivo is not yet edited by base conversion, but is edited by nucleotide insertion (unpublished data). The experiment that demonstrated this latter point is outlined schematically in Fig. 3. Under our digestion conditions, both dinucleotide bulges and dinucleotide mismatches within an RNA-DNA hybrid are susceptible to S1 nuclease cleavage (Visomirski-Robic and Gott, 1995 and unpublished data). Because two of the C to U changes affect adjacent nucleotides, the extent of editing at these sites can be estimated by annealing mitochondrial RNA to ssDNA complementary to either the edited or unedited sequence and determining the relative sensitivity to digestion with S1 nuclease. Because run-on transcription in isolated mitochondria is limited to 250–350 nucleotides, only the 3′ ends of nascent RNAs are labeled (Visomirski-Robic and Gott, 1997a). Thus, when ssDNA probes longer than 350 nucleotides are used in S1 protection experiments, regions of nascent RNA synthesized both in vivo (unlabeled) and in vitro (labeled) can be as-

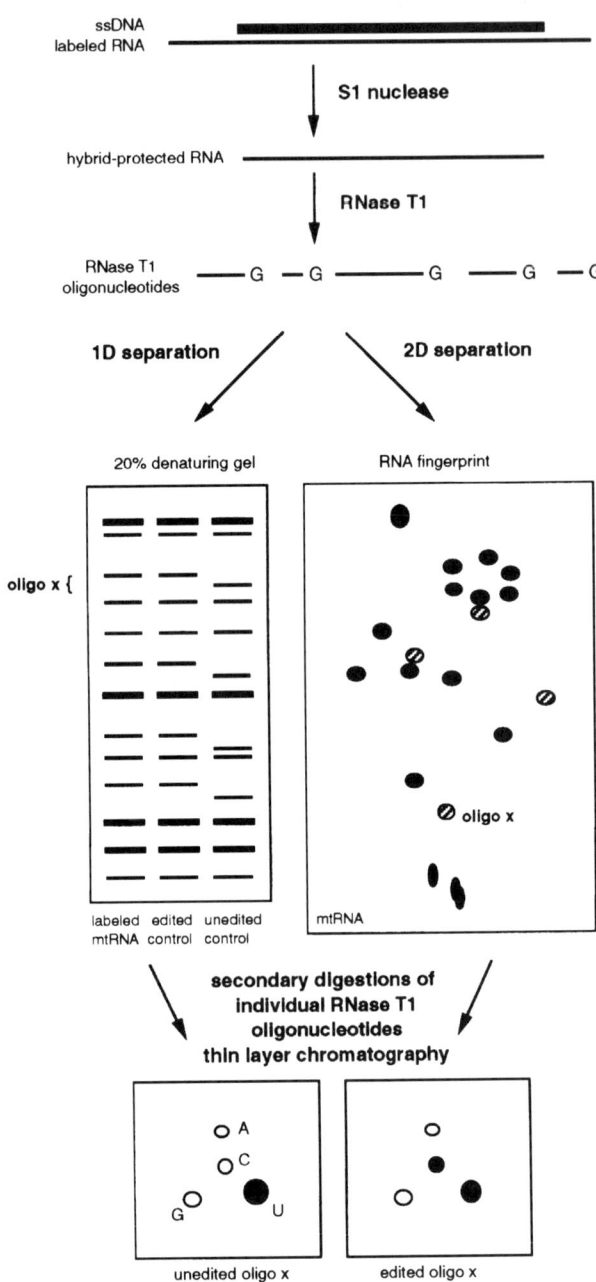

Figure 2. Analysis of labeled RNAs synthesized in vitro. Run-on transcripts are labeled in isolated mitochondria with [α-^{32}P] NTP prior to isolation of mitochondrial RNA. Total RNA is then hybridized to ssDNA corresponding to the region of interest and digested with S1 nuclease. Hybrid-protected RNAs are gel-purified and digested with RNase T$_1$. T$_1$ oligonucleotides are separated in one dimension (on 15–20% acrylamide denaturing gels) or in two dimensions via RNA fingerprinting (low-pH gel and homochromatography). Black spots indicate oligonucleotides which are present in both edited and unedited controls. Oligonucleotides whose mobility is affected by insertional or substitutional editing are indicated by hatched or shaded spots, respectively. Individual bands or spots can be isolated and analyzed further with additional nucleases, if desired. In the example shown, an [α-^{32}P]ATP-labeled T$_1$ oligonucleotide, x [CpUpApUp(c)pApCpUpApUpGp, where (c)

indicates the inserted C residue and underlining indicates labeled nucleotides] was digested with RNase T$_2$ and the resulting 3′ monophosphates (Np) were separated in two dimensions via thin-layer chromatography. The labeled phosphates are transferred to the nucleotide 5′ of each A residue. Examples of each of these techniques can be found in Visomirski-Robic and Gott (1995, 1997a, 1997b).

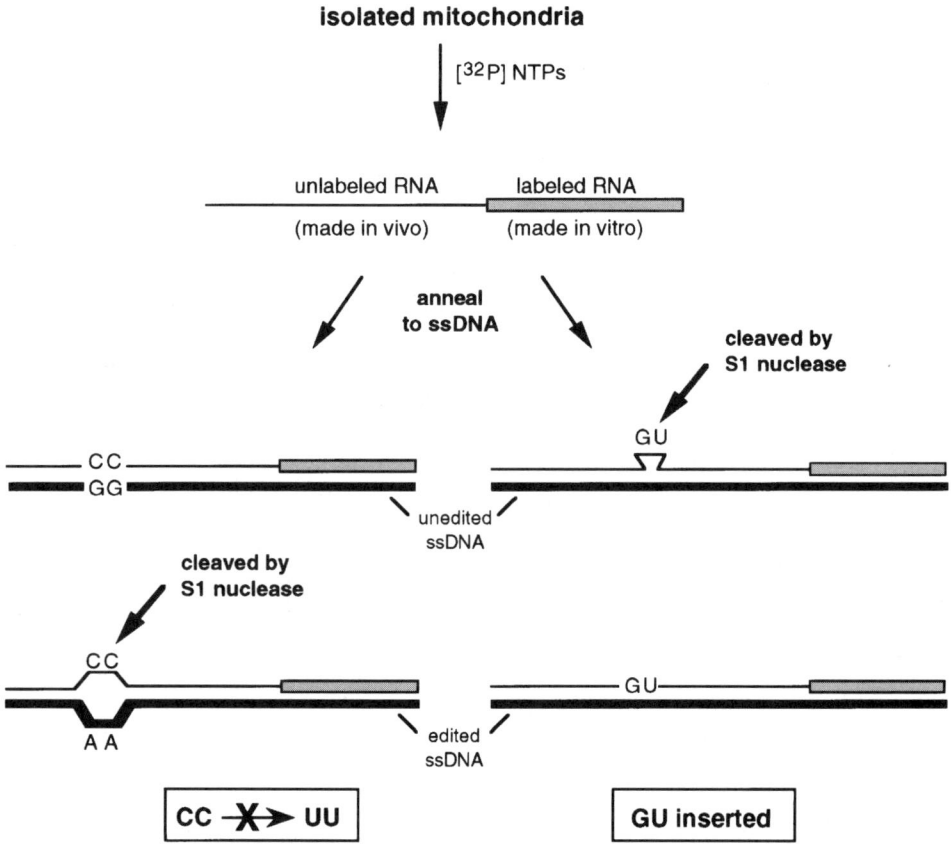

Figure 3. Schematic illustration of S1 nuclease protection experiments used to determine whether nascent RNAs synthesized in vivo are edited by C to U changes and dinucleotide insertions. The site of adjacent C to U changes is cleaved when nascent RNA is annealed to ssDNA corresponding to edited sequence, but fully protected by the unedited sequence. The opposite result is observed at sites of dinucleotide insertion, indicating that these two types of editing occur at different times in vivo.

sayed for dinucleotide insertions or adjacent base changes. As expected from previous in vitro studies (Visomirski-Robic and Gott, 1995), regions of the nascent *coI* mRNA that had been synthesized prior to mitochondrial isolation were efficiently edited at the dinucleotide insertion sites, but showed no evidence for base substitution at the sites of adjacent C to U changes. These data indicate that nucleotide insertions and base changes in *Physarum* mitochondria occur at different times relative to RNA synthesis, and are therefore likely to be mechanistically distinct.

Association Between Insertional Editing and Transcription

Analysis of newly synthesized RNAs produced in isolated mitochondria indicated that insertional editing is extremely efficient and highly accurate under standard labeling conditions because no unedited or misedited RNAs are detected (Visomirski-Robic and Gott, 1995). Manipulation of these conditions has allowed mechanistic questions to be addressed, including the issue of when editing occurs relative to transcription. At low nucleotide concentrations, transcribing RNA polymerases stall at certain sites along the mitochondrial DNA template, allowing the identification and isolation of individual nascent RNA species (Visomirski-Robic and Gott, 1997a). These RNAs clearly are still associated with the mitochondrial RNA polymerase, as they can be extended on the addition of limiting nucleotide. Interestingly, analysis of these RNA species has demonstrated that editing is already complete at sites located within 14–20 nucleotides of the 3' end of the growing RNA chain (Visomirski-Robic and Gott, 1997a). Thus, the editing activity can function in very close proximity to the transcriptional machinery. It should be noted that positions closer than 14 nucleotides from the site of transcription were not examined in these experiments; therefore, nucleotide insertion may actually occur at sites much closer to, or even at, the 3' end of newly synthesized RNAs (Visomirski-Robic and Gott, 1997a).

Mitochondrial labeling reactions can also be altered such that transcription proceeds in the absence of editing under defined conditions. For example, if the concentration of CTP is reduced to extremely low levels (~200 nM), RNAs that are not edited, or only partially edited, at C insertion sites can be synthesized in isolated mitochondria (Fig. 4) (Visomirski-Robic and Gott, 1997b). Thus, despite the close association between these processes, transcription is not obligatorily linked to insertional editing in *Physarum* mitochondria. Editing at other types of insertion sites, including the single U, UA, GU, and (somewhat surprisingly) CU insertions (see below), is not affected, since these sites are edited to the same extent under low CTP and standard "editing" conditions. Similarly, if the concentration of UTP is held very low, effects on U insertions are observed, but nonencoded C residues are added efficiently. The accuracy of transcription is unaffected by either condition, indicating that insertional editing is more sensitive to nucleotide concentration than is RNA synthesis (Visomirski-Robic and Gott, 1997b).

The establishment of conditions under which unedited and partially edited RNAs can be generated also made it possible to test whether C residues can be inserted posttranscriptionally (Visomirski-Robic and Gott, 1997b). In these experiments, labeled RNAs that were unedited at C insertion sites were first synthesized under limiting CTP concentrations in isolated mitochondria. These reactions were then chased with high levels of unlabeled nucleotides under conditions that normally yield edited molecules, and the labeled RNAs were analyzed to determine whether they had been processed by C insertion. Importantly, upon incubation under editing conditions, nucleotide insertion into previously synthesized, unedited RNA was not observed, even though newly transcribed RNA made during the "chase" reaction was fully edited (Fig. 4) (Visomirski-Robic and Gott, 1997b). Thus, the editing apparatus is unable to add nucleotides posttranscriptionally, providing strong evidence that insertional editing and transcription are closely associated in *Physarum* mitochondria (Visomirski-Robic and Gott, 1997a, 1997b).

Single Versus Dinucleotide Insertions

The existence of dinucleotide insertions in *Physarum* mitochondria has given rise to speculation regarding (i) whether the two nucleotides are added singly or as a unit, (ii) whether there is an order of addition at dinucleotide insertion sites, and (iii) whether the same editing activity is utilized for both single and dinucleotide insertions. Initial data from restriction enzyme and sequence analysis of RT-PCR amplification products from mitochondrial RNA suggested that the dinucleotides were inserted as a unit. In these experiments, only fully edited or completely unedited molecules were observed at all dinucleotide sites examined, even in the context of molecules that had been edited by single C insertions (described in Miller et al., 1993b). However, analysis of RNAs synthesized in isolated mitochondria has demonstrated that nucleotides can be added individually at dinucleotide sites, at least under limiting nucleotide conditions (Visomirski-Robic and Gott, 1997b). For example, when the concentration of UTP is very low in mitochondrial labeling reactions, most of the newly synthesized *coI* mRNA contains only a C at the CU insertion site. Similar evidence for single nucleotide addition has also been observed at the UA insertion site within the *coI* mRNA under these conditions, although the identity of the added nucleotide has not yet been determined in these experiments (unpublished data). Unexpectedly, however, when CTP is severely limiting, most of the *coI* mRNA is fully edited by CU insertion (Visomirski-Robic and Gott, 1997b). Although the order of nucleotide addition could not be addressed in this experiment, the greater frequency of C insertion at the CU site is consistent with the insertion of a C prior to U addition at this site. Finally, under standard mitochondrial labeling conditions, dinucleotides are efficiently inserted, even at sites close to the growing end of the RNA chain (Visomirski-Robic and Gott, 1997a, and unpublished data). Thus, dinucleotide insertions are also closely

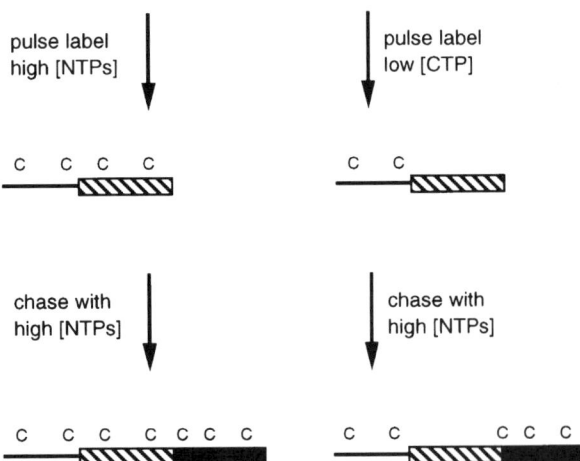

Figure 4. Coupling of transcription and editing in *Physarum*. Editing by C insertion (indicated by C's) is significantly reduced when RNAs are synthesized in the presence of low levels of CTP in isolated mitochondria. Once unedited RNA is made, it remains unedited even when incubated under "editing" conditions. RNA synthesized downstream of unedited regions is edited under these conditions. See Visomirski-Robic and Gott (1997b) for data and experimental details. Horizontal lines, RNA made in vivo; hatched bars, RNA made during pulse-labeling; solid bars, RNA made during chase.

associated with transcription in *Physarum* mitochondria, strongly suggesting that a single editing activity performs both single and dinucleotide insertions in *Physarum*.

Implications for the Mechanism of Insertional Editing

The data from in vitro labeling experiments have a number of implications regarding the mechanism of editing in *Physarum*. First, the available evidence suggests that insertional editing in *Physarum* is normally linked to transcription (Visomirski-Robic and Gott, 1997b). In this regard, *Physarum* editing is clearly different from nucleotide insertion in kinetoplastids (see Chapter 21), which occurs posttranscriptionally (Seiwert and Stuart, 1994), and may be somewhat similar to that in the paramyxoviruses, which involves the cotranscriptional addition of one or more guanosine residues by the viral polymerase (Vidal et al., 1990) (see Chapter 23).

Second, the pattern of editing in nascent *Physarum* RNAs is consistent with an overall 5' to 3' polarity of addition. The insertional activity appears to proceed unidirectionally because, as previously discussed, once unedited RNA is made, it is not processed further in isolated mitochondria (Fig. 4). In contrast, editing in kinetoplastids occurs in an overall 3' to 5' direction and individual editing sites may in fact be subject to multiple rounds of processing (Maslov and Simpson, 1992; Corell et al., 1993; Sugisaki and Takanami, 1993) (see Chapter 21).

Third, nucleotide insertions appear to occur independently of one another in *Physarum*. Edited sites can be flanked by unedited or partially edited sites under limiting NTP concentrations in vitro and, once nucleotide concentrations are restored to high levels, newly synthesized RNA downstream of the unedited region is edited efficiently (Visomirski-Robic and Gott, 1997b). These data also suggest that the editing machinery does not stall under limiting nucleotide conditions, but instead proceeds beyond unedited sites.

Finally, since editing has been observed within 14 nucleotides of the site of transcription (Visomirski-Robic and Gott, 1997a), RNA sequences more than 15 nucleotides downstream of editing sites cannot be required for specifying *Physarum* editing sites, as they would not have been synthesized at the time of nucleotide insertion. This indicates that extensive secondary structures involving downstream RNA sequences are not required for *Physarum* editing.

Taken together, the data are most consistent with a unique mode of insertional editing in *Physarum* in which transcription and nontemplated nucleotide addition are physically and/or mechanistically coupled. The lack of unedited regions in RNAs labeled in isolated mitochondria under standard, non-limiting nucleotide concentrations (Visomirski-Robic and Gott, 1995) and the proximity of editing to the site of transcription (Visomirski-Robic and Gott, 1997a), coupled with the inability of the editing activity to insert nucleotides into previously synthesized RNAs (Visomirski-Robic and Gott, 1997b), argue that there is a limited "window of opportunity" for editing in *Physarum* mitochondria. These features will need to be taken into account in generating plausible editing models.

POSSIBLE EDITING MECHANISMS

Parallels with Other Systems

While there is presently no single, unifying model for insertional editing in *Physarum*, there are at least precedents in other systems for many of the activities that might be involved in nucleotide insertion. At a minimum, the editing machinery must (i) recognize the editing site, (ii) identify the correct nucleotide(s) to insert, and (iii) catalyze the reaction. In addition, as editing and transcription appear to be linked, some means of stopping and starting transcription may also be needed. The purpose of this section is to identify analogous activities in other systems to suggest possible avenues of investigation and to aid in the development of explicit editing models.

Site selection

Given the proximity of nucleotide insertion to the active site of transcription, it is likely that the *Physarum* mitochondrial RNA polymerase (or another element associated with the transcription complex) plays some role, direct or indirect, in the editing process. Preliminary evidence from experiments with isolated *Physarum* mitochondria suggests that the rate at which the transcription complex moves along the DNA template can affect the extent of editing at a given site (unpublished data). This observation, along with data demonstrating that there is only a limited period in which RNAs can be edited (Visomirski-Robic and Gott, 1997b), suggests that the elongating transcription complex may be involved in signal recognition in *Physarum* editing. Template-dependent pausing of elongation complexes is used to regulate a number of other eukaryotic genes, including the *Drosophila* heat shock (Rasmussen and Lis, 1995) and mammalian c-*myc* gene (Krumm et al., 1995), and is needed for recognition of transcription termination signals (Landick, 1997). In the case of paramyxoviral editing, it has been postulated that pausing of the polymerase in the vicinity of the editing site is required for addition of nontemplated G residues (Jacques and Kolakofsky, 1991), and a similar strategy could be employed in *Physarum* mitochondria.

At least some of the information regarding the location of editing sites must be contained within the editing substrate itself, and could conceivably involve either primary sequences or higher-order structures (Fig. 5). While RNA sequences more than 14 nucleotides downstream of *Physarum* editing sites are not required for nucleotide insertion (Visomirski-Robic and Gott, 1997a), the region near the 3′ end of the nascent RNA that is to be edited is likely to interact with the editing machinery in some manner. Nascent RNAs are known to play an important role in pausing, transcription termination, attenuation, and antitermination (Fig. 5) (Das, 1992; Greenblatt et al., 1993; Lu et al., 1996; Chan et al., 1997), often via the formation of structures such as RNA hairpins (Platt, 1998). Potential RNA hairpins exist near some, but not all, *Physarum* editing sites (Visomirski-Robic, 1997), but the role of sequences and/or structures 5′ of insertional editing sites has not been systematically tested as yet.

Because the insertional editing activity is likely to be associated with elongating transcription complexes, sequences within the DNA template could also play a role in *Physarum* editing (Fig. 5). For example, the bacteriophage λ antiterminator Q protein recognizes a region of the DNA upstream of the pR′ terminator (the qut site [Yarnell and Roberts, 1992]) and modifies the *E. coli* RNA polymerase to allow read-through. There are also precedents for protein contacts with the single-stranded DNA within the transcription bubble, which affect both promoter clearance (Ring and Yarnell, 1996; Roberts and Roberts, 1996) and λQ protein-mediated antitermination (Ring and Roberts, 1994). Any of these potential contacts could play a role in *Physarum* editing. Additionally, the composition of the RNA-DNA hybrid, and the sequence and structure of downstream DNA can influence polymerase movement (Landick, 1997), playing, for example, a critical role in transcription termination. Bending of the DNA can cause polymerase pausing (Kerppola and Kane, 1990), and it is possible that DNA structure (or perhaps modification) could be used to mark the position of editing sites. Finally, editing sites could be localized by binding sites for "roadblock" molecules such as the mitochondrial termination factor mTERF (Daga et al., 1993) and the pol I terminator protein Reb1p (Lang et al., 1994). These proteins facilitate termination by binding to the DNA downstream of transcription terminators, effectively halting the progression of the RNA polymerase. Binding of roadblock molecules specific for editing sites would have to be reversible, however, to allow transcription to resume.

Specificity

The basis of the precision with which nucleotides are inserted into *Physarum* mitochondrial RNAs is still a mystery. One possibility is the existence of an as-yet-unidentified template acting either in *cis* or in *trans*. In the paramyxoviruses and Ebola virus, insertional editing is "pseudotemplated," i.e., the same

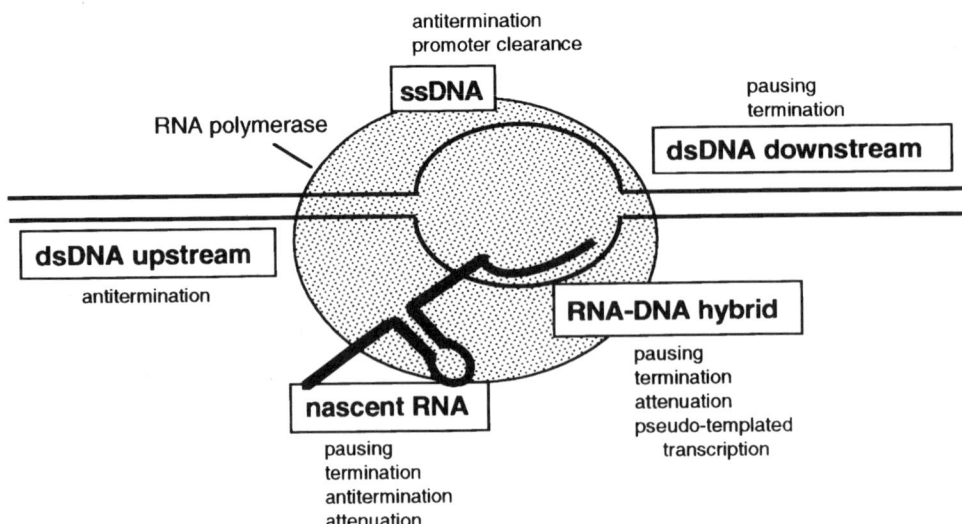

Figure 5. Location of potential *cis*-acting signals in *Physarum* editing. Precedents for roles in other processes for each of these regions are indicated (see references in text).

region of the template is read more than once (Jacques and Kolakofsky, 1991). For example, the editing site within the P gene of the paramyxoviruses falls within a 3–6 nt homopolymer C run on the template (see Chapter 23). Upon pausing of the viral RNA polymerase at the editing site, the nascent RNA is thought to realign on the template in an upstream direction. Subsequent extension of the RNA results in the addition of one or more pseudotemplated G's (Kolakofsky et al., 1993). Interestingly, this phenomenon is not restricted to viral editing sites, as pseudo-templated transcription is used in head and tail formation in viral RNAs (Jacques and Kolakofsky, 1991) (see Chapter 23) and prokaryotic transcriptional attenuation (Liu et al., 1994; Qi and Turnbough, 1995), and it may be an integral part of the transcription termination process (Jeong et al., 1996). Only approximately one-third of the added C residues in *Physarum* are inserted next to an encoded C, and few of the remaining sites are compatible with slippage of the nascent RNA relative to the DNA template (Mahendran et al., 1991; Gott et al., 1993; Mahendran et al., 1994). Thus, polymerase "stuttering" could account for the identity of only about 30% of inserted nucleotides in *Physarum* mitochondria.

In the mitochondria of *Acanthamoeba*, most tRNAs are edited by base substitutions at their 5' ends (Burger et al., 1995). These changes are likely to be the result of the deletion of encoded nucleotides and insertion of others (see Chapter 16), and are probably templated by the sequence of the 3' side of the acceptor stem (Lonergan and Gray, 1993). Editing of tRNAs in land snail mitochondria may also be "guided" by *cis*-acting sequences, but in this case only A residues are added to the 3' side of the acceptor stem (Yokobori and Paabo, 1995). In addition, intron sequences are used to direct both modification of some eukaryotic tRNAs (Johnson and Abelson, 1983; reviewed in Grosjean et al., 1997) and A to I editing in mRNAs (see Chapter 19). No introns have been found in *Physarum* mitochondrial genes, but nucleotide insertions at some sites within *Physarum* rRNAs and tRNAs could be templated by base pairing to a previously transcribed region of the RNA. Approximately half of the C insertion sites within the small subunit rRNA in *Physarum* mitochondria are paired when folded into the conserved secondary structure (Mahendran et al., 1994) and all six of the C residues added to *Physarum* tRNAs fall within base-paired regions (see Chapter 16). The pattern of nucleotide insertion is not a simple one, however, with added nucleotides falling within sequences 5' of potential guiding regions in some structures and 3' of potential templates in others. Obvious candidates for internal guide-like sequences are not present near most editing sites in *Physarum* mitochondrial mRNAs and, given the proximity of transcription and editing, a role for distant *cis*-acting elements seems unlikely. This does not preclude the involvement of transient, more proximal structures in determining the site or specificity of editing, however.

An editing "template" could also be provided in *trans*. Both pseudouridine formation (Ganot et al., 1997; Ni et al., 1997; Smith and Steitz, 1997) and ribose methylation (Bachellerie et al., 1995; Kiss-Laszlo et al., 1996; Tycowski et al., 1996) in rRNAs are targeted by specific small nucleolar RNAs (snoRNAs), which bind to the site of modification (see Chapter 13). Similarly, base pairing between individual gRNAs and the sequence just downstream of editing sites targets the editing site for endonucleolytic cleavage and directs the insertion of additional U residues into kinetoplast RNAs (Blum et al., 1990) (see Chapter 21). A strictly analogous mechanism is unlikely in *Physarum*, because much of the RNA immediately 3' of editing sites is likely to be involved in an RNA-DNA duplex in transcription complexes (Visomirski-Robic and Gott, 1997a). This would not necessarily preclude the use of a template that pairs with sequences 5' of editing sites, however. Thus far, individual guide-like RNA molecules have not been found in *Physarum* mitochondria (unpublished data and D. Miller, personal communication). However, a template associated with the editing activity could provide a plausible alternative. This guide-like molecule could involve an external nucleic acid template analogous to the RNA subunit of telomerase (Greider and Blackburn, 1989), which uses base pairing between its RNA subunit and the end of the DNA substrate to identify the appropriate register and generate the correct telomeric repeat (Gilley and Blackburn, 1996; Hammond et al., 1997). In this scenario, the insertion of each of the four nucleotides into *Physarum* RNAs would require a different alignment of the RNA with the template and some means of identifying the correct pairing.

Other sources of specificity that do not involve standard base pairing are also possible. Base specificity could reside, for example, in the nucleotide donor in the editing reaction. This possibility was suggested in some models of kinetoplastid editing (Blum et al., 1991; Cech, 1991; Sollner-Webb, 1991), in which the oligo(U) tail of the gRNA was proposed as the source of the added uridines. The fact that each of the four nucleotides is added to *Physarum* RNAs complicates this possibility, however. Precise nucleotide addition can also be directed by protein binding sites, such as those used by tRNA terminal transferase to add or restore the CCA ends of tRNAs (Masiakski and Deutscher, 1980). Again, the correct register is maintained in the final product, allowing the addition of specific nucleotides at the appropriate positions. Even small nucleic acid molecules can bind nucleotides with relatively high affinity and specificity (Sassanfar and Szostak, 1993; Ekland and Bartel, 1996), leaving open the possibility that the nucleotide to be inserted could be designated by either a bound nucleic acid or the single-stranded DNA within the transcription bubble, most likely via non-Watson–Crick interactions (Dieckmann et al., 1996). Finally, the identity of the added nucleotide could be specified through the use of separate catalytic activities for individual sites. In this case, the specificity of the reaction would be determined at the level of site recognition by the individual catalyst and its preference for a given nucleotide substrate.

Catalytic mechanisms

Nonencoded nucleotides are known to be added very close to the growing end of the RNA chain in *Physarum* mitochondria (Visomirski-Robic and Gott, 1997a), and it is possible that they are added concurrently with RNA synthesis. Although transcription normally is templated, many RNA polymerases are capable of adding nucleotides in an nontemplated or pseudotemplated manner, as described above (Jacques and Kolakofsky, 1991). Bacteriophage T7 RNA polymerase, for example, will add additional residues at the Ebola virus editing site (Volchkov et al., 1995) and at homopolymer tracts within other templates (Macdonald et al., 1993). In addition, this enzyme adds one or more untemplated nucleotides to the 3′ termini of transcripts at a relatively high frequency (Milligan et al., 1987). These characteristics are of particular interest because mitochondrial polymerases generally are single subunit enzymes with homology to the bacteriophage RNA polymerases (Masters et al., 1987; Cermakian et al., 1996). Other polymerases may also provide useful precedents for editing activities. For instance, DNA polymerases demonstrate higher fidelity than RNA polymerases due to their proofreading function. Misincorporated nucleotides are removed from the 3′ end of the newly synthesized DNA strand by a separate "editing" domain within the polymerase (Joyce and Steitz, 1994). An analogous activity capable of catalyzing the reverse reaction could be responsible for the addition of a nontemplated nucleotide to *Physarum* mitochondrial RNAs.

Recent experiments on *E. coli* transcription complexes have demonstrated that RNA polymerase can slide "backward" (Komissarova and Kashlev, 1997; Nudler et al., 1997), making the 3′ ends of nascent RNAs accessible to other proteins. Therefore, addition of extra nucleotides to the 3′ termini of nascent RNAs could potentially be carried out by *trans*-acting factors other than the mitochondrial RNA polymerase. In kinetoplasts, a terminal uridylyl transferase (TUTase) catalyzes the addition of U tails to gRNAs and is thought to be the enzyme responsible for uridine addition during insertional editing (Bakalara et al., 1989; reviewed in Stuart et al., 1997). A terminal transferase activity capable of using each of the four ribonucleotides as substrates has been detected in *Physarum* mitochondrial extracts (unpublished data). However, it is not yet known whether this activity is due to one or more enzymes, nor is there any evidence indicating whether this activity is involved in insertional editing. Other enzymes that add nucleotides to the 3′ ends of nucleic acids include poly(A)polymerase, which specifically adds adenosine residues to eukaryotic mRNAs (Wahle, 1995); terminal deoxynucleotidyl transferase, which can catalyze the addition of any of the four nucleotides to DNA substrates (Ratliff, 1981); tRNA nucleotidyltransferase (Deutscher, 1990); and telomerase (Autexier and Greider, 1994).

An obvious difficulty with editing models involving nontemplated 3′ end addition is that it is not clear how transcription could resume from an unpaired nucleotide and still maintain the degree of accuracy that is observed in *Physarum* mitochondria (Gott et al., 1993). Problems associated with elongation from a nontemplated residue cease to be an issue with editing models involving internal addition of nonencoded nucleotides. As first pointed out for kinetoplastid editing (Blum et al., 1990, 1991; Cech, 1991), two mechanistically distinct models of internal nucleotide insertion are possible: cleavage-ligation and transesterification. Both involve breaking the phosphodiester backbone, nucleotide addition, and rejoining, but differ significantly in the enzymatic mechanism and the site of nucleotide addition. A cleavage-ligation mechanism for *Physarum* editing would most likely involve an enzymatic cascade utilizing enzyme activities similar to those described for the trypanosomatids (see Chapter 21). As described above, a wide range of activities could be responsible for nucleotide addition prior to religation. Residues would most likely be added to the 3′ end of the 5′ cleavage product, as is observed in kinetoplastid editing (Kable et al., 1996), but could potentially be added at the 5′ end of the 3′ product (see Chapter 16) (Sollner-Webb, 1991).

There are several catalytic functions that could serve as models for an internal cleavage activity in *Physarum* mitochondria. For instance, polyadenylation reactions require site-specific cleavage of the

RNA to generate the 3' hydroxyl for poly(A)addition. Interestingly, in vivo the enzymes responsible for cleavage and polyadenylation work in concert, although the two reactions can be uncoupled from one another in vitro (Wahle, 1995). Cleavage could also be catalyzed by activities associated with the transcription complex. The elongation factors GreA, GreB, and TFIIS (Borukhov et al., 1993; Reines, 1994) allow the completion of transcripts associated with arrested elongation complexes by stimulating internal cleavage of the nascent RNA by the RNA polymerase (Surratt et al., 1991; Rudd et al., 1994). In these reactions the 3' cleavage product is lost, permitting extension of the new 3' end. If an analogous activity were used in a cleavage-ligation pathway of *Physarum* editing, both cleavage products would have to remain bound to the transcription complex to permit nucleotide addition and subsequent ligation. Retention of the 3' cleavage product would also allow the mitochondrial RNA polymerase to resume RNA synthesis at a perfectly paired 3' nucleotide. *Physarum* mitochondrial preparations have high levels of associated nucleolytic activity (unpublished data), but it should be possible to determine whether editing site-specific endonuclease and RNA ligase activities are present in these organelles.

Internal addition of nonencoded residues to *Physarum* RNAs could also occur via transesterification reactions similar to those used in the splicing of group I, group II, and nuclear pre-mRNA introns (reviewed in Cech, 1987; Sharp, 1994). Each of these reactions involves two sequential nucleolytic attacks at specific sites along the phosphodiester backbone by appropriately positioned 2' or 3' hydroxyl groups. In *Physarum*, the first transesterification reaction could involve the 3' hydroxyl of either an RNA cofactor, as proposed in models of kinetoplastid editing (Blum et al., 1991; Cech, 1991; Sollner-Webb, 1991), or a free nucleotide, as in the splicing of group I introns (Cech et al., 1981). A second transesterification reaction, involving attack of the 3' OH of the upstream cleavage product just 5' of the added residue (or at the α-β bond of the triphosphate of the "free" nucleotide), would result in the production of RNAs containing an extra, nonencoded nucleotide. Editing in isolated *Physarum* mitochondria is sensitive to the concentration of NTPs (Visomirski-Robic and Gott, 1997b), suggesting that the free nucleotide pool is the immediate precursor for editing. However, the unexpected enrichment of unlabeled C residues at certain editing sites in the *coI* mRNA (see discussion in Visomirski-Robic and Gott, 1997b) would be more consistent with nucleotides donated by another RNA molecule.

During splicing, the juxtaposition of reactive groups relies principally on RNA-RNA interactions (reviewed in Moore et al., 1993; Nilsen, 1994). However, in *Physarum* mitochondria the particular conformation of the editing substrate and attacking nucleophile could be accomplished through RNA-protein interactions, perhaps involving the mitochondrial polymerase and/or associated factors. Indeed, the requirement for such an RNA-protein scaffold could potentially explain the association between transcription and editing (Visomirski-Robic and Gott, 1997b). It should be emphasized, however, that there is currently no experimental evidence that insertional editing in *Physarum* goes through this or any of the other pathways mentioned above.

Substitutional Editing Mechanisms

As described previously, it appears that substitutional editing occurs by a mechanism separate from that responsible for nucleotide insertions in *Physarum* mitochondria. Base changes have been observed in a variety of other organisms, and it seems clear that multiple pathways can be used to achieve this end (see Chapters 16–19). The in vitro data suggest that it is likely that the apparent C to U changes observed in the *Physarum* *coI* mRNA are due to a post-transcriptional mechanism (Visomirski-Robic and Gott, 1995, 1997b). While it is tempting to speculate that these changes occur through hydrolytic deamination events similar to those that occur in plant mitochondria (Araya et al., 1992; Yu and Schuster, 1995) (see Chapter 17) and apolipoprotein B mRNA in mammalian nuclei (Johnson et al., 1993) (see Chapter 18), labeling studies to confirm the presence of U at these positions within the RNA or to directly demonstrate the deamination of a C to a U have not yet been possible due to the lack of an in vitro substitutional editing system.

The means by which sites of base substitution are selected in *Physarum* mitochondria is also currently unknown. Three of the four sites are clustered and could fall within the loop of a single RNA hairpin (Visomirski-Robic, 1997). The significance of this stem-loop is uncertain, however, as no analogous structure has been identified for the fourth C to U site. No sequence resembling the "mooring" sequence that lies 3' of the editing site within the apolipoprotein B mRNA (Backus and Smith, 1991) (see Chapter 18) is found in the vicinity of any of the four C to U changes in the *coI* mRNA, nor have other common downstream sequence elements been identified. In plants, sequences immediately upstream of editing sites are required for C to U changes in both chloroplasts (Bock et al., 1996; Chaudhuri and Maliga, 1996) and mitochondria (see Hanson et al., 1996; Maier et al., 1996) (Chapter 17). Presumably these sequences exert their effects at the level of RNA, although the exact nature of the signals within these upstream regions has not yet been elucidated. Thus, it is possible that synthesis of RNA from a specific upstream region could be necessary for editing at C to U sites in *Physarum* mitochondria.

Finally, *trans*-acting factors required for C to U changes in *Physarum* have yet to be defined. Nuclear mutations that affect base substitutions in plant mitochondria and chloroplasts have been identified (Lu and Hanson, 1992; Zeltz et al., 1993) (Chapter 16) and similar genetic approaches could be used to identify factors required for base substitutions in *Physarum*. Alternatively, molecular methods such as those used to identify RNA adenosine deaminases (ADARs [see Chapter 19]) or the APOBEC-1 enzyme required for apoB mRNA editing (Teng et al., 1993) could prove useful in this context.

OUTSTANDING QUESTIONS AND CONCLUDING REMARKS

Editing in the slime mold *P. polycephalum* is a very complex process that involves the insertion of each of the four nucleotides and specific base changes. Much of the mitochondrial genome of *Physarum* has yet to be sequenced, and it is possible that additional types of editing events may yet be discovered upon characterization of the remaining genes and their RNA products. For instance, insertion of only 5 of the 16 possible dinucleotides has been observed thus far (see Table 1); the presence or absence of other combinations may help to define the signals that specify an insertion site. Additional cases of base substitutions may also be found, creating a larger pool from which to define consensus sequences and/or secondary structures involved in editing.

The means by which these precise alterations occur are not understood, although the available data suggest that at least two separate mechanisms are likely. Insertional editing occurs by a novel mechanism that is either cotranscriptional or tightly linked to transcription, while base substitutions appear to be posttranscriptional. In both cases, however, questions regarding the manner in which editing sites are identified and the composition of the editing machinery remain unanswered. As discussed above, *cis*-acting elements could involve a primary recognition site, a sequence with base-pairing potential, or structural features of the RNA or template. *trans*-acting factors will also be required, and potential roles for these might include detecting or providing editing signals, aiding in conformational changes, templating added nucleotides, or serving as a nucleotide donor for inserted residues, as well as actual catalytic functions.

The mechanism of editing in *Physarum* mitochondria also remains to be elucidated. Many possible models can be envisioned, and further work will be needed to distinguish between them. For example, the connection between transcription and insertional editing needs to be clarified, as the experiments carried out thus far have not addressed whether nucleotides are added internally or at the 3' end of the RNA as it is being transcribed. Addition at the 3' end would suggest direct involvement of the mitochondrial RNA polymerase in nucleotide insertion. Conversely, internal addition would suggest that *Physarum* editing occurs via either a cleavage-ligation pathway or transesterification reactions.

It is also not known whether the same activity that recognizes insertional editing sites catalyzes the addition of nonencoded nucleotides. For instance, it is possible that the *Physarum* mitochondrial RNA polymerase recognizes an editing site and pauses to allow the editing activity to function. Subsequent nucleotide insertion could require the use of a separate or conformationally altered catalytic site on the polymerase itself or a distinct *trans*-acting factor(s). In the latter case, the transcription complex could serve as a scaffold to bind or position the substrate molecules. Alternatively, other components may bind to or otherwise specify an editing site, triggering nucleotide addition by either the polymerase or an associated factor.

Finally, the fact that each of the nucleotides can act as a substrate for the *Physarum* insertional editing machinery adds a further level of mechanistic complexity. The ultimate product of a given mitochondrial gene will be influenced by both the selection of the site and the identity of the inserted residue, making nucleotide choice an important element in the fidelity of gene expression. While the high degree of accuracy demonstrated by the editing activity argues in favor of a nucleic acid template (Maizels and Weiner, 1988; Benne, 1989), there is currently no evidence that an editing template exists, nor is it clear how one might interact with the editing substrate. Specific nucleotide binding sites in proteins could also provide such discrimination (Deutscher, 1990), but the proteins required for editing have yet to be identified.

Recent experiments using isolated mitochondria have offered glimpses into these mechanistic issues, and phylogenetic studies now in progress may shed light on the nature of the editing signals. Although numerous precedents for some of the activities potentially involved in *Physarum* editing exist, a model that accounts for all of the unusual features of editing in *Physarum* mitochondria has not yet been presented. Ultimately, proposed mechanisms will have to explain how each of the activities is coordinated as a whole to generate the incredible precision of the *Physarum* editing process.

Acknowledgments. We thank the members of the Case Western Reserve University (CWRU) Center for RNA Molecular Biology, the Cleveland RNA Club, the CWRU transcription journal club, and the CWRU Molecular Microbiology group for stimulating discussions and Rob Benne and Henri Grosjean for helpful suggestions. This work was supported by grants to J.M.G. from the National Science Foundation (MCB-9630672) and the National Institutes of Health (GM/OD54663).

REFERENCES

Araya, A., C. Domec, D. Begu, and S. Litvak. 1992. An in vitro system for the editing of ATP synthase subunit 9 mRNA using wheat mitochondrial extracts. *Proc. Natl. Acad. Sci. USA* **89:** 1040–1044.

Autexier, C., and C. W. Greider. 1994. Functional reconstitution of wild-type and mutant *Tetrahymena* telomerase. *Genes Dev.* **8:**563–575.

Bachellerie, J. P., B. Michot, M. Nicoloso, A. Balakin, J. W. Ni, and M. J. Fournier. 1995. Antisense snoRNAs: a family of nucleolar RNAs with long complementarities to rRNA. *Trends Biochem. Sci.* **20:**261–264.

Backus, J. W., and H. C. Smith. 1991. Apolipoprotein B mRNA sequences 3' of the editing site are necessary and sufficient for editing and editosome assembly. *Nucleic Acids Res.* **19:** 6781–6786.

Bakalara, N., A. M. Simpson, and L. Simpson. 1989. The *Leishmania* kinetoplast-mitochondrion contains terminal uridylyltransferase and RNA ligase activities. *J. Biol. Chem.* **264:** 18679–18686.

Benne, R. 1989. RNA-editing in trypanosome mitochondria. *Biochim. Biophys. Acta* **1007:**131–139.

Blum, B., N. Bakalara, and L. Simpson. 1990. A model for RNA editing in kinetoplastid mitochondria: "guide" RNA molecules transcribed from maxicircle DNA provide the edited information. *Cell* **60:**189–198.

Blum, B., N. R. Strum, A. M. Simpson, and L. Simpson. 1991. Chimeric gRNA-mRNA molecules with oligo(U) tails covalently linked at sites of RNA editing suggest that U addition occurs by transesterification. *Cell* **65:**543–550.

Bock, R., M. Hermann, and H. Kossel. 1996. In vivo dissection of *cis*-acting determinants for plastid RNA editing. *EMBO J.* **15:** 5052–5059.

Borukhov, S., V. Sagitov, and A. Goldfarb. 1993. Transcript cleavage factors from *E. coli*. *Cell* **72:**459–466.

Burger, G., I. Plante, K. M. Lonergan, and M. W. Gray. 1995. The mitochondrial DNA of the amoeboid protozoan, *Acanthamoeba castellanii*. Complete sequence, gene content and genome organization. *J. Mol. Biol.* **245:**522–537.

Byrne, E. M., G. J. Connell, and L. Simpson. 1996. Guide RNA-directed uridine insertion RNA editing in vitro. *EMBO J.* **15:** 6758–6765.

Casey, J. L., K. F. Bergmann, T. L. Brown, and J. L. Gerin. 1992. Structural requirements for RNA editing in hepatitis delta virus: evidence for a uridine-to-cytidine editing mechanism. *Proc. Natl. Acad. Sci.* **89:**7149–7153.

Cech, T. R. 1987. The chemistry of self-splicing RNA and RNA enzymes. *Science* **236:**1532–1539.

Cech, T. R. 1991. RNA editing: world's smallest introns? *Cell* **64:** 667–669.

Cech, T. R., A. J. Zaug, and P. J. Grabowski. 1981. In vitro splicing of the ribosomal RNA precursor of Tetrahymena: involvement of a guanosine nucleotide in the excision of the intervening sequence. *Cell* **27:**487–496.

Cermakian, N., T. M. Ikeda, R. Cedergren, and M. W. Gray. 1996. Sequences homologous to yeast mitochondrial and bacteriophage T3 and T7 RNA polymerases are widespread throughout the eukaryotic lineage. *Nucleic Acids Res.* **24:** 648–654.

Chan, C. L., D. G. Wang, and R. Landick. 1997. Multiple interactions stabilize a single paused transcription intermediate in which hairpin to 3' end spacing distinguishes pause and termination pathways. *J. Mol. Biol.* **268:**54–68.

Chaudhuri, S., and P. Maliga. 1996. Sequences directing C to U editing of the plastid psbL mRNA are located within a 22 nucleotide segment spanning the editing site. *EMBO J.* **15:** 5958–5964.

Corell, R. A., J. E. Feagin, G. R. Riley, T. Strickland, J. A. Guderian, P. J. Myler, and K. Stuart. 1993. *Trypanosoma brucei* minicircles encode multiple guide RNAs which can direct editing of extensively overlapping sequences. *Nucleic Acids Res.* **21:** 4313–4320.

Daga, A., V. Micol, D. Hess, R. Aebersold, and G. Attardi. 1993. Molecular characterization of the transcription termination factor from human mitochondria. *J. Biol. Chem.* **268:**8123–8130.

Das, A. 1992. How the phage lambda N gene product suppresses transcription termination: communication of RNA polymerase with regulatory proteins mediated by signals in nascent RNA. *J. Bacteriol.* **174:**6711–6716.

Deutscher, M. P. 1990. Ribonucleases, tRNA nucleotidyltransferase, and the 3' processing of tRNA. *Prog. Nucleic Acid Res. Mol. Biol.* **39:**209–237.

Dieckmann, T., E. Suzuki, G. K. Nakamura, and J. Feigon. 1996. Solution structure of an ATP-binding RNA aptamer reveals a novel fold. *RNA* **2:**628–640.

Driscoll, D. M., J. K. Wynne, S. C. Wallis, and J. Scott. 1989. An in vitro system for the editing of apolipoprotein B mRNA. *Cell* **58:**519–525.

Ekland, E. H., and D. P. Bartel. 1996. RNA-catalysed RNA polymerization using nucleoside triphosphates. *Nature* **382:** 373–376.

Ganot, P., M.-L. Bortolin, and T. Kiss. 1997. Site-specific pseudouridine formation in preribosomal RNA is guided by small nucleolar RNAs. *Cell* **89:**799–809.

Gilley, D., and E. H. Blackburn. 1996. Specific RNA residue interactions required for enzymatic functions of *Tetrahymena* telomerase. *Mol. Cell. Biol.* **16:**66–75.

Gott, J. M., L. M. Visomirski, and J. L. Hunter. 1993. Substitutional and insertional RNA editing of the cytochrome *c* oxidase subunit 1 mRNA of *Physarum polycephalum*. *J. Biol. Chem.* **268:** 25483–25486.

Greenblatt, J., J. R. Nodwell, and S. W. Mason. 1993. Transcriptional antitermination. *Nature* **364:**401–406.

Greider, C. W., and E. H. Blackburn. 1989. A telomeric sequence in the RNA of *Tetrahymena* telomerase required for telomere repeat synthesis. *Nature* **337:**331–337.

Grosjean, H., Z. Szweykowska-Kulinska, Y. Motorin, F. Fasiolo, and G. Simos. 1997. Intron-dependent enzymatic formation of modified nucleosides in eukaryotic tRNAs: a review. *Biochimie* **79:**293–302.

Hammond, P. W., T. N. Lively, and T. R. Cech. 1997. The anchor site of telomerase from *Euplotes aediculatus* revealed by photo-cross-linking to single- and double-stranded DNA primers. *Mol. Cell. Biol.* **17:**296–308.

Hanson, M. R., C. A. Sutton, and B. W. Lu. 1996. Plant organelle gene expression: altered by RNA editing. *Trends Plant Sci.* **1:** 57–64.

Jacques, J.-P., and D. Kolakofsky. 1991. Pseudo-templated transcription in prokaryotic and eukaryotic organisms. *Genes Dev.* 5:707–713.

Jeong, S. W., W. H. Lang, and R. H. Reeder. 1996. The yeast transcription terminator for RNA polymerase I is designed to prevent polymerase slippage. *J. Biol. Chem.* 271:16104–16110.

Johnson, D. F., K. S. Poksay, and T. L. Innerarity. 1993. The mechanism for apo-B mRNA editing is deamination. *Biochem. Biophys. Res. Commun.* 195:1204–1210.

Johnson, P. F., and J. Abelson. 1983. The yeast tRNAtyr gene intron is essential for correct modification of its tRNA product. *Nature* 302:681–687.

Joyce, C. M., and T. A. Steitz. 1994. Function and structure relationships in DNA polymerases. *Annu. Rev. Biochem.* 63:777–822.

Kable, M. L., S. D. Seiwert, S. Heidmann, and K. Stuart. 1996. RNA editing: a mechanism for gRNA-specified uridylate insertion into precursor mRNA. *Science* 273:1189–1195.

Kerppola, T. K., and C. M. Kane. 1990. Analysis of the signals for transcription termination by purified RNA polymerase II. *Biochemistry* 29:269–278.

Kiss-Laszlo, Z., Y. Henry, J.-P. Bachellerie, M. Caizergue-Ferrer, and T. Kiss. 1996. Site-specific ribose methylation of preribosomal RNA: a novel function for small nucleolar RNAs. *Cell* 85:1077–1088.

Kolakofsky, D., J. Curran, T. Pelet, and J.-P. Jacques. 1993. Paramyxovirus P gene mRNA editing, p. 105–123. In R. Benne (ed.), *RNA Editing: the Alteration of Protein Coding Sequences of RNA.* Ellis Horwood Limited, New York, N.Y.

Komissarova, N., and M. Kashlev. 1997. Transcriptional arrest: *Escherichia coli* RNA polymerase translocates backward, leaving the 3' end of the RNA intact and extruded. *Proc. Natl. Acad. Sci. USA* 94:1755–1760.

Krumm, A., L. B. Hickey, and M. Groudine. 1995. Promoter-proximal pausing of RNA polymerase II defines a general rate-limiting step after transcription initiation. *Genes Dev.* 9:559–572.

Landick, R. 1997. RNA polymerase slides home: pause and termination site recognition. *Cell* 88:741–744.

Lang, W. H., B. E. Morrow, Q. Ju, J. R. Warner, and R. H. Reeder. 1994. A model for transcription termination by RNA polymerase I. *Cell* 79:527–534.

Liu, C. G., L. S. Heath, and C. L. Turnbough. 1994. Regulation of pyrBI operon expression in *Escherichia coli* by UTP-sensitive reiterative RNA synthesis during transcriptional initiation. *Genes Dev.* 8:2904–2912.

Lonergan, K. M., and M. W. Gray. 1993. Editing of transfer RNAs in *Acanthamoeba castellanii* mitochondria. *Science* 259:812–816.

Lu, B., and M. R. Hanson. 1992. A single nuclear gene specifies the abundance and extent of RNA editing of a plant mitochondrial transcript. *Nucleic Acids Res.* 20:5699–5703.

Lu, Y., R. J. Turner, and R. L. Switzer. 1996. Function of RNA secondary structures in transcriptional attenuation of the *Bacillus subtilis* pyr operon. *Proc. Natl. Acad. Sci. USA* 93:14462–14467.

Macdonald, L. E., Y. Zhou, and W. T. McAllister. 1993. Termination and slippage by bacteriophage T7 RNA polymerase. *J. Mol. Biol.* 232:1030–1047.

Mahendran, R., M. R. Spottswood, and D. L. Miller. 1991. RNA editing by cytidine insertion in mitochondria of *Physarum polycephalum*. *Nature* 349:434–438.

Mahendran, R., M. S. Spottswood, A. Ghate, M.-L. Ling, K. Jeng, and D. L. Miller. 1994. Editing of the mitochondrial small subunit rRNA in *Physarum polycephalum*. *EMBO J.* 13:232–240.

Maier, R. M., P. Zeltz, H. Kossel, G. Bonnard, J. M. Gualberto, and J. M. Grienenberger. 1996. RNA editing in plant mitochondria and chloroplasts. *Plant Mol. Biol.* 32:343–365.

Maizels, N., and A. Weiner. 1988. In search of a template. *Nature* 334:469–470.

Masiakski, P., and M. P. Deutscher. 1980. Dissection of the active site of rabbit liver tRNA nucleotidyltransferase. *J. Biol. Chem.* 255:11240–11246.

Maslov, D. A., and L. Simpson. 1992. The polarity of editing within a multiple gRNA-mediated domain is due to formation of anchors for upstream gRNAs by downstream editing. *Cell* 70:459–467.

Masters, B. S., L. L. Stohl, and D. A. Clayton. 1987. Yeast mitochondrial RNA polymerase is homologous to those encoded by bacteriophages T3 and T7. *Cell* 51:89–99.

Miller, D., R. Mahendran, M. Spottswood, H. Costandy, S. Wang, M. L. Ling, and N. Yang. 1993a. Insertional editing in mitochondria of *Physarum*. *Semin. Cell Biol.* 4:261–266.

Miller, D., R. Mahendran, M. Spottswood, M. Ling, S. Wang, N. Yang, and H. Costandy. 1993b. RNA editing in mitochondria of *Physarum polycephalum*, p. 87–103. In R. Benne (ed.), *RNA Editing: the Alteration of Protein Coding Sequences of RNA.* Ellis Horwood, New York, N.Y.

Milligan, J. F., D. R. Groebe, G. W. Witherell, and O. C. Uhlenbeck. 1987. Oligoribonucleotide synthesis using T7 RNA polymerase and synthetic DNA templates. *Nucleic Acids Res.* 15:8783–8798.

Moore, M. J., C. C. Query, and P. A. Sharp. 1993. Splicing of precursors to mRNA by the spliceosome, p. 303–357. In R. F. Gesteland and J. F. Atkins (ed.), *The RNA World.* Cold Spring Harbor Laboratory Press, Plainview, N.Y.

Ni, J., A. L. Tien, and M. J. Fournier. 1997. Small nucleolar RNAs direct site-specific synthesis of pseudouridine in ribosomal RNA. *Cell* 89:565–573.

Nilsen, T. W. 1994. RNA-RNA interactions in the spliceosome: unraveling the ties that bind. *Cell* 78:1–4.

Nudler, E., A. Mustaev, E. Lukhtanov, and A. Goldfarb. 1997. The RNA-DNA hybrid maintains the register of transcription by preventing backtracking of RNA polymerase. *Cell* 89:33–41.

Platt, T. 1998. RNA structure in transcription elongation, termination, and antitermination, p. 541–575. In R. W. Simons and M. Grunberg-Manago (ed.), *RNA Structure and Function.* Cold Spring Harbor Laboratory Press, Cold Spring Harbor, N.Y.

Qi, F. X., and C. L. Turnbough. 1995. Regulation of codBA operon expression in *Escherichia coli* by UTP-dependent reiterative transcription and UTP-sensitive transcriptional start site switching. *J. Mol. Biol.* 254:552–565.

Rasmussen, E. B., and J. T. Lis. 1995. Short transcripts of the ternary complex provide insight into RNA polymerase II elongational pausing. *J. Mol. Biol.* 252:522–535.

Ratliff, R. L. 1981. Terminal deoxynucleotidyltransferase, p. 105–118. In P. D. Boyer (ed.), *The Enzymes.* Academic Press, New York, N.Y.

Reines, D. 1994. Nascent RNA cleavage by transcription elongation complexes, p. 263–278. In R. C. Conaway and J. W. Conaway (ed.), *Transcription: Mechanisms and Regulation.* Raven Press, Ltd., New York, N.Y.

Ring, B. Z., and J. W. Roberts. 1994. Function of a nontranscribed DNA strand site in transcription elongation. *Cell* 78:317–324.

Ring, B. Z., and W. S. Yarnell. 1996. Function of E. coli RNA polymerase sigma factor sigma(70) in promoter-proximal pausing. *Cell* 86:485–493.

Roberts, C. W., and J. W. Roberts. 1996. Base-specific recognition of the nontemplate strand of promoter DNA by E. coli RNA polymerase. *Cell* 86:495–501.

Rudd, M. D., M. G. Izban, and D. S. Luse. 1994. The active site of RNA polymerase II participates in transcript cleavage within arrested ternary complexes. *Proc. Natl. Acad. Sci. USA* **91:** 8057–8061.

Rundquist, B. A. 1995. RNA editing throughout the life cycle of *Physarum polycephalum*. M.S. thesis. Case Western Reserve University, Cleveland, Ohio.

Rundquist, B. A., and J. M. Gott. 1995. RNA editing of the *coI* mRNA throughout the life cycle of *Physarum polycephalum*. *Mol. Gen. Genet.* **247:**306–311.

Sassanfar, M., and J. W. Szostak. 1993. An RNA motif that binds ATP. *Nature* **364:**550–553.

Seiwert, S. D., and K. Stuart. 1994. RNA editing: transfer of genetic information from gRNA to precursor mRNA in vitro. *Science* **266:**114–117.

Sharp, P. A. 1994. Split genes and RNA splicing. *Cell* **77:**805–815.

Smith, C. M., and J. A. Steitz. 1997. Sno storm in the nucleolus: new roles for myriad small RNPs. *Cell* **89:**669–672.

Sollner-Webb, B. 1991. RNA editing. *Curr. Opin. Cell Biol.* **3:** 1056.

Stuart, K., T. E. Allen, S. Heidmann, and S. D. Seiwert. 1997. RNA editing of kinetoplastid protozoa. *Microbiol. Mol. Biol. Rev.* **61:**105–120.

Sugisaki, H., and M. Takanami. 1993. The 5' terminal region of the apocytochrome b transcript in *Crithidia fasciculata* is successively edited by two guide RNAs in the 3' to 5' direction. *J. Biol. Chem.* **268:**887–891.

Surratt, C. K., S. C. Milan, and M. J. Chamberlin. 1991. Spontaneous cleavage of RNA in ternary complexes of *Escherichia coli* RNA polymerase and its significance for the mechanism of transcription. *Proc. Natl. Acad. Sci. USA* **88:**7983–7987.

Teng, B., C. F. Burant, and N. O. Davidson. 1993. Molecular cloning of an apolipoprotein B messenger RNA editing protein. *Science* **260:**1816–1819.

Tycowski, K. T., C. M. Smith, M. D. Shu, and J. A. Steitz. 1996. A small nucleolar RNA requirement for site-specific ribose methylation of rRNA in Xenopus. *Proc. Natl. Acad. Sci. USA* **93:** 14480–14485.

Vidal, S., J. Curran, and D. Kolakosky. 1990. A stuttering model for paramyxovirus P mRNA editing. *EMBO J.* **9:**2017–2022.

Visomirski-Robic, L. M. 1997. Examination of insertional editing in *Physarum polycephalum* in an isolated mitochondrial system: a potential coupling of editing and transcription. Ph.D. thesis. Case Western Reserve University, Cleveland, Ohio.

Visomirski-Robic, L. M., and J. M. Gott. 1995. Accurate and efficient insertional RNA editing in isolated *Physarum* mitochondria. *RNA* **1:**681–691.

Visomirski-Robic, L. M., and J. M. Gott. 1997a. Insertional editing of nascent mitochondrial RNAs in *Physarum*. *Proc. Natl. Acad. Sci. USA* **94:**4324–4329.

Visomirski-Robic, L. M., and J. M. Gott. 1997b. Insertional editing in isolated *Physarum* mitochondria is linked to RNA synthesis. *RNA* **3:**821–837.

Visomirski-Robic, L. M., and J. M. Gott. Unpublished data.

Visomirski-Robic, L. M., C. Webb, and J. M. Gott. Unpublished data.

Volchkov, V. E., S. Becker, V. A. Volchkova, V. A. Ternovoj, A. N. Kotov, S. V. Netesov, and H. D. Klenk. 1995. GP mRNA of Ebola virus is edited by the Ebola virus polymerase and by T7 and vaccinia virus polymerases. *Virology* **214:**421–430.

Wahle, E. 1995. 3'-end cleavage and polyadenylation of mRNA precursors. *Biochim. Biophys. Acta* **1261:**183–194.

Yarnell, W. S., and J. W. Roberts. 1992. The phage lambda gene Q transcription antiterminator binds DNA in the late gene promoter as it modifies RNA polymerase. *Cell* **69:**1181–1189.

Yokobori, S., and S. Paabo. 1995. Transfer RNA editing in land snail mitochondria. *Proc. Natl. Acad. Sci. USA* **92:**10432–10435.

Yu, W., and W. Schuster. 1995. Evidence for a site-specific cytidine deamination reaction involved in C to U RNA editing of plant mitochondria. *J. Biol. Chem.* **270:**18227–18233.

Zeltz, P., W. R. Hess, K. Neckermann, T. Borner, and H. Kossel. 1993. Editing of the chloroplast *rpoB* transcript is independent of chloroplast translation and shows different patterns in barley and maize. *EMBO J.* **12:**4291–4296.

Zheng, H., T.-B. Fu, D. Lazinski, and J. Taylor. 1992. Editing on the genomic RNA of human hepatitis delta virus. *J. Virol.* **66:** 4693–4697.

Chapter 23

Cotranscriptional Paramyxovirus mRNA Editing: a Contradiction in Terms?

DANIEL KOLAKOFSKY AND STÉPHANE HAUSMANN

A SHORT PRIMER IN PARAMYXOVIRUS RNA SYNTHESIS

Paramyxoviruses contain a nonsegmented single-stranded RNA genome of ca. 15 kb. The genome is of negative polarity (complementary to the viral mRNAs) and is never found free, but is encapsidated by the viral N protein in the form of a helical nucleocapsid. This structure, rather than naked RNA, serves as the template for the synthesis of both the monocistronic mRNAs and the full-length antigenome (the intermediate in genome replication) by the viral polymerase (a complex of the phosphoprotein P and the large protein L). The linear RNA genome contains ca. 6 genes or transcriptional units, flanked by short noncoding leader and trailer regions. Each mRNA transcription unit begins on a conserved gene start sequence and ends on a conserved polyadenylation/termination sequence (Lamb and Kolakofsky, 1996).

Virus replication begins intracellularly with mRNA synthesis (primary transcription), where the polymerase starts at the 3′ end of the (−) template and sequentially produces the leader RNA and mRNAs by stopping and restarting at each of the junctions. The mRNAs contain poly(A) tails of several hundred nucleotides, and the start of this tail maps to the U_{4-7} stretch within the conserved termination sequence of the [−] template (for example, 3′ UNAUUCUUUUU for Sendai virus). The poly(A) tail presumably is made by polymerase slippage and reiterative copying of the U_{4-7} stretch (McGeoch, 1979; Rose, 1980; Schubert et al., 1980; Robertson et al., 1981; Giorgi et al., 1983; Gupta and Kingsbury, 1984), and this process is referred to as polymerase "stuttering." The strictly processive transcriptase is thought to start the next mRNA only after terminating poly(A) synthesis. During antigenome synthesis, the next stage in virus replication, ostensibly the same viral polymerase copies the same (−) template, but now reads through all the junctions to make an exact full-length complementary copy. However, unlike the mRNAs, the antigenomes are assembled into nucleocapsids concomitant with their synthesis, as genome synthesis and assembly are coupled. (Antigenomes presumably also function as templates only in the assembled form.) This coupling of antigenome synthesis and assembly is thought to play a key role in whether the junctional stop and restart signals are obeyed. For a more complete review, see Lamb and Kolakofsky (1996).

PARAMYXOVIRUS P GENE mRNA EDITING

The *Paramyxovirinae* currently are organized in three genera: the paramyxovirus genus (including Sendai virus [SeV], human parainfluenza virus type 1 [hPIV1], and human and bovine PIV3); morbilliviruses [e.g., measles virus and the distemper viruses]; and rubulaviruses (e.g., mumps virus and simian virus 5 {SV5}]). The first rubulavirus P gene sequence determined correctly was that of SV5 (Thomas et al., 1988). This yielded the surprising result that there was no single open reading frame (ORF) large enough to code for the SV5 P protein. Rather, the gene contained two separate ORFs at each end, which briefly overlapped in the middle (Fig. 1a, top line). mRNA transcribed from cDNAs that were exact copies of the SV5 P gene could be translated into the shorter V protein encoded by the 5′ proximal ORF, but not into the longer P protein. After a period of

---

Daniel Kolakofsky and Stéphane Hausmann • Department of Genetics and Microbiology, University of Geneva School of Medicine, CMU, 9 Ave. de Champel, CH-1211 Geneva, Switzerland.

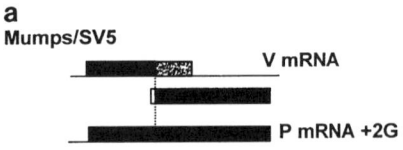
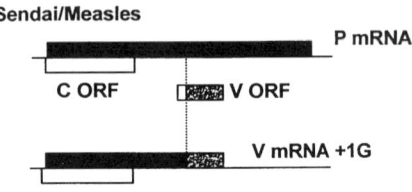
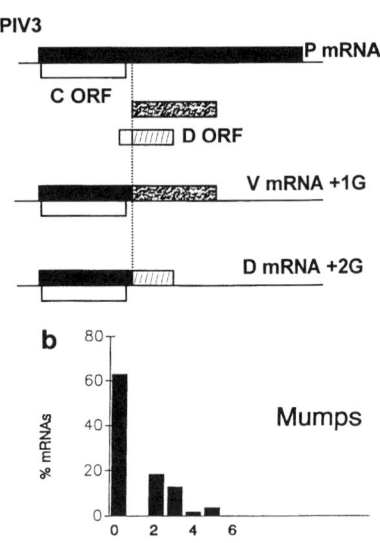
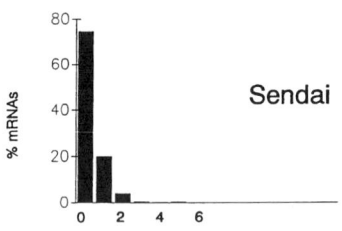
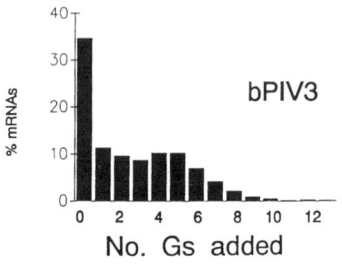

Figure 1. (a) Schematic representation of the paramyxovirus P gene mRNAs. The mRNAs are indicated as horizontal lines, and the ORFs are indicated as boxes. For each group, the upper line shows the mRNA that is an exact copy of the gene, and the beginning of the ORF box indicates the ribosomal start codon. When more than one ORF box is attached to the line, they are accessed by alternate initiation codons. The boxes below indicate alternate downstream ORFs that are accessed by G insertions in the mRNAs. The positions of the insertions are shown by the dotted vertical lines. The three possible ORFs are indicated by different shading. (b) Distribution of G insertions in various paramyxovirus P gene mRNAs. The distribution of the number of Gs inserted in the indicated paramyxovirus P gene mRNAs is shown as a bar graph. The mumps virus data is from Paterson and Lamb (1990) and was determined by sequencing 54 mRNA clones. The Sendai and bPIV3 data is from Pelet et al. (1991) and was determined by a primer extension method directly on the mRNA population. "No. Gs added" refers to the unedited mRNA.

frustration, the solution to this dilemma became clear when Thomas et al. (1988) found a second population of these mRNAs that had two G residues inserted within a short run of Gs where the two ORFs overlapped (dotted line, Fig. 1a). This fuses the separate ORFs at each end into a single continuous unit, and mRNAs prepared from the latter clones translated into P (but not V) proteins. A similar but complementary situation was independently uncovered for the measles virus P gene (Cattaneo et al., 1989) (Fig. 1a). In short order, other paramyxovirus P genes also were found to use this unusual mechanism (transcriptional frameshifting/alternative elongation/mRNA editing) to expand their repertoire of P gene products (Vidal et al., 1990a; Pelet et al., 1991; Galinski et al., 1992), as predicted (Thomas et al., 1988, Cattaneo et al., 1989).

All these viral P genes (with the notable exception of that of hPIV1) contain an A_nG_n purine run (Fig. 2a) at the start of a highly conserved internal, overlapping V ORF. The alternate base-pairing possibilities of this "slippery sequence" (including G:U bonds) are used to synthesise *trans*-frame proteins (that is, proteins whose N- and C-terminal portions are encoded by separate ORFs, like the human immunodeficiency virus [HIV] gag-pol polyprotein [Kollmus et al., 1996]). This situation is analogous to ribosomal frameshifting, which forms the HIV gag-pol polyprotein. However, for paramyxovirus mRNA editing, the alternate base pairing occurs during mRNA synthesis, via pseudotemplated transcription (Jacques and Kolakofsky, 1991). As a result, mRNAs with expanded G runs are transcribed from these genes in addition to those that are faithful copies of their templates, and the number of G insertions that occur for each virus group mirrors their requirements to switch between the in-frame and out-of-frame ORFs (reviewed in Kolakofsky et al., 1993). For the morbilliviruses and SeV, which require a +1G insertion to access the V ORF from the genome-encoded

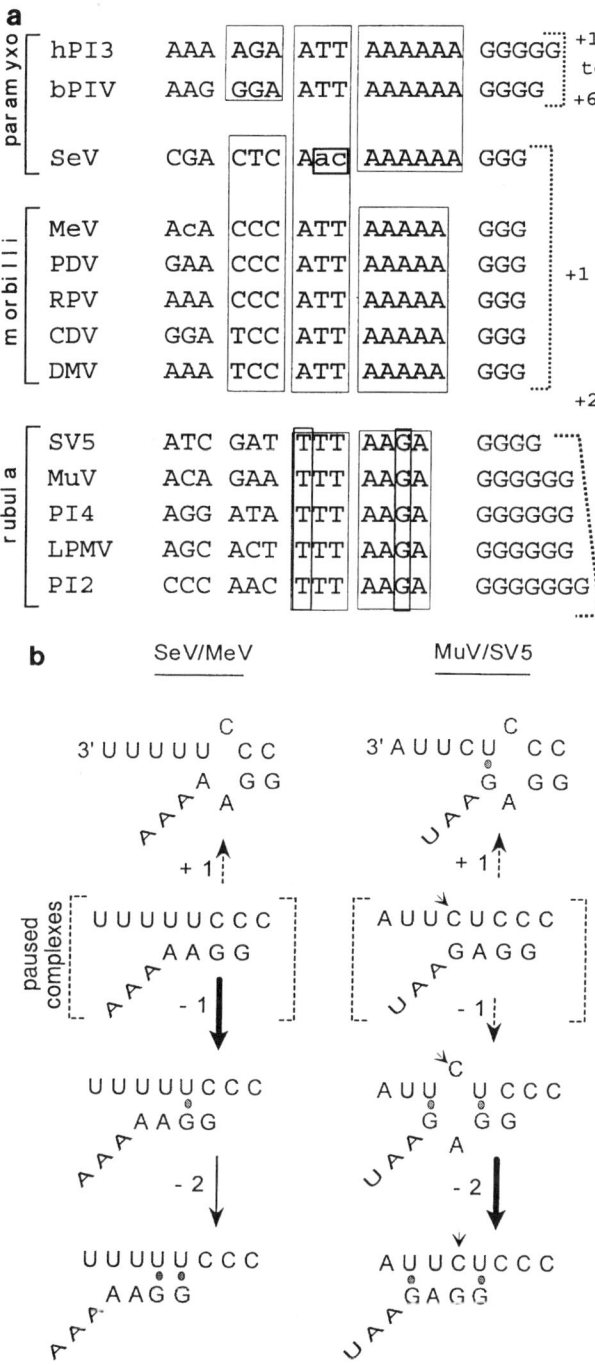

Figure 2. (a) Comparison of the paramyxovirus editing sites. The sequences are written as (+) RNA, 5' to 3', and are grouped into the three genera of the subfamily *Paramyxovirinae*. Spaces have been introduced to emphasize the different elements of the sequence. The short G run that is expanded on mRNA editing is shown on the right, together with the pattern of G insertions that occurs for each group (dotted brackets). Note that the A run that precedes the G run is the only part of this *cis*-acting sequence which is strictly conserved according to genera. Also note that the second A residue upstream of the rubulavirus G run is replaced by a G (highlighted with a rectangle), which presumably accounts for why rubulaviruses insert a minimum of two G residues (Vidal et al., 1990b). The shaded boxes indicate sequence conservations. When the boxed lower case "ac" is changed to TT as in the other members of the genus *Paramyxovirus*, SeV now edits its mRNA

P ORF (Fig. 1a), a single G is added as the predominant insertional event (Fig. 1b), generating a V mRNA coding for the *trans*-frame V protein. Two Gs, however, are also inserted at a much-reduced frequency, and this leads to a protein called W, which is essentially the N-terminal part of the P protein up to the editing site (this downstream ORF is quickly closed by a stop codon). For the rubulaviruses, in contrast, P is the *trans*-frame protein and these P genes require a 2G insertion at the editing site to access the remainder of the P ORF from the genome-encoded V ORF. Two G's are added here at high frequency when insertions occur (Fig. 1b). For hPIV3 and bovine PIV3, which contain an additional (D) ORF at this position, one to six G residues are inserted at roughly equal frequencies to accommodate the expression of all three overlapping ORFs (P, V, and D) (Pelet et al., 1991).

Paramyxovirus RNA synthesis is restricted to the cytoplasm, and these viruses fend for themselves in all aspects of mRNA synthesis. Given that they formed their poly(A) tails by the unusual mechanism of polymerase stuttering, this mechanism was immediately suggested as one that could also account for the unusual G insertions. Experimental evidence in favor of a stuttering mechanism came from work with SeV, a member of the remaining paramyxovirus genus. SeV mRNAs made in vitro with purified virions contained the same pattern of G insertions as those found intracellularly, indicating that the process was carried out by viral gene products. More tellingly, when mRNA was examined from cells doubly infected with SeV and a vaccinia virus expressing an almost identical P gene, the P mRNA expressed from vaccinia virus DNA was completely unedited, indicating that the editing activity could not function in *trans* (Vidal et al., 1990a).

like PIV3. (b) Realignment possibilities at the paramyxovirus mRNA editing sites. RNA synthesis complexes of the template (top strand, written 3' to 5') and nascent chain (5' to 3', bottom strand) containing four base pairs are shown, for the SeV group and morbilliviruses (left side) and rubulaviruses (right side). The polymerase (not shown) whose catalytic site contains the 3' end of the nascent (bottom) chain, is proposed to pause after incorporation opposite the middle template C residue (second level, in dotted brackets) and the nascent chain to realign on the template, allowing for U:G pairs (highlighted with a shaded circle in between) but not A:C pairs (which are shown looping out). The frequencies of the realignments that occur during natural virus infections are indicated by the strengths of the arrows. The numbers indicate the realignment of the nascent chain: minus as upstream, and plus as downstream, that is, the opposite of the subsequent insertions. For rubulaviruses, the −1 realignment is prevented by the marked (arrowhead) template C residue (which is a U in the SeV/MeV group). Once bypassed by a −2 shift (bottom panel), this template C aligns only with G's in further shifts. MeV, measles virus; MuV, mumps virus.

A COMPETITIVE KINETIC MODEL FOR PARAMYXOVIRUS POLYMERASE STUTTERING

It has been known for some time that *Escherichia coli* RNA polymerase can respond to intrinsic signals in the template DNA (and nascent RNA), which divert a fraction of the transcription complexes from the path of rapid chain elongation, for example, to pause, or to terminate the chain (see Landick, 1997). These processes are among the best-studied examples of transcriptional choice, and thus have the most detailed models. By analogy to *E. coli* RNA polymerase (von Hippel and Yager, 1992), the paramyxovirus transcription elongation complex has two choices at any template position I; it can extend the nascent chain by one nucleotide to form a transcript that is $I + 1$ residues in length, or it can be induced by features of the template or nascent chain sequence (perhaps assisted by protein factors) to form a processively unstable complex (Fig. 3). At this point (the first branchpoint), there is a significant probability that the active site together with its nascent chain is repositioned one or two residues upstream of the template C that has just been copied, such that one (paramyxoviruses and morbilliviruses) or two (rubulaviruses) of these C's are copied a second time when nucleotide addition recurs, resulting in pseudotemplated G insertions in the mRNA (stuttering). How the base pairs between the nascent chain and template are presumably broken and reformed in the realignment is unclear. Processively unstable complexes would form at only two types of template sites, polyadenylation sites at the ends of mRNA transcription units (3' UNAUUC<u>UUUUU</u> for Sendai virus) and the editing sites within the P genes (3' UUG-UUUUUU<u>CCC</u> for Sendai virus) which are both very rich in pyrimidines.

In a competitive kinetic model, the elongation and stuttering processes are characterized by specific overall rate constants ($k_{forward}$ and $k_{stutter}$) at each template position. The magnitude of these rate constants is expected to depend on template and nascent chain sequences and, in particular, on the base-pairing possibilities of the realigned upstream sequences. For *E. coli* RNA polymerase, where the average step time for addition of the next nucleotide is ~30 ms (at typical nonterminator positions), the activation barrier height to elongation was calculated at +16 kcal/mol, whereas the barrier to termination at these positions is thought to be >+30 kcal/mol (von Hippel and Yager, 1992). Hence, transcription occurs with an infinitesimal probability of spontaneous termination at nonterminator positions. At terminator positions, however, the length of time the polymerase pauses at this site is roughly equal to that for nascent chain release (1 to 10 s), and the barriers to elongation and termination are roughly equal (~+18 kcal/mol). By analogy, we assume that the barrier to stuttering at stuttering sites is somehow lowered (and that to strictly templated elongation is raised), such that a significant fraction of the elongation complexes are realigned upstream on the template before the next nucleotide is added.

In paramyxovirus infections, a variable fraction of the P gene mRNAs (20–60%) are edited, that is, they contain one or more pseudotemplated G insertions at the editing site. The efficiency of editing (EE) is then defined as the number of mRNAs with insertions at a given site devided by the total number of mRNAs. In the competitive kinetic (Eyring) model presented in Fig. 4, the editing efficiency of a transcription complex located at I will be

$$EE_I = (1 + e^{-\Delta(\Delta G^*)/RT})^{-1} \qquad (1)$$

where ΔG^* is the activation free energy, $\Delta(\Delta G^*)$ is the difference in heights of the two activation barriers of Fig. 4, R is the gas constant, and T is the temperature. EE_I thus ranges from 0 to 1 at each template position, as a function of the (position dependent) difference in heights of the activation barriers to elongation and stuttering. It would be 0 at nonstuttering positions, 0.2 to 0.6 at various editing sites, and very close to 1 at polyadenylation sites. As previously noted for the elongation-termination decision of *E. coli* RNA polymerase, equation 1 is based on general kinetic principles, and is independent of any particular structural model of the transcription complex at (in this case) an editing site.

The competitive kinetic model made two predictions for the *E. coli* RNA polymerase elongation-termination decision (von Hippel and Yager, 1992), which also should apply to stuttering. First, stuttering should be possible only at sites where the elongation complex is destabilized to such an extent that the heights of the elongation and stuttering barriers are roughly equal. Because of the large difference in barrier heights at nonstuttering positions, the position of stuttering sites should be strongly determined and the elongation-stuttering decision should have the character of a binary switch. This is of interest because neither form of polymerase stuttering normally occurs during antigenome synthesis where the viral polymerase is switched to the replication mode. Second, editing efficiencies should be easily modified (or regulated) at editing sites, because relatively small changes in either kinetic or thermodynamic compo-

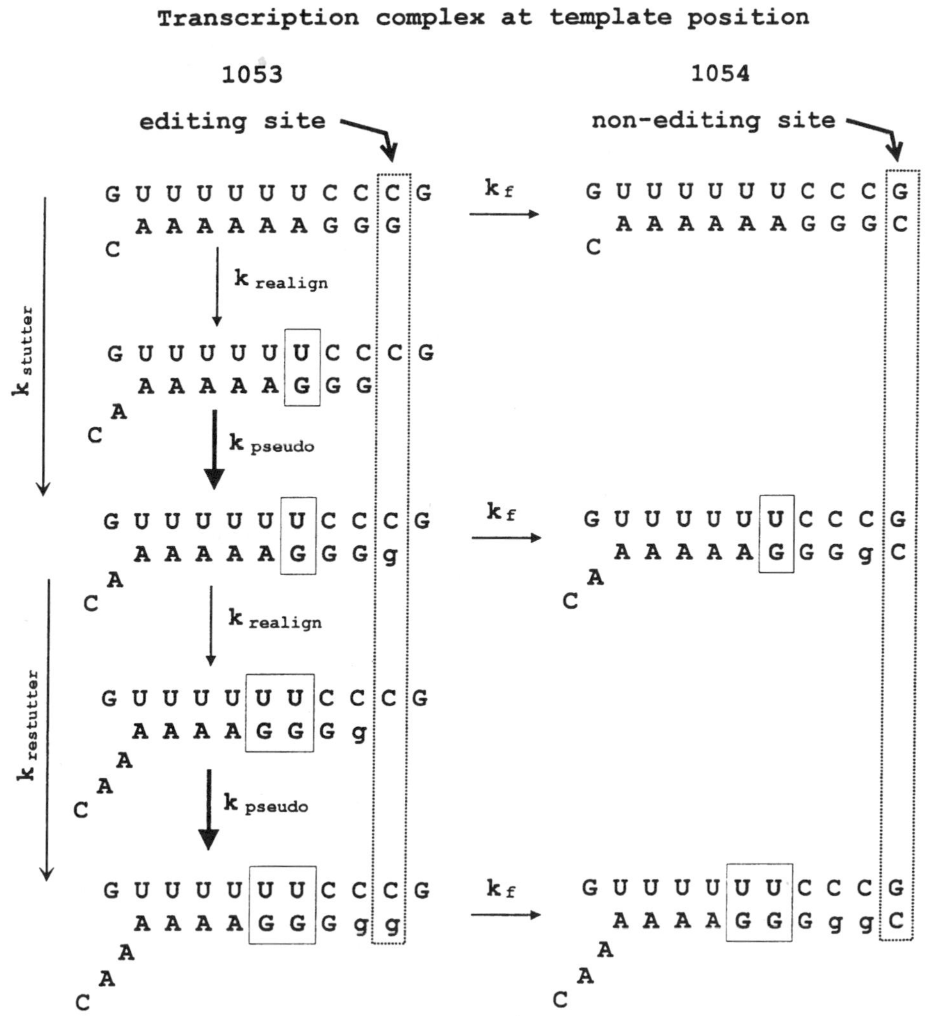

Figure 3. A stuttering model for paramyxovirus mRNA editing. The putative RNA:RNA hybrid between the polypyrimidine tract of the (−) template (top strand, written 3′ to 5′) and the polypurine run of the nascent mRNA chain (bottom strand, written 5′ to 3′) is shown, when the active site of the transcription complex (indicated by the long vertical box) is successively at template positions 1053 and 1054, representing editing (stuttering) and nonediting (nonstuttering) sites, respectively. At an editing site, the transcription complex has the choice of realigning its nascent RNA chain upstream on the template before the next nucleotide is added. This is because the minimum requirement that the realigned hybrid be nearly as stable as its predecessor has been met, because permissible U:G pairs are the only non-Watson–Crick pairs whose formation is required for the helical stack to remain unbroken. We assume that the rate constant for pseudotemplated addition of the G residue is much greater than that for realignment, and thus that $k_{stutter}$ is mostly determined by $k_{realign}$. Having stuttered once at template position 1053, the transcription complex is back to where it started from and has the same choice, but $k_{restutter}$ seems to be determined differently from $k_{stutter}$ (see text). Escape from this process requires the active site to move on to a site where realignment of the nascent chain upstream is no longer possible, because one or more critical base pairs will not be reformed.

nents of the activation energy barriers will produce large changes in EE_I due to the exponential form of equation 1.

Experimental evidence for a competitive kinetic model for paramyxovirus mRNA editing comes from studying mRNA synthesis with purified SeV virions. In these cell-free reactions, one G is inserted at a frequency of ca. 20%, and more than one G is added to ~5% of the mRNAs. Partial substitution for the GTP in the reaction mix with ITP (expected to destabilize to the nascent chain:template G:C pairs at the editing site), and very low concentrations of GTP (expected to increase the step time for elongation at the editing site), significantly increased the frequency of both single and multiple G insertions (Vidal et al., 1990b).

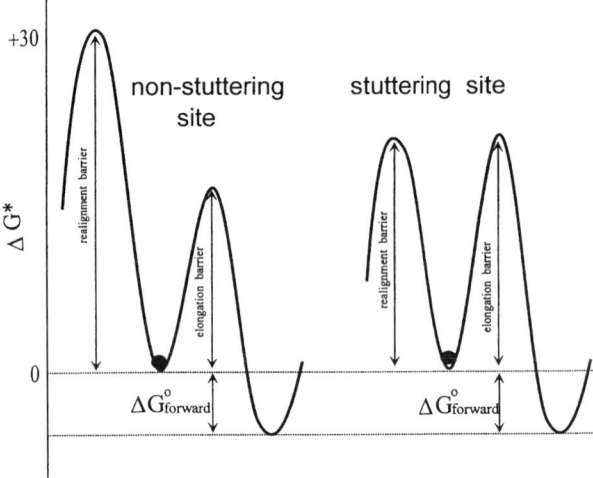

Figure 4. Thermodynamic reaction pathway of the transcription elongation complex. A schematic diagram of the relative activation barriers heights for elongation and stuttering (realignment) at stuttering and nonstuttering sites is shown. The black dot represents the transcription elongation complex. Barriers heights are given as ΔG^*, the free energy of activation. Note that the thermodynamic free energy of extending the chain from I to $I + 1$ ($\Delta G^\circ_{forward}$) is shown as being the same for stuttering and nonstuttering sites.

STRUCTURAL MODELS AND THERMODYNAMIC CONSEQUENCES

A structural model has been proposed to explain how certain RNA sequences destabilize the elongation complex, resulting in the equalization of the elongation and stuttering barriers at stuttering sites (Vidal et al., 1990b). The transcription complex is proposed to pause at the editing site I, and for the 3' end of the nascent mRNA, which is base paired to its template over a small number of residues, to be realigned by either one (paramyxoviruses and morbilliviruses) or two (rubulaviruses) positions upstream. The exact number of base pairs is unclear. As shown in Fig. 3, a minimum of 5 base pairs is required to explain how rubulaviruses realign by two positions rather than by one position (Fig. 2b), but the entire mRNA polypurine run is shown in a hybrid, similar to the eight base pairs thought to be formed in the E. coli RNA polymerase transcription complex (reviewed in Landick, 1997). At heteropolymeric nonstuttering sites, there is a large thermodynamic barrier to this realignment because very few stable base pairs would be expected to reform on realignment. We have referred to the polypyrimidine run of the conserved cis-acting editing signal as a "slippery" sequence, precisely because realignment of the 3' end of the mRNA (upstream) in this case does not interrupt the helical stack of base pairs (if U:G pairs are permitted; boxed in Fig. 3). The barrier to upstream realignment ($\Delta G^*_{backward}$) at editing sites will thus be much reduced relative to those at nonediting sites, such that $k_{stutter}$ now becomes similar in magnitude to $k_{forward}$, and stuttering competes with strictly templated elongation.

The cis-acting editing signal appears to control two distinct mechanistic branchpoints (Fig. 3). In the first, the transcription complex arrives at a site that allows realignment of the nascent chain upstream, and this decelerates the rate of strictly templated nucleotide addition ($k_{forward}$). If the nucleoside triphosphate is nevertheless added, this allows the polymerase to move past the potential site of editing, at least in part because the 3' base of the nascent chain at residue 1054 will no longer be paired to the template on realignment (Fig. 3). This branchpoint determines the overall efficiency of editing. However, if a stutter occurs, the polymerase is back to where it started (the only difference is that there is now a U:G pair in the template:nascent chain hybrid), and it can escape either to strictly templated elongation or to stutter again. This is ostensibly a second branchpoint because different viruses behave differently at the juncture. For all the viruses except human and bovine PIV3, the tendency to escape after an initial round of stuttering is high because mRNAs with additional G insertions are less and less frequent (Fig. 2b). For human and bovine PIV3, in contrast, 1 to 6 G's are added with roughly equal frequency; hence, the tendency to escape is low after an initial round of stuttering. The second branchpoint determines the number of G insertions after stuttering has commenced. There is some experimental evidence that $k_{stutter}$ and $k_{restutter}$ represent separate parts of this reaction pathway, as ITP substitution for GTP (and very low concentrations of GTP) in in vitro reactions increases the fraction of mRNAs with one G inserted ca. 2-fold (from 20 to 40%), but increases the fraction of mRNAs with >1 G insertions ca. 10-fold (from 2 to 20%) (Vidal et al., 1990b).

As previously mentioned, different paramyxoviruses insert a different distribution of G residues into their P gene mRNAs during the editing process, which is matched to the particular ORF organization of the different P genes (Fig. 1a). This variable pattern of G insertion could have been due only to the different polymerase (and nucleocapsid) proteins of each virus, which would respond differently to similar pause and realignment signals. For example, the SeV and bovine PIV3 N, P, and L proteins are sufficiently different that they cannot be individually exchanged for RNA synthesis, they function only as a matched set. However, in spite of these differences in viral N, P, and L proteins, they appear to respond very similarly to these stutter and restutter editing signals because the simple substitution for the SeV editing region (which contains a U_6C_3 polypyrimidine tract) with the bPIV3 region (with a U_6C_4 polypyrimidine tract) causes the SeV transcription complex to edit its mRNA like PIV3; that is, 1 to 6 G's are now inserted at approximately equal frequency, rather than in a strongly decreasing frequency (a decrease of roughly 75% at each step). In previous experiments, the determinants for this difference in $k_{restutter}$ were mapped to the sequences upstream of the polypyrimidine tract or the presence of four rather than three C's in the polypyrimidine tract (Jacques et al., 1994). More recently, we have found that the presence of four rather than three Cs in the polypyrimidine tract has no effect on the distribution of G insertions, which are determined solely by the upstream sequences, and in particular whether the dinucleotide 3' UG (SeV) or 3' AA (PIV3) (complement boxed in Fig. 2a) precedes the 3' $U_6C_{3\ or\ 4}$ polypyrimidine tract (Hausmann et al., unpublished data). Although the structural basis for how the upstream dinucleotide determines the frequency of restuttering ($k_{restutter}/k_{forward}$) is unclear, the finding that the two bases upstream of the tract can determine the relative rates of restuttering and escape suggests that signal is complex, and at least bipartite.

COTRANSCRIPTIONAL PARAMYXOVIRUS mRNA EDITING: A CONTRADICTION IN TERMS?

The title of this volume, *Modification and Editing of RNA*, has grown from the realization that the large number of processes that alter RNA structure often also alter the information content of these molecules by modifying the readout of the genetic code either directly or indirectly. "Editing," however, implies a posttranscriptional event, such as the addition of uridines to trypanosome mitochondrial cryptogene mRNAs to convert them to functional mRNAs (see Chapter 21), or the hydrolytic deamination of a given fraction of specific adenosines to inosines (which are then deciphered as guanosines) in certain nerve cell receptor mRNAs, which alters the ratio of polypeptides with different electrophysiological properties (see Chapter 19). For "strict constructionists" whose language is dominated by the edit menus of word-processing programs (for example, "Undo"), the term applies only to something that has been made one way and then changed for the purpose of altering (or correcting) the information content. The *American College Dictionary*, however, defines "edit" as "to supervise or direct the preparation of a newspaper, magazine, etc.," that is, something which transmits information (or mRNA in the context of biology).

Strictly speaking, whether the process that is the subject of this review may be called editing is open to interpretation. It has been referred to in several ways, for example, transcriptional frameshifting, because the consequences are identical to those of ribosomal frameshifting, or alternative elongation, because of the analogy with how different downstream ORFs are joined to a common upstream ORF in pre-mRNAs by alternative splicing. It has also been referred to as cotranscriptional mRNA editing even though the G insertions occur during mRNA synthesis by polymerase stuttering, because the informational consequences of the G insertions are so similar to those of other forms of insertional mRNA editing. However, it is the transcriptional process itself that is edited here, which leads to edited mRNAs. The insertion of C residues into *Physarum polycephalum* mitochondrial cryptogenes also occurs during mRNA synthesis, but the mechanism here is less clear (see Chapter 22).

Acknowledgment. We thank Hans Geiselmann (Geneva, Switzerland) for helpful discussions.

REFERENCES

Cattaneo, R., K. Kaelin, K. Baczko, and M. A. Billeter. 1989. Measles virus editing provides an additional cysteine-rich protein. *Cell* 56:759–764.

Galinski, M. S., R. M. Troy, and A. K. Banerjee. 1992. RNA editing in the phosphoprotein gene of the human parainfluenza virus type 3. *Virology* 186:543–550.

Giorgi, C., B. M. Blumberg, and D. Kolakofsky. 1983. Sendai virus contains overlapping genes expressed from a single mRNA. *Cell* 35:829–836.

Gupta, M., and D. Kingsbury. 1984. Complete sequences of the intergenic and mRNA start signals in the Sendai virus genome: homologies with the genome of vesicular stomatitis virus. *Nucleic Acids Res.* 12:3829–3841.

Hausmann, S., et al. Unpublished data.

Jacques, J.-P., and D. Kolakofsky. 1991. Pseudo-templated transcription in prokaryotic and eukaryotic organisms. *Genes Dev.* 5:707–713.

Jacques, J.-P., S. Hausmann, and D. Kolakofsky. 1994. Paramyxovirus mRNA editing leads to G deletions as well as insertions. *EMBO J.* 13:5496–5503.

Kolakofsky, D., J. Curran, T. Pelet, and J.-P. Jacques. 1993. Paramyxovirus P gene mRNA editing, p. 105–123. *In* R. Benne (ed.) *RNA Editing, the Alteration of Protein Coding Sequences of RNA.* Ellis Horwood.

Kollmus, H., M. W. Hentze, and H. Hauser. 1996. Regulated ribosomal frameshifting by an RNA-protein interaction. *RNA* 2:316–323.

Lamb, R. A., and D. Kolakofsky. 1996. *Paramyxoviridae*: the viruses and their replication. *In* B. Fields, D. Knipe, P. Howley, et al. (ed.), *Fields Virology*, 3rd ed. Raven Press, New York, N.Y.

Landick, R. 1997. RNA polymerase slides home: pause and termination site recognition. *Cell* 88:741–744.

McGeoch, D. J. 1979. Structure of the gene N:gene NS intercistronic junction in the genome of vesicular stomatitis virus. *Cell* 17:673–681.

Paterson, R. G., and R. A. Lamb. 1990. RNA editing by G-nucleotide insertion in mumps virus P-gene mRNA transcripts. *J. Virol.* 64:4137–4145.

Pelet, T., J. Curran, and D. Kolakofsky. 1991. The P gene of bovine parainfluenza virus 3 expresses all three reading frames from a single mRNA editing site. *EMBO J.* 10:443–448.

Robertson, J. S., M. Schubert, and R. A. Lazzarini. 1981. Polyadenylation sites for influenza virus mRNA. *J. Virol.* 38:157–163.

Rose, J. K. 1980. Complete intergenic and flanking gene sequences from the genome of VSV. *Cell* 19:415–421.

Schubert, M., J. D. Keene, R. C. Herman, and R. A. Lazzarini. 1980. Site on the vesicular stomatitis virus genome specifying polyadenylation and the end of the L gene mRNA. *J. Virol.* 34:550–559.

Thomas, S. M., R. A. Lamb, and R. G. Paterson. 1988. Two mRNAs that differ by two nontemplated nucleotides encode the amino coterminal proteins P and V of the paramyxovirus SV5. *Cell* 54:891–902.

Vidal, S., J. Curran, and D. Kolakofsky. 1990a. Editing of the Sendai virus P/C mRNA by G insertion occurs during mRNA synthesis via a virus-encoded activity. *J. Virol.* 64:239–246.

Vidal, S., J. Curran, and D. Kolakofsky. 1990b. A stuttering model for paramyxovirus P mRNA editing. *EMBO J.* 9:2017–2022.

von Hippel, P. H., and T. D. Yager. 1992. The elongation-termination decision in transcription. *Science* 255:809–812.

Chapter 24

Intracellular Locations of RNA-Modifying Enzymes

B. Edward H. Maden

During the biosynthesis of eukaryotic RNA, many posttranscriptional nucleoside modifications occur. The proportion of nucleosides that are modified in eukaryotic RNA is generally higher than in bacterial or archaeal RNA. Eukaryotic cells also differ from bacterial and archaeal cells in possessing an elaborate internal structure comprising nucleus, cytoplasm, endoplasmic reticulum and cytoskeleton, as well as organelles (mitochondria, as well as chloroplasts in plants and algae) that are of endosymbiotic evolutionary origin.

In this chapter I consider the subcellular locations in which the modifications to eukaryotic RNA occur. As with other steps in RNA processing, nucleoside modifications are predominantly nuclear events. However, some modifications occur late in the RNA processing pathways. It is not always easy to distinguish between nuclear and cytoplasmic locations for some of these "late" modification events, and evidence on some examples is discussed. I also examine some modifications that occur on mitochondrial RNA; these are of considerable interest in the context of protein targeting. In many eukaryotes, including mammals, all mitochondrial RNAs are encoded by organellar DNA and are transcribed locally, whereas the modifying enzymes are encoded by the nuclear genome and must be directed to the mitochondria. In various other eukaryotes some mitochondrial tRNAs are imported, as will be briefly discussed.

This chapter focuses mainly on tRNA and rRNA. Most tRNA molecules are built to a common general design, giving rise to the well-known cloverleaf secondary structure and "inverted L" tertiary structure (crystal structures have been determined at atomic resolution in only a few instances; see Dirheimer et al., 1995, for a review). The common design holds for most of the different tRNAs within a given species and for most tRNAs across the known phylogenetic spectrum, shaping the tRNA molecules for their common role in protein synthesis. Nevertheless, within their common plan there is much variety of detail between individual tRNAs.

As will be discussed, some of the modifications in tRNAs are also phylogenetically widespread. The fundamental biochemistry of these phylogenetically widespread modifications was most likely established early in the common ancestral lineage, before the misty dawn of eukaryotic evolution (see Chapter 29 by Cermakian and Cedergren for a discussion of the antiquity of some modifications and Doolittle, 1996, and Sogin et al., 1996, for recent commentaries on the unresolved question of eukaryotic origins). Eukaryotes retained those ancient modifications, arranged for them to happen in appropriate subcellular locations in the emerging cellular organization, and over time also progressively expanded the eukaryotic repertoire of tRNA modifications. Gene transfers from endosymbiotic bacteria may have contributed to the expanding repertoire. This is a plausible historical outline against which specific details of tRNA modifications in eukaryotes may be considered.

Ribosomes were as fundamental to common ancestral life as was tRNA, both coevolving as the translational machinery. Extensive vestiges of that ancient history are retained in present day rRNA from all organisms, affording the basis of the rRNA-based universal phylogeny (Woese, 1987; Woese et al., 1990). However, in contrast to tRNA, rRNAs from the deepest phylogenetic branches do not typically carry similar nucleoside modifications. In particular, eukaryotes have developed a distinctive pattern of rRNA modifications, with a high proportion of 2'-O-methyl ribose nucleosides and pseudouridines. Small nucleolar RNAs (snoRNAs) play a central role in determining this pattern, as has recently been discovered. These exciting developments are briefly summarized here and are described in detail in Chapter 12 by

---

B. Edward H. Maden • School of Biological Sciences, University of Liverpool, Crown Street, Liverpool L69 7ZB, United Kingdom.

Ofengand and Fournier and Chapter 13 by Bachellerie and Cavaillé.

mRNA modification in eukaryotes is reviewed by Bokar and Rottman in Chapter 10. Modifications to small nuclear RNAs (snRNAs) and snoRNAs are covered by Massenet et al. in Chapter 11. The intracellular locations of these events are briefly reviewed in this chapter.

In this chapter the terms "eukaryotic" and "eukaryotes" are applied in the context of RNA molecules that are synthesized in the cell nucleus and (for tRNA, rRNA, and mRNA) carry out their functions in the cytoplasm. Mitochondrial RNA molecules are designated as such.

tRNA

tRNA molecules contain the greatest abundance of modified nucleosides in relation to their size of any RNAs, and also the greatest variety of nucleoside modifications. Table 1 gives a simplified compilation of positions that are frequently modified; the data are summarized from Grosjean et al. (1995). (The standard cloverleaf structure of tRNA with nucleoside numbering is shown in Appendix 1.) The tabular display, while not encompassing all sites and kinds of modifications that occur in tRNA molecules, does highlight at a glance several noteworthy trends.

First, certain modifications occur at the indicated positions only in eukaryotic tRNAs. (In some instances, chemically similar modifications occur in other positions in prokaryotic tRNAs.) Second, many other modifications are common to eukaryotic and bacterial tRNAs at corresponding positions in the molecules. Third, a few modifications are common to eukaryotic and archaeal tRNAs at corresponding nucleoside positions (m_2^2G26 is an example.) Fourth, many, but not all, modifications in eukaryotic cytoplasmic tRNAs also occur in mitochondrial tRNAs.

The following working hypotheses might account for these observations. The genes for many eukaryotic tRNA modification enzymes arose first in bacteria or archaea or the common ancestral lineage (as briefly discussed in the beginning of this chapter; see also Chapter 29 by Cermakian and Cedergren). A smaller number of genes for tRNA modification enzymes arose specifically in some eukaryotes. Many eukaryotic tRNA modification enzymes have access to mitochondrial tRNAs.

Examples are accumulating in support of these inferences, as will be discussed below. If it is assumed meanwhile that the inferences are at least partly correct, then it follows that a brief sketch of the biosynthesis of some tRNA modifications as elucidated initially in prokaryotes is relevant to eukaryotic tRNA. [For an earlier detailed review of genetics, biosynthesis, and function of tRNA modifications, mainly in bacteria, see Björk (1995). For further discussions of some aspects of functions and genetics of tRNA modifications, see several other chapters in this book.]

Prokaryotic and Bacteriophage tRNAs

Several studies were undertaken in the 1970s on the temporal order of nucleoside modifications during the maturation of bacterial and bacteriophage tRNAs, and on the related question of whether the modifications occur preferentially on pre-tRNA or on endonucleolytically processed tRNA. Schaefer et al. (1973), studying pre-tRNA^{Tyr} and tRNA^{Tyr}, found that the pre-tRNA contains Ψ and small amounts of m^5U and i^6A, and no Gm. The modifications m^5U54, $\Psi55$, and $\Psi40$ (modern numbering) were produced in vitro both on pre-tRNA and on substrate that had been endonucleolytically processed (in vitro) into tRNA, the latter being the better substrate. A fractionated enzyme preparation produced only $\Psi55$ and no $\Psi40$ on pre-tRNA, $\Psi40$ being formed only on the cleaved substrate. This was early evidence for multiple Ψ synthase activities. Sakano et al. (1974) found in a temperature-sensitive mutant for pre-tRNA processing that the m^5U54, $\Psi55$, and D modifications are present in several precursor tRNAs, whereas Gm18 is absent. Ciampi et al. (1977) reinvestigated Ψ formation with reference to $\Psi40$ in the anticodon stem. The enzyme responsible for Ψ formation in this region in several tRNAs is called pseudouridine synthase I or TruA and is encoded by the *truA* (formerly *hisT*) gene. (For nomenclature of pseudouridine synthases and their genes, see Chapter 12 by Ofengand and Fournier and also Koonin, 1996.) Ciampi et al. (1977), in contrast to Schaefer et al. (1973), found that $\Psi40$ can in fact be formed on pre-tRNA.

Thus, the picture that emerged from the studies, although not clear in all details, was that some modifications in the T and D loops, including m^5U54, $\Psi55$, and D, but excluding Gm18, occur early during tRNA maturation, possibly on pre-tRNA in vivo, whereas Gm18 occurs later, with cleaved tRNA as the obligatory substrate, and the timing of $\Psi40$ and probably i^6A synthesis is intermediate.

Eukaryotic tRNAs

The early 1980s saw major developments in characterizing the pathways for eukaryotic tRNA maturation and modification by use of cloned tRNA genes and *Xenopus* assay systems. Two tRNAs in particular were studied by these methods, yeast tRNA^{Tyr}

Table 1. Frequently modified sites in tRNA[a]

Position	Archaea	Bacteria	Eukaryotes (cytoplasmic)	Mitochondria
8		$\underline{s^4U}$		
9			m^1G	m^1A, m^1G
10	m^2G, m_2^2G		m^2G	m^2G
12			ac^4C	ac^4C
13	Ψ	Ψ	Ψ	(Ψ)
15	$\underline{G}^{*b}$			
16		D	D	D
17		D	D	D
18		(Gm)	Gm	Gm
20		D	D	D
20a		D	D	D
20b			D, acp^3U	acp^3U
22		$\underline{m^1A}$		
26	m^2G, m_2^2G		m^2G, m_2^2G	m^2G, m_2^2G
27			Ψ	Ψ
28			Ψ	Ψ
31			Ψ	Ψ
32	Cm	s^2C, Cm, Ψ, Um	Cm, m^3C, Um, Ψ	Ψ
34[c]	Cm, ac^4C	I	I	
	mo^5U	Cm, ac^4C, k^2C	Cm	(Cm)
		Q, Gm	Q, **manQ, galQ**, Gm	(Q)
		mnm^5U, mnm^5s^2U	mcm^5s^2U	
		$cmnm^5U$, $cmnm^5Um$, $cmnn^5s^2U$, mo^5U, cmo^5U	$cmnm^5Um$	($cmnm^5U$)
37[d]	t^6A, m^1G	m^2A, m^6A		m^6A
		i^6A, ms^2i^6A	i^6A	i^6A, ms^2i^6A
		t^6A, ms^2t^6A	t^6A, ms^2t^6A	t^6A
		m^6t^6A	m^6t^6A	
			$\underline{m^1I}$	
		m^1G	m^1G, $\underline{vW, o^2vW}$	m^1G
38		Ψ	Ψ, $\underline{m^5C}$	Ψ
39	Ψ	Ψ	Ψ	Ψ
40	Ψ	(Ψ)	$\underline{m^5C}$	
46		m^7G	$\underline{m^7G}$	(m^7G)
47			$\underline{D}$	
48	m^5C		m^5C	m^5C
49	m^5C		m^5C	m^5C
54	$m^1Ψ$, Ψ	m^5U	m^5U	m^5U
55	Ψ	Ψ	Ψ	Ψ
56	$\underline{Cm}$			
57	$\underline{m^1I}$			
58	(m^1A)	(m^1A)	m^1A	(m^1A)

[a] The data are compiled from Grosjean et al. (1995), and are based (Steinberg et al., 1993) upon 59 archaeal tRNA sequences (mostly *Halobacter*), 130 bacterial sequences (mostly *E. coli*, *B. subtilis*, and *Mycoplasma*), 195 eukaryotic cytoplasmic sequences (a variety of single cell organisms, animals and plants) and 81 mitochondrial sequences. Generally sites at which modified nucleosides occur in at least 25% of sequences are included; at several sites the modification frequencies are much higher, such as Ψ55, which occurs in nearly all tRNA molecules. Modifications that occur only in one major lineage (e.g., eukaryotes) are shown in bold and underlined. Modifications that occur frequently in one lineage but infrequently in other lineages are shown in parentheses. Several sites not listed in the table are modified in some tRNAs but in fewer than 25% of available sequences; for example, Ψ occurs at one or two of the anticodon positions 34–36 in several eukaryotic tRNAs (see the text). For a pictorial display of all tRNA modification sites, with numerical frequencies, see Grosjean et al. (1995).
[b] Archaeosine.
[c] Wobble nucleoside. I results from deamination of A. Q and derivatives occur in multicellular but not unicellular eukaryotes.
[d] Nucleoside flanking anticodon.

by microinjection of the gene into *Xenopus* oocyte nuclei (Melton et al., 1980; Nishikura and De Robertis, 1981) and *Xenopus* tRNA^{Met} by transcription in a *Xenopus* cell-free system (Koski and Clarkson, 1982).

The yeast tRNA^{Tyr} primary transcript comprises 108 nucleotides (nt), containing a 5' leader, short 3' trailer and an intron, and lacks the terminal CCA. Processing starts with stepwise partial removal of the 5' leader (104-nt and 97-nt intermediates). Then in a 92-nt intermediate the 5' leader has been fully removed and the 3' trailer has also been removed and replaced by CCA, but the 14-nt intron is still present. Finally the 14-nt intron is removed to generate the mature-length tRNA (78 nt). All of these processing and splicing steps occur in the nucleus. Each processing intermediate was examined for its content of modified nucleosides; the findings for several nucleoside modifications were reported by Melton et al. (1980), and an essentially complete analysis for some 18 modifications was presented by Nishikura and De Robertis (1981). Figure 1 summarizes the findings.

The 104-nt intermediate already contains five modifications, including all three that are present in the T loop of mature tRNA (position 54 being predominantly Ψ rather than m^{5}U), m^{5}C48 at the base of the variable and T loops, and Ψ35. The last is close to the intron and becomes the middle nucleoside of the anticodon after splicing. The 92-nt precursor contains 11 additional modifications, including several D residues, two m^{2}G residues, m$_2^2$G26, and Ψ39 also close to the intron. Finally, after intron processing the hypermodified nucleosides Q34 and i^{6}A37 in the anticodon loop are formed.

The transcription and processing of a yeast tRNA gene in a *Xenopus* oocyte comprise a heterologous system. In fact, most of the nucleoside modifications in tRNA^{Tyr} are common to yeast and vertebrates. However, there are a few differences. At the sites of difference, *Xenopus* oocytes modified the yeast transcript according to the vertebrate pattern; in particular, Gm18, characteristic of yeast tRNA^{Tyr}, was not produced in the oocyte system, whereas m^{2}G6 and Q34, characteristic of vertebrate tRNA^{Tyr}, were produced (Fig. 1).

Because splicing is a nuclear event, it can be concluded that all of the modifications that take place before splicing, that is, all except the formation of Q34 and i^{6}A37, are also nuclear events. The temporal order of modifications, starting in the T loop and ending with hypermodifications in the anticodon region, is similar to that elucidated earlier (in less detail) for bacterial tRNAs as outlined previously.

To study transcription and processing in a system in which all components originated from a single (vertebrate) species, Koski and Clarkson (1982) utilized the cloned gene for *Xenopus* initiator tRNA^{Met} and a *Xenopus* cell-free system. The tRNA primary transcript contains short 5' leader and 3' trailer sequences, and lacks the CCA end. Unlike the tRNA^{Tyr} transcript it does not contain an intron. It also lacks any U residues in the D loop and contains A instead of U at position 54. Thus, several modifications that are found in most tRNAs cannot occur in this tRNA. Nevertheless, eight modified nucleosides are present in *Xenopus* initiator tRNA^{Met}. Seven of these were found in the in vitro transcripts, of which six were present (to varying extents) in the primary transcript and in a partly processed precursor. These were m^{1}G9, m^{2}G10, m^{2}G26, m^{7}G46, D47, and m^{1}A58. The modification t^{6}A37 in the anticodon loop was

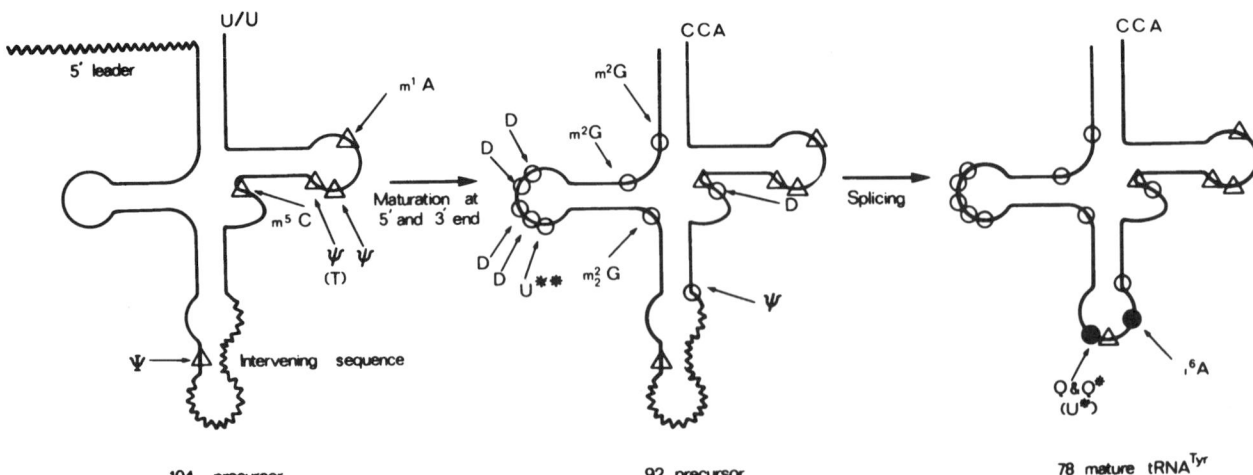

Figure 1. Sequential modifications to yeast tRNA^{Tyr} precursors in *Xenopus* oocytes. Modifications which occur on the 104-nt precursor, the 92-nt precursor, and the 78-nt mature tRNA are indicated. Reprinted from Nishikura and De Robertis (1981) with permission.

found only in mature-length tRNA. The modification m^{5}C48 was not found. Thus, apart from the absence of m^{5}C48 the findings resembled qualitatively those for tRNA^{Tyr}, with most modifications occurring on nuclear precursors but the modification at position 37, t^{6}A in this case, being found only on the mature-size tRNA.

Introns and tRNA modifications

Ψ occurs at position 35 in eukaryotic tRNAs^{Tyr} generally, and the presence of the intron is required for this modification to take place (Johnson and Abelson, 1983; Choffat et al., 1988; van Tol and Beier, 1988). m^{5}C34 formation in yeast pre-tRNA^{Leu} is also dependent on the presence of an intron (Strobel and Abelson, 1986), as are Ψ34 and Ψ36 formation in yeast minor pre-tRNA^{Ile} (Szweykowska-Kulinska et al., 1994) and m^{5}C40 formation in yeast pre-tRNA^{Phe} (Jiang et al., 1997). A short stem-loop substrate containing the intron was also a substrate for m^{5}C40 formation of yeast pre-tRNA^{Phe}. Thus, all of these modifications on anticodon stem-loop nucleosides are nuclear events. By contrast, the intron in yeast pre-tRNA^{Phe} completely hindered the formation of Cm32, Gm34, and m^{1}G37. The remaining modifications, m$^2_2$G26, Ψ39, m^{7}G46, m^{5}C49, m^{5}U54, and Ψ55, were unaffected or only slightly affected by the presence of an intron. On the basis of such experiments Grosjean et al. (1997) have classified tRNA modifications according to whether an intron is required or is inhibitory (that is, the absence of an intron is required) or is immaterial for the modification to occur (Table 2); several modifications at positions 34 and 37 require the absence of an intron, but this is not a general rule.

Nuclear or cytoplasmic locations of late modifications?

tRNA ligase, a component of the tRNA splicing machinery, is located at the inner nuclear membrane, most likely at the nuclear pore (Clark and Abelson, 1987). Completion of splicing is followed by transport of tRNA to the cytoplasm. Therefore, modifications that do not occur before splicing, such as Q34 and i^{6}A37 on tRNA^{Tyr}, might be expected a priori to occur in the cytoplasm. By analogy, other hypermodifications at positions 34 and 37 might be expected to occur in the cytoplasm even for tRNAs that do not undergo splicing, for example, t^{6}A37 in Xenopus tRNA^{Met}. As was previously noted, t^{6}A37 is indeed present only in the mature-length transcript (Koski and Clarkson, 1982), suggesting a cytoplasmic location for this modification. However, one cannot exclude the possibility that these late modifications occur while tRNA is being transported through the nuclear pore or immediately after traversing the nuclear pore complex. Efforts have been made to resolve this problem both by microinjection techniques and, in some instances, by analyzing the intracellular distribution of the respective enzymes.

An early study (Carbon et al., 1982) on a chimeric yeast tRNA^{Asp}, generated by in vitro enzymatic manipulations so as to uniquely label G34 at the 3' phosphate and then injected into the cytoplasm of Xenopus oocytes, showed that modification of G34 to Q34 and further modification to manQ (Haumont et al., 1987) occur in Xenopus cytoplasm. (As previously mentioned, these are Xenopus-specific modifications and do not occur in yeast.) Likewise, another chimeric yeast tRNA^{Asp}, engineered to contain U34, was modified by Xenopus cytoplasm to yield mcm^{5}U and mcm^{5}s^{2}U at position 34. In a further study, an in vitro manipulated tRNA^{Arg} with A34 was found to be correctly modified to inosine 34 after injection into Xenopus oocyte cytoplasm (Fournier et al., 1983). Recently, the enzyme responsible for inosine 34 formation in the anticodon of eukaryotic tRNAs has been characterized (Auxilien et al., 1996). Enzyme activity was detectable in cytoplasmic or S100 extracts from yeast and several other species. The enzyme was partly purified from yeast, and a single enzyme appears to catalyze I34 formation in

Table 2. Positive or negative intron dependence and intron insensitivity to modified nucleoside formation in eukaryotic tRNAs^a

Presence of intron required	Absence of intron required	Insensitive to presence of intron
m^{5}C34	Gm18	m^{2}G6
Ψ34	Cm32 (a)	m^{1}G9
Ψ35	Ψ32	m^{2}G10
Ψ36	Gm34 (a)	D16, 17, 20, 20a
m^{5}C40 (a)	Q34, galQ34	m$^2_2$G26 (a)
	m^{1}G37 (a)	Ψ27
	Y37	mcm^{5}s^{2}U34
	i^{6}A37	Ψ39 (a)
	Um44	m^{7}G46 (a)
		D47
		m^{5}C48
		m^{5}C49 (a)
		m^{5}U54 (a)
		Ψ55 (a)
		m^{1}A58

^{a} Adapted from Grosjean et al. (1997), with permission. Primary source references are in Grosjean et al. (1997). Several of the findings been confirmed in a recent study on yeast tRNA^{Phe} (Jiang et al., 1997) and are indicated by "(a)" in this table. Several of the modifications in column 3 normally occur early on pre-tRNA, e.g., modifications in and near the T loop, as discussed elsewhere in the text; in these instances some of the respective pre-tRNAs do, and others do not, contain an intron.

the seven yeast tRNAs that contain this modification. (See Grosjean et al., 1996a, for a review of inosine and 1-methyl inosine in tRNA.)

i^6A and t^6A formation at position 37 has also been studied. i^6A37 occurs in tRNAs that read codons starting with uridine. The enzyme responsible, N^6-isopentenyl-pyrophosphate:tRNA isopentenyl transferase (also called IPPT, or Mod5p in yeast), is encoded in yeast by the *MOD5* gene. Two forms of the protein are produced: one form is targeted to the mitochondria but also occurs in the cytoplasm; the other form occurs in the nucleus and cytoplasm (Boguta et al., 1994). Therefore, from the cellular distribution of the enzyme it cannot be excluded that some i^6A formation might take place in the nucleus.

t^6A37 formation has been shown to occur in *Xenopus* cytoplasm; T7 polymerase transcripts of yeast initiator tRNA^{Met} and tRNA^{Val} were used as substrates (Morin et al., 1998). In this analysis numerous variant tRNA transcripts were also used to probe the "tRNA architecture" requirements for t^6A formation.

tRNA architecture

Related to the temporal order and subcellular sites of the tRNA modification reactions is the question of which features of the tRNA architecture are recognized by the modifying enzymes. This question has been addressed by several authors, recently in detailed studies by Grosjean and coworkers, and also by Qian and Björk (1997) in a specific case study of m^1G37.

Grosjean et al. (1996b) microinjected unmodified transcripts of the gene for yeast tRNA^{Asp}, separately microinjected some 21 structural variants of this tRNA into the cytoplasm of *Xenopus* oocytes, and examined which modifications took place. The three-dimensional structure of yeast tRNA^{Asp} is known, and therefore it was possible to rationally design mutants of different degrees of structural severity with respect to the tRNA architecture. As in the earlier study of yeast tRNA^{Tyr} described above, *Xenopus* oocytes modified yeast tRNA^{Asp} according to the *Xenopus* pattern. Thus, 13 modifications occurred on the wild-type tRNA^{Asp} transcript, of which the following are normally present in yeast tRNA^{Asp}: $\Psi13$, D17, D20, $\Psi32$, m^1G37, m^5C49, m^5U54, and $\Psi55$. Five additional modifications are "*Xenopus* specific": m^2G6, m^2G26, manQ34 (as also previously studied by Carbon et al., 1982, and Haumont et al., 1987), $\Psi40$, and m^1A58.

The formation of some modified nucleosides was extremely sensitive to small perturbations in the three-dimensional structure, including the anticodon arm nucleosides m^1G37 and $\Psi40$ and the D arm nucleoside $\Psi13$. Others were insensitive to change in the overall three-dimensional structure and took place on small stem-loop substrates: m^5U54 (confirming an earlier in vitro study by Gu and Santi, 1991, and further studied by Becker et al., 1997a), as well as $\Psi55$ and m^1A58 in the T loop, and m^2G6 in the amino acid acceptor stem. m^2G26 and manQ34 showed intermediate sensitivities to the overall structure. The authors propose that tRNA-modifying enzymes be classified into two major groups according to their sensitivities to the overall tRNA architecture. Comparable studies have been extended to examine the structural requirements for I34 formation (Auxilien et al., 1996) and t^6A37 formation (Morin et al., 1998). Both of these modifications require the global L-shaped tRNA archietcture; I34 formation is also sensitive to structural perturbation and some sequence changes in the anticodon loop and stem.

Figure 2 is a recent summary of the dependence of several tRNA modifications upon features of tRNA

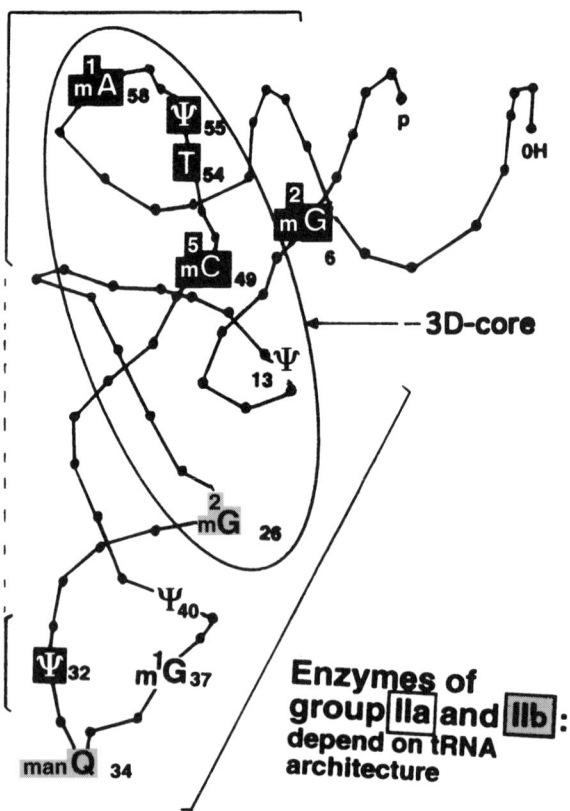

Figure 2. Classification of enzymes catalyzing tRNA modifications into two groups according to their sensitivities to the tRNA architecture. Enzymes of class I recognize fragments of tRNA structure, whereas enzymes of class II require the major features of the tRNA architecture to be intact. Reprinted from Grosjean et al. (1996b) with permission; see that reference for further details.

architecture, showing the classification of enzymes into groups I (insensitive to overall architecture) and II (sensitive to overall architecture). It is striking that some modifications that normally occur earliest in tRNA maturation, particularly those in the T loop, are also the least sensitive to overall tRNA architecture. Conversely, some of those that occur later are very sensitive to the overall architecture.

The study of Qian and Björk (1997) utilized a different experimental in vivo system, an isoacceptor of tRNA^{Pro} from *Salmonella typhimurium* that also contains m^{1}G37. The authors demonstrated by an ingenious mutagenic technique that this modification is also extremely sensitive to small perturbations in the three-dimensional structure. This result from a prokaryotic system is therefore in striking agreement with that for m^{1}G in eukaryotic yeast tRNA^{Asp}. This in turn leads to the expectation that an essentially intact tRNA architecture will be found to be prerequisite for m^{1}G37 formation in any tRNA from any phylogenetic source and, perhaps more generally, that some of the "rules" for architecture recognition by tRNA-modifying enzymes will be found to be evolutionarily highly conserved.

To summarize available data from the several experimental systems previously outlined, nucleoside modifications to tRNA follow a temporal sequence whose broad features may be phylogenetically conserved. Most of the modifications to eukaryotic tRNA take place in the nucleus; the final ones, particularly at positions 34 and 37, probably occur after or during transport of tRNA to the cytoplasm. There is a tendency for enzymes that modify their target sites earliest in the temporal sequence, upon early precursors to tRNA in eukaryotes, to be least sensitive to the overall tRNA architecture. Conversely, in at least some instances the later acting enzymes require an essentially complete tRNA architecture on which to act. Moreover, special eukaryotic recognition mechanisms exist for those modifications for which the presence of an intron is required. Such intron requiring modifications may be integral to the nucleocytoplasmic maturation sequence of some tRNAs, as the following example indicates.

Functional relationship of certain modification enzymes to nuclear pore proteins

PUS1 (Simos et al., 1996) is a gene encoding a yeast enzyme that catalyzes pseudouridine formation at several different locations in tRNA molecules: Ψ27, Ψ28, Ψ34, and Ψ36, of which Ψ34 and Ψ36 are formed in intron-containing pre-tRNAs (see Table 2). The gene was identified genetically through its interaction with *NSP1*, which encodes the nuclear pore protein Nsp1p, and *LOS1*, which encodes the nuclear protein Los1p (Shen et al., 1993) that is required for efficient pre-tRNA splicing. The genes interact in such a way that lack of all three genes causes synthetic lethality, whereas lack of any one or two of them is not lethal (Simos et al., 1996). These results suggest that *PUS1* is important for nucleocytoplasmic transport of tRNA, either because the tRNA needs to be modified or possibly because the protein Pus1p contributes in some other way to the transport process, for example as a transport protein. Searches for further such synthetic lethal combinations may afford a powerful approach to identifying interactions between modifications and transport.

Mitochondrial tRNAs

Mitochondria contain their own DNA that encodes certain key mitochondrial functions. There are numerous phylogenetic variations on the structural organization of the mitochondrial genome, concerning, for example, some of the functions that are encoded and the presence or absence of introns. However, mitochondrial DNA generally encodes its own rRNA, and mitochondria from most sources except trypanosomes encode all or most of the tRNA molecules that are required for translation. In addition, several integral inner membrane proteins of electron transport and oxidative phosphorylation are encoded. However, hundreds of other mitochondrial proteins are encoded by the nuclear genome, and are synthesized on cytoplasmic ribosomes before being transported into the mitochondria. Moreover, there are numerous examples from intermediary metabolism and nucleic acid enzymology in which different nuclear genes encode separate mitochondrial and nucleocytoplasmic isoenzymes, such as aconitase, several other enzymes of the citric acid cycle, DNA and RNA polymerases, aminoacyl-tRNA synthases and ribosomal proteins (briefly reviewed in Hopper et al., 1982). Here, however, I will be concerned only with mitochondrial tRNA-modifying enzymes. A distinctive mode of organellar targeting exists for some of these enzymes, as will be discussed.

As indicated in Table 1, mitochondrial tRNAs contain many of the nucleoside modifications that also occur in cytoplasmic tRNAs. (There are some exceptions, in particular, absence of inosine and low abundance of some other modifications at the wobble position.) Although only a few mitochondrial tRNA-modifying enzymes are currently characterized, all of them must be encoded by nuclear DNA. This follows from the fact that in several fully sequenced mitochondrial genomes all the DNA encodes other functions. Two mitochondrial tRNA-modifying enzymes in yeast have been intensively studied both biochemically and genetically (Martin and Hopper, 1994), and have become models for studies of mitochondrial targeting. Two further enzymes have been studied more recently (Lecointe et al., 1997; Becker et al., 1997b).

N^2,N^2-dimethylguanosine tRNA methyltransferase (Trm1p) is encoded by the nuclear gene *TRM1*. Mutation of this single gene results in m_2^2G deficiency in both cytoplasmic and mitochondrial tRNAs (Smolar and Svensson, 1974; Hopper et al., 1982). The gene was isolated and shown to encode the methyltransferase (Ellis et al., 1986). Sequence analysis revealed ATG codons at amino acid positions 1 and 17 (Ellis et al., 1987). The ATG for translational initiation is selected by transcription; some transcripts start just upstream of ATG1, allowing translational initiation at this codon. Other transcripts start between ATG1 and ATG17, allowing translation only from the latter codon. Proteins starting at ATG1 are more efficiently imported into mitochondria than are proteins starting at ATG17 (Ellis et al., 1989). Removal of nucleotides encoding the first 48 amino acids abolishes mitochondrial import (Rose et al., 1992). Thus, mitochondrial import of the enzyme is determined within the N-terminal region, with the full-length translation product especially favoring import.

The other destination of the enzyme is the nucleus, and is determined by a nuclear targeting signal, almost certainly amino acids 95 to 102 (KKSKKKRC) (Li et al., 1989). This is evidently the dominant targeting signal in the shorter translation product. The enzyme appears to localize near the nuclear periphery (Rose et al., 1992, 1995), consistent with m_2^2G formation being among the later modifications in the nuclear maturation pathway of cytoplasmic tRNAs. (This modification occurs on the 92-nt nuclear precursor to tRNA^{Tyr}; see Nishikura and De Robertis, 1981.)

IPPT (Mod5p), encoded by *MOD5*, has been mentioned above. This enzyme, like Trm1p, exists in two isoforms encoded by a single gene, and, also like Trm1p, differs in length at the N terminus according to which of two initiator ATG codons is used. Use of ATG codon 1 produces a longer protein; this is directed preferentially into mitochondria with some accumulation also in the cytoplasm. Use of ATG codon 12 produces a shorter protein that accumulates in both the nucleus and the cytoplasm (Boguta et al., 1994). Further experiments with fusion proteins indicated that amino acids 1–11, although necessary, are insufficient for mitochondrial targeting; hence, some of the targeting information is downstream from ATG12 (Boguta et al., 1994). Clearly, a complex interplay of amino acid sequences with intracellular recognition elements determines the partition of this enzyme between different cellular destinations.

The mechanism of sorting has been further dissected for Mod5p; *trans*-acting mutants defective in delivery of the enzyme to mitochondria were identified. One class of mutants implicates the actin cytoskeleton in delivery to the mitochondria; another class of mutants implicates mRNA 3' ends or protein synthesis as being important for protein distribution in *S. cerevisiae* (Zoladek et al., 1997).

There is also a single gene, *TRM2*, responsible for m^5U54 formation on cytoplasmic and mitochondrial tRNAs (Hopper et al., 1982). This gene is less fully characterized than *TRM1* and *MOD5*, although as previously mentioned, the enzyme has been assayed with cytoplasmic tRNA^{Asp} and variants as substrates (Becker et al., 1997a).

Recently, two other nuclear gene products in yeast, both of them pseudouridine synthases, have been characterized and shown to modify cytoplasmic and mitochondrial tRNAs. One of these, formerly called Deg1p, is encoded by the gene *DEG1* and catalyzes the formation of $\Psi38$ and $\Psi39$ in the anticodon stem-loop of several tRNAs (Lecointe et al., 1998). This enzyme is homologous to the *E. coli* pseudouridine synthase TruA, mentioned previously. The yeast enzyme, now renamed Pus3p, localizes both in the nucleus and in the cytoplasm as shown by immunofluorescence microscopy. Disruption of the DEG1 results in the failure of $\Psi38$ and $\Psi39$ synthesis in both cytoplasmic and mitochondrial tRNAs, indicating mitochondrial targeting of the gene product.

The other recently discovered pseudouridine synthase with dual specificity for cytoplasmic and mitochondrial tRNAs is encoded by the yeast gene *YNL292w*. This gene shows sequence homology to the *E. coli* tRNA:pseudouridine-55 synthase; the gene product was characterized and shown to be responsible for catalysis of synthesis of the (almost ubiquitous) $\Psi55$ in both cytoplasmic and mitochondrial tRNAs (Becker et al., 1997b), and was named Pus4p.

The known tRNA-modifying enzymes with dual specificity for cytoplasmic and mitochondrial tRNAs, and their subcellular locations, are summarized in Table 3.

Finally, the addition of CCA ends to cytoplasmic and mitochondrial tRNAs, although not a modification, is of interest in this context. The addition is brought about by the enzyme ATP (CTP):tRNA nucleotidyltransferase, encoded by the gene *CCA1*. Translation start signals ATG1 and ATG10 direct the enzyme to the mitochondria, whereas a translational start at ATG18 results in exclusively nucleocyto-

Table 3. Known tRNA-modifying enzymes with dual specificities for cytoplasmic and mitochondrial tRNAs

Enzyme name and origin[a]	Nucleoside(s) modified	Site of action[b]			Reference(s)
		N	C	M	
N^2,N^2-G-methyltransferase (yeast *TRM1*)	m^2_2G26	+	−	+	Rose et al., 1995
Isopentenyl transferase (yeast *MOD5*)	i^6A37	+?	+	+	Boguta et al., 1994
(Uridine-5)methyltransferase (yeast *TRM2*)	m^5U54	+	−	+	Hopper et al., 1982; Becker et al., 1997a
Ψ-Synthase 1 (yeast *PUS1*)	Ψ27, 28, 34, 36	+	−	?	Simos et al., 1996
Ψ-Synthase 3 (yeast *PUS3*)	Ψ38, 39	+	−	+	Lecointe et al., 1998
Ψ-Synthase 4 (yeast *PUS4*)	Ψ55	+	−	+	Becker et al., 1997b

[a] The names of the genes are in italics. The yeast enzyme designations are commonly derived from the gene names (e.g., Trm1p).
[b] N, nuclear; C, cytoplasmic; M, mitochondrial. A plus sign indicates the normal physiological sites of action of the enzymes. This is still unresolved for the isopentenyl transferase (Mod5p), as discussed in the text.

plasmic location of the enzyme (Chen et al., 1992; Martin and Hopper, 1994). Because CCA ends are essential to RNA function, yeast strains lacking the ability to add CCA ends to mitochondrial tRNAs are respiration deficient and cannot grow on nonfermentable carbon sources.

The common denominator in *TRM1*, *MOD5*, and *CCA1* is that, in each case, a single nuclear gene programs cytoplasmic ribosomes to deliver products to mitochondria and to cytoplasmic or nuclear compartments by appropriate choice between alternative translational initiation codons, with the longer protein in each instance being directed preferentially to mitochondria. (See summary in Fig. 3.) The term "sorting isozymes" has been coined for isoenzymes derived from single genes and with dual (or multiple) intracellular destinations (Martin and Hopper, 1994). It seems likely that several further enzymes with dual mitochondrial and cytoplasmic or nuclear targets will turn out to be sorting isozymes, including the pseudouridine synthase Pus1p, for which two gene products differing in length were demonstrated by immunoprecipitation (Simos et al., 1996), as well as Pus3p and Pus4p, and further enzymes catalyzing mitochondrial tRNA modifications. Such a phenomenon would account for the fairly high degree of similarity between eukaryotic cytoplasmic and mitochondrial tRNA modification patterns (see Table 1). Conversely, enzymes such as the A34 deaminase, which catalyzes the A to I conversion in cytoplasmic but not in mitochondrial tRNAs (Auxilien et al., 1996; Grosjean et al., 1996a), would be expected not to possess mitochondrial targeting signals.

Mitochondrial tRNA import and editing

Although in mammals all mitochondrial tRNAs are encoded by the mitochondrial genome, the situ-

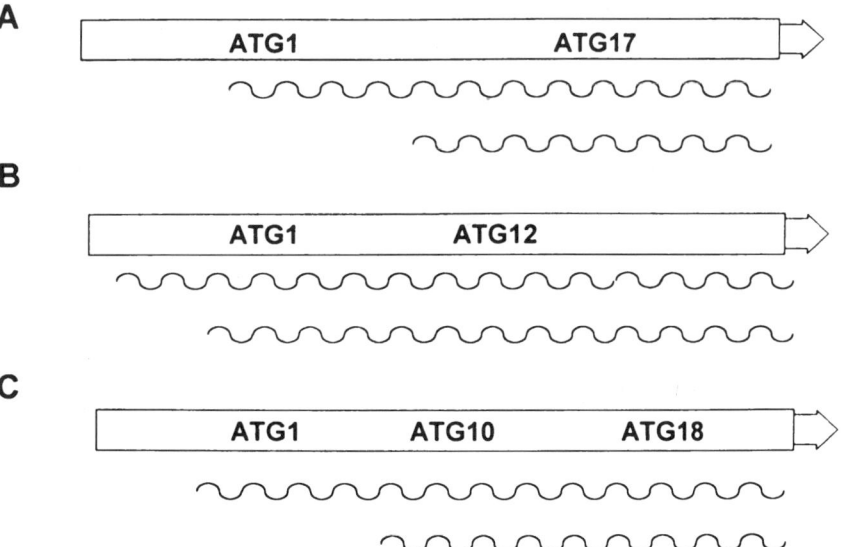

Figure 3. Schematic representation of translation start sites (in-frame ATG codons) and transcription start sites (starts of wavy lines) in the yeast genes *TRM1* (A), *MOD5* (B), and *CCA1* (C). Reproduced from Martin and Hopper (1994) with permission.

ation is less straightforward in many other eukaryotes. For example, in *S. cerevisiae* one of three lysine isoacceptors, tRNA-KI, with the anticodon CUU, is encoded by the nuclear genome and is distributed between cytoplasm and mitochondria (Tarassov and Martin, 1996, and references therein). Mitochondria of various higher plants produce some of their functional tRNAs by (separately or in combinations) editing of mitochondrially transcribed tRNAs, replacement (during the course of evolution) of some native mitochondrial tRNA genes by chloroplast-derived tRNA genes, and import of nucleus-encoded tRNAs from the cytoplasm (Dietrich et al., 1996, and references therein). In many unicellular eukaryotes, especially trypanosomes, the import of nucleus-encoded tRNAs into mitochondria is quite prevalent (Simpson et al., 1989; Hancock and Hajduk, 1990; Lye et al., 1993; Schneider et al., 1994a, 1994b; Hauser and Schneider, 1995). In *Trypanosoma brucei*, some nuclear genes encode tRNA species that are unique to the cytoplasm, others encode tRNAs that are unique to the mitochondria and still others encode tRNAs that are shared between cytoplasm and mitochondria (Hancock and Hajduk, 1990).

Editing of mitochondrial tRNAs is also fairly prevalent across a broad range of species, from unicellular organisms such as *Acanthamoeba* spp., the slime mold *Physarum polycephalum*, and a variety of land plants and animals (see Chapter 16 by Price and Gray, Chapter 17 by Marchfelder et al., and Chapter 22 by Gott and Visomirski-Robic).

To summarize, across the range of eukaryotes a great diversity of processes contributes to providing mitochondria with a full range of functional tRNAs.

rRNA

rRNA differs from tRNA in several obvious respects. It is much bigger, and it comprises both the framework of the ribosome and parts of the working machinery, notably in decoding and peptidyl transfer. In these respects it is more complicated than tRNA. However, in any given cell or organism rRNA is homogenous or nearly so, with all ribosomes performing the same function in protein synthesis. In this respect rRNA is simpler than tRNA, because a set of at least 30 distinct tRNA molecules is required for translation of the genetic code (somewhat fewer in mitochondria).

Eukaryotic pre-rRNA is transcribed by RNA polymerase I from tandem arrays of multiple rRNA genes, located on a single chromosome or, at most, a few chromosomes, whereas pre-tRNAs are transcribed by RNA polymerase III from numerous different chromosomal locations. The sites of rRNA transcription are at the heart of the nucleolus, a specialized region of the nucleus in which processing of pre-rRNA into preribosomes also occurs. Nucleoli organize around the multiple tandemly linked and transcribed rRNA genes. If rRNA genes are artificially dispersed, as on a multicopy plasmid, nucleolar components become dispersed through the nucleus, but ribosome biogenesis still occurs (Nierras et al., 1997).

There are more than 200 modifications per human ribosome and about 110 per ribosome in the yeast *Saccharomyces carlsbergensis*. However, the overall frequency of modification is lower than in tRNA: about 2 to 3% of nucleosides in eukaryotic rRNAs are modified, compared with 15 to 20% in eukaryotic tRNAs. Moreover, the diversity of modifications in eukaryotic rRNA molecules is much less than for tRNA. Most of the modifications of eukaryotic rRNA are of just two kinds: 2'-O-ribose methyl groups and Ψ. There are only about ten methylated bases and other modifications. These data are summarized for representative vertebrates and for *Saccharomyces cerevisiae* in Table 4. The corresponding data for *E. coli* rRNA are also included, showing the much smaller numbers of 2'-O-methyl groups and Ψ and the larger numbers of base methyl groups in this prokaryotic rRNA than in the eukaryotes. (See Table 5 for the base-modified nucleosides in eukaryotic rRNAs and Limbach et al., 1994, and Appendix 1 for further details on the phylogenetic distribution of modified nucleosides in rRNAs, especially among prokaryotes.) Nearly all of the modified nucleosides have been located in the respective rRNA sequences in Table 4.

The earliest studies on eukaryotic rRNA modification were by Lane and colleagues on wheat rRNA, as summarized in Chapter 1. The subsequent history of the field was described by Maden (1990a) and in summary form by Maden et al. (1995). A detailed review of the state of knowledge of modifications in relation to eukaryotic rRNA sequences and secondary structure up to 1990 was given by Maden (1990b).

The field has recently been transformed by discoveries of how the correct nucleosides are recognized for modification, specifically for 2'-O-methylation and pseudouridylation. snoRNAs are at the heart of the recognition processes. These developments are detailed in Chapter 12 by Ofengand and Fournier and Chapter 13 by Bachellerie and Cavaillé. A short commentary is essential in the context of this chapter and now follows.

2'-O-methyl nucleosides

Before the complete sequences of any rRNAs were known, considerable information had already

Table 4. Numbers of modified nucleosides in representative rRNAs[a]

Species and rRNA	2′-O methyl	Base methyl	Ψ	Other	Total	Total no. of nucleosides	% of nucleosides modified
Human							
SSU	40	5	~36	≥1	~82	1,869	4.4
5.8S	2		2		4	157	2.5
LSU	63–65	5	55	?	~124	5,025	2.5
Total	105–107	10	~93	1–2?	~210	7,051	3.0
Mouse							
SSU	40	5	~36	≥1	~82	1,869	4.0
5.8S	2		2		4	157	2.5
LSU	63–65	5	57	?	~126	4,712	2.7
Total	105–107	10	~95	1–2?	~212	6,738	3.1
Xenopus laevis							
SSU	33	5	~44	≥1	~83	1,826	4.5
5.8S	2		2		4	160	2.5
LSU	62–64	5	~52	?	121	4,110	2.9
Total	97–99	10	~98	1–2?	~208	6,096	3.4
Saccharomyces carlsbergensis							
SSU	18	4	14	≥1	37	1,798	2.1
5.8S			1		1	160	0.6
LSU	37	~5	30	?	~72	3,393	2.1
Total	55	~9	45	1–2?	~110	5,351	2.1
Escherichia coli							
SSU	1	10	1	?	~12	1,542	0.8
LSU	3	14	9	?	~26	2,904	0.9
Total	4	24	10	?	~38	4,446	0.85

[a] Most of the data were summarized in Maden (1990b) and are from primary sources cited therein. The Ψ numbers have been refined from recent sequence-specific analyses, summarized in Chapter 12 by Ofengand and Fournier. The yeasts *S. carlsbergensis* and *S. cerevisiae* are closely related and may have identical rRNA modifications. The *E. coli* data are from Cunningham et al. (1990) and from Chapter 12 by Ofengand and Fournier.

Table 5. Base-modified nucleosides in eukaryotic rRNA[a]

rRNA	Modification	Base no. in: Human	Base no. in: S. carlsbergensis
SSU	m¹acp³Ψ	1248	1187
	m⁷G	1639	1573
	m⁶A	1832	
	m⁶₂A	1850, 1851	1779, 1780
	ac⁴C	Unlocated	Unlocated
LSU	m¹A	1302	643
	m¹A		2140
	m⁵C	3751	2276
	m³U		~2631
	m⁶A	4179 or 4180	?
	m³U		~2840
	Cm	Ca. 4403	
	m³U	4490	

[a] Base numbering is as in Maden (1990b). Note that slightly different numbering systems for LSU rRNAs exist in the literature, due partly to sequencing ambiguities but also to small length heterogeneities (i.e., length variations between individual genes within species) in eukaryotic expansion segments. Nucleotide numbers on the same line for human and yeast rRNAs denote homologous positions in the secondary structure; nucleotide numbers on different lines denote nonhomologous positions. ac⁴C in SSU rRNA (Thomas et al., 1978) is currently unlocated. The two yeast m³U residues shown here were erroneously shown as Um in Maden (1990b).

been obtained on 2′-O-methylation in rRNA and pre-rRNA from representative eukaryotes. This information came from exploitation of the following facts and techniques, separately or in combination. RNA can be specifically labeled in its methyl groups by growth of methionine-requiring cells (animal cells, or yeast cells with a deficiency in methionine biosynthesis) in the presence of methyl-labeled methionine; 2′-O-methylation confers resistance to hydrolysis by alkali and by some enzymes on the adjacent phosphodiester bond; fingerprints of rRNA methylation patterns can readily be obtained by enzymic digestion of methyl-labeled RNA followed by two-dimensional electrophoretic fractionation of the products; methyl-labeled pre-rRNA can be obtained from various eukaryotic sources in sufficient quantity for analysis in this way.

Almost all of the 2′-O-methyls in rRNA were found to be present in the primary pre-rRNA transcript, 45S RNA in mammals and 35S RNA in yeast. 2′-O-methylation appeared to be closely linked to transcription, because blocking transcription by actinomycin D led to almost immediate cessation of 2′-O-methyl labeling (Maden and Salim, 1974); in brief pulse-labeling experiments some methylation was detected on nascent 45S pre-rRNA, at least in mammalian cells (Greenberg and Penman, 1966; Liau and Hurlbert, 1975). (One exception to the rule of early 2'-O-methylation was found, at a single site in LSU rRNA containing the remarkable modified triplet Um-Gm-Ψ, in a conserved loop near the peptidyl transfer region; there is appreciable delay in methylation of this Gm, although this still occurs on pre-rRNA [Eladari et al., 1977].)

It was also evident from the diversity of products in RNase T_1 digests of rRNA that a very large number of different primary sequences serve as recognition sites for 2'-O-methylation (Maden and Salim, 1974; Klootwijk and Planta, 1973). This conclusion was confirmed when the methylated oligonucleotide data were combined with the complete rRNA sequences, derived from the genes, to reveal the distribution of methyl groups along the rRNA sequences. All of the methyl group locations were identified in SSU rRNA and most of the methyl group locations were identified in LSU rRNA from various vertebrates (Maden, 1986, 1988), and most methyl groups in yeast rRNA were also placed (Raue et al., 1988). These complete sequence data also revealed that methylation is concentrated within conserved core regions of rRNA and does not occur in the eukaryotic expansion segments.

From the fact that all of the numerous 2'-O-ribose methylations occur on nucleolar pre-rRNA, it was evident that the enzymic machinery for rRNA 2'-O-methylation must be located in the nucleolus. What recognition processes could lead to more than one hundred 2'-O-methylations on diverse sequences on pre-rRNA almost as soon as it comes off the DNA template? Could a single enzyme specifically recognize all the sites, or is there a different enzyme for every site, or does the answer lie somewhere in-between? One nucleolar 2'-O-methylase had been partly characterized, with specificity for a unique 2'-O-methyl triplet, Am-Gm-Cm, near the middle of (mammalian) LSU rRNA (Segal and Eichler, 1991; Eichler, 1994). The nucleolar protein fibrillarin was known to play a global though undefined role in 2'-O-methylation (Tollervey et al., 1993). Otherwise, no progress had been made until very recently on the molecular recognition of the sites for of 2'-O-methylation.

It is easy to see in retrospect that something was missing from the picture. What that something was is one of the most spectacular and exciting recent developments in the field of RNA modification, and is detailed in Chapter 13 by Bachellerie and Cavaillé. In brief, Kiss-László et al. (1996) and Nicoloso et al. (1996) have characterized numerous snoRNAs, each with the following properties: short sequence elements called boxes C and D, and a long (12 or more nt) stretch of perfect complementarity to an rRNA tract containing a 2'-O-methylation site. Experimental evidence based on deletion of genes for these snoRNAs and after manipulation of critical sites in their nucleotide sequences has established that the snoRNAs serve as "guides" for adding the 2'-O-methyl groups at precise locations on the newly synthesized rRNA (Kiss-László, 1996; Cavaillé et al., 1996).

The enzymology of 2'-O-methylation of pre-rRNA has not yet been resolved, but clearly the new findings on the role of snoRNAs afford the potential of great simplification. It is possible in principle that a single 2'-O-methylase recognizes many if not all of the transient complexes of snoRNAs with their cognate rRNA sequences, methylating the target nucleosides in pre-rRNA. Fibrillarin probably functions indirectly by complexing with box C and D snoRNPs.

Since many of the 2'-O-methyl groups occur in various kinds of folded regions within the rRNA secondary structure (Maden, 1990b), it can be inferred that the interactions with the snoRNAs that specify the methylation sites must occur before rRNA folding becomes stabilized. This is consistent with the fact that 2'-O-methylations occur extremely rapidly upon pre-rRNA probably on the nascent transcript (Greenberg and Penman, 1966; Liau and Hurlbert, 1975). Moreover, the early timing of 2'-O-methylation upon pre-rRNA is in marked contrast with the late timing of 2'-O-methylations during tRNA maturation, with intron removal being a prerequisite for 2'-O-methylation of Gm18, Cm32, Gm34 and Um44 in tRNA (see Table 2). This is consistent with different recognition processes for 2'-O-methylation of rRNA and tRNA. Whereas rRNA is 2'-O-methylated through interactions with guide snoRNAs, 2'-O-methylations of tRNA most likely involve direct enzymic recognition of features of the tRNA architecture.

Lastly, it should be mentioned that to date snoRNAs have been discovered matching about 50% of the numerous 2'-O-methylation sites in mammalian rRNA (listed in Maden and Hughes, 1997, and by Bachellerie and Cavaillé in Chapter 13). The possibility that a few 2'-O-methylation sites are recognized by different mechanisms cannot yet be excluded. It will be interesting, for example, to discover whether or not there is a guide snoRNA for the delayed methylation of Gm in the LSU Um-Gm-Ψ sequence (Eladari et al., 1977) (Gm4459 in the numbering system of Maden and Hughes, 1997, and in Chapter 13).

Pseudouridines

Ψ is also an abundant modified nucleoside in eukaryotic rRNA (Table 4) and has been known for many years to be present in nucleolar pre-rRNA (Jeanteur et al., 1968; Brand et al., 1979). Until recently technical difficulties precluded the complete mapping of all the Ψ residues in any eukaryotic rRNA, although some progress was made with mammalian 18S rRNA (Choi and Busch, 1978; Maden and Wakeman, 1988). Recently, new methodology based on reverse transcription has permitted the mapping of all Ψ residues at sequence resolution in the rRNAs of several eukaryotes and prokaryotes (Ofengand and Bakin, 1997; see Chapter 12 by Ofengand and Fournier for details and further references). To date this method of analysis has not yet been applied to the whole of any pre-rRNA. Therefore, there has not yet been a formal bookkeeping check between all of the Ψ residues in rRNA and pre-rRNA such as was achieved for methylation (Maden and Salim, 1974). However, it is clear from oligonucleotide data in some instances (Brand et al., 1979), and can be inferred in many other instances, that most if not all pseudouridines are introduced to pre-rRNA in the nucleolus.

Meanwhile, among several important findings from the mapping work on mature rRNA was abundant evidence that Ψ occurs in a wide variety of local primary and secondary structure environments in rRNA. Thus, the recognition problem for Ψ formation seemed to be as complicated as it had appeared to be for 2′-O-methylation before the discovery of D box snoRNAs. Might there be a similar solution? As is fully described in Chapter 12 by Ofengand and Fournier, Ψ formation in eukaryotic pre-rRNA is indeed mediated by a novel class of snoRNAs, called H/ACA snoRNAs (Ni et al., 1997; Ganot et al., 1997).

The following aspects of H/ACA snoRNAs are important from the standpoint of this chapter. First, like snoRNA-mediated 2′-O-methylation, snoRNA-mediated pseudouridylation is also necessarily a nucleolar process. Second, although the precise modes of interaction between the two classes of snoRNAs (D box and H/ACA snoRNAs) and their target sites on pre-rRNA differ, both require that the pre-rRNA target be partly unfolded with respect to the final rRNA conformation. Therefore, early pseudouridylation is to be expected on pre-rRNA, as already discussed for 2′-O-methylation. Third, it is not yet known how many Ψ residues are specified by H/ACA box snoRNAs, but the findings by Ganot et al. (1997) of snoRNAs for the two Ψ sites in 5.8S rRNA, for the hypermodified m^{1}acp^{3}Ψ site in SSU rRNA, and for many other sites in SSU and LSU rRNA make it likely that this mechanism for Ψ site selection is a general one. Fourth, the operation of this mechanism for Ψ site selection in pre-rRNA by transient interaction with snoRNAs renders it possible that a single Ψ synthase catalyzes the formation of most or all Ψ residues in eukaryotic pre-rRNA, a similar prediction as for 2′-O-methylation.

In fact, the nucleolar protein Gar1 plays an essential role in pseudouridylation, probably not as the actual Ψ synthase but as a protein component of H/ACA snoRNP functional complexes (Bousquet-Antonelli et al., 1997). A candidate protein, Cbf5p, for the actual Ψ synthase has very recently been identified in yeast by Lafontaine et al. (1998), as outlined in Chapter 15 by Lafontaine and Tollervey.

Base modifications

Only about ten methylated bases occur in eukaryotic rRNA (Table 5). The base methyls of human (HeLa cell) LSU rRNA are detectable in 45S rRNA (Maden and Salim, 1974). Therefore these modifications must occur in the nucleolus. However, the base methyls of SSU rRNA are not present in 45S rRNA or in 20S rRNA, the immediate nucleolar precursor to cytoplasmic 18S (SSU) rRNA (Maden et al., 1972). Moreover, methyl label continues to be incorporated into the SSU base methylation sites for at least several minutes after rRNA transcription has been blocked by actinomycin D (Zimmerman, 1968; Salim and Maden, 1973; Maden and Salim, 1974). These findings operationally classify the SSU base methylations as late. From these HeLa cell data it is not clear whether these late methylations of SSU rRNA occur immediately before or just after SSU rRNA leaves the nucleus in the form of small preribosomal subunits (discussed by Maden and Salim, 1974), and no further work appears to have been done with a mammalian system (except for m^{1}acp^{3}Ψ [see below]).

The category of late or secondary methylations has been further studied with yeast (Brand et al., 1977). Table 5 shows that yeast LSU rRNA contains two m^{3}U residues that are distinct from the one m^{3}U in mammalian LSU rRNA. These two m^{3}U residues are not present in the primary transcript but are present in 29S RNA, the immediate nucleolar precursor to LSU rRNA. Therefore, these are delayed, but nucleolar, methylation events. (Also absent from the primary transcript but present in 29S is Um-Gm-Ψ; this observation by Brand et al., 1977, was contemporary with the report by Eladari et al., 1977, in which delayed methylation of the Gm was established as the cause of absence of the respective oligonucleotide from the RNA fingerprint.)

Turning to SSU rRNA in yeast, m^{7}G, which occurs at the homologous site in yeast and HeLa cells, is also absent from the yeast primary transcript, but is present in the nucleolar precursor (18S in the nomenclature of Brand et al. and 20S in later literature) to cytoplasmic SSU rRNA. Therefore, m⁷G formation appears to be a delayed nucleolar event in yeast.

Only one of the methylation sites in eukaryotic SSU rRNA is similarly methylated in prokaryotic rRNA. This is the well-known site containing two adjacent dimethyladenosines in the highly conserved 3'-terminal helix loop. Despite its high evolutionary conservation, methylation at this site is not essential for viability in *E. coli*; a mutant lacking the ability to methylate the two adjacent adenosines is viable and resistant to the antibiotic kasugamycin (Helser et al., 1971), showing mild impairment in some steps of protein synthesis (van Knippenberg, 1986). As previously noted for HeLa cells, the sequence was found to be unmethylated at this site in the yeast nucleolar precursor to cytoplasmic SSU rRNA, and it was concluded that the adenosine dimethylations occur after transport of 18S pre-rRNA molecules to the cytoplasm but before the final maturation to SSU rRNA (Brand et al., 1977; mature SSU rRNA is termed 17S in their nomenclature).

Recently the gene for the corresponding dimethyltransferase in yeast, *DIM1*, has been cloned and has been shown to be essential in this eukaryotic organism (Lafontaine et al., 1994). Depletion experiments (Lafontaine et al., 1995) have shown that the protein (Dim1p) plays a nucleolar role in pre-SSU rRNA processing, in a capacity that appears to be in addition to the methylating activity, since cleavages on the path to pre-SSU rRNA are inhibited in the absence of Dim1p. Moreover, dimethylation can be induced to occur upon earlier nucleolar pre-rRNA precursors than 20S in pre-rRNA mutants where cleavage in the internal transcribed spacer is delayed. Taken together, the findings imply that Dim1p may normally be associated with pre-rRNA both in the nucleolus and immediately after transport to the cytoplasm. Lafontaine et al. (1995) propose that "binding of Dim1p to the pre-ribosomal particle is monitored to ensure that only dimethylated pre-rRNA molecules are processed to (mature) 18S rRNA." This topic is reviewed further in Chapter 15.

One other base modification deserves comment: the hypermodified nucleoside designated m¹acp³Ψ, about two thirds of the way towards the 3' end of SSU rRNA (Table 5). This nucleoside was first characterized in rRNA by Saponara and Enger (1974). It occurs in eukaryotes from *S. carlsbergensis* to vertebrates. Its biosynthesis involves conversion of U to Ψ, then methylation at N1, and finally aminocarboxypropylation at N3, with both of the latter substituents derived from methionine (via *S*-adenosylmethionine [AdoMet]) by separate reactions. The biosynthetic sequence was characterized both for yeast and HeLa cells (Brand et al., 1978); the first two steps, Ψ formation and N1 methylation, occur in the nucleolus. It was concluded that the final step is cytoplasmic because it occurs immediately before the final maturation to SSU rRNA in yeast, which is a cytoplasmic event. Of particular interest is the recent finding of an H/ACA snoRNA that matches this hypermodification site, snR35 in yeast (Ganot et al., 1997). Evidently, the pseudouridylation step involves a standard snoRNA recognition.

Mitochondrial rRNA

It has been known for many years that mitochondrial rRNA contains extremely few modified nucleosides (Dubin, 1974; Klootwijk et al., 1975; Lambowitz and Luck, 1976). These early findings, obtained by the fingerprinting of mitochondrial rRNA, have been confirmed by newer methodology, including mass spectrometry by Sirum-Connolly et al. (1995) (see also Chapter 14 by Mason).

These findings for mitochondrial rRNA modification contrast with those for mitochondrial tRNA. Mitochondrial tRNAs contain many similar modifications to cytoplasmic tRNAs, due in at least some instances to sorting isoenzymes for a given modification being targeted both to mitochondria and to the nucleus or cytoplasm, as previously described. By contrast, the recognition processes for nucleolar pre-rRNA 2'-O-methylation and pseudouridylation, involving in most cases a different snoRNA for each modification site, together with the nucleolar proteins fibrillarin (for 2'-O-methylation) and Gar1p (for pseudouridylation), do not lend themselves to access to mitochondria. Instead, mitochondria have their own very limited set of rRNA-modifying enzymes. As already discussed, mitochondrial DNA does not encode any of its RNA-modifying enzymes. Therefore, the few mitochondrial rRNA-modifying enzymes must be encoded in the nuclear genome, synthesized on cytoplasmic ribosomes, and targeted specifically to the mitochondria. The gene for one of these enzymes, *PET56*, encodes a methylase that introduces one of only two 2'-O-methyl groups into yeast mitochondrial LSU rRNA (Gm2270); this event seems to be essential for large ribosomal subunit assembly (Sirum-Connolly and Mason, 1993; see also Chapter 14).

mRNA

Modification of eukaryotic mRNA is discussed in Chapter 10 by Bokar and Rottman. Here it is sufficient to note the cellular locations of the events. The three main modification events, m⁷G cap formation, 2′-O-methylation of one or two nucleosides next to the cap, and formation of m⁶A at some internal sites, are all nuclear events. Cap formation occurs cotranscriptionally on nascent mRNA. The minimum mRNA chain length that is accessible for capping by the vaccinia virus capping enzyme complex is ≥31 nt, and it is proposed that the capping complex may interact with RNA polymerase II, thereby gaining access to the 5′ end of nascent mRNA as soon as it emerges from the polymerase (Hagler and Shuman, 1992; reviewed in Shuman, 1995).

At least one m⁶A modification has been reported to occur within an intron, clearly establishing a nuclear location for this class of events (Chapter 10; see also Rottman et al., 1994). The m⁶A methyltransferase is an extremely large complex with components of 30, 200, and 875 kDa. This size and complexity are unusual for methyltransferases and raise the interesting possibility that the components may be involved in other steps in nuclear RNA processing such as mRNA splicing or transport (for recent developments see Chapter 10).

An increasing number of instances are coming to light where a particular nucleoside in a given mammalian mRNA is edited by deamination: cytosine to uridine, as in apolipoprotein B mRNA (see Chapter 18 by Chang et al.), or adenosine to inosine, as in mRNA for glutamate receptor channels (Chapter 19 by Rueter and Emeson; see also Bass, 1997). In at least some of these editing events the target nucleoside for deamination is within a double-stranded RNA region in which one strand is contributed by intron material (Bass, 1997). It can be concluded that editing on such intron-containing complexes is necessarily nuclear in location. As described in Chapter 18, apolipoprotein B mRNA is substantially edited on the prespliced pre-mRNA. Therefore, in this instance also, editing is predominantly a nuclear event. It can be inferred that mammalian mRNA editing in general occurs in the nucleus. The more complicated mRNA editing that takes place in organelles of plants and in mitochondria of organisms such as trypanosomes and *Physarum polycephalum* is discussed in Chapter 17 by Marchfelder et al., Chapter 21 by Hajduk and Sabatini, and Chapter 22 by Gott and Visomirski-Robic.

snRNAs AND snoRNAs

snRNA and snoRNA modification is discussed in Chapter 11 by Massenet et al. As for mRNA, comments relevant to the cellular locations of these events will suffice here.

The snRNAs of the major spliceosomal complex are termed U1, U2, U4, U5, and U6. Of these, U1, U2, U4, and U5 are transcribed by RNA polymerase II. Like pre-mRNA transcripts, they are cotranscriptionally capped with m⁷G in the nucleus. Upon assembly with cap-binding proteins, each of these RNAs is transported to the cytoplasm, where assembly with further proteins occurs. Next, further cap nucleoside methylation occurs to yield the trimethylated m^{2,2,7}G cap before the ribonucleoprotein (RNP) complexes are reimported into the nucleus.

U6 RNA is transcribed by RNA polymerase III and has a γ-methyl phosphate at its 5′ end. This RNA, in contrast to U1, U2, U4, and U5, does not exit from the nucleus during snRNP assembly.

The spliceosomal RNAs are also modified internally by 2′-O-methylation and pseudouridylation, with the Ψ residues being particularly numerous (13 Ψs) in mammalian U2 RNA. Pseudouridylation of snRNAs has been studied in vitro by incubating unmodified transcripts with HeLa cell extracts. Cytoplasmic (S100) extracts that are suitable for reconstitution of snRNP particles (Patton et al., 1987; Kleinschmidt et al., 1989) also contain enzymatic activity for site-specific formation of Ψ residues in U1, U2, and U5 RNAs (Patton, 1993, 1994a; Patton et al., 1994; see also Patton, 1994b, for a review). However, one of the three Ψ residues in U5 RNA, Ψ53, is not formed by cytoplasmic S100 extract, being formed instead by a nuclear extract; it is proposed that Ψ53 is modified in the nucleus after reimport of the RNP from the cytoplasm (Patton, 1994a).

The U4 and U6 snRNAs form an intermolecular complex. Each molecule contains three Ψ residues, of which all three in U4 RNA and two in U6 RNA are in regions of intermolecular contacts. Because U6 RNA does not exit from the nucleus, the Ψ modifications in U6 RNA must be nuclear events. Moreover, these modifications require interaction with U4 RNA (Zerby and Patton, 1996). Conversely, the modification of U4 RNA requires U6 RNA (Zerby and Patton, 1997) and thus must occur in the nucleus.

Experiments in which 5-fluorouracil-substituted transcripts were used as inhibitors of Ψ formation in each of the previously discussed studies led to the conclusion that several different Ψ synthases are involved in the modification of these snRNAs.

To summarize, several Ψ residues in mammalian snRNAs are most likely formed during the cytoplasmic phase of snRNP assembly, although Ψ53 formation in U5 RNA and formation of the Ψ residues in U6 and probably U4 RNAs are nuclear events. The cytoplasmic events are in contrast to pseudouridylation of tRNA and rRNA, which, as described earlier, occur on the precursor RNAs within the nucleus and nucleolus.

The snoRNAs comprise a very numerous array; a few of them, U3, U8, U14, and U22, are involved in the endonucleolytic cleavage reactions of pre-rRNA, whereas the rest serve as guides for 2'-O-methylation and pseudouridylation of pre-rRNA, as outlined earlier and fully described in Chapter 12 by Ofengand and Fournier and Chapter 13 by Bachellerie and Cavaillé. Some of the snoRNAs are independently transcribed by RNA polymerase II. Others, including the great majority of guide RNAs in vertebrates, are transcribed as parts of introns within the pre-mRNAs of protein-coding genes, and are then liberated during pre-mRNA processing. The snoRNAs that are independently transcribed by RNA polymerase II, exemplified by U3 and U8 RNAs, receive an m^7G cap that subsequently becomes further methylated into trimethyl $m^{2,2,7}G$. In contrast to the spliceosomal RNAs, the trimethylation of U3 and U8 RNAs occurs in the nucleus, probably because of the retention of these RNAs in the nucleus by a mechanism that involves the conserved box D. Occurrence of trimethylation of the caps of spliceosomal snRNAs in the cytoplasm and of U3 and other snoRNAs in the nucleus suggests that the N^2-dimethyltransferase for these cap modifications may be distributed between nucleus and cytoplasm.

For a more detailed account of biosynthetic and functional aspects of snRNA and snoRNA modifications and for more extensive references to the primary literature on the subject, see Chapter 11 by Massenet et al.

CONCLUDING COMMENTS

The principal findings, conclusions, and inferences reviewed in this chapter may now be summarized.

In cytoplasmic tRNA, most modifications occur in the nucleus upon pre-tRNA, before intron removal in those pre-tRNAs that contain introns. Late modifications include hypermodifications at positions 34 and 37, as well as several 2'-O-methylations. They probably occur in close conjunction with exit of the tRNA through the nuclear pore to the cytoplasm.

Mitochondrial tRNAs contain many of the same modifications as cytoplasmic tRNAs. However, mitochondrial DNA does not encode its own RNA-modifying enzymes. Instead, the enzymes are imported from the cytoplasm. In at least some instances the same nuclear gene encodes sorting isozymes for cytoplasmic tRNA and mitochondrial tRNA, with intracellular targeting being determined in part by the presence or absence of a mitochondrial targeting signal at the N terminus of the protein. In many eukaryotes some mitochondrial tRNAs are imported, and mitochondrial tRNA editing is also prevalent in many eukaryotes.

In cytoplasmic rRNA, the great majority of modifications occur on pre-rRNA in the nucleolus. Nearly all 2'-O-methylations occur rapidly upon the primary transcript, in contrast to the late occurrence of tRNA 2'-O-methylations. Pseudouridines are also introduced early. Guide snoRNAs specify the modification sites for both of these classes of modifications, and this process of site selection demands that the pre-rRNA is not yet stably folded into the mature rRNA conformation. A few modifications, including some base methylations, occur later, though still in the nucleolus. The two dimethyladenosine modifications and the final step in $m^1acp^3\Psi$ synthesis probably occur in the cytoplasm in close conjunction with transport from the nucleolus. The adenosine dimethylase plays an essential role in small ribosomal subunit maturation. This role is probably distinct from, though related to, its dimethylase activity.

Mitochondrial rRNA possesses very few modified nucleosides, but again the modifying enzymes are nucleus encoded. Because the guide snoRNAs that are responsible for so many modifications to pre-rRNA of cytoplasmic ribosomes are confined to the nucleolus, there is no sharing of this machinery with that of mitochondrial ribosome biogenesis. This is in contrast to the sorting isozymes whereby single genes direct isozymes for modifications to cytoplasmic and mitochondrial tRNAs.

In mRNA, capping, other modifications, and editing of nucleus encoded mRNA occur in the nucleus, probably before pre-mRNA splicing.

In snRNA, molecules undergo a complex maturation pathway with nuclear and cytoplasmic phases (except for U6 RNA) before returning to the nucleus as snRNPs. Some modifications, including several pseudouridylations, most probably occur in the cytoplasmic phase of maturation, whereas others occur in the final nuclear phase. In snoRNA, many of these RNAs have not yet been fully characterized for nucleoside modifications. However, the molecules do not leave the nucleus; therefore, any modifications must occur in the nucleus or nucleolus.

This overview is not complete or final. It has been assembled from two major lines of investigation, both of which are very actively ongoing. First, relevant RNA species have been examined for their contents of modified nucleosides. For tRNA, a few representative molecules have been studied in detail, with reference to mature tRNA, to precursor tRNA transcribed in vivo from microinjected genes, and to in vitro-transcribed unmodified tRNAs and their pre-tRNAs, tRNA mutants, and fragments. Representative systems have included yeast tRNA^{Tyr}, yeast tRNA^{Asp}, yeast tRNA^{Phe} and *Xenopus* tRNA^{Met}. These have proved extremely informative for understanding several aspects of tRNA modification, including the cellular locations of several steps. Future studies of other tRNAs will add substantially to the overall picture, in connection with correlations between tRNA architecture and the subcellular sites of modifications, for example.

The second line of investigation has been the isolation of modifying enzymes and their genes. Thus far only a small number of genes for eukaryotic RNA-modifying enzymes has been cloned. These include the six tRNA-modifying enzymes listed in Table 3, the adenosine dimethylase Dim1p, and the mitochondrial rRNA methylase Pet56p. This is likely to be an area of rapid expansion, especially in yeast now that the complete genome is known and many open reading frames are awaiting identification.

Availability of the genes and the means of manipulating their expression is of great importance, as stressed for several examples in this chapter, including *PUS1* in the section on cytoplasmic tRNA, *TRM1* and *MOD5* in the section on mitochondrial tRNA and *DIM1* in the section on cytoplasmic rRNA. The enzymes encoded by each of these genes are of particular interest from the standpoint of intracellular disribution. More generally, there is great potential for refined localization of enzymes by immunocytochemical methods, as exemplified by Trm1p, whose location is near to the inside of the nuclear membrane (Rose et al., 1995). It would be of interest to know whether enzymes that act earlier in the tRNA modification pathway are located diffusely through the nucleus, or whether they might be found localized in speckles, comparable to those described for spliceosomal components (Carmo-Fonseca et al., 1991a, 1991b). For tRNAs, it is conceivable that many molecules that undergo similar modifications, for example the m^{5}U54 and Ψ55 modifications, might be channeled to a specific nuclear location for this purpose. To sum up, characterization of further modifying enzymes will undoubtedly lead to insights in several areas: their mode of interaction with their substrates, their intracellular trafficking, identification of targeting sequences for their intracellular destinations, and observations on precise localization by fluorescence and confocal microscopy.

Until now, many of the developments in eukaryotic RNA modification have been made with traditional organisms such as *S. cerevisiae, Xenopus laevis,* and humans. There are sound experimental and biomedical reasons for these choices, and compelling reasons why yeast, with its great potential for genetic analysis, will remain at the forefront. Nevertheless, there is a broad potential for work with other eukaryotes, notably plants, in which some of the earliest work on rRNA modification was done (see Chapter 1 by Lane), and some evolutionarily deeply rooted unicellular eukaryotes. These not only include parasites of medical importance, but many of them also afford approaches to unravelling the origins and early evolution of eukaryotic RNA modifications.

Finally, although this chapter has concentrated on eukaryotes, in which the questions of intracellular locations and organization are fairly obvious, it should not be excluded that some level of intracellular organization of RNA-modifying enzymes could also occur in prokaryotes, for example by multienzyme interactions between RNA-modifying enzymes and other enzymes of RNA function, such as aminoacyl-tRNA synthases.

Acknowledgments. Work in the author's laboratory on rRNA and modified nucleosides has been supported by the Medical Research Council and the Wellcome Trust. I thank John Hughes and Luminita Paraoan for critical discussions and Beryl Foulkes for the careful typing of the manuscript. Finally, I thank Henri Grosjean and Rob Benne for their work and suggestions as editors.

REFERENCES

Auxilien, S., P. F. Crain, R. W. Trewyn, and H. Grosjean. 1996. Mechanism, specificity and general properties of the yeast enzyme catalyzing the formation of inosine-34 in the anticodon of transfer RNA. *J. Mol. Biol.* 262:437–458.

Bass, B. L. 1997. RNA editing and hypermutation by adenosine deamination. *Trends Biochem. Sci.* 22:157–162.

Becker, H. F., Y. Motorin, M. Sissler, C. Florentz, and H. Grosjean. 1997a. Major identity determinants for enzymatic formation of ribothymidine and pseudouridine in the TΨ-loop of yeast tRNAs. *J. Mol. Biol.* 274:505–518.

Becker, H. F., Y. Motorin, R. J. Planta, and H. Grosjean. 1997b. The yeast gene YNL292w encodes a pseudouridine synthase (Pus4) catalyzing the formation of Ψ55 in both mitochondrial and cytoplasmic tRNAs. *Nucleic Acids Res.* 25:4493–4499.

Björk, G. R. 1995. Biosynthesis and function of modified nucleosides, p. 165–205. *In* D. Söll and U. L. RajBhandary (ed.), *tRNA: Structure, Biosynthesis and Function.* ASM Press, Washington, D.C.

Boguta, M., L. A. Hunter, W.-C. Shen, E. C. Gillman, N. C. Martin, and A. K. Hopper. 1994. Subcellular locations of MOD5 proteins: mapping of sequences sufficient for targeting to mitochondria and demonstration that mitochondrial and nuclear isoforms comingle with the cytosol. *Mol. Cell. Biol.* 14:2298–2306.

Bousquet-Antonelli, C., Y. Henry, J-P. Gélugne, M. Caizergues-Ferrer, and T. Kiss. 1997. A small nucleolar RNP protein is required for pseudouridylation of eukaryotic ribosomal RNAs. *EMBO J.* 16:4770–4776.

Brand, R. C., J. Klootwijk, T. J. M. van Steenbergen, A. J. de Kok, and R. J. Planta. 1977. Secondary methylation of yeast ribosomal precursor RNA. *Eur. J. Biochem.* 75:311–318.

Brand, R. C., J. Klootwijk, R. J. Planta, and B. E. H. Maden. 1978. Biosynthesis of a hypermodified nucleotide in *Saccharomyces carlsbergensis* 17S and HeLa cell 18S ribosomal ribonucleic acid. *Biochem. J.* **169**:71–77.

Brand, R. C., J. Klootwijk, C. P. Sibum, and R. J. Planta. 1979. Pseudouridylation of yeast ribosomal precursor RNA. *Nucleic Acids Res.* **7**:121–134.

Carbon, P., E. Haumont, S. De Henau, G. Keith, and H. Grosjean. 1982. Enzymatic replacement in vitro of the first anticodon base of yeast tRNA^{Asp}: application to the study of tRNA maturation in vivo, after microinjection into frog oocytes. *Nucleic Acids Res.* **10**:3715–3732.

Carmo-Fonesca, M., D. Tollervey, R. Pepperkok, S. M. L. Barabino, A. Merdes, C. Brunner, P. D. Zamore, M. R. Green, E. Hurt, and A. I. Lamond. 1991a. Mammalian nuclei contain foci which are highly enriched in components of the pre-mRNA splicing machinery. *EMBO J.* **10**:195–206.

Carmo-Fonesca, M., R. Pepperkok, B. S. Sproat, W. Ansorge, M. S. Swanson, and A. I. Lamond. 1991b. In vivo detection of snRNP-rich organelles in the nuclei of mammalian cells. *EMBO J.* **10**:1863–1873.

Cavaillé, J., M. Nicoloso, and J.-P. Bachellerie. 1996. Targeted ribose methylation of RNA in vivo directed by tailored antisense RNA guides. *Nature* **383**:732–735.

Chen, J. Y., P. B. M. Joyce, C. L. Wolfe, M. C. Steffen, and N. C. Martin. 1992. Cytoplasmic and mitochondrial tRNA nucleotidyl transferase activities are derived from the same gene in the yeast *Saccharomyces cerevisiae*. *J. Biol. Chem.* **267**:14879–14883.

Choffat, Y., B. Suter, R. Behra, and E. Kubli. 1988. Pseudouridine modification in the tRNA^{Tyr} anticodon is dependent on the presence, but independent of the size and sequence, of the intron in eukaryotic tRNA^{Tyr} genes. *Mol. Cell. Biol.* **8**:3332–3337.

Choi, Y. C., and H. Busch. 1978. Modified nucleotides in T1 RNase oligonucleotides of 18S ribosomal RNA of the Novikoff hepatoma. *Biochemistry* **17**:2551–2560.

Ciampi, M. S., F. Arena, and R. Cortese. 1977. Biosynthesis of pseudouridine in the in vitro transcribed tRNA^{Tyr} precursor. *FEBS Lett.* **77**:75–82.

Clark, M. W., and J. Abelson. 1987. The subcellular localization of tRNA ligase in yeast. *J. Cell Biol.* **105**:1515–1526.

Cunningham, P. R., C. J. Weitzmann, D. Nègre, J. G. Sinning, V. Frick, K. Nurse, and J. Ofengand. 1990. In vitro analysis of the role of rRNA in protein synthesis: site-specific mutation and methylation, p. 243–253. *In* W. E. Hill, A. Dahlberg, R. A. Garrett, P. B. Moore, D. Schlessinger, and J. R. Warner (ed.), *The Ribosome: Structure, Function and Evolution*. American Society for Microbiology, Washington, D.C.

Dietrich, A., I. Small, A. Cosset, J. H. Weil, and L. Maréchel-Drouard. 1996. Editing and import: strategies for providing plant mitochondria with a complete functional set of transfer RNAs. *Biochimie* **78**:518–529.

Dirheimer, G., G. Keith, P. Dumas, and E. Westhof. 1995. Primary, secondary and tertiary structures of tRNAs, p. 93–126. *In* D. Söll and U. L. RajBhandary (ed.), *tRNA: Structure, Biosynthesis, and Function*. ASM Press, Washington, D.C.

Doolittle, W. F. 1996. Some aspects of the biology of cells and their possible evolutionary significance, p. 1–21. *In* D. M. Roberts, P. Sharp, G. Anderson, and M. A. Collins (ed.), *Evolution of Microbial Life*. Cambridge University Press, Cambridge, England.

Dubin, D. 1974. Methylated nucleotide content of mitochondrial ribosomal RNA from hamster cells. *J. Mol. Biol.* **84**:257–273.

Eichler, D. C. 1994. Characterization of a nuclear 2'-O-methyltransferase and its involvement in the methylation of mouse precursor ribosomal RNA. *Biochimie* **76**:1115–1122.

Eladari, E., A. Hampe, and F. Galibert. 1977. Nucleotide sequence neighbouring a late modified guanylic residue within the 28S ribosomal RNA of several eukaryotic cells. *Nucleic Acids Res.* **4**:1759–1767.

Ellis, S. R., M. J. Morales, J. M. Li, A. K. Hopper, and N. C. Martin. 1986. Isolation and characterization of the *TRM1* locus, a gene essential for the N^2,N^2-dimethylguanosine modification of both mitochondrial and cytoplasmic tRNA in *Saccharomyces cerevisiae*. *J. Biol. Chem.* **261**:9703–9709.

Ellis, S. R., A. K. Hopper, and N. C. Martin. 1987. Amino-terminal extension generated from an upstream AUG codon is not required for mitochondrial import of yeast N^2,N^2-dimethylguanosine-specific tRNA methylase. *Proc. Natl. Acad. Sci. USA* **84**:5172–5176.

Ellis, S. R., A. K. Hopper, and N. C. Martin. 1989. Amino-terminal extension generated from an upstream AUG codon increases the efficiency of mitochondrial import of yeast N^2,N^2-dimethylguanosine-specific tRNA methyltransferase. *Mol. Cell. Biol.* **9**:1611–1620.

Fournier, M., E. Haumont, S. De Henau, J. Gangloff, and H. Grosjean. 1983. Post-transcriptional modification of the wobble nucleotide in anticodon-substituted yeast tRNA$^{Arg}_{II}$ after microinjection into *Xenopus laevis* oocytes. *Nucleic Acids Res.* **11**:707–718.

Ganot, P., M.-L. Bortolin, and T. Kiss. 1997. Site-specific pseudouridine formation in preribosomal RNA is guided by small nucleolar RNAs. *Cell* **89**:799–809.

Greenberg, H., and S. Penman. 1966. Methylation and processing of ribosomal RNA in HeLa cells. *J. Mol. Biol.* **21**:527–535.

Grosjean, H., S. Auxilien, F. Constantinesco, C. Simon, Y. Corda, H. F. Becker, D. Foiret, A. Morin, Y. X. Jin, M. Fournier, and J. L. Fourrey. 1996a. Enzymatic conversion of adenosine to inosine and N^1-methylinosine: a review. *Biochimie* **78**:488–501.

Grosjean, H., J. Edqvist, K. B. Stråby, and A. R. Giegé. 1996b. Enzymatic formation of modified nucleosides in tRNA: dependence on tRNA architecture. *J. Mol. Biol.* **255**:67–85.

Grosjean, H., M. Sprinzl, and S. Steinberg. 1995. Posttranscriptionally modified nucleosides in transfer RNA: their locations and frequencies. *Biochimie* **77**:139–141.

Grosjean, H., Z. Szweykowska-Kulinska, Y. Motorin, F. Fasiolo, and G. Simos. 1997. Intron-dependent enzymic formation of modified nucleosides in eukaryotic tRNAs: a review. *Biochimie* **79**:293–302.

Gu, X., and D. V. Santi. 1991. The T-arm of tRNA is a substrate for tRNA (m^5U54)-methyltransferase. *Biochemistry* **30**:2999–3002.

Hagler, J., and S. Shuman. 1992. A freeze-frame view of eukaryotic transcription during elongation and capping of nascent mRNA. *Science* **255**:983–986.

Hancock, K., and S. L. Hajduk. 1990. The mitochondrial tRNAs of *Trypanosoma brucei* are nuclear encoded. *J. Biol. Chem.* **265**:19208–19215.

Haumont, E., L. Droogmans, and H. Grosjean. 1987. Enzymatic formation of queuosine and of glycosyl queuosine in yeast tRNAs microinjected into *Xenopus laevis* oocytes: the effect of the anticodon loop sequence. *Eur. J. Biochem.* **168**:219–225.

Hauser, R., and A. Schneider. 1995. tRNAs are imported into mitochondria of *Trypanosoma brucei* independently of their genomic context and genetic origin. *EMBO J.* **14**:4212–4220.

Helser, T. L., J. E. Davies, and J. E. Dahlberg. 1971. Change in methylation of 16S ribosomal RNA associated with mutation to kasugamycin resistance in *Escherichia coli*. *Nat. New Biol.* **233**:12–14.

Hopper, A. K., A. H. Furukawa, H. D. Pham, and N. C. Martin. 1982. Defects in modification of cytoplasmic and mitochondrial transfer RNAs are caused by single nuclear mutations. *Cell* **28**:543–550.

Jeanteur, P., F. Amaldi, and G. Attardi. 1968. Partial sequence analysis of ribosomal RNA from HeLa cells. II Evidence for sequences of non-ribosomal type in 45S and 32S ribosomal precursors. *J. Mol. Biol.* 33:757–775.

Jiang, H.-Q., Y. Motorin, Y.-X. Jin, and H. Grosjean. 1997. Pleiotropic effects of intron removal on base modification pattern of yeast tRNA^{Phe}: an in vitro study. *Nucleic Acids Res.* 25:2694–2701.

Johnson, P. F., and J. Abelson. 1983. The yeast tRNA^{Tyr} gene intron is essential for correct modification of its tRNA product. *Nature* 302:681–687.

Kiss-László, Z., Y. Henry, J.-P. Bachellerie, M. Caizergues-Ferrer, and T. Kiss. 1996. Site-specific ribose methylation of preribosomal RNA: a novel function for small nucleolar RNAs. *Cell* 85:1077–1088.

Kleinschmidt, A. M., J. R. Patton, and T. Pederson. 1989. U2 small nuclear RNP assembly *in vitro*. *Nucleic Acids Res.* 17:4817–4828.

Klootwijk, J., and R. J. Planta. 1973. Analysis of the methylation sites in yeast ribosomal RNA. *Eur. J. Biochem.* 39:325–333.

Klootwijk, J., I. Klein, and L. A. Grivell. 1975. Minimal post-transcriptional modification of yeast mitochondrial ribosomal RNA. *J. Mol. Biol.* 97:337–350.

Koonin, E. V. 1996. Pseudouridine synthases: four families of enzymes containing a putative uridine-binding motif also conserved in dUTPases and dCTP deaminases. *Nucleic Acids Res.* 24:2411–2415.

Koski, R. A., and S. G. Clarkson. 1982. Synthesis and maturation of *Xenopus laevis* methionine tRNA gene transcripts in homologous cell-free extracts. *J. Biol. Chem.* 257:4514–4521.

Lafontaine, D., J. Delcour, A.-L. Glasser, J. Desgrès, and J. Vandenhaute. 1994. The DIM1 gene responsible for the conserved m$_2^6$Am$_2^6$A dimethylation in the 3' terminal loop of 18S rRNA is essential in yeast. *J. Mol. Biol.* 241:492–497.

Lafontaine, D., J. Vandenhaute, and D. Tollervey. 1995. The 18S rRNA dimethylase Dim1p is required for pre-ribosomal RNA processing in yeast. *Genes Dev.* 9:2470–2481.

Lafontaine, D. L. J., C. Bousquet-Antonelli, Y. Henry, M. Caizergues-Ferrer, and D. Tollervey. 1998. The box H+ACA snoRNAs carry Cbf5p, the putative rRNA pseudouridine synthase. *Genes Dev.* 12:527–537.

Lambowitz, A. M., and D. J. L. Luck. 1976. Studies on the *Poky* mutant of *Neurospora crassa*: fingerprint analysis of mitochondrial ribosomal RNA. *J. Biol. Chem.* 251:3081–3095.

Lecointe, F., G. Simos, A. Sauer, E. C. Hurt, Y. Motorin, and H. Grosjean. 1998. Characterization of the yeast protein Deg1 as pseudouridine synthase (Pus3) catalyzing the formation of Ψ38 and Ψ39 in the tRNA anticodon loop. *J. Biol. Chem.* 273:1316–1323.

Li, J. M., A. K. Hopper, and N. C. Martin. 1989. N^{2}N^2-dimethylguanosine-specific tRNA methyltransferase contains both nuclear and mitochondrial targeting signals in *Saccharomyces cerevisiae*. *J. Cell Biol.* 109:1411–1419.

Liau, M. C., and R. B. Hurlbert. 1975. The topographical order of 18S and 28S ribosomal ribonucleic acids within the 45S precursor molecule. *J. Mol. Biol.* 98:321–332.

Limbach, P. A., P. F. Crain, and J. A. McCloskey. 1994. Summary: the modified nucleosides of RNA. *Nucleic Acids Res.* 22:2183–2196.

Lye, L. F., D. H. T. Chen, and Y. Suyama. 1993. Selective import of nuclear-encoded tRNAs into mitochondria of the protozoan *Leishmania tarentolae*. *Mol. Biochem. Parasitol.* 58:233–246.

Maden, B. E. H. 1986. Identification of the locations of the methyl groups in 18S ribosomal RNA from *Xenopus laevis* and man. *J. Mol. Biol.* 189:681–699.

Maden, B. E. H. 1988. Locations of methyl groups in 28S rRNA of *Xenopus laevis* and man: clustering in the conserved core of molecule. *J. Mol. Biol.* 201:289–314.

Maden, B. E. H. 1990a. The modified nucleotides in ribosomal RNA of man and other eukaryotes, p. B265–301. *In* C. W. Gehrke and C. T. Kuo (ed.), *Chromatography and Modification of Nucleosides*, part B. *Biological Roles and Functions of Modification*. Elsevier, New York, N.Y.

Maden, B. E. H. 1990b. The numerous modified nucleotides in eukaryotic ribosomal RNA. *Prog. Nucleic Acid Res. Mol. Biol.* 39:241–303.

Maden, B. E. H., and J. M. X. Hughes. 1997. Eukaryotic ribosomal RNA: the recent excitement in the nucleotide modification problem. *Chromosoma* 105:391–400.

Maden, B. E. H., and M. Salim. 1974. The methylated nucleotide sequences in HeLa cell ribosomal RNA and its precursors. *J. Mol. Biol.* 88:133–164.

Maden, B. E. H., and J. A. Wakeman. 1988. Pseudouridine distribution in mammalian 18S ribosomal RNA: a major cluster in the central region of the molecule. *Biochem. J.* 249:459–464.

Maden, B. E. H., M. Salim, and D. F. Summers. 1972. Maturation pathway for ribosomal RNA in the HeLa Cell nucleolus. *Nat. New Biol.* 237:5–9.

Maden, B. E. H., M. E. Corbett, P. A. Heeney, K. Pugh, and P. M. Ajuh. 1995. Classical and novel approaches to the detection and localization of the numerous modified nucleotides in eukaryotic ribosomal RNA. *Biochimie* 78:22–29.

Martin, N. C., and A. K. Hopper. 1994. How single genes provide tRNA processing enzymes to mitochondria, nuclei and the cytosol. *Biochimie* 76:1161–1167.

Melton, D. A., E. M. De Robertis, and R. Cortese. 1980. Order and intracellular location of the events involved in the maturation of a spliced tRNA. *Nature* 284:143–148.

Morin, M., S. Auxilien, B. Senger, R. Tewari, and H. Grosjean. 1998. Structural requirements for enzymatic formation of threonylcarbamoyladenosine (t^{6}A) in tRNA: an *in vivo* study with *Xenopus* oocytes. *RNA* 4:24–36.

Ni, J., A. L. Tien, and M. J. Fournier. 1997. Small nucleolar RNAs direct site-specific synthesis of pseudouridine in ribosomal RNA. *Cell* 89:565–573.

Nicoloso, M., L.-H. Qu, B. Michot, and J.-P. Bachellerie. 1996. Intron-encoded, antisense small nucleolar RNAs: the characterization of nine novel species points to their direct role as guides for the 2'-O-ribose methylation of rRNAs. *J. Mol. Biol.* 260:178–195.

Nierras, C. R., S. W. Liebman, and J. R. Warner. 1997. Does *Saccharomyces* need an organized nucleolus? *Chromosoma* 105:444–451.

Nishikura, K., and E. M. De Robertis. 1981. RNA processing in microinjected *Xenopus* oocytes. Sequential addition of base modifications in a spliced transfer RNA. *J. Mol. Biol.* 145:405–420.

Ofengand, J., and A. Bakin. 1997. Mapping to nucleotide resolution of pseudouridine residues in large subunit ribosomal RNAs from representative eukaryotes, prokaryotes, archaebacteria, mitochondria and chloroplasts. *J. Mol. Biol.* 266:246–268.

Patton, J. R. 1993. Multiple pseudouridine synthase activities for small nuclear RNAs. *Biochem. J.* 290:595–600.

Patton, J. R. 1994a. Formation of pseudouridine in U5 small nuclear RNA. *Biochemistry* 33:10423–10427.

Patton, J. R. 1994b. Pseudouridine formation in small nuclear RNAs. *Biochimie* 76:1129–1132.

Patton, J. R., R. J. Patterson, and T. Pederson. 1987. Reconstitution of the U1 small nuclear ribonucleoprotein particle. *Mol. Cell. Biol.* 7:4030–4037.

Patton, J. R., M. R. Jacobson, and T. Pederson. 1994. Pseudouridine formation in U2 small nuclear RNA. *Proc. Natl. Acad. Sci. USA* 91:3324–3328.

Qian, Q., and G. R. Björk. 1997. Structural requirements for the formation of 1-methylguanosine in vivo in tRNA$^{Pro}_{GGG}$ of *Salmonella typhimurium*. *J. Mol. Biol.* 266:283–296.

Raué, H. A., J. Klootwijk, and W. Musters. 1988. Evolutionary conservation of structure and function of high molecular weight ribosomal RNA. *Prog. Biophys. Mol. Biol.* 51:77–129.

Rose, A. M., H. G. Bedford, W. C. Shen, C. L. Greer, A. K. Hopper, and N. C. Martin. 1995. Location of N2,N2-dimethylguanosine-specific tRNA methyltransferase. *Biochimie* 77:45–53.

Rose, A. M., P. B. Joyce, A. K. Hopper, and N. C. Martin. 1992. Separate information required for nuclear and subnuclear localization: additional complexity in localizing an enzyme shared by mitochondria and nuclei. *Mol. Cell. Biol.* 12:5652–5658.

Rottman, F. M., J. A. Bokar, P. Narayan, M. E. Shambaugh, and R. Ludwiczak. 1994. N^6-adenosine methylation in mRNA: substrate specificity and enzyme complexity. *Biochimie* 76:1109–1114.

Sakano, H., Y. Shimura, and H. Ozeki. 1974. Selective modification of nucleosides of tRNA precursors accumulated in a temperature sensitive mutant of *Escherichia coli*. *FEBS Lett.* 48:117–121.

Salim, M., and B. E. H. Maden. 1973. Early and late methylations in HeLa cell ribosome maturation. *Nature* 244:334–336.

Saponara, A. G., and M. D. Enger. 1974. The isolation from ribonucleic acid of substituted uridines containing α-aminobutyrate moieties derived from methionine. *Biochim. Biophys. Acta* 349:61–77.

Schaefer, K. P., S. Altman, and D. Söll. 1973. Nucleotide modification in vitro of transfer RNA^{Tyr} of *Escherichia coli*. *Proc. Natl. Acad. Sci. USA* 70:3626–3630.

Schneider, A., J. Martin, and N. Agabian. 1994a. A nuclear encoded tRNA of *Trypanosoma brucei* is imported into mitochondria. *Mol. Cell. Biol.* 14:2317–2322.

Schneider, A., K. P. McNally, and N. Agabian. 1994b. Nuclear-encoded mitochondiral tRNAs of *Trypanosom brucei* have a modified cytidine in the anticodon loop. *Nucleic Acids Res.* 22:3699–3705.

Segal, D. M., and D. C. Eichler. 1991. A nucleolar 2'-O-methyltransferase: specificity and evidence for its role in the methylation of mouse 28S precursor ribosomal RNA. *J. Biol. Chem.* 266:24385–24389.

Shen, W.-C., D. Selvakumar, D. R. Stanford, and A. K. Hopper. 1993. The *Saccharomyces cerevisiae* LOS1 gene involved in pre-tRNA splicing encodes a nuclear protein that behaves as a component of the nuclear matrix. *J. Biol. Chem.* 268:19436–19444.

Shuman, S. 1995. Capping enzyme in eukaryotic mRNA synthesis. *Prog. Nucleic Acid Res. Mol. Biol.* 50:101–129.

Simos, G., H. Tekotte, H. Grosjean, A. Segref, K. Sharma, D. Tollervey, and E. C. Hurt. 1996. Nuclear pore proteins are involved in the biogenesis of functional tRNA. *EMBO J.* 15:2270–2284.

Simpson, A. M., Y. Suyama, H. Dewes, D. Campbell, and L. Simpson. 1989. Kinetoplastid mitochondria contain functional tRNAs which are encoded in nuclear DNA and also contain small minicircle and maxicircle transcripts of unknown function. *Nucleic Acids Res.* 17:5427–5445.

Sirum-Connolly, K., and T. L. Mason. 1993. Functional requirement of a site-specific ribose methylation in ribosomal RNA. *Science* 262:1886–1889.

Sirum-Connolly, K., J. M. Peltier, P. F. Crain, J. A. McCloskey, and T. L. Mason. 1995. Implications of a functional large ribosomal RNA with only three modified nucleotides. *Biochimie* 77:30–39.

Smolar, N., and I. Svensson. 1974. Transfer RNA methylating activity of yeast mitochondria. *Nucleic Acids Res.* 1:707–718.

Sogin, M. L., J. D. Silberman, G. Hinkle, and H. G. Morrison. 1996. Problems with molecular diversity in the eukarya, p. 167–184. *In* D. M. Roberts, P. Sharp, G. Anderson, and M. A. Collins (ed.), *Evolution of Microbial Life*. Cambridge University Press, Cambridge, England.

Steinberg, S., A. Misch, and M. Sprinzl. 1993. Compilation of tRNA sequences and sequences of tRNA genes. *Nucleic Acids Res.* 21:3011–3015.

Strobel, M. C., and J. Abelson. 1986. Effect of intron mutations on processing and function of *Saccharomyces cerevisiae* SUP53 tRNA in vitro and in vivo *Mol. Cell. Biol.* 6:2663–2673.

Szweykowska-Kulinska, Z., B. Senger, G. Keith, F. Fasiolo, and H. Grosjean. 1994. Intron-dependent formation of pseudouridines in the anticodon of *Saccharomyces cerevisiae* minor tRNA^{Ile}. *EMBO J.* 13:4636–4644.

Tarassov, I. A., and R. P. Martin. 1996. Mechanisms of tRNA import into yeast mitochondria: an overview. *Biochimie* 78:502–510.

Thomas, G., J. Gordon, and H. Rogg. 1978. N^4-acetylcytidine, a previously unidentified labile component of the small subunit of eukaryotic ribosomes. *J. Biol. Chem.* 253:1101–1105.

Tollervey, D., H. Lehtonen, R. Jansen, H. Kern, and E. C. Hurt. 1993. Temperature-sensitive mutations demonstrate roles for yeast fibrillarin in pre-rRNA processing, pre-rRNA methylation, and ribosome assembly. *Cell* 72:443–457.

Van Knippenberg, P. H. 1986. Structural and functional aspects of the N^6,N^6dimethyladenosines in 16S ribosomal RNA, p. 412–424. *In* B. Hardesty and G. Kramer (ed.), *Structure, Function and Genetics of Ribosomes*. Springer-Verlag, New York, N.Y.

Van Tol, H., and H. Beier. 1988. All human tRNA^{Tyr} genes contain introns as a prerequisite for pseudouridine biosynthesis in the anticodon. *Nucleic Acids Res.* 16:1951–1966.

Woese, C. R. 1987. Bacterial evolution. *Microbiol. Rev.* 51:221–271.

Woese, C. R., O. Kandler, and M. L. Wheelis. 1990. Towards a natural system of organisms: proposals for the domains Archaea, Bacteria, and Eucarya. *Proc. Natl. Acad. Sci. USA* 87:4576–4579.

Zerby, D. B., and J. R. Patton. 1996. Metabolism of pre-messenger RNA splicing co-factors: modification of U6 RNA is dependent on its interaction with U4 RNA. *Nucleic Acids Res.* 24:3583–3589.

Zerby, D. B., and J. R. Patton. 1997. Modification of human U4 RNA requires U6 RNA and multiple pseudouridine synthases. *Nucleic Acids Res.* 25:4808–4815.

Zimmerman, E. F. 1968. Secondary methylation of ribosomal ribonucleic acid in HeLa cells. *Biochemistry* 7:3156–3164.

Zoladek, T., G. Vaduva, L. A. Hunter, M. Boguta, B. D. Go, N. C. Martin, and A. K. Hopper. 1997. Mutations altering the mitochondrial-cytoplasmic distribution of Mod5p implicate the actin cytoskeleton and mRNA 3' ends and/or protein synthesis in mitochondrial delivery. *Mol. Cell. Biol.* 15:6884–6894.

Chapter 25

Genetics and Regulation of Base Modification in the tRNA and rRNA of Prokaryotes and Eukaryotes

MALCOLM E. WINKLER

The regulation of base modification in tRNA and rRNA molecules is not well understood in any organism at either the level of enzyme activity or gene control. The goal of this chapter is to describe some of the better-understood systems, primarily in bacteria and yeast, that exhibit regulation of modification enzyme activity or gene expression. This chapter is not meant to be exhaustive, especially because several excellent reviews (Björk 1995a, 1995b, 1996; Agris, 1996; Martin and Hopper, 1994; Grosjean et al., 1997) have recently appeared covering many aspects of the regulation, functions, and formation of base modifications in tRNA and rRNA molecules. In addition, the enzymology and functions of base modifications in tRNA and rRNA are reviewed elsewhere in this volume (see Chapter 6 by Auffinger and Westhof, Chapter 2 by Grosjean et al., Chapter 8 by Garcia and Goodenough-Lashua, Chapter 9 by Romier et al., Chapter 12 by Ofengand and Fournier, Chapter 14 by Mason, Chapter 24 by Maden, Chapter 26 by Björk and Rasmuson, and Chapter 27 by Curran; also see Appendix 3 by Garcia and Goodenough-Lashua). Rather than present what is known about every modification, this chapter attempts to bring out the common themes about regulation that are emerging from ongoing studies of the modification process. This chapter also attempts to identify knowledge gaps and questions that need to be addressed in future studies. Genetic and biochemical approaches used to identify tRNA modification genes are presented first, using bacteria and yeast as model systems. Regulation of modification in bacteria is then described. The chapter ends with a discussion of several modes of regulation that are unique to eukaryotic cells. The base modifications in this chapter are depicted in Fig. 1 in the order in which they are discussed. The genetic organization and known modes of regulation of bacterial and eukaryotic modification genes are compiled in Table 1. Appendix 4 by Leung et al. contains the chromosome locations and database accession numbers of the known tRNA and rRNA modification genes in a wide range of organisms. Prospects for future studies are included at the end of each section.

IDENTIFICATION OF tRNA AND rRNA MODIFICATION GENES IN BACTERIA AND YEAST BY GENETIC AND BIOCHEMICAL APPROACHES

The analysis of the regulation of RNA modification depends on the identification of genes and the development of enzyme assays for protein purification. Until fairly recently, these two goals were linked because substrates for modification enzymes were prepared from mutants that lacked the function of a given modification enzyme. The bulk tRNA or rRNA isolated from such mutants specifically lacked certain base modifications, which could be added by the enzymes in crude extracts or purified protein preparations from wild-type strains (for example, see Cortese et al., 1974; Leung et al., 1997).

Modification genes were classically identified in bacteria by mutations that decreased tRNA suppression properties, affected translational regulation (especially of operons controlled by attenuation), affected properties of translation such as antibiotic resistance, or decreased the amounts of specific modifications detectable in biochemical screens. The presence of modified bases, especially in the anticodon stem-loop, can profoundly alter the decoding properties of tRNA molecules (reviewed in Chapter

---

Malcolm E. Winkler • Department of Microbiology and Molecular Genetics, University of Texas Houston Medical School, Houston, Texas 77030-1501.

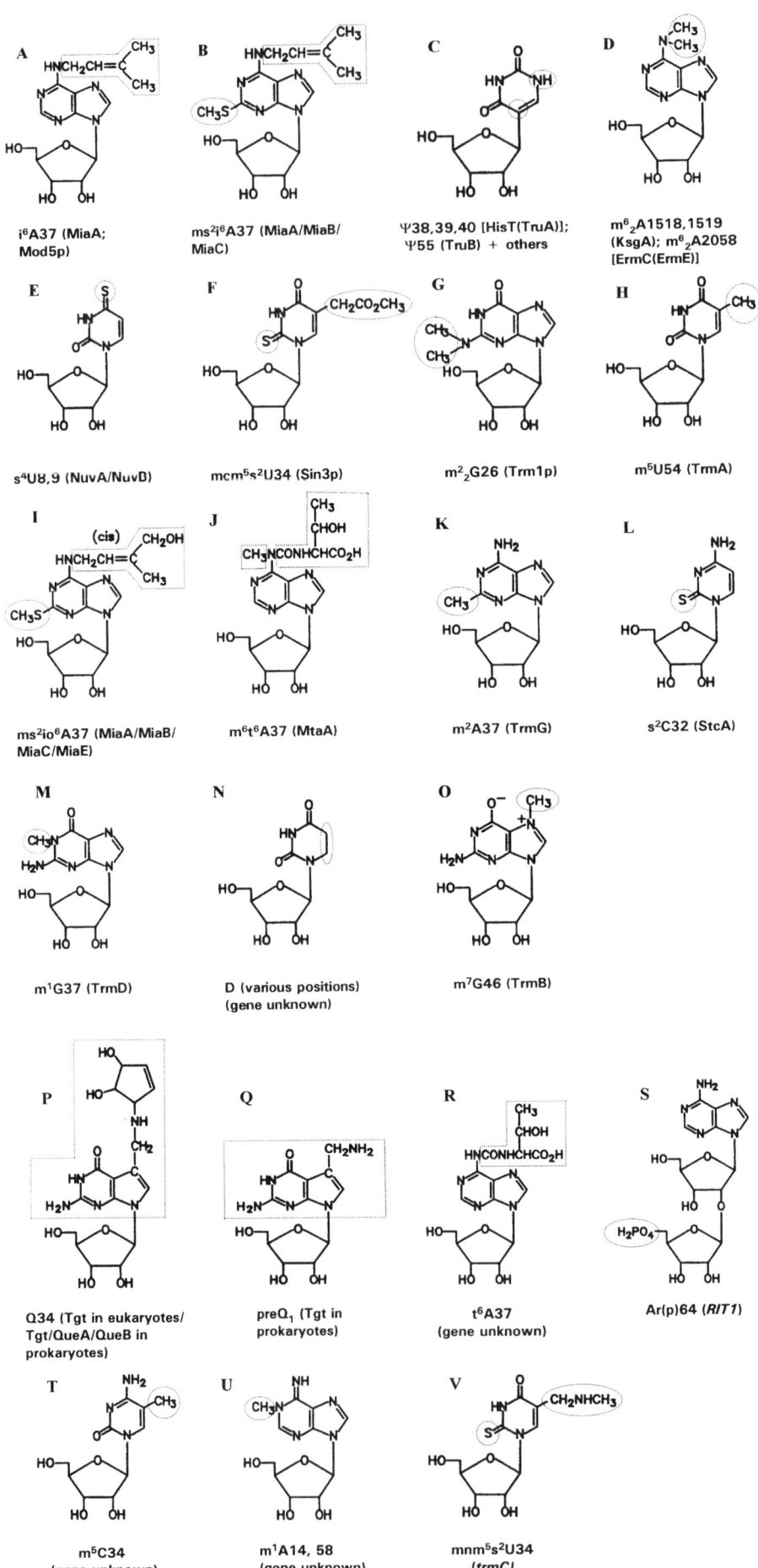

27 by Curran) (Björk, 1995a, 1995b, 1996). For example, fully modified wild-type tRNA species misread certain *lacZ*(UGA) nonsense codons sufficiently to produce a Lac⁺ (red color) phenotype on MacConkey lactose indicator plates. This misreading is prevented and a Lac⁻ (white color) phenotype results in strains containing *miaA* mutations, which mediate formation of i⁶A37 as the first step in ms²i⁶A37 in about 12% of all bacterial tRNA molecules (Fig. 1A and B) (Connolly and Winkler 1989, 1991; Petrullo et al., 1983). Likewise, the *su9* suppressor tRNA effectively suppresses the *lacZ*(UGA) nonsense mutation and gives high levels of β-galactosidase production (Connolly and Winkler, 1989, 1991; Petrullo et al., 1983). This nonsense suppression is significantly reduced in *miaA* mutants (Connolly and Winkler 1989, 1991; Petrullo et al., 1983). The efficiency of nonsense codon suppression has been exploited to analyze the effects of tRNA modification on decoding and codon context in bacteria and eukaryotes (see Björk 1995a, 1995b; Hagervall et al., 1990; Hopper, 1990; Hopper and Martin, 1992). The difficulty in using tRNA suppression to identify modification genes is knowing which nonsense codon-suppressor pair will give an antisuppressor phenotype when a certain modification is missing in a mutant. For example, *miaB* mutations, which cause a defect in ms² addition to i⁶A37 (Fig. 1B), decrease suppression of *lacZ*(UAG) codons by *supF30* tRNA more strongly than that of *supD10* or *supJ60* tRNA (Esberg and Björk, 1995), and *miaB* mutations do not influence *su9* suppression of a *lacZ*(UGA) codon (Leung and Winkler, 1997). It is perhaps no coincidence that the original *miaB* mutation was screened as an antisuppressor of *supF30*-mediated readthrough of UAG codons (Esberg and Björk, 1995).

Several base modification mutants were selected as regulators of bacterial operons that are controlled by attenuation mechanisms, which couple leader peptide translation efficiency to transcription termination (reviewed by Winkler, 1996; Landick et al., 1996). For example, selection of increased histidine (*his*) operon expression in the presence of the analogs 3-amino-1,2,4-triazole and 1,2,4-triazole-3-alanine led to the isolation of mutations that affected tRNA^His biosynthesis, including *hisT* (alternatively called *truA*; see Appendix 4 by Leung et al.) mutants defective in pseudouridine (Ψ) 38, 39, and 40 formation (Fig. 1C) (reviewed by Winkler, 1996). Similarly, *miaA* mutations defective in ms²i⁶A37 formation (Fig. 1A and B) were initially found as upregulators of tryptophan (*trp*) operon attenuation (Yanofsky and Soll, 1977; Landick et al., 1996). Conversely, base modification mutants often are more resistant or sensitive to a variety of metabolic analogs than their wild-type parent strains (Ericson and Björk, 1986; Esberg and Björk, 1995). However, it is not usually possible to predict these patterns of resistance and sensitivity to metabolic analogs and incorporate them into genetic selections for specific modification mutants.

Besides affecting translation control mechanisms, the absence of some base modifications alters properties of the translational apparatus. For example, *ksgA* mutations, which mediate formation of m₆²A1518 and 1519 in the 3'-end of 16S rRNA (Fig. 1D), were first identified as causing resistance to the antibiotic kasugamycin (see references in van Gemen et al., 1987). The *ermC* (*ermE*) gene of gram-positive bacteria encodes a methyltransferase that forms m₆²A2058 in 23S rRNA (Vester and Douthwaite, 1994). The presence of this modification imparts resistance to macrolide antibiotics, such as erythromycin (reviewed by Bechhofer 1990; Vellanoweth, 1993). *nuvA* and *nuvC*, which encode the enzymes that form s⁴U8 (Fig. 1E), were identified by resistance to near-UV light (de Araujo and Favre, 1986; Favre et al., 1986). Near-UV light cross-links s⁴U8 to nearby C13 in tRNA. The cross-linked tRNA does not function in translation, which in turn causes a growth delay that prevents induction of the SOS regulon and SOS mutagenesis (de Araujo and Favre, 1986; Favre et al., 1986). Thus, s⁴U8 is thought to act as an "antiphotomutageneic device" (de Araujo and Favre, 1986). However, these antibiotic or near-UV resistances are specific properties of *ksgA, ermC*, or *nuvA/nuvC* mutants or expression and cannot be used to identify other modification mutants.

Figure 1. Structures of the modified bases in tRNA and rRNA discussed in this chapter. Abbreviations for the modified bases and the enzymes that catalyze the modifications, which are enclosed in dotted lines, are indicated. Structures were redrawn or modified from Limbach et al. (1995). Positions refer to eubacterial or eukaryotic tRNA unless rRNA is indicated. (A) i⁶A37, N^6-Isopentenyladenosine; (B) ms²i⁶A37, 2-methylthio-N^6-isopentenyladenosine; (C) Ψ (various positions), pseudouridine; (D) m₆²A2058 (in 23S rRNA) or 1518 and 1519 in 16S rRNA, N^6,N^6-dimethyladenosine; (E) s⁴U8, 4-thiouridine; (F) mcm⁵s²U34, 5-methoxycarbonylmethyl-2-thiouridine; (G) m₂²G26, N^2,N^2-dimethylguanosine; (H) m⁵U54, ribosylthymine; (I) ms²io⁶A37, 2-methylthio-N^6-(*cis*-hydroxyisopentenyl)adenosine; (J) m⁶t⁶A37, N^6-methyl-N^6-threonylcarbamoyladenosine; (K) m²A37, 2-methyladenosine; (L) s²C32, 2-thiocytidine; (M) m¹G37, 1-methylguanosine; (N) D, 5,6-dihydrouridine (various positions); (O) m⁷G46, 7-methylguanosine; (P) Q34, queuosine; (Q) preQ₁34, 7-aminomethyl-7-deazaguanosine; (R) t⁶A37, N^6-threonylcarbamoyladenosine; (S) Ar(p)64, 2'-O-ribosyladenosine (phosphate); (T) m⁵C34, 48, 49, 5-methylcytidine; (U) m¹A14, 58, 1-methyladenosine; (V) mnm⁵s²U34, 5-methylaminomethyl-2-thiouridine.

Table 1. Summary of known modes of regulation of tRNA and rRNA modification genes in prokaryotes and eukaryotes[a]

Operon or gene[b]	Modification enzyme encoded	Growth rate regulation	Presence of FIS binding site	Effect of FIS on gene expression	Presence of stringent discriminator	Stringent control	Additional forms of regulation
Prokaryotes							
trmA	m⁵U54 tRNA methyltransferase	Yes (transcription and enzyme activity levels)	Yes	Unknown	Yes	Yes	
tgt-queA	tgt, Q34 tRNA transglycosylase; queA, epoxy-queuosine-34 tRNA synthase	Unknown	Yes	Yes (induction)	Yes	Unknown	
rpsP-21K-trmD-rplS	m¹G37 tRNA methyltransferase	Yes (transcription but not protein amount or enzyme activity)	Yes	Unknown	No	Yes (transcription but not enzyme activity)	Differential, noncoordinate regulation by internal mRNA regulatory structures that inhibit TrmD translation
amiB-mutL-miaA-hfq-hflX-hflK-hflC	i⁶A37 tRNA prenyltransferase	Unknown	Yes	No effect on protein amount	Yes	Unknown	Heat shock transcription; transcript destabilization by Hfq and RNase E; enzyme activity inhibited in vivo by nucleotide di- and triphosphates
surA-pdxA-ksgA-apaG-apaH	m⁶₂A1518 and 1519 16S rRNA dimethyltransferase	Yes (transcription of the surA-pdxA-ksgA cotranscript and from the P_{ksgA} internal promoter)	None in P_{ksgA} internal promoter; presumed in upstream promoter	Yes (induction)	Upstream promoter not located yet; yes in P_{ksgA} internal promoter	No (upstream transcription decreased by amino acid limitation, but not ppGpp dependent; P_{ksgA} internal promoter not stringently regulated)	Translational coupling; translational repression

CHAPTER 25 • REGULATION OF BASE MODIFICATION 445

Table 1. *Continued*

Operon or gene[b]	Modification enzyme encoded	Growth rate regulation	Presence of FIS binding site	Effect of FIS on gene expression	Presence of stringent discriminator	Stringent control	Additional forms of regulation
ermC (*ermE*)	m_2^6A2058 23S rRNA dimethyltransferase	Transcription is constitutive	NA[c]	NA	NA	NA	Induction by erythromycin by antiattenuation mechanism; mRNA stabilization; translational repression
pdxB-usg (*asd'*)-*hisT* (*truA*)-*dedA*	Ψ38, 39, 40 tRNA synthase I	Yes (transcription and enzyme activity levels)	Yes	Unknown	Yes	Unknown	Translational coupling
trmB	m^7G46 tRNA methyltransferase	No	Sequence of gene not identified yet				
trmC	$cmnm^5U34 \rightarrow nm^5s^2U34 \rightarrow mmm^5s^2U34$ tRNA methyltransferase	No	Sequence of gene not identified yet			No	
Eukaryotes							
TGT	Q34 tRNA transglycosylase	NA	NA	NA	NA	NA	Enzyme activity stimulated by phosphorylation by protein kinase C
TRM1	m_2^2G26 tRNA dimethyltransferase	NA	NA	NA	NA	NA	Transcription from upstream and internal promoters; two isozymes are synthesized by translation from the different length transcripts
MOD5	i^6A37 tRNA prenyltransferase	NA	NA	NA	NA	NA	Transcription from two upstream promoters; two isozymes are synthesized by translation from two different in-frame AUG start codons in the longer transcript; context and mRNA structure influence AUG selection in the bifunctional transcript

[a] See text and figures for details and references.
[b] In the order in which they are discussed in the text. Modification genes in multifunctional bacterial operons are indicated in boldface type.
[c] NA, not applicable.

Early biochemical screens for mutants lacking specific tRNA and rRNA modification enzymes were performed by in vitro modification assays or thin-layer chromatography of bases (see Björk, 1995b). In general, bacteria were treated with doses of powerful mutagens, such as nitrosoguanidine, sufficient to cause multiple mutations in each cell. Surviving mutated bacteria were pooled into small groups, and total RNA was extracted from each group. To identify tRNA and rRNA methyltransferase mutants (e.g., *trmA*⁻ [Fig. 1H], *trmD*⁻ [Fig. 1M], *trmB*⁻ [Fig. 1O], and *trmC*⁻ [Fig. 1V]), the extracted RNA was assayed in vitro for its ability to accept radiolabeled methyl groups when mixed with a crude extract of wild-type bacteria (Marinus et al., 1975; Björk and Isaksson, 1970; Björk and Kjellin-Straby, 1978). To identify mutants defective in Q biosynthesis (Fig. 1P and Q), ^{32}P-labeled total RNA was digested with nucleases, and the labeled bases were resolved by thin-layer chromatography (Noguchi et al., 1982; Okada et al., 1978). These biochemical screening approaches were extremely tedious and were effective in identifying only a limited number of tRNA and rRNA modification genes (Björk, 1995b).

Reduced efficiency of tRNA suppressor function was also used to identify yeast mutants defective in tRNA modification, such as *MOD5* or *SIN1* mutants lacking i⁶A37 (Fig. 1A) and *SIN3* mutants lacking mcm⁵s²U34 (Fig. 1F) (reviewed by Hopper, 1990; Hopper and Martin, 1992). Less direct biochemical approaches, such as screening of mutagenized cells for methyl-deficient tRNA or detection of altered tRNA mobility during polyacrylamide gel electrophoresis, were used to identify yeast *TRM1* or *TRM2* mutants lacking m$_2^2$G26 (Fig. 1G) or m⁵U54 (Fig. 1H), respectively (Hopper, 1990). A novel genetic approach to finding modification genes in yeast recently emerged from studies of functional interactions of nuclear pore proteins (nucleoporins) (Simos et al., 1996). A synthetic lethal genetic screen was used to find new mutations that were lethal in a conditional *NSP1* mutant, which lacks a functional nucleoporin involved in formation of nuclear pore complexes (Wimmer et al., 1992). Some of the new synthetic lethal mutations were in genes encoding other subclasses of nucleoporins (Wimmer et al., 1992). Attempts to complement mutations in other groups led to the cloning of genes involved in tRNA biogenesis, including *LOS1*, which is required for pre-tRNA splicing, and *PUS1*, which encodes the intranuclear intron-dependent Pus1p (Ψ34, Ψ36) synthase (Fig. 1C), which is homologous to *E. coli* HisT (TruA) (Simos et al., 1996). Furthermore, the triple combination of *NSP1 PUS1 LOS1* mutations was lethal, and the double *LOS1 PUS1*, but not the single *LOS1* or *PUS1*, mutant was defective in tRNA suppressor function (Simos et al., 1996). Together, these results suggest a functional interaction between enzymes involved in tRNA biogenesis, including certain tRNA modification enzymes, and the nuclear pore complex (Simos et al., 1996). They also suggest that powerful synthetic lethal approaches may be used to identify other genes that mediate the formation of certain tRNA modifications in yeast.

Three recent developments should lead to identification of the more than 50% of modification genes that remain unknown in *Escherichia coli, Salmonella typhimurium*, and yeast. First, the complete genomes of several organisms, including *E. coli* and *Saccharomyces cerevisiae*, are now known (see Appendix 4 by Leung et al.); therefore, it is a matter of finding which unassigned open reading frames mediate tRNA and rRNA base modification. The example of tRNA and rRNA pseudouridine synthases implies that some families of modifying enzymes do not contain readily recognizable motifs that can help assign function (Nurse et al., 1995; Wrzesinski et al., 1995). On the other hand, once a modification gene is identified in one organism, it will often be possible to find its homologs in other organisms with sequenced genomes (see Appendix 4 by Leung et al.). Second, the availability of relatively rapid and reliable high-pressure liquid chromatography (HPLC) methods to analyze total base contents in bulk tRNA (Buck et al., 1983; Gehrke et al., 1982) makes it much more feasible than before to screen for mutants missing certain tRNA modifications. For example, *miaE*, which hydroxylates ms²i⁶A37 to ms²io⁶A37 (Fig. 1I) in aerobically grown *S. typhimurium*, was located by pooling *E. coli* transformants containing an *S. typhimurium* plasmid genomic library and assaying by HPLC for the appearance of ms²io⁶A37, which is not normally present in *E. coli* (Persson and Björk, 1993). Other modification genes have been identified by a combination of luck and HPLC screening, including *mtaA* (m⁶t⁶A37), *trmG* (m²A37), and *stcA* (s²C32) (Fig. 1J to L) (Björk, 1995b, 1996). Third, it is now possible to synthesize in vitro RNA molecules by phage RNA polymerases (Sampson and Uhlenbeck, 1988) for use as substrates in schemes to purify modification enzymes. For example, a highly successful, systematic initiative is now under way to use reverse genetics to identify and clone the tRNA and rRNA pseudouridine synthases of *E. coli* (Chapter 8 by Garcia and Goodenough-Lashua and Chapter 12 by Ofengand and Fournier). This approach still requires the design and purification of suitable RNA substrates and the development of protein purification schemes using extracts of wild-type (nonoverexpressing) cells. Substrate design often requires considerable innovation, such as for the purification of the Ψ516 synthase (Fig. 1C), whose preferred substrate is a fragment of 16S rRNA complexed with 30S ribosomal proteins (Chapter 12 by Ofengand and Fournier; Ofengand et al., 1995).

Prospects

New genetic approaches combined with HPLC screening and reverse genetics will likely lead to the identification of the remaining tRNA and rRNA modification genes in bacteria and yeast. Newly identified modification genes can be rapidly targeted for mutagenesis in bacteria and yeast to determine whether their functions are essential and which cellular processes they mediate. The identification of these genes will greatly speed future investigations of the biochemical properties, cellular localization, and physical interactions of tRNA and rRNA modification enzymes. Identification of additional modification genes will also lead to a fuller understanding of the genetic regulation and function of tRNA and rRNA modification. The regulation of tRNA and rRNA enzymes and genes and the manipulation of tRNA modification as a genetic tool to gain information about fundamental biological processes, such as isozyme sorting, are described further in this chapter.

REGULATION OF tRNA AND rRNA MODIFICATION IN BACTERIA

Two separate, but related, issues need to be considered about the regulation of tRNA and rRNA modification in bacteria. The first issue is whether the cellular capacity for modification is regulated to maintain fully modified RNA molecules in cells growing under different physiological conditions. This issue is related to the amounts and catalytic capacity of tRNA and rRNA modification enzymes and the regulation of the genes that encode these enzymes. The second issue is whether the kinds or levels of modifications change in a given RNA molecule in cells grown under different physiological conditions. This issue would imply that changes in modification levels could serve as active regulatory devices.

Amounts and Catalytic Capacity of tRNA Modification Enzymes Are Sometimes Limited in Bacterial Cells

Regulation of rRNA and tRNA amounts in bacteria

To understand the possible need to regulate the cellular RNA modification capacity, the regulation of rRNA and tRNA amounts in bacteria needs to be briefly reviewed. The most complete information about this topic is available for *E. coli* (Bremer and Dennis, 1996), which is used as the model bacterial system in most of this chapter. However, a fast-growing, facultative anaerobe, such as *E. coli*, is undoubtedly not an appropriate model for all bacteria. The cellular amounts of ribosomes and tRNA are strongly regulated in *E. coli* and related enteric bacteria growing exponentially at different rates (Bremer and Denis, 1996). For example, the number of ribosomes and tRNA molecules is about 11-fold greater in *E. coli* growing fast at 2.5 doublings per h than slowly at 0.6 doublings per h. Over this range, the fast-growing cells have a volume about 3-fold greater than that of the slow-growing cells (Bremer and Dennis, 1996); consequently, the cellular concentration of ribosomes and tRNA molecules is about 3-fold greater in fast-growing cells than in slow-growing bacteria. To maintain this difference in concentration, the rate of accumulation of ribosomes and tRNA is about 45-fold greater in fast-growing than slow-growing exponential cells (calculated from $d[N]/dt = [(\ln 2)/(\text{cell doubling time})] \times [N]_{\text{exponential cells}}$, where N is the cellular amount of ribosomes of tRNA) (Bremer and Dennis, 1996). Therefore, there is a strong potential need to regulate cellular modification processes to match the changes in the rates of rRNA and tRNA synthesis. In addition, the rates of rRNA and tRNA synthesis are strongly downregulated by the stringent response in cells subjected to partial nitrogen or amino acid limitation (Gourse et al., 1996). Again, there is a potential need to coordinate modification processes with the change in rRNA and tRNA synthesis rates. The mechanisms that regulate rRNA and tRNA synthesis are considered further in the following sections.

Amounts and catalytic capacity of modification enzymes

With a few exceptions described in a later section, tRNA and rRNA are thought to be fully modified in wild-type bacteria during balanced growth (see Buck et al., 1983). A large excess of amounts and activities of modification enzymes would imply that tRNA and rRNA modification is not regulated in cells under different growth conditions. On the other hand, limiting enzyme amounts and activities would imply a need to regulate modification potential to accommodate the changes in the rates of stable RNA synthesis in cells grown under different conditions (Dong et al., 1996; Gourse et al., 1996; Keener and Nomura, 1996). Although still incomplete, two lines of investigation suggest that the amounts and activi-

Titration of modification capacity by tRNA overexpression

Several determinations of amounts of modified bases in tRNA molecules extracted from cells overexpressing a cloned tRNA species have suggested that modification enzymes can be readily saturated in vivo. When tRNA$_1^{Leu}$ was overexpressed 4-fold to 7-fold in *E. coli*, hypomodified (that is, partially modified) tRNA$_1^{Leu}$ species were detected that lacked m^{1}G37 (Fig. 1M) and Ψ38 and Ψ40 (Fig. 1C) (Wahab et al., 1993). Curiously, tRNA^{Tyr}, which was not overexpressed in these bacteria, was fully modified with pseudouridine in its anticodon stem and loop. These results were interpreted to mean that the *trmD*-encoded (m^{1}G37) methyltransferase and *hisT*(*truA*)-encoded (Ψ38, 39, 40) synthase could be titrated and that less abundant tRNA^{Tyr} is preferentially pseudouridinylated compared to tRNA$_1^{Leu}$ (Wahab et al., 1993). This interpretation makes the important, but unproven, assumption that all of the overexpressed tRNA$_1^{Leu}$ folded correctly in vivo and could serve as a substrate for the TrmD and HisT modification enzymes.

Overexpression of tRNA^{Phe} by about 11-fold in *E. coli* resulted in the accumulation of tRNA^{Phe} species that lacked ms^{2}i^{6}A37 or i^{6}A37 (Fig. 1A and B) but contained its normal distribution of other modifications (Wilson and Roe, 1989). In this case, hypomodified tRNA^{Phe} lacking ms^{2}i^{6}A37 was charged with phenylalanine by tRNA^{Phe} synthetase to the same extent as fully modified tRNA^{Phe}. This result and the presence of the other modifications normally found implied that the hypomodified tRNA^{Phe} was folded correctly. These findings are consistent with titration of a limited amount of the MiaA prenyltransferase that synthesizes i^{6}A37 in the anticodon of tRNA^{Phe}.

On the other hand, overexpression of tRNA^{Asp} by 15-fold in *E. coli* did not appear to titrate the activities of several tRNA modification enzymes, including those that synthesize 5,6-dihydrouridine (Fig. 1N), Ψ (Fig. 1C), s^{4}U8 (Fig. 1E), m^{2}A37 (Fig. 1K), m^{7}G46 (Fig. 1O), m^{5}U54 (Fig. 1H), and queuosine (Q34; Fig. 1P) (Martin et al., 1993). It is noteworthy that tRNA^{Asp} does not contain m^{1}G37 (Fig. 1M) or ms^{2}i^{6}A37 (Fig. 1B), whose formation was titrated by overexpressing other tRNA species. Another caution in interpreting tRNA overexpression data comes from experiments in which a heterologous tRNA^{Trp} species from *Bacillus subtilis* was overexpressed in *E. coli* (Xue et al., 1993). Modification of the *B. subtilis* tRNA^{Trp} resembled that expected for the *E. coli* host, with the exception that i^{6}A37 (Fig. 1A) was present instead of ms^{2}i^{6}A37 (Fig. 1B). Although this undermodification could indicate titration of the MiaB enzyme, it could just as easily indicate titration of the iron-dependent cofactor or some other metabolite required for ms^2 formation (Xue et al., 1993). Nevertheless, taken together, these tRNA overexpression experiments suggest that some tRNA modification activities may be limiting and more readily titrated in vivo than others.

Cellular amounts and kinetics of tRNA modification enzymes

As many as 36 and 45 different rRNA and tRNA modification enzymes, respectively, may be present in cells of enteric bacteria such as *E. coli* and *S. typhimurium* (Björk, 1995a, 1996). Until recently, few of these enzymes had been purified. However, this situation is rapidly changing with the biochemical characterization of several modification enzymes, including the TrmA (m^{5}U54) methyltransferase (Fig. 1H) (Gu et al., 1996; Gu and Santi 1991; Kealey and Santi, 1995), the TrmD (m^{1}G37) methyltransferase (Fig. 1M) (Holmes et al., 1995, 1992), the Tgt (preQ$_1$34 in prokaryotes or Q34 in eukaryotes) transglycosylases (Fig. 1P and Q) (Chong et al., 1995; Curnow and Garcia, 1995; Morris et al., 1995; Romier et al., 1996), the MiaA (i^{6}A37) prenyltransferase (Fig. 1A) (Leung et al., 1997; Moore and Poulter, 1997), the HisT (TruA) (Ψ38, 39, 40) synthase (Fig. 1C) (Kammen et al., 1988), the TruB (Ψ55) synthase (Fig. 1C) (Nurse et al., 1995), and several additional tRNA and rRNA pseudouridine synthases (see Chapters 8 and 12). Cellular modification capacities can be estimated from the number of enzyme molecules per cell, the k_{cat}^{app} turnover number, and the rate of tRNA synthesis.

The generalization has been made that tRNA modification enzymes are invariably present in low amounts in bacterial cells (Björk, 1995a). However, this generalization is not completely correct. Estimates of cellular amounts of tRNA modification enzymes were made by comparing the specific activities of purified enzymes with those of the enzymes in crude cellular extracts. This analysis makes the assumptions that the purified enzymes are completely active and that there is no inhibition of activity in crude extracts. These assumptions have not usually been verified. Rough estimates of enzyme amounts can also be made from the yields from enzyme purification schemes (Okada and Nishimura, 1979), but active enzyme is undoubtedly lost in these schemes. The early impression that certain modification enzymes, such as the TrmA (m⁵U54) methyltransferase (Fig. 1H) (Ny et al., 1988) and Tgt (preQ₁34) transglycosylase (Fig. 1Q) (Okada and Nishimura, 1979), are present in low amounts needs to be rechecked by direct methods, such as quantitative Western immunoblotting or two-dimensional polyacrylamide gel electrophoresis (see Leung et al., 1997; Wikström and Björk, 1988). The TrmD (m¹G37) methyltransferase (Fig. 1M) was found by two-dimensional gel analysis to be expressed at moderate amounts of about 260 monomers per gene equivalent (Wikström and Björk, 1988). As a standard of comparison, the enzymes that synthesize tryptophan in *E. coli* are present at about 1,000 copies per cell when the wild-type *trp* operon is fully derepressed (Morse and Yanofsky, 1969, 1968). Even by using activity comparisons, the HisT (TruA) (Ψ38, 39, 40) synthase (Fig. 1C) was estimated to be a moderately abundant bacterial enzyme at about 400 to 800 monomers per cell (Kammen et al., 1988). A recent determination by quantitative Western immunoblotting showed that the MiaA (i⁶A37) prenyltransferase (Fig. 1A) is also a moderately abundant enzyme at about 650 monomers per cell (≈ 1 μM) in bacteria growing exponentially in enriched minimal-glucose medium (Leung et al., 1997).

MiaA may need to be present in comparatively high cellular amounts, because its activity is strongly competitively inhibited for its prenyl substrate, dimethylallyl diphosphate (alternatively called Δ²-isopentenyl pyrophosphate), by nucleotide di- and triphosphates (Leung et al., 1997). The relative abundance of MiaA and its high kinetic activity for tRNA modification ($k_{cat}^{app} = 0.44$ s⁻¹) are probably necessary to overcome this inhibition and to just barely modify the tRNA synthesized in vivo (Leung et al., 1997). Little is known about the inhibition or regulation of the activities of other bacterial RNA modification enzymes by compounds present in cells. Regulation patterns suggest that the activity of the TrmD (m¹G37) methyltransferase (Fig. 1M) is regulated in vivo by an unknown molecule (Wikström and Björk, 1988). Besides tRNA methylation, the TrmD protein may also carry out a function required for the rapid growth of bacterial cells (Persson et al., 1995). Likewise, the TrmA (m⁵U54) methyltransferase (Fig. 1H) seems to have another essential activity or function besides tRNA modification. *trmA* null mutations cannot be constructed except in a *trmA*⁺ merodiploid strain (Persson et al., 1992); however, this essential function is not tRNA modification because viable *trmA* mutants were isolated that completely lack modification activity (Persson et al., 1992). The TrmA (m⁵U54) methyltransferase covalently binds to a fragment of the 3'-end of 16S rRNA, but this binding is not thought to catalyze or affect the modification activity of the enzyme (Ny et al., 1988). Regulation of the activity of the mammalian Tgt (Q34) transglycosylase (Fig. 1P) by activated protein kinase C is described below in the section on eukaryotic modification.

Another crucial consideration is whether tRNA modification enzymes are inhibited by nonsubstrate tRNA species. The issue of what positive and negative determinants in tRNA allow modification enzymes to distinguish substrates from similarly shaped nonsubstrates is a topic of considerable current interest (see Chapter 8 by Garcia and Goodenough-Lashua). The activities of the TrmD (m¹G37) methyltransferase (Fig. 1M) and the HisT (TruA) (Ψ38, 39, 40) synthase (Fig. 1C) are inhibited by nonsubstrate tRNAs in in vitro assays (Redlak et al., 1997). In contrast, the activities of other tRNA modification enzymes, such as the MiaA (i⁶A37) prenyltransferase (Fig. 1A), do not seem to be affected by nonsubstrate tRNAs (Leung et al., 1997). Modification enzymes that are inhibited by nonsubstrate tRNA species may need to possess more favorable kinetic parameters or be present in greater cellular amounts than those that are not inhibited.

Consideration of the kinetic parameters of other purified tRNA modification enzymes, whose activities are not known to be regulated, suggests that they are not present in excess in vivo. The TrmA (m⁵U54) methyltransferase (Fig. 1H) modifies every eubacterial tRNA species (Gu et al., 1996; Gu and Santi, 1991; Kealey and Santi, 1995). Consequently, TrmA must make approximately 2.5×10^3 modifications per min per cell to keep up with the total tRNA synthesized in bacteria growing with a 54-min doubling time (see Leung et al., 1997, for calculation). However, the k_{cat}^{app} for the TrmA methylation reaction is ≈ 2.9 per min (Gu et al., 1996), which means that ≈ 860 TrmA monomers per cell would need to be present to keep tRNA fully modified with m⁵U54. Estimates based on specific activities suggest that there are only about 200 TrmA molecules per cell in bacteria growing in minimal-glucose medium (Ny et al., 1988). Similarly, the Tgt (preQ₁34) transglycosylase (Fig. 1P) modifies about 25% of total tRNA species (Curnow et al., 1993). Reported values of the k_{cat}^{app} of the Tgt enzyme for guanine exchange into tRNA range from 0.3 to 2.4 per min (Reuter and Ficner, 1995; Curnow and Garcia, 1995). Therefore, assuming that the rate of guanine exchange reflects that of preQ₁34 incorporation, Tgt would need to be present at a moderate to an abundant amount between ≈ 260 and 2,000 monomers per cell to meet this load. Tgt does not appear to be an abundant *E. coli* protein (Okada and Nishimura, 1979). These considerations make the key assumptions that the purified modification enzymes are completely active in vitro and that kinetic parameters determined in vitro can be applied directly to the in vivo environment within growing cells. Nevertheless, these calculations of kinetic capacity support the notion that tRNA modification enzymes are not present in excess in cells so that their amounts or activities might be modulated in cells grown under different conditions.

Prospects

Comprehensive new information about the cellular amounts, kinetics, and regulation of activity of tRNA and rRNA modification enzymes should emerge in the next few years. These studies are being hastened by the ease of purifying large quantities of modification enzymes by affinity-tag chromatography (for example, see Leung et al., 1997; Moore and Poulter, 1997) for antibody production and enzymological characterization. It already seems apparent that some tRNA modification enzymes are not present in excess in bacterial cells and may be maintained in quantities that are just sufficient to fully modify their tRNA substrates. However, it does not follow that bacterial modification enzymes are necessarily present in low cellular amounts, because the activities of some tRNA modification enzymes are inhibited by nonsubstrate tRNA species and by compounds other than their substrates and products.

Diverse Structures and Regulation of Genes That Encode Bacterial tRNA and rRNA Modification Enzymes

The approximately 80 tRNA and rRNA modification genes in *E. coli* and *S. typhimurium* are scattered throughout the bacterial chromosome (Appendix 4 by Leung et al.; Björk 1995a, 1995b, 1996). Some modification genes, such as *trmA* (m^{5}U54 methyltransferase; Fig. 1H), occur as monocistronic operons (Fig. 2; Table 1) (Lindström et al., 1985). Some genes that mediate steps in the same modification pathway, such as *tgt* and *queA* (preQ$_1$34 insertion and Q34 biosynthesis; Fig. 1P and Q), are grouped together into polycistronic operons (Fig. 3; Table 1) (Pogliano and Beckwith, 1994; Slany and Kersten, 1992). However, many modification genes are grouped into multifunctional operons, which have also been referred to as complex operons or superoperons. Some of these multifunctional operons, such as the one containing *trmD* (m^{1}G37 methyltransferase; Fig. 1M), contain other genes involved in translation (ribosome subunits in the case of *trmD*) (Fig. 4; Table 1) (Byström and Björk, 1982; Byström et al., 1983). Other multifunctional operons, such as the ones containing *miaA* (i^{6}A37 tRNA prenyltransferase; Figs. 1A and 5) or *ksgA* (m$^6_2$A1518 and 1519 rRNA dimethyltransferase; Fig. 1D and 6), group together modification genes and genes without obvious direct links to translation, such as *amiB* (cell wall hydrolysis) and *mutL* (DNA mismatch repair) for *miaA* and *pdxA* (pyridoxal 5′-phosphate biosynthesis) for *ksgA* (Table 1) (Connolly and Winkler, 1989, 1991; Roa et al., 1989; Tsui et al., 1996; Tsui and Winkler, 1994; Tsui et al., 1994; van Gemen et al., 1987). The transcriptional and translational organization of these multifunctional operons is often exceedingly complicated compared with those of the *lac* and *trp* operon paradigms. This section starts with a brief summary of the regulation of rRNA and tRNA operons in *E. coli*. Next there are descriptions of tRNA modification genes, especially *trmA* and *tgt-queA*, which are the best candidates for coordinate regulation with stable RNA synthesis. The section continues with the regulation of the complex, multifunctional *trmD* and *miaA* tRNA modification operons. The section ends with regulation of the *ermC* and *ksgA* rRNA modification operons.

Mechanisms that regulate stable RNA synthesis in *E. coli*

An obvious speculation is that the expression of tRNA and rRNA modification genes might be coordinated with that of rRNA and tRNA operons. The simplest version of this hypothesis is that the same transcriptional mechanisms that regulate rRNA and tRNA operons also regulate the expression of RNA modification genes. However, this hypothesis is complicated by the diverse organization of tRNA genes and the many mechanisms that regulate rRNA and tRNA synthesis in enteric bacteria, such as *E. coli*, which for the sake of discussion is again used as a model.

tRNA genes are located in three different arrangements in *E. coli*: as spacers in rRNA operons, in tRNA operons containing one or more tRNA species, and in mixed operons that specify tRNA species and proteins (reviewed by Inokuchi and Yamao, 1995). The transcriptional regulation of rRNA operon expression is reasonably well understood (Fig. 7; reviewed by Gourse et al., 1996). Briefly, rRNA operons contain two promoters. Maximum transcription from the major P1 and minor P2 promoters depends on the presence of Up elements located immediately upstream from the −35 box in the promoters (Fig. 7). The Up elements contact the carboxyl-terminal region of the α subunit of RNA

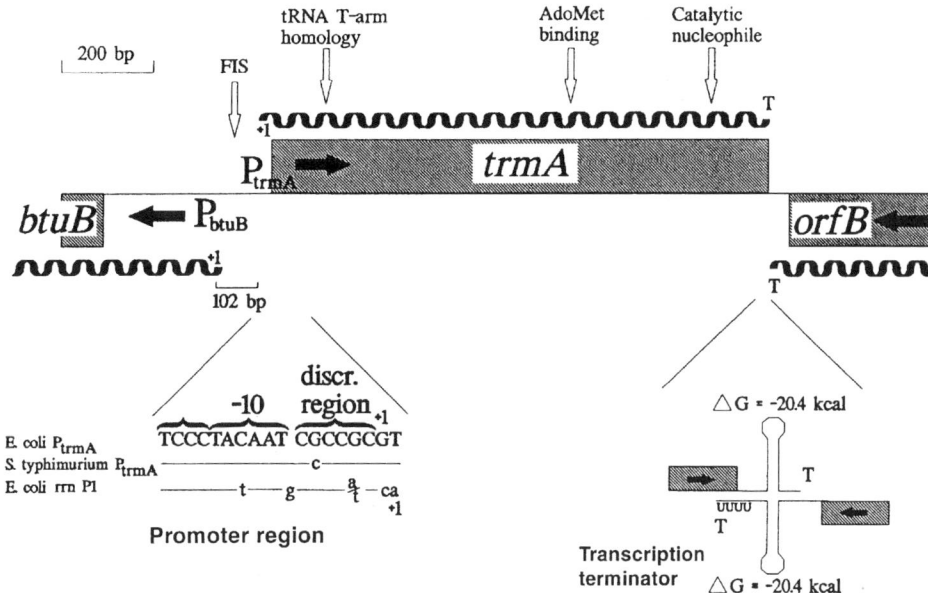

Figure 2. Structure of the monocistronic *trmA* operon (m⁵U54 methyltransferase; Fig. 1H) in *E. coli* (89.65 min) and *S. typhimurium*. The locations of the FIS protein binding site and stringent discriminator in the P_{trmA} promoter and the bifunctional transcription terminator structure at the ends of *trmA* and *orfB* (*yijD*) are indicated. The divergently transcribed *btuB* gene encodes an outer membrane protein involved in vitamin B_{12} uptake. See text for additional details. Adapted from Björk (1995a).

polymerase and thereby stimulate transcription initiation. Transcription from the P1 rRNA promoters is also positively regulated by the binding of the FIS (factor for inversion stimulation) regulatory protein upstream from the Up elements (Fig. 7) (Finkel and Johnson, 1992). Transcription of the seven rRNA operons of *E. coli* is downregulated by the stringent response during limitation for amino acids. Stringent downregulation, which is mediated directly or indirectly by the nucleotide ppGpp, has been correlated with a specific discriminator sequence located between the −10 and +1 positions of rRNA promoters (see consensus in Fig. 2) (Gourse et al., 1996; Inokuchi and Yamao, 1995). Transcription of rRNA operons is also positively regulated in cells growing with increasing growth rates over certain ranges. Neither stringent nor growth rate control depends on FIS binding or the Up elements. Growth rate control involves the core P1 promoter, especially the −35 region, and likely is brought about by a feedback mechanism that coordinates rRNA synthesis with translation capacity. Finally, transcription of the long rRNA operon primary precursor transcripts requires antitermination mechanisms that operate after transcription has initiated (Fig. 7) (Gourse et al., 1996).

In contrast to rRNA operons, FIS protein directly or indirectly mediates positive growth rate regulation of some, but not all, of the tRNA species

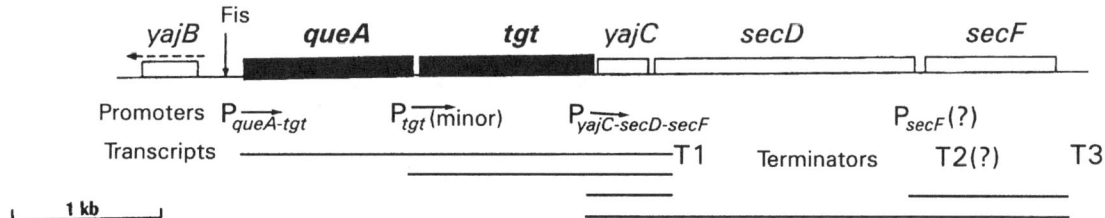

Figure 3. Structure of the *E. coli* (9.17 min) and *S. typhimurium queA-tgt* locus, which catalyzes Q34 biosynthesis (Fig. 1P). *tgt* encodes the preQ₁34 transglycosylase that inserts preQ₁ into tRNA (Fig. 1Q), and *queA* encodes epoxy-queuosine synthase, which catalyzes a step in the conversion of the inserted preQ₁34 to Q34 (see Slany and Kersten, 1992). The locations of promoters and terminators are indicated along with transcripts detected on Northern blots (Pogliano and Beckwith, 1994; Slany and Kersten, 1992). The relationship between the *queA-tgt* operon and the downstream *yajC, secD,* and *secF* genes is discussed in the text. Adapted from Björk (1995a), based on data from Pogliano and Beckwith (1994) and Slany and Kersten (1992).

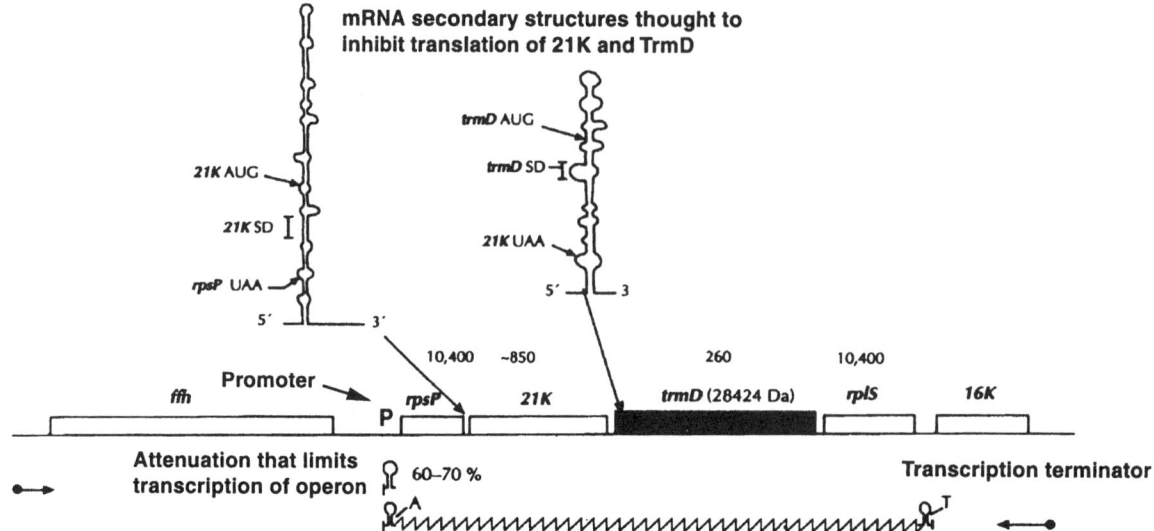

Figure 4. Structure of the multifunctional *trmD* operon (m¹G37 methyltransferase; Fig. 1M) of *E. coli* (59.10 min) and *S. typhimurium*. The four genes of the operon *rpsP* (ribosomal protein S16), *21K* (function unknown), *trmD*, and *rplS* (ribosomal protein L19) are transcribed into the single long mRNA indicated. The locations of the promoter (P), Rho-factor independent attenuator before the first (*rpsP*) gene, and terminator (T) at the end of the operon are indicated along with folded secondary structures in the mRNA that are thought to inhibit translation of the 21K and TrmD proteins. Control of the adjacent *ffh* (protein component of signal recognition protein) and *yfiB* (*16K*) (nonessential protein) genes is separate from that of the *trmD* operon. See text for additional details. Adapted from Björk (1995a).

located outside of rRNA operons (Emilsson and Nilsson, 1995; Nilsson and Emilsson, 1994). This pattern of FIS stimulation was most obvious in bacteria growing at their fastest rates (Emilsson and Nilsson, 1995). FIS binding has been detected upstream of the *tyrT* and *thrU-tufB* tRNA operons (see references in Nilsson et al., 1992). On the other hand, accumulation of some tRNA species is unaffected or actually increased in *fis* mutants compared to their *fis*⁺ parent (that is, FIS appeared to act as a negative regulator) (Nilsson and Emilsson, 1994). Thus, it is possible that FIS, which is produced in exponentially growing but not stationary cells (Nilsson et al., 1992), modulates the composition of cellular tRNA pools in bacteria growing at different rates (Emilsson and Nilsson, 1995, Nilsson and Emilsson, 1994). However, not all growth rate control of tRNA operons is mediated by FIS. For example, growth-rate control of *leuV* (tRNA$_1^{Leu}$) transcription seems to require only the core promoter region, but not FIS protein (Bauer et

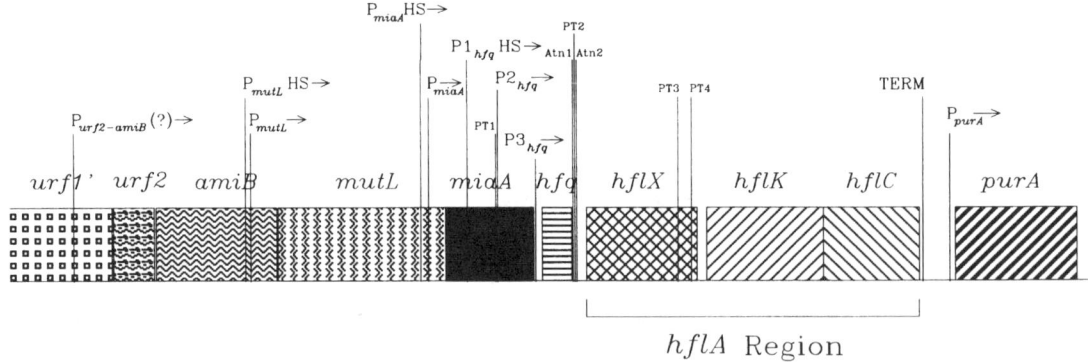

Figure 5. Structure of the multifunctional *miaA* operon (i⁶A37 prenyltransferase; Fig. 1A) in *E. coli* K-12 (94.75 min). The figure is drawn to scale. Besides *miaA*, the multifunctional operon, which has no intercistronic spaces between *urf1*, *urf2*, *amiB*, and *miaA*, includes *amiB* (cell wall amidase), *mutL* (DNA mismatch repair), *hfq* (RNA chaperon global regulator), and the *hflA* region (protease). The locations of multiple standard Eσ⁷⁰-specific promoters (P), heat shock Eσ³²-specific promoters (P-HS), transcript processing sites (PT), transcriptional attenuators (Atn), and the transcription terminator at the end of the operon (TERM) are indicated. The Hfq chaperone is thought to act as a negative regulator of MiaA expression by destabilizing the *miaA* transcript. See text for additional details. Adapted from Tsui et al. (1996).

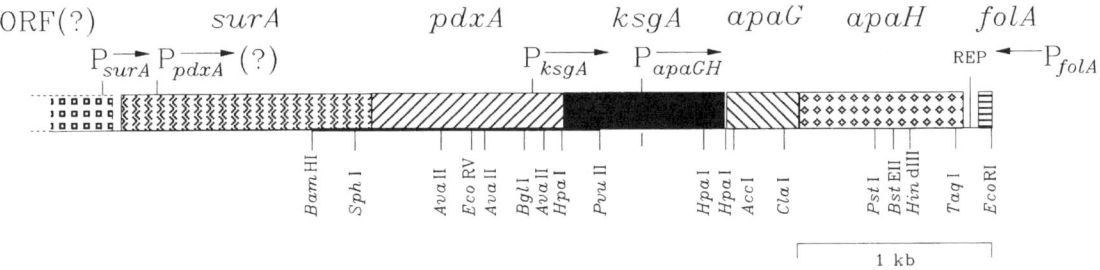

Figure 6. Structure of the multifunctional *ksgA* operon (m$_2^6$A1518 and 1519 dimethyltransferase; Fig. 1D) of *E. coli* (1.11 min). Besides *ksgA*, the operon consists of *surA* (peptidyl-prolyl *cis-trans* isomerase), *pdxA* (pyridoxine 5′-phosphate ring closure), *apaG* (unknown function), and *apaH* (diadenosine [AppppA] tetraphosphatase). The positions of the mapped P$_{ksgA}$ and P$_{apaGH}$ promoters are indicated. The locations of the P$_{surA}$ and P$_{pdxA}$ promoters are approximate. About 50% of *ksgA* transcription originates from promoters upstream from P$_{ksgA}$. See text for other details. Adapted from Roa et al. (1989), based on Tormo et al. (1990) and van Gemen et al. (1987).

al., 1993), as is the case for rRNA P1 promoters (Gourse et al., 1996).

Similar to rRNA, the synthesis of bulk tRNA molecules is stringently controlled (reviewed by Cashel et al., 1996). Many tRNA operon promoters contain discriminator sequences that may play a role in stringent downregulation (Inokuchi and Yamao, 1995; Bauer et al., 1993). Stringent response has been demonstrated directly for several individual tRNA operons (Cashel et al., 1996; Inokuchi and Yamao, 1995). In addition, several tRNA promoters contain upstream regions that are necessary for optimal promoter strength. In the case of the *leuV* (tRNA$_1^{Leu}$) promoter, this upstream region contains an Up element similar to the one found in rRNA P1 promoters (Bauer et al., 1993). Recently, it has become possible to measure the cellular amounts of 44 of the 46 tRNA species in *E. coli* growing at different rates (Dong et al., 1996). These studies revealed an elaborate physiological mechanism that decreases the abundance of all tRNA species normalized to that of ribosomes with increasing growth rates between 0.4 and 2.5 doublings per h. This mechanism leads to a correlation between the biased distribution of tRNA species and the biased usage of corresponding codons (Dong et al., 1996). In addition, some tRNA species are specified by multiple genes, and gene dosage likely plays some role in the mechanism that sets

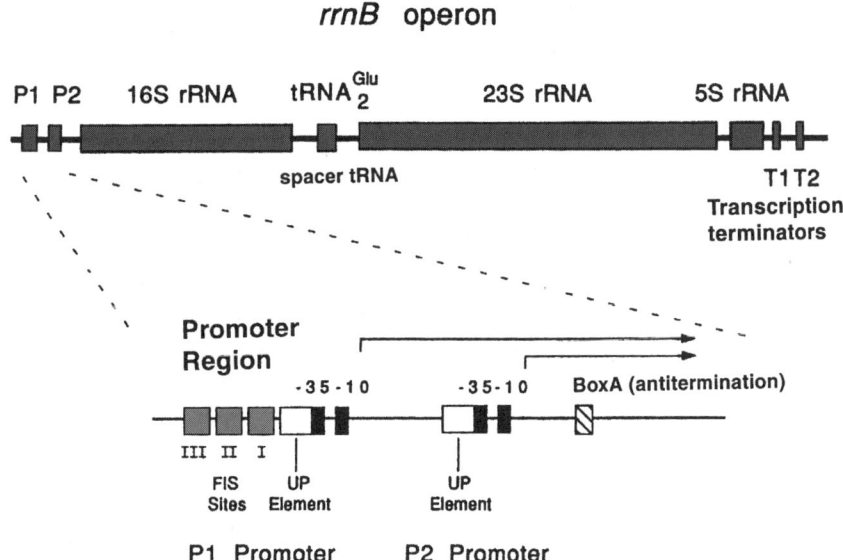

Figure 7. Structure and transcriptional control of the *rrnB* rRNA operon of *E. coli* (89.74 min). The upper figure shows the structure of the entire operon, which extends from the P1 and P2 promoters to the T1 and T2 transcription terminators and includes genes for 16S rRNA, tRNA$_2^{Glu}$, 23S rRNA, and 5S rRNA. The lower figure shows the FIS binding sites and Up elements that contribute to expression from the P1 and P2 promoters and the box A element that functions in transcription antitermination of the nontranslated operon. See text and Gourse et al. (1996) for additional details. Adapted from Gourse et al. (1996).

tRNA amounts at different growth rates (Inokuchi and Yamao, 1995). Finally, it needs to be noted that the tRNA genes of other bacteria are organized differently from those of E. coli and, consequently, may be regulated differently. For example, many of the tRNA genes of gram-positive bacteria, such as Bacillus subtilis and Staphylococcus aureus, are organized in large clusters that are associated with rRNA operons (reviewed by Green and Vold, 1993).

Coordinate control of modification gene expression with stable RNA synthesis

Of the approximately 80 tRNA and rRNA modification genes in E. coli and S. typhimurium, growth rate and stringent control have been demonstrated for only one, trmA (m^{5}U54 methyltransferase [Fig. 1H; Table 1]) (Gustafsson et al., 1991). The trmA gene (Fig. 2) and the hisT (truA) (Ψ38, 39, 40 synthase [Fig. 1C]) genes are the only tRNA modification genes currently known whose transcript and polypeptide amounts and enzyme activities are positively regulated by growth rate (Tsui et al., 1991). The trmA promoter is preceded by a putative FIS binding site (Fig. 2), but it is not known whether FIS regulates trmA transcription in vitro or in vivo. The amount of transcript and polypeptide produced from the trmD gene (m^{1}G37 methyltransferase; Fig. 1M) increases with cell growth rate (Fig. 4) (Wikström and Björk, 1988); however, curiously, the TrmD (m^{1}G37) methyltransferase activity does not correspondingly increase (Table 1) (Ny and Björk, 1980). This result implies that trmD gene expression is regulated positively by growth rate, but the enzyme activity is regulated negatively so that nearly the same enzyme specific activity is maintained under all growth conditions. The nature of this compensatory negative regulation is unknown. The main promoter of the miaA gene (i^{6}A37 prenyltransferase [Fig. 1A and 5]) is preceded by several putative FIS binding sites (Tsui and Winkler, 1994); however, the cellular amount of MiaA protein was the same in a fis null mutant and its fis$^+$ parent (Table 1) (Leung and Winkler, 1997). It is not yet known whether a compensatory mechanism, such as the one operating at rRNA P1 promoters (Gourse et al., 1996), increases miaA transcription in fis mutants.

The major promoter region of the queA-tgt operon (Q34 biosynthesis [Fig. 1P and Q and 3]) was shown by band shift and footprinting assays to bind FIS protein strongly in vitro (Table 1) (Slany and Kersten, 1992); however, it has not been established whether queA-tgt operon expression is positively regulated by growth rate. FIS protein stimulated expression of a queA-lacZ transcriptional fusion by a modest 2-fold in vivo (Slany and Kersten, 1992); however, it is not known whether the amounts of the actual queA-tgt transcript and gene products are stimulated by FIS in vivo. The tgt gene is transcribed from a minor internal promoter that occurs at the end of the queA reading frame, but this minor internal promoter may contribute less than 10% of tgt transcripts in vivo (Fig. 3) (Slany and Kersten, 1992). Initially, it was thought that queA and tgt were cotranscribed with the downstream yajC and secD reading frames (Reuter et al., 1991; Slany and Kersten, 1992), which encode a dispensable membrane protein of unknown function and a protein required for efficient protein export, respectively (Fig. 3) (Pogliano and Beckwith, 1994). However, further structural characterization of this locus showed the presence of individual queA-tgt-yajC, yajC, and yajC-secD-secF transcripts (Pogliano and Beckwith, 1994). Thus, queA and tgt are cotranscribed with yajC, but not secD and secF, and the majority of yajC transcripts arise from an internal promoter located near the end of the tgt reading frame (Fig. 3). Multifunctional operons are considered later in this chapter in the sections on the trmD, miaA, and ksgA operons.

Besides growth rate control, trmA is the only modification gene known thus far to be stringently controlled (Table 1) (Gustafsson et al., 1991; Ny and Björk, 1977). This control was detected at the level of enzyme activity (Ny and Björk, 1977). Although corresponding downregulation of trmA transcript amount was not demonstrated, it seems likely because chloramphenicol addition to cultures prevented stringent control, implying a dependence on translation of trmA mRNA (Ny and Björk, 1977). The trmA promoter region contains a discriminator region similar to those found in stringently controlled genes (Gustafsson et al., 1991). Promoters that drive queA-tgt, miaA, and hisT (truA) transcription contain discriminators (Table 1) (Slany and Kersten, 1992), but it is unknown whether these genes are stringently controlled. As with growth rate control, the regulation of the trmD gene is somewhat paradoxical with respect to stringent control. trmD is in a multifunctional operon whose transcription is stringently controlled (Fig. 4) (Wikström and Björk, 1988). Yet the rate of synthesis and enzyme activity of the TrmD (m^{1}G37) methyltransferase did not change during amino acid limitation (Table 1) (Wikström and Björk, 1988). Again, there appears to be a compensatory mechanism that maintains TrmD (m^{1}G37) methyltransferase activity at the same level under different physiological conditions. Finally, the activities of certain other modification genes, such as trmC, are not stringently controlled (Table 1) (Ny and Björk, 1977).

In summary, the notion that the same mechanisms regulate stable RNA and modification enzyme synthesis is attractive, but largely untested except for *trmA*. Even for *trmA*, data about mechanisms are incomplete; for most other modification genes, there are simply no data about regulation. Moreover, there is an additional complication for monocistronic *trmA*. *trmA* is required for *E. coli* viability, but this essential function does not seem to be m^5U54 methylation (Persson et al., 1992). Thus, it is possible that the essential function of *trmA* may not be related to stable RNA modification or translation and needs to be regulated in ways that are not coordinated with stable RNA synthesis. Likewise, *trmD* is suspected to have a second function not related to m^1G37 methylation (Persson et al., 1995), which may partly account for the lack of growth rate and stringent control.

Noncoordinate regulation of the TrmD (m^1G37) methyltransferase with ribosomal proteins S16 and L19

The *trmD* gene is located in the middle of a multifunctional operon that contains the *rpsP* and *rplS* genes, which encode ribosomal proteins S16 and L19, and the *21K* gene, which encodes a 21-kDa polypeptide of unknown function (Fig. 4) (Byström and Björk, 1982; Byström et al., 1983; Persson et al., 1995). The structure and regulation of the *trmD* operon are notable for several properties. The four genes are transcribed into a single 2,100-nucleotide (nt) transcript in the order *rpsP-21K-trmD-rplS* (Fig. 4) (Byström et al., 1989). No internal promoters or transcript processing sites were detected over the length of the operon, and different segments of the 2,100-nt transcript decayed with comparable chemical half-lives of around 4 min (Byström et al., 1989). A Rho-factor independent transcription terminator is located between the operon promoter and the first (*rpsP*) coding region; this terminator reduces the full-length transcripts synthesized in vitro by about 67%, but it was not determined whether this terminator functions as an attenuator in vivo (Byström et al., 1989). The amount of *trmD* operon transcript is positively regulated by growth rate and negatively regulated by the stringent response in bacteria limited for amino acids or amino-acid-charged tRNA (Byström et al., 1989).

Expression of the *21K* and *trmD* genes is strongly differentially regulated at the posttranscriptional level, compared to that of the two ribosomal proteins. In exponentially growing bacteria, the ribosomal proteins are present in about 10,000 monomers per cell, whereas 21K and TrmD are present in 850 and 260 monomers per gene equivalent, 12- and 40-fold differences, respectively (Wikström and Björk, 1988). Part of this differential regulation may simply reflect less efficient ribosome binding sites of the *21K* and *trmD* genes compared to those of the *rpsP* and *rplS* genes. Several other possible hypotheses were ruled out for this differential regulation. The decreased amounts of 21K and TrmD proteins cannot be accounted for by differences in protein stabilities (Wikström and Björk, 1989), dropoff of translating ribosomes, or decreased translation of rare codons (Wikström et al., 1992). In addition, fusion studies pinpointed a negative control element quite far (50–125 nt) into the 21K protein reading frame (Wikström and Björk, 1989; Wikström et al., 1992). This segment of mRNA was postulated to form an extensive paired secondary structure that blocks the upstream Shine-Dalgarno and AUG start codon of the *21K* gene (Fig. 4) (Wikström and Björk, 1989; Wikström et al., 1992). Extensive mutagenesis of this region relieved the block in translational initiation by creating alternative secondary structures that freed the *21K* ribosome binding site (Wikström et al., 1992). These studies further demonstrated that the translation stop codon of *rpsP* is sufficiently upstream so that terminating ribosomes do not strongly disrupt the secondary structure blocking *21K* translation (Wikström et al., 1992). However, moving the *rpsP* stop codon closer to the *21K* start codon increased expression of *21K-lacZ* translational fusions, and combinations of mutations that altered RNA secondary structures and moved the *rpsP* stop codon closer increased *21K* expression by as much as 150-fold (Wikström et al., 1992).

Less-detailed experiments suggested that a similar negative translational element decreases *trmD* expression relative to that of *rpsP*, *21K*, and *rplS* (Wikström et al., 1992). The negative regulation of *trmD* translation seems to be independent from that of the *21K* gene, since *21K* Up mutations did not lead to increased *trmD* translation (Wikström et al., 1992). The negative regulator of *trmD* translation is weaker than that of *21K*, and it is likely that a large factor in the relatively low translation of *trmD* is a poor ribosome binding site. Finally, transcription of the *trmD* operon is negatively regulated by stringent response (Byström et al., 1989). The rate of synthesis of the S16 and L19 ribosomal proteins drops in parallel to the decrease in transcription; however, the rates of synthesis and amounts of the 21K and TrmD (m^1G37) methyltransferase remain unaffected by stringent control (Wikström and Björk, 1988). The posttranscriptional mechanism that maintains 21K and TrmD protein amounts during amino acid limitation is unknown.

miaA and the organization and function of complex multifunctional operons in E. coli

The *miaA* gene (i^{6}A37 prenyltransferase; Fig. 1A) is in an exceedingly complicated multifunctional operon that includes *amiB*, *mutL*, *hfq*, and the *hflA* region (Fig. 5). The *amiB* gene encodes an amidase involved in cell wall hydrolysis (Tsui et al., 1994). The end of the *amiB* reading frame overlaps that of *mutL*, which plays crucial roles in DNA mismatch repair (Fig. 5) (Connolly and Winkler, 1991; Tsui et al., 1994). Likewise, there is no intercistronic space between *mutL* and *miaA* (Connolly and Winkler, 1991; Tsui et al., 1994). Downstream of *miaA* is *hfq*, which encodes a recently discovered RNA binding protein that acts as a pleiotropic regulator (Tsui et al., 1996, 1994, 1997). There is a short intercistronic space of 86 bp between *miaA* and *hfq*. *hfq* is followed in the operon by the three-gene *hflA* region (*hflX*, *hflK*, and *hflC*), which encodes a protease that inhibits formation of bacteriophage lambda lysogens (Tsui et al., 1996, 1994). The operon does not continue beyond *hflC* (Tsui et al., 1996, 1994).

Because spaces between coding regions do not exist at many places, promoters are found in the reading frames of upstream genes throughout this multifunctional operon (Fig. 5). To accommodate this gene arrangement, the apparent promoter strengths increase from the beginning to the end of the operon (Tsui et al., 1996; Tsui and Winkler, 1994; Tsui et al., 1994). *miaA* is transcribed from two promoters (P$_{miaA}$ and P$_{miaA}$HS) and as a cotranscript with the upstream *mutL* repair gene (Tsui et al., 1996; Tsui and Winkler, 1994). P$_{miaA}$ is a relatively strong promoter that is transcribed by the standard Eσ^{70} RNA polymerase (Tsui et al., 1996; Tsui and Winkler, 1994). However, transcripts from the Eσ^{70}-dependent P$_{miaA}$ promoter have the unusual property that they increase significantly in amount in cells subjected to heat shock (Tsui et al., 1996). P$_{miaA}$HS is transcribed by Eσ^{32} RNA polymerase during heat shock, but much less transcript seems to originate at P$_{miaA}$HS than P$_{miaA}$ (Tsui et al., 1996). The continued transcription of *miaA* suggested that *miaA* function might be required at higher physiological temperatures. This expectation was confirmed by the finding that *miaA* mutants grown at 37°C in rich medium fail to form colonies at 44°C on plates containing rich medium (Tsui et al., 1996). Thus, under this condition, *miaA* function is essential for bacterial viability. Suggestively, there are indications that rRNA P1 promoters may also be regulated by heat shock (Gourse et al., 1996).

The *miaA* reading frame contains two additional promoters (P1$_{hfq}$HS and P2$_{hfq}$) that drive the downstream *hfq* and *hflA*-region genes (Fig. 5) (Tsui et al., 1996). A third promoter (P3$_{hfq}$) is located immediately downstream of *miaA* (Fig. 5) (Tsui et al., 1996). P1$_{hfq}$HS is transcribed by Eσ^{32} RNA polymerase in cells grown at all temperatures, and the amount of P1$_{hfq}$HS transcript increases upon heat shock. P2$_{hfq}$ and P3$_{hfq}$ are transcribed by standard Eσ^{70} RNA polymerase. The *hfq* and *hflA*-region transcripts are processed in vivo at several discrete sites (PT1 to 4 [Fig. 5]), and two transcription attenuators located between *hfq* and the *hflA* region reduce *hflA*-region transcription significantly (Atn 1 and 2 [Fig. 5]).

An unusual feature of all the promoters in the *mutL-miaA-hfq* multifunctional operon is that they occur far upstream of the genes they transcribe (Fig. 5). This arrangement means that each transcript from the operon contains a relatively long leader region. For example, transcription from P$_{miaA}$, P$_{miaA}$HS, P1$_{hfq}$HS, and P2$_{hfq}$ result in leaders of 271 nt, 201 nt, 891 nt, and 488 nt, respectively, before the starts of the *miaA* or *hfq* coding regions (Tsui et al., 1996; Tsui and Winkler, 1994). Whether these long leaders play regulatory roles or are themselves translated is presently unknown. However, there are strong precedents for transcript leader regions playing roles in attenuation and other regulatory mechanisms (see references in Tsui and Winkler, 1994).

Why complex multifunctional operons exist in enteric bacteria, such as *E. coli* and *S. typhimurium*, is presently the topic of considerable speculation. It is possible that such gene arrangements are merely accidental. However, the *amiB-mutL-miaA* grouping is conserved in *Haemophilus influenzae* (Fleischmann et al., 1995). *hfq* is located apart from *miaA* in *H. influenzae*, but this arrangement may reflect a different role in global regulation played by Hfq protein in *E. coli* and *H. influenzae* (Fleischmann et al., 1995). For example, Hfq is thought to act as a positive regulator of translation of the RpoS (stationary-phase) sigma factor in *E. coli* (Muffler et al., 1996); however, *H. influenzae* does not seem to contain an *rpoS* gene (Fleischmann et al., 1995). It is unlikely that *miaA* is grouped with these other genes according to function, because none of the other genes in the operon is directly involved in modification, tRNA maturation, or translation.

Coregulation is a more compelling reason for the grouping of *miaA* with these other genes in *E. coli*. One reason to group *miaA* and *hfq* together in *E. coli* (Fig. 5) may be to allow regulation of MiaA cellular amount by the Hfq pleiotropic regulatory protein. Hfq is an RNA binding protein that has been likened to a chaperone (Muffler et al., 1997). The Hfq protein autoregulates its own expression at the level of mRNA stability of transcripts initiated at the P$_{miaA}$

and P1$_{hfq}$HS promoters (Tsui et al., 1997). At the same time, Hfq negatively regulates cellular MiaA amounts (Tsui et al., 1997). In exponentially growing or stationary-phase cells, an *hfq* null mutant contains about 2.5- or 6-fold more MiaA, respectively, than its *hfq*$^+$ parent (Tsui et al., 1997). Lack of functional Hfq completely reverses the 2- to 3-fold drop in MiaA amount in cells that enter stationary phase (Tsui et al., 1997). Downregulation of MiaA amount correlates with the decrease in tRNA synthesized as cells enter stationary phase (Huisman et al., 1996; Leung et al., 1997). Hfq may decrease the stability of the P$_{miaA}$ and P1$_{hfq}$ transcripts by blocking translation and thereby exposing normally protected mRNA regions to nucleases or by directly presenting the mRNA to nucleases.

A second reason for grouping *miaA* between *mutL* and *hfq* may be structural. On the one hand, *mutL* is really a separate gene in a different operon from *miaA*, because so much more *miaA* transcript originates from P$_{miaA}$ than by cotranscription from P$_{mutL}$ (Fig. 5) (Tsui et al., 1996; Tsui and Winkler, 1994; Tsui et al., 1994). On the other hand, *mutL* and *miaA* are very much members of a classical operon, because every *mutL* transcript contains *miaA* as a downstream cis-acting structural element. Likewise, every *miaA* transcript contains *hfq*, even though *hfq* itself is expressed independently. The cotranscripts that result from such arrangements may possess certain stability features that ensure proper expression of the upstream gene. Another important related, but unresolved, issue is whether MiaA and Hfq are expressed with equal efficiency from transcripts that initiate at different promoters.

Considerably more work needs to be done to discover other regulators of *miaA* expression. FIS (Table 1) and other global regulators, including the HN-S and IHF nucleoid-associated proteins and RpoS (sigma factor S), did not influence the cellular amount of MiaA in exponentially growing cells (Leung and Winkler, 1997). Genetic approaches are in progress using transcriptional and translational fusions between *miaA* and *lacZ* to screen for other regulators of *miaA*. Finally, the possible role of mRNA processing in regulating *miaA* expression needs to be analyzed further. Transcripts initiated at P$_{miaA}$ appear to be processed by RNase E, which is a major essential endonuclease that starts the breakdown of many different mRNA species and processes rRNA transcripts in *E. coli* (see references in Tsui and Winkler, 1994). The amount of P$_{miaA}$ transcript was increased about 10-fold in an *rnaE*(Ts) mutant shifted to a nonpermissive temperature compared to the *rnaE*$^+$ parent (Tsui and Winkler, 1994). This result suggests a model in which RNase availability may regulate the amount of MiaA during growth transitions. For example, when bacteria are upshifted from poor to rich medium, there is a burst of rRNA and tRNA transcription (see Tsui and Winkler, 1994). The increased amount of pre-rRNA will begin to titrate away the cellular amount of RNase E, which is autoregulated by cleaving its own transcript. A temporary drop in the amount of free RNase E may increase the amount of *miaA* transcript and thereby increase the amount of MiaA when cells are increasing their amounts of tRNA substrates. This model has not yet been directly tested, and RNA processing has not yet been examined for other genes that encode tRNA and rRNA modification enzymes.

ermC and *ksgA* and possible feedback regulation of rRNA modification genes

The regulation of the ErmC dimethyltransferase, which forms m$_2^6$A2058 (Fig. 1D) in 23S rRNA in *B. subtilis* and other gram-positive bacteria, has been studied extensively (Fig. 8) (reviewed by Bechhofer 1990; de Smit and van Duin, 1990; Vellanoweth, 1993). In the absence of erythromycin, the amount of ErmC protein is negligible and A2058 is present in 23S rRNA. Upon exposure to erythromycin, the amount of ErmC is strongly induced, and A2058 is modified to m$_2^6$A2058, which renders translation resistant to the antibiotic. Transcription of *ermC* is thought to be constitutive and does not change when erythromycin is present (see Bechhofer, 1990; Vellanoweth, 1993). Induction of ErmC expression occurs by an antiattenuation mechanism (Gryczan et al., 1980; Horinouchi and Weisblum, 1980). The *ermC* transcript contains a leader segment that precedes the *ermC* reading frame, and the *ermC* leader encodes a small polypeptide (Fig. 8). In the absence of erythromycin, this translated leader RNA folds into a secondary structure that sequesters the *ermC* ribosome binding site and effectively prevents translation of *ermC* (upper panels of Fig. 8). Erythromycin causes the ribosome to stall during translation of the leader peptide. The stalled ribosome blocks formation of the inhibiting RNA secondary structures and allows translation initiation of ErmC (lower panel of Fig. 8). The stalled ribosome also is thought to increase the stability of the *ermC* transcript, thereby amplifying expression of the *ermC* gene (see Bechhofer, 1990; Vellanoweth, 1993).

There are two mechanisms that autoregulate *ermC* expression. The attenuation mechanism itself is necessarily autoregulating. As erythromycin-resistant ribosomes accumulate, they no longer stall in the *ermC* leader, attenuation is restored, and expression of ErmC is reduced (Narayanan and Dubnau, 1987).

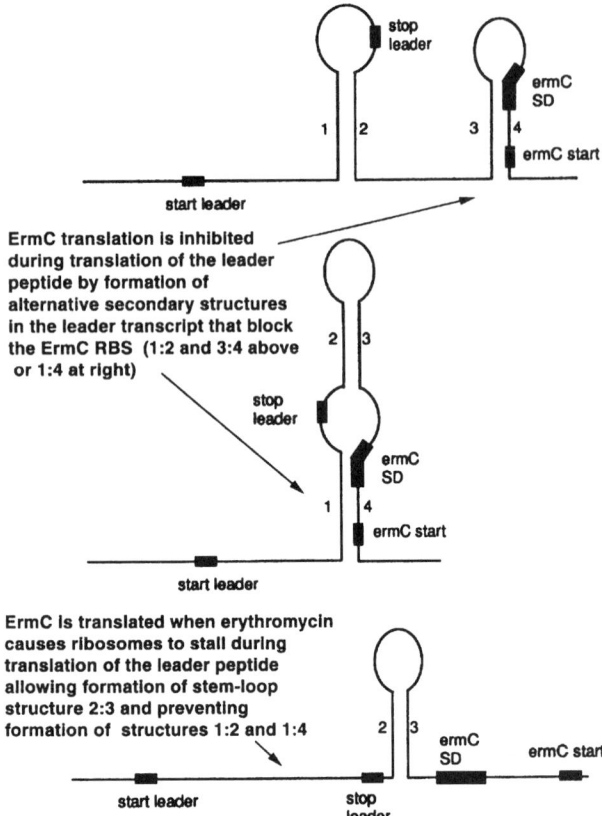

Figure 8. Antiattenuation model for induction of *ermC* operon (m$_2^6$A2058 dimethyltransferase; Fig. 1D) expression by the antibiotic erythromycin in *B. subtilis* and other gram-positive bacteria. The top two alternative mRNA secondary structures, which sequester the *ermC* ribosome binding site (RBS), form when the *ermC* leader transcript is translated in the absence of erythromycin. The bottom structure, which allows *ermC* translation, forms when ribosomes stall during translation of the leader peptide in the presence of erythromycin. See text for additional details. Adapted from Vellanoweth (1993). Start leader, translation start codon of leader peptide; stop leader, translation stop codon of leader peptide; *ermC* SD, Shine-Dalgarno sequence preceding *ermC*; *ermC* start, *ermC* translation start codon.

In addition, the ErmC protein binds near the 5'-end of the *ermC* transcript and acts as a translational repressor of its own expression (Bechhofer, 1990; Breidt and Dubnau, 1990; Vellanoweth, 1993). Translational repression is common for operons that encode ribosomal proteins (reviewed by Keener and Nomura, 1996), and it is noteworthy that it also occurs for this modification gene.

There is some evidence that expression of the *ksgA* gene is also translationally autoregulated. *ksgA* encodes a second dimethyltransferase that adds m$_2^6$A to positions 1518 and 1519 near the 3'-end of 16S rRNA (Fig. 1D; Table 1). *ksgA* is located in a complex multifunctional operon that includes *surA* (peptidyl-prolyl *cis-trans* isomerase), *pdxA* (pyridoxine 5'-phosphate ring closure), *apaG* (unknown function), and *apaH* [diadenosine (AppppA) tetraphosphatase] (Fig. 6) (Blanchin-Roland et al., 1986; Roa et al., 1989; Rouviere and Gross, 1996; Tormo et al., 1990; van Gemen et al., 1987, 1989). About 50% of *ksgA* transcripts are cotranscripts with *surA-pdxA*; the other 50% of *ksgA* transcripts begin at the P$_{ksgA}$ promoter located within *pdxA* about 150 bp upstream of the start of the *ksgA* reading frame (Fig. 6) (Roa et al., 1989). This relatively long *ksgA* leader may play a regulatory role in translational coupling, as discussed in the following section. New results suggest that *ksgA* transcription is positively regulated by growth rate by increases in the amounts of both the *surA-pdxA-ksgA* cotranscript and the transcript initiated at the P$_{ksgA}$ internal promoter (Pease et al., submitted for publication). Growth rate control of *surA-pdxA-ksgA* cotranscription is regulated by FIS, whereas transcription from P$_{ksgA}$ is not (Pease et al., submitted for publication). Thus, the total amount of *ksgA* transcript is induced by FIS, but the effect is dampened by the lack of response of the P$_{ksgA}$ promoter. Likewise, *surA-pdxA-ksgA* cotranscription is downregulated in a ppGpp-independent manner by amino acid limitation; however, P$_{ksgA}$ transcription does not respond to amino acid limitation (Pease et al., 1997), again dampening the overall decrease on *ksgA* transcription.

Preliminary evidence suggests that translation of KsgA is coupled to translation of the upstream reading frame whose stop codon overlaps the KsgA start codon (van Gemen et al., 1987). This reading frame turned out to be *pdxA* in the cotranscript (Fig. 6) (Roa et al., 1989). Thus, if translational coupling is required for KsgA expression from transcripts initiated at P$_{ksgA}$, then there must be translation of a peptide from the P$_{ksgA}$ leader transcript. Translational coupling may be a feature of other tRNA and rRNA modification genes. Preliminary experiments suggested that expression of the *hisT* (*truA*) gene (Ψ38, 39, 40 synthase; Fig. 1C) requires translational coupling in minicells (Arps et al., 1985; Marvel et al., 1985). Overlap between the *mutL* and *miaA* reading frames has also been noted (Fig. 5) (Connolly and Winkler, 1991; Tsui et al., 1994).

Experiments using a *ksgA-lacZ* translational fusion suggested that the amount of KsgA is autoregulated by translational repression (van Gemen et al., 1989). This conclusion was partly corroborated by experiments showing that KsgA synthesized in a coupled in vitro transcription–translation system is reduced by addition of purified unlabeled KsgA protein. Furthermore, KsgA was shown to bind specifically to an RNA fragment that contains the *ksgA* ribosome binding site and to an RNA fragment containing the segment normally modified in 16S rRNA (van Gemen et al., 1989). Further analysis of *ksgA* autoregulation has not been reported.

Prospects

The structure, expression, and regulation have been investigated for only a handful of the many tRNA and rRNA modification genes that occupy as much as 1% of the genetic material in *E. coli* and *S. typhimurium* (Björk, 1995a, 1995b, 1996). A systematic study of the regulation of bacterial modification genes is long overdue. As a group, the bacterial modification genes are diverse (Table 1; Fig. 2 to 8). The expression of some modification genes seems to be coordinated with the synthesis of stable RNA, whereas some are not coordinated (Table 1). Thus, the regulation of tRNA and rRNA modification genes will likely fall into groups without a single general mechanism. Moreover, several different mechanisms may be employed to achieve the same patterns of regulation for different modification genes. Finally, concerted study of these unusual genes and operons should reveal new essential cellular functions, indicate unique modes of gene regulation, and offer insights into the evolution of operons and chromosome structure in bacteria.

tRNA AND rRNA MODIFICATION LEVELS AS A REGULATORY DEVICE

An attractive hypothesis is that tRNA and rRNA modification levels might change in bacteria subjected to certain growth or stress conditions (Persson, 1993). Altered modification levels could then affect the expression of genes that are controlled by translational mechanisms, such as attenuation (Landick et al., 1996; Winkler, 1996). With three exceptions, there currently is no strong evidence that stable RNA modification levels change in wild-type bacteria in a concerted fashion to elicit regulation. The emphasis of this assertion is on wild type, since mutations in modification genes or in biosynthetic genes that influence modification levels show a variety of marked phenotypes (Chapter 26 by Björk and Rasmuson; Björk 1995a, 1995b, 1996). The two regulatory issues are whether tRNA undermodification occurs in wild-type bacteria and whether the tRNA undermodification that does occur causes significant changes in the regulation of specific genes. The three exceptions are methylation of 23S rRNA by ErmC in gram-positive bacteria (above), ms²-methylthiolation of i⁶A37 in *E. coli* and *S. typhimurium* (Buck and Griffiths, 1982; Esberg and Björk, 1995), and o⁶-hydroxylation of ms²i⁶A37 in *S. typhimurium*, but not *E. coli* (Persson, 1993; Persson and Björk, 1993). The amount of ErmC methyltransferase is regulated by antiattenuation and autoregulation so that m$_2^6$A2058 (Fig. 1D) will be present in 23S rRNA only when cells are exposed to erythromycin (Table 1; Fig. 8) (Vellanoweth, 1993). Methylthiolation of i⁶A37 to form ms²i⁶A37 (Fig. 1A and 1B) is catalyzed by the MiaB and MiaC enzyme activities (Buck and Griffiths, 1982; Esberg and Björk, 1995). The MiaB thiotransferase requires iron for activity, and even moderate limitation of *E. coli* and *S. typhimurium* for iron, such as growth in MOPS minimal medium lacking trace elements, leads to i⁶A37 undermodification (Buck and Griffiths, 1982; Esberg and Björk, 1995; Winkler, 1997). i⁶A37 undermodification has been correlated with a modest increase in the biosynthesis of the iron chelator enterochelin (Buck and Ames, 1984), but the physiological role, if any, of i⁶A37 undermodification in iron-limited cells awaits further investigation. The MiaE hydroxyltransferase forms ms²io⁶A37 (Fig. 1I) from ms²i⁶A37 (Fig. 1B) in *S. typhimurium* grown in the presence of molecular oxygen. Although *miaE* mutants are not slowed in transitions between aerobic and anaerobic growth (Persson, 1993; Persson and Björk, 1993), lack of ms²io⁶A37 impairs intermediary metabolism in *S. typhimurium* (Persson, 1993; Persson and Björk, 1993). Curiously, *E. coli*, which lacks the MiaE enzyme, does not show similarly impaired metabolism and must regulate certain key metabolic genes differently than *S. typhimurium*.

Several other early reports suggested changes in tRNA modification levels in stressed bacteria (Chase et al., 1974; Kitchingman and Fournier, 1977; Thomale and Nass, 1978; Turnbough et al., 1979). For the most part, these observations have not been pursued further. In addition, any apparent changes in the amounts of modified bases resolved by HPLC from cells grown under different conditions must be interpreted cautiously because of the complex physiological regulation of individual tRNA species (Dong et al., 1996; Emilsson and Nilsson, 1995; Nilsson and Emilsson, 1994). For example, the relative amount of t⁶A37 (Fig. 1R) in tRNA increases as *E. coli* cells progress from early exponential to stationary phase in minimal-salts-glucose medium (Winkler, 1997). This observation could explain the apparent discrepancy that *S. typhimurium* cells seemed to contain less t⁶A37 than *E. coli* cells (Buck et al., 1983); these comparisons were done with tRNA preparations from exponentially growing *S. typhimurium* and stationary-phase *E. coli*. However, it is not possible to ascribe the increase in t⁶A37 content to a change in modification level without knowing how the amounts of the tRNA species that contain t⁶A37 change during this growth transition.

Prospects

The hypothesis that tRNA and rRNA modifications act as regulatory devices has not been generally established and needs to be demonstrated convincingly for several model modifications. The undermodification of i^6A37 and ms^2i^6A37 in cells limited for iron and oxygen, respectively, may be the leading candidates to test this hypothesis (Buck and Ames, 1984; Buck and Griffiths, 1982; Persson, 1993; Persson and Björk, 1993). The genes whose expression is affected by these undermodifications need to be identified by two-dimensional gel analyses or genetic methods using random lacZ fusions. After these targets are identified, it will be possible to access whether the magnitude of the effects of undermodification on gene expression are sufficiently large to readily indicate physiological significance.

REGULATION OF tRNA AND rRNA MODIFICATION IN EUKARYOTIC CELLS

The same considerations about enzyme amounts, catalytic capacities, and gene control discussed above for bacteria apply to the regulation of tRNA and rRNA modification in eukaryotic cells. Less is now known about these topics in eukaryotic cells compared to prokaryotes, but this situation is changing rapidly. Moreover, several aspects of tRNA and rRNA modification that occur in eukaryotic but not prokaryotic cells have been studied in detail. A number of eukaryotic tRNA and rRNA modification enzymes are being purified and enzymatically characterized (Chapter 2 by Grosjean et al., Chapter 8 by Garcia and Goodenough-Lashua, Chapter 12 by Ofengand and Fournier, and Chapter 24 by Maden). The involvement of snoRNAs as trans-acting guides in the modification of eukaryotic rRNAs is discussed elsewhere (Chapter 12 by Ofengand and Fournier, Chapter 13 by Bachellerie and Cavaillé, Chapter 14 by Mason, and Chapter 24 by Maden) and is not discussed here.

Several studies suggest that certain tRNA modification enzymes are not present in excess in yeast and some kinds of mammalian cells. Antibodies against peptides have been prepared and used to locate the yeast Mod5p (i^6A37) prenyltransferase (Fig. 1A) and Trm1p (m_2^2G26) dimethyltransferase (Fig. 1G) isozymes in different subcellular compartments. Indirect immunofluorescence and fractionation experiments suggest that the Mod5p and Trm1p modification enzymes are present in low abundance and may not be present in excess (Gillman et al., 1991; Rose et al., 1995). Consistent with this interpretation, overexpression of the Trm1p-II isozyme fails to saturate the sites that localize it to the periphery of the cell nucleus (Rose et al., 1995; Martin and Hopper, 1994). Nonetheless, the amounts of these isozymes in different subcellular compartments have not been reported. Overexpression of initiator $tRNA_i^{Met}$ in a yeast mutant in which the genes for elongator $tRNA_m^{Met}$ were deleted titrated the 2'-O-ribosyl phosphate [Ar(p)64] transferase encoded by RIT1 (Fig. 1S) (Aström and Byström, 1994; Aström et al., 1993). The resulting undermodified $tRNA_i^{Met}$ functioned in both translation initiation and elongation (Aström and Byström, 1994; Aström et al., 1993). Thus, the Rit1p [Ar(p)64] ribosyl phosphate transferase does not seem to be present in excess in yeast. In aerobically grown HeLa cells, insertion of free-Q base (Fig. 1P) into tRNA was incomplete during the most active phase of cell proliferation (Langgut and Reisser, 1995). Sufficient amounts of free-Q base seemed to be transported into these growing cells, and the incomplete modification suggested that tRNA synthesis outpaced insertion of free-Q base by the Tgt (Q34) transglycosylase in this cell line growing under these conditions (Langgut and Reisser, 1995). Comparatively low activity of the Tgt (Q34) transglycosylase may also partly explain the incomplete Q34 modification in tRNA found in rapidly growing tumor cells (see Chapter 26 by Björk and Rasmuson; Dirheimer et al., 1995; Langgut and Reisser, 1995; Morgan et al., 1996; Morris et al., 1995).

A considerable amount is known about the regulation of the enzyme activity of the mammalian Tgt (Q34) transglycosylase (Fig. 1P) by the protein kinase C signal transduction pathway, the sequential modification as tRNA molecules progress from the nucleus to the cytoplasm, and the isozyme sorting of the Mod5p and Trm1p modification enzymes to different subcellular compartments. These three topics are briefly discussed in the remainder of this chapter, with an emphasis on genetic approaches to isozyme sorting in yeast. End maturation and the removal of introns from cytoplasmic and mitochondrial precursor tRNA are beyond the scope of this chapter and are reviewed for yeast by Hopper and Martin (1992). Processing of eukaryotic ribosomal RNA is reviewed by Eichler and Craig (1994). The modification of eukaryotic mRNA is discussed in Chapter 10 by Bokar and Rottman and is not covered here. Chapter 24 by Maden further describes the intracellular localization of the enzymes that modify tRNA, rRNA, mRNA, snRNA, and snoRNA in eukaryotic cells.

Regulation of Mammalian Tgt (Q34) Transglycosylase Enzyme Activity

Free-Q base (Fig. 1P) is provided to mammals as a nutritional requirement instead of being synthesized de novo as in *E. coli* and other bacteria (see Langgut and Reisser, 1995; Morris et al., 1995). Q34 modification by the Tgt transglycosylase has been implicated in cellular differentiation, development, control of energy metabolism, regulation of protein synthesis during reentry into the cell cycle, and hypoxic (low oxygen) stress management (see Chapter 26 by Björk and Rasmuson and references in Langgut and Reisser, 1995; Morris et al., 1995). Q34 hypomodifcation (partial modification) has frequently been found in neoplastic tissues and transformed tumor cell lines (see Chapter 26 by Björk and Rasmuson; Langgut and Reisser, 1995; Morris et al., 1995). Two recent reports suggest an important link between the control of Q34 transglycosylation and the protein kinase C (PKC) signal transduction pathway involved in regulating cell growth (Langgut and Reisser, 1995; Morris et al., 1995). At one level, uptake of free-Q base is positively regulated by the PKC pathway; inducers of PKC activity stimulate free-Q base uptake, whereas inhibitors decrease uptake (Langgut and Reisser, 1995; Morris et al., 1995). Chronic exposure of cultured cells to phorbol esters, which is thought to downregulate PKC activity, resulted in severe Q34 deficiency in tRNA; however, uptake of free-Q base was still sufficiently high to suggest that the activity of the Tgt enzyme itself was regulated by the PKC pathway (Langgut and Reisser, 1995; Morris et al., 1995). This hypothesis was confirmed in HeLa cells grown under hypoxic conditions (Langgut and Reisser, 1995). Measurements of free-Q base and Q34-modified tRNA showed that free-Q base was available in these cells, but was not incorporated into tRNA. Addition of serum factors or polypeptide growth factors known to stimulate PKC activity led to rapid Q34 modification of tRNA in hypoxic HeLa cells (Langgut and Reisser, 1995).

Further studies of purified enzymes revealed that eukaryotic Tgt (Q34) transglycosylase (Fig. 1P) exists as a heterodimer (Morris et al., 1995), whereas the prokaryotic Tgt (preQ$_1$34) transglycosylase (Fig. 1Q) is a monomer (Reuter and Ficner, 1995). Furthermore, PKC directly phosphorylates the regulatory subunit of the mammalian Tgt (Q34) transglycosylase heterodimer (Fig. 9). Unphosphorylated Tgt is an ≈104-kDa heterodimer consisting of an ≈60-kDa regulatory subunit and an ≈35-kDa catalytic subunit (Morris et al., 1995). Phosphorylation of the 60-kDa regulatory subunit is thought to cause dissociation of the heterodimer and release of an active catalytic subunit. Phosphatases present in Tgt enzyme preparations removed the phosphate ester group from the regulatory subunit and decreased Tgt enzyme activity, presumably by allowing heterodimer formation (Morris et al., 1995). Consistent with this model (Fig.

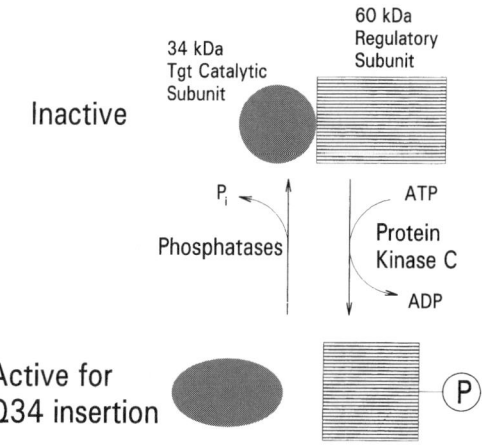

Figure 9. Model for regulation of Tgt (Q34) transglycosylase activity by phosphorylation by protein kinase C (PKC) in mammalian cells. The unphosphorylated form of the Tgt enzyme (top) exists as a heterodimer containing a catalytic and a regulatory subunit. Phosphorylation of the regulatory subunit by PKC (bottom) causes the subunits to dissociate releasing the active catalytic subunit. Based on Morris et al. (1995).

9), Tgt enzyme activity was restored by the addition of PKC and ATP to these preparations (Morris et al., 1995). This restoration of Tgt enzyme activity was blocked or stimulated by inhibitors or activators of PKC, respectively (Morris et al., 1995), which did not directly affect Tgt enzyme activity in the absence of PKC and ATP. Thus, according to this model, phosphatases or factors that decrease PKC activity, such as chronic exposure to phorbol esters, lead to heterodimer formation, inhibition of Tgt (Q34) transglycosylase activity, and hypomodification for Q34 in cellular tRNA (Langgut and Reisser, 1995; Morris et al., 1995). Conversely, stimulation of PKC activity will maintain uptake and full incorporation of free-Q base into tRNA.

Sequential Modification Pathways in Eukaryotic Cells

Sequential modification and cellular location of rRNA, tRNA, and mRNA modifying enzymes in eukaryotic cells are reviewed in Chapter 24 by Maden, and only a few points relevant to the regulation of tRNA modification are made here. An early classic study demonstrated that the modification of eukaryotic tRNA occurs by a sequential pathway (Nishikura and De Robertis, 1981). In this study, a gene specifying an intron-containing tRNA^{Tyr} from yeast was injected into *Xenopus* oocytes, and modification patterns were followed in pre-tRNA^{Tyr} and mature tRNA^{Tyr}. Certain modifications, such as Ψ35 (Fig. 1C), m^{5}C49 (Fig. 1T), m^{5}U54 (Fig. 1H), Ψ55 (Fig. 1C), and m^{1}A (Fig. 1U), were added to the primary tRNA^{Tyr} transcript before end processing or intron removal. Other modifications were added at the same time as end processing. Finally, some modifications, such as Q34 (Fig. 1P) and i^{6}A37 (Fig. 1A), were added only after intron removal. Thus, end processing and intron removal act to order the modification process in eukaryotic cells.

A recent broad analysis of the tRNA modification process led to the formulation of the working hypothesis that tRNA modification enzymes may be roughly grouped into two classes (Grosjean et al., 1996). Class I enzymes modify bases in the amino acid acceptor minihelix, consisting of the m^{5}U54-Ψ55-C and acceptor stem microhelices, and do not tend to depend heavily on tRNA tertiary structure for substrate recognition. Class II enzymes modify bases in the anticodon minihelix, consisting of the D and anticodon microhelices, and generally depend on correct tRNA folding and tertiary structure. Interestingly, the enzymes that modify primary tRNA transcripts in oocyte injection experiments fall into class I (Grosjean et al., 1996). It was further postulated that this first set of modifications directs the efficient folding of the pre-tRNA molecules so that they can be acted on by the class II enzymes (Grosjean et al., 1996). Thus, the tRNA processing pathway in eukaryotes may optimize tRNA folding.

Introns have been shown to act as both positive and negative determinants of tRNA modification and thereby to further order the modification pathway (reviewed by Grosjean et al., 1997). Formation of certain modifications, such as Ψ34 and Ψ36 (Fig. 1C) in yeast tRNA^{Ile}, requires the intron sequence rather than the pre-tRNA or mature tRNA. Likewise, formation of Ψ35 (Fig. 1C) in tRNA^{Tyr} and m^{5}C34 (Fig. 1T) in tRNA^{Leu} depends on the presence of introns (see Grosjean et al., 1997). It has been postulated that the introns in these tRNA molecules act as *cis*-acting internal guide sequences (Grosjean et al., 1997) analogous to the *cis*-acting sequences found in certain RNA editing enzymes (see Chapter 21 by Hajduk and Sabatini) and the *trans*-acting snoRNA sequences involved in rRNA modification (see Chapter 13 by Bachellerie and Cavaillé). In contrast, the presence of introns seems to block Q34 (Fig. 1P) and i^{6}A37 (Fig. 1A) formation, which suggests that these are among the last modifications added. This sequential pathway necessarily implies that some modification enzymes are localized to the nucleus, where end maturation and splicing occur, whereas enzymes that modify mature tRNA species after splicing may be primarily cytoplasmic (see Chapter 24 by Maden) (Martin and Hopper, 1994). In this regard, it is noteworthy that the intron-dependent Pus1p (Ψ34, Ψ36) synthase (Fig. 1C) functionally interacts with the nuclear pore complex by the genetic criterion of synthetic lethality (Simos et al., 1996). Thus, the nuclear modification enzymes may be an integral component of a larger molecular machine involved in tRNA maturation and export (Simos et al., 1996). Isozyme sorting is considered in the next section.

There are several qualifications about sequential modification. First, late-acting modification enzymes do not necessarily depend on the presence of previously added modified bases for specificity or activity. This property has been demonstrated best for the purified bacterial Tgt (preQ$_1$34) transglycosylase (Fig. 1Q) (Curnow et al., 1993) and MiaA (i^{6}A) prenyltransferase (Fig. 1A) (Leung et al., 1997), but it will likely also apply to some eukaryotic modification enzymes. Consistent with this notion, bacteria and yeast mutants containing singly inactivated tRNA modification genes accumulate mature tRNA species lacking only the corresponding modification (Connolly and Winkler, 1989; Esberg and Björk, 1995; Hopper and Martin, 1992); other modifications seem to be fully present. Second, some modification enzymes will modify both pre-tRNA and mature tRNA, although the pre-tRNA is the substrate in vivo. For example, m$^2_2$G26 (Fig. 1G) is readily added in vitro to mature tRNA species isolated from *TRM1* mutants of yeast (see Hopper and Martin, 1992). Third, it may not always be possible to extrapolate the tRNA recognition properties of a tRNA modification enzyme from one organism to another, even if the modification enzymes are evolutionary homologs.

Genetic Approaches to Isozyme Sorting

The yeast Trm1p (m$^2_2$G26) dimethyltransferase (Fig. 1G) and Mod5p (i^{6}A37) prenyltransferase (Fig. 1A) were among the first models for how isozymes are sorted among the nucleus, cytoplasm, and mitochondria of eukaryotic cells. This topic was reviewed recently (Martin and Hopper, 1994) and will only be described briefly here. An exciting extension of this work is the use of genetic methods to dissect the sorting process itself.

Two isozymes of Trm1p and Mod5p are synthesized in yeast by two different mechanisms (Fig. 10) (see references in Martin and Hopper, 1994). The single *TRM1* gene has two different transcription start points (Fig. 10A). The Trm1p-I isozyme with 16 additional residues at its amino-terminal end is translated from the first AUG start codon in the longer transcript, which is less abundant than the shorter transcript (Fig. 10). The longer Trm1p-I isozyme is targeted to mitochondria, but the 16-amino-acid terminal extension alone is not necessary for targeting (Fig. 10B). Instead, the terminal extension increases

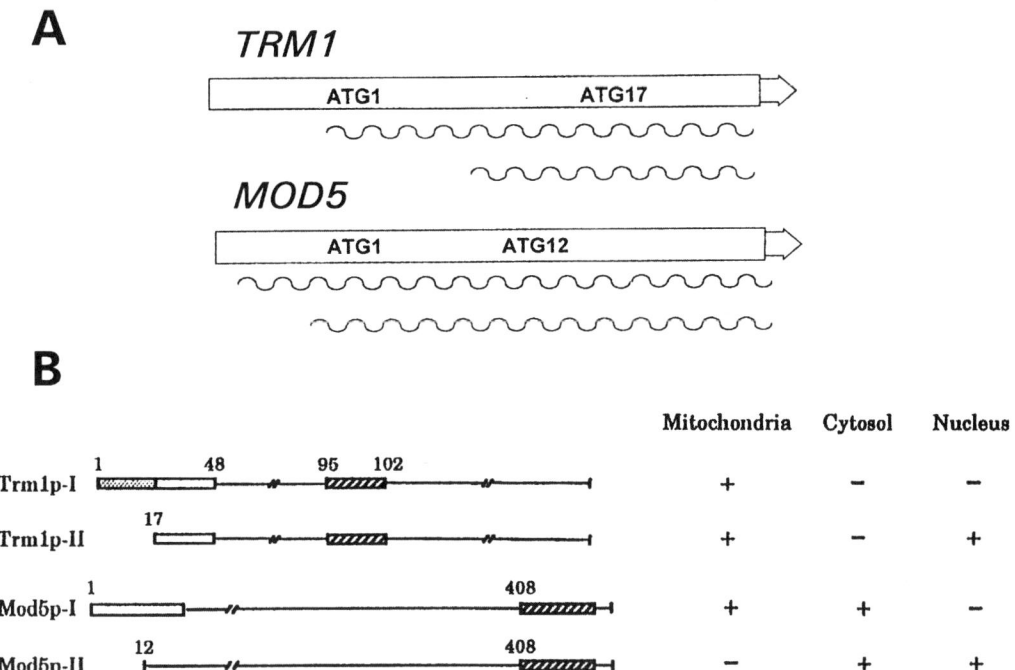

Figure 10. Differential expression of yeast *TRM1* (m$_2^2$G26 dimethyltransferase; Fig. 1G) and *MOD5* (i^{6}A37 prenyltransferase; Fig. 1A) that leads to isozyme sorting in yeast. (A) Transcription and translation start points in *TRM1* and *MOD5*. *TRM1* is transcribed from two different promoters. The larger isozyme (Trm1p-I) is translated from ATG1 in the longer transcript, and the smaller isozyme (Trm1p-II) is translated from ATG17, which is the first start codon in the shorter transcript. *MOD5* is also transcribed from two promoters, but both transcripts contain the ATG1 and ATG12 translation start codons. The larger (Mod5p-I) and smaller (Mod5p-II) isozymes are translated from ATG1 and ATG12, respectively, in the larger bifunctional transcript. See text and Martin and Hopper (1994) for additional details. (B) Signals for targeting and isozyme distribution of the *TRM1* and *MOD5* gene products in yeast. The amino acid sequences of the isozymes are indicated by the lines. Open boxes, regions sufficient for efficient targeting to mitochondria; shaped box in Trm1p-I, region that improves import into mitochondria; hatched boxes, nuclear targeting or localization signals. + and −, present and absent, respectively, in the indicated subcellular compartments. See text and Martin and Hopper (1994) for additional details. Adapted from Martin and Hopper (1994).

the efficiency of mitochondrial import. The signal for mitochondrial targeting is located within the first 48 amino acids of the longer Trm1p-I isozyme (Fig. 10B), which is not processed during import into the mitochondrial matrix. The shorter Trm1p-II isozyme is translated from the shorter *TRM1* transcript starting at an AUG codon that serves as an internal in-frame methionine codon in the longer *TRM1* transcript (Fig. 10A). The shorter Trm1p-II isozyme is a nuclear peripheral, and not an integral membrane, protein that is targeted by the internal segment between amino acids 70 and 213, which includes a nuclear targeting signal found in other eukaryotic nuclear proteins (Fig. 10B) (Martin and Hopper, 1994; Rose et al., 1995). Protein fusions between Trm1p and β-galactosidase were distributed aberrantly throughout the nucleus, suggesting that the native conformation of the Trm1p protein may be necessary for proper targeting to the periphery of the nucleus. Finally, mutations in the nuclear targeting sequence lead to increased accumulation of Trm1p-II in the cytoplasm, but not in mitochondria (Martin and Hopper, 1994; Rose et al., 1995). This result suggests that the distribution of Trm1p-II between the nucleus and mitochondria (Fig. 10B) is not simply caused by competition for targeting information in these two subcellular compartments (Martin and Hopper, 1994; Rose et al., 1995).

Two isozymes of Mod5p are also synthesized, but by a different mechanism than the Trm1p isozymes. Two in-frame AUG translational start codons are present within two different *MOD5* transcripts with 5'-ends about 14 and 4 nt upstream from the first AUG start codon (Fig. 10A) (see references in Martin and Hopper, 1994). About 10% and 90% of Mod5p present in yeast cells arise from the first and second AUG, respectively. The longer transcript is bifunctional in that both AUG translation start codons are used to encode the two Mod5p isozymes. The first AUG is not contained in a consensus eukaryotic ribosome binding site, but mutations that improve the first AUG ribosome binding site increase the amount of the longer Mod5p-I isozyme at the expense of the shorter Mod5p-II species (Fig. 10). Besides AUG context, secondary structures in the RNA leader preceding the first AUG seem to play a role in determining which AUG is used for translation initiation. The 11 additional residues of the longer Mod5p-I isozyme are necessary but not sufficient for mitochondrial import, and amino acids 12 to 21, which are present in both isozymes, are part of the mitochondrial targeting signal (Fig. 10B). Unexpectedly, the longer Mod5p-I isozyme is also present in the cytoplasm. Another unanticipated result is that the shorter Mod5p-II isozyme is present in the cytoplasm and the nucleus. A cytoplasmic location was anticipated, because i^6A37 is added late after pre-tRNA processing and splicing. A perfect nuclear targeting signal is located in a domain at the carboxyl-terminal end of yeast Mod5p (Fig. 10B) that is not present in prokaryotic MiaA homologs (Connolly and Winkler, 1991; Dihanich et al., 1987). Thus, isozymes of Mod5p are present in the nucleus, cytoplasm, and mitochondria of yeast.

The isozyme sorting of Mod5p has been incorporated into a powerful general selection for mutants deficient in mitochondrial import (Zoladek et al., 1995). Three mutations were introduced into yeast Mod5p-I contained on a single-copy plasmid: the second AUG was changed to a non-initiation ATT (Ile) codon, and the two lysine codons at positions 14 and 15 were changed to arginine codons to increase the hydrophobic moment of the mitochondrial targeting signal. This plasmid was then introduced into a *mod5* disruption mutant. The net result of the mutations was to increase transport of the mutant Mod5p-I into the mitochondria and to deplete the cytoplasmic pool of Mod5p so much that suppressor tRNAs, such as $tRNA^{Tyr}_{UAA}$, were undermodified for i^6A37 (Fig. 1A). Undermodification decreases the efficiency of suppressor tRNAs, and the Mod5p triple mutant no longer showed suppression of a nonsense codon in the *lys2* gene. Mutants with second-site mutations that reduced transport of the mutant Mod5p-I into mitochondria were selected at low temperatures (23°C) for growth in the absence of lysine. This LYS+ phenotype resulted from accumulation of the engineered mutant Mod5p-I in the cytoplasm, i^6A37 modification of the suppressor $tRNA^{Tyr}_{UAA}$, and suppression of the nonsense mutation in *lys2*. Mutants containing impaired mitochondrial protein import would be expected to show defects in respiration. Therefore, the LYS+ mutants were screened for lack of growth on the nonfermentable carbon source, glycerol, at 34°C and 38°C. This scheme led to the identification of four complementation groups of *mdp* (mitochondrial down protein) import mutants.

mdp2 mutations were in the *VRP1* gene that affects the actin cytoskeleton (Zoladek et al., 1995), *mdp3* mutations were in the *PAN1* gene that encodes a protein that interacts with mRNA 3'-ends and affects the initiation of protein synthesis (Zoladek et al., 1995), and *mdp1* mutations were in the *RSP5* gene that encodes ubiquitin-protein ligase (Zoladek et al., 1997). Together, these results implicate the ubiquitin system, the actin cytoskeleton, and protein synthesis in protein import into mitochondria and the subcellular sorting of proteins in eukaryotic cells (Zoladek et al., 1997, 1995).

Prospects

Studies of tRNA and rRNA modification in eukaryotic cells have focused on determining the sequential order and subcellular locations of modification enzymes (see Chapter 24 by Maden), isozyme sorting, the RNA determinants required by the modification enzymes (see Chapter 2 by Grosjean et al.), and the functions of modifications in tRNA and rRNA (see Chapter 6 by Auffinger and Westhof, Chapter 12 by Ofengand and Fournier, and Chapter 14 by Mason). The new work on regulation and subcellular localization described above raises several important issues. Are the activities of eukaryotic modification enzymes besides the Tgt (Q34) transglycosylase (Langgut and Reisser, 1995; Morris et al., 1995) regulated by phosphorylation by protein kinase C? Is the expression of eukaryotic modification genes or the activities of eukaryotic modification enzymes linked to signal transduction pathways other than the PKC pathway? To what extent can the many well-documented cases of tumor-induced and developmentally induced changes in tRNA modification patterns (see Chapter 26 by Björk and Rasmuson) be explained in terms of regulation by signal transduction pathways? What type of physical interaction is implied by the functional interaction between the Pus1p Ψ34, Ψ36 synthase and proteins involved in tRNA biogenesis, such as Los1p, and nucleoporins, such as Nsp1p (Simos et al., 1996)? Are tRNA and rRNA modification enzymes generally found in complexes in the nucleus, cytoplasm, and mitochondria of eukaryotic cells? What role, if any, do tRNA modification levels play in the import of tRNA into mitochondria? Base modification does not appear to be essential for tRNA import into mitochondria (reviewed by Tarassov and Martin, 1996), but base modification does affect the efficiency of the aminoacylation of some tRNA species (Sylvers et al., 1993), and aminoacylation is required for efficient import of tRNA into mitochondria (Tarassov and Martin, 1996). Finally, little is known about the recycling of modified bases and whether there are specific degradation systems that actively remove modified bases from rRNA and tRNA (McLennan, 1975). The biochemical and molecular biological framework to address these issues currently exists.

Genetically tractable model systems, such as yeast, also offer considerable promise to answer key questions about the regulation of tRNA and rRNA modification in eukaryotic cells. The transcription and translation start points for certain yeast tRNA modification genes, such as *MOD5* and *TRM1*, are available (Fig. 10A) (Martin and Hopper, 1994), and it should be possible to determine whether transcription or translation of these genes is regulated in cells grown under different conditions. The use of synthetic lethal strategies to identify functional interactions of the Pus1p (Ψ34, 36) synthase (Fig. 1C) and components of the nuclear pore complex (Simos et al., 1996; Wimmer et al., 1992) and the application of i^{6}A37 modification (Fig. 1A) to dissect isozyme sorting (Zoladek et al., 1997, 1995) are discussed above. Another recent example is the use of a genetic screen to identify mutations that allow yeast initiator tRNA$_i^{Met}$ to function in translation elongation (Aström and Byström 1994; Aström et al., 1993). These studies led to the identification of the *RIT1* gene, which encodes the Ar(p)64 transferase that forms 2'-O-ribosyl phosphate at A64 of tRNA$_i^{Met}$ (Fig. 1S). Repeated recovery of *rit1* mutants in independent screens reinforced the conclusion that the 2'-O-ribosyl phosphate A64 modification is the major negative determinant that prevents tRNA$_i^{Met}$ from binding to the eEF-1α used in translation elongation (Aström and Byström, 1994; Aström et al., 1993). Future genetic strategies should provide insights into the complicated functional and physical interactions that seem to be part of the modification process in eukaryotic cells. Finally, comparisons of the properties of yeast and bacterial null mutants deficient in tRNA and rRNA modification enzymes should lead to a better understanding of the synthesis and diverse functions of modified bases. For example, it was recently reported that yeast m$_2^6$A dimethyltransferase, designated Dim1p (see Fig. 1D), is required for viability, whereas its KsgA homolog is not thought to be essential in bacteria (Lafontaine et al., 1994; Leveque et al., 1990). The basis is unknown for this surprising difference in what superficially appear to be conserved rRNA modifications and enzymes, but it may reflect fundamental differences in the rRNA biogenesis pathways in prokaryotic and eukaryotic cells (Lafontaine et al., 1994).

Acknowledgments. I thank the members of my laboratory for helping to gather the information discussed in this chapter and H.-C. T. Tsui, H.-C. E. Leung, Y. Chen, R. Benne, and G. Garcia for comments. I especially thank Henri Grosjean for insightful, critical suggestions on earlier versions of this chapter. This work is supported by grant MCB-9420416 from the U.S. National Science Foundation and grant R01-GM37561 from the National Institute of General Medical Sciences.

REFERENCES

Agris, P. F. 1996. The importance of being modified: roles of modified nucleosides and Mg^{2+} in RNA structure and function. *Prog. Nucleic Acid Res. Mol. Biol.* 53:79–129.

Arps, P. J., C. C. Marvel, B. C. Rubin, D. A. Tolan, E. E. Penhoet, and M. E. Winkler. 1985. Structural features of the *hisT* operon of *Escherichia coli* K-12. *Nucleic Acids Res.* 13:5297–5315.

Aström, A. U., and A. S. Byström. 1994. Rit1, a tRNA backbone-modifying enzyme that mediates initiator and elongator discrimination. *Cell* 79:535–546.

Aström, S. U., U. Pawel-Rammingen, and A. S. Byström. 1993. The yeast initiator tRNA^{Met} can act as an elongator tRNA^{Met} *in vivo*. *J. Mol. Biol.* 233:43–58.

Bauer, B. F., R. M. Elford, and W. M. Holmes. 1993. Mutagenesis and functional analysis of the *Escherichia coli* tRNA^{Leu} promoter. *Mol. Microbiol.* 7:265–273.

Bechhofer, D. H. 1990. Triple post-transcriptional control. *Mol. Microbiol.* 4:1419–1423.

Björk, G. R., and L. A. Isaksson. 1970. Isolation of mutants of *Escherichia coli* lacking 5-methyluracil in transfer ribonucleic acid or 1-methylguanine in ribosomal RNA. *J. Mol. Biol.* 51:83–100.

Björk, G. R., and K. Kjellin-Straby. 1978. General screening procedure for RNA modificationless mutants: isolation of *Escherichia coli* strains with specific defects in RNA methylation. *J. Bacteriol.* 133:499–507.

Björk, G. R. 1995a. Biosynthesis and function of modified nucleosides, p. 165–205. In D. Söll and U. RajBhandary (ed.), *tRNA: Structure, Biosynthesis, and Function*. ASM Press, Washington, D.C.

Björk, G. R. 1995b. Genetic dissection of synthesis and function of modified nucleosides in bacterial transfer RNA. *Prog. Nucleic Acid Res. Mol. Biol.* 50:263–338.

Björk, G. R. 1996. Stable RNA modification, p. 861–886. In F. C. Neidhardt, R. Curtiss III, J. L. Ingraham, E. C. C. Lin, K. B. Low, B. Magasanik, W. S. Reznikoff, M. Riley, M. Schaechter, and H. E. Umbarger (ed.), Escherichia coli *and* Salmonella: *Cellular and Molecular Biology*, 2nd ed. ASM Press, Washington, D.C.

Blanchin-Roland, S., S. Blanquet, J.-M. Schmitter, and G. Fayat. 1986. The gene for *Escherichia coli* diadenosine tetraphosphatase is located immediately clockwise to *folA* and forms an operon with *ksgA*. *Mol. Gen. Genet.* 205:515–522.

Breidt, F., and D. Dubnau. 1990. Identification of *cis*-acting sequences required for translational autoregulation of the *ermC* methylase. *J. Bacteriol.* 172:3661–3668.

Bremer, H., and P. P. Dennis. 1996. Modulation of chemical composition and other parameters of the cell by growth rate, p. 1553–1570. In F. C. Neidhardt, R. Curtiss III, J. L. Ingraham, E. C. C. Lin, K. B. Low, B. Magasanik, W. S. Reznikoff, M. Riley, M. Schaechter, and H. E. Umbarger (ed.), Escherichia coli *and* Salmonella: *Cellular and Molecular Biology*, 2nd ed. ASM Press, Washington, D.C.

Buck, M., and B. N. Ames. 1984. A modified nucleotide in tRNA as a possible regulator of aerobiosis: synthesis of *cis*-2-methylthioribosylzeatin in tRNA of *Salmonella*. *Cell* 36:523–531.

Buck, M., M. Connick, and B. N. Ames. 1983. Complete analysis of tRNA-modified nucleosides by high-performance liquid chromatography: the 29 modified nucleosides of *Salmonella typhimurium* and *Escherichia coli* tRNA. *Anal. Biochem.* 129:1-13.

Buck, M., and E. Griffiths. 1982. Iron mediated methylthiolation of tRNA as a regulator of operon expression in *Escherichia coli*. *Nucleic Acids Res.* 10:2609-2624.

Byström, A. S., and G. R. Björk. 1982. The structural gene (*trmD*) for the tRNA (m^{1}G) methyltransferase is part of a four polypeptide operon in *Escherichia coli* K-12. *Mol. Gen. Genet.* 188:447-454.

Byström, A. S., K. J. Hjalmarsson, P. M. Wikström, and G. R. Björk. 1983. The nucleotide sequence of an *Escherichia coli* operon containing genes for the tRNA(m^{1}G)methyltransferase, the ribosomal proteins S16 and L19 and a 21-K polypeptide. *EMBO J.* 2:899-905.

Byström, A. S., A. von Gabain, and G. R. Björk. 1989. Differentially expressed *trmD* ribosomal protein operon of *Escherichia coli* is transcribed as a single polycistronic mRNA species. *J. Mol. Biol.* 208:575-586.

Cashel, M., D. R. Gentry, V. J. Hernandez, and D. Vinella. 1996. The stringent response, p. 1458-1496. *In* F. C. Neidhardt, R. Curtiss III, J. L. Ingraham, E. C. C. Lin, K. B. Low, B. Magasanik, W. S. Reznikoff, M. Riley, M. Schaechter, and H. E. Umbarger (ed.), Escherichia coli *and* Salmonella: *Cellular and Molecular Biology*, 2nd ed. ASM Press, Washington, D.C.

Chase, R., G. M. Tener, and I. C. Gillam. 1974. Changes in levels of amino acid acceptors in tRNA from *Escherichia coli* grown under various conditions. *Arch. Biochem. Biophys.* 163:306-317.

Chong, S., A. W. Curnow, T. J. Huston, and G. A. Garcia. 1995. tRNA-guanine transglycosylase from *Escherichia coli* is a zinc metalloprotein. Site-directed mutagenesis studies to identify the zinc ligands. *Biochemistry* 34:3697-3701.

Connolly, D. M., and M. E. Winkler. 1989. Genetic and physiological relationships among the *miaA* gene, 2-methylthio-N^6-(Δ^2-isopentenyl)-adenosine tRNA modification, and spontaneous mutagenesis in *Escherichia coli* K-12. *J. Bacteriol.* 171:3233-3246.

Connolly, D. M., and M. E. Winkler. 1991. Structure of *Escherichia coli* K-12 *miaA* and characterization of the mutator phenotype caused by *miaA* insertion mutations. *J. Bacteriol.* 173:1711-1721.

Cortese, R., H. O. Kammen, S. J. Spengler, and B. N. Ames. 1974. Biosynthesis of pseudouridine in transfer ribonucleic acid. *J. Biol. Chem.* 249:1103-1108.

Curnow, A. W., and G. A. Garcia. 1995. tRNA-guanine transglycosylase from *Escherichia coli*. Minimal tRNA structure and sequence requirements for recognition. *J. Biol. Chem.* 270:17264-17267.

Curnow, A. W., F. Kung, K. A. Koch, and G. A. Garcia. 1993. tRNA-guanine transglycosylase from *Escherichia coli*: gross tRNA structural requirements for recognition. *Biochemistry* 32:5239-5346.

de Araujo, A. C., and A. Favre. 1986. Near ultraviolet damage induces the SOS responses in *Escherichia coli*. *EMBO J.* 5:175-179.

de Smit, M. H., and J. van Duin. 1990. Control of procaryotic translational initiation by mRNA secondary structure. *Prog. Nucleic Acid Res. Mol. Biol.* 38:1-35.

Dihanich, M. E., D. Najarian, R. Clark, E. C. Gillman, N. C. Martin, and A. K. Hopper. 1987. Isolation and characterization of *MOD5*, a gene required for isopentenylation of cytoplasmic and mitochondrial tRNAs of *Saccharomyces cerevisiae*. *Mol. Cell. Biol.* 7:177-184.

Dirheimer, G., W. Baranowski, and G. Keith. 1995. Variations of tRNA modifications, particularly of their queuine content in higher eukaryotes. Its relation to malignancy grading. *Biochimie* 77:99-103.

Dong, H. J., L. Nilsson, and C. G. Kurland. 1996. Co-variation of tRNA abundance and codon usage in *Escherichia coli* at different growth rates. *J. Mol. Biol.* 260:649-663.

Eichler, D. C., and N. Craig. 1994. Processing of eukaryotic ribosomal RNA. *Prog. Nucleic Acid Res. Mol. Biol.* 49:197-239.

Emilsson, V., and L. Nilsson. 1995. Factor for inversion stimulation-growth rate regulation of serine and threonine tRNA species. *J. Biol. Chem.* 270:16610-16614.

Ericson, J. U., and G. R. Björk. 1986. Pleiotropic effects induced by modification deficiency next to the anticodon of tRNA from *Salmonella typhimurium* LT2. *J. Bacteriol.* 166:1013-1021.

Esberg, B., and G. R. Björk. 1995. The methylthio group (ms^2) of N^6-(4-hydroxyisopentenyl)-2-methylthioadenosine (ms^{2}io^{6}A) present next to the anticodon contributes to the decoding efficiency of the tRNA. *J. Bacteriol.* 177:1967-1975.

Favre, A., V. Chams, and A. C. de Araujo. 1986. Photosensitized UVA light induction of the SOS response in *Escherichia coli*. *Biochimie* 68:857-864.

Finkel, S. E., and R. C. Johnson. 1992. The Fis protein: it's not just for DNA inversion anymore. *Mol. Microbiol.* 6:3257-3265.

Fleischmann, R. D., M. D. Adams, O. White, R. A. Clayton, E. F. Kirkness, A. R. Kerlavage, C. J. Bult, J. F. Tomb, B. A. Dougherty, J. M. Merrick, et al. 1995. Whole-genome random sequencing and assembly of *Haemophilus influenzae* Rd. *Science* 269:496-512.

Gehrke, C. W., K. C. Kuo, R. A. McCune, and K. O. Gerhardt. 1982. Quantitative enzymatic hydrolysis of tRNAs. Reversed-phase high-performance chromatography of tRNA nucleosides. *J. Chromatogr.* 230:297-308.

Gillman, E. C., L. B. Slusher, N. C. Martin, and A. K. Hopper. 1991. *MOD5* translation initiation sites determine N^6-isopentenyladenosine modification of mitochondrial and cytoplasmic tRNA. *Mol. Cell. Biol.* 11:2382-2390.

Gourse, R. L., T. Gall, M. X. Bartlett, J. A. Appleman, and W. Ross. 1996. rRNA transcription and growth rate-dependent regulation of ribosome synthesis in *Escherichia coli*. *Annu. Rev. Microbiol.* 50:645-677.

Green, C. J., and B. S. Vold. 1993. tRNA, tRNA processing, and aminoacyl-tRNA synthetases, p. 683-698. *In* A. L. Sonenshein, J. A. Hoch, and R. Losick (ed.), Bacillus subtilis *and Other Gram-Positive Bacteria: Biochemistry, Physiology, and Molecular Genetics*. ASM Press, Washington, D.C.

Grosjean, H., J. Edqvist, K. B. Straby, and R. Giegé. 1996. Enzymatic formation of modified nucleosides in tRNA: dependence on tRNA architecture. *J. Mol. Biol.* 255:67-85.

Grosjean, H., Z. Szweykowska-Kulinska, Y. Motorin, F. Fasiolo, and G. Simos. 1997. Intron-dependent enzymatic formation of modified nucleosides in eukaryotic tRNAs: a review. *Biochimie* 79:293-302.

Gryczan, T. J., G. Grandi, J. Hahn, R. Grandi, and D. Dubnau. 1980. Conformational alteration of mRNA structure and the posttranscriptional regulation of erythromycin-induced drug resistance. *Nucleic Acids Res.* 8:6081-6097.

Gu, D., K. M. Ivanetich, and D. V. Santi. 1996. Recognition of the T-arm of tRNA by tRNA (m^{5}U54)-methyltransferase is not sequence specific. *Biochemistry* 35:11652-11659.

Gu, X., and D. V. Santi. 1991. The T-arm of tRNA is a substrate for tRNA (m^{5}U54)-methyltransferase. *Biochemistry* 30:2999-3002.

Gustafsson, C., P. H. R. Lindström, T. G. Hagervall, K. B. Esberg, and G. R. Björk. 1991. The *trmA* promoter has regulatory features and sequence elements in common with the rRNA P1 promoter family of *Escherichia coli*. *J. Bacteriol.* 173:1757-1764.

Hagervall, T. G., J. U. Ericson, K. B. Esberg, J. Li, and G. R. Björk. 1990. Role of tRNA modification in translational fidelity. *Biochim. Biophys. Acta* **1050**:263–266.

Holmes, W. M., C. Andraos-Selim, and M. Redlak. 1995. tRNA-m^{1}G methyltransferase interactions: touching bases with structure. *Biochimie* **77**:62–65.

Holmes, W. M., C. Andraos-Selim, I. Roberts, and S. Z. Wahab. 1992. Structure requirements for tRNA methylation. Action of *Escherichia coli* tRNA(guanosine-1)methyltransferase on tRNA$_1^{Leu}$ structural variants. *J. Biol. Chem.* **267**:13440–13445.

Hopper, A. K. 1990. Genetic methods for study of trans-acting genes involved in processing of precursors of yeast cytoplasmic transfer RNAs. *Meth. Enzymol.* **181**:400–421.

Hopper, A. K., and N. C. Martin. 1992. Processing of yeast cytoplasmic and mitochondrial precursor tRNAs, p. 99–142. *In* E. W. Jones, J. R. Pringle, and J. R. Broach (ed.), *The Molecular and Cellular Biology of the Yeast Saccharomyces*. Cold Spring Harbor Press, Plainview, N.Y.

Horinouchi, S., and B. Weisblum. 1980. Posttranscriptional modification of mRNA conformation: mechanism that regulates erythromycin-induced resistance. *Proc. Natl. Acad. Sci USA* **77**:7079–7083.

Huisman, G. W., D. A. Siegle, M. M. Zambrano, and R. Kolter. 1996. Morphological and physiological changes during stationary phase, p. 1672–1682. *In* F. C. Neidhardt, R. Curtiss III, J. L. Ingraham, E. C. C. Lin, K. B. Low, B. Magasanik, W. S. Reznikoff, M. Riley, M. Schaechter, and H. E. Umbarger (ed.), *Escherichia coli* and *Salmonella*: *Cellular and Molecular Biology*, 2nd ed. ASM Press, Washington, D.C.

Inokuchi, H., and F. Yamao. 1995. Structure and expression of prokaryotic tRNA genes, p. 17–30. *In* D. Söll and U. L. RajBhandary (ed.), *tRNA: Structure, Biosynthesis, and Function*. ASM Press, Washington, D.C.

Kammen, H. O., C. C. Marvel, L. Hardy, and E. E. Penhoet. 1988. Purification, structure, and properties of *Escherichia coli* tRNA pseudouridine synthase I. *J. Biol. Chem.* **263**:2255–2263.

Kealey, J. T., and D. V. Santi. 1995. Stereochemistry of tRNA(m^{5}U54)-methyltransferase catalysis: 19F NMR spectroscopy of an enzyme-FUraRNA covalent complex. *Biochemistry* **34**:2441–2446.

Keener, J., and M. Nomura. 1996. Regulation of ribosome synthesis, p. 1417–1431. *In* F. C. Neidhardt, R. Curtiss III, J. L. Ingraham, E. C. C. Lin, K. B. Low, B. Magasanik, W. S. Reznikoff, M. Riley, M. Schaechter, and H. E. Umbarger (ed.), *Escherichia coli* and *Salmonella*: *Cellular and Molecular Biology*, 2nd ed. ASM Press, Washington, D.C.

Kitchingman, G. R., and M. J. Fournier. 1977. Modification-deficient transfer ribonucleic acids from relaxed control *Escherichia coli*: structures of the major undermodified phenylalanine and leucine transfer tRNAs produced during leucine starvation. *Biochemistry* **16**:2213–2220.

Lafontaine, D., J. Delcour, A.-L. Glasser, J. Desgres, and J. Vandenhaute. 1994. The DIM1 gene responsible for the conserved m$_2^6$Am$_2^6$A dimethylation in the 3'-terminal loop of 18S rRNA is essential in yeast. *J. Mol. Biol.* **241**:492–497.

Landick, R., C. L. Turnbough, and C. Yanofsky. 1996. Transcription attenuation, p. 1263–1286. *In* F. C. Neidhardt, R. Curtiss III, E. C. C. Lin, K. B. Low, B. Magasanik, J. L. Ingraham, W. S. Reznikoff, M. Riley, M. Schaechter, and H. E. Umbarger (ed.), *Escherichia coli* and *Salmonella*: *Cellular and Molecular Biology*, 2nd ed. ASM Press, Washington, D.C.

Langgut, W., and T. Reisser. 1995. Involvement of protein kinase C in the control of tRNA modification with queuine in HeLa cells. *Nucleic Acids Res.* **23**:2488–2491.

Leung, H. C., Y. Chen, and M. E. Winkler. 1997. Regulation of substrate recognition by the MiaA tRNA prenyltransferase modification enzyme of *Escherichia coli* K-12. *J. Biol. Chem.* **272**:13073–13083.

Leung, H.-C. E., and M. E. Winkler. 1997. Unpublished observation.

Leveque, F., S. Blanchin-Roland, G. Fayat, P. Plateau, and S. Blanquet. 1990. Design and characterization of *Escherichia coli* mutants devoid of Ap$_4$N-hydrolase activity. *J. Mol. Biol.* **212**:319–329.

Limbach, P. A., P. F. Crain, S. C. Pomerantz, and J. A. McCloskey. 1995. Appendix 1: structures of modified nucleosides, p. 551–555. *In* D. Söll and U. L. RajBhandary (ed.), *tRNA: Structure, Biosynthesis, and Function*. ASM Press, Washington, D.C.

Lindström, P. H. R., D. Stüber, and G. R. Björk. 1985. Genetic organization and transcription from the gene (*trmA*) responsible for the synthesis of tRNA (uracil-5)-methyltransferase by *Escherichia coli*. *J. Bacteriol.* **164**:1117–1123.

Marinus, M. G., N. R. Morris, D. Söll, and T. C. Kwong. 1975. Isolation and partial characterization of three *Escherichia coli* mutants with altered transfer ribonucleic acid methylases. *J. Bacteriol.* **122**:257–265.

Martin, F., G. Eriani, S. Eiler, D. Moras, G. Dirheimer, and J. Gangloff. 1993. Overproduction and purification of native and queuine-lacking *Escherichia coli* tRNA^{Asp}. *J. Mol. Biol.* **234**:965–974.

Martin, N. C., and A. K. Hopper. 1994. How single genes provide tRNA processing enzymes to mitochondria, nuclei and the cytosol. *Biochimie* **76**:1161–1167.

Marvel, C. C., P. J. Arps, B. C. Rubin, H. O. Kammen, E. E. Penhoet, and M. E. Winkler. 1985. *hisT* is part of a multigene operon in *Escherichia coli* K-12. *J. Bacteriol.* **161**:60–71.

McLennan, B. D. 1975. Enzymatic demodification of transfer RNA species containing N^6(delta^2-isopentenyl)adenosine. *Biochem. Biophys. Res. Commun.* **65**:345–351.

Moore, J. A., and C. D. Poulter. 1997. *Escherichia coli* dimethylallyl diphosphate:tRNA dimethylallyltransferase: a binding mechanism for recombinant enzyme. *Biochemistry* **36**:604–614.

Morgan, C. J., F. L. Merrill, and R. W. Trewyn. 1996. Defective transfer RNA-queuine modification in C3H10T1/2 murine fibroblasts transfected with oncogenic *ras*. *Cancer Res.* **56**:594–598.

Morris, R. C., B. J. Brooks, P. Eriotou, D. F. Kelly, S. Sagar, K. L. Hart, and M. S. Elliott. 1995. Activation of transfer RNA-guanine ribosyltransferase by protein kinase C. *Nucleic Acids Res.* **23**:2492–2498.

Morse, D. E., and C. Yanofsky. 1969. Amber mutants of the *trpR* regulatory gene. *J. Mol. Biol.* **44**:185–193.

Morse, D. E., and C. Yanofsky. 1968. The internal low-efficiency promoter of the tryptophan operon of *Escherichia coli*. *J. Mol. Biol.* **38**:447–451.

Muffler, A., D. Fischer, and R. Hengge-Aronis. 1996. The RNA-binding protein HF-I, known as a host factor for phage Q$_\beta$ RNA replication, is essential for RpoS translation in *Escherichia coli*. *Genes Dev.* **10**:1143–1151.

Muffler, A., D. D. Traulsen, D. Fischer, R. Lange, and R. Hengge-Aronis. 1997. The RNA-binding protein HF-I plays a global regulatory role which is largely, but not exclusively due to its role in expression of the sigma^S subunit of RNA polymerase in *Escherichia coli*. *J. Bacteriol.* **179**:297–300.

Narayanan, C. S., and D. Dubnau. 1987. An in vitro study of the translational attenuation model of *ermC* regulation. *J. Biol. Chem.* **262**:1756–1765.

Nilsson, L., and V. Emilsson. 1994. Factor for inversion stimulation-dependent growth rate regulation of individual tRNA species in *Escherichia coli*. *J. Biol. Chem.* **269**:9460–9465.

Nilsson, L., H. Verbeek, E. Vijgenboom, C. van Druen, A. Vanet, and L. Bosch. 1992. FIS-dependent *trans* activation of stable

RNA operons of *Escherichia coli* under various growth conditions. *J. Bacteriol.* 174:921–929.

Nishikura, K., and E. M. De Robertis. 1981. RNA processing in microinjected *Xenopus* oocytes. Sequential addition of base modifications in the spliced transfer RNA. *J. Mol. Biol.* 145:405–420.

Noguchi, S., T. Nishimura, Y. Hirota, and S. Nishimura. 1982. Isolation and characterization of an *Escherichia coli* mutant lacking tRNA-guanine transglycosylase. Function and biosynthesis of queuosine in tRNA. *J. Biol. Chem.* 257:6544–6550.

Nurse, K., J. Wrzesinski, A. Bakin, B. G. Lane, and J. Ofengand. 1995. Purification, cloning, and properties of the tRNA pseudouridine 55 synthase from *Escherichia coli*. *RNA* 1:102–112.

Ny, T., and G. R. Björk. 1980. Growth rate-dependent regulation of transfer ribonucleic acid(5-methyluridine) methyltransferase in *Escherichia coli* B/r. *J. Bacteriol.* 141:67–73.

Ny, T., and G. R. Björk. 1977. Stringent regulation of the synthesis of a transfer ribonucleic acid biosynthetic enzyme: transfer ribonucleic acid (m^5U) methyltransferase from *Escherichia coli*. *J. Bacteriol.* 130:635–641.

Ny, T., P. H. R. Lindström, T. G. Hagervall, and G. R. Björk. 1988. Purification of transfer RNA (m^5U54)-methyltransferase from *Escherichia coli*. Association with RNA. *Eur. J. Biochem.* 177:467–475.

Ofengand, J., A. Bakin, J. Wrzesinski, K. Nurse, and B. G. Lane. 1995. The pseudouridine residues of ribosomal RNA. *Biochem. Cell. Biol.* 73:915–924.

Okada, N., and S. Nishimura. 1979. Isolation and characterization of a guanine insertion enzyme, a specific tRNA transglycosylase from *Escherichia coli*. *J. Biol. Chem.* 254:3061–3066.

Okada, N., S. Noguchi, S. Nishimura, T. Ohgi, T. Goto, P. F. Crain, and J. A. McCloskey. 1978. Structure determination of a nucleoside Q precursor isolated from *E. coli*: 7-(aminomethyl)-7-deazaguanosine. *Nucleic Acids Res.* 5:2289–2296.

Pease, A. J., B. R. Roa, K. A. Betchel, and M. E. Winkler. Growth rate-dependent regulation of the *Escherichia coli pdxA* and *pdxB* genes, encoding proteins essential for pyridoxal phosphate coenzyme biosynthesis. Submitted for publication.

Persson, B. C. 1993. Modification of tRNA as a regulatory device. *Mol. Microbiol.* 8:1011–1016.

Persson, B. C., and G. R. Björk. 1993. Isolation of the gene (*miaE*) encoding the hydroxylase involved in the synthesis of 2-methylthio-*cis*-ribozeatin in tRNA of *Salmonella typhimurium* and characterization of mutants. *J. Bacteriol.* 175:7776–7785.

Persson, B. C., G. O. Bylund, D. E. Berg, and P. M. Wikström. 1995. Functional analysis of the *ffh-trmD* region of the *Escherichia coli* chromosome by using reverse genetics. *J. Bacteriol.* 177:5554–5560.

Persson, B. C., C. Gustafsson, D. E. Berg, and G. R. Björk. 1992. The gene for a tRNA modifying enzyme, m^5U54-methyltransferase, is essential for viability in *Escherichia coli*. *Proc. Natl. Acad. Sci. USA* 89:3995–3998.

Petrullo, L. A., P. J. Gallagher, and D. Elseviers. 1983. The role of 2-methylthio-N^6-isopentenyladenosine in readthrough and suppression of nonsense codons in *Escherichia coli*. *Mol. Gen. Genet.* 190:289–294.

Pogliano, K. J., and J. Beckwith. 1994. Genetic and molecular characterization of the *Escherichia coli secD* operon and its products. *J. Bacteriol.* 176:804–814.

Redlak, M., C. Andraosselim, R. Giegé, C. Florentz, and W. M. Holmes. 1997. Interaction of tRNA with tRNA (guanosine-1) methyltransferase-binding specificity determinants involves the dinucleotide G(36)PG(37) and tertiary structure. *Biochemistry* 36:8699–8709.

Reuter, K., and R. Ficner. 1995. Sequence analysis and overexpression of the *Zymomonas mobilis tgt* gene encoding tRNA-guanine transglycosylase: purification and biochemical characterization of the enzyme. *J. Bacteriol.* 177:5284–5288.

Reuter, K., R. Slany, F. Ullrich, and H. Kersten. 1991. Structure and organization of *Escherichia coli* genes involved in biosynthesis of the deazaguanine derivative queuine, a nutrient factor for eukaryotes. *J. Bacteriol.* 173:2256–2264.

Roa, B. B., D. M. Connolly, and M. E. Winkler. 1989. Overlap between *pdxA* and *ksgA* in the complex *pdxA-ksgA-apaG-apaH* operon of *Escherichia coli* K-12. *J. Bacteriol.* 171:4767–4777.

Romier, C., K. Reuter, D. Suck, and R. Ficner. 1996. Crystal structure of tRNA-guanine transglycosylase: RNA modification by base exchange. *EMBO J.* 15:2850–2857.

Rose, A. M., H. G. Belford, W. C. Shen, C. L. Greer, A. K. Hopper, and N. C. Martin. 1995. Location of N^2,N^2-dimethylguanosine-specific tRNA methyltransferase. *Biochimie* 77:45–53.

Rouviere, P. E., and C. A. Gross. 1996. SurA, a periplasmic protein with peptidyl-prolyl isomerase activity, participates in the assembly of outer membrane porins. *Genes Dev.* 10:3170–3182.

Sampson, J. R., and O. C. Uhlenbeck. 1988. Biochemical and physical characterization of an unmodified yeast phenylalanine transfer RNA transcribed *in vitro*. *Proc. Natl. Acad. Sci. USA* 85:1033–1037.

Simos, G., H. Tekotte, H. Grosjean, A. Segref, K. Sharma, D. Tollervey, and E. C. Hurt. 1996. Nuclear pore proteins are involved in the biogenesis of functional tRNA. *EMBO J.* 15:2270–2284.

Slany, R. K., and H. Kersten. 1992. The promoter of the *tgt/sec* operon in *Escherichia coli* is preceded by an upstream activation sequence that contains a high affinity FIS binding site. *Nucleic Acids Res.* 16:4193–4198.

Sylvers, A. A., K. C. Rogers, M. Shimizu, E. Ohtsuka, and D. Söll. 1993. A 2-thiouridine derivative in $tRNA^{Glu}$ is a positive determinant for aminoacylation by *Escherichia coli* glutamyl-tRNA synthetase. *Biochemistry* 32:3836–3841.

Tarassov, I. A., and R. P. Martin. 1996. Mechanisms of tRNA import into yeast mitochondria: an overview. *Biochimie* 78:502–510.

Thomale, J., and G. Nass. 1978. Alterations of the intracellular concentration of aminoacyl-tRNA synthetases and isoaccepting tRNAs during amino-acid limited growth in *Escherichia coli*. *Eur. J. Biochem.* 85:407–418.

Tormo, A., M. Almiron, and R. Kolter. 1990. *surA*, an *Escherichia coli* gene essential for survival in stationary phase. *J. Bacteriol.* 172:4339–4347.

Tsui, H. C. T., G. Feng, and M. E. Winkler. 1996. Transcription of the *mutL* repair, *miaA* tRNA modification, *hfq* pleiotropic regulator, and *hflA* region protease genes of *Escherichia coli* K-12 from clustered $E\sigma^{32}$-specific promoters during heat shock. *J. Bacteriol.* 178:5719–5731.

Tsui, H. C. T., H. C. E. Leung, and M. E. Winkler. 1994. Characterization of broadly pleiotropic phenotypes caused by an *hfq* insertion mutation in *Escherichia coli* K-12. *Mol. Microbiol.* 13:35–49.

Tsui, H. C. T., and M. E. Winkler. 1994. Transcriptional patterns of the *mutL-miaA* superoperon of *Escherichia coli* K-12 suggest a model for posttranscriptional regulation. *Biochimie* 76:1168–1177.

Tsui, H.-C. T., P. J. Arps, D. M. Connolly, and M. E. Winkler. 1991. Absence of *hisT*-mediated tRNA pseudouridylation results in a uracil requirement that interferes with *Escherichia coli* K-12 cell division. *J. Bacteriol.* 173:7395–7400.

Tsui, H.-C. T., G. Feng, and M. E. Winkler. 1997. Negative regulation of *mutS* and *mutH* repair gene expression by the Hfq and RpoS global regulators of *Escherichia coli* K-12. *J. Bacteriol.* 179:7476–7487.

Tsui, H.-C. T., G. Zhao, G. Feng, H.-C. E. Leung, and M. E. Winkler. 1994. The *mutL* gene of *Escherichia coli* K-12 forms a superoperon with a gene encoding a new cell-wall amidase. *Mol. Microbiol.* **11:**189–202.

Turnbough, C. L., R. J. Neill, R. Landsberg, and B. N. Ames. 1979. Pseudouridylation of tRNAs and its role in regulation in *Salmonella typhimurium*. *J. Biol. Chem.* **254:**5111–5119.

van Gemen, B., H. J. Koets, C. A. M. Plooy, J. Bodlaender, and P. H. van Knippenberg. 1987. Characterization of the *ksgA* gene of *Escherichia coli* determining kasugamycin sensitivity. *Biochimie* **69:**841–848.

van Gemen, B., J. Twisk, and P. H. van Knippenberg. 1989. Autogenous regulation of the *Escherichia coli ksgA* gene at the level of translation. *J. Bacteriol.* **171:**4002–4008.

Vellanoweth, R. L. 1993. Translation and its regulation, p. 699–712. *In* A. L. Sonenshein, J. A. Hoch, and R. Losick (ed.), Bacillus subtilis *and Other Gram-positive Bacteria: Biochemistry, Physiology, and Molecular Genetics*. ASM Press, Washington, D.C.

Vester, B., and S. Douthwaite. 1994. Domain V of 23S rRNA contains all the structural elements necessary for recognition by the ErmE methyltransferase. *J. Bacteriol.* **176:**6999–7004.

Wahab, S. Z., K. O. Rowley, and W. M. Holmes. 1993. Effects of tRNA$_1^{Leu}$ overproduction in *Escherichia coli*. *Mol. Microbiol.* **7:**253–263.

Wikström, P. M., and G. R. Björk. 1988. Noncoordinate translation-level regulation of ribosomal and nonribosomal protein genes in the *Escherichia coli trmD* operon. *J. Bacteriol.* **170:**3025–3031.

Wikström, P. M., and G. R. Björk. 1989. A regulatory element within a gene of a ribosomal protein operon of *Escherichia coli* negatively controls expression by decreasing the translational efficiency. *Mol. Gen. Genet.* **219:**381–389.

Wikström, P. M., L. K. Lind, D. E. Berg, and G. R. Björk. 1992. Importance of mRNA folding and start codon accessibility in the expression of genes in a ribosomal protein operon of *Escherichia coli*. *J. Mol. Biol.* **224:**949–966.

Wilson, R. K., and B. Roe. 1989. Presence of the hypermodified nucleotide N^6-(Δ^2-isopentenyl)-2-methylthioadenosine prevents codon misreading by *Escherichia coli* phenylalanyl-transfer RNA. *Proc. Natl. Acad. Sci. USA* **86:**409–413.

Wimmer, C., V. Doye, P. Grandi, U. Nehrbass, and E. C. Hurt. 1992. A new subclass of nucleoporins that functionally interact with nuclear pore protein NSP1. *EMBO J.* **11:**5051–5061.

Winkler, M. E. 1996. Biosynthesis of histidine, p. 485–505. *In* F. C. Neidhardt, R. Curtiss III, E. C. C. Lin, K. B. Low, B. Magasanik, J. L. Ingraham, W. S. Reznikoff, M. Riley, M. Schaechter, and H. E. Umbarger (ed.), Escherichia coli *and* Salmonella: *Cellular and Molecular Biology*, 2nd ed. ASM Press, Washington, D.C.

Winkler, M. E. 1997. Unpublished observation.

Wrzesinski, J., A. Bakin, K. Nurse, B. G. Lane, and J. Ofengand. 1995. Purification, cloning, and properties of the 16S RNA pseudouridine 516 synthase from *Escherichia coli*. *Biochemistry* **34:**8904–8913.

Xue, H., A. L. Glasser, J. Desgres, and H. Grosjean. 1993. Modified nucleotides in *Bacillus subtilis* tRNA^{Trp} hyperexpressed in *Escherichia coli*. *Nucleic Acids Res.* **21:**2479–2486.

Yanofsky, C., and L. Soll. 1977. Mutations affecting tRNA^{Trp} and its charging and their effect on regulation of transcription at the attenuator of the tryptophan operon. *J. Mol. Biol.* **113:**663–677.

Zoladek, T., A. Tobiasz, G. Vaduva, M. Boguta, N. C. Martin, and A. K. Hopper. 1997. *MDP1*, a *Saccharomyces cerevisiae* gene involved in mitochondrial/cytoplasmic protein distribution, is identical to the ubiquitin-protein ligase gene *RSP5*. *Genetics* **145:**595–603.

Zoladek, T., G. Vaduva, L. A. Hunter, M. Boguta, B. D. Go, N. C. Martin, and A. K. Hopper. 1995. Mutations altering the mitochondrial-cytoplasmic distribution of Mod5p implicate the actin cytoskeleton and mRNA 3' ends and/or protein synthesis in mitochondrial delivery. *Mol. Cell. Biol.* **15:**6884–6894.

Chapter 26

Links between tRNA Modification and Metabolism and Modified Nucleosides as Tumor Markers

GLENN R. BJÖRK AND TORGNY RASMUSON

tRNA MODIFICATION AS A REGULATORY DEVICE: A HYPOTHESIS

Versatility of tRNA

tRNA is a key component in translation. In this capacity, the various tRNA species decode at different speeds and fidelity that depend on the availability of the tRNA species, of the codon choice, and of the codon context. Thus, there is a multitude of variability in the action of the various tRNA species, and this variability depends on many components of the translation apparatus, one of which is the degree of tRNA modification. However, besides its function in translation, which is its main function in the cell, tRNA also participates in cellular reactions that are not directly connected to translation.

Some tRNA may act as a cofactor in enzymatic reactions. The first precursor to porphyrins in many bacteria and organelles is glutamate-1-semialdehyde (GSA), which is formed after the ligation of Glu to tRNA^{Glu}. The NADP dependent reduction of Glu-tRNA^{Glu} to GSA and deacylated RNA^{Glu} is catalyzed by the glutamyl-tRNA reductase. Free tRNA^{Glu} is released and can be recharged to Glu-tRNA^{Glu}, and can then either be used once more in the synthesis of GSA or recycled into protein synthesis (Verkamp et al., 1995). The synthesis of Gln and Asn occurs after misacylation of their cognate tRNA species by Glu and Asp, respectively. Thus, a tRNA dependent transamidation of Glu-tRNA^{Gln} and Asp-tRNA^{Asn} results in Gln-tRNA^{Gln} and Asn-tRNA^{Asn}, respectively (Wilcox and Nirenberg, 1968; Schön et al., 1988; Curnow et al., 1996). In the synthesis of selenocysteine (SeCys), the conversion of Ser to SeCys also occurs following serine ligation to the tRNA (reviewed in Low and Berry, 1996). Moreover, formylation of initiator t-RNA^{Met} in *Escherichia coli* occurs after the acylation of methionine to the tRNA. A specific tRNA^{Gly} species that does not participate in protein synthesis is involved in the synthesis of peptidoglycan (Stewart et al., 1971). Aminoacylated tRNAs (aa-tRNAs) participate in the synthesis of aminoacyl phophatidylglycerol (Nesbitt and Lennarz, 1968), and aa-tRNAs participate in the terminal addition of amino acid residues to proteins, thereby targeting the protein to degradation (reviewed in Söll, 1993). In the transcription machinery of *E. coli*, fMet-tRNA$_i^{Met}$ interacts with the RNA polymerase (Pongs and Ulbrich, 1976), which suggests that this tRNA species may influence the transcription activity in some way. Moreover, transcription antitermination of bacterial mRNAs coding for periplasmic components of the high-affinity transport system for leucine is sensitive to availability of Leu-tRNA and the presence of Ψ in the anticodon region (Quay et al., 1978; Williamson and Oxender, 1992). It is suggested that the Leu-tRNA may alter the activity or expression of a *trans*-acting factor that inhibits premature termination (Williamson and Oxender, 1992). Also, yeast glyceraldehyde-3-phosphate dehydrogenase may, in addition to being a key enzyme in glycolysis, have another function as a transporter of tRNAs from the nucleus to the cytoplasm (Singh and Green, 1993), although this view has been challenged (Boelens et al., 1995). Various retroviruses use tRNAs as primers for reverse transcriptase, and the modification status of these tRNAs is critical (Isel et al., 1993). Thus, there are many cases in which a specific tRNA species acts as a cofactor in an enzymatic reaction. More extensive discussions of the various functions of tRNA have recently been published (Söll, 1993; Ibba et al., 1997).

Glenn R. Björk • Department of Microbiology, Umeå University, S-90187 Umeå, Sweden. **Torgny Rasmuson** • Department of Oncology, Umeå University, S-90187 Umeå, Sweden.

tRNA Modification as a Potential Regulatory Device

tRNA modification or tRNA modifying enzymes may act as regulatory links between translation and various parts of the intermediary and central metabolism, or in other metabolic reactions important for the adaptation of the cell to various changes in the environment. There are four ways in which such a regulatory link may be manifested.

The degree of tRNA modification influences translation and regulates gene expression

The efficiency of a tRNA species to read a certain codon is influenced by its concentration in the cell, the fraction of the tRNA population that is aminoacylated, and the degree of modification. The modified nucleosides play a pivotal role in improving the decoding capacity of the tRNA, but also in decreasing the codon sensitivity, influencing the codon choice, and maintaining the reading frame. Moreover, the impact of a specific modified nucleoside is dependent on the tRNA species (see reviews by Björk, 1992, 1996, 1995a, 1995b; Li et al., 1997) (also see Chapter 27). Thus, the degree of modification of a tRNA may influence the translation in a very profound way.

One specific modified nucleoside may be part of more than one tRNA species. Because the tRNA modifying enzymes are recognizing various structures and sequences in the tRNA, the synthesis of one specific modified nucleoside may occur with different rates depending on which tRNA is used as a substrate. Therefore, the individual tRNAs may harbor various degrees of modification during different physiological statuses of the cell. Because the individual tRNA genes are regulated differently as a function of growth rate (Emilsson and Kurland, 1990), these genes may respond to physiological signals other than the growth rate. Also, since some of the tRNA modifying enzymes are not regulated as tRNA (Ny et al., 1980), there may be physiological conditions that induce an imbalance between the level of tRNA modifying enzymes and their substrates, the tRNAs (also see Chapter 25). Such a condition may therefore induce partial hypomodification of some specific tRNA species. Thus, there are reasons to believe that each tRNA may not always be fully modified and that this degree of hypomodification may be tRNA species dependent.

As stated previously, the presence of modified nucleosides influences various activities of the tRNA. Therefore, the degree of modification may be a regulatory device by influencing translation of a specific mRNA. If so, a regulatory device must show some specificity; that is, it should act at a specific cellular reaction and not in a general way. There are several ways to implement the required specificity. First, as stated previously, there are reasons to believe that the deficiency of a modified nucleoside may not be the same for all tRNA species where it is normally occurring. Therefore, certain physiological conditions may determine that a particular modified nucleoside is not present in one tRNA species but is present in another species. Thus, a specific hypomodification may be induced that will generate a specific regulatory response by the influence that this modified nucleoside has on the decoding property of the tRNA. Second, experiments have shown that an element of specificity may be implemented by the codon context sensitivity of the unmodified tRNA, by codon choice, or by a certain sequence that is sensitive to frameshift. For example, the lack of ms^2io^6A37 renders the tRNA more sensitive than the wild-type tRNA to the 3' nucleoside (Ericson and Björk, 1991). Moreover, the impact of a specific modified nucleoside is dependent on the tRNA it is part of (Li et al., 1997). Some sequences are more prone to frameshifting, which is when the ribosome changes its reading frame either to the left (−1 frameshifting) or to the right (+1 frameshifting) (Farabaugh, 1996). Modified nucleosides may prevent such translational errors (Björk et al., 1989; Hagervall et al., 1993; Schwartz and Curran, 1997; Qian and Björk, 1997). Third, some modified nucleosides are influencing the aminoacylation of tRNA (reviewed in Björk, 1995b); therefore, the degree of modification may regulate the level of charging of a subpopulation of the tRNA. Thus, there are several ways in which the degree of tRNA modification influences the efficiency and quality of translation to have a specific effect on gene expression. One example of how a specific sequence causes a desired regulatory response is provided by the mechanisms of transcriptional and translational attenuation in bacteria. In these mechanisms, the rate with which the ribosome traverses a specific set of clustered codons in the leader mRNA regulates the expression of the operon. If the tRNAs reading these clustered codons are undermodified, as, for example, in *miaA* or *hisT* mutants, the rate of translation will decrease and the expression of the operon will increase. Thus, the level of modification of the tRNAs reading these specific codons will influence the rate with which they are read. Note that in bacteria, translation of one gene in an operon may also influence the translation of a downstream gene either by the mechanism of translational coupling or by influencing the attenuation of transcription, as in the case of expression of the *pyrE* gene (Bonekamp et al., 1984). Hypomodified tRNA reduces the efficiency of translation, resulting in a lower level of a potential regulatory protein for which concentration in the cell is critical and for which changes in its level may result in a cascade of changes in the metabolism. Whereas attenuation only operates in bacteria, the latter mechanism may also operate in eukaryotes, with their nuclei, which separate transcription and translation into separate compartments. Moreover, the programmed frameshifting sites, which are present in both bacteria and eukaryotes, are certain structures in mRNA that enhance frameshifting; it is well documented that this regulatory device is used in several viral systems and by cellular genes (Farabaugh, 1996). Indeed, tRNA modification influences in bacteria the efficiency of frameshifting both at −1 (Brierley et al., 1997) and at +1 (Li et al., 1997; Schwartz and Curran, 1997; Qian and Björk, 1997) programmed frameshifting sites. Thus, even at such sites, the level of tRNA modification in bacteria may have a profound influence on gene expression. Further study will determine if this is also true in eukaryotic systems. However, lack of hypermodified nucleosides in the anticodon loop of tRNAs in retrovirus-infected cells is correlated with the participation of these modification-deficient tRNAs in the frameshifting event to produce the *gag-pol* peptide required for virus growth (Hatfield et al., 1989). Thus, the "shifty" tRNAs that mediate the frameshifting event in these eukaryotic systems may be hypomodified isoacceptors.

tRNA modification may influence the reactions in which tRNA acts as a cofactor

The fact that tRNA also participates in various enzymatic reactions as a cofactor leads us to hypothesize that tRNA modification may also be critical for the rate of catalysis of the enzymatic step. Thus, contrary to the more expected impact on translation by tRNA modification (as previously stated), this hypothesis also postulates a role of tRNA modification in the nonconventional activities of the tRNA.

Bacterial tRNA modifying genes are part of complex operons to ensure coordination to cellular metabolism

Many of the genes encoding bacterial tRNA modifying enzymes are part of complex operons in which apparently unrelated genes are present (see Chapter 25). Three examples will be mentioned here.

The *miaA* gene is part of an operon that shows extremely complex regulation (Tsui et al., 1996) (see Fig. 5 in Chapter 25). The operon includes also the *mutL* gene, which is located just upstream from the *miaA* gene. The translation of the *mutL* and *miaA* genes is translationally coupled, which results in a constant ratio of these two peptides under certain physiological conditions (Connolly and Winkler, 1989, 1991). Thus, the expression of these two proteins is coregulated not only at the transcriptional level but also at the translational level. Mutations in the *mutL* gene cause increased frequencies of GC→AT and AT→GC transitions. Although the primary function of the *miaA* protein is to synthesize i^{6}A37 in tRNA, mutations in the *miaA* gene also increase the rate of GC→TA transversions, that is, a distinct mutator specificity compared to that induced by mutations in the *mutL* gene. The mechanism behind the *miaA* mediated mutator phenotype is not known. However, it is conceivable that the *miaA* operon is designed to coordinate the synthesis of the MiaA and MutL proteins to achieve a balance between them for an adequate DNA metabolism and to set the frequency of spontaneous mutations as a response to physiological changes. This may be an advantage for the cell under certain stress conditions because a higher mutation rate provides a selective advantage (Tröbner and Piechocki, 1984).

The *trmD* operon consists of four genes of which the first and the last in the transcriptional unit encode ribosomal proteins (r-proteins) S16 and L19, respectively (Byström et al., 1983). The second gene (*rimM*) encodes a 21-kDa protein involved in ribosome maturation (Bylund et al., 1997) and the third gene, *trmD*, encodes the tRNA(m^{1}G37)methyltransferase. This apparent misplacement of a gene encoding a tRNA modifying enzyme in a ribosomal protein operon (Björk, 1985) is conserved among several eubacteria (Li and Björk, 1997). This is also true for the distantly related *Mycoplasma* spp. (Li and Björk, 1997), although the *rimM* counterpart is absent in the *trmD* operon of these organisms (Fraser et al., 1995). The transcriptional regulation is the same for all four genes in the *trmD* operon (Byström et al., 1989), although regulation at the translation level differs (Wikström and Björk, 1988). The gene organization of the *trmD* operon may ensure a coordinated ribosome assembly as suggested earlier (Björk, 1985) by coregulating the synthesis of the r-proteins S16 and L19 with that of RimM. Whether the TrmD protein facilitates ribosome assembly is presently unknown. Alternatively, its expression ensures an efficient and correct translation that may be sensed by mechanisms that regulate the synthesis and assembly of ribosomes (Condon et al., 1995; Gourse et al., 1996).

The *truA* (*hisT*) gene encodes the tRNA(Ψ38, 39,40)synthetase, which catalyzes the formation of Ψ in positions 38, 39, and 40 in about half of the tRNA species present in *E. coli*. This gene is part of a complex operon, conserved among several enterobacterial species, that contains genes with apparently unrelated functions (Arps and Winkler, 1987). Upstream from the *hisT* gene, the *pdxB* gene is involved in the synthesis of pyridoxal phosphate, which is a coenzyme in many aminotransferase reactions and therefore is involved in the synthesis of many amino acids, including histidine. The range by which the histidine operon is regulated is very substantial and it is obtained by a transcriptional attenuator that monitors the efficiency of translation of seven histidine codons in the leader *his*-mRNA (Johnston et al., 1980). The efficiency of tRNA^{His} is very dependent on the presence of Ψ38,39; lack of these modified nucleosides results in a derepression of the *his* operon almost as large as that observed when the attenuator has been deleted (Lewis and Ames, 1972). Therefore, it is reasonable to assume that coregulation of the *pdxB* and *hisT* genes would help to adjust the pyridoxine (vitamin B$_6$) supply relative to the demand required by the histidine operon expression. It is known that the synthesis of several other pyridoxal phosphates requiring amino acid biosynthetic enzymes is regulated by an attenuation mechanism that is sensitive to tRNAs with Ψ in the anticodon stem and loop. Therefore, the coregulation of *pdxB* and *hisT* expression may be a way to globally regulate amino acid metabolism (Arps and Winkler, 1987).

Thus, genes for tRNA modifying enzymes may be part of complex operons to ensure coregulated synthesis of key enzymes in various metabolic pathways.

tRNA modifying enzymes have more functions than to modify tRNA

It was recently discovered that the tRNA-(m^{5}U54)methyltransferase encoded by the *trmA* gene has two functions (Persson et al., 1992). One function is to catalyze the formation of m^{5}U54 in tRNA, and the other is an essential but to date unknown function. This latter function may be related to the fact that a subpopulation of the TrmA peptide is covalently bound to 16S rRNA (Gustafsson and Björk, 1993). Interestingly, the yeast tRNA(m^{5}U54)-methyltransferase also has two functions: one catalyzes the formation of m^{5}U54 in tRNA and the other may be a DNA endonuclease activity (von Pawel-Rammingen, 1997). The tRNA(Ψ)synthetase from yeast is part of a multiprotein complex that is involved in the transport and splicing of tRNA (Simos et al., 1996). In this complex the tRNA(Ψ)synthetase may be important not only to catalyze the formation of Ψ in tRNA but also, in conjunction with the other proteins, to transport tRNA through the nuclear pore and promote the splicing event. Interestingly, the yeast tRNA(m$_2^2$G26)methyltransferase is also located at the nuclear pore (Rose et al., 1995) and therefore may also have a function other than to methylate the precursor tRNA, for example, the transportation of tRNA from the nucleus to the cytoplasm. Thus, some tRNA modifying enzymes may have more than one function, and this may be an alternative way to link tRNA modification to specific parts of cellular metabolism.

The degree of tRNA modification therefore may be a regulatory device, acting by any of these four mechanisms. If so, there should be physiological conditions that induce hypomodification of tRNA. Moreover, genetically induced hypomodification should induce metabolic effects that could be explained by any of the previously mentioned mechanisms. In the following sections we will discuss evidence that there are indeed several physiological conditions that induce hypomodification. Moreover, some genetically induced defects in tRNA modification also cause changes in the intermediary and central metabolism. Therefore, at least two of the basic requirements of the hypothesis are fulfilled.

METABOLICALLY INDUCED CHANGES IN tRNA MODIFICATION

Hypomodification of tRNA can be caused by growth conditions in which some precursors to modified nucleosides are limiting or by adding antibiotics (reviewed in Björk, 1995b). However, bacteria in balanced growth but in nutrient-limiting conditions also induce specific undermodification; for example, growth at the limiting isoleucine concentration induced a specific deficiency of a subset of modified nucleosides (Thomale and Nass, 1978). Clearly, these physiological stress situations inhibit the synthesis of a subset of modified nucleosides, among which the synthesis of the ms^{2}i(o)^{6}A37 seems to be especially sensitive. Also, yeast in logarithmic growth but just before entering the stationary phase accumulates methyl-deficient tRNA (Kjellin-Stråby and Phillips, 1968). Growth rate as such seems to influence the synthesis of some modified nucleosides. Whereas the level of mnm^{5}s^{2}U34 in *E. coli* tRNA$^{Glu}_{mnm5s2UUC}$ is invariant of growth rate, the level of s^{4}U8,9 in some but not all tRNA species decreases with increasing growth rate (Emilsson et al., 1992). Whether this occurrence of hypomodification is due to imbalance between the capacity to modify and the amount of tRNA is not known. However, it is known that some modifying enzymes are present in limiting amounts. Therefore, the formation of the corresponding modified nucleosides may be especially sensitive to small imbalances between the enzyme and its substrate.

(See further discussion on this item in Chapter 25.) Moreover, in *E. coli*, the formation of ms^2i^6A37 and Q34 is growth phase dependent (Bartz et al., 1970; Singhal et al., 1981) as is the formation of s^2U derivatives in *Agrobacterium tumefaciens* (Chackalaparampil and Cherayil, 1981). Interestingly, the level of queuosine (Q) in *E. coli* decreases in stationary phase, suggesting that the Q-tRNA turns over with a higher rate than other tRNAs (developmental changes in Q are discussed later in this chapter). Thus, various growth conditions may influence the level of tRNA modification in specific ways; this seems to be true for both bacteria and yeast.

Some stress conditions for bacteria induce deficiency in modified nucleosides in tRNA. Iron limitations result in a deficiency of the ms^2 group of $ms^2i(o)^6A37$ (Wettstein and Stent, 1968; Rosenberg and Gefter, 1969; Griffiths and Humphreys, 1978). The conversion of GTP to the precursor $preQ_1$ in the synthesis of Q does not occur when iron is limiting (Kersten and Kersten, 1990). A deficiency in the ms^2 group of $ms^2i(o)^6A37$ stimulates the transport of aromatic amino acids and increases the frequency of GC→TA transversions, as does a mutation in the *miaA* gene (Buck and Griffiths, 1981; Connolly and Winkler, 1989, 1991). The uptake of iron is dependent on the concentration of enterochelin, the synthesis of which starts with chorismic acid. This metabolite is also the starting point for the synthesis of folate and two other vitamins, as well as the aromatic amino acids. Moreover, the level of chorismic acid in bacteria also regulates the synthesis of cmo^5U34/$mcmo^5U34$ (Björk, 1980; Hagervall et al., 1990). The synthesis of mnm^5s^2U34 in *E. coli* and m^5U54 in gram-positive organisms is dependent on folate (Delk and Rabinowitz, 1975; Kersten et al., 1975; Taya and Nishimura, 1977). Therefore, the level of chorismic acid is critical for the synthesis of several modified nucleosides [$mcmo^5U34$, cmo^5U34, $ms^2i(o)^6A37$, mnm^5s^2U34, and Q in both gram-negative and gram-positive bacteria, and m^5U54 in gram-positive bacteria]. Consequently, the degree of tRNA modification, set by the availability of chorismic acid, may then regulate several parts of the central and intermediary metabolism.

The modified nucleoside mnm^5s^2U34 is converted to mnm^5se^2U34 in bacteria when selenium is available in the growth medium (Ching et al., 1985). When no selenium is available, only mnm^5s^2U34 is found. After the addition of SeO_4^{2-}, the level of mnm^5se^2U34 in tRNA is increased and there is a corresponding decrease of mnm^5s^2U34. At 1.0 μM SeO_4^{2-}, there are equal amounts of these two modified nucleosides in tRNA. When the SeO_4^{2-} concentration is increased further, the level of mnm^5se^2U34 is decreased and that of mnm^5s^2U34 is increased. Thus, the concentration of SeO_4^{2-} in the growth medium sets the ratio between mnm^5se^2U34 and mnm^5s^2U34 in tRNA. Since these two modified nucleosides have a slightly different codon choice (Wittwer and Ching, 1989), the ratio of these modified nucleoside can be expected to influence the efficiency of translation. As discussed later in this chapter, selenium in the diet of mice also influences the ribose 2′-O-methylation of the wobble nucleoside in $tRNA^{(Ser)Sec}_{mcm5UmCA}$ (Diamond et al., 1993).

Some modified nucleosides in bacterial tRNA may act as sensors for various stress conditions. Upon near-UV irradiation, the s^4U8 is cross-linked to the nearby C13 (Favre et al., 1969; Thomas and Favre, 1975). This cross-linking does not occur in a mutant (*nuvA*) strain lacking s^4U8, and such a strain is more easily killed by near-UV light than the wild type (Kramer et al., 1988). Prior exposure of the cell to near-UV light, which induces the s^4U8-C13 cross-link, antagonizes the mutagenic effect of 254-nm UV light and inhibits subsequent induction of the SOS response (Favre et al., 1985). The latter system inhibits cell division and induces several DNA repair systems. The fact that this photoprotection by near-UV light requires the s^4U8 (Thomas and Favre, 1980; Favre et al., 1985; Caldeira de Araujo and Favre, 1986) strongly suggests that this modified nucleoside acts as a sensor for near-UV light and mediates the response to protect the cell from such a stress.

The facultative bacteria sustain a substantial stress when they move between the presence and absence of oxygen. Upon such stress, the synthesis of two modified nucleosides, Q and ms^2io^6A37, is sensitive to the oxygen tension in the culture. The hydroxylation of ms^2i^6A37 to ms^2io^6A37 in *Salmonella typhimurium* requires oxygen (Buck and Ames, 1984). This bacterium synthesizes vitamin B_{12} only under anaerobic conditions (Jeter et al., 1984) and this vitamin is required for the conversion of epoxy-Q (oQ) to Q (Frey et al., 1988). Therefore, the tRNA of this bacterium contains oQ when the bacterium is grown in the presence of oxygen, provided that vitamin B_{12} is not added to the growth medium, but has Q in its tRNA when grown anaerobically. Thus, when bacteria are shifted between anaerobic and aerobic conditions, the tRNA modification changes; these changes may be sensors for aerobiosis (Buck and Ames, 1984). However, the *miaE* mutant of *S. typhimurium*, which lacks the hydroxyl group and has ms^2i^6A37 instead of ms^2io^6A37, shows no defects when transferred between anaerobic and aerobic conditions, suggesting that this hydroxyl group is not essential for such transitions.

In the bacterium *A. tumefaciens* the synthesis of s^2U derivatives is dependent on both the growth phase and the temperature (Chackalaparampil and Cherayil, 1981). Also, in the extreme thermophile *Thermus thermophilus* the thiolation reaction forming m^5s^2U54 from m^5U54 is linearly correlated with the growth temperature; this results in a linear correlation to the melting temperature of the tRNA (Watanabe et al., 1976). Moreover, it has been observed that in thermophilic bacteria the content of Gm18 increases with increasing temperature (Agris et al., 1973; Kumagai et al., 1980). This temperature regulation is not because of derepression of the synthesis of tRNA(Gm18)methyltransferase, but rather because of an increased activity of the enzyme (Kumagai et al., 1980). Thus, in some bacteria the temperature regulates the synthesis of some modified nucleosides.

In summary, there are several documented examples of how metabolic changes induce specific undermodifications of tRNA; thus, one requirement of the previously discussed hypothesis is fulfilled. Since such induced changes in tRNA modification have been observed in bacteria, yeast, and mice, the hypothesis may be valid in both eubacteria and eukaryotes. According to the hypothesis, such hypomodified tRNA may in turn serve as regulatory mediators. If so, modification-deficient tRNA should induce some specific changes in the metabolism.

MUTATIONS THAT CAUSE tRNA MODIFICATION DEFICIENCY MAY ALSO CAUSE ALTERATIONS IN THE CENTRAL OR INTERMEDIARY METABOLISM

tRNA Modifications: Links to and Influences on Central and Intermediary Metabolism

In bacteria, an unexpected link between tRNA modification and intermediary metabolism is the requirement of chorismic acid or its derivative (X in the reaction below) in a previously unknown metabolic pathway to synthesize uridine-5-oxyacetic acid (cmo^5U34) and its methyl ester ($mcmo^5U34$) (Björk, 1980; Hagervall et al., 1990). Therefore, all mutants defective in the common aromatic amino acid pathway, which is common to the synthesis of the three aromatic amino acids and enterochelin, folate, menaquinone, and ubiquinone, also lack these two modified uridines in position 34 (wobble position). As discussed above, enterochelin is a key metabolite in the transport of iron, folate is key in C1 metabolism, and ubiquinone and menaquinone are key in aerobic and anaerobic respiration, respectively. The biosynthetic pathway for the synthesis of $(m)cmo^5U34$ in *E. coli* and in *S. typhimurium* may be as follows:

$$U34^{Chor\ or\ X} \to ho^5U^{AdoMet} \to mo^5U34 \to cmo^5U34^{AdoMet} \to mcmo^5U34$$

Only one of the two carbon atoms in the side chain of cmo^5U34 originates from S-adenosyl-L-methionine (AdoMet) (Björk, 1980). The synthesis of its methyl ester also uses AdoMet as a methyl donor (Pope et al., 1978). Thus, the synthesis of $(m)cmo^5U34$ is sensitive to the levels of chorismic acid and AdoMet. Accordingly, the level of chorismic acid controls the synthesis of cmo^5U34. Derivatives of 5-hydroxyuridines (xo^5U) are thought to expand the intrinsic wobble capacity of U in such a way that, in addition to reading codons ending with A and G, it also reads codons ending with U (Yokoyama et al., 1985). These kinds of derivatives are found in *E. coli* tRNAs reading all four codons in a codon family. In the Ala family, there are two tRNA species, $tRNA^{Ala}_{GGC}$ and $tRNA^{Ala}_{cmo5UGC}$, that read the four Ala codons. One of the tRNA species, $tRNA^{Ala}_{cmo5UGC}$, has cmo^5U34 and is suggested to read GCU, GCA, and GCG. The $tRNA^{Ala}_{GGC}$, which has G34, reads GCU and GCC. Recently, an *E. coli* strain was constructed such that the gene (*ala2*) that encodes the G34-containing $tRNA^{Ala}_{GGC}$ was deleted. This mutant therefore has only the cmo^5U34-containing $tRNA^{Ala}_{cmo5UGC}$ species. Since this mutant is viable, the $tRNA^{Ala}_{cmo5UGC}$ is able to read all four Ala codons (Gabriel et al., 1996) when the isoacceptor $tRNA^{Ala}_{GGC}$ is absent. As previously stated, chorismic acid is required for the synthesis of cmo^5U34. This metabolite is synthesized from erythrose-4-phosphate and phosphoenolpyruvate, and one of the intermediates is shikimic acid (SA). The *aroD* gene product catalyzes one of the steps upstream of the synthesis of SA. Therefore, one can manipulate the level of cmo^5U34 in the tRNA by growing an *aroD* mutant in the presence or absence of shikimic acid, which results in presence or absence of cmo^5U34 in the tRNA (Björk, 1980; Hagervall et al., 1990). The *aroD* mutation was introduced into a strain ($\Delta ala2$; kindly obtained from W. McClain) which possesses only $tRNA^{Ala}_{cmo5UGC}$. This double mutant ($\Delta ala2\ aroD$) was shown to require shikimic acid for growth (Björk, 1997). Therefore, the $tRNA^{Ala}_{cmo5UGC}$ requires the cmo^5 modification to read efficiently all four Ala codons. Thus, the cmo^5 modification of $tRNA^{Ala}_{cmo5UGC}$ extends the wobble capacity not only to U-ending codons but also to the C-ending codons. The presence of the cmo^5 modification may be critical under certain conditions at which isoacceptor tRNAs that read the U/C-ending codons are low in abundance. Therefore, the level of chorismic acid controls the level of cmo^5U34, which in turn may regulate the synthesis of certain proteins in the cell.

The modified nucleoside ms²io⁶A37 in tRNA may be synthesized in vivo in *E. coli* and in *S. typhimurium* in the following steps:

$$A37^{miaA} \rightarrow i^6A37^{miaB,miaC} \rightarrow ms^2i^6A37^{miaE} \rightarrow ms^2io^6A37$$

Judged from the chemical reactions involved, at least four enzymatic activities are required for the synthesis of ms²io⁶A37. The corresponding genetic loci are designated *miaA*, -*B*, -*C*, and -*E*. The last step in the synthetic pathway for ms²io⁶A37 is dependent on the presence of molecular oxygen (O_2) (Buck and Ames, 1984). This step is present in most eubacteria but not in *E. coli*, since this organism lacks the *miaE* gene (Persson and Björk, 1993). The tRNA(ms²io⁶A37)hydroxylase is present under anaerobic conditions, although the hydroxylation reaction does not occur under these conditions (Buck and Ames, 1984). It was suggested that ms²io⁶A37 may function as a regulator of aerobiosis. Unexpectedly, the *miaE* mutant of *S. typhimurium* is unable to grow aerobically on the citric acid cycle (CAC) intermediates fumarate, succinate, and malate (Fig. 1) (Persson et al., 1997). This inability to grow on these dicarboxylic acid intermediates of the CAC is dependent on the state of modification in position 37 of those tRNAs normally having ms²io⁶A37, and not to a second function of the tRNA(ms²io⁶A37)hydroxylase. Therefore, the oxygen tension, which controls the hydroxylation reaction (Buck and Ames, 1984), regulates the ability of *S. typhimurium* to grow on these CAC intermediates. Interestingly, an *miaA* mutation in *E. coli* or in *S. typhimurium* induces changes in the central metabolism, since the synthesis of the enzyme 6-phosphogluconate dehydrogenase, which is encoded by the *gnd* gene and is part of the pentose-phosphate cycle (Fig. 1), is derepressed in an *miaA1* mutant. The *miaA1* mediated effect is specific, since the synthesis of another enzyme, the *zwf* gene product (Fig. 1), in the pentose-phosphate cycle is not affected (Jones et al., 1990). Moreover, a mutation in the *miaA* gene of *E. coli* also increases the oxidation of various carbon compounds and some amino acids, such as proline and glutamine (Tsui et al., 1994). Thus, the degree of modification of ms²io⁶A37 is critical for maintaining a proper balance within the central and intermediary metabolism. It has also been noted that the *trmD3* mutant of *S. typhimurium*, which lacks m¹G37 in its tRNA, also has a reduced ability to grow on these CAC intermediates (Fig. 1) (Nilsson and Björk, 1997). Moreover, the *trmD3* mutant and a *hisT* mutant, which lacks Ψ in positions 38, 39, and 40, show increased ability to oxidize some carbon compounds (Li and Björk, 1995). These results are consistent with the suggestion that the degree of modification in tRNA may be a regulatory device acting at various steps in the central and intermediary metabolism.

1-Methylguanosine (m¹G37) is present in position 37 (next to and 3' of the anticodon) in tRNAs reading codons CUN(Leu), CCN(Pro) and CGG(Arg) in tRNAs from all organisms (where N is any of the four major nucleosides; besides being found in position 37, m¹G is present in position 9 in archaic, eukaryotic, mitochondrial, and plant tRNAs, but not in eubacterial tRNAs). A mutant of *S. typhimurium* has been isolated that is defective in the synthesis of m¹G37 due to a mutation (*trmD3*) in the structural gene (*trmD*) for the tRNA(m¹G37)methyltransferase (Björk et al., 1989). The *trmD3* mutation causes the cell to grow 24% slower than the wild type in MOPS-glucose medium, with a corresponding reduction of the polypeptide chain elongation rate (Li and Björk, 1995). Moreover, the mutation increases +1 frameshifting (Björk et al., 1989; Hagervall et al., 1993) and reduces the aa-tRNA selection in a tRNA-dependent manner (Li et al., 1997). Thus, the presence of m¹G37 promotes an efficient translation and prevents +1 frameshifting. A wild-type *S. typhimurium* strain is sensitive to the addition of adenine to the medium. This Ade sensitivity is relieved by the addition of thiamine because these two metabolites share a common intermediate (aminoimidazole ribotide [AIR] [Fig. 1]) in their synthesis (Newell and Tucker, 1968); the addition of Ade represses the synthesis of AIR, resulting in the reduced synthesis of thiamine. However, the *trmD3* mutant is resistant to Ade (Nilsson and Björk, 1997), suggesting that it is able to synthesize thiamine in an adequate amount although the purine pathway is repressed. Other results also support this conclusion. Therefore, in the biosynthetic pathway of thiamine, there must be a gene for which the expression is sensitive to the level of m¹G37 in tRNA, indicating that efficient translation is a prerequisite for adequate synthesis of thiamine. Therefore, these results suggest that a correct level of m¹G37 in tRNA may be a regulatory device in the synthesis of thiamine. Moreover, the *trmD3* mutant is extremely sensitive to the amino acid analog nitrotyrosine, which is not caused by an increased uptake of the analog or a decreased synthesis of tyrosine or tryptophan (the nitrotyrosine is transported by a tryptophan specific permease) (Li and Björk, 1995). It was suggested that the *trmD3* mutation may influence the C1 metabolism because wild-type cells are also sensitive to nitrotyrosine in the presence of some metabolites known to influence the C1 metabolism.

The above-mentioned examples of how undermodified tRNA may influence metabolism have been

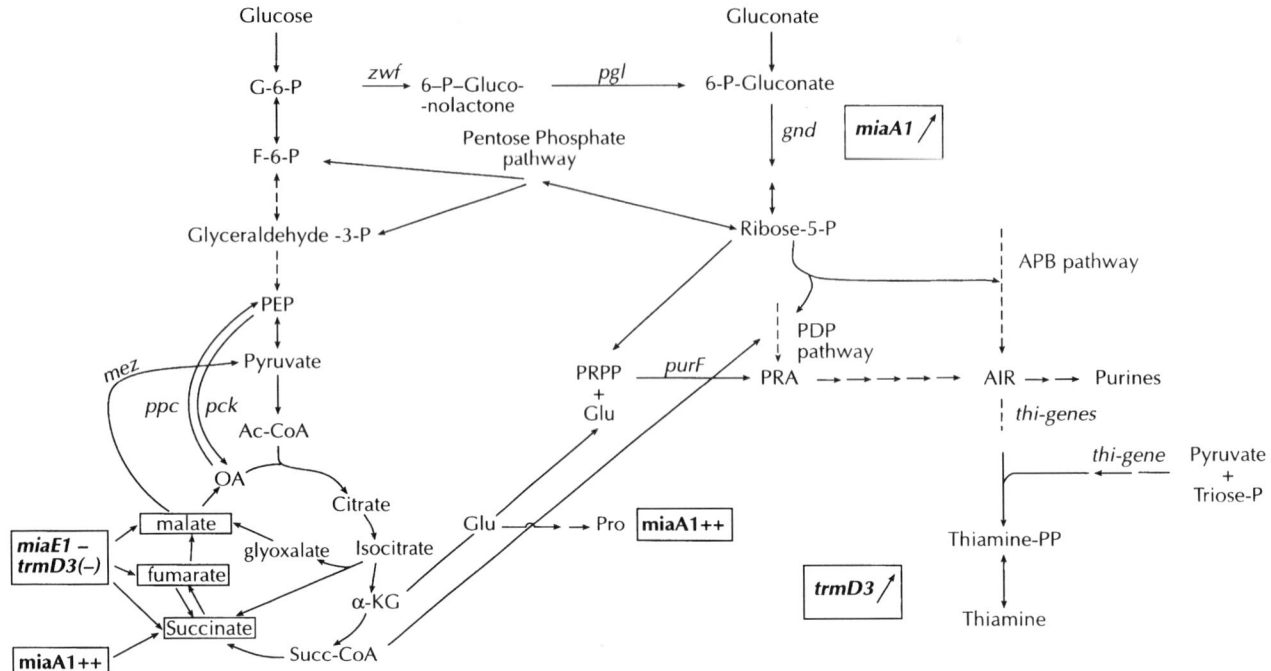

Figure 1. Hypomodification of tRNA alters central and intermediary metabolism in bacteria. Mutants (*trmE*) defective in the hydroxylation of ms^{2}i^{6}A37 to ms^{2}io^{6}A37 in *S. typhimurium* are unable (denoted −) and the *trmD3* mutant defective in the synthesis of m^{1}G37 has a reduced ability [denoted (−)] to grow on the indicated CAC cycle intermediates. An *miaA1* mutant of *E. coli*, which lacks ms^{2}i^{6}A37 in its tRNA, grows better than the wild type (denoted ++; wild type is assumed to be +) on succinate or proline as a carbon source. Moreover, the Gnd enzyme is derepressed about twofold (denoted ↗) in the same mutant. Mutation in the *trmD* gene, which results in lack of m^{1}G37 in the tRNA, apparently derepresses the synthesis of thiamine (denoted ↗). See text for references.

obtained with bacterial mutants. However, a mutant of yeast that lacks the m$_2^2$G26 has a reduced ability to ferment ethanol (Stråby et al., 1997). Thus, whereas links between tRNA modification and metabolism are well documented in bacteria, such links also exist in eukaryotes.

tRNA Modification: Influences on Pathogenicity of Bacteria

tRNAs specific for Tyr, His, Asn, and Asp from eubacteria and eukaryotes, except yeast, contain queuosine (Q34) in the wobble position. The synthesis is complex and may proceed as follows in *E. coli* (several biosynthetic steps may be involved, although only one arrow is shown):

In eukaryotes, some of which do not synthesize queuine, the fully modified base queuine is incorporated. The addition of hexoses occurs only when Q is part of the tRNA. Therefore, the synthesis of Q in eukaryotic tRNAs may consist of only two steps: (i) the insertion of queuine and (ii) GDP-man and GDP-gal dependent transglycosylation. The latter reactions occur only with tRNA^{Asp} and tRNA^{Tyr}, respectively (Okada et al., 1977).

Q, or its glycosylated derivatives manQ and galQ, may slightly affect the aminoacylation of some tRNA species but not others (reviewed in Björk, 1996, 1995b). The structure of Q suggests coding properties that are similar to those of G, although it has been shown that the presence of Q influences the codon choice and misreading of the stop codon UAG (reviewed in Björk, 1996, 1995b). The *E. coli* mutant

$$GTP \rightarrow \text{base of preQ}_0 \xrightarrow{queB} \text{base of preQ}_1$$
$$\searrow$$
$$tgt \rightarrow tRNA_{preQ_134} \xrightarrow{queA} tRNA_{oQ34} \xrightarrow{B12} tRNA_{Q34}$$
$$\nearrow$$
$$tRNAs_{GUN}$$

(*tgt*), which completely lacks Q in its tRNA, grows as well as the wild type, and the only phenotype induced by the *tgt* mutation is a slight effect on the survival in stationary phase (Noguchi et al., 1982). Therefore, the lack of Q does not seem to have a strong effect on the cell physiology. However, a *tgt* mutation in *Shigella flexneri* reduces the expression of the virulence associated genes by 50% (Durand et al., 1994). It was further shown that the primary effect induced by the *tgt* mutation is a posttranscriptional reduction of the expression of the *virF* gene. The VirF protein is encoded on the large virulence plasmid and regulates positively the expression of two other genes, *virG* and *virB*, involved in the virulence of the bacterium. The *virG* gene encodes a 116-kDa surface-exposed membrane protein and is essential for the spreading of invading bacteria. The *virB* gene encodes a positive regulator for the *ipa* operon that encodes three invasive associated proteins. Moreover, the VirB protein also regulates the expression of three other gene clusters (regions 3, 4, and 5), the products of which are essential for the virulence of the bacteria (see review by Salyers and Whitt, 1994). Thus, the target gene, *virF*, for the *tgt*-mediated reduced virulence of *S. flexneri* is the first regulatory gene in a regulatory cascade that governs the degree of virulence of this bacterium. One interpretation is that the lack of Q in any of the tRNAs specific for Tyr, His, Asn, or Asp influences the translation of *virF* mRNA, resulting in a reduced synthesis of the VirF regulator. Although there is experimental evidence that Q in tRNA may influence the efficiency and fidelity of translation, the reported effects are small or require a special context in the mRNA (codon choice or UAG suppression). Supporting the interpretation that Q deficiency affects translation of *virF* mRNA is the fact that an *miaA* mutation, which results in lack of ms^{2}i^{6}A37 in tRNA, strongly impairs the expression of the *virF* gene and the other virulence genes of *S. flexneri* (Durand et al., 1997). Moreover, the result for the *miaA*-mediated effect is also a posttranscriptional reduction of the *virF* gene expression. Because there is extensive experimental evidence that lack of ms^{2}i^{6}A37 reduces the efficiency of the coding ability of tRNA (up to 99% in some codon contexts [Bouadloun et al., 1986]), this result supports and extends the conclusion that the translation of the *virF* mRNA is sensitive to the degree of tRNA modification. Of the tRNAs containing Q34, only the two tRNA^{Tyr} species also contain ms^{2}i^{6}A37. Therefore, the common denominator for the poor translation of the *virF* mRNA may be a Tyr codon, which is placed in a codon context sensitive to the presence of Q34 and ms^{2}i^{6}A37. Clearly, tRNA modification is important for the ability of *S. flexneri* to act as a pathogen.

In the plant pathogen *A. tumefaciens*, a mutation in the *miaA* gene also induces a reduced virulent phenotype (Gray et al., 1992). A Tn5 insertion in the *miaA* gene reduces two- to threefold the expression of four virulence-associated genes and fivefold the expression of one gene. One explanation for the reduced expression of the virulence genes could be that the lack of ms^{2}i^{6}A37 in tRNA reduces the translation of a regulatory gene (for example, *virG* in *A. tumefaciens*, not related to *virG* in *S. flexneri*), resulting in a reduced cascade synthesis of the downstream virulence-associated proteins. The fact that virulence in both *Shigella* and *Agrobacterium* is sensitive to tRNA modification suggests that translation of a regulatory gene in the virulence cascade is sensitive to the degree of modification of tRNA. Since the virulence of a bacterium is in many cases regulated by a cascade of virulence-associated genes, the effect of an incorrect or an inefficient translation of an upstream regulatory gene may be amplified by the regulatory cascade network, and may result in a dramatic effect on the virulence phenotype. Thus, the degree of tRNA modification may be a powerful regulatory device in such systems.

tRNA Modification: Influences on Resistance to Antibiotics

Tetracyclines are broad-spectrum antibiotics that are effective against both gram-positive and gram-negative bacteria. This group of antibiotics is known to prevent aminoacyl-tRNA binding to the A site. However, no mutation in any of the components of the 30S subunit has been identified that confers tetracycline resistance, and the precise mechanism of the action of tetracycline is not known. The tetracycline resistant determinant tet(M) was identified in *Streptococcus* (Burdett et al., 1982), and has been shown to be expressed in a variety of bacteria, including *E. coli*. The Tet(M) protein interacts with the protein machinery to render it resistant to tetracycline (Burdett, 1991). It shows homology with GTPases that participate in protein synthesis, especially to EF-G but also to a lesser extent to EF-Tu (Burdett, 1991). However, it cannot substitute for any of these elongation factors and it is unlikely that Tet(M) is a functional homolog of either EF-G or EF-Tu (Burdett, 1996). It is assumed that Tet(M) acts in conjunction with EF-G at a step before the A site binding of aa-tRNA to release the antibiotic bound nearby (Burdett, 1996). A mutation in the *miaA* gene, which results in lack of ms^{2}i^{6}A37 in the tRNA, renders the cell sensitive to tetracycline in the presence of the Tet(M) protein (Burdett, 1993). This suggests that ms^{2}i^{6}A37 may have a role in the action of tetracycline to inhibit protein synthesis or that it is involved in the mechanism with which Tet(M) protein circumvents the action of the antibiotic. If the action of Tet(M) protein is to stabilize the binding of aa-tRNA to the ribosome in the presence of the antibiotic, the lack of this modified nucleoside, which is known to affect the anticodon–codon interaction (reviewed in Björk, 1996, 1995b), may create a situation on the ribosome that this protein is unable to fulfill such a task. This should render the cell sensitive to the antibiotic even though the Tet(M) protein is present. Although the precise mechanism of how a lack of ms^2i^6A37 acts under this situation is not known, the results show that tRNA modification may influence the action of various antibiotics.

DEVELOPMENTAL AND TUMOR-INDUCED CHANGES IN tRNA MODIFICATION

Queuine and Q-Containing tRNA in Development and Tumorigenesis

Queuosine in position 34 of tRNAs specific for Tyr, His, Asn, and Asp is synthesized in bacteria by the insertion of the base of $preQ_1$ by tRNA-(Q34)transglycosylase (the TGT-enzyme). This incorporated base is then modified to Q. The bacterial enzyme only inserts the base of $preQ_1$ and not queuine (the base of Q), whereas the eukaryotic enzyme inserts queuine. Whereas bacteria can synthesize the precursor $preQ_0$, eukaryotic organisms are unable to make queuine and have to rely on the presence of the base in the diet. Queuine can therefore be considered a vitamin. The presence of Q in eukaryotic tRNAs depends partly on the presence and the activity of the TGT-enzyme and also on the availability of queuine. It was early recognized (see recent review by Dirheimer et al., 1995) that tRNAs from malignant tissues were lacking Q (Okada et al., 1978; Nishimura, 1983; Katze et al., 1983). Because some of the tumors analyzed were slow in growing but still lacked Q in their tRNA (Okada et al., 1978), lack of Q in tRNA cannot merely be a consequence of the fast growth of the tumor cells. The Q deficiency is probably caused by a deficiency in the uptake of queuine caused by low availability of queuine in the neighborhood of the tumor cells. However, this may not be the only reason, since injection of a vast excess of queuine to Ehrlich ascites tumor-bearing mice only partially restored the Q content in tRNA (Katze and Beck, 1980). Protein kinase C (PKC) is normally activated by diacylglycerol, which is generated by the cleavage of phosphatidylinositol-4,5-bisphosphate (PIP_2), but this activation can also be mimicked by phorbolesters. PKC phosphorylates serine or threonine residues in various proteins in a cell-dependent manner. The activity of PKC influences the activity of plasma membranes, ion channels, and the transcription of specific genes. Because PKC modulates the uptake of queuine (Elliott and Crane, 1990a), Q-deficiency in neoplastic tissues may be a consequence of a reduced uptake, which is caused not only by the availability of queuine but also by the activity of the specific queuine uptake system. Since interferons, which are induced by viral infection and decreased PKC activity (Ito et al., 1988), also inhibit queuine uptake (Elliott and Crane, 1990b), viral infection may induce the synthesis of Q-deficient tRNA caused by a poor uptake of queuine. Moreover, PKC activity is also essential for maintaining the activity of the TGT enzyme (Morris et al., 1995). The substrate for the PKC-catalyzed phosphorylation is the 60-kDa TGT subunit, which upon phosphorylation associates with the 34.5-kDa catalytic subunit and forms the 100-kDa active TGT enzyme. Therefore, the PKC-dependent phosphorylation of TGT enzyme is pivotal for the activity and is an important regulatory component of Q modification of the tRNA. Thus, the reason that tRNA in tumor tissues lacks Q is partly because of poor uptake of queuine. However, queuine also influences the PKC signaling pathway, which besides influencing the uptake of queuine also modulates the activity of the TGT enzyme. These results tie tRNA metabolism to the signal transduction pathway control of gene expression and tumor promotion.

Warburg (1956) first noted that glycolysis is increased in tumor cells, that is, the cell metabolism of tumors is changed into a more anaerobic condition, resulting in increased fermentative metabolism and decreased respiratory metabolism. Indeed, transformation of some oncogenes, such as ras, increases aerobic glycolysis (Racker et al., 1985). Whereas addition of queuine to queuine-deficient nontransformed cells stimulates proliferation, the addition of queuine stimulated the proliferation of some transformed cells but inhibited others (Langgut et al., 1993b). This differential response to the addition of queuine is correlated to the glycolytic activity of the cell; that is, queuine stimulates transformed cells under aerobic conditions but inhibits the proliferation in transformed cells grown under oxygen limitation. However, Q content in tRNA was not correlated to the proliferation activity. Therefore, growth modulation by queuine is not directly related to the absence or presence of Q in tRNA, but instead depends on the aerobic or glycolytic state of the cell and on the availability of queuine. This non-tRNA-mediated queuine response may be attributed to its manner of modulating the mitogenic signaling initiated by epidermal growth factor (EGF), which activates the autophosphorylation of its receptor. The phosphorylated EGF receptor induces a signaling pathway that activates expression of c-*fos*. The modulation of the EGF signaling pathway by queuine is correlated by its enhancement of the autophosphorylation of the EGF receptor (Langgut et al., 1993a). In HeLa cells, queuine stimulates the proliferation at aerobic conditions, whereas it inhibits the proliferation at hypoxic conditions (Langgut et al., 1990; Reisser et al., 1994). The activity of one of the various lactate dehydrogenase isoenzymes, LDH-A_4, is low at aerobic condition compared to hypoxic conditions. The addition of queuine further decreases the activity of the enzyme at aerobic conditions and stimulates the activity at hypoxic conditions. The effect is posttranslational because the level of the peptide and the level of its mRNA do not change upon addition of queuine (Reisser et al., 1994). However, the activity of another isoenzyme, LDH_k, decreases at hypoxic condition. Therefore, the presence of queuine rather drastically changes the activity of the various lactate dehydrogenase isoenzymes. These results suggest that the level of queuine in the cell is an important factor in adapting the cell to different oxygen tension.

Various peptide growth factors activate complex signaling pathways in multicellular organisms. EGF and platelet-derived growth factor (PDGF) are recognized by surface located receptors with intrinsic tyrosine kinase activity (reviewed in Fantl et al., 1993). The addition of EGF stimulates the growth of HeLa cells; this stimulation of proliferation is antagonized by addition of queuine to the medium. However, PDGF mediated stimulation is enhanced by the presence of queuine (Langgut, 1995). It is known that binding these growth factors to their receptors results in the activation of distinct protein kinases, and these activities were influenced by the presence of queuine in the growth medium. These results suggest that queuine may modulate various signaling pathways through its influence on the interaction of the growth factor with its receptor (Langgut, 1995). However, the question was not addressed as to whether the queuine effect is through Q-containing tRNA.

The slime mold *Dictyostelium discoideum* normally lives on bacteria, which it engulfs and digests. However, pure cultures of it can be maintained on a defined medium. This slime mold shows a morphologically distinct cell cycle in which the amoeba-like single cell aggregates into a slug upon starvation, and the life cycle culminates into a fungus-like fruiting body, which is differentiated into a stalk and a head. The signal to aggregate is cyclic AMP, which is excreted upon starvation by one of the amoeba-like cells, which will be the starting point for the aggregation to a slug. *D. discoideum* contains Q in the same subset of tRNAs as other organisms. A pure culture of *D. discoideum* grows equally well without queuine in the growth medium. However, if the amoebae are starved for a nutrient, they will not differentiate into a stalk cell unless queuine is present in the medium (Schachner and Kersten, 1984). Concomitantly, the level of lactate decreases and the population of the various NAD-dependent D-(−)-lactate dehydrogenases changes (Schachner et al., 1984). It was suggested that Q-containing tRNA in some way is involved in the metabolism of lactate, which is reminiscent to the previously mentioned cases that neoplastic transformed cells also are characterized by changes in lactate metabolism. (Note that yeast, in which anaerobic metabolism does not end in lactate but in ethanol, does not have Q in its tRNA). Moreover, the stability of $tRNA^{Tyr}$ and $tRNA^{Asp}$, which both normally contain Q34, is dependent on the presence of queuine in the medium (Ott and Kersten, 1985). Therefore, it seems as if queuine as such or Q-containing tRNAs play a pivotal role in the development of *D. discoideum*.

In summary, these examples and others [Friend leukemia cells (Lin et al., 1980) and plant cells (Beier et al., 1987)] suggest that queuine in the medium or Q-containing tRNAs play an important role in the development of various organisms and in tumorigenesis.

Organ Specificity and tRNA Methylation

Analyses of the tRNA modification patterns in various organs have revealed certain organ specificity. There are six isoacceptors of $tRNA^{Ser}$ in the rat brain. Three of them are also found in other rat organs, such as the liver. The brain-specific species lack Gm present in the DHU loop (Gm18) (Rogg et al., 1977). Interestingly, another ribose methylated nucleoside, mcm^5Um34, which is present in the wobble position of $tRNA^{(Ser)Sec}$, is also differentially synthesized in various organs and this response is dependent on the presence of selenium in the diet (Diamond et al., 1993). The $tRNA^{(Ser)Sec}_{mcm^5UmCA}$ is first aminoacylated with serine, which is later converted to selenocysteine (Sec). The Sec-$tRNA^{(Sec)Sec}_{mcm^5UmCA}$ decodes a UGA in the mRNA, which results in a Sec-containing peptide. This selenium-dependent formation of mcm^5Um34 is especially pronounced in liver, muscle, and kidney, and to a lesser degree in brain, whereas the formation of it is independent of selenium in testis. Indeed, only 6% of this tRNA species contain mcm^5Um in testis irrespective of the concentration of selenium in the diet. The ribose methylation reduces wobble capacity of U and results in reduced ability to read G-ending codons (Kawai et al., 1992). Thus, the ribose methylation will ensure the reading of UGA in this case, and will prevent any misreading of UGG. The ribose methylation will therefore influence the codon choice of this tRNA and consequently influence the efficiency of translation of some mRNAs. Although the mechanism of the selenium-dependent methylation is not known, it demonstrates that tRNA modification may in some instances be organ specific and influenced by the environment and the type of diet of the organism.

MODIFIED NUCLEOSIDES AS TUMOR MARKERS

Early Clinical Observations

The modification of RNA takes place at the polynucleotide level and is catalyzed by specific enzymes. As previously mentioned, there are conditions in bacteria (Bartz et al., 1970; Singhal et al., 1981) and in slime mold (Ott and Kersten, 1985) in which increased turnover of tRNA occurs. Because no degradation pathway is known for the modified nucleosides, the presumption is that they will accumulate in the cytoplasm and be excreted into the growth medium. In multicellular organisms, the modified nucleosides are excreted into the urine, and the level is most likely a reflection of the turnover of RNA in the organisms. The most frequently observed modified nucleoside is pseudouridine (Ψ), which is present in most of the different RNA species in the cell. Already in 1959, Adler and Gutman (1959) showed that Ψ is excreted in human urine, and later Dlugajczyk and Eiler (1966) presented evidence for the lack of catabolism of Ψ in humans. Increased urinary excretion of methylated purines was demonstrated in patients with leukemia (Adams et al., 1960; Park et al., 1962), and similar results were found in transplanted tumors in rats (Mandel et al., 1966). tRNA methyltransferases from neoplastic tissue differ both quantitatively and qualitatively from those of normal tissue (Craddock, 1970; Borek, 1971; Viale, 1971; Murphy et al., 1976). The increased urinary excretion of modified nucleosides in cancer patients is mainly because of increased turnover of tRNA rather than cell death (Borek et al., 1977).

Modified Nucleosides in Animal Model Systems

Rats transplanted with Yoshida tumors had increased urinary excretion of m^{7}Gua and 1-methylhypoxanthine (Hanski and Stehlik, 1980), and rats with Morris hepatoma had increased excretion of five modified nucleosides (Lin et al., 1984). Both the level and number of RNA catabolites increased after a lag period, despite early rapid tumor growth. The late appearance of the increase of excretion might indicate that the source of RNA catabolites is the host tissue rather than the tumor. In rats transplanted with osteogenic sarcomas, the number of elevated nucleosides increased parallel to tumor volume, but with a great variability between animals (Lin et al., 1987). Furthermore, there was a reduction of the number of elevated nucleosides following tumor resection. Moreover, excretion of Ψ and other modified nucleosides decreased following chemotherapy in nude mice transplanted with human mesothelioma (Buhl et al., 1985).

Following intracutaneous injections in mice with 3-methylcholanthrene, twelve modified nucleosides and bases were analyzed in urine (Thomale and Nass, 1982). The excretion of most nucleosides and bases increased before the tumors were palpable. Thus, increased excretion of modified nucleosides is an early sign of tumor development. X-ray induced malignant lymphoma in mice resulted in a pathologic elevation of excretion of eight modified nucleosides several weeks before the lymphomas were diagnosed (Thomale et al., 1984). However, they also found two mice with pathologically increased excretion without any signs of focal neoplastic lesions.

In summary, these animal studies on transplanted, induced, and spontaneous tumors and other studies (Russo et al., 1984; Esposito et al., 1985) show that in most cases an increased level of modified nucleosides occurs in urine or serum before any sign of tumor can be observed clinically. This increased level of modified nucleosides most likely reflects an increased turnover of stable RNA.

Physiological Excretion of Modified Nucleosides

The excretion of normal and modified nucleobases in healthy subjects from 0–16 years of age is correlated with the growth velocity (Schöch et al., 1981). Among healthy probands from birth to the age of 40 years the excretion of Ψ, C, A, U, m^{1}G, and m$_2^2$G was highest in neonates and the levels decreased steadily during the first 10 years of life (Müller-Wickop et al., 1986). During adolescence there was a slower decrease, and after the age of 20 years the excretion pattern was fairly stable. Most studies indicate that women have a slightly higher excretion than men (Speer et al., 1979; Müller-Wickop et al., 1986; Rasmuson et al., 1991), although others found no difference between men and women (Itoh et al., 1993). The serum level of Ψ in adults seems to be unaffected by age or sex (Higley et al., 1982).

In summary, these results suggest that in healthy individuals the level of excreted modified nucleosides is independent of age among adults. Children have a higher level than adults; this decreases with age. Whether differences exist in the level of modified nucleosides in urine between sexes is not clear.

Modified Nucleosides in Nonneoplastic Diseases

The excretion of modified nucleosides was studied in a large group of pre- and dysmature neonates (Van Acker et al., 1993), but the excretion pattern did not differ between patients and healthy neonates. Mills et al. (1982) analyzed urinary nucleoside excretion in a small group of patients with severe combined immunodeficiency (SCID). High excretion of modified nucleosides was observed, possibly as a consequence of chronic infections.

The occurrence of small compounds like nucleosides that are excreted in urine should be influenced by renal function. Indeed, Ψ is accumulated in uremic plasma, and elevated serum Ψ levels were reduced as a result of peritoneal dialysis and reduced even more after hemodialysis (Asatoor, 1968; Bernert et al., 1988; Schoots et al., 1988; Struijk et al., 1991).

Elevated urinary excretion of modified nucleosides was also found in patients with AIDS and AIDS-related complex (Borek et al., 1986; Fischbein et al., 1987). Moreover, the levels of Ψ and m_2^2G correlated with the degree of lymphadenopathy (Fischbein et al., 1987). There was also an inverse relation between Ψ, m^1A and m_2^2G and the percentages of total T-lymphocytes, T-suppressor lymphocytes, and the number of natural killer cells, possibly as an effect of chronic viral infection. In a study of human immunodeficiency virus (HIV)-positive intravenous heroin addicts, Intrieri et al. (1996) found that serum Ψ was an independent predictor of progression to AIDS in Centers for Disease Control and Prevention (CDC) stage A2 HIV-infected patients.

In summary, increased excretion of modified nucleosides is found in children with severe combined immunodeficiency and in patients with AIDS and AIDS-related complex. In renal insufficiency Ψ is accumulated in plasma.

Modified Nucleosides in Malignant Diseases

Solid tumors

Gastrointestinal tract tumors. The frequency of elevated urinary excretion of Ψ varied from 55% in cancer of the esophagus (Masuda et al., 1993) to 38% in gastric, 45–65% in colonic, and 29–38% in rectal cancers (Higley et al., 1982; Rasmuson et al., 1984). Furthermore, 89% of the patients with colorectal cancer had elevated excretion of at least one modified nucleoside (Holstege et al., 1986). In all these studies there was a tendency, although not statistically significant, toward higher excretion in more advanced stages of the disease.

Masuda et al. (1993) analyzed the expression of Ψ and m^1A with immunohistochemistry in esophageal cancer and found overexpression in tumor compared to normal esophageal epithelium. They also found a correlation between Ψ and m^1A levels and clinical stage, but only in advanced disease (stages III–IV).

In hepatocellular carcinoma (HCC), elevated urinary excretion of Ψ in 70% of the patients was observed (Tamura et al., 1988, 1986a). The levels were significantly higher in patients with HCC compared to patients with cirrhosis or hepatitis or healthy controls. They also demonstrated a reduction of Ψ excretion after arterial embolization of the tumor. Serum Ψ in HCC was studied by Amuro et al. (1988), who found elevated levels in 52% compared to 7% in patients with cirrhosis. No correlation between serum Ψ and tumor diameter was found. To separate cirrhotic ascites from ascites caused by HCC, Castaldo et al. (1996) analyzed Ψ in ascitic fluid and serum in patients with HCC and cirrhosis. Ascitic and serum Ψ levels were significantly higher in HCC patients than in cirrhosis patients.

Breast cancer. In breast cancer, Tormey et al. (1980) found correlations to disease stage for Ψ, m^1I and m_2^2G, but not in response to chemotherapy. In a smaller study Vreken and Tavenier (1987), analyzing six modified nucleosides in urine, showed that 88% of the patients had at least one abnormally elevated nucleoside, and 69% had at least two. Rasmuson et al. (1987) found elevated levels of Ψ in 36% of patients with metastatic breast carcinoma, compared to 7% in patients free from disease after primary surgery. Despite the association between urinary Ψ and disease stage, no prognostic information for relapse or survival seems to exist.

Respiratory tract tumors. In bronchogenic carcinoma, Ψ is the most frequently analyzed modified nucleoside, but Waalkes et al. (1982) also analyzed m^1A, m^1I, m^2G and m_2^2G, and McEntire et al. (1989) analyzed a total of 29 different modified nucleosides. Excretion was generally higher in small-cell lung cancer (SCLC) patients than in patients with other histologic types (non-small-cell lung cancer [NSCLC]) (Tamura et al., 1986b; Lu et al., 1994). The disease stage in SCLC is usually separated in limited disease (LD) or extensive disease (ED), and significantly higher levels of nucleosides were found in ED patients than in LD patients (Waalkes et al., 1982; Tamura et al., 1987; Lu et al., 1994). Chemotherapy is the treatment of choice in SCLC, and in several studies the level of Ψ in serum or urine varied parallel to the clinical response to chemotherapy (Waalkes et al., 1982; Tamura et al., 1986b; Lu et al., 1994). Furthermore, both Waalkes et al. (1982) and Rasmuson et al. (1983b) showed that the modified nucleosides were prognostic for survival.

Urologic tumors. Urinary excretion of four modified nucleosides in patients with renal cell carcinoma was only slightly elevated (Koshida et al., 1985). However, Kvist et al. (1990) found that 87% of patients with this tumor had elevated excretion of Ψ. Furthermore, the level of Ψ in urine from the ureter on the tumor side was almost twice as high as that in urine in the bladder, where it is diluted by urine from the other kidney. This difference indicates a tumor association. In another study on renal cell carcinoma (Rasmuson et al., 1991), urinary excretion of Ψ was elevated in 41% of the patients; the level correlated to tumor grade, disease stage, and tumor diameter. Disease stage and Ψ excretion were independent prognostic factors for survival.

In summary, increased excretion of modified nucleosides is documented in patients with tumors from several organ systems. The highest frequencies are observed in patients with colonic, hepatocellular, and small-cell lung cancer. There seems to be a correlation to disease stage, and in some tumors the excretion gives prognostic information.

Hematological Malignancies

Leukemia. The role of modified nucleosides in leukemia was reviewed by Trewyn and Grever (1986). Analysis of Ψ and methylated guanines in chronic myeloproliferative syndromes revealed elevated levels of these modified nucleosides compared to controls (Hogan et al., 1970; Schöch et al., 1979). Furthermore, Ψ and methylated bases correlated with white cell count. Nielsen and Killman (1983) found that 84% of the patients with acute and chronic myelogenous leukemia had elevated Ψ excretion, and a trend toward correlation between Ψ and the percentage of marrow blast cells was observed. In acute lymphatic and myelogenous leukemia the excretion of m^1I and m_2^2G was most frequently elevated, and patients with lymphatic leukemia had higher levels than those with myelogenous leukemia (Heldman et al., 1983a, 1983b, 1983c). In chronic myelogenous leukemia the best discriminators of disease activity were Ψ, m^1I, and m_2^2G. A note of caution must be raised since a few falsely elevated levels of modified nucleosides were found in patients with urinary tract infections.

Itoh et al. (1992) analyzed urinary Ψ and m^1A in patients with different types of leukemia, and found elevated levels in 77% and 62%, respectively. A correlation between nucleoside level and treatment response was also observed. Pane et al. (1993) assessed serum Ψ in acute leukemia. Patients with acute lymphatic leukemia had higher levels than those with acute myelogenous leukemia. Furthermore, serum Ψ level was prognostic for survival.

Malignant lymphoma and myelomatosis. In two studies, elevated urinary excretion of Ψ, m^1Gua and m^7Gua was found in ~50% of patients with Hodgkin's disease (Pinkard et al., 1972; Cooper et al., 1977). Pseudouridine was associated with negative prognostic indicators such as the presence of B-symptoms and atypical histiocytes, advanced disease stage, and tRNA methylase activity. In a group of patients with Hodgkin and non-Hodgkin lymphomas, a correlation between disease stage and therapy response to nucleoside excretion was observed (Rasmuson et al., 1983a).

In a large study of Ψ in patients with myelomatosis, Sörensen et al. (1985) found that the Ψ excretion was significantly higher in patients compared to healthy controls. Elevated Ψ was also correlated to a worse prognosis. A weak correlation between urinary Ψ and serum β_2-microglobulin was also found. However, when prognosis was evaluated, stratified of serum β_2-microglobulin level, Ψ added little as a prognostic factor.

Recently, Rasmuson and Björk (1995) presented a long-term follow-up study on Ψ excretion in patients with malignant lymphomas classified according to the Kiel system. Pseudouridine correlated to disease stage in high-grade malignant (HGM) lymphoma and also to the presence of B symptoms. The prognostic value for survival was settled in a multivariate analysis for patients with non-Hodgkin's lymphoma.

In summary, elevated excretion of modified nucleosides is frequently found in patients with hematological malignant diseases. Particularly high levels are observed in acute leukemia and in high-grade malignant lymphoma. Association with disease stage and prognosis is observed, and as a result of therapy, reduced excretion is found.

CONCLUSION AND PERSPECTIVES

The level of tRNA modification is changed by metabolic stress conditions and by entry into different developmental stages. There are many known examples in which read-through or frameshifting can act as a regulatory device. Moreover, the efficiency of translation also regulates gene expression. The level of tRNA modification may strongly affect such regulatory devices since it is well established that modified nucleosides in tRNA influence not only the efficiency of the tRNA but also the codon context sensitivity of it and reading frame maintenance. Moreover, some tRNA modifying enzymes may have a function other than to modify tRNA. Therefore, tRNA modification or the corresponding enzymes may be sensors for various environmental stress or developmental signals, which direct the cellular response to such changes. We have reviewed experimental facts supporting the hypothesis that tRNA modification may be a global regulatory device and link translation with central and intermediary metabolism. However, thus far a complete regulatory circuit has not been established. Improved knowledge of the function of the modified nucleosides and of the molecular mechanism with which translation and metabolism are linked may in the future reveal such regulatory circuits and facilitate our understanding as to why they are operating. If so, a better understanding of the dynamics and interplay between translation and metabolism will be obtained and will help us to understand the cellular response that occurs at various physiological stresses and during development.

Thus far, no degradation pathway for modified nucleosides has been identified in bacteria or in yeast. Since modified nucleosides cannot be incorporated into the growing RNA chain, one would expect that upon degradation of tRNA or rRNA in bacteria and yeast, such modified nucleosides would be found in the cytoplasm or excreted into the growth medium. However, since tRNA and rRNA are very stable, the pool of modified nucleosides is most likely very low if it exists at all. Since tRNA is known to turn over under certain starvation conditions (Bartz et al., 1970; Singhal et al., 1981; Ott and Kersten, 1985), one would expect an increased pool of the modified nucleosides at these conditions since they cannot be reused in RNA synthesis. If so, either these free modified nucleosides may just be excreted as waste products or they may play a functional role during these starvation conditions. Note that a bacterium or yeast cell spends a considerable part of its life cycle at such extreme conditions in nature. It would not be surprising if the free modified nucleosides are not merely a waste product but have an overlooked role. Interestingly, queuine in higher organisms may have such a role, since it influences the physiology of the cell in a tRNA independent way (see above). Unfortunately, no investigation has been made to analyze the pool of free modified nucleosides at conditions when one can suspect that stable RNA is degraded. Moreover, no report addressing the possibility of an alternative function has been presented besides the action of queuine. In multicellular organisms, there is turnover of stable RNA, and modified nucleosides are found intact in rather large amounts in serum and in the urine. Therefore, urinary excretion of modified nucleosides reflects the turnover of RNA. Most of the RNA, which contains modified nucleosides in the multicellular eukaryotes, is stable and participates in translation (rRNA and tRNA) or in splicing (the various U RNAs). The latter contain very high levels of Ψ for which function has not been established. The level of modified nucleosides in serum and urine is therefore a reflection of the need for these RNA species in translation and in maturation of mRNA, provided that when not in high demand these unused RNAs are degraded. Thus, poor translation may induce degradation of these RNAs, resulting in increased excretion of modified nucleosides, which may be a biological marker for an imbalance between metabolism and translation—a condition that may be induced in tumorigenesis. Indeed, the excretion of modified nucleosides, both in animal model systems and in clinical materials, is increased in individuals with most malignant tumors, especially those with high proliferative rates, in nonneoplastic diseases such as AIDS, and in some infectious diseases. However, no specific pattern of excretion of modified nucleosides has been associated with a certain illness, although this may still be the case since in most studies only Ψ in either urine or serum has been monitored. Although an elevated level of excreted modified nucleoside is not a specific marker of malignancy, several studies indicate that analysis of excretion of modified nucleosides could be used to monitor the effect of therapy and as a prognostic indicator.

Acknowledgments. This work was supported by grants from the Swedish Cancer Society (project 680 to G.R.B.), the Swedish Natural Science Council (project BU-2930 to G.R.B.), and from the Lions Cancer Research Foundation (to T.R.). We thank Kerstin Stråby and Jerome Durand for critical readings of the manuscript.

REFERENCES

Adams, W. S., F. Davis, and M. Nakatani. 1960. Purine and pyrimidine excretion and leukemic subjects. *Am. J. Med.* 212: 726–734.

Adler, M., and A. B. Gutman. 1959. Uridine isomer (5-ribosyluracil) in human urine. *Science* 130:862–863.

Agris, P. F., H. Koh, and D. Söll. 1973. The effect of growth temperatures on the *in vivo* ribose methylation of *Bacillus stearothermophilus* transfer RNA. *Arch. Biochem. Biophys.* 154: 277–282.

Amuro, Y., H. Nakaoka, S. Shimomura, M. Fujikura, T. Yamamoto, S. Tamura, T. Hada, and K. Higashino. 1988. Serum pseudouridine as a biochemical marker in patients with hepatocellular carcinoma. *Clin. Chim. Acta* 178:151–158.

Arps, P. J., and M. E. Winkler. 1987. An unusual genetic link between vitamin B_6 biosynthesis and tRNA pseudouridine modification in *Escherichia coli* K-12. *J. Bacteriol.* 169:1071–1079.

Asatoor, A. M. 1968. Retention of pseudouridine and 4-amino-5-imidazole carboxamide in uraemia. *Clin. Chim. Acta* 20:407–411.

Bartz, J., D. Söll, W. J. Burrows, and F. Skoog. 1970. Identification of the cytokinin-active ribonucleosides in pure *Escherichia coli* tRNA species. *Proc. Natl. Acad. Sci. USA* 67:1448–1453.

Beier, H., U. Zech, E. Zubrod, and H. Kersten. 1987. Queuine in plants and plant tRNAs: difference between embryonic tissue and mature leaves. *Plant Mol. Biol.* 8:345–353.

Bernert, J. T., Jr., C. J. Bell, J. Guntupalli, and W. H. Hannon. 1988. Pseudouridine is unsuitable as an endogenous renal clearance marker. *Clin. Chem.* 34:1011–1017.

Björk, G. R. 1980. A novel link between the biosynthesis of aromatic amino acids and transfer RNA modification in *Escherichia coli*. *J. Mol. Biol.* 140:391–410.

Björk, G. R. 1985. *E. coli* ribosomal protein operons: the case of the misplaced genes. *Cell* 42:7–8.

Björk, G. R. 1992. The role of modified nucleosides in tRNA interactions, p. 23–85. *In* D. L. Hatfield, B. J. Lee, and R. M. Pirtle (ed.), *Transfer RNA in Protein Synthesis*. CRC Press, Boca Raton, Fla.

Björk, G. R. 1995a. Genetic dissection of synthesis and function of modified nucleosides in bacterial transfer RNA. *Prog. Nucleic Acid Res. Mol. Biol.* 50:263–338.

Björk, G. R. 1995b. Biosynthesis and function of modified nucleosides in tRNA, p. 165–205. *In* D. Söll and U. L. RajBhandary (ed.), *tRNA: Structure, Biosynthesis, and Function*. ASM Press, Washington, D.C.

Björk, G. R. 1996. Stable RNA modification, p. 861–886. *In* F. C. Neidhardt, R. Curtiss III, J. L. Ingraham, E. C. C. Lin, K. B. Low, B. Magasanik, W. S. Reznikoff, M. Riley, M. Schaechter, and H. E. Umbarger (ed.), *Escherichia coli and Salmonella: Cellular and Molecular Biology*, 2nd ed. ASM Press, Washington, D.C.

Björk, G. R. 1997. Unpublished results.

Björk, G. R., P. M. Wikström, and A. S. Byström. 1989. Prevention of translational frameshifting by the modified nucleoside 1-methylguanosine. *Science* 244:986–989.

Boelens, W. C., I. Palacios, and I. W. Mattaj. 1995. Nuclear retention of RNA as a mechanism for localization. *RNA* 1:273–283.

Bonekamp, F., K. Clemmesen, O. Karlström, and K. F. Jensen. 1984. Mechanism of UTP-modulated attenuation at the *pyrE* gene of *Escherichia coli*: an example of operon polarity control through the coupling of translation to transcription. *EMBO J.* 3:2857–2861.

Borek, E. 1971. Introduction. *Cancer Res.* 31:596–597.

Borek, E., B. S. Baliga, C. W. Gehrke, C. W. Kuo, S. Belman, W. Troll, and T. P. Waalkes. 1977. High turnover rate of transfer RNA in tumor tissue. *Cancer Res.* 37:3362–3366.

Borek, E., O. K. Sharma, F. L. Buschman, D. L. Cohn, K. A. Penley, F. N. Judson, B. S. Dobozin, C. R. Horsburgh, Jr., and C. H. Kirkpatrick. 1986. Altered excretion of modified nucleosides and β-aminoisobutyric acid in subjects with acquired immunodeficiency syndrome or at risk for acquired immunodeficiency syndrome. *Cancer Res.* 46:2557–2561.

Bouadloun, F., T. Srichaiyo, L. A. Isaksson, and G. R. Björk. 1986. Influence of modification next to the anticodon in tRNA on codon context sensitivity of translational suppression and accuracy. *J. Bacteriol.* 166:1022–1027.

Brierley, I., M. R. Meredith, A. J. Bloys, and T. G. Hagervall. 1997. Expression of a coronavirus ribosomal frameshifting signal in *Escherchia coli*: influence of tRNA anticodon modification on frameshifting. *J. Mol. Biol.* 270:360–373.

Buck, M., and B. N. Ames. 1984. A modified nucleotide in tRNA as a possible regulator of aerobiosis: synthesis of *cis*-2-methylthioribosylzeatin in the tRNA of *Salmonella*. *Cell* 36:523–531.

Buck, M., and E. Griffiths. 1981. Regulation of aromatic amino acid transport by tRNA: role of 2-methylthio-N^6-(delta^2-isopentenyl)-adenosine. *Nucleic Acids Res.* 9:401–414.

Buhl, L., C. Dragsholt, P. Svendsen, E. Hage, and M. R. Buhl. 1985. Urinary hypoxanthine and pseudouridine as indicators of tumor development in mesothelioma-transplanted nude mice. *Cancer Res.* 45:1159–1162.

Burdett, V. 1991. Purification and characterization of Tet(M), a protein that renders ribosomes resistant to tetracycline. *J. Biol. Chem.* 266:2872–2877.

Burdett, V. 1993. tRNA modification activity is necessary for Tet(M)-mediated tetracycline resistance. *J. Bacteriol.* 175:7209–7215.

Burdett, V. 1996. Tet(M)-promoted release of tetracycline from ribosomes is GTP dependent. *J. Bacteriol.* 178:3246–3251.

Burdett, V., J. Inamine, and S. Rajagopalan. 1982. Multiple tetracycline resistance determinants in *Streptococcus*, p. 155–158. *In* D. Schlessinger (ed.), *Microbiology—1982*. American Society for Microbiology, Washington, D.C.

Bylund, G. O., L. C. Wipemo, L. A. C. Lundberg, and P. M. Wikström. 1998. RimM and RbfA are essential for efficient processing of 16S rRNA in *Escherichia coli*. *J. Bacteriol.* 180:73–82.

Byström, A. S., K. J. Hjalmarsson, P. M. Wikström, and G. R. Björk. 1983. The nucleotide sequence of an *Escherichia coli* operon containing genes for the tRNA(m^1G)methyltransferase, the ribosomal proteins S16 and L19 and a 21-K polypeptide. *EMBO J.* 2:899–905.

Byström, A. S., A. von Gabain, and G. R. Björk. 1989. Differentially expressed *trmD* ribosomal protein operon of *Escherichia coli* is transcribed as a single polycistronic mRNA species. *J. Mol. Biol.* 208:575–586.

Caldeira de Araujo, A., and A. Favre. 1986. Near ultraviolet DNA damage induces the SOS responses in *Escherichia coli*. *EMBO J.* 5:175–179.

Castaldo, G., M. Intrieri, G. Calcagno, L. Cimino, G. Budillon, L. Sacchetti, and F. Salvatore. 1996. Ascitic pseudouridine discriminates between hepatocarcinoma-derived ascites and cirrhotic ascites. *Clin. Chem.* 42:1843–1846.

Chackalaparampil, I., and J. D. Cherayil. 1981. Changes in 2-thiouridine derivatives of transfer ribonucleic acid of Agrobacterium tumefaciens. *Biochem. Int.* 2:121–128.

Ching, W. M., L. Tsai, and A. J. Wittwer. 1985. Selenium-containing transfer RNAs. *Curr. Top. Cell Regul.* 27:497–507.

Condon, C., C. Squires, and C. L. Squires. 1995. Control of rRNA transcription in *Escherichia coli*. *Microbiol. Rev.* 59:623–645.

Connolly, D. M., and M. E. Winkler. 1989. Genetic and physiological relationships among the *miaA* gene, 2-methylthio-N^6-(Δ^2-isopentenyl)-adenosine tRNA modification, and spontaneous mutagenesis in *Escherichia coli* K-12. *J. Bacteriol.* 171:3233–3246.

Connolly, D. M., and M. E. Winkler. 1991. Structure of *Escherichia coli* K-12 *miaA* and characterization of the mutator phenotype caused by *miaA* insertion mutations. *J. Bacteriol.* 173:1711–1721.

Cooper, I. A., G. R. Wray, and T. L. Murphy. 1977. Urinary excretory patterns of tRNA degradation products: a marker of Hodgkin's cell metabolism? *Eur. J. Cancer* 13:1309–1312.

Craddock, V. M. 1970. Transfer RNA methylases and cancer. *Nature* 228:1264–1268.

Curnow, A. W., M. Ibba, and D. Söll. 1996. tRNA-dependent asparagine formation. *Nature* 382:589–590.

Delk, A. S., and J. C. Rabinowitz. 1975. Biosynthesis of ribosylthymine in the transfer RNA of *Streptococcus faecalis*: a folate-dependent methylation not involving S-adenosylmethionine. *Proc. Natl. Acad. Sci. USA* 72:528–530.

Diamond, A. M., I. S. Choi, P. F. Crain, T. Hashizume, S. C. Pomerantz, R. Cruz, C. J. Steer, K. E. Hill, R. F. Burk, J. A. McCloskey, and D. L. Hatfield. 1993. Dietary selenium affects methylation of the wobble nucleoside in the anticodon of selenocysteine tRNA(:Ser:Sec). *J. Biol. Chem.* 268:14215–14223.

Dirheimer, G., W. Baranowski, and G. Keith. 1995. Variations in trna modifications, particularly of their queuine content in higher eukaryotes—its relation to malignancy grading. *Biochimie* 77:99–103.

Dlugajczyk, A., and J. J. Eiler. 1966. Lack of catabolism of 5-ribosyluracil in man. *Nature* 212:611–612.

Durand, J. M., N. Okada, T. Tobe, M. Watarai, I. Fukuda, T. Suzuki, N. Nakata, K. Komatsu, M. Yoshikawa, and C. Sasakawa. 1994. *vacC*, a virulence-associated chromosomal locus of *Shigella flexneri*, is homologous to *tgt*, a gene encoding tRNA-guanine transglycosylase (Tgt) of *Escherichia coli* K-12. *J. Bacteriol.* 176:4627–4634.

Durand, J. M. B., G. R. Björk, A. Kuwae, M. Yoshikawa, and C. Sasakawa. 1997. The modified nucleoside 2-methylthio-N^6-isopentenyladenosine in tRNA of *Shigella flexneri* is required for expression of virulence genes. *J. Bacteriol.* 179:5777–5782.

Elliott, M. S., and D. L. Crane. 1990a. Protein kinase C modulation of queuine uptake in cultured human fibroblasts. *Biochem. Biophys. Res. Commun.* 171:393–400.

Elliott, M. S., and D. L. Crane. 1990b. Interferon induced inhibition of queuine uptake in cultured human fibroblasts. *Biochem. Biophys. Res. Commun.* 171:384–392.

Emilsson, V., and C. G. Kurland. 1990. Growth rate dependence of transfer RNA abundance in *Escherichia coli*. *EMBO J.* 9:4359–4366.

Emilsson, V., A. K. Näslund, and C. G. Kurland. 1992. Thiolation of transfer RNA in *Escherichia coli* varies with growth rate. *Nucleic Acids Res.* 20:4499–4505.

Ericson, J. U., and G. R. Björk. 1991. tRNA anticodons with the modified nucleoside 2-methylthio-N^6-(4-hydroxyisopentenyl)-adenosine distinguish between bases 3' of the codon. *J. Mol. Biol.* 218:509–516.

Esposito, F., T. Russo, R. Ammendola, A. Duilio, F. Salvatore, and F. Cimino. 1985. Pseudouridine excretion and transfer RNA primers for reverse transcriptase in tumors of retroviral origin. *Cancer Res.* 45:6260–6263.

Fantl, W. J., D. E. Johnson, and L. T. Williams. 1993. Signalling by receptor tyrosine kinases. *Annu. Rev. Biochem.* 62:453–481.

Farabaugh, P. J. 1996. Programmed translational frameshifting. *Microbiol. Rev.* 60:103–134.

Favre, A., E. Hajnsdorf, K. Thiam, and A. Caldeira de Araujo. 1985. Mutagenesis and growth delay induced in *Escherichia coli* by near-ultraviolet radiations. *Biochimie* 67:335–342.

Favre, A., M. Yaniv, and A. M. Michelson. 1969. The photochemistry of 4-thiouridine in *Escherichia coli* t-RNA$_1^{Val}$. *Biochem. Biophys. Res. Commun.* 37:266–271.

Fischbein, A., J. G. Bekesi, S. Solomon, E. Borek, and O. K. Sharma. 1987. Modified nucleosides in patients with acquired immune deficiency syndrome (AIDS) and individuals at high risk of AIDS: correlation with lymphadenomegaly and immunological parameters. *Cancer Detect. Prev. Suppl.* 1:589–596.

Fraser, C. M., J. D. Gocayne, O. White, M. D. Adams, R. A. Clayton, R. D. Fleischmann, C. J. Bult, A. R. Kerlavage, G. Sutton, J. M. Kelley, J. L. Fritchman, J. F. Weidman, K. V. Small, M. Sandusky, J. Fuhrmann, D. Nguyen, T. R. Utterback, D. B. Saudek, and C. A. Phillips. 1995. The minimal gene complement of Mycoplasma genitalium. *Science* 270:397–403.

Frey, B., J. McCloskey, W. Kersten, and H. Kersten. 1988. New function of vitamin B_{12}: cobamide-dependent reduction of epoxyqueuosine to queuosine in tRNAs of *Escherichia coli* and *Salmonella typhimurium*. *J. Bacteriol.* 170:2078–2082.

Gabriel, K., J. Schneider, and W. H. McClain. 1996. Functional evidence for indirect recognition of G.U in tRNA(Ala) by alanyl-tRNA synthetase. *Science* 271:195–197.

Gourse, R. L., T. Gaal, M. S. Bartlett, J. A. Appleman, and W. Ross. 1996. rRNA transcription and growth rate-dependent regulation of ribosome synthesis in *Escherichia coli*. *Annu. Rev. Microbiol.* 50:645–677.

Gray, J., J. Wang, and S. B. Gelvin. 1992. Mutation of the *miaA* gene of *Agrobacterium tumefaciens* results in reduced *vir* gene expression. *J. Bacteriol.* 174:1086–1098.

Griffiths, E., and J. Humphreys. 1978. Alterations in tRNAs containing 2-methylthio-N^6-(delta^2-isopentenyl)-adenosine during growth of enteropathogenic *Escherichia coli* in the presence of iron-binding proteins. *Eur. J. Biochem.* 82:503–513.

Gustafsson, C., and G. R. Björk. 1993. The tRNA-(m^{5}U54)-methyltransferase of *Escherichia coli* is present in two forms *in vivo*, one of which is present as bound to tRNA and to a 3'-end fragment of 16 S rRNA. *J. Biol. Chem.* 268:1326–1331.

Hagervall, T. G., Y. H. Jönsson, C. G. Edmonds, J. A. McCloskey, and G. R. Björk. 1990. Chorismic acid, a key metabolite in modification of tRNA. *J. Bacteriol.* 172:252–259.

Hagervall, T. G., T. M. Tuohy, J. F. Atkins, and G. R. Björk. 1993. Deficiency of 1-methylguanosine in tRNA from *Salmonella typhimurium* induces frameshifting by quadruplet translocation. *J. Mol. Biol.* 232:756–765.

Hanski, C., and G. Stehlik. 1980. Increased concentration of 7-methylguanine and 1-methylhypoxanthine in urine of rats bearing Yoshida tumour. *Cancer Lett.* 9:339–343.

Hatfield, D., Y. X. Feng, B. J. Lee, A. Rein, J. G. Levin, and S. Oroszlan. 1989. Chromatographic analysis of the aminoacyl-tRNAs which are required for translation of codons at and around the ribosomal frameshift sites of HIV, HTLV-1, and BLV. *Virology* 173:736–742.

Heldman, D. A., M. R. Grever, J. S. Miser, and R. W. Trewyn. 1983a. Relationship of urinary excretion of modified nucleosides to disease status in childhood acute lymphoblastic leukemia. *JNCI* 71:269–273.

Heldman, D. A., M. R. Grever, C. E. Speicher, and R. W. Trewyn. 1983b. Urinary excretion of modified nucleosides in chronic myelogenous leukemia. *J. Lab. Clin. Med.* 101:783–792.

Heldman, D. A., M. R. Grever, and R. W. Trewyn. 1983c. Differential excretion of modified nucleosides in adult acute leukemia. *Blood* 61:291–296.

Higley, B., J. De Mello, Jr., D. J. Oakes, and G. R. Giles. 1982. Urinary pseudouridine/creatinine ratio as an indicator of gastrointestinal cancer. *Br. J. Surg.* 69:699–701.

Hogan, A., A. Creuss-Callaghan, and J. J. Fennelly. 1970. Studies of pseudouridine changes in chronic lymphatic leukemia during therapy. *Ir. J. Med. Sci.* 3:505–511.

Holstege, A., M. Pauw, R. Haring, R. Kirchner, J. Pausch, and W. Gerok. 1986. Die Wertigkeit einer erhohten Urinausscheidung modifizierter Nukleoside als Tumormarker beim Kolonkarzinom. *Verh. Dtsch. Ges. Inn. Med.* 92:114–120.

Ibba, M., A. W. Curnow, and D. Söll. 1997. Aminoacyl-tRNA synthesis: divergent routes to a common goal. *Trends Biochem. Sci.* 22:39–42.

Intrieri, M., G. Calcagno, G. Oriani, F. Pane, F. Zarrilli, P. T. Cataldo, M. Foggia, M. Piazza, F. Salvatore, and L. Sacchetti. 1996. Pseudouridine and 1-ribosylpyridin-4-one-3-carboxamide (PCNR) serum concentrations in human immunodeficiency virus type 1-infected patients are independent predictors for AIDS progression. *J. Infect. Dis.* **174**:199–203.

Isel, C., R. Marquet, G. Keith, C. Ehresmann, and B. Ehresmann. 1993. Modified nucleotides of transfer RNA$_3^{Lys}$ modulate primer template loop-loop interaction in the initiation complex of hiv-1 reverse transcription. *J. Biol. Chem.* **268**:25269–25272.

Ito, M., Y. Takami, F. Tanabe, S. Shigeta, K. Tsukui, and Y. Kawade. 1988. Modulation of protein kinase C activity during inhibition of tumor cell growth by IFN-beta and -gamma. *Biochem. Biophys. Res. Commun.* **150**:126–132.

Itoh, K., S. Aida, S. Ishiwata, S. Sasaki, N. Ishida, and M. Mizugaki. 1993. Urinary excretion of modified nucleosides, pseudouridine and d1-methyladenosine, in healthy individuals. *Clin. Chim. Acta* **217**:221–223.

Itoh, K., T. Konno, T. Sasaki, S. Ishiwata, N. Ishida, and M. Misugaki. 1992. Relationship of urinary pseudouridine and 1-methyladenosine to activity of leukemia and lymphoma. *Clin. Chim. Acta* **206**:181–189.

Jeter, R. M., B. M. Olivera, and J. R. Roth. 1984. *Salmonella typhimurium* synthesizes cobalamin (vitamin B$_{12}$) de novo under anaerobic growth conditions. *J. Bacteriol.* **159**:206–213.

Johnston, H. M., W. M. Barnes, F. G. Chumley, L. Bossi, and J. R. Roth. 1980. Model for regulation of the histidine operon of *Salmonella*. *Proc. Natl. Acad. Sci. USA* **77**:508–512.

Jones, W. R., G. J. Barcak, and R. E. Wolf, Jr. 1990. Altered growth-rate-dependent regulation of 6-phosphogluconate dehydrogenase level in *hisT* mutants of *Salmonella typhimurium* and *Escherichia coli*. *J. Bacteriol.* **172**:1197–1205.

Katze, J. R., and W. T. Beck. 1980. Administration of queuine to mice relieves modified nucleoside queuosine deficiency in Ehrlich ascites tumor tRNA. *Biochem. Biophys. Res. Commun.* **96**:313–319.

Katze, J. R., W. T. Beck, C. S. Cheng, and J. A. McCloskey. 1983. Why is tumor tRNA hypomodified with respect to Q nucleoside? *Recent Results Cancer Res.* **84**:146–159.

Kawai, G., Y. Yamamoto, T. Kamimura, T. Masegi, M. Sekine, T. Hata, T. Iimori, T. Watanabe, T. Miyazawa, and S. Yokoyama. 1992. Conformational rigidity of specific pyrimidine residues in tRNA arises from posttranscriptional modifications that enhance steric interaction between the base and the 2'-hydroxyl group. *Biochemistry* **31**:1040–1046.

Kersten, H., and W. Kersten. 1990. Biosynthesis and function of queuine and queosine tRNAs, p. B69–B108. *In* C. W. Gehrke and K. C. T. Kuo (ed.), *Chromatography and Modification of Nucleosides*, part B. *Biological Roles and Function of Modification*. Elsevier, Amsterdam, The Netherlands.

Kersten, H., L. Sandig, and H. H. Arnold. 1975. Tetrahydrofolate-dependent 5-methyluracil-tRNA transferase activity in *B. subtilis*. *FEBS Lett.* **55**:57–60.

Kjellin-Stråby, K., and J. H. Phillips. 1968. Studies on microbial ribonucleic acid. VI. Appearance of methyl-deficient transfer ribonucleic acid during logarithmic growth of *Saccharomyces cerevisiae*. *J. Bacteriol.* **96**:760–767.

Koshida, K., J. Harmenberg, U. Stendahl, B. Wahren, E. Borgstrom, L. Helstrom, and L. Andersson. 1985. Urinary modified nucleosides as tumor markers in cancer of the urinary organs or female genital tract. *Urol. Res.* **13**:213–218.

Kramer, G. F., J. C. Baker, and B. N. Ames. 1988. Near-UV stress in *Salmonella typhimurium*: 4-thiouridine in tRNA, ppGpp, and ApppGpp as components of an adaptive response. *J. Bacteriol.* **170**:2344–2351.

Kumagai, I., K. Watanabe, and T. Oshima. 1980. Thermally induced biosynthesis of 2'-O-methylguanosine in tRNA from an extreme thermophile, *Thermus thermophilus* HB27. *Proc. Natl. Acad. Sci. USA* **77**:1922–1926.

Kvist, E., K. E. Sjölin, and J. Iversen. 1990. Excretion patterns of pseudouridine and beta-aminoisobutyric acid in patients with tumours of the upper urinary tract. *Scand. J. Urol. Nephrol.* **24**:287–292.

Langgut, W. 1995. Regulation of signalling by receptor tyrosine kinases in HeLa cells involves the q-base. *Biochem. Biophys. Res. Commun.* **207**:306–311.

Langgut, W., T. Reisser, and H. Kersten. 1990. Queuine modulates growth of HeLa cells depending on oxygen availability. *Biofactors* **2**:245–249.

Langgut, W., T. Reisser, H. Kersten, and S. Nishimura. 1993a. Modulation of epidermal growth factor receptor activity and related responses by the 7-deazaguanine derivative, queuine. *Oncogene* **8**:3141–3147.

Langgut, W., T. Reisser, S. Nishimura, and H. Kersten. 1993b. Modulation of mammalian cell proliferation by a modified tRNA base of bacterial origin. *FEBS Lett.* **336**:137–142.

Lewis, J. A, and B. N. Ames. 1972. Histidine regulation in *Salmonella typhimurium*. XI. The percentage of transfer RNA^{His} charged *in vivo* and its relation to the repression of the histidine operon. *J. Mol. Biol.* **66**:131–142.

Li, J. N., and G. R. Björk. 1995. 1-Methylguanosine deficiency of tRNA influences cognate codon interaction and metabolism in *Salmonella typhimurium*. *J. Bacteriol.* **177**:6593–6600.

Li, J.-N., and G. R. Björk. 1997. Unpublished results.

Li, J.-N., B. Esberg, J. F. Curran, and G. R. Björk. 1997. Three modified nucleosides present in the anticodon region influence the *in vivo* aa-tRNA selection differently and in a tRNA dependent manner. *J. Mol. Biol.* **271**:209–221.

Lin, W., J. W. Mackenzie, and I. Clark. 1984. Excretion of RNA catabolites by rats with hepatoma transplants. *Cancer Lett.* **22**:187–192.

Lin, W., J. W. Mackenzie, and I. Clark. 1987. Effect of transplanted osteogenic sarcoma on urinary RNA catabolites. *Cancer Lett.* **35**:47–57.

Lin, V. K., W. R. Farkas, and P. F. Agris. 1980. Specific changes in Q-ribonucleoside containing transfer RNA species during Friend leukemia cell erythroid differentiation. *Nucleic Acids Res.* **8**:3481–3489.

Low, S. C., and M. J. Berry. 1996. Knowing when not to stop: selenocysteine incorporation in eukaryotes. *Trends Biochem. Sci.* **21**:203–208.

Lu, J. Y., R. S. Lai, L. L. Liang, H. C. Wang, and T. I. Lin. 1994. Evaluation of urinary pseudouridine as a tumor marker in lung cancer. *J. Formos. Med. Assoc.* **93**:25–29.

Mandel, L. R., P. R. Srinivasan, and E. Borek. 1966. Origin of urinary methylated purines. *Nature* **209**:586–588.

Masuda, M., T. Nishihira, K. Itoh, M. Mizugaki, N. Ishida, and S. Mori. 1993. An immunohistochemical analysis for cancer of the esophagus using monoclonal antibodies specific for modified nucleosides. *Cancer* **72**:3571–3578.

McEntire, J. E., K. C. Kuo, M. E. Smith, D. L. Stalling, J. W. Richens, R. W. Zumwalt, C. W. Gehrke, and B. W. Papermaster. 1989. Classification of lung cancer patients and controls by chromatography of modified nucleosides in serum. *Cancer Res.* **49**:1057–1062.

Mills, G. C., F. C. Schmalstieg, R. J. Koolkin, and R. M. Goldblum. 1982. Urinary excretion of purines, purine nucleosides, and pseudouridine in immunodeficient children. *Biochem. Med.* **27**:37–45.

Morris, R. C., B. J. Brooks, P. Eriotou, D. F. Kelly, S. Sagar, K. L. Hart, and M. S. Elliott. 1995. Activation of transfer rna-guanine ribosyltransferase by protein kinase c. *Nucleic Acids Res.* **23**:2492–2498.

Müller-Wickop, J., H. Lorenz, K. Winkler, and N. Erb. 1986. The age dependency of the creatinine-related concentration of ribonucleosides in human urine. *J. Clin. Chem. Clin. Biochem.* 24: 993–999.

Murphy, T. L., I. A. Cooper, G. W. Wray, P. N. Ironside, and J. Matthews. 1976. Transfer RNA and transfer RNA methylase activity in spleens of patients with Hodgkin's disease and histiocytic lymphoma. *J. Natl. Cancer Inst.* 56:215–219.

Nesbitt, J. A. I., and W. J. Lennarz. 1968. Participation of aminoacyl transfer ribonucleic acid in aminoacyl phosphatidylglycerol synthetase. I. Specificity of lysyl phosphatidylglycerol synthetase. *J. Biol. Chem.* 243:3088–3095.

Newell, P. C., and R. G. Tucker. 1968. Precursors of the pyrimidine moiety of thiamine. A new route of pyrimidine biosynthesis involving purine intermediates. *Biochem. J.* 106:279–287.

Nielsen, H. R., and S.-A. Killman. 1983. Urinary excretion of β-aminoisobutyrate and pseudouridine in acute and chronic myeloid leukemia. *JNCI* 71:887–891.

Nilsson, K., and G. R. Björk. 1997. Unpublished results.

Nishimura, S. 1983. Structure, biosynthesis, and function of queuosine in transfer RNA. *Prog. Nucleic Acid Res. Mol. Biol.* 28: 49–73.

Noguchi, S., Y. Nishimura, Y. Hirota, and S. Nishimura. 1982. Isolation and characterization of an *Escherichia coli* mutant lacking tRNA-guanine transglycosylase. Function and biosynthesis of queuosine in tRNA. *J. Biol. Chem.* 257:6544–6550.

Ny, T., J. Thomale, K. Hjalmarsson, G. Nass, and G. R. Björk. 1980. Non-coordinate regulation of enzymes involved in transfer RNA metabolism in *Escherichia coli*. *Biochim. Biophys. Acta* 607:277–284.

Okada, N., N. Shindo-Okada, and S. Nishimura. 1977. Isolation of mammalian tRNAAsp and tRNATyr by lectin-Sepharose affinity column chromatography. *Nucleic Acids Res.* 4:415–423.

Okada, N., N. Shindo-Okada, S. Sato, Y. H. Itoh, K. Oda, and S. Nishimura. 1978. Detection of unique tRNA species in tumor tissues by *Escherichia coli* guanine insertion enzyme. *Proc. Natl. Acad. Sci. USA* 75:4247–4251.

Ott, G., and H. Kersten. 1985. Differential turnover of tRNAs of the queuosine family in *Dictyostelium discoideum* and its possible role in regulation. *Biol. Chem. Hoppe-Seyler* 366:69–76.

Pane, F., M. Savoia, G. Fortunato, A. Camera, B. Rotoli, F. Salvatore, and L. Sacchetti. 1993. Serum pseudouridine in the diagnosis of acute leukaemias and as a novel prognostic indicator in acute lymphoblastic leukaemia. *Clin. Biochem.* 26: 513–520.

Park, R. W., J. F. Holland, and A. Jenkins. 1962. Urinary purines in leukemia. *Cancer Res.* 22:469–477.

Persson, B. C., and G. R. Björk. 1993. Isolation of the gene (*miaE*) encoding the hydroxylase involved in the synthesis of 2-methylthio-*cis*-ribozeatin in tRNA of *Salmonella typhimurium* and characterization of mutants. *J. Bacteriol.* 175:7776–7785.

Persson, B. C., C. Gustafsson, D. E. Berg, and G. R. Björk. 1992. The gene for a tRNA modifying enzyme, m^{5}U54-methyltransferase, is essential for viability in *Escherichia coli*. *Proc. Natl. Acad. Sci. USA* 89:3995–3998.

Persson, B. C., O. Olafsson, H. Lundgren, L. Hederstedt, and G. R. Björk. 1997. Unpublished results.

Pinkard, K. J., I. A. Cooper, R. Motteram, and C. N. Turner. 1972. Purine and pyrimidine excretion in Hodgkin's disease. *J. Natl. Cancer Inst.* 49:27–38.

Pongs, O., and N. Ulbrich. 1976. Specific binding of formylated initiator-tRNA to *Escherichia coli* RNA polymerase. *Proc. Natl. Acad. Sci. USA* 73:3064–3067.

Pope, W. T., A. Brown, and R. H. Reeves. 1978. The identification of the tRNA substrates for the supK tRNA methylase. *Nucleic Acids Res.* 5:1041–1057.

Qian, Q., and G. R. Björk. 1997. Unpublished results.

Quay, S. C., R. P. Lawther, G. W. Hatfield, and D. L. Oxender. 1978. Branched-chain amino acid transport regulation in mutants blocked in tRNA maturation and transcriptional termination. *J. Bacteriol.* 134:683–686.

Racker, E., R. J. Resnick, and R. Feldman. 1985. Glycolysis and methylaminoisobutyrate uptake in rat-1 cells transfected with *ras* or *myc* oncogenes. *Proc. Natl. Acad. Sci. USA* 82:3535–3538.

Rasmuson, T., G. R. Björk, L. Damber, L. Jacobsson, A. Jeppsson, T. Stigbrand, and G. Westman. 1987. Tumor markers in mammary carcinoma. An evaluation of carcinoembryonic antigen, placental alkaline phosphatase, pseudouridine and CA-50. *Acta Oncol.* 26:261–267.

Rasmuson, T., and G. R. Björk. 1995. Urinary excretion of pseudouridine and prognosis of patients with malignant lymphoma. *Acta Oncol.* 34:61–67.

Rasmuson, T., G. R. Björk, L. Damber, S. E. Holm, L. Jacobsson, A. Jeppsson, B. Littbrand, T. Stigbrand, and G. Westman. 1983a. Evaluation of carcinoembryonic antigen, tissue polypeptide antigen, placental alkaline phosphatase, and modified nucleosides as biological markers in malignant lymphomas. *Recent Results Cancer Res.* 84:331–343.

Rasmuson, T., G. R. Björk, L. Damber, S. E. Holm, L. Jacobsson, A. Jeppsson, T. Stigbrand, and G. Westman. 1983b. Tumor markers in bronchogenic carcinoma. An evaluation of carcinoembryonic antigen, tissue polypeptide antigen, placental alkaline phosphatase and pseudouridine. *Acta Radiol.* 22: 209–214.

Rasmuson, T., G. R. Björk, L. Damber, S. E. Holm, L. Jacobsson, A. Jeppsson, T. Stigbrand, and G. Westman. 1984. Tumor markers in colorectal carcinoma. An evaluation of carcinoembryonic antigen, tissue polypeptide antigen, placental alkaline phosphatase and pseudouridine. *Acta Radiol.* 23:27–32.

Rasmuson, T., G. R. Björk, S. O. Hietala, R. Stenling, and B. Ljungberg. 1991. Excretion of pseudouridine as an independent prognostic factor in renal cell carcinoma. *Acta Oncol.* 30:11–15.

Reisser, T., W. Langgut, and H. Kersten. 1994. The nutrient factor queuine protects HeLa cells from hypoxic stress and improves metabolic adaptation to oxygen availability. *Eur. J. Biochem.* 221:979–986.

Rogg, H., P. Müller, G. Keith, and M. Staehelin. 1977. Chemical basis for brain-specific serine transfer RNAs. *Proc. Natl. Acad. Sci. USA* 74:4243–4247.

Rose, A. M., H. G. Belford, W. C. Shen, C. L. Greer, A. K. Hopper, and N. C. Martin. 1995. Location of n-2,n-2-dimethylguanosine-specific trna methyltransferase. *Biochimie* 77:45–53.

Rosenberg, A. H., and M. L. Gefter. 1969. An iron-dependent modification of several transfer RNA species in *Escherichia coli*. *J. Mol. Biol.* 46:581–584.

Russo, T., A. Colonna, F. Salvatore, F. Cimino, S. Bridges, and C. Gurgo. 1984. Serum pseudouridine as a biochemical marker in the development of AKR mouse lymphoma. *Cancer Res.* 44: 2567–2570.

Salyers, A. A., and D. D. Whitt. 1994. *Bacterial Pathogenesis: a Molecular Approach*. ASM Press, Washington, D.C.

Schachner, E., H. J. Aschhoff, and H. Kersten. 1984. Specific changes in lactate levels, lactate dehydrogenase patterns and cytochrome b559 in *Dictyostelium discoideum* caused by queuine. *Eur. J. Biochem.* 139:481–487.

Schachner, E., and H. Kersten. 1984. Queuine deficiency and restoration in *Dictyostelium discoideum* and related early developmental changes. *J. Gen. Microbiol.* 130:135–144.

Schöch, G., M. Garbrecht, G. Heller Schöch, H. Baisch, and W. Leifer. 1979. Die Ausscheidung von normalen und modifizierten Nucleobasen im Urin bein chronischen myeloproliferativen Syndromen. *Blut* 38:391–396.

Schöch, G., H. Lorenz, G. Heller-Schöch, H. Baisch, and P. Clemens. 1981. Die Altersabhangigkeit der normalen und modifizierten Nucleobasen im Urin als Ausdruck der Wachstumsgeschwindigkeit. *Monatsschr. Kinderheilkd.* **129**:29–33.

Schön, A., C. G. Kannangara, S. Gough, and D. Söll. 1988. Protein biosynthesis in organelles requires misaminoacylation of tRNA. *Nature* **331**:187–190.

Schoots, A. C., P. G. Gerlag, A. W. Mulder, J. A. Peeters, and C. A. Cramers. 1988. Liquid-chromatographic profiling of solutes in serum of uremic patients undergoing hemodialysis and chronic ambulatory peritoneal dialysis (CAPD); high concentrations of pseudouridine in CAPD patients. *Clin. Chem.* **34**:91–97.

Schwartz, R, and J. F. Curran. 1997. Analyses of frameshifting at UUU-pyrimidine sites. *Nucleic Acids Res.* **25**:2005–2011.

Simos, G., H. Tekotte, H. Grosjean, A. Segref, K. Sharma, D. Tollervey, and E. C. Hurt. 1996. Nuclear pore proteins are involved in the biogenesis of functional tRNA. *EMBO J.* **15**:2270–2284.

Singh, R., and M. R. Green. 1993. Sequence-specific binding of transfer RNA by glyceraldehyde-3-phosphate dehydrogenase. *Science* **15**:365–368.

Singhal, R. P., R. A. Kopper, S. Nishimura, and N. Shindo-Okada. 1981. Modification of guanine to queuine in transfer RNAs during development and aging. *Biochem. Biophys. Res. Commun.* **99**:120–126.

Söll, D. 1993. Transfer RNA: an RNA for all seasons, p. 157–184. *In* R. F. Gesteland and J. F. Atkins (ed.), *The RNA World*. Cold Spring Harbor Laboratory Press, Cold Spring Harbor, N.Y.

Sörensen, S. H., D. A. Brown, E. H. Cooper, K. A. Kelly, and I. C. MacLennan. 1985. Urinary pseudouridine excretion in myelomatosis. *Br. J. Cancer* **52**:863–866.

Speer, J., C. W. Gehrke, K. C. Kuo, T. P. Waalkes, and E. Borek. 1979. tRNA breakdown products as markers for cancer. *Cancer* **44**:2120–2123.

Stewart, T. S., R. J. Roberts, and J. L. Strominger. 1971. Novel species of tRNA. *Nature* **230**:36–38.

Stråby, K. B., M. Eriksson, A. Housseini, K. Spångberg, and M. Sterner. 1997. A possible link between tRNA modification and stress response in the yeast *Saccharomyces cerevisiae*. *Yeast* **13**:S1–S274.

Struijk, D. G., A. C. Schoots, L. H. Koole, H. J. van der Reijden, G. C. Koomen, R. T. Krediet, and L. Arisz. 1991. Transport kinetics of pseudouridine during hemodialysis and continuous ambulatory peritoneal dialysis. *J. Lab. Clin. Med.* **118**:74–80.

Tamura, S., Y. Amuro, T. Nakano, J. Fujii, Y. Moriwaki, T. Yamamoto, T. Hada, and K. Higashino. 1986a. Urinary excretion of pseudouridine in patients with hepatocellular carcinoma. *Cancer* **57**:1571–1575.

Tamura, S., J. Fujii, T. Nakano, T. Hada, and K. Higashino. 1986b. Urinary pseudouridine as a tumor marker in patients with small cell lung cancer. *Clin. Chim. Acta* **154**:125–132.

Tamura, S., H. Fujioka, T. Nakano, Y. Amuro, T. Hada, N. Nakao, and K. Higashino. 1988. Urinary pseudouridine as a biochemical marker in the diagnosis and monitoring of primary hepatocellular carcinoma. *Am. J. Gastroenterol.* **83**:841–845.

Tamura, S., H. Fujioka, T. Nakano, T. Hada, and K. Higashino. 1987. Serum pseudouridine as a biochemical marker in small cell lung cancer. *Cancer Res.* **47**:6138–6141.

Taya, Y., and S. Nishimura. 1977. Purification and properties of the tRNA methylase specific for synthesis of 5-methylaminomethyl-2-thiouridine, p. 251–257. *In* F. Salvatore, E. Borek, V. Zappia, H. G. Williams-Ashman, and F. Schlenk (ed.), *Biochemistry of Adenosylmethionine*. Columbia University Press, New York, N.Y.

Thomale, J., A. Luz, and G. Nass. 1984. Excretion of modified nucleosides during development of malignant lymphomas in mice after whole body irradiation. *J. Cancer Res. Clin. Oncol.* **108**:302–307.

Thomale, J., and G. Nass. 1978. Alteration of the intracellular concentration of aminoacyl-tRNA synthetases and isoaccepting tRNAs during amino-acid limited growth in *Escherichia coli*. *Eur. J. Biochem.* **85**:407–418.

Thomale, J., and G. Nass. 1982. Elevated urinary excretion of RNA catabolites as an early signal of tumor development in mice. *Cancer Lett.* **15**:149–159.

Thomas, G., and A. Favre. 1975. 4-Thiouridine as the target for near-ultraviolet light induced growth delay in *Escherichia coli*. *Biochem. Biophys. Res. Commun.* **66**:1454–1461.

Thomas, G., and A. Favre. 1980. 4-Thiouridine triggers both growth delay induced by near-ultraviolet light and photoprotection. *Eur. J. Biochem.* **113**:67–74.

Tormey, D. C., T. P. Waalkes, and C. W. Gehrke. 1980. Biological markers in breast carcinoma—clinical correlations with pseudouridine, N^2,N^2-dimethylguanosine, and 1-methylinosine. *J. Surg. Oncol.* **14**:267–273.

Trewyn, R. W., and M. R. Grever. 1986. Urinary nucleosides in leukemia: laboratory and clinical applications. *Crit. Rev. Clin. Lab. Sci.* **24**:71–93.

Tröbner, W., and R. Piechocki. 1984. Competition between isogenic *mutS* and *mut+* populations of *Escherichia coli* K12 in continuously growing cultures. *Mol. Gen. Genet.* **198**:175–176.

Tsui, H.-C. T., G. Feng, and M. E. Winkler. 1996. Transcription of the *mutL* repair, *miaA* tRNA modification, *hfq* pleiotropic regulator, and *hflA* region protease genes of *Escherichia coli* K-12 from clustered $E\sigma^{32}$-specific promoters during heat shock. *J. Bacteriol.* **178**:5719–5731.

Tsui, H. C. T., H. C. E. Leung, and M. E. Winkler. 1994. Characterization of broadly pleiotropic phenotypes caused by an *hfq* insertion mutation in *Escherichia coli* K-12. *Mol. Microbiol.* **13**:35–49.

Van Acker, K. J., F. J. Eyskens, R. M. Verkerk, and S. S. Scharpe. 1993. Urinary excretion of purine and pyrimidine metabolites in the neonate. *Pediatr. Res.* **34**:762–766.

Verkamp, E., A. M. Kumar, A. Lloyd, O. Martins, N. Stange-Thomann, and D. Söll. 1995. Glutamyl-tRNA as an intermediate in glutamate conversions, p. 545–550. *In* D. Söll and U. L. RajBhandary (ed.), *tRNA: Structure, Biosynthesis, and Function*. ASM Press, Washington, D.C.

Viale, G. L. 1971. Transfer RNA and transfer RNA methylase in human brain tumors. *Cancer Res.* **31**:605–608.

von Pawel-Rammingen, U. 1997. Biosynthesis and function of initiator tRNA in *Saccharomyces cerevisiae*. Ph.D. thesis. Umeå Universitet, Umeå, Sweden.

Vreken, P., and P. Tavenier. 1987. Urinary excretion of six modified nucleosides by patients with breast carcinoma. *Ann. Clin. Biochem.* **24**:598–603.

Waalkes, T. P., M. D. Abeloff, D. S. Ettinger, K. B. Woo, C. W. Gehrke, K. C. Kuo, and E. Borek. 1982. Modified ribonucleosides as biological markers for patients with small cell carcinoma of the lung. *Eur. J. Cancer Clin. Oncol.* **18**:1267–1274.

Warburg, O. 1956. On the origin of cancer cells. *Science* **123**:309–314.

Watanabe, K., M. Shinma, T. Oshima, and S. Nishimura. 1976. Heat-induced stability of tRNA from an extreme thermophile, *Thermus thermophilus*. *Biochem. Biophys. Res. Commun.* **72**:1137–1144.

Wettstein, F. O., and G. S. Stent. 1968. Physiologically induced changes in the property of phenylalanine tRNA in *Escherichia coli*. *J. Mol. Biol.* **38**:25–40.

Wikström, P. M., and G. R. Björk. 1988. Noncoordinate translation-level regulation of ribosomal and nonribosomal protein genes in the *Escherichia coli trmD* operon. *J. Bacteriol.* **170**:

3025–3031.

Wilcox, M., and M. Nirenberg. 1968. Transfer RNA as a cofactor coupling amino acid synthesis with that of protein. *Proc. Natl. Acad. Sci. USA* **61:**229–236.

Williamson, R. M., and D. L. Oxender. 1992. Premature termination of in vivo transcription of a gene encoding a branched-chain amino acid transport protein in *Escherichia coli*. *J. Bacteriol.* **174:**1777–1782.

Wittwer, A. J., and W. M. Ching. 1989. Selenium-containing tRNA^{Glu} and tRNA^{Lys} from *Escherichia coli:* purification, codon specificity and translational activity. *Biofactors* **2:**27–34.

Yokoyama, S., T. Watanabe, K. Murao, H. Ishikura, Z. Yamaizumi, S. Nishimura, and T. Miyazawa. 1985. Molecular mechanism of codon recognition by tRNA species with modified uridine in the first position of the anticodon. *Proc. Natl. Acad. Sci. USA* **82:**4905–4909.

Chapter 27

Modified Nucleosides in Translation

JAMES F. CURRAN

Cellular physiology is fundamentally dependent on the functions of our translational apparatus, and these functions are dependent on modified nucleosides. Cells are composed primarily of protein, and cellular growth and maintenance are strictly dependent on the synthesis of this protein mass (Alberts et al., 1994). Cells devote a large fraction of their resources to protein synthesis. The translational apparatus is a major fraction of the cell, and translation is the most costly macromolecular biosynthetic process, costing more than 10 times as much NTP as RNA and DNA synthesis. Clearly, translational accuracy and efficiency are important to cellular physiology. Modified nucleosides contribute to genetic translation in several ways. Base modifications at the wobble position of tRNA (position 34) can change the patterns of hydrogen bond donors and acceptors, and thus directly change base pairing specificity. Nucleoside modifications in the anticodon region can also affect translational specificity and efficiency by less direct mechanisms. By modulating anticodon arm conformation, modifications can affect tRNA:message complex structure and stability. These modulations can increase translational efficiency and prevent frameshifting.

This chapter examines the translational functions of modified nucleosides in the anticodon arm of tRNA. There are modifications at other positions within tRNA, but our knowledge of translational effects is limited to the modifications in the anticodon region. Emphasis is placed on the effects that the loss of specific modifications has on the activities of tRNA. A common theme is that modifications generally affect decoding efficiency, but not all tRNAs may be equally affected by the loss of the same nucleoside modification. This complexity of function undoubtedly reflects the complexity of the translational apparatus and processes. There are a number of other recent, relevant reviews on the biosynthesis and function of modified nucleosides in tRNA. Björk (1995a, 1995b, 1995c) has published several recent, thorough reviews on the genetics, biochemistry and function of modified nucleosides. Agris (1996) discusses the many ways in which modifications can affect the physicochemical properties of bases and reviews physical studies of model oligonucleotides. Yokoyama and Nishimura (1995) provide detailed descriptions of the decoding properties of anticodon region nucleotides with emphases on how nucleoside structure affects decoding specificity. Another very useful reference is the paper by Lim (1994) in which molecular modeling was used to predict the decoding properties of all nucleotides at position 34 in tRNA. Though the work in the paper is theoretical, it does explain the decoding spectra of modified nucleosides at position 34. This work should serve as a useful reference for the possible structural features of anticodon:codon complexes that can guide the experimentation and interpretation of functional studies. Chapter 5 of this volume reviews the structural and biophysical properties of modified nucleosides. Other related works include those by Osawa et al. (1992), who provide interesting perspectives on the evolution of the genetic code, and by Ikemura (1992), who gives an interesting discussion of the relationships between codon usage and tRNA content.

A BRIEF VIEW OF GENETIC TRANSLATION

Before we consider the effects of modifications on translation, it is helpful to review certain aspects of the decoding process. A detailed review of translation is beyond the scope of this chapter, and I will focus on the roles of the anticodon arm. Readers wanting more thorough discussions should consult

---

James F. Curran • Department of Biology, Wake Forest University, P.O. Box 7325, Winston-Salem, North Carolina 27109.

the many excellent recent reviews on protein synthesis. These include the chapters by Yarus and Smith (1995) on the functions of tRNA during translation and by Kurland et al. (1995) on ribosomal accuracy. Thompson (1988) describes the evidence that supports the widely accepted proofreading model for aminoacyl-tRNA selection, and Nierhaus (1993, 1996) proposes an interesting, hypothetical mechanism for aminoacyl-tRNA selection that does not require proofreading. Other useful works are the reviews on ribosomes and translation by Green and Noller (1997) and by Hardesty et al. (1991).

The structural features of the anticodon arm are shown in Fig. 1. The anticodon is at bottom right, with the bases projecting back and toward the right. The anticodon arm plays several roles during translation. First, aminoacyl-tRNA (aa-tRNA) selection requires that it bind tightly to the triplet in the ribosomal A (aminoacyl) site. Next, stable pairing to the message helps maintain the reading frame during and after translocation to the P (peptidyl) site. In addition, once in the P site, the anticodon arm interacts with the tRNAs that attempt to read the next codon in the A site (Smith and Yarus, 1989). These functions are dependent on features throughout the anticodon arm of tRNA. Obviously, base pairing between the anticodon and the message is paramount. The codon bases at positions one and two require Watson–Crick pairing to tRNA bases 36 and 35, respectively. These positions are rarely modified in tRNA; an exception that is described later in this section is pseudouridine (Ψ) at position 35 (center of anticodon) of a yeast tRNA. The third position of the codon can be read by non-Watson–Crick (wobble) base pair geometries. One consequence is that individual tRNAs can read more than one codon. In these wobble pairs, the third codon nucleotide is fixed in the standard A conformation (Lim and Venclovas, 1992; Yokoyama and Nishimura, 1995), while the anticodon nucleoside 34 assumes a conformation that is stabilized by two hydrogen bonds between the paired bases. In 1966, Crick published his famous "wobble hypothesis" (Crick, 1966), which predicts the acceptable wobble base pairs. Crick reasoned that base pairs whose structures are similar to Watson–Crick might be incorporated at the wobble position without disrupting the rest of the anticodon: codon complex. Representatives of the predicted, allowable structures are shown in the left half of Fig. 2. These predictions have been shown to be largely correct, and the few exceptions are discussed in the following paragraphs.

tRNAs contain a wide variety of modified nucleosides at the wobble position. By altering the pattern of hydrogen bond donors and acceptors, modifications can directly affect translation. The adenosine to inosine (I) modification, for example, allows base pairing of three bases (C, U, and A) and thus extends wobbling. Instead of altering hydrogen bonding groups, many modifications affect wobbling by altering nucleoside conformation. Certain of those modifications increase wobbling, while others restrict it. We can therefore think of specific modifications as being either determinants or antideterminants for the translation of specific codons.

The structure and stability of the entire anticodon arm is important for decoding. The anticodon is at the 5' end of a stacked helix that is continuous with the anticodon stem (see Fig. 1). This helix positions the anticodon for interaction with the message. Sequences near the anticodon, including modified nucleosides, are highly correlated with the base at position 36, which reads the first base of the cognate codon(s) (Yarus, 1982). Mutations at these positions can dramatically decrease decoding efficiency (Yarus et al., 1986). There are numerous examples of modified nucleosides near the anticodon that facilitate aa-tRNA selection and reading frame maintenance, presumably by modulating the structure and stability of tRNA:message complexes. Position 37, which is immediately 3' to the nucleoside that decodes the first codon base, is very frequently modified. Modifications here can strongly increase decoding efficiency, and may prevent misreading (Jukes, 1973). A common theme is that base modifications at position 37 increase anticodon stability, at least in part by increasing base stacking (Jukes, 1973; Nishimura, 1972). Modified nucleosides at other positions in the anticodon arm also increase anticodon arm stability and translational efficiency. In addition, modifications at position 37 may even help determine which tRNA nucleotides function as the "anticodon." At least one modification deficiency at position 37 (m^1G deficiency) may prevent G37 from pairing with the message (Pieczenik, 1980), and the absence of m^1G can cause frameshifting (Björk et al., 1989; Hagervall et al., 1993).

Though the anticodon region is important, other factors also contribute to translation. For example, aa-tRNAs undergo conformational changes during ribosomal selection (Yarus and Smith, 1995), and those changes can depend on tRNA sequences far from the anticodon (Curran et al., 1995; Schultz and Yarus, 1994a, 1994b, 1989b, 1989c). Unfortunately, we do not fully understand how the anticodon arm contributes to these structural dynamics. Message sequences near codons (codon context) can also affect decoding efficiency and accuracy (Bossi and Roth, 1980; Buckingham, 1994; Lim, 1997; Miller and Albertini, 1983; Parker, 1989; Precup and Parker, 1987; Salser,

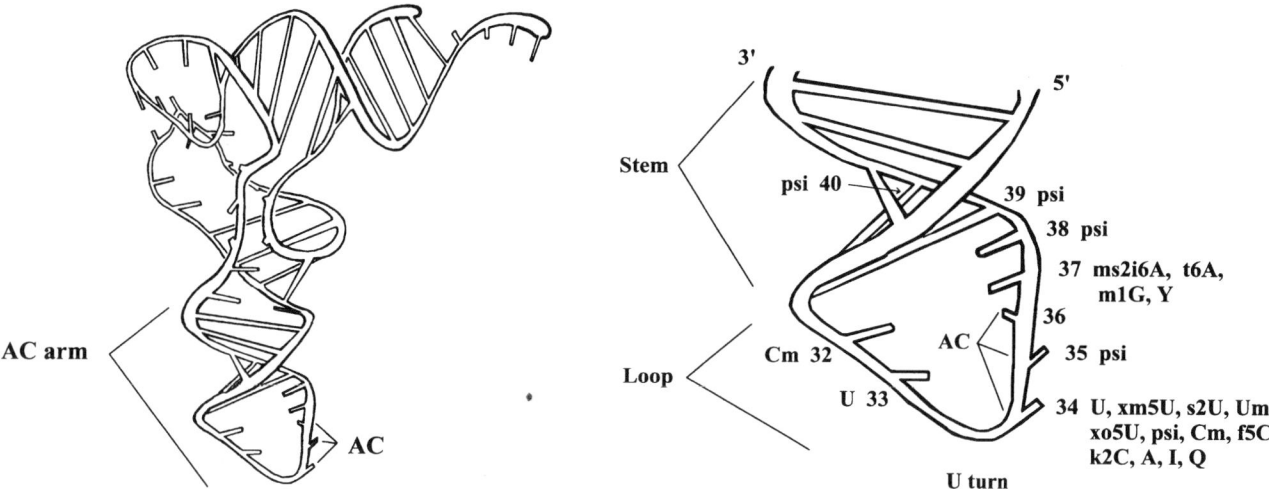

Figure 1. Schematic structure of the tRNA anticodon arm. The ribbon structure of yeast tRNA^{Phe} is shown on the left. The anticodon arm (AC arm) is the bottom quarter of the molecule, and the anticodon nucleotides (AC) are indicated at the bottom. A detailed view of the AC arm is shown on the right. The stem includes the five base pairs on the top, and the loop contains seven nucleotides. The anticodon nucleotides are 36, 35, and 34, which read codon bases 1, 2, and 3, respectively. The "U turn" in the backbone occurs between the highly conserved U33 and the 5′ end of the anticodon. All of the nucleotides discussed in the text are indicated. (Adapted from Saenger, 1984.)

1969; Tate and Mannering, 1996; Yarus and Curran, 1992). At least some context effects may be due to interactions between the nascent polypeptide and the ribosome (Gu et al., 1994; Lovett and Rogers, 1996). The mechanisms by which these extrinsic factors affect decoding are largely unknown and, therefore, unpredictable. Finally, it seems clear that tRNAs in the P and A ribosomal sites interact, but the structural details of the interactions between anticodon arms are not yet clear (see Lim et al., 1992). All of this suggests that specific modified nucleosides in the anticodon region may not affect all tRNAs equally and that effects may vary among message sites. Recent observations suggest that this phenomenon may be rather common. Further work on the translational mechanism is needed to fully understand the roles of modified nucleosides in this important cellular process.

EXPERIMENTAL APPROACHES

This section discusses the effects of modified nucleosides at various positions in the anticodon arm. There are data on the translational effects of a subset of the modified nucleosides that occur within the anticodon arm. All of the nucleosides discussed here are indicated in Fig. 1. The structures of the modified nucleosides can be found in Appendix 1 of this book. For each modification, I briefly discuss its known or predicted physical effects on anticodon loop structure, which are determined by physical methods and molecular modeling. I then discuss the functional data on decoding properties, which are determined primarily from in vivo and in vitro protein synthesis assays.

The physical methods and models are useful because they can indicate intrinsic physical-chemical properties of RNAs that may determine in vivo function. However, because they are measured out of the context of the cellular ribosome, these systems provide models for what may occur in vivo. The in vivo assays neatly account for all of the cellular conditions missing from in vitro studies. However, most in vivo assays monitor a specific translational phenotype, such as nonsense suppression. Therefore, unless large numbers of assays are employed, in vivo studies may not provide the molecular detail needed for full understanding of the underlying physical processes. Therefore, our understanding of the roles of modified nucleosides is most complete for cases in which complementary physical and decoding functional data exist.

Data on the structural and physical effects of modified nucleosides are obtained through molecular modeling studies and various spectroscopic methods, including circular dichroism (CD) and nuclear magnetic resonance (NMR) (Agris et al., 1992; Agris, 1996; Agris et al., 1997; Davis and Poulter, 1991; Davis, 1995; Varani and Tinoco, 1991; Watanabe, 1980; Watanabe et al., 1994; Yokoyama et al., 1979; Yokoyama et al., 1985). Structural studies are reviewed in Chapter 5 of this volume. One method that has been particularly useful for understanding tRNA

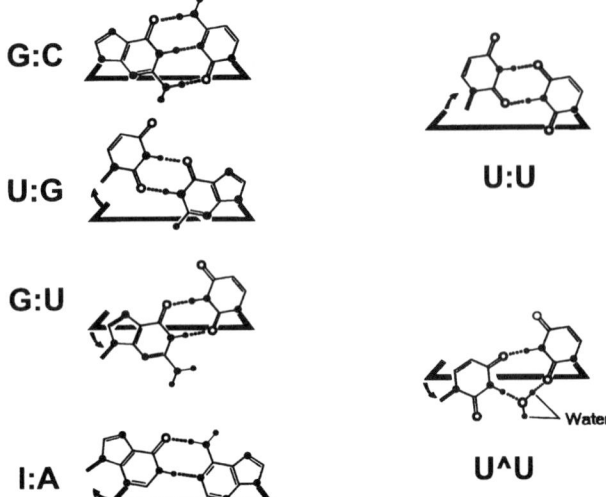

Figure 2. Examples of base pair geometries that may occur at the wobble position. On the left are representatives of the four structures predicted by Crick (1966) to function in decoding. For each base pair, the codon nucleotide is on the right and is fixed in the standard 3'-endo, A conformation (see the discussions in Li and Venclovas, 1992; Yokoyama and Nishimura, 1995). The anticodon nucleotide may assume a "wobble" conformation. The boldface brackets indicate the positions of the N-glycosidic bonds for the Watson–Crick base pairs. Anticodon nucleotide wobble deviations from Watson–Crick geometry are indicated by curved arrows. At the top left is the G:C base pair, which is representative of the Watson–Crick base pairs (G:C, C:G, A:U, U:A, and I:C), all of which are geometrically equivalent. Just below are the U:G and G:U base pairs, in which the anticodon bases are slightly shifted into the major and minor grooves, respectively. The I:U base pair is geometrically equivalent to the G:U base pair. At the bottom left is the "long wobble" I:A base pair. On the right are two hypothetical schemes for base pairing between uridines. At the top right is the "short wobble" conformation favored by Yokoyama and Nishimura (1995). This base pair requires a 2'-endo conformation for the anticodon uridine. At the bottom right is the conformation preferred by Lim (1995) in which a water molecule in the minor groove bridges between the uridine bases. This base pair does not require a 2'-endo conformation, but it does require a significant rotation of the anticodon base about the N-glycosidic bond (propeller twist). The propeller twist is not depicted in this two-dimensional diagram. (Adapted from Saenger, 1984.)

function is Grosjean's assay (Grosjean et al., 1976, 1978, 1985; Grosjean and Houssier, 1990; Houssier and Grosjean, 1985; Houssier et al., 1988; Vacher al., 1984) for duplex stability using tRNAs paired with complementary anticodons. Here, tRNAs are allowed to pair, and their stability is monitored by measuring rates of dissociation following a temperature jump. Comparisons of many pairs of tRNAs, including species that lack certain modifications, are very useful for understanding how those modifications affect anticodon function. A variety of studies monitor the effects of nucleoside modification with in vitro translation systems. Such systems can be supplemented with tRNAs that differ in their modification states (e.g., Takai et al., 1996) to determine how nucleoside modifications affect tRNA translational function.

Many of the in vivo assays are performed with *Escherichia coli* and *Salmonella typhimurium* mutants that fail to perform specific modifications (reviewed in Björk, 1995a). There are also limited data from eukaryotic systems. Commonly used in vivo assays include measurements of growth rates and polypeptide elongation rate. The latter is determined from measurements of the time required to complete synthesis of β-galactosidase following induction of the *lac* operon (Ericson and Björk, 1986; Palmer et al., 1983). Useful information is also obtained from assays for the deattenuation of amino acid biosynthetic operons. In these systems, expression of the genes for the biosynthetic enzymes is dependent on ribosomal progression through codons specific for the amino acid products of the operon encoded by a leader message sequence (Barnes, 1978; Johnston et al., 1980; Lee and Yanofsky, 1977; reviewed in Landick et al., 1996). Factors that decrease ribosomal rates can cause the expression (deattenutation) of the downstream genes. If modification-deficient tRNAs deattenuate a specific operon, then one concludes that the modification deficiency decreases translational efficiency at the regulatory codons in the leader sequence. The previously described assays have all been used to show that modification deficiencies can reduce the rate of translation.

Nonsense suppression assays are also commonly used (Eggertsson and Söll, 1988; Murgola, 1995). Here, nonsense (termination) codons located within the coding regions of reporter genes are read by mutant tRNAs that have anticodons complementary to nonsense codons (suppressor tRNAs). This suppression of the nonsense mutation allows for the production of the protein, and the level of production is an indication of the translational activity of the suppressor tRNA. The absence of certain modifications can reduce the translational efficiencies of suppressor tRNAs. Missense reading assays (Parker and Friesen, 1980) test for the ability of tRNAs to misread codons. The absence of certain modifications can affect the abilities of tRNAs to misread.

Frameshifting assays are commonly used to determine whether a modification contributes to reading frame maintenance (Björk et al., 1989; Brierley et al., 1997; Hagervall et al., 1993a, 1993b; Qian and Björk, personal communication; Schwartz and Curran, 1997). High levels of frameshifting are related to weak tRNA:message pairing (Brierley et al., 1992; Curran, 1993; Dinman et al., 1991; Fu and

Parker 1994; Schwartz and Curran 1997; Tsuchihashi and Brown, 1992), which allows for slippage. In addition, the absence of specific modifications may cause tRNAs to read out of frame. Effects on frameshifting are detected by assays for suppression of frameshift mutations and by assays for programmed frameshifting. Suppression of a frameshift mutation requires that a translational frameshift occur at or near the site of the mutation to restore the reading frame, allowing the production of a normal (or nearly normal) protein. Because background gene function is usually very low for frameshift mutants, frameshift suppression can be extremely sensitive. Often, however, the codon at which the suppressing frameshift occurs is not apparent, and polypeptide sequencing is necessary to determine the site of the shift. The peptide sequence may suggest which tRNAs are responsible.

Assays at programmed frameshift sites have the advantage that the sites and mechanisms of the frameshifts are known. Two kinds of programmed frameshifts have been used. I will provide only brief sketches here; there are excellent recent reviews of programmed frameshifting (Atkins and Gesteland, 1995; Farabaugh, 1996; Gestland and Atkins, 1996). The "simultaneous slippage" (Jacks et al., 1988; Weiss et al., 1989) frameshifts occur at seven-nucleotide message sequences of the general type: X XXY YYZ (ten Dam, 1990). Here, frameshifting occurs when the two tRNAs paired at the last two message triplets (that is, XXY YYZ) in the ribosomal P and A sites simultaneously slip leftward by one nucleotide. In bacteria, frameshifting may occur by the simultaneous slippage of two tRNAs and perhaps also by slippage of the single tRNA paired at the second codon (Farabaugh, 1996; Horsfield et al., 1995; Yelverton et al., 1994). The use of these programmed frameshifts to probe the effects of undermodification on tRNA function has begun recently (Atkins and Gesteland, 1995; Brierley et al., 1997).

The other programmed frameshift used in the study of modified nucleosides is the *E. coli prfB* (RF2) frameshift (Craigen et al., 1985), which is used to give information about the rate of aa-tRNA selection. At this message site, frameshifting occurs by a single tRNA slipping rightward at a specific codon (Fig. 2). tRNA slippage competes with normal translation of the codon 3' of the slippery site (Craigen and Caskey, 1986; Curran and Yarus, 1988, 1989; Sipley and Goldman, 1993) such that slow translation of the 3' codon increases frameshifting and vice versa. To assay for effects of modifications on rates of aa-tRNA selection, *lacZ*/RF2 fusions are constructed such that β-galactosidase production is dependent on programmed frameshifting (Curran and Yarus, 1988, 1989). Site-directed mutagenesis is used to place various codons 3' of the frameshift site, and the effects of modified nucleosides on aa-tRNA selection at those codons are estimated by comparing the β-galactosidase activities in strains that do and do not modify the corresponding tRNAs (Li et al., 1997; Qian et al., 1997). If, for example, higher frameshift-dependent β-galactosidase activities are observed in a modification-deficient strain, then one concludes that the absence of that modification decreases the rate of aa-tRNA selection at that codon.

MODIFIED NUCLEOSIDES IN THE ANTICODON

tRNA position 34 reads the third codon base, and is the only decoding position for which non-Watson–Crick, or wobble, base pairing geometries are permissible. This position is also commonly modified on either the base or the ribose. Some modifications have obvious effects on the decoding spectrum by altering the array of H bonding groups. In other cases, modifications alter the relative affinities for codons by altering anticodon loop structure. In addition to changing the decoding ranges of tRNA, these modifications can also improve translational efficiencies at cognate codons.

Uridine at Position 34

Uridine at the wobble position is almost always modified, and the nature of the modification is correlated with the base or bases read. I begin with a discussion of unmodified U34 because models for decoding by the modified forms are extended from those for unmodified U. In his classic paper on the wobble hypothesis Crick (1966) predicted that U should only read A and G. He considered that U:C and U:U wobble pairs might be stable, but omitted them from his pairing rules because such pairing might cause translational error. We now know that tRNAs with U34 can indeed read codons ending with all four bases (Andachi et al., 1989; Heckman et al., 1980; Osawa et al., 1992; Sibler et al., 1986). tRNAs with unmodified U34 occur in mitochondria and in *Mycoplasma*, and those tRNAs read all four codons in family codon boxes (Sprinzl et al., 1996). These are streamlined genetic systems in which the numbers of tRNAs have been greatly reduced during evolution. Thus it seemed possible that the "extended wobbling" by U34 was a quirk of these simplified systems. However, experiments have shown that extended wobbling by U34 can also occur in more complex genetic systems. For example, *E. coli* ribosomes also allow the decoding of all four GGN codons by the *M. capricolum* tRNA^{Gly} or by the *E. coli* tRNA$_1^{Gly}$ mutated to have the *Mycoplasma* tRNA^{Gly} anticodon loop (Lustig et al., 1993).

Two mutually exclusive classes of molecular models for U:pyrimidine pairs in anticodon:codon complexes have been proposed. Neither model completely explains all of the translational phenotypes of modified and unmodified tRNAs. However, they are good working hypotheses that may guide further experimentation. The model developed by Yokoyama and Nishimura (1985, 1995) and their coworkers suggests "short wobble" U:pyrimidine pairs in which the backbones of the two strands are closer to each other than in standard Watson–Crick pairs (Fig. 2, top right). This requires that the U34 ribose assume a 2′-endo configuration instead of the 3′-endo conformation needed for Watson–Crick base pairing. The structure in Fig. 2 is for the U:U short wobble pairs; in order for the U:C pair to assume a similar structure, the C must be protonated at the N3 position (pK, ~4.2). Thus, in this model U:U pairs are more stable than U:C pairs, unless the ribosomal environment somehow raises the pK of C.

In the alternate model proposed by Lim (1992, 1994), base pairing between anticodon U and either U or C occurs through a water bridge, that is, U$^\wedge$U and U$^\wedge$C (U$^\wedge$U is shown in Fig. 2, bottom right). Bridged U$^\wedge$C and U$^\wedge$U pairs have been observed in crystal structures of model RNAs (Baeyens et al., 1995; Holbrook et al., 1991). These bridged pairs do not require a "short" 2′-endo configuration, but do require a significant (~35°) propeller twist of the anticodon U such that it does not stack strongly onto the rest of the anticodon. The bridging water molecule can make a total of four hydrogen bonds, including two to solvent, and Lim predicts that these bonds compensate for the weakened stacking. In summary, the principal characteristics of these two models are a 2′-endo short wobble (Yokoyama and Nishimura, 1995) and alternately a water bridge between bases (Lim, 1994). Studies of model RNAs have shown that both short wobbles (Schejter et al., 1982; Westhof et al., 1985) and water bridges (Baeyens et al., 1995; Holbrook et al., 1991) can occur in pyrimidine:pyrimidine pairs and that conformation (Baeyens et al., 1995; Lietzke et al., 1996) and base pair stability (Wu et al., 1995) are highly context dependent. Because the relationships of those model RNAs to anticodon:codon complexes are not yet clear, they do not distinguish the two decoding models. Thus, both models are viable, and it is even possible that both might occur.

U34 Modifications Associated with Restricted Wobbling

The A- and G-ending codons of the split codon boxes are frequently read by tRNAs that have a modified U34. To maintain accurate reading of the genetic code, these tRNAs should not read pyrimidine-ending codons. tRNAs with these modified U's may also be relatively inefficient at G-ending codons because, in many cases (perhaps always in eukaryotes), cells maintain a tRNA with C34 to specifically read the G-ending codon (Sprinzl et al., 1996). The common modifications in these tRNAs are xm^{5}U34 species (xmnm^{5}U34 and xmcm^{5}U34) and s^2 (or se^2), and some tRNAs have Um34 (Sprinzl et al., 1996). Because these modifications are almost ubiquitously associated with tRNAs that should not read U- and C-ending codons, they may contribute to "restricted wobbling." For example, mitochondria and *Mycoplasma*, which as previously described use tRNAs with unmodified U34 in family boxes, use tRNAs with xm^{5}U34 to read A- and G-ending codons in split codon boxes (Andachi et al., 1989; Martin et al., 1990; Tanaki et al., 1991). This correlation between modification and restricted wobbling suggests, but does not prove, that these modifications contribute to restricted wobbling. These modifications may increase translational efficiency at A-ending codons because mutations that affect modification can weaken nonsense suppression and reduce the growth rates of yeast (Grossenbacher et al., 1986; Heyer et al., 1984).

Presumably, tRNAs that read the A- and G-ending codons of split boxes do not frequently misread the corresponding U- and C-ending codons. However, explicit information on this point is very limited. Working with an *E. coli* system, Parker and coworkers have made the only direct assay for misreading by an appropriate tRNA. They determined that tRNA^{Lys}, which contains mnm^{5}s^{2}U34 and normally reads AAA and AAG, misreads the asparagine codons AUU and AUC at frequencies of only about 3×10^{-3} and 3×10^{-4}, respectively (Precup and Parker, 1987). These values flank the estimated average error rate (Kurland et al., 1995; Parker, 1989). The difference in the error rates of these codons is correlated with a strong codon bias in highly expressed *E. coli* genes. The relatively error-prone AUU occurs much less frequently than AUC in those genes (Dalphin et al., 1997).

There is evidence that tRNAs that have xm^5 and s^2 modifications at U34 translate their G-ending codons relatively inefficiently, as might be expected if wobbling were restricted. For example, in Grosjean's paired tRNA system, the s^{2}U:A pair is much more stable than an unmodified U:A pair (Houssier et al., 1988), and the s^{2}U:G pair is unstable (Grosjean et al., 1978). In addition, *E. coli* has tRNAs for Gln, Lys, and Glu that contain xm^{5}s^{2}U34, and all of them prefer their A-ending codons over their G-ending codons in vitro (Agris et al., 1973; Lustig et al., 1981). Another example occurs for *E. coli* tRNA$_4^{Arg}$, which decodes AGA better than AGG in vivo (Spanjaard et al., 1990). In addition, AAG is more frameshift prone than AAA at the *E. coli dnaX* programmed frameshift site (Tsuchihashi and Brown, 1992). Assuming that weak tRNA:message pairing increases tRNA slippage, then pairing is weaker at the AAG codon. For lysine and glutamate, which have single tRNAs to decode their respective codons, greater tRNA:message stability is correlated with codon usage in *E. coli*. In this organism, highly expressed genes strongly prefer the A-ending codon over the G-ending codon (Dalphin et al., 1997).

In eukaryotes, modified U34 may translate G-ending codons very inefficiently, if at all. Here, split box codons ending in G and A are read by two tRNAs, one with modified U34 and another with C34 (Sprinzl et al., 1996). Guthrie and Abelson (1982) suggested that the C34-containing tRNAs might be necessary because the G-ending codons may not be translated by tRNAs that have modified U34. Consistent with this idea, it was subsequently shown that the yeast tRNA$_{CUG}^{Gln}$ gene is essential (Weiss and Friedberg, 1986), which shows that the glutamine tRNA with a modified U34 cannot adequately translate the CAG glutamine codon.

The molecular models of both Lim (1994) and Yokoyama and Nishimura (1995) predict that the xm^5, s^2 and Um modifications restrict wobbling, but the two models propose different mechanisms. Perhaps because of the complexity of the decoding process, the models do not fully explain all of the data on decoding. Nonetheless, the models are very good working hypotheses that may be modified with further information about tRNA structure and function. I present the structural predications about the s^2 and Um modifications together because, as in all A-related helices, the uridine 2'-OH and 2-carbonyl are in close proximity such that substitutions at either position may interfere with the other. This steric interference can restrict nucleoside conformation. Lim argues that either the 2-thio group or 2'-O-methyl group will prevent the propeller twist that he predicts is necessary to read pyrimidines with the water bridge. Additionally, both of these groups interfere with the hydrogen bonding between a water molecule and the 2-amino group of the G. In their model, Yokoyama and Nishimura predict that either adduct stabilizes the 3'-endo conformation and inhibits the formation of the 2'-endo conformations needed for wobble pairing (Kawai et al., 1992). Thus, both models predict that either adduct will restrict wobbling. The physical studies on dinucleotides containing modified U by Agris and coworkers (Agris et al., 1992; Sierzputowska-Gracz et al., 1987; Smith et al., 1992) showed that the s^2 group restricts nucleoside movement, favoring the 3'-endo form, which would promote Watson–Crick pairing. These data are consistent with both models. In the Yokoyama model it is this restriction of conformation that restricts wobbling. In the Lim model U34 pairs with uridine in a 3'-endo form with or without the s^2 modification.

The other common class of modifications associated with restricted wobbling by U34 is the xm^{5}U class (for example, mnm^{5}U, mcm^{5}U, and cmcm^{5}U). Lim predicts that all 5-methyl substitutions will inhibit reading of pyrimidines. The rationale is that the propeller twist needed to read pyrimidines places the 5-methyl group such that it will disrupt favorable interactions between a solvent molecule and functional groups in the major groove. Such solvent molecules are observed in RNA double helices (Dock-Bregeon et al., 1989; Holbrook et al., 1991). Because the loss of those interactions would not be sufficiently compensated by the xm^{5}U:pyrimidine base pair interaction, Lim (1994) considers such reading unlikely. In their model, Yokoyama and Nishimura (1995) argue that the xm^{5}U adducts interact favorably with the 5' phosphate to stabilize the 3'-endo conformation at the expense of the 2'-endo conformation. The 3'-endo conformation would favor Watson–Crick pairing. In addition, they predict that the amino group of mnm^5 adduct will interact favorably with the 2'-OH of U33 to help stabilize the "U-turn," which defines the 5' end of the anticodon helix (see Fig. 1; Holbrook et al., 1978). An interaction that stabilizes the U-turn might facilitate decoding. Yokoyama and Nishimura further suggest that the xm^{5}U, s^{2}U, and Um modifications have additive effects, such that multiple modifications should more strongly inhibit pairing to pyrimidines (1995). This can explain why so many nucleosides have more than one modification, xm^5 and either s^2 or Um, for example.

Both the Lim and Yokoyama-Nishimura models predict that xm^5 modifications will favor decoding of A-ending codons at the expense of G-ending codons. Those predictions are in general agreement with the codon preferences exhibited by various xm^{5}U34-containing tRNAs described previously. However, there are observations on the decoding efficiencies and frameshifting properties of mnm^5-deficient tRNAs that are not explained by these models. For example, a very recent study of frameshifting in *E. coli* also suggested that the mnm^5 group increases tRNA:message stability at both A- and G-ending codons. tRNA^{Lys} normally contains mnm^{5}s^{2}U34, but the *E. coli trmE* mutation prevents synthesis of the mnm^5 adduct (Elseviers et al., 1984). Brierley et al. (1997) showed that frameshifting increases at variants of the infectious bronchitis virus programmed frameshift site that include either of the lysine codons at the frameshift point. Assuming that these increases in frameshifting are related to decreases in tRNA: message stability, then loss of the mnm^5 adduct weakens tRNA:message interaction at both lysine codons. However, both models predict that loss of the mnm^5 adduct should weaken pairing to A and strengthen pairing to G.

Nonsense suppression assays have also given data that are not fully compatible with the molecular models. The mnm^5 group increases translational efficiency at A-ending codons (Elseviers et al., 1984; Hagervall et al., 1984), as might be expected from both the Lim and Yokoyama and Nishimura models. However, mnm^5 increases translational efficiency even more strongly at G-ending codons (Elseviers et al., 1984; Hagervall et al., 1984), and this effect is not explained by either model. Interestingly, this pattern is observed for three mutations that block different steps in the synthesis of mnm^5 and that, therefore, have different modifications. Suppressors with these modified nucleosides have relative suppression efficiencies in the order mnm^{5}s^{2}U (wild type) > cmnm^{5}s^{2}U (*trmC1*) > nm^{5}s^{2}U (*trmC2*) > s^{2}U (*trmE*). Thus, the wild type has the apparent optimum modification, and it is actually intermediate in size (Hagervall et al., 1984). Together, these data suggest that the mnm^5 group may play some specific structural/functional role for increasing translational efficiency that is not anticipated by the Lim and Yokoyama and Nishimura models.

Pseudouridine (Ψ) at position 34 occurs in a minor isoleucine tRNA from yeast (reported in Yokoyama and Nishimura, 1995). The tRNA reads AUA, and presumably does not frequently read the AUG methionine codon. Thus, it is expected that Ψ34 restricts wobbling. Possible molecular explanations are provided by physical studies. Ψ is found at several positions in various tRNAs. It was found that regardless of the position, the N^1 imino proton of Ψ exchanges slowly with solution. These data suggest that the imino proton is hydrogen bonded and that this bond does not depend on the sequence context (Griffey et al., 1985). It was suggested that a water molecule hydrogen bonded between the N^1 imino group and the sugar:phosphate backbone could explain these data (Davis and Poulter, 1991). Recently, comparisons of the crystal structures of the modified and unmodified forms of *E. coli* tRNA$_2^{Gln}$ indicate that such water bridges occur between the N^1 imino groups and the 5' phosphates of Ψ at three positions within the tRNA (Arnez and Steitz, 1994). Ψ also increases base stacking (Davis, 1995). Thus, it is possible that a water bridge and increased base stacking stabilize the Watson–Crick conformation and thus inhibit decoding of AUG by this tRNA.

Recent observations show that restricted U wobbling can be achieved without U34 modification. The *Mycoplasma* tRNA^{Gly} that has an unmodified U34 reads all four glycine codons (GGN), and an *E. coli* tRNA$_1^{Gly}$ mutated to have the same anticodon loop also reads all of the glycine codons (Lustig et al., 1993). However, this extended wobbling by the *E. coli/Mycoplasma* hybrid tRNA requires a cytosine at position 32 (two bases 5' of the anticodon). If C32 is mutated to U, then the decoding spectrum is strongly restricted in the order A > G >> U and C. Though the molecular mechanism for this restriction is not known, the decoding pattern might be sufficient to provide accurate reading of NNR codons within split boxes. Furthermore, *Halobacterium volcanii* (an archaeon) has a tRNA$_{UAA}^{Leu}$ with an unmodified U34 (Gupta, 1984). This tRNA apparently reads the UUA and UUG leucine codons, but not the UUU and UUC phenylalanine codons. These observations show that modification of U34 may not be a prerequisite to decoding restriction. Furthermore, xm^{5}U34 modifications are not always associated with the need for restricted wobbling. Eukaryotes commonly use tRNAs that have xm^{5}U-type modifications in family boxes (Sprinzl et al., 1996), in which the third codon position is irrelevant to accuracy. It seems likely that the xm^5 modifications are used in those tRNAs to increase translational efficiency at A-ending codons.

In summary, tRNAs that read the A- and G-ending codons in split boxes contain a variety of uridine modifications that are associated with and contribute to restricted wobbling. Features other than uridine modifications can also restrict wobbling. Certain modifications also increase translational efficiency.

U34 Modifications in Family Boxes

Most genetic systems use at least two tRNAs in family boxes. One of those tRNAs has G34, and should read codons ending in C and U. In eukaryotes, A- and G-ending codons are read by tRNAs that have xm^{5}U34 and C34, respectively. Inosine 34-containing tRNAs are also common in eukaryotes, and those tRNAs read C- and U-ending codons and possibly A-ending codons, though I:A decoding is inefficient. Bacteria commonly use tRNAs that have xo^{5}U-type uridine at position 34 (x = CH$_3$ or CH$_2$COOH) to read codons within family boxes, but not in split boxes (Sprinzl et al., 1996; Yokoyama and Nishimura, 1995). Unlike the xm^5 modifications, which may help restrict wobbling, the xo^5 adducts may facilitate wobbling. tRNAs with xo^{5}U34 have been observed to decode triplets ending in A, G, and U, but not C (Ishikura et al., 1971).

Yokoyama and Nishimura (1995) predict that these modifications facilitate pairing at U- and G-ending codons by making it easier for the modified base to assume the 2′-endo conformation needed for wobbling. Lim (1994) attributes extended wobbling to a different structural property. Though all substitutions at position 5 tend to inhibit the large propeller twist needed for reading uridine with a water bridge, the 5-oxygen atom of xo^{5}U compensates for this loss by forming a hydrogen bond to a water molecule fixed in the major groove. Lim also predicts that xo^{5}U pairs to A and G will not be hindered because a large propeller twist is not required for U:A and U:G pairs. Thus, both models predict that xo^{5}U34-modified tRNAs should read codons ending in A, G and U.

A recent study from the Yokoyama lab (Takai et al., 1996) addresses the decoding properties of cmo^{5}U34 in *E. coli* tRNA$_1^{Ser}$, which normally reads UCA, UCG, and UCU. They compared the decoding efficiencies of the normal, modified tRNA to those of an in vitro transcript that lacks all nucleoside modifications, including cmo^{5}U34. The lack of modifications dramatically reduces the efficiency of the tRNA at UCU and UCG, while the decoding of UCA is only slightly reduced. Neither the modified nor the unmodified tRNA can read UCC. Their major conclusion is that the modification enhances reading of UCU and UCG, as predicted by their structural models (Yokoyama et al., 1985; Yokoyama and Nishimura, 1995). This interpretation is reasonable, though it remains formally possible that the translational phenotypes are also affected by the absence of other modifications, including two in the anticodon loop (Cm34 and ms^{2}i^{6}A37).

There is evidence that the xo^{5}U34 modifications may generally increase translational efficiency. *E. coli* aro mutants do not make xo^{5}U34 modifications (Björk, 1980; Hagervall et al., 1990), and show a slightly reduced growth rate (Björk, 1995a). These data suggest that the xo^5 modifications increase translational efficiency, but it is not known whether this results from enhanced wobbling.

In summary, the xo^{5}U34 modifications may increase translational efficiency, particularly at U- and G-ending codons, but more work is necessary to test these hypotheses.

Modified Cytidines at Position 34

Several tRNAs contain modified C34. Cm34 allows the reading primarily of G-ending codons, but at least some tRNAs with this nucleoside can also recognize A-ending codons. For example, many bacteria and chloroplasts use single tRNAs to read both the UUG and the UUA leucine codons, and those tRNAs have Cm34 (Sprinzl et al., 1996). However, there are cases for which Cm:A decoding would result in translational error. Many eukaryotes and archaea have tRNA^{Met}, which has Cm34. These tRNAs should only read AUG because AUA is an isoleucine codon. Interestingly, in *E. coli*, a high-frequency translational error involving a Cm:A mispair does indeed occur. *E. coli* has a single tryptophan tRNA that has Cm34 and which decodes the UGG tryptophan codon. However, this tRNA also reads through the UGA stop codon at frequencies of a few percent (Hirsh and Gold, 1971). There are no mutants that fail to make this 2′-O-methylation, so it is not known whether the modification affects the frequency of this translational error. In any case, because the apparent decoding specificity of Cm appears to vary among genetic systems and tRNAs, other unknown features of tRNAs must strongly contribute to specificity.

In the "universal" genetic code, the isoleucine codon group is unique in that it includes three (AUU, AUC, and AUA) of the four codons in a family box but does not include the related G-ending codon (AUG). In eukaryotes, the isoleucine codons are commonly read by a single tRNA that has inosine at position 34. In bacteria, organelles, and certain archaea, the isoleucine codons are read by two tRNAs, one that contains G34 for reading AUU and AUC and another that contains a C34 modified with lysidine (or a derivative) at position 2 for reading AUA. The k^{2}C34 modification apparently switches base pairing specificity of C34 from G to A (Muramatsu et al., 1988). It is possible that the decoding of AUA by this tRNA is inefficient because both the codon (Dalphin et al., 1997) and the corresponding tRNA (Ikemura, 1981) are rare in *E. coli*. Finally, the scarcity of isoleucine tRNAs that have modified U may be related to the inherently ambiguous nature of decoding with U34; the repertoire of modified uridines may not include nucleosides that can adequately prevent the decoding of AUG by isoleucine tRNAs.

Muramatsu and colleagues (1988) propose two alternate base pairing schemes for k^{2}C34:A, depending on the tautomeric state of for k^{2}C34. The imino form for the modified bases allows pairing in a Watson–Crick geometry, and the amino form pairs with A with an U:G wobble geometry. Lim (1994) proposes a base pair structure similar to the amino form of Yokoyama and Nishimura, but with a significant propeller twist to allow hydrogen bonding between the amino groups of the bases and water molecules in the major groove.

Recently, f⁵C34 and the related f⁵Cm34 have been identified in bovine mitochondrial methionine (Takemoto et al., 1995) and bovine cytoplasmic leucine (Païs de Barros et al., 1996) tRNAs, respectively. The leucine tRNA corresponds to UUG and possibly UUA, while the methionine tRNAs correspond to AUG and possibly AUA (in animal mitochondrial genetic code, AUA and AUG are both specific for methionine). The roles of these modifications are not clear. They are expected to restrict wobbling, and an f⁵C34-containing tRNA^{Met} can only decode AUG in an *E. coli*-based in vitro translation system (Takemoto et al., 1995). However, no tRNAs specific for the A-ending codons have yet been identified. It is possible that these modifications prevent misreading of the corresponding U- and C-ending codons or that they increase translational efficiency at the G-ending codons.

Adenosine and Inosine at Position 34

Adenosine 34 is almost never observed in mature tRNA; instead, A34 in primary transcripts is usually deaminated to inosine (base = hypoxanthine), which is structurally similar to guanosine except that the base lacks the 2-amino group. tRNAs with I34 are widely distributed taxonomically, occurring in all bacteria and eukarya for which representative sets of tRNAs have been characterized (Sprinzl et al., 1996). However, I34 has not been found in archaea. Bacteria have single I34-containing tRNAs (Sprinzl et al., 1996) that are specific for the CGU, CGC, and CGA arginine codons. Eukaryotes commonly use I34-containing tRNAs to decode various family boxes and the isoleucine codon group (AUU, AUC, and AUA). From his structural analyses, Crick (1966) predicted that I34-containing tRNAs should translate three codons ending in C, U, and A. The three base pairs with inosine have different structures. I:C is Watson–Crick, and I:U has the same geometry as a standard G:U pair. In contrast, I:A is a purine:purine pair, and it might assume a "long wobble" conformation (Crick, 1966) (Fig. 2).

Crick predicted that A:G and A:A would not occur during decoding. However, yeast mitochondrial tRNA$^{Arg}_{ACG}$ (Sibler et al., 1986) and *Mycoplasma* tRNA$^{Thr}_{AGU}$ (Andachi et al., 1989) apparently read all four members of their respective codon family boxes with unmodified A34. In addition, Boren et al. (1993) found that an *E. coli* tRNA$^{Gly}_I$ mutated to have A34 can read all four glycine codons in *E. coli* extracts. Why then is A modified to I in family boxes? Structural modeling by Lim (1995) suggests that A can indeed read all four bases, though the significant propeller twist required to read C, G, and A will weaken the anticodon:codon complexes. However, Lim suggests that the most significant problem occurs after translocation of the tRNA:message complex to the P site. There, the 6-amino group of anticodon A34 (in all A:X pairs) is predicted to interfere with reading the next codon in the A site. Thus, the deamination of A to make I relieves this potential problem. However, this prediction depends on the relative positions of the ribosomal P and A sites, which have not been determined (Lim et al., 1992).

Whether decoding with the I:A pair is efficient, or indeed whether it occurs at all, has been widely debated. In the experiments that elucidated the genetic code, ribosomes programmed with the CGA triplet bound arginine tRNA (Söll et al., 1965). In those experiments, tRNA binding occurred at the ribosomal P site, which does not necessarily have the same specificity as the actual decoding site (the A site). There is evidence that suggests that *Saccharomyces cerevisiae* may have difficulty translating with I:A. In this yeast, the A-ending codons in family boxes have two nominally cognate isoacceptors, one with inosine and one with a modified uridine in the first position (Guthrie and Abelson, 1982; Sprinzl et al., 1996). This apparent duplication of specificity may be necessary because inosine-containing tRNAs are not sufficiently active (Guthrie and Abelson, 1982) at NNA codons. An experimental challenge to the prediction that inosine can recognize adenosine was provided by Munz et al. (1981), who showed that strains of the yeast *Schizosaccharomyces pombe* are not viable if the only tRNA available to decode UCA contains I34. Because the UCA codon is relatively abundant in yeast (1.6% of codons [Dalphin et al., 1997]), the average gene contains several UCA codons. Thus, the growth inhibition may be due to wholesale inhibition of translation by the inability to efficiently translate this common codon by using an I34-containing tRNA.

On the other hand, inosine-containing tRNAs must translate A-ending codons in some organisms. Extensive characterizations of the tRNA repertoires of the bacteria *E. coli* (Komine et al., 1990), *Micrococcus luteus* (Kano et al., 1991), and *Mycoplasma capricolum* (Andachi et al., 1989) and the yeast *Candida cylindracea* (Suzuki et al., 1994) show that I34-containing tRNAs are the only adapters available to read certain NNA codons. However, Ikemura (1992) noticed that A-ending codons putatively read with inosine are rare in virtually all organisms. This apparently universal form of codon bias may reflect general functional deficiencies for decoding with the I:A base pair. Indeed, this particular codon bias may even be an essential genomic feature. As previously noted, fission yeast is not viable if the only tRNA available to translate the common UCA codon is an I34-containing isoacceptor.

Recent work from my lab (Curran, 1995) strongly suggested that decoding with I:A is indeed very inefficient, causing problems during two successive ribosomal cycles. We studied the in vivo decoding properties of *E. coli* tRNA$_{ICG}^{Arg}$, which is the only tRNA in *E. coli* available to read the arginine codons CGU, CGC, and CGA. The CGA codon is read with an I:A wobble pair. We showed that relative to CGU and CGC, the CGA codon is slow to accept aa-tRNA in the RF-2 based frameshift competition assay. In addition, relative to CGU and CGC, CGA is a poor 5′ context for amber suppression, which suggests that the bulky I:A pair may interfere with translation of the next codon. We also showed that a cluster of CGA codons causes a strong polar effect on the expression of a downstream *lacZ* gene. Presumably, slow translation of the CGA cluster allows ribosomes to lag behind the RNA polymerase, invoking *rho*-mediated termination (Guillerez et al., 1991; Stanssens et al., 1986). These experimental observations can explain the dramatic bias against CGA in *E. coli* genes (Dalphin et al., 1997).

I previously described evidence that I:A may interfere with translation of the next codon. Recently, Lim (1995) suggested that this interference may be due to a steric hindrance of the ribosomal A site by the large I:A wobble pair in the P site. While it must be true that the A and P sites are in close proximity, their relative positions are not known (Lim et al., 1992). Thus, Lim's model remains a useful hypothesis at this point. Determination of the relative positions of the decoding sites will help explain this and other interactions between tRNAs on the ribosome (Yarus and Smith, 1995).

Queuosine at Position 34

Many bacteria and eukaryotes use tRNAs with Q34 to read NAY codons. In certain metazoans, Q is further modified by glycosylation (Sprinzl et al., 1996). These modifications do not directly affect the base pairing face of G, though they may slightly alter the relative affinities of the base for C and U. Because the NAY codons all occur within split boxes, it is important that these tRNAs do not read A- and G-ending codons. Whether the Q modifications prevent such misreading is not completely clear. Consistent with the idea that Q may prevent misreading are findings that tyrosine tRNAs that have G34 suppress amber codons, but tRNA^{Tyr} cannot suppress if G34 is modified to Q (Beier et al., 1984; Bienz and Kubli, 1981). It is possible that the Q modifications severely restrict the ability of the base to assume a *syn* conformation (Yokoyama and Nishimura, 1995), which could inhibit the reading of G-ending codons though a *syn* Q:G wobble pair. On the other hand, the Q modifications might not be necessary to prevent widespread misreading because an *E. coli tgt* mutant, which fails to make the Q base, has no severe translational phenotype and does not affect exponential growth (Noguchi et al., 1982). The only noticeable phenotype of the *tgt* mutant is that it does not survive as well as the *tgt*$^+$ strain during prolonged stationary culture (Noguchi et al., 1982). Though it cannot be excluded that the stationary phase effect is caused by an unknown, secondary mutation (Björk, 1995; Frey et al., 1989), it is intriguing that certain frameshift errors are much more common in the stationary phase (Barak et al., 1996; Fu and Parker, 1994; Schwartz and Curran, 1997). It thus seems possible that Q34 minimizes some sort of translational error in the stationary phase.

The modification G→Q tends to shift the base pairing preference from C to U in *Xenopus* oocytes (Meier et al., 1985) and *E. coli* (Harada and Nishimura, 1972). Further, in Grosjean's paired tRNA assay (Grosjean et al., 1978), Q:U is about 3-fold more stable than G:U, and Q:C is slightly less stable than G:C. Under the assumption that frameshifting is more likely for weak anticodon:codon complexes, these results are also in agreement with two observations at programmed frameshift sites. At the programmed frameshift site in MMTV, AUC is more frameshift prone than is AUU (Chamorro et al., 1992). These data are consistent with the idea that a relatively weak Q:C pair at the AUC codon facilitates the frameshift. A similar observation was made in an *E. coli* system. Brierley et al. (1997) compared frameshift frequencies of variants of the infectious bronchitis virus programmed frameshift site in strains of *E. coli* that either make or do not make the Q modification (*tgt*$^+$ and *tgt*$^-$). With either of the asparagine codons at the frameshift point, frameshifting is rather weak in *E. coli*. Nonetheless, in the *tgt*$^-$ host, frameshifting doubles at AUU and decreases by half at AUC. These data imply that Q stabilizes pairing to U and destabilizes pairing to C. These data are also consistent with the suggestion that programmed frameshifting can be modulated by the modification state of the slippery tRNAs (Hatfield et al., 1989).

This apparent preference for U-ending codons is not reflected in codon usage in *E. coli*. In highly expressed *E. coli* genes, U- and C-ending codons in split boxes always show a preference for the C-ending triplet (Dalphin et al., 1997). This rule holds for the NAY codons, and is very strong for the asparagine codon pair (Dalphin et al., 1997). It is possible that the C-ending codons are preferred for reasons that are more significant than the small effects observed on tRNA:message stability. One possibility is that the U-ending codons may be more frequently misread by the tRNAs specific to the NAA and NAG codons in the same codon box. As described above, tRNA^{Lys} misreads the asparagine AAU codon about ten times more often than it misreads AAC (Precup and Parker, 1987).

Pseudouridine in the Center of the Anticodon

In eukaryotic cytoplasmic tyrosine tRNAs, the U35 present in the primary transcript is converted to pseudouridine (Ψ). As described above for Ψ34, Ψ35 may stabilize base pairing in the Watson–Crick geometry. There are functional data that show that Ψ does stabilize anticodon:codon interactions. Johnson and Abelson (1983) show that the suppression efficiency of the yeast nonsense suppressor derived from tRNA^{Tyr} (anticodon U-Ψ-A) is decreased if Ψ35 is replaced with U35. Similarly, readthrough of UAG and UAA stop codons in tobacco mosaic virus by tRNA^{Tyr} (anticodon G-Ψ-U) is reduced by replacement of Ψ35 with U35 (Zerfass and Beier, 1992). These data show that Ψ35 increases translational efficiency. Unexpectedly, then, Grosjean et al. (1978) found that yeast tRNA^{Tyr} forms a relatively unstable complex when paired with a central Ψ:A pair to E. coli tRNA$_1^{Val}$. The reason for this apparent instability of the Ψ:A pair in this assay is not known.

MODIFIED NUCLEOSIDES IN THE 3′ SIDE OF THE ANTICODON LOOP

The nucleotide 3′ of the anticodon (that is, at position 37; see Fig. 1) is highly correlated with the base at position 36 (Yarus, 1982), which reads the first codon base. Position 37 is virtually always a purine, and is usually modified. Modifications often occur at more than one position on the base. These generalizations hold for all three living domains and the subcellular organelle systems (Grosjean et al., 1995). There is a great deal of information about the effects of many of the common modified nucleosides on decoding.

A common theme is that these modified bases may stabilize cognate anticodon:codon interactions, primarily through increased base stacking (Jukes, 1973; Nishimura, 1972). This increased stability is associated with increased translational efficiency and in some cases with enhanced reading frame maintenance. Increased stability of Watson–Crick pairs at the first codon position would also tend to reduce miscoding errors (Jukes, 1973). Modifications may also prevent base 37 from pairing with the message (Pieczenik, 1980). Because such use of an "out-of-phase anticodon" can lead to frameshift errors, these modifications may contribute to reading frame maintenance. Another possibility is that base modifications may prevent pairing across the anticodon loop. There is a strong correlation between potential Watson–Crick base pairing within the anticodon loop of primary transcripts and base modifications that would inhibit such pairing in mature tRNA (Dao et al., 1994). There is evidence that a base pair between positions 32 and 38, which are on opposite ends of the anticodon loop, may strongly inhibit decoding. Yarus et al. (1986) found that an Su7 derivative that has C32 and G38 has an extraordinarily low translational efficiency. This inefficiency can be relieved by mutations at either position that disrupt the putative base pair across the anticodon loop.

ms^{2}i(o)^{6}A37

The modified nucleosides ms^{2}i^{6}A37 and ms^{2}io^{6}A37 [collectively referred to as ms^{2}i(o)^{6}A37] are present in most bacterial tRNAs that decode UNN codons. It has been proposed that these modifications may stabilize anticodon:codon interactions, effectively compensating for the relatively weak A:U base pair at the first codon position (Nishimura, 1972). In Salmonella typhimurium, mutations in four genes affect the synthesis of this hypermodified nucleoside, and some of the effects of these mutations on translation have been characterized in S. typhimurium and in E. coli (reviewed in Björk, 1995b, 1995c). The miaA product catalyzes the first step, the 6-isopentenylation. This adduct is required for subsequent modification steps, so miaA mutants have completely unmodified A37. Addition of the 2-methylthio adduct is performed by the products of the miaB and miaC genes. Therefore, miaB and miaC mutants have the 6-isopentenyl group, but not the 2-methylthio group. In S. typhimurium, the hydroxylation of the isopentenyl group is the last step, carried out by the miaE gene product and only under aerobic growth conditions. Interestingly, this gene is missing in E. coli, and E. coli tRNAs are not hydroxylated. Expression of the S. typhimurium miaE gene in E. coli results in tRNA hydroxylation, with no obvious growth or translational phenotype. Also, in S. typhimurium, a miaE mutation does not affect amber suppression. Together, these results suggest that hydroxylation does not affect translational function. It is very curious, however, that a *Salmonella miaE* mutant is unable to grow aerobically with certain citric acid cycle intermediates as carbon sources (Persson and Björk, 1993). The physiological basis for this is unknown, but this phenomenon is not apparently related to the translational activities of the unmodified tRNAs.

The translational properties of tRNAs from *miaA* mutants, which lack both the 6-isopentenyl and 2-methylthio groups, have been characterized both in vivo and in vitro. In general, ms^{2}i(o)^{6}A37-deficient tRNAs are inefficient in translation. For example, elongation rates are decreased. The average translational cycle is slowed about fourfold in vivo (Ericson and Björk, 1986), and poly(U) translation by tRNA^{Phe} is reduced in vitro (Diaz et al., 1986). In addition, nonsense suppressors are generally weakened (Bouadloun et al., 1986; Ericson and Björk, 1986; Petrullo et al., 1983), and depending on context, the deficiency can be as great as 99% (Björnsson and Isaksson, 1993). Unmodified suppressors are generally more sensitive to message context (Bouadloun et al., 1986; Ericson and Björk, 1991). Not only does *miaA* reduce cognate translation, but it also reduces the readthrough of UGA codons by tRNA^{Trp} (Bouadloun et al., 1986; Petrullo et al., 1983) and the errant reading of UGG by tRNA^{Cys} (Bouadloun et al., 1986). Furthermore, ms^{2}i^{6}A can also increase initiation efficiency of a cleverly designed variant of *E. coli* initiator tRNA^{fMet} (Mangroo et al., 1995). Wild-type initiator tRNA has the CAU anticodon and is not modified at A37. Mutation of the anticodon to make it complementary to the amber codon (anticodon CUA) causes A37 to acquire the ms^{2}i^{6}A modifications. This variant tRNA initiates translation at an amber codon, but this ability is decreased about 10-fold in an *miaA* strain (Mangroo et al., 1995). All of these data indicate that ms^{2}i^{6}A deficiencies generally make tRNAs poor substrates for protein synthesis. These modifications may also prevent misreading, as proposed by Jukes (1973), because undermodifed *E. coli* tRNA^{Phe} misreads CUU leucine codons in an in vitro protein synthesis assay (Wilson and Roe, 1989).

The absence of these modifications in *E. coli* tRNA^{Trp} weakens the bond of the anticodon:anticodon complex to tRNA$^{Pro}_{xo5UGG}$ by eightfold (Vacher et al., 1984). This corresponds to a decrease in stability of about 1.3 kcal/mol, which could readily be explained by increased base stacking due to the hydrophobic ms^{2}i^{6}A modifications (Ericson and Björk, 1991; Vacher et al., 1984). However, the major and minor serine tRNAs, both of which naturally lack these modifications at A37 (Grosjean et al., 1985), form stable complexes with the complementary glycine tRNA (Houssier and Grosjean, 1985). Apparently, other features in the anticodon arm compensate for the lack of these modifications in those tRNAs. Those serine tRNAs have two features that might contribute. First, these are the only serine tRNAs that have a three-purine anticodon (GGA) (Grosjean et al., 1985). It is possible that base stacking by these three purines stabilizes pairing by the anticodon sufficiently for normal function. These tRNAs also have an unusual G:Ψ base pair at positions 30:40 near the bottom of the anticodon stem (Grosjean et al., 1985) (Fig. 1). This unusual base pair is likely to affect anticodon arm structure because it is a strong antideterminant for isopentenylation at A37 (Björk, 1995b; Motorin et al., 1997). The precise effect of the G:Ψ pair on anticodon arm structure is not known, but this unusual base pair may help position the GGA anticodon for favorable duplex interaction.

The paired tRNA assay suggests that that the loss of these modifications may not affect all tRNAs equally, and there are translational data to support this possibility. Bouadloun et al. (1986) found that amber suppressors are affected to different degrees by the *miaA* mutation. The suppression efficiencies of the amber suppressors derived from tRNA$^{Ser}_2$ (Su1) and tRNA^{Tyr} (Su3) are reduced about 10-fold by an *miaA* mutation, but that of the suppressor derived from tRNA$^{Leu}_5$ (*supJ*) is reduced only about 2-fold. The loss of ms^{2}i^{6}A may even affect different tRNAs at different steps during the ribosomal cycle. Diaz and Ehernberg (1991) showed that ms^{2}i^{6}A-deficient tRNA^{Phe} reacts normally during initial selection at the ribosome, but is then strongly proofread. Consistent with the first observation, the RF2-based frameshift assay (Fig. 3) showed that the rates of ribosomal selection of Phe-tRNA^{Phe} are not decreased in an *mia1* host (Li et al., 1997). However, the aa-tRNA selection rates at UCG (Ser) and UGC (Cys) codons are all affected by the *mia1* mutation (Li et al., 1997). Together, these data suggest that the serine and cysteine tRNAs may be sensitive at least during initial tRNA selection, although tRNA^{Phe} is only affected at a later stages.

The translational effects of *miaA* have been extensively characterized with nonsense suppression assays, and the RF2-based frameshift assay is usually performed with sense tRNAs. To correlate these two assays, Li et al. (1997) examined rates of aa-tRNA selection of suppressor (*supF*, derived from tRNA^{Tyr}) at amber (UAG) codons placed at the frameshift point of the RF2-based frameshift assay. The nucleotide 3' to the amber codon was randomized so that context effects on the rate of aa-tRNA selection could be studied (Pedersen and Curran, 1991). The *miaA* mutation inhibits the suppressor in all contexts, but the

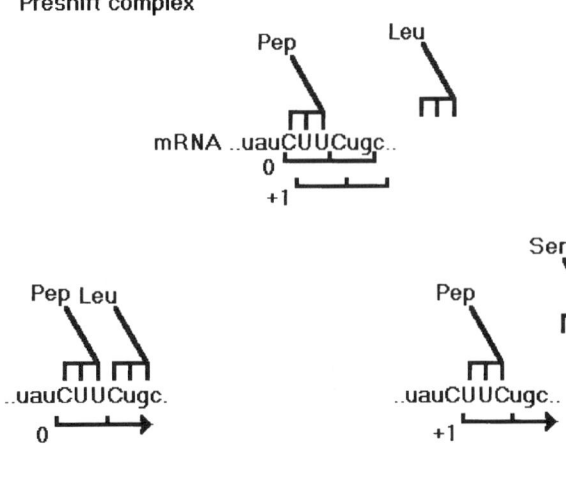

Figure 3. Frameshifting competes with normal translation at the RF2-programmed frameshift site. Shown is an example of an RF2 variant used to measure the relative rate of aa-tRNA selection at the leucine CUG codon. Prior to the frameshift (top) the peptidyl-tRNA is base paired to message CUU in the ribosomal P site. The next triplet, CUG, is available for decoding in the A site. Normal translation occurs if tRNA^{Leu} binds productively at the A site, fixing the "0" reading frame. Alternatively, frameshifting occurs when the peptidyl-tRNA slips one nucleotide rightward onto UUC, fixing the "+1" reading frame. (For mechanistic details of the RF2-programmed frameshift, see Atkins and Gesteland, 1995; Curran and Yarus, 1988; Farabaugh, 1996; Gesteland and Atkins, 1996; and Weiss et al., 1988.)

effect is especially strong at the "C" context. Thus, these data show that in this assay, as in the standard suppression assay, *miaA* increases context sensitivity.

The absence of the ms^{2}io^{6}A modifications also makes tRNA^{Phe} more likely to frameshift at UUU-Y sites (Qian and Björk, personal communication; Schwartz and Curran, 1997). Fu and Parker (1994) showed that UUU-Y sites are frameshift-prone because tRNA^{Phe} frequently slips rightward from the UUU codon onto the overlapping UUY triplet. We followed up on that observation and found that the *miaA* mutation increased frameshifting about two-fold (Schwartz and Curran, 1997). Qian and Björk (personal communication) also found that *miaA* increases frameshifting at UUU-Y, but by a factor of 20. The reason why the two studies gave different magnitudes for the *miaA* effect is not clear, but it is apparent that in the normal tRNA^{Phe}, the ms^{2}io^{6}A modifications prevent frameshifting. The most likely mechanism is tRNA:message slippage caused by the weakened duplex.

The *miaB* mutation causes loss of the 2-methylthio group, but A37 retains the 6-isopentenyl adduct. The ms^2 group increases duplex stability in the paired tRNA assay (Houssier and Grosjean,

1985). However, lack of the ms^2 has relatively minor effects on protein synthesis. There is no detectable decrease in growth rate, and the rate of synthesis of the β-galactosidase enzyme is not affected (Esberg and Björk, 1995). The *miaB* mutation does weaken nonsense suppressors (Esberg and Björk, 1995), but not to the extent that the *miaA* mutation does. In the RF2-based frameshift assay, *miaB* has much smaller effects on rates of aa-tRNA selection than *miaA* does at tyrosine, tryptophan, and cysteine codons (Li et al., 1997). Context effects on the rates of aa-tRNA selection were also studied at an amber codon placed at the shift point of the RF2-based frameshift assay. At three of the contexts (A, C, and G), *miaB* caused a minor reduction in aa-tRNA selection. However, at the C context, the reduction was about fivefold, which is the same as that observed for the *miaA* mutation. Thus, it appears as though both the ms^2 and io^6 adducts are important for tRNA function and that at most sites the i^{6}A adduct is more important. This could be because of the greater stacking potential of the large io^6 group (Esberg and Björk, 1995; Li et al., 1997). However, the ms^2 group is also very important at C contexts.

m^{1}G37

m^{1}G37 is present in the tRNAs that read CUN (leucine), CCN (proline) and CGN (arginine) codons of prokaryotes. The modification is predicted to have two effects on decoding. First, the methyl adduct at the N1 position of G prohibits base pairing with Watson–Crick geometry. This might prevent out-of-phase reading with a shifted or expanded anticodon (Pieczenik, 1980). Second, because the 1-methyl group increases base hydrophobicity, it may increase base stacking (Nishimura, 1972), which could increase decoding efficiency. There is evidence that this modification may have both of these functions in translation.

Björk et al. (1989) attempted to recover defective alleles of the corresponding tRNA methylase gene (*trmD*) by selecting for weakened suppression by the frameshift suppressor *sufJ*, which normally contains this methylation. The anticipated antisuppressor phenotype was not found, but they instead recovered a mutant (*trmD3*) of this gene that can independently suppress certain frameshift mutations. Various phenotypes of this mutant have been observed. Mutations in the methylase and mutations in tRNA^{Pro} that inhibited methylation (Qian and Björk, 1997) suppressed frameshift mutations. Mutation *trmD3* decreased growth rates (~10–25%) in various media (Li and Björk, 1995). Assuming that only the codons read by these tRNAs are affected, this corresponds to an average decrease of about threefold for the rates of translation of those codons. Apparently, however, codons read by these tRNAs are not equally affected by the loss of m¹G. The *leu* attenuator, which is sensitive to the rates of translation of four consecutive leucine codons (CUA CUA CUA CUC; Carter et al., 1985) whose tRNAs normally contain m¹G, is not deattenuated by the *trmD3* mutation (Li and Björk, 1995). Other data suggest that translation of these particular leucine codons may be much less sensitive to *trmD3* than are other codons. Using the RF2-based frameshift competition assay (Fig. 3), Li et al. (1997) showed that the rate of aa-tRNA selection at CUA was decreased about 30% and that at CUC was unaffected. In contrast, the rates at leucine codons CUU and CUG were decreased about 35%. The rates at proline codons CCC, CCA, and CCG and the arginine codon CGG were all decreased even more severely (50–90%). These results suggest that the m¹G group can affect anticodon arm structure and decoding function, but effects are highly dependent on the broader tRNA sequence and on the codon.

Frameshifting caused by the *trmD3* mutation has been studied in some detail. Peptide sequence analyses showed that proline is inserted at certain frameshift sites (Hagervall et al., 1993b). Leucine tRNAs may also frameshift, but at levels too low to generate sufficient product for peptide sequencing. Suitable sites for frameshifting by arginine tRNAs were not present in the suppression windows used. Thus far, several +1 frameshift mutants are suppressed. No suppression of three −1 mutations was detected.

Frameshifting at proline codons occurs at all four CCC-N sites (Hagervall et al., 1993b). Because peptide sequencing shows that CCC-A and CCC-U are decoded as "proline," this may be formally considered four-base translation. More than one molecular explanation seems possible. For example, the unmodified tRNA may read with a four-base anticodon, using G37 to read the 5′ C at CCC-N. Consistent with this idea is the observation that frameshifting does not occur at ACC-U (Hagervall et al., 1993b), which shows that elimination of the complementarity between the G37 and the first base of the message quadruplet eliminates frameshifting. However, this cannot be conclusive because the absence of frameshift suppression at ACC-U could also result from poor competition with the wild-type threonyl-tRNA, which decodes ACC as a triplet. Thus, it is also possible that the unmodified tRNA binds to CCC in the "0" frame at the A site and that the weakened tRNA:message complex realigns subsequent to aa-tRNA selection. Distinguishing these possibilities will be important for understanding whether m¹G prevents frameshifts by preventing out-of-phase pairing or by stabilizing the complex in the "0" phase. In principle, these models could be distinguished by the effect of the modification on duplex stability. If the methyl group stabilizes duplexes, its absence will weaken them. In contrast, if the methyl group prevents pairing, its absence will increase duplex stability by allowing the formation of a fourth base pair.

t⁶A37 and Derivatives

Virtually all organisms use tRNAs that have a t⁶A derivative 3′ to the anticodon to read ANN codons (Grosjean et al., 1995; Sprinzl et al., 1996). This adduct may stabilize the U:A base pair at the first codon position (Nishimura, 1972), and may also prevent misreading at the first position with a U:G base pair, as suggested by Jukes (1973). Starvation of *E. coli* cells for threonine can lead to as much as half of the relevant tRNAs being unmodified (Miller et al., 1976). tRNA$^{Ile}_{GAU}$ isolated from threonine-starved cells does not function well in in vitro protein synthesis or in binding to programmed ribosomes. These findings are in apparent contrast to the results of Weissenbach and Grosjean (1981), who used several assays to show that t⁶A deficiency has negligible effects on yeast tRNA$^{Arg}_{III}$ function. These workers showed that binding to programmed ribosomes is essentially unaffected and that binding to a cognate triplet or to a tRNA with a complementary anticodon was decreased by a factor of only about two. They concluded that the t⁶A modification does increase duplex stability slightly, probably through base stacking. The difference between their results and those of Miller et al. (1976) could be explained either by the *E. coli* tRNA having an additional, unknown deficiency in threonine-starved cells, or because like many other modifications t⁶A may not affect all tRNAs equally.

Recently, Qian et al. (1997) have isolated the first mutation affecting synthesis of a t⁶A derivative. This mutation, *E. coli tta1*, prevents the further N⁶-methylation of t⁶A modified tRNAs. The mutant retains t⁶A, but lacks the additional N⁶-methyl group found on specific t⁶A-modified tRNAs. Several in vivo assays show that the loss of this methyl group has little or no effect on tRNA function. For example, growth rate, polypeptide chain elongation rate, and the rate of aa-tRNA selection as measured with the RF2-based competition assay are completely unaffected. The only noticeable effect is a twofold deattenuation of the *thr* operon. The leader region of this operon encodes eight threonine codons (Lynn et al., 1987) that are read by tRNAs that normally contain this methyl group (Sprinzl et al., 1996). It is not clear why this effect is not evident in the bulk protein synthesis, but it may be related to the clustering of the threonine codons in the leader or an exquisite sensitivity for this attenuation regulatory mechanism. Further understanding of the role of this ubiquitous t⁶A37 will be facilitated by mutations that affect threonylation.

The Y Base at Position 37

Eukaryotic tRNA$_{AAG}^{Phe}$ species often have the large Y base at position 37 (Sprinzl et al., 1996). In paired tRNA complexes, the Y base increases the stability of tRNAs paired through their anticodons by severalfold (Grosjean et al., 1976). It has been suggested that this modified nucleoside increases base stacking (Nishimura, 1972), which might stabilize the relatively weak pairing to the phenylalanine codons (Grosjean et al., 1976; Grosjean and Houssier, 1990; Jukes 1973), especially to UUU. One consequence of this stabilization may be reduced frameshifting. In *E. coli*, the ms²i⁶A modifications increase tRNA:message stability, probably by increased stacking, and in this bacterium, tRNA^{Phe} becomes more frameshift-prone with the loss of the ms²io⁶A37 modifications (Qian and Björk, personal communication; Schwartz and Curran, 1997). Hatfield et al. (1989) correlated the undermodification of tRNAs with retroviral infection, and noted that phenylalanine codons are commonly included in eukaryotic viral programmed frameshift sites (ten Dam et al., 1990). Thus, it is possible that absence of the Y base during retroviral infection contributes to programmed frameshifting during viral gene expression. Direct tests of the idea that tRNA undermodification affects programmed frameshifting are under way (Atkins and Gesteland, 1995; Brierley et al., 1997).

Pseudouridine on the 3' Side of the Anticodon Arm

Pseudouridine (Ψ) occurs at position 38 at the 3' end of the anticodon loop, and at positions 39 and 40 in the 3' side of the anticodon stem. Nearly half of the tRNAs in *E. coli* contain one or more of these modifications; essentially every U at these positions in the primary transcript is converted to Ψ (Sprinzl et al., 1996). Mutations in *hisT* fail to perform these modifications (Singer et al., 1972). As described previously, Ψ either at the wobble position or in the center of the anticodon increases translational efficiency probably by stabilizing anticodon:codon pairing. NMR analyses show that Ψ39 at the base of the anticodon stem increases the stability of the anticodon arm of *E. coli* tRNA^{Phe} (Davis and Poulter, 1991). Pseudouridine in the 3' side of the anticodon arm generally increases translational efficiency, perhaps because of stabilization of the anticodon region.

Mutations in the gene that converts U to Ψ (*hisT*) were isolated by selection for deregulation (deattenuation) of the *his* operon (Roth et al., 1966). The *hisT* mutation also deattenuates the *ilv* and *leu* operons (Cortese et al., 1974; Rizzino et al., 1977), and decreases the translational rate of the *lacZ* gene (Palmer et al., 1983). These observations show that absence of Ψ can decrease translational efficiency, but do not indicate which tRNAs or positions within tRNAs are responsible.

Various tRNAs have Ψ at different positions. Comparisons of the effects of the *hisT* mutation on the functions of various tRNAs suggest that Ψ is most influential at position 38, which is at the 3'-most position of the anticodon loop (Fig. 1). Amber suppressor Su2 (*supE*), which is derived from tRNA$_2^{Gln}$, contains Ψ at positions 38 and 39. In a *hisT* strain, the suppression efficiency of Su2 drops about 20-fold (Bossi and Roth, 1980). In contrast, suppressor Su3 (*supF*), which is derived from tRNA^{Tyr}, is less than 2-fold affected by *hisT* (Hagervall et al., 1990). Because this tRNA is modified only at position 39, the data from these two suppressors suggest that Ψ at position 38 may be more important to suppression efficiency. Other evidence suggesting the relative importance of Ψ at position 38 comes from studies of misreading errors. In these assays, cells are starved for a specific amino acid, and the rates of misreading of the starved codons are measured (Parker, 1982; Parker and Friesen, 1980). The *hisT* mutation strongly reduces starvation-induced misreading of histidine codons by glutaminyl-tRNA, which suggests that the absence of Ψ at positions 38 and 39 weakens misreading by the glutaminyl-tRNAs. In contrast, starvation-induced misreading of asparagine codons by tRNA^{Lys} is not decreased by *hisT*. However, the lysine tRNA normally has Ψ only at position 39. Together, these data suggest that Ψ may not be as important at position 39 as it is at position 38.

In the RF2-programmed frameshift assay, Ψ in the anticodon region is important to aa-tRNA selection of the leucine tRNAs that act at the CUN codons. For these tRNAs, the severity of the effect of the *hisT* mutation depends on the number of Ψ normal present in the anticodon arm. For example, the rates of aa-tRNA selection at CUU and CUC, which are read by a tRNA that has Ψ only at position 38 (tRNA$_2^{Leu}$), are only modestly affected by the *hisT* mutation. In contrast, *hisT* strongly decreases the rate of aa-tRNA selection at CUA, which is read by a tRNA having Ψ at positions 38 and 39, and at CUG, which is read primarily by a tRNA that has Ψ at positions 38 and 40. In this assay Ψ40 does not appear to contribute to the rate of aa-tRNA selection of proline tRNAs. Proline codons CCU and CCC are read by tRNA$_2^{Pro}$, which has Ψ40, and the *histT* mutation does not affect aa-tRNA selection at those codons.

In summary, though these data were gathered with different assays and different tRNAs, they suggest the provisional hypotheses that Ψ is most important at position 38 and that effects may be cumulative when Ψ occurs at more than one position.

Modified Nucleosides on the 5′ Side of the Anticodon Loop

The 5′ side of the anticodon loop contains two nucleotides (positions 32 and 33) that are important for anticodon loop structure and function. Base 33 immediately 5′ of the anticodon is virtually always an unmodified U. Uridine 33 helps to stabilize the "U-turn" in the anticodon loop, which helps define the 5′ end of the anticodon (Holbrook et al., 1978) (Fig. 1). Substitution of any base for U33 decreases the translational efficiency of amber suppressor Su7 by at least 10-fold (Yarus et al., 1986). Base 32 is always a pyrimidine, and it is frequently modified (Grosjean et al., 1995; Stanssens et al., 1986). The common modifications are Um, Cm, s^{2}C, and Ψ. Structurally, all of these modifications are predicted to restrict nucleotide flexibility (see the section on U34 modifications associated with restricted wobbling). Rigidity here could help stabilize the anticodon.

There is strong evidence that base identity at position 32 is important to translational efficiency. Amber suppressor Su7 normally has Cm32. A mutational analysis showed that substitution of either purine at position 32 reduces suppression efficiency more than 10-fold (Yarus et al., 1986). Position 32 can also affect the decoding range of tRNA. Boren et al. (1993) observed that an *E. coli* tRNA^{Gly} variant with an unmodified U34 can translate all four GGN codons if base 32 is cytosine; however, if the base 32 is uracil, then only GGA is efficiently decoded. The molecular mechanisms for these effects of the base at position 32 are not known, but are presumably related to effects on the microstructure of the anticodon loop.

It therefore seems entirely possible that nucleoside modifications at position 32 affect translational efficiency; however, there are no studies that explicitly test for the translational effects of modification deficiency at this position. Yarus et al. (1986) attempted to distill such an effect from their analyses of base substitution mutations in the anticodon arm of Su7. They examined over 30 mutants that vary several-hundred-fold in their translational efficiencies. The wild-type suppressor has Cm32, and the various anticodon arm mutants have different levels of this modification. They did not observe a significant relationship between translational efficiency and the extent of Cm modification. However, a minor role for the 2′-methyl group may have been obscured by the dramatic effects caused by many of the base substitutions.

An Interaction between mnm^{5}s^{2}U34 and t^{6}A37 May Affect tRNA^{Lys} Function

Lysine tRNAs from all genetic systems commonly have xm^{5}U34-type and t^{6}A37-type modifications. In bacteria, this single lysine tRNA reads both the AAA and AAG lysine codons. In *E. coli*, two interesting biological properties of tRNA^{Lys}, loss from the ribosome as peptidyl-tRNA (Heurgue-Hamard et al., 1996) and programmed frameshifting in *dnaX* (Tsuchihashi and Brown, 1992), are thought to result from relatively weak tRNA:message interaction.

Physical studies suggest that the *E. coli* tRNA^{Lys} may have an unusual anticodon loop structure that may contribute to the unusual translational properties. This tRNA has the modified nucleosides mnm^{5}s^{2}U34 and t^{6}A37 in the anticodon loop. In other tRNAs, the base at position 34 is stacked onto the remainder of the anticodon where it is poised for interaction with the message (Fig. 1). This conformation also makes the nucleoside accessible to certain reagents in solution. Several observations suggest that nucleoside 34 in tRNA^{Lys} may be relatively inaccessible to biochemical probes. First, the phosphodiester linkage between nucleotides 34 and 35 is unusually refractory to digestion by single strand-specific RNases (Harada and Dahlberg, 1975; Watanabe, 1980). In addition, the s^2 group at nucleoside 34 is relatively slowly oxidized during exposure to H$_2$O$_2$ (Agris et al., 1997; Watanabe et al., 1994). Further evidence that the s^2 group is in an unusual conformation comes from near-UV CD spectral analyses, which illuminate thiouridine nucleosides (Watanabe, 1980). The near-UV CD spectrum of native tRNA^{Lys} is significantly different from those of the mnm^{5}s^{2}U nucleotide, denatured tRNA^{Lys}, and native tRNA^{Glu} (which also has mnm^{5}s^{2}U nucleotide but unmodified A37) (Watanabe et al., 1994). These data suggest that the mnm^{5}s^{2}U34 residue is in a different chemical environment in native tRNA^{Lys} than it is in the other molecules. Finally, the anticodon loop of tRNA^{Lys} may not bind a Mg^{2+} ion (Agris et al., 1997), which also distinguishes it from those of other tRNAs (Holbrook et al., 1978; Saenger, 1984).

NMR analyses of synthetic pentanucleotides based on the tRNA^{Lys} anticodon loop provide a structural explanation for these unusual chemical and decoding properties. Agris et al. (1997) showed that the

s^2U base is not stacked onto the rest of the anticodon, but is instead rotated up into the loop where it interacts with t^6A37. Analyses of control oligonucleotides showed that this interaction requires the modifications at both 34 and 37. This unusual conformation can explain why nucleotide 34 is less accessible to solvent. This conformation would also weaken anticodon:codon interaction. First, the anticodon stack is distorted, which should inhibit pairing with the message. In addition, the absence of the Mg^{2+} ion in the anticodon loop may also weaken decoding because Mg^{2+} may help shape the anticodon stack and shield the net negative charge of the anticodon region (Agris, 1996; Holbrook et al., 1978; Saenger, 1984). Thus, although these measurements were made on a synthetic fragment, this structure provides a provisional explanation for all of the physical biological properties noted for *E. coli* tRNA^{Lys}. Furthermore, Agris et al. (1996) noted that these same structural features may extend to lysine tRNAs from other organisms that have xm^5U34-type and t^6A37-type modifications.

Other Modified Nucleosides That Affect Translation

Modifications outside the anticodon region can also affect protein synthesis. One modification restricts yeast initiator tRNA (tRNA$_i^{Met}$) from interacting with elongation factor EF1α during the elongation phase of protein synthesis. The initiator tRNAs from plants and fungi have 2'-phosphoribosyl adenosine at position 64 (Glasser et al., 1991; Sprinzl et al., 1996). This phosphate group is not present in elongator tRNAs. Initiator tRNAs from yeast and wheat germ that contain this phosphate do not interact strongly with either yeast EF1α·GTP or *E. coli* EF-Tu·GTP complexes. However, removal of the phosphate allows interaction with those factors, and the unmodified tRNAs can function as elongator tRNAs in vivo (Forster et al., 1993). Thus, the phosphoribosyl group acts as an antideterminant for elongation (Forster et al., 1993).

There are many other modified bases at other positions in tRNA, including those that are used to name the D-arm (dihydrouridine) and the T-Ψ-C arm (thymidine and pseudouridine). Very little is known about the functions of the modifications in protein synthesis.

Nucleoside modifications in messenger RNA and ribosomal RNA can also affect protein synthesis. These include the caps of eukaryotic messages and specific base modifications in certain messages. The molecular biology of these modifications is described in Chapter 10. rRNA modifications are largely simple methylations, Ψ and dihydrouridine. Pseudouridine in rRNA is discussed in Chapter 12. These modifications are clustered in conserved, functionally important regions, such as the decoding sites on the small subunit and the peptidyl transferase center on the large subunit (Kowalak et al., 1994; Lane et al., 1995; Ofengand and Bakin, 1997). Thus, they are likely important for proper structure and function of those regions. However, the functions of these modifications in translation are not yet understood.

CONCLUSION AND OUTLOOK

Cellular physiology is largely "protein synthesis." This cardinal process is fundamentally dependent on the proper tRNA:message interactions within the ribosomal coding sites. Virtually all tRNAs contain modified nucleosides within the anticodon region, and it has become abundantly clear that they contribute to translation in a number of ways (Table 1). Through their effects on nucleoside chemical properties, modifications can affect anticodon loop structure and function. The most common modifications occur at position 34, which reads the third codon nucleotide, and at position 37, which is immediately 3' of the anticodon. At position 34, modifications such as inosine and k^2C completely alter base pairing specificity by changing the patterns of H bond donors and acceptors. Other modifications at position 34 can affect base pairing specificity or range by altering preferred nucleoside conformations. Examples include xm^5U34, s^2U34, and $Um34$, which

Table 1. Summary of the effects of modified nucleosides on translation

Position(s)	Modification(s)	Principal effect(s) of translation
32	Um, Cm, Ψ	Unknown
34	xm^5U, Um	Restricts wobbling, reads A > G
	xo^5U	Enhances wobbling, reads A, G, and U
	I (modified A)	Decodes U, C and A, but decoding of A is inefficient
	Q (modified G)	Minor effects on the decoding of U and C
35	Ψ	Enhances reading of A and inhibits wobbling
37	$ms^2i(o)^6A$, m^1G	Strengthens anticodon:codon interaction, increases rates of aa-tRNA selection, prevents frameshifting
	Y	Strengthens anticodon:codon interaction
38–40	Ψ	Increases translational efficiency, especially at position 38

appear to restrict wobble pairing by U34. By inhibiting base pairing to U- and C-ending codons, these modifications contribute directly to translational accuracy in split codon boxes. In addition, the xm^5 modifications may also increase decoding efficiency at cognate codons. In contrast, the xo^{5}U34 modifications found in many bacterial tRNAs may increase wobbling by improving the interactions of U34 with G- and U-ending codons. Another position that is commonly modified is position 37, which is immediately 3' of the anticodon. Modifications here generally stabilize anticodon:codon pairing, and can have very large effects on decoding efficiency and accuracy. The absence of modifications can cause frameshifting by increasing tRNA:message slippage or by allowing base pairing between the unmodified base and the message. Modifications at other positions in the anticodon arm also affect decoding efficiency, but are less well studied.

It is very interesting that nucleoside modifications differentially affect various tRNAs. The xm^5 and s^{2}U modifications at U34 are generally associated with restriction of wobbling and may contribute to prevention of reading U- and C-ending codons. However, these modifications are occasionally found in tRNAs that read codons in family boxes. In addition, there are tRNAs specific for A- and G-ending codons that do not have modified U34. Further, certain modifications affect the translational efficiencies of only a subset of the tRNAs that contain them. It is becoming clear that these examples are far too common to be considered "exceptions." Undoubtedly, this complexity of phenotype reflects the inherent complexity of the translational process.

It is clear that features of tRNA outside of the anticodon contribute strongly to translational efficiency and accuracy, but the mechanisms are largely unknown. The variability of the effects of modified nucleosides among tRNAs may provide insight into these other mechanisms. tRNAs that function normally despite the loss of a particular modification may have other features that compensate for the missing modification. The identification and study of such features will contribute greatly to our understanding of the fundamentally important translational process.

Our ability to study the roles of modified nucleosides has been enhanced in recent years. Improved methods for identifying modifications greatly facilitate their study (Limbach et al., 1995) (see Chapter 3). Model RNAs that include modified nucleosides are extremely useful for predicting preferred structures and conformations (Agris et al., 1997; Yokoyama and Nishimura, 1992; see Chapters 4 and 5). In addition, recent advances in computer modeling will continue to be helpful for making structural predictions that can guide experimentation and interpretation (Lin, 1994). Furthermore, we have accumulated a wide array of probes of the decoding functions in vivo (Björk et al., 1989; Brierley et al., 1997; Curran, 1995; Ericson and Björk, 1986; Hagervall et al., 1993a, 1993b; Li et al., 1997) and with in vitro protein synthesis assays (Takai et al., 1996; Takemoto et al., 1995). Finally, these studies are facilitated by clever selections for mutants and rapid characterizations of modification-deficient tRNAs (Björk et al., 1989; Heyer et al., 1984; Qian and Björk, 1997). The future looks very promising for the continued study and understanding of the roles that modified nucleosides play in the fundamentally important translational process.

Acknowledgments. I thank Glenn Björk, Ian Brierley, Franco Fasiolo, Valery Lim, and Qiang Qian for communications of unpublished data. I also thank the editors for their help. My research is supported by NIH grant GM-52643.

REFERENCES

Agris, P. F., D. Söll, and T. Seno. 1973. Biological function of 2-thiolation in *Escherichia coli* glutamic acid transfer ribonucleic acid. *Biochemistry* 12:4331–4337.

Agris, P. F., H. Sierzputowska-Gracz, W. Smith, A Malkiewicz, S. Sochacka, and B. Nawrot. 1992. Thiolation of uridine carbon-2 restricts the motional dynamics of the transfer RNA wobble position. *J. Am. Chem. Soc.* 114:2652–2656.

Agris, P. F. 1996. The importance of being modified: roles of modified nucleosides and Mg^{2+} in RNA structure and function. *Prog. Nucleic Acid Res. Mol. Biol.* 53:79–129.

Agris, P. F., R. Guenther, P. S. Ingram, M. M. Basti, J. W. Stuart, E. Sochacka, and A. Malkiewicz. 1997. Unconventional structure of tRNA$^{Lys}_{SUU}$ anticodon explains tRNA's role in bacterial and mammalian ribosomal frameshifting and primer selection by HIV-1. *RNA* 3:420–428.

Alberts, B., D. Bray, J. Lewis, M. Raff, K. Roberts, and J. D. Watson. 1994. *Molecular Biology of the Cell*. Garland Press, New York, N.Y.

Andachi, T., F. Yamao, A. Muto, amd S. Osawa. 1989. Codon recognition pattern as deduced from sequences of the complete set of transfer RNA species in *Mycoplasma capricolum*. *J. Mol. Biol.* 209:37–54.

Arnez, J. G., and T. A. Steitz. 1994. Crystal structure of unmodified tRNA^{Gln} complexed with glutamyl-tRNA synthetase and ATP suggests a possible role for pseudo-uridines in stabilization of RNA structure. *Biochemistry* 33:7560–7567.

Atkins, J. F., and R. F. Gesteland. 1995. Discontinuous triplet reading with or without re-pairing by peptidyl-tRNA, p. 471–490. In D. Söll and U. RajBhandary (ed), *tRNA: Structure, Biosynthesis, and Function*. American Society for Microbiology, Washington, D.C.

Baeyens, K. J., H. L. De Bont, and S. R. Holbrook. 1995. Structure of an RNA double helix including uracil-uracil base pairs in an internal loop. *Nat. Struct. Biol.* 2:52–62.

Barak, Z., J. Gallant, D. Lindsley, B. Kwieciszewski, and D. Heidel. 1996. Enhanced ribosomal frameshifting in stationary phase cells. *J. Mol. Biol.* 263:140–148.

Barnes, W. M. 1978. DNA sequence from the histidine control region: seven histidine codons in a row. *Proc. Natl. Acad. Sci. USA* 75:4281–4285.

Beier, H., M. Barciszewska, and H. Sickinger. 1984. The molecular basis for the differential translation of TMV RNA in tobacco protoplasts and wheat germ extracts. *EMBO J.* 3:1091–1096.

Bienz, M., and E. Kubli. 1981. Wild-type tRNA$_G^{Tyr}$ reads the TMV stop codon, but the Q base-modified tRNA$_Q^{Tyr}$ does not. *Nature* 294:188–190.

Björk, G. R. 1980. A novel link between the biosynthesis of aromatic amino acids and transfer RNA modification in *Escherichia coli*. *J. Mol. Biol.* 140:391–410.

Björk, G. R., P. M. Wikström, and A. S. Byström. 1989. Prevention of translational frameshifting by the modified nucleoside 1-methylguanosine. *Science* 244:986–989.

Björk, G. R. 1995a. Genetic dissection of synthesis and function of modified nucleosides in bacterial transfer RNA. *Prog. Nucleic Acid Res. Mol. Biol.* 50:263–338.

Björk, G. R. 1995b. Biosynthesis and function of modified nucleosides, p. 165–205. *In* D. Söll and U. L. RajBhandary (ed.), *tRNA: Structure, Biosynthesis and Function.* American Society for Microbiology, Washington, D.C.

Björk, G. R. 1995c. Stable RNA modification, p. 861–886. *In* F. C. Neidhardt, R. Curtiss III, J. L. Ingraham, E. C. C. Lin, K. B. Low, B. Magasanik, W. S. Reznikoff, M. Riley, M. Schaechter, and H. E. Umbarger (ed.), Escherichia coli *and* Salmonella*: Cellular and Molecular Biology,* 2nd ed. ASM Press, Washington, D.C.

Björnsson, A., and L. A. Isaksson. 1993. UGA codon context which spans three codons. Reversal by ms^{2}i^{6}A37 in tRNA, mutation in *rpsD*(S4) or streptomycin. *J. Mol. Biol.* 232:1017–1029.

Borén, T., P. Elias, T. Sammuelsson, C. Claesson, M. Barciszewska, C. W. Gehrke, K. C. Kuo, and F. Lustig. 1993. Undiscriminating codon reading with adenosine in the wobble position. *J. Mol. Biol.* 230:739–749.

Bossi, L., and J. R. Roth. 1980. The influence of codon context on genetic code translation. *Nature* 286:123–127.

Bouadloun, F., T. Srichaiyo, L. A. Isakson, and G. R. Björk. 1986. Influence of modification next to the anticodon in tRNA on codon context sensitivity of translational suppression and accuracy. *J. Bacteriol.* 166:102–1027.

Brierley, I., A. J. Jenner, and S. C. Inglis. 1992. Mutational analysis of the "slippery-sequence" component of a coronavirus ribosomal frameshifting signal. *J. Mol. Biol.* 227:463–479.

Brierley, I., M. R. Meredith, A. J. Bloys, and T. G. Hagervall. 1997. Expression of a coronavirus ribosomal frameshift signal in *Escherichia coli*: influence of tRNA anticodon modification on frameshifting. *J. Mol. Biol.* 271:1–14.

Buckingham, R. H. 1994. Codon context and protein synthesis: enhancements of the genetic code. *Biochimie* 76:351–354.

Carter, P. W., D. L. Weiss, H. L. Weith, and J. M. Calvo. 1985. Mutations that convert the four leucine codons of the *Salmonella typhimurium leu* leader to four threonine codons. *J. Bacteriol.* 162:943–949.

Chamorro, M., N. Parkin, and H. E. Varmus. 1992. An RNA pseudoknot and an optimal heptameric shift site are required for highly efficient ribosomal frameshifting on a retroviral messenger RNA. *Proc. Natl. Acad. Sci. USA* 89:713–717.

Cortese, R., R. Landsberg, R. A. Haar, H. E. Umbarger, and B. N. Ames. 1974. Pleiotropy of *hisT* mutants blocked in pseudouridine synthesis in tRNA: leucine and isoleucine-valine operons. *Proc. Natl. Acad. Sci. USA* 71:1857–1861.

Craigen, W. J., R. G. Cook, W. P. Tate, and C. T. Caskey. 1985. Bacterial peptide chain release factors: conserved primary structure and possible frameshift regulation of release factor 2. *Proc. Natl. Acad. Sci. USA* 82:3616–3620.

Craigen, W. J., and C. T. Caskey. 1986. Expression of peptide chain release factor 2 requires high-frequency frameshift. *Nature* 322:272–275.

Crick, F. H. C. 1966. Codon-anticodon pairing: the wobble hypothesis. *J. Mol. Biol.* 19:548–555.

Curran, J. F., and M. Yarus. 1988. Use of tRNA suppressors to probe regulation of Escherichia coli release factor 2. *J. Mol. Biol.* 203:75–83.

Curran, J. F., and M. Yarus. 1989. Rates of aminoacyl-tRNA selection at 29 sense codons *in vivo*. *J. Mol. Biol.* 209:65–77.

Curran, J. F. 1993. Analyses of effects of tRNA:message stability on frameshift frequency at the Escherichia coli RF2 programmed frameshift site. *Nucleic Acids Res.* 21:1837–1843.

Curran, J. F. 1995. Decoding with the A:I wobble pair is inefficient. *Nucleic Acids Res.* 23:683–688.

Curran, J. F., E. S. Poole, W. P. Tate, and B. L. Gross. 1995. Selection of aminoacyl-tRNAs at sense codons: the size of the tRNA variable loop determines whether the immediate 3' nucleotide to the codon has a context effect. *Nucleic Acids Res.* 23:4104–4108.

Dalphin, M. E., C. M. Brown, P. A. Stockwell, and W. P. Tate. 1997. The translational signal database, TransTerm: more organisms, complete genomes. *Nucleic Acids Res.* 25:246–247.

Dao, V., R. Guenther, A. Malkiewicz, B. Nawrot, E. Sochacka, A. Kraszewski, K. Everett, and P. F. Agris. 1994. Ribosome binding of DNA analogs of tRNA requires base modifications and supports the "extended anticodon." *Proc. Natl. Acad. Sci. USA* 91:2125–2129.

Davis, D. R., and C. D. Poulter. 1991. ^{1}H-^{15}N NMR studies of *Escherichia coli* tRNA^{Phe} from *hisT* mutants: a structural role for pseudouridine. *Biochemistry* 30:4223–4231.

Davis, D. R. 1995. Stabilization of RNA stacking by pseudouridine. *Nucleic Acids Res.* 23:5020–5026.

Diaz, I., M. Ehrenberg, and C.-G. Kurland. 1986. How do combinations of *rpsL*$^-$ and *miaA*$^-$ generate streptomycin dependence? *Mol. Gen. Genet.* 208:373–376.

Diaz, I., and M. Ehrenberg. 1991. ms^{2}i^{6}A deficiency enhances proofreading in translation. *J. Mol. Biol.* 222:1161–1171.

Dinman, J. D., T. Icho, and R. B. Wickner. 1991. A −1 ribosomal frameshift in a double-stranded RNA virus of yeast forms a *gag-pol* fusion protein. *Proc. Natl. Acad. Sci. USA* 88:174–178.

Dock-Bregeon, A. C., B. Chevrier, A. Podjarny, J. Johnson, J. S. deBear, G. R. Gough, P. T. Gilham, and D. Moras. 1989. Crystallographic structure of an RNA helix [U(UA)$_6$A]$_2$. *J. Mol. Biol.* 209:459–474.

Eggertsson, G., and D. Söll. 1988. Transfer ribonucleic acid-mediated suppression of termination codons in *Escherichia coli*. *Microbiol. Rev.* 52:354–374.

Elseviers, D., L. A. Petrullo, and P. J. Gallagher. 1984. Novel *E. coli* mutants deficient in biosynthesis of 5-methylaminomethyl-2-thiouridine. *Nucleic Acids Res.* 12:3521–3534.

Ericson, J. U., and G. R. Björk. 1986. Pleiotropic effects induced by modification deficiency next to the anticodon of tRNA from *Salmonella typhimurium* LT2. *J. Bacteriol.* 166:1013–1021.

Ericson, J. U., and G. R. Björk. 1991. tRNA anticodons with the modified nucleoside 2-methylthio-N^6-(hydroxyisopentenyl)-adenosine distinguish between bases 3' of the anticodon. *J. Mol. Biol.* 218:509–516.

Esberg, B., and G. R. Björk. 1995. The methylthio group (ms^2) of N6-(4-hydroxyisopentenyl)-2-methylthioadenosine (ms^{2}io^{6}A) present next to the anticodon contributes to the decoding efficiency of tRNA. *J. Bacteriol.* 177:1967–1975.

Farabaugh, P. J. 1996. Programmed translational frameshifting. *Microbiol. Rev.* 60:103–134.

Förster, C., K. Chakraburtty, and M. Sprinzl. 1993. Discrimination between initiation and elongation of protein biosynthesis in yeast: identity assured by a nucleotide modification in the initiator tRNA. *Nucleic Acids Res.* **21:**5679–5683.

Frey, B., G. Jänel, U. Michelsen, and H. Kersten. 1989. Mutations in the *Escherichia coli fnr* and *tgt* genes: control of molybdate reductase activity and the cytochrome *d* complex by *fnr. J. Bacteriol.* **171:**1524–1530.

Fu, C., and J. Parker. 1994. A ribosomal frameshifting error during translation of the *argI* mRNA of *Escherichia coli. Mol. Gen. Genet.* **243:**434–441.

Gesteland, R. F., and J. F. Atkins. 1996. Recoding: dynamic reprogramming of translation. *Annu. Rev. Biochem.* **65:**741–768.

Glasser, A.-L., J. Desgres, J. Heitzler, C. W. Gehrke, and G. Keith. 1991. O-Ribosyl-phosphate purine as a constant modified nucleotide located at position 64 in cytoplasmic initiator tRNAs^{Met} of yeasts. *Nucleic Acids Res.* **19:**5199–5203.

Green, R., and H. F. Noller. 1997. Ribosomes and translation. *Annu. Rev. Biochem.* **66:**679–716.

Griffey, R. H., D. Davis, Z. Yamaizumi, S. Nishimura, A. Bax, B. Hawkins, and C. D. Poulter. 1985. ^{15}N-labeled *Escherichia coli* tRNA^{fMet}, tRNA^{Glu}, tRNA^{Tyr}, and tRNA^{Phe}. Double resonance and two-dimensional NMR of N^1-labelled pseudouridine. *J. Biol. Chem.* **260:**9734–9741.

Grosjean, H., D. G. Söll, and D. M. Crothers. 1976. Studies of the complex between tRNAs with complementary anticodons. I. Origins of enhanced affinity between complementary triplets. *J. Mol. Biol.* **103:**499–519.

Grosjean, H. J., S. DeHenau, and D. M. Crothers. 1978. On the physical basis for ambiguity in genetic coding interactions. *Proc. Natl. Acad. Sci. USA* **75:**610–614.

Grosjean, H., K. Nicoghosian, E. Haumont, D. Söll, and R. Cedergren. 1985. Nucleotide sequences of two serine tRNAs with a GGA anticodon: the structure-function relationships in the serine family of tRNAs. *Nucleic Acids Res.* **13:**5697–5706.

Grosjean, H., and C. Houssier. 1990. Codon recognition: evaluation of the effects of modified bases in the anticodon loop of tRNA using the temperature-jump relaxation method. *J. Chromatogr. Library* **45:**A555–A595.

Grosjean, H., M. Sprinzl, and S. Steinberg. 1995. Posttranscriptionally modified nucleosides in transfer RNA: their locations and frequencies. *Biochimie* **77:**139–141.

Grossenbacher, A.-M., B. Stadelmann, W.-D. Heyer, P. Thuriaux, and J. Kohli. 1986. Antisuppressor mutations and sulfur-carrying nucleosides in transfer RNAs of *Schizosaccharomyces pombe. J. Biol. Chem.* **261:**16351–16355.

Gu, Z., R. Harrod, E. J. Rogers, and P. S. Lovett. 1994. Properties of a pentapeptide inhibitor of peptidyltransferase that is essential for *cat* gene regulation by translational attenuation. *J. Bacteriol.* **176:**6238–6244.

Guillerez, J., M. Gazeau, and M. Dreyfus. 1991. In the *Escherichia coli lacZ* gene the spacing between the translating ribosome is sensitive to the efficiency of translation initiation. *Nucleic Acids Res.* **19:**6743–6750.

Gupta, R. 1984. *Halobacterium volcanii* tRNAs: identification of 41 tRNAs covering all amino acids, and the sequences of 33 class 1 tRNAs. *J. Biol. Chem.* **259:**9461–9471.

Guthrie, C., and J. Abelson. 1982. Organization and expression of tRNA genes in Saccharomyces cerevisiae, p. 487–528. *In* J. N. Strathern, E. W. Jones, and J. R. Broach (ed.), *The Molecular Biology of the Yeast* Saccharomyces: *Metabolism and Gene Expression.* Cold Spring Harbor Laboratory Press, Cold Spring Harbor, N.Y.

Hagervall, T. G., J. U. Ericson, B. Esberg, J.-N. Li, and G. R. Björk. 1984. Undermodification in the first position of the anticodon of *supG*-tRNA reduces translational efficiency. *Mol. Gen. Genet.* **196:**194–200.

Hagervall, T. G., Y. H. Jönsson, C. G. Edmonds, J. A. McCloskey, and G. R. Björk. 1990. Chorismic acid, a key metabolite in modification of tRNA. *J. Bacteriol.* **172:**252–259.

Hagervall, T. G., B. Esberg, J.-N. Li, T. M. F. Tuohy, J. F. Atkins, J. F. Curran, and G. R. Björk. 1993a. Functional aspects of three nucleosides, Ψ, ms^{2}io^{6}A, and m^{1}G, present in the anticodon loop of tRNA, p. 67–78. *In* K. Nierhaus (ed.), *The Translational Apparatus.* Plenum Press, New York, N.Y.

Hagervall, T. G., T. M. F. Tuohy, J. F. Atkins, and G. R. Björk. 1993b. Deficiency of 1-methylguanosine in tRNA from *Salmonella typhimurium* induces frameshifting by quadruplet decoding. *J. Mol. Biol.* **232:**756–765.

Harada, F., and S. Nishimura. 1972. Possible anticodon sequences of tRNA^{His}, tRNA^{Asn}, and tRNA^{Asp} from *Escherichia coli* B. Universal presence of nucleoside Q in the first position of the anticodon of three transfer ribonucleic acids. *Biochemistry* **13:**300–306.

Harada, F., and J. E. Dahlberg. 1975. Specific cleavage of tRNA by nuclease S1. *Nucleic Acids Res.* **2:**865–871.

Hardesty, B., O. W. Odom, and J. Czworkowski. 1990. Movement of tRNA through ribosomes during peptide elongation, p. 366–372. *In* W. E. Hill, A. Dahlberg, R. A. Garrett, P. B. Moore, D. Schlessinger, and J. R. Warner (ed.), *The Ribosome: Structure, Function, and Evolution.* American Society for Microbiology, Washington, D.C.

Hatfield, D., Y.-X. Feng, B. J. Lee, A. Rein, J. G. Levin, and S. Oroszlan. 1989. Chromatographic analysis of the aminoacyl-tRNAs which are required for translation of codons at and around the ribosomal frameshift sites of HIV, HTLV-1, and BLV. *Virology* **173:**736–742.

Heckman, J. E., J. Sarnoff, B. Alzner-DeWeerd, S. Yin, and U. L. RajBhandary. 1980. Novel features in the genetic code and codon reading patterns in *Neurospora crassa* mitochondria based on sequences of six mitochondrial tRNAs. *Proc. Natl. Acad. Sci. USA* **77:**3159–3163.

Heurgue-Hamard, V., L. Mora, G. Guarneros, and R. H. Buckingham. 1996. The growth defect in *Escherichia coli* deficient in peptidyl hydrolase is due to starvation for Lys-tRNA^{Lys}. *EMBO J.* **15:**2826–2833.

Heyer, W.-D., P. Thuriaux, J. Kohli, P. Ebert, H. Kersten, C. Gehrke, K. C. Kuo, and P. F. Agris. 1984. An antisuppressor mutation of *Schizosaccharomyces pombe* affects posttranscriptional modification of the "wobble" U base in the anticodon of tRNAs. *J. Biol. Chem.* **259:**2856–2862.

Hirsh, D., and L. Gold. 1971. Translation of the UGA triplet *in vitro* by tryptophan transfer RNAs. *J. Mol. Biol.* **58:**459–468.

Holbrook, S. R., J. L. Sussman, R. W. Warrant, and S.-H. Kim. 1978. Crystal structure of yeast phenylalanine transfer RNA. II. Structural features and functional implications. *J. Mol. Biol.* **123:**631–660.

Holbrook, S. R., C. Cheong, I. Tinoco, and S.-H. Kim. 1991. Crystal structure of an RNA double helix incorporating a track of non-Watson–Crick base pairs. *Nature* **353:**579–581.

Horsfield, J. A., D. N. Wilson, S. A. Mannering, F. M. Adamski, and W. P. Tate. 1995. Prokaryotic ribosomes recode the HIV-1 *gag-pol* −1 frameshift sequence by an E/P site post-translocation simultaneous slippage mechanism. *Nucleic Acids Res.* **23:**1487–1494.

Houssier, C., and H. Grosjean. 1985. Temperature jump relaxation studies on the interactions between transfer RNAs with complementary anticodons: the effect of modified bases adjacent to the anticodon triplet. *J. Biomol. Struct. Dyn.* **3:**387–408.

Houssier, C., P. Degrée, K. Nicoghosian, and H. Grosjean. 1988. Effect of uridine dethiolation in the anticodon triplet of tRNA^{Glu}

on its association with tRNA^Phe. *J. Biomol. Struct. Dyn.* 5: 1259–1266.

Ikemura, T. 1981. Correlation between the abundance of *Escherichia coli* transfer RNAs and the occurrence of the respective codons in its protein genes: a proposal for a synonymous codon choice that is optimal for the *E. coli* translational system. *J. Mol. Biol.* 151:389–409.

Ikemura, T. 1992. Correlation between codon usage and tRNA content in microorganisms, p. 87–111. *In* D. L. Hatfield, B. J. Lee, and R. M. Pirtle (ed.), *Transfer RNA in Protein Synthesis.* CRC Press, Boca Raton, Fla.

Ishikura, H., Y. Yamada, and S. Nishimura. 1971. Structure of serine tRNA from *Escherichia coli.* I. Purification of serine tRNA's with different codon responses. *Biochim. Biophys. Acta* 228:471–481.

Jacks, T., H. D. Madhani, F. R. Masiarz, and H. E. Varmus. 1988. Signals for ribosomal frameshifting in the Rous sarcoma virus *gag-pol* region. *Cell* 55:447–458.

Johnson, P. F., and J. Abelson. 1983. The yeast tRNA^Tyr gene intron is essential for correct modification of its tRNA product. *Nature* 302:681–687.

Johnston, H. M., W. M. Barnes, F. G. Chumley, L. Bossi, and B. N. Ames. 1980. Model for regulation of the histidine operon of *Salmonella. Proc. Natl. Acad. Sci. USA* 77:508–512.

Jukes, T. H. 1973. Possibilities for the evolution of the genetic code from a preceding form. *Nature* 246:22–26.

Kano, A., Y. Andachi, T. Ohama, and S. Osawa. 1991. Novel anticodon composition of transfer RNAs in *Micrococus luteus,* a bacterium with a high genomic G+C content. Correlation with codon usage. *J. Mol. Biol.* 221:387–401.

Kawai, G., Y. Yamamoto, T. Kamimura, T. Masegi, M. Sekine, T. Hata, T. Iimori, T. Watanabe, T. Miyazawa, and S. Yokoyama. 1992. Conformational rigidity of specific pyrimidine residues in tRNA arises from posttranscriptional modifications that enhance steric interaction between the base and the 2'-hydroxyl group. *Biochemistry* 31:1040–1046.

Komine, Y., T. Andachi, A. Inokuchi, and H. Ozeki. 1990. Genomic organization and physical mapping of the transfer RNA genes in *Escherichia coli* K12. *J. Mol. Biol.* 212:579–598.

Kowalak, J. A., E. Breunger, and J. A. McCloskey. 1994. Posttranscriptional modification of the central loop of domain V in *Escherichia coli* 23 S ribosomal RNA. *J. Biol. Chem.* 270:17758–17764.

Kurland, C. G., D. Hughes, and M. Eherenberg. 1995. Limitations of translational accuracy, p. 979–1004. *In* F. C. Neidhardt, R. Curtiss III, J. L. Ingraham, E. C. C. Lin, K. B. Low, B. Magasanik, W. S. Reznikoff, M. Riley, M. Schaechter, and H. E. Umbarger (ed.), Escherichia coli *and* Salmonella: *Cellular and Molecular Biology,* 2nd ed. ASM Press, Washington, D.C.

Landick, R., C. L. Turnbough, Jr., and C. Yanofsky. 1996. Transcription attenuation, p. 1263–1286. *In* F. C. Neidhardt, R. Curtiss III, J. L. Ingraham, E. C. C. Lin, K. B. Low, B. Magasanik, W. S. Reznikoff, M. Riley, M. Schaechter, and H. E. Umbarger (ed.), Escherichia coli *and* Salmonella: *Cellular and Molecular Biology,* 2nd ed. ASM Press, Washington, D.C.

Lane, B. G., J. Ofengand, and M. W. Gray. 1995. Pseudouridine and O2'-methylated nucleosides. Significance of their selective occurrence in rRNA domains that function in ribosome-catalyzed synthesis of the peptide bonds in proteins. *Biochimie* 77:7–15.

Lee, F., and C. Yanofsky. 1977. Transcription termination at the *trp* operon attenuators of *Escherichia coli* and *Salmonella typhimurium*: RNA secondary structure and regulation of termination. *Proc. Natl. Acad. Sci. USA* 74:4365–4369.

Li, J.-N., and G. R. Björk. 1995. 1-Methylguanosine deficiency of tRNA influences cognate codon interaction and metabolism in *Salmonella typhimurium. J. Bacteriol.* 177:6593–6600.

Li, J.-N., B. Esberg, J. F. Curran, and G. R. Björk. 1997. Three modified nucleosides present in the anticodon stem and loop influence the *in vivo* aa-tRNA selection in a tRNA-dependent manner. *J. Mol. Biol.* 271:209–221.

Lietzke, S. E., C. L. Barnes, J. A. Berglund, and C. E. Kundrot. 1996. The structure of an RNA dodecamer shows how tandem U-U base pairs increase the range of stable RNA structures and the diversity of recognition sites. *Structure* 4:917–930.

Lim, V., Č. Venclovas, A. Spirin, R. Brimacombe, P. Mitchell, and F. Müller. 1992. How are tRNAs and mRNA arranged in the ribosome? An attempt to correlate the stereochemistry of the tRNA-mRNA interaction with constraints imposed by the ribosomal topography. *Nucleic Acids Res.* 20:2627–2637.

Lim, V. I., and Č. Venclovas. 1992. Codon-anticodon pairing. A model for interacting codon-anticodon duplexes located at the ribosomal A- and P-sites. *FEBS Lett.* 313:133–137.

Lim, V. I. 1994. Analysis of the action of wobble nucleoside modifications on codon-anticodon pairing within the ribosome. *J. Mol. Biol.* 240:8–19.

Lim, V. I. 1995. Analysis of the action of the wobble adenine on codon reading within the ribosome. *J. Mol. Biol.* 252:277–282.

Lim, V. I. 1997. Analysis of interactions between the codon-anticodon duplexes within the ribosome: their role in translation. *J. Mol. Biol.* 266:877–890.

Limbach, P. A., P. F. Crain, and J. A. McCloskey. 1995. Characterization of oligonucleotides and nucleic acids by mass spectrometry. *Curr. Opin. Biotechnol.* 6:96–102.

Lovett, P. S., and E. J. Rogers. 1996. Ribosome regulation by the nascent peptide. *Microbiol. Rev.* 60:366–385.

Lustig, F., P. Elias, T. Axberg, T. Samuelsson, I. Tittawella, and U. Lagerkvist. 1981. Codon reading and translational error. Reading of the glutamine and lysine codon during protein synthesis *in vitro. J. Biol. Chem.* 256:2635–2643.

Lustig, F., T. Borén, C. Claesson, C. Simonsson, M. Barciszewska, and U. Lagerkvist. 1993. The nucleotide in position 32 of the tRNA anticodon loop determines ability of anticodon UCC to discriminate among glycine codons. *Proc. Natl. Acad. Sci. USA* 90:3343–3347.

Lynn, S. P., W. S. Burton, T. J. Donahue, R. M. Gould, R. I. Gumport, and J. F. Gardner. 1987. Specificity of the attenuation response of the threonine operon of *Escherichia coli* determined by the threonine and isoleucine codons in the leader transcript. *J. Mol. Biol.* 194:59–69.

Mangroo, D., P. A. Limbach, J. A. McCloskey, and U. L. RajBhandary. 1995. An anticodon sequence mutant of *Escherichia coli* initiator tRNA: possible importance of a newly acquired base modification next to the anticodon on its activity in initiation. *J. Bacteriol.* 177:2858–2862.

Martin, R. P., A. P. Sibler, C. W. Gehrke, K. Kuo, C. G. Edmonds, J. A. McCloskey, and G. Dirheimer. 1990. 5-[((Carboxymethyl)-amino]methyl)]uridine is found in the anticodon of yeast mitochondrial tRNAs recognizing two-codon families ending in a purine. *Biochemistry* 29:956–959.

Meier, F., B. Suter, H. Grosjean, G. Keith, and E. Kubli. 1985. Queuosine modification of the wobble base in tRNA^His influences 'in vivo' decoding properties. *EMBO J.* 4:823–827.

Miller, J. H., and A. M. Albertini. 1983. Effects of surrounding sequence on the suppression of nonsense codons. *J. Mol. Biol.* 164:59–71.

Miller, J. P., Z. Hussein, and M. P. Schweizer. 1976. The involvement of the anticodon adjacent modified nucleoside N-9-(-D-ribofuranosyl)-purine-6-carbamoyl-threonine in the biological function of *E. coli* tRNA^Ile. *Nucleic Acids Res.* 3:1185–1201.

Motorin, Y., G. Bec, R. Tewari, and H. Grosjean. 1997. Transfer RNA recognition by the *Escherichia coli* Δ^2-isopentenyl-pyrophosphate:tRNA Δ^2-isopentenyl transferase: dependence on the anticodon arm structure. *RNA* 3:721–733.

Munz, P., U. Leupold, P. Agris, and J. Kohli. 1981. In vivo decoding rules studied in *Schizosaccharomyces pombe* are at variance with *in vitro* data. *Nature* 294:187–188.

Muramatsu, T., S. Yokoyama, N. Horie, A. Matsuda, T. Ueda, Z. Yamaizumi, Y. Kuchino, S. Nishimura, and T. Miyazawa. 1988. A novel lysine-substituted nucleoside in the first position of the anticodon of minor isoleucine tRNA from *Escherichia coli*. *J. Biol. Chem.* 263:9261–9267.

Murgola, E. J. 1995. Translational suppression: when two wrongs DO make a right, p. 491–509. *In* D. Söll and U. L. RajBhandary (ed.), *tRNA: Structure, Biosynthesis, and Function*. American Society for Microbiology, Washington, D.C.

Nierhaus, K. H. 1993. Solution of the ribosome riddle: how the ribosome selects the correct aminoacyl-tRNA out of 41 similar contestants. *Mol. Microbiol.* 9:661–669.

Nierhaus, K. H. 1996. An elongation factor turn-on. *Nature* 379:491–492.

Nishimura, S. 1972. Minor components in transfer RNA: their characterization, location and function. *Prog. Nucleic Acid Res. Mol. Biol.* 12:49–85.

Noguchi, S., Y. Nishimura, Y. Hirota, and S. Nishimura. 1982. Isolation and characterization of an *Escherichia coli* mutant lacking tRNA-guanine transglycosylase. *J. Biol. Chem.* 257:6544–6550.

Ofengand, J., and A. Bakin. 1997. Mapping to nucleotide resolution of pseudouridine residues in large subunit ribosomal RNAs from representative eukaryotes, prokaryotes, archaebacteria, mitochondria and chloroplasts. *J. Mol. Biol.* 266:246–268.

Osawa, S., T. H. Jukes, K. Watanabe, and A. Muto. 1992. Recent evidence for evolution of the genetic code. *Microbiol. Rev.* 56:229–264.

Païs de Barros, J.-P., G. Keith, C. E. Adlouni, A.-L. Glasser, G. Mack, G. Dirheimer, and J. Degrès. 1996. 2'-O-methyl-5-formylcytidine (f^{5}Cm), a new modified nucleotide at the 'wobble' position of two cytoplasmic tRNAsLeu(NAA) from bovine liver. *Nucleic Acids Res.* 24:1489–1496.

Palmer, D. T., P. H. Blum, and S. W. Artz. 1983. Effect of the *hisT* mutation of *Salmonella typhimurium* on translation elongation rate. *J. Bacteriol.* 153:357–363.

Parker, J. 1982. Specific mistranslation in *hisT* mutants of *Escherichia coli*. *Mol. Gen. Genet.* 190:405–409.

Parker, J., and J. D. Friesen. 1980. "Two out of three" codon reading leading to mistranslation *in vivo*. *Mol. Gen. Genet.* 177:439–445.

Parker, J. 1989. Errors and alternatives in reading the universal genetic code. *Microbiol. Rev.* 53:273–298.

Pedersen, W. T., and J. F. Curran. 1991. Effects of the nucleotide 3' to an amber codon on ribosomal selection rates of suppressor tRNA and release factor 1. *J. Mol. Biol.* 219:231–241.

Persson, B., and G. R. Björk. 1993. Isolation of the gene (*miaE*) encoding the hydroxylase involved in the synthesis of 2-methylthio-*cis*-ribozeatin in tRNA of *Salmonella typhimurium* and characterization of mutants. *J. Bacteriol.* 175:7776–7785.

Petrullo, L. A., P. J. Gallagher, and D. Elseviers. 1983. The role of 2-methylthio-N^6-isopentenyladenosine in readthrough and suppression of nonsense codons. *Mol. Gen. Genet.* 190:289–294.

Pieczenik, G. 1980. Predicting coding function from nucleotide sequence or survival of "fitness" of tRNA. *Proc. Natl. Acad. Sci. USA* 77:3539–3543.

Precup, J., and J. Parker. 1987. Missense misreading of asparagine codons as a function of codon identity and context. *J. Biol. Chem.* 262:11351–11355.

Qian, Q., and G. R. Björk. 1997. Structural requirements for the formation of 1-methylguanosine *in vivo* in tRNA$^{Pro}_{GGG}$ of Salmonella typhimurium. *J. Mol. Biol.* 266:283–296.

Qian, Q., J. F. Curran, and G. R. Björk. 1997. Unpublished data.

Qian, Q., and G. R. Björk. Personal communication.

Rizzino, A., M. Mastanduno, and M. Freundlich. 1977. Partial derepression of the isoleucine-valine enzymes during methionine starvation in *Salmonella typhimurium*. *Biochim. Biophys. Acta* 475:267–275.

Roth, J. R., D. N. Anton. and P. E. Hartman. 1966. Histidine regulatory mutations in *Salmonella typhimurium*. I. Isolation and general properties. *J. Mol. Biol.* 22:305–323.

Saenger, W. 1984. *Principles of Nucleic Acid Structure*. Springer-Verlag, New York, N.Y.

Salser, W. 1969. The influence of the reading context upon the suppression of nonsense codons. *Mol. Gen. Genet.* 105:125–130.

Schejfer, E., S. Roy, V. Sanchez, and A. G. Redfield. 1982. Nuclear Overhauser effect study of yeast tRNA$^{Val}_1$: evidence for uridine:pseudouridine pairing. *Nucleic Acids Res.* 10:8297–8305.

Schultz, D. W., and M. Yarus. 1994a. tRNA structure and ribosomal function. I. tRNA nucleotide 27–43 mutations enhance first position wobble. *J. Mol. Biol.* 235:1381–1394.

Schultz, D. W., and M. Yarus. 1994. tRNA structure and ribosomal function. II. Interaction between anticodon helix and other tRNA mutations. *J. Mol. Biol.* 235:1395–1405.

Schwartz, R. S., and J. F. Curran. 1997. Analyses of frameshifting at UUU-pyrimidine sites. *Nucleic Acids Res.* 25:2005–2011.

Sibler, A. P., G. Dirheimer, and R. P. Martin. 1986. Codon reading patterns in *Saccharomyces cerevisiae* based on sequences of mitochondrial tRNAs. *FEBS Lett.* 194:131–138.

Sierzputowska-Gracz, H., E. Sochacka, A. Malkiewicz, K. C. Kuo, C. W. Gehrke, and P. F. Agris. 1987. Chemistry and structure of modified uridines in the anticodon, wobble position of transfer RNA are determined by thiolation. *J. Am. Chem. Soc.* 109:7171–7177.

Singer, C. E., G. R. Smith, R. Cortese, and B. N. Ames. 1972. Mutant tRNA^{His} ineffective in repression and lacking two pseudouridine modifications. *Nat. New Biol.* 238:72–74.

Sipley, J., and E. Goldman. 1993. Increased ribosomal accuracy increases a programmed translational frameshift in *Escherichia coli*. *Proc. Natl. Acad. Sci. USA* 90:2315–2319.

Smith, D., and M. Yarus. 1989a. tRNA-tRNA interactions within cellular ribosomes. *Proc. Natl. Acad. Sci. USA* 86:4397–4401.

Smith, D., and M. Yarus. 1989b. Transfer RNA structure and coding specificity. I. Evidence that a D-arm mutation reduces tRNA dissociation from the ribosome. *J. Mol. Biol.* 206:489–501.

Smith, D., and M. Yarus. 1989c. Transfer RNA structure and coding specificity. II. A D-arm tertiary interaction that restricts coding range. *J. Mol. Biol.* 206:503–511.

Smith, W., H. Sierzputowska-Gracz, S. Sochacka, A Malkiewicz, and P. F. Agris. 1992. Chemistry and structure of modified uridine dinucleosides are determined by thiolation. *J. Am. Chem. Soc.* 114:7989–7997.

Söll, D., E. Ohtsuka, D. S. Jones, R. Lohrmann, H. Hayatsu, S. Nishimura, and H. G. Khorana. 1965. Studies on polynucleotides. XLIX. Stimulation of the binding of aminoacyl-tRNA's to ribosomes by ribotrinucleotides and a survey of codon assignments for 20 amino acids. *Proc. Natl. Acad. Sci. USA* 54:1378–1385.

Spanjaard, R. A., K. Chen, J. R. Walker, and J. van Duin. 1990. Frameshift suppression at tandem AGA and AGG codons by cloned tRNA genes: assigning a codon to *argU* tRNA and T4 tRNA^{Arg}. *Nucleic Acids Res.* 18:5031–5036.

Sprinzl, M., C. Steegborn, F. Hübel, and S. Steinberg. 1996. Compilation of tRNA sequences and sequences of tRNA genes. *Nucleic Acids Res.* 24:68–72.

Stanssens, P., E. Remaut, and W. Fiers. 1986. Inefficient translation initiation causes premature transcription termination in the *lacZ* gene. *Cell* 44:711–718.

Suzuki, T., T. Ueda, T. Yokogawa, K. Nishimura, and K. Watanabe. 1994. Characterization of serine and leucine tRNAs in an asporogenic yeast *Candida cylindracea* and evolutionary implications of genes for tRNA$_{CAG}^{Ser}$ responsible for translation of a non-universal genetic code. *Nucleic Acids Res.* 22:115–123.

Takai, K., H. Takaku, and S. Yokoyama. 1996. Codon-reading specificity of an unmodified form of *Escherichia coli* tRNA$_1^{Ser}$ in cell-free protein synthesis. *Nucleic Acids Res.* 24:2894–2899.

Takemoto, C., T. Koike, T. Yokogawa, L. Benkowski, L. L. Spremulli, T. A. Ueda, K. Nishikawa, and K. Watanabe. 1995. The ability of bovine mitochondrial transfer RNA Met to decode AUG and AUA codons. *Biochimie* 77:104–108.

Tanaka, R., Y. Andachi, and A. Muto. 1991. Evolution of tRNAs and tRNA genes in *Acholeplasma laidlawii*. *Nucleic Acids Res.* 19:6787–6792.

Tate, W. P., and S. A. Mannering. 1996. Three, four or more: the translational stop signal at length. *Mol. Microbiol.* 21:213–219.

ten Dam, E., C. Pleij, and L. Bosch. 1990. RNA pseudoknots: translational frameshifting and readthrough of viral RNAs. *Virus Genes* 4:121–136.

Thompson, R. C. 1988. EFTu provides an internal kinetic standard for translational accuracy. *Trends Biochem. Sci.* 13:91–93.

Tsuchihashi, Z., and P. O. Brown. 1992. Sequence requirements for efficient translational frameshifting in the *Escherichia coli dnaX* gene and the role of an unstable interaction between tRNA^{Lys} and an AAG lysine codon. *Genes Dev.* 6:511–519.

Vacher, J., H. Grosjean, C. Houssier, and R. H. Buckingham. 1984. The effect of point mutations affecting *Escherichia coli* tryptophan tRNA on anticodon-anticodon interactions and on UGA suppression. *J. Mol. Biol.* 177:329–342.

Varani, G., and I. Tinoco. 1991. RNA structure and NMR spectroscopy. *Q. Rev. Biophys.* 24:479–532.

Watanabe, K. 1980. Reactions of 2-thioribothymidine and 4-thiouridine with hydrogen peroxide in transfer ribonucleic acids from *Thermus thermophilus* and *Escherichia coli* as studied by circular dichroism. *Biochemistry* 19:5542–5549.

Watanabe, K., N. Hayashi, A. Oyama, K. Nishikawa, T. Ueda, and K. Miura. 1994. Unusual anticodon loop structure found in E. coli lysine tRNA. *Nucleic Acids Res.* 22:79–87.

Weiss, R. B., D. M. Dunn, A. E. Dahlberg, J. F. Atkins, and R. F. Gesteland. 1988. Reading frame switch caused by base-pair formation between the 3' end of 16S rRNA and the mRNA during elongation of protein synthesis in *Escherichia coli*. *EMBO J.* 7:1503–1507.

Weiss, R. B., D. M. Dunn, M. Shuh, J. F. Atkins, and R. F. Gesteland. 1989. *E. coli* ribosomes re-phase on retroviral frameshift signals at rates ranging from 2 to 50 percent. *New Biol.* 1:159–169.

Weiss, W. A., and E. C. Friedberg. 1986. Normal yeast tRNA$_{CAG}^{Gln}$ can suppress amber codons and is encoded by an essential gene. *J. Mol. Biol.* 192:725–735.

Weissenbach, J., and H. Grosjean. 1981. Effect of threonylcarbamoyl modification (t^{6}A) in yeast tRNA$_{III}^{Arg}$ on codon-anticodon and anticodon-anticodon interactions. *Eur. J. Biochem.* 116:207–213.

Westhof, E., P. Dumas, and D. Moras. 1985. Crystallographic refinement of yeast aspartic acid transfer RNA. *J. Mol. Biol.* 184:119–145.

Wilson, R. K., and B. A. Roe. 1989. Presence of the hypermodified nucleotide N^6-(Δ^2-isopentenyl)-2-methylthioadenosine prevents codon misreading by *Escherichia coli* phenylalanyl-transfer RNA. *Proc. Natl. Acad. Sci. USA* 86:409–413.

Wu, M., J. A. McDowell, and J. H. Turner. 1995. A periodic table of symmetric tandem mismatches in RNA. *Biochemistry* 34:3204–3211.

Yarus, M. 1982. Translational efficiency of transfer RNA's: uses of an extended anticodon. *Science* 218:646–652.

Yarus, M., S. W. Cline, P. Wier, L. Breeden, and R. C. Thompson. 1986. Actions of the anticodon arm in translation on the phenotypes of RNA mutants. *J. Mol. Biol.* 192:235–255.

Yarus, M., and J. F. Curran. 1992. The translational context effect, p. 319–365. *In* D. A. Hatfield, B. J. Lee, and R. M. Pirtle (ed.), *Transfer RNA in Protein Synthesis*. CRC Press, Boca Raton, Fla.

Yarus, M., and D. Smith. 1995. tRNA on the ribosome: a waggle theory, p. 443–469. *In* D. Söll and U. L. RajBhandary (ed.), *tRNA: Structure, Biosynthesis, and Function*. American Society for Microbiology, Washington, D.C.

Yelverton, E., D. Lindsley, P. Yamauchi, and J. Gallant. 1994. The function of a ribosomal frameshifting signal from human immunodeficiency virus-1 in *Escherichia coli*. *Mol. Microbiol.* 11:303–313.

Yokoyama, S., Z. Yamaizumi, S. Nishimura, G. Kawai, and T. Miyazawa. 1979. ^{1}H NMR studies on the conformational characteristics of 2-thiopyrimidine nucleotides found in transfer RNAs. *Nucleic Acids Res.* 6:2611–2626.

Yokoyama, S., T. Watanabe, K. Murao, H. Ishikura, Z. Yamaizumi, S. Nishimura, and T. Miyazawa. 1985. Molecular mechanism of codon recognition by tRNA species with modified uridine in the first position of the anticodon. *Proc. Natl. Acad. Sci. USA* 82:4905–4909.

Yokoyama, S., and S. Nishimura. 1995. Modified nucleosides and codon recognition, p. 207–233. *In* D. Söll and U. L. RajBhandary (ed.), *tRNA: Structure, Biosynthesis, and Function*. American Society for Microbiology, Washington, D.C.

Zerfass, K., and H. Beier. 1992. Pseudouridine in the anticodon GΨA of plant cytoplasmic tRNA^{Tyr} is required for UAG and UAA suppression in the TMV-specific context. *Nucleic Acids Res.* 20:5911–5918.

Chapter 28

Importance of Modified Nucleotides in Replication of Retroviruses, Plant Pararetroviruses, and Retrotransposons

ROLAND MARQUET

Retroviruses are RNA viruses that infect not only mammals but also birds, fish, and insects. Retrotransposons are widespread transposable elements that are closely related to retroviruses. Pararetroviruses, including caulimoviruses and badnaviruses, are DNA viruses that infect plants. As for all retroids, the replication cycle of retroviruses, retrotransposons, and pararetroviruses includes a reverse transcription step (for a review, see Marquet et al., 1995, and references therein). Reverse transcription of the retroviral genome takes place in the cytoplasm of the infected cell (Fig. 1). The resulting provirus is integrated into the host genome, and is transcribed and translated by the cellular machinery (Fig. 1). The replication cycle of retrotransposons is very similar. However, because they do not code for envelope proteins, the virus-like particles are not released in the extracellular medium (Gabriel and Boeke, 1993). Even though pararetroviruses have a DNA genome, they replicate through an RNA intermediate that is reverse transcribed in the cytoplasm. Unlike retroviruses and retrotransposons, the viral DNA is maintained in the nucleus as a circular, covalently closed, minichromosome (Bonneville and Hohn, 1993).

Unlike other retroids, retroviruses, pararetroviruses, and most retrotransposons use a specific tRNA isoacceptor to prime reverse transcription (Marquet et al., 1995; Levin, 1997). The genomic RNAs of these retroids contain a primer binding site (PBS) located close to their 5' ends, which is usually complementary to the 3' end of a specific tRNA species of the host cell. In retroviruses, the PBS is 18 nt in length, while it can be significantly shorter in plant pararetroviruses (8 to 12 nt) and some retrotransposons (8 to 18 nt). Human immunodeficiency virus type 1 (HIV-1), the etiological agent of AIDS, like all immunodeficiency viruses, utilizes tRNA$_3^{Lys}$ as its replication primer, while the unique avian tRNA^{Trp} isoacceptor and two tRNA^{Pro} isoacceptors differing by two nucleotides in the anticodon loop are the primers of most avian and murine retroviruses, respectively (Fig. 2). tRNA$_i^{Met}$ is used by plant pararetroviruses and some retrotransposons (Fig. 2), while other retrotransposons utilize tRNA^{Asn}, tRNA^{Lys}, tRNA^{Arg}, tRNA^{Ser} or tRNA^{Leu} as primers (Marquet et al., 1995).

The primer tRNA is encapsidated and annealed to the PBS during formation and maturation of the viral (or virus-like) particles (Fig. 1) (Marquet et al., 1995). The conversion of the genomic RNA into double-stranded DNA is achieved in the infected cell by the RNA- and DNA-dependent DNA polymerase encoded by the retroid (reverse transcriptase [RT]), through a complex series of steps (Fig. 3) (Gilboa et al., 1979). During the initiation of minus-strand strong-stop DNA synthesis, tRNA is used as a primer (step I in Fig. 3), while during the plus-strand strong-stop DNA synthesis, the 3' end of the tRNA is used as a template to generate the plus copy of the PBS (step III in Fig. 3). In the next sections, I will summarize our knowledge concerning the importance of modified nucleotides in the selection and annealing of the primer tRNA and in the reverse transcription process.

In retroviruses and retrotransposons, the viral enzymes, including RT, are synthesized as Gag-Pol precursors, while the pol gene of cauliflower mosaic virus (CaMV; the caulimovirus prototype) is expressed separately from the gag gene (Bonneville and Hohn, 1993). An independent expression of pol was also recently demonstrated in human foamy virus (Yu et al., 1996). Synthesis of the Gag-Pol precursor requires either "readthrough" suppression of a termi-

---

Roland Marquet • Unité Propre de Recherche 9002 du Centre National de la Recherche Scientifique, Institut de Biologie Moléculaire et Cellulaire, 15 rue R. Descartes, F-67084 Strasbourg Cedex, France.

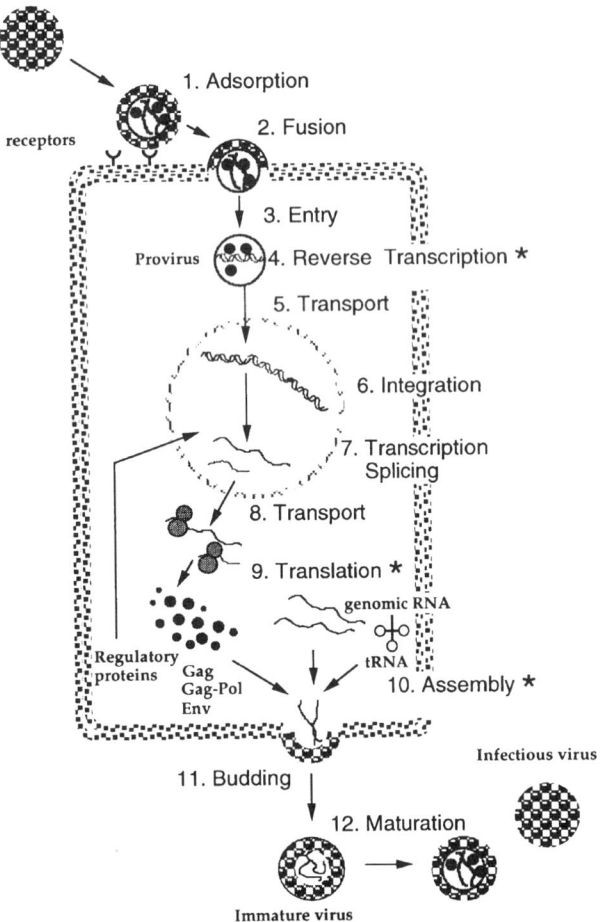

Figure 1. The retroviral replication life cycle. Reverse transcription, integration, and maturation require viral enzymes encoded by the *pol* gene. The steps that are affected by modified nucleotides are indicated by asterisks.

nation codon or ribosomal frameshifting in the 3' (−1) or 5' (+1) direction. The efficiency of readthrough suppression and ribosomal frameshifting might be strongly affected by the modified nucleotides in the anticodon loop of the tRNAs involved in these events.

Finally, not only the modified nucleotides of tRNAs, but also those present in the genomic RNA itself might affect the replication of retroviruses and retrotransposons. The identification of such modifications and their possible roles are described in the last section of this chapter.

IMPORTANCE OF MODIFIED NUCLEOTIDES FOR PRIMER SELECTION AND ANNEALING

Selective Encapsidation of Primer tRNAs

The processes of tRNA selection and annealing in pararetroviruses and retrotransposons remain poorly understood. Most of our knowledge comes from studies on avian, murine and human retroviruses. During viral assembly, the dimeric genomic RNA (reviewed in Paillart et al., 1996) is encapsidated together with a variety of cellular RNAs, including 4S, 7–9S, 18S, and 28S RNAs (Bishop et al., 1970; Erikson and Erikson, 1970; Emmanoil-Ravicovitch et al., 1973; Faras et al., 1973; Sawyer and Dahlberg, 1973) (Fig. 1). The 4S RNA fraction has been identified as tRNAs, some of which were used as primers for reverse transcription (Harada et al., 1975, 1979). In all retroviruses, the tRNA species encapsidated into the viral particles represents a se-

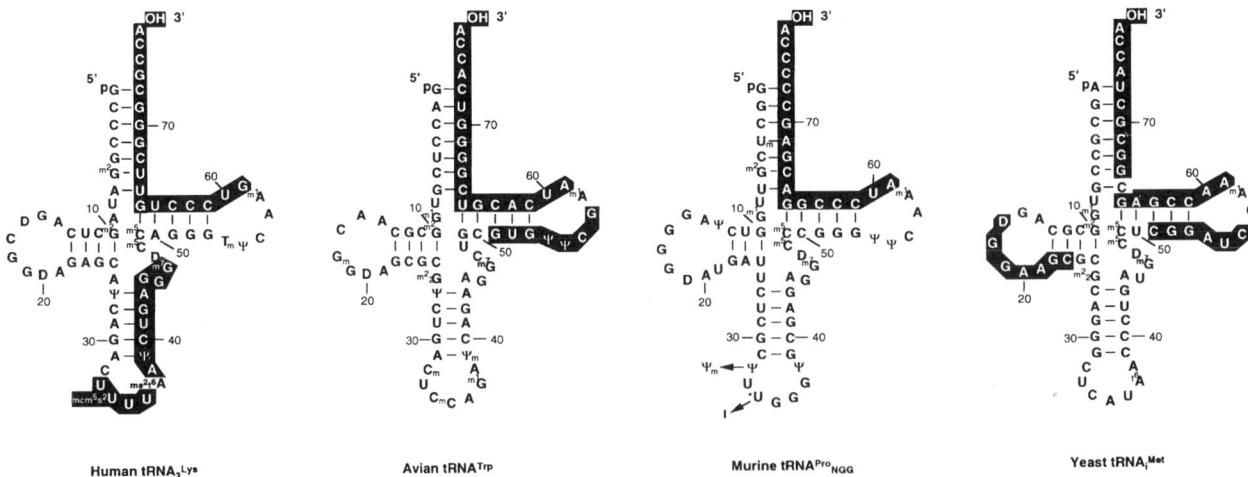

Figure 2. Secondary structure of some of the tRNAs used as primers by retroviruses, retrotransposons, and pararetroviruses. The nucleotides that have been proposed to interact with the genomic RNA of retroids (see text) are indicated in white on a black background. In tRNA^{Pro}, *U is an unidentified modified U; in tRNA$_i^{Met}$, A^r stands for O-β-ribosyl-(1″-2′)-adenosine-5″ phosphate. The differences between the two tRNA^{Pro} isoacceptors that are used as primers by murine leukemia virus are indicated.

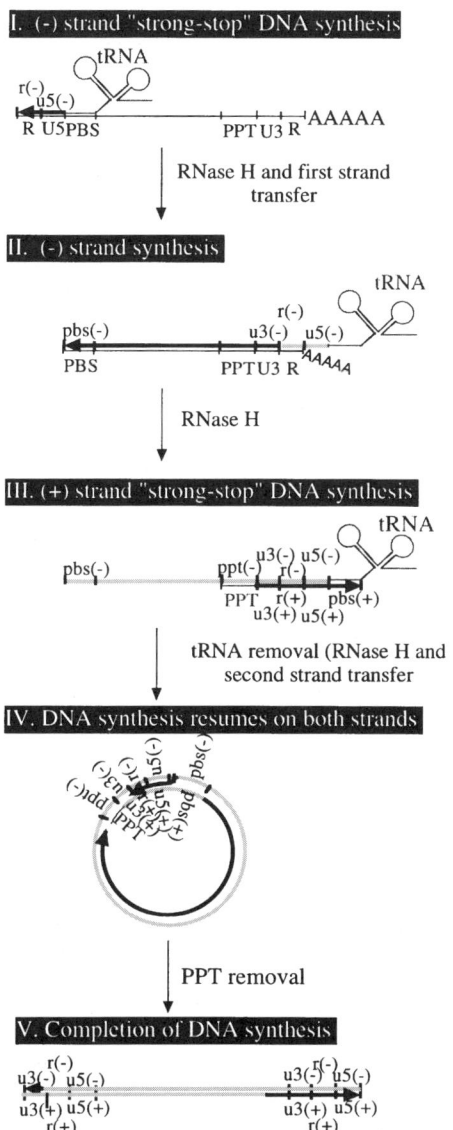

Figure 3. Schematic representation of the process of reverse transcription. RNA and DNA are represented by thin and thick lines, respectively. The DNA region that is being synthesized during each step is shown in black. Capital and lowercase letters are used to indicate regions in the RNA (PBS, U5, ...) and in the DNA [u3(−), u3(+), . . .], respectively. The − and + refer to the minus and plus DNA strands, respectively. U3 and U5 are unique sequences present at the 3' or 5' end of the genomic RNA, respectively, while R is a repeated sequence present at both ends of the genomic RNA. PPT is the polypurine tract. In step I, reverse transcription is initiated from the 3' hydroxyl end of the primer tRNA annealed to the PBS, and proceeds until RT reaches the 5' end of the template. The corresponding cDNA is named minus-strand strong-stop DNA. After degradation of the 5' end of the template by RNase H and the first-strand transfer, synthesis of the minus strand resumes (step II). The plus-strand synthesis is initiated from the PPTs that are generated by RNA regions resistant to the RNase H degradation (step III). Depending on the retrovirus, one or two PPTs may be used as primers. When the polymerizing plus-strand DNA chain reaches the end of the minus-strand DNA template, RT copies the primer tRNA until it is stopped by a modified nucleoside. After the second-strand transfer, which leads to a circular intermediate, DNA synthesis resumes on both strands (step IV). The

lected subpopulation of the host cell tRNAs. The enrichment of the primer tRNA varies considerably from one retrovirus to the other. Compared to the cytoplasm of the infected cells, tRNA^{Trp} is enriched 20-fold in the avian retroviruses, where it constitutes the most abundant tRNA species (Waters and Mullin, 1977). In HIV-1, isoacceptors 1, 2, and 3 of tRNA^{Lys} are also enriched by 10- to 20-fold. Together, they represent 60% of the encapsidated tRNA species (Kleiman et al., 1991; Jiang et al., 1992). In comparison, the fourfold enrichment of tRNA^{Pro} in murine leukemia virus (MLV) is relatively modest. The differences in the efficiency of tRNA packaging might be related to different selection mechanisms. In avian retroviruses (Peters and Hu, 1980) and HIV-1 (Mak et al., 1994, 1997), the RT coding region of the *pol* gene is required for selective packaging of the primer tRNA, while this is not in the case of MLV (Levin and Seidman, 1981; Fu et al., 1997). Parallel in vitro studies showed that RTs from avian retroviruses (Panet et al., 1975; Hizi et al., 1977) and HIV-1 (Barat et al., 1989; Richter-Cook et al., 1992; Zakharova et al., 1995) tightly bind their primer tRNA, while MLV RT does not (Panet and Berliner, 1978).

Selection of the Primer tRNA Is Affected by Modified Nucleotides

Initial reports indicated that unmodified synthetic tRNA$_3^{Lys}$ binds to HIV-1 RT with a 10-fold reduced affinity, as compared to natural, fully modified tRNA$_3^{Lys}$ (Barat et al., 1989, 1991). These results suggested that the posttranscriptional modifications of tRNA$_3^{Lys}$ were required for efficient selection of the primer tRNA. However, they were not confirmed by more recent studies (Thrall et al., 1996). On the other hand, chromatographic analysis of the tRNAs incorporated in HIV-1 revealed multiple forms of tRNA$_3^{Lys}$, while a single form was found in uninfected and infected cells (Li et al., 1996). The differences existing between these subspecies were not identified but might correspond to different levels of posttranscriptional modifications. Indeed, there are several examples where the level and the type of modifications (especially hypermodifications) depend on the growth conditions of the cell and its malignancy grading (see Chapter 26 by Björk and Rasmuson) (Dirheimer et al., 1995). Obviously, studying the importance of modified nucleosides of tRNA$_3^{Lys}$ for its incorporation into HIV-1 particles deserves further effort. However, such studies are hampered by the

---

nicks and overhangs are repaired by the host enzymes to yield regular double-stranded DNA (step V).

lack of an in vitro system mimicking the encapsidation process.

The importance of posttranscriptional modifications in the packaging of the primer tRNA was clearly demonstrated in the case of avian retroviruses (Keith and Heyman, 1990). The unique chicken tRNA^{Trp} isoacceptor could be separated into two subspecies differing by a posttranscriptional modification of nucleotide 7 in the acceptor stem, which is either G or m^{2}G (Fig. 2). Interestingly, only the nonmethylated species is encapsidated in avian retroviruses (Keith and Heyman, 1990). Thus, it is likely that methylation of G to m^{2}G at position 7 can prevent the recognition of tRNA^{Trp} by the viral proteins involved in the selective packaging of the primer tRNA. It was previously observed that RT from avian retroviruses can bind 100% of tRNA^{Trp} extracted from the viral particles, but only 30% of cellular tRNA^{Trp} (Panet et al., 1975). Furthermore, tRNA^{Trp} isolated from viral particles and cellular tRNA^{Trp} differ in their suppression efficiencies of the UGA codon (Cordell et al., 1980). These results indicated that the packaging machinery of retroviruses can be extremely sensitive to minor differences in tRNAs.

tRNA Annealing to Genomic RNA

The role of modified nucleotides in the annealing of the primer tRNA to the PBS has not been explicitly studied. However, it is well-known that the posttranscriptional modifications of tRNAs stabilize their secondary and tertiary structures (Sampson and Uhlenbeck, 1988; Hall et al., 1989; Perret et al., 1990) (see also Chapter 5 by Davis). Since annealing of the 3′-terminal 18 nucleotides of the primer tRNA to the PBS requires opening of the acceptor and TΨC stems, one may speculate that the modified nucleosides increase the energy barrier of tRNA binding to the viral RNA. More detailed discussions of tRNA/viral RNA annealed complexes follow in the next sections.

IMPORTANCE OF MODIFIED NUCLEOTIDES FOR INITIATION OF REVERSE TRANSCRIPTION

Avian Retroviruses

Even though schematic drawings such as Fig. 3 usually represent the genomic and primer RNAs interacting solely through base pairing of the PBS with the 3′ end of the tRNA, it was proposed more than 20 years ago that the TΨC loop of the primer tRNA^{Trp} of Rous sarcoma virus (RSV) might potentially interact with a sequence upstream of the PBS (Haseltine et al., 1976). The existence and potential role of this additional interaction was tested both in vitro and in vivo by combining genetic and functional studies (Cobrinik et al., 1988, 1991; Aiyar et al., 1992, 1994). These studies indicated that this interaction is most probably involved in the annealing of the primer tRNA to the viral RNA or in the very first polymerization steps (Fig. 2). These authors suggested that similar interactions may indeed exist in several other retroviruses, including MLV, HIV-1 and HIV-2 (Aiyar et al., 1992). In the absence of chemical or enzymatic probing data, intermolecular rearrangements that might take place in the viral RNA or in tRNA^{Trp} during formation of the primer/template complex were not identified. Furthermore, the possible role of modified nucleotides in the stabilization of the extended viral RNA/tRNA^{Trp} interactions was not investigated. Indeed, it is likely that Ψ_{54} and Ψ_{55} stabilize the interaction of the TΨ arm with the viral RNA (see Chapter 5 by Davis and Chapter 6 by Auffinger and Westhof).

HIV-1

Structural studies

The conformation of the binary complex formed by the genomic RNA of HIV-1 and its replication primer tRNA$_3^{Lys}$ was tested by extensive in vitro chemical and enzymatic probing (Isel et al., 1993, 1995). Chemical probing was used to test the conformation of the viral RNA by determining the accessibility of every base at one of its Watson–Crick positions and the N-7 position of purines (Ehresmann et al., 1987). The modified bases were identified by a primer extension assay using RT from avian myeloblastosis virus (AMV). The primer extension assay does not work on the natural tRNA$_3^{Lys}$ due to the presence of naturally occurring posttranscriptional modifications, especially m^{1}A$_{58}$, that block elongation of reverse transcriptase (Hagenbüchle et al., 1978; Wittig and Wittig, 1978; Youvan and Heast, 1981; Ehresmann et al., 1987). With natural tRNAs, structural probing was limited to the cases where specific cleavage could be induced at the chemically modified nucleotides (for example, probing of N-3 position of C's, N-7 position of G's, and N-7 position of A's). Nucleotides accessible to the chemical probes were identified by using 3′ or 5′ end-labeled tRNAs. Nucleotides cleaved by RNases could be monitored either by primer extension or by using end-labeled RNAs. Thus, nucleases were useful to test both viral RNA and tRNA$_3^{Lys}$ (Isel et al., 1993, 1995).

By comparing the conformational changes induced in the viral RNA upon hybridization of either tRNA$_3^{Lys}$ or an 18-mer oligodeoxyribonucleotide whose sequence was complementary to the PBS, it was apparent that the intermolecular primer/template interactions were not limited to the annealing of the 3' end of tRNA$_3^{Lys}$ to the HIV-1 PBS (Isel et al., 1993). Indeed, very complex interactions took place between the two RNA molecules. Analysis of the complete probing data set allowed construction of the secondary structure model depicted in Fig. 4 (Isel et al., 1995). Apart from the interaction between the PBS and the 18 nt at the 3' end of tRNA$_3^{Lys}$ (helix 7F), a 14-nt sequence corresponding to the anticodon loop, the 3' part of the anticodon stem and part of the variable loop of tRNA$_3^{Lys}$ interacted with two nonadjacent sequences located upstream of the PBS (helices 3E, 5D, and 6C). The structure of tRNA$_3^{Lys}$ and that of the viral RNA were strongly modified during formation of the binary complex (Fig. 2 and 4). Only the secondary structure of the dihydrouridine arm was not affected (Fig. 4), while the 5' part of the acceptor stem became base paired with the 5' part of the TΨC stem (Fig. 4).

Quite unexpectedly, it was found that this highly organized and complex secondary structure could not be formed when the natural tRNA$_3^{Lys}$ was replaced by its in vitro transcribed counterpart, which lacks all posttranscriptional modifications. In fact, the only intermolecular interaction taking place with this synthetic primer was formation of the PBS helix (Isel et al., 1993). The modified nucleotides that could affect the stability of intermolecular helices are mcm^{5}s^{2}U$_{34}$ (S$_{34}$), ms^{2}t^{6}A$_{37}$ (R$_{37}$), Ψ$_{39}$ and m^{7}G$_{46}$ (Fig. 4). Indeed, it was shown that specific dethiolation of mcm^{5}s^{2}U$_{34}$ of tRNA$_3^{Lys}$ by hydrogen peroxide strongly, but not completely, destabilized these helices. The accessibility of the corresponding nucleotides in the viral RNA toward chemical probes was dramatically increased, especially in the case of A$_{166}$, which is paired with mcm^{5}s^{2}U$_{34}$ (Isel et al., 1993). The stabilizing effect of 2-thiolation has been previously demonstrated for synthetic polynucleotides (Mazumdar et al., 1974; Scheit and Faerber, 1975), as well as for tRNAs. In thermophilic bacteria, significant thermostabilization is provided by m^{5}s^{2}U$_{54}$, which replaces the "classical" m^{5}U$_{54}$ (Davanloo et al., 1979; Horie et al., 1985). The 2-thioketo group has also been shown to stabilize the anticodon–anticodon interaction between tRNA^{Gln} (anticodon mnm^{5}s^{2}UUC) and tRNA^{Phe} (anticodon G$_m$AA) (see Chapter 7 by Grosjean et al.) (Houssier et al., 1988). Additional candidates for the stabilization of the extended tRNA$_3^{Lys}$/HIV-1 RNA interactions are ms^{2}t^{6}A$_{37}$ and Ψ$_{39}$. Indeed, the stabilizing effects of t^{6}A derivatives (Chapter 7 by Grosjean et al.) (Houssier and Grosjean, 1985) and Ψ residues (see Chapter 5 by Davis and Chapter 6 by Auffinger and Westhof) are well documented.

Information concerning the three-dimensional structure of the primer/template was obtained by psoralen cross-linking: it was shown that helix 4 could be cross-linked to the anticodon of tRNA$_3^{Lys}$ (Skripkin et al., 1996). However, this cross-link was not obtained when helix 6C was destabilized either by a mutation in the viral RNA or by using synthetic tRNA$_3^{Lys}$ (Skripkin et al., 1996). These results confirmed the importance of the modified nucleotides of tRNA$_3^{Lys}$, especially mcm^{5}s^{2}U$_{34}$, in the stabilization of the secondary and tertiary structures of the HIV-1 primer/template complex.

Importance of tRNA$_3^{Lys}$/HIV-1 RNA interactions for viral replication

It was proposed that the interaction of the anticodon loop of tRNA$_3^{Lys}$ with the AAAAUG sequence (Fig. 4), which is also located in a loop (termed the A-rich loop) in the free form of the viral RNA (Baudin et al., 1993), might be important for the selection

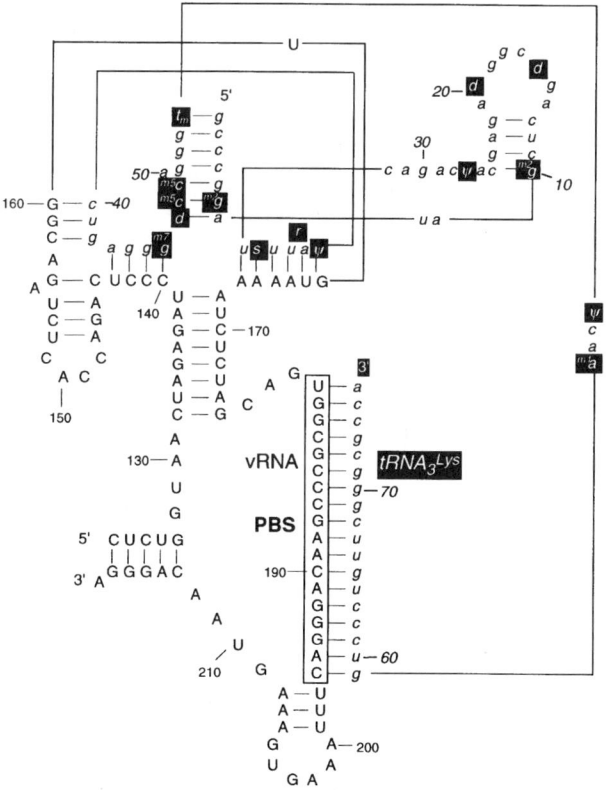

Figure 4. Secondary structure model of the viral RNA/tRNA$_3^{Lys}$ (template/primer) complex. The sequences of the viral RNA and tRNA$_3^{Lys}$ are indicated by upper- and lowercase letters, respectively. The modified nucleotides of tRNA$_3^{Lys}$ are highlighted and the PBS of the viral RNA is boxed. Adapted from Isel et al. (1995).

or annealing of the primer tRNA (Isel et al., 1993). Indeed, it was shown in cell culture that a tRNA$_3^{Lys}$ in which the mcm^{5}s^{2}UUU anticodon was replaced by CUA was encapsidated in the HIV-1 particles (Huang et al., 1994), but was not used as a primer (Huang et al., 1996), probably because of a defect in the annealing with the viral RNA. Conversely, deletion of the A-rich loop results in diminished levels of infectivity and reduced synthesis of viral DNA (Liang et al., 1997). This loop was progressively restored during long-term culture of the mutant virus (Liang et al., 1997).

Further support for the importance of the extended interactions between tRNA$_3^{Lys}$ and HIV-1 RNA arose from experiments in which the HIV-1 PBS was altered to make it complementary to other tRNAs. Replication of the mutated virus was dramatically reduced and reversion to the wild-type PBS was rapidly observed (Li et al., 1994; Das et al., 1995; Wakefield et al., 1995). However, viruses that stably maintained a PBS complementary to tRNA^{His} (Wakefield et al., 1996) or tRNA^{Met} (Kang et al., 1997) could be obtained if the A-rich loop in the viral RNA was also mutated to be complementary to the anticodon loop of these tRNAs. The modified nucleotides present in the anticodon loop of these tRNAs are not the same as those found in tRNA$_3^{Lys}$: the anticodon loop sequence is ΨUGUGGC in tRNA^{His} and CUC$_m$AUt^{6}AA in elongator tRNA^{Met} (Sprinzl et al., 1998). Thus, stabilization of the tRNA and viral RNA interactions by the modified nucleotides, rather than the exact nature of modified nucleotides in the anticodon loop, was important for viral replication. For instance, tRNA^{His}, which has a very G/C-rich anticodon arm, supported HIV-1 replication even though the only modified nucleotide in its anticodon is Ψ_{32}.

Functional studies

The initiation of HIV-1 reverse transcription has been studied in detail in vitro (Isel et al., 1996; Lanchy et al., 1996). It was shown that when an oligodeoxyribonucleotide complementary to the PBS was used as primer, cDNA synthesis started in the elongation mode, without any initiation phase, unlike cDNA synthesis from natural tRNA$_3^{Lys}$, which started in a different mode termed initiation (Isel et al., 1996). The initiation of reverse transcription was observed to be a specific process, requiring the homologous RT (Isel et al., 1996). Indeed, among the RTs that naturally use tRNA$_3^{Lys}$, only HIV-1 RT was able to efficiently initiate HIV-1 reverse transcription from this primer (Arts et al., 1996b). In contrast, cDNA synthesis from the oligodeoxyribonucleotide primer was efficient with heterologous RTs (Isel et al., 1996). Amino acids in the so-called "primer grip" region of HIV-1 RT that are required for extension of RNA, but not DNA, have been recently identified by alanine-scanning mutagenesis (Powell et al., 1997). Furthermore, deletion of the C terminus of HIV-1 RT inhibits extension of tRNA$_3^{Lys}$ but not of a DNA primer (Arts et al., 1996a). The initiation and elongation modes could also be discriminated by their sensitivity to manganese ions: elongation was inhibited by Mn^{2+}, while initiation was not (Isel et al., 1996).

The efficiency of initiation of HIV-1 reverse transcription also heavily relies on the presence of the posttranscriptional modifications of tRNA$_3^{Lys}$. It was shown that, while the natural tRNA was a very efficient primer, its in vitro synthesized counterpart was the least efficient among all tested primers (Isel et al., 1996). The low level of extension of the synthetic tRNA, compared to the natural one, was not because of a reduction of the polymerization rate of RT in the preformed complex of initiation of reverse transcription. Instead, it was because of the fact that RT did not specifically recognize the synthetic tRNA bound to the viral RNA, while it did efficiently recognize the natural tRNA$_3^{Lys}$/viral RNA complex (Lanchy et al., 1996). To test whether the modified nucleotides of tRNA$_3^{Lys}$ were important for efficient initiation of reverse transcription by themselves, or whether they had only an indirect role through formation of the extended interactions between the natural primer and the viral RNA, a mutated viral RNA preventing those interactions was used as template. Initiation of reverse transcription of the mutant template using the natural tRNA$_3^{Lys}$ as primer was as efficient as that of the wild-type viral RNA, thus indicating that the posttranscriptional modifications of the primer are important per se (Isel et al., 1996). However, disruption of the extended primer/template interactions resulted in a defect in the transition between the initiation and elongation phases, producing an accumulation of short initiation products (Isel et al., 1996). Thus, in the natural HIV-1 primer/template, the modifications of tRNA$_3^{Lys}$ were directly involved in the initiation of reverse transcription and, indirectly, through stabilization of tRNA/viral RNA interactions, in the transition between the initiation and elongation modes.

Murine Leukemia Virus

The solution structure of the binary complex formed between the genomic RNA of MLV and either natural or synthetic tRNA^{Pro} was recently studied by chemical and enzymatic probing (Fossé et al., 1998). Unlike in HIV-1, the intermolecular interaction is limited to the annealing of the 18 3′ terminal nucleotides of the natural tRNA^{Pro} to the viral PBS. Surprisingly, extended interactions take place between MLV RNA and synthetic tRNA^{Pro}. In the absence of modified nucleotides, the five nucleotides at the 5′ end of tRNA^{Pro} (GGCUC) are bound to the genomic RNA, upstream from the PBS (Fossé et al., 1998) (Fig. 2). The modified nucleotides of natural tRNA^{Pro} that prevent this interaction were not identified. Nucleotides Um$_4$ and m^{2}G$_6$ might be important in this respect. Because in the complex formed between natural tRNA^{Pro} and MLV RNA the $_1$GGCUC$_5$ sequence of the primer forms an intramolecular interaction with $_{51}$GGGΨΨ$_{55}$ in the TΨC arm, the pseudouridines at positions 54 and 55 might also prevent the extended primer/template interactions. It is tempting to speculate that in the MLV initiation complex of reverse transcription, the modified bases prevent misfolding of the primer and template. Indeed, it has been postulated that the role of m$_2^2$G in cytosolic tRNAs may be to prevent their folding into the unusual mitochondrial tRNA pattern (Steinberg and Cedergren, 1995). However, because the priming efficiencies of natural and synthetic tRNA^{Pro} have not been compared, this remains only an attractive hypothesis.

The efficiencies of tRNA^{Pro} and tRNA^{Trp} in priming reverse transcription of RNA from MLV and RSV, respectively, were compared to the efficiencies of 18-mer oligoribonucleotides complementary to the PBS of these viruses (in the presence of the homologous RT) (Isel et al., 1997). In both viral systems, the first nucleotides were added with equal efficiency to the oligoribonucleotides and the natural primers. These results indicated that the modified nucleotides of tRNA^{Pro} and tRNA^{Trp} are not crucial from the initiation of reverse transcription of MLV and RSV. However, in both systems, synthesis of the minus-strand strong-stop DNA was more efficient when the natural tRNA was used as primer. Thus, the modified nucleotides of tRNA^{Pro} and tRNA^{Trp} affect either the transition from the initiation to the elongation phase or elongation itself (Isel et al., 1997).

Ty1 Retrotransposon

Enzymatic probing (Friant et al., 1996) and sequence comparison (Friant et al., 1997) revealed that interactions between tRNA$_i^{Met}$ and the genomic RNA of the yeast Ty1 retrotransposon are not limited to the annealing of the 3′ terminal 10 nt of the tRNA with the PBS. In addition, three short sequences located downstream from the PBS interact with the TΨ and D arms of tRNA$_i^{Met}$ (Friant et al., 1996) (Fig. 2). Nucleotides 59–65 of the primer interact with a complementary sequence located 6–12 nt downstream from the PBS. By stacking on the PBS helix, this interaction could extend the double-stranded region up to 17 base pairs, a size similar to that of the PBS helix of retroviruses (Friant et al., 1996). In vivo and in vitro experiments indicated that the extended interactions between the primer tRNA$_i^{Met}$ and the Ty1 RNA template could play a role in the reverse transcription process (Wilhelm et al., 1994; Keeney et al., 1995). In particular, these interactions may favor the annealing of tRNA$_i^{Met}$ (Friant et al., 1996). As for avian retroviruses, these structural studies were conducted with an in vitro transcribed primer tRNA (Friant et al., 1996). The possible role of modified nucleotides in the stabilization of the extended template RNA/tRNA$_i^{Met}$ interactions was not investigated. However, it is likely that D$_{16}$ might destabilize the interaction between the D loop and the genomic RNA (Fig. 2) (see Chapter 5 by Davis).

IMPORTANCE OF MODIFIED NUCLEOTIDES FOR SYNTHESIS OF PLUS-STRAND STRONG-STOP DNA

Choice of the Primer and Template

During synthesis of minus-strand strong-stop DNA, tRNA is used as the primer and viral RNA is used as the template (Fig. 3, step I). On the contrary, during synthesis of plus-strand strong-stop DNA, the 3′ end of the tRNA is used as the template and the growing DNA chain is used as the primer (Fig. 3, step III). Does the nature of the nucleic acid hybridized to the tRNA (RNA or DNA) govern the primer usage of RT, or are other factors involved? To solve this question, short oligodeoxyribonucleotides and oligoribonucleotides were hybridized to the 3′ end of tRNA$_3^{Lys}$. These oligonucleotides were designed in such a way that both the annealed oligonucleotides and tRNA$_3^{Lys}$ could theoretically be used either as primer or as template by HIV-1 RT (Yusupova et al., 1996). When oligodeoxyribonucleotides were hybridized to the natural or synthetic tRNA$_3^{Lys}$, HIV-1 RT used exclusively the DNA strand as primer. On the contrary, when the oligoribonucleotides were annealed to natural tRNA$_3^{Lys}$, HIV-1 RT selected tRNA as primer. Thus, the chemical nature of the ribose-phosphate backbone (DNA versus RNA) of the nucleic acid bound to tRNA$_3^{Lys}$ appeared to be the main determinant of the primer choice by HIV-1 RT. However, when an oligoribonucleotide was hybridized to in vitro-transcribed tRNA$_3^{Lys}$, the oligonucleotide and synthetic tRNA were indiscriminately used both as primer and as template. Again, these results indicate that the posttranscriptional modifications of tRNA$_3^{Lys}$ contribute to the primer selection by HIV-1 RT (Yusupova et al., 1996).

Termination of Plus-Strand Strong-Stop DNA Synthesis

Eighteen-nucleotide-long PBS

According to the generally accepted scheme of reverse transcription, only the 3' end of the primer tRNA is copied during synthesis of the plus-strand strong-stop DNA (Fig. 3, step III). Baltimore and coworkers proposed that the m^{1}A$_{58}$ of tRNA^{Pro} acted as a termination signal during synthesis of the plus-strand strong-stop DNA, allowing only the 18 nt at the 3' end of the tRNA to be reverse transcribed during MLV replication (Gilboa et al., 1979). Indeed, the primer tRNAs used by all retroviruses and some long terminal repeat retrotransposons that have an 18-nt PBS complementary to their 3' ends have an m^{1}A at position 58 (Holzschu et al., 1995; Marquet et al., 1995). These tRNAs include tRNA^{Pro}, tRNA^{Trp}, tRNA$_3^{Lys}$, tRNA$_{1,2}^{Lys}$, tRNA^{Gln}, tRNA^{His}, tRNA^{Asn}, tRNA^{Arg} and tRNA^{Ser}. It was shown that m^{1}A$_{58}$ efficiently stops reverse transcription of tRNA$_3^{Lys}$ by HIV-1 RT in vitro (Ben-Artzi et al., 1996; Yusupova et al., 1996; Burnett and McHenry, 1997). Furthermore, when A$_{58}$ was not modified, DNA synthesis could not continue after the second-strand transfer (Fig. 3, step IV) because the 3' end of the overextended plus-strand strong-stop DNA was not complementary to the minus-strand retroviral DNA (Burnett and McHenry, 1997).

PBS complementary to tRNA$_i^{Met}$

Plant pararetroviruses and several retrotransposons use tRNA$_i^{Met}$ as primer (Marquet et al., 1995). Surprisingly, these retroids differ in the length of their PBSs; for example, they comprise 8 nucleotides in Ty3, 10 in Ty1 and Ty2, and 14 in Ta (a retrotransposon from *Arabidopsis thaliana*), which are complementary to the 3' end of tRNA$_i^{Met}$. Since even different retrotransposons from yeast differ in the length of the PBS, this length is obviously not dictated only by the posttranscriptional modifications of tRNA$_i^{Met}$. In yeasts, the modified nucleotide closest to the 3' end of tRNA$_i^{Met}$ is an O-β-ribosyl-(1''-2')-purine-5'' phosphate at position 64 (Keith et al., 1990; Glasser et al., 1991). If RT were blocked just before this nucleotide, one would expect a 12-nt-long PBS. This ambiguity was solved by several recent studies. Analysis of the plus-strand, strong-stop DNA isolated from Ty1 and Ty3 indicated that, in the most abundant product, the last 12 bases at the 3' end of tRNA$_i^{Met}$ were reverse transcribed (Lauermann et al., 1996; Wilhelm et al., 1997). However, genetic studies suggested that, unlike the retroviruses, the primer tRNA sequence was not inherited during Ty1 reverse transcription and that full-length plus-strand strong-stop DNA was not a direct intermediate in Ty1 replication (Lauermann and Boeke, 1994; Lauermann et al., 1996). Thus, minor less-than-full-length, plus-strand molecules could be the active intermediates in Ty reverse transcription (Wilhelm et al., 1997). However, a generation of these shorter replication intermediates probably also relied somehow on the presence of the modified purine at position 64. Indeed, retrotransposition of the Ty1 element was reduced in a yeast strain deficient in 2'-O-ribosyl phosphate transferase that modifies tRNA$_i^{Met}$ at position 64 (Astrom and Bystrom, 1994).

Other PBSs

In some cases, the PBS of a retrotransposon is complementary to an internal part of the primer tRNA, rather than to its 3' end. The PBSs of the *Drosophila* retrotransposon *copia* (14 nt) and of Ty5 (13 nt) are complementary to the anticodon arm of tRNA$_i^{Met}$ (Marquet et al., 1995). The primer isolated from *copia* virus-like particles corresponds to the 5' 39 nt of *Drosophila* tRNA$_i^{Met}$ (Kikuchi et al., 1986). This tRNA fragment might be generated by aberrant RNase P cleavage of the full-length tRNA (Kikuchi et al., 1990). In this case, the PBS is complementary to nucleotides 27 to 40 of tRNA$_i^{Met}$. During synthesis of the plus-strand strong-stop DNA, RT has apparently to reverse transcribe the t^{6}A$_{37}$ present in the anticodon loop of the tRNA, and the terminator of reverse transcription is most probably m^{2}G$_{26}$ (Kikuchi et al., 1986). Even though the presence of t^{6}A$_{37}$ has not been absolutely demonstrated, the absence of cleavage by RNase U$_2$ was indicative of a modified adenosine at this position. In *Drosophila* retroelement mdg3, the primer has been proposed to be tRNA$^{Leu(CAA)}$, analogous to yeast tRNA$_3^{Leu}$, that lacks its 3'-terminal 5 nucleotides (Saigo, 1986). It is presently unknown how the 3' end of this primer is generated. Termination of plus-strand strong-stop DNA synthesis might be due to N-1 methylation of residue A$_{58}$, although this has not been demonstrated experimentally (Robinson and Davidson, 1981). (Note that there is a nucleotide inversion in the figure of Saigo, 1986.)

IMPORTANCE OF MODIFIED NUCLEOTIDES FOR SYNTHESIS OF THE Gag-Pol PRECURSOR

In cauliflower mosaic virus (CaMV), and most probably in the other plant pararetroviruses, RT is expressed independently of the structural proteins (Bonneville and Hohn, 1993). In most retroviruses and retrotransposons, however, the viral enzymes, including RT, are synthesized as Gag-Pol precursors. This fusion protein offers an efficient strategy to incorporate the viral enzymes into the viral particles (for reviews, see Mesnard and Lebeurier, 1991; Levin et al., 1993b). An exception is human foamy virus, in which an independent expression of *pol* was recently demonstrated (Enssle et al., 1996; Yu et al., 1996). Synthesis of the Gag-Pol precursor usually requires either "readthrough" suppression of a termination codon or ribosomal frameshifting in the 3' (-1) or 5' ($+1$) direction. These mechanisms allow overexpression of the major internal proteins of the virus, as compared to the viral enzymes (including RT, protease, and, in retroviruses, integrase). An exception is the *Schizosaccharomyces pombe* retrotransposon Tf1, in which the Gag and Pol proteins are derived from a single large open reading frame (Levin et al., 1993a). In this case, the virus-like particles contain excess Gag protein relative to integrase because of a regulated degradation process (Atwood et al., 1996). In the *Drosophila melanogaster* retrotransposon *copia*, the Gag-Pol precursor also results from translation of a single open reading frame of the genome-length RNA and overexpression of the structural proteins is obtained by translation of a spliced subgenomic RNA (Brierley and Flavell, 1990; Yoshioka et al., 1990).

Readthrough Suppression of Termination Codons

In mammalian type C retroviruses [including MLV, feline leukemia virus (FeLV), baboon endogenous virus (BaEV), gibbon ape leukemia virus (GaLV), and spleen necrosis virus (SNV)], the *pol* sequences are located downstream from the UAG *gag* termination codon, with no independent *pol* initiation codon (Levin et al., 1993b). Purification and sequencing of MLV protease (the first enzyme encoded by the MLV *pol* gene) revealed that this protein is translated from an mRNA containing the UAG termination codon by suppression of this termination codon by a glutamine tRNA (Yoshinaka et al., 1985).

Since the UAG codon is efficiently suppressed (~5%) when located at the *gag-pol* junction of type C retroviral RNAs (Jamjoom et al., 1977), but not when found at the end of the cellular coding sequences (Capone et al., 1986; Sedivy et al., 1987), the viral mRNA must contain a readthrough signal. This signal was first localized within 99 codons surrounding the MLV *gag-pol* junction (Panganiban, 1988). Sequence comparison of the *gag-pol* junction of type C retroviruses revealed that their genomic RNAs can all potentially fold into a pseudoknot structure located 8 nt downstream of the *gag* termination codon (ten Dam et al., 1990). Not only is the pseudoknot structure required for efficient readthrough suppression (Wills et al., 1991; Feng et al., 1992), but also most of the sequence of the spacer region and of the pseudoknot region is of crucial importance (Honigman et al., 1991; Feng et al., 1992; Wills et al., 1994).

The possibility that virus infection may affect the population of suppressor tRNA^{Gln} has been considered by several groups. It has been proposed that the suppressor tRNA is a minor glutamine tRNA having a UmUG anticodon (Kuchino et al., 1987). This minor tRNA^{Gln} differs from the major tRNA^{Gln} isoacceptor (anticodon CUG) at positions 4, 34, and 68, and occurs in hyper- and hypomodified forms (Kuchino et al., 1987). It was reported that the concentration of this tRNA is increased in MLV-infected cells (Kuchino, 1987). However, this result was not confirmed by a chromatographic analysis, which showed that the separation profile of the different tRNA^{Gln} isoacceptors was unaffected by MLV infection (Feng et al., 1989). The latter observation is consistent with the fact that the UAG codon at the *gag-pol* junction is suppressed with equal efficiency in infected and uninfected cells (Panganiban, 1988). Similarly, in vitro suppression of the MLV UAG codon at the *gag-pol* junction was equally stimulated by tRNAs extracted from infected and uninfected murine cells (Feng et al., 1989).

Minus One Ribosomal Frameshifting

Ribosomal frameshifting in the 5' (-1) direction is the strategy used by most retroviruses (except type C retroviruses) to synthesize the Gag-Pol precursor (for reviews, see Hatfield et al., 1992; Levin et al., 1993b; Brierley, 1995). Minus one ribosomal frameshifting is not restricted to retroviruses or to other viruses (for a review, see Brierley, 1995) but also takes place in prokaryotic and eukaryotic genes (for a review, see Chapter 27 by Curran). Indeed, some retroviruses use two such events to complete the precursor synthesis: one between the *gag* and *pro* (encoding protease) genes and the second between the *pro* and *pol* (encoding RT) genes (see Table 1). Ribosomal -1 frameshifting was directly demonstrated by sequencing of the transframe peptide corresponding to the *gag-pro* overlap of mouse mammary tumor

Table 1 −1 frameshift signals in retroviruses

Retrovirus[a]	Gene overlap	Frameshift signal
BLV	gag-pro	A AAA AAC
BLV	pro-pol	U UUA AAC
EIAV	gag-pol	A AAA AAC
FIV	gag-pol	G GGA AAC
gypsy	gag-pol	U UUU UUA
HIV-1	gag-pol	U UUU UUA
HIV-2	gag-pol	U UUU UUA
HTLV-I	gag-pro	A AAA AAC
HTLV-I	pro-pol	U UUA AAC
HTLV-II	gag-pro	A AAA AAC
HTLV-II	pro-pol	U UUA AAC
MMTV	gag-pro	A AAA AAC
MMTV	pro-pol	G GAU UUA
MPMV	gag-pro	G GGA AAC
MPMV	pro-pol	A AAU UUU
SIV	gag-pol	U UUU UUA
RSV	gag-pol	A AAA AAC
SRV-1	gag-pro	G GGA AAC
SRV-1	pro-pol	A AAU UUU
SRV-2	gag-pro	G GGA AAC
SRV-2	pro-pol	A AAU UUU
STLV-I	gag-pro	A AAA AAC
STLV-I	pro-pol	U UUA AAC
Visna virus	gag-pol	G GGA AAC

[a] BLV, bovine leukemia virus; EIAV, equine infectious anemia virus; FIV, feline immunodeficiency virus; HIV, human immunodeficiency virus; HTLV, human T-cell leukemia virus; MMTV, mouse mammary tumor virus; MPMV, Mason-Pfizer monkey virus; SIV, simian immunodeficiency virus; RSV, Rous sarcoma virus; SRV, simian retrovirus; STLV, simian T-cell leukemia virus. References to the viral sequences can be found in Hatfield et al., (1992). The frameshift signal is represented as codons in the gag (0) frame.

virus (MMTV) (Hizi et al., 1987), the *gag-pol* overlaps of Rous sarcoma virus (RSV) (Jacks et al., 1988a) and HIV-1 (Jacks et al., 1988b), and the *pro-pol* overlap of human T-cell leukemia virus (HTLV-I) (Nam et al., 1993).

Analysis of the *gag-pol*, *gag-pro* and *pro-pol* overlaps revealed that the frameshift site is located at the 3' end of a heptanucleotide containing two homopolymeric triplets (XXXYYYZ) (Table 1) (for reviews see Hatfield et al., 1992; Levin et al., 1993b; Voytas and Boeke, 1993; Brierley, 1995). It is usually assumed that frameshifting takes place at this heptanucleotide by simultaneous slippage of two adjacent ribosome bound tRNAs from the zero reading frame (X XXY YYZ) to the −1 frame (XXX YYY) (Table 1 and Fig. 5 A) (Jacks et al., 1988a). After the −1 shift, the two tRNAs are bound to the viral mRNA by at least two of the three anticodon positions (Fig. 5A) (Jacks et al., 1988a). After transfer of the peptidyl-tRNA from the A to the P ribosomal site, translation continues normally in −1 frame. These results are in keeping with the sequencing results of the transframe peptide (Hizi et al., 1987; Jacks et al., 1988a, 1988b; Nam et al., 1993), which indicated that frameshift occurs at the second codon of the heptanucleotide signal (X XXY YYZ).

In addition to the slippery sequence, efficient −1 frameshifting requires an RNA secondary structure located 6 to 10 nt downstream (Fig. 5A) (Hatfield et al., 1992; Levin et al., 1993b; Brierley, 1995). This RNA structure is either a stem-loop or, most frequently, a pseudoknot. There is no direct link between the slippery sequence and the downstream structure. For instance, even though the same heptanucleotide sequence is found at the *gag-pro* junctions of MMTV and HTLV-II, a pseudoknot and a stem-loop structure are used by the former and the latter virus, respectively. The RNA structure downstream of the slippery sequence probably stimulates frameshifting by slowing down the translating ribosomes and pausing them over the slippery sequence (Jacks et al., 1988a).

Only three different codons (AAC, UUA, and UUU) are found at the 3' end of the −1 frameshift signal of retroviruses (Table 1). Three additional codons are known to allow efficient −1 frameshifting in eukaryotes: AAA, AAU, and UUC (see Chapter 27 by Curran) (Brierley, 1995). Except for tRNA^{Leu} (decoding UUA), the tRNAs decoding these codons all have a hypermodified base at position 34 or 37: Q in tRNA^{Asn} (decoding AAC and AAU), Y in tRNA^{Phe} (decoding UUU and UUC), and mcm^{5}s^{2}U in tRNA^{Lys} (decoding AAA). The posttranscriptional modifications of the tRNAs involved in frameshifting have been analyzed in infected cells (Hatfield et al., 1989). Interestingly, it was found that most of the tRNA^{Phe} lacks Y base in HIV-1-infected cells, and most of the tRNA^{Asn} lacks Q base in cells infected with bovine leukemia virus (BLV), HTLV-I and simian retrovirus (SRV) (Hatfield et al., 1989). This observation led to the proposal that frameshifting might require hypomodified tRNA^{Asn}, tRNA^{Phe}, and tRNA^{Lys} and, thus, that virus-induced hypomodification of tRNA might enhance frameshifting (Hatfield et al., 1992). This hypothesis is apparently not supported by recent studies. In CD4$^+$ human lymphoid cells (Cassan et al., 1994), as well as in CD4-expressing human 293 cells (Reil et al., 1994), the frameshifting efficiency at the HIV-1 *gag-pol* junction is not affected by HIV-1 infection. However, the HIV-1 *gag-pol* junction is probably not the best model to test this hypothesis, since the last codon in the 0 frame is read by a naturally hypomodified tRNA^{Leu}. Expression of the wild-type and mutated ribosomal frameshift signals of the coronavirus infectious bronchitis virus (IBV) in *Escherichia coli* allowed testing of the influence of tRNA anticodon modification on the frameshift efficiency (Brierley et al., 1997). Lack of Q base in tRNA^{Asn} (anticodon QUU) did not significantly affect

CHAPTER 28 • MODIFIED NUCLEOSIDES IN VIRAL AND TRANSPOSON REPLICATION

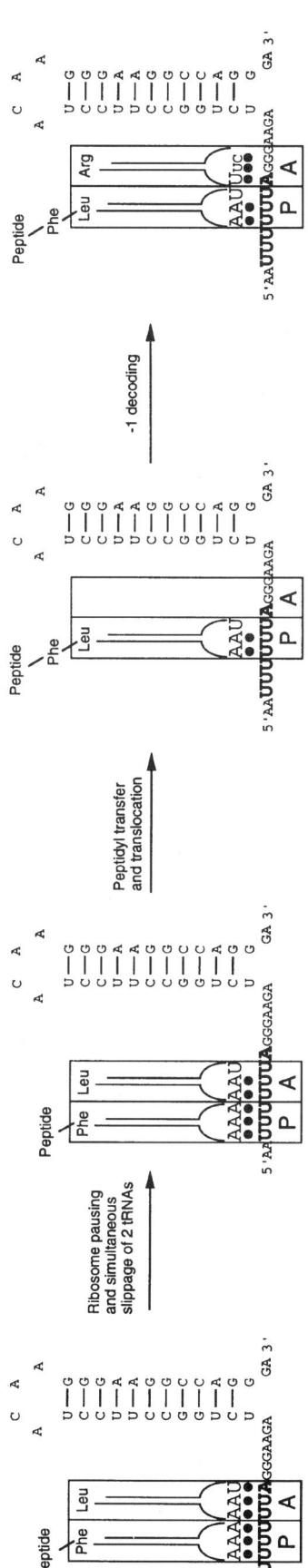

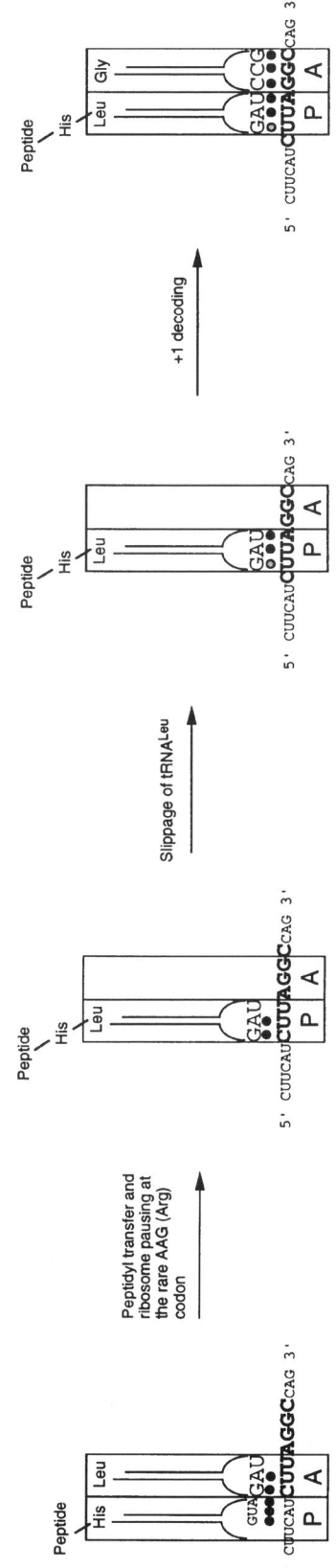

Figure 5. Ribosomal frameshifting. The −1 (A) and +1 (B) frameshiftings are illustrated with HIV-1 and Ty1, respectively. The A and P ribosomal sites are schematically represented.

frameshifting at UUUAAAC and UUUAAAU slippery sequences. Among several mutants of tRNA^{Lys} (anticodon mnm^{5}s^{2}UUU), only the one with anticodon s^{2}UUU significantly increased frameshifting at UUUAAAA (but not at UUUAAAG). The data support a role for the amino group of mnm^5 in shaping the anticodon structure (Brierley et al., 1997). Since the modifications of U$_{34}$ are not the same, it is unclear whether these conclusions can be extended to eukaryotes.

Plus One Ribosomal Frameshifting

In yeast Ty retrotransposons, the *gag* and *pol* genes are separated by a frameshift in the 3' (+1) direction (for reviews, see Hatfield et al., 1992; Voytas and Boeke, 1993). Two different +1 frameshift mechanisms exist. Unlike in retroviruses, ribosomal pausing is achieved by the presence of a rare codon downstream of the frameshift site (Hatfield et al., 1992; Voytas and Boeke, 1993). In Ty1, Ty2, and probably Ty4, frameshifting is mediated by the CUUAGGC heptanucleotide (Fig. 5B) (Belcourt and Farabaugh, 1990). After binding of tRNA$^{Leu(UAG)}$ (an unusual leucine isoacceptor that recognizes all six leucine codons) to the CUU codon, ribosomes pause in front of the AAG codon, because of the low availability of tRNA$^{Arg(CUU)}$. Ribosome stalling allows realignment of tRNA$^{Leu(UAG)}$ on the UUA-Leu codon and decoding of the downstream message in the +1 frame (Fig. 5B) (Belcourt and Farabaugh, 1990). Indeed, translational frameshifting increases 3-fold to 17-fold in cells lacking tRNA$^{Arg(CUU)}$ and decreases in cells overproducing it (Kawakami et al., 1993).

In Ty3, a +1 frameshift occurs within the GCGAGUU sequence (Farabaugh et al., 1993; Vimaladithan and Farabaugh, 1994). After binding of tRNA$^{Ala(CGC)}$ to the GCG codon and peptidyl transfer, translational pausing is induced by the low availability of the tRNA^{Ser} specific for the AGU codon. Ribosome stalling then allows +1 frameshifting by misplacement of valyl-tRNA^{Val} on the GUU codon in the ribosomal A site, without slippage of the peptidyl-tRNA$^{Ala(CGC)}$ in the P site (Farabaugh et al., 1993; Vimaladithan and Farabaugh, 1994). The particular features of peptidyl-tRNA$^{Ala(CGC)}$ that allow out-of-frame binding of valyl-tRNA^{Val} are currently unknown, but could involve modified nucleotides (Vimaladithan and Farabaugh, 1994).

IMPORTANCE OF POSTTRANSCRIPTIONAL MODIFICATIONS OF VIRAL RNA

Methylations

In addition to having 5'-methylated cap structures, many eukaryotic mRNAs contain internal methylated bases, mostly m^{6}A (Banerjee, 1980). Methylation of internal adenosines occurs in the nucleus very rapidly after synthesis of the precursor mRNA, before splicing (Revel and Groner, 1978; Chen-Kiang et al., 1979). Thus, RNAs from retroviruses and retrotransposons, which are transcribed from the integrated provirus, are also targets for the cellular methylase.

Indeed, RNA extracted from RSV particles contains an average of 12 to 15 m^{6}As (Furuichi et al., 1975; Stoltzfus and Dimock, 1976; Beemon and Keith, 1977). These methylated bases are not randomly distributed; they are found exclusively in the 3' half of the genomic RNA (Beemon and Keith, 1977). Thirteen methylated sites were identified by fingerprinting, paper electrophoresis, and thin-layer chromatography at nucleotides 6394, 6447, 6507, 6718, 7414, 7424, 7890, 8014, 8339, 8366, 8413, 8485, and 8633 (Kane and Beemon, 1985; Csepany et al., 1990). The modified sites are located in the *src* and *env* genes and usually correspond to the consensus sequence RGACU (Beemon, 1985; Csepany et al., 1990).

Attempts at understanding the function of m^{6}A have been made in which competitive inhibitors of *S*-adenosylmethionine were used to block internal modifications (Stoltzfus and Dane, 1982). These authors observed a buildup of unspliced viral RNA and a decreased level of spliced subgenomic RNAs when they used cycloleucine as methylation inhibitor, and they suggested a role for m^{6}A in mRNA splicing. However, they did not address the relationship between methylation and RNA processing directly, and secondary effects of the drug on this and other cellular functions cannot be ruled out. Using a different approach, Beemon and coworkers mutated −1, +2 or +3 positions of the consensus sequence to prevent methylation of A$_{7417}$, A$_{7425}$, A$_{8014}$ and A$_{8485}$, without substituting for these adenines (Kane and Beemon, 1987; Csepany et al., 1990). Even though these mutations prevented methylation, neither viral RNA processing nor viral replication was affected (Kane and Beemon, 1987; Csepany et al., 1990). However, it cannot be excluded that methylation of other sites might be functionally important.

Deaminations and the Editing Process

Another modification that can affect retroviral RNA is adenosine deamination. A double-stranded, RNA-dependent adenosine deaminase, originally found in *Xenopus* oocytes (Bass and Weintraub, 1987; Rebagliati and Melton, 1987) (see Chapter 19 by Rueter and Emeson), exists in the nuclei of all somatic cells examined (Wagner et al., 1990). Indeed,

Hajjar and Linial (1995) reported an avian recombinant provirus containing a short region of A-to-G hypermutation. The hypermutated region corresponded to complementary sequences present in the recombinant virus. Thus, this RNA duplex was a substrate for double-stranded, RNA-dependent adenosine deaminase that converted A's to I's, which were further converted to G's during reverse transcription (Hajjar and Linial, 1995). A similar mechanism might explain the appearance of mutant spleen necrosis viruses (Kim et al., 1996) and an avian retrovirus (Felder et al., 1994) with A-to-G hypermutations. In the latter case, as much as 48% of the A residues in the U3 region of the long terminal repeats were mutated to G residues. Thus, adenosine deamination might contribute to the high variability of retroviruses and contribute to the generation of hypervariants (Felder et al., 1994; Hajjar and Linial, 1995; Kim et al., 1996).

Adenosine deamination may also affect other retroviral functions. Indeed, adenosine deamination was observed when HIV-1 RNA was injected in the nuclei of *Xenopus* oocytes (Sharmeen et al., 1991). Modification was observed in the transactivation response (TAR) site, a sequence found at the 5' ends of the genomic and messenger HIV-1 RNAs that adopts a stem-loop structure. TAR contains a binding site for the transactivating protein Tat. Surprisingly, a single adenosine located in the Tat binding site was deaminated in a Tat-dependent manner. Wild-type TAR RNA was deaminated only in the presence of Tat. However, mutants of TAR RNA that do not bind Tat were equally modified in the presence and absence of Tat (Sharmeen et al., 1991). The role of this Tat-dependent RNA modification is not known. Since no A-to-G mutation is observed in the HIV-1 DNA provirus that would correspond to the A-to-I modification of RNA in TAR, the deaminated RNA is not reverse transcribed. Instead, adenosine deamination might direct the modified RNA toward the translation machinery and prevent its encapsidation. Indeed, in in vitro translation and *Xenopus* microinjection assays, TAR has been shown to inhibit the translation of downstream genes. In *Xenopus* oocytes, the inhibitory effect of TAR is relieved by Tat (Braddock et al., 1989).

CONCLUSIONS AND PERSPECTIVES

Modified nucleotides might be critical for a number of steps in the replication life cycle of retroviruses, plant pararetroviruses, and retrotransposons. The importance of modified nucleotides of the primer tRNA for encapsidation and reverse transcription is well established. In contrast, the roles of modifications of the viral RNA remain speculative. It is important to realize that no general conclusion that would be valid for all retroviruses, plant pararetroviruses, and retrotransposons can be drawn regarding the importance of modified nucleotides. Each virus or transposon represents a particular case. This is probably because of the pronounced ability of retroids to rapidly evolve, which results from the error prone reverse transcription process.

Human retroviruses cause several fatal diseases, including AIDS (HIV-1 and HIV-2), leukemia (HTLV-I and HTLV-II), and possibly multiple sclerosis (Perron et al., 1997). In light of our current knowledge, it is not unrealistic to speculate that some of the RNA modifying enzymes may be suitable targets against these pathogens. Obviously, defining the exact roles of modified nucleotides in the replication life cycles of plant pararetroviruses, retrotransposons, and especially retroviruses deserves more attention.

Acknowledgments. I thank Bernard Ehresmann, Chantal Ehresmann, Catherine Isel, Gérard Keith, Jean-Marc Lanchy, Stuart F. J. Le Grice, E. Skripkin, and G. Yusupova for their contribution to the original work summarized in this chapter, and J. Stephen Lodmell for critical reading of the manuscript. The help of Henri Grosjean in the elaboration of the manuscript is greatly acknowledged. Research on the initiation of HIV-1 reverse transcription was supported by the Agence Nationale de Recherches sur le SIDA (ANRS).

REFERENCES

Aiyar, A., D. Cobrinik, Z. Ge, H. J. Kung, and J. Leis. 1992. Interaction between retroviral U5 RNA and the TΨC loop of the tRNA^{Trp} primer is required for efficient initiation of reverse transcription. *J. Virol.* 66:2464–2472.

Aiyar, A., Z. Ge, and J. Leis. 1994. A specific orientation of RNA secondary structure is required for initiation of reverse transcription. *J. Virol.* 68:611–618.

Arts, E. J., M. Ghosh, P. S. Jacques, B. Ehresmann, and S. F. J. Le Grice. 1996a. Restoration of tRNA$_3^{Lys}$-primed (−) strand DNA synthesis to an HIV-1 reverse transcriptase mutant with extended tRNAs. Implications for retroviral replication. *J. Biol. Chem.* 271:9054–9061.

Arts, E. J., S. R. Stetor, X. G. Li, J. W. Rausch, K. J. Howard, B. Ehresmann, T. W. North, B. M. Wohrl, R. S. Goody, M. A. Wainberg, and S. F. J. Le Grice. 1996b. Initiation of (−) strand DNA synthesis from tRNA$_3^{Lys}$ on lentiviral RNAs: implications of specific HIV-1 RNA-tRNA$_3^{Lys}$ interactions inhibiting primer utilization by retroviral reverse transcriptases. *Proc. Natl. Acad. Sci. USA* 93:10063–10068.

Astrom, S. U., and A. S. Bystrom. 1994. Rit1, a tRNA backbone-modifying enzyme that mediates initiator and elongator tRNA discrimination. *Cell* 79:535–546.

Atwood, A., J. H. Lin, and H. L. Levin. 1996. The retrotransposon Tf1 assembles virus-like particles that contain excess Gag relative to integrase because of a regulated degradation process. *Mol. Cell. Biol.* 16:338–346.

Banerjee, A. K. 1980. 5'-terminal cap structure in eucaryotic messenger ribonucleic acids. *Microbiol. Rev.* 44:175–205.

Barat, C., S. F. J. Le Grice, and J. L. Darlix. 1991. Interaction of HIV-1 reverse transcriptase with a synthetic form of its replication primer, tRNA^{Lys,3}. *Nucleic Acids Res.* 9:751–757.

Barat, C., V. Lullien, O. Schatz, G. Keith, M. T. Nugeyre, F. Grüninger-Leitch, F. Barré-Sinoussi, S. F. J. Le Grice, and J. L. Darlix. 1989. HIV-1 reverse transcriptase specifically interacts with the anticodon domain of its cognate primer tRNA. *EMBO J.* 8:3279–3285.

Bass, B. L., and H. Weintraub. 1987. A developmentally regulated activity that unwinds RNA duplexes. *Cell* 48:607–613.

Baudin, F., R. Marquet, C. Isel, J. L. Darlix, B. Ehresmann, and C. Ehresmann. 1993. Functional sites in the 5' region of human immunodeficiency virus type-1 RNA form defined structural domains. *J. Mol. Biol.* 229:382–397.

Beemon, K., and J. Keith. 1977. Localization of N^6-methyladenosine in the Rous sarcoma virus genome. *J. Mol. Biol.* 113:165–179.

Belcourt, M. F., and P. J. Farabaugh. 1990. Ribosomal frameshifting in the yeast retrotransposon Ty: tRNAs induce slippage on a 7 nucleotide minimal site. *Cell* 62:339–352.

Ben-Artzi, H., J. Shemesh, E. Zeelon, B. Amit, L. Kleiman, M. Gorecki, and A. Panet. 1996. Molecular analysis of the second template switch during reverse transcription of the HIV RNA template. *Biochemistry* 35:10549–10557.

Bishop, J. M., W. E. Levinson, N. Quintrell, D. Sullivan, H. Fanshier, and J. Jackson. 1970. The low molecular weight RNAs of Rous sarcoma virus. I. The 4S RNA. *Virology* 42:182–195.

Bonneville, J. M., and T. Hohn. 1993. A reverse transcriptase for cauliflower mosaic virus: state of the art, 1992, p. 357–390. *In* A. M. Skalka, and S. P. Goff (ed.), *Reverse Transcriptase*. Cold Spring Harbor Laboratory Press, Cold Spring Harbor, N.Y.

Braddock, M., A. Chambers, W. Wilson, M. P. Esnouf, S. E. Adams, A. J. Kingsman, and S. M. Kingsmann. 1989. HIV-1 TAT "activates" presynthesized RNA in the nucleus. *Cell* 58:269–279.

Brierley, C., and A. J. Flavell. 1990. The retrotransposon *copia* controls the relative levels of its gene products post-transcriptionally by differential expression from its two major mRNAs. *Nucleic Acids Res.* 18:2947–2951.

Brierley, I. 1995. Ribosomal frameshifting on viral RNAs. *J. Gen. Virol.* 76:1885–1892.

Brierley, I., M. R. Meredith, A. J. Bloys, and T. G. Hagervall. 1997. Expression of a coronavirus ribosomal frameshift signal in *Escherichia coli*: influence of tRNA anticodon modification on frameshifting. *J. Mol. Biol.* 270:360–373.

Burnett, B. P., and C. S. McHenry. 1997. Posttranscriptional modification of retroviral primers is required for late stages of DNA replication. *Proc. Natl. Acad. Sci. USA* 94:7210–7215.

Capone, J. P., J. M. Sedivy, P. A. Sharp, and U. L. RajBhandary. 1986. Introduction of UAG, UAA, and UGA nonsense mutations at a specific site in the *Escherichia coli* chloramphenicol acetyltransferase gene: use in measurement of amber, ochre, and opal suppression in mammalian cells. *Mol. Cell. Biol.* 6:3059–3067.

Cassan, M., N. Delaunay, C. Vaquero, and J. P. Rousset. 1994. Translational frameshifting at the Gag-Pol junction of human immunodeficiency virus type 1 is not increased in infected T-lymphoid cells. *J. Virol.* 68:1501–1508.

Chen-Kiang, S., J. R. Nevins, and J. E. Darnell. 1979. N^6-methyladenosine in adenovirus type 2 nuclear RNA is conserved in the formation of messenger RNA. *J. Mol. Biol.* 22:327–353.

Cobrinik, D., A. Aiyar, Z. Ge, M. Katzman, H. Huang, and J. Leis. 1991. Overlapping retrovirus U5 sequence elements are required for efficient integration and initiation of reverse transcription. *J. Virol.* 65:3864–3872.

Cobrinik, D., L. Soskey, and J. Leis. 1988. A retroviral RNA secondary structure required for efficient initiation of reverse transcription. *J. Virol.* 62:3622–3630.

Cordell, B., F. M. DeNoto, J. F. Atkins, R. F. Gesteland, J. M. Bishop, and H. M. Goodman. 1980. The forms of $tRNA^{Trp}$ found in avian sarcoma virus and uninfected chicken cells have structural identity but functional distinctions. *J. Biol. Chem.* 255:9358–9368.

Csepany, T., A. Lin, C. J. Baldick, Jr., and K. Beemon. 1990. Sequence specificity of mRNA N^6-adenosine methyltransferase. *J. Biol. Chem.* 265:20117–20122.

Das, A. T., B. Klaver, and B. Berkhout. 1995. Reduced replication of human immunodeficiency virus type 1 mutants that use reverse transcription primers other than the natural $tRNA_3^{Lys}$. *J. Virol.* 69:3090–3097.

Davanloo, P., M. Sprintzl, K. Watanabe, M. Albani, and K. Kersten. 1979. Role of ribothymidine in the thermal stability of transfer RNA as monitored by proton magnetic resonance. *Nucleic Acids Res.* 6:1571–1581.

Dirheimer, G., W. Baranowski, and G. Keith. 1995. Variations in tRNA modifications, particularly of their queuine content in higher eukaryotes. Its relation to malignancy grading. *Biochimie* 77:99–103.

Ehresmann, C., F. Baudin, M. Mougel, P. Romby, J. P. Ebel, and B. Ehresmann. 1987. Probing the structure of RNA in solution. *Nucleic Acids Res.* 15:9109–9128.

Emmanoil-Ravicovitch, R., C. J. Larsen, M. Bazilier, J. Robin, J. Pris, and M. Boiron. 1973. Low-molecular-weight RNAs of murine sarcoma virus: comparative studies of free and 70S RNA-associated components. *J. Virol.* 12:1625–1627.

Enssle, J., I. Jordan, B. Mauer, and A. Rethwilm. 1996. Foamy virus reverse transcriptase is expressed independently from the Gag protein. *Proc. Natl. Acad. Sci. USA* 93:4137–4141.

Erikson, E., and R. L. Erikson. 1970. Isolation of amino acid acceptor RNA from purified avian myeloblastosis virus. *J. Mol. Biol.* 52:387–390.

Farabaugh, P. J., H. Zhao, and A. Vimaladithan. 1993. A novel programed frameshift expresses the POL3 gene of retrotransposon-Ty3 of yeast—frameshifting without transfer RNA slippage. *Cell* 74:93–103.

Faras, A. J., A. C. Garapin, W. E. Levinson, J. M. Bishop, and H. M. Goodman. 1973. Characterization of the low-molecular-weight RNAs associated with the 70S RNA of Rous sarcoma virus. *J. Virol.* 12:334–342.

Felder, M.-P., D. Laugier, B. Yatsula, P. Dezélée, G. Calothy, and M. Marx. 1994. Functional and biological properties of an avian variant long terminal repeat containing multiple A to G conversions in the U3 sequence. *J. Virol.* 68:4759–4767.

Feng, Y.-X., D. L. Hatfield, A. Rein, and J. G. Levin. 1989. Translational readthrough of the murine leukemia virus *gag* gene amber codon does not require virus-induced alteration of tRNA. *J. Virol.* 63:2405–2410.

Feng, Y.-X., H. Yuan, A. Rein, and J. G. Levin. 1992. Bipartite signal for readthrough suppression in murine leukemia virus mRNA: an eight-nucleotide purine-rich sequence immediately downstream of the *gag* termination codon followed by an RNA pseudoknot. *J. Virol.* 66:5127–5132.

Fossé, P., M. Mougel, G. Keith, E. Westhof, B. Ehresmann, and C. Ehresmann. 1998. Modified nucleotides of $tRNA^{Pro}$ restrict interactions in the binary primer/template complex of M-MLV. *J. Mol. Biol.* 275:731–746.

Friant, S., T. Heyman, O. Poch, M. Wilhelm, and F. X. Wilhelm. 1997. Sequence comparison of the Ty1 and Ty2 elements of the yeast genome supports the structural model of the $tRNA_i^{Met}$-Ty1 RNA reverse transcription initiation complex. *Yeast* 13:639–645.

Friant, S., T. Heyman, M. L. Wilhelm, and F. X. Wilhelm. 1996. Extended interactions between the primer $tRNA_i^{Met}$ and genomic

RNA of the yeast Ty1 retrotransposon. *Nucleic Acids Res.* 24: 441–449.

Fu, W., B. A. Ortiz-Conde, R. J. Gorelick, S. H. Hughes, and A. Rein. 1997. Placement of tRNA primer on the primer-binding site requires *pol* gene expression in avian but not murine retroviruses. *J. Virol.* 71:6940–6946.

Furuichi, Y., A. J. Shatkin, E. Stavnezer, and J. M. Bishop. 1975. Blocked, methylated 5′-terminal sequence in avian sarcoma virus RNA. *Nature* 257:618–620.

Gabriel, A., and J. D. Boeke. 1993. Retrotransposon reverse transcription, p. 275–328. *In* A. M. Skalka and S. P. Goff (ed.), *Reverse Transcriptase*. Cold Spring Harbor Laboratory Press, Cold Spring Harbor, N.Y.

Gilboa, E., S. W. Mitra, S. Goff, and D. Baltimore. 1979. A detailed model of reverse transcription and tests of crucial aspects. *Cell* 18:93–100.

Glasser, A. L., J. Desgres, J. Heitzler, C. W. Gehrke, and G. Keith. 1991. O-ribosyl-phosphate purine as a constant modified nucleotide located at position 64 in cytoplasmic initiatior tRNAs(Met) of yeasts. *Nucleic Acids Res.* 19:5199–5203.

Hagenbüchle, O., M. Santer, J. A. Steitz, and R. J. Mans. 1978. Conservation of the primary structure at the 3′ end of 18 S rRNA from eucaryotic cells. *Cell* 13:551–563.

Hajjar, A. M., and M. L. Linial. 1995. Modification of retroviral RNA by double-stranded RNA adenosine deaminase. *J. Virol.* 69:5878–5882.

Hall, K. B., J. R. Sampson, O. C. Uhlenbeck, and A. G. Redfield. 1989. Structure of an unmodified tRNA molecule. *Biochemistry* 28:5794–5801.

Harada, F., G. G. Peters, and J. E. Dahlberg. 1979. The primer tRNA for Moloney murine leukemia virus DNA synthesis. Nucleotide sequence and aminoacylation of tRNA^{Pro}. *J. Biol. Chem.* 254:10979–10985.

Harada, F., R. C. Sawyer, and J. E. Dalhberg. 1975. A primer ribonucleic acid for initiation of *in vitro* Rous sarcoma virus deoxyribonucleic acid synthesis. *J. Biol. Chem.* 250:3487–3497.

Haseltine, W. A., D. G. Kleid, A. Panet, E. Rothenberg, and D. Baltimore. 1976. Ordered transcription of RNA tumor viruses genomes. *J. Mol. Biol.* 106:109–131.

Hatfield, D. L., Y.-X. Feng, B. J. Lee, A. Rein, J. G. Levin, and S. Oroszlan. 1989. Chromatographic analysis of the aminoacyl-tRNAs which are required for translation of codons at and around the ribosomal frameshift sites of HIV, HTLV-I and BLV. *Virology* 173:736–742.

Hatfield, D. L., J. G. Levin, A. Rein, and S. Oroszlan. 1992. Translation suppression in retroviral gene expression. *Adv. Virus Res.* 41:193–239.

Hizi, A., L. E. Henderson, T. D. Copeland, R. C. Sowder, C. V. Hixson, and S. Oroszlan. 1987. Characterization of mouse mammary tumor virus gag-pro gene products and the ribosomal frameshift site by protein sequencing. *Proc. Natl. Acad. Sci. USA* 84:7041–7045.

Hizi, A., J. P. Leis, and W. K. Joklik. 1977. The RNA dependent DNA polymerase of avian sarcoma virus B77: binding of viral and nonviral ribonucleic acids to the α, β, and $\alpha\beta$ forms of the enzyme. *J. Biol. Chem.* 252:6878–6884.

Holzschu, D. L., D. Martineau, S. K. Fodor, V. M. Vogt, P. R. Bowser, and J. W. Casey. 1995. Nucleotide sequence and protein analysis of a complex piscine retrovirus, walleye dermal sarcoma virus. *J. Virol.* 69:5320–5331.

Honigman, A., D. Wolf, S. Yaish, H. Falk, and A. Panet. 1991. *Cis*-acting RNA sequences control the Gag-Pol translation readthrough in murine leukemia virus. *Virology* 183:313–319.

Horie, N., M. Hara-Yokoyama, S. Yokoyama, K. Watanabe, Y. Kuchino, S. Nishimura, and T. Miyazawa. 1985. Two tRNAI^{leu} species from an extreme thermophile, Thermus thermophilus HB8: effect of 2-thiolation of ribothymidine on the thermostability of tRNA. *Biochemistry* 24:5711–5715.

Houssier, C., P. Degée, K. Nicoghosian, and H. Grosjean. 1988. Effect of uridine dethiolation in the anticodon triplet of tRNA(Glu) on its association with tRNA(Phe). *J. Biomol. Struct. Dyn.* 5:1259–1266.

Houssier, C., and H. Grosjean. 1985. Temperature jump relaxation studies on the interactions between transfer RNAs with complementary anticodons. The effect of modified bases adjacent to the anticodon triplet. *J. Biomol. Struct. Dyn.* 3:387–399.

Huang, Y., J. Mak, Q. Cao, Z. Li, M. A. Wainberg, and L. Kleiman. 1994. Incorporation of excess wild-type and mutant tRNA$_3^{Lys}$ into human immunodeficiency virus type 1. *J. Virol.* 68:7676–7683.

Huang, Y., A. Shalom, Z. Li, J. Wang, J. Mak, M. A. Wainberg, and L. Kleiman. 1996. Effects of modifying the tRNA$_3^{Lys}$ anticodon on the initiation of human immunodeficiency virus type 1 reverse transcription. *J. Virol.* 70:4700–4706.

Isel, C., C. Ehresmann, G. Keith, B. Ehresmann, and R. Marquet. 1995. Initiation of reverse transcription of HIV-1: secondary structure of the HIV-1 RNA/tRNA$_3^{Lys}$ (template/primer) complex. *J. Mol. Biol.* 247:236–250.

Isel, C., C. Ehresmann, G. Keith, B. Ehresmann, and R. Marquet. 1997. Two step synthesis of (−) strong-stop DNA by avian and murine reverse transcriptases *in vitro*. *Nucleic Acids Res.* 25:545–552.

Isel, C., J. M. Lanchy, S. F. J. Le Grice, C. Ehresmann, B. Ehresmann, and R. Marquet. 1996. Specific initiation and switch to elongation of human immunodeficiency virus type 1 reverse transcription require the post-transcriptional modifications of primer tRNA$_3^{Lys}$. *EMBO J.* 15:917–924.

Isel, C., R. Marquet, G. Keith, C. Ehresmann, and B. Ehresmann. 1993. Modified nucleotides of transfer-RNA$_3^{Lys}$ modulate primer/template loop-loop interaction in the initiation complex of HIV-1 reverse transcription. *J. Biol. Chem.* 268:25269–25272.

Jacks, T., H. D. Madhani, F. R. Masiarz, and H. E. Varmus. 1988a. Signals for ribosomal frameshifting in the Rous sarcoma virus *gag-pol* junction. *Cell* 55:447–458.

Jacks, T., M. D. Power, F. R. Masiarz, P. A. Luciw, P. J. Barr, and H. E. Varmus. 1988b. Characterization of ribosomal frameshifting in HIV-1 *gag-pol* expression. *Nature* 331:280–283.

Jamjoom, G. A., R. B. Naso, and R. B. Arlinghaus. 1977. Further characterization of intracellular precursor polyproteins of Rauscher leukemia virus. *Virology* 78:11–34.

Jiang, M., J. Mak, M. A. Wainberg, M. A. Parniak, E. Cohen, and L. Kleiman. 1992. Variable transfer RNA content in HIV-1(IIIB). *Biochem. Biophys. Res. Commun.* 185:1005–1015.

Kane, S. E., and K. Beemon. 1987. Inhibition of methylation at two internal N^6-methyladenosine sites caused by GAC to GAU mutations. *J. Biol. Chem.* 262:3422–3427.

Kane, S. E., and K. Beemon. 1985. Precise localization of m6A in Rous sarcoma virus RNA reveals clustering of methylation sites: implications for RNA processing. *Mol. Cell. Biol.* 5:2298–2306.

Kang, S. M., Z. J. Zhang, and C. D. Morrow. 1997. Identification of a sequence within U5 required for human immunodeficiency virus type 1 to stably maintain a primer binding site complementary to tRNA^{Met}. *J. Virol.* 71:207–217.

Kawakami, K., S. Pande, B. Faiola, D. P. Moore, J. D. Boeke, P. J. Farabaugh, J. N. Strathern, Y. Nakamura, and D. J. Garfinkel. 1993. A rare tRNA-Arg(CCU) that regulates Ty1 element ribosomal frameshifting is essential for retrotransposition in Saccharomyces cerevisiae. *Genetics* 135:309–320.

Keeney, J. B., K. B. Chapman, V. Lauermann, D. F. Voytas, S. U. Astrom, U. von Pawel-Rammingen, A. Bystrom, and J. D.

Boeke. 1995. Multiple molecular determinants for retrotransposition in a primer tRNA. *Mol. Cell. Biol.* 15:217–226.

Keith, G., A. L. Glasser, J. Desgres, K. C. Kuo, and C. W. Gehrke. 1990. Identification and structural characterization of O-β-ribosyl-(1″-2′)-adenosine-5″-phosphate in yeast methionine initiator tRNA. *Nucleic Acids Res.* 18:5989–5993.

Keith, G., and T. Heyman. 1990. Heterogeneities in vertebrate tRNAs^{Trp}. Avian retroviruses package only as a primer the tRNA^{Trp} lacking modified m^{2}G in position 7. *Nucleic Acids Res.* 18:703–710.

Kikuchi, Y., Y. Ando, and T. Shiba. 1986. Unusual priming mechanism of RNA-directed DNA synthesis in *copia* retrovirus-like particles of *Drosophila*. *Nature* 323:824–826.

Kikuchi, Y., N. Sasaki, and Y. Ando-Yamagami. 1990. Cleavage of tRNA within the mature tRNA sequence by the catalytic RNA of RNase P: implication for the formation of the primer tRNA fragment for reverse transcription in *copia* retrovirus-like particles. *Proc. Natl. Acad. Sci. USA* 87:8105–8109.

Kim, T., R. A. Mudry, C. A. I. Rexode, and V. K. Pathak. 1996. Retroviral mutation rates and A-to-G hypermutations during different stages of retroviral replication. *J. Virol.* 70:7594–7602.

Kleiman, L., S. Gaudry, F. Boulerice, M. A. Wainberg, and M. A. Parniak. 1991. Incorporation of tRNA into normal and mutant HIV-1. *Biochem. Biophys. Res. Commun.* 174:1272–1280.

Kuchino, Y., H. Beier, N. Akita, and S. Nishimura. 1987. Natural UAG suppressor glutamine tRNA is elevated in mouse cells infected with Moloney murine leukemia virus. *Proc. Natl. Acad. Sci. USA* 84:2668–2672.

Lanchy, J. M., C. Ehresmann, S. F. J. Le Grice, B. Ehresmann, and R. Marquet. 1996. Binding and kinetic properties of HIV-1 reverse transcriptase markedly differ during initiation and elongation of reverse transcription. *EMBO J.* 15:7178–7187.

Lauermann, V., and J. D. Boeke. 1994. The primer tRNA sequence is not inherited during ty1 retrotransposition. *Proc. Natl. Acad. Sci. USA* 91:9847–9851.

Lauermann, V., K. Nam, J. Trambley, and J. D. Boeke. 1996. Plus-strand strong-stop DNA synthesis in retrotransposon Ty1. *J. Virol.* 69:7845–7850.

Levin, H. L. 1997. It's prime time for reverse transcriptase. *Cell* 88:5–8.

Levin, H. L., D. C. Weaver, and J. D. Boeke. 1993a. Novel gene expression mechanism in a fission yeast retroelement: Tf1 proteins are derived from a single primary translation product. *EMBO J.* 12:4885–4895.

Levin, J. G., D. L. Hatfield, S. Oroszlan, and A. Rein. 1993b. Mechanisms of translational suppression used in the biosynthesis of reverse transcriptase. *In* A. M. Skalka, and S. P. Goff (ed.), *Reverse Transcriptase*. Cold Spring Harbor Laboratory Press, Cold Spring Harbor, N.Y.

Levin, J. G., and J. G. Seidman. 1981. Effect of polymerase mutations on packaging of primer tRNA^{Pro} during murine leukemia virus assembly. *J. Virol.* 38:403–408.

Li, X. G., J. Mak, E. J. Arts, Z. X. Gu, L. Kleiman, M. A. Wainberg, and M. A. Parniak. 1994. Effects of alterations of primer-binding site sequences on human immunodeficiency virus type 1 replication. *J. Virol.* 68:6198–6206.

Li, Z., A. Shalom, Y. Huang, J. Mak, E. Arts, M. A. Wainberg, and L. Kleiman. 1996. Multiple forms of tRNA(Lys3) in HIV-1. *Biochem. Biophys. Res. Commun.* 227:530–540.

Liang, C., X. Li, L. Rong, P. Inouye, Y. Quan, L. Kleiman, and M. A. Wainberg. 1997. The importance of the A-rich loop in human immunodeficiency virus type 1 reverse transcription and infectivity. *J. Virol.* 71:5750–5757.

Mak, J., M. Jiang, M. A. Wainberg, M.-L. Hammarskjöld, D. Rekosh, and L. Kleiman. 1994. Role of Pr160$^{gag-pol}$ in mediating the selective incorporation of tRNA^{Lys} into human immunodeficiency virus type 1 particles. *J. Virol.* 68:2065–2072.

Mak, J., A. Khorchid, Q. Cao, Y. Huang, I. Lowy, M. A. Parniak, V. R. Prasad, M. A. Wainberg, and L. Kleiman. 1997. Effects of mutations in Pr160(gag-pol) upon tRNA$_3^{Lys}$ and Pr160(gag-pol) incorporation into HIV-1. *J. Mol. Biol.* 265:419–431.

Marquet, R., C. Isel, C. Ehresmann, and B. Ehresmann. 1995. tRNAs as primer of reverse transcriptase. *Biochimie* 77:113–124.

Mazumdar, S. K., W. Saenger, and K. H. Scheit. 1974. Molecular structure of poly-2-thiouridylic acid, a double helix with nonequivalent polynucleotide chains. *J. Mol. Biol.* 85:213–229.

Mesnard, J. M., and G. Lebeurier. 1991. How do viral reverse transcriptases recognize their RNA genome? *FEBS Lett.* 287:1–4.

Nam, S. H., T. D. Copeland, M. Hatanaka, and S. Oroszlan. 1993. Characterization of ribosomal frameshifting for expression of *pol* gene products of human T-cell leukemia virus type I. *J. Virol.* 67:196–203.

Paillart, J. C., R. Marquet, E. Skripkin, C. Ehresmann, and B. Ehresmann. 1996. Dimerization of retroviral genomic RNAs: structural and functional implications. *Biochimie* 78:639–653.

Panet, A., and H. Berliner. 1978. Binding of tRNA to reverse transcriptase of RNA tumor viruses. *J. Virol.* 26:214–220.

Panet, A., W. A. Haseltine, D. Baltimore, G. Peters, F. Harada, and J. E. Dahlberg. 1975. Specific binding of tryptophan transfer RNA to avian myeloblastosis virus RNA-dependent DNA polymerase (reverse transcriptase). *Proc. Natl. Acad. Sci. USA* 72:2535–2539.

Panganiban, A. T. 1988. Retroviral *gag* gene amber codon suppression is caused by an intrinsic *cis*-acting component of the viral mRNA. *J. Virol.* 62:3574–3580.

Perret, V., A. Garcia, J. Puglisi, H. Grosjean, J. P. Ebel, C. Florentz, and R. Giege. 1990. Conformation in solution of yeast tRNA(Asp) transcripts deprived of modified nucleotides. *Biochimie* 72:735–743.

Perron, H., J. A. Garson, F. Bedin, F. Beseme, G. Paranhos-Baccala, F. Komurian-Pradel, F. Mallet, P. W. Tuke, C. Voisset, J. L. Blond, B. Lalande, J. M. Seigneurin, and B. Mandrand. 1997. Molecular identification of a novel retrovirus repeatedly isolated from patients with multiple sclerosis. *Proc. Natl. Acad. Sci. USA* 94:7583–7588.

Peters, G., and J. Hu. 1980. Reverse transcriptase as the major determinant for selective packaging of tRNAs into avian sarcoma virus particles. *J. Virol.* 36:692–700.

Powell, M. D., M. Ghosh, P. S. Jacques, K. J. Howard, S. F. J. Le Grice, and J. G. Levin. 1997. Alanine-scanning mutations in the "primer grip" of p66 reverse transcriptase result in selective loss of RNA priming activity. *J. Biol. Chem.* 272:13262–13269.

Rebagliati, M. R., and D. A. Melton. 1987. Antisense RNA injections in fertilized frog eggs reveal an RNA duplex unwinding activity. *Cell* 48:599–605.

Reil, H., M. Höxter, D. Moosmayer, G. Pauli, and H. Hauser. 1994. CD4 expressing human 293 cells as a tool for studies in HIV-1 replication: the efficiency of translational frameshifting is not altered by HIV-1 infection. *Virology* 205:371–375.

Revel, M., and Y. Groner. 1978. Post-transcriptional and translational controls of gene expression in eukaryotes. *Annu. Rev. Biochem.* 47:1079–1126.

Richter-Cook, N. J., K. J. Howard, N. M. Cirino, B. M. Wöhrl, and S. F. J. Le Grice. 1992. Interaction of transfer RNA(Lys-3) with multiple forms of human immunodeficiency virus reverse transcriptase. *J. Biol. Chem.* 267:15952–15957.

Robinson, R. R., and N. Davidson. 1981. Analysis of a *Drosophila* tRNA gene cluster: two tRNA^{Leu} genes contain intervening sequences. *Cell* 23:251–259.

Saigo, K. 1986. A potential primer for reverse transcription of *mdg3*, a *Drosophila copia*-like element, is a leucine tRNA lacking its 3' terminal 5 bases. *Nucleic Acids Res.* **14**:4370.

Sampson, J. R., and O. C. Uhlenbeck. 1988. Biochemical and physical characterization of an unmodified yeast phenylalanine transfer RNA transcribed in vitro. *Proc. Natl. Acad. Sci. USA* **85**:1033–1037.

Sawyer, R. C., and J. E. Dahlberg. 1973. Small RNAs of Rous sarcoma virus: characterization by two dimensional polyacrylamide gel electrophoresis and fingerprint analysis. *J. Virol.* **12**:1226–1237.

Scheit, K. H., and P. Faerber. 1975. The effects of thioketo substitution upon uracil-adenine interactions in polyribonucleotides. *Eur. J. Biochem.* **50**:549–555.

Sedivy, J. M., J. P. Capone, U. L. RajBhandary, and P. A. Sharp. 1987. An inducible mammalian amber suppressor: propagation of a poliovirus mutant. *Cell* **50**:379–389.

Sharmeen, L., B. Bass, N. Sonenberg, H. Weintraub, and M. Groudine. 1991. Tat-dependent adenosine-to-inosine modification of wild-type transactivation response RNA. *Proc. Natl. Acad. Sci. USA* **88**:8096–8100.

Skripkin, E., C. Isel, R. Marquet, B. Ehresmann, and C. Ehresmann. 1996. Psoralen crosslinking between human immunodeficiency virus type 1 RNA and primer tRNA$_3^{Lys}$. *Nucleic Acids Res.* **24**:509–514.

Sprinzl, M., C. Horn, M. Brown, A. Ioudovitch, and S. Steinberg. 1998. Compilation of tRNA sequences and sequences of tRNA genes. *Nucleic Acids Res.* **26**:148–153.

Steinberg, S., and R. Cedergren. 1995. A correlation between N2-dimethylguanosine presence and alternate tRNA conformers. *RNA* **1**:886–891.

Stoltzfus, C. M., and R. W. Dane. 1982. Accumulation of spliced avian retrovirus mRNA is inhibited in S-adenosylmethionine-depleted chicken embryo fibroblasts. *J. Virol.* **42**:918–931.

Stoltzfus, C. M., and K. Dimock. 1976. Evidence for methylation of B77 avian sarcoma genome RNA subunits. *J. Virol.* **18**:586–595.

ten Dam, E. B., C. W. A. Pleij, and L. Bosch. 1990. RNA pseudoknots: translational frameshifting and readthrough on viral RNA. *Virus Genes* **4**:121–136.

Thrall, S. H., J. Reinstein, B. M. Wohrl, and R. S. Goody. 1996. Evaluation of human immunodeficiency virus type 1 reverse transcriptase primer tRNA binding by fluorescence spectroscopy: specificity and comparison to primer/template binding. *Biochemistry* **35**:4609–4618.

Vimaladithan, A., and P. J. Farabaugh. 1994. Special peptidyl-tRNA molecules can promote translational frameshifting without slippage. *Mol. Cell. Biol.* **14**:8104–8116.

Voytas, D. F., and J. D. Boeke. 1993. Yeast retrotransposons and transfer RNAs. *Trends Genet.* **9**:421–427.

Wagner, R. W., C. Yoo, L. Wrabetz, J. Kamholtz, J. Buchhalter, N. F. Hassan, K. Khalili, S. U. Kim, B. Perussia, F. A. McMorris, and K. Nishikura. 1990. Double-stranded RNA unwinding activity is detected ubiquitously in primary tissues and cell lines. *Mol. Cell. Biol.* **10**:5586–5590.

Wakefield, J. K., S.-M. Kang, and C. D. Morrow. 1996. Construction of a type 1 human immunodeficiency virus that maintains a primer binding site complementary to tRNA^{His}. *J. Virol.* **70**:966–975.

Wakefield, J. K., A. G. Wolf, and C. D. Morrow. 1995. Human immunodeficiency virus type 1 can use different tRNAs as primers for reverse transcription but selectively maintains a primer binding site complementary to tRNA$_3^{Lys}$. *J. Virol.* **69**:6021–6029.

Waters, L. C., and B. C. Mullin. 1977. Transfer RNA in RNA tumor virus. *Prog. Nucleic Acid Res. Mol. Biol.* **20**:131–160.

Wilhelm, M., T. Heyman, S. Friant, and F. X. Wilhelm. 1997. Heterogeneous terminal structure of Ty1 and Ty3 reverse transcripts. *Nucleic Acids Res.* **25**:2161–2166.

Wilhelm, M., F. X. Wilhelm, G. Keith, B. Agoutin, and T. Heyman. 1994. Yeast Ty1 retrotransposon: the minus-strand primer binding site and a *cis*-acting domain of the Ty1 RNA are both important for packaging of primer tRNA inside virus-like particles. *Nucleic Acids Res.* **22**:4560–4565.

Wills, N. F., R. F. Gesteland, and J. F. Atkins. 1991. Evidence that a downstream pseudoknot is required for translational readthrough of the Moloney murine leukemia virus gag stop codon. *Proc. Natl. Acad. Sci. USA* **88**:6991–6995.

Wills, N. M., R. F. Gesteland, and J. F. Atkins. 1994. Pseudoknot-dependent read-through of retroviral *gag* termination codons: importance of sequences in the spacer and loop 2. *EMBO J.* **13**:4137–4144.

Wittig, B., and S. Wittig. 1978. Reverse transcription of tRNA. *Nucleic Acids Res.* **5**:1165–1178.

Yoshinaka, Y., I. Katoh, T. D. Copeland, and S. Oroszlan. 1985. Murine leukemia virus protease is encoded by the *gag-pol* gene and is synthesized through suppression of an amber termination codon. *Proc. Natl. Acad. Sci. USA* **82**:1618–1622.

Yoshioka, K., H. Honma, M. Zushi, S. Kondo, S. Togashi, T. Miyake, and T. Shiba. 1990. Virus-like particle formation of *Drosophila copia* through autocatalytic processing. *EMBO J.* **9**:535–541.

Youvan, D. C., and J. E. Heast. 1981. A sequence from Drosophila melanogaster 18S rRNA bearing the conserved hypermodified nucleoside amΨ: analysis by reverse transcription and high-performance liquid chromatography. *Nucleic Acids Res.* **9**:1723–1741.

Yu, S. F., D. N. Baldwin, S. R. Gwynn, S. Yendapalli, and M. L. Linial. 1996. Human foamy virus replication: a pathway distinct from that of retroviruses and hepadnaviruses. *Science* **271**:1579–1582.

Yusupova, G., J. M. Lanchy, G. Yusupov, G. Keith, S. F. J. Le Grice, C. Ehresmann, B. Ehresmann, and R. Marquet. 1996. Primer selection by HIV-1 reverse transcriptase on RNA-tRNA$_3^{Lys}$ and DNA-tRNA$_3^{Lys}$ hybrids. *J. Mol. Biol.* **261**:315–321.

Zakharova, O. D., L. Tarrago-Litvak, M. Fournier, M. L. Andreola, M. N. Repkova, A. G. Venyaminova, S. Litvak, and G. A. Nevinsky. 1995. Interaction of primer tRNA$_3^{Lys}$ with the p51 subunit of human immunodeficiency virus type 1 reverse transcriptase: a possible role in enzyme activation. *FEBS Lett.* **361**:287–290.

Chapter 29

Modified Nucleosides Always Were: an Evolutionary Model

NICOLAS CERMAKIAN AND ROBERT CEDERGREN

THE EVOLUTIONARY FRAMEWORK

The RNA world hypothesis of the origin of cellular life on earth is a good example of how interpreting a process or a structure—in this case RNA catalysis—within an evolutionary framework can have profound consequences for subsequent directions in biology. The origin of introns is an even better example for our purposes because it has been approached in different ways; hypotheses generated by phylogenetic analyses of present-day introns have coexisted, not always very happily, with chemical/biological rationalizations of intron origins. Since we will use both phylogenetic arguments and chemical rationalizations in analyzing the origin of modified nucleotides in biological systems, the differences between them must be clear. Phylogenetic analyses start at the present day and make use of the organismal distribution of traits to infer the past, that is, properties of the most recent common ancestor. Chemical and biological rationalizations of the past are based on the extrapolation of certain molecular properties, which can be logically linked to a process or a structure present during the period of emerging life forms. These extrapolations require neither intermediates nor ancestors, the plausibility being judged solely on the coherence of the proposal.

The differences between these two approaches are brought into focus when one considers that intron splicing is a phenomenon that can be invoked in an undeniably rational manner to explain the formation of modern genes (Gilbert, 1978). On the other hand, no one has been able to demonstrate a wide-enough distribution of introns to support hypotheses of a prebiotic origin. The controversy surrounding the origin of introns is thus fundamentally due to a conflict between the inferences of the two methodologies.

In general, when one speaks of a modified nucleoside, minor nucleosides are meant, even though this certainly does not mean minor in the sense of function (see Chapter 1 by Lane). Under this definition, the spectrum of "modified" nucleosides is expanded out of the realm of posttranscriptional modifications of RNA and leads to the consideration of the origin of nucleosides and their derivatives in a more global manner. Within our definition, it seems perfectly clear that modified nucleosides did not have an origin any different from that of the "basic nucleosides" or the plethora of other small organic compounds that were found in the primordial soup. Indeed, it was proposed some time ago that the numerous nucleotide-derived cofactors and prosthetic groups had an ancient origin and could even be viewed as relics of the early involvement of RNA molecules in catalysis of various cellular reactions (Lazcano, 1994a). These hypotheses also suggest that during the emergence of the key cellular components these nucleoside components in RNA were already present. We believe, therefore, that the issue is not the origin of modified nucleotides—they always were—but rather how and why they remain in RNA and DNA molecules. We do, however, bow to consensus opinion in this book and use the term modified nucleoside.

THE PHYLOGENETICS OF MODIFIED NUCLEOSIDES

The tried and true methodology in evolutionary studies is the determination of the organismal distribution of a given trait. Then, by using the principle of parsimony, the origins of a property generally can be ascribed to the most recent ancestor of the two most distant organisms possessing the trait. In the

---

Nicolas Cermakian and Robert Cedergren • Département de Biochimie, Université de Montréal, Montréal, Québec, Canada H3C 3J7.

case of the modified nucleosides, we are interested not only in the organism in which they are found but also in the RNA species and the position occupied by modified nucleosides.

Modified nucleosides occur in rRNAs, mRNAs (eukaryotic and viral), small nuclear RNAs, and other small RNAs (see Appendix 1). Thus, we can infer that modified nucleosides are present in most known RNA species. As can be seen in Appendix 1, most of the modified nucleosides that have been compiled and studied are found in tRNAs. Here, we evaluate the distribution of modified nucleosides in certain RNA species of the three domains of life (archaea, bacteria, and eukaryotes).

The phylogenetic distribution of modified nucleosides in tRNA shows that more than half (60 out of 96) are found only in either eukaryotes, bacteria, or archaea, whereas 22 are found in two domains—7 are common to eukaryotes and archaea, 7 are common to bacteria and archaea, and 8 are common to eukaryotes and bacteria. However, these data do not allow us to infer the presence of modified nucleotides before the divergence of the three domains.

Fourteen modifications are present in all three domains (15, including queuosine derivatives) and of these several are found at comparable positions in tRNAs. Table 1 is derived from the information provided in Appendices 1 and 5 and Grosjean et al. (1995b). First, most of the conserved modified nucleosides are in or near the anticodon loop (positions 34 and 37–39). They are among the simplest modifications with the exception of queuosine (Q) and the threonylated residue t^6A37. In fact, these conserved modifications (that is, pseudouridine [Ψ] and methylated nucleotides) are the most frequent modifications in tRNAs and also in other RNAs (Appendices 1 and 5). Another striking point is that two of the modified nucleosides of Table 1, m^1G37 and t^6A37, are nearly ubiquitous in tRNAs reading codons beginning by a C or an A, respectively (Björk, 1995). Therefore, based on the distribution of modified nucleosides, those of Table 1 would have been present in tRNA at the point at which the eukaryote/archaea branch diverged from the bacterial lineage (Fig. 1). We may assume that the enzymatic systems required to modify the nucleosides also were present at that time, although the precise mechanisms of their formation may have differed from the modern enzymatic processes for their biosynthesis. Note that these modifications inferred to have been present before the divergence of the three domains are likely to form a minimal set, because certain modifications could have been lost in one domain or another. The distribution of modified nucleosides showing presence in two of the three domains supports this conjecture.

It is interesting that two of the most widespread tRNA modifications of Table 1, pseudouridylation and 2′-O-methylation, are also the most frequent modifications in RNAs. In fact, Ψ and Cm are among the only four modified nucleosides that are found in RNAs from all three organismal domains (Table 2). These observations strengthen our point that these modifications are very ancient. At the very least, therefore, there is strong phylogenetic evidence for the early appearance of modified nucleosides.

THE DEEP PAST

The RNA world hypothesis states that at some point in early evolution, living organisms relied entirely on RNA for both the storage of genetic information and catalysis; they would have been devoid of encoded proteins and DNA (Gilbert, 1986; Darnell and Doolittle, 1986; Cedergren and Grosjean, 1987). The intertwining of the RNA world and RNA catalysis has led to novel hypotheses, experiments and discoveries in almost every biological process that involves RNA. The study of RNA processing and translation, as well as in vitro selection of RNA molecules with a wide range of catalytic activities, provides additional support for the notion of an all-RNA world (Lazcano, 1994b). RNA catalysis was such a perfect solution to the genotype/phenotype dichotomy in origin-of-life theories that practically all modern scenarios are based on some form of RNA catalysis. As a result, many underpinnings of this hypothesis have unfortunately managed to escape detailed critical evaluation.

Among the issues that we believe to be worthy of thought is the purity of the RNA in the RNA

Table 1. Modified nucleosides found at corresponding positions in tRNAs from eukaryotes, bacteria, and archaea[a]

Position in tRNA	Modified nucleoside[b]
13	Ψ[c]
34	Cm
34	Q[d]
37	t^6A
37	m^1G[e]
38	Ψ
39	Ψ[f]
55	Ψ[g]
58	m^1A[c]

[a] Data taken from Appendices 1 and 5 and Grosjean et al. (1995b).
[b] Abbreviations as in Appendix 1.
[c] Not found in chloroplasts.
[d] There are different variants of Q depending on base substituents (see Appendix 1).
[e] Modification found in more than 80% of tRNAs with a G37.
[f] Modification found in more than 80% of tRNAs with a U39.
[g] Modification found in more than 90% of tRNAs with a U55.

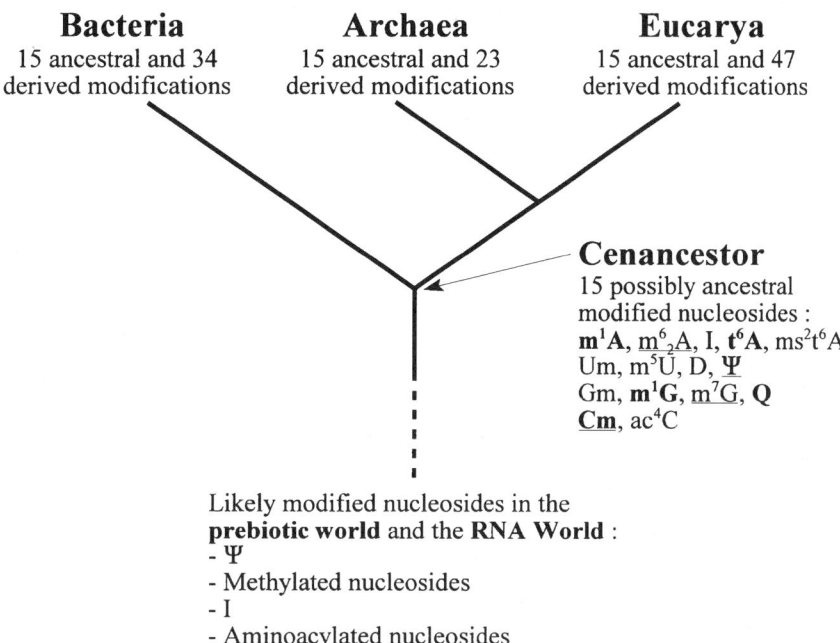

Figure 1. Nucleoside modifications likely to have been occurring in the cenancestor and in the RNA world. The upper part of the figure presents the results of the phylogenetic analysis of modified nucleotides. The data are from Table 1 and Appendix 1. The modified nucleosides found at corresponding positions in tRNAs from the three domains (Table 1) are in bold. The modified nucleosides found in rRNAs of all three domains are underlined (Table 2). Derived modifications are considered to have occurred later in evolution. The lower part of the figure refers to the discussion about the prebiotic world and RNA world. See the text for more details.

world. The emergence of the RNA world would have taken place in a molecular soup composed of massive numbers of small chemicals. Even though molecular complexity can be a precondition to the origin of life in some theories (Kauffman, 1993), chemically this mixture is a nightmare (Joyce and Orgel, 1993). How do the poorly represented purines and pyrimidines in this diluted soup get together with the poorly represented ribose to form nucleosides? More importantly, how could the chemistry of these precursors distinguish so well between the "nonmodified" and the "modified" bases and riboses? In summary, how does a self-replicating, presumably polymeric RNA arise de novo from a complex mixture of nucleotides and derivatives thereof (Joyce and Orgel, 1993)? It seems much more plausible that primordial "RNA" was more a representative of the variety of nucleoside-like compounds that existed in the soup than a pure ribonucleotide species.

A possible criticism of the above scenario would be if modified nucleotides somehow blocked the catalytic activity of RNA upon which the emergence of an RNA world depended. However, work in our laboratory has shown that mixed ribo- and deoxyribo-oligonucleotides can perform catalytic functions (Perreault et al., 1990; Yang et al., 1992; Chartrand et al., 1995). Based on this work, it was proposed that such mixed polymers could have preceded the RNA world or served as a bridge between the RNA world and the present-day DNA-protein world. This possibility was presented in some depth by Bussière and Perreault (1995). Since those early experiments, many nucleic acid catalysts have been found, either by using modified nucleosides in chemical synthetic schemes (Usman et al., 1996) or, in the case of catalytic DNAs, by in vitro selection (Cuenoud and Szostak, 1995). Rather than thinking that the idea of mixed polymers is ill founded, we consider it surprising that such little critical thought has been given to the state of RNA produced in the primordial soup.

At the present time, the mechanism of abiotic polymerization of RNA (here, we use RNA to signify

Table 2. Conserved modified nucleosides in rRNAs from eukaryotes, bacteria, and archaea[a]

Modified nucleoside[b]	rRNA species[c]		
	Archaea	Bacteria	Eukaryotes
m^6_2A	16S	16S/23S	18S
Cm	5S	23S	5.8S/17S/18S/28S
m^7G	16S	?[d]	5S/17S/18S/26S/28S
Ψ	16S	16S/23S	5S/5.8S/18S/28S

[a] Data taken from Limbach et al. (1994).
[b] Abbreviations as in Appendix 1.
[c] Refers to the rRNAs containing the modified nucleoside.
[d] The precise position of the modified nucleoside has not been determined.

any polymer involving a sugar, a nitrogen base, and a phosphate residue) or their precursors is controversial (Joyce and Orgel, 1993; Szathmáry, 1997); therefore, we can only imagine how modified nucleosides might have become part of the early polymers. First, nucleic acids could have been produced by condensation of nucleotides (as they are today, albeit in an indirect manner), or by copolymerization of nucleosides and phosphates. Alternate schemes would involve addition of a nitrogen base to a sugar-phosphate backbone or, less likely, insertion of a base or a nucleoside into a preformed polyphosphate backbone. These three possible, but nonexhaustive, pathways relate in different ways to how modified nucleosides could be incorporated during synthesis of the polymer.

The condensation reaction of nucleosides takes place at the 3' and 5' positions of the sugar; thus, unless some intramolecular cyclic intermediate would be formed, this reaction could not discriminate with regard to the substituents on the nitrogen base of the nucleotide. If this condensation were templated by a complementary strand, discrimination would be possible, but only in the case where the nitrogen base could not form the proper hydrogen bonds with the template strand (Ferris et al., 1996). Clearly this property would tend to eliminate some modified (and unmodified) nucleotides, but many would still be good candidates. For example, no effective discrimination could be imagined between uracil and thymine derivatives. Modifications on the sugar residue such as 2'-O-methylation of a ribose should not affect condensation either, unless the 2'-OH was the primary nucleophile. Therefore, a condensation mechanism could only be partially discriminatory if a template strand was involved in polymerization.

The alternate synthetic pathway, the addition of a nitrogen base to a sugar-phosphate backbone, is subject to the reactivity of the ring nitrogen involved in formation of the glycosidic bond. The synthesis of a nucleoside would involve the nucleophilic attack of the ring nitrogen (either N3 or N9) on the C1' carbonyl carbon; therefore, the better the nucleophile, the more likely the reaction. Normally, methyl groups would be considered electron donating, and therefore depending on their ring position would provide a higher nucleophilicity to the ring nitrogen and therefore a greater probability of reaction. Electron withdrawing substituents on the ring would lower the nucleophilicity. This second way of incorporating bases (eventually modified bases) into RNA can be illustrated by current mechanisms involved in some tRNA modifications. For instance, G is replaced by queuine (Q) at wobble position 34 in many tRNAs, by a mechanism known as transglycosylation,

in which the base is replaced without cleavage of the phosphodiester backbone (see Chapter 9 by Romier et al.). Although Q is a hypermodified nucleotide, the transglycosylation reaction per se is rather simple; indeed, the enzyme that performs the base transfer does not require any additional energy source (see Chapter 9 by Romier et al.). Also significant here is the discovery that abasic hammerhead ribozymes can be rescued by the addition of exogenous bases, which restore the activity by inserting in the cavity created by the absence of the normal base (Peracchi et al., 1996). Finally, the insertion of a preformed nucleoside to a polymeric phosphate should not be sensitive to the structure of the nitrogen base for the same reasons as we invoked above for the condensation reaction.

The incorporation of modified nucleosides into nucleic acid polymers is therefore quite plausible, and the use of a particular nucleoside may depend only on its concentration in the primordial soup. Again, considering the chemistry of nucleosides, it would seem that the moderately modified nucleosides would be present in the highest concentrations in the primordial soup along with the standard nucleosides (Fig. 1). Such modified nucleosides would likely include the sugar or base methylated nucleosides; inosine (I), which can be obtained easily by deamination of A (see Chapter 19 by Rueter and Emeson) (Grosjean et al., 1995a); pseudouridine (Ψ), which could theoretically be generated from U via an isomerization pathway; and possibly the aminoacylated nucleosides, but not the hypermodified nucleotides generally found in the anticodon loop of tRNAs.

A MODEL OF THE EMERGENCE OF MODIFIED NUCLEOSIDES

Up to this point, we have made a case for the presence and the incorporation of modified nucleosides into RNA under prebiotic conditions. In parallel, the phylogenetic evidence suggests that some simple modified nucleosides existed at the point when the major organismal groups diverged. The question is now: Are we able to provide a reasonable scenario to bridge the one- or two-billion-year gap between the primordial soup and the divergence of major living forms? Fortunately, the use of the phylogenetic data not only provides evidence for the early existence of modified nucleosides in RNA molecules, but also allows us to focus on only those modifications that appear to be the oldest.

At the beginning, a minimum of distinctive chemical characteristics (see the previous section) would have promoted the more or less random incorporation of nucleosides and modified nucleosides into RNA. In an evolutionary sense, this situation is acceptable but hardly ideal because the information on the position and the nature of the modification would be lost at each generation in the absence of some way to store this information. Therefore, rather early in the emergence of more sophisticated systems, the position and the nature of the modified nucleoside would have to be encoded. A hint as to how this could have happened has been provided by recent experimental results showing that modifications of ribosomal RNAs are addressed or guided by certain small nucleolar RNAs (see Chapter 12 by Ofengand and Fournier and Chapter 13 by Bachellerie and Cavaillé). Intriguingly, the modifications that we have predicted to be most prevalent in the primordial soup, and demonstrated to be the most widespread from phylogenetic analysis, are also those that have an RNA involvement in their synthesis. It must be kept in mind, however, that RNA-guided rRNA modification has so far been found only in eukaryotes.

We propose, therefore, that at some early stage small RNA molecules were used to direct the positions at which modified nucleosides were to be present. At first, it could be imagined that a modified nucleoside was incorporated during synthesis of the RNA, thanks to the presence of small RNA molecules that added new discriminatory properties to the template-copying mechanism. Later, the modifications would have been products of posttranscriptional modifications. This eventuality would be required when the exceedingly more discriminatory protein polymerases took over the role of RNA synthesis. Here again the known chemistry of the modifications is in agreement with the interpretation that we make. The modern synthesis of the methylated nucleosides is assured by a methyl transferase and S-adenosylmethionine, the methyl donor (Appendix 3). S-Adenosylmethionine, the product of a simple addition of an amino acid to adenosine (i.e., a modified nucleoside), likely existed at the time when cellular systems were emerging. The presence of an RNA to direct the positioning of the methyl group raises the possibility that an RNA once had a methyl transferase activity in this process. Pseudouridine is simply an isomer of uridine and could also be the product of an RNA-catalyzed reaction. It is also possible that these guide RNAs were internal. Perhaps the complex present-day recognition of modification sites in tRNA (see, for example, Grosjean et al., 1996) is an example of how *cis*-guidance could occur, with the guide RNA being intramolecular or within an intron (Grosjean et al., 1997). *cis*-guidance has been proposed in the case of modification of bacterial rRNA (see Chapter 12 by Ofengand and Fournier). An interesting parallel can be made here with RNA editing (see also below); in trypanosomatid mitochondria, insertion and deletion of U are mediated by guide RNA acting in *trans* through base pairing (see Chapter 21 by Hajduk and Sabatini); moreover, it has been proposed that editing could be in some cases mediated by *cis*-acting guide RNA, for example, in tRNAs (see Chapter 16 by Price and Gray) or in glutamate neuroreceptor editing (Herbert, 1996). These RNA-mediated RNA modification systems would then have given way to the presumed, modern protein-enzyme modification system.

The driving force for the presence of modified nucleosides could have been in rendering superior properties to the RNA. First, they could confer thermal or pH stability on RNA molecules, which would otherwise have been rather fragile and reactive (Conrad et al., 1995). Such a stability advantage has also been discussed in the context of mixed RNA-DNA polymers (Yang et al., 1992; Bussière and Perreault, 1995). Second, the presence of a variety of chemical groups on modified nucleosides could have expanded the functionality of early ribozymes, by modulating the recognition of other nucleic acids, or by allowing a greater range of catalytic activities. An example of the modulation of nucleic acid interactions is the influence of modified nucleosides in the evolution of the genetic code, through altering codon-anticodon interactions (Cedergren et al., 1986; Osawa et al., 1992). Enlargement of the range of catalytic activities is illustrated by the case of Ψ, which is much more chemically versatile than U (see Chapter 1 by Lane and Chapter 12 by Ofengand and Fournier) (Lane et al., 1995; Ofengand and Bakin, 1997). Third, modified nucleosides could modify RNA base pairing as in the case of m^2_2G (Steinberg and Cedergren, 1996) and m^1A (Helm et al., 1997), which may prevent alternative base pairing in tRNAs. Furthermore, they could be involved in regulation of RNA activity by modulating the turnover of these molecules; in present-day RNAs, modified nucleosides can be recognition signals for specific nucleases (Masaki et al., 1997). Finally, several hydrophobic isopentenyl-containing modified nucleosides known as cytokinins, apart from their occurrence in tRNAs, are present in a free state in a variety of organisms where they exert regulatory functions (Björk, 1995). Whatever the evolutionary mechanism by which these new properties arise, it may be impossible to go back once established.

The research on editing and on nucleoside modification that has been carried out during the last decade, and which is reviewed in detail in other chapters in this book, brought to light various analogies between these two phenomena. One striking example is the use of guide RNAs both in rRNA modification (see Chapter 12 by Ofengand and Fournier and Chapter 13 by Bachellerie and Cavaillé) and in some editing reactions (see Chapter 21 by Hajduk and Sabatini). Another point of convergence of editing and modification studies is the deamination of bases from A to I and from C to U, which are generated via similar mechanisms (see Chapter 19 by Rueter and Emeson and Chapter 20 by Carter). In fact, in some cases it is not clear whether we should speak of editing or modification (see Chapter 16 by Price and Gray; Covello and Gray, 1993), especially in the case of the production of I by deamination of A: its occurrence is called editing when the mRNA for some glutamate receptors (GluR) is concerned (Herbert, 1996) but is called base modification in the case of tRNA (Appendix 1). Also, U formed by an editing phenomenon involving enzymatic deamination of an encoded C can be viewed as a modified nucleoside (see Appendix 1).

These parallels between RNA editing and modification prompted us to examine the former through the same evolutionary approach as the latter. In fact, an evolutionary scenario for the origin and retention of editing in RNA molecules parallels that of nucleoside modifications. Editing might have first occurred in an undirected and random way through spontaneous chemical base transformations. We have already discussed, for example, the likelihood of base deamination in the primordial soup. Thereafter, editing of RNA molecules might have been selected for some advantage, for example, the capacity of forming more stable double-stranded structures, as in the case of tRNA editing (see Chapter 16 by Price and Gray), or assisting early, less-discriminatory RNA polymerases (Conrad et al., 1995). This would have favored the establishment of a way to specifically edit sites in RNA, for example, through guide RNAs in *trans* (see Chapter 21 by Hajduk and Sabatini) or in *cis* (see Chapter 16 by Price and Gray).

In spite of this rationalization of what might have happened in the deep past, there is no phylogenetic evidence that RNA editing is very ancient. In fact, there is general agreement that all present-day manifestations of editing are likely to have arisen recently, in particular lineages of eukaryotes (see Chapter 17 by Marchfelder et al. and Chapter 21 by Hajduk and Sabatini) (Covello and Gray, 1993; Scott, 1995; Weiner and Maizels, 1990). This hypothesis is based on the fact that phenomena grouped under the banner "editing" are diverse in effects and molecular mechanisms, and are each restricted to a group of organisms or subcellular compartments: for instance, RNA-guided U insertion/deletion in trypanosome mitochondria (see Chapter 21 by Hajduk and Sabatini), C-to-U and U-to-C transition by deamination in plant organelles (Freyer et al., 1997; Simpson, 1990; Yu and Schuster, 1995), intron-guided A-to-I deamination in GluR mRNA (see Chapter 19 by Rueter and Emeson), or editing by various mechanisms in the amino acid acceptor stem of tRNAs of different lineages (see Chapter 16 by Price and Gray). On the other hand, no editing mechanism is known to be widespread, and editing is apparently absent in archaea and bacteria (Appendix 2).

This situation is in sharp contrast to nucleotide modifications, for which, as has been previously discussed, both chemical rationalization and phylogenetic evidence point toward an ancient, prebiotic origin. The situation regarding what we know about the origin of editing is rather analogous to the case of introns. Whereas RNA splicing fit well in the RNA world hypothesis, and can help to explain many observations pertaining to the early evolution of RNA and protein domains, no phylogenetic evidence, based on the known distribution of introns, clearly supports an early appearance of these entities (Hurst and McVean, 1996).

CONCLUSION

We have proposed a stepwise chronology for the early origin and the maintenance of modified nucleosides in RNA. A similar model could be established for RNA editing, which is similar to RNA modification in many of its present-day manifestations; however, contrary to the case of modifications, no phylogenetic evidence supports an early occurrence of editing. Our model involves first a random incorporation of modified nucleosides, which later gave way to an RNA-guided modification system. In the case of tRNAs, there is no evidence for present-day RNA guiding, although such a system could be imagined early in the evolution of replicating cells. In the case of the rRNAs, aspects of the RNA-guided system are still present, as evidenced by the startling results demonstrating a role for the small nucleolar RNAs in the maturation of ribosomal RNA. The current evidence of the enzymes involved in RNA-guided modification leaves the door open to speculation that RNA has a greater role in this process than simply guiding modification. Could it be that remnants of the modification system of the RNA world still exist?

Acknowledgments. We are particularly grateful to Michael Gray for a critical reading of this chapter. We thank the Natural Science and Engineering Research Council of Canada for support. R.C. is the Richard Ivey Professor of the Canadian Institute for Advanced Research. N.C. has a predoctoral fellowship from the Fonds pour la Formation de Chercheurs et l'Aide à la Recherche du Québec.

REFERENCES

Björk, G. R. 1995. Biosynthesis and function of modified nucleosides, p. 165–205. *In* D. Söll and U. L. RajBhandary (ed.), *tRNA: Structure, Biosynthesis, and Function*. American Society for Microbiology, Washington, D.C.

Bussière, F., and J.-P. Perreault. 1995. On the road to a DNA-protein world. *RNA* 1:451–452.

Cedergren, R., and H. Grosjean. 1987. On the primacy of primordial RNA. *BioSystems* 20:175–180.

Cedergren, R., H. Grosjean, and B. Larue. 1986. Primordial reading of genetic information. *BioSystems* 19:259–266.

Chartrand, P., S. C. Harvey, G. Ferbeyre, N. Usman, and R. Cedergren. 1995. An oligodeoxyribonucleotide that supports catalytic activity in the hammerhead ribozyme domain. *Nucleic Acids Res.* 23:4092–4096.

Conrad, F., A. Hanne, R. K. Gaur, and G. Krupp. 1995. Enzymatic synthesis of 2'-modified nucleic acids: identification of important phosphate and ribose moieties in RNase P substrates. *Nucleic Acids Res.* 23:1845–1853.

Covello, P. S., and M. W. Gray. 1993. On the evolution of RNA editing. *Trends Genet.* 9:265–268.

Cuenoud, B., and J. W. Szostak. 1995. A DNA metalloenzyme with DNA ligase activity. *Nature* 375:611–614.

Darnell, J. E., and W. F. Doolittle. 1986. Speculations on the early course of evolution. *Proc. Natl. Acad. Sci. USA* 83:1271–1275.

Ferris, J. P., A. R. Hill, R. Liu, and L. E. Orgel. 1996. Synthesis of long prebiotic oligomers on mineral surfaces. *Nature* 381:59–61.

Freyer, R., M. Kiefer-Meyer, and H. Kössel. 1997. Occurence of plastid RNA editing in all major lineages of land plants. *Proc. Natl. Acad. Sci. USA* 94:6285–6290.

Gilbert, W. 1978. Why genes in pieces? *Nature* 271:501.

Gilbert, W. 1986. The RNA world. *Nature* 319:618.

Grosjean, H., F. Constantinesco, D. Foiret, and N. Benachenhou. 1995a. A novel enzymatic pathway leading to 1-methylinosine modification in *Haloferax volcanii* tRNA. *Nucleic Acids Res.* 23:4312–4319.

Grosjean, H., M. Sprinzl, and S. Steinberg. 1995b. Posttranscriptionally modified nucleosides in transfer RNA: their locations and frequencies. *Biochimie* 77:139–141.

Grosjean, H., J. Edqvist, K. B. Stråby, and R. Giegé. 1996. Enzymatic formation of modified nucleosides in tRNA: dependence on tRNA architecture. *J. Mol. Biol.* 255:67–85.

Grosjean, H., Z. Szweykowska-Kulinska, Y. Motorin, F. Fasiolo, and G. Simos. 1997. Intron-dependent enzymatic formation of modified nucleosides in eukaryotic tRNAs: a review. *Biochimie* 79:293–302.

Helm, M., H. Brulé, F. Degoul, C. Cepanec, J. P. Leroux, R. Giegé, and C. Florentz. 1997. The presence of a modified nucleotide is required for the cloverleaf folding of a human mitochondrial tRNA. Oral presentation 3-13 of the 17th International tRNA Workshop, May 10–15, Kazusa Akademia Center, Chiba, Japan.

Herbert, A. 1996. RNA editing, introns and evolution. *Trends Genet.* 12:6–9.

Hurst, L. D., and G. T. McVean. 1996. A difficult phase for intron-early. *Curr. Biol.* 6:533–536.

Joyce, G. F., and L. E. Orgel. 1993. Prospects for understanding the origin of the RNA world, p. 1–25. *In* R. F. Gesteland and J. F. Atkins (ed.), *The RNA World*. Cold Spring Harbor Laboratory Press, Cold Spring Harbor, N.Y.

Kauffman, S. A. 1993. *The Origins of Order. Self-Organization and Selection in Evolution*. Oxford University Press, New York, N.Y.

Lane, B. G., J. Ofengand, and M. W. Gray. 1995. Pseudouridine and O2'-methylated nucleosides. Significance of their selective occurrence in rRNA domains that function in ribosome-catalyzed synthesis of the peptide bonds in proteins. *Biochimie* 77:7–15.

Lazcano, A. 1994a. The transition from nonliving to living, p. 61–69. *In* S. Bengtson (ed.), *Early Life on Earth*. Columbia University Press, New York, N.Y.

Lazcano, A. 1994b. The RNA world, its predecessors, and its descendants, p. 70–80. *In* S. Bengtson (ed.), *Early Life on Earth*. Columbia University Press, New York.

Limbach, P. A., P. F. Crain, and J. A. McCloskey. 1994. Summary: the modified nucleosides of RNA. *Nucleic Acids Res.* 22:2183–2196.

Mazaki, H., T. Ogawa, K. Tomita, T. Ueda, K. Watanabe, and T. Uozumi. 1997. Colicin E5: a novel ribonuclease targeting a specific group of tRNAs. Oral presentation 2-1 of the 17th International tRNA Workshop, May 10–15, Kazusa Akademia Center, Chiba, Japan.

Ofengand, J., and A. Bakin. 1997. Mapping to nucleotide resolution of pseudouridine residues in large subunit ribosomal RNAs from representative eukaryotes, prokaryotes, archaebacteria, mitochondria and chloroplasts. *J. Mol. Biol.* 266:246–268.

Osawa, S., T. H. Jukes, K. Watanabe, and A. Muto. 1992. Recent evidence for evolution of the genetic code. *Microbiol. Rev.* 56:229–264.

Peracchi, A., L. Beigelman, N. Usman, and D. Herschlag. 1996. Rescue of abasic hammerhead ribozymes by exogenous addition of specific bases. *Proc. Natl. Acad. Sci. USA* 93:11522–11527.

Perreault, J.-P., T. F. Wu, B. Cousineau, K. K. Ogilvie, and R. Cedergren. 1990. Mixed deoxyribo- and ribo-oligonucleotides with catalytic activity. *Nature* 344:565–567.

Scott, J. 1995. A place in the world for RNA editing. *Cell* 81:833–836.

Simpson, L. 1990. RNA editing—a novel genetic phenomenon? *Science* 250:512–513.

Szathmáry, E. 1997. The first two billion years. *Nature* 387:662–663.

Steinberg, S., and R. Cedergren. 1996. A correlation between N2-dimethylguanosine presence and alternate tRNA conformers. *RNA* 1:886–891.

Usman, N., L. Beigelman, and J. A. McSwiggen. 1996. Hammerhead ribozyme engineering. *Curr. Opin. Struct. Biol.* 6:527–533.

Weiner, A. M., and N. Maizels. 1990. RNA editing: guided but not templated? *Cell* 61:917–920.

Yang, J. H., N. Usman, P. Chartrand, and R. Cedergren. 1992. Minimum ribonucleotide requirement for catalysis by the RNA hammerhead domain. *Biochemistry* 31:5005–5009.

Yu, W., and W. Schuster. 1995. Evidence for a site-specific cytidine deamination reaction involved in C to U RNA editing of plant mitochondria. *J. Biol. Chem.* 270:18227–18233.

Modification and Editing of RNA
Edited by Henri Grosjean and Rob Benne
© 1998 ASM Press, Washington, D.C.

Appendix 1: Chemical Structures and Classification of Posttranscriptionally Modified Nucleosides in RNA

YURI MOTORIN AND HENRI GROSJEAN

The compilation in this appendix shows the chemical structures, symbols, and common names of the 95 modified ribonucleosides isolated as of August 1997 from different cellular RNAs (essentially tRNA, rRNA, snRNA, and mRNA) from various organisms (A, archaea; B, eubacteria; E, eukaryotes). They are classified into five groups according to their parent ribonucleosides: adenosine, uridine, guanosine, cytidine, and pseudouridine (the "fifth nucleoside" [see Chapter 1 by Lane]. All names and symbols are the conventional ones and are defined by Limbach et al. (1994), except for archaeosine, for which the abbreviation gQ is used (Agris, 1996). Another symbol for archaeosine used by Crain and McCloskey (1997) is G^+. The one-letter code (sometimes a symbol or numeral) used in the tRNA bank by Sprinzl et al. (1998) is also indicated in parentheses after each name. Numbering of atoms in the purine and pyrimidine rings is also given. Arrows in the diagrams indicate the atoms involved in hydrogen bond formation as donors or acceptors of protons in classical Watson-Crick base pairs. Most of the data come from the reference paper by Limbach et al. (1994), where additional information can be found (*Chemical Abstracts* registry number and index names, initial literature citations for structural characterization, occurrence and chemical synthesis of modified ribonucleosides, and useful notes). Detailed identification information for each modified ribonucleoside can also be obtained on the World Wide Web at http://medstat.med.utah.edu/RNAmods/RNAmods.shtml (see Crain and McCloskey, 1997). Additional information on the occurrence and frequencies of each modified ribonucleoside in tRNAs is given in Grosjean et al. (1995) and Sprinzl et al. (1998). For snRNAs, the reference paper by Gu and Reddy (1997) should be consulted.

Acknowledgments. We are indebted to J. McCloskey (University of Utah) and M. Sprinzl (University of Bayreuth) for their advice and for providing data. This work was supported by grants to H.G. from the CNRS, ARC, and ANRS in France.

REFERENCES

Agris, P. F. 1995. The importance of being modified: roles of modified nucleosides and Mg^{2+} in RNA structure and function. *Prog. Nucleic Acid Res. Mol. Biol.* 53:79–129.

Crain, P. F., and J. A. McCloskey. 1997. The RNA modification database. *Nucleic Acids Res.* 25:126–127.

Grosjean, H., M. Sprinzl, and S. Steinberg. 1995. Posttranscriptionally modified nucleosides in tRNA: their locations and frequencies. *Biochimie* 77:139–141.

Gu, J., and R. Reddy. 1997. Small RNA database. *Nucleic Acids Res.* 25:98–101.

Limbach, P. A., P. F. Crain, and J. A. McCloskey. 1994. Summary: the modified nucleosides of RNA. *Nucleic Acids Res.* 22:2183–2196.

Sprinzl, M., C. Horn, M. Brown, A. Ioudovitch, and S. Steinberg. 1998. Compilation of tRNA sequences and sequences of tRNA genes. *Nucleic Acids Res.*, 26:148–153.

Yuri Motorin and Henri Grosjean • Laboratoire d'Enzymologie et Biochimie Structurales du CNRS, Bâtiment 34, 1 av. de la Terrasse, 91198 Gif-sur-Yvette, France.

Modified pseudouridines

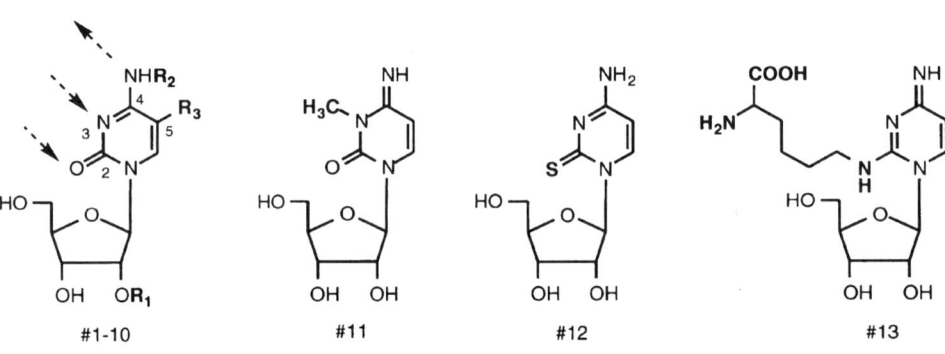

#	Symbol	Name (one letter code in Sprinzl's tRNA bank)	Source	R_1	R_3	R_4	Nr in Limbach, 1994
1	Ψ	pseudouridine (P)	A,B,E	H	H	H	50
2	Ψm	2'-O-methylpseudouridine (Z)	A,E	CH_3	H	H	56
3	m^1Ψ	1-methylpseudouridine (])	A	H	H	CH_3	55
4	m^3Ψ	3-methylpseudouridine	B	H	CH_3	H	-
5	m^1acp^3Ψ	1-methyl-3-(3-amino-3-carboxypropyl)pseudouridine	E	H	acp	CH_3	86

Modified cytidines

#	Symbol	Name (one letter code in Sprinzl's tRNA bank)	Source	R_1	R_2	R_3	Nr in Limbach, 1994
1	Cm	2'-O-methylcytidine (B)	A,B,E	CH_3	H	H	22
2	m^4C	N^4-methylcytidine	E	H	CH_3	H	82
3	m^4Cm	N^4,2'-O-dimethylcytidine	B	CH_3	CH_3	H	83
4	ac^4C	N^4-acetylcytidine (M)	A,B,E	H	$-COCH_3$	H	24
5	ac^4Cm	N^4-acetyl-2'-O-methylcytidine	A	CH_3	$-COCH_3$	H	27
6	m^5C	5-methylcytidine (?)	A,E	H	H	CH_3	21
7	m^5Cm	5,2'-O-dimethylcytidine	A	CH_3	H	CH_3	26
8	hm^5C	5-hydroxymethylcytidine	E	H	H	$-CH_2OH$	84
9	f^5C	5-formylcytidine (>)	E	H	H	-CHO	25
10	f^5Cm	2'-O-methyl-5-formylcytidine (°)	E	CH_3	H	-CHO	-
11	m^3C	3-methylcytidine (')	B,E	-	-	-	20
12	s^2C	2-thiocytidine (%)	A,B	-	-	-	23
13	k^2C	lysidine (})	B,E	-	-	-	28

APPENDIX 1 · CHEMICAL STRUCTURES OF MODIFIED NUCLEOSIDES 545

Modified uridines

3-amino-3-carboxypropyl (acp)

#1-6, 8-29   #7   #30-31

#	Symbol	Name (one letter code in Sprinzl's tRNA bank)	Source	R$_1$	R$_2$	R$_3$	R$_4$	Nr in Limbach, 1994
1	Um	2'-O-methyluridine (J)	A,B,E	CH$_3$	O	H	H	53
2	s^{2}U	2-thiouridine (2)	A,E	H	S	H	H	57
3	s^{2}Um	2-thio-2'-O-methyluridine	A	CH$_3$	S	H	H	60
4	m^{3}U	3-methyluridine	B	H	O	CH$_3$	H	85
5	m^{3}Um	3,2'-O-dimethyluridine	E	CH$_3$	O	CH$_3$	H	92
6	acp^{3}U	3-(3-amino-3-carboxypropyl)uridine (X)	B	H	O	acp	H	61
7	s^{4}U	4-thiouridine (4)	A,B	-	-	-	-	58
8	m^{5}U	ribosylthymine (T)	A,B,E	H	O	H	CH$_3$	52
9	m^{5}Um	5, 2'-O-dimethyluridine (\)	B,E	CH$_3$	O	H	CH$_3$	54
10	m^{5}s^{2}U	5-methyl-2-thiouridine (F)	A,B	H	S	H	CH$_3$	59
11	ho^{5}U	5-hydroxyuridine	B,E	H	O	H	OH	62
12	mo^{5}U	5-methoxyuridine (5)	B	H	O	H	-OCH$_3$	63
13	cmo^{5}U	uridine 5-oxyacetic acid (V)	B	H	O	H	-OCH$_2$COOH	64
14	mcmo^{5}U	uridine 5-oxyacetic acid methyl ester	B	H	O	H	-OCH$_2$COOCH$_3$	65
15	cm^{5}U	5-carboxymethyluridine	E	H	O	H	-CH$_2$COOH	87
16	mcm^{5}U	5-methoxycarbonylmethyluridine (1)	E	H	O	H	-CH$_2$COOCH$_3$	68
17	mcm^{5}Um	5-methoxycarbonylmethyl-2'-O-methyluridine	E	CH$_3$	O	H	-CH$_2$COOCH$_3$	69
18	mcm^{5}s^{2}U	5-methoxycarbonylmethyl-2-thiouridine (3)	E	H	S	H	-CH$_2$COOCH$_3$	70
19	ncm^{5}U	5-carbamoylmethyluridine (&)	E	H	O	H	-CH$_2$CONH$_2$	75
20	ncm^{5}Um	5-carbamoylmethyl-2'-O-methyluridine (~)	E	CH$_3$	O	H	-CH$_2$CONH$_2$	76
21	chm^{5}U	5-(carboxyhydroxymethyl)uridine	E	H	O	H	-CH(OH)COOH	66
22	mchm^{5}U	5-(carboxyhydroxymethyl)uridinemethyl ester (,)	E	H	O	H	-CH(OH)COOCH$_3$	67
23	nm^{5}s^{2}U	5-aminomethyl-2-thiouridine	B	H	S	H	-CH$_2$NH$_2$	71
24	mnm^{5}U	5-methylaminomethyluridine ({)	B	H	O	H	-CH$_2$NHCH$_3$	72
25	mnm^{5}s^{2}U	5-methylaminomethyl-2-thiouridine (S)	B	H	S	H	-CH$_2$NHCH$_3$	73
26	mnm^{5}se^{2}U	5-methylaminomethyl-2-selenouridine	A,B	H	Se	H	-CH$_2$NHCH$_3$	74
27	cmnm^{5}U	5-carboxymethylaminomethyluridine (!)	B,E	H	O	H	-CH$_2$NHCH$_2$COOH	77
28	cmnm^{5}Um	5-carboxymethylaminomethyl-2'-O-methyluridine (})	B	CH$_3$	O	H	-CH$_2$NHCH$_2$COOH	78
29	cmnm^{5}s^{2}U	5-carboxymethylaminomethyl-2-thiouridine ($)	B	H	S	H	-CH$_2$NHCH$_2$COOH	79
30	D	dihydrouridine (D)	A,B,E	H	O	H	H	51
31	m^{5}D	dihydroribosylthymine	E	H	O	H	CH$_3$	93

Modified adenosines

#	Symbol	Name (one letter code in Sprinzl's tRNA bank)	Source	R_1	R_2	R_3	R_4	Nr in Limbach, 1994
		Adenosine derivatives						
1	Am	2'-O-methyladenosine (:)	A,E	CH_3	H	H	H	4
2	m^2A	2-methyladenosine (/)	B	H	CH_3	H	H	2
3	m^6A	N^6-methyladenosine (=)	A,B,E	H	H	H	CH_3	3
4	m^6_2A	N^6,N^6-dimethyladenosine	A,B,E	H	H	CH_3	CH_3	80
5	m^6Am	$N^6,2'$-O-dimethyladenosine	E	CH_3	H	H	CH_3	88
6	m^6_2Am	$N^6,N^6,2'$-O-trimethyladenosine	E	CH_3	H	CH_3	CH_3	89
7	ms^2m^6A	2-methylthio-N^6-methyladenosine	B	H	CH_3-S-	H	CH_3	5
8	i^6A	N^6-isopentenyladenosine (+)	E	H	H	H	Dimethylallyl	6
9	ms^2i^6A	2-methylthio-N^6-isopentenyladenosine (*)	B	H	CH_3-S-	H	Dimethylallyl	7
10	io^6A	N^6-(cis-hydroxyisopentenyl)adenosine (`)	E	H	H	H	cis-hydroxy-methyl-methylallyl	8
11	ms^2io^6A	2-methylthio-N^6-(cis-hydroxyisopentenyl)-adenosine	E	H	CH_3-S-	H	cis-hydroxy-methyl-methylallyl	9
12	g^6A	N^6-glycinylcarbamoyladenosine	E	H	H	H	-CONHCH$_2$COOH	10
13	t^6A	N^6-threonylcarbamoyladenosine (6)	A,B,E	H	H	H	threonylcarbamoyl	11
14	m^6t^6A	N^6-methyl-N^6-threonylcarbamoyladenosine (E)	B	H	H	CH_3	threonylcarbamoyl	13
15	ms^2t^6A	2-methylthio-N^6-threonylcarbamoyl-adenosine ([)	A,B,E	H	CH_3-S-	H	threonylcarbamoyl	12
16	hn^6A	N^6-hydroxynorvalylcarbamoyladenosine	A,B	H	H	H	hydroxynorvalyl-carbamoyl	14
17	ms^2hn^6A	2-methylthio-N^6-hydroxynorvalylcarbamoyl-adenosine	A,B	H	CH_3-S-	H	hydroxynorvalyl-carbamoyl	15
18	Ar(p)	2'-O-ribosyladenosine (phosphate) (^)	E	2'-O-ribosyl	H	H	H	16
19	m^1A	1-methyladenosine (")	A,B,E	-	-	-	-	1
		Inosine derivatives						
20	I	inosine (I)	A,B,E	H	H	-	-	17
21	Im	2'-O-methylinosine	E	CH_3	H	-	-	81
22	m^1I	1-methylinosine (O)	A,E	H	CH_3	-	-	18
23	m^1Im	1,2'-O-dimethylinosine	A	CH_3	CH_3	-	-	19

Dimethylallyl (isopentenyl)

cis-hydroxymethyl-methylallyl (cis-hydroxyisopentenyl)

threonylcarbamoyl

hydroxynorvalylcarbamoyl

2'-O-ribosyl

Modified guanosines

#1-7   #8-10   #11-16   #17-23

#	Symbol	Name (one letter code in Sprinzl's tRNA bank)	Source	R_1	R_2	R_3	R_4	Nr in Limbach, 1994
		Guanosine derivatives						
1	Gm	2'-O-methylguanosine (#)	A,B,E	CH_3	H	H	H	32
2	m^1G	1-methylguanosine (K)	A,B,E	H	H	H	CH_3	29
3	m^2G	N^2-methylguanosine (L)	A,E	H	CH_3	H	H	30
4	m^2_2G	N^2,N^2-dimethylguanosine (R)	A,E	H	CH_3	CH_3	H	33
5	m^2Gm	N^2,2'-O-dimethylguanosine	A	CH_3	CH_3	H	H	34
6	m^2_2Gm	N^2,N^2,2'-O-trimethylguanosine (I)	A	CH_3	CH_3	CH_3	H	35
7	Gr(p)	2'-O-ribosylguanosine (phosphate)	E	2-O-ribosyl	H	H	H	36
		m^7-Guanosine derivatives						
8	m^7G	7-methylguanosine (7)	A,B,E	-	H	H	-	31
9	$m^{2,7}G$	N^2,7-dimethylguanosine	E	-	CH_3	H	-	90
10	$m^{2,2,7}G$	N^2,N^2,7-trimethylguanosine	E	-	CH_3	CH_3	-	91
		Wyosine derivatives						
11	imG	wyosine	E	-	-	-	H	41
12	mimG	methylwyosine	A	-	-	-	CH_3	42
13	OHyW*	undermodified hydroxywybutosine	E	-	-	-	S1	40
14	yW	wybutosine (Y)	E	-	-	-	S2	37
15	OHyW	hydroxywybutosine	E	-	-	-	β-hydroxy-S2	39
16	o_2yW	peroxywybutosine (W)	E	-	-	-	β-peroxy-S2	38
		7-deazaguanosine derivatives						
17	Q	queuosine (Q)	B,E	-	-	-	S3	43
18	oQ	epoxyqueuosine	B	-	-	-	epoxy-S3	44
19	galQ	galactosyl-queuosine (9)	E	-	-	-	gal-S3	45
20	manQ	mannosyl-queuosine (8)	E	-	-	-	man-S3	46
21	$preQ_0$	7-cyano-7-deazaguanosine	B	-	-	-	-CN	47
22	$gQ (G^+)$	Archaeosine (alternate name 7-formamidino-7-deazaguanosine) (())	A	-	-	-	-C(=NH)NH$_2$	49
23	$preQ_1$	7-aminomethyl-7-deazaguanosine	B	-	-	-	$-CH_2NH_2$	48

S1   S2   ß-hydroxy-S2   ß-peroxy-S2   2'-O-ribosyl

S3   epoxy-S3   man-S3   gal-S3

APPENDIX 1 • CHEMICAL STRUCTURES OF MODIFIED NUCLEOSIDES

Distribution of modified nucleotides by RNA type

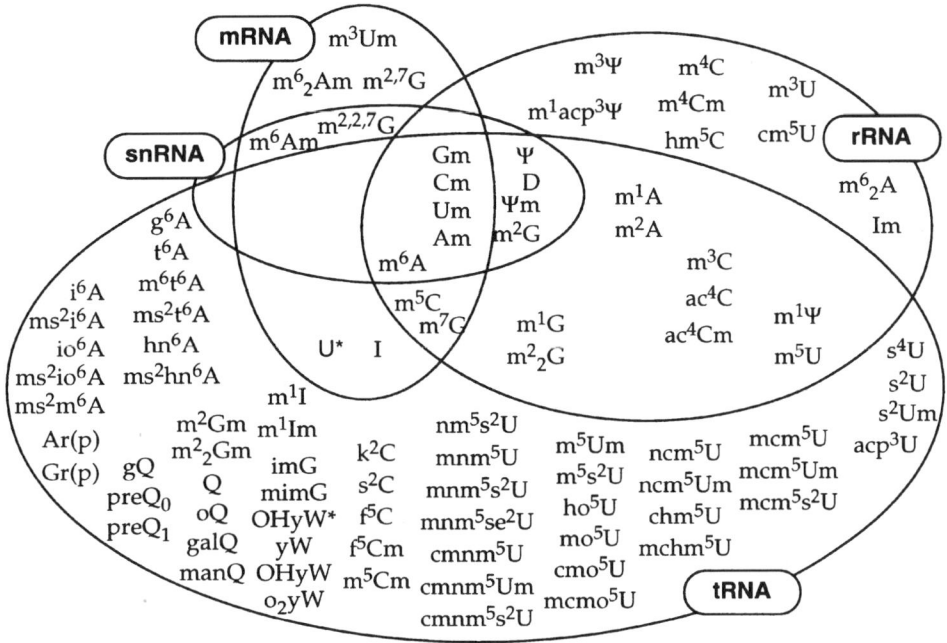

Phylogenetic distribution

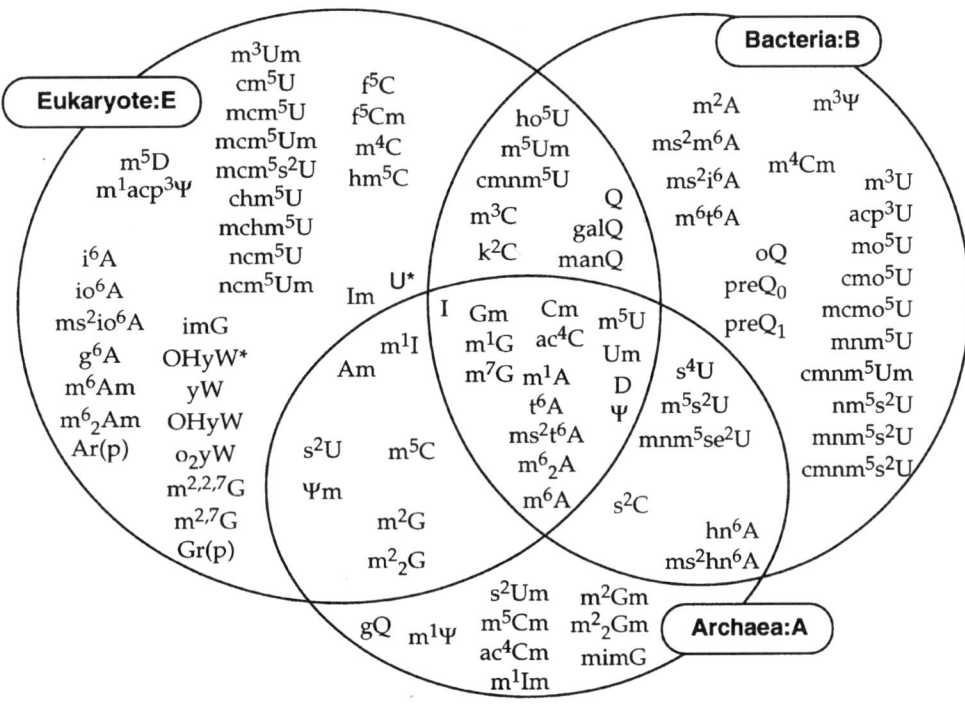

U* corresponds to the uridine that is formed by enzymatic deamination of the encoded cytosine by postranscriptional editing phenomena

Appendix 2: RNA Editing Types and Characteristics

ROB BENNE AND DAVE SPEIJER

On the following two pages, Table A1 shows the types and characteristics of RNA editing.

REFERENCES

Novo, F. J., A. Kruszewski, K. D. MacDermot, G. Goldspink, and D. C. Gorecki. 1995. Editing of human alpha-galactosidase RNA resulting in a pyrimidine to purine conversion. *Nucleic Acids Res.* 23:2636–2640.

Ojala, D., J. Montoya, and G. Attardi. 1981. tRNA punctuation model of RNA processing in human mitochondria. *Nature* 290:470–474.

Rajput, B., J. Ma, and I. K. Vijaj. 1994. Structure and organization of mouse GlcNac-1-phosphate transferase gene. *J. Biol. Chem.* 269:9590–9597.

Sharma, P. M., M. Bowman, S. L. Madden, F. J. Rauscher III, and S. Sukumar. 1994. RNA editing in the Wilms' tumor susceptibility gene, WT1. *Genes Dev.* 8:720–731.

Volchkov, V. E., S. Becker, V. A. Volchkova, V. A. Ternovoj, A. N. Kotov, S. V. Netesov, and H. D. Klenk. 1995. GP mRNA of Ebola virus is edited by the Ebola virus polymerase and by T7 and vaccinia virus polymerases. *Virology* 214:421–430.

Rob Benne and Dave Speijer • Department of Biochemistry, Academic Medical Center, University of Amsterdam, Meibergdreef 15, 1105 AZ Amsterdam, The Netherlands.

Table A1. RNA editing types and characteristics[a]

Type[b]	Organism (genome)	Transcript(s)[c]	cis-acting element(s)	trans-acting factor(s)	Mechanism	Reference[d]
Insertion/deletion editing						
U insertion/deletion	Kinetoplastids (mt)	mRNAs	Anchoring sequence	gRNAs, TUTase, RNA ligase, endonuclease, U exonuclease, other factors	Cleavage, TUTase or U exonuclease action, and ligation	Chapter 21
Mostly C insertion; also U, UA, AA, CU, GU, and GC	*Physarum polycephalum* (mt)	mRNAs, tRNAs, rRNA	?	?	Linked to transcription	Chapter 22
G insertion	Paramyxoviruses (v)	P mRNA	Slippery sequence	Viral polymerase	Pseudotemplated transcription	Chapter 23
A insertion	Ebola viruses (v)	GP mRNA	Slippery sequence	Viral polymerase	Pseudotemplated transcription	Volchov et al., 1995; see Chapter 23
3'-terminal A addition	Vertebrates (mt)	mRNAs	Flanking tRNA structure	Endonuclease, TATase	Cleavage and TATase action	Ojala et al., 1981
3'-terminal A addition[e]	Metazoan animals (mt)	tRNAs	3'-overlapping tRNA?	Endonuclease, TATase?	Cleavage and TATase action	Chapter 16
C to A, A to G, U to G, and U to A	*Acanthamoeba castellanii*, chytridiomycete fungi (mt)	tRNAs	Internal guide sequence?	Endo- or exonuclease, nucleotidyltransferase?	Replacement of first three 5' nt	Chapter 16
Modification editing						
C to U	Land plants	mRNAs (mt, cp), tRNAs (mt), rRNAs (mt)	Flanking sequence	?	C deamination	Chapter 17
	Mammals (n)	mRNAs, apoB (Gln → stop), NF1 (Arg → stop)	Mooring sequence, efficiency and AU-rich elements	C deaminase (APOBEC-1), other factors	C deamination	Chapter 18
	Physarum polycephalum (mt)	cox1 mRNA	?	?	?	Chapter 22
	Marsupials (mt)	tRNA (Gly → Asp anticodon)	?	?	?	Chapter 16

APPENDIX 2 • EDITING TYPES AND CHARACTERISTICS   553

			Flanking sequence			
U to C	Land plants (mt, cp)	mRNAs	?	?	U amination?	Chapter 17
	Mammals (n)	WT1 mRNA (Leu → Pro)	?	?	?	Sharma et al., 1994
A to I	Mammals (n)	mRNAs, GluR-B, -C, -D, -5, -6, 5-HT$_{2c}$R	dsRNA structure	dsRNA A deaminases (ADAR1 and -2)	A deamination	Chapter 19
	Human hepatitis delta virus (v)	Antigenome (stop → Trp)	dsRNA structure	ADAR1	A deamination	Chapter 19
A to I?[f]	Squids (n)	Kv2 K$^+$ channel mRNA	?	?	A deamination?	Chapter 19
	Drosophila melanogaster (n)	4f-rnp mRNA	?	?	A deamination?	Chapter 19
G to A[d]	Mice (n)	GPT mRNA (Cys → Tyr)	?	?	Base or nt replacement?	Rajput et al., 1994
U to A	Humans (n)	α-Galactosidase mRNA (Phe → Tyr)	?	?	Base or nt replacement?	Novo et al., 1995

[a] ADAR, A deaminase for RNA; apoB, apolipoprotein B; cox1, cytochrome *c* oxidase subunit 1; GluR, glutamate receptor; GP, glycoprotein; GPT, GlcNac-1-phosphate transferase; 5-HT$_{2c}$R, serotonin 2C receptor; NF1, neurofibromatosis type 1 gene; TATase, terminal adenylyltransferase; TUTase, terminal uridylyltransferase; WT1, Wilms' tumor susceptibility gene 1; cp, chloroplast; ds, double stranded; mt, mitochondrial; n, nuclear; nt, nucleotide; v, viral; ?, uncertain or unknown.
[b] Insertion/deletion editing is defined as an editing process during which phosphodiester bonds are made and/or broken, resulting in almost all cases in an edited RNA in which the number of nucleotides has changed. In modification editing, the identity of a nucleotide is changed by base conversion or substitution without disruption of the phosphodiester backbone. The status of G-to-A and U-to-A editing in two mammalian RNAs is unclear: it could be the result of either base substitution (in which case it would qualify as modification editing) or nucleotide replacement (in which case it should be considered insertion/deletion editing).
[c] For modification editing, the translational consequences of editing have been indicated for RNAs that contain a single editing site.
[d] For other references, please check the corresponding chapters.
[e] In platypuses, A's and C's are added.
[f] A comparison of genomic and cDNA sequences shows multiple apparent A-to-G changes, suggesting that the RNAs are edited by A-to-I conversion (see Chapter 19). The presence of I's in edited RNAs has not yet been established, however.
[g] This (putative) editing event has been reported for the NIH/SW mouse strain. Whether it occurs in other strains and species is unknown.

Appendix 3: General Properties of RNA-Modifying and -Editing Enzymes

GEORGE A. GARCIA AND DEEANNE M. GOODENOUGH-LASHUA

Table A1 summarizes the properties of RNA-modifying and -editing enzymes.

REFERENCES

Aschhoff, H. J., H. Elten, H. H. Arnold, G. Mahal, W. Kersten, and H. Kersten. 1976. 7-Methylguanine specific tRNA-methyltransferase from *Escherichia coli*. *Nucleic Acids Res.* **3**:3109–3122.

Becker, H. F., Y. Motorin, M. Sissler, C. Florentz, and H. Grosjean. 1997. Major identity determinants for enzymatic formation of ribothymidine and pseudouridine in the TΨ-loop of yeast tRNAs. *J. Mol. Biol.* **274**:505–518.

Bokar, J. A., M. E. Rath-Shambaugh, R. Ludwiczak, P. Narayan, and F. Rottman. 1994. Characterization and partial purification of mRNA N^6-adenosine methyltransferase from HeLa cell nuclei. Internal mRNA methylation requires a multisubunit complex. *J. Biol. Chem.* **269**:17697–17704.

Byström, A. Personal communication.

Cimino, F., C. Traboni, A. Colonna, P. Izzo, and F. Salvatore. 1981. Purification and properties of several transfer RNA methyltransferases from *S. typhimurium*. *Mol. Cell. Biochem.* **36**:95–104.

Colonna, A., G. Ciliberto, R. Santamaria, F. Cimino, and F. Salvatore. 1983. Isolation and characterization of a tRNA (guanine-7)-methyltransferase from *Salmonella typhimurium*. *Mol. Cell. Biochem.* **52**:97–106.

Ensinger, M. J., and B. Moss. 1976. Modification of the 5′ terminus of mRNA by an RNA (guanine-7)-methyltransferase from HeLa cells. *J. Biol. Chem.* **251**:5283–5291.

Glick, J. M., V. M. Averyhart, and P. S. Leboy. 1978. Purification and characterization of two tRNA-(guanine)-methyltransferases from rat liver. *Biochim. Biophys. Acta* **518**:158–171.

Glick, J. M., and P. S. Leboy. 1977. Purification and properties of tRNA (adenine-1)-methyltransferase from rat liver. *J. Biol. Chem.* **252**:4790–4795.

Grosjean, H. Personal communication.

Herbert, A., K. Lowenhaupt, J. Spitzner, and A. Rich. 1995. Chicken double-stranded RNA adenosine deaminase has apparent specificity for Z-DNA. *Proc. Natl. Acad. Sci. USA* **92**:7550–7554.

Hurwitz, J., M. Gold, and M. Arders. 1964. The enzymatic methylation of ribonucleic acid and deoxyribonucleic acid. IV. The properties of the soluble ribonucleic acid methylating enzymes. *J. Biol. Chem.* **239**:3474–3482.

Izzo, P., and R. Gantt. 1977. Partial purification and characterization of an N2-guanine RNA methyltransferase from chicken embryos. *Biochemistry* **16**:3576–3581.

Kim, U., T. L. Garner, T. Sanford, D. Speicher, J. M. Murray, and K. Nishikura. 1994. Purification and characterization of double-stranded RNA adenosine deaminase from bovine nuclear extracts. *J. Biol. Chem.* **269**:13480–13489.

Lecointe, F., G. Simons, A. Sauer, E. Hurt, Y. Motorin, and H. Grosjean. 1998. Characterization of yeast protein DegI as pseudouridine synthase (Pus3) catalyzing the formation of Ψ38 and Ψ39 in tRNA anticodon loop. *J. Biol. Chem.* **273**:1316–1323.

Locht, C., J. L. Beaudart, and J. Delcour. 1983. Partial purification and characterization of mRNA (guanine-7) methyltransferase from the yeast Saccharomyces cerevisiae. *Eur. J. Biochem.* **134**:117–121.

Martin, S. A., and B. Moss. 1975. Modification of RNA by mRNA guanylyltransferase and mRNA (guanine-7)-methyltransferase from vaccinia virions. *J. Biol. Chem.* **250**:9330–9335.

Martin, S. A., E. Paoletti, and B. Moss. 1975. Purification of mRNA guanylyltransferase and mRNA (guanine-7)-methyltransferase from vaccinia virions. *J. Biol. Chem.* **250**:9322–9329.

Mutzel, R., D. Malchow, D. Meyer, and H. Kersten. 1986. tRNA (adenine-N1)-methyltransferase from *Dictyostelium discoideum*. Purification, characterization and developmental changes in activity. *Eur. J. Biochem.* **160**:101–108.

Nau, F., G. Pham-Coeur-Joly, and J. M. Dubert. 1983. A study of some molecular and kinetic properties of two tRNA methyltransferases from mouse plasmocytoma. *Eur. J. Biochem.* **130**:261–268.

O'Connell, M. A., A. Gerber, and W. Keller. 1997. Purification of human double-stranded RNA-specific editase 1 (hRED1) involved in editing of brain glutamate receptor B pre-mRNA. *J. Biol. Chem.* **272**:473–478.

Ofengand, J. Personal communication.

Persson, B., C. Gustafsson, D. Berg, and G. Björk. 1992. The gene for a tRNA modifying enzyme, m^5U54-methyltransferase, is essential for viability in *Escherichia coli*. *Proc. Natl. Acad. Sci. USA* **89**:3995–3998.

Poldermans, B., L. Roza, and P. H. Van Knippenberg. 1979. Studies on the function of two adjacent N6,N6-dimethyladenosines near the 3′ end of 16 S ribosomal RNA of *Escherichia coli*. III. Purification and properties of the methylating enzyme and methylase-30 S interactions. *J. Biol. Chem.* **254**:9094–9100.

Rottman, F. M., J. A. Bokar, P. Narayan, M. E. Shambaugh, and R. Ludwiczak. 1994. N^6-adenosine methylation in mRNA: substrate specificity and enzyme complexity. *Biochimie* **76**:1109–1114.

Samuelsson, T., and M. Olsson. 1990. Transfer RNA pseudouridine synthases in *Saccharomyces cerevisiae*. *J. Biol. Chem.* **265**:8782–8787.

Sipe, J. E., W. M. Anderson, Jr., C. N. Remy, and S. H. Love. 1972. Characterization of S-adenosylmethionine: ribosomal ribonucleic acid-adenine (N^6-) methyltransferase of *Escherichia coli* strain B. *J. Bacteriol.* **110**:81–91.

Taylor, M. J., and R. Grant. 1979. Partial purification and characterization of a ribonucleic acid N2-guanine methyltransferase associated with avian myeloblastosis virus. *Biochemistry* **18**:5253–5258.

George A. Garcia and DeeAnne M. Goodenough-Lashua • College of Pharmacy, University of Michigan, Ann Arbor, Michigan 48109-1065.

Table A1. General properties of RNA-modifying and -editing enzymes[a]

Function, enzyme, and organism(s)	Quaternary structure	Molecular mass (kDa)	Cofactor(s) and cosubstrate(s)[b]	Ion requirement(s)	pH optimum	Site(s) of modification	Truncated RNAs as substrates	Reference(s)
Methylation								
tRNA (cytosine-5-)-methyltransferase (EC 2.1.1.29, m^{5}C)								
Escherichia coli			AdoMet			tRNA		Hurwitz et al., 1964
Mouse		110–140	AdoMet	Monovalent cation	8–9	tRNA		Nau et al., 1983
Human (HeLa cell)		72	AdoMet		7.3	tRNA-48/49	No	Chapter 8
tRNA (guanine-N^1-)-methyltransferase (EC 2.1.1.31, m^{1}G)								
E. coli		31	AdoMet	Mg^{2+}, Ca^{2+}, Mn^{2+} (stimulatory)	8–8.5	tRNA-37	Yes	Chapter 8; Hurwitz et al., 1964
Salmonella typhimurium		50	AdoMet			tRNA		Cimino et al., 1981
Rat		83	AdoMet	Putrescine	8	tRNA		Glick et al., 1978
tRNA (guanine-N^2-)-methyltransferase (EC 2.1.1.32, m^{2}G)								
Rat		69	AdoMet	Spermidine	8	tRNA-10		Glick et al., 1978
Chicken		77	AdoMet	Mg^{2+}, NH$_4^+$ (stimulatory)	8–8.5	tRNA-10		Izzo and Gantt, 1977
tRNA (guanine-N^7-)-methyltransferase (EC 2.1.1.33, m^{7}G)								
E. coli		10–30	AdoMet	Mg^{2+} (stimulatory)	7.5–8	tRNA-46		Hurwitz et al., 1964; Aschhoff et al., 1976
S. typhimurium	Monomeric	25–30	AdoMet	Mono- or divalent cation		tRNA		Colonna et al., 1983
tRNA (guanosine-2'-O-)-methyltransferase (EC 2.1.1.34, Gm)								
Thermus thermophilus			AdoMet			tRNA-18		Chapter 8
E. coli		25.3	AdoMet			tRNA-18	Yes	Chapter 8; Persson et al., 1992
tRNA (uracil-5-)-methyltransferase (EC 2.1.1.35, m^{5}U)								
E. coli		42	AdoMet	Monovalent cation	8, 8.5–9[c]	tRNA-54	Yes	Chapter 8; Hurwitz et al., 1964
Yeast		64.2	AdoMet			tRNA-54		Chapter 8; Byström, personal communication
Bacillus subtilis			5,10-Methylene THF, FADH$_2$			tRNA-54		Chapter 8

APPENDIX 3 • PROPERTIES OF MODIFYING AND EDITING ENZYMES 557

Enzyme	Subunit	MW	Cofactor	Ions	pH	Substrate	Other	Reference
tRNA (adenine-N^1-)-methyltransferase (EC 2.1.1.35, m^1A)								
T. thermophilus		60	AdoMet	Mg^{2+}		tRNA-58	Yes	Chapter 8
Tetrahymena pyriformis		210	AdoMet			tRNA-58	Yes	Chapter 8
Dictyostelium discoideum			AdoMet	Monovalent cation		tRNA-58		Mutzel et al., 1986
Mouse		200–230	AdoMet	Monovalent cation	7.3	tRNA-58		Nau et al., 1983
Rat		95	AdoMet	Putrescine, NH_4^+	8	tRNA-58		Glick and Leboy, 1977
rRNA (adenine-N^6-)-methyltransferase (EC 2.1.1.48, m^6A), *E. coli*			AdoMet	Mg^{2+}, monovalent (stimulatory)	7.5–8.2	23S rRNA, 16S rRNA		Sipe et al., 1972
mRNA (guanine-N^7-)-methyltransferase (EC 2.1.1.56, m^7G [cap])								
Vaccinia virus		95	AdoMet	Mg^{2+}	Broad (around 7)	mRNA		Martin and Moss, 1975; Martin et al., 1975
Yeast		49	AdoMet		7	mRNA		Locht et al., 1983
Human (HeLa cell)		56	AdoMet		7.2–7.8	mRNA		Ensinger and Moss, 1976
mRNA (nucleoside 2′-O-)-methyltransferase (EC 2.1.1.57, Nm [cap]), vaccinia virus	Monomeric	39	AdoMet	No divalent cations	7.5	mRNA (cap)		Chapters 8 and 13
mRNA (2′-O-methyladenosine-N^6-)-methyltransferase (EC 2.1.1.62, m^6Am), human (HeLa cell)			AdoMet		7.25	mRNA		Chapter 13
rRNA (adenosine-2′-O-)-methyltransferase (EC 2.1.1.66, Am), *Staphylococcus aureus*		35–38	AdoMet		7.5–7.6	23S rRNA		Chapter 13
tRNA (guanine-N^2,N^2-)-methyltransferase (EC 2.1.1.–, m_2^2G)								
S. cerevisiae	α_4?	63	AdoMet			tRNA-26		Chapter 8
T. pyriformis		200–250	AdoMet					Chapter 8
Avian myeloblastosis virus		220	AdoMet (stimulatory)	Mg^{2+}, NH_4^+	7.6–7.9	tRNA		Taylor and Gantt, 1979
mRNA (adenosine-N^6-)-methyltransferase (EC 2.1.1.–, m^6A), human (HeLa cell)	$\alpha\beta\gamma$	1,105 (30, 200, 875)	AdoMet			mRNA		Bokar et al., 1994; Rottman et al., 1994

Table A1. Continued

Function, enzyme, and organism(s)	Quaternary structure	Molecular mass (kDa)	Cofactor(s) and cosubstrate(s)[b]	Ion requirement(s)	pH optimum	Site(s) of modification	Truncated RNAs as substrates	Reference(s)
rRNA (adenosine-N^6,N^6)-methyltransferase (EC 2.1.1.–, m_2^6A)								
E. coli	Monomeric	30	AdoMet	Mg^{2+}	7.8	16S rRNA-1518/1519		Poldermans et al., 1979
S. cerevisiae	Monomeric	35.9	AdoMet			18S rRNA-1779/1780		Chapter 15
rRNA (cytosine-5-)-methyltransferase (EC 2.1.1.–, m^5C), *E. coli*		48.3	AdoMet			16S rRNA-967		Ofengand, personal communication
rRNA (guanine-N^2-)-methyltransferase (EC 2.1.1.–, m^2G), *E. coli*		37.6	AdoMet			16S rRNA-1207		Ofengand, personal communication
Nucleolar 2′-O-methyltransferase (EC 2.1.1.–, Nm), mouse (tumor)		50–150, 130[c]	AdoMet	None	7.5	Cellular, nucleolar, and 45S rRNA		Chapters 8 and 13
Deamination								
tRNA adenosine deaminase (EC 3.5.4.4, I), yeast		75 46	None None	Mg^{2+}		tRNA-34 tRNA-37	No	No Chapter 8; Grosjean, personal communication
Double-stranded RNA adenosine deaminase (EC 3.5.4.–, I)								
Xenopus laevis		120		Zn^{2+} (catalytic?)	6.5–8	Double-stranded RNA		Chapters 8, 19, and 20
Chicken (lung)		140						Herbert et al., 1995
Bovine (liver)		83, 88, 93						Kim et al., 1994
Bovine (thymus)	Monomeric	100, 116		Zn^{2+} (catalytic ?)		mRNA		Chapters 8, 19, and 20
Human (HeLa cell)		90						O'Connell et al., 1997
mRNA cytidine deaminase (EC 3.5.4.–, C to U)				Zn^{2+} (catalytic ?)				Chapter 20 and Appendix 2

Isomerization

Enzyme	MW (kDa)	Substrate	Cofactor/Ion	pH	Site	Other	Reference
Pseudouridine synthase I (EC 5.4.99.12, Ψ)							
E. coli	31		Monovalent cation or Mg^{2+}	9	tRNA-38, -39, -40	No	Chapters 8 and 12
S. cerevisiae	51		Zn^{2+} (structural)		tRNA-38, -39, not -40	No	Chapter 12; Lecointe, 1998
	62				tRNA-27, -28, -34, -35, -36	Yes	Chapters 8 and 12
Bovine (thymus)	180	None			tRNA-38, -39, -40		Chapter 8
tRNA pseudouridine 13 synthase (EC 5.4.99.12, Ψ), yeast			K^+ or NH_4^+	5.5	tRNA-13		Samuelsson and Olsson, 1990
tRNA pseudouridine 55 synthase (EC 5.4.99.12, Ψ)							
E. coli	35, 40		No Mg^{2+}		tRNA-55	Yes	Chapter 12
Yeast	45, 55		NaCl, Mg^{2+}	7.5	tRNA-55	Yes	Chapter 12; Becker, 1997
16S pseudouridine 516 synthase (EC 5.4.99.–, Ψ), E. coli	25.9				16S rRNA-516		Chapter 12
23S pseudouridine 2605 synthase (EC 5.4.99.–, Ψ), B. subtilis	26				23S rRNA-2605		Chapter 12
Dual-specificity pseudouridine synthase (EC 5.4.99.–, Ψ), E. coli	24.9		No Mg^{2+}		tRNA-32, 23S rRNA-746	Yes	Chapters 8 and 12

Alkylation

Enzyme	MW (kDa)	Substrate	Cofactor/Ion	pH	Site	Other	Reference
Dimethylallyl diphosphate tRNA dimethylallyltransferase (EC 2.5.1.8, i^6A)							
E. coli	33.5, 55[c]	Dimethylallyl pyrophosphate	Divalent cation, Mg^{2+}	8, 6.5–9[c]	tRNA-37		Chapters 8 and 25
Lactobacillus acidophilus		Dimethylallyl pyrophosphate	Mg^{2+}	7.5–8	tRNA-37		Chapter 8
Yeast	57–63	Dimethylallyl pyrophosphate	Mg^{2+}	7.5–8	tRNA-37		Chapter 8
Zea mays L.	Monomeric	Dimethylallyl pyrophosphate	Mg^{2+}	7.8	tRNA-37		Chapter 8
N-(Purin-6-ylcarbamoyl)-threonine synthetase (t^6A), E. coli	22, 50–60[c]	L-Threonine, ATP, bicarbonate	Mg^{2+}	7.7–8.2	tRNA-37	No	Chapter 8

Table A1. Continued

Function, enzyme, and organism(s)	Quaternary structure	Molecular mass (kDa)	Cofactor(s) and cosubstrate(s)[b]	Ion requirement(s)	pH optimum	Site(s) of modification	Truncated RNAs as substrates	Reference(s)
2′-O-Ribosylphosphate transferase [A/Gr(P)], yeast		57	PRPP			tRNA-64		Chapter 8
Hypermodification								
tRNA-guanine transglycosylase (EC 2.4.2.29, Q)								
E. coli	α, α_3	42.5	PreQ$_1$	Mg^{2+}, Zn^{2+} (structural)	6–8	tRNA-34	Yes	Chapters 8 and 9
Zymomonas mobilis	Monomeric	55	PreQ$_1$	Mg^{2+}, Zn^{2+} (structural)		tRNA-34		Chapters 8 and 9
Wheat (germ)	α_2	140	Queuine	Mg^{2+} or Na$^+$	7.6	tRNA-34		Chapter 8
Rat (liver)	Monomeric	80	Queuine, preQ$_1$	None		tRNA-34		Chapter 8
Rabbit	$\alpha\beta$	103 (43, 60)	Queuine	Monovalent cation		tRNA-34		Chapter 8
Bovine	$\alpha\beta$	96 (33, 66)	Queuine	None		tRNA-34		Chapter 8
tRNA-guanine transglycosylase (EC 2.4.2.29, archaeosine), Haloferax volcanii		78	PreQ$_0$			tRNA-15	No	Chapter 8
S-adenosylmethionine tRNA ribosyl transferase-isomerase (EC 2.4.2.-, Q), E. coli	Monomeric	39	AdoMet			tRNA-34	Yes	Chapters 8 and 9

[a] This table contains entries only for those modification enzymes which have been isolated and at least partially characterized. There are no entries where only a gene and/or modification activity has been reported. The enzymes are categorized by type of reaction catalyzed and then listed numerically by EC number within each category. Blank entries indicate that no report was found for a given property.
[b] AdoMet, S-adenosylmethionine; THF, tetrahydrofolate; FADH$_2$, reduced flavin adenine dinucleotide; PRPP, phosphoribosylpyrophosphate.
[c] Multiple entries indicate that different values were reported by different investigators.

Modification and Editing of RNA
Edited by Henri Grosjean and Rob Benne
© 1998 ASM Press, Washington, D.C.

Appendix 4: Genetic Locations and Database Accession Numbers of RNA-Modifying and -Editing Enzymes

HON-CHIU EASTWOOD LEUNG, TORD G. HAGERVALL, GLENN R. BJÖRK, AND
MALCOLM E. WINKLER

Tables A1 and A2 (following pages) show the genetic locations, designations, and database accession numbers of RNA-modifying and -editing enzymes.

Hon-Chiu Eastwood Leung and Malcolm E. Winkler • Department of Microbiology and Molecular Genetics, University of Texas Houston Medical School, Houston, Texas 77030-1501.   **Tord G. Hagervall and Glenn R. Björk** • Department of Microbiology, University of Umeå, S-90187 Umeå, Sweden.

Table A1. RNA modification

Base modification[a]	Type(s) of RNA[b]	Organism[c]	Former and present genetic symbol(s)	Suggested new genetic symbol for bacterial genes[d]	Nucleotide sequence accession no. (source)[e]	Genome contig	Map position or ORF[f]	SWISS-PROT or protein sequence accession no. (source)[e]
s⁴U8, s⁴U9	tRNA	Ec	nuvA	sfuA	NA[g]	NA	~9.0 min	NA
	tRNA	Ec	nuvC	sfuC	NA	NA	~44.0 min	NA
	tRNA	St	nuvA	sfuA	NA	NA	~9.0 min	NA
Gm18	tRNA	Ec	spoU	trmH	AE000442 (B)	ECD080.00	82.37 min	P19396 (D)
m²₂G26	tRNA	Sc	TRM1	TRM1	Z48758 (B)	NA	YDR120c	P15565 (D)
s²C32	tRNA	Ec	stcA	stcA	NA	NA	26-33 min	NA
s²U34 (in mnm⁵s²U34)	tRNA	Ec	asuE, ycfC, trmU	mnmA	AE000213 (B)	ECD025.00	25.67 min	P25746 (A)
mnm⁵s²U34	tRNA	Hi	ycfC	mnmA	HIU32747 (B)	NA	NA	P44796 (D)
	tRNA	Ec	trmC	mnmC	NA	NA	~50.0 min	NA
	tRNA	Ec	trmF	mnmG	NA	NA	~83.0 min	NA
cmnm⁵s²U34 (in mnm⁵s²U34)	tRNA	Ec	trmE, thdF	mnmE	AE000447 (B)	ECD080.00	83.71 min	P25522 (A)
	tRNA	Hi	HI1002	mnmE	U32781 (B)	NA	NA	NA
	tRNA	Pp	50K, thdF	mnmE	X62540 (B)	NA	NA	P25755 (D)
mnm⁵Se²U34 (in mnm⁵Se²U34)	tRNA	Ec	selD	selD	M30184 (B)	ECD035.00	39.76 min	P16456 (A)
	tRNA	Hi	selD	selD	U32705 (B)	NA	NA	A64054 (D)
	tRNA	Rc	selD	selD	X78346 (B)	NA	NA	1122427 (D)
	tRNA	Dd	selD	selD	U68248 (B)	NA	1.495	1657427 (D)
	tRNA	Dm	selD	selD	AJ000672 (B)	NA	2R50D/E	2292838 (D)
	tRNA	Hs	selD	selD	U34044 (B)	NA	Xq13	P49903 (D)
PreQ₁ (in Q34)	tRNA	Ec	tgt	tgt	M63939 (B)	ECD005.00	9.17 min	P19675 (A)
	tRNA	Sfl	vacC	tgt	D26469 (B)	NA	NA	599588 (D)
	tRNA	Hi	tgt	tgt	U32710 (B)	NA	NA	P44594 (D)
	tRNA	Zm	tgt	tgt	49283 (B)	NA	NA	P28720 (D)
	tRNA	Mj	tgt	tgt	U67495 (B)	NA	NA	D64354 (D)
oQ34 (in Q34)	tRNA	Ec	queA	queA	M37702 (B)	ECD005.00	9.14 min	P21516 (A)
	tRNA	Hi	queA	queA	U32710 (B)	NA	NA	P44595 (D)
	tRNA	Ss	queA	queA	D64004 (B)	NA	NA	1001768 (D)
	tRNA	Hp	queA	queA	AE000613 (B)	NA	NA	2314205 (D)

APPENDIX 4 • LOCATIONS OF MODIFYING AND EDITING ENZYMES   563

Q34	tRNA	Ce	tgt	tgt	Z73899 (B)	NA	Chromosome IV	1340027 (D)
m¹G37	tRNA	Ec	trmD	trmD	X01818 (B)	ECD055.00	59.10 min	P07020 (A)
	tRNA	St	trmD	trmD	X74933 (B)	NA	57.87 min	P36245 (C)
	tRNA	Mt	trmD	trmD	Z74024 (B)	NA	NA	Q10797 (D)
	tRNA	Sm	trmD	trmD	L23334 (B)	NA	NA	P36244 (D)
	tRNA	Mg	trmD	trmD	U39731 (B)	NA	NA	P47683 (D)
	tRNA	Hi	trmD	trmD	U32705 (B)	NA	NA	P43912 (D)
i⁶A37 (in ms²i⁶A37)	tRNA	Ec	miaA	miaA	M63655 (B)	ECD090.00	94.75 min	P16384 (A)
	tRNA	Ss	miaA	miaA	D90911 (B)	NA	NA	1653199 (D)
	tRNA	Hi	miaA	miaA	U32692 (B)	NA	NA	P44495 (D)
	tRNA	Mt	miaA	miaA	Z98209 (B)	NA	NA	2292961 (D)
	tRNA	Atu	miaA	miaA	M83532 (B)	NA	NA	P38436 (D)
	tRNA	St	miaA	miaA	SG10233 (C)	NA	95.01 min	P37724 (C)
	tRNA	Ml	miaA	miaA	U00019 (B)	NA	NA	P46811 (D)
	tRNA	Hp	miaA	miaA	AE000642 (B)	NA	NA	2314590 (D)
i⁶A37 (in i⁶A37)	tRNA	Sc	MOD5	MOD5	M15991 (B)	NA	YOR274w	P07884 (D)
	tRNA	Ce	MOD5	MOD5	U13642 (B)	NA	Chromosome III	532094 (D)
ms²i⁶A37	tRNA	Ec	f474, miaB	miaB	AE000170 (B)	ECD010.00	14.97 min	1786882 (D)
	tRNA	St	miaB	miaB	NA	NA	~15 min	NA
	tRNA	Hi	HI0019	miaB	U32687 (B)	NA	NA	1572963 (D)
ms²io⁶A37 (in ms²io⁶A37)	tRNA	St	miaE	miaE	X73368 (B)	NA	97.2 min	Q08015 (C)
m²A37	tRNA	Ec	trmG	trmG	NA	NA	11–26 min	NA
m⁶⁶A37	tRNA	Ec	mtaA	tsaA	NA	NA	4.6 min	NA
Inosine 37	tRNA	Sc	HRA400, SLA2	tad1	Z72765 (B)	NA	YGL243w	P53065 (D)
Ψ27, Ψ28, Ψ34, Ψ36	tRNA	Sc	PUS1	PUS1	X80673 (B)	NA	YPL212c	S65231 (D)
					Z73568 (B)			
Ψ38, Ψ39	tRNA	Sc	DEG1	PUS3	D44600 (B)	NA	YFL001w	P31115 (D)
Ψ38, Ψ39, Ψ40	tRNA	Ec	hisT	truA	M15542 (B)	ECD050.00	52.43 min	P07649 (A)
	tRNA	Hi	hisT	truA	U32837 (B)	NA	NA	P45291 (D)
	tRNA	Ss	hisT	truA	D90905 (B)	NA	NA	1652401 (D)
	tRNA	Bs	hisT	truA	NA	NA	NA	JT0612 (D)
	tRNA	Bb	hisT	truA	Y09141 (B)	NA	NA	1665842 (D)
	tRNA	Mpn	hisT	truA	AE000061 (B)	NA	NA	S73961 (D)
	tRNA	Mg	hisT	truA	U39695 (B)	NA	NA	B64220 (D)
	tRNA	Hp	hisT	truA	AE000553 (B)	NA	NA	2313463 (B)
	tRNA	Mj	hisT	truA	U67608 (B)	NA	NA	A64509 (D)

564 LEUNG ET AL.

Table A1. Continued

Base modification[a]	Type(s) of RNA[b]	Organism[c]	Former and present genetic symbol(s)	Suggested new genetic symbol for bacterial genes[d]	Nucleotide sequence accession no. (source)[e]	Genome contig	Map position or ORF[f]	SWISS-PROT or protein sequence accession no. (source)[e]
m⁵U54	tRNA	Ec	trmA	trmA	M57568 (B)	ECD085.00	89.65 min	P23003 (A)
	tRNA	Hi	trmA	trmA	U32766 (B)	NA	NA	P31812 (D)
	tRNA	Ng	trmA	trmA	M90807 (B)	NA	NA	P55134 (D)
	tRNA	St	trmA	trmA	M57569 (C)	NA	89.65 min	P22038 (C)
Ψ55	tRNA	Ec	truB, yhbA, P35	truB	42221 (B)	ECD065.00	71.32 min	P09171 (D)
	tRNA	Hi	truB	truB	U32809 (B)	NA	NA	P45142 (D)
	tRNA	Sc	PUS4	PUS4	Z71568 (B)	NA	YNL292w	P48567 (D)
(p)rA64	tRNA	Sc	RIT1	RIT1	Z49704 (B)	NA	YMR283c	P23796 (D)
Ψ32, Ψ746	tRNA, rRNA	Ec	rluA, yabO	rluA	AE000116 (B)	ECD000.00	1.29 min	P39219 (D)
Ψ516	rRNA	Ec	rsuA, bcr	rsuA	AE000308 (B)	ECD045.00	49.07 min	1042177 (D)
Ψ2605	rRNA	Bsu	rluB, ypuL	rluB	L09228 (B)	NA	206.8°	P35159 (D)
m¹G745	rRNA	Ec	rrmA, yebH	rrmA	AE000276 (B)	ECD040.00	41.04 min	P36999 (D)
m⁵C967	rRNA	Ec	rsmB, fmu	rsmB	X77091 (B)	ECD065.00	73.98 min	P36929 (D)
m²G1207	rRNA	Ec	yjjT, rsmC	rsmC	AE000507 (B)	ECD090.00	99.22 min	P39406 (D)
m²₂A1518, m⁶₂A1519	rRNA	Ec	ksgA	ksgA	M11054 (B)	ECD000.00	1.11 min	P06992 (D)
m⁶₂A1779, m⁶₂A1780	rRNA	Sc	DIM1	RRM1	L26480 (B)	NA	YPL266w	P41819 (D)
m²G2270	rRNA	Sc	PET56	RRM2	L19947 (B)	NA	YOR201c	P25270 (D)
methylation	RNA	Bf	ermFU	ermFU	M62487 (B)	NA	NA	S34413 (D)
Gm	rRNA	Mo	fmrO	fmrO	D13171 (B)	NA	NA	Q08325 (D)
Gm	rRNA	Mr	grm	rmg	M55521 (B)	NA	NA	P24619 (D)
Gm	rRNA	Mpu	grm	rmg	M55520 (B)	NA	NA	P24618 (D)
mnA methylation	rRNA	Se	ermE	ermE	X51891 (B)	NA	NA	1333825 (D)
	rRNA	Sfr	ermSF	ermSF	M19269 (B)	NA	NA	153253 (D)
Ψ (position unknown)[b]	rRNA	Ec	sfhB	sfhB	AE000346 (B)	ECD055.00	58.93 min	P33643 (D)
	rRNA	Ec	yceC	yceC	AE000209 (B)	ECD020.00	24.66 min	P23851 (D)
	rRNA	Ec	yciL	yciL	AE000225 (B)	ECD025.00	28.56 min	P37765 (D)
	rRNA	Ec	yjbC	yjbC	AE000475 (B)	ECD085.00	91.14 min	P32684 (D)
	rRNA	Ec	yqcB	yqcB	1255723 (B)	ECD060.00	62.98 min	Q46918 (D)
mRNA capping	mRNA	VV	N2L	N2L	L22579 (B)	NA	NA	P33808 (D)

	SV	ORF H2L	ORF H2L			
mRNA				L22012 (B)	NA	Q08512 (D)
mRNA	VC	D1	D1	M15058 (B)	NA	P20980 (D)
<u>mRNA</u>	VWR	D12	D12	X03729 (B)	NA	P04318 (D)
ssRNA	TVCV	NA	NA	U03387 (B)	NA	514836 (D)

[a] In order of positions in RNA molecules. The chemical moiety added by the enzyme is underlined.

[b] Underlining indicates true assignments for which substrate specificities of enzymes were confirmed experimentally; entries are putative assignments based solely on amino acid alignments. ssRNA, single-stranded RNA.

[c] Atu, *Agrobacterium tumefaciens*; Bb, *Borrelia burgdorferi*; Bfr, *Bacteroides fragilis*; Bs, *Bacillus* sp.; Bsu, *Bacillus subtilis*; Ce, *Caenorhabditis elegans*; Dd, *Dictyostelium discoideum*; Dm, *Drosophila melanogaster*; Ec, *Escherichia coli*; Hi, *Haemophilus influenzae*; Hp, *Helicobacter pylori*; Hs, *Homo sapiens*; Mg, *Mycoplasma genitalium*; Mj, *Methanococcus jannaschii*; Ml, *Mycobacterium leprae*; Mo, *Micromonospora olivasterospora*; Mr, *Micromonospora rosea*; Mpn, *Mycoplasma pneumoniae*; Mpu, *Micromonospora purpurea*; Mt, *Mycobacterium tuberculosis*; Ng, *Neisseria gonorrhoeae*; Pp, *Pseudomonas putida*; Rc, *Rhodobacter capsulatus*; Sc, *Saccharomyces cerevisiae*; Se, *Saccharopolyspora erythraea*; Sfl, *Shigella flexneri*; Sfr, *Streptomyces fradiae*; Sm, *Serratia marcescens*; Sp, *Schizosaccharomyces pombe*; Ss, *Synechocystis* sp.; St, *Salmonella typhimurium*; SV, swinepox virus; TVCV, turnip vein-clearing virus; VC, vaccinia virus strain Copenhagen; VV, variola virus; VWR, vaccinia virus strain WR; Zm, *Zymomonas mobilis*.

[d] Based on names of modified bases instead of phenotypes of modificationless mutants or other historical designations.

[e] A, *E. coli* Database Collection homepage (http://susi.bio.uni-giessen.de/ecdc.html); B, National Center for Biotechnology Information homepage, Nucleotides query (http://www.ncbi.nlm.nih.gov/Entrez/); C, K. E. Sanderson, A. Hessel, S.-L. Liu, and K. E. Rudd, "The genetic map of *Salmonella typhimurium*, edition VIII," p. 1903–1999, *in* F. C. Neidhardt et al. (ed.), *Escherichia coli* and *Salmonella: Cellular and Molecular Biology*, 2nd ed., ASM Press, Washington, D.C., 1996; D, National Center for Biotechnology Information homepage, Proteins query (http://www.ncbi.nlm.nih.gov/Entrez/).

[f] ORF, open reading frame.

[g] NA, not available.

[h] Taken from Table 6 of Chapter 12 by Ofengand and Fournier.

566 LEUNG ET AL.

Table A2. RNA editing

RNA editing type	Type of RNA[a]	Former and present genetic symbol(s)	Organism[b]	Enzyme	Nucleotide sequence accession no.[c]	Map position or ORF[d]	SWISS-PROT or protein sequence accession no.[e]
G insertion	mRNA	L	SeV F	RNA polymerase β subunit	X58886	NA[f]	Q06996
	mRNA	L	SeV Z	RNA polymerase β subunit	M69046	NA	P27566
	mRNA	L	RV	RNA polymerase β subunit	Z30697	NA	P41357
	mRNA	L	SPIV41	RNA polymerase β subunit	X64275	NA	P35341
	mRNA	L	PDV	RNA polymerase β subunit	D10371	NA	P35946
	mRNA	L	MeV AIK-C	RNA polymerase β subunit	S58435	NA	P35975
	mRNA	L	CDV	RNA polymerase β subunit	L13195	NA	P24658
	mRNA	L	SV5	RNA polymerase β subuit	D13868	NA	Q03396
	mRNA	L	MuV	RNA polymerase β subunit	D10575	NA	P30929
	mRNA	L	SeV E	RNA polymerase β subuit	D00053	NA	P06829
	mRNA	L	HPIV1	RNA polymerase β subunit	X03614	NA	P06447
	mRNA	L	HPIV2	RNA polymerase β subuit	X57559	NA	P26676
	mRNA	L	HPIV3	RNA polymerase β subunit	M21649	NA	P12577
	mRNA	L	NDV	RNA polymerase β subunit	X05399	NA	P11205
	mRNA	L	MeV E	RNA polymerase β subunit	M20865	NA	P12576
	mRNA	L	HRSV	RNA polymerase β subunit	M75730	NA	P28887
A insertion	mRNA	L	EV	RNA-directed RNA polymerase	L11365	NA	Q05318
	mRNA	L	MV	RNA-directed RNA polymerase β subunit	Z29337	NA	Q03039

APPENDIX 4 • LOCATIONS OF MODIFYING AND EDITING ENZYMES   567

C to U	mRNA	Mouse	*APOBEC-1*	ApoB mRNA editing protein	U22264	Chromosome 6	P51908
	mRNA	Rat	*APOBEC-1*	ApoB mRNA editing protein	L07114	NA	P38483
	mRNA	Hs	*APOBEC-1*	ApoB mRNA editing protein	L26234	12p13.1–p13.2	P41238
	mRNA	Rabbit	*APOBEC-1*	ApoB mRNA editing protein	U10695	NA	P47855
A to I	dsRNA	Rat	*RED1, rADAR2*	dsRNA-specific editase I	U43534	NA	P51400
	dsRNA	Hs	*hRED1, ADARB1, hADAR2*	dsRNA-specific editase	X99383	21q22.3	1707504
	dsRNA	Hs	*DRADA2a, hADAR2a*	dsRNA-specific adenosine deaminase	U76420	21q22.3	2039298
	dsRNA	Hs	*DRADA2b, hADAR2b*	dsRNA-specific adenosine deaminase	U76421	21q22.3	2039300
	dsRNA	Hs	*DRADA2c, hADAR2c*	dsRNA-specific adenosine deaminase	U76422	21q22.3	2039302
	dsRNA	Xla	*dsRAD-1, xADAR1.1*	dsRNA-specific adenosine deaminase	U88065	NA	1932813
	dsRNA	Xla	*dsRAD-2, xADAR1.2*	dsRNA-specific adenosine deaminase	U88066	NA	1932815
	dsRNA	Rat	*RED2, ADARB2*	dsRNA-specific adenosine deaminase	U74586	NA	1814270
	dsRNA	Rat	*DRADA, rADAR1*	dsRNA-specific adenosine deaminase	U18942	NA	P55266
	dsRNA	Hs	*DRADA, hADAR1*	dsRNA-specific adenosine deaminase	U10439	NA	P55265
	dsRNA	Chicken	*cADAR1*	dsRNA-specific adenosine deaminase homolog	NA	NA	1050971

[a] Underlining indicates true assignments for which substrate specificities of enzymes were confirmed experimentally. Other entries are putative assignments based solely on amino acid alignments.
[b] CDV, canine distemper virus strain Onderstepoort; EV, Ebola virus; HPIV1, -2, and -3, human parainfluenza virus types 1, 2, and 3, respectively; HRSV, human respiratory syncytial virus; Hs, *Homo sapiens*; MeV A and E, measles virus strains AIK-C and Edmonston, respectively; MuV, mumps virus strain Miyahara vaccine; MV, Marburg virus strain Popp; NDV, Newcastle disease virus; PDV, phocine distemper virus; RV, rinderpest virus strain RBOK; SeV E, F, and Z, Sendai virus strains Enders, Fushimi, and Z, respectively; SPIV41, simian parainfluenza virus 41; SV5, simian virus 5 strain 21004-WR; Xla, *Xenopus laevis*.
[c] National Center for Biotechnology Information homepage, Nucleotides query (http://www.ncbi.nlm.nih.gov/Entrez/).
[d] ORF, open reading frame.
[e] National Center for Biotechnology Information homepage, Proteins query (http://www.ncbi.nlm.nih.gov/Entrez/).
[f] NA, not available.

Appendix 5: Location and Distribution of Modified Nucleotides in tRNA

PASCAL AUFFINGER AND ERIC WESTHOF

Several excellent compilations and reviews devoted to modified nucleotides in tRNAs have been published (see, for example, Agris, 1996; Björk, 1995; Dirheimer, 1983; Dirheimer et al., 1995; Grosjean et al., 1982; Grosjean et al., 1995). The large number of known tRNA sequences, as well as the information on their modified nucleotides, makes tRNA one of the most interesting classes of RNA molecules from which phylogenetic information on modified and unmodified nucleotides can be extracted.

The present compilation gives an overview of the location and distribution of modified nucleotides in tRNAs from the 1998 release of the tRNA database containing 546 tRNA sequences and 2,726 sequences of tRNA genes (Sprinzl et al., 1998). Figure A1 shows the conventional numbering of tRNA nucleotides. Figure A2 shows the distribution of modified and unmodified nucleotides from the 546 tRNA sequences. Figure A3 shows the proportion of each of the four A, G, C, and T nucleotides from the 2,726 sequences of tRNA genes. Figure A4 details the distribution and location of the modified nucleotides which have been detected in several positions in tRNA. Table A1 groups all the modified nucleotides found at the hypermodified positions 34 and 37 in the anticodon loop (see also Appendix 6).

In this compilation, all tRNA sequences present in the tRNA database have been taken into account independently of the organisms from which they have been determined (as long as they could be fitted in the canonical tRNA alignment). A classification of the distribution of modified nucleotides by type of organism can be found in the compilation by Grosjean et al. (1995). The different tRNA categories in which the modified nucleotides are found are indicated in Fig. A4 and Table A1 in order to show their origin. For consistency, we considered only the modifications present in the most recent compilation of sequences (Sprinzl et al., 1998). For example, the modification Gr(p), which is present on the ribose 2'OH of the T helix (Glasser et al., 1991) but is not listed in the tRNA database, is not shown.

As in any database, a few errors resulting from typographical or sequencing errors may still be present. If an error is suspected, the reader is prompted to check the original references listed by Sprinzl et al. (1998) (http://www.uni-bayreuth.de/departments/biochemie/trna/index.html).

Although it is expected that the distribution of nucleotides from tRNA genes represented in Fig. A3 will not substantially change with the addition of new sequences of tRNA genes, given the large number of gene sequences (2,726) included in this compilation, the actual view of the distribution of modified nucleotides will probably evolve with the inclusion of new tRNA sequences.

Figure A2 displays an asymmetric distribution of the modified nucleotides, which are found more often at the 5' end of helices (residues 8 to 10 for the D stem, residues 26 to 28 and 37 to 39 for the anticodon stem, and residues 47 to 49 for the T stem). It is interesting that several positions have never been observed to be modified (residues 2, 5, and 70 and the CCA end of the acceptor stem; residues 11, 23, and 24 in the D stem; residues 33, 42, and 43 in the anticodon hairpin; residue 45 in the variable loop; and residues 59, 62, and 63 in the T loop and stem. Besides the well-known biases in nucleotide distribution at several positions (e.g., U at position 8, G at position 10, A at position 14, R at position 15, and Y at position 48) dictated by conserved tertiary base pairs, it is worth remarking that the central base pairs of the acceptor stem are among the only base pairs with an equal partition of the four Watson-Crick combinations (Fig. A3). However, while at the 5'-end there is a strong bias for an R1-Y72 base pair, there is a near absence of C7-G66 base pairs at the other end. Interestingly, the base pairs in the acceptor stem contain the most frequently observed identity elements (along with the bases of the anticodon loop) for the aminoacyl-tRNA synthetases (Giegé et al., in press). The central base pairs in the T and anticodon stems also present an almost equal distribution

---

Pascal Auffinger and Eric Westhof • Institut de Biologie Moléculaire et Cellulaire du CNRS, 15 rue René Descartes, 67084 Strasbourg Cedex, France.

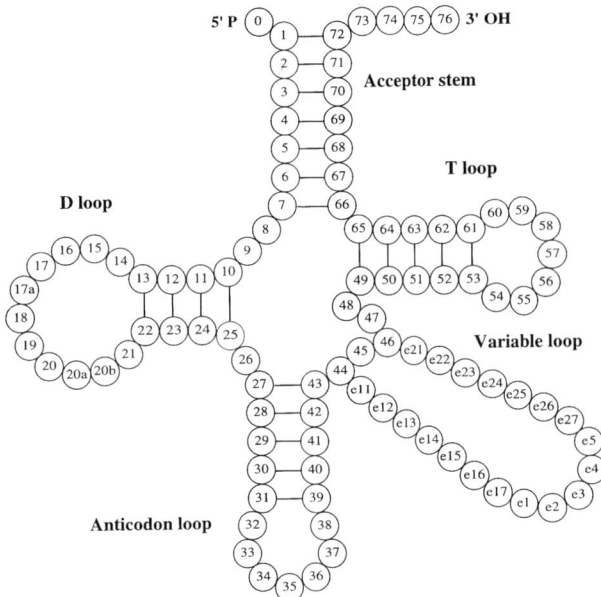

Figure A1. Cloverleaf structure of tRNA with conventional numbering of nucleotides. For a full description of the secondary structure representation, see Sprinzl et al. (1998).

among the four Watson-Crick base pairs. The location and distribution of modified nucleotides are detailed in Fig. A4. As expected, dihydrouridine residues are mainly found in the D loop (but surprisingly also at position 47). The high proportion of Ψ residues in the anticodon hairpin is noticeable (see Chapter 6). Positions 34 and 37 are highly modified, and thus the detailed distribution of natural and modified nucleotides at these positions is given in Table A1 (see also Appendix 6). From this table, it appears that modifications occurring at position 37 never occur at position 34 and vice versa. Also, while adenines are strongly avoided at position 34, they constitute 44% of the nucleotides at position 37. Inosine (I) is the only modified adenosine known to be present at position 34. The three other nucleotides, G, C, and U, are present in almost equal proportions at position 34. However, while 10 and 8% of G and C nucleotides, respectively, are modified at position 34, about 19% of U nucleotides are modified at that location. In conclusion, it is apparent that the expression "rare nucleotides" is a misnomer.

Acknowledgments. We thank Philippe Dumas, Henri Grosjean, and Gérard Keith for help and comments.

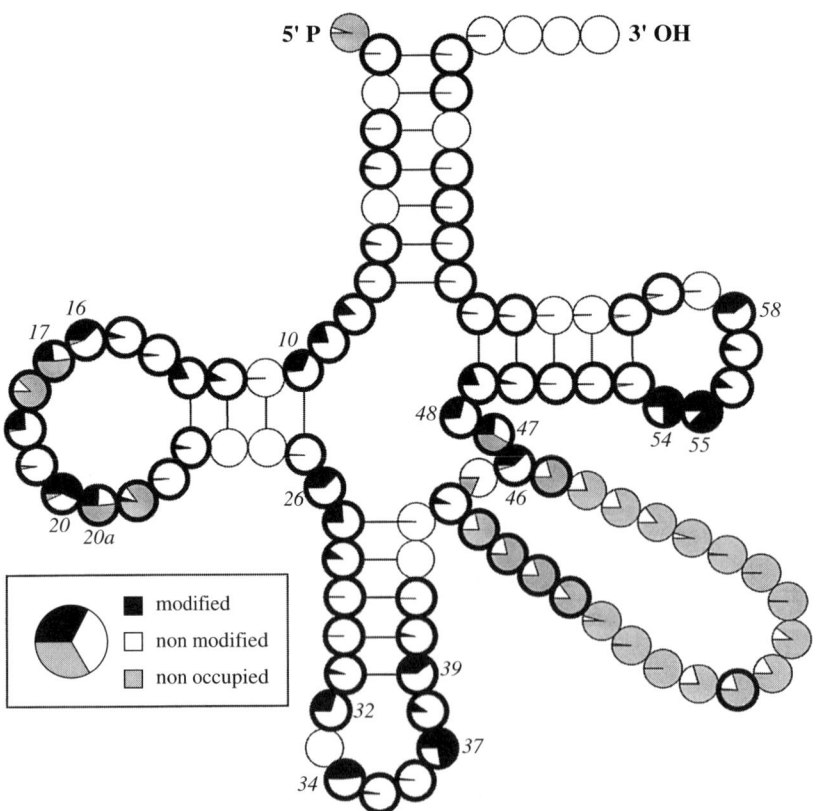

Figure A2. Distribution of modified and unmodified nucleotides in tRNA from the 546 tRNA sequences listed in the 1998 release of the tRNA database (Sprinzl et al., 1998). The circles surrounded by a bold line mark the sites where modified nucleotides have been listed in the tRNA database. The numbered positions are those where more than 25% of the nucleotides are modified; they are arbitrarily considered hypermodified.

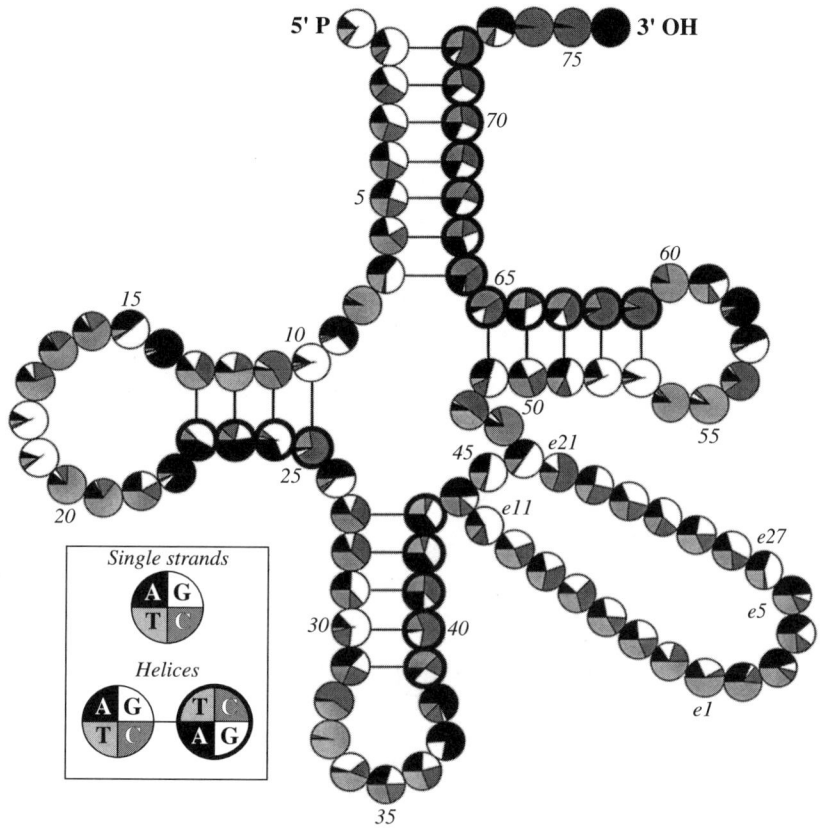

Figure A3. Distribution of the four nucleotides, A, G, C, and T, found in tRNA genes from the 2,726 tRNA gene sequences listed in the 1998 release of the tRNA database (Sprinzl et al., 1998). The circles surrounded by a bold line have been flipped in order to show the base pair complementarity in the stems. Note that the nucleotide positions in the variable loop as well as some positions in the D loop are not occupied in all tRNA sequences. The proportions of occupied and unoccupied sites are indicated in Fig. A2.

REFERENCES

Agris, P. F. 1996. The importance of being modified: roles of modified nucleosides and Mg^{2+} in RNA structure and function. *Prog. Nucleic Acid Res. Mol. Biol.* **53**:79–129.

Björk, G. 1995. Biosynthesis and function of modified nucleosides, p. 165–205. *In* D. Söll and U. L. RajBhandary (ed.), *tRNA: Structure, Biosynthesis, and Function.* ASM Press, Washington, D.C.

Dirheimer, G. 1983. Chemical nature, properties, location and physiological and pathological variations of modified nucleosides in tRNAs. *Recent Results Cancer Res.* **84**:15–46.

Dirheimer, G., G. Keith, P. Dumas, and E. Westhof. 1995. Primary, secondary, and tertiary structures of tRNAs, p. 93–126. *In* D. Söll and U. L. RajBhandary (ed.), *tRNA: Structure, Biosynthesis, and Function.* ASM Press, Washington, D.C.

Giegé, R., M. Sissler, and C. Florentz. Structural and mechanistic aspects in tRNA identity expression. *Nucleic Acids Res.*, in press.

Glasser, A. L., J. Desgres, J. Heitzler, J., C. W. Gehrke, and G. Keith. 1991. O-ribosyl-phosphate purine as a constant modified nucleotide located at position 64 in cytoplasmic initiator tRNAs^{Met} of yeasts. *Nucleic Acids Res.* **19**:5199–5203.

Grosjean, H., R. J. Cedergren, and W. McKay. 1982. Structure in tRNA data. *Biochimie* **64**:387–397.

Grosjean, H., M. Sprinzl, and S. Steinberg. 1995. Posttranscriptionally modified nucleosides in transfer RNA: their locations and frequencies. *Biochimie* **77**:139–141.

Sprinzl, M. C. Horn, M. Brown, A. Ioudovitch, and S. Steinberg. 1998. Compilation of tRNA sequences and sequences of tRNA genes. *Nucleic Acids Res.* **26**:148–153.

Szweykowska-Kulinska, Z., B. Senger, G. Keith, F. Fasiolo, and H. Grosjean. 1994. Intron-dependent formation of pseudouridines in the anticodon of *Saccharomyces cerevisiae* minor tRNA^{Ile}. *EMBO J.* **13**:4636–4644.

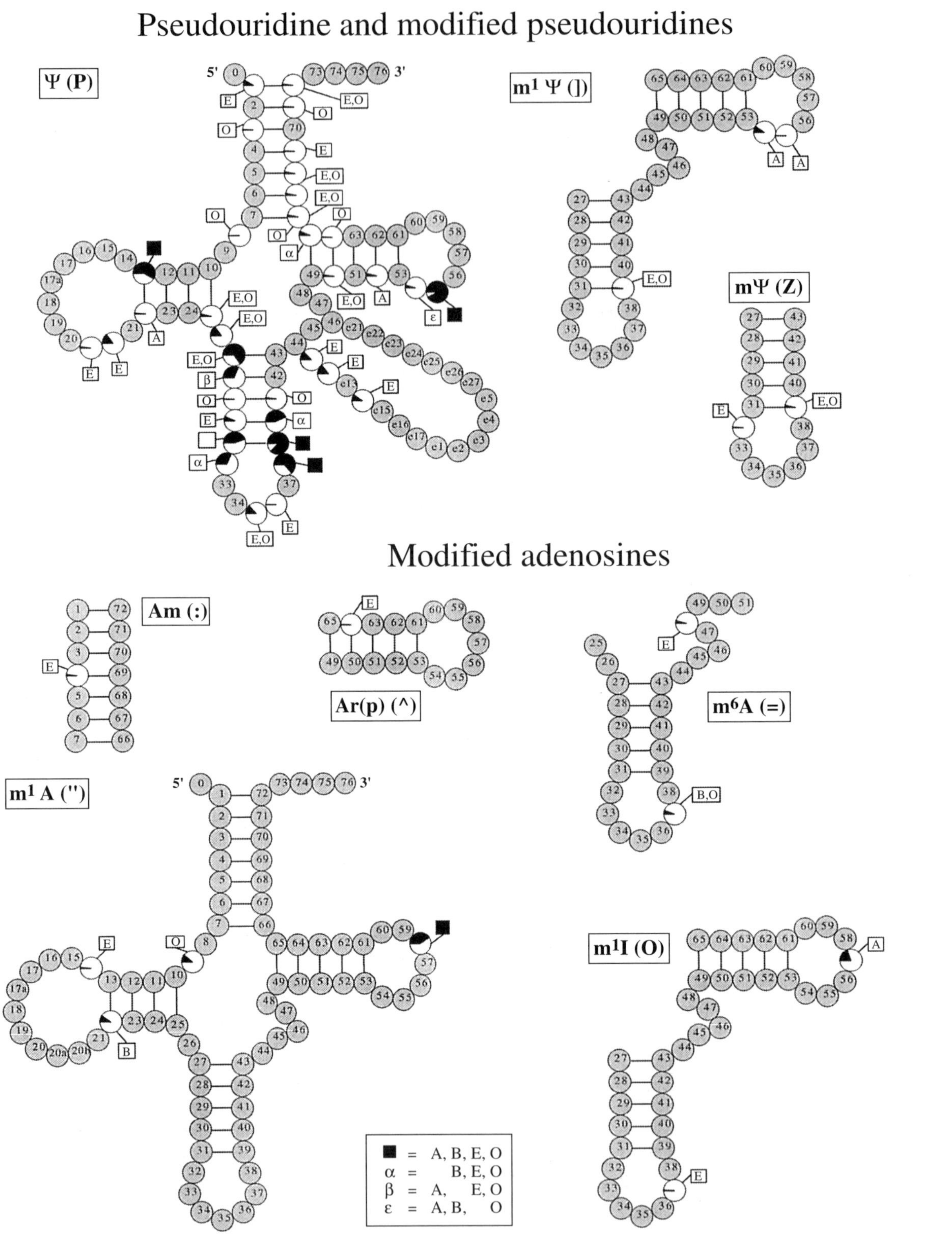

Figure A4. Location and distribution of modified nucleotides. Symbols for modified nucleotides are as in Appendix 1. The letters A, B, and E refer to the three domains (A, archaea [59 sequences]; B, eubacteria [133 sequences]; E, eukaryotes [212 sequences]), while O includes the remaining 142 mitochondrial, chloroplastic, and viroid tRNA sequences. Black squares indicate modifications occurring in all four domains (A, B, E, and O); α, β, and ε indicate modifications occurring in the (B, E, O), (A, E, O), and (A, B, O) subdomains, respectively. The percentage of occurrence of a specific modified base relative to the total number of nucleotides of the same class (A, G, C, or U) is indicated by the black portion of the white disks.

Modified uridines

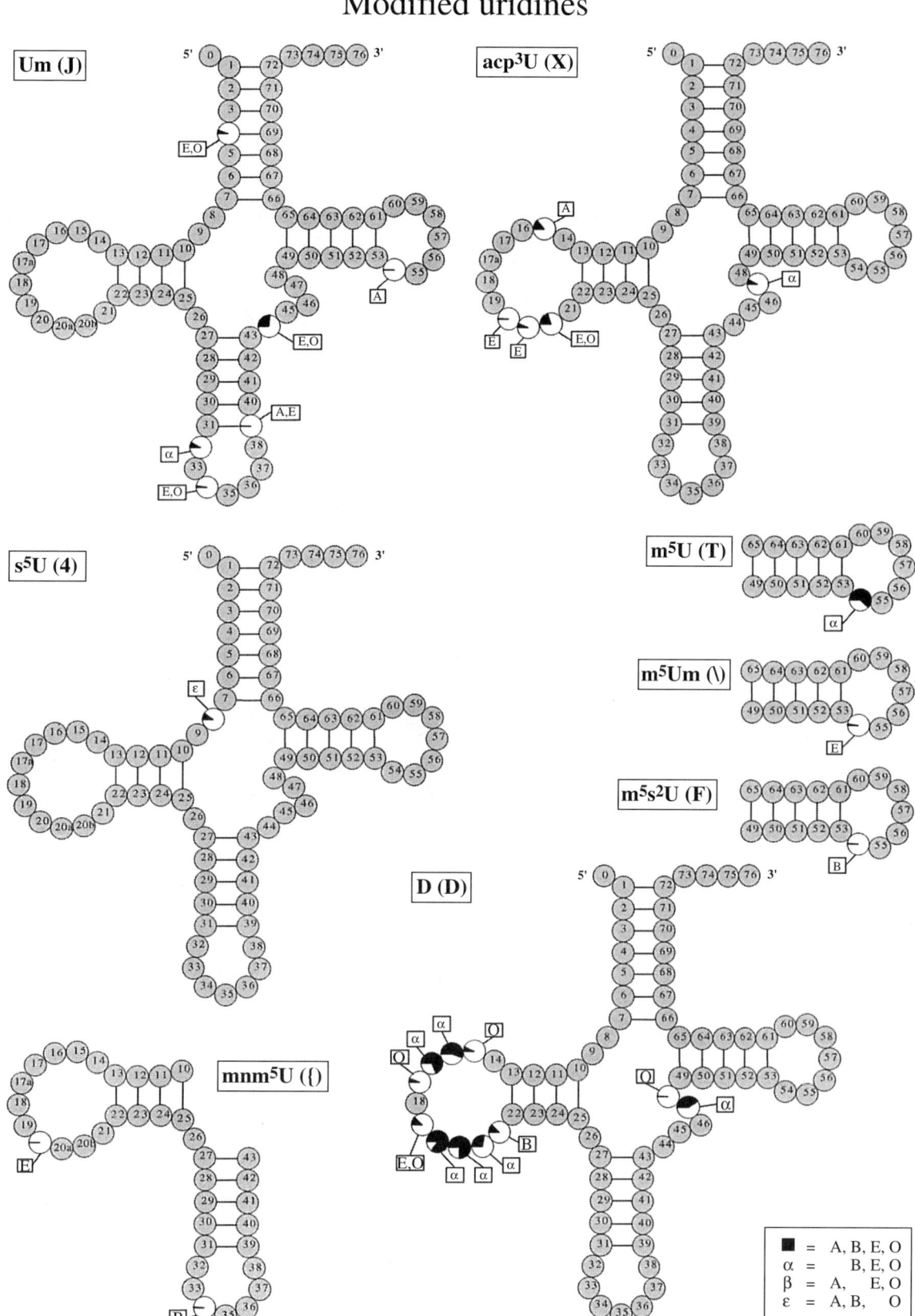

Figure A4. *Continued*

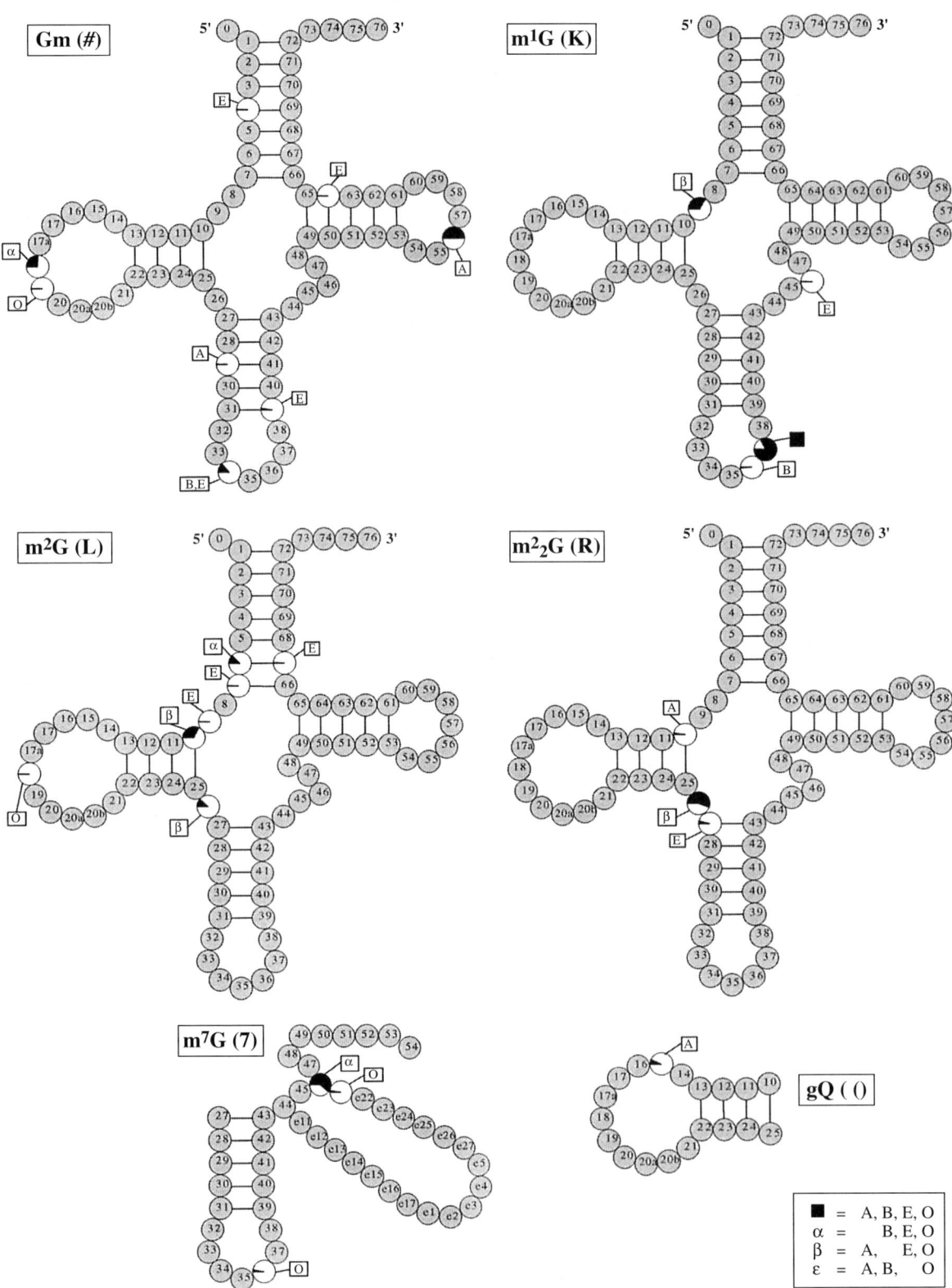

Figure A4. *Continued*

Modified cytidines

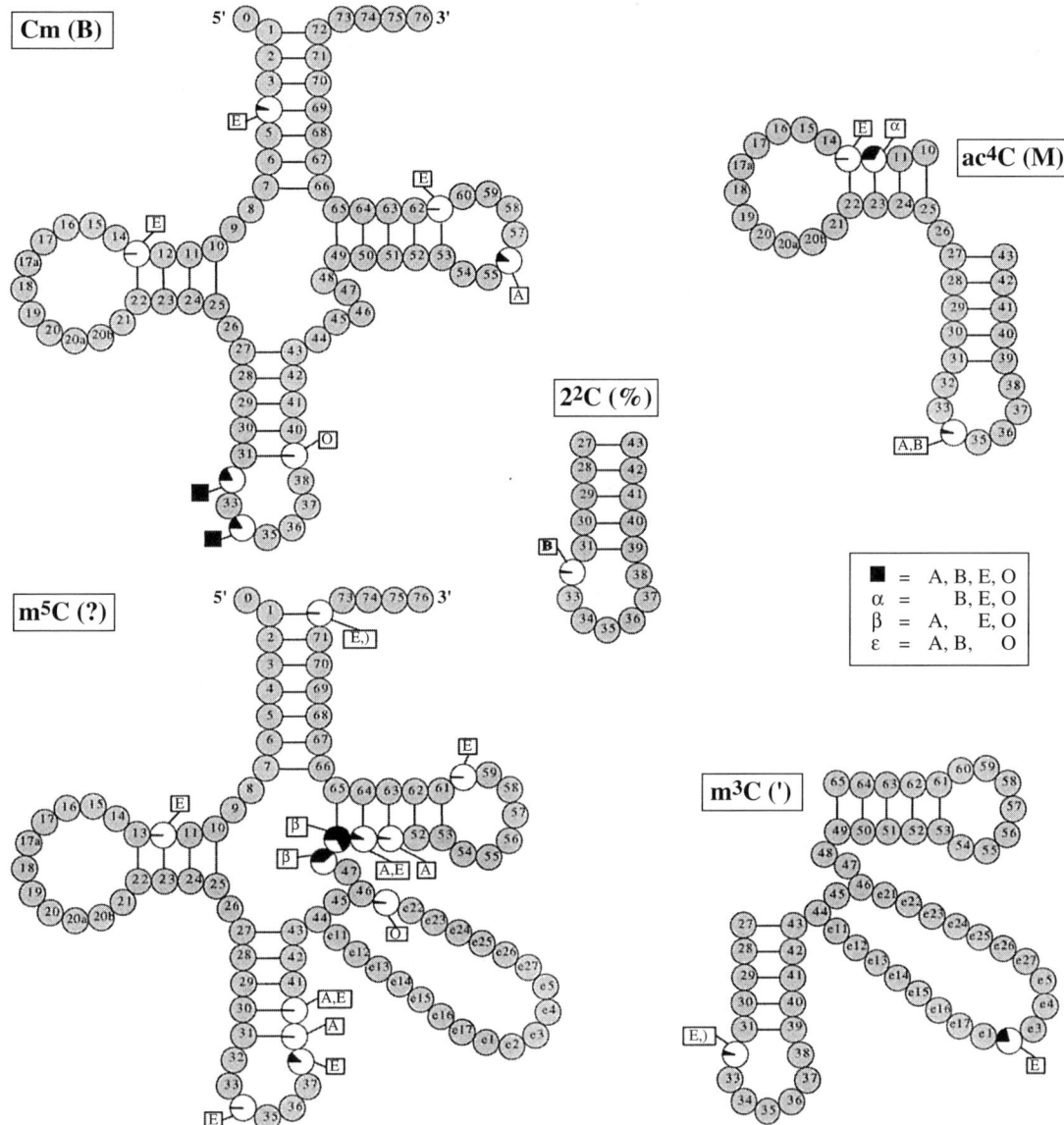

Figure A4. *Continued*

Table A1. Distribution of natural and modified nucleotides at tRNA positions 34 and 37

Name and no.[a]	Symbol[b]	Code[c]	Position 34				Position 37			
			No. of tRNAs[d]	Relative %[e]	Absolute %[f]	Source(s)[g]	No. of tRNAs	Relative %	Absolute %	Source(s)
Adenosine	A	A	2	4	1	B, O	145	36	26	A, B, E, O
	Unknown A	H	1	2	1	B	29	7	5	B, E, O
2	m^2A	/					13	3	2	B, E, O
3	m^6A	=					24	6	4	B, O
8	i^6A	+					29	7	5	B, E, O
9	ms^2i^6A	*					24	6	4	B, E, O
10	io^6A	`					5	1	<1	E
13	t^6A	6					114	29	21	A, B, E, O
14	m^6t^6A	E					7	2	1	B, E, O
15	ms^2t^6A	[					3	<1	<1	B, E
20	I	I	46	94	8	B, E, O				
22	m^1I	O					6	2	1	E
	Total		49	100	9		401	100	44	
Guanosine	G	G	116	68	21	A, B, E, O	2	1	<1	E
	Unknown G	;	3	2	<1	A, E	5	3	1	B, E
1	Gm	#	23	14	4	B, E				
2	m^1G	K					116	83	21	A, B, E, O
14	yW	Y					7	5	1	E
16	o^2yW	W					11	8	2	E
17	Q	Q	20	12	4	B, E, O				
19	galQ	9	3	2	1	E				
20	manQ	8	4	2	<1	E				
	Total		169	100	31		141	100	26	
Cytidine	C	C	110	73	20	A, B, E, O	2	100	<1	B
	Unknown C	<	5	3	<1	E, O				
1	Cm	B	24	16	4	A, B, E, O				
4	ac^4C	M	7	5	1	A, B				
6	m^5C	?	1	<1	<1	E				
9	f^5C	>	1	<1	<1	E				
10	f^5Cm	∞	1	<1	<1	E				
13	k^2C	}	3	2	<1	B				
	Total		152	100	28		2	100	<1	
Uridine	U	U	56	36	10	A, B, E, O	1	100	<1	E
	Unknown U	N	46	29	8	A, B, E, O				
1	Um	J	4	2	<1	E, O				
2	s^2U	2	1	<1	<1	E				
12	mo^5U	5	6	4	1	A, B				
13	cmo^5U	V	5	3	<1	E				
16	mcm^5U	1	3	2	<1	E				
18	mcm^5s^2U	3	6	4	1	E				
19	ncm^5U	&	4	2	<1	E				
22	$mchm^5U$	,	1	<1	<1	E				
24	nmn^5U	{	2	1	<1	B				
25	mnm^5s^2U	S	6	4	1	E, O				
27	$cmnm^5U$	!	9	6	2	E, O				
28	$cmnm^5Um$	)	4	2	<1	B, E				
29	$cmnm^5s^2U$	$	2	1	<1	B				
1[h]	Ψ	P	1	<1	<1	B				
	Total		156	100	29		1	100	<1	

[a] For the numbering, name, and chemical structure of the modified nucleotides, see Appendix 1. Only the nucleotides detected at position 34 and 37 are indicated.
[b] See also Appendix 1.
[c] One-letter or one-symbol code used in the tRNA database by Sprinzl et al. (1998).
[d] Number of tRNAs reported in the tRNA database in which these modified nucleotides have been located.
[e] Number of nucleotides of a specific type relative to the total number of nucleotides of its class (i.e. A, G, C, or U).
[f] Number of nucleotides of a specific type relative to the total number of nucleotides found at that position (the 1998 release of the tRNA database contains 546 tRNA sequences).
[g] Origin of the tRNA sequences in which the modified nucleotides are reported (A, archaea; B, eubacteria; E, eukaryotes; O, others [chloroplastic, mitochondrial, and viroid tRNA sequences]).
[h] The sequence of *Saccharomyces cerevisiae* minor tRNA[Ile] containing the rare ΨAΨ anticodon (Szweykowska-Kulinska et al., 1994) has been added to the present compilation. See also Fig. 2 of Chapter 6.

Modification and Editing of RNA
Edited by Henri Grosjean and Rob Benne
© 1998 ASM Press, Washington, D.C.

Appendix 6: Modified Nucleosides at Positions 34 and 37 of tRNAs and Their Predicted Coding Capacities

GLENN R. BJÖRK

This appendix summarizes the presence of modified nucleosides at positions 34 (the wobble position) and 37 (adjacent to and 3' of the anticodon) in tRNAs (Fig. A2 to A5). Data are compiled from Sprinzl et al. (1998). The predicted coding capacities of the various modified nucleosides are based on Crick's wobble hypothesis (Crick, 1966), on the coding rules for xm^5s^2U and xo^5U derivatives (Yokoyama et al., 1985), on the restrictive coding features of 2'-O-ribose methylated derivatives (Kawai et al., 1992), and on the inefficient interaction of inosine (I) in the wobble position with A (Curran, 1995). These as well as other considerations are summarized in Table 2 of Björk (1995). It should be pointed out, however, that in many instances the predicted coding capacities have not been experimentally verified and should be taken as predictions rather than facts.

Acknowledgments. This work was supported by grants from the Swedish Natural Science Research Council (project B-BU2930) and from the Swedish Cancer Society (project 680). I thank Tord Hagervall, Umeå University, for comments on the manuscript.

REFERENCES

Björk, G. R. 1995. Biosynthesis and function of modified nucleosides, p. 165–205. *In* D. Söll and U. L. RajBhandary (ed.), *tRNA: Structure, Biosynthesis, and Function.* ASM Press, Washington, D.C.

Crick, F. H. C. 1966. Codon-anticodon pairing. The wobble hypothesis. *J. Mol. Biol.* 19:548–555.

Curran, J. F. 1995. Decoding with the A-I wobble pair is inefficient. *Nucleic Acids Res.* 23:683–688.

Hjalmarsson, K. J., and M. Wikström. Unpublished results.

Inagaki, Y., A. Kojima, Y. Bessho, H. Hori, T. Ohama, and S. Osawa. 1995. Translation of synonymous codons in family boxes by Mycoplasma capricolum tRNAs with unmodified uridine or adenosine at the first anticodon position. *J. Mol. Biol.* 25:486–492.

Kawai, G., Y. Yamamoto, T. Kamimura, T. Masegi, M. Sekine, T. Hata, T. Iimori, T. Watanabe, T. Miyazawa, and S. Yokoyama. 1992. Conformational rigidity of specific pyrimidine residues in tRNA arises from posttranscriptional modifications that enhance steric interaction between the base and the 2'-hydroxyl group. *Biochemistry* 31:1040–1046.

Kuchino, Y., N. Hanyu, F. Tashiro, and S. Nishimura. 1985. Tetrahymena thermophila glutamine tRNA and its gene that corresponds to UAA termination codon. *Proc. Natl. Acad. Sci. USA* 82:4758–4762.

Pais de Barros, J. P., G. Keith, C. El Adlouni, A. L. Glasser, G. Mack, G. Dirheimer, and J. Desgres. 1996. 2'-O-methyl-5-formylcytidine (f^5 C), a new modified nucleotide at the 'wobble' of two cytoplasmic tRNAs Leu (NAA) from bovine liver. *Nucleic Acids Res.* 24:1489–1496.

Sprinzl, M., C. Horn, M. Brown, A. Ioudovitch, and S. Steinberg. 1998. Compilation of tRNA sequences and sequences of tRNA genes. *Nucleic Acids Res.* 26:148–153.

Takai, K., N. Horie, Z. Yamaizumi, S. Nishimura, T. Miyazawa, and S. Yokoyama. 1994. Recognition of UUN codons by two leucine tRNA species from Escherichia coli. *FEBS Lett.* 344:31–34.

Takeoto, C., T. Koike, T. Yokogawa, L. Benkowski, L. L. Spremulli, T. A. Ueda, K. Nishikawa, and K. Watanabe. 1995. The ability of bovine mitochondrial transfer RNA(met) to decode AUG and AUA codons. *Biochimie* 77:104–108.

Yokoyama, S., T. Watanabe, K. Murao, H. Ishikura, Z. Yamaizumi, S. Nishimura, and T. Miyazawa. 1985. Molecular mechanism of codon recognition by tRNA species with modified uridine in the first position of the anticodon. *Proc. Natl. Acad. Sci. USA* 82:4905–4909.

	U	C	A	G	
U	UUU, UUC } Phe (F) UUA, UUG } Leu (L)	UCU, UCC, UCA, UCG } Ser (S)	UAU, UAC } Tyr (Y) UAA, UAG } STOP	UGU, UGC } Cys (C) UGA STOP UGG Trp (W)	U C A G
C	CUU, CUC, CUA, CUG } Leu (L)	CCU, CCC, CCA, CCG } Pro (P)	CAU, CAC } His (H) CAA, CAG } Gln (Q)	CGU, CGC, CGA, CGG } Arg (R)	U C A G
A	AUU, AUC, AUA } Ile (I) AUG Met (M)	ACU, ACC, ACA, ACG } Thr (T)	AAU, AAC } Asn (N) AAA, AAG } Lys (K)	AGU, AGC } Ser (S) AGA, AGG } Arg (R)	U C A G
G	GUU, GUC, GUA, GUG } Val (V)	GCU, GCC, GCA, GCG } Ala (A)	GAU, GAC } Asp (D) GAA, GAG } Glu (E)	GGU, GGC, GGA, GGG } Gly (G)	U C A G

Figure A1. The universal genetic code. The amino acids are abbreviated by the standard three-letter code and, in parentheses, by the one-letter code.

Glenn R. Björk • Department of Microbiology, Umeå University, S-90 187 Umeå, Sweden.

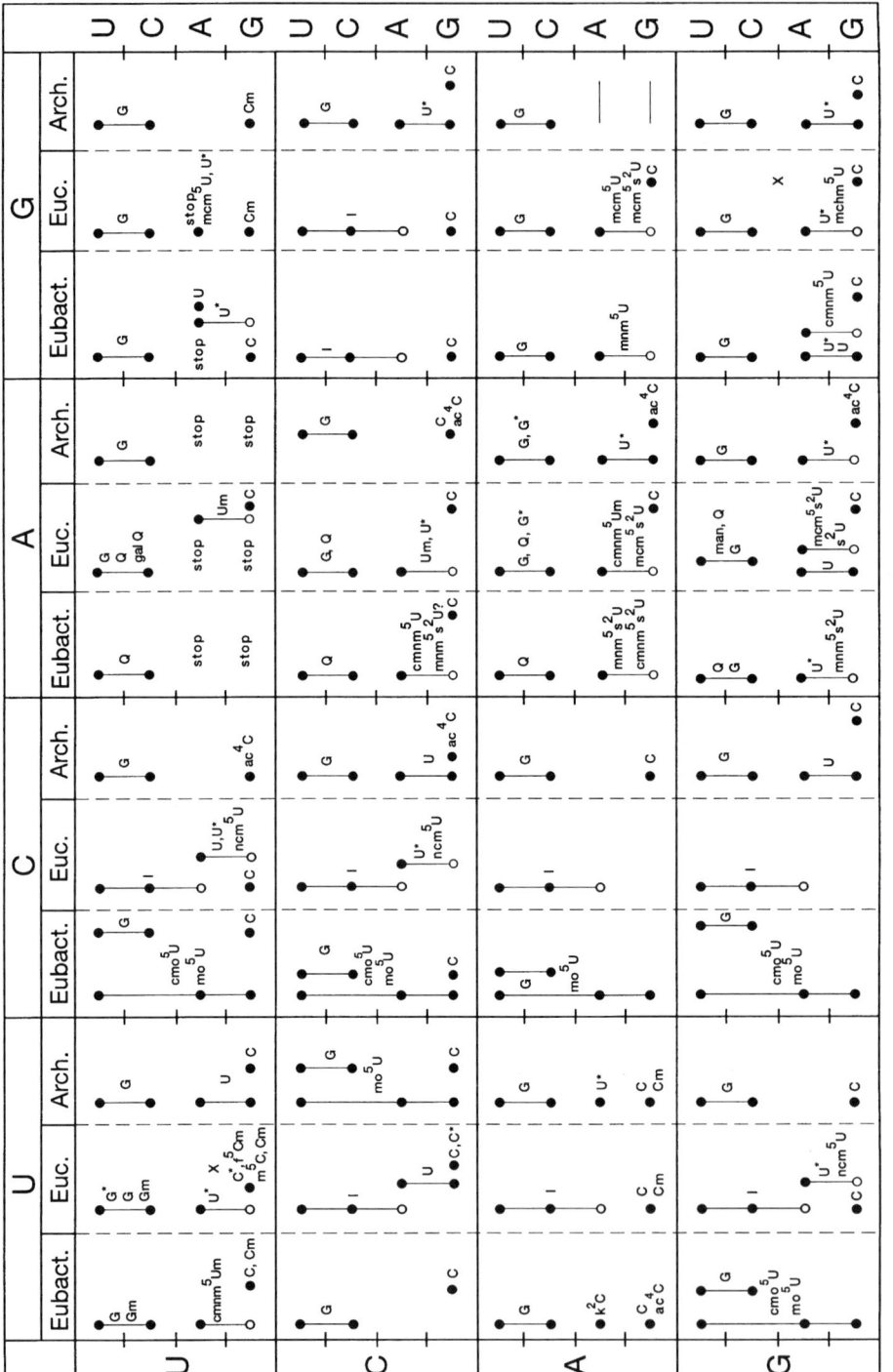

Figure A2. Presence of modified nucleosides in position 34 (the wobble position) and their predicted coding capacities in cytoplasmic tRNAs. The figure is modified from Björk (1995). The circles indicate predicted coding preferences exhibited by the various tRNAs containing the indicated modified nucleosides. Closed circles indicate efficient interaction, and open circles indicate less-well-recognized codons (e.g., *Escherichia coli* tRNA$^{Lys}_{mnm5s2UUU}$ decodes AAA efficiently but AAG less well [Yokoyama et al., 1985]; tRNA$^{Ala}_{cmo5U}$ reads GCG, GCA, and GCU [Yokoyama et al., 1985]; *E. coli* tRNA$^{Arg}_{ICG}$ reads CGU and CGC efficiently and CGA inefficiently [Curran, 1995]; and so forth for other tRNAs). The f^{5}C34 reads only codons ending in G (Takemoto et al., 1995), and the predicted codon preference of f^{5}Cm is likewise restricted to codons ending in G (Pais de Barros et al., 1996). An asterisk indicates an unknown modified derivative of the indicated nucleoside, an X indicates an unknown nucleoside, and a question mark after a nucleoside indicates uncertainties in the identification. tRNA$^{Leu}_{cmnm5UmAA}$ in *E. coli* contains a modified A34 according to Sprinzl et al. (1998), but this modification has been shown to be cmnm^{5}Um (Takai et al., 1994). Eubact., eubacteria, Euc., eucarya; Arch., archaea.

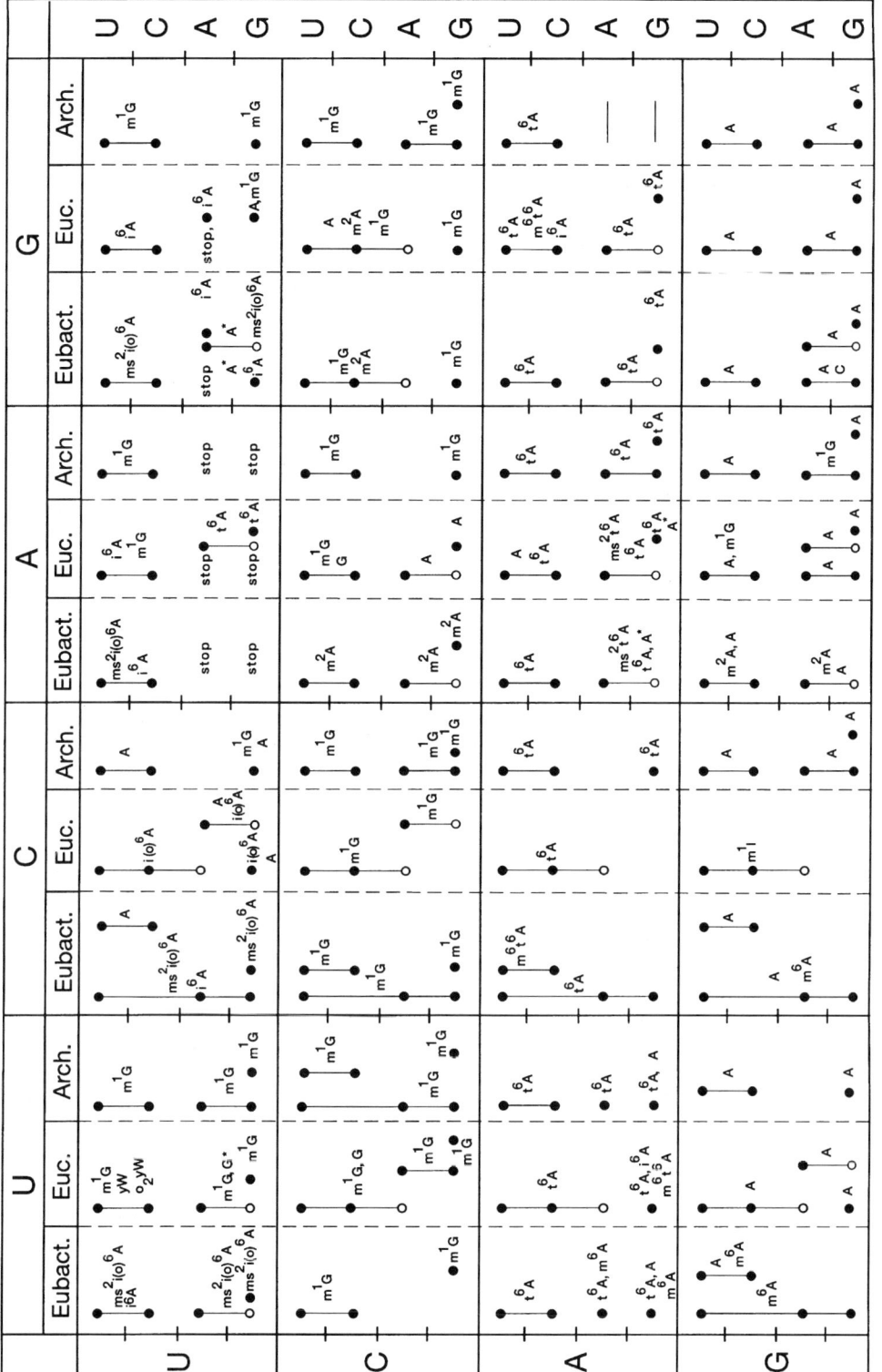

Figure A3. Presence of modified nucleosides in position 37 (adjacent and 3' to the anticodon) in cytoplasmic tRNAs. The coding capacities of tRNAs are indicated as described in the legend to Fig. A2. The figure is modified from Björk (1995). ms²i(o)⁶A37 indicates that in most eubacteria, the indicated tRNAs contain ms²io⁶A37, whereas tRNAs from some eubacteria, like *E. coli*, contain ms²i⁶A37. Two glutamine tRNAs from *Tetrahymena thermophila* have t⁶A37 and C34 or Um34 and recognize the stop codon UAA (Kuchino et al., 1985). The three *E. coli* tRNA^Leu_3, which read codons starting with C, have an unidentified modified G in position 37. However, mutations in the *trmD* gene, which is the structural gene for the tRNA(m¹G37) methyltransferase, affect the elution properties of these three tRNA^Leu species in *E. coli* and *Salmonella typhimurium* (Hjalmarsson and Wikström, unpublished results). Therefore, the modified G must be m¹G or a derivative of m¹G. Eubact., eubacteria; Euc., eucarya; Arch., archaea. Asterisks indicate unknown modified derivatives of the indicated nucleosides.

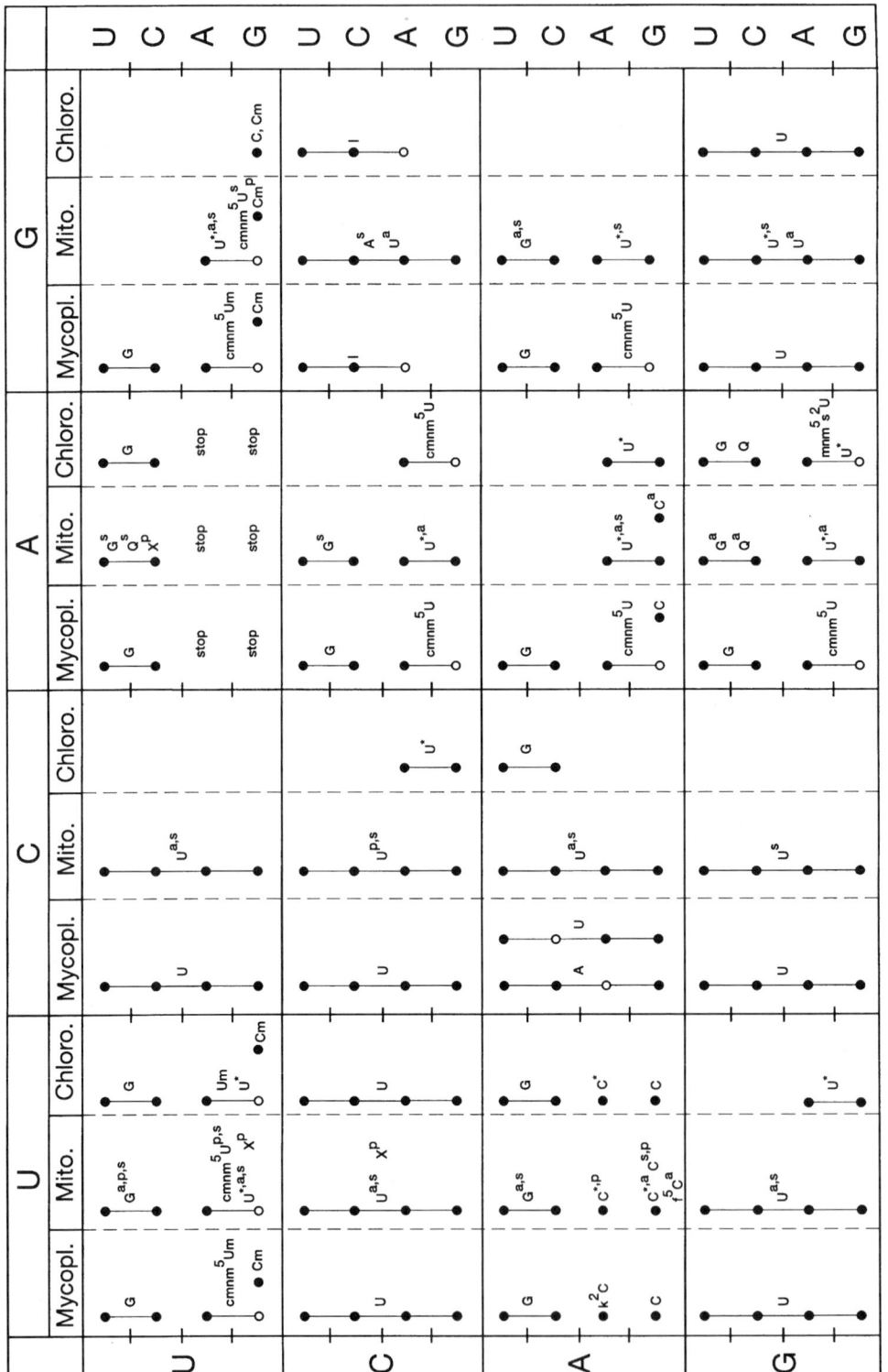

Figure A4. Modified nucleosides in the wobble position of tRNA from a *Mycoplasma* sp. (Mycopl.), mitochondria (Mito.), and chloroplasts (Chloro.). The coding capacities of tRNAs are indicated as described in the legend to Fig. A2. Of the two tRNA^{Thr} species from *Mycoplasma*, one contains A34 and the other contains U34. tRNA$^{Thr}_{AGU}$ reads ACU, ACC, and ACG efficiently and ACA less well, whereas tRNA$^{Thr}_{UGU}$ reads ACU, ACA, and ACG efficiently and ACC less well (Inagaki et al., 1995). Superscripts: a, animal mitochondria; p, plant mitochondria; s, single-celled organismal mitochondria. Asterisks indicate unknown modified derivatives of the indicated nucleosides.

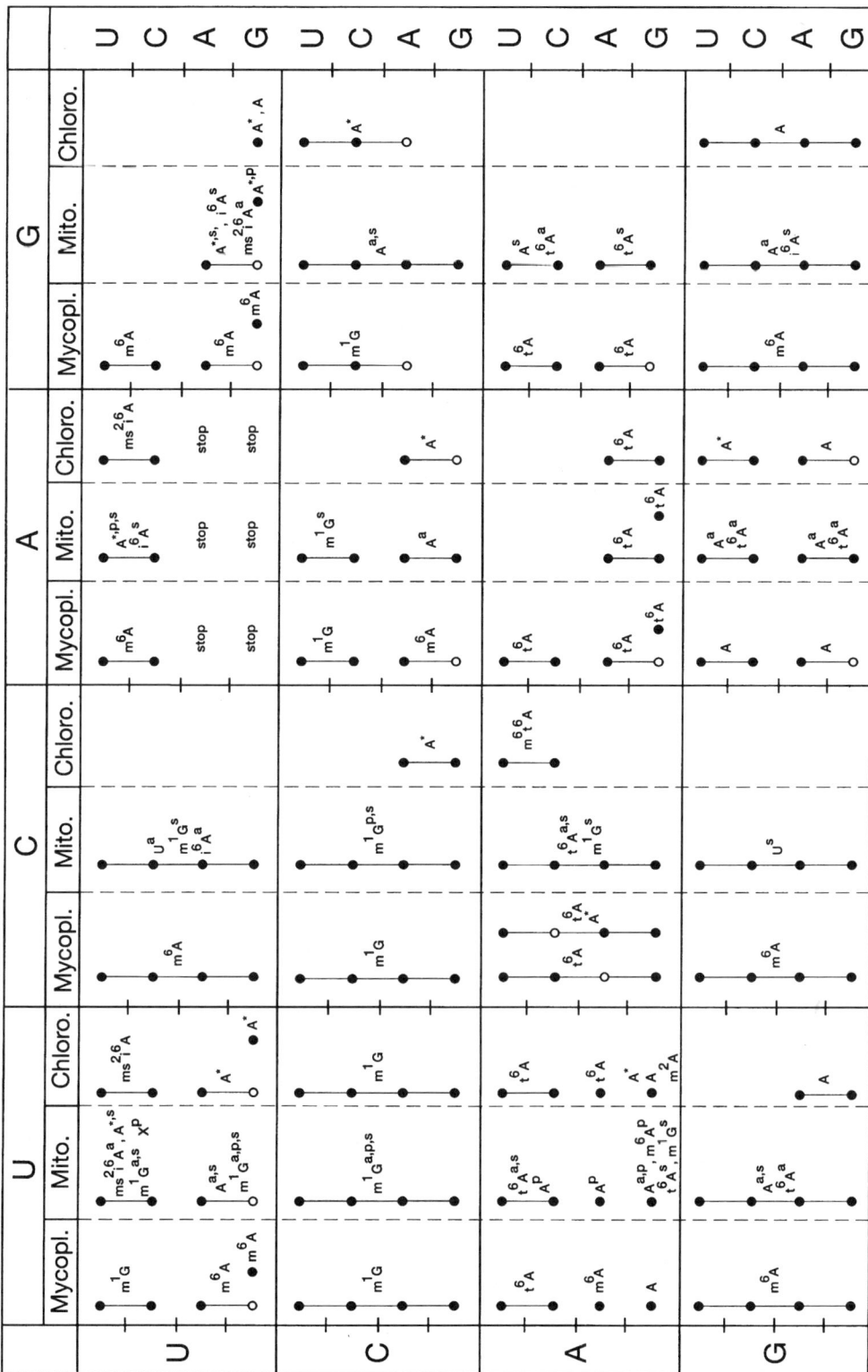

Figure A5. Modified nucleosides in position 37 (adjacent and 3' to the anticodon) in tRNAs from a *Mycoplasma* sp. (Mycopl.), mitochondria (Mito.), and chloroplasts (Chloro.). The coding capacities of tRNAs are indicated as described in the legends to Fig. A2 and A4. Superscripts: a, animal mitochondria; p, plant mitochondria; s, single-celled organismal mitochondria. Asterisks indicate unknown modified derivatives of the indicated nucleosides.

INDEX

A site, topography, 76–77
ACA box, 243
Acanthamoeba, mitochondrial tRNA editing, 293–295
Acellular slime mold, *see Physarum*
N^4-Acetylcytidine, conformational and thermodynamic effects, 95
N^4-Acetyl-2′-O-methylcytidine, conformational and thermodynamic effects, 95
Active site, *E. coli* cytidine deaminase, 366–367
ADAR, *see* Adenosine deaminase, acting on RNA
Adenosine deaminase, 147
    acting on RNA (ADAR), 352–357, 363–375
        family of ADAR enzymes, 356–357
        dsRNA specific, 352–355
            editase 1, 355–356
        lack of role in RNA editing, 372–373
        tRNA:adenosine-34 deaminase, 146–147, 155
N^6-Adenosine methyltransferase, mRNA, from HeLa cells, 192–198
    characterization, 192–198
    characterization of MT-B, 197–198
    GenBank and EST database homology searches, 192–196
    Northern blot analysis of MT-A70 expression, 196–197
    purification and cDNA cloning of MT-A, 192–194
    subnuclear localization of MT-A70, 197
Adenosine-to-inosine conversion, 372–373
    mRNA, 343–361
        adenosine deaminases, 352–357
        glutamate receptor subunits, 344–348
        hepatitis delta virus, 348–349
        RNA substrates, 344–352
        RNA substrates with multiple editing sites, 351–352
        serotonin 2C receptor, 349–351
    prospects, 373–374
Adenosylmethionine, 136
Aerobiosis, sensors for, 475–476
Affinity labeling, 60–62
    azidonucleotides, 61
    convertible nucleotides, 61–62
    halonucleotides, 61
    ribose modifications, 62
    thionucleotides, 60–61
AIDS, 483
AIDS-related complex, 483
Alkylation, 150–154, 162

Allen, Frank, 4
Allosteric activation, *E. coli* cytidine deaminase, 370–371
Amides, tRNA, 10–11
Aminoacyl-tRNA synthetases
    interaction with tRNA, 72
    tRNA-mRNA interactions in control of gene expression, 126
Anomeric effects, modified nucleosides, 89–90
Antibiotic resistance, bacteria, tRNA modification and, 479–480
Anticodon
    carboxylate salts, esters, and amides in and near, 10–11
    modifications
        A34 and I34, 502–503
        C34, 501–502
        pseudouridine, 504
        queuosine at position 34, 503–504
        U34, 497–498
        U34 modifications and restricted wobble, 498–500
        U34 modifications in family boxes, 500–501
Anticodon loop, 493
    modifications in 3′ side, 504–510
        m^1G37, 506–507
        $ms^2i(o)^6A37$, 504–506
        pseudouridine, 508–509
        t^6A37, 507
        Y base at position 34, 508
    pseudouridine, 104–106
Anticodon-anticodon interactions
    between tRNAs, 113–133
    features of anticodon-anticodon complex, 114–117
    model, 122–123
    modified nucleotides in
        nucleotides 3′-adjacent to anticodon, 117–118
        role for wobble base, 118–122
    temperature-jump relaxation studies, 114–117
Antisense RNA
    control of ColE1 plasmid replication, 126–127
    snoRNA, 264–265
        box C/D, 262
apobec-1, 330–332, 364, 371–372
    annulment of expression, pathophysiology, 338–339
    developmental, hormonal, and nutritional control of mRNA expression, 336–337
    E. coli cytidine deaminase as model, 368–369
    evolution, 332–334
    hepatic overexpression, pathophysiology, 337–338

mechanism of action, 332–334
Apolipoprotein B
  mRNA editing, 325–342
    annulment of apobec-1 expression, 338–339
    apobec-1, 330–332
    developmental, hormonal, and nutritional control, 336–337
    discovery, 326
    evolution and mechanism of apobec-1, 332–334
    hepatic overexpression of apobec-1, 337–338
    physiological consequences, 325–326
    sequence specificity, 327–330
    species and tissue specificity, 326
    structure of editing enzyme complex, 334–336
    subcellular location, 326–327
  physiology of apolipoprotein B-containing lipoproteins, 325
  structure of ApoB-100 and ApoB-48, 325–326
Archaebacteria
  pseudouridine in rRNA, 246
  tRNA-guanine transglycosylase, 171, 178–179
Archaeosine
  formation, 157, 160, 169
  function, 169
  location, 169
Atomic mutagenesis, 59
Avian retrovirus, initiation of reverse transcription, 520
Azidonucleotides, 61

Bacteria, see also Prokaryotes
  complex multifunctional operons, 456–457
  tRNA editing, 292–293
  tRNA modifications
    antibiotic resistance and, 479–480
    pathogenicity and, 478–479
Bacteriophage, tRNA modifications, 422
Base composition analysis, 31–32
Base stacking
  correlation with sugar conformation, 88
  effect of modified nucleosides, 90
Base-methylated nucleosides, 6–9
  formation, 6
Biolistic technique, detecting enzyme activity, 39
Box C/D antisense snoRNA, 262
Box H/ACA snoRNA, 243
  direct pseudouridine site selection, 243–244
Breast cancer, modified nucleosides as tumor markers, 483

5-Carbamoylmethyl-2′-O-methyluridine, formation, 143–144
Carbamoyl-phosphate synthetase, 153
Carboxylate salts, tRNA, 10–11
Cell electroporation technique, detecting enzyme activity, 38–39
Cgf5p, yeast, 242
  component of snoRNPs, 285–286
Chemical synthesis, RNA substrates, 28–29
Chloroplasts, see also Plant organelles
  pseudouridine in rRNA, 246

Chytridiomycete fungi, mitochondrial tRNA editing, 293–295
Cloning, prokaryotic pseudouridine synthases, 238–239
Codon recognition, by tRNA on ribosome, 123–125
ColE1 plasmid, replication, antisense RNA control, 126–127
Conformation, modified nucleosides, 85–102
Convertible nucleotides, affinity labeling, 61–62
Crithidia fasciculata, see Kinetoplastid protozoa
Cross-linking, 76–78
Cytidine deaminase, 147
  acting on RNA, 363–375
  E. coli, 364
    allosteric activation, 370–371
    asymmetry in substrate binding, 370–371
    composite, dimeric configuration, 368–369
    configuration of substrate and product complexes, 367
    hydrolytic deamination, 366
    inaccessible active site and editing specificity, 366–367
    model for apobec-1, 368–369
    peptide mimic of RNA substrate, 369–370
    structural reaction profile, 366–367
  evolution, 365–366
Cytidine deaminase superfamily, 364–365
Cytidine-to-uridine editing, 368–369
  prospects, 373–374
  signatures, 371–372

Deamination, 145–147, 162
  viral RNA, 528–529
Developmental changes, tRNA modification, 480–482
Developmental control
  apobec-1 mRNA expression, 336–337
  mitochondrial biogenesis in kinetoplastid protozoa, 380–381
  mRNA editing, 389–390
Didelphis virginiana, mitochondrial tRNA editing, 295–296
Dihydrouridine
  conformational and thermodynamic effects, 94
  formation, 154–155
Dim1p, 281–285
Dimerization, HIV-1 RNA, 127–129
N^6,N^6-Dimethyladenosine, 281–285
N^2,N^2-Dimethylguanosine, formation, 142–143
Dimethylase, rRNA, 281–285
Dimethyltransferase, 462–464
D-loop, tRNA, interaction with TΨ-loop, 125
DNA, plus-strand strong-stop, see Plus-strand strong-stop DNA
DNA ligase, T4, incorporation of modified nucleotides into RNA, 71–72
Double-stranded RNA-specific adenosine deaminase, 352–355
  editase 1, 355–356

E site, topography, 76–77
Editosome complex, 330–332

structure, 334–336
Electroporation technique, detecting enzyme activity, 38–39
Electrospray ionization mass spectrometry, 48–49
Escherichia coli
   cytidine deaminase, 364
      allosteric activation, 370–371
      asymmetry in substrate binding, 370–371
      composite, dimeric configuration, 268–269
      configuration of substrate and product complexes, 367
      hydrolytic deamination, 366
      inaccessible active site and editing specificity, 366–367
      model for apobec-1, 368–369
      peptide mimic of RNA substrate, 369–370
      structural reaction profile, 366–367
   miaA gene, 456–457
   reconstitution of 50S ribosomal subunits, 276
Esters, tRNA, 10–11
Eukaryotes, *see also specific eukaryotes*
   guide RNA, 242–243
   pseudouridine biosynthesis, 242–246
   pseudouridine synthetases, 245–246
   RNA-modifying enzymes
      isozyme sorting, 462–464
      subcellular location, 421–440
   Tgt transglycosylase, 460–461
   tRNA-guanine transglycosylase, 171, 176–178
Eukaryotic mRNA
   methylated nucleosides
      enzymes in cap methylation, 185–186
      within cap structure, 184–185
      modifications, 435
         inosine, 186
         N^6-methyladenosine, 187–192
         5-methylcytidine, 186–187
         5′-terminal cap, 183–184
         within internal regions, 186–192
Eukaryotic rRNA
   modifications, 430–434
   prospects, 464–465
   pseudouridine, 242–246
   regulation of modification, 460–465
   ribose methylation, 255–272
Eukaryotic snoRNA, modifications, 435–436
Eukaryotic snRNA, modifications, 435–436
Eukaryotic tRNA
   modifications, 422–430
   prospects, 464–465
   regulation of modification, 460–465
   sequential modification pathways, 461–462
Evolution
   apobec-1, 332–334
   cytidine deaminase, 365–366
   deep past, 536–538
   mitochondrial tRNA editing, 301–302
   modified nucleosides, 535–541
   origin-of-life theories, 535–541
   RNA editing in plants, 317–318
   rRNA, 421
   tRNA, 421
   tRNA-guanine transglycosylases, 179–180
Excretion, modified nucleosides, 482–483
Expressed sequence tag database, search for homology to mRNA N^6-adenosine methyltransferase, 192–196

Family boxes, tRNAs in, 500–501
Feedback regulation, rRNA modification genes, 457–458
5-Formylcytidine
   conformational and thermodynamic effects, 94–95
   formation, 143–144
Frameshifting, ribosomal, Gag-Pol precursor synthesis, 525–528
Frameshifting assay, 496–497
Functional group analysis, 72–74
   interaction of HIV-1 Tat protein with TAR RNA, 72–74
   interaction of tRNAs with aminoacyl-tRNA synthetases, 72
   reagents, 62–63

Gag-Pol precursor, synthesis
   minus one ribosomal frameshifting, 525–528
   modified nucleotides and, 525–528
   plus one ribosomal frameshifting, 528
   readthrough suppression of termination codons, 525
Gastrointestinal tract tumor, modified nucleosides as tumor markers, 483
3′-Gauche effect, modified nucleosides, 89–90
GenBank, search for homology to mRNA N^6-adenosine methyltransferase, 192–196
Gene expression
   analysis, consequences of RNA editing, 318
   effect of tRNA modification, 472–473
   regulation through mRNA editing, 388–389
Genetics, RNA modification in prokaryotes, 450
Glutamate receptor subunits, adenosine-to-inosine conversion in mRNA, 344–348
Guanine transglycosylase, *see* tRNA-guanine transglycosylase
(Guanine-7)-methyltransferase, 185
Guide RNA, 268–269, 539
   eukaryotic, 242–243
   mitochondrial mRNA editing in kinetoplastid protozoa, 381–384
   primary structure, 381
   prokaryotic, 241–242
   pseudouridine in rRNA, 244–245
   snoRNA, 255–272
      box H/ACA snoRNA, 245
      discovery, 255–256
      evolution, 286–287
      as RNA chaperone, 267–268

H box, 243
Halonucleotides, 61
HeLa cells, mRNA N^6-adenosine methyltransferase, 192–198

Hepatitis delta virus, adenosine-to-inosine conversion in mRNA, 348–349
Heterologous RNA, as RNA substrate, 29–30
Historical perspectives on modified nucleosides, 1–20
  base-methylated nucleosides, 6–9
  general perspectives and prospects, 11–14
  history of RNA research, 1–3
  hypermodification, 9–11
  perspectives on future, 13–14
  pseudouridine, 3–6
  sugar-methylated nucleosides, 6–9
  targeting sites of scission and modification, 9
HIV-1, see human immunodeficiency virus type 1
Hormonal control, of apobec-1 mRNA expression, 336–337
Human immunodeficiency virus type 1
  initiation of reverse transcription, 520–522
    functional studies, 522
    lysine tRNA, 521–522
    structural studies, 520–521
  RNA
    dimerization, 127–129
    interaction with human tRNA, 125–126
    interaction with lysine tRNA, 521–522
  Tat protein, interaction with TAR RNA, 73–74
Hydrogen bonds, intermolecular, between polynucleotides, 2
Hypermodified nucleosides, 9–11
  formation, 157–161
  rRNA, 434
  tRNA, 10
    carboxylate salts, esters, and amides, 10–11

In vitro recombinant RNA molecules, 24–28
Inosine, see also Adenosine-to-inosine conversion
  eukaryotic mRNA, 186
  formation, 145–147
Intron, 202–203
  AT-AC, splicing, 217–220
  GT-AG, splicing, 206–219
  origin of, 535
  RNA editing in plant organelles, 309
  self-splicing, 75
  tRNA modifications and, 425
Intron-encoded snoRNA, 261–262
Iron limitation, 475
Isomerization, 147–150, 162
Isopentenyl pyrophosphate isomerase, 151
Isopentenyl transferase, 151–152
N^6-Isopentenyladenosine
  conformational and thermodynamic effects, 97
  formation, 150–152
Isopentenylation, 150–152
Isozyme sorting, genetic approaches, 462–464

Kinetoplastid, mitochondrial genome, 378–380
Kinetoplastid protozoa
  mitochondrial biogenesis, developmental regulation, 380–381
  mitochondrial mRNA editing, 377–393
    assembly of editing complex, 385
    cleavage-ligation mechanism, 382–384
    components of editing complex, 385–388
    direction of pre-mRNA editing, 381–382
    functions, 388–390
    gRNA, 381–384
    gRNA/mRNA complex, 385–386
    mitochondrial RNPs in, 384–385, 387–388
    phylogenic distribution, 377–378
    proteins of editing complex, 386–387
    transesterification and enzymatic models, 382

Land plants, mitochondrial tRNA editing, 296–299
Leishmania tarentolae, see Kinetoplastid protozoa
Leukemia, modified nucleosides as tumor markers, 484
Lipoprotein, see Apolipoprotein B
Liver, overexpression of apobec-1, pathophysiology, 337–338
Loop-loop interactions, 113–133
Lymphoma, modified nucleosides as tumor markers, 484
Lysidine, conformational and thermodynamic effects, 97
Lysine tRNA, 509–510
  HIV-1 RNA interactions, 521–522
  translation, 509–510

Malignant lymphoma, modified nucleosides as tumor markers, 484
Mammalian cells, Tgt transglycosylase, 460–461
Mass spectrometry
  electrospray ionization, 48–49
  ionization methods, 48
  modified nucleosides, 47–57
    characterization of nucleosides in mixtures, 50–51
    collision-induced dissociation of nucleoside ions, 51–52
    future prospects, 55–56
    LC/ESI-MS, 49–50
    mapping nucleosides in oligonucleotides from RNA, 52–53
    mapping nucleosides in rRNA, 53–54
    mapping nucleosides in tRNA, 54–55
    oligonucleotide sequencing, 55
    RNA hydrolysates, 49–50
    tandem, 51–52
Matrix-assisted laser desorption ionization, 48
Maxicircle, 378–379
Messenger RNA (mRNA)
  eukaryotic, 435
    enzymes in methylation, 185–186
    inosine, 186
    N^6-methyladenosine, 187–192
    5-methylcytidine, 186–187
    modified nucleosides in 5'-terminal cap, 183–184
    modified nucleosides within internal regions, 186–192
  mapping pathway through ribosome, 76
  modifications, 5' terminus, 8
  paramyxovirus, P gene, 413–418
  splicing, 75
  subcellular location of RNA-modifying enzymes, 434–435

tRNA interactions, control of aminoacyl-tRNA
    synthetase gene expression, 126
Messenger RNA (mRNA) cap
  methylated nucleosides, 184–186
  modified nucleosides, 183–184
Messenger RNA (mRNA) editing
  adenosine-to-inosine conversion, 343–361
    adenosine deaminases, 352–357
    glutamate receptor subunits, 344–348
    hepatitis delta virus, 348–349
    RNA substrates, 344–352
    RNA substrates with multiple editing sites, 351–352
    serotonin 2C receptor, 349–351
  apolipoprotein B, 325–342
    annulment of apobec-1 expression, 338–339
    apobec-1, 330–332
    developmental, hormonal, and nutritional control, 336–337
    discovery, 326
    evolution and mechanism of apobec-1, 332–334
    hepatic overexpression of apobec-1, 337–338
    physiological consequences, 325–326
    sequence specificity, 327–330
    species and tissue specificity, 326
    structure of editing enzyme complex, 334–336
    subcellular location, 326–327
  developmental regulation, 389–390
  mitochondria of kinetoplastid protozoa, 377–393
    assembly of editing complex, 385
    cleavage-ligation mechanism, 382–384
    components of editing complex, 385–388
    direction of pre-mRNA editing, 381–382
    functions, 388–390
    gRNA, 381–384
    gRNA/mRNA complex, 385–386
    mitochondrial RNPs in, 384–385, 387–388
    phylogenic distribution, 377–378
    proteins of editing complex, 386–387
    transesterification and enzymatic models, 382
  mitochondria of *Physarum*, 395–411
  paramyxovirus
    cotranscriptional, 413–420
    structural models, 418–419
    thermodynamic consequences, 418–419
  plant organelles
    base conversions proceed without order in mRNA, 313
    biochemistry of nucleotide alterations, 313–315
    noncoding regions, 309
    partially edited mRNA, 310–312
    translational consequences, 310
  regulation of gene expression, 388–389
Metabolism
  induction of changes in tRNA modification, 474–476
  links with tRNA modification, 471–491
Metazoan animals, mitochondrial tRNA editing, 299–300
2′-O-Methyl nucleosides, rRNA, 430–432
5-Methyl pyrimidines, conformational and thermodynamic effects, 91
1-Methyladenosine, formation, 141–142

N^6-Methyladenosine, *see also* N^6-Adenosine
    methyltransferase, mRNA
  eukaryotic mRNA, 187–190
    function, 189–192
    in vitro system for sequence specific formation, 188–189
    mutation of methyladenosine sites, 189–190
    sequence specificity in vivo, 187–188
    studies using methylation inhibitors, 190–192
  formation, 142–143
2-O-Methyladenosine-N^6-methyltransferase, 185–186
1-Methyl-3-(3-amino-3-carboxylpropyl)pseudouridine, 249
5-Methylaminomethyl-2-selenouridine, formation, 155–157
5-Methylaminomethyl-2-thiouridine
  conformational and thermodynamic effects, 95–96
  formation, 155–157
5-Methylaminomethyluridine, conformational and thermodynamic effects, 95–96
2′-O-Methylase, 281
Methylation, 136–144, 162
  base-methylated nucleosides, 6–9
  on carbon, 136–141
  enzymes in mRNA cap methylation, 185–186
  on oxygen, 143–144
  on primary nitrogen, 142–143
  ribose in rRNA, 255–272
    catalysis of 2′-O-ribose methylation, 266–267
    dependence on cognate antisense snoRNA, 264–265
    identifying patterns, 256–257
    patterns in eukaryotes, 257–259
    perspectives, 269–270
    prokaryotes, 259, 266
    ribosome biogenesis and, 259–261
    snoRNA/rRNA canonical duplex, 265–266
    snoRNA/rRNA duplexes at methylation sites, 262–264
  rRNA, 433–434
  sugar-methylated nucleosides, 6–9
  on tertiary nitrogen, 141–142
  tRNA, organ specificity, 481–482
  viral RNA, 528
2′-O-Methylation, conformational and thermodynamic effects, 90
Methylation inhibitors, studies of methyladenosine in mRNA, 190–192
5-Methylcytidine
  conformational and thermodynamic effects, 91
  eukaryotic mRNA, 186–187
  formation, 136–141
2′-O-Methylcytidine, formation, 143–144
2′-O-Methyl-N^2,N^2-dimethylguanosine, conformational and thermodynamic effects, 97
1-Methylguanosine
  anticodon loop, 506–507
  formation, 141–142
7-Methylguanosine, formation, 141–142
N^2-Methylguanosine, formation, 142–143
2′-O-Methylguanosine

formation, 143–144
  yeast mitochondrial rRNA, 274
1-Methylinosine, formation, 141–142, 155
1-Methylpseudouridine, 109, 249
3-Methylpseudouridine, 249
2′-O-Methylpseudouridine, 109, 249
2-Methylthio-N^6-(cis-hydroxyisopentenyl)adenosine, anticodon loop, 504–506
2-Methylthio-N^6-isopentenyladenosine, anticodon loop, 504–506
2-Methylthio-N^6-methyladenosine, formation, 145
5-Methyl-2-thiouridine, conformational and thermodynamic effects, 91
Methyltransferase
  AdoMet dependent, 136
  (guanine-7)-methyltransferase, 185
  m^6A, 142
  m^1A58, 141
  m^1G37 methyltransferase, 141
  tRNA methyltransferase, S-adenosylmethionine dependent, 161
  tRNA-(5-methylaminomethyl-2-thiouridine)-methyltransferase, 156
2′-O-Methyltransferase, 143–144, 278
  cap specific, 185–186
2′-O-Methyluridine, yeast mitochondrial rRNA, 274
miaA gene, 456–457
Minicircle, 378–380
Minus one ribosomal frameshifting, Gag-Pol precursor synthesis, 525–528
Mitochondria, see also Plant organelles
  biogenesis, developmental control in kinetoplastid protozoa, 380–381
  Pet56p, 285
Mitochondrial genome, kinetoplastids, 378–380
Mitochondrial mRNA editing, kinetoplastid protozoa, 377–393
  assembly of editing complex, 385
  cleavage-ligation mechanism, 382–384
  components of editing complex, 385–388
  direction of pre-mRNA editing, 381–382
  functions, 388–390
  gRNA, 381–384
  gRNA/mRNA complex, 385–386
  mitochondrial RNPs in, 384–385, 387–388
  phylogenic distribution, 377–378
  proteins of editing complex, 386–387
  transesterification and enzymatic models, 382
Mitochondrial ribonucleoprotein complex
  enzymatic activities, 387–388
  identification, 384–385
  in mitochondrial mRNA editing, 384–385
Mitochondrial RNA editing, Physarum, 395–411
  biological function, 395
  catalytic mechanisms, 406–407
  characteristics, 396–398
  insertional editing, 398–401, 403
    transcription and, 401–402
  isolated mitochondria, 399–400
  mechanism, 400–408

methods used to study, 398–400
parallels with other systems, 403–407
prospects, 408–409
single versus dinucleotide insertions, 402–403
site selection, 403–404
specificity, 404–406
substitutional editing, 398–401, 407–408
Mitochondrial rRNA
  modified nucleosides, 434
  pseudouridine, 246
  yeast
    conservation of RNA-modifying enzymatic activities, 277–278
    Gm2251, 274
    LSU, 273–280
    modified nucleotides, 273–280
    modified nucleotides in peptidyl transferase center, 273–276
    prospects, 278–279
    pseudouridine, 274–276
    Um2552, 274
Mitochondrial tRNA editing, 293–300
  Acanthamoeba, 293–295
  chytridiomycete fungi, 293–295
  D. virginiana, 295–296
  evolution, 301–302
  import and editing, 429–430
  land plants, 296–299
  metazoan animals, 299–300
  P. polycephalum, 295
  plants, 310
  RNA-modifying enzymes, 427–429
Modified nucleosides
  biophysical properties, 85–102
  chemical structures, 543–549
  classification, 543–549
  conformation, 85–102
    anomeric and 3′-gauche effects, 89–90
    base stacking effects, 90
    effect of specific modifications, 90–97
    stereoelectronic effects, 88–89
    steric effects, 88–89
  detection, 47–57
  evolution, 535–541
  historical perspective on, 1–20
    base-methylated nucleosides, 6–9
    general perspectives and prospects, 11–14
    history of RNA research, 1–3
    hypermodification, 9–11
    pseudouridine, 3–6
    sugar-methylated nucleosides, 6–9
  mass spectrometry, 47–57
  NMR analysis, 85–88
    calculation of thermodynamic parameters from, 87–88
    correlation of sugar conformation with base stacking, 88
    measurable parameters, 85–86
  nonneoplastic disease, 483
  phylogenetics, 535–536

physiological excretion, 482
structure analysis, 47–57
synthesis of plus-strand strong-stop DNA, 523–524
thermodynamic effects of specific modifications, 90–97
translation, 493–516, see also Translation
tumor markers, see Tumor markers
Modified nucleotides
  affinity labeling, 60–62
  functional group analysis, reagents for, 62–63
  incorporation into RNA, 59–84
    chemical cleavage methods, 67–69
    chemical synthesis, 63–65
    covalent joining of RNA molecules, 70
    DNA-dependent RNA polymerase, 65–67
    enzymatic cleavage methods, 69
    at internal positions, 67–70
    labeling 3′ end, 67
    perspectives, 78–79
    recombinant RNA techniques, 67–72
    semisynthetic methods, 69–70
    synthetic methods, 63–67
    T4 DNA ligase, 71–72
    T4 RNA ligase, 70–71
    template-dependent chemical ligation, 72
  localizing in RNA, 30–36
    base composition analysis, 31–32
    direct sequencing, 30–31
    nearest-neighbor analysis, 32–33
    primer extension sequencing analysis, 33–36
  in replication, 517–533
  in RNA loop-loop interaction, 113–133
Molecular dynamics simulation, 103–110
mRNA, see Messenger RNA
Murine leukemia virus, initiation of reverse transcription, 522–523
Mutations
  strains defective for given nucleotide modification, 29
  that affect rRNA modifications, 12–13
  that affect tRNA modifications, 11–12
Myelomatosis, modified nucleosides as tumor markers, 484

Nearest-neighbor analysis, 32–33
NMR analysis, modified nucleosides, 85–88
  calculation of thermodynamic parameters, 87–88
  correlation of sugar conformation with base stacking, 88
  measurable parameters, 85–86
Nonsense suppression assay, 496
Northern blot analysis, mRNA N^6-adenosine methyltransferase, 196–197
Nuclear pore proteins, 427
Nucleoside(s), modified, see Modified nucleosides
Nucleoside deaminase, 363–375
Nucleotides
  intramolecular covalent bonds in, 1–2
  modified, see Modified nucleotides
Nutritional control, apobec-1 mRNA expression, 336–337

Oligonucleotide sequencing, mass spectrometry, 55
Oocyte microinjection technique, detecting enzyme activity, 36–38
Operon, tRNA modification genes in bacteria, 473–474
Opossum, mitochondrial tRNA editing, 295–296
Organ specificity, tRNA methylation, 481–482
Organelles, see also Chloroplasts; Mitochondria
  plant, see Plant organelles
  pseudouridine in rRNA, 246
Origin-of-life theories, 535–541
Oxidation, 162

P gene, paramyxovirus, mRNA editing, 413–418
P site, topography, 76–77
Paramyxovirus
  mRNA editing
    cotranscriptional, 413–420
    P gene, 413–418
    structural models, 418–419
    thermodynamic consequences, 418–419
  polymerase stuttering, kinetic model, 416–418
  RNA synthesis, 413
Pararetrovirus, replication, modified nucleotides in, 517–533
Pathogenicity, bacteria, tRNA modification and, 478–479
PBS, see Primer binding site
Peptide mimic, of RNA substrate of cytidine deaminase, 369–370
Peptidyl transfer center, rRNA
  coclustering of type of modification, 5–6
  pseudouridine, 5
Peptidyl transferase center, yeast mitochondrial rRNA, modified nucleotides, 273–276
PET56, yeast, 276–277, 281
  mutation that suppresses requirement for, 277
Pet56p, mitochondrial rRNA-modifying enzyme, 285
Phenolic extraction, RNA, 2–3
Phylogenetics, modified nucleosides, 535–536
Physarum, RNA editing in mitochondria, 395–411
  biological function, 395
  catalytic mechanisms, 406–407
  characteristics, 396–398
  insertional editing, 398–401, 403
    transcription and, 401–402
  isolated mitochondria, 399–400
  mechanism, 400–408
  methods used to study, 398–400
  parallels with other systems, 403–407
  prospects, 408–409
  single versus dinucleotide insertions, 402–403
  site selection, 403–404
  specificity, 404–406
  substitutional editing, 398–401, 407–408
  tRNA, 295
Plant organelles
  mitochondrial tRNA editing, 296–299, 310
  RNA editing, 307–323
    base conversions proceed without order in mRNA, 313
    biochemistry of nucleotide alteration, 313

C to U, 307–313
  consequences for analysis of gene expression, 318
  evolution, 317–318
  experimental difficulties, 315
  functional advantage, 316–320
  in introns, 309
  mechanism, 313–316
  noncoding regions, 309
  partially edited mRNA, 310–312
  potential applications, 320
  pseudogenes, 312–313
  selection of nucleotide to be edited, 315–316
  silent editing in reading frames, 310
  similar processes in mitochondria and plastids, 319–320
  tissue-specific differences, 318–319
  translational consequences, 310
  U to C, 307–313
Plant pararetrovirus, replication, modified nucleotides in, 517–533
Plus one ribosomal frameshifting, Gag-Pol precursor synthesis, 528
Plus-strand strong-stop DNA synthesis
  choice of primer and template, 523–524
  modified nucleosides and, 523–524
  termination, 524
Polymerase stuttering, paramyxovirus, kinetic model, 416–418
Polynucleotides, intermolecular hydrogen bonds, 2
Pre-mRNA, 201–203
  complexing with snRNA, 9
  direction of editing, 381–382
Prenyltransferase, 462–464
  i^6A37, 456–457
Pre-rRNA processing
  Cbf5p, 285–286
  Dim1p, 281–285
  Pet56p, 285
  prospects, 287
  regulatory aspects, 281–288
Primer binding site (PBS), 517, 520
  complementary to $tRNA_i^{met}$, 524
  eighteen nucleotide long, 524
Primer extension sequencing analysis, 33–36
Primer tRNA, 517–533
  annealing to genomic RNA, 520
  plus-strand strong-stop DNA synthesis, 523–524
  selection, effect of modified nucleotides, 519–520
  selective encapsidation, 518–519
Primordial soup, 537
Prokaryotes, see also Bacteria
  gRNA, 241–242
  pseudouridine biosynthesis, 238–242
Prokaryotic rRNA
  amounts and catalytic capacity of modifying enzymes, 447–448
  modification levels as regulatory device, 459–460
  prospects, 450
  regulation of amounts of RNA, 447
  regulation of modification, 447
  regulation of synthesis, 450–454
  ribose methylation, 259, 266
  rRNA modification genes, 441–447
    coordination of gene expression and RNA synthesis, 454–455
    feedback regulation, 457–458
    regulation, 450
  structure of modifying enzymes, 450
Prokaryotic tRNA
  amounts and catalytic capacity of modifying enzymes, 447–450
  editing, 292–293
  modification genes, 441–447
    coordination of gene expression and RNA synthesis, 454–455
    in operons, 473–474
    regulation, 450
  modification levels as regulatory device, 459–460
  modifications, 422
    regulation, 447
  prospects, 450
  regulation of amounts of RNA, 447
  regulation of synthesis, 450–454
  structure of modifying enzymes, 450
  titration of modification capacity by tRNA overexpression, 448
  tRNA-guanine transglycosylase, 170–171
    mechanism of base exchange, 174–176
    $preQ_1$ binding by, 171–173
    three-dimensional structure, 171
    tRNA binding and recognition by, 173–174
Protein(s), in snoRNA-directed pseudouridine synthesis, 245–246
Protozoa, kinetoplastid, see Kinetoplastid protozoa
Pseudo-base pair, 105–106
Pseudogenes, RNA editing in plant organelles, 312–313
Pseudouridine
  anticodon of tRNA, 504
  conformational and thermodynamic effects, 92–94
  formation, 6, 147–150
  Frank Allen's laboratory, 4
  historical perspective, 3–6
  rRNA, 5, 229–253, 432–433
    archaea, 246
    biosynthesis in eukaryotes, 242–246
    biosynthesis in organelles, 246
    biosynthesis in prokaryotes, 238–242
    correlation with other modified nucleosides, 235–236
    essential residues in LSU, 248–249
    essential residues in SSU, 248
    formation with gRNA, 244–245
    location, 229–238
    location in LSU RNA, 231–232
    location in SSU RNA, 231
    methodology, 229–230
    modified pseudouridines, 249
    number, 229–238
    number in LSU RNA, 230–231
    number in SSU RNA, 230

peptidyl transfer center, 5
prospects, 249–250
role in LSU, 247–248
role in ribosome activity, 247–248
role in ribosome biosynthesis, 246–247
role in SSU, 247
structural environment, 236–238
yeast mitochondrial, 274–276
snoRNA-directed synthesis, 242–246
tRNA, 103–112
anticodon loop, 508–509
functional roles, 107–109
modified pseudouridines, 109
position 13, 107
position 27, 107
position 32 of anticodon hairpin, 104–106
position 34, 107, 109
position 35, 107–108
position 36, 107–109
position 39, 106–107
position 55, 107
pseudo-base pair, 105–106
specific hydration, 104–105
Pseudouridine synthase, 147–150, 278, 281
prokaryotic
cloning, 238–239
multiple specificity, 240–241
RNA recognition by, 239–240
Pseudouridine synthetase, 474
eukaryotic, 245–246

Queuine
formation, 157, 169
function, 169
in development and tumorigenesis, 480–481
location, 169
Queuosine
conformational and thermodynamic effects, 97
formation, 157–160
tRNA, 292, 480–481, 503–504

Readthrough suppression, termination codons, 525
Recombinant RNA techniques, incorporation of modified nucleotides into RNA, 67–72
Reduction, 154–155, 162
Replication
ColE1 plasmid, antisense RNA control, 126–127
modified nucleotides in, 517–533
Respiratory tract tumor, modified nucleosides as tumor markers, 483–484
Retroids, replication, modified nucleotides in, 517–533
Retrotransposon
initiation of reverse transcription, 523
replication, modified nucleotides in, 517–533
Retrovirus
avian, initiation of reverse transcription, 520
replication, modified nucleotides in, 517–533
Reverse transcription, *see also* Primer tRNA
avian retrovirus, 520
HIV-1, 520–522

initiation, modified nucleotides in, 520–523
murine leukemia virus, 522–523
Ty1 retrotransposon, 523
Ribonucleoprotein complex, mitochondrial
enzymatic activities, 387–388
identification, 384–385
in mRNA editing, 384–385
Ribose methylase, rRNA, 266–267
Ribose methylation, rRNA, 255–272
catalysis of 2'-O-ribose methylation, 266–267
dependence on cognate antisense snoRNA, 264–265
identifying patterns, 256–257
patterns in eukaryotes, 257–259
perspectives, 269–270
prokaryotes, 259, 266
ribosome biogenesis and, 259–261
snoRNA/rRNA canonical duplex, 265–266
snoRNA/rRNA duplexes at methylation sites, 262–264
Ribose methyltransferase, PET56, 276–277
Ribose modifications, affinity labeling, 62
Ribosomal frameshifting
minus one, Gag-Pol precursor synthesis, 525–528
plus one, Gag-Pol precursor synthesis, 528
Ribosomal proteins
ribosome assembly, 247
S16 and L19, 455
Ribosomal RNA (rRNA)
eukaryotes, 255–272, 430–434
prospects, 464–465
regulation of modification, 460–465
evolution, 421
hypermodified nucleosides, 434
mapping modified nucleosides by mass spectrometry, 53–54
methylated bases, 433–434
2'-O-methyl nucleosides, 430–432
mitochondrial, modified nucleosides, 434
modifications
biochemical and genetic changes, 12–13
genetics and regulation, 441–469
regulatory aspects, 281–288
3' terminus, 7–8
Pet56p, 285
Physarum mitochondria, 395–411
prokaryotes
amounts and catalytic capacity of modifying enzymes, 447–448
coordination of modification gene expression and RNA synthesis, 454–455
feedback regulation of modification genes, 457–458
modification genes, 441–447
modification levels as regulatory device, 459–460
prospects, 450
regulation of amounts of RNA, 447
regulation of modification, 447
regulation of modification genes, 450
regulation of synthesis, 450–454
structure of modifying enzymes, 450
pseudouridine, 5, 229–253, 432–433
archaea, 246

biosynthesis in eukaryotes, 242–246
biosynthesis in organelles, 246
biosynthesis in prokaryotes, 238–242
correlation with other modified nucleosides, 235–236
essential residues in LSU, 248–249
essential residues in SSU, 248
formation with guide RNA, 244–245
location, 229–238
location in LSU RNA, 231–232
location in SSU RNA, 231
methodology, 229–230
modified pseudouridines, 249
number, 229–238
number in LSU RNA, 230–231
number in SSU RNA, 230
peptidyl transfer center, 5
prospects, 249–250
role in LSU, 247–248
role in ribosome activity, 247–248
role in ribosome biosynthesis, 246–247
role in SSU, 247
structural environment, 236–238
ribose methylation, 255–272
catalysis of 2'-O-ribose methylation, 266–267
dependence on cognate antisense snoRNA, 264–265
identifying patterns, 256–257
patterns in eukaryotes, 257–259
perspectives, 269–270
prokaryotes, 259, 266
ribosome biogenesis and, 259–261
snoRNA/rRNA canonical duplex, 265–266
snoRNA/rRNA duplexes at methylation sites, 262–264
subcellular location of RNA-modifying enzymes, 430–434
yeast, 281–288
modification genes, 441–447
yeast mitochondria
conservation of RNA-modifying enzymatic activities, 277–278
Gm2251, 274
LSU, 273–280
modified nucleotides, 273–280
modified nucleotides in peptidyl transferase center, 273–276
prospects, 278–279
pseudouridine, 274–276
Um2552, 274
Ribosome(s)
biosynthesis, 246–247
assembly of RNA and proteins, 247
folding of RNA, 247
formation of tertiary structure, 247
processing of RNA, 246–247
ribose methylation in rRNA and, 259–261
codon recognition by tRNA, 123–125
mapping pathway of mRNA through, 76
peptidyl transfer site, 1
reconstitution, *E. coli* 50S subunit, 276

topography of A, P, and E sites, 76–77
2'-O-Ribosyladenosinephosphate, formation, 154
2'-O-Ribosylguanosinephosphate, formation, 154
2'-O-Ribosylphosphate, 154
Ribosylthymine
conformational and thermodynamic effects, 91
formation, 136–141
RNA, *see also* Guide RNA; Messenger RNA; Ribosomal RNA; Small nuclear RNA; Small nucleolar RNA; Transfer RNA
functional group analysis, *see* Functional group analysis
history of RNA research, 1–3
in vitro recombinant molecules, 24–28
in vitro transcription, construction of DNA template, 24–25
purification
phenolic extraction, 2–3
salt precipitation, 2–3
sequencing
direct approach, 30–31
nearest-neighbor analysis, 32–33
primer extension sequencing analysis, 33–36
RNA catalysis, 535, 537
RNA chaperones, 267–268
RNA deaminase, 146–147
RNA editing, *see also specific types of RNA*
evolution, 538–540
nucleoside deamination and base conversion, 363–364
Physarum mitochondria, 395–411
biological function, 395
catalytic mechanisms, 406–407
characteristics, 396–398
insertional editing, 398–401, 403
insertional editing and transcription, 401–402
isolated mitochondria, 399–400
mechanism, 400–408
parallels with other systems, 403–407
prospects, 408–409
single versus dinucleotide insertions, 402–403
site selection, 403–404
specificity, 404–406
substitutional editing, 398–401, 407–408
plant organelles, 307–323
C to U, 307–313
functional advantage, 316–320
mechanism, 313–316
U to C, 307–313
potential applications, 320
pseudogenes, 312–313
silent editing in reading frames, 310
tissue-specific differences, 319
types and characteristics, 551–553
"RNA enzymes," 4
RNA genes, runoff transcripts as RNA substrates, 23–24
RNA hydrolysate, detection of modified nucleosides by LC/ESI-MS, 49–50
RNA ligase, T4, incorporation of modified nucleotides into RNA, 70–71
RNA modification genes
bacterial operons, 473–474

feedback regulation, 457–458
prokaryotes, 450
    coordination of gene expression and RNA synthesis, 454–455
RNA polymerase, DNA-dependent, incorporation of modified nucleotides into RNA, 65–67
RNA splicing, biochemical mechanisms, 74–76
RNA substrates
    adenosine-to-inosine conversion in mRNA, 344–352
        with multiple editing sites, 351–352
    chemical synthesis, 28–29
    for RNA-modifying and RNA-editing enzymes, 22–30
    from mutants defective in nucleotide modification, 29
    in vitro recombinant RNA molecules, 24–28
    natural heterologous RNA, 29–30
    peptide mimic, 369–370
    perspectives, 39–40
    runoff transcripts of RNA genes, 23–24
RNA world hypothesis, 535–541
RNA-editing enzymes
    database accession numbers, 561–567
    deaminases, 363–375
    detecting activity, 36–39
        biolistic technique, 39
        cell electroporation technique, 38–39
        oocyte microinjection technique, 36–38
        using transformed cells, 39
    general properties, 555–560
    genetic locations, 561–567
    identification, 21–46
    identification of new modifications, 40
    in vitro versus in vivo assays, 40–41
    localizing modified nucleotides in RNA, 30–36
    mechanisms, 125–168
    number, 21–22
    perspectives, 39–41
    prospects, 161–162
    RNA substrate, 22–30
RNA-modifying enzymes
    conservation in yeast, 277–278
    database accession numbers, 561–567
    detecting activity, 36–39
        biolistic technique, 39
        cell electroporation technique, 38–39
        oocyte microinjection technique, 36–38
        using transformed cells, 39
    functional relationship to nuclear pore proteins, 427
    general properties, 555–560
    genetic locations, 561–567
    identification, 21–46
    identification of new modifications, 40
    in vitro versus in vivo assays, 40–41
    isozyme sorting, 462–464
    localizing modified nucleotides in RNA, 30–36
    mechanisms, 125–168
    multiple functions, 474
    number, 21–22
    perspectives, 39–41
    prokaryotes, amounts and catalytic capacity, 447–450
    prospects, 161–162

RNA substrate, 22–30
RNA-guided site selection, 268–269
subcellular locations, 421–440
rRNA, see Ribosomal RNA
RUMT reaction, 136–141
Runoff transcripts of RNA genes, as RNA substrates, 23–24

Salt precipitation, RNA, 2–3
Selenation, 156–157
Selenium, 156–157, 475
Selenocysteine tRNA, 291
Serotonin 2C receptor, adenosine-to-inosine conversion in mRNA, 349–351
Severe combined immunodeficiency, 483
Silent RNA editing, 310
Slime mold, see Physarum
Small nuclear RNA (snRNA), 201
    box H/ACA, 242–245
    complexing with pre-mRNA, 9
    in splicing of AT-AC introns, 217–220
    in splicing of GT-AG introns, 206–219
    modifications, 5' terminus, 8
    posttranscriptional modifications, 206–220
        detection, 203–204
        internal, 211–216, 220
        in spliceosome assembly and function, 216–220
        terminal, 219–220
        5'-terminal structure and nuclear export, 206–211
        5'-terminal structure and nuclear import, 211
    spliceosomal, 206–220
    subcellular locations of RNA-modifying enzymes, 435–436
Small nucleolar ribonucleoprotein particles (snoRNPs), 201
    Cbf5p, 285–286
Small nucleolar RNA (snoRNA), 201, 278, see also Guide snoRNA
    antisense, 264–265
    box C/D, 262
    box H/ACA, 243
        in pseudouridine site selection, 243–244
    complexing with pre-mRNA, 9
    intron encoded, 261–262
    posttranscriptional modifications, 204–206
        detection, 203–204
        internal, 206
        5' terminal, 204–206
    as RNA chaperones, 267–268
    snoRNA/rRNA duplexes at methylation sites, 262–266
    subcellular location of RNA-modifying enzymes, 435–436
snoRNA, see Small nucleolar RNA
snoRNPs, see Small nucleolar ribonucleoprotein particles
snRNA, see Small nuclear RNA
Species specificity, apolipoprotein B mRNA editing, 326
Spliceosome, 201–203
    assembly and function, role of snRNA, 216–220
    intermolecular interactions within, 77–78
SRM1-1, 277

Stereoelectronic effects, 88–90
Steric effects, modified nucleosides, 88–89
Stress response, 475
Subcellular location
　apolipoprotein B mRNA editing, 326–327
　RNA-modifying enzymes, 421–440
Sugar conformation, correlation with base stacking, 88
Sugar-methylated nucleosides, 6–9
　formation, 6
Sulfurtransferase system, tRNA, 145

Tandem mass spectrometry, 51–52
TAR, see trans-activation response region
Tat protein, HIV-1, interaction with TAR RNA, 73–74
Temperature-jump relaxation method, 114
　features of anticodon-anticodon complex, 114–117
Termination codons, readthrough suppression, 525
TGT, see tRNA-guanine transglycosylase
Thiolation, 145, 162
Thionucleotides, 60–61
2-Thiouridine
　conformational and thermodynamic effects, 91
　formation, 145
4-Thiouridine
　conformational and thermodynamic effects, 91–92
　formation, 145
Threonylation, 152–154
N^6-Threonylcarbamoyladenosine
　anticodon loop, 507
　conformational and thermodynamic effects, 96
　formation, 152–154
Thymidylate synthase, 136–137
Tissue specificity
　apolipoprotein B mRNA editing, 326
　RNA editing in plant organelles, 318–319
TΨ-loop, tRNA, interaction with D-loop, 125
trans-activation response region (TAR), interaction of HIV-1 Tat protein with TAR RNA, 73–74
Transcription
　cotranscriptional paramyxovirus mRNA editing, 413–420
　insertional RNA editing and, 401–402
　regulation in prokaryotes, 450–454
Transfer RNA (tRNA)
　anticodon-anticodon interactions, 113–133
　carboxylate salts, esters, and amides in and near anticodons, 10–11
　codon recognition on ribosome, 123–125
　complementary to primer binding site, 524
　editing, 289–305
　　detection and quantification, 290–291
　　false starts, 291–292
　　prokaryotes, 292–293
　editing in mitochondria, 293–300
　　Acanthamoeba, 293–295
　　chytridiomycete fungi, 293–295
　　D. virginiana, 295–296
　　evolution, 301–302
　　land plants, 296–299
　　metazoan animals, 299–300

　　P. polycephalum, 295
　　plants, 310
　eukaryotes, 422–430
　　prospects, 464–465
　　regulation of modification, 460–465
　　sequential modification pathways, 461–462
　evolution, 421
　human, interaction with HIV-1 RNA, 125–126
　hydration and dynamics, 103–112
　hypermodification, 10
　hypomodification, 474–476
　interaction between D-loop and TΨ-loop, 125
　interaction with aminoacyl-tRNA synthetases, 72–73
　lysine, 509–510, 521–522
　mapping modified nucleosides by mass spectrometry, 54–55
　methylation, organ specificity, 481–482
　modification(s)
　　antibiotic resistance in bacteria and, 479–480
　　bacterial pathogenicity and, 478–479
　　bacteriophage, 422
　　developmental changes in, 480–482
　　effect on reactions in which tRNA acts as cofactor, 473
　　effect on translation and gene expression, 472–473
　　genetics and regulation, 441–469
　　introns and, 425
　　late modifications, 425–426
　　links to metabolism, 471–491
　　location and distribution, 569–576
　　metabolically induced changes in, 474–476
　　perspectives, 484–485
　　position 34, 577–581
　　position 37, 577–581
　　as regulatory device, 471–474
　　relationship of modifying enzymes to nuclear pore proteins, 427
　　3′ terminus, 7–8
　　5′ terminus, 8
　　tRNA architecture and, 7, 426–427
　　tumor-induced changes in, 480–482
　modification genes
　　bacterial operons, 473–474
　　mutations that affect modifications, 11–12
　　yeast, 441–447
　modifications in anticodon, 493
　　A34 and I34, 502–503
　　C34, 501–502
　　pseudouridine, 504
　　queuosine at position 34, 503–504
　　U34, 497–498
　　U34 modification in family boxes, 500–501
　　U34 modifications and restricted wobbling, 498–500
　modifications in 3′ side of anticodon loop, 504–510
　　m^1G37, 506–507
　　$ms^2i(o)^6A37$, 504–506
　　pseudouridine, 508–509
　　t^6A37, 507
　　Y base at position 37, 508
　modifications outside anticodon region, 510

modified nucleosides as tumor markers, 482–484
  animal models, 482
  hematological malignancies, 484
  physiological excretion, 482–483
  solid tumors, 483–484
modified nucleosides in nonneoplastic disease, 483
modified nucleosides in translation, 493–516
molecular dynamics simulation, 103–110
mRNA interactions, control of aminoacyl-tRNA synthetase gene expression, 126
multiple functions of modifying enzymes, 474
Physarum mitochondria, 395–411
primer, *see* Primer tRNA
prokaryotes, 422
  amounts and catalytic capacity of modifying enzymes, 447–450
  coordination of modification gene expression and RNA synthesis, 454–455
  modification genes, 441–447
  modification levels as regulatory device, 459–460
  prospects, 450
  regulation of amounts of RNA, 447
  regulation of modification, 447
  regulation of modification genes, 450
  regulation of synthesis, 450–454
  structure of modifying enzymes, 450
  titration of modification capacity by tRNA overexpression, 448
pseudouridine, 103–112
  functional roles, 107–109
  modified pseudouridines, 109
  position 13, 107
  position 27, 107
  position 32 of anticodon hairpin, 104–106
  position 34, 107, 109
  position 35, 107–108
  position 36, 107–109
  position 39, 106–107
  position 55, 107
  pseudo-base pair, 105–106
  specific hydration, 104–105
queuosine, 292, 480–481
selenocysteine, 291
subcellular location of RNA-modifying enzymes, 422–430
$tRNA_i^{Met}$, 524
versatility, 471
Transformed cells, detecting enzyme activity, 39
Transglycosylation, 162
Translation
  effect of tRNA modification, 472–473
  modified nucleosides in, 493–516
    experimental approaches, 495–497
    nucleosides in anticodon, 497–504
    outlook, 510–511
  mRNA editing in plant organelles, 310
  review, 493–495
Transposon, replication, modified nucleotides in, 517–533
Trisubstituted pyrophosphates, 72

trmA gene, 474
trmD gene, 455
TrmD methylase, noncoordinate regulation with ribosomal proteins, 455
tRNA, *see* Transfer RNA
tRNA-guanine transglycosylase (TGT), 157–160
  amino acid sequences, 176–178
  archaebacterial, 171, 178–179
  base exchange, 169–182
  eukaryotic, 171, 176–178
    regulation, 460–461
  evolutionary relationships, 179–180
  prokaryotic, 170–171
    mechanism of base exchange, 174–176
    $preQ_1$ binding by, 171–173
    three-dimensional structure, 171
    tRNA binding and recognition by, 173–174
  prospects for future, 179–180
Trypanosoma brucei, *see* Kinetoplastid protozoa
Tumor markers, modified nucleosides, 482–484
  animal models, 482
  clinical observations, 482
  hematological malignancies, 484
  physiological excretion, 482–483
  solid tumors, 483–484
Tumor-induced changes, tRNA modification, 480–482
Ty1 retrotransposon, initiation of reverse transcription, 523

Urinalysis, modified nucleosides, 482–484
Urologic tumor, modified nucleosides as tumor markers, 484

Viral replication, modified nucleotides in, 517–533
Viral RNA
  deaminations and editing process, 528–529
  methylation, 528
  modifications, 528–529

Wobble pairing, 498–500
  in anticodon-anticodon interactions, 118–122
  expansion by 5-modified uridines, 96
  restriction by 5-modified uridines, 95–96
Wobble position, 577–581
Wyosine
  conformational and thermodynamic effects, 97
  formation, 160–161

Y base, anticodon loop, 508
Yeast, *see also* Eukaryotes
  Cbf5p, 242, 285–286
  Dim1p, 281–285
  mitochondrial rRNA
    conservation of RNA-modifying enzymes, 277–278
    Gm2251, 274
    LSU, 273–280
    modified nucleotides, 273–280
    modified nucleotides in peptidyl transferase center, 273–276
    prospects, 278–279

pseudouridine, 274–276
Um2552, 274
PET56, 276–277, 281
    mutation that suppresses requirement for, 277

Pet56p, 285
rRNA, 281–288
rRNA modification genes, 441–447
tRNA modification genes, 441–447